The Geometrid Moths of Europe

Volume 5

Peder Skou & Pasi Sihvonen

In memory of the most eminent lepidopterist of our time, and a dear friend,

Professor Niels Peder Kristensen

2nd March 1943 to 6th December 2014.

Niels provided a lot of help to us during the preparation of this volume.

The Geometrid Moths of Europe

Axel Hausmann (ed.)

Volume 5

Peder SKOU & Pasi SIHVONEN

Subfamily Ennominae I

(Abraxini, Apeirini, Baptini, Caberini, Campaeini, Cassymini, Colotoini, Ennomini, Epionini, Gnophini (part), Hypochrosini, Lithinini, Macariini, Prosopolophini, Theriini and 34 species of uncertain tribus association)

BRILL

Editor: Axel Hausmann, Zoologische Staatssammlung München, Germany.

Linguistic editor: Martin Corley, Faringdon, United Kingdom.

Colour plates photographed and compiled by: Igor Kostjuk, Kiev, Ukraine.

Printed by: Vinderup Bogtrykkeri ApS

This publication should be cited as: Skou, P. & Sihvonen, P. (2015): Ennominae I. – In A. Hausman (ed.): The Geometrid Moths of Europe 5: 1-657. Brill, Leiden.

Date of publication: 31.07.2015

Authors's addresses:

Peder Skou
Aamosen 1
5762 Vester Skerninge
Denmark
E-mail: info@apollobooks.dk

Pasi Sihvonen
Vanhatie 16
02880 Veikkola
Finland
and
Research Affairs
University of Helsinki
P.O. Box 33 (Yliopistonkatu 4)
FI-00014 University of Helsinki, Finland
E-mail: pasi.sihvonen@helsinki.fi

Front cover: Ennomos fuscantaria (Haworth, 1809). Denmark: Veksø, Sjælland, 13.ix.2013. Photo by Bjarne Skule, Denmark.

Numerous photos of differential external features shown in Text-figs modified from Igor Kostjuk's plate photographs. Axel Hausmann wrote the paragraphs for the genetic data sections.

The work of Pasi Sihvonen was supported by Societas Entomologica Helsingforsiensis (visit to Zoologische Staatssammlung München, München, Germany), SYNTHESYS project, which is financed by European Community Research Infrastructure Action under the FP7 Integrating Activities Programme (visit to the Natural History Museum, London, United Kingdom) and Apollo Books, Vester Skerninge, Denmark (visit to Zoologische Staatssammlung München, München, Germany).

Library of Congress Control Number: 2015943542
ISBN 978-90-04- 25220-2 (hardback, Vol. 5)
ISBN 978-90-04-26573-8 (e-book)

Contents

Preface

Ennominae are the most species-rich geometrid subfamily with some 10,700 species having been described worldwide, so far (SCOBLE & HAUSMANN 2007), and the number of undescribed species will amount to at least 7,000 additional species, judging from the numerous unnamed genetic clusters in the Global Campaign of DNA Barcoding (iBOL Lepidoptera).

The present fifth volume of The Geometrid Moths of Europe, covering roughly the first half of European Ennominae, is a product of team-work. PEDER SKOU prepared adult plates, distribution maps and for the species accounts wrote the parts on distribution, phenology, biology, habitat and remarks where relevant. PASI SIHVONEN dissected the genitalia and prepared the genitalia plates, text-figures and tables, wrote introduction, systematic account (including systematic position of the Ennominae within the Geometridae, monophyly of the Ennominae, Ennominae systematics, classification of European Ennominae genera into tribes), methodological notes, on species and subspecies, and for the species accounts the parts on taxonomy, external characters and abdomen, variation, male genitalia, female genitalia, similar species and remarks where relevant. In a few details, the style of this fifth volume differs from that of the previous volumes, changes (e.g. in the way to present nomenclature, in the length of morphological descriptions, in the style of photographing genitalia etc.) have been made when they seemed to be justified scientifically.

This book – like the other volumes of the series – provides an updated, valuable, modern tool for species identification. We hope, that it will stimulate the re-examination of all existing collections, in order to improve quality of identification. Parallel to this volume, a continuous update of the European checklist is published (www.faunaeur.org; www.zsm.mwn.de/lep/gme.htm), containing digital illustrations and additional data. The site of the Geometridae research initiative Forum Herbulot (www.herbulot.de) provides a list of family-group names and further information on geometridology.

At the present stage (October 2014), we have generated, in the framework of the Global Geometrid Campaign of the Barcode of Life project, DNA barcodes of more than 920 (out of 988) European geometrid species (cf. HAUSMANN 2011; HAUSMANN *et al.* 2013). As in the case of Volume 3 we are happy that we could include this large amount of genetic information into our analyses and we thank PAUL HEBERT, SUJEEVAN RATNASINGHAM and their teams (Biodiversity Institute of Ontario, Guelph, Canada; cf. Acknowledgements) for providing us the related research and finance for DNA sequencing (chemicals, salaries).

The editor thanks the authors for preparing the book in such a diligent and competent way and is confident that the readers will be similarly satisfied with this volume.

Axel Hausmann (*editor*), Munich, 15.1.2015

Abstract

Volume 5 of the book series *The Geometrid Moths of Europe* presents the first part of the revision of the European taxa of the subfamily Ennominae, covering 141 species, classified into 59 genera. The current volume treats the tribes Abraxini (4 species), Apeirini (1 species), Baptini (2 species), Caberini (3 species), Campaeini (7 species), Cassymini (7 species), Colotoini (2 species), Ennomini (6 species + 13 species of potential Ennomini association), Epionini (3 species + 5 species of potential Epionini association), Gnophini (23 selected species, remaining species are dealt with in vol. 6), Hypochrosini (1 species + 2 species of potential Hypochrosini association), Lithinini (1 species + 5 species of potential Lithinini association), Macariini (40 species), Prosopolophini (3 species), Theriini (4 species) and 9 species of uncertain tribus association. The high number of species with uncertain or potential tribus associations demonstrates the need for a higher-level Ennominae revision in global context. Potential systematic positions of the uncertain genera are discussed. Volume 6 treats the remaining European Ennominae taxa with focus on the Gnophini (rest) and the Boarmiini.

The species are presented with their full name, reference to the original description, fully referenced junior synonyms, (potentially) unavailable names, description, distribution, phenology, biology including host plants, habitat, and differential features from similar species. Further, for each species the following information is provided: colour photographs of the adults, greyscale photographs of the male genitalia, including the everted vesica, greyscale photographs of the female genitalia, and distribution map for Europe and adjacent countries. Genetic data is provided for the majority of species. 151 text-figures illustrate and explain diagnostic characters and other morphological structures.

The following taxonomic changes are proposed.

One tribus name is downgraded to synonymy: Aleucini DJAKONOV, 1936 (type genus *Aleucis* GUENÉE, 1845) is downgraded to synonymy with Theriini HERBULOT, 1963 syn. n. Aleucini is a senior synonym, but without usage in the recent literature, thus we propose not to apply the Aleucini in favour of stability of nomenclature, as Theriini is largely accepted.

Two genus names are downgraded to synonymy: *Atomorpha* STAUDINGER, 1901, is proposed to be a junior synonym of *Isturgia* HÜBNER, 1823, syn. n., *Megaspilates* WARREN, 1894, is proposed to be a junior synomym of *Aspitates* TREITSCHKE, 1825, syn. n.

17 new species and subspecies level synonymies are proposed: *Abraxas pantaria* subspecies *urganda* (STOLL, 1781) is downgraded to synonymy with *Abraxas pantaria* (LINNAEUS, 1767) syn. n., *Apeira syringaria* subspecies *druentaria* LERAUT, 2009, is downgraded to synonymy with *Apeira syringaria* (LINNAEUS, 1758) syn. n., *Chariaspilates formosaria* subspecies *pannonicus* VOJNITS, 1977, is downgraded to synonymy with *Chariaspilates formosaria* (EVERSMANN, 1837) syn. n., *Enanthyperythra legataria* subspecies *transcaucasica* WEHRLI, 1937 is downgraded to synonymy with *Enanthyperythra legataria* (HERRICH-SCHÄFFER, 1852) syn. n., *Ennomos autumnaria* subspecies *sareptana* WEHRLI, 1940 is downgraded to synonymy with *Ennomos autumnaria* (WERNEBURG, 1859) syn. n., *Ennomos quercaria* subspecies *dryadaria* RAMBUR, 1866 is downgraded to synonymy with *Ennomos quercaria* (HÜBNER, 1813) syn. n., *Isturgia spodiaria* subspecies *mizanensis* (WEHRLI, 1940) is downgraded to synonymy with *Isturgia spodiaria* (LEFÈBVRE, 1832) syn. n., *Heliomata glarearia* subspecies *aurearia* (STAUDER, 1920) is downgraded to synonymy with *Heliomata glarearia*

(DENIS & SCHIFFERMÜLLER, 1775) syn. n., *Heliomata glarearia* subspecies *schawerdae* (STAUDER, 1920) is downgraded to synonymy with *Heliomata glarearia* (DENIS & SCHIFFERMÜLLER, 1775) syn. n., *Isturgia murinaria* subspecies *uralica* (WEHRLI, 1937) is downgraded to synonymy with *Isturgia murinaria* (DENIS & SCHIFFERMÜLLER, 1775) syn. n., *Macaria liturata* subspecies *deceptans* (STAUDINGER, 1901) is downgraded to synonymy with *Macaria liturata* (CLERCK, 1759) syn. n., *Narraga fasciolaria* subspecies *danubialis* MOUCHA & POVOLNY, 1957 is downgraded to synonymy with *Narraga fasciolaria* (HUFNAGEL, 1767) syn. n., *Narraga (Naragula) tessularia* subspecies *kasyi* MOUCHA & POVOLNY, 1957 is downgraded to synonymy with *Narraga tessularia* (METZNER, 1845) syn. n., *Perconia strigillaria* subspecies *cretaria* (EVERSMANN, 1844) is downgraded to synonymy with *Perconia strigillaria* (HÜBNER, 1787) syn. n., *Plagodis pulveraria* subspecies *gadmensis* (RÄTZER, 1890) is downgraded to synonymy with *Plagodis pulveraria pulveraria* (LINNAEUS, 1758) syn. n., *Odontopera graecarius* (BANG-HAAS, 1910) is downgraded to synonymy with *Odontopera bidentata* (CLERCK, 1759) syn. n.

Status of nine taxa are revised: *Stegania ochrearia* BANG-HAAS, 1910 is downgraded to a subspecies of *Stegania trimaculata* (VILLERS, 1789) as *Stegania trimaculata ochrearia* (BANG-HAAS, 1910) stat. n., *Stegania oranaria* (WEHRLI, 1930), originally described as a variation of *Stegania trimaculata* (VILLERS, 1789), but treated as a separate species by SCOBLE (1999), is downgraded back to a variation of *Stegania trimaculata* subsp. *ochrearia* BANG-HAAS, 1910 new combination and stat. rev., *Stegania postrecta* WEHRLI, 1930, originally described as a variation of *Stegania trimaculata* (VILLERS, 1789), but treated as a separate species by SCOBLE (1999), is downgraded back to variation of *Stegania trimaculata* subsp. *ochrearia* BANG-HAAS, 1910 new combination and stat. rev., *Narraga cappadocica* HERBULOT, 1943, originally described as a subspecies of *Narraga nelvae* (ROTHSCHILD, 1912) but treated as a separate species in recent publications, is downgraded back to a subspecies of *Narraga nelvae* (ROTHSCHILD, 1912) as *Narraga nelvae cappadocica* HERBULOT, 1943, stat. rev., *Ennomos duercki* REISSER, 1958, is downgraded to a subspecies of *Ennomos quercaria* (HÜBNER, 1813) as *Ennomos quercaria duercki* REISSER, 1958 stat. n., *Ennomos freidbergi* HAUSMANN, 1997, is downgraded to a subspecies of *Ennomos quercaria* (HÜBNER, 1813) as *Ennomos quercaria freidbergi* HAUSMANN, 1997 stat. n., *Ennomos lissochila* (PROUT, 1929) is downgraded to a subspecies of *Ennomos quercaria* (HÜBNER, 1813) as *Ennomos quercaria lissochila* (PROUT, 1929) stat. n., *Crocallis cypriaca* FISCHER, 2003 is downgraded to a subspecies of *Crocallis tusciaria* (BORKHAUSEN, 1793) as *Crocallis tusciaria cypriaca* FISCHER, 2003 stat. n., *Dyscia govii* FIUMI & FLAMIGNI, 2013, is downgraded to a subspecies of *Dyscia innocentaria* (CHRISTOPH, 1885) as *Dyscia innocentaria govii* FIUMI & FLAMIGNI, 2013 stat. n.

Seven new combinations are proposed: *Liodes homochromata* MABILLE, 1869 is transferred from *Liodesina* WEHRLI, 1938 to *Isturgia* HÜBNER, 1823, comb. n., *Atomorpha punctistrigaria* CHRISTOPH, 1897 is transferred from *Atomorpha* STAUDINGER, 1901 to *Isturgia* HÜBNER, 1823, comb. n., *Atomophora* [sic!] *falsaria* ALPHÉRAKY, 1892 is transferred from *Atomorpha* STAUDINGER, 1901 to *Isturgia* HÜBNER, 1823, comb. n., *Fidonia heydemanni* CHRISTOPH, 1885, is transferred from *Atomorpha* STAUDINGER, 1901 to *Isturgia* HÜBNER, 1823, comb. n., *Stegania mabillearia* LUCAS, 1907 is transferred from *Atomorpha* STAUDINGER, 1901 to *Isturgia* HÜBNER, 1823, comb. n., *Deuteronomos ifranaria* RUNGS, 1941 is transferred from subspecies of *Ennomos quercaria* (HÜBNER, 1813) to subspecies of *Ennomos erosaria* (DENIS & SCHIFFERMÜLLER), 1775, comb. n., *Phalaena mundataria* STOLL, 1782 is transferred from *Megaspilates* WARREN, 1894 to *Aspitates* TREITSCHKE, 1825, comb. n.

Status of two taxa are uncertain: Status of *Ennomos effractaria* FREYER, 1841 is uncertain, being potentially a junior synonym of *E. fuscantaria* (HAWORTH, 1809). Following SIHVONEN *et al.* (2014), taxonomic combination and status of *Hemithea squalidaria* O. G. COSTA, 1848, recently classified in genus *Hylaea* HÜBNER, 1822, is uncertain, being potentially valid at species level and potentially belonging to another genus in the subfamily Geometrinae.

One species is excluded from the European fauna: *Chiasmia syriacaria* (STAUDINGER, 1871), due to the absence of authentic material.

The current volume is the first monograph in the Geometridae where the vesica of the male genitalia was everted for all included species. It is shown to be an additional and valuable source of taxonomical information: the vesicas are structurally similar in numerous species within certain genera, providing diagnostic genus-level features (for instance in *Macaria* CURTIS, 1826), yet the vesicas may vary considerably even between closely related species, providing diagnostic species-level features (for instance in *Stegania* GUENÉE, 1844).

Data from papers published up to 31.12.2014 (with few exceptions in early 2015), that have come to our knowledge, are included in the manuscript.

Acknowledgements

We are indebted to the following persons, both professional and amateur lepidopterists and other experts, who have helped us to complete this book. Their invaluable help is greatly appreciated.

LEIF AARVIK, Norway: Providing access to the Norwegian distribution data.

HENRY BARLOW, Malaysian Nature Society & Southdene Sdn. Bhd., Malaysia: permission to reprint the Ennominae cladogram shown in Text-fig. 3.

JEROME BARBUT, France: Provision of photographs and taxonomic information.

KNUD BECH, Denmark: Loan of material.

Prof. Dr. STOYAN BESHKOV, Bulgaria: Provision of intensive faunistic data, biological information of many different kinds. Cooperation in DNA barcoding.

ROLF BLÄSIUS, Germany: Provision of faunistic data.

DANIEL BOLT, Switzerland: Provision of intensive faunistic data.

ULF BUCHSBAUM, Germany: Provision of faunistic data.

Dr. FEZA CAN, Turkey: Provision of literature, faunistic data.

JOHN CHAINEY, United Kingdom: Loan of material, taxonomic discussions and support.

MARTIN CORLEY, United Kingdom: Linguistic editing of the manuscript, provision of biological information, provision of intensive faunistic data.

Dr. JORGE DANTART, Spain: Provision of literature and intensive faunistic data.

Dr. LEONARDO DAPPORTO, Italy: Provision of intensive faunistic data.

Dr. JEREMY DEWAARD, Canada: Identification of barcoded specimens.

THOMAS DRECHSEL, Germany Provision of faunistic and biological data.

Dr. SVEN ERLACHER, Germany: Provision of faunistic data (Saxonia, Thuringia), cooperation in DNA barcoding.

IMRE FAZEKAS, Hungary: Provision of faunistic data.

RALF FIEBIG: Provision of faunistic data.

HEINZ FISCHER, Germany: Provision of literature.

GABRIELE FIUMI, Italy: Loan of material, taxonomic discussions.

EGBERT FRIEDRICH, Germany: Provision of biological data and intensive faunistic data.

DIETER FRITSCH, Germany: Provision of intensive faunistic data, informations on biology, proof reading of the colour plate legends, taxonomic discussions, loan of material.

Dr. JÖRG GELBRECHT, Germany: Information on immature stages and distribution, taxonomic discussions.

BARRY GOATER, United Kingdom: Provision of faunistic data.

Dr. STANISLAV GOMBOC, Slovenia: Provision of biological and intensive faunistic data.

GUIDO GOVI, Italy: Loan of material, taxonomic discussions.

PENNY HALE, Spain: Provision of faunistic data (southern Spain), cooperation in DNA barcoding.

Dr. AXEL HAUSMANN, Germany: Editor, provision and writing the genetic data chapters, providing the necessary infrastructure for the adult moth photography, loan of material, kind support to the authors' visits at Zoologische Staatssammlung in Munich, provision of all kinds of information including taxonomic discussions and support.

Dr. PAUL HEBERT, Canada: Kind cooperation in DNA barcoding, initiation of the global geometrid campaign of DNA barcoding, sequencing of 20,000 geometrid species and 2,500 geometrid type specimens by late 2014, covering almost all species of this volume.

JEAN-PAUL HERZET, Belgium: Provision of faunistic and habitat information.

Dr. JEREMY HOLLOWAY, Great Britain: Taxonomic discussions and support.

HENRY HOLMBERG, Finland: Loan of material.

MARTIN HONEY, United Kingdom: Loan of specimens, taxonomic discussions, linguistic guidance and support.

Dr. PETER HUEMER, Austria: Provision of faunistic data, extensive cooperation in DNA barcoding, systematic discussions and habitat informations.
Dr. TIES HUIGENS, The Netherlands: Providing permission to use Dutch distribution data gathered by De Vlinderstichting.
CARSTEN HVIID, Denmark: Provision of habitat data.
LARS IMBY, Sweden: Provision of intensive faunistic data.
Dr. MARCO INFUSINO, Italy: Provision of faunistic data (Sicily), loan and donation of material. Cooperation in DNA barcoding.
Prof. Dr. PREDRAG JAKSIC, Serbia: Provision of literature.
URMAS JÜRIVETE, Estonia. Provision of intensive faunistic data.
Dr. LAURI KAILA, Finland: Loan of material.
OLE KARSHOLT, Denmark: Numerous discussions related to Ennominae, access to the ZMUC collection, provision of literature.
GERGELY KATONA, Hungary: Provision of taxonomic information.
CHRISTOPH KAYSER, Germany: Provision of biological data.
Dr. IAN KITCHING, United Kingdom: Taxonomic discussions and support.
Dr. TONI KOREN, Croatia: Provision of faunistic data.
Dr. IGOR KOSTJUK, Ukraine: Photography and preparation of colour plates, taxonomic discussions, provision of literature, biological data and loan of material.
Prof. Dr. NIELS PEDER KRISTENSEN, Denmark: Numerous discussions related to Ennominae, access to the ZMUC collection.
ECKARD KRÜGER, Germany: Provision of faunistic data.
Dr. MARTIN KRÜGER, South Africa: Taxonomic discussions.
JAAKKO KULLBERG, Finland: Provision of biological data, loan of material.
Dr. DON LAFONTAINE, Canada: Identification of barcoded specimens.
Dr. BERNARD LANDRY, Switzerland: Provision of material.
Dr. ZDENEK LASTUVKA, Czech Republic: Provision of intensive faunistic and biological data.
GYULA M. LÁSZLÓ, Hungary: Provision of intensive faunistic data, provision of biological data.
MICHAEL LEIPNITZ, Germany. Provision of biological data, photographs, taxonomic discussions.
Dr. ANTOINE LÉVÊQUE, France: provision of faunistic data, cooperation in DNA barcoding.
MAURIZIO LUPI, Italy: Provision of faunistic and biological data.
Dr. HANS LÖBEL, Germany: Provision of faunistic data.
Dr. ADAM MALKIEWICZ, Poland: Provision of intensive faunistic data.
Dr. REINHARD MASCHLER, Germany: Provision of faunistic data.
Dr. IVA MIHOCI, Croatia: Provision of intensive faunistic data, cooperation in DNA barcoding.
Dr. VLADIMIR MIRONOV, Russia: Provision of intensive faunistic data, taxonomic discussions and support.
PHILIPPE MOTHIRON, France: For providing us access to information on his web site Lepinet (www.lepinet.fr).
Dr. MARKO MUTANEN, Finland: Provision of faunistic data, literature, scientific discussions. Extensive cooperation in DNA barcoding.
Dr. BERND MÜLLER, Germany: Provision of intensive faunistic data and numerous biological informations.
KARI NUPPONEN, Finland: Provision of intensive faunistic data, numerous biological informations, loan of material, taxonomic discussions.
Dr. ERLING ÓLAFSSON, Iceland: Provision of faunistic data.
Mag. SIEGFRIED ORTNER, Austria: Provision of intensive faunistic data.
Dr. ERKI ÕUNAP, Estonia: Loan of material.
ARTTU PAARLAHTI, Finland: Technical expertise with distribution maps.
Dr. PAOLO PARENZAN, Italy: Revision of manuscript, extensive provision of taxonomic and faunistic data and discussions.
GERGELY PETRÁNYI, Hungary: Provision of intensive faunistic data.

Dr. ALEXANDER N. POLTAVSKY, Russia: Provision of literature, biological data and intensive faunistic data.
WILLY DE PRINS, Belgium/United Kingdom: Providing access to Belgian distributional data, provision of taxonomic data.
NORBERT PÖLL, Austria: Provision of intensive faunistic and biological data, loan of material and much other help during the preparation of the manuscript. Cooperation in DNA barcoding.
FRANZ PÜHRINGER, Austria: Cooperation in DNA barcoding.
MAIJA-LEENA REUNANEN, Finland: Translation of German literature.
Dr. HARTMUT RIETZ, Germany: Provision of intensive faunistic data, provision of biological data.
LÁSZLÓ RONKAY, Hungary: Provision of taxonomic information, taxonomic discussions.
Prof. NILS RYRHOLM, Sweden: Provision of intensive faunistic data, provision of biological data.
PAUL SAMMUT, Malta: Provision of faunistic data and food plant informations.
LAURENT SARI, Finland: Translation of French literature.
FRANTIŠEK ŠARŽIK, Slovakia: Permission to use a photograph of *Artiora evonymaria* (Denis & Schiffermüller) as a Text-fig.
Dr. NICOLAJS SAVENKOVS, Latvia: Provision of intensive faunistic data.
Dr. STEFANO SCALERCIO, Italy: Provision of literature and intensive faunistic data, loan and donation of material. Cooperation in DNA barcoding.
STEFANO SCIVOLA, Italy: Provision of biological data.
Dr. ANDREAS SEGERER, Germany: Provision of faunistic data. Cooperation in DNA barcoding.
HARRI SIHVONEN, Finland: Loan of material, provision of taxonomic data and scanning of slides.
BJARNE SKULE, Denmark: Provision of habitat data and front cover photograph.
MANFRED SOMMERER, Germany: Provision of faunistic data, loan and donation of extensive material. Cooperation in DNA barcoding.
Dr. DIRK STADIE, Germany: Provision of faunistic, biological and taxonomic information, taxonomic discussions.
HERMANN STAUDE, South Africa: Provision of taxonomic data.
Dr. ALEKSANDAR STOJANOVIĆ, Serbia: Provision of literature and biological data.
Dr. DIETER STÜNING, Germany: Loan of material, taxonomic information, provision of literature, translation of German literature, taxonomic and distribution discussions, identification of barcoded vouchers, cooperation in DNA barcoding and support.
GIEDRIUS SVITRA, Lithuania: Providing access to Lithuanian distributional data.
Dr. LEVENTE SZÉKELY, Romania: Provision of intensive faunistic and taxonomic data.
IVARS SULCS, Latvia: Provision of intensive faunistic data.
Dr. ANDREY VALENTINOVICH SVIRIDOV, Russia: Provision of taxonomic information.
CLAUDE TAUTEL, France: Provision of intensive faunistic data as well as biological information.
Dr. ROBERT TRUSCH, Germany: Provision of literature, loan of material, taxonomic discussions.
ANGUS TYNER, Ireland: Provision of intensive faunistic data.
Dr. DRAGAN VAJGAND, Serbia: Provision of biological data, literature and intensive faunistic data.
THIERRY VARENNE, France: Provision of biological data.
Dr. JAAN VIIDALEPP, Estonia: Loan of material, provision of information, taxonomic discussions.
HARTMUT WEGNER, Germany: Provision of intensive faunistic data, provision of biological information.
Dr. MIKKO YLIKANGAS, Finland: Translation of Russian literature.
Dr. JOSEP YLLA I ULLASTRE, Spain: Provision of biological data and intensive faunistic data.
Dr. NORBERT ZAHM, Germany: Provision of intensive faunistic data, biological data and literature. Cooperation in DNA barcoding.

We owe special thanks to AXEL HAUSMANN, PAOLO PARENZAN and MARTIN CORLEY for their exhaustive and constructive scientific review of the whole manuscipt.

The work of PASI SIHVONEN was supported by Societas Entomologica Helsingforsiensis (visit to Zoologische Staatssammlung München, München, Germany) and SYNTHESYS project http://www.synthesys.info/. which is financed by European Community Research Infrastructure Action under the FP7 Integrating Activities Programme (visit to the Natural History Museum, London, United Kingdom).

The analyses of genetic data (DNA barcoding) were supported by grants of the Bavarian ministry of science (project BFB), by the German ministry of science (BMBF; project GBOL) to Axel Hausmann, and by Genome Canada (Ontario Genomics Institute) in the framework of the International Barcode of Life (iBOL) iniative, WG 1.9 http://ibol.org to PAUL D. N. HEBERT. The genetic analyses have received considerable support from PAUL D. N. HEBERT and his competent teams at the Biodiversity Institute of Ontario (BIO) and the Canadian Centre for DNA Barcoding (CCDB University of Guelph). The data management and analysis system BOLD was provided by SUJEEVAN RATNASINGHAM.

AXEL HAUSMANN edited and reviewed the manuscript, wrote the paragraphs for the genetic data sections, provided the facilities for adult plate photography, loaned material, engaged in numerous taxonomic discussions and provided continuous support during the preparation of the book. AXEL's invaluable help and support is greatly appreciated.

IGOR KOSTJUK photographed nearly all adult specimens and arranged the numerous individual photographs into the final adult colour plates. We thank IGOR for the high-quality work and for the patience he showed when specimens were rearranged over and over again before the final plates were ready. Also the precise numbering of all 714 specimens on the colour plates has been done by IGOR. We are most thankful for all the assistance and help IGOR has provided us.

The Entomology Library (SARAH SWORDER) and the Picture Library (GRAHAM SMITH) of the Natural History Museum, London, UK, are thanked for locating and reproducing the image of *H. squalidaria*. Text-Fig. 131 is reproduced under license from the Natural History Museum. Publishing company Brill (MICHIEL THIJSSEN) is thanked for covering the photography and reproduction fees for the image.

PEDER SKOU wishes to thank his partner ANNE-GRETE KLAUSEN for joining him on much field work that has produced much informations for this volume. Further PEDER wishes to thank PASI SIHVONEN for having joined him as co-author. Without PASI's brilliant overview and great working capacity this volume could have turned out to be a never-ending story. Finally PEDER wishes to thank FINN ANDERSEN, MICHAEL KAVIN, DANNY NILSSON and BJARNE SKULE, who all have joined him on one or more field trips.

PASI SIHVONEN is very grateful to PEDER SKOU for asking him to co-author this book with him. It has been a priviledge to learn from his vast knowledge of the Geometridae. PASI wants to present his special thanks to his wife, KAROLIINA REUNANEN, and to their children EERO and SIIRI, for enduring with much understanding the approximately seven year period over which this book was prepared. PASI prepared the book outside office hours, during evenings, weekends and holidays.

Introduction

The geometrid subfamily Ennominae includes about 46% of all Geometridae, currently over 10,700 species are considered valid globally (SCOBLE *et al.* 1995, MINET & SCOBLE 1999, SCOBLE 1999, SCOBLE & HAUSMANN 2007). Approximately 300 Ennominae species occur in Europe (MÜLLER 1996, HAUSMANN 2001), roughly half of them are dealt with in this volume.

The Ennominae are, like most other geometrids, mostly delicate and slender-bodied, medium to large Geometridae. A number of robust-bodied groups exist, often with adults that fly in autumn. The wings are often finely fasciated and patterns are continued on hindwings also. External appearance of both sexes are generally similar, sexual dimorphism is sometimes apparent in size differences (wing span), in wing colouring and in structure of the antennae, those of males are often more expanded than those of females. The Ennominae comprises a high number of species with short-winged or brachypterous females, sometimes the wings are absent altogether (aptery) (YAMAMOTO & SATO 2007, WAHLBERG *et al.* 2010). The Ennominae are predominantly active at night, only a few species in Europe are diurnal, among those are *Isturgia famula* (ESPER, 1787), *Pseudopanthera macularia* (LINNAEUS, 1758), and species of genera *Psodos* TREITSCHKE, 1825 and *Glacies* MILLIÈRE, 1869. Some tropical diurnal groups like *Milionia* WALKER, 1854 and related genera in the Australian tropics and South-East Asia are particularly brightly coloured (HOLLOWAY 1994). In Africa a small group of cycad-feeding moths are also colourful and diurnal (for instance STAUDE 2001, STAUDE & SIHVONEN 2014), their Ennominae association has been supported only recently (SIHVONEN *et al.* 2015). The Ennominae larvae normally have prolegs present on segments A6 and A10, as in other Geometridae, thus causing the typical looping in the movement of the larvae. There are a number of Ennominae taxa, especially in Southern Hemisphere, which have prolegs more or less developed also on segments A3-A5 (YOUNG 2004, SIHVONEN *et al.* 2015). The European *Campaea margaritaria* (LINNAEUS) and *Gerinia honoraria* (DENIS & SCHIFFERMÜLLER) have an additional, rudimentary pair of legs on the eight segment (PORTER 2010, LEPIFORUM 2013). The age of the subfamily Ennominae has been estimated to be about 41 million years, whereas the lineage leading to the Geometridae diverged from the Uraniidae+Sematuridae about 72 million years ago, during the late Cretaceous (WAHLBERG *et al.* 2013, fig. S1).

The Ennominae are cosmopolitan in distribution, and species of the subfamily are found in all biogeographical realms except Antarctica. South America has the highest number of species (SCOBLE *et al.* 1995). The subfamily is more diverse at low latitudes and altitudes, in the tropics they show their greatest species diversity in the lower montane zone at around 1000 m above sea level (HOLLOWAY 1984, 1994, HOLLOWAY *et al.* 1990, BREHM & FIEDLER 2003, AXMACHER *et al.* 2004); although in Ecuador no decline in Ennominae species diversity was observed along an altitude gradient from 1000 m up to 2700 m (BREHM, SUESSENBACH & FIEDLER 2003). Purely open habitat species are rare, although a number of species respond positively to forest disturbance and are probably associated with plant taxa characteristic of regenerating forest (HOLLOWAY 1994).

In Europe the proportion of Ennominae species within the Geometridae is relatively low, approximately 30%, if compared with global diversity, where their proportion is nearly half of the total species number of Geometridae (SCOBLE 1999, HAUSMANN 2001, BREHM & FIEDLER 2003). Host plant data for the Ennominae are rather poor, particularly for the tropical species, but it appears that the Ennominae larvae feed prevalently on leaves of trees and shrubs, not so much on leaves of herbaceous plants and apparently not at all on monocotyledons (SKOU 1986, BREHM 2002). Polyphagy appears to be rather widespread, but some specialism to particular plant family can be found at tribal level, for example the Macariini and some Cassymini show preference for Fabaceae (HOLLOWAY *et al.* 2001, SCOBLE & KRÜGER 2002), Lithinini show preference for ferns (RINDGE 1986), the Neotropical Palyadini show strong preference for Myrsinaceae (SCOBLE 1995, BREHM 2002) and the southern African Diptychini for Cycadaceae, Stangeriaceae and Zamiaceae, ie. cycads (STAUDE & SIHVONEN 2014). Needles of conifers are also commonly consumed, examples of such species are found in many tribes (BREHM 2002). Several Ennominae species are considered pests (KUDLER 1978, ZHANG 1994), *Erannis defoliaria* (CLERCK, 1759) and *Bupalus piniaria* (LINNAEUS, 1758) are perhaps the most well-known. Also *Abraxas grossulariata* (LINNAEUS, 1758) was earlier considered a pest on gooseberry and currant bushes but since the middle of the 20th century its adundance has decreased and it has become extinct in many parts of Europe (HAUSMANN 2001). In Malaysia the adult Ennominae favoured savoury feeding almost exclusively when sweet (fermenting banana) and savoury (rotting prawns) baits were offered simultaneously. The same was observed throughout the Lepidoptera (HOLLOWAY *et al.* 2013).

Taxonomy of the western Palaearctic Lepidoptera is perhaps the most extensively studied of all biogeographic realms. HAUSMANN (2001) has estimated that at least 95% of the existing geometrid fauna of Europe is probably known (see also HAUSMANN, HASZPRUNAR & HEBERT 2011). This high ratio can be explained by characteristics of the fauna, which is rather poor compared to tropical regions, in addition to high number of both amateur and professional lepidopterists investigating that fauna, and there has been a long tradition for such studies.

Despite that, comprehensive modern revisionary works that would cover the complete Geometridae fauna of Europe, including the Ennominae, are missing. Formerly widely used reference books on the European Geometridae (CULOT 1917, CULOT 1919, PROUT 1912–16, PROUT 1934–39, WEHRLI 1939–1954) are still used for identification, but unfortunately they lack illustrations of anatomical details, including the genitalia, which were not routinely illustrated at that time. Further, these monographs have outdated nomenclature and numerous taxonomic changes have occurred since then. In addition to these works, there are numerous, but regionally restricted, identification guides and monographs available on the European Geometridae. Those include for instance the fauna of northern Europe (SKOU 1986), Finland (MIKKOLA *et al.* 1985, 1989), Germany and Central Europe (FORSTER & WOHLFAHRT 1981, KOCH 1984, EBERT 2001), Poland (BLESZYNSKI 1960-66, BUSZKO 2000, MALKIEWICZ 2012), Spain (REDONDO & GASTÓN 1999), the Iberian Peninsula (REDONDO *et al.* 2009), Great Britain (BROOKS 1991) Italy (FLAMIGNI *et al.* 2007) and France (ROBINEAU 2007). A recent addition is the pocket-size field guide on the European Geometridae (LERAUT 2009). The field guide is generally very helpful as an identification guide, although not all European species are illustrated on the plates. LERAUT also proposed numerous taxonomic changes, but this aspect is dealt with in a debatable manner. The major problem is the apparent lack of scientific approach: in many cases the author does not provide information on the material upon which the conclusions are drawn. This means that the analyses are not verifiable. Furthermore, in most cases the conclusions are drawn without examination of type specimens. For details see SIHVONEN (2009).

The *Geometrid Moths of Europe* series aims to present a modern synthesis covering the European fauna. The series will be published in six volumes. Thus far four volumes have been published (HAUSMANN 2001, 2004, MIRONOV 2003, HAUSMANN & VIIDALEPP 2012). The

Ennominae will be dealt with in the volumes 5 and 6. The present volume, vol. 5, covers about half of the European Ennominae fauna.

Systematic position of the Ennominae within the Geometridae

The Geometridae as a whole are diagnostic (e.g. HOLLOWAY 1997, MINET & SCOBLE 1999, RAJAEI *et al.* 2015), but monophylies of the subfamilies have been weakly demonstrated, and the evolutionary relationships among all major groupings have remained uncertain (e.g. HAUSMANN 2001). The biggest obstacle in obtaining a more robust classification has been the lack of global view. The systematic position of the Ennominae within the Geometridae has therefore been controversial and varied. HOLLOWAY (1997) proposed a tentative phylogeny for the Geometridae, recognising nine subfamilies. In the morphology based analysis the Orthostixinae were hypothesized to be sister-taxon to the Ennominae.

Almost all recent analyses that have examined higher-level relationships of the Geometridae have been based on molecular data. In the analysis of ABRAHAM *et al.* (2001), which was based on a limited selection of northern European taxa, the Archiearinae were recovered as sister to the Ennominae *sensu lato* clade, which included also the Geometrinae. In a detailed examination of southern hemisphere Ennominae by YOUNG (2006), which was based on molecular data from the 28S D2 ribosomal nuclear gene fragment, the Ennominae were found as the most derived group but as not monophyletic because the Alsophilinae (*Alsophila* HÜBNER, 1816) were nested within it. Young concluded that *Alsophila* is probably a derived boarmiine (Ennominae) with some unusual features that are likely adaptations to an extreme environment. YAMAMOTO & SOTA (2007), who analysed phylogenetic relationships on the grounds of two mitochodrial and two nuclear gene sequences, recovered a clade containing the Geometrinae and Oenochrominae as the sister taxa to the Ennominae. Again, the Alsophilinae were nested within the Ennominae, thus agreeing with the results of YOUNG (2006). WAHLBERG *et al.* (2010), who studied the evolution of female flightlessness among the Ennominae, performed a phylogenetic analysis of morphological and molecular characters. The results supported the monophyly of the Ennominae (including *Alsophila* of the traditional subfamily Alsophilinae). They recovered the Geometrinae as sister to the Ennominae, the Ennominae being the most derived family of the Geometridae.

SIHVONEN *et al.* (2011) performed the most extensive molecular analysis to date to examine the geometrid affinities in a global context. The analysis was based on eight genes, and the maximum likelihood analyses included 164 taxa. The analysed taxa represented all previously recognised subfamilies and nearly 90% of the recognised tribes (Text-fig. 1). The Ennominae were found to be the sister-taxon to the clade containing the Geometrinae and Oenochrominae sensu stricto + Desmobathrinae (tribe Eumeleini). The Ennominae were recovered as a monophyletic lineage, including the Alsophilinae and Orthostixinae, thus agreeing with the earlier results (YOUNG 2006, YAMAMOTO & SOTA 2007, WAHLBERG *et al.* 2010). MUTANEN *et al.* (2010) and REGIER *et al.* (2013) studied the phylogenetic relationships of the order Lepidoptera. The overall pattern of the Geometridae relationships in both studies, although represented by rather few taxa only, agreed with the results of SIHVONEN *et al.* (2011): the Geometrinae were recovered as the sister taxon to the Ennominae, the latter being the most derived lineage of the Geometridae. In SIHVONEN *et al.* (2011) Sematuridae + Epicopeiidae were recovered as the sister-taxon to the Geometridae. Other recent studies have recovered Uraniidae (REGIER *et al.* 2013, RAJAEI *et al.* 2015) or Uraniidae + Sematuridae (MUTANEN *et al.* 2010) as the sister-taxon to the Geometridae (cf Text-fig. 1).

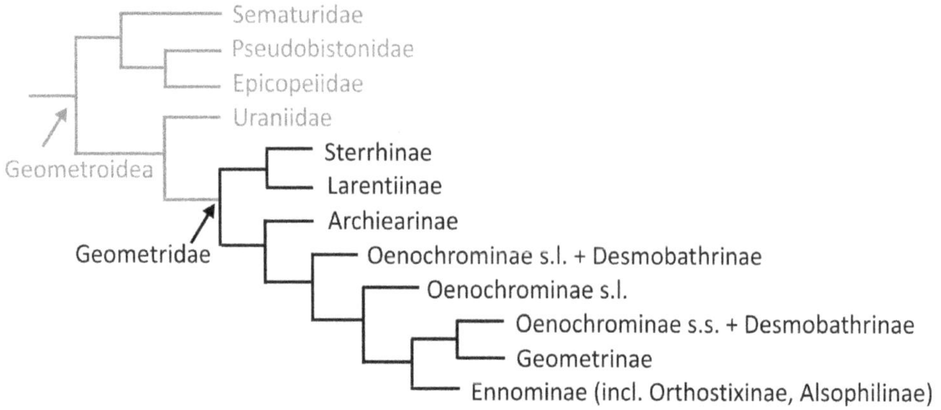

Text-fig 1. Simplified overview of the Geometroidea relationships, redrawn from various sources. The relationships of the Geometridae subfamilies (shown as black) are according to SIHVONEN *et al.* (2011), those of Sematuridae, Pseudobistonidae, Epicopeiidae and Uraniidae (shown as grey) are according to REGIER *et al.* (2013) and RAJAEI *et al.* (2015). The newly described Pseudobistonidae MINET, RAJAEI & STÜNING has been classified as the sister-taxon to the Epicopeiidae (RAJAEI *et al.* 2015). The branch lengths are presented uniformly to show potential relationships, they do not indicate genetic distances. The relationships resulting from various studies are still not fully concordant, see for instance MUTANEN *et al.* (2010) for an alternative hypothesis of the basal groups.

Monophyly of the Ennominae

Based on several molecular studies (YOUNG 2006, YAMAMOTO & SOTA 2007, WAHLBERG *et al.* 2010, SIHVONEN *et al.* 2011), there is growing evidence that the Ennominae are a monophyletic entity. The morphology based definition of the subfamily is still weak. The traditionally held diagnostic feature for the Ennominae is the loss (or reduction to a fold) of vein M2 on the hindwing (HOLLOWAY 1994, MINET & SCOBLE 1999, BELJAEV 2006a, BELJAEV 2006b), which is the only morphological feature that is shared by all major tribal groupings within the subfamily (Text-fig. 2). The loss of vein M2 is not exclusive as there are several taxa with apparent reversal of this character, for instance the Neotropical Melanolophiini (PITKIN 2002), and the genera *Alsophila* HÜBNER, 1825, *Epirranthis* HÜBNER, 1823 and *Naxa* WALKER, 1856 (SIHVONEN *et al.* 2011). Additional less definitive features are the posteriorly directed transverse comb or setae that are located on the third abdominal sternite (HOLLOWAY 1997, HOLLOWAY *et al.* 2001) and a spiny, mushroom-like signum in the corpus bursa of the female genitalia. The presence of setae seems to be correlated with a hair-pencil, which is usually concealed in a deep groove on the tibia of the hindleg. Unlike in many other Geometridae subfamilies, the tympanal organs vary widely in Ennominae (COOK & SCOBLE 1992, our observations). There are some modifications, for instance the occurrence of a lobe on the cavus wall, which potentially is a unique character within the Ennominae (Text-figs 3–5).

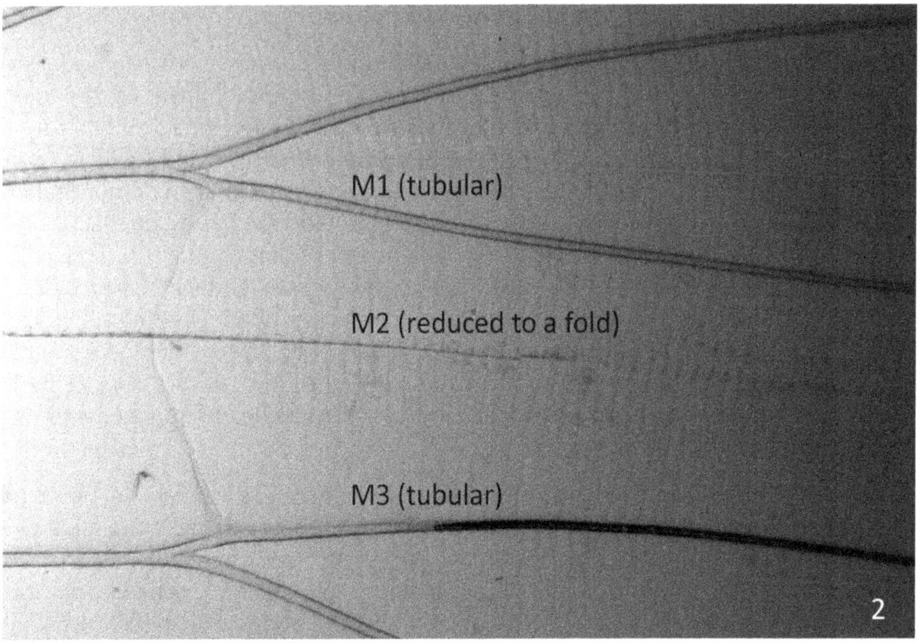

Text-fig. 2. Morphological synapomorphy of the Ennominae, the loss (or reduction to a fold) of hindwing vein M2. Detail of *Macaria liturata* (CLERCK, 1759), Finland: Kirkkonummi, 16.vi.2008, slide PS1227 (coll. P. SIHVONEN).

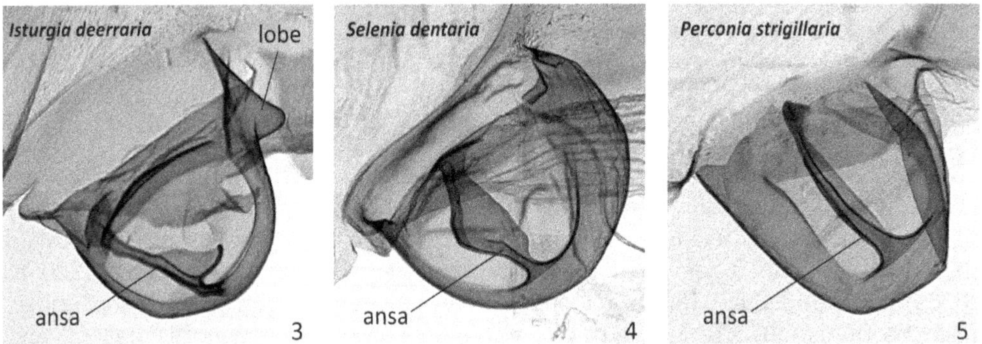

Text-figs 3–5. The Ennominae tympanal organs vary widely, some of the structures are potentially unique within the Ennominae. Text-fig. 3. *Isturgia deerraria* (WALKER, 1861) (slide PS1836, Malta, ex. larva 13.x.2007, coll. SAMMUT). Asymmetric extension of the ansa is potentially restricted to a few species in genus *Isturgia* HÜBNER, 1823 whereas the lobe on the cavus wall is found widely in the Ennominae but not in other subfamilies, see also Text-fig. 144: *Onychora agaritharia* (DARDOIN, 1842). Text-fig. 4. *Selenia dentaria* (FABRICIUS, 1775) (slide PS1298, Finland, 12–13.v.1994, coll. P. SIHVONEN). Bottle-shaped ansa, found widely across the Ennominae. Text-fig. 5. *Perconia strigillaria* (HÜBNER, 1787), (slide PS1350, Finland, 21.vi.1999, coll. P. SIHVONEN). Narrow ansa, restricted to few Ennominae lineages.

Ennominae systematics

The species level taxonomy of regional faunas within the Ennominae has been extensively studied but the current state of the higher classification of the Ennominae, i.e. grouping of species into genera, and the genera into tribes and to other higher categories, is currently poorly resolved. This is likely a result of the great number of species and extensive homoplasy in morphology, particularly in the genitalic characters, as has been demonstrated by difficulties encountered by many authors elsewhere in Lepidoptera also (e.g. MILLER 1991, KRISTENSEN 1991, MINET & SCOBLE 1999). Secondly, many studies have focused on regional faunas. A consequence of the regional classifications has been a tendency towards describing new genera on an *ad hoc* basis, i.e. new genera are described when new species are identified that do not fit the existing system. Many generic concepts within the Ennominae appear to have originated in this way, for example in the Macariini (SCOBLE & KRÜGER 2002). Currently the over 10,000 Ennominae species are classified in approximately 1100 genera (PITKIN 2002).

Characters that have been used in higher classification of the Ennominae, i.e. which have been considered phylogenetically informative, are mostly derived from structures of the adult. Among those are for example the absence or presence and structure of the male forewing fovea (HOLLOWAY 1994), structures of the anellar area of the male genitalia, particularly the labides (BELJAEV 2006a, BELJAEV 2006b), the division of the valve (HOLLOWAY 1994), arrangement of setae on the male third abdominal segment (HOLLOWAY 1994), structure and size of tympanal organs (HOLLOWAY 1994), modifications of the posterior margin of male abdominal sternite eight (PITKIN 2002, SCOBLE & KRÜGER 2002), structure of the uncus (SCOBLE & KRÜGER 2002), presence or absence of the coremata (HOLLOWAY 1994, PITKIN 2002) and the form of the signum (HOLLOWAY 1994, PITKIN 2002). Characters of the immature stages in a phylogenetic context are yet to be fully exploited. Those that have been used, include for instance the number of cremaster hooks of the pupae (FORBES 1948, MCGUFFIN 1981, HOLLOWAY 1994), other pupal characters (PATOČKA 2003, 2004, 2006), number of larvae prolegs (MCQUILLAN 1985) and number of setae on the A6 proleg of larvae (DUGDALE 1961, HEITZMAN 1985, POOLE 1987).

During recent years, two morphology based hypotheses have been presented that attempted to depict the potential tribal relationships of the Ennominae. HOLLOWAY's (1997) hypothesis is shown in Text-fig. 6. He recognised 21 tribes, and numerous relationships are unsolved. When this is compared against the hypotheses suggested by BELJAEV (2006a, 2006b) (not illustrated), who recognised 24 tribes, one genus group and five supertribes, it is apparent that there is no consensus which tribal names have diagnostic meaning and how they are related. YAMAMOTO & SOTA (2007), WAHLBERG *et al.* (2010) and ÕUNAP *et al* (2011) examined the Ennominae relationships using molecular data. Their analyses were not aimed to resolve the Ennominae relationships as such, but all analyses concluded that there are two main lineages in the Ennominae, these were termed the ennomines and boarmiines in WAHLBERG *et al.* (2010). These two main lineages were confirmed in a molecular analysis by SIHVONEN *et al.* (2011) (Text-fig. 7).

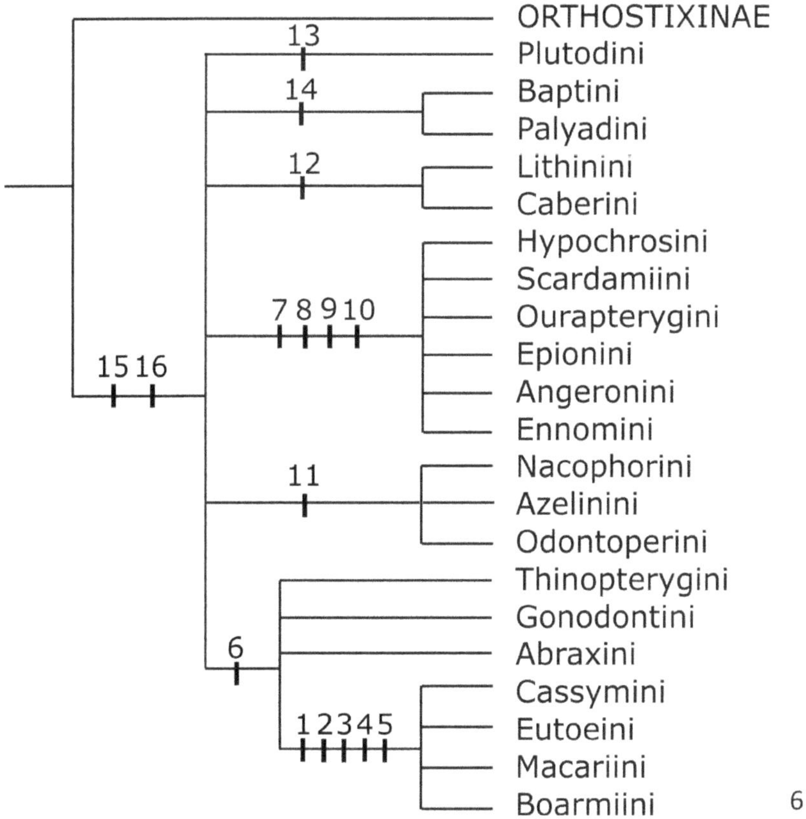

Text-fig. 6. Morphology based, tentative phylogeny of the Ennominae after HOLLOWAY (1997), reprinted with permission from Malaysian Nature Society & Southdene Sdn. Bhd. The tentative synapomorphic characters (not necessarily unique) have been added by us, according to the descriptive texts in HOLLOWAY (1994). Our character interpretations have been confirmed by J. HOLLOWAY (pers. comm.). Note that not all the tribes depicted in the figure occur in Europe. 1. Cremaster of pupa divergently bifid. 2. Male forewing foveate in high proportion of genera (fovea is structurally different in each, illustrated in Holloway 1994). 3. Presence of cucullate anterior apical portion of the valve supporting a concentration of the setae. 4. Tendency for the walls of the polygonal cells in eggs ornamenting the surface to be heavy and for the cells to be aligned in longitudinal rows. 5. Presence in many genera of transverse comb of posteriorly directed setae on the male third abdominal segment. HOLLOWAY (1994) notes that setae are also present in the Abraxini, Gonodontini, Lithinini and Ourapterygini, thus making it unclear whether the comb is a feature of the Ennominae ground plan, and therefore its consistent absence in a tribe is a derived feature, or whether it can be treated as a synapomorphy for the tribes that possess it. 6. Socii much reduced or absent. 7. Loss of setal comb. 8. Presence of strong furca. 9. Collared eggs. 10. Additional SV setae on the A6 proleg. 11. Multiple setae (16) on the proleg on A6. 12. Columnar arrangement of egg polygonal cells, leading to the walls separating adjacent rows forming longitudinal ridges. 13. Signum absent. 14. Broad, central zone of long setae becoming denser and often longer towards the apex and ventral margin of the valve. 15. Hindwing vein M2 not tubular. 16. Signum single, often mushroom-shaped, spined or dentate.

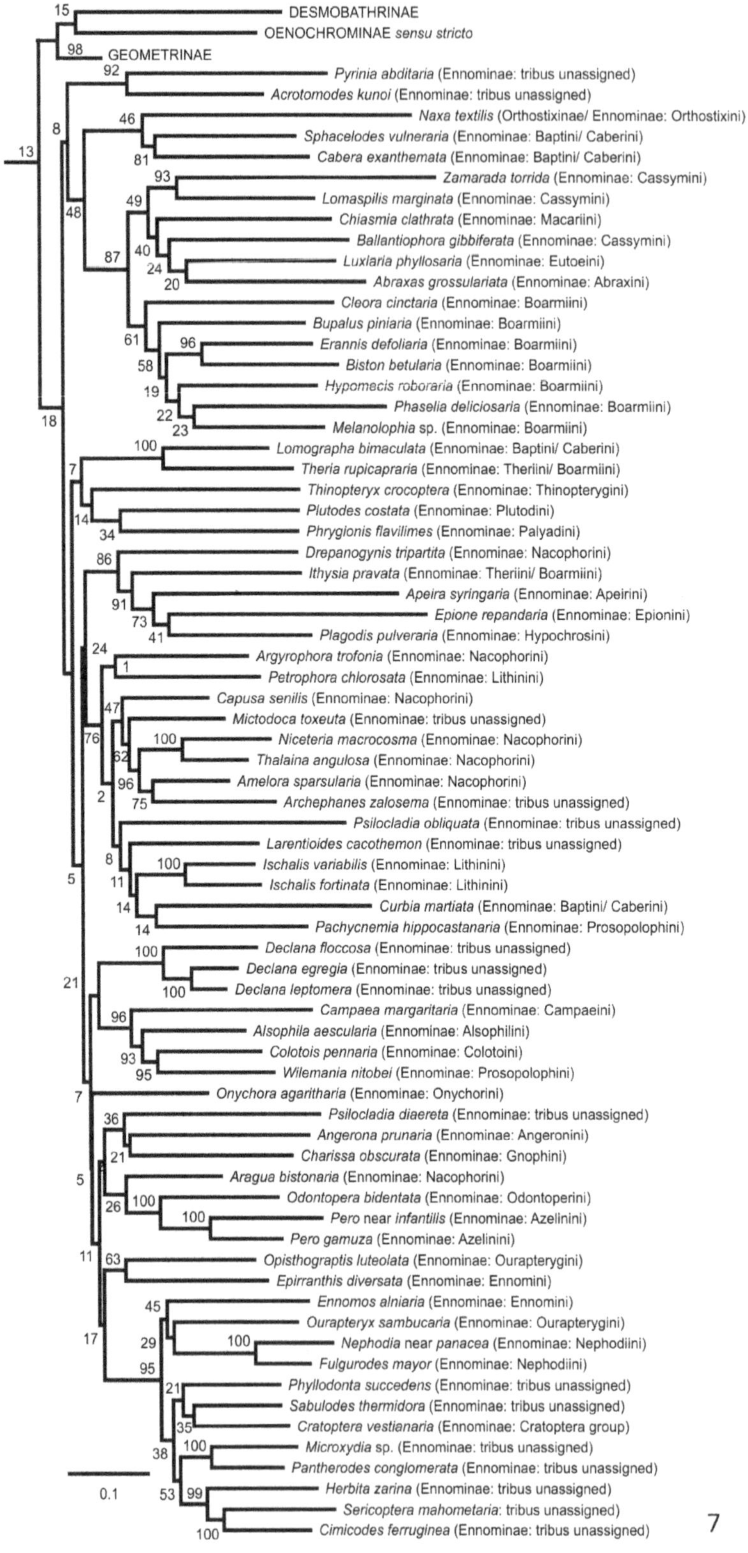
DESMOBATHRINAE
OENOCHROMINAE sensu stricto
GEOMETRINAE
Pyrinia abditaria (Ennominae: tribus unassigned)
Acrotomodes kunoi (Ennominae: tribus unassigned)
Naxa textilis (Orthostixinae/ Ennominae: Orthostixini)
Sphacelodes vulneraria (Ennominae: Baptini/ Caberini)
Cabera exanthemata (Ennominae: Baptini/ Caberini)
Zamarada torrida (Ennominae: Cassymini)
Lomaspilis marginata (Ennominae: Cassymini)
Chiasmia clathrata (Ennominae: Macariini)
Ballantiophora gibbiferata (Ennominae: Cassymini)
Luxiaria phyllosaria (Ennominae: Eutoeini)
Abraxas grossulariata (Ennominae: Abraxini)
Cleora cinctaria (Ennominae: Boarmiini)
Bupalus piniaria (Ennominae: Boarmiini)
Erannis defoliaria (Ennominae: Boarmiini)
Biston betularia (Ennominae: Boarmiini)
Hypomecis roboraria (Ennominae: Boarmiini)
Phaselia deliciosaria (Ennominae: Boarmiini)
Melanolophia sp. (Ennominae: Boarmiini)
Lomographa bimaculata (Ennominae: Baptini/ Caberini)
Theria rupicapraria (Ennominae: Theriini/ Boarmiini)
Thinopteryx crocoptera (Ennominae: Thinopterygini)
Plutodes costata (Ennominae: Plutodini)
Phrygionis flavilimes (Ennominae: Palyadini)
Drepanogynis tripartita (Ennominae: Nacophorini)
Ithysia pravata (Ennominae: Theriini/ Boarmiini)
Apeira syringaria (Ennominae: Apeirini)
Epione repandaria (Ennominae: Epionini)
Plagodis pulveraria (Ennominae: Hypochrosini)
Argyrophora trofonia (Ennominae: Nacophorini)
Petrophora chlorosata (Ennominae: Lithinini)
Capusa senilis (Ennominae: Nacophorini)
Mictodoca toxeuta (Ennominae: tribus unassigned)
Niceteria macrocosma (Ennominae: Nacophorini)
Thalaina angulosa (Ennominae: Nacophorini)
Amelora sparsularia (Ennominae: Nacophorini)
Archephanes zalosema (Ennominae: tribus unassigned)
Psilocladia obliquata (Ennominae: tribus unassigned)
Larentioides cacothemon (Ennominae: tribus unassigned)
Ischalis variabilis (Ennominae: Lithinini)
Ischalis fortinata (Ennominae: Lithinini)
Curbia martiata (Ennominae: Baptini/ Caberini)
Pachycnemia hippocastanaria (Ennominae: Prosopolophini)
Declana floccosa (Ennominae: tribus unassigned)
Declana egregia (Ennominae: tribus unassigned)
Declana leptomera (Ennominae: tribus unassigned)
Campaea margaritaria (Ennominae: Campaeini)
Alsophila aescularia (Ennominae: Alsophilini)
Colotois pennaria (Ennominae: Colotoini)
Wilemania nitobei (Ennominae: Prosopolophini)
Onychora agaritharia (Ennominae: Onychorini)
Psilocladia diaereta (Ennominae: tribus unassigned)
Angerona prunaria (Ennominae: Angeronini)
Charissa obscurata (Ennominae: Gnophini)
Aragua bistonaria (Ennominae: Nacophorini)
Odontopera bidentata (Ennominae: Odontoperini)
Pero near infantilis (Ennominae: Azelinini)
Pero gamuza (Ennominae: Azelinini)
Opisthograptis luteolata (Ennominae: Ourapterygini)
Epirranthis diversata (Ennominae: Ennomini)
Ennomos alniaria (Ennominae: Ennomini)
Ourapteryx sambucaria (Ennominae: Ourapterygini)
Nephodia near panacea (Ennominae: Nephodiini)
Fulgurodes mayor (Ennominae: Nephodiini)
Phyllodonta succedens (Ennominae: tribus unassigned)
Sabulodes thermidora (Ennominae: tribus unassigned)
Cratoptera vestianaria (Ennominae: Cratoptera group)
Microxydia sp. (Ennominae: tribus unassigned)
Pantherodes conglomerata (Ennominae: tribus unassigned)
Herbita zarina (Ennominae: tribus unassigned)
Sericoptera mahometaria: tribus unassigned)
Cimicodes ferruginea (Ennominae: tribus unassigned)
0.1

7

⇦ Text-fig. 7. Modified overview of the RAxML maximum likelihood analysis by SIHVONEN *et al.* (2011), only the Ennominae are shown. Parentheses describe tribus associations of the analysed taxa prior to the analysis. Numbers above branches indicate bootstrap values.

Classification of European Ennominae genera into tribes

The European Ennominae fauna as a whole has not previously been classified into tribes, but there is some data available on regional faunas that give an indication of classification on that level. HERBULOT (1962–1963 classified the French Ennominae fauna into 15 tribes, VIIDALEPP (1996) arranged the Ennominae species that occur in the European part of Russia into 17 tribes, KULLBERG *et al.* (2001), following VIIDALEPP (1996), arranged the much less species-diverse Finnish Ennominae fauna into 14 tribes and VIVES MORENO (1994) classified the fauna of Iberian Peninsula into 15 tribes. MINET & SCOBLE (1999) recognised in the global context about 20 Ennominae tribal names with some substantiation, about 10 of which occur in Europe. HOLLOWAY (1994) surveyed the literature very extensively and came up with 63 family-group names for possible usage within the Ennominae, including synonyms. Forum HERBULOT (2014), which is an international scientific community with research focused on Geometridae, has attempted to create a synthesis of all available information. According to FORUM HERBULOT (2014), there are 27 valid tribes in the Ennominae.

Below we have tried to diagnose the Ennominae tribes that occur in Europe and to list their constituent genera to give an overview of the current classification. It is apparent that much more research is needed to have the relationships clarified and to understand which are their constituent genera. Morphological differences between the tribes are often minimal, potential synapomorphies are often general and of limited diagnostic value. Therefore, we have started to wonder if the current, rather detailed tribus classification bears historical burden that has not been critically questioned. An indication of that was provided by SIHVONEN *et al.* (2011), who performed an extensive molecular analysis to examine the geometrid affinities in a global context. Phylogenetic relationships of numerous Ennominae tribes and their constituent taxa, as traditionally conceived, were questioned (Text-fig. 7). Thus, it seems apparent that some tribus level synonymies may be necessary.

Abraxini WARREN, 1893. Putatively about 190 species, most in East Asia (HOLLOWAY 1994), one genus in the Nearctic (*Ligdia* GUENÉE, 1858) (FERGUSON 2008). Potential synapomorphies: Valva divided: its dorsal arm digitate, apex weakly setose: its ventral arm with wide apex, apex sometimes dentate. Coremata absent. Male 8th sternite unmodified. Fovea in male forewing absent. Chaetosemata normal. Thorax yellowish. Abdomen with complex array of dark maculae. Wings white, bases often yellow (HOLLOWAY 1994). Extent of Abraxini is unclear (HOLLOWAY 1994, CHOI *et al.* (1998). Abraxini, Macariini, Cassymini and non-European Eutoeini are structurally very similar. BELJAEV (2006a, 2006b) has proposed a broad concept of Macariini, including the above-mentioned four tribes. These four tribes have a setal comb on the 3rd male abdominal sternite (but absent in *A. grossulariata* (LINNAEUS), our observation) and the socii are absent (HOLLOWAY 1994). SIHVONEN *et al.* (2011) analysed the Abraxini, Macariini, Eutoeini and Cassymini in a broad molecular phylogenetic framework (Text-fig. 7). The four tribes grouped together in a single clade, but the material is too limited to draw conclusions on how these species-rich taxa may be related. Selected literature: INOUE (1972, 1984, 1994), HOLLOWAY (1994), HOLLOWAY *et al.* (2001), PITKIN (2002), STÜNING & HAUSMANN (2002), FERGUSON (2008). Included European genera: *Abraxas, Ligdia.*

Apeirini STEKOLNIKOV & KUZNETZOV, 1982. Definition is restricted to genus *Apeira* (GISTL, 1848) only. Ten species in the Palaearctic region. Potential synapomorphies (homologies are difficult to interpret): Uncus absent. Gnathos arms separate, massive. Valvae directed downwards. Two separate sclerotisations ventrad of juxta. Sclerotised sterigma reaches papillae anales (our observations). Pupa cremaster with two apical and numerous (about 30) minute, curved setae (PATOČKA & TURČANI 2005). Potentially related taxa include *Epione* (Epionini) and *Plagodis* (potentially Hypochrosini) (SIHVONEN *et al.* 2011). Included European genus: *Apeira*.

Baptini FORBES, 1948. The geographical extent and number of included species in the tribe is unclear. Neotropical taxa are reviewed by PITKIN (2002), Nearctic taxa (only genus *Lomographa* HÜBNER, 1825) by MCGUFFIN (1981) and the fauna of Borneo by HOLLOWAY (1994). No convincing synapomorphies for the tribe have been identified (HOLLOWAY *et al.* 2001, PITKIN 2002), except potentially the swelling of the base of vein Sc in the male hindwing and the gnathos being weakly developed or absent in the male genitalia as in the Caberini. Tribes Caberini (type genus is *Cabera* TREITSCHKE, 1825) and Baptini FORBES, 1948 (type genus is *Lomographa* HÜBNER, 1825) have been considered to be closely related (ABRAHAM *et al.* 2001), and Baptini was proposed to be a junior synonym of the Caberini by PITKIN (2002) and later treated as a subtribe of Caberini (FORUM HERBULOT 2014). The close relationship was also supported by a DNA analysis by YAMAMOTO and SOTA (2007) but opposed in another DNA study that included considerably more taxa and genes (SIHVONEN *et al.* 2011). In the latter *Lomographa* grouped together with *Theria* HÜBNER, 1825 (Theriini) and these two taxa were not related to the Caberini. Included European genus: *Lomographa*.

Caberini DUPONCHEL, 1845. According to FORUM HERBULOT (2014), the concept includes also Erastriini HERRICH-SCHÄFFER, 1845, Brotini GROTE, 1882, Deiliniini WARREN, 1894, Catopyrrhini WARREN, 1894 and Sphacelodini FORBES, 1948. The geographical extent and number of included species in the tribe is unclear. Neotropical taxa are reviewed by PITKIN (2002), Nearctic taxa by MCGUFFIN (1981), and the fauna of Borneo by HOLLOWAY (1994). VIIDALEPP (1996), who considered European material also, included in the tribe ten genera and 22 species in the Palaearctic region. No convincing synapomorphies for the tribe have been identified (HOLLOWAY *et al.* 2001, PITKIN 2002), except potentially the swelling of the base of vein Sc in the male hindwing and the gnathos being weakly developed or absent in the male genitalia as in the Baptini. Tribes Caberini (type genus is *Cabera* TREITSCHKE, 1825) and Baptini FORBES, 1948 (type genus is *Lomographa* HÜBNER, 1825) have been considered to be closely related (ABRAHAM *et al.* 2001), and Baptini was proposed to be a junior synonym of the Caberini by PITKIN (2002) and later treated as a subtribe of Caberini (FORUM HERBULOT 2014). The close relationship was also supported by a DNA analysis by YAMAMOTO and SOTA (2007) but opposed in another DNA study that included considerably more taxa and genes (SIHVONEN *et al.* 2011). In the latter *Cabera* (type genus of the Caberini) and the mainly Neotropical *Sphacelodes* GUENÉE 1858 (type genus of Sphacelodini FORBES, 1948, synonymised with the Caberini in PITKIN (2002)) were recoved as sister taxa, whereas *Lomographa* grouped together with *Theria* HÜBNER, 1825 (Theriini) and these two taxa were not related to the Caberini. Included European genus: *Cabera*.

Campaeini FORBES, 1948. Metrocampini TUTT, 1896 is an older name but it is based on *Metrocampa* BRUAND, 1846, a junior objective synonym of *Campaea* LAMARCK, 1816 (FLETCHER 1979, SCOBLE 1999). A small tribe in the Palaearctic, Oriental and Nearctic regions, containing putatively four genera and less than ten species (VIVES MORENO 1994, VIIDALEPP 1996, LERAUT 1997, SCOBLE 1999). PITKIN (2002) excluded the Neotropical taxa from the Campaeini, those were earlier placed in genus *Hylaea* HÜBNER, 1822. Potential synapomorphies (our observations): Wings green. Subapical spine in the ventral margin of valva (potentially unique). Dorsal lobe/extension at the base of valva. Narrow uncus. Absence of socii. Absence of gnathos. Elongated saccus. Apophyses anteriores absent or reduced in length. Corpus bursae spinose. Included European genera: *Campaea, Gerinia, Hylaea, Adalbertia, Pungeleria.*

Cassymini HOLLOWAY, 1994. Total number of species is uncertain. The tribe is most species-rich in the Oriental tropics, but also numerous species occur in Africa (genus *Zamarada* MOORE, 1887 with over 200 species) (FLETCHER 1974, HOLLOWAY 1994) and in the Neotropics (24 genera and approximately 70 species) (PITKIN 2002). Potential synapomorphies: Valva divided: its dorsal arm long, slender, sinuous or angled, apex usually with a spine; its ventral arm with narrow apex, with two weak projections. Coremata well developed (absent in European species). Male 8th sternite unmodified. Fovea in male forewing present in some genera. Chaetosemata normal. Pupa with bifid cremaster (HOLLOWAY 1994). Abraxini, Macariini, Cassymini and non-European Eutoeini are structurally very similar. BELJAEV (2006a, 2006b) has proposed a broad concept of Macariini, including the above-mentioned four tribes. SIHVONEN *et al.* (2011) analysed the Abraxini, Macariini, Eutoeini and Cassymini in a broad molecular phylogenetic framework (Text-fig. 7). The four tribes grouped together in a single clade, but the material is too limited to draw conclusions on how these species-rich taxa may be related. Selected literature: HOLLOWAY (1994), SCOBLE & KRÜGER (2002), FERGUSON (2008). Included European genera: *Lomaspilis, Stegania.*

Colotoini WEHRLI, 1940. Extent of the tribe, and thus the number of species is unclear, see for instance VIIDALEPP (1996). Colotoini is potentially restricted to the Palaearctic region, and the genera *Colotois* HÜBNER, 1823 and *Dasycorsa* PROUT, 1915 have three and one species, respectively. Potential synapomorphies: Posterior margin of frons with lateral setae (Text-figs 8–9). Hindtibia spurs of unequal length. Uncus often bilobed, asymmetrical. Gnathos absent. Valva divided, dorsal arm setose. Juxta narrow at base, posterior margin concave. Male 8th sternite unmodified. Ductus bursae weakly funnel-shaped.

In molecular phylogenetic frameworks *Colotois pennaria* (LINNAEUS, 1761) grouped together with East-Asian *Descoreba simplex* BUTLER, 1878 and to a larger node also containing *Wilemania nitobei* NITOBE, 1907 (YAMAMOTO & SOTA 2007), and with *W. nitobei* (SIHVONEN *et al.* 2011), type species of the monotypic tribe Wilemanini. The morphology does not unambiguously support inclusion of *Wilemania* PROUT, 1928 in the Colotoini, it is dissimilar, and it has a number of unique features, most notably the stout setae on abdominal tergites 2–4 in both sexes. *Colotois* HÜBNER and *Dasycorsa* PROUT are morphologically similar to *Compsoptera* BLANCHARD (type genus of Prosopolophinil). The potential relationships of the Apochimini, Wilemanini and the Prosopolophini in relation to the Colotoini should be studied in an analytical context. Included European genera: *Colotois*, *Dasycorsa.*

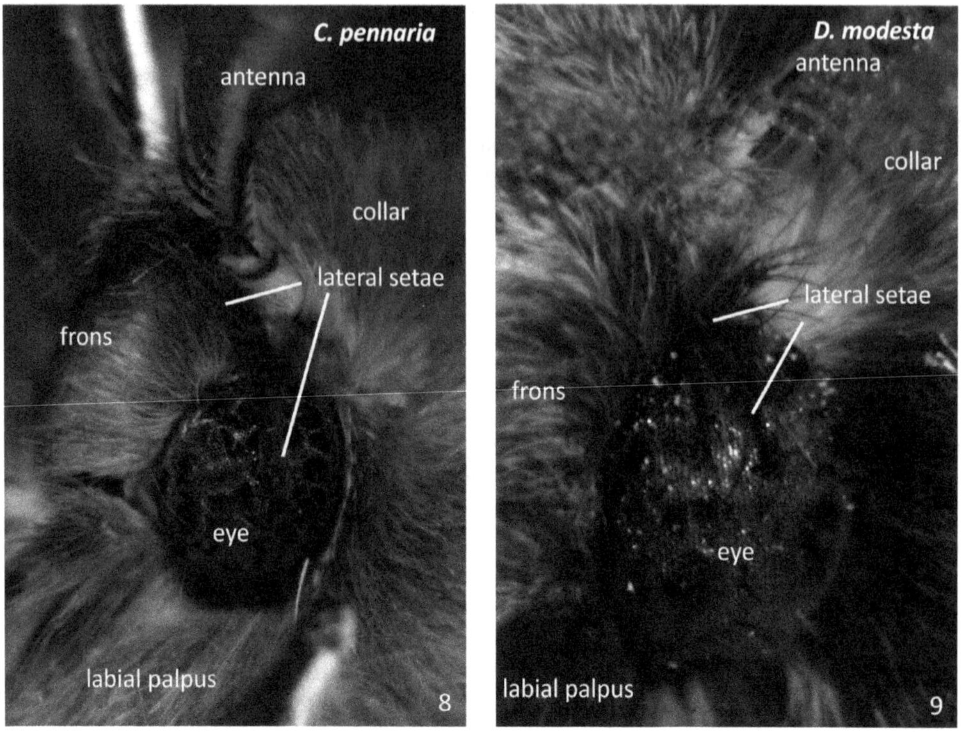

Text-figs 8–9. *Colotois pennaria* (LINNAEUS, 1761) and *Dasycorsa modesta* (STAUDINGER, 1879), showing the posterior margin of the frons with lateral setae that curve over the compound eye. Text-fig 8. *C. pennaria* male (Finland, 10–19.ix.1995). Text-fig 9. *D. modesta* male (Romania, 26.iv.1964). The structure is found in the Colotoini genera *Colotois* HÜBNER, 1823 and *Dasycorsa* PROUT, 1915, and also in the (related?) *Apochima flabellaria* (HEEGER, 1838) (type genus of the Apochimini) and the non-European *Wilemania nitobei* NITOBE, 1907 (type genus of the Wilemaniini). *A. flabellaria* is treated in the Geometrid Moths of Europe, vol. 6, and *W. nitobei* occurs in East-Asia.

Ennomini DUPONCHEL, 1845. The tribe, as defined by BELJAEV (2008a, 2008b), includes also Ourapterygini and altogether 148 genera. 141 of those occur in the Nearctic region, seven of those occur in the Palaearctic region, and in Europe only two genera are included, *Ennomos* TREITSCHKE and *Ourapteryx* LEACH. Potential synapomorphies: Vinculum with a membranous area on its ventral side. Paired muscle m3 attached distally to the medial invagination of the basal portion of the juxta (BELJAEV 2008a, 2008b). The first character is not unique because it is absent in several *Ennomos* species, and present elsewhere in the Ennominae, for instance in *Selenia*, and potentially also in *Cepphis*, *Epione* and *Pseudopanthera* (our observation). In previous works the European Ennomini have had a wider compass, with four (LERAUT 1997), six (HAUSMANN, MIRONOV & VIIDALEPP 2011), 14 (VIVES MORENO 1994) or 19 genera (VIIDALEPP 1996). Prosopolophini has been suggested to be sister taxon to the Ennomini (BELJAEV 2008a). In a molecular phylogenetic analysis of the world Geometridae (SIHVONEN *et al.* 2011) *Ennomos* (Ennomini) and *Ourapteryx* (Ourapterygini) grouped next to each other but the Prosopolophini, represented in the molecular analysis by *Pachycnemia* and non-European *Wilemania*, were not related to the Ennomini. BELJAEV's work (2008a) was not analytical, the mentioned characters were weighted subjectively, and he did not provide information where the excluded, former Ennomini genera should be classified. It is possible that the Ennomini, as traditionally delimited, has included a series of remotely related genera that have been united by superficial habitus rather than diagnosable synapomorphies. In a molecular, phylogenetic analysis of limited European taxa, ÕUNAP *et al.* (2011) concluded that *Ennomos*, *Opisthograptis* and *Crocallis* grouped together but *Selenia*, a genus frequently classified in the Ennomini, was not included in the same cluster of taxa. Further research is needed. Included European genera: *Ennomos*. Ennomini of uncertain association: *Selenia, Artiora, Odontopera, Crocallis, Ourapteryx.*

Epionini BRUAND, 1846. Species diversity, distribution and number of included genera, other than *Epione*, are uncertain. Scardamiini WARREN, 1894, with the type genus *Scardamia* GUENÉE, 1858, and Hypochrosiini GUENÉE, 1857, with the type genus *Hypochrosis* GUENÉE, 1858, as diagnosed and illustrated in HOLLOWAY (1994) appear closely related. Potential synapomorphies: Consistent absence of setal comb on male 3rd sternite, perhaps associated with loss of tibial hair pencil. Looping of the vinculum to support lateral coremata. Presence of strong furca. The proposed synapomorphies are very general. *Cepphis* and *Epione* are structurally similar, perhaps indicating close relationship. References: HOLLOWAY (1994). Included European genera: *Epione, Therapis*. Epionini of uncertain association: *Cepphis, Pseudopanthera, Eilicrinia.*

Gnophini DUPONCHEL, 1845. Species diversity, distribution and number of included genera are uncertain. Forum HERBULOT (2014) lists Dasydini DUPONCHEL, 1845 and Sionini DUPONCHEL, 1845 as junior synonyms. Based on the morphology of examined European taxa it seems apparent to us that the Angeronini FORBES, 1948, which also includes the Aspitatini DUPONCHEL, 1845 (see MCGUFFIN 1981, HOLLOWAY 1994) could potentially be brought into synonymy with the Gnophini. In a DNA-based maximum likelihood analysis *Angerona* (type genus of Angeronini) and *Charissa* (Gnophini) were recovered as sister taxa (SIHVONEN *et al.* 2011). A similar result was obtained by YAMAMOTO & SOTA (2007), where *Angerona* and *Chariaspilates* were recovered as sister taxa. The mainly Nearctic Angeronini genera *Euchlaena* HÜBNER, *Lytrosis* HULST and *Xantholype* WARREN are also similar (illustrated for instance in MCGUFFIN 1981). VIIDALEPP *et al.* (2007) proposed the following as potential Gnophini synapormorphies: Uncus short and birdhead-shaped. Absence of fovea on forewing. Absence of setae on male abdominal sternite A3. Costal projection on valva usually present, bearing a spine or group of spines. Juxta is often H-shaped with short basal and longer distal appendages. Pupal cremaster with setae D1, SD1, L1, D2 present and setae D2 separated. Included European genera in Geometrid Moths of Europe, vol. 5 (numerous other taxa are included in vol. 6): *Hypoxystis, Angerona, Siona, Chariaspilates, Aspitates, Dyscia, Perconia.*

Hypochrosini GUENÉE, 1858. About 270 species, classified into about 15 genera. The geographical extent is unclear but it contains taxa at least in the Nearctic (as Anagogini in MCGUFFIN 1987), the Palaearctic and in the Oriental regions (HOLLOWAY 1994). Most taxa occur in the Oriental region. Potential synapomorphies: Furca originates from pocket between valves. Other proposed synapomorphies are general: Furca often asymmetrical. Saccus not elongated. HOLLOWAY (1994) proposed that the Anagogini FORBES, 1948 is a junior synonym of the Hypochrosini GUENÉE, 1858. Type species of the Anagogini is *Phalaena pulveraria* (LINNAEUS, 1758) from Europe and type of the Hypochrosini is *Hypochrosis sternaria* GUENÉE, 1858 from India. Morphological differences between the Hypochrosini,

Ourapterygini and Lithinini, as diagnosed in HOLLOWAY (1994), and the Epionini, are small. HOLLOWAY *et al.* (2001) proposed that probably the furca in the Hypochrosini and in the Ourapterygini are not homologous. In a molecular, phylogenetic analysis of limited European taxa only, ÕUNAP *et al.* (2011) concluded that *Cepphis*, *Selenia*, *Plagodis* and *Pseudopanthera* grouped together. In a molecular, phylogenetic analysis of the world Geometridae, the Hypochrosini, Ourapterygini and Lithinini did not group together (SIHVONEN *et al.* 2011). Included European genera: *Heterolocha*. Hypochrosini of uncertain association: *Plagodis*.

Lithinini FORBES, 1948. About 100–150 species, classified into about 30 genera, cosmopolitan distribution (WEINTRAUB & SCOBLE 2004, PARRA & HERNANDEZ 2010). Potential synapomorphies: Fern-feeding common. 6th abdominal segment of larvae with six or more SV setae (characteristally Ennominae have three or four setae). One or more of segments A6–A8 with extremely long setae L1. Other proposed synapomorphic characters are rather general: discal spot present, postmedial band or row of spots present on veins on ventral surface of hindwings, costa of valvae usually curved upwards to tegumen, transtilla triangular, anellus with pair of setose processes, furca usually with apical spine or cluster of setae, vesica with small number of large cornuti, corpus bursae long, signum with elongate ridge or sclerotised strip (HOLLOWAY 1994, HOLLOWAY *et al.* 2001, WEINTRAUB & SCOBLE 2004). PARRA & FERNANDEZ (2010) have analysed synapomorphic characters for the Neotropical Lithinini genera. Commonly the larvae drop or jump off the host plant when disturbed (MCGUFFIN 1987, WEINTRAUB & SCOBLE 2004). We have studied morphology of the Lithinini extensively from a global perspective, but not by employing analytical approach, and confirm that the above-mentioned morphological features are rather diagnostic for the group, yet the differences between the Nacophorini and the Lithinini are small, both tribes apparently being part of a larger complex (see SIHVONEN *et al.* 2015). Particularly the triangular transtilla, the postero-lateral processes in the anellus at the end of the juxta, which often are slender and terminate in a cluster of spines or cristae, are typical. The latter are often generally termed furca, but PITKIN (2002) proposed a neutral term anellar processes to be used, and restricted the term furca to the Ourapterygini. In the female genitalia the long, elongated corpus bursae, signum with horizontal ridge and weakly dentate margins, and anterior end of the corpus bursae with appendix bursae separated by a constriction, are diagnostic. *Pachycnemia* STEPHENS, 1829 agrees morphologically with the listed characters, and the close association is supported by the molecular data also (see below). *Pachycnemia* STEPHENS, 1829 is the type genus of the Pachycnemiini KIRBY, 1903. Thus, if the Lithinini and Pachycnemiini were brought into synonymy, the latter has seniority. We do not propose synonymisation until a thorough analytical study on the subject has been carried out, because potentially other tribus-level names are involved (see text under *Pachycnemia*) and also SIHVONEN *et al.* (2015). Based on the literature, the systematic position of the Lithinini is unclear. RINDGE (1986) proposed the Lacariini ORFILA & SCHAJOVSKOY, 1959, as a junior subjective synonym of the Lithinini. HOLLOWAY (1997), based on morphology, proposed that the Lithinini and Caberini may be closely related. YOUNG (2006, 2008) concluded that the Australian Nacophorini and the Lithinini have paired processes in the anellus in the male genitalia and the molecular evidence supports the unification of these two tribes. The Lithinini was represented by *Idiodes* GUENÉE, 1858 that is speciose in Africa, the Oriental region and Australia. Based on our morphological survey of numerous taxa in the Lithinini and the Nacophorini, we agree with the conclusions of YOUNG (2006, 2008). SIHVONEN *et al.* (2011), based on DNA, concluded that the Caberini was not related. Instead, the Lithinini and Nacophorini and their constituent taxa formed a group whose taxa were intermingled. Within that group, the New Zealand *Ischalis* WALKER, 1863 and the Palaearctic *Petrophora* HÜBNER, 1811, did not group together (Text-fig 7). Selected literature: RINDGE (1986), MCGUFFIN (1987), HOLLOWAY (1994), WEINTRAUB & SCOBLE (2004), PARRA & HERNANDEZ (2010). Included European genus: *Petrophora*. Lithinini of uncertain association: *Pachycnemia, Perigune*.

Macariini GUENÉE, 1858. About 700 species, the tribe as a whole has global distribution (SCOBLE & KRÜGER 2002). KRÜGER (2001) noted that the Afrotropical region today holds the highest number of basal members of the clade suggesting a late Gondwanaland origin of the tribe in Africa. Only 15 described species are listed for Australia (MCQUILLAN & EDWARDS 1996). No single character defines the tribe, but diagnostic features include one or more of the following (SCOBLE & KRÜGER 2002): Chaetosemata extend across head, either completely or partly. Uncus with pair of enlarged setae (horns) at or near apex. Valva divided; its dorsal arm rather straight, widest at apex, apex setose; its ventral arm with round or elongated apex. Coremata absent. Male 8th sternite often modified. Fovea in male forewing present. Abraxini, Macariini, Cassymini and non-European Eutoeini are structurally very similar. BELJAEV (2006a, 2006b) has proposed a broad concept of Macariini, including the above-mentioned four tribes. SIHVONEN *et al.* (2011) analysed the Abraxini, Macariini, Eutoeini and Cassymini in a broad molecular phylogenetic framework (Text-fig 7). The four tribes grouped together in a single clade, but the material is too limited to draw conclusions how these species-rich taxa may be related. KRÜGER (2001) recognised two groups within the Macariini: The *Platypepla* group of genera and other Macariini. Both are present in Europe, but only two small genera (*Acanthovalva, Narraga*) fall within the *Platypepla* group. Larvae are frequently encountered on legumes (SCOBLE & KRÜGER 2002), deciduous trees and conifers (SEPPÄNEN 1970, FERGUSON 2008). Pupal cremaster well-developed and typically bifurcate with truncated or pointed tips frequently of differing length (KRÜGER 2001). Selected literature: KRÜGER (2001), SCOBLE & KRÜGER (2002). Included European genera: *Acanthovalva, Narraga, Heliomata, Isturgia, Itame, Neognopharmia, Macaria, Chiasmia, Digrammia, Rhoptria, Ortaliella*.

Prosopolophini WARREN, 1894. Palaearctic tribe, currently with one genus and three species only. Potential synapomorphies: Male antennae bipectinate, pectinations very long. Collar with dorsally raised setae, giving iroquois-like appearance (Text-fig 145). Hindleg tibia of both sexes with 2+2 spurs, which are of unequal length (outer spur shorter). Gnathos absent. Valva divided, dorsal arm setose. Juxta sclerotised plate, dorsal margin often bilobed. Vesica with 'backwards' directed cornuti. Signum roundish, ventral margin dentate.

Colotois HÜBNER (type genus of Colotoini) and *Dasycorsa* PROUT (Colotoini) are morphologically similar to *Compsoptera* BLANCHARD (type genus of Prosopolophinil). The potential relationships should be studied in an analytical context.

FORUM HERBULOT (2014) suggests that Pachycnemiini KIRBY, 1903 is a junior synonym of Prosopolophini. We have examined *Pachycnemia* STEPHENS and it is morphologically dissimilar to *Compsoptera* BLANCHARD (type genus of Prosopolophini). The dissimilarity is also supported by the DNA, in a maximum likelihood analysis it grouped to a cluster of species consisting mostly, but not exclusively, of the Lithinini (SIHVONEN *et al.* 2011). We place *Pachycnemia* in a group of genera with uncertain tribus association. According to BELJAEV (2008a) the Ennomini and the Prosolophini may be sister taxa. As explained above, the Ennominae as a whole needs thorough, analytical higher-level revision. Such analysis could potentially clarify the relationships of the Prosopolophini to other taxa. Other genera associated with the Colotoini and Prosopolophini in the literature (for instance VIIDALEPP 1996, BELJAEV 2008a), must be checked against *Compsoptera jourdanaria* (Serres) and allied taxa. Included European genera: *Compsoptera*.

Theriini HERBULOT, 1963. The concept also includes Cheimatobiini TUTT, 1896, which is a senior synonym, being based on *Cheimatobia* STEPHENS, 1829. We also include in the Theriini the Aleucini DJAKONOV, 1936: 484, new synonym (DJAKONOV 1936). The Aleucini is also a senior synonym, because it is based on the Palaearctic *Aleucis* GUENÉE, 1845. Aleucini has been without usage in the recent literature, apart from a mention (as Aleucini) in BELJAEV (2008b). Therefore we propose not to apply the Aleucini, following the same reasoning as in the Cheimatobiini by FORUM HERBULOT (2014), as Theriini is largely accepted (see also HOLLOWAY 1994). *Aleucis* and *Theria* HÜBNER, 1825 are structurally very similar, see the adult and genitalia plates. Theriini is a small tribe, restricted to the Palaearctic region. VIIDALEPP (1996) included in it five genera and six species. Of those, *Ithysia* HÜBNER, 1825 is potentially misplaced (see SIHVONEN *et al.* 2011). Potential synapomorphies, in addition to the adult activity in winter months and early spring (our observations): Gnathos vestigial. Juxta granulate (laterally). Papillae anales with scoop-shaped setae (floricomous ovipositor) (Text-figs 118–121). Apophyses short, stout. Female tympanal organs very small. Pupa cremaster with two stout, straight setae (PATOČKA & TURČANI 2005). Included European genera: *Aleucis, Theria*.

Genera of uncertain tribus association, See remarks under each genus: *Lignyoptera, Opisthograptis, Toulgoetia, Enanthyperythra, Eumera, Asovia, Lhommeia, Onychora*.

Against this background, which shows that there is no consensus in the classification, we have made the decision to arrange the higher taxa and species within the genera largely, but not solely, according to conventional and practical aspects. The order therefore does not necessarily reflect the phylogenetic relationships of the taxa. As a basis for the arrangement of tribes in this monograph has been the *Distributional Checklist of European Geometridae* by MÜLLER (1996). This work is complemented by the works of HOLLOWAY (1994, 1997), in which he diagnoses a large number of Ennominae tribes and discuss their relationships (Text-fig 6). Naturally, we have also consulted carefully major revisionary works such as those by MCGUFFIN (1977, 1981, 1987), VIIDALEPP (1996), SCOBLE & KRÜGER (2002), PITKIN (2002), BELJAEV (2006a, 2006b), FERGUSON (2008), SIHVONEN *et al.* (2011, whose Ennominae results shown are shown in Text-fig 7), to name a few examples, and have tried to apply the knowledge from those works into present context.

Methodological notes

In this book series (see HAUSMANN 2001) terminology of the wing pattern (Text-fig 10) and wing venation (Text-figs 11, 12) largely follows that of MCGUFFIN (1977). In the Ennominae the wings often bear only two transverse lines, therefore when there is only one transverse line between the basal and medial area, it is termed antemedial line, and when there is only one transverse line between the medial area and terminal area, it is called postmedial line.

The specimens shown in the adult plates were photographed individually, removed from the background and edited in Adobe Photoshop (version 11.0). The single images were compiled into plates with Photoshop and drop shadows were added to final plates.

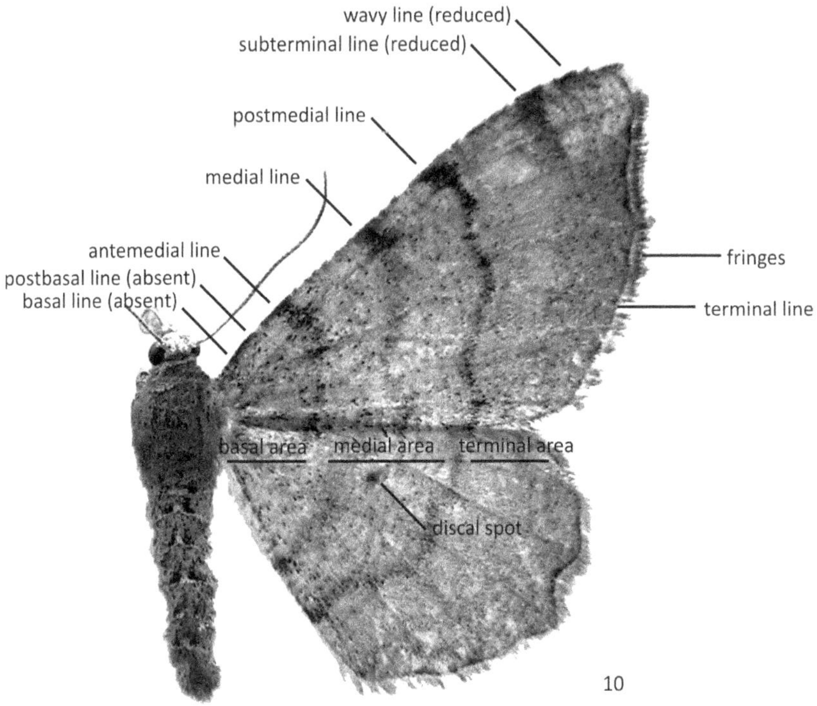

Text-fig 10. Terminology of wing pattern, illustrated on *Macaria liturata* (CLERCK, 1759) male (Finland, 2.vii.1994).

The genitalia and abdomen were prepared following the methods described by HARDWICK (1950). The male aedeagus is shown both as vesica uneverted, to allow comparison with older literature, and as vesica everted. The vesica of species with small aedeagus was everted via the caecum that was cut open by placing the aedeagus inside a hypodermic syringe (SIHVONEN 2001), or in species with large aedeagus, such as *Ennomos* species, by following the method described by LAFONTAINE & MIKKOLA (1998). A small incision was cut in the female corpus bursae before it was mounted on a slide in Euparal to avoid the corpus bursae from shrinking. Terminology for the male genitalia (see the first male genitalia plate) and female genitalia (see the first female genitalia plate) follows KLOTS (1970) and KRISTENSEN (2003), and, in cases where homologies of structures were difficult to identify, descriptive terms were used instead. All taxa shown in the genitalia plates were mounted in euparal and majority of those were photographed in ventral view with a Leica DM1000 microscope and integrated Leica DF295 digital camera. Numerous taxa were photographed in 2–6 images of different depth of focus and combined into single images using image-stacking software CombineZP (HADLEY 2012). About 800 slides of various structures were prepared for this book. Original images were cleaned and edited with Adobe Photoshop (version 11.0) and final plates were compiled with CorelDraw (version X5).

For the DNA analyses, one or two legs were removed from each dried specimen and stored in an individual tube, which in most cases contained ethanol. DNA extraction, amplification, and sequencing of the barcode region of the mitochondrial cytochrome c oxidase I

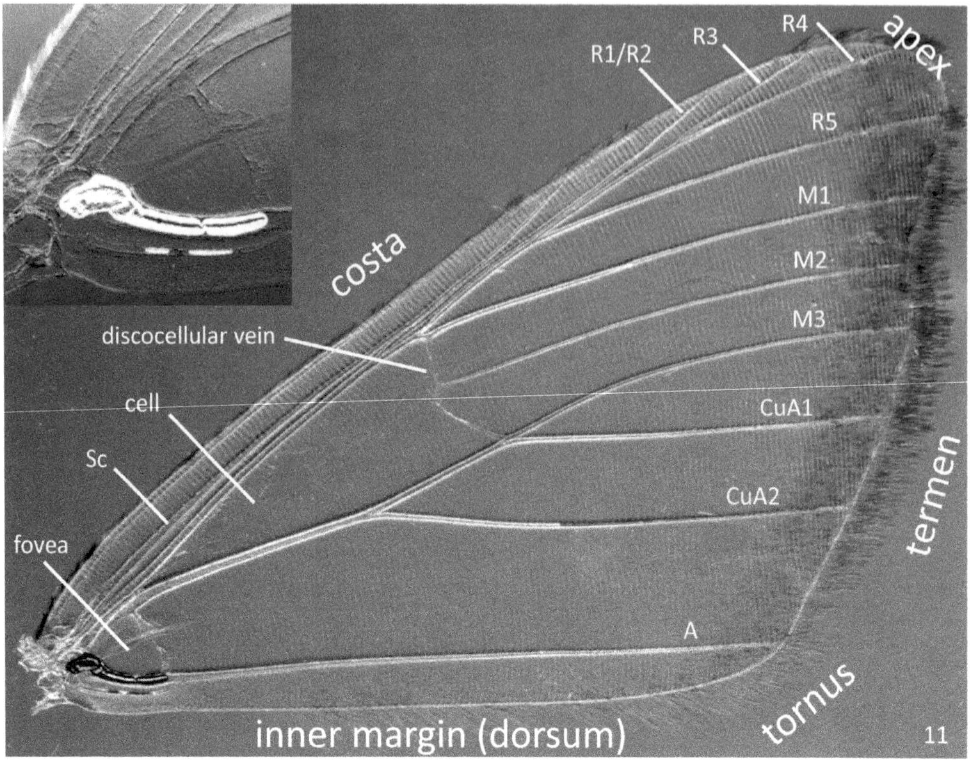

Text-fig 11. Terminology of forewing venation, with fovea enlarged, illustrated on *Macaria liturata* (CLERCK, 1759) male (Finland: Savonlinna, 20–25.ix.1993, slide PS1117, coll. P. SIHVONEN). The fovea, which is a blister-like small area at the base of the forewing upperside, occurs in the males of many Ennominae, it is only rarely seen in females. Abbreviations of the forewing veins are: Sc=subcostalis, R1–R5=radials, M1–M3=medial veins, CuA1–CuA2=cubitan veins and A=anal vein. In many species R1 of forewing turns back into R2 between the cell and costa, forming a small cell called the areole or accessory cell (the figured species lacks this structure). Some Ennominae species have up to three areoles.

(COI) gene (658 base pairs) were carried out in the Canadian Centre for DNA Barcoding, Ontario, Canada, using standard high throughput protocols (IVANOVA *et al.* 2006, DEWAARD *et al.* 2008), those are described at CCDB (2013). Sequence divergence within and between species was calculated using the Kimura 2-parameter model (KIMURA 1980) and the neighbour-joining algorithm (SAITOU & NEI 1987), as implemented in BOLD (http://www.boldsystems.org/). The Barcode Index Number (BIN), an algorithm of BOLD database (RATNASINGHAM & HEBERT 2007), was used to verify (sub-)species identifications since the clusters have been shown to perform high concordance with (sub-)species when combined with a morphological analysis (RATNASINGHAM & HEBERT 2013).

Nomenclature and citations of type specimens and type localities for genus and species entries are given in the same format as in, and largely following, SCOBLE (1999). The taxa are arranged in alphabetical order, and, if the current interpretation of its taxonomic status differs from the original status, the latter is given (Text-fig. 13). Synonyms are indented.

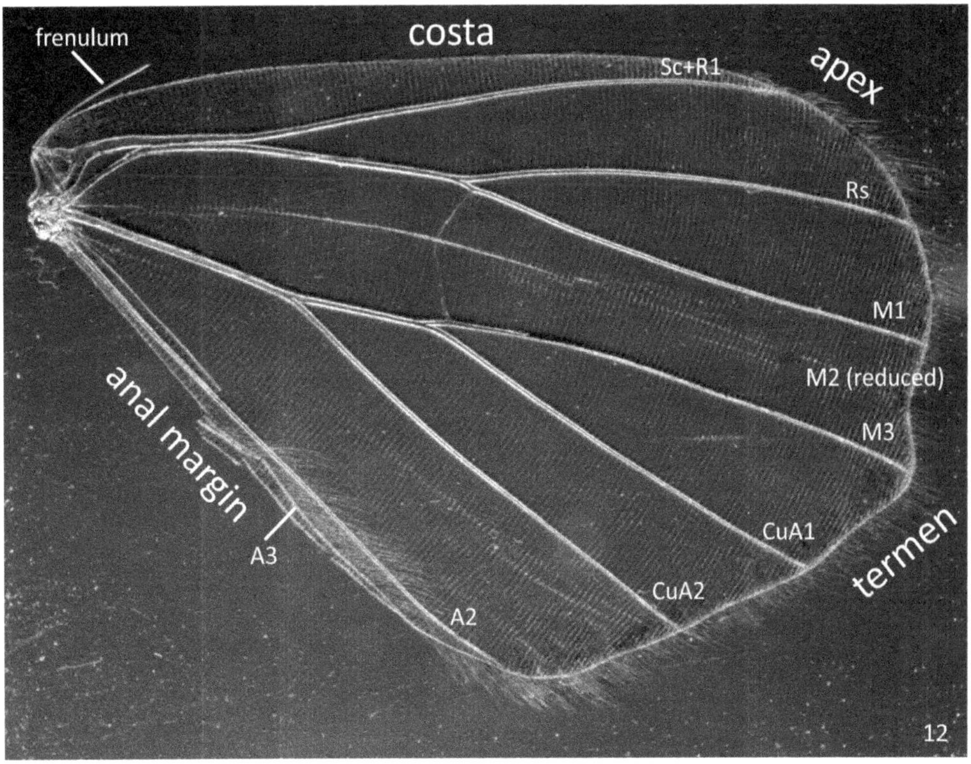

Text-fig 12. Terminology of hindwing venation, illustrated on *Macaria liturata* (CLERCK, 1759) male (Finland: Kirkkonummi, 15.vi.2008, slide PS1227, coll. P. SIHVONEN). Abbreviations of the hindwing veins are: Sc=subcostalis, R1=radial, Rs=radial sector, M1–M3=medial veins, CuA1–CuA2=cubitan veins and A=anal veins.

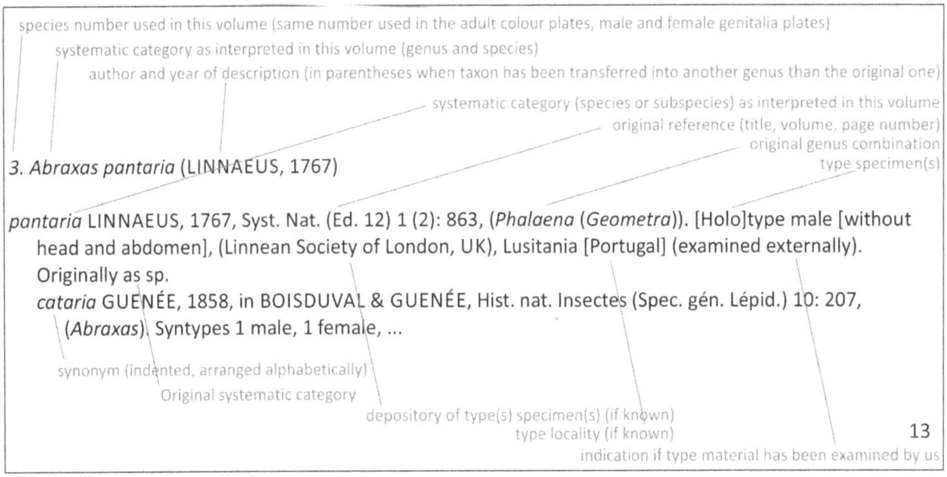

Text-fig. 13. The format of nomenclatural citations is given in the same format as in, and largely following, SCOBLE (1999). Synonyms are indented, arranged alphabetically. If relevant, reference(s) for additional information are given at the end of each entry.

No extensive survey was carried out to find and list all unavailable names (see ICZN 1999, Articles 10.2, 13.1). A large number of intrasubspecific names have been introduced in the European Ennominae, therefore only a selection of the names described in the comprehensive publications are listed, for instance (Prout 1912–1916, Wehrli 1939–1954, Lempke 1970, Fletcher 1979, Vives Moreno 1994, Leraut 2009, The Global Lepidoptera Names Index (Anonymous 2013)), supplemented with some unavailable names published more recently.

Distribution maps were prepared by compiling data from collection specimens. These data were supplemented by adding data from recent faunistic inventories of United Kingdom (Hill *et al.* 2010) Norway (Aarvik *et al.* 2000), Finland (Huldén *et al.* 2000, Kullberg *et al.* 2001), the Netherlands (Anonymous 2012), Austria (Huemer & Malicky 2009, Kerschbaum & Pöll 2010), Spain (Redondo *et al.* 2009), Italy (Flamigni *et al.* 2007) and south-western Germany (Ebert 2001). These maps were further supplemented by sending out blank maps of each species to numerous experienced lepidopterists across Europe. Data from these sources were drawn together, and final maps were sent out one more time to same lepidopterists for approval and for additional data. Distribution maps show the hypothetical resident distribution area (grey), combined with localities from which specimens have been examined (black dots, about 30 000 data points). Unless otherwise mentioned in the maps, old records are marked with open dots, when the species has probably disappeared from that locality. Doubtful records are marked with question marks.

Larval food plants and information on habitats are based on the authors' own original data, literature data, and personal communications with numerous lepidopterists. For first hand verified food plant data, we use the expression "found on" while for literature data we use "recorded on". Vertical distribution refers mainly to the analysis of collection label data, with personal communications and literature data used as an additional source.

Phenology is mainly based on collection label data, while literature data are used as an additional source. It is important to note that the flight period may vary greatly annually and geographically. Therefore single outlying records are usually omitted.

On the legends to the colour plates the label data – localities – are given in original language whilst country names are in English, and with reference to the collection where the specimen is deposited.

The colour names used in the species descriptions were taken from the Wikipedia article *List of Colors* (Anonymous 2011a). In several instances, descriptive colour names were used also, taken from familiar contexts.

The book is a joint undertaking and both authors bear mutual responsibility for its content. For practical reasons, and to avoid duplication during preparation, the following division of labour was established. Peder Skou prepared adult plates, compiled data and extrapolated the distribution areas for distribution maps and for the species accounts wrote the parts on distribution and abundance, phenology, biology, habitat and remarks where relevant. Pasi Sihvonen dissected the genitalia and prepared the genitalia plates, text-figures and tables, wrote the introduction, systematic account (including systematic position of the Ennominae within the Geometridae, monophyly of the Ennominae, Ennominae systematics, classification of European Ennominae genera into tribes), methodological notes, on species and subspecies, and for the species accounts, the parts on taxonomy, external characters and abdomen, variation, male genitalia, female genitalia, similar species and remarks where relevant.

On species and subspecies

To delimit the studied taxa, we have used the phylogenetic species concept of NIXON & WHEELER (1990), which is more applicable than the biological species concept of MAYR (1942, 1963) when working with museum material. This is true particularly in cases where information on immature stages and biology is unknown or uncertain. The morphological data was supported by molecular data from the CO1 gene, the DNA barcode, which was available for the majority of examined taxa, and data on ecology and biology.

In many instances there was no unambiguous way to interpret the data (whether phenotypical, ecological, morphological, molecular or other) to decide whether a taxon should be considered valid at species or subspecies level. The DNA barcodes were in this regard a valuable source of new information, but in some instances they did not provide a straightforward answer either. All such cases have been mentioned in the text. When making subjective taxonomic decisions (hypotheses), we used all available information, including existing knowledge on genetic variation in DNA barcodes in the European Geometridae (HAUSMANN *et al.* 2013). HAUSMANN *et al.* (2013) found a mean genetic distance of 8.8% when congeneric species were considered and the mean nearest neighbour divergence was 4.4%. Some taxonomically problematic cases in the context of this volume include, for instance:

- *Isturgia sparsaria* (HÜBNER) – *Isturgia messapiaria* (SOHN-RETHEL). We regard these as separate species on the basis of external and tentative genitalia differences and different ecological niches, DNA barcodes diverge by only 0.15%. Alternatively *I. messapiaria* could have been treated as a subspecies of *I. sparsaria* or even a junior synonym of it.
- *Neognopharmia stevenaria* (BOISDUVAL). We regard this as one species despite genetic divergence of Spanish specimens by 4.8%, but no correlating differences were observed in the external appearance nor in the genitalia. Alternatively a new taxon could have been described on the basis of DNA differences.
- *Crocallis tusciaria* subsp. *cypriaca* FISCHER. We regard this as a subspecies on the grounds of similar genitalia and allopatric occurrence, despite DNA barcodes diverging by 6.2% from the main cluster. Alternatively taxon *cypriaca* could have been considered valid at species level.
- *Crocallis auberti* OBERTHÜR. We regard this as one species despite populations from southern Spain diverging genetically from those of Sicily by 5.6%. No correlating differences were observed in the external appearance nor in the genitalia. Alternatively a new taxon could have been described on the basis of DNA differences.

It became apparent during the course of preparation of the book that many subspecies, particularly those described before the early 1940s, prior to the time when the more revisionary and comparative taxonomy arose (for instance GASTON *et al.* 1995), were described by using criteria that do not correspond to the modern understanding of the subspecies concept. According to MAYR & ASHLOCK (1991: 430), the subspecies may be defined as being "an aggregate of local populations of a species inhabiting a geographic subdivision of the range of the species and differing taxonomically from other populations of the species." Taxonomists are most likely to give subspecies names to geographically isolated and morphologically distinctive populations of a species. As noted by BLACKWELDER (1967), the subspecies cannot be known until the species limits are known, and until that particular species has been well enough studied that variants or clinal variations that might possibly be described as subspecies can be recognised. This demanding work is beyond the scope of the book, and examination of type specimens or a limited number of museum samples often cannot answer these questions. Therefore the subspecies, particularly the non-European subspecies, are largely reproduced here as they appear in the works of MÜLLER (1996) and SCOBLE (1999), unless the taxa have been revised more recently.

We estimate that a large number of the subspecies, which are considered valid among the European Ennominae at present, will turn out to be invalid with a closer study. We welcome revisions attempting to look further into this matter. We also ask our fellow lepidopterists to consider the above-mentioned guidelines of MAYR & ASHLOCK (1991) and BLACKWELDER (1967), before formally describing subspecies.

Abbreviations

BECH	Private collection of KNUD BECH, Ølsted, Denmark
BMNH	The Natural History Museum, London, United Kingdom
FIEBIG	Private collection of RALF FIEBIG, Roßleben, Germany
FLAMIGNI	Private collection of CLAUDIO FLAMIGNI, Forli, Italy
GOVI	Private collection of GUIDO GOVI, Forli, Italy
HONEY	Private collection of MARTIN HONEY, London, United Kingdom
IPPE	Estonian University of Life Sciences, Tartu, Estonia
LSSU	Laboratory of Animal Systematics and Faunistics, Samara University, Russia
MSNM	Museo Civico di Storia Naturale Milano, Italy
MÜLLER	Private collection of BERND MÜLLER, Berlin, Germany
NUPPONEN	Private collection of KARI and TIMO NUPPONEN, Espoo, Finland
POLTAVSKY	Private collection of ALEXANDER POLTAVSKY, Rostov-on-Don, Russia
RUCKDESCHEL	Private collection of WALTER RUCKDESCHEL, Übersee, Germany
SAMMUT	Private collection of PAUL SAMMUT, Rabat, Malta
H. SIHVONEN	Private collection of HARRI SIHVONEN, Mäntsälä, Finland
P. SIHVONEN	Private collection of PASI SIHVONEN, Veikkola, Finland
SMNK	Staatliches Museum für Naturkunde, Karlsruhe, Germany
SKOU	Private collection of PEDER SKOU, Vester Skerninge, Denmark
TAUTEL	Private collection of CLAUDE TAUTEL, Paris, France
ZFMK	Zoologisches Forschungsinstitut und Museum A. Koenig, Bonn, Germany
ZMH	Zoological Museum, University of Helsinki, Finland
ZISP	Zoological Institute, Russian Academy of Sciences, St Petersburg, Russia
ZOLOTUHIN	Private collection of VLADIMIR ZOLOTUKHIN, Ulyanovsk, Russia
ZMKU	Zoological Museum of Kiev, Ukraine
ZSM	Zoologische Staatssammlung München, Germany

ab.	aberration
BIN	Barcode Index Number
coll.	collection
f.	form
pers. comm.	personal communication
var.	variation

Systematic Account

Subfamily **Ennominae** DUPONCHEL, 1845

Abraxas LEACH, 1815

Abraxas LEACH, [1815] 1830, Brewster's Edinburgh Encyclopaedia 9: 134. Type species: *Phalaena grossulariata* LINNAEUS, 1758 [Europe]. Genus synonyms are after SCOBLE (1999).

Callispilus AGASSIZ, 1847. [Emendation of *Calospilos* HÜBNER].

Calospilos HÜBNER, 1825. Type species: *Phalaena ulmata* FABRICIUS, 1775. England. [Junior synonym of *Phalaena sylvata* SCOPOLI, 1763. Italy]

Chooreechillum LUCAS, 1901. Type species: *Chooreechillum distitans* LUCAS, 1901. Australia. [Junior synonym of *Potera flavimacula* WARREN, 1896. Australia].

Dextridens WEHRLI, 1934. Type species: *Abraxas sinopicaria* WEHRLI, 1934. China.

Diceratodesia WEHRLI, 1935. Type species: *Abraxas pusilla* BUTLER, 1880. India.

Empriononyx WEHRLI, 1936. Type species: *Zerene leopardina* KOLLAR, 1848. India. [Unnecessary replacement name for *Priononyx* WEHRLI, 1935]

Isostictia WEHRLI, 1934. Type species: *Abraxas picaria* MOORE, 1868. India.

Mesohypoleuca WEHRLI, 1935. Type species: *Abraxas metamorpha* WARREN, 1893. India.

Omophyseta WARREN, 1894. Type species: *Abraxas triseriaria* HERRICH-SCHÄFFER, 1850-1858. Java.

Potera MOORE, 1879. Type species: *Potera marginata* MOORE, 1879. Burma.[Junior synonym of *Abraxas triseriaria* HERRICH-SCHÄFFER, 1855]

Priononyx WEHRLI, 1935. Type species: *Zerene leopardina* KOLLAR, 1844. India. [Junior homonym of *Priononyx* DAHLBOM 1843-1845 (Hymenoptera)]

Rhabdotaedoeagus WEHRLI, 1935. Type species: *Abraxas martaria* GUENÉE, [1858]. India.

Silabraxas SWINHOE, 1900. Type species: *Abraxas lobata* HAMPSON, 1895. India. [Junior synonym of *Abraxas capitata* WARREN, 1894]

Spilota HÜBNER, 1822. Type species: *Phalaena grossulariata* LINNAEUS, 1758. [Junior homonym of *Spilota* BILLBERG, 1820 (Coleoptera) and junior objective synonym of *Abraxas* LEACH, 1815]

Spinuncus WEHRLI, 1935. Type species: *Abraxas celidota* WEHRLI, 1931. China.

Trimeresia WEHRLI, 1935. Type species: *Abraxas miranda* BUTLER, 1878. Japan.

Zerene TREITSCHKE, 1825. Type species: *Phalaena grossulariata* LINNAEUS, 1758. [Junior homonym of *Zerene* HÜBNER, 1816 (Lepidoptera: Pieridae) and junior objective synonym of *Abraxas* LEACH, 1815]

Zirene UNGER, 1856 [Emendation of *Zerene* TREITSCHKE]

UNAVAILABLE NAMES (name used in suprageneric sense): *Spilota* HÜBNER, 1818 (see FLETCHER 1979). – (included in work rejected for nomenclatural purposes): *Spilote* HÜBNER, 1806 (see FLETCHER 1979). – (incorrect original spellings): *Spilote* HÜBNER, 1822; *Spitole* HÜBNER, 1822 (see FLETCHER 1979). – (incorrect subsequent spellings): *Xerene* STEPHENS, 1829; *Lerene* HAAN, 1830; *Zerena* GUÉRIN-MÉNEVILLE, 1843 (see FLETCHER 1979).

DIVERSITY AND DISTRIBUTION: *Abraxas* contains about 190 species, mostly in the east Palaearctic and the Oriental regions, a few species in Australia (SCOBLE 1999).

EXTERNAL CHARACTERS AND ABDOMEN: Large moths, wingspan often exceeding 40 mm, wings white or whitish, decorated with black, dark brown, yellow-orange and blueish spots and blotches, which are present particularly at forewing base, forewing tornus and hindwing anal margins. Abdomen yellow-orange, decorated with black markings. Tympanal organs small. Male 3rd sternite often setose. Male 8th sternite unmodified. CHOI *et al.* (1998) diagnosed the genus by the following synapomorphies: male forewing fovea is absent, forewing basal part is prominent, hindtibia is expanded, and the abdomen is distinct by black dots.

MALE GENITALIA: Uncus basally tapered, sometimes with basal processes or large flap-like process, or uncus absent altogether. Tegumen and vinculum forming broad, elongate structure. Gnathos ring-shaped, incomplete ring-shaped or absent. Juxta wide plate, sclerotised basally, with pair of pit-like invaginations. Valva divided. Costal arm of valva distinct, rod-shaped, sometimes basally expanded and medially curved. Ventral part of valva prominent, margins variably dentate. Aedeagus slender with long caecum. Vesica tubular, sometimes expanded at base, cornuti are absent. CHOI *et al.* (1998) diagnosed the genus by the following synapomorphies: ampulla [of male valva] is with a dentate process, harpe [of male valva] is with a process, aedeagus has long caecum, and cornuti are absent. Tympanic cavi have been noted to be relatively small (COOK & SCOBLE 1992, STÜNING & HAUSMANN 2002).

FEMALE GENITALIA: Papillae anales rather large, sometimes sclerotised laterally. Lamella postvaginalis irregular, degree of sclerotisation variable. Lamella antevaginalis prominent plate, sclerotised. Ductus bursae rather long, wide, sclerotised. Corpus bursae large, elongated, mostly membranous. Signum stellate. CHOI *et al.* (1998) did not diagnose the genus by any female genitalia characters.

BIOLOGY: The larvae of European species are polyphagous, they have been found for instance on *Prunus padus*, *Ribes* species, *Ulmus* and *Salix* (SEPPÄNEN 1970, EBERT 2003) and recorded on *Betula* (FLAMIGNI *et al.* 2007). Adults are nocturnal but easily flushed up from vegetation during daytime.

IMMATURE STAGES: Larvae of the European *Abraxas* species are conspicuous, banded longitudinally with yellow, white and black, or more irregularly patterned in the same colours (AIZPÚRUA 2002, EBERT 2003). Pupae of the European species are illustrated in AIZPÚRUA (2002). Cremaster hooks are equal in size, four pairs in *A. grossulariata* (PATOČKA & TURČANI 2005).

REMARKS: Often two subgenera have been recognised, i.e. *Abraxas* and *Calospilos* (e.g. INOUE 1972, 1984, 1995). Most species of the subgenus *Abraxas* have a fairly punctate, delicate pattern, they lack a strong blotch at forewing base and the gnathos is present as a ring. Species of the subgenus *Calospilos* have the wing pattern strong, blotchy, with a strong blotch at forewing base and gnathos is absent (HOLLOWAY 1994). Female genitalia of both subgenera are similar. We follow recent literature and have not recognised the mentioned subgenera. Instead we consider *Calospilos* a junior synonym of *Abraxas* (e.g. VIIDALEPP 1996, SCOBLE 1999, STÜNING & HAUSMANN 2002). Systematic position: Abraxini (HOLLOWAY 1994), but see 'Classification of European Ennominae genera into tribes'.

1. Abraxas grossulariata (LINNAEUS, 1758)

grossulariata grossulariata LINNAEUS, 1758, Syst. Nat. (Ed. 10) 1: 525, (*Phalaena* (*Geometra*)). Syntypes, including putative lectotype male (Linnean Society of London, UK), Europe (examined externally).

cuneifera WARREN, 1895, Novit. Zool. 2: 125, (*Abraxas*). Holotype female.

grossulariata memorabilis INOUE, 1946, Bull. lepidopt. Soc. Japan 1 (1): 12, (*Abraxas*). Syntype(s) (The Natural History Museum, London, UK), Korea: Kaijo (examined externally).

grossulariata conspurcata BUTLER, 1878, Ann. Mag. nat. Hist. (5) 1: 440, (*Abraxas*). Syntype(s) (The Natural History Museum, London, UK), Japan: Yokohama. Originally as sp., Valid at subspecies rank according to SCOBLE (1999), (examined externally).

grossulariata dsungarica WEHRLI, 1939, in SEITZ, Gross-Schmett. Erde 4 (Suppl.): 280, pl. 21: e, (*Abraxas*). Syntype(s), [Central Asia]: Ala-Tau, Sajan Mts, 1500 m. Valid at subspecies rank according to SCOBLE (1999).

grossulariata minor HERZ, 1905, Ezheg. zool. Muz. 9: 353: pl. 1, fig. 17, (*Abraxas*). Syntypes, including at least 4 males, 3 females, Korea; [Russia]: Amur region; Ussuri; Bureja Mountains; Schilka Islands, Ust-Strielka. Originally as var., valid at subspecies rank according to SCOBLE (1999).

grossulariata notabilis WEHRLI, 1939, in SEITZ, Gross-Schmett. Erde 4 (Suppl.): 281, pl. 21: e, (*Abraxas*). Syntype(s), China: Kansu (Lanchowfu), Tanchow; Kansu (south-west), 2000-3000 m.; Yangtse; Ginfu-shan, 900 m. Valid at subspecies rank according to SCOBLE (1999).

grossulariata ribesata STAUDINGER, 1892, Dt. ent. Z. Iris 5: 161, (*Abraxas*). Holotype, Central Asia: Transalai. Valid at subspecies rank according to SCOBLE (1999).

INTRASUBSPECIFIC NAMES, POTENTIALLY UNAVAILABLE: *aderdoniensis* RAYNOR, 1923 (ab.); *actinotata* RAYNOR, 1920 (ab.); *albipalliata* RAYNOR, 1909 (ab.); *albispatiata* RAYNOR, 1909 (ab.); *albomarginata* RAYNOR, 1903 (ab.); *albovarleyata* PORRITT, 1917 (ab.); *alona* THIERRY-MIEG, 1910 (ab.); *angustifasciata* RAYNOR, 1920 (ab.); *antemarginata* RAYNOR, 1923 (ab.); *ardana* THIERRY-MIEG, 1910 (ab.); *aureofasciata* PORRITT, 1920 (ab.); *aureomarginata* COCKAYNE, 1951 (ab.); *aureopicta* COCKAYNE, 1951 (ab.); *aurivestita* COCKAYNE, 1934 (ab.); *axantha* RAYNOR, 1903 (ab.); *basicaerulea* HUTCHINSON, 1969 (ab.); *brevifasciata* LEMPKE, 1970 (f.); *candida* RAYNOR, 1903 (ab.); *centralipuncta* RAYNOR, 1910 (ab.); *chalcobares* RAYNOR, 1907 (ab.); *chalcostrata* COCKAYNE, 1949 (ab.); *chalcozona* RAYNOR, 1903 (ab.); *chrysostrota* RAYNOR, 1909 (ab.); *cliftoniana* RAYNOR, 1907 (ab.); *cockayni* HUTCHINSON, 1974 (ab.); *continua* LEMPKE, 1951 (f.); *crocea* RAYNOR, 1920 (ab.); *croceolineata* LEMPKE, 1970 (f.); *cuneata* RAYNOR, 1909 (ab.); *cupreofasciata* RAYNOR, 1909 (ab.); *deleta* COCKERELL, 1889 (ab.); *depauperata* HUTCHINSON, 1974 (ab.); *diluta* LEMPKE, 1951 (f.); *dohrnii* KOENIG, 1883 (ab.); *eppingensis* HUTCHINSON, 1974 (ab.); *eremodisca* RAYNOR, 1920 (ab.); *exquisita* RAYNOR, 1919 (ab.) (primary homonym(?) of *exquisita* PORRITT, 1919); *exquisita* PORRITT, 1919 (ab.) (primary homonym(?) of *exquisita* RAYNOR, 1919); *exquisita-aenae* RAYNOR, 1920 (ab.); *flavipalliata* RAYNOR, 1909 (ab.); *flavofasciata* HUENE, 1901 (ab.) (=synonym of *dohrnii* KOENIG, 1883 according to The Global Lepidoptera Names Index (ANONYMOUS 2013)); *fractifasciata* RAYNOR, 1920 (ab.); *formosa* COCKAYNE, 1951 (ab.); *fulvapicata* RAYNOR, 1903 (ab.); *hazeleignensis* RAYNOR, 1903 (ab.); *igneofasciata* RAYNOR, 1909 (ab.); *impunctifasciata* AUSLOW, 1919 (ab.); *infrabifasciata* RAYNOR, 1909 (ab.); *infrafasciata* RAYNOR, 1909 (ab.); *intrafasciata* RAYNOR, 1909; *infraguttata* COCKAYNE, 1949 (ab.); *inframaculata* STOVIN, 1939 (ab.); *iochalca* RAYNOR, 1909 (ab.); *juncta* LEMPKE, 1970 (f.); *latifasciata* RAYNOR, 1920 (ab.); *latilimbata* COCKAYNE, 1951 (ab.); *leucomelaina* COCKAYNE, 1951 (ab.); *leucosticta* RAYNOR, 1920 (ab.); *lilacifasciata* COCKAYNE, 1949 (ab.); *lunulata* PORRITT, 1920 (ab.); *luteavarleyata*

PORRITT, 1920 (ab.); *lacteanigra* RAYNOR, 1909 (ab.); *lacteasparsa* RAYNOR, 1907 (ab.); *lacticolor* RAYNOR, 1903 (ab.) (=synonym of *dohrnii* KOENIG, 1883 according to The Global Lepidoptera Names Index (ANONYMOUS 2013)); *lutescens* LEMPKE, 1951 (f.); *luteovenata* LEMPKE, 1951 (f.); *magnipuncta* LEMPKE, 1951 (ab.); *malmundariense* DONCKIER, 1881 (ab.); *mediofasciata* LEMPKE, 1951 (f.); *melanapicata* PORRITT, 1921 (ab.); *melanoneura* COCKAYNE, 1951 (ab.); *melanozona* RAYNOR, 1903 (ab.); *mixta* PORRITT, 1920 (ab.); *nanata* LAMBILLION, 1907 (ab.); *nigra* RAYNOR, 1909 (ab.); *nigralbata* COCKAYNE, 1951 (ab.); *nigricostata* RAYNOR, 1909 (ab.); *nigroapicata* RAYNOR, 1923 (ab.) (=synonym of *melanapicata* PORRITT, 1921 according to The Global Lepidoptera Names Index (ANONYMOUS 2013); *nigrocaerulea* RAYNOR, 1909 (ab.); *nigrocincta* AUSLOW, 1919 (ab.); *nigrocingulata* COCKAYNE, 1946 (ab.); *nigrocretacea* RAYNOR, 1923 (ab.); *nigrofasciata* RAYNOR, 1909 (ab.); *nigrofusa* RAYNOR, 1920 (ab.); *nigrolineata* RAYNOR, 1923 (ab.); *nigrolutea* RAYNOR, 1907 (ab.); *nigropalliata* COCKAYNE, 1951 (ab.); *nigroradiata* REBEL, 1904 (ab.) (=synonym of *malmundariense* DONCKIER, 1881 according to The Global Lepidoptera Names Index (ANONYMOUS 2013)); *nigrosparsata* RAYNOR, 1903 (ab.); *nigrotincta* RAYNOR, 1910 (ab.); *nigrovarleyata* PORRITT, 1920 (ab.); *nigrovelata* COCKAYNE, 1939 (ab.); *nigrovenata* RAYNOR, 1909 (ab.); *obscura* COCKERELL, 1889 (ab.) (=synonym of *varleyata* PORRITT, 1920 according to The Global Lepidoptera Names Index (ANONYMOUS 2013); *odersfeltia* PORRITT, 1920 (ab.); *ovalidisca* COCKAYNE, 1951 (ab.); *paucisignata* LEMPKE, 1951 (f.); *pernigrata* THIERRY-MIEG, 1910 (ab.); *perpallida* HUTCHINSON, 1974 (ab.); *plurifasciata* RAYNOR, 1920 (ab.); *postfimbriata* COCKAYNE, 1951 (f.); *postmarginata* RAYNOR, 1923 (ab.); *privata* COCKAYNE, 1939 (ab.); *pulchra* RAYNOR, 1919 (ab.); *pulchra-aena* RAYNOR, 1920 (ab.); *pura* LEMPKE, 1970 (f.); *quasivarleyata* HUTCHINSON, 1969 (ab.); *radiata* RAYNOR, 1909 (ab.); *radioreversa* COCKAYNE, 1951 (ab.); *raynori* PORRITT, 1920 (ab.); *rubrolutea* RAYNOR, 1909 (ab.); *sebaria* COCKAYNE, 1951 (ab.); *semilutea* RAYNOR, 1910 (ab.); *seminigra* COCKAYNE, 1939 (ab.); *semiviolacea* RAYNOR, 1903 (ab.); *sexstrigata* COCKAYNE, 1951 (ab.); *sparsatahazeleighensis* PORRITT, 1920 (ab.); *sparsatavarleyata* PORRITT, 1920 (ab.); *striata* COCKAYNE, 1952 (ab.); *subangulata* RAYNOR, 1923 (ab.); *subviolacea* RAYNOR, 1903 (ab.); *supralutea* RAYNOR, 1920 (ab.); *triangulata* HUTCHINSON, 1969 (ab.); *triumvirorum* HUTCHINSON, 1974 (ab.); *unilineata* COCKAYNE, 1952 (ab.); *varleyata* PORRITT, 1920 (ab.); *vauata* PORRITT, 1920 (ab.); *venusta* COCKAYNE, 1951 (ab.); *xantha* RAYNOR, 1920 (ab.).

EXTERNAL CHARACTERS AND ABDOMEN: Wingspan ♂ 34–42 mm, ♀ 35–43 mm. Wings white, forewing base and subterminal lines yellow-orange, surrounded by large blackish spots. Fringes chequered black and white. Blackish spots weaker on hindwings, yellow areas usually absent. Discal spots present on all wings, elongated, blackish. Wings coloured below as above, except yellow almost absent. Frons and collar black. Antennae unicolorous dark, filiform in both sexes. Thorax yellow, dorsal part black. Hindtibia with 2+2 spurs. Abdomen with two rows of black dots ventrally, one dorsally and laterally, additional lateral spots on segments 3–4. Tympanal organs small. Sternites and tergites 3–8 of both sexes undifferentiated.

VARIATION: Size of black spots variable, sometimes fused and may cover most of basal and medial area, melanic specimens may be almost uniform blackish.

MALE GENITALIA: Uncus short, basally with wing-like screlotised projections. Gnathos arms fused, forming narrow ring. Juxta large, medially elongated plate. Valva divided. Dorsal arm of valva narrow, curved ventrally, setose, narrowing to apex, apex with several small thorns. Ventral arm of valva wide, sclerotised, equally wide from base to apex, apex margin straight.

Costa with two strong projections, weakly dentate, at base round knob. Ventral margin with weak sclerotisation at halfway. Saccus roundish, with very narrow medial element. Aedeagus narrow, slightly bent ventrally. Caecum long, with keel-shaped membrane. Apex with one sclerotised tooth. Vesica slightly expanded near base, without cornutus.

FEMALE GENITALIA: Papillae anales large, setose, basally sclerotised. Apophyses straight, narrow. Lamella antevaginalis sclerotised, lunular. Lamella postvaginalis irregular, invaginated medially. Antrum large, long, cone-shaped, sclerotised. Ductus bursae long, membranous, prominent. Corpus bursae oval, membranous, striated and weakly sclerotised around signum. Signum large, weakly stellate, on ventral side.

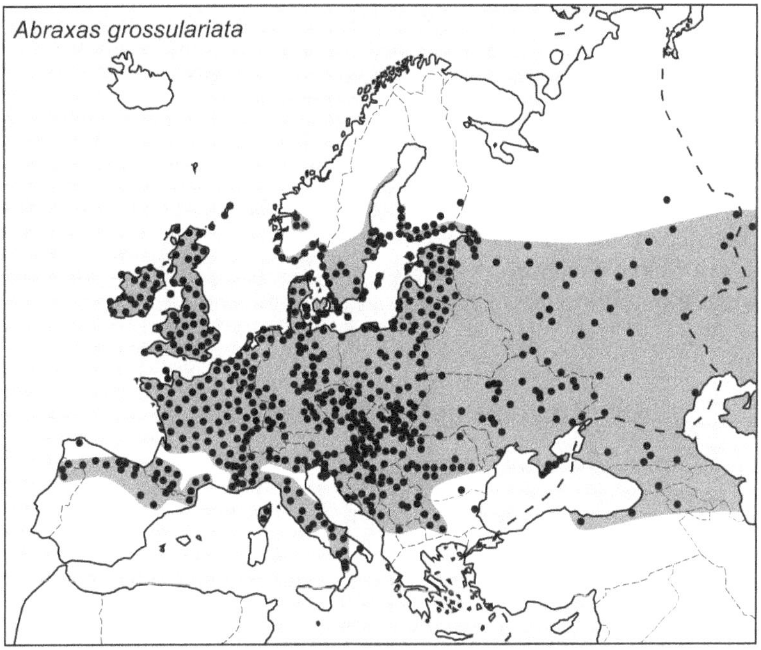

DISTRIBUTION AND ABUNDANCE: Eurasiatic. Widely distributed in Europe from southern Fennoscandia to northern Portugal and Spain, southern Italy including Corsica and most of the Balkans and from the British Isles including Ireland through central and eastern Europe to south-eastern European Russia and the Urals. The species has earlier been reported from Malta (BORG 1932), but its presence there has never been verified (P. SAMMUT pers. comm.). – Outside Europe in northern Turkey, Caucasus and through southern Siberia into China, Far East Russia and Japan (INOUE 1977). – *A. grossulariata* was formerly a common species that was often found in larger numbers, but in recent decades it has become much less common in a large part of its European distribution area and it is now rarely seen in larger numbers (for instance HAUSMANN 2001, HORSFIELD & MACDONALD 2004).

PHENOLOGY: Univoltine. In northern Europe it flies from early July and until late August (SKOU 1984 & 1986) and in southern Europe slightly earlier from June to August (FLAMIGNI *et al.* 2007). Overwintering as larva (SKOU 1984, 1986). The adults can be found in the vegetation during daytime but are not often flushed up from the vegetation. Caterpillars from late April until mid-June in central Europe. Hibernation takes place as an egg. The moths fly mainly at night and come readily to light. In Britain the species can often be seen flying in the late afternoon well before sunset (M. Corley pers. comm.).

BIOLOGY: Polyphagous. The larva has been found on *Ribes nigrum* (B. MÜLLER pers. comm.), on *R. uva-crispa*, *R. rubrum* (H. WEGNER pers. comm.), *Salix*, *Prunus padus* and *Euonymus europaeus* (EBERT 2003). Further it has been recorded on *Ribes uva-crispa*, *R. rubrum*, *Crataegus*, *Prunus spinosa* (SKOU 1984 & 1986), *Ribes alpinum* (FLAMIGNI *et al.* 2007), rarely also on a number of other deciduous trees and bushes such as *Corylus avellana* and *Fagus sylvatica* (AIZPÚRUA 2002). In the Hebrides the main food plant is *Calluna* (SOUTH 1961). In Portugal on *Vaccinium myrtillus* (M. Corley pers. comm.). Pupation takes place between spun leaves of the food plant.

HABITAT: The species occurs in most places where its food plants grow, especially in gardens and parks and in forests along rivers. It occurs from sea level and in the Alps up to 1900 m and in the central Appenines up to 1500 m (FLAMIGNI *et al.* 2007).

SIMILAR SPECIES: There are no similar species in Europe, but *A. sylvata* and *A. pantaria* show some resemblance.

GENETIC DATA: Genetically homogeneous in Europe (n=12, from United Kingdom and Finland to southern Italy), but one specimen from Finland diverging by 5.0%. In Asia, from Turkey to eastern China, genetically heterogeneous (n=13), with three other BINs. Nearest species: several Asian congenerics at 7–8% distance. Nearest species in Europe: *Selenia lunularia* (8.2%).

2. *Abraxas sylvata* (SCOPOLI, 1763)

sylvata sylvata SCOPOLI, 1763, Ent. Carniolica: 220, fig. 546, (*Phalaena*). Syntype(s), lost. ([Slovenja (north-west)]: Carniolia).

pantherata HÜBNER, 1790, Beitr. Gesch. Schmett.: 2 (47), pl. 2, fig. N, (*Ph[alaena] Geom[etra]*). Syntype(s), lost, Ukraine.

ulmaria HÜBNER, 1799, Samml. Eur. Schmett. 5 Geometrae (1): pl. 16, fig. 85, (*Geometra*). Syntype(s), lost, Europe.

ulmata FABRICIUS, 1775, Syst. Ent.: 632, (*Phalaena*). Syntype(s), England.

sylvata ijimai INOUE, 1955, Kontyu 22 (3/4): 75, (*Abraxas*). Holotype male (The Natural History Museum, London, UK), Japan: Kushiro, Shibecha, Futatsu-yama. Valid at subspecies rank according to SCOBLE (1999) (examined externally).

sylvata intensa COCKERELL, 1906, Nature, Lond. 73: 341, (*Abraxas*). Syntype(s), Japan. Originally as var., valid at subspecies rank according to SCOBLE (1999).

sylvata ischna WEHRLI, 1939, in SEITZ, Gross-Schmett. Erde 4 (Suppl.): 284, pl. 21: g, (*Abraxas*). Syntype(s), [Russia]: Eastern Sajan territories mountains, 2000 m. Valid at subspecies rank according to SCOBLE (1999).

sylvata microtate WEHRLI, 1931, Mitt. dt. ent. Ges. 2 (7): 107, (*Abraxas*). Holotype male (Zoologische Forschungsinstitut und Museum Alexander Koenig, Bonn, Germany), [China]: Manchuria, Kirin [Jilin], Mauschan. Originally as sp., valid at subspecies rank according to SCOBLE (1999).

INTRASUBSPECIFIC NAMES, POTENTIALLY UNAVAILABLE: *argyphea* COCKAYNE, 1946 (ab.); *bifasciata* HANNEMANN, 1919 (ab.); *circumnotata* COCKAYNE, 1952 (ab.); *confluens* HANNEMANN, 1919 (ab.); *fasciata* LEMPKE, 1951 (f.); *glomerata* COCKAYNE, 1934 (ab.); *guttata* HANNEMANN, 1919 (ab.); *intermedia* TUTT, 1897 (ab.); *magnimaculata* LEMPKE, 1970 (f.); *nigrescens* HANNEMANN, 1919 (ab.); *nigrocellata* LEMPKE, 1970 (f.); *obscura* TUTT, 1897 (ab.); *pantarioides* SPITZ, 1909 (ab.); *plumbea* REBEL, 1910 (ab.) (=synonym of *suffusa* TUTT, 1897 according to The Global Lepidoptera Names Index (ANONYMOUS 2013)); *purissima* COCKAYNE, 1952 (ab.); *radiata* HANNEMANN, 1919 (ab.); *reducta* HANNEMANN, 1917 (ab.); *rufomaculata* LEMPKE, 1951 (f.); *seminigra* HANNEMANN, 1919 (ab.); *striata* LEMPKE, 1970 (f.); *suffusa* TUTT, 1897 (ab.); *transversa* TUTT, 1897 (ab.).

EXTERNAL CHARACTERS AND ABDOMEN: Wingspan ♂ 31–40 mm, ♀ 34–41 mm. Wings whitish, forewing markings grey except forewing base and inner margin near tornus with brown to yellow blotch, often with grey to blue centre. Subterminal line expressed by oval grey dots. Fringes grey. Grey dots weaker on hindwings. Subterminal line expanded in anal margin to brown to yellow blotch with grey to blue centre. Discal spots normally present on all wings, elongated, grey. Wings coloured below as above, except yellow almost absent. Frons dark brown to black. Collar yellowish. Antennae unicolorous grey, filiform in both sexes. Thorax yellow, dorsally dark brown. Hindtibia with 2+2 spurs. Abdomen yellow, with two rows of black dots ventrally and dorsally, one row laterally and additional lateral spots on segments 5–8. Tympanal organs small. Male 3rd sternite with several stout setae, arranged in one row, otherwise sternites and tergites 3–8 of both sexes undifferentiated.

VARIATION: Size of grey spots variable, sometimes fused and may cover most of forewing terminal area.

MALE GENITALIA: Uncus short, apex expanded. Juxta small plate, weakly setose. Valva divided. Dorsal part of valva setose, narrowing to apex, apex with one small thorn. Ventral part of valva wide, slightly narrowing to apex, costa with numerous small, sharp teeth. Ventral margin with short, acute, separate expansion halfway. Saccus roundish. Aedeagus narrow, straight. Caecum short, apex with sclerotised tooth. Vesica expanded at base, without cornutus.

FEMALE GENITALIA: Papillae anales soft, setose. Lamella antevaginalis sclerotised, lunular. Lamella postvaginalis strongly projected medially. Antrum long, wide, tubular. Ductus bursae wide, irregularly sclerotised. Corpus bursae large, round, membranous. Signum large, stellate, on dorsal side. KAMEDA (2004) has described abnormal female genitalia.

DISTRIBUTION AND ABUNDANCE: Eurasiatic. Widely distributed in Europe from southern Fennoscandia to northernmost Iberian Peninsula, the southern Alps and the northern Balkans and from the British Isles including Ireland in the west and through central and eastern Europe to southern European Russia and the Ural mountains. – Outside Europe in Turkey and Caucasus and further through southern Siberia to China, Far East Russia and Japan. – A common species in Europe, but mainly found in limited number of specimens, but now and then *A. sylvata* can occur in larger numbers. In some areas, such as in Denmark, where Dutch elm disease in recent decades has destroyed the majority of the elm trees, *A. sylvata* seems to have declined similarly. However, in the most recent years, the species is again beginning to be more common.

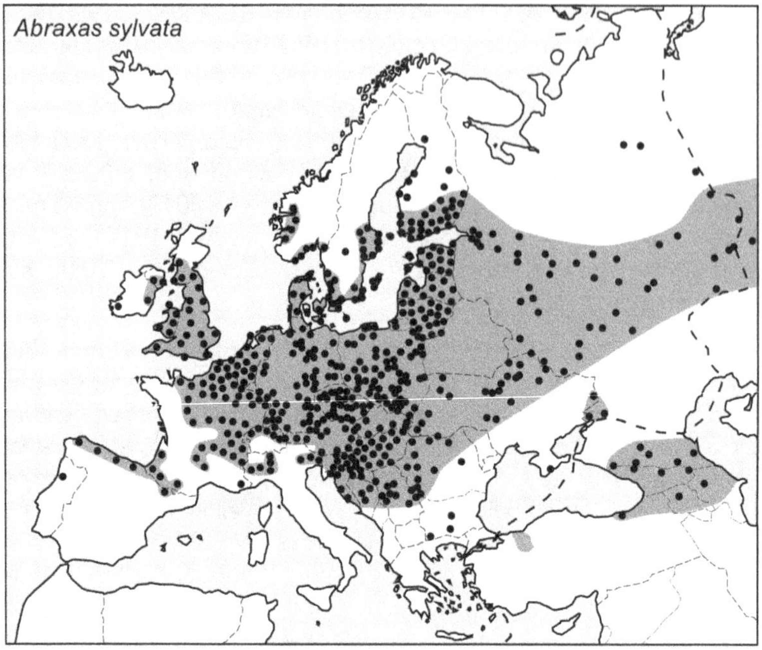

BIOLOGY: Polyphagous. The larva is mainly found on *Ulmus glabra*, and also on other *Ulmus* species. Further it has been found on *Prunus padus* and *Alnus glutinosa* (H. WEGNER pers. comm.), on *Corylus avellana*, *Prunus padus*, *P. avium* and *Viburnum lantana* (EBERT 2003) and it has been recorded on *Betula*, *Fagus*, *Platanus*, *Rhamnus* and *Frangula* (FLAMIGNI *et al.* 2007).

HABITAT: The species can be found in most places where the food plants occur such as forests, gardens and parks. In southern Germany especially in damp and shady valleys with dense forests (LEIPNITZ pers. comm.) and most common in moist birch forest with *Prunus padus* (A. HAUSMANN pers. comm.). In northern Italy it is found up to 1000 m above sea level (FLAMIGNI *et al.* 2007) and in the Pyrenees up to 1200 m (REDONDO *et al.* 2009).

SIMILAR SPECIES: Weakly marked specimens of *A. sylvata* may be similar to *A. pantaria*, but the frons is dark brown to blackish in *A. sylvata* (yellowish in *A. pantaria*), fringes are grey (white), spots in the abdomen are black (often with yellowish centre), male 3rd sternite has stout setae (setae are absent), lamella postvaginalis is strongly projected medially and ductus bursae is rather wide (lamella postvaginalis is weakly projected medially and ductus bursae is narrower).

GENETIC DATA: Genetically homogeneous from western Europe to the Iran (n=18, from seven countries). Nearest species in Europe: *A. pantaria* (7.5%), several other congenerics from Asia at 5–7% distance.

3. *Abraxas pantaria* (LINNAEUS, 1767)

pantaria LINNAEUS, 1767, Syst. Nat. (Ed. 12) 1 (2): 863, (*Phalaena* (*Geometra*)). [Holo]type male [without head and abdomen], (Linnean Society of London, UK), Lusitania [Portugal] (examined externally). Originally as sp.

cataria GUENÉE, [1858], in BOISDUVAL & GUENÉE, Hist. nat. Insectes (Spec. gén. Lépid.) 10: 207, (*Abraxas*). Syntypes 1 male, 1 female, Gibraltar. Originally as sp. See remarks.

pantata DUPONCHEL, 1830, in GODART & DUPONCHEL, Hist. nat. Lépid. Papillons Fr. 8 (1): 244, pl. 187, fig. 3, (*Zerene*). Emendation of *pantaria* LINNAEUS.

urganda STOLL, 1782, in CRAMER, Uitlandsche Kapellen (Papillons exot.) 4: 162, 252 (index), pl. 372, fig. B, (*Phal*[*aena*] *Geomet*[*ra*]. Syntype(s) male(s), [type locality not given] (examined externally, based on original description). Originally as sp., treated valid at subspecies rank by SCOBLE (1999), downgraded here from subspecies rank (new synonym), quoting absence of differential features, falling within intraspecific variation. See Text-fig 14 and Remarks.

INTRASUBSPECIFIC NAMES, POTENTIALLY UNAVAILABLE: *astiasoi* AGENJO (ab.?), 1968; *dafaucei* AGENJO, 1968 (ab.?); *romanyki* AGENJO, 1968 (ab.?).

EXTERNAL CHARACTERS AND ABDOMEN: Wingspan ♂ 31–33 mm, ♀ 32–41 mm. Wings whitish yellow, forewing markings weak, grey, except base and inner margin near tornus with small brown-grey blotch, often with grey to blue centre. Subterminal line expressed by roundish, small grey dots, strongest near costa. Fringes white. Markings weak on hindwings, subterminal line expanded in anal margin to brown to yellow blotch with weak grey to blue centre. Discal spots normally weak, visible on forewings only. Wings coloured below as above, except markings weaker and yellow normally absent. Frons yellowish. Antennae greyish, those of male weakly setose, those of female filiform. Thorax yellow, dorsally dark brown. Hindtibia with 2+2 spurs. Abdomen yellow, with rows of black dots dorsally, laterally and ventrally, often with yellow centre. Tympanal organs small. Sternites and tergites of 3–8 of both sexes undifferentiated.

VARIATION: Grey and brown-grey markings may be absent in forewing costa and in subterminal line.

MALE GENITALIA: Uncus short, apex expanded. Juxta small plate, weakly setose. Valva divided. Dorsal part of valva setose, narrowing to apex, apex with one small thorn. Ventral part of valva wide, slightly narrowing to apex, costa margin with several small, sharp teeth. Ventral margin with weak, acute sclerotisation at halfway. Saccus roundish. Aedeagus narrow, straight. Caecum short, apex with sclerotised tooth. Vesica constricted at base, then expanded, without cornutus.

FEMALE GENITALIA: Papillae anales soft, setose. Lamella antevaginalis sclerotised, lunular. Lamella postvaginalis weakly projected medially. Antrum long, narrow, tubular. Ductus bursae wide, weakly sclerotised. Corpus bursae rather small, round, membranous. Signum large, stellate, on dorsal side.

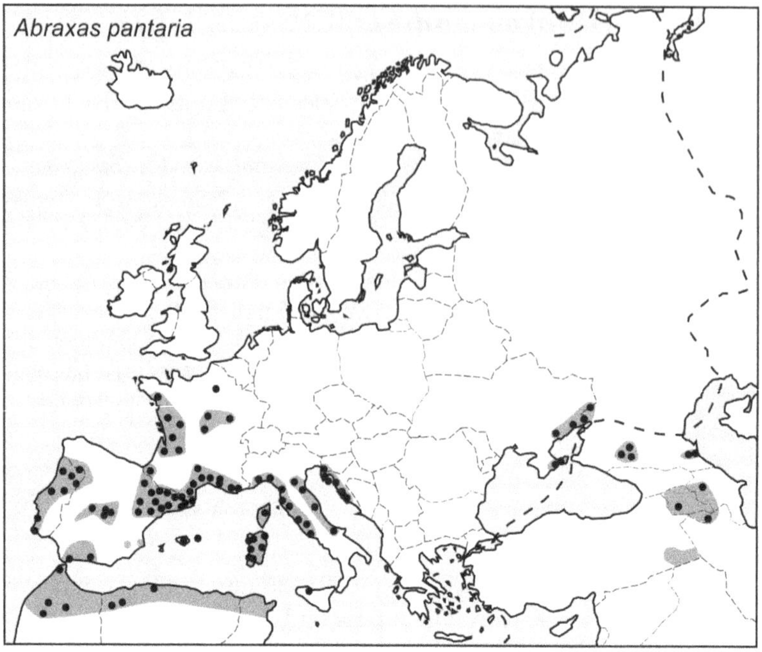

DISTRIBUTION AND ABUNDANCE: Mediterranean (South-European-Westasiatic). Disjunct, in Europe in western and central part of the Mediterranean region and in south-eastern Ukraine and southern European Russia. – Outside Europe the species occurs in the Caucasus region and in Morocco and Algeria. Reported from Turkey (VIIDALEPP 1996, KOÇAK & KEMAL 2009). – In Europe a somewhat local but relatively common species that in general occurs in limited numbers. However, mass occurrences are known and the species can be so numerous that it is a pest.

PHENOLOGY: Uni- or bivoltine. In Spain apparently in two generations, at least at lower altitudes, from late April to late September. In Italy in general univoltine, but in some localities bivoltine (FLAMIGNI *et al.* 2007). In Turkey also in one generation from late June and until mid-August (ÖZBEK & CALMASUR 2010). Hibernation takes place as a pupa (ÖZBEK & CALMASUR 2010). The adult moths fly at night and come frequently to light.

BIOLOGY: Monophagous. The caterpillar has been found on *Fraxinus excelsior* (ÖZBEK & CALMASUR 2010) and it has been recorded on *F. angustifolia* (REDONDO *et al.* 2009) and *Ulmus* (AIZPÚRUA 2002). Pupation takes place in the ground (ÖZBEK & CALMASUR 2010).

HABITAT: The species occurs in habitats where the food plants grow, both in open bush land and in more closed forests. In Spain it is found from sea level up to 1300 m (REDONDO *et al.* 2009).

SIMILAR SPECIES: See *A. sylvata*.

GENETIC DATA: Genetically homogeneous (Spain, France, Croatia). Nearest species in Europe: *A. sylvata* (7.5%), several other congenerics from Asia at 6–7% distance.

REMARKS: Taxon *cataria* (GUENÉE) is treated in the recent literature as a form or an aberration (PROUT 1912–1916, WEHRLI 1939–1954, VIVES MORENO 1994, FLAMIGNI *et al.* 2007), apart from SCOBLE (1999), who considered it valid at subspecies level. Taxon *cataria* is almost without grey-brown markings (apart from tornus and forewing base) and specimens are generally smaller (plate 1, fig. 3d, illustrated also in PROUT 1912–1916, FLAMIGNI *et al.* 2007). Potentially these specimens represent partial second generation, because many individuals are observed in July, August and September, but not completely because in Sardinia most of the specimens caught in August belong to the nominal form (CAO pers. comm. *in* FLAMIGNI *et al.* 2007). We consider *cataria* to fall within intraspecific variation, quoting the absence of differential features and sympatric occurrence with nominal taxon.

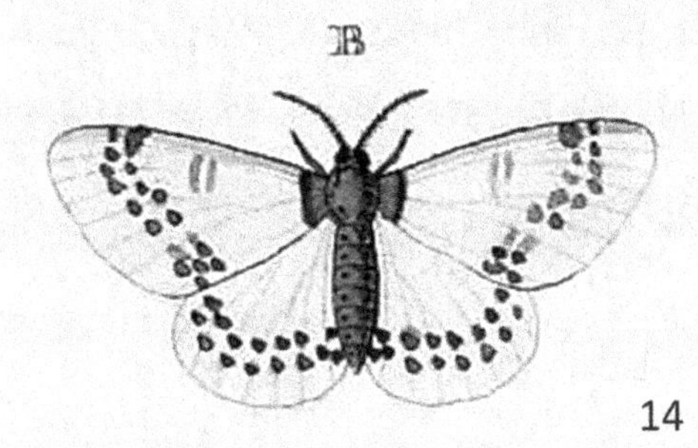

14

Text-fig. 14. The original illustration of *Phal.[aena] Geomet.[ra] urganda* STOLL, 1781, plate 372, fig. B (STOLL 1781). We consider the specimen to fall within intraspecific variation, and treat it as a junior synonym of *A. pantaria* (LINNAEUS). See text for details. Figure is reproduced under Internet Archive's terms of use from the Open Library https://openlibrary.org/books

Taxon *urganda* (STOLL) has, according to the original, hand-drawn picture a distinct postmedial line that is made up of double row of grey spots (Text-fig. 14). The illustrated specimen is a male, with setose antennae (STOLL 1781, p. 25: 'Antennes plumacees'). We consider the specimen to fall within intraspecific variation, and treat it as a junior synonym of *A. pantaria*. We have not located a mention of the taxon in the recent literature, apart from SCOBLE (1999), who considered it valid at subspecies level.

Ligdia GUENÉE, 1858

Ligdia GUENÉE, [1858], Hist. nat. Insectes (Spec. gén. Lépid.) 10: 209. Type species: *Geometra adustata* DENIS & SCHIFFERMÜLLER, 1775 [Austria]. Genus synonymy is after SCOBLE (1999).

Harpicosta WEHRLI, 1936. Type species: *Ligdia extratenebrosa* WEHRLI, 1936. China.

DIVERSITY AND DISTRIBUTION: *Ligdia* contains 11 species, mainly in Africa and in the Palaearctic regions, including Japan and China, a few species possibly reaching Oriental region and one species in the Nearctic.

EXTERNAL CHARACTERS AND ABDOMEN: Moderate sized moths, wingspan often about 25 mm, wings whitish, markings black, brown, grey and yellow-orange. Forewing base with prominent dark area. Fovea in male forewing absent. Tympanal organs normal sized. Male 3rd sternite setose, 8th sternite unmodified. CHOI *et al.* (1998) diagnosed the genus by three external synapomorphies: forewing basally prominent, abdomen distinct by black dots, and forewing subterminal band dorsally prominent.

MALE GENITALIA: Uncus triangular. Gnathos narrow ring, medially elongated. Valva divided. Dorsal arm of valva long, curved. Ventral arm of valva wide, distally bifurcate. Cornuti present. CHOI *et al.* (1998) diagnosed the genus by one synapomorphy: gnathos triangular.

FEMALE GENITALIA: Ductus bursae wide, long, sclerotised. Corpus bursae elongated, membranous. Signum stellate.

BIOLOGY: Larvae of *L. adustata* have been found feeding on *Euonymus europaeus* (Celastraceae) (EBERT 2003), and the North American *L. wagneri* FERGUSON & ADAMS, 2008 potentially on *E. obovatus* (FERGUSON 2008). Adults are nocturnal.

IMMATURE STAGES: *Ligdia adustata* has green larvae, with weak and sparse lateral and dorsal red-brown markings. Tint of yellow may also be present (EBERT 2003). Cremaster hooks are absent, but cremaster sides are obtuse-angled in middle in *L. adustata* (PATOČKA & TURČANI 2005).

REMARKS: *Ligdia* is considered closely related to *Abraxas*, the relationship being supported by similar wing pattern, including the dark blotch at forewing base and in termen, the absence of fovea in forewing and several other features (HOLLOWAY 1994, CHOI *et al.* 1998). The inclusion of *Ligdia* in the Abraxini was tentative (STÜNING pers. comm. in HOLLOWAY 1994), but it was adopted in the later checklists without comments. *Ligdia* is in many features intermediate between the Abraxini and Cassymini, but majority of characters are more similar to the Abraxini. Also VIVES MORENO (1994), FERGUSON (2008) and HAUSMANN *et al.* (2011) included *Ligdia* in the Abraxini. Systematic position: Abraxini.

4. *Ligdia adustata* (DENIS & SCHIFFERMÜLLER, 1775)

adustata DENIS & SCHIFFERMÜLLER, 1775, Ankündung syst. Werkes Schmett. Wienergegend: 114, (*Geometra*). Syntype(s), lost, [Austria]: Vienna district.

inspersata STAUDINGER, 1879, Horae soc. ent. Rossicae XIV (4): 443, (*Abraxas*). Syntype(s), [Turkey, Mugla Province]: Kerasdere [Karadere].

INTRASUBSPECIFIC NAMES, POTENTIALLY UNAVAILABLE: *alba* LEMPKE, 1970 (f.); *extincta* HANNEMANN, 1917 (ab.); *obscura* LEMPKE, 1951 (f.); *plumbosa* COCKAYNE, 1950 (ab.); *postumbrata* LEMPKE, 1970 (f.); *suffusa* PROUT, 1915 (ab.).

EXTERNAL CHARACTERS AND ABDOMEN: Wingspan of both sexes 24–27 mm. Wings bone-white. Forewing basal area dark, medial line weakly expressed as a row of elongated dark markings, postmedial line suffused with orange-brown. Terminal area dark except apex and and tornus, which are whiteish-yellow. Terminal line dark, markings elongated and interrupted. Fringes dusty white, slightly chequered below apex. Hindwings dusty white, weakly decorated, dark areas in wing base and in terminal area absent. Wavy line distinct, white. Terminal line distinct between apex and termen only. Discal spots weak. Wings more yellowish below, markings weaker. Frons and collar dark brown. Antennae mixed brown and whitish, weakly fasciculate in male, filiform in female. Thorax dark brown. Hindtibia with 2+2 spurs. Abdomen light brown. Tympanal organs large. Male 3rd sternite with several stout setae, arranged in multiple rows, otherwise sternites and tergites 3–8 of both sexes undifferentiated.

VARIATION: Ground colour mostly bone-white, occasionally suffused with grey. Terminal area usually dark violet-brown, occasionally light brown.

MALE GENITALIA: Uncus wide, setose. Gnathos fused ventrally, with minute lateral arms. Juxta triangular. Valva divided. Dorsal part of valva narrow, curved ventrally, setose, apex acute. Ventral part of valva wide, bilobed, costa terminating in acute apex, ventral margin folded partially, setose. Saccus wide, round. Aedeagus narrow. Caecum short, round, apex with one sclerotised tooth. Vesica narrow tube, with several basal cornuti.

FEMALE GENITALIA: Papillae anales, soft, setose. Apophyses straight, long, narrow. Lamella antevaginalis bowl-shaped. Lamella postvaginalis small median sclerotisation. Antrum wide. Ductus bursae wide, sclerotised, striated and slightly angled. Corpus bursae elongated, membranous. Signum large, stellate, on dorsal side.

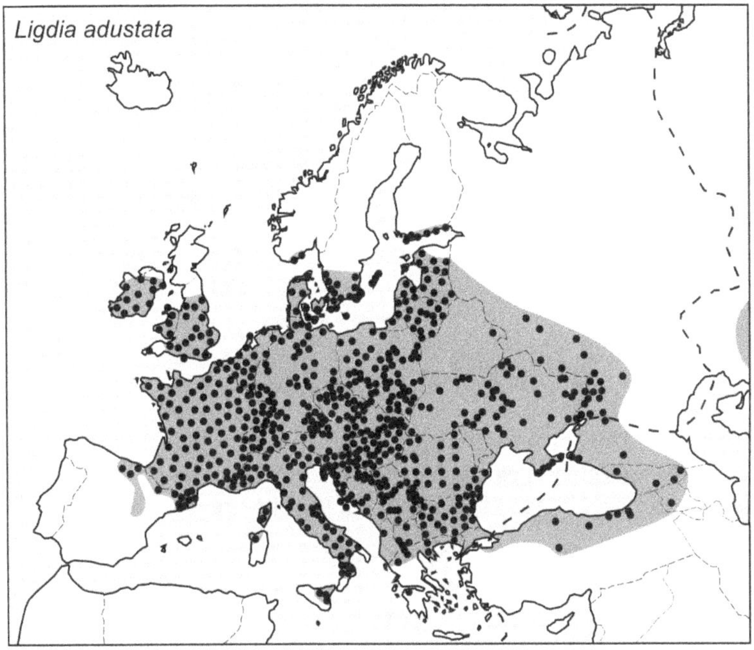

DISTRIBUTION AND ABUNDANCE: European-Westasiatic. In Europe from southern Fennoscandia to northern Spain, southern Italy including Corsica, Sardinia and Sicily and to northern Greece and from Britain including Ireland through central and eastern Europe and southern European Russia to southern Urals. – Outside Europe *L. adustata* is found in Turkey, the Caucasus region, north-western Kazakhstan and Turkmenistan. – In Europe a common species, but usually it occurs in limited numbers. However, in some areas where the food plant is planted in gardens and parks, it can be one of the most abundant Geometrid species.

PHENOLOGY: Mainly bivoltine. In two overlapping generations from mid-May till late August in northern Europe, in Italy from March to September or October and with only one generation at higher altitudes. Caterpillars in May to July and August to October, depending on latitude. Hibernation takes place as a pupa. The moth flies at night and comes readily to light, but is also occasionally flushed up during daytime.

BIOLOGY: Monophagous. The larva has repeatedly been found feeding on *Euonymus europaeus*, the only known food plant. The pupation takes place between spun leaves.

HABITAT: In all habitat types where the food plant occurs: clearings in forests, forest edges, parks, and gardens. Found up to 1600 m (FLAMIGNI *et al.* 2007). In the Pyrenees up to 1300 m (REDONDO *et al.* 2009).

SIMILAR SPECIES: There are no similar species in Europe. In Korea and Japan *L. japonaria* LEECH, 1897 is somewhat similar externally (adult and genitalia are illustrated in CHOI *et al.* (1998)), and in North America *L. wagneri* FERGUSON & ADAMS 2008 (adult and genitalia are illustrated in FERGUSON (2008).

GENETIC DATA: Genetically comparatively homogeneous in Europe and North Africa (n=36, from 13 countries), but one specimen from Hungary diverging by 2.8%. Nearest species in Europe: *Chemerina caliginearia* (RAMBUR, 1833) (9.0%). West Asian *Ligdia coctata* GUENÉE, 1858 and several other congenerics at 10–11% distance.

Lomaspilis HÜBNER, 1825

Lomaspilis HÜBNER, [1825] 1816, Verz. bekannter Schmett.: 306. Type species: *Phalaena marginata* LINNAEUS, 1758 [Europe] by subsequent designation by GROTE, 1902, Insektenbörse 19: 314. Genus synonyms are after SCOBLE (1999).

Lomatospilus AGASSIZ, 1847. [Emendation of *Lomaspilis* HÜBNER, 1825]

Poecilophasia STEPHENS, 1831. Type species: *Phalaena marginata* LINNAEUS, 1758.

DIVERSITY AND DISTRIBUTION: *Lomaspilis* contains three species, all occur in the Palaearctic region.

EXTERNAL CHARACTERS AND ABDOMEN: Small moths, wings brownish-black and white. Thorax and abdomen dark, smooth and glossy. Male hindleg slightly dilated. Antenna of both sexes filiform. Male 3rd sternite with group of setae, 8th sternite unmodified.

MALE GENITALIA: Uncus tapering. Gnathos ring-shaped. Juxta medially elongated, triangular plate. Valva divided. Dorsal arm of valva long, narrow, curved ventrally, hook-shaped. Ventral arm of valva wide at base, densely setose apically. Vesica large, with apical diverticulum, cornuti grouped together. CHOI *et al.* (1998) recognised two synapomorphies: valva is with a medially invaginated costa and the cornuti are spinular.

FEMALE GENITALIA: Papillae anales large. Sclerotisations around ostium bursae weak. Ductus bursae short, sclerotised laterally. Corpus bursae pear-shaped, caudal part wide, striated and sclerotised, cephalad part membranous. Signum asteroid-shaped, often located at cephalad end of corpus bursae. CHOI *et al.* (1998) recognised two synapomorphies: signum is present, and it is asteroid-shaped.

BIOLOGY: Larvae of *Lomaspilis* species have been found for instance on *Salix*, *Populus*, *Corylus* (HULDÉN *et al.* 2000), *L. opis* has been recorded on *Betula* (SEPPÄNEN 1970) and *L. bithynica* on *Populus tremula* (GELBRECHT *et al.* 2004). Adults are nocturnal but easily flushed up from vegetation during daytime.

IMMATURE STAGES: Larva of *L. marginata* is green, with minute dark green dorsal spots. Intersegmental areas are slightly paler (EBERT 2003). Pupa has a bifid cremaster in *L. marginata* (PATOČKA & TURČANI 2005).

REMARKS: The diagnostic features of the European *Lomaspilis* species are reviewed by GELBRECHT *et al.* (2004). Inclusion of *Lomaspilis* in the Cassymini was proposed rather recently (D. STÜNING pers. comm. in HOLLOWAY 1994), it is also adopted elsewhere (HAUSMANN, MIRONOV & VIIDALEPP 2011). Systematic position: Cassymini.

5. *Lomaspilis marginata* (LINNAEUS, 1758)

marginata LINNAEUS, 1758, Syst. Nat. (Ed. 10) 1: 527, (*Phalaena* (*Geometra*)). Syntypes, including putative lectotype (Linnean Society of London, UK) (examined externally).

maculata HÜBNER, 1786, Beitr. Gesch. Schmett. 1 (1): 24, pl. (1)3, fig. P, (*Ph*[*alaena*] *Geom*[*etra*]). Syntype(s) female(s), lost, [Germany]: Augsburg.

marginaria HÜBNER, 1799, Samml. Eur. Schmett. 5 Geometrae (1): pl. 15, fig. 80, (*Geometra*). Syntype(s), lost, Europe. Emendation(?) of *marginata* LINNAEUS.

naevaria HÜBNER, 1799, Samml. Eur. Schmett. 5 Geometrae (1): pl, 15, fig. 79, (*Geometra*). Syntype(s), lost, Europe.

pollutaria HÜBNER, 1799, Samml. Eur. Schmett. 5 Geometrae (1): pl. 15, fig. 77, (*Geometra*). Syntype(s), lost, Europe.

staphylaeata SCOPOLI, 1763, Ent. Carniolica: 221, fig. 548, (*Phalaena*). Syntype(s), [Slovenja (north-west)]: Carniolia.

INTRASUBSPECIFIC NAMES, POTENTIALLY UNAVAILABLE: *albociliata* HÖRHAMMER, 1923 (ab.); *albomarginata* OSTHELDER, 1931 (ab.); *andrearia* LAMBILLION, 1919 (ab.); *artoni* BURRAU, 1950 (ab.?); *brunnescens* LEMPKE, 1951 (f.); *conflua* STRAND, 1900 (ab.); *demarginata* FUCHS, 1899 (ab.) (=synonym of *pollutaria* HÜBNER, [1799] according to The Global Lepidoptera Names Index (ANONYMOUS 2013)); *diluta* COCKAYNE, 1950 (ab.); *discocellularis* STRAND, 1919 (ab.); *dumeei* JOANNIS, 1912 (ab.); *hjordisi* LINGONBLAD, 1946 (ab.); *hortulata* SULZER, 1761 (f.) (=synonym of *pollutaria* HÜBNER, [1799] according to The Global Lepidoptera Names Index (ANONYMOUS 2013)); *huenei* STRAND, 1904 (ab.) (=synonym of *mediofasciata* HÖFNER, 1897 according to The Global Lepidoptera Names Index (ANONYMOUS 2013)); *lacticolor* LEMPKE, 1951 (f.); *limbata* HORMUZAKI, 1899 (ab.) (=synonym of *pollutaria* HÜBNER, [1799] according to The Global Lepidoptera Names Index (ANONYMOUS 2013)); *mediofasciata* HÖFNER, 1899 (ab.); *mediofasciata* HOYNINGEN-HUENE, 1901 (ab.) (=homonym of *mediofasciata* HÖFNER, 1899); *naevata* HÜBNER, 1790; *nigrofasciata* SCHÖYEN, 1882 (ab.); *nigrosparsata* COCKAYNE, 1946 (ab.); *nigro-unicolorata* HOVER, 1904 (ab.); *pollutaria* HÜBNER, [1799]; *postalbata* LEMPKE, 1951 (f.); *punctaria* AGENJO, 1968 (ab.?); *semialbata* MELLAERTS, 1928 (ab.?); *staphyleata* HOYNINGEN-HUENE, 1901 (ab.); *subdeleta* COCKERELL, 1889 (ab.); *suffusa* PROUT, 1915 (ab.); *virgulata* AGENJO, 1968 (ab.?); wendlandtiata Fuch, 1899 (=junior synonym of *subdeleta* COCKERELL, 1889). PROUT (1912–1916) and WEHRLI (1939–1954) list further intrasubspecific names.

EXTERNAL CHARACTERS AND ABDOMEN: Wingspan ♂ 23–27 mm, ♀ 21–25 mm. Wings silky glossy, white, blackish or brownish black patterns of fore and hindwings generally as in the other two *Lomaspilis* species. Forewing basal spot elongated, medial band constricted, often interrupted at CuA2 or reduced to one costal spot. Terminal band concave between M3 and CuA1. Fringes unicolorous blackish. Dark bands continued on hindwings except in basal area. Medial band often reduced and terminal band concave between M3 and CuA1. Discal spots normally absent. Wings coloured below as above. Frons and collar brown. Antennae brown, those of male weakly setose, female antennae filiform. Thorax brown. Male and female hindtibia with 2+2 spurs. Abdomen concolorous with thorax, scattered with white scales ventrally. Tympanal organs large. Male 3rd sternite with several stout setae, arranged in single or few rows, otherwise sternites and tergites 3–8 of both sexes undifferentiated.

VARIATION: Shape and extensiveness of dark areas varies quite a lot, particularly in the medial band, and to a lesser degree in the terminal area.

MALE GENITALIA: Uncus evenly tapering to round tip. Gnathos ring-shaped. Juxta elongated medially. Valva divided. Dorsal arm of valva long, narrow, curved ventrally, hook-shaped, apex acute. Ventral arm of valva simple, distally round, margin of costa setose, medially deeply invaginated. Saccus elongated, sometimes with a pair of medial wrinkles. Aedeagus straight. Caecum round. Vesica large, expanded, elongate, cylindrical. Cornuti grouped together, their number (8–14) and shape variable.

FEMALE GENITALIA: Papillae anales soft, setose. Apophyses posteriores longer than apophyses anteriores. Lamella antevagnialis membranous. Lamella postvaginalis small plate. Ductus bursae laterally sclerotised near antrum, rather narrow, anterior part of ductus bursae wide, sclerotised, striated. Corpus bursae elongated or round, large sac, sclerotised, partly striated. Signum stellate, often on ventral side of corpus bursae.

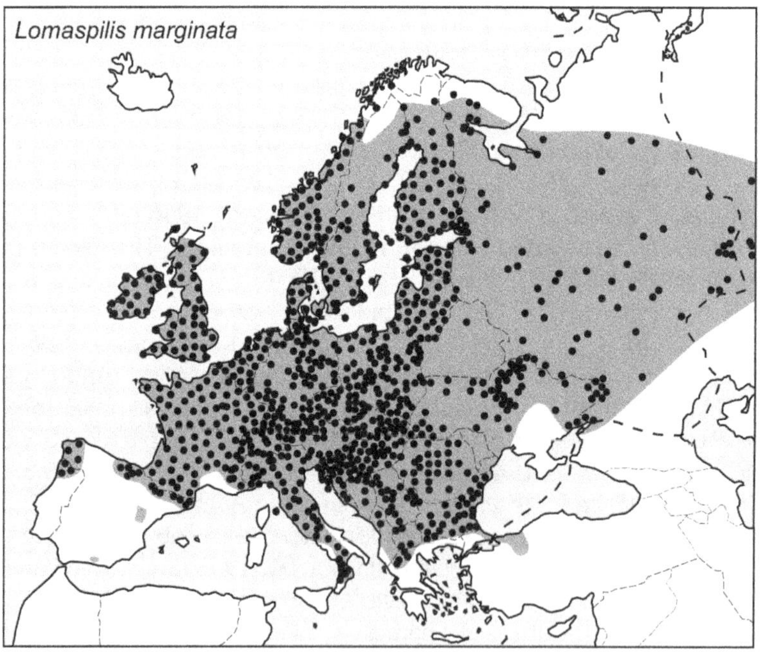

DISTRIBUTION AND ABUNDANCE: Eurasiatic. In Europe from northern Fennoscandia to the Iberian Peninsula, Italy and northern Greece and from the British Isles through central and eastern Europe to southern European Russia and the Ural mountains. – Outside Europe widely distributed through Siberia, China, Far East Russia to Korea, Japan, the Kurile Islands and

Sakhalin. It has been reported from Turkey (KOÇAK & KEMAL 2009), but no specimens have been traced. – In Europe a common species that sometimes occurs in larger numbers.

PHENOLOGY: Uni- or bivoltine. In northern Europe usually from late May to mid-August in two overlapping generations, in southern Germany already from late April (EBERT 2003), in Italy from March-April to September or October (FLAMIGNI *et al.* 2007) and in northern Spain from April to early August (REDONDO *et al.* 2009). At higher altitudes in only one generation from mid-June to mid-August. Hibernation takes place as a pupa. The adults fly up easily by day when disturbed and fly actively at night when they come readily to light.

BIOLOGY: Oligo- or polyphagous. The larva has been found on *Populus*, *Salix* and *Corylus* (HULDÉN *et al.* 2000). H. WEGNER (pers. comm.) has found caterpillars on *S. aurita* and *S. cinerea* from June to September. As rare food plants also *Quercus* and *Fagus sylvatica* are mentioned (EBERT 2003). FLAMIGNI *et al.* (2007) describe the caterpillar as polyphagous on broadleaved trees. Pupation takes place in the ground.

HABITAT: In deciduous forests, brush, hedges, moors, along water courses, in parks and gardens. In Fennoscandia it is found up to 300 m above sea level and up to approximately 2000 m in the Alps and the Pyrenees.

SIMILAR SPECIES (EXTERNAL CHARACTERS): *L. bithynica* and *L. opis* (Text-figs 15–17). External, rather diagnostic features include the colour of dark patterns on wings: brownish black (*L. marginata*), blueish black to black (*L. bithynica*), mostly black (*L. opis*). Forewing basal spot shape: elongate (*L. marginata*), trapezoid, slightly convave (*L. bithynica*) or oval (*L. opis*). Orientation and pattern of medial band: symmetry axis of forewing medial band towards inner margin (*L. marginata*), symmetry axis towards tornus (*L. bithynica* and *L. opis*). Medial band variable, but can be reduced to one costal spot on forewings and nothing on hindwings (*L. marginata*), medial band variable but often continuous on forewings (*L. bithynica*), medial band variable but rarely continuous on forewings (*L. opis*). Shape and intensity of dark sports varies considerably, particularly in the medial band, and to a lesser degree in the terminal area in *L. marginata*. The markings vary less in *L. bithynica* and *L. opis*.

SIMILAR SPECIES (GENITALIA CHARACTERS): According to GELBRECHT *et al.* (2004), there is only one male genitalia character that is reliable in species identification, the form of the vesica. Our findings support the views of GELBRECHT *et al.* (2004). The others listed below must be treated with caution. Male genitalia: saccus with pair of wrinkles in middle (*L. marginata, L. opis*), saccus without wrinkles (*L. bithynica*). Form of vesica: small diverticulum at base of the ductus ejaculatorius absent (*L. marginata, L. opis*), diverticulum present (*L. bithynica*). Distal end of vesica approximately straight (*L. marginata*), slightly expanded (*L. bithynica, L. opis*). GELBRECHT *et al.* (2004) list as reliable female genitalia characters the width of ostium bursae and the shape of bursa copulatrix. We have failed to find reliable differences in the ostium width, but agree that corpus bursae shapes are often diagnostic. Bursa copulatrix is pear-shaped (*L. marginata, L. opis*) or elongated (*L. bithynica*). Position of the signum is variable, but it is in most instances on ventral side of the bursa copulatrix in *L. marginata* and *L. bithynica*, at cephalad end in *L. opis*.

GENETIC DATA: Genetically comparatively homogeneous in Europe (n=35, from ten countries). Nearest species: *L. bithynica* (0.9%).

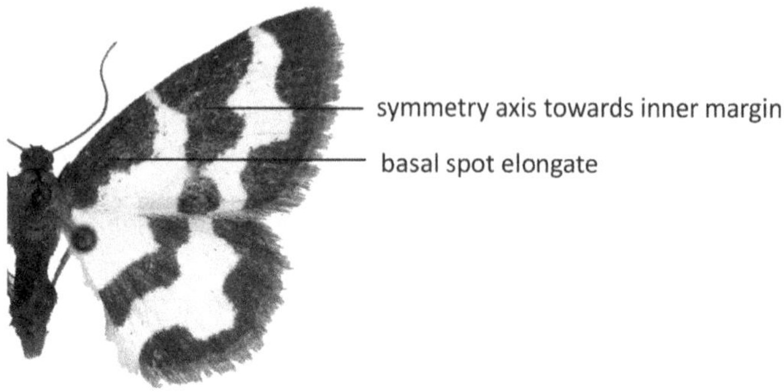

L. marginata 15

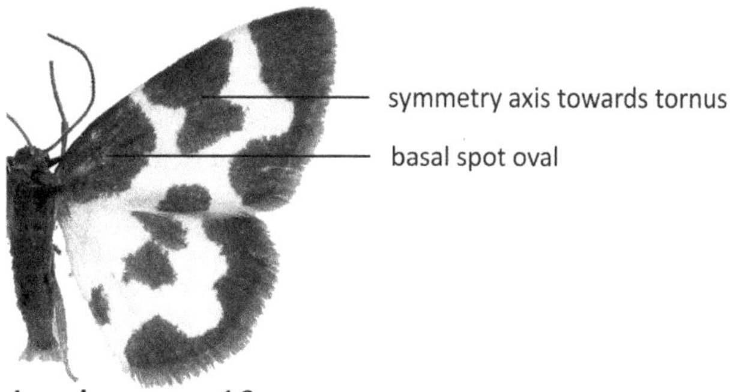

L. opis 16

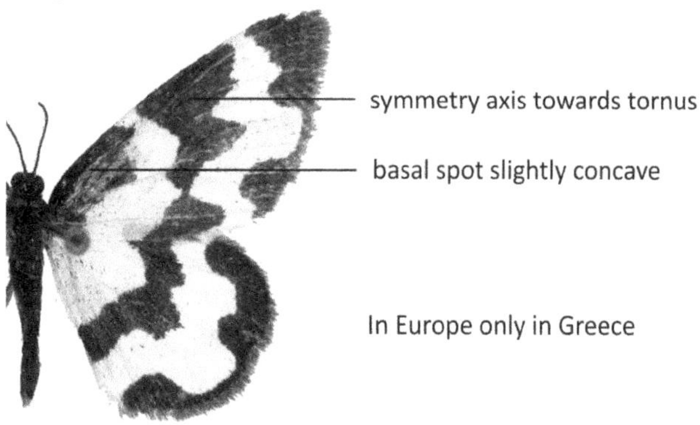

L. bithynica 17

Text-figs 15–17. Diagnostic characters (indicated) of European *Lomaspilis* species. Text-fig. 15: *L. marginata* (LINNAEUS, 1758) (Denmark, 12.vi.1976). Text-fig. 16: *L. opis* BUTLER, 1878 (Finland, 12.vi.1973). Text-fig. 17: *L. bithynica* WEHRLI, 1954 (Turkey, 5.vii.1983).

6. ***Lomaspilis opis*** BUTLER, 1878

opis opis BUTLER, 1878, Ann. Mag. nat. Hist. (5) 1: 442, (*Lomaspilis*). Syntype(s) (The Natural History Museum, London, UK), [Japan]: Yokohama (examined externally).

opis amurensis HEDEMANN, 1881, Horae Soc. ent. ross. 16: 44 (*Abraxas*). Syntypes 42, Russia: Amur. Originally as var. of *Phalaena (Geometra) marginata* LINNAEUS, 1758, valid at subspecies rank according to SCOBLE (1999).

opis nigrita HEYDEMANN, 1936, Ent. Z. Frankf. a.M. 49: 510, figured, (*Lomaspilis*). Syntypes male(s), female(s), Finland: Mikkeli. Valid at subspecific rank (SCOBLE 1999) (types not found, original description (HEYDEMANN 1936) illustrates adults and genitalia of both sexes, these are examined).

INTRASUBSPECIFIC NAMES, POTENTIALLY UNAVAILABLE: *kumakurai* INOUE, 1956 (ab.).

EXTERNAL CHARACTERS AND ABDOMEN, based on European subsp. ***nigrita***: Wingspan ♂ 21–24 mm, ♀ 21–25 mm. Fore and hindwing colour and patterns generally as in other two *Lomaspilis* species, see Description of *L. marginata* above, but with the following differences (Text-figs. 15–17): basal and medial spots mostly black, basal spot oval, symmetry axis of forewing medial band towards tornus, medial band variable but rarely continuous on forewings.

VARIABLILITY: Shape and extensiveness of dark areas varies little, patterns rather uniform.

MALE GENITALIA: Generally as in the other two *Lomaspilis* species, see *L. marginata* above, but with following differences. Saccus is triangular, with medial wrinkles, small diverticulum at base of ductus ejaculatorius is absent and distal end of vesica is slightly expanded.

FEMALE GENITALIA: Generally as in other two *Lomaspilis* species, see Description of genitalia characters in *L. marginata* above, but with following differences: Corpus bursae is pear-shaped, signum is often positioned at cephalad end of the corpus bursae.

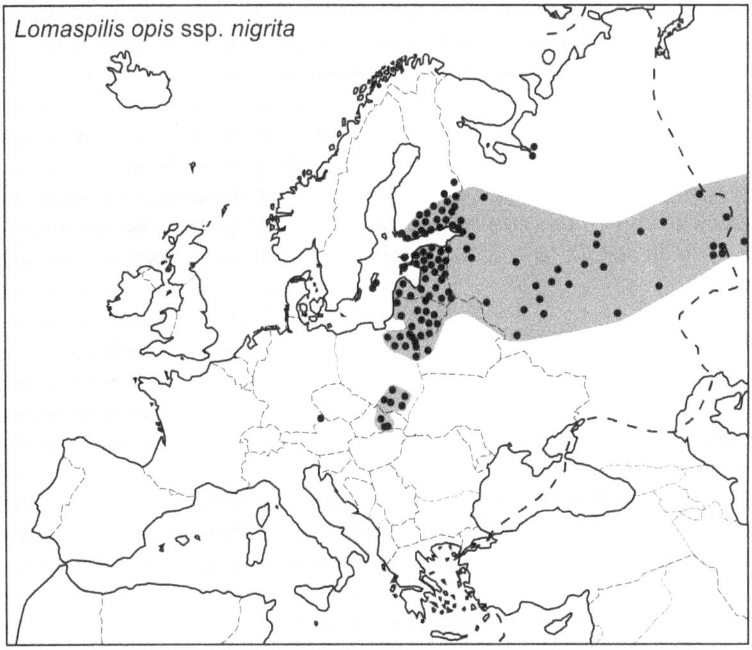

DISTRIBUTION AND ABUNDANCE: Eurasiatic. In Europe (subsp. *nigrita* HEYDEMANN 1944) from south-eastern Finland, the Baltic countries, eastern Poland and Slovakia through Russia to southern Urals. In south-easternmost Germany an isolated occurrence (RITT *et al.* 2011). – Outside Europe in Far East Russia, Sakhalin Island, southern Kuriles Islands and Japan subsp. *opis*). Reported erroneously from Turkey by KOÇAK & KEMAL (2009), but their data are undoubtedly for *L. bithynica*. – In Europe a relatively rare and local species.

PHENOLOGY: Univoltine. The species occurs in southern Finland from mid-June to late July and in eastern Poland from early June to early July (MALKIEWICZ 2012). Caterpillars from June to late August or September. Overwintering takes place as a pupa. The moths fly at night when they are readily attracted to light.

BIOLOGY: Monophagous. The caterpillar has been recorded on *Betula* (HULDÉN 2000). Reared on *Betula pendula* (M. LEIPNITZ pers. comm.). Pupation takes place in the ground.

HABITAT: In Finland the species occurs locally in open birch forests. In Poland in moist deciduous or mixed forests, birch growth and thickets. In most of the European distribution area only in lowlands. In Slovakia up to 900 m (Z. LÁSTŮVKA pers. comm.). In southern Urals up to 1000 m (K. NUPPONEN pers. comm.).

SIMILAR SPECIES: See *L. marginata* above.

GENETIC DATA: Genetically identical in Europe (n=8, from Finland, Germany and Poland). Nearest species: *L. marginata* (1.9%) and *L. bithynica* (1.7%).

7. *Lomaspilis bithynica* WEHRLI, 1939

bithynica WEHRLI, 1939, in SEITZ, Gross-Schmett. Erde 4 (Suppl.): 291, pl. 22: c, (*Lomaspilis*). Holotype female, Asia Minor [Turkey]: Bithynia, Boli [Bolu] (Zoologische Forschungsinstitut und Museum Alexander Koenig, Bonn, Germany). Originally as f. (examined externally)

EXTERNAL CHARACTERS AND ABDOMEN: Wingspan ♂ 23–24 mm, ♀ 21–25 mm. Fore and hindwing colour and patterns generally as in other two *Lomaspilis* species, see Description of *L. marginata* above, but with the following differences (Text-figs. 15–17): basal and median spots are black with bluish suffusion, terminal band is pure black, basal spot is trapezoid and slightly concave between Cu and anal veins, medial band is variable but often continuous on forewings.

VARIABLILITY: Shape and extensiveness of dark areas varies little, patterns rather uniform.

MALE GENITALIA: Generally as in other two *Lomaspilis* species, see Description of *L. marginata* above, but with the following differences. Saccus is triangular, without medial wrinkles, vesica has a small globular diverticulum at the base of the ductus ejaculatorius.

FEMALE GENITALIA: Generally as in other two *Lomaspilis* species, see Description of *L. marginata* above, but with the following differences. The corpus bursae is elongated and the signum is located on ventral side of the corpus bursae.

DISTRIBUTION AND ABUNDANCE: Pontic. In Europe only two specimens have been recorded from the vicinity of Grevena in northern Greece (GELBRECHT *et al.* 2004). – Outside Europe apparently with limited distribution in northern and north-eastern Turkey and the Caucasus region. Not listed from Turkey by KOÇAK & KEMAL (2009). *L. bithynica* was only recognised as a valid species recently (GELBRECHT *et al.* 2004), hence the limited information.

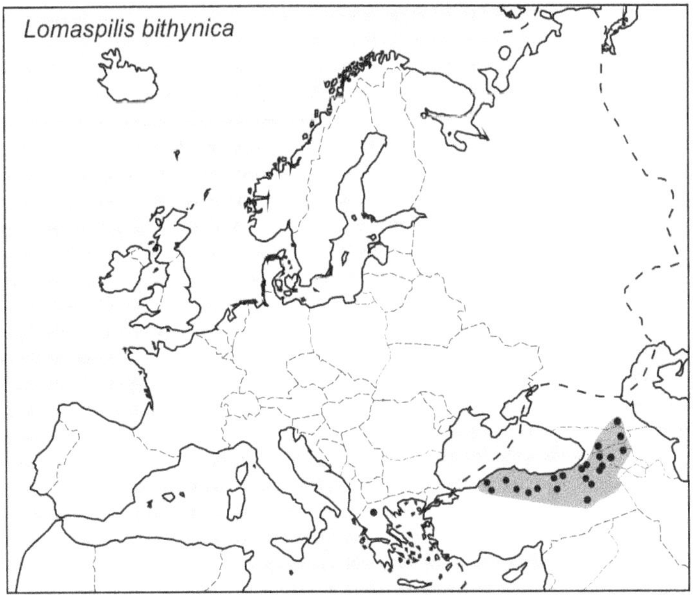

PHENOLOGY: Univoltine. The moths fly from late May to late July. Under breeding conditions a few specimens of a second generation emerged in September and early October. In rearing the species overwintered as pupa (M. LEIPNITZ pers. comm.). The moths fly at night and come to light (GELBRECHT *et al.* 2004).

BIOLOGY: Monophagous. The species has been recorded on *Populus tremula* (GELBRECHT *et al.* 2004), the only known food plant so far.

HABITAT: In Turkey in deciduous forests. In Turkey *L. bithynica* has been found between 800 and 2300 m above sea level while the only two European specimens were recorded at 600 m altitude (GELBRECHT *et al.* 2004).

REMARKS: This species was searched for by PEDER SKOU in 2010 where it has been recorded from near Grevena in northern Greece, but without success and no suitable habitats with *Populus tremula* seem to be present there. The two European specimens have undoubtedly been found near Grevena, but their occurrence there is difficult to explain and the species should be searched for both near Grevena and in suitable habitats in northern Greece and in Macedonia and Bulgaria.

SIMILAR SPECIES: See *L. marginata* above.

GENETIC DATA: In Turkey and Georgia diverging to two separate BINs at 2.8% distance (n=7) between each other, and at 0.9 and 2.7% from *L. marginata*.

REMARKS: GELBRECHT *et al.* (2004) reinstated *L. bithynica* as valid at species level, and we share their view. LERAUT (2009) treated *L. bithynica* as a subspecies of *L. opis* without analytical argumentation.

Steganía GUENÉE, 1845

Stegania GUENÉE, [1845] 1844, Cat. méth. Lépid. Eur.: 270. Type species: *Geometra permutataria* HÜBNER, [1799] 1796, a junior subjective synonym of *Stegania trimaculata* (VILLERS, 1789) (Gallia [France]). Genus synonyms are after SCOBLE (1999).

Calomicta GUMPPENBERG, 1887. Type species: *Stegania dalmataria* GUENÉE, [1858]. Dalmatia. [Unnecessary replacement name for *Eustegania* GUMPPENBERG, 1887]

Eustegania GUMPPENBERG, 1887. Type species: *Stegania dalmataria* GUENÉE, [1858]. Dalmatia.

Terpnomicta LEDERER, 1853. [Unnecessary replacement name for *Stegania* GUENÉE, 1845]

UNAVAILABLE NAMES (incorrect subsequent spelling): *Stegalia* DESMAREST, 1847 (see FLETCHER 1979).

DIVERSITY AND DISTRIBUTION: *Stegania* contains 10 species, mainly in the Palaearctic, a few species in North Africa and one species is described from East Bali. There are four species in Europe.

EXTERNAL CHARACTERS AND ABDOMEN: Wings of European species are light yellow or ochreous, markings of transverse and terminal lines and veins are different shades of brown. Thorax and abdomen light yellow to ochreous. Male 3rd sternite with or without setae, 8th sternite slightly triangular.

MALE GENITALIA: Genitalia are small and generally similar to each other in all European species, whereas the aedeagi, and vesicas in particular, are large and complex. The male vesica offers most reliable, diagnostic, species-level features. Uncus round, weakly bifurcate, or with two hooks. Juxta wide plate, not elongated medially. Valva divided. Dorsal part of valva narrow, hook-shaped (i. e., acutely bent ventrally). Ventral part of valva simple, costa medially invaginated, apex weakly setose. Saccus elongated medially. Male 8th tergite triangular. Aedeagus large, with several curved, separated cornuti. Vesica large, complex, with one or two distinct diverticula.

FEMALE GENITALIA: Distal part of the corpus bursae long, straight or slightly bent, sclerotised and striated. Signum stellate.

BIOLOGY: Three European species have been recorded on *Populus* (EBERT 2003).

IMMATURE STAGES: Larva of *S. trimaculata* and *S. dilectaria* are illustrated in EBERT (2003). They are long, slender, green, green-brown or brown in colour with elongated dorsal markings present or absent. Larva and pupa of *S. trimaculata* are in illustrated in AIZPÚRUA (2003). Cremaster of pupa is elongated and shortly bifurcate (PATOČKA & TURČANI 2005).

REMARKS: *Heterostegane* HAMPSON, 1893 was considered a junior synonym of *Stegania* GUENÉE, 1845 by LERAUT (1997). *Heterostegane* HAMPSON is mainly an Old World tropical genus, but a few species extend into the Palaearctic. Facies of *Heterostegane* are similar to *Stegania*, but the two genera differ in the genitalia structures (HOLLOWAY 1994). Inclusion of *Stegania* in the Cassymini was established rather recently (HOLLOWAY 1994). Systematic position: Cassymini.

8. *Steganía cararia* (HÜBNER, 1790)

cararia cararia HÜBNER, 1790, Beitr. Gesch. Schmett. 2 (3) 77, pl. (3)4, fig. X, (*Ph[aleana] Geom[etra]*). Syntype(s) female(s), lost.

cararia lungtanensis WEHRLI, 1939, in SEITZ, Gross-Schmett. Erde 4 (Suppl.): 294, pl 22: fig. f, (*Lomographa*). Syntypes 1 male, 3 females (The Natural History Museum, London, UK), [China]: South East China, Province Kiangsu, Lungtan, near Nanking (examined externally). Valid at subspecies rank according to SCOBLE (1999).

EXTERNAL CHARACTERS AND ABDOMEN: Wingspan of both sexes 19–23 mm. Wings straw-yellow, densely stippled with brown, most pronounced in forewing costa and basal area, only subterminal line visible, angled, dividing terminal area into three distinct cells. Terminal line continuous, brown. Fringes chequered with straw-yellow and brown. Pattern of hindwings the same. Discal spots present in all wings, elongated, brown. Wings coloured below as above, but less stippled with brown. Frons and collar stippled with brown, antennae similarly coloured. Male antennae weakly fasciculate, female antennae filiform. Thorax straw-yellow to brown. Hindtibia with 2+2 spurs. Abdomen concolorous with thorax and wings. Tympanal organs large. Male 3rd sternite with row of small setae and 3–4 intersegmental membrane expanded. Male 8th tergite triangular, otherwise abdominal segments of both sexes undifferentiated.

VARIATION: There is very little variation, density of the dark markings varies little.

MALE GENITALIA: Uncus triangular, apex with two horns. Gnathos arms fused, granular medially. Juxta simple, wide plate. Tegumen wide. Valva divided. Dorsal arm of valva long, narrow, acutely bent ventrally, hook-shaped, apex acute. Ventral part of valva simple, distally tapering and setose. Dorsal margin (costa) invaginated, ventral margin (sacculus) approximately straight. Saccus triangular. Aedeagus wide, large compared to genitalia, caecum round. Vesica large, with narrow apical diverticulum, one prominent cornutus near apex and row of sclerotised, short cornuti at base. Ductus ejaculatorius opens from apex.

FEMALE GENITALIA: Papillae anales soft, setose, elongated. Both apophyses narrow, straight. Apophyses posteriores about twice the length of apophyses anteriores. Lamella antevaginalis narrow, weakly sclerotised, lunular. Lamella postvaginalis small medial plate. Ductus bursae short, wide, sclerotised laterally. Distal part of corpus bursae wide, sclerotised. Base of ductus seminalis enlarged, coiled. Medial part of corpus bursae heavily sclerotised, spinose and striated. Corpus bursae large, elongated sac, weakly striated. Signum prominent, stellate.

DISTRIBUTION AND ABUNDANCE: Eurasiatic. In Europe from northernmost Spain and France through northern Italy and central and eastern Europe up to Estonia and eastwards through Russia. – Outside Europe through southern Siberia and northern China to Far East Russia. – In Europe a locally distributed and not common species that is rather rarely found in numbers.

PHENOLOGY: Univoltine. From early June to late July in south-western Germany (EBERT 2003). In Poland the species flies from early June to late July (MALKIEWICZ 2012). According to FLAMIGNI *et al.* (2007), in some localities bivoltine with flight time both earlier and later compared to places with only one generation. The adult moth comes to light.

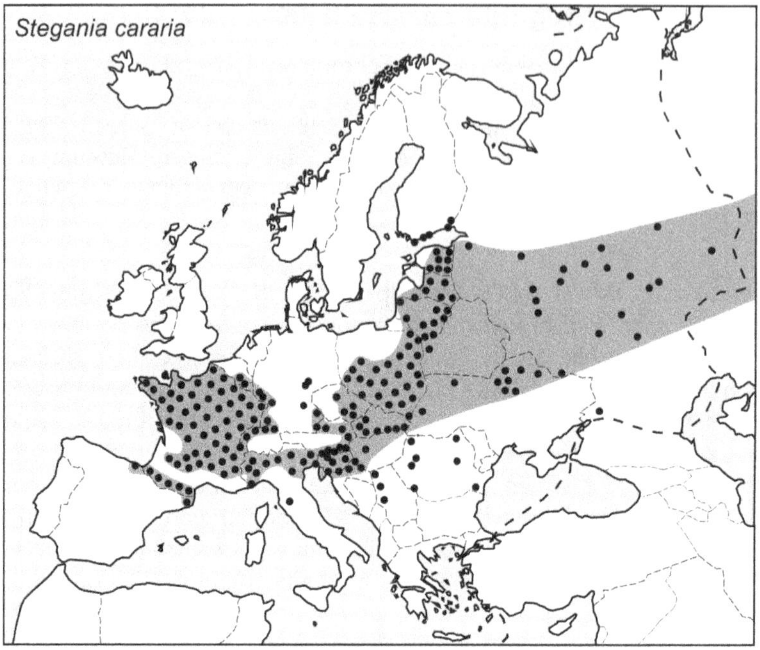

BIOLOGY: Monophagous. The caterpillar is recorded on *Populus* species. EBERT (2003) mentions a caterpillar found on *P. canadensis*.

HABITAT: Mainly in forests along rivers and other forests on moist ground, but also in *Populus* plantations. The species occurs from sea level up to 1200 m in northern Italy (FLAMIGNI *et al.* 2007) and up to 800 m in the Pyrenees (REDONDO *et al.* 2009).

SIMILAR SPECIES: There are no similar species in Europe. *Lomographa* (*Heterostegane*) *minax* PROUT, 1931, currently combined with *Heterostegane* HAMPSON, 1893 (SCOBLE 1999) from W. China is externally very similar to *S. cararia*. We have examined the (holo)type male in the BMNH and the superficial original description (PROUT 1931), but not the genitalia. WEHRLI (1939–1954) considered it conspecific with *S. cararia*.

GENETIC DATA: Genetically homogeneous from Germany and Finland to Russia (n=5, from four countries). Nearest species: An undescribed species of *Lomaspilis* from China (7.9%).

REMARKS: CHOI *et al.* (1998) combined *S. cararia* HÜBNER with *Heterostegane* HAMPSON, which is mainly an Old World tropical genus, but a few species extend into the Palaearctic. Facies of *Heterostegane* are similar to *Stegania*, but the two genera differ in the genitalia structures (HOLLOWAY 1994).

9. *Stegania trimaculata* (VILLERS, 1789)

trimaculata trimaculata VILLERS, 1789, Linn. ent. 2: 384, (*Phal*[*aena] Geom*[*etra*]). Syntype(s), Gallia [France].

albicaria BRUAND, 1846, Mém. Soc. Emul. doubs 2 (2) (3/4): 137, (*Stegania*). Syntypes, [France]: Jura, Dampierre; Bourgogne, Nuits.

cognataria LEDERER, 1853, Verh. zoo. –bot. Ver. Wien 3: 261, (*Terpnomicta*). Syntype(s), Europe [aus Südfrankreich].

permutaria commutaria GUENÉE, [1858], in BOISDUVAL & GUENÉE, Hist. nat. Insectes (Spec. gén. Lépid.) 10: 48, (*Stegania*). Syntypes 6, France (central).

permutataria HÜBNER, [1799], Samml. Eur. Schmett. 5 Geometrae (1): pl. 17, fig. 92, (*Geometra*). Syntype(s), lost, Europe.

trimaculata ochrearia BANG-HAAS, 1910, Dt. ent. Z. Iris 24: 47, pl. 3, fig. 19, (*Stegania*). Holotype male, Tunisia: Ain Draham (examined externally). Originally as sp., downgraded here to subspecies rank (new status), quoting allopatric occurrence (new status). See remarks.

permutaria xanthinaria OBERTHÜR, 1923, Études Lépid. comp. 20: 231, pl. 561, fig. 4839, (Stegania). Syntype(s), Tunisia. Originally as var. Synonyny after SCOBLE (1999).

oranaria WEHRLI, 1930, Int. ent. Z. 23 (37/38): 438, (Lomographa). Syntypes 13 males, 4 females (of which 1 male, 1 female are labeled as paratypes (The Natural History Museum, London, UK, examined externally), [Algeria]: Oran, Aflou. Originally as var. of *trimaculata* (VILLERS), but treated as a separate species by SCOBLE (1999), is downgraded back to variation of *Stegania trimaculata* subsp. *ochrearia* BANG-HAAS, quoting absence of diagnostic features (new combination and status revised). See remarks.

postrecta WEHRLI, 1930, Int. ent. Z. 23 (37/38): 438, (*Lomographa*). Syntypes, [Algeria]: [Oran, Aflou?]. Originally as var. of *trimaculata* (VILLERS), but treated as a separate species by SCOBLE (1999), is downgraded back to variation of *Stegania trimaculata* subsp. *ochrearia* BANG-HAAS, quoting absence of diagnostic features (new combination and status revised). See remarks.

UNAVAILABLE NAMES (incorrect subsequent spelling): *permutaria* DESMAREST, 1848 (see FLETCHER 1979) – Intrasubspecific names, potentially unavailable: *bermeja* RIBBE, 1912 (ab.); *cadahiai* AGENJO, 1968 (f.).

EXTERNAL CHARACTERS AND ABDOMEN: Wingspan ♂ 19–21 mm, ♀ 20–23 mm. Dimorphic species. Nominotypical form unicolorous light yellow, medial and postmedial lines narrow, brown, expanded on costa forming sdots. Postmedial line rather straight, before it turns towards costa. Subterminal line represented by brown spot on costa. Terminal line brown, cilia uniform light yellow. On hindwings only postmedial line and terminal lines visible, brown. Discal spots weak, elongated, brown. Wings coloured below as above, but lighter. Labial palpi short, brown. Frons ochre-yellow, near upper margin suffused brown. Collar light yellow to brown. Antennae straw-yellow, mixed with brown. Male antennae bipectinate, female antennae filiform. Thorax light yellow to brown. Male and female hindtibia with 2+2 spurs. Abdomen concolorous with thorax and wings. Tympanal organs large. Male 3rd sternite without setae, 8th tergite triangular. Otherwise abdominal segments of both sexes undifferentiated.

VARIATION: Form *cognataria* LEDERER ochre-yellow, suffused with brown, veins also brown, thus forming weakly reticulate pattern. Medial and postmedial lines distinct, brown. Subterminal line represented by enlarged brown spot on costa, another dark brown area near tornus. Terminal line brown, continuous, cilia weakly chequered brown and yellow. Hindwings similarly stippled brown, veins brown. Postmedial line distinct. Discal spots strong, brown, elongated. Wings coloured below as above, but lighter. Density of the brown markings varies a little.

Subsp. ***ochrearia*** variable in external appearance, from ochreous-yellow with grey-brown medial lines and partially grey-brown terminal area (*ochrearia*) to densely irrorated with grey-brown and strong medial lines, forming reticulate pattern, to pale straw with narrow medial lines and partially grey-brown terminal area (*postrecta*), to pale straw with narrow medial lines and forewing costa with three spots, to pale straw, almost without markings except curved brown streak near forewing costa (*oranaria*) (Text-figs 18–21). Locally one colour form may be more common than the others; ochreous-yellow specimens are common in Tunisia, whereas specimens almost without any markings are commonly found in Algeria, and medial lines are often visible in Moroccan specimens. See remarks.

18 19 20 21

Text-figs 18–21. External variation of *Stegania trimaculata* subsp. *ochrearia* BANG-HAAS, 1910 in North Africa. Text-fig. 18. Male, Algeria: Sebdou, 16.viii.1918, genitalia slide PS1964 (coll. BMNH). Text. fig. 19. Female, Algeria: Hammam Righa, 14.v.1917, slide PS1965 (coll. BMNH). Text-fig. 20. Male, Tunisia: Ain Draham, 6–19.viii.1911, slide PS1971 (coll. BMNH). Text-fig. 21. Female, Tunisia: Ain Draham, 7–16.ix.1911, slide PS1972 (coll. BMNH).

MALE GENITALIA: Generally all European *Stegania* species have similar male genitalia (see *S. cararia*), but *S. trimaculata* has the following minor, diagnostic features. Uncus apex with two acute teeth, sometimes almost absent. Anterior margin of juxta plate concave. Vesica is diagnostic: apical diverticulum coiled, with large cornutus, second cornutus medially. Ductus ejaculatorius opens from medial part.

FEMALE GENITALIA: Papillae anales soft, setose. Apophyses short, narrow. Lamella antevaginalis membranous. Lamella postvaginalis small, posterior margin invaginated medially. Ductus bursae short, membranous. Distal part of corpus bursae long, straight, sclerotised, striated. Anterior part of corpus bursae round sac. Signum stellate.

DISTRIBUTION AND ABUNDANCE: Sub-Mediterranean. In Europe from Germany, southern Netherlands, to the Iberian Peninsula, southern France including Corsica and southern Italy including Sardinia and Sicily. In Russia an old record from Volga-Don region (MIRONOV *et al.* 2008). Due to the large distance to the nearest occurrence of the species, this record may be a result of mis-labelling. In northern Greece an isolated record (S. ORTNER pers. comm.). – Outside Europe the species occurs in Morocco, Algeria and Tunisia (subsp. *ochrearia*). According to VIIDALEPP (1996) also known from Turkmenistan, Uzbekistan and Turkey. However, due to the western distribution, occurrence in Turkmenistan and Uzbekistan are unlikely and occurrence in Turkey may be due to confusion with *S. dilectaria*. KOÇAK & KEMAL (2009) listed *S. trimaculata* from Bursa and Kirikkale provinces, but these records are also most likely due to confusion with *S. dilectaria*. – In Europe a rather common species that is mainly found in limited numbers.

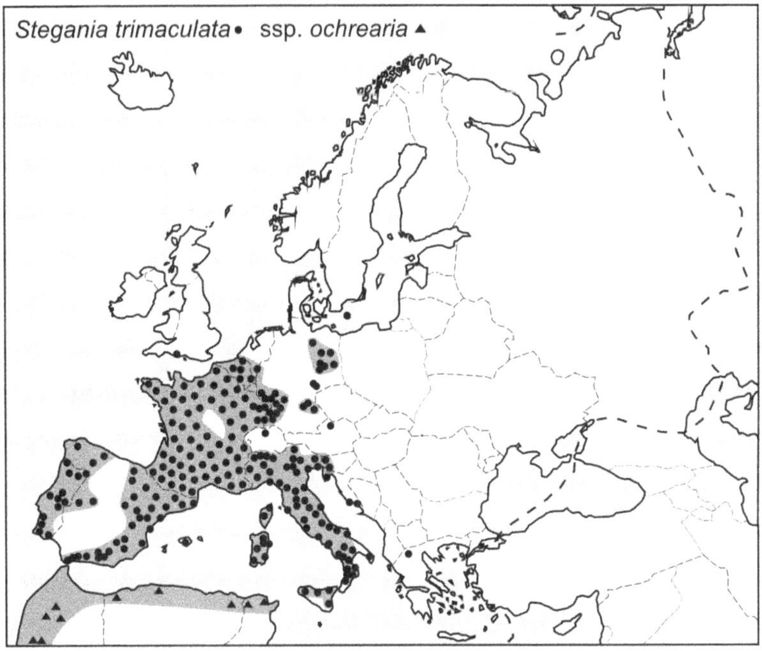

PHENOLOGY: Bivoltine. In south-western Germany from late April to mid-June and from late July to mid-August (EBERT 2003), and in Poland from early May to late July and from early August to late September (MALKIEWICZ 2012). In southern Spain and southern Italy from mid-March to late September. In southern Spain the species has also been recorded in October (P. HALE pers. comm.). The larvae occur from June to September (FAJCIK 2003) in southern central Europe, in Italy and Spain possibly later. In Sardinia also in late May (M. LEIPNITZ pers. comm.). The adults fly at night and come frequently to light.

BIOLOGY: Oligophagous. The larva has been found on *Populus*, especially *P. alba*, *P. nigra* and *P. canadensis* (EBERT 2003). In Sardinia on *Salix purpurea* and other *Salix* species (M. LEIPNITZ pers. comm.). In Morocco on *Salix pedicellata* (RUNGS 1981). Reared on *Populus tremula* and *P. canadensis* (E. FRIEDRICH pers. comm.). Based on rearings, pupation takes place on or in the ground (M. LEIPNITZ pers. comm.).

HABITAT: Mainly along rivers but also in towns and forests with plantations of the food plant. Recorded up to 1400 m above sea level both in Italy (FLAMIGNI *et al.* 2007) and Spain (coll. P. SKOU).

SIMILAR SPECIES: *S. trimaculata* f. *cognataria* and *S. dilectaria* (HÜBNER) can be identified by differences in the shape of forewing postmedial line. It is nearly parallel with termen in *S. trimaculata* f. *cognataria* before it turns towards costa, in *S. dilectaria* it is distinctly undulated or concave before it turns towards costa (Text-figs 22–23).

GENETIC DATA: Genetically homogeneous from Europe to North Africa (n=22, from six countries). Nearest species: *Stegania dalmataria* (5.5%).

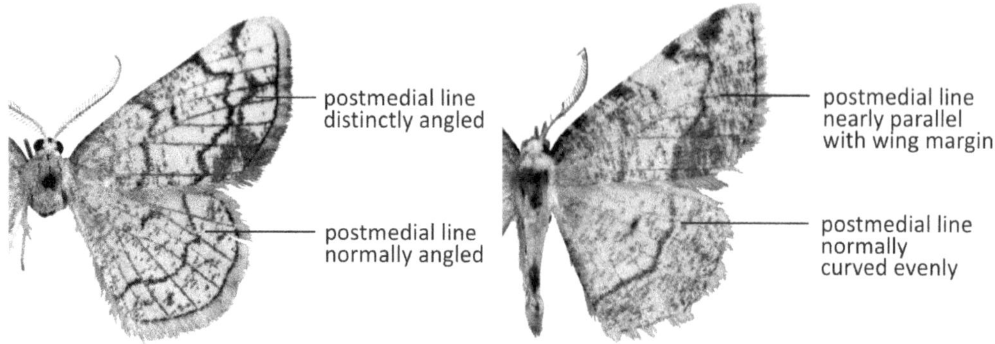

Text-figs 22–23. Diagnostic characters (indicated) of *Steganía* species. Text-fig. 22. *S. dilectaria* (HÜBNER, 1790) (Hungary, 5.v.1948). Text-fig. 23. *S. trimaculata* (VILLERS, 1789) f. *cognataria* (France, 23.vii.1992).

REMARKS: *Stegania ochrearia* BANG-HAAS, 1910 from Tunisia has been considered valid at species level (WEHRLI 1939–1954, SCOBLE 1999), being characterised by ochreous yellow wings, distinct grey-brown spot near forewing apex and partially grey-brown terminal area. WEHRLI (1939–1954) also considered it valid at species level, writing (p. 295, our translation from German): 'The male reproductive organ proves a close relationship with *trimaculata*, but valva and uncus are different'. We have examined specimens, which are externally similar to the *ochrearia* holotype, from the type locality of *ochrearia* in Tunisia (Ain Draham), and also from Algeria, including both male and female genitalia (dissected specimens in the coll. BMNH), but have not noticed differences in the mentioned structures, or elsewhere in the genitalia including the vesica, which is diagnostic between the European species (see the Genitalia plate). Although the current location or fate of the *ochrearia* holotype is unknown (not mentioned in the original description (BANG-HAAS, 1910)), the identity of the taxon is not in doubt thanks to a good grey-scale photograph of the holotype in the original description (plate 3, fig. 19). The holotype is an unusually contrasting specimen, but it falls within variation of *S. trimaculata* (VILLERS) in North Africa. We therefore downgrade *Stegania ochrearia* BANG-HAAS, 1910 to synonymy with *Stegania trimaculata* (VILLERS, 1789), but we consider the North African populations valid at subspecies level because they are allopatric and external appearances are partially different from the nominal subspecies (see Text-figs 18–21). *S. trimaculata* is genetically homogeneous from Europe to North Africa.

Both *Stegania oranaria* (WEHRLI, 1930) and *Stegania postrecta* (WEHRLI, 1930) were described from Algeria as variations of *S. trimaculata* (VILLERS). Adults are illustrated in WEHRLI (1939–1954). We have examined specimens from North Africa (Algeria, Tunisia), including the genitalia and vesica (dissected specimens in the coll. BMNH), whose external appearances range from pale straw, almost without markings (*oranaria*) to contrasting, partially reticulate (*postrecta*). We agree with WEHRLI's original intention (1930) and consider these taxa to fall within variation of *S. trimaculata*. SCOBLE (1999) treated these two taxa valid at species level.

10. Steganía dilectaria (HÜBNER, 1790)

dilectaria dilectaria HÜBNER, 1790, Beitr. Gesch. Schmett. 2(3): 76, pl. (3) 4, fig. W, (*Ph[alaena] Geom[etra]*). Syntype(s) male(s), lost, [Austria]: Vienna.

commutataria HÜBNER, [1819], Samml. Eur. Schmett. 5 Geometrae (1): pl. 98, fig. 505, (*Geometra*). Syntype(s), lost, Europe.

dilectaria trimaculoides WEHRLI, 1931, Mitt. münch. ent. Ges. 21 (2): 44, (*Lomographa*). Syntypes 70, [Turkey]: Marasch [Maras]. Valid at subspecies rank according to SCOBLE (1999).

UNAVAILABLE NAMES (manuscript name): *taurica* WEHRLI (In Zoologische Staatssammlung, Munich, Germany. One male labelled as cotype of *Lomographa trimaculata* f. *taurica*, mentioned under *Lomographa dilectaria* in WEHRLI 1939 (1939–1954)).

EXTERNAL CHARACTERS AND ABDOMEN: Wingspan 21–24 mm. Wings straw-yellow, weakly stippled with brown, veins also brown thus forming weakly reticulate pattern. Medial and postmedial lines distinct, brown. Postmedial line distinctly undulated or concave before it turns towards costa. Subterminal line angled, not expanded in costa. Terminal area dark near tornus. Terminal line brown, continuous, fringes weakly chequered yellow and brown. Hindwings similarly stippled brown, veins brown. Postmedial line distinct, weakly undulating, turning towards thorax near costa. Subterminal line sometimes weakly developed. Discal spots distinct, brown, lunular. Wings coloured below as above but lighter. Frons brown, ventral margin straw-yellow. Labial palpi and collar straw-yellow. Antennae brownish, male antennae bipectinate, female antennae filiform. Thorax light brown. Hindtibia with 2+2 spurs. Abdomen concolorous with thorax and wings. Tympanal organs large. Male 3rd sternite without setae, 8th tergite triangular. Otherwise abdominal segments of both sexes undifferentiated.

VARIATION: There is very little variation. Forewing terminal area may be reticulate or more evenly brown.

Subsp. ***trimaculoides*** is without reticulate pattern, paler, and the appearance is similar to *trimaculata*. Subsp. *trimaculoides* has postmedial line with distinct angles on both wings (angles smaller or absent in *trimaculata*).

MALE GENITALIA: Generally all European *Stegania* species have similar male genitalia, but vesica offers diagnostic characters. Basal part granulate, apical diverticulum curved, pointing forward, with large cornutus. Second cornutus medially. Ductus ejaculatorius opens from medial part.

FEMALE GENITALIA: Papillae anales soft, setose. Apophyses short, narrow. Lamella antevaginalis narrow ring. Lamella postvaginalis small, covered densely with short setae, weakly invaginated medially. Ductus bursae short, membranous. Distal part of corpus bursae long, sclerotised, slightly bent, striated, weakly turned along its longitudinal axis. Corpus bursae roundish. Signum stellate.

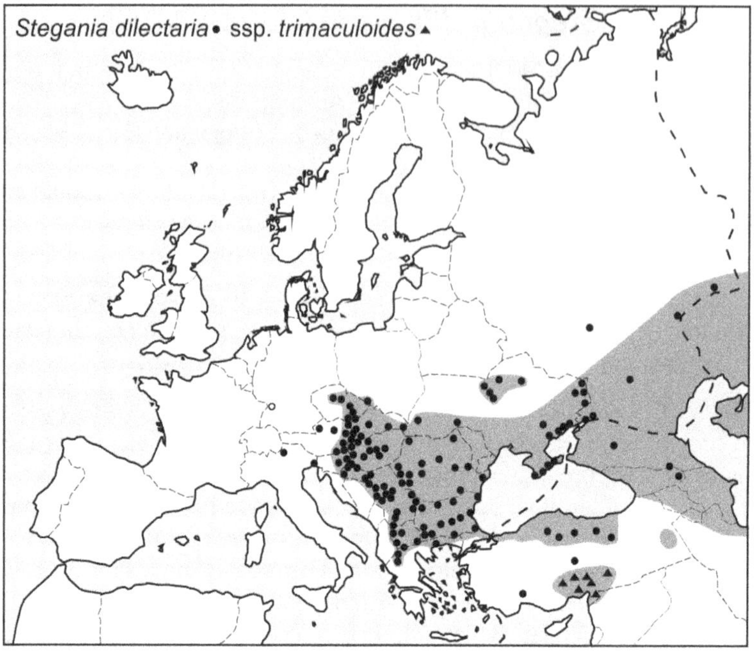

DISTRIBUTION AND ABUNDANCE: East-European-Westasiatic. In Europe from the Czech Republic and eastern Austria through most of the Balkans to northern Greece, in central and southern Ukraina and south-eastern European Russia to the southern Urals. A few specimens by Karlsruhe in south-western Germany in 1901 (EBERT 2003) and a few scattered records in Italy and Austria. – Outside Europe in Turkey (in southern Turkey subsp. *trimaculoides*), Georgia, Azerbaijan, northern Iran and parts of central Asia. – In most countries local and occurring in limited numbers.

PHENOLOGY: Uni- or bivoltine. From mid May to mid-June and from mid-July to late August. In the Czech Republic from mid-May to mid-June and from mid-July to late August (Z. LAŠTŮVKA pers. comm.). In Romania from mid-April to mid-May and from late July to late August (SZÉKELY 2011). In southern Urals from mid-June to mid-July (K. NUPPONEN pers. comm.). Hibernation takes place as a pupa (SZÉKELY 2011, MALKIEWICZ 2012). The moths fly at night and come to light.

BIOLOGY: Monophagous. Caterpillars have been recorded on *Populus* from June to September (FAJCIK 2003). B. MÜLLER (pers. comm.) has reared the larvae on *Populus nigra*. In rearings, pupation takes place in the ground (M. LEIPNITZ pers. comm.).

HABITAT: In forests along rivers, in *Populus* plantations and in parks. In open grassland, along roads and small rivers with bushes and small trees of *Populus* (M. LEIPNITZ pers. comm.). At lower altitudes (FAJCIK 2003). In Greece up to 1400 m and in Turkey up to 1500 m (coll. P. SKOU).

SIMILAR SPECIES: See *Stegania trimaculata* f. *cognataria* (Text-fig. 23).

GENETIC DATA: No data available.

11. Stegania dalmataria GUENÉE, 1858

dalmataria dalmataria GUENÉE, [1858], in BOISDUVAL & GUENÉE, Hist. nat. Insectes (Spec. gén. Lépid.) 10: 47, (*Stegania*). Holotype male (Zoologisches Forschungsinstitut und Museum Alexander Koenig, Bonn, Germany), (examined externally]. [Holotype is without locality data, See remarks].

arenaria STAUDINGER, 1892, Dt. ent. Z. Iris 5: 162, (*Stegania*). Syntypes 4 m, Central Asia [Kyrgyzstan]: Issyk-Kul. Originally as var.

dalmataria altaica WEHRLI, 1939, in SEITZ, Gross-Schmett. Erde 4 (Suppl.): 295, pl. 22: g, (*Lomographa*). Holotype female, [Russia]: Altai (not examined), paratype male (The Natural History Museum, London, UK) (examined externally). Valid at subspecies rank according to SCOBLE (1999).

EXTERNAL CHARACTERS AND ABDOMEN: Wingspan of both sexes 21–24 mm. Wings beige-brown in European specimens, heavily stippled with fine dark grey scales, creating striated appearance, most pronounced in forewing costa, and basal areas in fore and hindwings. Antemedial line weak, dark. Subterminal line distinct in males, with dark suffusion, angled, often fading away before costa, terminal area darker. Subterminal line weakly expressed in females, and contrast between terminal area and rest of wing barely visible. Terminal line fine. Fringes same colour as wings. Pattern of hindwings same, subterminal line angled at M3. Discal spots present in all wings, weaker in females, lunular. Wings more uniform below, stippled appearance missing. Subterminal line wide, suffused. Frons brown, mixed with yellow. Collar straw-yellow. Antennae straw-yellow, bipectinate in males, filiform in females. Thorax same colour as wings. Hindtibia with 2+2 spurs. Tympanal organs large. Male 3rd sternite without setae, 8th tergite triangular. Otherwise abdominal segments of both sexes undifferentiated.

VARIATION: The external variation of the species is quite remarkable between the isolated places where material has been examined: European specimens (Volgograd, Ural, Kazakhstan) are beige-brown with brown and grey markings, Asian specimens are straw-yellow, markings light brown, with tint of violet (Kirghizstan) or pale grey with brown and rust-brown markings (Tadzikistan). Subterminal line often suffused in males, more distinct in females.

MALE GENITALIA: Generally all European *Stegania* species have similar male genitalia, but *S. dalmataria* has the following minor difference: Saccus elongated (not triangular). Vesica is diagnostic: apical diverticulum large, curved, with large cornutus, second cornutus medially. Ductus ejaculatorius opens from medial part.

FEMALE GENITALIA: Generally as in *S. dilectaria*, but *S. dalmataria* has the following minor differences: distal part of the corpus bursae is not turned along its longitudinal axis, its angle is closer to the corpus bursae and the corpus bursae is slightly smaller.

DISTRIBUTION AND ABUNDANCE: East-European-Westasiatic. In Europe only known from south-eastern European Russia and southern Urals. – Outside Europe in Georgia, Armenia and Azerbadjan, central Asia, south-western Siberia and the Altai mountains. Reputedly recorded from Turkey (VIIDALEPP 1996, KOÇAK & KEMAL 2009), the last mentioned without any province name, but no Turkish specimens have been traced. – In Europe only few both older and more recent records from south-eastern European Russia.

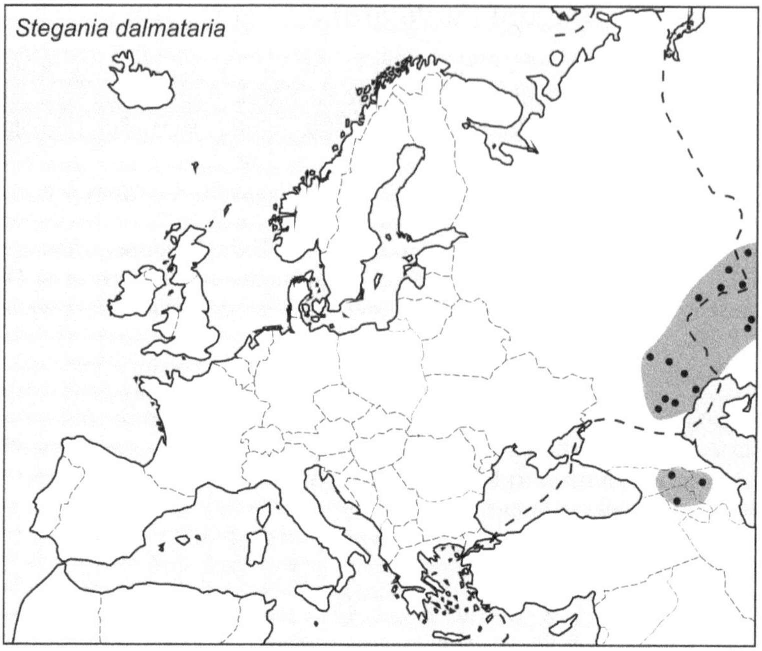

PHENOLOGY: Probably bivoltine. Scarce data from south-easternmost Europe from late May to early and late July. When material from central Asia is included the flight time appears to be from late April to late August, possibly with a gap between the generations in late June. The moth is active by night and comes to light (K. NUPPONEN pers. comm.). The hibernation takes place as a pupa (T. DRECHSEL pers. comm.).

BIOLOGY: The larva has been reared on *Populus* × *canadensis*. The pupation takes place between two leaves. Three generations were obtained by rearing (T. DRECHSEL pers. comm.). According to K. NUPPONEN (pers. comm.) the larval host plant is probably *Salix*.

HABITAT: Riversides and wet riverside meadows with plenty of *Salix*. In western Kazakhstan also in moist hollows in sandy areas where *Salix caspica* is frequent (K. NUPPONEN pers. comm.). Also found in a river valley in Kazakhstan where plenty of *Populus* trees grow (T. DRECHSEL pers. comm.). In Europe a lowland species found from 50 m up to 220 m. In central Asia the species has been found up to 2000 m.

SIMILAR SPECIES: The beige-grey colour of wings makes the species easy to idenfity.

GENETIC DATA: Nearest species: *Stegania trimaculata* (5.5%).

REMARKS: GUENÉE (1858, p. 48) mentions in the original description that the type locality of *dalmataria* GUENÉE, 1858 is 'Dalmatie'. The holotype male, illustrated in WEHRLI (1939–1954), and deposited in Zoologisches Forschungsinstitut und Museum Alexander Koenig, Bonn, Germany has three labels, but all are without locality data (D. STÜNING pers. comm.). Dalmatia, which is mostly in present-day Croatia, is, according to present knowledge about 3000 km outside the known distribution of *dalmataria*, and the holotype is similar in appearance to specimens from Volgograd (Sarepta) and European part of Kazakhstan. For these reasons, we consider Dalmatia an erroreous type locality, the holotype potentially originating from south-eastern European Russia. We illustrate a male in the adult plate

(Figure 11a), labelled as ‚Dalm.‘ (in Zoologische Staatssammlung, Munich, Germany), but consider the locality data of this specimen also dubious because the specimen is externally similar to south-western Russian specimens. Possibly the locality data was added to the specimen later on. LERAUT (2009) noted that *S. arenaria* STAUDINGER occurs sympatrically with the nominate subspecies. Therefore he considers it to be valid at species level, unlike SCOBLE (1999), who considered it a junior synonym of *dalmataria* GUENÉE.

Macariini: The *Platypepla* group of genera

The group is diagnosed by KRÜGER (2001) and it is defined by two likely synapomorphies: the complete or near-complete separation of costa and sacculus in the male genitalia and the partial or total covering of the corpus bursae wall in denticles or spines in the female genitalia. The cremaster of pupa with lateral filamentous processes (illustrated in KRÜGER 2001) may prove to be a further synapomorphy. Most species are yellowish-ochre or dark grey or brown. Included European genera in the group are *Acanthovalva* KRÜGER and *Narraga* WALKER.

Acanthovalva KRÜGER, 2001

Acanthovalva KRÜGER, 2001, Bull. nat. Hist. Mus. Lond. (Ent.) 70 (1): 33, 31 (key). Type species: *Geometra inconspicuaria* HÜBNER, [1819]. There are no genus synonyms.

DIVERSITY AND DISTRIBUTION: *Acanthovalva* contains six species in the Afrotropical region including Madagascar and the Palaearctic. Potentially one additional species in the Nearctic region and one in the Oriental region should be combined to *Acanthovalva* (KRÜGER 2001). One species in Europe.

EXTERNAL CHARACTERS AND ABDOMEN: Small to medium-sized, wings grey to greyish-brown to orange, wings rounded. Three transverse lines dark, wings suffused, otherwise markings simple. Male antennae bipectinate with long rami, forming plumose appearance, female antennae ciliate or very finely serrated. Male 3rd sternite without setae, 8th sternite modified in numerous taxa.

MALE GENITALIA: Uncus with numerous strong spines. Gnathos absent. Valva bilobed, connected only at base. Dorsal arm of valva with spines, particularly its ventral margin. Ventral arm of valva wide. Vesica without cornuti, often with small diverticulum (see Table 1).

FEMALE GENITALIA: Region around ostium bursae variably sclerotised. Corpus bursae round or elongated. Corpus bursae densely or sparsely spinose, or with small signum (see Table 1).

BIOLOGY: The larva of *A. inconspicuaria* has been recorded on *Trifolium* (REDONDO *et al.* 2009). Species are found for instance in open habitats, secondary forests and savannas. Most species are nocturnal, *A. focularia* (GEYER, 1837), which is known from South Africa and Namibia, is also diurnal (KRÜGER 2001).

IMMATURE STAGES: Data available for *A. inconspicuaria* only, egg resembling the standard *Chiasmia*-type in shape but more rounded (KRÜGER 2001). No data available on pupa.

REMARKS: *Acanthovalva* and *Narraga* are considered closely related (KRÜGER 2001). Systematic position: Macariini (KRÜGER 2001, SCOBLE & KRÜGER 2002).

12. *Acanthovalva inconspicuaria* (HÜBNER, 1819)

inconspicuaria inconspicuaria HÜBNER, [1819], Samml. Eur. Schmett. 5. Geometrae (1): pl. 97, fig. 500, (*Geometra*). Syntype(s), lost, [Europe]. Synonyms are after KRÜGER (2001).

cinerascens BUTLER, 1875, Ann. Mag. Nat. Hist. (4) 16: 418, (*Acidalia*). Holotype female (The Natural History Museum, London, UK), [South Africa]: Natal (examined externally).

cinnamomaria ROTHSCHILD, 1914, Novit. zool. 21: 353, (*Tephrina*). Holotype female (The Natural History Museum, London, UK), Algeria (central): Guelt-es-Stel (examined externally).

osyraria GUENÉE, [1858], in BOISDUVAL & GUENÉE, Hist. nat. Insectes (Spec. gén. Lépid.) 10: 148, (*Selidosema*). Holotype male, [Spain]: Andalusia.

perturbata BASTELBERGER, 1908, Jb. Nassau. Ver. Naturk. 61: 74, (*Tephrina*). Holotype male (Naturwissenschaftliche Sammlung, Museum Wiesbaden, Germany), German South-West Africa [Namibia]: Gochas.

pumicaria LEDERER, 1855, Verh. zool. –bot. Ver. Wien 5 (Abh.): 213, pl. 3, fig. 6, (*Eubolia*). Syntypes 3 males, [Eastern Mediterranean, Beirut (Lebanon)] (examined externally, based on original description). See remarks.

inconspicuaria tsekubia WEHRLI, 1940, in SEITZ, Gross-Schmett. Erde 4 (Suppl.): 401, pl. 32: e, (*Tephrina*). Syntypes 6 males, China: Szechuan (Sichuan), Tseku. [Originally as subsp. of *pumicaria*]. See remarks.

EXTERNAL CHARACTERS AND ABDOMEN: Wingspan of both sexes 21–24 mm (spring generation), 17–18 mm (summer generation), no data on November specimens (REDONDO *et al.* 2009). Wings grey to grey-brown to brown, markings weak and dark brown. Transverse lines faint, weakly undulating, present as weak spots in costa. Medial line most pronounced, also in hindwing. Wings often heavily irrorated with dark brown. Terminal area often darkest. Terminal line faint, more or less continuous, dark brown. Cilia lighter than wings. Discal spots present in all wings, distinct, elongated. Wings below lighter, transverse lines weaker, densely striated with greyish brown. Frons and collar whitish-grey, antennae similarly coloured dorsally. Male antennae bipectinate with long rami, forming plumose appearance, female antennae weakly fasciculate. Thorax and abdomen same colours as wings. Hindtibia of both sexes 2+2 spurs, pairs often of unequal length. Tympanal organs medium-sized. Male 8th sternite slightly triangular, posterior margin concave medially, extensions small and round. Otherwise abdominal segments of both sexes undifferentiated.

VARIATION: Transverse lines and discal spots may be absent, and ground colour varies from grey to grey-brown to brown.

MALE GENITALIA: Uncus prominent, setose, apex acute. Base of uncus spinose dorsally. Gnathos absent. Juxta minute. Tegumen sclerotised, prominent plate. Pedunculi of tegumen wide, notable plates. Valva massive, divided deeply. Dorsal arm of valva wide, with ventral lobe at 2/3 length. Its ventral margin spinose, medial area setose and apex wide, round. Ventral arm of valva wide, triangular, margins upturned and setose. Saccus elongated, membranous. Aedeagus long, narrow, curved. Caecum long, apex spoon-shaped. Vesica with finger-like diverticulum near base, expanded and angled in its middle part. Sclerotisations and cornuti absent.

FEMALE GENITALIA: Papillae anales large, slightly elongated. Apophyses straight, long, narrow. Lamella antevaginalis horizontally sclerotised ridge with small, round, sclerotised plate medially. Lamella postvaginalis narrow, horizontally sclerotised ridge with membranous, curved sutura more caudally. Caudal part of ductus bursae sclerotised, wide, tapering.

Cephalad part of ductus bursae membranous, short, narrow. Caudal part of corpus bursae slightly elongated. Corpus bursae roundish, spinose all over.

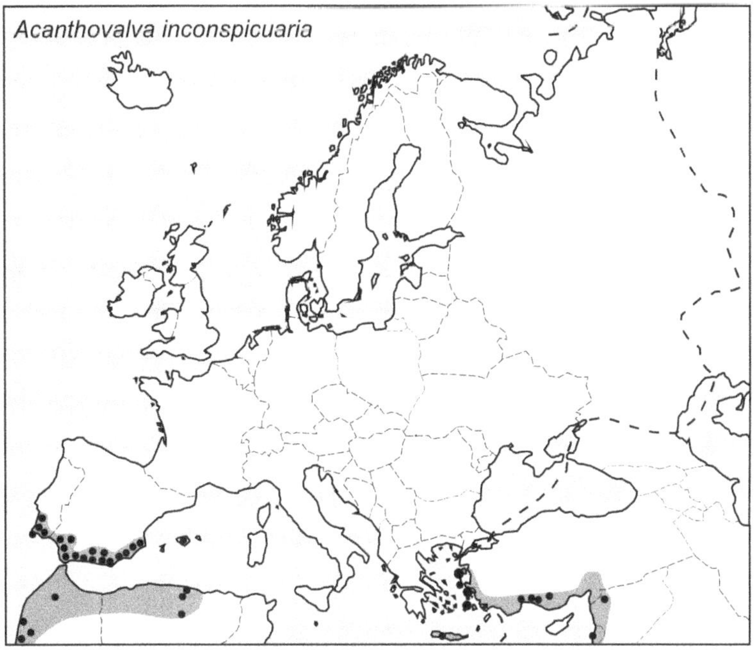

DISTRIBUTION AND ABUNDANCE: Palaeotropical. In Europe disjunct occurence in the south and south-east of the Iberian Peninsula and on the Greek islands of Crete, Kos, Samos and Mytilini. – Outside Europe widely distributed through Africa to South Africa and in Asia in southern Turkey, Cyprus, through the Middle East and Arabia to Pakistan. – In Spain a common species on suitable habitats where it often can be abundant, in the Greek islands apparently more scarce.

PHENOLOGY: Bi- or trivoltine in Europe. In the Iberian Peninsula from early April to mid-October in several and apparently overlapping generations, in Portugal to early November (M. CORLEY pers. comm.). In Samos in two generations from mid-May to early June and from late August to mid November (FRITSCH *et al.* 2014). The moths fly at night and both males and females come readily to light. It also flies by daytime, but only in sunshine, short distances within a very limited area (M. LEIPNITZ pers. comm.). In Morocco the larva has been found in early April at 600 m altitude (H. RIETZ pers. comm.), but apart from that, larval time appears to be unknown.

BIOLOGY: Monophagous(?). The larva has been recorded on *Trifolium* (REDONDO *et al.* 2009). Despite a number of rearing attempts from eggs with a variety of plants from the locality, such as *Erica scoparia*, *E. arborea*, *Ulex*, *Trifolium* and *Lavendula*, the caterpillars refused all plant species offered to them (M. LEIPNITZ pers. comm.).

HABITAT: A rather large variety of open lowland habitats and it has been recorded from sea level up to 650 m and according to REDONDO *et al.* (2009) even up to 2000 m in Sierra Nevada in southern Spain. In Morocco at 1900 m (N. PÖLL pers. comm.). Probably records from these high altitudes are wind blown specimens.

SIMILAR SPECIES: There are no similar species in Europe. *Isturgia catalaunaria* (GUENÉE) (see text under *catalaunaria*) and *I. spodiaria* (LEFÈBVRE) are externally somewhat similar but the genitalia are distinctly different.

GENETIC DATA: Genetically heterogeneous, with populations from Spain corresponding well to those from Iran, Yemen and Morocco to western Africa and Kenya in the South (n=22). Populations from Greece to Turkey, Lebanon and Israel diverging by 1.7%, populations from South Africa by 3.6%, the latter probably not conspecific. Nearest species: several non-congeneric and non-related species from other continents at 8.5–9% distance (long branch attraction), *A. focularia* (GEYER, 1837) from South Africa at 12.5% distance and probably not closely related either.

REMARKS: We follow KRÜGER (2001), who combined *inconspicuaria* (HÜBNER), with *Acanthovalva* KRÜGER, 2001. REDONDO & GASTÓN (1999, p. 9) combined *inconspicuaria* with *Tephrina* GUENÉE, 1858. SCOBLE (1999) provisionally combined *inconspicuaria* with putative '*Semiothisa*'. KRÜGER (2001) noted that WEHRLI (1940) described large examples of what he considered to be *inconspicuaria* from west China as subsp. *tsekubia* WEHRLI. KRÜGER has examined a topotypical male of *tsekubia* (M. KRÜGER genitalia slide No. 6 (Zoologisches Forschungsinstitut und Museum Alexander Koenig, Bonn, Germany) and concluded that *tsekubia* is not conspecific, although closely related (KRÜGER 2001). VIIDALEPP (1996) treated *pumicaria* (LEDERER, 1855) valid at species level. In the absence of authentic material, we follow earlier authors (FLETCHER 1978, SCOBLE 1999, KRÜGER 2001) and consider it a junior synonym of *inconspicuaria* (HÜBNER, 1819). The original hand-drawn black-and-white illustration does not allow unambiguous identification (LEDERER 1855, pl. 3, fig. 6).

Narraga WALKER, 1861

Narraga WALKER, 1861, List Specimens lepid. Insects Colln Br. Mus. 23: 948 (key). Type species: *Geometra cebraria* HÜBNER, [1799] 1796, a junior synonym of *Phalaena fasciolaria* HUFNAGEL, 1767, Europe. Genus synonymy is after HODGES *et al.* (1983) and SCOBLE & KRÜGER (2002).

Fernaldella HULST, 1896. Type species: *Fidonia fimetaria* GROTE & ROBINSON, 1870. USA.

DIVERSITY AND DISTRIBUTION: *Narraga* contains provisionally eight species, and it is Holarctic in distribution. Species occur in Europe, mostly in central and southern Europe, and in North Afica, Asia Minor, China, and North America (SCOBLE & KRÜGER 2002).

EXTERNAL CHARACTERS AND ABDOMEN: Small moths. Wings brown, tessellated, usually with light yellow to whitish markings on both upperside and underside. Hindwing underside and forewing apex lighter, markings sharper. Fringes chequered. Fovea absent. Chaetosemata extended across head as broad band. Antenna of male bipectinate. Male 8th sternite emarginate in European species.

MALE GENITALIA: Uncus pointed, horns lacking but setae well-developed. Gnathos arms wide, medial element prominent, upturned. Valva deeply divided, but not completely separated. Dorsal arm of valva widest at apex, setose. Ventral arm of valva wider, apex may be bent ventrally. Saccus often prominent. Vesica with or without sclerotisations (see Table 1).

Table 1. Diagnostic characters of the European Macariini genera. External characters are in several instances too variable to provide diagnostic features above species-groups levels. Numerous characters, as presented in the table, are not unique and the morphological differences between the genera are small. Particularly the diagnostic characters of *Isturgia* HÜBNER, 1823 are diffuse. Data is based on KRÜGER (2001), SCOBLE & KRÜGER (2002), and our own observations.

Genus	Selected diagnostic characters	Number of European species
Acanthovalva	Uncus with numerous strong spines dorsally. Sclerite below uncus absent. Gnathos absent. Male 8th sternite slightly concave. Signum absent, corpus bursae spinose.	1
Narraga	Uncus setose dorsally. Sclerite below uncus absent. Gnathos prominent. Male 8th sternite emarginate. Signum absent, or with 1 or 2 small signa.	3
Heliomata	Uncus setose dorsally. Sclerite below uncus absent. Juxta with cross medially. Male 8th sternite not differentiated. Male hindtibia swollen and with hair pencil. Signum stellate.	1
Isturgia	Uncus setose dorsally. Sclerite below uncus absent. Ventral arm of valva wide, somewhat triangular or pointed. Male 8th sternite undifferentiated, weakly emarginate or with small projections. Lamella postvaginalis narrow, lunular plate. Signum stellate or absent.	16
Itame	Uncus setose dorsally. Sclerite below uncus absent. Ventral arm of valva rounded. Male 8th sternite with small projections. Signum stellate.	1
Neognopharmia	Uncus setose dorsally, with medial extension. Ventral arm of valva dentate. Male 8th sternite with small projections. Papillae anales with modified setae. Signum stellate.	1
Macaria	Uncus horns present, moderate or small. Sclerite below uncus absent. Ventral arm of valva rounded or weakly elongated. Male 8th sternite deeply cleft. Lamella postvaginalis narrow, lunular plate. Signum stellate.	11
Chiasmia	Uncus horns present, spine-like. Sclerite below uncus present. Ventral arm of valva often flat. Male 8th sternite excavated. Lamella postvaginalis prominent plate. Signum stellate or absent.	3
Digrammia	Uncus horns with 2–4 pairs of strong, subapical setae. Sclerite below uncus absent. Ventral arm of valva often with acute apical process. Lamella postvaginalis bilobed, 'winged'. Signum absent.	1
Rhoptria	Uncus densely setose. Ventral arm of valva wide, flat. Male 8th sternite weakly concave, not otherwise modified. Vesica with several long cornuti. Posterior part of corpus bursae long, curved, striated. Signum stellate or absent.	2
Ortaliella	Uncus densely setose. Ventral arm of valva narrow, its dorsal margin with sclerotised process. Juxta downturned. Male 8th sternite weakly concave, not otherwise modified. Vesical with several short cornuti. Papillae anales, apophyses posteriores, apophyses anteriores and 9th segment extremely long. Signum stellate.	1

FEMALE GENITALIA: Lamella postvaginalis often irregular, horizontally sclerotised ridge. Posterior part of ductus bursae sclerotised, long, often funnel-shaped. Corpus bursae without signum, or if present, with one or two small signa having large denticulate projections (see Table 1).

BIOLOGY: Larvae have been recorded feeding on Compositae (Asteraceae) (SCOBLE & KRÜGER 2002): *N. fasciolaria*, *N. tessularia* (FORSTER & WOHLFAHRT 1981) and *N. nelvae* have been found on *Artemisia* (GELBRECHT 1998, REDONDO *et al.* 2009), North American *N. fimetaria* is recorded from *Gutierrezia* species and *N. georgiana* from *Chrysoma pauciflosculosa* (FERGUSON 2008).

IMMATURE STAGES: Cremaster of pupa rather short, bifurcate, or with two additional and very short projections giving serrate appearance (PATOČKA & TURČÁNI 2005).

REMARKS: FERGUSON (2008) classified the three North American species in the genus *Fernaldella* Hulst, 1896. We agree with the view of SCOBLE & KRÜGER (2002), and consider *Fernaldella* a junior synonym of *Narraga*. This is supported by tessellated wing pattern, excavated valva in the male genitalia, signum structure and association of larva with Compositae (Asteraceae, *Artemisia* in European species), among other things (SCOBLE & KRÜGER 2002). Wings are held folded over the back at rest (KRÜGER 2011). FERGUSON (2008) reports the same in North American species (as *Fernaldella*) and with only the pattern of the underside of the hindwing and tip of the forewing showing. *Narraga* species may be both diurnal and nocturnal. Systematic position: Macariini (KRÜGER 2001, SCOBLE & KRÜGER 2002).

13. *Narraga fasciolaria* (HUFNAGEL, 1767)

fasciolaria fasciolaria HUFNAGEL, 1767, Berlin Mag. 4 (5): 516, (*Phalaena*). Syntype(s), lost, [Germany]: Berlin region.

cebraria HÜBNER, [1799], Samml. Eur. Schmett. 5 Geometrae (1): pl. 24, fig. 129, (*Geometra*). Syntype(s), lost, Europe.

fasciolaria danubialis MOUCHA & POVOLNY, 1957, Cas. ceské Spol. ent. 54 (3): 220, pl. 2, figs. 9–13; pl. 5, figs 39, 40, (*Narraga*). Holotype male (The Natural History Museum, London, UK), [North-East] Hungary: Nagy Nyir (examined externally). Originally as subspecies, treated valid at subspecies rank by SCOBLE (1999), downgraded here from subspecies rank (new synonym), quoting absence of disjunct differential features, falling within intraspecific variation.

zebraria DUPONCHEL, 1829, in Godart & Duponchel, Hist. nat. Lépid. Papillons Fr. 7 (2): 428, pl. 166, fig. 1, (*Fidonia*). Emendation of *cebraria* HÜBNER.

fasciolaria fumipennis PROUT, 1915, in SEITZ, The Macrolepidoptera of the World 4: 397, (*Narraga*). Holotype male (The Natural History Museum, London, UK), [China]: Peking. Valid at subspecies rank according to SCOBLE (1999) (examined externally).

fasciolaria turkestanica MOUCHA & POVOLNY, 1957, Cas ceské Spol. ent 54(3): 222, pl. 1, figs 6, 7; pl. 5, fig. 37, (*Narraga*). Holotype male (The Natural History Museum, London, UK), E Turkestan: Valley Malaya River, Almatinka, 4000 ft. Valid at subspecies rank according to SCOBLE (1999) (examined externally).

INFRASUBSPECIFIC NAMES, POTENTIALLY UNAVAILABLE: *subtusochrea* AGENJO, 1956 (f.).

EXTERNAL CHARACTERS AND ABDOMEN: Wingspan 17–21 mm in both sexes. Wings ochreous brown in males, ochreous areas more prominent in females. Wide brown medial lines forming a tessellated pattern. Postmedial line fused to brown terminal area near tornus. Terminal line brown, fringes chequered with ochreous and brown. Lines continued on hindwings except antemedial line weak. Discal spots rarely visible. Wing patterns repeated below, hindwings and forewing apex paler, markings in those areas sharper. Frons and collar whitish brown. Antennae chequered whitish brown, male antennae bipectinate, female antennae filiform. Thorax whitish brown. Hindtibia with 2+2 spurs. Abdomen whitish brown. Tympanal organs large, ansa narrow at base, central flap round, apex with weak hammer-headed tip. Posterior margin of male 8th sternite deeply emarginate. Other sternites and tergites of both sexes undifferentiated.

VARIATION: Amount of brown markings variable from bicolorous ochreous brown to almost uniform dark brown.

MALE GENITALIA: Uncus short, stout, weakly setose, apex bare. Gnathos arms wide, fused ventrally, forming small upright hook. Juxta membranous. Valva deeply divided. Dorsal arm of valva long, narrow, straight, weakly channel-shaped, apex round, setose. Ventral arm of valva wide, apex with round ventral extension. Sacculus with sclerotised projection (harpe?). Saccus small, weakly elongated medially. Aedeagus straight, narrower towards apex. Caecum short, apex weakly sclerotised and round. Vesica simple tube, without diverticula and sclerotisations.

FEMALE GENITALIA: Papillae anales soft, setose. Apophyses straight, narrow. Lamella postvaginalis distinct sclerotised band, small tooth medially. Ventral margin of ostium bursae slightly irregular. Ductus bursae narrow, sclerotised. Posterior part of corpus bursae narrow, membranous, apical part round sac. One or two small signa with denticulate projections.

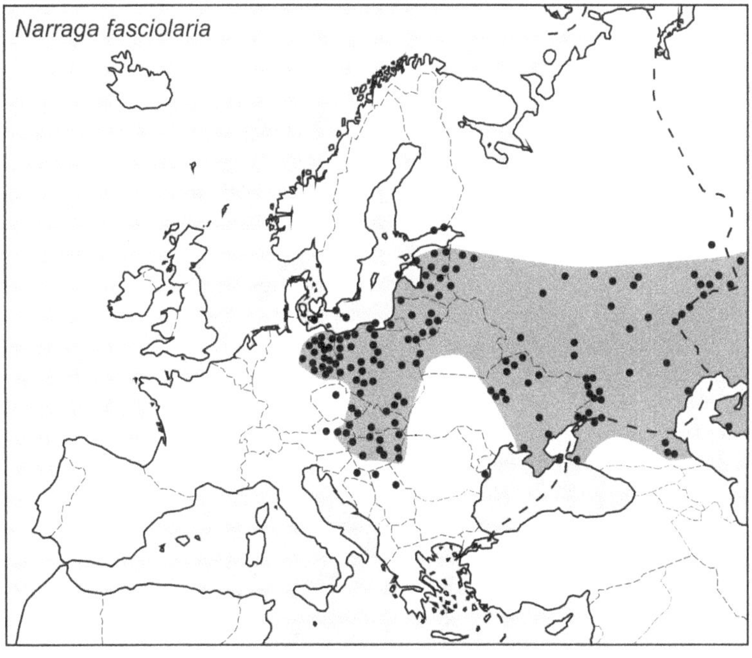

DISTRIBUTION AND ABUNDANCE: Palaearctic. From central Europe including eastern Germany to the east through a large part of eastern Europe including the north-westernmost part of the Balkans to south-eastern European Russia and the southern Urals. North of the Baltic Sea, the species is found as a rare migrant in Finland, Sweden and Denmark. – Outside Europe known from the Urals through central Asia and southern Siberia, the Altai mountains and Mongolia to northern China, Far East Russia, Korea and Japan. The species has been recorded from Turkey by RIEMIS (1994) and (KOÇAK & KEMAL 2009), but according to GELBRECHT (1998) this is probably due to a misidentification of *N. nelvae* subsp.

cappadocica. All *Narraga* specimens we have been able to trace from Turkey have turned out to be *N. nelvae* subsp. *cappadocica*. – In most of the European distribution area a locally occurring species that is often common in its localities.

PHENOLOGY: Bivoltine. In Germany the species flies in two overlapping generations from mid-May to late August, 2nd generation only partial in Germany (GELBRECHT 1998). Further south from early May. The species is most abundant in late June and in July (GELBRECHT 1998). In rearings the vast majority of pupae hibernated and emerged the year after (B. MÜLLER pers. comm.). This could indicate that the second generation is only partial. Larvae from late June to October. Hibernation takes place as a pupa (H. WEGNER pers. comm.). The adults are easily flushed up by day and fly also actively by day. K. NUPPONEN (pers. comm.) reports that the moths fly around at sunset and at dusk, but they are not really diurnal. At night they come to light, but not so often (GELBRECHT 1998). M. LEIPNITZ (pers. comm.) reports that the moths fly only in sunshine. In Denmark, where the species only occurs as a migrant, all recorded specimens have come to light and no specimens have been seen flying by day.

BIOLOGY: Monophagous. Larva has been found on *Artemisia campestris* (GELBRECHT 1998). The pupation takes place in the ground (M. LEIPNITZ pers. comm.).

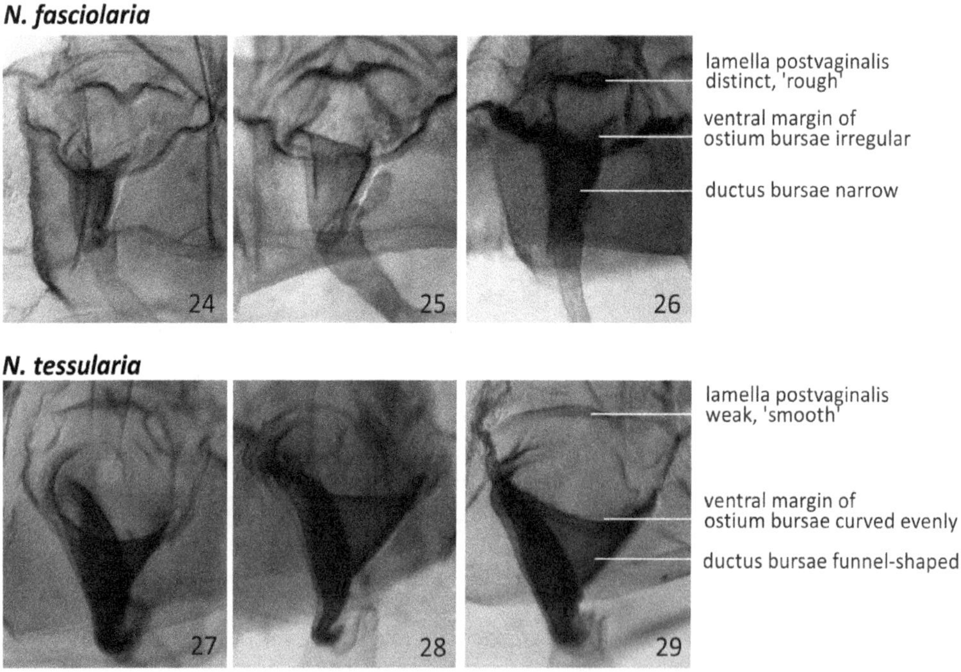

Text-figs 24–29. Diagnostic female genitalia characters (indicated) of *Narraga fasciolaria* (HUFNAGEL, 1767) and *N. tessularia* (METZNER, 1845). Text-figs 24–26. *N. fasciolaria*. 24. Russia: SW-Altai, Katun valley, 1200 m, 22–27.vi.1983, slide PS1353 (coll. ZMH). 25. Germany: Zehnedick, 20.v.1949, slide PS1176 (coll. ZMH). 26. Germany: Zehnedick, 28.v.1956, slide PS1178 (coll. ZMH). Text-figs 27–29 *N. tessularia*. 27. Austria: Illmitz am Neusiedler-see, ex larva, July 1956, slide PS1181 (coll. ZMH). 28. Russia: Tuva republic, E. Tannu-Ola mts, 1250–2000 m, 17.vi.1995, slide PS1354 (coll. ZMH). 29. Austria: Illmitz, ex larva 1982, slide PS1203 (coll. SKOU).

HABITAT: A xerothermophilous species occurring in dry sandy places such as fallow fields, edges of sandpits, forest fire breaks, along roads and railroads and military training areas (GELBRECHT 1998). In Slovakia up to 500 m (Z. LAŠTŮVKA pers. comm.) and in south-eastern Kazakhstan found at 800 m altitude. In Turkey at 2500 m (K. NUPPONEN pers. comm.).

SIMILAR SPECIES. All European *Narraga* species are externally similar, therefore we recommend the genitalia to be used in the identification. Both male and female genitalia structures are distinct (see genitalia plates and Text-figs 24–26). Females of *N. tessularia* are often small and clearly contrasting brown-white, perhaps being the only *Narraga* specimens that can be reliably identified on the basis of external features alone.

GENETIC DATA: Genetically homogeneous from Germany (n=2) to north-eastern China (n=13; subsp. *fumipennis* Prout, 1915), but populations of Samara and Kazakhstan diverging by 1.9% (subsp. *turkestanica*?). Nearest species: *Narraga tessularia* (6.3%).

REMARKS: When this book was prepared, it became evident that in collections *N. fasciolaria* and *N. tessularia* were mixed. We recommend that genitalia be used in the identification of European *Narraga* species so that distribution of these taxa may be known in better detail.

14. *Narraga nelvae* (ROTHSCHILD, 1912)

nelvae nelvae ROTHSCHILD, 1912, Novit. zool. 19: 126, (*Fidonia*). Syntypes including 1 male, 4 females (The Natural History Museum, London, UK), Algeria: Batna (examined externally).

nelvae cappadocica HERBULOT, 1943, Bull. Soc. ent. Fr. 48 (6): 84, (*Narraga*). Holotype male (Muséum National d'Histoire Naturelle, Paris, France), Asia Minor [Turkey]: Caesarea (examined, including genitalia). Originally as subsp. of *nelvae*, treated valid at species rank for instance in VIIDALEPP (1996) and SCOBLE (1999). Revised here to the original status at subspecies level, quoting allopatric occurrence (stat. rev.).

nelvae catalaunica HERBULOT, 1943, Bull. Soc. ent. Fr. 48 (6): 83, (*Narraga*). Holotype male (Zoologische Staatssammlung, Munich, Germany), [Spain]: Catalonia, Province de Lerida, Juneda (examined externally).

EXTERNAL CHARACTERS AND ABDOMEN, description is based on the European specimens, subsp. ***catalaunica*** from Spain (Text-fig. 31): Wingspan 16–21 mm in both sexes. Wing pattern similar to *N. fasciolaria* (see above), with the following minor, quantitative differences. In *N. nelvae* subsp. *catalaunica* female markings are sharper. Suffused, brown discal spots are normally present in all wings. Fringes chequered whitish and brown. Anterior margin of male 8th sternite weakly concave medially, posterior margin deeply concave, membrane between 8th and 9th segments with two membranous pockets. Other sternites and tergites of both sexes undifferentiated.

Subsp. ***nelvae*** (Text-fig. 30) slightly smaller (16–19 mm). Externally similar to European populations (*N. nelvae* subsp. *catalaunica*), but the male genitalia capsule and the aedeagus are considerably smaller than those in the Spanish and Turkish populations. Aedeagus apex occasionally with minute dentate sclerotisations. Female genitalia are similar in all subspecies.

Subsp. ***cappadocica*** (Text-fig. 32) slightly smaller (16–19 mm). Externally similar to nominal subspecies, the male genitalia capsule and the aedeagus are considerably larger than in the North African populations, and about same size as those in the Spanish specimens.

Aedeagus apex often with small dentate sclerotisations. Female genitalia are similar in all subspecies, that of *N. nelvae* subsp. *cappadocica* (as *N. cappadocica*) is illustrated in LERAUT (2009). Hostplant of *N. nelvae* subsp. *cappadocica* is an *Artemisia* species from the *A. absinthium* group (J. GELBRECHT pers. comm.).

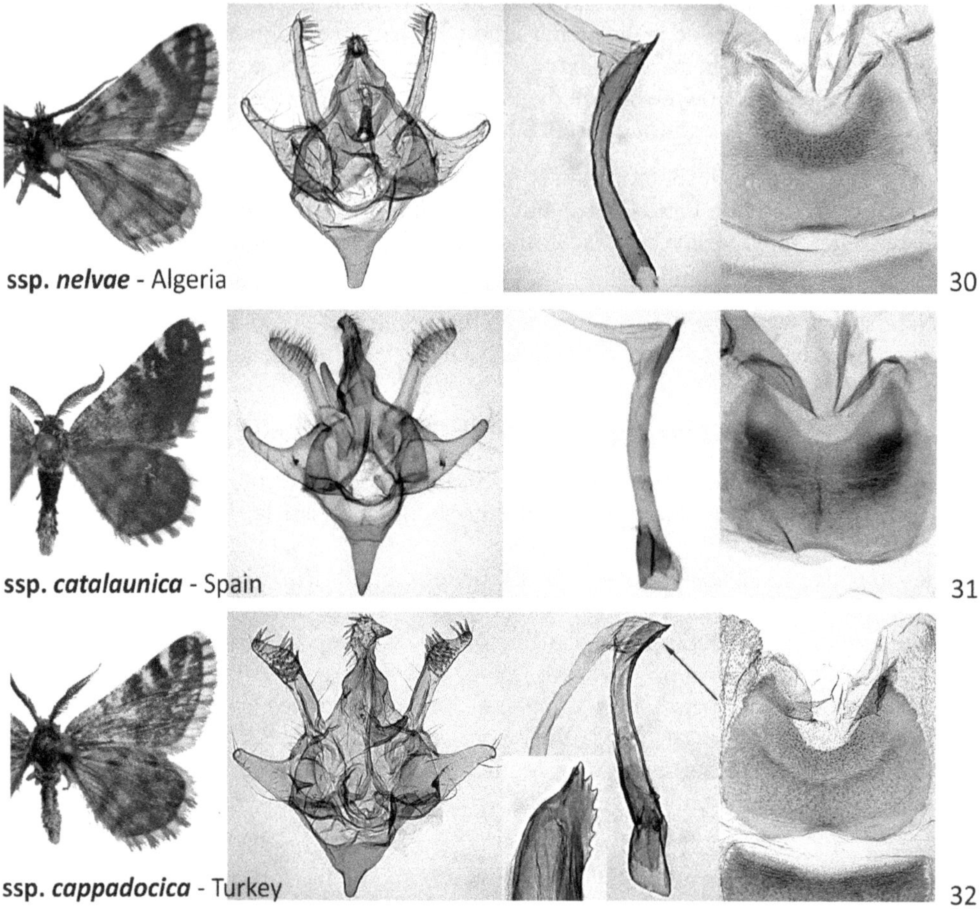

Text-figs. 30–32. External appearance and genitalia of *Narraga nelvae* (ROTHSCHILD, 1912) in Algeria (subsp. *nelvae* (ROTHSCHILD, 1912)), Spain (subsp. *catalaunica* HERBULOT, 1943) and Turkey (subsp. *cappadocica* HERBULOT, 1943). North African specimens are the smallest, including the genitalia capsule and the aedeagus. Turkish specimens often have small dentate sclerotisation in the aedeagus apex (shown enlarged in insert), whereas these are smaller in North African specimens and often absent in Spanish specimens. Length of saccus is variable, even within each subspecies. Female genitalia are similar in all subspecies. Text-fig. 30. subsp. *nelvae*. Adult and all genitalia figures from Algeria: Lambessa [Lambaesa], 1902, slide PS1652 (coll. ZSM). Text-fig. 31. subsp. *catalaunica*. Adult: Spain, prov. Murcia, Bolnuevo, 15.iv.2001 (coll. SKOU); genitalia: Spain: Granada, 17.vii.1980, slide PS1481 (coll. SKOU); vesica and 8th sternite: Spain: Granada, 800 m, 27.v.1988, slide PS1204 (coll. SKOU). Text-fig. 32. subsp. *cappadocica*. Adult and aedeagus spines: Turkey: Kayseri, 25.v.1997, slide PS2003 (coll. NUPPONEN); genitalia and vesica: Turkey: Pontisches Gebirge, Ilgaz, ca. 800 m, 23–25.vi.1992, slide PS1790 (coll. MÜLLER); 8th sternite: Turkey: Pontisches Gebirge, Ilgaz, ca. 800 m, 6–7.vii.1992, slide PS1720 (coll. MÜLLER). Figures are not to scale.

VARIATION: Amount of brown markings slightly variable. Light brown markings on forewing costa normally clearly visible but they may be suffused and less contrasting.

MALE GENITALIA: Uncus well-developed, forming downwards turning hook, setose, apex acute. Gnathos arms wide, fused ventrally, forming upright hook. Juxta weakly developed, mostly membranous. Valva deeply divided. Dorsal arm of valva long, apex spoon-shaped, setose. Ventral arm of valva wider at base, narrowing to apex, apex roundish, margins sparsely setose, sclerotised tooth midventrally. Saccus wide at base, elongated medially. Aedeagus long, narrow, evenly curved, caecum very short, apex sclerotised and acute, occasionally with minute dentate sclerotisations. Vesica widest at base, weakly curved, without diverticula and sclerotisations.

FEMALE GENITALIA: Papillae anales soft, setose. Apophyses straight, narrow. Lamella postvaginalis large, weakly sclerotised, lunular plate. Posterior margin of ostium bursae very prominent, sclerotised, often with shallow invagination medially. Ductus bursae curved, sclerotised, long. Corpus bursae oval, membranous, weak. Signum absent.

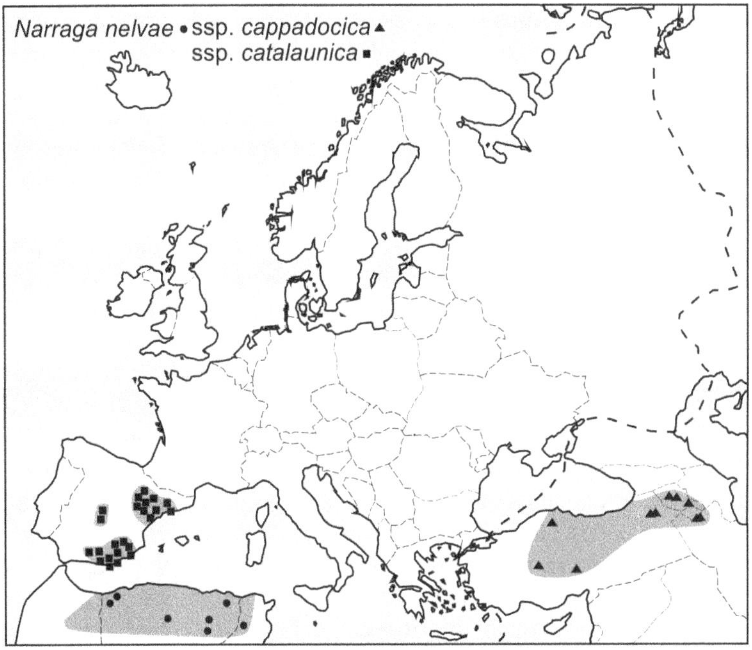

DISTRIBUTION AND ABUNDANCE: Mediterranean. In Europe only in Spain (subsp. *catalaunica*). – Outside Europe in Turkey and Armenia (subsp. *cappadocica*) and in North Africa in Morocco, Algeria and Tunisia (subsp. *nelvae*). – In Spain a somewhat local species, but where it occurs it is often found in numbers.

PHENOLOGY: Bi- or trivoltine. The adult moths fly from mid-April until late September in two or three overlapping generations. The hibernation probably takes place as pupa. In rearings pupation takes place in the ground (M. LEIPNITZ pers. comm.). K. NUPPONEN (pers. comm.) reports that the moths fly around at sunset and at dusk, but he does not regard the species to be to be truely day flying. G. KING (pers. comm.) has repeatedly seen the moths flying actively by day. The moths are also active at night and come readily to light, often in numbers.

BIOLOGY: Monophagous. The caterpillar has been found feeding on *Artemisia herba-alba* (G. KING pers. comm.). The pupation takes place in the ground (M. LEIPNITZ pers. comm.).

HABITAT: Xerothermophilous. Open localities with low and scattered vegetation. From sea level and mainly up to about 800 m altitude, in central Spain up to 2000 m according to REDONDO *et al.* (2009), in Turkey up to 1550 m. Possibly the species is not resident at such high altitudes but is migrating altitudinally.

SIMILAR SPECIES. All European *Narraga* species are externally similar, therefore we recommend that genitalia be used for identification. Both male and female genitalia structures are distinct (see genitalia plates and Text-figs 24–32).

GENETIC DATA: Genetically heterogenous, two barcoded specimens from Spain (subsp. *catalaunica*; including the barcoded holotype) diverging from three Turkish specimens (subsp. *cappadocica*) by 5.9%. Nearest species: *Narraga tessularia* (9.0%).

REMARKS: *N. nelvae* is a widespread species with three putative subspecies, requiring an extensive integrative (morphology, biology, DNA) revision. The discovery that the Spanish and Turkish specimens are genetically considerably divergent (5.9%) may require, if supported by future materials also, these taxa to be considered valid at species level. Current genetic material is too limited (no material analysed from North Africa), therefore we consider the North African (subsp. *nelvae*), the Spanish (subsp. *catalaunica*) and the Turkish (subsp. *cappadocica*) populations valid at subspecies level. The populations are, according to current knowledge, allopatric and the external appearance and the genitalia morphology of the populations are without any distinct differences, except that the male genitalia capsule and the aedeagus in North African populations are considerably smaller than in the Spanish and Turkish populations. REDONDO *et al.* (2009, p. 43) concluded that subsp. *catalaunica* 'does not show any significant differences from other specimens from North Africa, and this is very likely to be just a variant form, if a common one, of the species, and lies within the range of geographic variation of the species [*nelvae*]'. LERAUT (2009) raised *N. catalaunica* HERBULOT, to species rank by small quantitative differences in the male genitalia. He did not discuss or illustrate the female genitalia. LERAUT (2009) did not indicate on what material the conclusions were based on, and whether any type material was examined. Based on our observations, including examination of the type specimens, the indicated quantitative differences are minute, and depend on how the cover glass has been placed on top of the genitalia. *Narraga nelvae* subsp. *cappadocica* was originally described as valid at subspecies level, we reinstate here the original combination at subspecies level, following HERBULOT (1943).

The distribution of *N. nelvae* is unusual, resembling that of the *Cinglis humifusaria* (EVERSMANN, 1837) and *C. andalusiaria* WAGNER, 1935 species complex (HAUSMANN 2004).

15. Narraga tessularia (METZNER, 1845)

tessularia tessularia METZNER, 1845, Stettin. ent. Ztg. 6 (6): 186, (*Geom*[*etra*]). Syntypes males, females, Algeria; Russia (southern). See remarks.

atromacularia HERRICH-SCHÄFFER, 1848, Syst. Bearbeitung Schmett. Eur. 3 (31): 91; ibidem (1846), pl. 35, figs. 216, 217 (non binominal); pl. 36, figs. 218-220 (non binominal); (1848), pl. 59, fig. 367 (non binominal), (*Fidonia*). Syntypes males, females, [Russia]: Ural.

baltearia FREYER, 1845, Neuere Beitr. Schmettkde 5 (79): 153, pl. 474, figs. 1, 2, (*Fidonia*). Syntype(s), Mai monat, Orenburg.

isabel AGENJO, 1956, Eos. Madr. 32: 22, pl. 1, figs. 1, 2, (*Narraga*). Holotype male (Museo Nacional de Ciencias Naturales, Madrid, Spain), Spain: Granada, de Puebla de Don Fadrique.

tessularia kasyi MOUCHA & POVOLNY, 1957, Cas. ceské Spol. ent. 54 (3): 230, pl. 2, fig. 16; pl. 3, figs. 17–22; pl. 6, fig. 49; text-figs 2, 4b, (*Narraga (Naragula)*). Holotype male (Národni Múzeum Prirodovedecké Múzeum, Prague, Czech Republic), Austria: Illmitz am Neusiedler See (examined, based on literature). Paratype male in the Natural History Museum, London, UK (examined externally). Originally as subspecies of *tessularia* METZNER, 1845, treated valid at subspecies rank also by SCOBLE (1999). Downgraded here from subspecies rank, quoting absence of diagnostic features (new synonym).

tessularia pannonica VOJNITS, 1977, Folia ent. hung. 30 (1): 170, (*Narraga*). Holotype (Termeszettudomanyi Muzeum Allattara (Hungarian Natural History Museum), Budapest, Hungary), [Hungary]: Ujszász. We follow HAUSMANN *et al.* (2011), and consider *pannonica* a junior synonym of *kasyi* MOUCHA & POVOLNY, 1957, quoting the absence of disjunct differential features. SCOBLE (1999) considers *pannonica* valid at subspecies rank. See remarks.

tessularia ilia WEHRLI, 1940, in SEITZ, Gross-Schmett. Erde 4 (Suppl.): 394, pl. 32: a, (*Narraga*). Syntypes 2 females, Central Asia [Kazakhstan]: Ili District. Originally as var. of *fasciolaria*. Valid at subspecies rank according to SCOBLE (1999).

EXTERNAL CHARACTERS AND ABDOMEN: Wingspan ♂ 17–19 mm, ♀ 13–17 mm. Wing pattern similar to *N. fasciolaria* (see above), with the following minor, quantitative differences (compare to *N. fasciolaria*). *N. tessularia* females are smaller and markings are more contrasting. Small, brown discal spots are normally present in all wings. Postmedial line often not fused to brown terminal area near tornus, particularly in females. Fringes chequered white and brown in females. Posterior margin of male 8th sternite concave with medial sclerotisation. Other sternites and tergites of both sexes undifferentiated.

VARIATION: Extent of brown markings less variable than in *N. fasciolaria*. Ground colour varies from whitish to yellowish brown in males, ground colour in females mostly whitish and contrasting.

MALE GENITALIA: Uncus short, stout, weakly setose. Gnathos arms wide, fused ventrally, forming small plate. Juxta weakly sclerotised plate. Valva divided. Dorsal arm of valva long, slightly curved, apex round and setose. Ventral arm of valva wide, apex with sharp ventral extension. Sacculus weakly setose. Saccus large, round. Aedeagus straight, caecum short, apex sclerotised and acute. Vesica narrow at base, with dentate cornutus, base of ductus ejaculatorius sclerotised.

FEMALE GENITALIA: Papillae anales soft, setose. Apophyses straight, narrow. Lamella postvaginalis weak sclerotised band. Ventral margin of ostium bursae curved evenly. Ductus bursae funnel-shaped, sclerotised (Text-figs 27–29). Corpus bursae elongated, pear-shaped sac. One small signum, shape variable, normally with sharp projections (Text-figs 33–35).

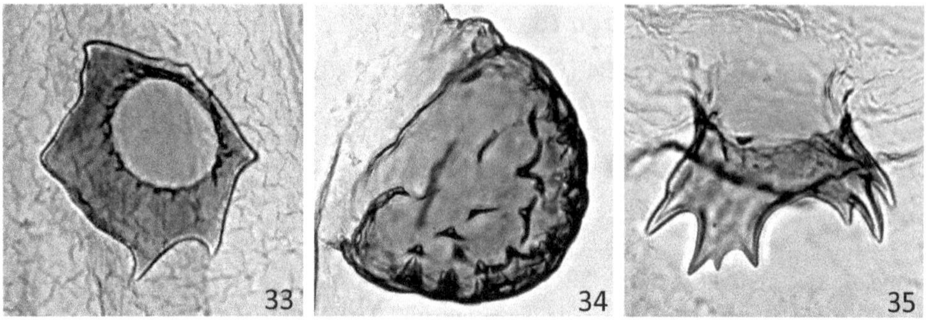

Text-figs 33–35. Variation in the signum shape in *Narraga tessularia* (METZNER, 1845). Text-fig. 33. Austria: N-Burgenland, ex larva vii.1956, slide PS1181 (coll. ZMH). Text-fig. 34. Austria: Burgenland, ex larva 1982 (abnormal?), slide PS1203 (coll. SKOU). Text-fig. 35. Russia: Tuva rep., E. Tannu-Ola Mountains, 1250m–2000m, 17.vi.1995, slide PS1354 (coll. ZMH).

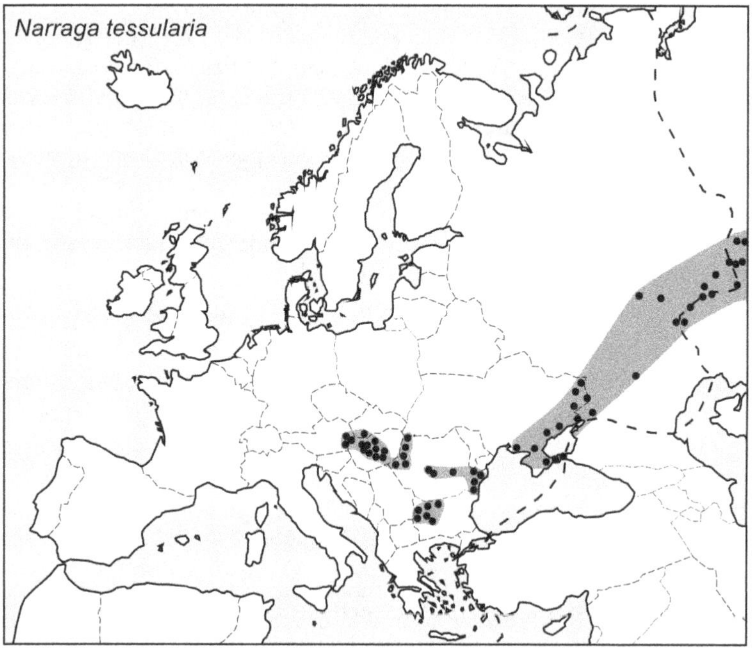

DISTRIBUTION AND ABUNDANCE: European-Westasiatic. In Europe in south-eastern central Europe, the Balkans, Ukraine, southern European Russia and southern Ural mountains. – Outside Europe in Kazakhstan, south-western Siberia, Altai and Tuva. – A very local species, but rather often it can be found in numbers at its localities.

PHENOLOGY: Bivoltine. The adult moths fly from mid-April to mid-June and from early July to late August. In Romania from early May to mid-June and again from mid-July to late September, most abundant from late July to late August (L. SZÉKELY pers. comm.). According to M. LEIPNITZ (pers. comm.) from late May to early July and from mid-August to late September. The species overwinters as a pupa, and in rearings most of the pupae hibernated and emerged the year after (B. MÜLLER pers. comm.). The moths fly actively by day, but come also to light. K. NUPPONEN (pers. comm.) reports that the moths fly around at sunset and at dusk, but he does not regard the species as really day flying. However, according to M. LEIPNITZ (pers. comm.) the moths are entirely day flying and they fly only in sunshine and within very limited areas with the food plant, which they hardly leave.

BIOLOGY: Monophagous. The larvae have been recorded on *A. campestris* (FAJCIK 2003), from mid-May to September. L. SZÉKELY (pers. comm.) reports to have found many larvae on *Artemisia maritima* in late October. Reared larvae have refused *A. campestris* (M. LEIPNITZ pers. comm.). KASY (1959) carried out a breeding experiment and concluded, that the species feeds entirely on *Artemisia maritima*. The pupation takes place in the ground (KASY 1959, M. LEIPNITZ pers. comm.).

HABITAT: Xerothermophilous. A steppe species with preference for halophilic conditions. In Romania strictly associated with salty steppe localities with *Artemisia maritima* (L. SZÉKELY pers. comm.), sandy steppe and grassland, halophilic meadows and on banks of salt lakes (M. LEIPNITZ pers. comm.).

SIMILAR SPECIES. All European *Narraga* species are externally similar, therefore we recommend that genitalia be used for identification. Both male and female genitalia structures are distinct (see genitalia plates and Text-figs 27–29).

GENETIC DATA: Genetically comparatively homogenous from Ukraine, Kazakhstan and Kyrgyzstan (n=3). Nearest species: *Narraga fasciolaria* (6.3%).

REMARKS: The description of *Geom*[*etra*]) *tessularia* METZNER, 1845 was based on specimens from Algeria and (southern) Russia. It is likely that the specimen(s) from Algeria are conspecific with *Narraga nelvae* (ROTHSCHILD, 1912), because *tessularia* is not known from North Africa. We do not know if the original METZNER'S type specimens still exist. If they do, and if a lectotype is to be designated, it should be from (southern) Russia.

Narraga isabel AGENJO, 1956, described from a male and a female, supposedly collected from Spain, has been found to be a synonym of *N. tessularia* (EXPÓSITO 2005). This was confirmed by examination of authentic material by REDONDO *et al.* (2009). LERAUT (2009) considered *N. isabel* a valid species 'resembling *N. tessularia* but having a different genitalia'. LERAUT did not provide any material or concrete description to substantiate his argument, and did not cite the above-mentioned references.

Specimens from Austria, treated valid at subspecies rank as *kasyi* MOUCHA & POVOLNY, 1957 (also in SCOBLE 1999), are without disjunct differences relative to nominal subspecies. Specimens are often less contrasting, but not exclusively, and ground colour is ochreous brown and resembling in appearance *N. fasciolaria*. We consider *kasyi* a junior subjective synonym of *tessularia* METZNER, 1845, the differences between nominal subspecies and *kasyi* are small and quantitative (VOJNITS 1977), and the Austrian populations are adjacent (not allopatric) to populations in Hungary and Slovakia.

The Other Macariini group of genera (*sensu* KRÜGER)

The group is diagnosed in KRÜGER (2001), but it is unclear if the included taxa are monophyletic. The group is diagnosed by two potential synapomorphies: incomplete separation of costa and sacculus in the male genitalia and the presence of an antrum bearing two vertical and slightly sclerotised bands. The antrum may be strongly modified, making it difficult to recognise the vertical bands (KRÜGER 2011). Included European genera are *Heliomata* GROTE & ROBINSON, *Isturgia* HÜBNER, *Itame* HÜBNER, *Neognopharmia* WEHRLI, *Macaria* CURTIS, *Chiasmia* HÜBNER, *Digrammia* GUMPPENBERG, *Rhoptria* GUENÉE and *Ortaliella* HAUSMANN.

Heliomata GROTE & ROBINSON, 1866

Heliomata GROTE & ROBINSON, 1866, Proc. ent. Soc. Philad. 6: 29. Type species: *Heliomata infulata* GROTE, 1863 (U.S.A.: Northern Virginia). Genus synonymy is after SCOBLE (1999).

Pepasmenoptera GUMPPENBERG, 1887. Type species: *Geometra glareāria* [DENIS & SCHIFFERMÜLLER], 1775. Austria.

UNAVAILABLE NAMES (incorrect subsequent spelling): *Haliommata* PACKARD, 1876 (see FLETCHER 1979).

DIVERSITY AND DISTRIBUTION: *Heliomata* contains three species, two of those are Nearctic, only one species in Europe (SCOBLE & KRÜGER 2002).

EXTERNAL CHARACTERS AND ABDOMEN: Small moths, wingspan 15–23 mm. The North American species and the only European species are very dissimilar externally. North American species (two): wings black, with white or yellow bands and with metallic blue scales. European species (one): wings light brown, transverse lines darker, suffused, cilia chequered. Male hindleg slightly swollen, with hair pencil. Male 3rd sternite with setae and posterior margin of 8th sternite straigth, not emarginated. SCOBLE & KRÜGER (2002) diagnosed the genus by the following characters: Chaetosemata widely separated, not extended. Antennae not bipectinate. Male 3rd sternite with distinctive pecten from sternal window, 8th sternite not emarginate (SCOBLE & KRÜGER 2002).

MALE GENITALIA: Uncus triangular, pointed, horns absent. Gnathos with central element well developed. Valva deeply excavated, sacculus narrow, curved, pointed. Juxta broad plate with distinctive creases forming a cross (see the male genitalia of *H. glarearia* (DENIS & SCHIFFERMÜLLER)). Vesica without cornuti or with group of minute cornuti (see Table 1).

FEMALE GENITALIA: Ductus bursae short. Corpus bursae long, posterior part elongated. Corpus bursae not greatly expanded. Signum single, moderate in size, stellate (see Table 1).

BIOLOGY: The larvae have been recorded feeding on numerous species of Fabaceae, for instance *Medicago*, *Trifolium*, *Lotus* (FLAMIGNI *et al.* 2007) and on *Robinia* (FERGUSON 2008). All known species are diurnal.

IMMATURE STAGES: Larvae are green, with narrow light lateral stripe. Both green and brown colour forms have been recorded for *H. glarearia* (illustrated in EBERT 2003) and *H. cycladata* (MCGUFFIN 1972). Cremaster shortly bifurcate in *H. glarearia* (PATOČKA & TURČANI 2005).

REMARKS: The European species *Geometra glarearia* DENIS & SCHIFFERMÜLLER, 1775 has traditionally been combined with the North American genus *Heliomata* (GROTE & ROBINSON) (e.g. MÜLLER 1996, SCOBLE 1999). *Geometra glarearia* is externally very distinct from its North American congeners (illustrated in MCGUFFIN 1972, SCOBLE & KRÜGER 2002, FERGUSON 2008). The male genitalia, however, particularly the distinct cross in the juxta leave no doubt as to its affinity with *Heliomata* (SCOBLE & KRÜGER 2002). VIIDALEPP (1996) has combined *Geometra glarearia* with *Semiothisa* HÜBNER, but it has been shown that *Semiothisa* is a Neotropical genus (SCOBLE & KRÜGER 2002). Earlier *Heliomata* was placed in the Abraxini (e.g. MCGUFFIN 1972), and the Cassymini association has been considered potential (HOLLOWAY 1994), but recently the Macariini association has been endorsed by the divided male valva (SCOBLE & KRÜGER 2002) and structures of the larva and pupa (FERGUSON 2008). Systematic position: Macariini.

16. *Heliomata glarearia* (DENIS & SCHIFFERMÜLLER, 1775)

glarearia [DENIS & SCHIFFERMÜLLER], 1775, Ankündung syst. Werkes Schmett. Wienergegend: 106, (*Geometra*). Syntype(s), lost, [Austria]: Vienna district.

glarearia aurearia STAUDER, 1920, Int. ent. Z. 14: 39, (*Chiasma* [sic!] (*Phasiane*)). Syntype(s), [Austria]: Wels. Originally as f. (STAUDER 1920), treated valid at subspecies rank by SCOBLE (1999), downgraded here from subspecies rank, quoting absence of diagnostic features (new synonym). See remarks.

glarearia schawerdae STAUDER, 1920, Int. ent. Z. 14: 38, (*Chiasma* [sic!] (*Phasiane*)). Syntypes, [Croatia?]: Mid Dalmatia; [Slovenia]: Innerkrain: Wippach. Originally as f. (STAUDER 1920), treated valid at subspecies rank by SCOBLE (1999), downgraded here from subspecies rank, quoting absence of diagnostic features (new synonym). See remarks.

UNAVAILABLE NAMES (incorrect subsequent spelling): *glarcaria* ECKSTEIN, 1923 (ECKSTEIN 1923). – (incorrect subsequent authorship): *glarearia* BRAHM, 1791. – (incorrect subsequent authorship): *aurearia* STAUDINGER, 1920 (The Global Lepidoptera Names Index (ANONYMOUS 2013); should be *aurearia* STAUDER, 1920)); *schawerdae* STAUDINGER, 1920 (The Global Lepidoptera Names Index (ANONYMOUS 2013); should be *schawerdae* STAUDER, 1920)). – Intrasubspecific names, potentially unavailable: *latestrigaria* DJAKONOV, 1926 (ab.); *lutea* GILMER, 1909 (ab.); *praepotentaria* SCHAWERDA, 1918 (ab.); *schawerdae* STAUDER, 1920 (f.); *uniformis* KAUTZ, 1923 (ab.); *virgata* DANNEHL, 1928 (f.). WEHRLI (1939–1954) lists further intrasubspecific names.

EXTERNAL CHARACTERS AND ABDOMEN: Wingspan 20-24 mm. Wings straw-yellow, medial lines weakly defined, brown. Antemedial and postmedial lines strongest but suffused. Subterminal line often fused with subterminal line. Terminal line brown, discontinuous. Fringes chequered yellow and brown. Lines continued on hindwings. Female markings tend to be sharper. Discal spots present, brown, sometimes difficult to identify. Wings coloured below as above. Frons and collar mixed straw yellow and brown. Male antennae fasciculate, female antennae filiform. Thorax straw-yellow and brown. Hindtibia with 2+2 spurs, male hindtibia swollen, with hair pencil. Abdomen concolorous with thorax and wings. Tympanal organs large, ansa bottle-shaped, tip minute. Male 3rd sternite with minute setae, posterior margin of 8th sternite undifferentiated.

VARIATION: Ground colour varies from dark straw yellow to whitish, markings may be very weak or cover most of the wing area. Numerous forms have been described.

MALE GENITALIA: Uncus triangular, pointed, horns absent. Gnathos arms fused, ventral element prominent. Juxta broad plate, distinctive cross medially. Valva deeply excavated. Dorsal arm of valva very large, its apex and ventral margin to half distance setose, round extension in ventral margin. Ventral arm of valva wide at base, pointed, apex sharp, its ventral margin weakly setose. Saccus wide, shallow. Aedeagus narrow, slightly curved, caecum short. Vesica expanded at base, surface slightly granular, without cornutus.

FEMALE GENITALIA: Papillae anales setose. Apophyses narrow, rather short. Lamella antevaginalis tiny, narrow sclerotised plate. Lamella postvaginalis not sclerotised. Ductus bursae short, wide. Base of ductus seminalis enlarged. Corpus bursae long, membranous sac, posterior part slightly grooved, anterior end slightly expanded. Signum small, multiple stellate.

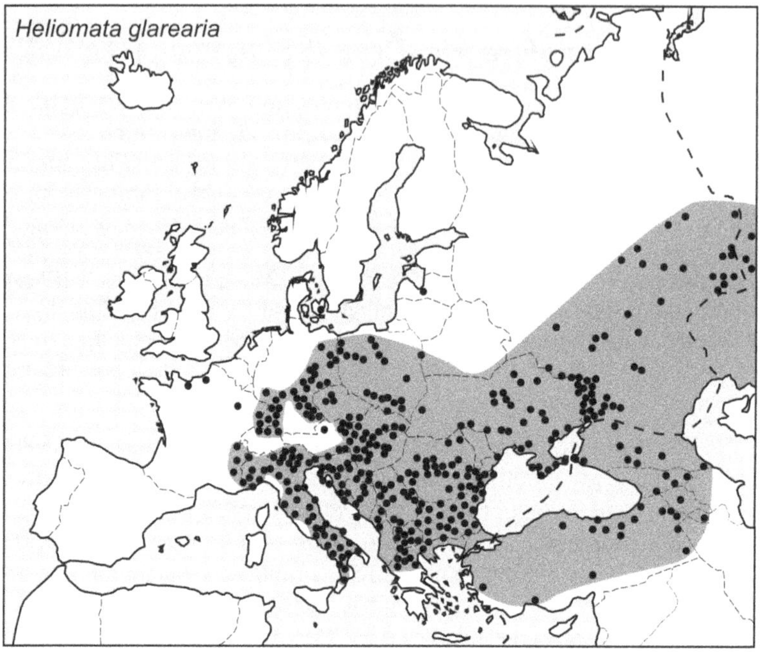

DISTRIBUTION AND ABUNDANCE: West-Palaearctic. In Europe from eastern France and Germany through much of central Europe, Italy, the Balkans to northern Greece, Ukraine and south-eastern European Russia to the Urals. – Outside Europe in central Asia, Caucasus, Transcaucasus and Turkey. – In Europe a rather common species that normally does not occur in large numbers.

PHENOLOGY: Probably bivoltine. The adults fly from late April to early September, apparently in two overlapping generations. In Poland from early May to late June and from mid-July to mid-August (MALKIEWICZ 2012). In southern Italy from early April to mid-September (A. HAUSMANN pers. comm.). The larva can be found from mid-May to early July and in August-September. The species hibernates as pupa. The species flies actively both by day and night, and comes readily to light.

BIOLOGY: Polyphagous. The larva is recorded on Fabaceae *Medicago sativa*, *M. lupulina*, *M. prostrata*, *Trifolium*, *Hippocrepis comosa* and *Lathyrus pratensis* (FLAMIGNI *et al.* 2007).

HABITAT: Thermophilous.The species occurs in many kinds of open, warm habitats, especially in grasslands from sea level up to 1400 m (FLAMIGNI *et al.* 2007). In Turkey up to 2000 m (coll P. SKOU). It seems to prefer calcareous soil. At higher altitudes the species often occurs on south- and south-western slopes (EBERT 2003).

SIMILAR SPECIES: Externally *H. glarearia* resembles *Ematurga atomaria* (LINNAEUS, 1758), especially the females are similar. Male antennae are fasciculate in *H. glarearia* (bipectinate in *E. atomaria*), female wings are straw-yellow, markings brown (white-grey in *E. atomaria)*. Also *Macaria carbonaria* (CLERCK, 1759) is similar but it is considerably smaller, black-and-white.

GENETIC DATA: Genetically homogeneous in Europe, but with two separate BINs at distances of 2.5–3% in Iran and from Turkey to central Asia. Nearest related species: North American *Heliomata cycladata* GROTE & ROBINSON, 1866 (10.8%).

REMARKS: Taxa *aurearia* STAUDER and *schawerdae* STAUDER (*schawerdai* according to WEHRLI (1939–54)) are downgraded here from subspecies rank (SCOBLE 1999) to synonymy. The diagnostic features fall within the variation of the species and the analysed CO1 genes from Croatian specimens nest together with other analysed European material (A. HAUSMANN pers. comm.). Earlier WEHRLI (1939–54) has questioned the distinct status of *aurearia*.

Isturgia HÜBNER, 1823

Isturgia HÜBNER, [1823] 1816, *Verz. bekannter Schmett.*: 297. Type species: *Geometra conspicuata* [DENIS & SCHIFFERMÜLLER], 1775, a junior subjective synonym of *G. limbaria* FABRICIUS, 1775, Austria. Genus synonyms are after SCOBLE & KRÜGER (2002), synonymy of *Atomorpha* with *Isturgia* is proposed here.

Atomorpha STAUDINGER, 1901 is transferred from the Gnophini (for instance VIIDALEPP 1996) to the Macariini and considered a junior subjective synonym of *Isturgia* HÜBNER, 1813, quoting the structural similarities (new synonym). Type species: *Atomophora falsaria* ALPHÉRAKY, 1892. [China:] Gan-Sou province. [Replacement name for *Atomophora* ALPHÉRAKY, 1892]. See remarks.

Atomophora ALPHÉRAKY, 1892. Type species: *Atomophora falsaria* ALPHÉRAKY, 1892. Junior homonym of *Atomophora* REUTER, 1884 (Hemiptera).

Bichroma GUMPPENBERG, 1887. Type species: *Noctua famula* ESPER, 1787. Europe.

Enconista LEDERER, 1853. Type species: *Fidonia perspersaria* DUPONCHEL, [1829], a junior subjective synonym of *miniosaria* DUPONCHEL, 1829; France: Montpellier.

Histurgia AGASSIZ, 1847 [unjustified emendation].

Tephrina GUENÉE, [1854]. Type species: *Geometra murinaria* [DENIS & SCHIFFERMÜLLER], 1775. Austria.

UNAVAILABLE NAMES (incorrect subsequent spelling): *Dichroma* GUMPPENBERG, 1887 (see VIVES MORENO 1994).

DIVERSITY AND DISTRIBUTION: *Isturgia* contains approximately 60 species, distribution area of the species cover the Palaearctic and mainland Oriental regions, including in the latter Tibet also. In addition many Afrotropical species, which are dealt with by KRÜGER (2001). If *Selidosema penthearia* GUENÉE, [1858], listed under "*Hypomecis*" HÜBNER, 1821 in SCOBLE 1999, is also included in *Isturgia*, then distribution area extends into Australia also.

EXTERNAL CHARACTERS AND ABDOMEN: Medium to large moths, wingspan approximately 25–40 mm. Wings yellow, orange, brown, grey or mixture of these. Often postmedial line distinct and terminal area darkest, this is repeated in hindwings. Discal spots small but distinct. Male antennae bipectinate. Male 3rd sternite setose occasionally. Posterior margin of male 8th sternite undifferentiated, weakly emarginated, or with small, occasionally dentate, projections.

MALE GENITALIA: Uncus pointed, setose, horns absent. Sclerite below uncus absent. Gnathos band-like, arms fused to form triangular and upturned element or straight, medial band. Valva deeply divided. Dorsal arm of valva long, narrow, often widest at apex, setose. Ventral arm of valva wide at base, apex round or sharply pointed. Saccus wide, often medially extended. Bullae tympani often with roundish lobe. Male genitalia rather uniform, differences between species often small (see Table 1).

FEMALE GENITALIA: Ductus bursae short, narrow, sclerotised laterally. Posterior part of corpus bursae elongated, anterior part expanded, sometimes with diverticulum. Signum stellate, branches short or long, located in anterior part (see Table 1).

BIOLOGY: Other than for *Salvia* (Labiatae), all larval records are on Fabaceae (SCOBLE & KRÜGER 2002). They have been recorded to feed on for instance *Cytisus*, *Genista*, *Retama*, *Calicotome*, *Acacia*, *Trifolium* and *Vicia* (SCOBLE & KRÜGER 2002, REDONDO *et al.* 2009). A few species fly in sunshine and rest with the wings closed over body (e.g. *I. limbaria*), but most are nocturnal.

IMMATURE STAGES: Larvae of *I. famula*, *I. limbaria* and *I. miniosaria* with whitish-yellow lateral stripe, *I. limbaria* has both green and brown forms (illustrated in EBERT 2003 and AIZPÚRUA 2002). Cremaster elongated and forked in *I. roraria* (PATOČKA & TURČANI 2005).

REMARKS: SCOBLE & KRÜGER (2002) brought together into *Isturgia* the externally dissimilar genera *Bichroma*, *Enconista* and *Tephrina* on the basis of their similar male genitalia structures. The three North-American monotypic genera *Eumaria*, *Trigrammia* and *Mellilla* were also considered similar but kept separate. FERGUSON (2008) mentions that these three genera could perhaps be congeneric to *Isturgia*. Selected literature: HAUSMANN (1990), SCOBLE & KRÜGER (2002), FERGUSON (2008). Systematic position: Macariini (KRÜGER 2001, SCOBLE & KRÜGER 2002).

Genus *Atomorpha* STAUDINGER agrees with the definition of *Isturgia* as diagnosed in SCOBLE & KRÜGER (2002). We therefore propose *Atomorpha* STAUDINGER, 1901, to be a junior synonym of *Isturgia* HÜBNER, 1823 and propose the following new genus combinations, these taxa are transferred from *Atomorpha* to *Isturgia* (see SCOBLE (1999) for more complete taxa entries, including synonymies).

- *falsaria* ALPHÉRAKY, 1892, in ROMANOFF, Mém. Lépid. 6: 66, (*Atomophora*). Lectotype male (Zoological Institute of Russian Academy of Sciences, St. Petersburg, Russia), [China:] Gan-Sou province (examined externally). Lectotype imago, non-type male genitalia and paralectotype female genitalia are illustrated in TROFIMOVA (2009). Not in Europe (new combination).
- *hedemanni* CHRISTOPH, 1885, in ROMANOFF, Mém. Lépid. 2: 121, pl. 6, fig. 3, (*Fidonia*). Holotype male (Zoological Institute of Russian Academy of Sciences, St. Petersburg, Russia), [Turkmenistan:] Ashkhabad (examined externally). Holotype imago and non-type male genitalia are illustrated in TROFIMOVA (2009). Not in Europe (new combination).
- *mabillearia* LUCAS, 1907, Bull. Soc. ent. Fr. 1907 (11): 180, pl. 18, fig. 8, (*Stegania*). Syntypes males, females (not located, not examined), Tunisia: Nefta; Kébili; Zarcine. *Isturgia mabillearia* (LUCAS) is presented by numerous specimens (>20 exx.) in the Natural History Museum, London, UK, but no types have been traced. PROUT (1912–16) considers *A. mabillearia* (LUCAS) a potentially 'local form' of *A. hedemanni*. Variable species, numerous forms have been described, which are discussed in WEHRLI (1939–54). Not in Europe (new combination).

- *punctistrigaria* CHRISTOPH, 1897, Dt. ent. Z. Iris 6: 95, (*Atomorpha*). Lectotype male (Zoological Institute of Russian Academy of Sciences, St. Petersburg, Russia), [Turkmenistan:] Ashkabad (examined externally). Lectotype male, paralectotype male genitalia and paralectotype female genitalia are illustrated in TROFIMOVA (2009). Present in Europe, see below (new combination).

17. *Isturgia famula* (ESPER, 1787)

famula famula ESPER, 1787, Die Schmett. 4 (34): pl. 106, fig. 4; (1789), ibidem, 4: 164, ((*Phalaena*) *Noctua*). Syntype(s) male(s), France: Lyon; Germany: Bavaria.

concordaria HÜBNER, [1799], Samml. Eur. Schmett. 5 Geometrae (1): pl. 24, fig. 126, (*Geometra*). Syntype(s), lost, Europe.

intricataria VILLERS, 1789, Linn. Ent. 2: 327, (*P*[*halaena*] *G*.[*eometra*]), syntype(s), S. France.

famula brunnea LE CERF, 1923, Bull. Soc. ent. Fr. 1923 (15): 198, (*Bichroma*). Holotype male (The Natural History Museum, London, UK), Morocco: Harcha. Valid at subspecies rank according to SCOBLE (1999). [originally as f].

INTRASUBSPECIFIC NAMES, POTENTIALLY UNAVAILABLE: *tenebrosa* KITT, 1925.

EXTERNAL CHARACTERS AND ABDOMEN: Wingspan 22–29 mm. Forewings ochre-yellow to whitish, medial lines and subterminal line brown, heavily suffused with brown. Subterminal line often incomplete, made up of brown spots. Fringes chequered brown and white. Hindwings orange, suffused with brown, particularly in terminal area, medial lines brown. Contrast between forewings and hindwings distinct. Discal spots present on all wings, larger on forewings. Forewings below orange-brown, subterminal line made up of large, suffused dark brown spots, terminal area and costa suffused with brown, diffuse whitish areas in terminal area. Hindwings below whitish, often whitish line(s) from base to terminal area, medial lines and veins brown, stippled with brown. Frons peculiar, protruding, flat on top, dorsal margin round, mixed black and white (Text-fig 36). Collar black and white. Antennae brown and white, bipectinate in males, filiform in females. Thorax black and white, thorax and legs covered with long hair. Hindtibia with 2+2 spurs. Abdomen dark, suffused with white. Tympanal organs rather large. Sternites and tergites 3–8 of both sexes undifferentiated.

VARIATION: Forewing whitish areas may be ochre-yellow and patterns may be sharply contrasting or suffused. Hindwing veins and medial line may be dark, thus forming a distinct reticulate pattern.

Subsp. ***brunnea*** smaller, wingspan 15–22 mm. Forewings darker, without whitish areas, and hindwings more uniform ochre-yellow, thus contrast between these pronounced. Hindwings below without, or with fragments of, whitish areas.

MALE GENITALIA: Uncus evenly tapering, setose dorsally. Gnathos arms wide, fused medially, forming triangular and upturned element. Juxta small plate. Valva divided. Dorsal arm of valva long, narrow, widest at apex, costa and apex setose. Ventral arm of valva wide, slightly pointed, few setae at apex. Saccus wide, slightly elongated medially. Aedeagus narrow, straight, caecum minute. Vesica slightly expanded at base, opens at 45 degree angle away from aedeagus, cornuti absent.

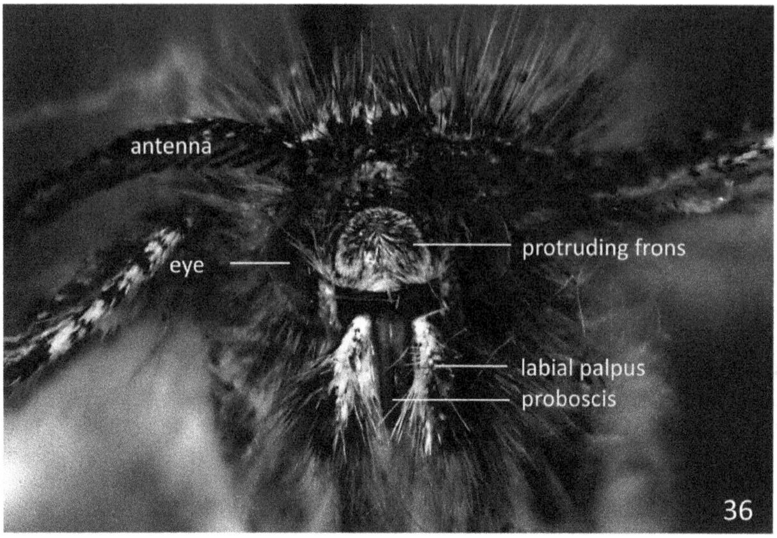

Text-fig 36. Frons of *Isturgia famula* (ESPER, 1787) male, Portugal: Beira Alta province, Viseu, 22–23.v.1971. The structure is peculiar, it is protruding, with a flat surface, and its dorsal margin is round. Both sexes have similar structure.

FEMALE GENITALIA: Papillae anales slightly elongated, setose. Apophyses straight, narrow. Lamella antevaginalis membranous. Lamella postvaginalis small, sclerotised plate. Ductus bursae short, sclerotised laterally. Corpus bursae elongated, anterior end expanded, membranous, thin. Signum small, stellate, with numerous points.

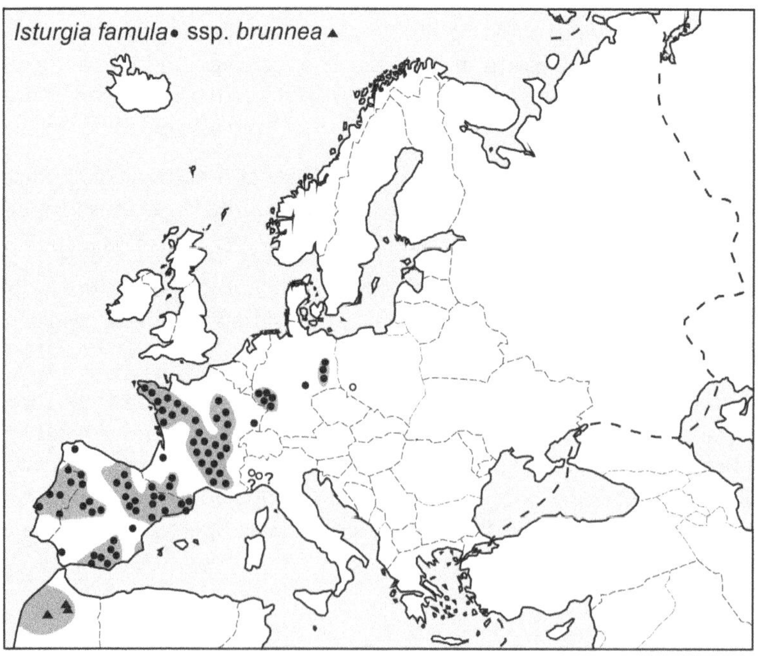

DISTRIBUTION AND ABUNDANCE: Sub-Mediterranean. In Europe scattered from Germany through France to Spain and Portugal. Old records from north-western Italy where the species is now extinct. In coll. ZMH an old specimen from Wroclaw in south-western Poland. – Outside Europe only occurring in Morocco (subsp. *brunnea*). – In Europe a rather local and not common species that is rarely found in numbers.

PHENOLOGY: Univoltine. In Spain the adults fly from late March until mid-June, at higher altitudes until mid-July. Further north later, from early May to early July. Caterpillars from April or May to late in the summer. The moths fly actively in sunshine in the morning and in the afternoon, but do not come to light.

BIOLOGY: Oligophagous. The larva is recorded on *Genista* and *Cytisus* (REDONDO *et al.* 2009) while in Germany it has entirely been found on *Cytisus scoparius* (EBERT 2003). WENCK (1950) describes a mass occurrence on *Genista pilosa*. The pupation takes place relatively deep in the ground (RÖDEL & TRUSCH 1997).

HABITAT: Xerothermophilous. Occurs in a number of different open and warm habitats from rather arid localities in Spain to heaths in Germany. In Spain up to 2300 m altitude in Sierra Nevada, at this altitude maybe due to altitudinal migration.

SIMILAR SPECIES: There are no similar species in Europe. *Athroolopha pennigeraria* (HÜBNER), *A. chrysitaria* (HÜBNER) and *Eurranthis plummistaria* (VILLERS) have somewhat reminiscent facies.

GENETIC DATA: Nearest species: *I. spodiaria* (6.1%), *I. roraria* (6.3%), and *I. miniosaria* (6.4%).

18. *Isturgia limbaria* (FABRICIUS, 1775)

limbaria limbaria FABRICIUS, 1775, Syst. Ent. 624, (*P.*[*halaena*]). Syntype(s) male(s) (The Natural History Museum, London, UK), England (examined externally).

auroraria HÜBNER, [1787], Beitr. Gesch. Schmett. 1 (2): 27, pl. 4, fig. Y, 1, 2, (*P.*[*halaena*] *G.*[*eometra*]. Syntype(s), lost.

circumdataria VILLERS, 1789, Linn. ent. 11: 330, pl. 6, fig. 10, (*Phal*[*aena*] *Geom*[*etra*]. Syntype(s), Europe, Gallia [France].

conspicuaria BORKHAUSEN, 1794, Natur. eur. Schmett. 5: 465, (*Phalaena Geometra*). syntype(s), Europe.

conspicuata [DENIS & SCHIFFERMÜLLER], 1775, Ankündung syst. werkes Schmett. Wienergegend: 316, (Geometra). Syntype(s), lost, [Austria]: Vienna district.

limbaria anzascaria STAUDINGER, 1892, Dt. ent. Z. Iris 5: 199, (*Fidonia*). Syntypes including 2 males, 2 females, [Italy]: N Piemont, Val d'Anzasca, bbei Macugnaga. Originally as var., valid at subspecies rank according to SCOBLE (1999).

limbaria delimbaria STAUDINGER, 1892, Dt. ent. Z. Iris 5: 198, (*Fidonia*). Syntype(s), [France]: Basses Alpes, near Castellane and Digne. Originally as var., valid at subspecies rank according to SCOBLE (1999).

limbaria nigrostriata HEYDEMANN, 1938, Ent. Z. Frankf. a. M. 52 (3): 23, pl. 1, figs 21, 22, (*Fidonia*). Holotype, [Germany]: Holstein. Originally as var., valid at subspecies rank according to SCOBLE (1999).

limbaria pedemontaria STAUDINGER, 1892, Dt. ent. Z. Iris 5: 198, (*Fidonia*). Syntypes including 2 males, 1 female, [Italy]: Piemonte; [France]: Alpes Maritimes. Originally as var., valid at subspecies rank according to SCOBLE (1999).

limbaria rablensis ZELLER, 1868, Verh. zool. -bot. Ges. Wien 18 (Abh.): 587, (*Fidon.*[*ia*]). Syntype(s), including male(s), [Alps]: Carinthia [Raibl in Oberkärnthen (Cave del Predil)] Originally as var., valid at subspecies rank according to SCOBLE (1999).

limbaria styriaca SCHWINGENSCHUSS, 1911, Verh. zool. -bot. Ges. Wien 61 (1/2): (46), (*Fidonia*). Syntypes including males and females, [Austria]: Steiermark, Poster, near Prebichl, 1400-1600 m. Valid at subspecies rank according to SCOBLE (1999).

INTRASUBSPECIFIC NAMES, POTENTIALLY UNAVAILABLE: *fumata* MATHEW, 1907 (ab.); *infuscata* THIERRY-MIEG, 1916 (ab.); *nigricaria* BUBACEK, 1915 (ab.); *postclara* LEMPKE, 1953 (f.); *postdemarginata* LEMPKE, 1953 (f.); *postnigrescens* LEMPKE, 1953 (f.); *quadripunctaria* FUCHS, 1899 (ab.).

EXTERNAL CHARACTERS AND ABDOMEN: Wingspan 19–26 mm, females often smaller. Wings yellow, speckled with dark brown scales. Forewing costa and terminal area on both wings dark brown. Fringes dark brown. Discal spots small, often absent. Wings coloured below as above, except hindwings more speckled, striated, and with radiating whitish lines from base. Frons, collar, thorax and abdomen mixed black and (light) yellow. Antennae black and white dorsally, bipectinate in male, filiform in female. Hindtibia with 2+2 spurs. Tympanal organs medium-sized, bullae tympani with roundish lobe (see genitalia plate). Sternites and tergites 3–8 of both sexes undifferentiated.

VARIATION: Ground colour varies from dark, golden yellow to light yellow, and the amount of dark brown scales is variable. Numerous subspecies and forms have been described on the basis of differences observed in the width and intensity of the terminal area, density of dark scales and differences in the whitish lines in ventral surface of hindwings. See remarks.

MALE GENITALIA: Uncus evenly tapering, setose dorsally. Gnathos arms wide, fused medially, forming triangular and upturned element. Juxta small plate. Valva divided. Dorsal arm of valva long, narrow, about equal width throughout, costa and apex setose. Ventral arm of valva wide, short, apex round with few setae. Saccus wide, slightly concave medially. Aedeagus narrow, rather straight, caecum minute, apex sharp. Vesica expanded at base, opens at 90 degree angle, cornuti absent.

FEMALE GENITALIA: Papillae anales slightly elongated, setose. Apophyses straight, narrow. Lamella antevaginalis narrow plate. Lamella postvaginalis small, shallow pocket. Ductus bursae short, sclerotised laterally. Corpus bursae elongated, posterior end slightly striated, anterior end expanded, membranous. Signum small, single stellate.

DISTRIBUTION AND ABUNDANCE: West-European. In Europe from Great Britain (now extinct), the Benelux countries, France including Corsica and north-easternmost Spain to parts of Germany, Austria, northern Italy, Slovenia. Austrian records for *Isturgia limbaria* and *I. roraria* are based on the Austrian distribution atlas (HUEMER & MALICKY 2009) and thus, like records from other areas, may change in the light of further studies. – The species has not been found outside Europe. – A locally distributed and not common species that is rarely found in numbers.

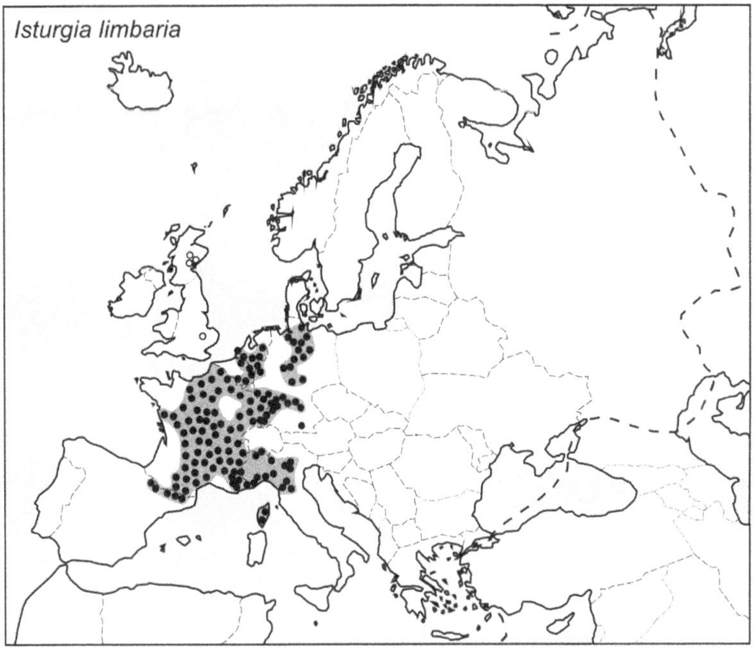

PHENOLOGY: Bivoltine. According to EBERT (2003) in two overlapping generations, in south-western Germany from late April to mid-August with an approximate separation between the two generations in late June. In northern Italy univoltine from May to July (FLAMIGNI *et al.* 2007), in Spain and France from late April to mid-August, probably also in two generations. Larva in May-June and in August-September. Hibernation takes place as a pupa. The moths fly actively by day in sunshine.

BIOLOGY: Oligophagous. The larva has been found on *Cytisus scoparius* (EBERT 2003) and it is also recorded on *Genista* (FLAMIGNI *et al.* 2007). The pupa overwinters (EBERT 2003).

HABITAT: Xerothermophilous. Warm sun-exposed slopes where the food plants occurs. In south-western Germany the species occurs up to 1000 m (EBERT 2003), in northern Italy from 200 to 2000 m (FLAMIGNI *et al.* 2007) and in the Pyrenees from 700 up to 2000 m (REDONDO *et al.* 2009).

SIMILAR SPECIES (Text-figs 37–40): *I. roraria* has terminal area light brown (dark brown in *I. limbaria*), wings are striated with brown scales (irrorated with brown scales) and ventral surface of hindwing is without whitish lines (radiating whitish lines from base). Male and female genitalia of *I. limbaria* and *I. roraria* are similar. See remarks.

GENETIC DATA: The *limbaria/roraria* species complex genetically splitting up into four BINs. *I. limbaria* from Germany and France to central Italy comparatively homogeneous but populations from South Tyrol (Alto Adige) diverging by 4.3%, see under following species for ambiguous populations. Nearest species: *I. roraria* (4.1%).

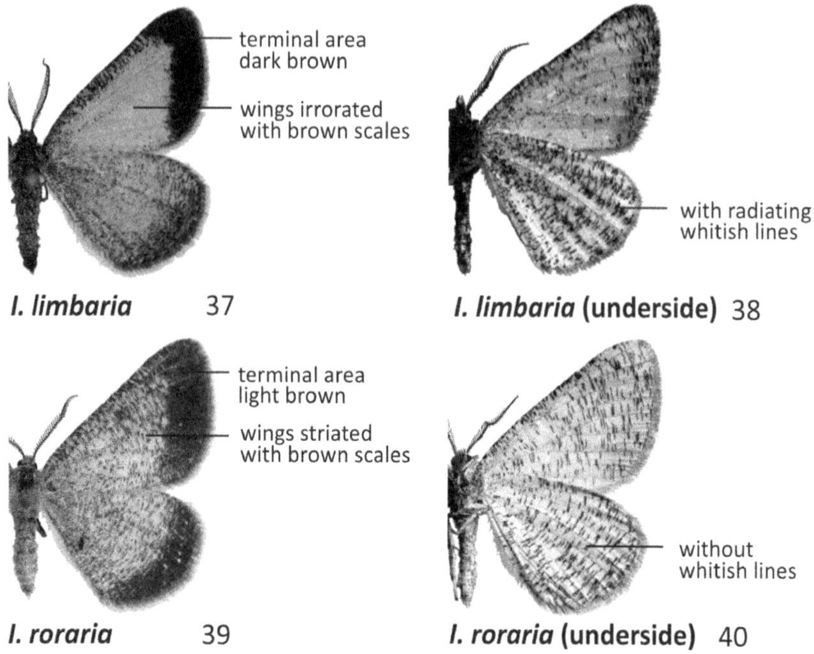

Text-figs 37–40. Diagnostic characters (indicated) of European *Isturgia* HÜBNER, 1823 species. Text-fig. 37: *I. limbaria* (FABRICIUS, 1775), above, France: Laubert, 9.vi.2007. Text-fig. 38. *I. limbaria*, below, Italy: Tuscany, Alpi-Apuane, 800 m, 5.vii.1999. Text-fig. 39: *I. roraria* (FABRICIUS, 1776), above, Slovakia: Remetské Hámre, 17.vii.1983. Text-fig. 40, *I. roraria*, below Germany, 2.vii.1933.

REMARKS: *Isturgia limbaria* (FABRICIUS) and *I. roraria* (FABRICIUS) form a species complex that needs a thorough detailed analysis, but it is beyond the scope of this work. Therefore we have repeated the subspecies here as they are presented in SCOBLE (1999). Identification through traditional means has proven difficult and this complex has been presented as an example of a species pair that hybridise occasionally in nature (HAUSMANN 2001). Preliminary genetic results from a distance model analysis have not offered a straightforward answer to the problem either. HAUSMANN *et al.* (2011) presented this as a possible case of local introgression. Intermediate populations between *I. limbaria* and *I. roraria* are known to occur in the southern Alps.

J. GELBRECHT (pers. comm.) has bred material of the complex extensively, apart from northern Italy, with the conclusion that larvae and pupae also have different structural features that correlate with the adult features. In his breeding experiments he also found out that several taxa in this complex hybridise.

Preliminary examination of tympanal organs suggests that the bullae tympani have a roundish lobe in *I. limbaria*, but not in *I. roraria* (see the male genitalia plate). The reliability of this character needs to be tested on more extensive material from the whole distribution range of the complex.

19. *Isturgia roraria* (FABRICIUS, 1776)

roraria FABRICIUS, [1776], Genera Insect. 285, (*Phalaena*). Syntype(s), Europe. See remarks.

adspersaria FABRICIUS, 1787, Mantissa Insect. 2: 189, (*Phalaena*). Syntype(s) (Zoological Museum, University of Copenhagen, Denmark), [Germany]: Saxony, Halle.

conspicuaria ESPER, [1801], Die Schmett. 5 (5): 124, pl. 24, figs. 2-4, (*Ph[alaena] Geom[etra]*). Syntype(s), [Germany]: Frankfurt. Junior primary homonym of *conspicuaria* Borkhausen, 1794.

spartariaria HÜBNER, [1799], Samml. Eur. Schmett. 5 Geometrae (1): pl. 22, fig. 116, (*Geometra*). Syntype(s), lost, Europe.

spartiaria TREITSCHKE, 1827, Schmett. Eur. 6 (1): 270, (*Fidonia*). Emendation of *spartiaria* HÜBNER?

INTRASUBSPECIFIC NAMES, POTENTIALLY UNAVAILABLE: *aequestriga* HIRSCHKE, 1910 (ab.).

EXTERNAL CHARACTERS AND ABDOMEN: Wingspan 20–26 mm. Wings and other structures similar to *I. limbaria* (see above), but with the following differences (Text-figs. 37–40): wings more speckled, scales brown. Wings coloured below as above, except more striated, and hindwings without radiating whitish lines from base. Tympanal organs medium-sized, bullae tympani without roundish lobe (see genitalia plate). Sternites and tergites 3–8 of both sexes undifferentiated.

MALE GENITALIA: Similar to *I. limbaria* (see above).

FEMALE GENITALIA: Similar to *I. limbaria* (see above), but lamella postvaginalis is often more sclerotised, posterior margin round, but the differences are not clear-cut.

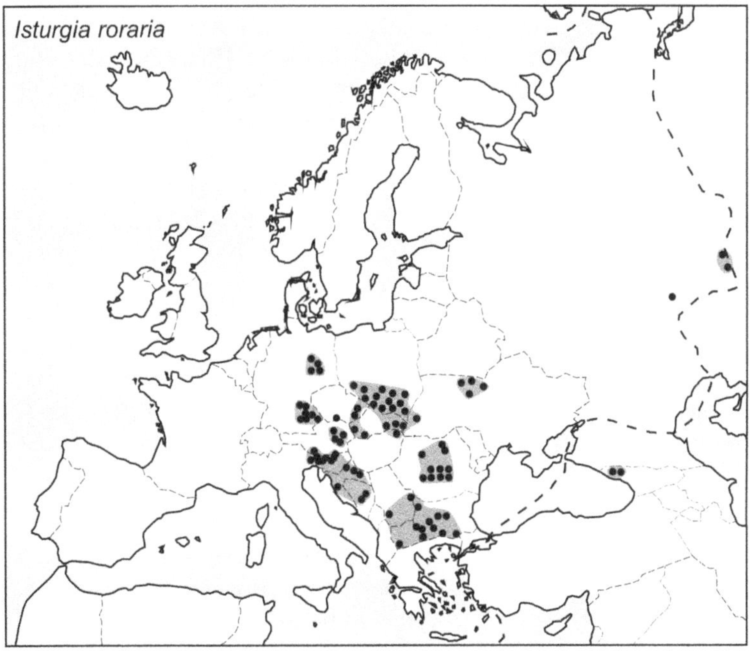

Distribution and abundance: European. The species occurs scattered from central and south-eastern Germany through eastern central Europe and parts of the Balkans and further in eastern European Russia and in the Ural mountains. The distribution is somewhat uncertain because of confusion with *Isturgia limbaria*. – Outside Europe only an isolated occurrence in Georgia. – A locally distributed and rather scarce species.

Phenology: Bivoltine. In southern Poland in two generations from late April to mid-June and from mid-July to late August in two generations (Malkiewicz 2012), in Austria from early May. Caterpillars from May to July and in August-September. The moths fly actively by day in the sunshine.

Biology: Oligophagous. The caterpillar has been found on *Cytisus scoparius* (E. Friedrich pers. comm.) and it is further recorded on *Genista tinctoria* (Buszko 2000).

Habitat: Xerothermophilous.Warm slopes, forest edges, clearings, open places with scattered bushes, heaths.

Similar species: *I. limbaria* (see above).

Genetic data: The *limbaria/roraria* species complex genetically splitting up into four BINs. *I. roraria* genetically homogeneous from Germany across north-easternmost Italy to Bulgaria, but with a separate BIN in Romania at a distance of 1.1%. Barcoded specimens from South Tyrol (see preceding species) and north-easternmost Italy were identified on BOLD (according to wing coloration) as *I. limbaria*, the existence of intermediate 'hybrid' populations was postulated and figured in GME volume 1 (Hausmann 2001; introduction).

Remarks: *Isturgia limbaria* (Fabricius) and *I. roraria* (Fabricius) form a species complex that needs a thorough detailed analysis, see *I. limbaria*.

Cleogene ostrogovichi Caradja, 1930, described from Romania, was classified in the genus *Crocota* (e.g. Scoble 1999) until a new, monotypic genus, *Palaeocrocota* was described by Stanescu (2002). Stanescu illustrated the holotype male of *P. ostrogovichi* (in the National Museum of Natural History, Bucharest, Romania], including its genitalia, which led D. Stüning (pers. comm.) to the conclusion that *P. ostrogovichi* is a melanic specimen of *I. roraria*, and should be treated as its junior synonym. D. Stüning (pers. comm.) has a manuscript in preparation and he will formally propose the synonymy.

20. Isturgia homochromata (Mabille, 1869), new combination

homochromata Mabille, 1869, Annls Soc. Ent. Fr. (4) 9 (1): 57, pl. 2, fig. 9, (*Liodes*). Holotype male (C. Herbulot Collection in Zoologische Staatssammlung, Munich, Germany), [France, Corsica:] Bastia (examined, including genitalia). Originally as sp. Transferred here from *Liodesina* Wehrli, 1938 to *Isturgia* Hübner, 1823, quoting structural similarities (new combination). See remarks.

External characters and abdomen: Wingspan 20 mm (1 male), female is unknown. Wings unicolorous brown, slightly shining, transverse lines not visible, hindwings paler. Fringes light brown. Discal spots absent. Wings coloured below as above. Frons, collar and thorax brown, abdomen brown, sparsely scattered with sand-coloured scales. Antennae unicolorous brown, male antennae bipectinate. Male hindtibia with 2+2 spurs. Tympanal organs medium-sized. Sternites and tergites 3–8 of male undifferentiated.

VARIATION: Unknown, only the holotype male is known.

MALE GENITALIA: Uncus evenly tapering, setose dorsally. Gnathos arms wide, bare, fused medially, forming triangular and upturned element. Juxta small plate. Valva divided. Dorsal arm of valva long, narrow, widest at apex, costa and apex setose. Ventral arm of valva wide, short, apex round with few setae. Saccus wide, elongated medially. Aedeagus narrow, slightly curved ventrally, caecum minute, apex sharp. Vesica narrow tube, slightly expanded at base, opens at 45 degree angle towards aedeagus, cornuti absent.

FEMALE GENITALIA: Unknown.

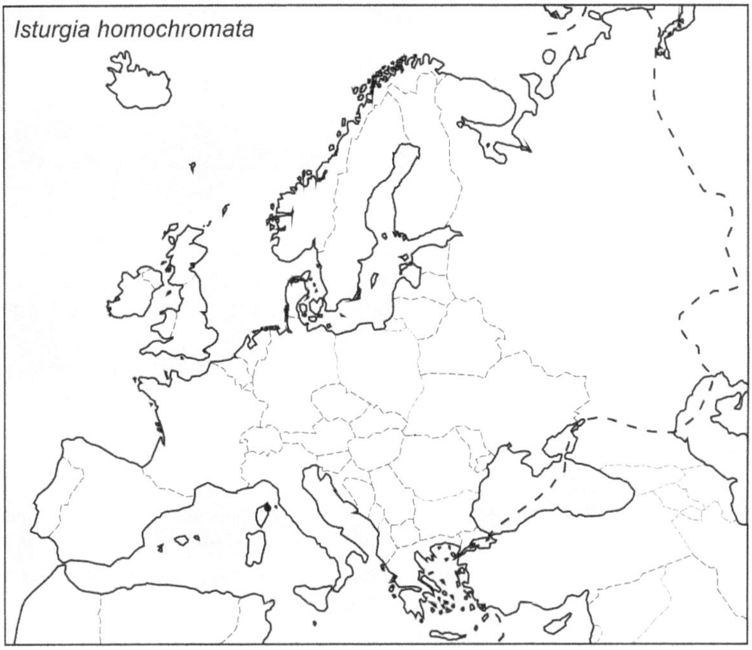

DISTRIBUTION AND ABUNDANCE: Endemic to Corsica. The only known specimen originates from the 1860's by the sea near Bastia, Corsica.

PHENOLOGY: The only known specimen was reared in May.

BIOLOGY: The only known specimen was found as a caterpillar, and according to MABILLE (1867), as far as he could remember, it came from *Genista corsica*.

HABITAT: Unknown, but if the above mentioned caterpillar was found near the coast, the habitat must have been Mediterranean macchia vegetation.

SIMILAR SPECIES: The non-related *Minoa murinata* (SCOPOLI, 1763) (Larentiinae) is usually unicolorous mouse-grey and forewing margin is more round (illustrated in HAUSMANN & VIIDALEPP 2012) (forewing brown, margin less round in *homochromata*).

GENETIC DATA: No data available.

REMARKS: Only one specimen of *Isturgia homochromata* (MABILLE) is known, the holotype male. *Liodes homochromata* MABILLE, classified recently in genus *Liodesina* WEHRLI, 1938 (e.g. SCOBLE 1999), agrees with the generic description of *Isturgia* as diagnosed in SCOBLE & KRÜGER (2002). Bipectinate male antenna and shape of the male genitalia, including the male 8th sternite, of *I. homochromata* is similar with *I. limbaria*, *I. roraria* and *I. famula*. Type species of the genus *Liodesina* is *L. mesatlantis* WEHRLI, 1938, described from Morocco. *L. mesatlantis* is also unicolorous brown, therefore externally similar to *L. homochromata*. WEHRLI (1938) described the male genitalia of *L. mesatlantis* (no illustration) and concluded that *Liodesina* is related to the genera *Epione* and *Lignyoptera*. We have dissected and examined both male (from Algeria: Guelt-es-Stel, 24.viii.1913, slide PS1267) and female genitalia (from Algeria: Prov. Oran, Sebdou, 8-12.ix.1918, slide PS1707/BMNH GEO 24001) of *L. mesatlantis* (material is in the Natural History Museum, London, UK) and conclude that they are structurally very similar to the European *Perigune* species.

21. *Isturgia murinaria* (DENIS & SCHIFFERMÜLLER, 1775)

murinaria murinaria [DENIS & SCHIFFERMÜLLER], 1775, Ankündung syst. Werkes Schmett. Wienergegend: 105, (*Geometra*). Syntype(s), lost, [Austria]: Vienna district.

buffonaria MILLIÈRE, 1867, Iconogr. Descr. Chenilles Lépid. inédits 2 (20): 368, pl. 89, fig. 2, (*Tephrina*). Type male, [France]: Hyéres. See remarks.

cineraria DUPONCHEL, 1829, in GODART & DUPONCHEL, Hist. nat. Lépid. Papillons Fr. 7 (2): 440, pl. 167, figs 1, 2, (*Fidonia*). Syntypes males, females, France (mid). See remarks.

disspersaria HÜBNER, [1823], Verz. bekannter Schmett.: 296, (*Perconia*). [Replacement name for *murinaria* [DENIS & SCHIFFERMÜLLER], 1775]

epimysaria OBRAZTSOV, 1934, Z. öst. Ent-Ver. 19 (11/12): 73, (*Tephrina*). Syntypes 7 males, 7 females, Ukraine (southern): Nikolajev; Park von Vessjolaja Bokovenjka; Ssemjonovka.

intractaria WALKER, [1863], List Specimens lepid. Insects Colln Br. Mus. 26: 1685, (*Aspilates*). Holotype female (The Natural History Museum, London, UK) (examined externally).

murinaria uralica WEHRLI, 1937, Amat. Papillons 8 (16/17): 245, (*Tephrina*). Syntype(s), [Russia]: Urals (south), 2 paratype females (The Natural History Museum, London, UK) (examined externally). Originally as subspecies of *murinaria* [DENIS & SCHIFFERMÜLLER], 1775, treated valid at subspecies rank also by SCOBLE (1999). Downgraded here from subspecies rank, quoting absence of diagnostic features (new synonym). See remarks.

myosaria ESPER, [1806], Die Schmett. 5 (10): pl. 49, fig. 5, (*Geom*[*etra*]). Syntype(s), [Europe].

nubecularia VILLERS, 1789, Linn. Ent. 2: 326, *Phal.*[*aena*] *Geom.*[*etra*]. Syntype(s), Europe.

respersaria BORKHAUSEN, 1794, Natur. eur. Schmett. 5: 221, (*Phalaena Geometra*). Syntype(s), Europe.

INTRASUBSPECIFIC NAMES, POTENTIALLY UNAVAILABLE: *impunctata* BERGMANN, 1955 (ab.).

EXTERNAL CHARACTERS AND ABDOMEN: Wingspan 22–28 mm. Wings light grey, markings brown, medial lines more pronounced in females. Basal and medial areas irrorated with brown, terminal area often darker. Antemedial and medial lines often weakly developed. Postmedial line often strongest, its interior side whitish, line slightly turning inwards near costa. Terminal line dark brown, discontinuous. Fringes chequered brown and whitish. Discal spots brown, elongated. Wings coloured below as above, except markings lighter and more striated. Frons, collar, thorax and abdomen mixed grey and brown. Antennae grey and brown dorsally, bipectinate in male, filiform in female. Hindtibia with 2+2 spurs. Tympanal organs medium-sized. Posterior margin of male 8th sternite weakly serrate, slightly concave medially. Other male and female sternites and tergites undifferentiated.

VARIATION: Markings vary from rather uniform brown to sparsely irrorated brown, in the latter light grey ground colour more visible and specimens more contrasting.

MALE GENITALIA: Uncus constricted at base, margins slightly expanded, apex triangular, setose dorsally. Gnathos arms narrow, fused medially, forming wide, horizontal ridge. Juxta small plate. Valva divided. Dorsal arm of valva long, narrow, slightly expanded at apex, costa and apex setose. Ventral arm of valva wide, its ventral margin weakly serrate, with few setae, apex sharp. Saccus wide, slightly concave medially. Aedeagus wide, straight, caecum absent, apex sharp. Vesica opens at about 90 degree angle, narrow tube, with large, roundish diverticulum, its surface with minute sclerotisations all over, cornuti absent.

FEMALE GENITALIA: Papillae anales slightly elongated, setose. Apophyses straight, narrow. Sinus vaginalis prominent, weakly sclerotised, striated. Lamella antevaginalis heart-shaped, sclerotised plate. Lamella postvaginalis sclerotised, concave medially. Ductus bursae short, wide, sclerotised laterally. Corpus bursae large, posterior end elongated, striated, narrow, anterior end very large, oval, membranous. Signum small, stellate.

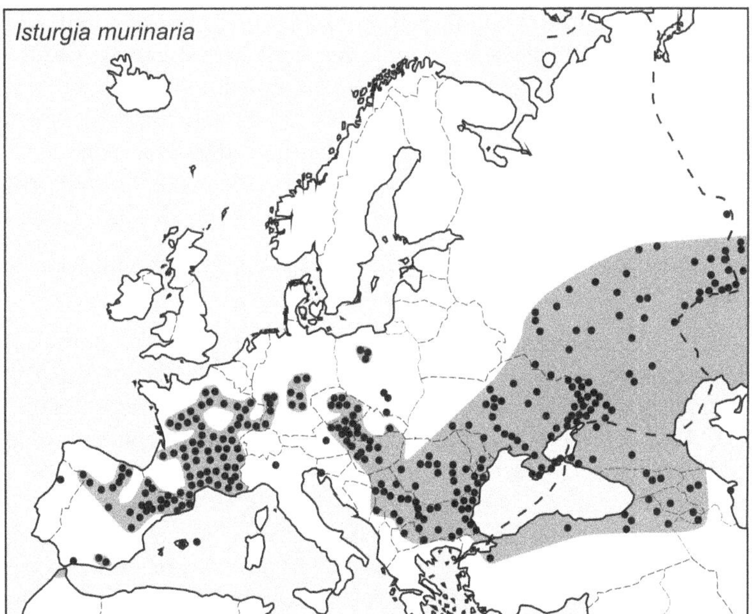

DISTRIBUTION AND ABUNDANCE: Eurasiatic. In Europe from the Iberian Peninsula and France through parts of central Europe, south-eastern Europe and Ukraine to European Russia and southern Urals. – Outside Europe distributed further to the east through central Asia and southern Siberia to Mongolia and to the south to Caucasus, Transcaucasus and Turkey. Also recorded from northernmost Morocco. – In many parts of Europe a common species that sometimes occurs in numbers.

PHENOLOGY: Bivoltine, but the second generation seems to be partial. From late March to late June and from mid-July to early September. In Poland from late April to mid-June and from mid-July to mid-August (MALKIEWICZ 2012). In Romania from late April to mid-June and from late July to mid-September (L. SZÉKELY pers. comm.). In France in late September (N. PÖLL pers. comm.). In Spain possibly trivoltine with specimens in late February and late March and until late September (J. YLLA I ULLASTRE pers. comm.). Larva in May-June and August to October. Moths are active at night and come readily to light.

BIOLOGY: Oligophagous on Fabaceae. The larva has been found on *Medicago sativa* and *Onobrychis viciifolia* (BERGMANN 1955) and was reared on *O. viciifolia* (E. FRIEDRICH pers. comm.). According to SZÉKELY (2011) also recorded on *Trifolium* and BERGMANN (1955) records also *Onobrychis sativa*. In rearings the larvae have accepted *Vicia* and *Medicago* as food plants (B. MÜLLER pers. comm.). In Germany most larvae of the first generation do not give rise to a second generation, but hibernate together with the offspring of the second generation as pupa. (EBERT 2003).

HABITAT: Open and warm habitats with scattered bushes and trees from sea level up to 1500 m (REDONDO *et al.* 2009).

SIMILAR SPECIES (Text-figs 41–46): *I. spodiaria* and *I. assimilaria* have postmedial line clearly turning inwards near costa and it is suffused in hindwing (postmedial line nearly parallel with wing margin and it is distinct in *I. murinaria*), fringes are uniform in *I. spodiaria* (chequered in *I. murinaria*). Male and female genitalia of *I. murinaria* are clearly different from *I. spodiaria* and *I. assimilaria*. *Isturgia berytaria* (STAUDINGER) has distinct forewing discal spot and postmedial line is angled towards wing base on inner margin (dorsum). Non-European *Isturgia hopfferaria* (STAUDINGER, 1879) from Transcaucasus, Turkey and further East (VIIDALEPP 1996), is similar but it is larger (wingspan up to 34 mm) and irroration is distinct. Male and female genitalia are distinct, allowing reliable identification. LERAUT (2009) illustrates adults.

GENETIC DATA: Genetically homogeneous from France, Spain to central Asia. Nearest species: *I. hopfferaria* (STAUDINGER, 1879) from Turkey (3.6%).

REMARKS: SCOBLE (1999) considers the taxon *buffonaria* MILLIÈRE, 1867, from France, valid at species level. Description is based on a single male (illustrated in PROUT 1912–1916), whose current depositary is unknown, and therefore its identity has caused confusion. WEHRLI (1939–1954) gives a detailed account of it, concluding that *buffonaria* is provisionally to be considered a small, aberrative form of *I. murinaria* (DENIS & SCHIFFERMÜLLER, 1775). Due to absence of authentic material, we follow WEHRLI (1939–1954). KRÜGER (2001) considers the taxon *cineraria* Duponchel, 1829, from France, valid at subspecies level. He mentions that (p. 73): 'The status of *cineraria* is not clear and it is probably of intrasubspecific rank, as suggested by PROUT (1915 [1912–1916]), and as listed in the BMNH card index. However, as I have been unable to examine North African material, I follow the most recent treatment by RUNGS (1981).'

We downgrade taxon *uralica* WEHRLI, 1937 from subspecies rank (WEHRLI 1937, SCOBLE 1999) to synonymy with *murinaria* DENIS & SCHIFFERMÜLLER, 1775, quoting absence of diagnostic features (new synonym). The specimens from southern Urals are characterised by paler ground colour, particularly on the inner margin of postmedial line, but such specimens fall within intraspecific variation. The specimens from southern Urals are also paler in some other species, for instance in *Perconia strigillaria* (HÜBNER, 1787). Both male genitalia (Russia: S-Ural, Cheliabinsk district, near Moskovo village, 26.v.1998, slide PS1571) and female genitalia (Russia: S-Ural, Cheliabinsk district, Ajat river near Nikolaevka village, 3.vii.1997, slide PS1912) of examined southern Urals specimens are similar with other examined specimens (Russia: Novosibirsk oblast; Austria; Ukraine).

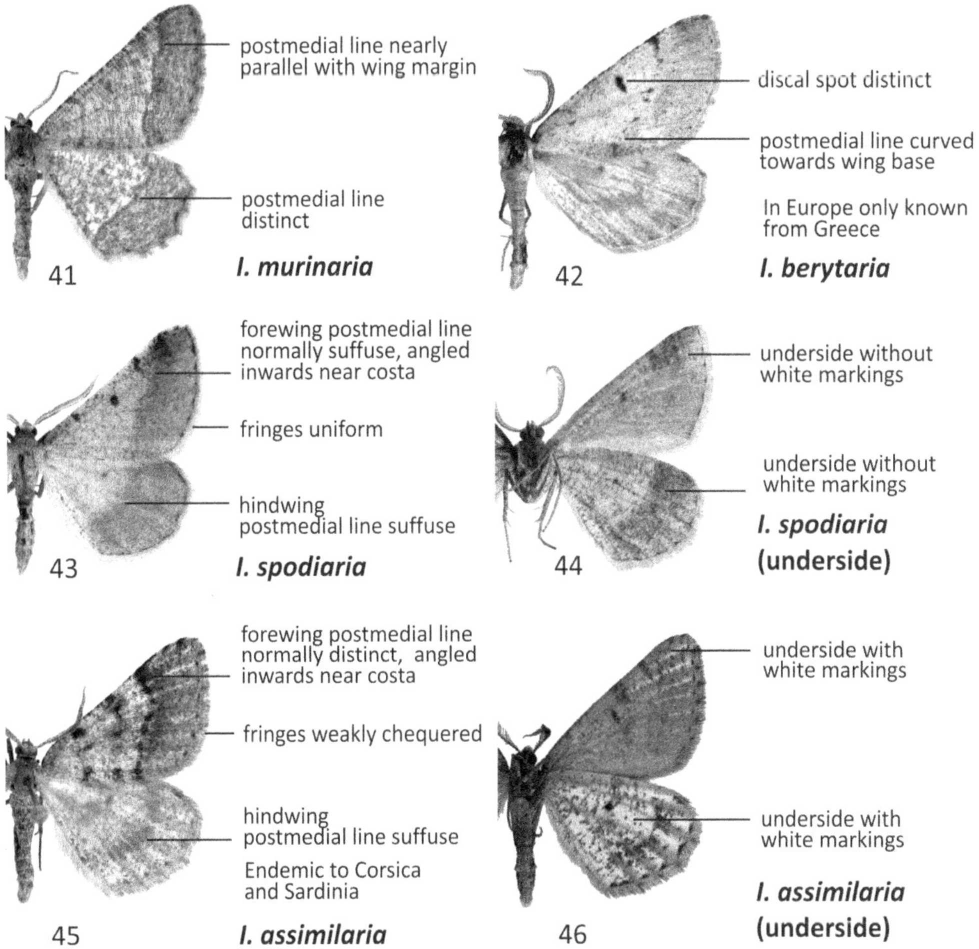

Text-figs 41–46. Diagnostic characters (indicated) of European *Isturgia* HÜBNER, 1823 species. Text-fig. 41: *I. murinaria* (DENIS & SCHIFFERMÜLLER, 1775), Spain, 22.v.1998. Text-fig. 42: *I. berytaria* (STAUDINGER, 1892), Greece, 3.x.1984. Text-fig. 43: *I. spodiaria* (LEFÈBVRE, 1832), Spain, 7.v.2000. Text-fig. 44. *I. spodiaria* (LEFÈBVRE, 1832) underside, Spain: Marbella, 22.ix.1987. Text-fig. 45: *I. assimilaria* (RAMBUR, 1833), Sardinia, 22.v.2004. Text-fig. 46. *I. assimilaria* (RAMBUR, 1833) underside, Italy: Sardinia, 19–20.v.2004.

22. *Isturgia catalaunaria* (GUENÉE, 1858)

catalaunaria GUENÉE, [1858], in BOISDUVAL & GUENÉE, Hist. nat. Insectes (Spec. gén. Lépid.) 10: 108, (*Psamatodes*). Holotype male, [Spain]: Catalonia [KRÜGER 2001, p. 76: not located in The Natural History Museum, London, UK or Muséum National d'Histoire Naturelle, Paris, France, and probably lost (not examined)].

catalaunaria ningwuana WEHRLI, 1940, in SEITZ, Gross-Schmett. Erde 4 (Suppl.): 400, pl. 32: e, (*Tephrina*). Syntypes 3 males, China: Szechuan (Sichuan), Ningwuan.

cogitata WALKER, [1863], List Specimens lepid. Insects Colln Br. Mus. 26: 1661, (*Panagra?*). Holotype male (The Natural History Museum, London, UK), South Africa (examined, including genitalia).

dataria WALKER, 1861, List. Specimens lepid. Insects Coll. Br. Nus. 23: 961, (*Tephrina*). Holotype male (Oxford University Museum of Natural History, Oxford, UK), South Africa: Cape.

defectaria WALKER, 1861, List Specimens lepid. Insects Colln Br. Mus 23: 962, (*Tephrina*). Holotype male (Oxford University Museum of Natural History, Oxford, UK), South Africa: Cape.

largificaria MÖSCHLER, 1887, Abh. senckenb. naturforsch. Ges. 15 (1): 95, fig. 20, (*Semiothisa*). Type, Gold Coast [Ghana]: Accra.

occupata WALKER, 1862, List Specimens lepid. Insects Colln Br. Mus. 24: 1071, (*Aspilates*). Lectotype male (The Natural History Museum, London, UK), South Africa (examined, including genitalia). [After KRÜGER (2001): The lectotype of *Aspilates occupata* (designated in KRÜGER 2001) is conspecific with *Isturgia catalaunaria* GUENÉE, [1858]. JANSE (1932: 238, 239) indicates that the type series was mixed and contained specimens of *deerraria* as well].

proxantharia WALKER, [1863], List Specimens lepid. Insects Colln Br. Mus. 26: 1679, (*Aspilates*). Lectotype male (The Natural History Museum, London, UK), South Africa: Knysna (examined, including genitalia), paralectotype female (The Natural History Museum, London, UK), South Africa: Cape Town (examined externally).

INTRASUBSPECIFIC NAMES, POTENTIALLY UNAVAILABLE: *proxaulkaria* WALKER, 1863 (DISTANT 1892).

EXTERNAL CHARACTERS AND ABDOMEN: Wingspan 22–28 mm. Wings light grey, densely stippled with dark grey. Antemedial and medial lines faint, suffuse. Postmedial line strongest, particularly on hindwings, brown, its outer margin suffused brown, frequently with row of small dark spots. Terminal area often darkest. Terminal line sharp, discontinous. Fringes grey. Discal spots sharp, small. Wings below brown, markings fainter, more striated. Frons and collar light brown, thorax and abdomen light grey, mixed with dark grey. Antennae mixed white and dark brown dorsally, bipectinate in male, filiform in female. Hindtibia with 2+2 spurs. Tympanal organs medium-sized, bullae tympani with roundish lobe. Posterior margin of male 8th sternite weakly sclerotised, slightly serrate. Other male and female sternites and tergites undifferentiated.

VARIATION: Wing colour varies from dark to light grey, terminal area often darker than rest of wings. Medial lines and discal spots can be weakly developed or absent.

MALE GENITALIA: Uncus short, margins slightly angled, setose dorsally. Gnathos arms wide, fused medially, forming triangular and upturned element. Juxta small plate. Valva divided. Dorsal arm of valva long, narrow, slightly expanded at apex, costa and apex setose. Ventral arm of valva wide, short, with longitudinal ridge near ventral margin, apex acute, ventral margin with few setae. Saccus slightly triangular. Aedeagus straight, caecum minute, with subapical keel, apex narrow and sharp. Vesica narrow tube, with large, elongated diverticulum, with numerous microcornuti in restricted area.

FEMALE GENITALIA: Papillae anales elongated, weakly striated, setose. Apophyses straight, narrow. Lamella antevaginalis and postvaginalis irregularly sclerotised, within membranous pocket. Ductus bursae medium length, wide, sclerotised laterally. Base of ductus seminalis sclerotised. Posterior part of corpus bursae narrow, striated, partly sclerotised. Anterior part of corpus bursae large, pear-shaped, membranous. Signum stellate, situated dorsally.

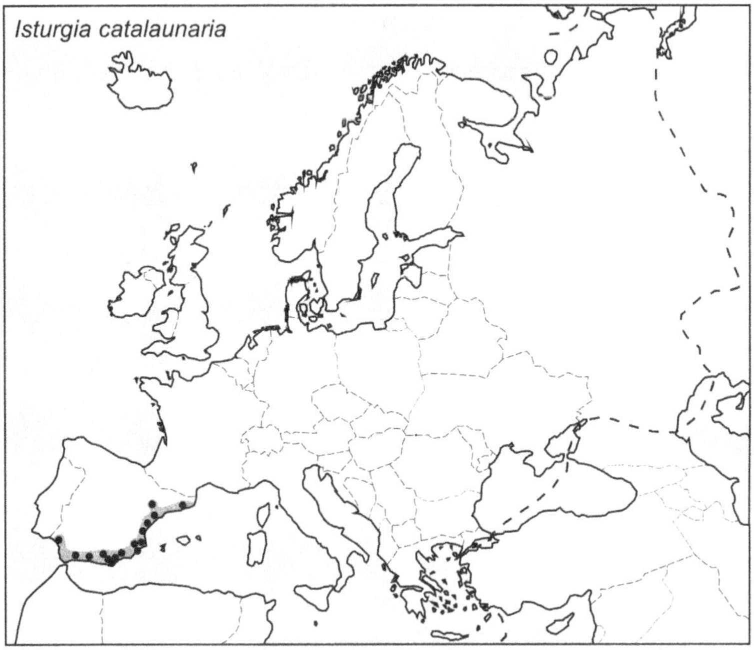

DISTRIBUTION AND ABUNDANCE: Palaeotropical. In Europe only in eastern and southern Spain, close to the coast. – Outside Europe the species is widely distributed in the Afrotropical region south of Sahara to South Africa and in Arabia (KRÜGER 2001). So far we have no records from North Africa. – In Spain a somewhat local species that may occur in numbers.

PHENOLOGY: Probably trivoltine. Generations overlapping from late March to early November. In Africa the species flies throughout the year (KRÜGER 2001). The moths fly at night and come readily to light.

BIOLOGY: Oligophagous on Fabaceae. REDONDO *et al.* (2009) records the caterpillar on *Vicia*. The caterpillar has been reared on *Dorycnium* (SPULER 1903–1910).

HABITAT: Xerothermophilous. Warm and dry places with sparse vegetation, relatively close to coasts, common in the semi-deserts in south-eastern Spain. The species has been found from sea level up to 400 m altitude.

SIMILAR SPECIES: *Acanthovalva inconspicuaria* is somewhat similar but smaller, forewing discal spot is larger and forewing medial lines form small dots on costa (wingspan larger, forewing discal spot smaller and forewing medial lines normally indistinct on costa in *catalaunaria*).

GENETIC DATA: Genetically homogeneous in Spain (n=7), well corresponding to one of the haplotypes from the Arabian Peninsula. Examined populations from ten African countries slightly diverging in two different BINs at distances of 0.6–1.2%. Nearest species: *I. univirgaria* (MABILLE, 1880) from Madagascar (6.1%).

23. *Isturgia deerraria* (WALKER, 1861)

deerraria WALKER, 1861, List Specimens lepid. Insects Colln Br. Mus. 23: 962, (*Tephrina*). Lectotype male (The Natural History Museum, London, UK), [South Africa] (examined, including genitalia). See *occupata* WALKER, 1862 under *Isturgia catalaunaria* (GUENÉE, 1858).

nemorivaga WALLENGREN, 1872, Ofvers. K. VeteskAkad. Förh. Stockh. 29 No. 3: 59, (*Tephrina*). Holotype female [After KRÜGER (2001): not traced in RMS [Naturhistoriska Riskmuseet, Stockholm, Sweden] or UZIL [Universitets Zoologiska Institut, Lund, Sweden] and probably lost]. [South Africa]: Caffraria interior.

INTRASUBSPECIFIC NAMES, POTENTIALLY UNAVAILABLE: *dissocia* WARREN, 1897 (ab.).

EXTERNAL CHARACTERS AND ABDOMEN: Wingspan 14–21 mm. Wings ochreous-grey, markings brown, basal and medial areas lighter, terminal area darker. Wings irrorated with brown. Basal and medial lines absent or weakly developed, suffused. Postmedial line strongest, brown, fine, straight or weakly curved, frequently outer margin bordered by one or several blackish marks. Terminal line sharp, discontinuous. Fringes ochreous-grey. Discal spots large, suffuse. Wings below ochreous-grey, markings light brown, suffuse, striated. Terminal area light brown, dark brown on forewing termen and hindwing terminal area. Terminal area with ochreous-grey spots on both wings. Fringes ochreous-grey. Frons and collar ochreous brown, thorax and abdomen lighter, mixed with brown. Antennae ochreous-grey dorsally, bipectinate in male, filiform in female. Hindtibia with 2+2 spurs. Tympanal organs medium-sized, bullae tympani with small roundish lobes. Posterior margin of male 8th sternite with two, sclerotised lobes, finely serrate, submargin area scobinate. Other male and female sternites and tergites undifferentiated.

VARIATION: Wing colour varies from ochreous-whitish to dark grey. Terminal area varies from light brown to dark brown. According to KRÜGER (2001), melanistic specimens are observed.

MALE GENITALIA: Uncus wide, margins slightly angled, densely setose dorsally. Gnathos arms narrow, fused medially, forming wide, horizontal ridge. Juxta small plate. Valva divided. Dorsal arm of valva long, narrow, margins and apex setose. Ventral arm of valva wide, apex with few setae. Its ventral margin finely serrated. Saccus weakly elongated. Aedeagus narrow, caecum minute, apex striated. Vesica opens at 45 degree angle. Vesica wide at base, with single lateral cornutus and numerous microcornuti on other side.

FEMALE GENITALIA: Papillae anales slightly elongated, setose. Apophyses long, narrow. Lamella antevaginalis unrecognisable. Lamella postvaginalis narrow, sclerotised plate. Ductus bursae medium length, wide, with longitudinal sclerotisations. Posterior part of corpus bursae elongated, partly sclerotised and striated, near ductus seminalis scobinate. Shape of anterior part of corpus bursae variable, normally elongated horizontally, membranous. Signum large, stellar arms long.

DISTRIBUTION AND ABUNDANCE: Palaeotropical. In Europe only in south-western Iberian Peninsula, Malta and Lampedusa. – Outside Europe widely distributed in Africa from the Mediterranean to South Africa and in the Middle East. – In Spain a rather rare species that generally is found as single specimens, in Malta apparently more common.

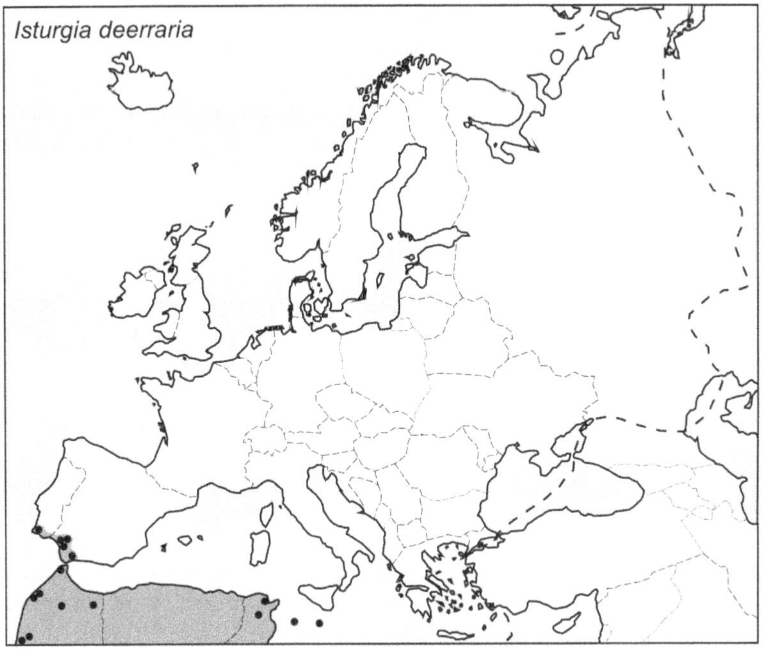

PHENOLOGY: In Europe probably bivoltine, in Malta possibly trivoltine. On the Iberian Peninsula specimens have been found in June and July and from September to December (REDONDO *et al.* 2009), in August (P. HALE pers. comm.) and in Malta a single specimen has been found in May and many in August, September and November. In North Africa also recorded in late January, mid-March, early May, mid-June to late August, mid-September to early October and mid- and late December. According to KRÜGER (2001) the adults are active throughout the year in the Afrotropical region. In the Arabian Peninsula also flying in the spring. The moths are active at night and come to light.

BIOLOGY: There are no reports of natural food plants in Europe. By rearing from eggs in Malta, the larvae accepted *Acacia karroo* as food plant, but ignored *A. cyanophylla* (P. SAMMUT pers. comm.). In Morocco recorded on various species of *Acacia* (RUNGS 1981). KRÜGER (2001) also records it on *Acacia* and furthermore on *Peltophorum*. By rearing in Malta the development of the species has been very short. The eggs hatch after five days, the larval stage is approximately 10 days and the pupa hatches after only approximately seven days (P. SAMMUT pers. comm.).

HABITAT: Xerothermohilous. In the Iberian Peninsula in open lowland habitats, including overgrown coastal dunes. According to REDONDO *et al.* (2009) up to 400 m altitude. In Africa in a large variety of habitats (KRÜGER 2001).

SIMILAR SPECIES: No similar species in Europe. Numerous similar species in Africa, particularly *I. sublimbata* (BUTLER, 1884), which is known from Middle-East, Arabia and East Africa, cannot be identified reliably without examination of the genitalia (KRÜGER 2001).

GENETIC DATA: Genetically homogeneous in Europe (n=6, from Spain, Portugal and Malta), Africa and Arabian Peninsula (n=73). Nearest species: *I. arizeloides* (KRÜGER, 2001) from South Africa (4.3%).

REMARKS: Until recently, *Isturgia deerraria* has been considered a subspecies of *Isturgia pulinda* (WALKER, 1860) from Sri Lanka. KRÜGER (2001) concluded, after examination of the type specimens (holotype males of *erraria* and *pulinda* are in the Natural History Museum, London, UK) that, although they are closely related, *I. pulinda* is not conspecific with the African *I. deerraria. I. deerraria* has only relatively recently been recorded from Europe (AGENJO 1974).

24. *Isturgia disputaria* (GUENÉE, 1858)

disputaria GUENÉE, [1858], in Boisduval & Guenée, Hist. nat. Insectes (Spec. gén. Lépid.) 10: 489, (*Eubolia?*). Holotype male (The Natural History Museum, London, UK), Egypt (examined, including genitalia).

curvifera WALKER, 1870, Entomologist 5: 154, (*Aspilates*). Syntype(s) male(s), Egypt: Cairo.

gazella BLACKMORE, 1872, Entomologist's mon. Mag. 8: 231, (*Eubolia*). Holotype male, Morocco (north-west): Tangier.

granitalis BUTLER, 1883, Proc. zool. Soc. Lond. 1883 (2): 171, (*Tephrina*). Lectotype female (The Natural History Museum, London, UK), India: Mhow (examined, including genitalia).

indotata WALKER, [1863], List Specimens lepid. Insects Colln Br. Mus. 26: 1681, (*Aspilates*). Holotype male (The Natural History Museum, London, UK), North Hindostan [India].

infrictaria SWINHOE, 1866, Proc. zool. Soc. Lond. 1885 (4): 862, (*Macaria*). Lectotype female (The Natural History Museum, London, UK), [India:] Poona (examined, including genitalia).

lithina BUTLER, 1883, Proc. zool. Soc. Lond. 1883 (2): 171, (*Tephrina*). Lectotype female (The Natural History Museum, London, UK), [Pakistan:] Kurrachee ([Karachi]; Solun; Umballa (examined, including genitalia).

martiniaria OBERTHÜR, 1876, Études ent. 1: 61, pl. 4, fig. 8 (*Fidonia*). Syntype(s) including male(s), Algeria: Biskra.

strenuata WALKER, [1863], List Specimens lepid. Insects Colln Br. Mus. 26: 1647, (*Macaria*). Lectotype male (The Natural History Museum, London, UK), [India]; Punjab (examined, including genitalia).

strenuataria BUTLER, 1880, Ann. Mag. nat. Hist. (5): 224, (*Tephrina*). [Emendation of *strenuata* WALKER].

subocellata WARREN, 1896, Novit. zool. 3: 413, (*Tephrina*). Holotype male (The Natural History Museum, London, UK), [Yemen:] Arabia, S[heik] Othman (examined, including genitalia).

trilineata KRÜGER, 1939, Ann. Mus. libico Stor. nat. 1: 355, pl. 17, figs 117–119, (*Tephrina*). Syntypes male(s), female(s), [Libya]: Beni Ulid; Tauorga; Uadi Zarzur; Gebel es-Soda; Zella. [originally as subspecies of *disputaria* GUENÉE].

zebrina BUTLER, 1883, Proc. zool. Soc. Lond. 1883 (2): 171, (*Tephrina*). Lectotype female (The Natural History Museum, London, UK), India: Mhow (examined, including genitalia).

INTRASUBSPECIFIC NAMES, POTENTIALLY UNAVAILABLE: *darmouini* DUMONT, 1932 (ab.); *hachia* DUMONT, 1932 (ab.); *nigrescente* D. LUCAS, 1948 (ab.?).

EXTERNAL CHARACTERS AND ABDOMEN: Wingspan 18–25 mm. Wings light ochreous-yellow, medial lines brown, weakly undulating. Antemedial and medial lines strongest in inner margin, medial line widest, postmedial line complete, its outer margin with light brown band. Terminal area darker. Terminal line dark, complete. Fringes unicolorous light. Discal spots dark, elongated, occasionally ring-shaped. Wings below lighter, markings more striated.

Females are less contrasting, forewing postmedial line weaker, particularly near inner margin. Frons, collar, thorax and abdomen light ochreous yellow. Antennae mixed white and brown dorsally, bipectinate in male, filiform in female. Male hindtibia slightly swollen, both sexes with 2+2 spurs. Tympanal organs large, bullae tympani with roundish lobe. Male 3rd sternite with setal comb, posterior margin of 8th sternite furcate, with acute tips, their inner margin finely serrated, subbasal area scobinated. Other male and female sternites and tergites undifferentiated.

VARIATION: Ground colour varies from light ochreous-yellow to almost uniform grey, females are less contrasting. Postmedial line strongest, other medial lines may be weaker, occasionally reduced to brownish dots in inner margin. Numerous forms have been described.

MALE GENITALIA: Uncus roundish, base slightly swollen, finely setose dorsally. Gnathos arms narrow, fused medially, forming wide, horizontal ridge. Juxta small plate. Valva divided. Dorsal arm of valva long, narrow, weakly expanded at apex, base, costa and apex setose. Ventral arm of valva wide, short, apex acute, with few setae. Saccus slightly triangular. Aedeagus long, narrow, straight, caecum minute, apex pointed, striated. Vesica narrow tube, opens at 45 degree angle, with two lateral cornuti. Long, slightly curved cornutus attached to lateral diverticulum, group of minute cornuti in more basal position.

FEMALE GENITALIA: Papillae anales slightly elongated, setose. Apophyses straight, narrow. Lamella antevaginalis and postvaginalis forming small, roundish plate around ostium bursae, dorsal part slightly lunular. Ductus bursae medium length, wide, sclerotised. Base of ductus seminalis expanded. Corpus bursae large, elongated, pear-shaped. Posterior end elongated, striated, weakly sclerotised, anterior end roundish, membranous. Signum on roundish lateral diverticulum, large, stellar arms long.

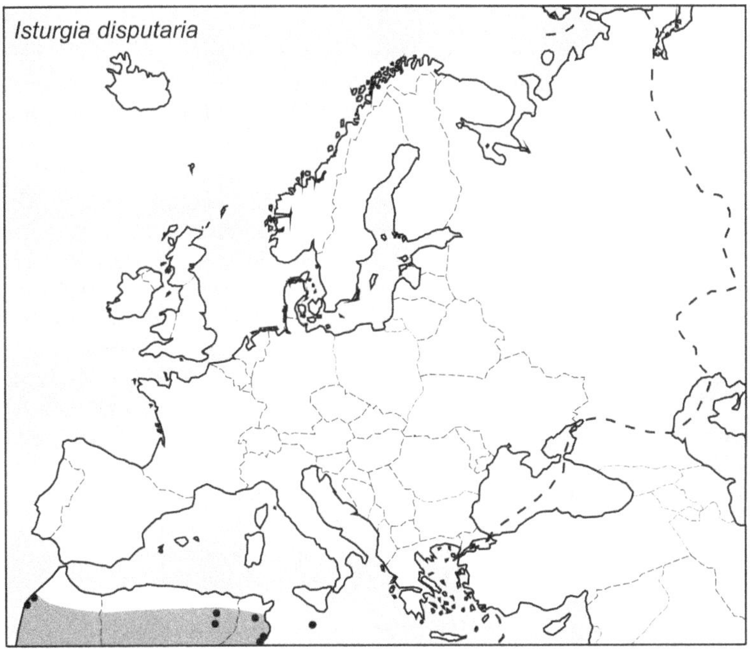

DISTRIBUTION AND ABUNDANCE: Palaeotropical. In Europe only known from a single specimen found in Malta. – Outside Europe the species is widely distributed in the whole of Africa including Madagascar, the Arabian Peninsula and in India (KRÜGER 2001). – In Malta only a single specimen in 2010 (CATANIA 2010).

PHENOLOGY: Multivoltine. The single specimen in Malta was found in late March. In Morocco in April and May (RUNGS 1981). In central Libya and Israel during the whole year, but most abundant in autumn and spring (A. HAUSMANN & N. PÖLL pers. comm.). The adult moths are active at night and come readily to light (N. PÖLL pers. comm.).

BIOLOGY: Monophagous. The species has been recorded feeding on *Acacia tortilis* (DUMONT 1932). It may, however, also feed on other species of *Acacia*.

HABITAT: Xerothermophilous. The species occurs in arid habitats from sea level and mainly at lower altitudes, in Morocco up to 1280 m (N. PÖLL pers. comm.), in Oman up to 900 m. In the Near East a typical species of oases and wadis (A. HAUSMANN pers. comm.).

SIMILAR SPECIES: *Digrammia rippertaria* (DUPONCHEL) lacks forewing medial shade near inner margin (present in *disputaria*). Genitalia of both sexes in these species allow easy identification.

GENETIC DATA: Genetically rather homogeneous from Morocco and Gambia to Pakistan (n=60, from 14 countries). Nearest species: *I. deerraria* (9.3%).

REMARKS: *Isturgia disputaria* was reported as new to the European fauna in 2010, and it is known from Malta only. *I. disputaria* is likely to be a migrant in Europe, at least the 2010 record was associated with strong southern wind (CATANIA 2010).

25. *Isturgia spodiaria* (LEFÈBVRE, 1832)

spodiaria LEFÈBVRE, 1832, Magasin Zool. Paris Class 9: unpaginated, pl. 8, (*Fidonia*). Syntypes 2 males, 2 females, [Italy]: Sicily, Val di Noto.

cerataria GUENÉE, [1858], in BOISDUVAL & GUENÉE, Hist. nat. Insectes (Spec. gén. Lépid.) 10: 146, (*Selidosema*). Holotype male (The Natural History Museum, London, UK), [Italy: Sicily]: Messina (examined externally).

mizanensis WEHRLI, 1940, in SEITZ, Gross-Schmett. Erde 4 (Suppl.): 396, pl. 32: b, (*Itame*). Holotype male (Zoologische Forschungsinstitut und Museum Alexander Koenig, Bonn, Germany), [Morocco]: Mrassine (examined, including genitalia, illustrated in Text-fig 51). Originally as subspecies of *Itame spodiaria* LEFÈBVRE, treated valid at subspecies level by SCOBLE (1999) (as *Isturgia spodiaria* subspecies *mizanensis*), downgraded here to synonymy, quoting absence of diagnostic features (new synonym). See remarks.

semicanaria FREYER, [1832], Neuere Beitr. Schmettkde 1 (13): 145, pl. 78, fig. 2, (*Fidonia*). Syntype(s) [not stated but apparently 7 females (KRÜGER 2001)], [Italy]: Sicily.

spodiaria gaetana WEHRLI, 1940, in SEITZ, Gross-Schmett. Erde 4 (Suppl.): 396, pl. 32: b, (*Itame*). Paratype male (The Natural History Museum, London, UK), Italy (southern/middle): Formia near Gaeta (examined externally). See remarks.

spodiaria plumbeolata AMARY, 1840, Esercit. accad. Aspir. Nat. Napoli 2 (1): 83, pl. 6, figs. 4a, b, (*Phidonia*). Syntype(s) male(s), [Italy]: Capri. [Originally as sp]. See remarks.

EXTERNAL CHARACTERS AND ABDOMEN: Wingspan 25–30 mm. Wings ash-grey, markings weak, often antemedial present as dot in costa, postmedial most visible, suffused, clearly turning inwards near costa. Terminal area darkest, suffused with brown. Terminal line weak, dark brown, discontinous. Fringes unicolorous ash-grey. Hindwings often without medial lines, terminal area darker. Discal spots large, brown, elongated in forewings, small in hindwings. Wings brown below, hindwing markings striated and more pronounced. Frons brownish, collar, thorax and abdomen ash grey, sparsely mixed with brown. Antennae white dorsally, bipectinate in male, filiform in female. Hindtibia with 2+2 spurs. Tympanal organs medium-sized, bullae tympani with roundish lobe. Sternites and tergites 3–8 undifferentiated.

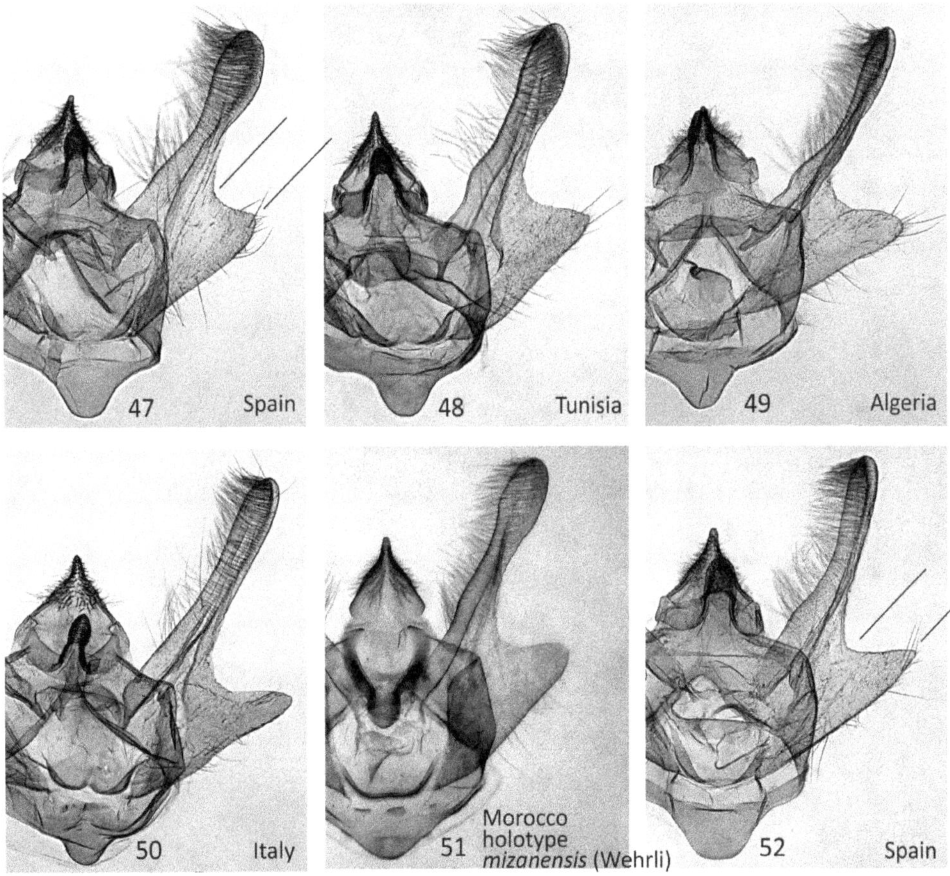

Text-figs 47–52. Intraspecific variation of the valva shape in *Isturgia spodiaria* (LEFÈBVRE, 1832). Apex of the ventral arm of valva varies from acute to round, as does the concavity between the dorsal and the ventral arms of valva (n=15). Shape of the valva is somewhat correlated with geographical locality, but not exclusively, demonstrated here by two Spanish specimens that are located at extreme ends of variation. Other male genitalia structures, including the aedeagus and vesica, are structurally more homogeneous. Text-fig. 47. Spain: Prov. Cádiz, 17–18.v.2008, slide PS1848 (coll. SKOU). Text-fig. 48. Tunisia: Hammamet, 12–18.iii.1973, PS1212 (coll. ZMH). Text-fig. 49. Algeria: Bône, May 1931, PS1858 (coll. ZFMK). Text-fig. 50. Italy: Formia, April 1910, PS1872/BMNH GEO 24846 (coll. BMNH). Text-fig. 51. Holotype, *Itame spodiaria mizanensis* WEHRLI, 1941, Morocco: Mrassine, April 1921 (coll. ZFMK). Text-fig. 52. Spain: Malaga, 22.ix.1987, slide PS1831 (coll. SKOU).

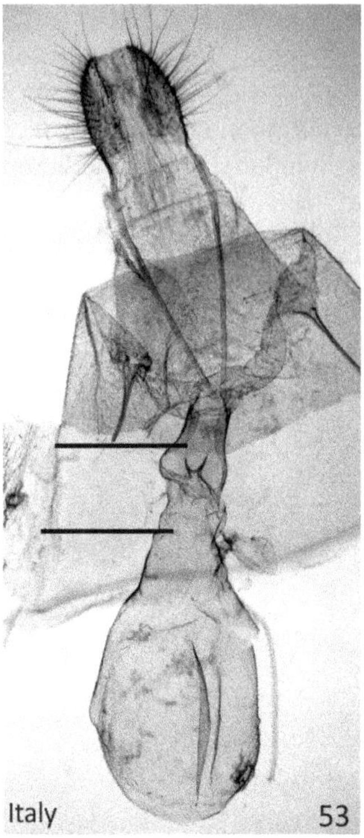
Italy
53
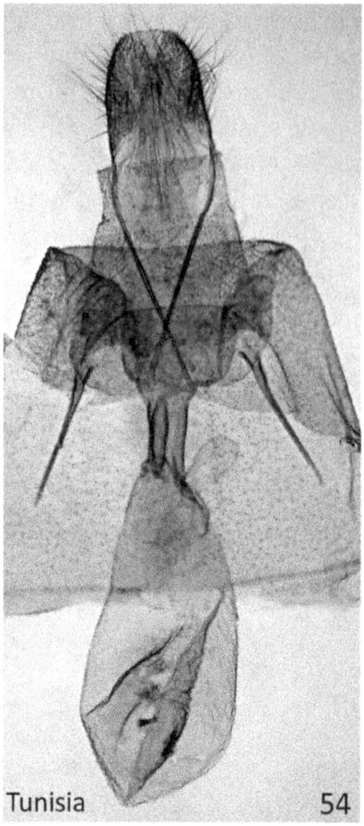
Tunisia
54
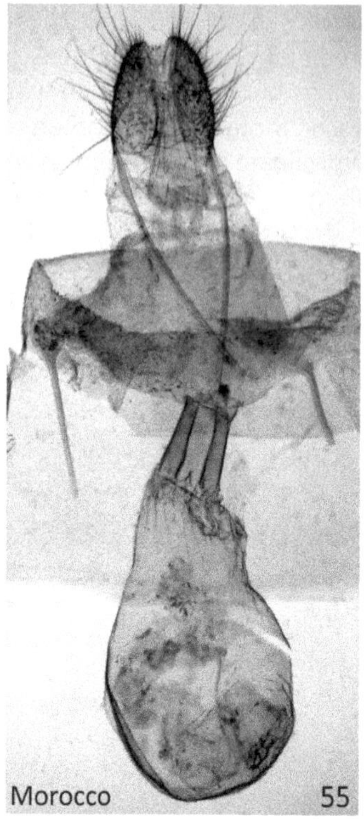
Morocco
55
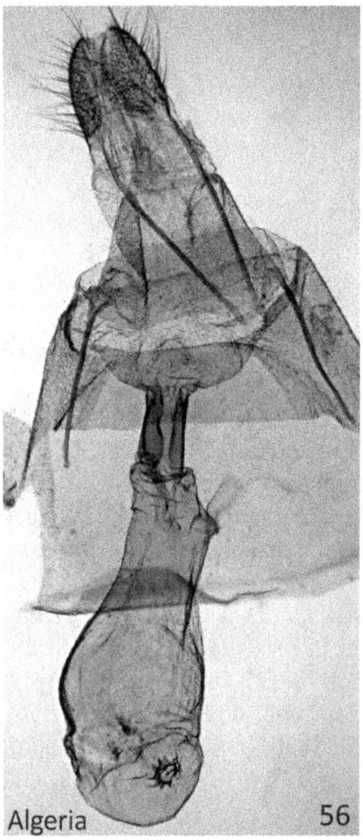
Algeria
56
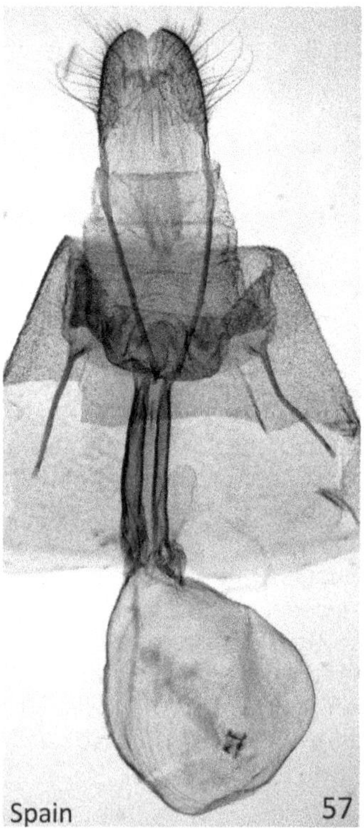
Spain
57
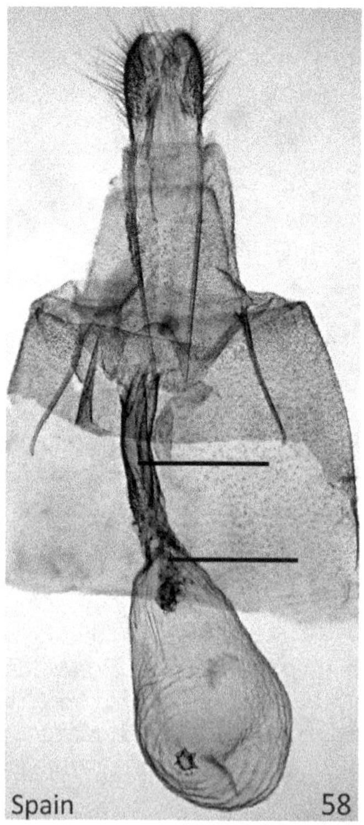
Spain
58

⇦ Text-figs 53–58. Variation of the female genitalia in *Isturgia spodiaria* (LEFÈBVRE, 1832), see also female genitalia plates. Length and shape of the ductus bursae and posterior part of the corpus bursae are variable (n=11). Structures somewhat correlate with geographical locality: Spanish specimens have the longest ductus bursae and the posterior part of ductus bursae is not expanded. Other specimens have shorter ductus bursae and posterior part of ductus bursae is expanded to varying degrees. Only two specimens were examined from Italy, the ductus bursae shape of those is unique when compared against other material. Text-fig. 53. Italy: Formia, April 1910, slide PS1873/BMNH GEO 24847 (coll. BMNH). Text-fig. 54. Tunisia: Hammamet, 4–11.iii.1975, PS1855 (coll. ZMH). Text-fig. 55. Morocco: Taddert, 1 700 m, March 1937, PS1875/BMNH GEO 24849 (coll. BMNH). Text-fig. 56. Algeria: Bône, May 1931, PS1860 (coll. ZFMK). Text-fig. 57. Spain: prov. Cadiz, 4–7.iii.2005, PS1832 (coll. SKOU). Text-fig. 58. Spain: Malaga, 16.xi.1980, PS1854 (coll. SKOU).

VARIATION: Ground colour can be different shades of grey, postmedial line often most visible but it can be absent. Tympanal organs vary geographically. Examined Tunisian specimens are without roundish lobe in the bullae tympani (n=4), whereas all other examined specimens (n=22) from Italy, Algeria, Morocco and Spain are with distinct roundish lobe. Wingspan variable, from 22 mm (smallest examined male, Morocco: Tangier, 25.x.1934, slide PS1874/BMNH GEO 24848, coll. BMNH) to 37 mm (largest examined female, Morocco: High Atlas, 1700 m, iii.1937, slide PS1875/BMNH GEO 24849, coll. BMNH). Text-figs 47–52 and 53–58 illustrate variation in the male and female genitalia.

MALE GENITALIA: Uncus evenly tapering, setose dorsally. Gnathos arms wide, fused medially, wide at base, forming triangular and upturned element, surface granulate. Juxta small plate. Valva divided. Dorsal arm of valva long, narrow, about equal width, costa and apex setose. Ventral arm of valva wide, short, apex round with few setae. Saccus small. Aedeagus narrow, slightly curved, caecum minute, apex sharp. Vesica expanded at base, with weak sclerotisations, opens at 90 degree angle, cornuti absent. Shape of male valva variable, See remarks and Text-figs 47–52.

FEMALE GENITALIA: Papillae anales slightly elongated, setose. Apophyses straight, narrow. Lamella antevaginalis membranous. Lamella postvaginalis tiny, lunular plate. Ductus bursae straight, short, sclerotised laterally. Base of ductus seminalis expanded. Corpus bursae membranous, posterior part wide, anterior part round. Signum tiny, weakly stellate, shape variable. Shape and length of ductus bursae variable, See remarks and Text-figs 53–58.

DISTRIBUTION AND ABUNDANCE: West-Mediterranean. In Europe disjunct distribution, only occurring in southern Spain, Sicily and southern and central Italy. – Outside Europe the species is found in Libya, Tunisia, Algeria, Morocco and Mauritania. – In Europe a locally common species that now and then is found in numbers.

PHENOLOGY: Bivoltine. In southern Spain from mid-March to early May and again from late September to late November. In southernmost Spain recorded in all months of the year except July (P. HALE pers. comm.). In Italy apparently with shorter flight times from April to early June and from September to late October (FLAMIGNI *et al.* 2007). At high altitudes in Morocco the species flies apparently also in July, perhaps only in one generation. The moths fly at night and come to light.

BIOLOGY: Oligophagous on Fabaceae. Recorded on *Calicotome* and *Genista* (REDONDO *et al.* 2009), also FLAMIGNI *et al.* (2007) record *Genista*.

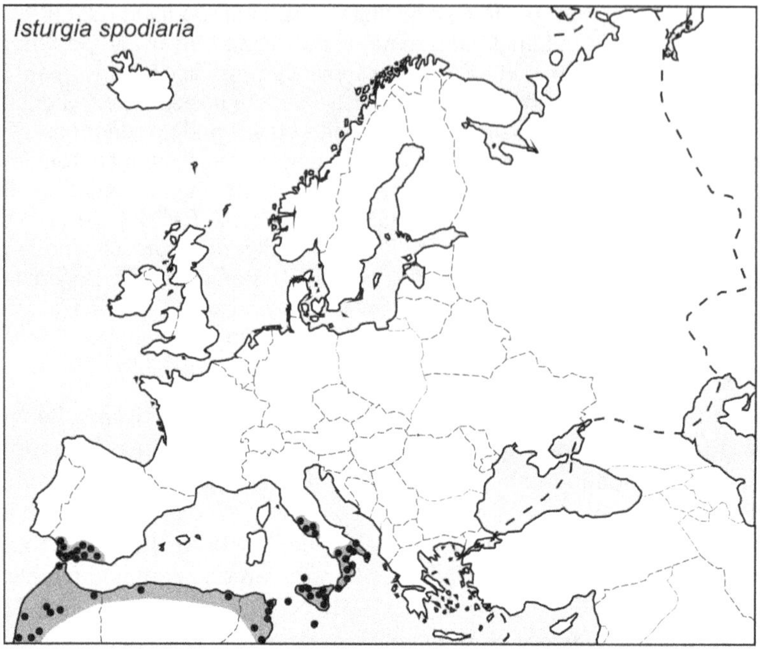

HABITAT: Xerothermophilous. In Europe the species occurs in a number of different habitat types such as overgrown coastal dunes, oak forests, macchia and other uncultivated places and it has been found from sea level up to 1350 m in Sicily, according to REDONDO *et al.* (2009) up to 1500 m in southern Spain. In the Atlas mountains in Morocco it has been recorded up to at least 2400 m altitude.

SIMILAR SPECIES: *I. murinaria*, *I. assimilaria* and *I. berytaria*. See *I. murinaria* and Text-figs 41–46. Gnathos in the male genitalia of *I. assimilaria* narrower at base (wider at base in *I. spodiaria*), vesica less expanded at base (vesica more expanded at base in *I. spodiaria*). Ductus bursae in the female genitalia of *I. assimilaria* shorter (ductus bursae longer in *I. spodiaria*).

GENETIC DATA: Populations from Italy and Tunisia diverging from those of Spain and northern Morocco by 6.9%, requiring investigation based on more extensive material. Nearest species: North African *I. teknaria* (POWELL & RUNGS, 1942) (6.9% from Italian populations) and *I. roraria* (7.0%).

REMARKS: An analysis of extensive material throughout the distribution area of *I. spodiaria* suggests that the external and genitalia characters somewhat correlate with geographical locality, but not exclusively. When more material was examined the correlations became weaker, see Text-figs 47–52 and 53–58. For this reason we follow FLAMIGNI *et al.* (2007) and consider the taxon *gaetana* WEHRLI, 1940 and *plumbeolata* AMARY, 1840 as synonyms of *I. spodiaria* (LEFÈBVRE, 1832) and downgrade taxon *mizanensis* WEHRLI, 1940 from Morocco to synonymy with *I. spodiaria* (LEFÈBVRE). TRUSCH & MÜLLER (2001) consider the Tunisian specimens to belong to *I. spodiaria* subsp. *spodiaria* (LEFÈBVRE), whereas KRÜGER (2001) and LERAUT (2009) treat all North African populations as *I. spodiaria* subsp. *mizanensis* (WEHRLI). The male genitalia show weaker correlation with geographical locality than the female genitalia. The latter form three rather distinct groups: Spain, North Africa and

Italy. The female genitalia of examined Italian specimens (n=2) are the most distinct, having short, wide, and anteriorly expanded ductus bursae (Text-fig. 53). This correlates with the genetic divergence, but not with the male genitalia, including the aedeagus length. More extensive genetic, ecological and life history material from throughout the distribution area needs to be analysed to resolve the issue.

26. *Isturgia assimilaria* (RAMBUR, 1833)

assimilaria RAMBUR, 1833, Annls Soc. ent. Fr. 2 (1): 34, pl. 2, figs 9-10, (*Fidonia*). Syntype(s), [France]: Corsica.

sardalta BYTINSKI-SALZ, 1934, Int. ent. Z. 28 (14): 168, (*Tephrina* (*Eubolia*)). Syntypes males, females, [Italy]: Sardinia (Gennargentu 1800 m).

INTRASUBSPECIFIC NAMES, POTENTIALLY UNAVAILABLE: *delineata* BYTINSKI-SALZ, 1934 (ab.).

EXTERNAL CHARACTERS AND ABDOMEN: Wingspan 24–26 mm. Wings grey, weakly irrorated with brown. Medial lines variably visible, strongest at costa, postmedial line strongest, clearly turning inwards near costa. Terminal area darkest, brownish. Terminal line distinct, dark brown, discontinuous. Fringes chequered grey and brown. Hindwings often without medial lines, terminal area darker. Discal spots suffused, brown. Forewings grey below, hindwings white, irrorated with brown, postmedial line brown, terminal line distinct, discontinuous. Frons, collar, thorax and abdomen grey, sparsely mixed with brown. Antennae grey dorsally, bipectinate in male, filiform in female. Hindtibia with 2+2 spurs. Tympanal organs large, meeting ventrally. Sternites and tergites 3–8 undifferentiated.

VARIATION: Ground colour can be different shades of grey, medial lines can be weak and suffused or dark and distinct. Forewings below may have whitish stripes in terminal area.

MALE GENITALIA: Generally as in *Isturgia spodiaria*, see entry above, but with following minor differences: Ventrally fused gnathos arms narrower at base. Saccus more elongated medially. Vesica less expanded at base, with very weak sclerotisation.

FEMALE GENITALIA: Generally as in *Isturgia spodiaria*, see entry above, but with following differences: Ductus bursae straight, shorter, sclerotised also ventrally. Posterior part of corpus bursae narrower, more elongated, anterior part less expanded.

DISTRIBUTION AND ABUNDANCE: Tyrrhenian. Endemic to Corsica and Sardinia where the species is rather common and often found in numbers.

PHENOLOGY: Univoltine. The species flies from early April until late June or early July, depending on altitude. Caterpillars from late May to late July (M. LEIPNITZ pers. comm.). The species hibernates as a pupa. Moths fly actively at sunset and again at night when they come to light.

BIOLOGY: Monophagous. The caterpillar is recorded on *Genista* (FLAMIGNI *et al.* 2007) from May to July. Also reared on *Genista*, the larval time is relatively short with only three weeks (B. MÜLLER pers. comm.). In rearings pupation on and in the ground (M. LEIPNITZ pers. comm.).

HABITAT: Occurs in a variety of habitats, dry as well as more moist and open land with scattered trees and bushes also in more dense woodland. The moths are found from sea level up to 1400 m on Corsica, in Sardinia up to 1800 m according to FLAMIGNI *et al.* (2007).

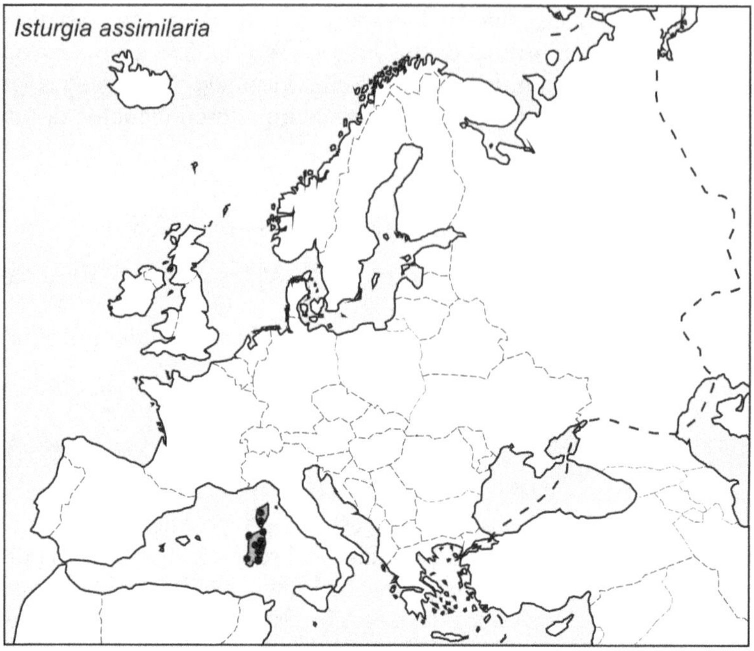

SIMILAR SPECIES: *I. murinaria*, *I. berytaria* and *I. spodiaria*, see *I. murinaria*, Text-figs 41–46, and descriptions of genitalia characters above.

GENETIC DATA: No data available.

27. *Isturgia miniosaria* (DUPONCHEL, 1829)

miniosaria miniosaria DUPONCHEL, 1829, in GODART & DUPONCHEL, Hist. nat. Lépid. Papillons Fr. 7 (2): 368, pl. 160, fig. 4, (*Boarmia*). Syntype(s), [France]: near Valenciennes.

miniosaria gabriellae SILVA CRUZ, 1978, Anais Fac. Ciěnc. Porto 61 (1-4): 56, fig. 1; text-fig. 1, (*Enconista*). Holotype male (Instituto Zoologia ,Dr Augusto Nobre', Porto, Portugal), Portugal: Lousado (Famalicão (examined, including genitalia).

miniosaria perspersaria DUPONCHEL, [1830], in GODART & DUPONCHEL, Hist. nat. Lépid. Papillons Fr. 7 (2): 458, pl. 169, figs. 1, 2, (*Fidonia*). Syntype(s), France: Montpellier.

duponcheli PROUT, 1915, in SEITZ, The Macrolepidoptera of the World 4: 407, pl. 23: l, (*Enconista*). [Unnecessary replacement name for *perspersaria* DUPONCHEL (SCOBLE 1999), but see LERAUT (2009)].

miniosaria rungsi LERAUT 2009, Moths of Europe, vol. 2, Geometrid Moths: 79, pl. 15, figs 1, 2, fig. 42b (*Isturgia*). Holotype male (Muséum National d´Histoire Naturelle, Paris, France), Morocco: Middle-Atlas, Gara Midelt. Paratype male (Muséum National d´Histoire Naturelle, Paris, France) from Morocco is illustrated in LERAUT (2009). LERAUT also illustrates a paratype female, but no female is designated as a type in the description.

INTRASUBSPECIFIC NAMES, POTENTIALLY UNAVAILABLE: *cretacea* SCHAWERDA, 1928 (ab.); *grisea* SCHAWERDA, 1928 (ab.); *infuscata* SCHWINGENSCHUSS, 1930 (ab.); *obscura* SCHAWERDA, 1928 (ab.); *rosescens* SCHAWERDA, 1928 (ab.).

EXTERNAL CHARACTERS AND ABDOMEN: Wingspan 32–40 mm. Wings whitish grey to beige, often heavily irrorated with grey, brown or blackish. Medial lines dark, slightly undulating, angled at costa. Terminal area occasionally darkest, particularly in hindwings. Terminal line blackish, discontinuous. Fringes unicolorous. Forewing margin rounded. Discal spots dark, elongated. Wings below as above, markings weaker and more diffuse. Frons, collar, thorax and abdomen same colour as wings, irrorated with dark scales. Antennae mixed grey and brown-blackish dorsally, bipectinate in male, filiform in female. Both sexes with 2+2 spurs. Tympanal organs medium-sized, bullae tympani with roundish lobe. Sternites and tergites 3–8 of both sexes undifferentiated.

VARIATION: One of the most variable European geometrid species. Ground colour varies from whitish to grey, brown, rust-brown to dark grey-brown. Medial lines can be distinct, blackish, brownish and weak, or nearly absent. Markings can be distinct, occasionally these are absent and wings are irrorated with with blackish-brown scales. Discal spots normally present, well developed, but can be absent. Numerous forms have been described. Several of these are discussed by WEHRLI (1939–54). Algerian specimens are uniform in appearance, forewings are beige with weak black antemedial and postmedial lines, hindwings are paler. See remarks.

Subsp. ***rungsi*** with wings beige-white to ash-grey and brick-brown, weakly irrorated with dark scales. Medial lines reduced to weak markings. Ventral arm of valva wide, apex more rounded than in nominate subspecies (illustrated in LERAUT 2009). Subspecies status in relation to *duponcheli* (PROUT) needs to be confirmed. See remarks.

MALE GENITALIA: Uncus triangular, setose dorsally. Gnathos arms wide, fused medially, forming triangular and upturned element, surface granulate. Juxta small plate. Valva divided. Dorsal arm of valva long, narrow, about equal width throughout, costa and apex setose. Ventral arm of valva wide, short, ventral margin slightly concave, apex slightly angled, with few setae. Saccus weakly elongated medially. Aedeagus long, narrow, slightly curved, caecum minute, apex short, acute. Vesica expanded at base, with numerous microcornuti, opens at 90 degree angle, large cornuti absent.

FEMALE GENITALIA: Papillae anales roundish, setose. Apophyses straight, apophyses anteriores approximately half length of apophyses posteriores. Lamella antevaginalis membranous. Lamella postvaginalis sclerotised, rounded-plate. Ductus bursae long, narrow, sclerotised laterally. Corpus bursae membranous, posterior part elongated, narrow, striated, anterior part larger, pear-shaped. Signum large, weakly stellate.

DISTRIBUTION AND ABUNDANCE: West-Mediterranean. In Europe in southern France and the Iberian Peninsula. Records from Croatia and South Tyrol (PROUT 1912–1916, and references repeated elsewhere in more recent literature) are erroneous (for instance Tyrol in FLAMIGNI *et al.* 2007, Croatia in HAUSMANN *et al.* 2011). – Outside Europe recorded from North Africa. However, distribution in North Africa requires further research. Subsp. *rungsi* occurs in Morocco, but numerous specimens labeled as *miniosaria* in various collections from elsewhere in North Africa need to be verified against other taxa in the species complex (see similar species and remarks). Therefore the examined specimens from North Africa are shown as question marks. – In Europe a common species that is often found in numbers, sometimes in large numbers.

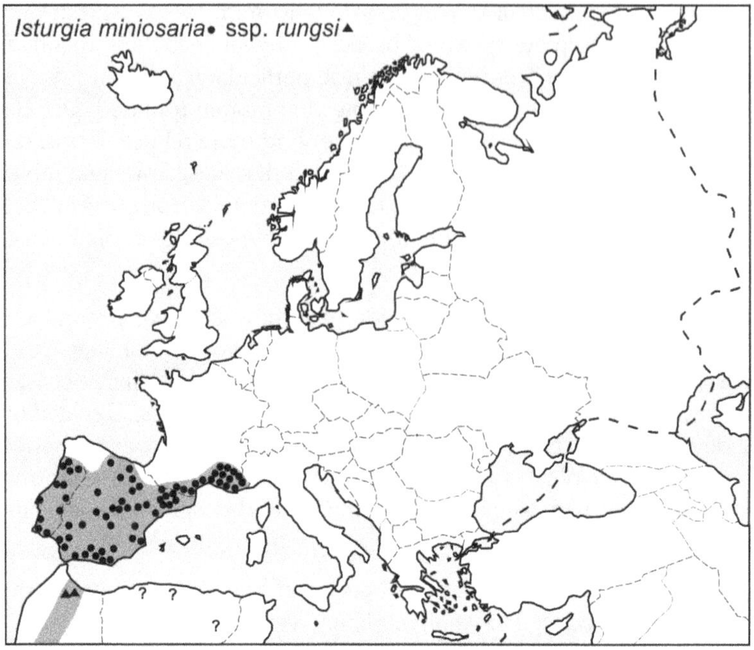

PHENOLOGY: Univoltine. The moths fly from late August or early September in northern Spain and until late November in southern Spain. Larvae in the spring. The moths are night active and come readily to light, often in numbers.

BIOLOGY: Oligophagous on Fabaceae. Caterpillars have been found on *Retama sphaerocarpa* (B. MÜLLER pers. comm.) and in Portugal on *Cytisus striatus* (M. CORLEY pers. comm.). Caterpillars have been found in numbers on *C. multiflorus* and *Pterospartum tridentatum* (MENDES 1909). The larva is further recorded on *Spartium*, *Genista* (REDONDO *et al.* 2009) and *Ulex* (AIZPÚRUA 2002).

HABITAT: The species occurs in a large variety of habitats from overgrown dunes by the sea and very warm and dry open low land places to localities with scattered bushes at middle altitudes and rocky slopes with patches of grass above the timber line in the higher mountains.

SIMILAR SPECIES: No similar species in Europe, particularly the large size of *I. miniosaria* is diagnostic. *Isturgia hausmanni* KRÜGER, 2001, *I. terminipuncta* KRÜGER, 2001, *I. tozeurensis* LERAUT, 2009 (See remarks), which are known from Tunisia, and *I. rubrior* (HAUSMANN, 1990) from Morocco, are similar (HAUSMANN 1990, KRÜGER 2001, LERAUT 2009). In the Iberian Peninsula the unrelated *Odontognophos perspersata* (TREITSCHKE, 1827) is slightly similar.

GENETIC DATA: The few examined specimens from northern, eastern and southern Spain (n=3) genetically heterogeneous, maximum variation 2.5%. Examined specimens from south-western Morocco strongly diverging by 6.1%. Nearest species (sister species *I. hausmanni* (KRÜGER, 2001) from Tunisia not yet barcoded). *I. tennoa* (PINKER, 1978) from the Canary Islands (8.5% from Spanish populations) and *I. berytaria* (6.7% from Moroccan populations).

REMARKS: The synonymy of *Enconista miniosaria* subsp. *gabriellae* SILVA CRUZ, 1978, with *Boarmia miniosaria* DUPONCHEL, 1829, was proposed by HAUSMANN & LENZ (1993). The Iberian specimens have been treated as subsp. *miniosaria* (ground colour varying from whitish to brown with blackish markings) and the non-Iberian specimens (for instance France, North Africa) have been treated as subsp. *duponcheli* (ground colour mostly beige-grey, irrorated with grey and markings weak) (PROUT 1912–16, KRÜGER 2001). We illustrate numerous specimens from the Iberian Peninsula to show that such distinct features do not correlate with geographical location of the specimen. Therefore we do not recognise subspecies in Europe. KRÜGER (2001), following PROUT (1912–16), treated the African populations as subsp. *duponcheli* (PROUT, 1915). LERAUT (2009) described Moroccan populations as subsp. *rungsi*, with a note that this subspecies probably lives in Algeria also. More extensive material throughout the distribution area needs to be analysed to solve the issue, particularly to find out the relationship between *duponcheli* (PROUT) and *rungsi* LERAUT in North Africa and whether the populations south and north of the Pyrenees are distinct at subspecies level.

Isturgia tozeurensis LERAUT, 2009 from south west Tunisia: Tozeur, which is part of the *Isturgia miniosaria* group of species, is potentially a junior synonym of *Isturgia hausmanni* KRÜGER, 2001 or *Isturgia terminipuncta* KRÜGER, 2001. Apparently LERAUT was not aware of these two Tunisian *Isturgia* species, which are very similar to *tozeurensis*, because they were not mentioned in the original description (LERAUT 2009, p. 80, figs 42, 43). Despite the similarities in external appearance, and in the genitalia, we will not propose synonymy here because LERAUT does not illustrate the holotype of *tozeurensis* (housed in Muséum National d'Histoire Naturelle, Paris, France) and the hand drawn genitalia illustrations of *tozeurensis* are too superficial to allow reliable identification with either *hausmanni* (more probable candidate based on the external appearance) or *terminipuncta*.

28. *Isturgia berytaria* (STAUDINGER, 1892)

berytaria (STAUDINGER, 1892), Dt. ent. Z. Iris 5 (1): 200, (*Halia*). Lectotype male, paralectotype female (Zoologisches Museum der Humboldt Universität, Berlin, Germany), [Lebanon]: near Beirut. Further paralectotypes in Zoologisches Museum der Humboldt Universität, Berlin, Germany (KRÜGER 2011). Originally described from 9 males, 2 females.

EXTERNAL CHARACTERS AND ABDOMEN: Wingspan 29–33 mm. Wings elongated, grey, glossy, antemedial and medial lines reduced to weak blackish spots on costa, postmedial line broken into small blackish dots. Terminal area weakly irrorated with black scales, subapical area brown on costa. Terminal line fine, discontinuous, fringes concolorous with wings. Forewing discal spots very prominent, blackish, hindwing discal spots smaller, lighter. Hindwings without markings, underside patterns shining weakly through, margin at termen weakly undulated. Wings below more brownish, particularly hindwings, medial lines blackish, wide, particularly in hindwings. Hindwing with thin white streak, running across wing. Frons, collar, thorax and abdomen concolorous with wings, weakly irrorated with blackish scales. Abdomen with lateral row of blackish spots. Antennae mixed grey and blackish dorsally, bipectinate in male, filiform in female. Hindtibia with 2+2 spurs. Tympanal organs medium-sized, male bullae tympani with cone-shaped lobe. Sternites and tergites 3–8 of both sexes undifferentiated.

Variation: Very little. The density of markings varies somewhat, postmedial line may be nearly absent or rather distinct.

Male genitalia: Uncus triangular, margins slightly expanded, setose dorsally. Gnathos arms wide, fused medially, forming triangular and upturned element, surface granulate. Juxta small plate. Valva divided. Dorsal arm of valva long, narrow, about equal width, costa and apex setose. Ventral arm of valva wide, short, ventral margin slightly concave, apex round, with few setae. Saccus weakly elongated medially. Aedeagus slightly curved, caecum minute, apex sharp. Vesica expanded at base, weakly sclerotised, opens at 90 degree angle, cornuti absent.

Female genitalia: Papillae anales slightly elongated, setose. Apophyses straight, narrow. Lamella antevaginalis and postvaginalis forming small, narrow ring around ostium bursae. Ductus bursae medium length, narrow, sclerotised laterally. Corpus bursae elongated, posterior end narrow, long, striated, anterior end larger, roundish. Signum circular, weakly stellate.

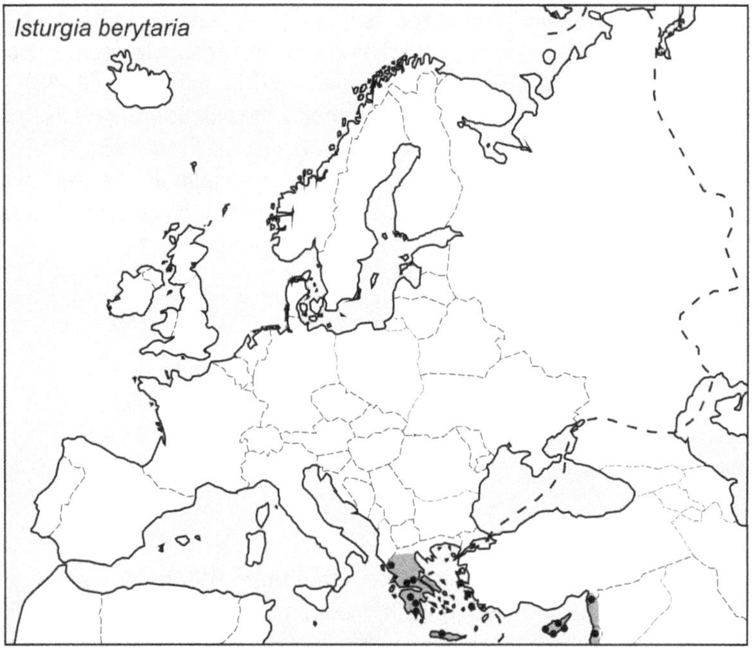

Distribution and abundance: East-Mediterranean. In Europe only known from mainland Greece and some of the Greek islands including Crete. – Outside Europe in Cyprus, Turkey, Lebanon, Israel and Jordan. – In Greece apparently very local, but where it occurs, it can often be found in numbers.

Phenology: Univoltine. The species flies late in the year from early October to early December. In Samos in October to November (Fritsch *et al.* 2014), in Cyprus until late December (N. Pöll pers. comm.). Larva from late March to late May (D. Fritsch pers. comm.). Probably the species hibernates as an egg. Night active and comes to light.

BIOLOGY: Oligophagous on Fabaceae. On Samos the larvae have been found on *Calicotome villosa* (D. FRITSCH & E. FRIEDRICH pers. comm.) and in Cyprus on *Calicotome spinosa* (B. MÜLLER pers. comm.) and *Genista fasselata* (WILTSHIRE 1948).

HABITAT: The moths have been found in open pine forest with a lot of *Calicotome* in between (B. MÜLLER pers. comm.). In Europe from sea level up to 780 m. In Jabal Moussa, Kasrouan in Lebanon it is rather common from 800 up to 1200 m where the typical habitat is semiopen dry woodland with *Quercus* and *Pinus* (J. KULLBERG pers. comm.).

SIMILAR SPECIES: *Isturgia spodiaria* has terminal area darker, postmedial line meets forewing termen at 2/3 of wing length and wings are roundish (terminal area about same colour as rest of wing, postmedial line meets forewing termen at 1/2 of wing length and wings are elongated in *I. berytaria*). See text-figs 41–46. Male genitalia are without distinct differences. Anterior part of corpus bursae is larger and more round in *I. spodiaria* (smaller and elongated in *I. berytaria*), signum is small, irregular in *I. spodiaria* (signum is large, circular in *I. berytaria*).

GENETIC DATA: Genetically somewhat heterogeneous from Greece to Turkey and Israel (n=5), maximum variation 1.7%. Nearest species: *I. miniosaria* (6.7%).

29. Isturgia sparsaria (HÜBNER, 1809)

sparsaria HÜBNER, [1809], Samml. Eur. Schmett. 5 Geometrae (1 & 2): pl. 63, fig. 326, 327, (*Geometra*). Syntype(s), lost, Europe.

acquiaria MILLIÈRE, 1875, Bull. Soc. ent. Fr. 1875: 166, (*Thamnonoma* (*Halia*)). Syntype(s) male(s), Italy: [Piemonte, Alessandria]: Acquit Terme ('Acqui-les-Bains').

nolaria HÜBNER, [1809], Samml. Eur. Schmett. 5 Geometrae (1 & 2): pl. 63, fig. 325, (*Cosymbia*). Syntype(s), lost, Europe.

tephrania BOISDUVAL, 1840, Genera Index meth. eur. Lepid.: 201, (*Eubolia*). [Unnecessary replacement name for *sparsaria* HÜBNER].

EXTERNAL CHARACTERS AND ABDOMEN: Wingspan 23–28 mm. Wings pale straw, weakly irrorated with brown. Antemedial line weak, angled at costa, medial line rudimentary, postmedial line strongest, narrow, sharp. Terminal area occasionally darker than rest of wing. Terminal line weakly dotted. Fringes concolorous with wings. Discal spots small, sharp. Wings below as above, postmedial line wider, more suffused. Frons, collar, thorax and abdomen concolorous with wings, weakly irrorated with brown. Antennae concolorous with wings dorsally, irrorated with brown, bipectinate in male, filiform in female. Hindtibia with 2+2 spurs. Tympanal organs medium-sized, bullae tympani with roundish lobe. Sternites and tergites 3–8 of both sexes undifferentiated.

VARIATION: Very little. Postmedial line may be weak or distinct, irroration varies slightly, and terminal area may be concolorous with wings or slightly darker.

MALE GENITALIA: Uncus triangular, setose dorsally. Gnathos arms wide, fused medially, forming triangular and upturned element, surface granulate. Juxta small plate. Valva divided. Dorsal arm of valva long, narrow, slightly expanded at apex, costa and apex setose. Ventral arm of valva wide, short, dorsal margin straight, apex angled, with few setae. Saccus weakly elongated medially. Aedeagus slightly curved, thick, caecum minute, apex sharp. Vesica slightly expanded at base, weakly sclerotised, opens at 135 degree angle, cornuti absent.

FEMALE GENITALIA: Papillae anales slightly elongated, setose. Apophyses straight, narrow. Lamella antevaginalis and lamella postvaginalis membranous. Ductus bursae short, wide, weakly sclerotised. Posterior end of corpus bursae elongated, weakly striated, anterior end large, roundish sac, expanded over half length of corpus bursae. Signum small, elongated, stellar arms rudiment.

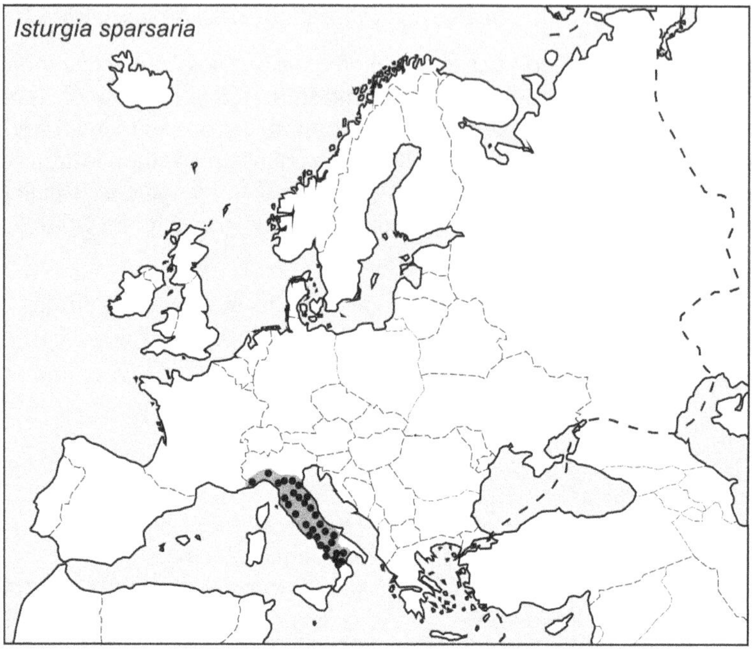

DISTRIBUTION AND ABUNDANCE: Central Mediterranean. Endemic to Italy where it is distributed throughout most of the Apennine mountain chain. The species is not rare, regionally it is seen in limited numbers (N. ZAHM pers. comm.), sometimes it is common in Basilicata and Calabria (A. HAUSMANN pers. comm.) and in some years it can be common in Emilia-Romagna (C. FLAMIGNI pers. comm.).

PHENOLOGY: Bivoltine. From late April to June and from early July to September. Flies at night and comes to light.

BIOLOGY: Unknown.

HABITAT: Apparently in a rather large range of habitats. The species is found from close to sea level up to 1600 m in southern Italy (FLAMIGNI *et al.* 2007), often on forested or semi-forested dry hills and slopes on calcareous ground (A. HAUSMANN pers. comm.).

SIMILAR SPECIES: *Isturgia messapiaria* (SOHN-RETHEL) is ochreous brown, irrorated with dark brown, and medial and postmedial lines are wide, brown, suffused. In male genitalia the costa of ventral arm of valva is slightly concave (straight in *I. sparsaria*). In female genitalia (only one *I. messapiaria* female was available for study, differences must be considered tentative) the corpus bursae is expanded at anterior end only (corpus bursae expanded over half the distance in *I. sparsaria*), the signum is round, stellar arms are short but recognisable (signum is elongated, stellar arms are vestigial in *I. sparsaria*).

GENETIC DATA: Genetically homogeneous in southern Italy (n=8), but one specimen from central Italy diverging by 2.0%. Nearest species: *I. messapiaria* (0.15%), almost barcode-sharing, but, so far, always diverging by at least one base pair.

REMARKS: *Isturgia sparsaria* has a wide ecological niche, ranging from sea level to 1600 metres, and occurring usually on rocky or stony, comparatively dry sites, often on limestone. *I. messapiaria* has a narrow ecological niche, it is strictly confined to hygrophilous riverside vegetation and swampy areas with moist meadows, without rocks, in the Sila Mountains in southern Italy at about 1200 up to 1500 metres. Given the differences found in the external characters, in the male and female genitalia, and in their ecological preferences, we consider *I. sparsaria* and *I. messapiaria* valid at species level but recognise that further studies are needed. See genetic data.

30. Isturgia messapiaria (SOHN-RETHEL, 1929)

messapiaria SOHN-RETHEL, 1929, Dt. ent. Z. Iris 43 (1): 21 pl. 2, figs 17, 18, (*Itame*). Syntypes 14 males, 6 females, Italy: Calabria, Sila Mountains, 600-1200 m. See remarks.

EXTERNAL CHARACTERS AND ABDOMEN: Wingspan 26–29 mm. Externally as *Isturgia sparsaria* (HÜBNER), see above, but with the following differences. Wings rusty, ochreous brown, irrorated with dark brown. Antemedial line weak, medial and postmedial lines wide, brown, suffused. Outer margin of postmedial line brown, terminal area often dark brown.

VARIATION: Very little. Females are slightly lighter in colour, less irrorated with brown. Terminal area may be concolorous with wings or slightly darker.

MALE GENITALIA: As in *Isturgia sparsaria* (HÜBNER), see above, but with the following differences: Ventral arm of valva wide, short, dorsal margin concave, apex elongated.

FEMALE GENITALIA: As in *Isturgia sparsaria* (HÜBNER), see above, but with the following differences (only one female of *I. messapiaria* was available for study, differences must be considered tentative): Corpus bursae expanded at anterior end only. Signum round, stellar arms short but recognisable.

DISTRIBUTION AND ABUNDANCE: Central Mediterranean. Endemic to southern Italy where the species only occurs in the Sila mountains in Calabria where it flies very locally. It is mainly found as single specimens, recently also in numbers (A. HAUSMANN pers. comm.).

PHENOLOGY: Univoltine. The species flies from early June to early August (FLAMIGNI *et al.* 2007). The moth flies at night and comes readily to light.

BIOLOGY: Unknown.

HABITAT: Hygrophilous. The species is found along water courses, on open damp meadows between deciduous or pine forests and in forested narrow valleys with *Alnus*, *Salix* and *Pinus* (A. HAUSMANN pers. comm.). According to FLAMIGNI *et al.* (2007) from 600 up to 1300 m altitude.

SIMILAR SPECIES: *Isturgia sparsaria* (HÜBNER), see entry above.

GENETIC DATA: Four examined specimens almost barcode-sharing with *I. sparsaria* (0.15%), but, so far, always diverging by at least one base pair.

REMARKS: SOHN-RETHEL (1929) did not indicate where the type specimens were deposited. We have examined several specimens, including those deposited in coll. FIORI, in Museo Civico di Storia Naturale, Milano, Italy, labelled Calabria, Fossiata, 6.VIII. 1921, A. FIORI. These specimens could potentially be syntypes. We do not designate lectotype or paralectotypes because *I. messapiaria* is easily identifiable by external characters, and identity of the taxon is not in doubt. *I. messapiaria* is also illustrated in FLAMIGNI *et al.* (2007). See remarks on *Isturgia sparsaria* above.

31. *Isturgia arenacearia* (DENIS & SCHIFFERMÜLLER, 1775)

arenacearia arenacearia [DENIS & SCHIFFERMÜLLER], 1775, Ankündung syst. Werkes Schmett. Wienergegend: 102, (*Geometra*). Syntype(s), lost, [Austria]: Vienna district.

arenacearia stena WEHRLI, 1940, in SEITZ, Gross-Schmett. Erde 4 (Suppl.): 399, pl. 32: d, (*Tephrina*). Syntypes including 1 male, 1 female, [Central Asia]: Tai-Shan, 1550 m. [Originally as f].

discriminaria WALKER, 1862, List Specimens lepid. Insects Colln Br. Mus. 24: 1073, (*Aspilates*). Holotype male (The Natural History Museum, London, UK), [China]: Shanghai (examined externally).

flavidaria EVERSMANN, 1852, Bull. Soc. imp. Nat. Moscou 25 (1): 162, (*Aspilates*). Syntype(s), [Russia]: Lower Volga.

imperviata WALKER, [1863], List Specimens lepid. Insects Colln Br. Mus. 26: 1682, (*Aspilates*). Holotype female (The Natural History Museum, London, UK), [China]: Shanghai (examined externally).

irradiata WALKER, 1862, List Specimens lepid. Insects Colln Br. Mus. 24: 1072, (*Aspilates*?). Syntypes 2 males (The Natural History Museum, London, UK), [China]: Shanghai (examined externally).

arenacearia mongolica VOJNITS, 1974, Annales hist.-nat. Mus. natn. hung. 66: 284, figs 1a, 2a, (*Tephrina*). Holotype male (Hungarian Natural History Museum, Budapest, Hungary), Mongolia: Central Aimak, 26 km E of Somon Lun, 1180 m. Paratype male (The Natural History Museum, London, UK), Mongolia. Valid at subspecies rank according to SCOBLE (1999) (examined externally).

INFRASUBSPECIFIC NAMES, POTENTIALLY UNAVAILABLE: *brunneo-violacea* DJAKONOV, 1926 (ab.).

EXTERNAL CHARACTERS AND ABDOMEN: Wingspan 24–28 mm. Wings light straw to yellowish grey to ochre-yellow, irrorated with grey. Antemedial and medial lines weak, postmedial line distinct, concave, inner margin light, outer margin shaded or broken into dark spots. Terminal area darker than rest of wing. Terminal line distinctly spotted. Fringes concolorous with wings. Discal spots small, dark, distinct. Wings below as above, paler, markings weaker. Frons, collar, thorax and abdomen concolorous with wings, weakly irrorated with brown. Antennae concolorous with wings dorsally, bipectinate in male, filiform in female. Hindtibia with 2+2 spurs. Tympanal organs small. Posterior margin of male 8th sternite with two round lobes, margins weakly serrate. Other male and female sternites and tergites undifferentiated.

VARIATION: Ground colour varies from light straw to yellowish grey to ochre-yellow. Antemedial and medial lines are normally weak, if visible, then suffused. Outer margin of postmedial line is normally shaded or broken into dark spots. Wings, particularly hindwing may be heavily irrorated with grey.

MALE GENITALIA: Uncus constricted at base, margins slightly expanded, apex triangular, setose dorsally. Gnathos arms narrow, fused medially, forming wide, horizontal ridge, surface granulate. Juxta small plate. Valva divided. Dorsal arm of valva slightly expanded at apex, costa and apex setose. Ventral arm of valva wide, long, narrow ridge near ventral margin, with few setae, apex sharp. Saccus wide, often concave medially. Aedeagus long, narrow, caecum minute, subapical area with small teeth, apex pointed. Vesica opens at 45 degree angle, narrow at base, with large, roundish diverticulum, its surface with minute sclerotisations all over and one large cornutus.

FEMALE GENITALIA: Papillae anales slightly elongated, setose. Apophyses straight, narrow. Sinus vaginalis prominent, sclerotised, lunular. Lamella antevaginalis narrow, round ring. Lamella postvaginalis sclerotised, concave medially. Ductus bursae short, narrow, sclerotised laterally. Corpus bursae large, posterior end elongated, striated, sclerotised, anterior end very large, oval, membranous. Signum small, stellate.

DISTRIBUTION AND ABUNDANCE: Eurasiatic-Palaearctic. In Europe from Corsica, Italy including Sardinia and Sicily and southernmost Switzerland to southern and eastern Poland, through south-eastern central Europe and the Balkans through Ukraine and southern European Russia to the Ural mountains. – Outside Europe in north-western and north-eastern Turkey, the Caucasus region, central Asia and further eastwards in southern Siberia, Mongolia, Korea, eastern China, the Amur region and Japan. – In Europe a common species that only rarely occurs in numbers. However, according to ATANASOVA *et al.* (2010) the species can occur as a serious pest on *Medicago sativa*.

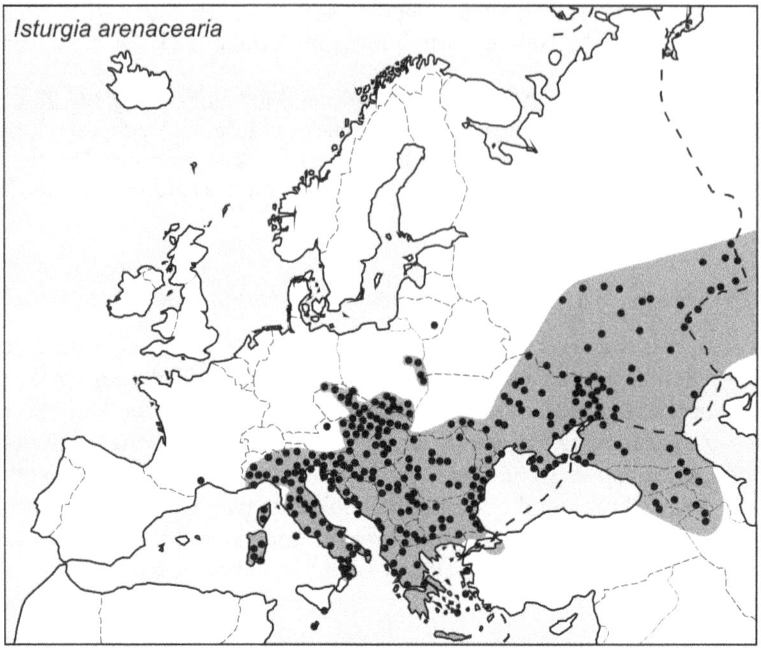

PHENOLOGY: Bivoltine. Generations apparently overlapping, from early May to late September. In central Asia also in late April. The moths fly at night and come to light.

BIOLOGY: Oligophagous on Fabaceae. The caterpillar has been recorded on *Melilotus alba* (FLAMIGNI *et al.* 2007), *Coronilla varia* (SPULER 1903–1910), *Medicago sativa* (ROBERTI 1964, ATANASOVA *et al.* 2010) and on *Glycine max* (ZANDIGIACOMO & DALLA MONTÀ 1982). In rearings the larva has been fed with *Vicia cracca* (WOLFSBERGER 1960).

HABITAT: Xerothermophilous. In warm and dry places, slopes, steppe, places with scattered bushes. In southern Europe apparently in a larger variety of habitats, probably due to the warmer climate. It has been found from sea level up to 900 m altitude (FLAMIGNI *et al.* 2007).

SIMILAR SPECIES: No similar species in Europe. Genitalia structures of *Isturgia murinaria* (DENIS & SCHIFFERMÜLLER) are somewhat similar.

GENETIC DATA: Genetically homogeneous from Europe to north-eastern China. Nearest species: *I. murinaria* (4.6%).

32. *Isturgia punctistrigaria* (CHRISTOPH, 1897), new combination

punctistrigaria CHRISTOPH, 1893, Dt. ent. Z. Iris 6 (1): 95, (*Atomorpha*). Lectotype male [Turkmenistan]: Ashkabad (examined externally), paralectotype male [Turkmenistan]: Ashkabad (genitalia examined), paralectotype female [Turkmenistan] Ashkabad (genitalia examined). All type specimens, 4 males and 3 females are in Zoological Institute of Russian Academy of Sciences, St. Petersburg, Russia). Lectotype is designated and illustrated in TROFIMOVA (2009). Originally as sp. Transferred here from *Atomorpha* STAUDINGER, 1901 to *Isturgia* HÜBNER, 1823, quoting structural similarities (new combination).

EXTERNAL CHARACTERS AND ABDOMEN: Wingspan 16–21 mm. Wings whitish grey, with tint of yellow, irrorated with brown. Two whitish areas radiating across forewings, thus medial lines broken into brown spots, incomplete, present on costa. Subterminal line faint, brown. Terminal line fine, normally discontinuous. Forewing fringes chequered whitish and brown, hindwing fringes whitish. Hindwings grey, terminal area darker, medial lines present as brown spots on anal margin. Forewing discal spots small, brown, hindwing discal spots absent. Wings below greyish, markings weak. Frons, collar, thorax and abdomen concolorous with wings, weakly irrorated with brown. Antennae concolorous with wings dorsally, fasciculate in male, filiform in female. Hindtibia with 2+2 spurs. Tympanal organs small. Posterior margin of male 8th sternite with two round lobes. Other male and female sternites and tergites undifferentiated.

VARIATION: Very little. Forewing ground colour varies from whitish grey to yellowish to moss-green. Hindwing colour varies from white to grey.

MALE GENITALIA: Uncus slightly elongated, setose dorsally. Gnathos arms wide, fused ventrally, forming triangular and upturned element, surface granulate. Juxta small plate. Valva divided. Dorsal arm of valva slightly expanded at apex, costa and apex setose. Ventral arm of valva wide, short, dorsal margin concave, apex roundish, with few setae. Saccus very wide, slightly concave. Aedeagus long, narrow, caecum minute, apex round. Vesica opens at 135 degree angle, wider at base, surface weakly granulate, with one large, dentate cornutus.

FEMALE GENITALIA: Papillae anales slightly elongated, setose. Apophyses straight, narrow. Lamella antevaginalis membranous. Lamella postvaginalis sclerotised, small, narrow plate. Posterior part of ductus bursae short, sclerotised laterally, narrower at anterior end. Anterior part of ductus bursae membranous, striated. Corpus bursae large, posterior end sclerotised laterally, narrow, elongated, anterior end large, roundish. Signum small, roundish, spinose all over, stellar arms short.

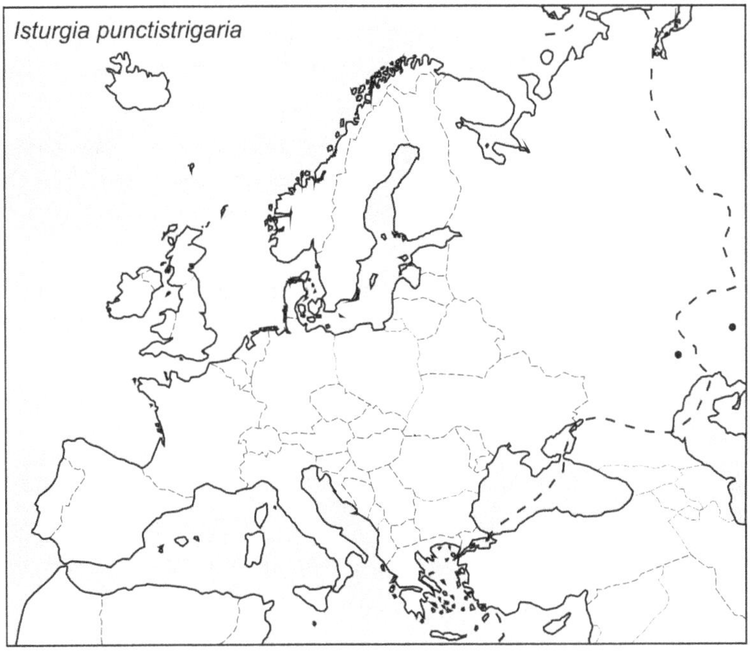

DISTRIBUTION AND ABUNDANCE: Turanian. Limited information is available for this species which is only known from three localities in Turkmenistan and five in Kazakhstan, one of them in the European part of Kazakhstan where only a single male specimen has been found so far (TROFIMOVA 2009).

PHENOLOGY: Probably bi- or trivoltine. The few known specimens have the following collecting data: mid- and late April, early, mid and late May, late July and early October. These data show undoubtedly a first generation from mid-April to late May. How the remaining dates should be interpretated should await further information. The moths fly at night and come to light (K. NUPPONEN pers. comm.).

BIOLOGY: Unknown.

HABITAT: Xerothermophilous, psammophilous. The few known localities where the species has been found are sand deserts or semi deserts with sparse vegetation (J. VIIDALEPP pers. comm.) and in sand dunes (K. NUPPONEN pers. comm.).

SIMILAR SPECIES: No similar species in Europe.

GENETIC DATA: No data available.

REMARKS: See genus treatment of *Isturgia* above, we propose *Atomorpha* STAUDINGER to be a junior synonym of *Isturgia* HÜBNER.

Itame HÜBNER, 1823

Itame HÜBNER, [1823] 1816, Verz. bekannter Schmett.: 299. Type species: *Geometra vincularia* HÜBNER, [1813] 1796. Europe. No genus synonyms.

UNAVAILABLE NAMES (incorrect subsequent spelling): *Itama* HÜBNER, 1826 (see FLETCHER 1979).

DIVERSITY AND DISTRIBUTION: *Itame* contains provisionally two species (*I. vincularia* and *I. teknaria* POWELL & RUNGS), which are found in southern France, Spain, Portugal, Algeria (PROUT 1915–16), Morocco and Tunisia.

EXTERNAL CHARACTERS AND ABDOMEN: Wings grey, markings brown to black. Outer margin of postmedial line with wide, brown band. Forewing discal spot prominent, right-angled. Male and female antennae bipectinate. Male 3rd sternite without setae, posterior margin of 8th sternite modified.

MALE GENITALIA: Uncus setose, horns absent. Gnathos arms fused, weakly upturned medially. Valva divided. Dorsal arm of valva long, narrow, setose. Ventral arm of valva rounded, with several narrow flap-like projections. Saccus bifurcate (see Table 1).

FEMALE GENITALIA: Sinus vaginalis, lamella antevaginalis and lamella postvaginalis forming sclerotised ring around ostium bursae. Signum weakly stellate (see Table 1).

BIOLOGY: The caterpillar of *I. vincularia* subsp. *vincularia* (HÜBNER) has been recorded feeding on *Rhamnus alaternus* (REDONDO *et al.* 2009).

IMMATURE STAGES: No description or illustration of egg, larva or pupa has been located.

REMARKS: Although a number of species have been included in *Itame*, both in the Palaearctic (e.g. MÜLLER 1996) and in the Nearctic regions (e.g. HODGES *et al.* (1983), SCOBLE & KRÜGER (2002) restricted the genus to the type species, and probably to *I. teknaria* POWELL & RUNGS. The other species have been recombined into various other genera. The strict definition of *Itame* by SCOBLE & KRÜGER (2002) was not adopted by LERAUT (2009). Systematic position: Macariini (KRÜGER 2001, SCOBLE & KRÜGER 2002).

33. *Itame vincularia* (HÜBNER, 1813)

vincularia vincularia HÜBNER, [1813], Samml. Eur. Schmett. 5 Geometrae (1): pl. 78, fig. 402, (*Geometra*). Syntype(s), lost, Europe.

vincularia latefasciata ROTHSCHILD, 1914, Novit. zool. 21 (4): 352, (*Itame*). Syntypes 23 males, 8 females (The Natural History Museum, London, UK), Algeria (central): Guelt-Es-Stel (examined externally). Valid at subspecies rank according to SCOBLE (1999). See remarks.

vincularia lycioidaria HERBULOT, 1957, Bull. mens. Soc. linn. Lyon 26 (7): 191, (*Itame*). Holotype male (Coll. Herbulot, in Zoologische Staatssammlung, Munich, Germany), Tunisia: Gafsa, Diebel (examined externally). Valid at subspecies rank according to SCOBLE (1999). See remarks.

vincularia mrassinaria OBERTHÜR, 1923, Études Lépid. comp. 20: 249, pl. 561, fig. 4838, (*Tephrina*). Holotype, Morocco: Mrassine (not located in the Natural History Museum, London, UK), paratype male (The Natural History Museum, London, UK), Morocco: Mrassine (examined externally). Originally as var., valid at subspecies rank according to SCOBLE (1999). See remarks.

atlantis PROUT, 1928, in LE CERF & TALBOT, Bull. Hill Mus. Witley 2 (2): 120, (*Itame*). Syntype female (The Natural History Museum, London, UK), Morocco: Great Atlas, Tenfecht, 3000–4000 ft (examined externally).

EXTERNAL CHARACTERS AND ABDOMEN: Wingspan 28–32 mm. Wings grey-brown, weakly irrorated with brown. Antemedial line weak, angled. Medial line rudimentary, present as dark dot on costa. Postmedial line narrow, slightly undulating, its outer margin with conspicuous, wide, chocolate-brown band, darkest on costa, interrupted and whitish at veins. Subterminal line whitish, undulating. Terminal area often darker, more irrorated with brown. Terminal line fine, discontinuous. Fringes concolorous with wings. Hindwings uniform grey-brown, postmedial and subterminal lines inconspicuous whitish bands. Forewing discal spots prominent, rectangular, hindwing discal spots weak, elongated. Wings below uniform grey, irrorated with brown. Frons, collar, thorax and abdomen concolorous with wings, weakly irrorated with brown. Antennae grey-brown dorsally, sparsely irrorated with brown, bipectinate in both sexes, pectinations shorter in female. Hindtibia with 2+2 spurs. Tympanal organs medium-sized, bullae tympani with small roundish lobe. Posterior margin of male 8th sternite with two round lobes. Other sternites and tergites of both sexes undifferentiated.

VARIATION: Ground colour varies from light grey to purplish grey and density of markings varies from intense to rather weak. Occasionally forewings may be heavily irrorated, appearing dark brown to blackish. Brown band between postmedial and subterminal lines in forewing can be rather uniform throughout the wing, normally strongest on costa, or reduced to a light-brown and orange spot on costa only. Forewing discal spot is normally prominent, rectangular, but can be considerably weaker.

Subsp. ***latefasciata*** on average with dark borders next to ante- and postmedial lines intermediate between subsp. *lycioidaria* and subsp. *mrassinaria*, brown on costa, weaker towards termen. Forewing discal spot brown, rather prominent. Genitalia not examined. See remarks.

Subsp. ***lycioidaria*** on average with dark borders next to ante- and postmedial lines faint, orange-brown. Forewing discal spot brown, not prominent. Genitalia not examined. See remarks.

Subsp. ***mrassinaria*** on average with dark borders next to ante- and postmedial lines well developed, dark brown. Forewing discal spot blackish, prominent. Aedeagus apex with two long, lateral prongs and one or two additional, minute, ventral prongs (see male genitalia plate). LERAUT (2009) raised *mrassinaria* OBERTHÜR to species rank on the basis of this minute difference, in addition to the slightly differently shaped margin of the ventral arm of valva. Our material indicates that the margin shape is dependent on the way the cover glass is placed on top of the genitalia and how well the valvae are spread, thus appearing as a preparation artefact. Furthermore, LERAUT (2009) appears to have mixed the aedeagi. The aedeagus below *vincularia* is apparently *mrassinaria* and vice versa. Female genitalia with margin of the lamella antevaginalis even, and signum small, stellar arms reduced (see female genitalia plate). See genetic data and Remarks.

MALE GENITALIA: Uncus constricted at base, triangular, setose dorsally. Gnathos arms wide, fused ventrally, forming weakly upturned element, surface granulate. Juxta large plate. Valva divided. Dorsal arm of valva long, narrow, weakly curved, costa and apex setose. Ventral arm of valva wide, round, with three narrow, high ridges, apex with few setae. Saccus large, deeply concave medially. Aedeagus stout, caecum wide, with two lobes, apex with two short, lateral prongs, occasionally with one additional, minute ventral prong. Vesica almost straight tube, subbasal area slightly expanded, with scale-like, minute sclerotisations. Dorsal diverticulum bearing single cornutus.

FEMALE GENITALIA: Papillae anales elongated, setose. Apophyses straight, narrow. Lamella antevaginalis and lamella postvaginalis strongly sclerotised, fused laterally. Lateral parts wide, wing-like. Margins of lamella antevaginalis and lamella postvaginalis concave. Ductus bursae short, wide, sclerotised. Base of ductus seminalis slightly expanded. Posterior end of corpus bursae same width as ductus bursae, sclerotised, densely ribbed and striated. Anterior end large, round, membranous. Signum small, stellar arms short but distinct.

DISTRIBUTION AND ABUNDANCE: West Mediterranean. In Europe only in southern France and in the Iberian Peninsula. – Outside Europe in north-western Africa where the species has been found in Tunisia (subsp. *lycioidaria*), Algeria (subsp. *latefasciata*) and Morocco (subsp. *latefasciata* and *mrassinaria*). – In south-western Europe a common species that can be found nearly anywhere and often in numbers. Within its flight time it is rarely missing at light.

PHENOLOGY: Bi- or trivoltine. The adult moths have been found from mid-March until early October, depending on altitude and latitude, in two or three more or less overlapping generations. Recorded in southernmost Spain from January to August and in October (P. HALE pers. comm.). The larva has been found in mid-October (M. CORLEY pers. comm.). The hibernation probably takes place as a pupa. The moth flies at night and comes readily to light.

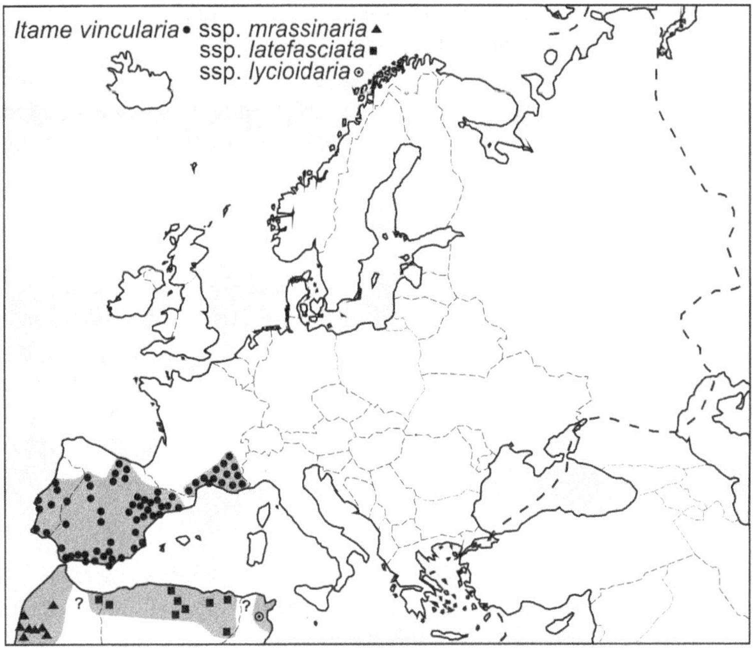

BIOLOGY: Monophagous. The caterpillar has been found on *Rhamnus alaternus* (M. CORLEY pers. comm.) and *R. lycioides* (M. LEIPNITZ pers. comm.). REDONDO *et al.* (2009) also records it on *R. alaternus*. It has been reared on *R. cathartica* (M. LEIPNITZ pers. comm.). In rearings pupation takes place between withered leaves (M. LEIPNITZ pers. comm.).

HABITAT: Xerothermophilous. The species occurs in many different dry habitats from sea level up to at least 1600 m, according to REDONDO *et al.* (2009), up to 2000 m in southern Spain.

SIMILAR SPECIES: According to the original, unillustrated description, *Itame teknaria* POWELL & RUNGS, 1942 from Morocco is similar but smaller. We have not had original material available for study, the holotype male has not been traced. LERAUT (2009) illustrates an *I. teknaria* male from Morocco, but the specimen does not fully agree with the original, written description.

GENETIC DATA: Genetically homogeneous in Europe (n=6, from France, Spain, Portugal). Nearest taxa: *I. vincularia* subsp. *mrassinaria* (OBERTHÜR, 1923) from Morocco (3.0%) and '*Itame vincularia* subsp. *lycioidaria*' Herbulot, 1957 from Tunisia, (4.5%). See remarks.

REMARKS: In North Africa three subspecies have been described, subsp. *latefasciata* ROTHSCHILD from Algeria, subsp. *lycioidaria* HERBULOT from Tunisia and subsp. *mrassinaria* OBERTHÜR from Morocco. Externally there are small quantitative differences that apparently correlate with minute quantitative differences in the shape and number of projections in the apex of male aedeagus, shape of margin of the lamella antevaginalis and the shape of signum (own observations), see genitalia plates. KRÜGER (2001) concluded that the North African taxa probably do not represent more than geographical races. LERAUT (2009) raised subsp. *mrassinaria* OBERTHÜR to species status on the basis of small differences in the external characters and the male genitalia. Unfortunately he did not indicate exact geographical position of the examined Moroccan specimen(s), and if any relevant type material was examined. There is a risk of potential confusion because subsp. *latefasciata* ROTHSCHILD may also occur in Middle Atlas mountains in Morocco (L. LEHMANN pers. comm.). We have only had access to a restricted amount of North African specimens and we consider our data inadequate to make formal changes to the current classification. More research is needed.

Neognopharmia WEHRLI, 1953

Neognopharmia WEHRLI, 1953, in SEITZ, Gross-Schmett. Erde 4 (Suppl.): 568. Type species: *Halia stevenaria* Boisduval, 1840, Russia (south); Spain. No genus synonyms. *Neognopharmia* WEHRLI is often stated to have been described in 1951. In that publication (WEHRLI 1951), there is only a mention of the subgenus name and no description. Proper description is given in WEHRLI (1939–1954, the page in question was published in 1953). This is also noted by RAJAEI *et al.* (2012), who noted that (p. 5): 'According to [ICZN 1999] article 13.6, a name described later than 1930 cannot be made available by "indication".'

DIVERSITY AND DISTRIBUTION: *Neognopharmia* has been considered a monotypic genus, including *N. stevenaria* only, but it is likely to include three additional species, *Gnophos cataleucaria* STAUDINGER, 1901 (from *Gnopharmia*), *Gnophos hoerhammeri* BRANDT, 1938 from *Gnopharmia* and a new species from Pakistan (NORBERT PÖLL, in prep., mentioned in RAJAEI *et al.* 2012). All species occur in the Palaearctic region.

EXTERNAL CHARACTERS AND ABDOMEN, based on European species: Wings brownish to grey, terminal area darker. Wing margins dentate, elongated at hindwing M3. Male forewing with fovea. Chaetosemata round. Male antennae bipectinate with long setae. Male 8th sternite modified, posterior margin with two round lobes.

MALE GENITALIA, based on European species: Uncus wide, apex weakly elongated, densely setose dorsally. Valva divided. Ventral arm of valva with sharp, tooth-like projections. Vesica with cornuti (see Table 1).

FEMALE GENITALIA, based on European species: Papillae anales with modified setae (floricomous ovipositor) (Text-fig. 59). Posterior part of corpus bursae very long, striated. Signum very large, stellate (see Table 1).

BIOLOGY: Larva of *N. stevenaria* has been found on *Crataegus azarolus* and *Prunus dulcis* (M. LEIPNITZ pers. comm.) and reared on the leaves of *Prunus spinosa* (LEPIFORUM 2013) and *Crataegus* (B. MÜLLER pers. comm.). Potentially closely related *Gnopharmia kasrunensis* WEHRLI, 1939 and *G. colchidaria* LEDERER, 1870 have been found on wild almond (*Prunus scoparia*) (RAJAEI 2010). All mentioned taxa share specialised setae on the ovipositor of the female (Text-fig. 59), and it is assumed that the specialised setae may be used to scrape dirt or minute fragments of bark over the eggs (SCOBLE & KRÜGER 2002, RAJAEI 2010). This hypothesis is supported by the fact that the papillae anales of many females have particles lodged among the hairs (SCOBLE & KRÜGER 2002). Similar modified setae are present in *Theria rupicapraria* (DENIS & SCHIFFERMÜLLER), *T. primaria* HAWORTH and in *Aleucis* GUENÉE.

Text-fig. 59. *Neognopharmia stevenaria* (BOISDUVAL, 1840), detail of the modified ovipositor setae (floricomous setae) in the papillae anales. Ukraine: Crimea, 10.vi.1996, slide PS1356 (coll. ZMH).

IMMATURE STAGES: Larva and pupa of *N. stevanaria* are illustrated in LEPIFORUM (2013). Larva is a twig-mimic, brown to yellowish-brown, with tint of violet. Each abdominal segment has minute, ring-shaped dot laterally. Pupa cremaster is elongated, apex is with one pair of weakly curved setae.

REMARKS: *Halia stevenaria* has been classified in various genus combinations, to *Gnophos* subgenus *Ctenognophos* PROUT, 1915 (e.g. PROUT 1912–16), to *Gnophos* subgenus *Neognopharmia* (WEHRLI 1951), to *Gnopharmia* subgenus *Neognopharmia* WEHRLI, 1953 (WEHRLI 1953), to *Neognopharmia* WEHRLI, 1953 (MÜLLER 1996, HAUSMANN 1993) and to *Gnopharmia* STAUDINGER, 1892 (e.g. VIVES MORENO 1994, LERAUT 1997, SCOBLE & KRÜGER 2002, HAUSMANN, MIRONOV & VIIDALEPP 2011). VIIDALEPP (1996) classified *stevenaria* to *Neognophina* WEHRLI, 1953, which apparently is referable to *Neognopharmia* WEHRLI, 1953. Subsequently, the chosen genus combination has resulted in *stevenaria* being combined to the Gnophini (VIIDALEPP 1996) or to the Macariini (LERAUT 1997). In most instances the genus combination has not been justified.

Numerous structures indicate that *Halia stevenaria* is a Macariini, a conclusion shared by RAJAEI *et al.* (2012) also. According to RAJAEI *et al.* (2012) *Gnopharmia* STAUDINGER (also a Macariini) and *Neognopharmia* WEHRLI are closely related, yet distinctive, thus they maintained *Neognopharmia* valid at genus level. These two genera share a number of characters, namely (after RAJAEI *et al.* 2012): the structure of the antennae (apical segments not pectinated, rami scaled dorsally, arising distally from the flagellomeres), the presence of a fovea on the male forewings, the structure of the legs (long epiphysis of forelegs, unmodified tibia of hindlegs), the absence of a setal comb on sternite 3, the presence and similar form of octavals, the form of the valva in male genitalia (separated into dorsal and ventral parts), and the presence of a floricomus. Future studies are needed to shed light on the systematics of these two genera, and whether they should be synonymised. Systematic position: Macariini.

34. Neognopharmia stevenaria (BOISDUVAL, 1840)

stevenaria BOISDUVAL, 1840, Gnera Index meth. eur. Lepid.: 187, (*Halia*). Syntype(s), Russia (south); Hispan. mer. [Spain].

lapidisaria FREYER, 1841, Neuere. Beitr. Schmettkde 4 (59): 113, pl. 353, fig. 3, (*Gnophos*). Syntype(s), [Russia]: Sarepta [Volgograd].

EXTERNAL CHARACTERS AND ABDOMEN: Wingspan of both sexes 27–33 mm. Wings ash-grey, forewing medial area irrorated with brown, markings brown to blackish. Medial lines forming three dark spots on costa, antemedial and medial lines otherwise weak, postmedial complete, slightly undulating. Terminal area darker, brownish, towards wing margin. Discal spots minute. Terminal line fine, discontinuous. Fringes brownish. Hindwing basal and medial area irrorated with brown, wing margin dentate, elongated at M3. Wings coloured below greyish brown, densely irrorated with brown, only postmedial line visible. Frons, collar, thorax and abdomen concolorous with wings. Antennae concolorous with wings dorsally, bipectinate in male, filiform in female. Hindtibia with 2+2 spurs. Tympanal organs medium-sized. Posterior margin of male 8th sternite with two round, sclerotised lobes, margin between them concave. Other sternites and tergites of both sexes undifferentiated.

VARIATION: Ground colour varies from whitish-grey to dark, brownish grey. Occasionally contrast between terminal area and basal parts of wing may be high. Degree of brown irroration variable. Frons normally concolorous with wings, but may be brown. See remarks and Text-figs. 60–64 and 65–68.

MALE GENITALIA (see also Text-figs 60–63): Uncus constricted at base, dorsal part round, setose, ventral part narrow, elongated, bare, apex blunt. Tegumen margin with round lobe. Gnathos arms wide, fused medially, forming weakly upturned element, surface granulate. Juxta large plate. Valva divided. Dorsal arm of valva long, rather wide, its dorsal margin slightly curved, costa and apex setose. Ventral arm of valva narrow, its ventral margin curved inwards and dentate, margin weakly setose. Saccus wide, round. Aedeagus long, narrow, curved. Caecum round, apex elongated, sharp. Vesica opens at approximately 90 degree angle, basal part wider, with long cornutus, occasionally with second small cornutus, apex of longer cornutus weakly dentate or brush-like. Apical part of vesica narrow.

Text-figs. 60–64. Morphological variation (adult, genitalia, aedeagus with everted vesica and 8th sternite) within *Neognopharmia stevenaria* (BOISDUVAL, 1840), compared with non-European *N. hoerhammeri* (BRANDT, 1941). The margin of the ventral part of valva is round in Spanish specimens (n=2), slightly angled or weakly curved inwards in eastern European specimens (n=5), and blunt-ended with single large tooth in *N. hoerhammeri* (n=1). Posterior margin of the 8th sternite is without lateral concavity in *N. hoerhammeri*. See genetic data and Remarks for details. Text-fig. 60. *N. stevenaria* adult, genitalia and vesica from Spain: Algarve, 11–12.v.1971, slide PS1355 (coll. ZMH) and 8th sternite from Spain: Huesca, 1900 m, 22.vii.1992, slide PS1562 (coll. ZMH). Text-fig. 61. *N. stevenaria* from Greece: Lakonia, 30.vi–3.vii.1985, all pictures from slide PS1953 (coll. SKOU). Text-fig. 62. *N. stevenaria* from Russia: North Caucasus, Kislovodsk, 9.vi.1997, all pictures from slide PS1957 (coll. SKOU). Text-fig. 63. *N. stevenaria* from Russia: South Ural, Orenburg Region, 1–4.vi.1998, all pictures from slide PS1958 (coll. SKOU). Text-fig. 64. *N. hoerhammeri* from Iran: Pireh, Fars, 6.v.1940, all pictures from slide PS1969/ BMNH GEO 24897 (coll. BMNH). ⇨

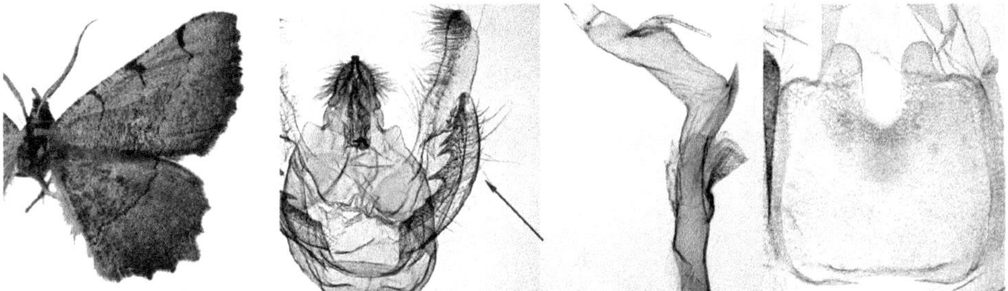

N. stevenaria - Spain 60

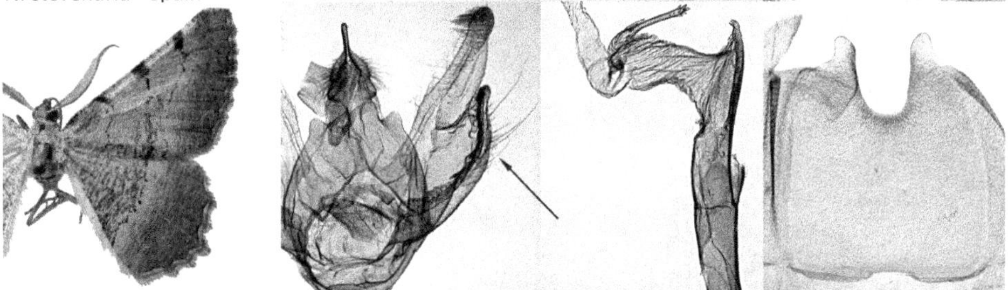

N. stevenaria - Greece 61

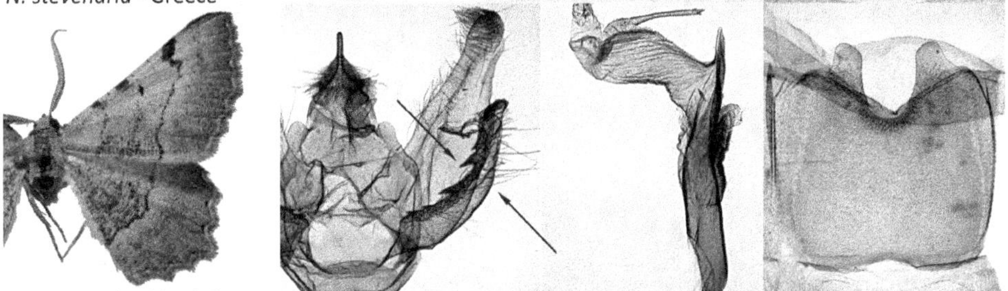

N. stevenaria - North Caucasus 62

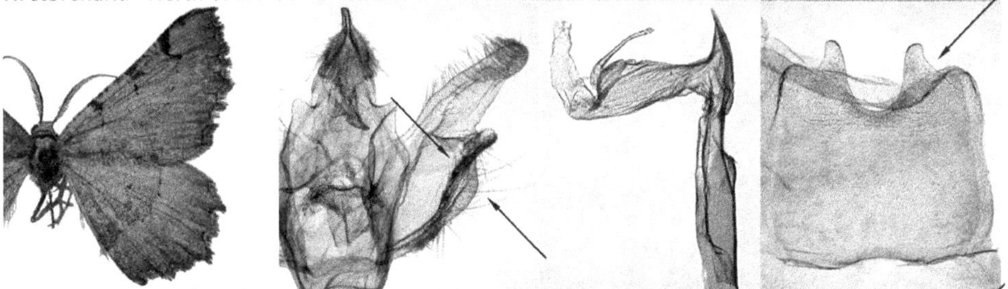

N. stevenaria - South Ural 63

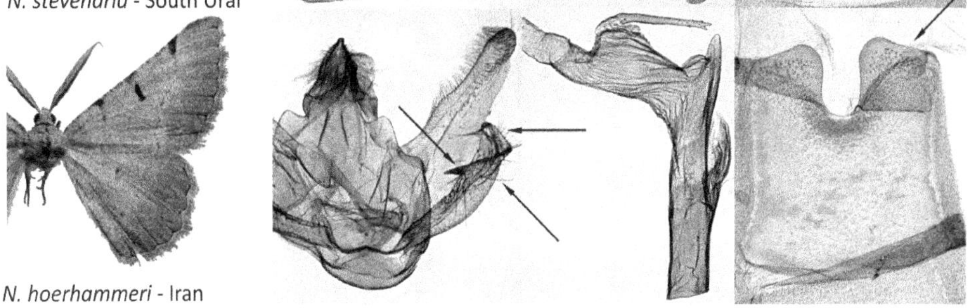

N. hoerhammeri - Iran 64

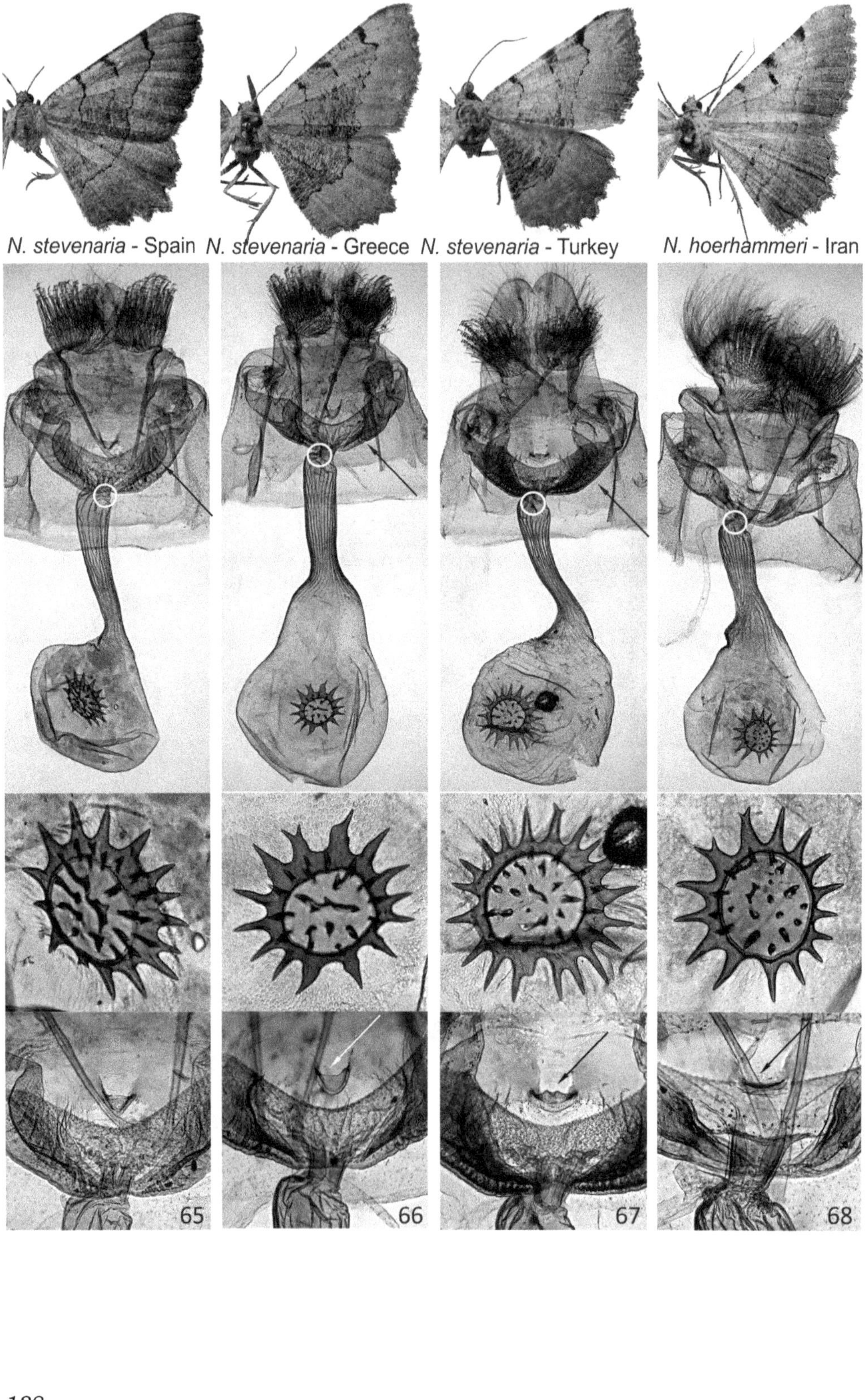
N. stevenaria - Spain
N. stevenaria - Greece
N. stevenaria - Turkey
N. hoerhammeri - Iran
65
66
67
68

⇦ Text-figs 65–68. Morphological variation (adult, genitalia, signum, ostium bursae) within *Neognopharmia stevenaria* (BOISDUVAL, 1840), compared with non-European *N. hoerhammeri* (BRANDT, 1941). Circle indicates the point of origin of ductus seminalis. The shape of anterior part of the corpus bursae varies slightly (n=4) and the medial sclerotisation of the lamella postvaginalis is with medial ridge in the Turkish specimen (n=1), compared to European specimens that have the structure evenly concave. Lamella antevaginalis is less sclerotised and the medial sclerotisation of the lamella postvaginalis is less distinct in *N. hoerhammeri*. See genetic data and Remarks for details. Text-fig. 65. *N. stevenaria* from Spain: Andalucia, 1.v.1979, all pictures from slide PS1952 (coll. SKOU). Text-fig. 66. *N. stevenaria* from Greece: Epiros, 17–31.v.1999, all pictures from slide PS1951 (coll. P. SIHVONEN). Text-fig. 67. *N. stevenaria* from Turkey: Yesilyurt, Malatya 1400m, 24.vii.1995, all pictures from slide PS1955 (coll. SKOU). Text-fig. 68. *N. hoerhammeri* from Iran: N-Fars, Komehr-Khan, c. 8500 ft, 9.vii.1950, all pictures from slide PS1970/BMNH GEO 24898 (coll. BMNH).

FEMALE GENITALIA (see also Text-figs 65–67): Papillae anales large, dorsal part setose, ventral part with membranous sac, containing modified, stout setae, widest at apex (Text-fig. 59). Apophyses short, stout. Lamella antevaginalis and lamella postvaginalis fused. Lamella antevaginalis channel-shaped, striated, ribbed, distinctly sclerotised laterally, membranous midventrally. Lamella postvaginalis small plate, margin deeply concave. Ductus bursae short, striated, sclerotised. Posterior part of corpus bursae long, curved, striated. Anterior part round, striated, membranous. Signum size variable, often medium-sized, always stellate.

DISTRIBUTION AND ABUNDANCE: Mediterranean-Turanian. In Europe disjunct. In the Iberian Peninsula and from north-easternmost Italy through the Balkans including Greece, Bulgaria and Romania to Ukraine, south-eastern European Russia and southern Urals. New for southern Urals Orenburg Region, Bylyaevka District, Donskoe Village (coll. P. SKOU). – Outside Europe in Turkey, Cyprus, Israel, Caucasus, Transcaucasus, northern Iraq and Iran. – In Europe a relatively common species that is mainly found in limited numbers.

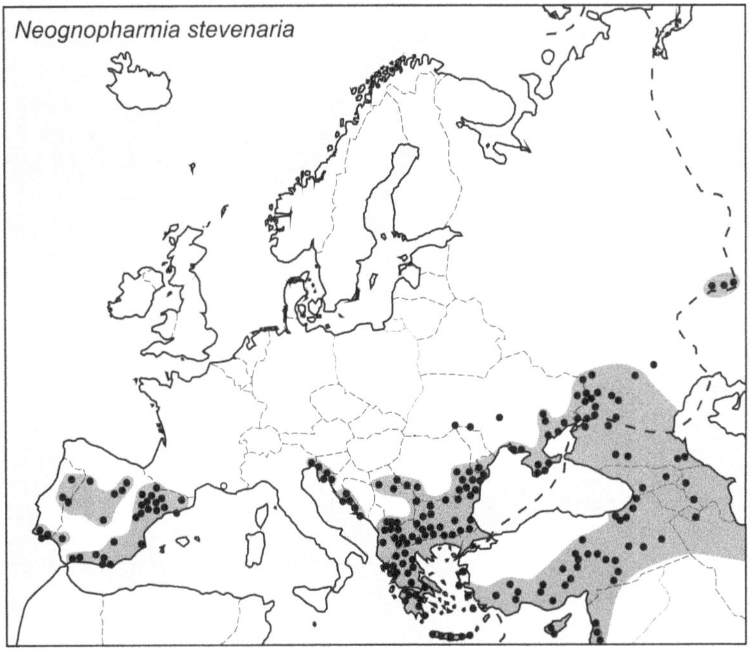

PHENOLOGY: Bivoltine. The moths fly from early April to late August, at warmer localities already from mid-March, in two overlapping generations. In Romania from early May to mid-June and from late July to mid-September (L. SZÉKELY pers. comm.). Overwintering takes place as pupa (M. LEIPNITZ pers. comm.). The moths fly at night and come to light.

BIOLOGY: Oligophagous on Rosaceae. In Cyprus the caterpillar has been found on *Crataegus azarolus* and *Prunus dulcis* (M. LEIPNITZ pers. comm.). Recorded on various herbaceous plants (SZÉKELY 2011). The caterpillar has been reared on leaves of *Crataegus* (B. MÜLLER pers. comm.) and *Prunus spinosa* (D. FRITSCH pers. comm.).

HABITAT: Xerothermophilous. In a variety of open land habitats, with low vegetation and scattered bushes, mainly in dry places and often in areas of abandoned cultivation. From sea level up to at least 1200 m. In the Iberian Peninsula up to 2000 m (REDONDO *et al.* 2009). In Turkey up to at least 1700 m and on Cyprus up to 1950 m.

SIMILAR SPECIES: No similar species in Europe. Non-European *Neognopharmia hoerhammeri* (BRANDT, 1941), from Iran (WEHRLI 1939–1954) is very similar (Text-figs 60–64 and 65–68). Posterior margin of male 8th sternite is without lateral concavity, the margin of the ventral part of valva is blunt-ended with single large tooth in *N. hoerhammeri* (8th sternite concave, valva margin round, valva with several teeth in *N. stevenaria*). Lamella antevaginalis is less sclerotised and medial sclerotisation of lamella postvaginalis is weak in *N. hoerhammeri* (lamella antevaginalis distinctly sclerotised and medial sclerotisation of lamella postvaginalis distinct, concave in *N. stevenaria*).

GENETIC DATA: Genetically heterogeneous in Europe, populations from Spain (n=4) diverging from those of south-eastern Europe to Kazakhstan and to the Eastern Mediterranean (n=12) by 4.8%. In the Middle East with two other BINs. Nearest species: *N. hoerhammeri* (BRANDT, 1941) from Iran (7.6%).

REMARKS: The genetic distance observed between analysed European specimens induced us to examine the genitalia across the Europe, to see if there would be diagnosable morphological differences that correlate with the genetic differences. We examined specimens from Spain, Ukraine, Greece, Turkey and Russia (South Ural and North Caucasus) (n=11). We failed to find distinct morphological differences between the examined specimens. There is some variation in the size and number of sclerotised teeth in the ventral arm of valva, shape of the valva margin (Text-figs 60–63), and size of the signum and shape of corpus bursae (Text-figs 65–63). A single female from Turkey has the medial sclerotisation of the lamella antevaginalis with a medial ridge. On the other hand, it is known that the genitalia in many Macariini genera, for instance in *Isturgia* HÜBNER, 1823 and *Macaria* CURTIS, 1826 are rather uniform and thus not very informative between species. This has also been illustrated in the related non-European genus *Gnopharmia* STAUDINGER, 1892 (RAJAEI *et al.* 2012, RAJAEI *et al.* 2013). We also examined *Neognopharmia hoerhammeri* (BRANDT, 1941), and although genetically 7.6% different from *N. stevenaria* (BOISDUVAL, 1840), the genitalia are similar to *N. stevenaria*, having only small quantitative differences (see Text-figs 64, 68).

Macaria CURTIS, 1826

Macaria CURTIS, 1826 September 1, Br. Ent. 3: 132. Type species: *Phalaena liturata* CLERCK, 1759 [Europe]. Genus synonyms are after SCOBLE (1999), SCOBLE & KRÜGER (2002) and PITKIN (2002).

Azata WALKER, 1860. Type species: *Azata idriasaria* WALKER, 1860.

Catastictis GUMPPENBERG, 1887. [Unnecessary replacement name for *Eufitchia* PACKARD, 1876].

Colutogyna LEDERER, 1853. [Unnecessary replacement name for *Pygmaena* BOISDUVAL].

Diastictis HÜBNER, [1823]. Type species: *Geometra artesiaria* [DENIS & SCHIFFERMÜLLER], 1775. Austria. [Junior homonym of DIASTICTIS Hübner (1818) (Lepidoptera: Pyralidae). Replaced by *Proutictis* BRYK]

Dysmigia WARREN, 1895. Type species: *Fidonia loricaria* EVERSMANN, 1837. [Tatarstan]: Kazan.

Elpiste GUMPPENBERG, 1887. Type species: *Halia tripunctaria* PACKARD, 1873. USA: California. [Junior synonym of *Tephrina lorquinaria* GUENÉE].

Epelis HULST, 1896. Type species: *Fidonia truncataria* WALKER, 1862. Canada.

Eufitchia PACKARD, 1876. Type species: *Abraxas ribearia* FITCH, 1848. USA: New York State.

Eupisteria BOISDUVAL, 1840. Type species: *Geometra quinquaria* HÜBNER, [1822]. Europe. [Junior subjective synonym of *Geometra brunneata* THUNBERG. Sweden].

Eutropa HÜBNER, [1831] 1825. Type species: *Eutropa distribuaria* HÜBNER, [1831] 1825. [USA]: Pennsylvania.

Fidonia HERRICH-SCHÄFFER, 1855. Type species: *Geometra pinetaria* HÜBNER, [1799]. Europe. [Junior homonym of *Fidonia* TREITSCHKE, 1825 (Geometridae)].

Gladela GROSSBECK, 1909. [Replacement name for *Sympherta* HULST, 1896].

Grammatophora STEPHENS, 1829. Type species: *Phalaena wauaria* LINNAEUS, 1758. Europe. [Replacement name for *Halia* DUPONCHEL, [1829], and a senior objective synonym of *Thamnonoma* LEDERER, 1853].

Halia DUPONCHEL, [1829]. Type species: *Phalaena wauaria* LINNAEUS, 1758. Sweden. [Junior homonym of *Halia* RISSO, 1826. Replaced by *Grammatophora* STEPHENS, 1829].

Lissoplaga WARREN (1894). Type species: *Fidonia edmondsi ni* BUTLER, 1882, *sensu* WARREN, 1894. Chile.

Parasemia HÜBNER, 1823. Type species: *Parasemia agnitaria* HÜBNER, 1823. Brasil. [Identity not established, SCOBLE *in* PITKIN (2002). Junior synonym of *Phal[aeana] Geomet[ra] gambarina* STOLL, 1781. Surinam. Junior homonym of *Parasemia* HÜBNER, 1816 (Lepidoptera: Arctiidae)]

Philobia DUPONCHEL, [1829]. Type species: *Phalaena notata* LINNAEUS, 1758, by original designation, Europe.

Physosteganía WARREN (1894). Type species: *Stegania pustularia* GUENÉE, [1858]. North America.

Prophasiane MCDUNNOUGH, 1939. Type species: *Semiothisa octolineata* HULST, 1887. USA: Arizona.

Proutictis BRYK, 1938. Type-species: *Geometra artesiaria* [DENIS & SCHIFFERMÜLLER], 1775. Austria. [Replacement name for *Diastictis* HÜBNER, [1823]]

Psamatodes GUENÉE, [1858]. Type species: *Psamatodes rimosata* GUENÉE, [1858]. Brazil.

Pseudoisturgia POVOLNY & MOUCHA, 1957. Type species: *[Phalaena] carbonaria* CLERCK, 1759. Sweden.

Pygmaena BOISDUVAL, 1840. Type species: *Geometra venetaria* HÜBNER, [1809] 1796. Europe.

Sciagraphia HULST, 1896. Type species: *Macaria granitata* GUENÉE, [1858]. [USA]: Pennsylvania.

Speranza CURTIS, 1828. Type species: *Speranza sylvaria* CURTIS, 1828. Scotland. [Junior subjective synonym of *S. brunneata* THUNBERG. Scotland].

Sympherta HULST, 1896. Type species: *Halia tripunctaria* PACKARD, 1873. USA: California. [Junior homonym of *Sympherta* FÖRSTER (Hymenoptera) and junior objective synonym of *Elpiste* GUMPPENBERG, 1887]

Thamnonoma LEDERER, 1853. Type species: *Phalaena wauaria* LINNAEUS, 1758. Sweden. [Replacement name for *Halia* DUPONCHEL, [1829], but a junior objective synonym of *Grammatophora* STEPHENS, 1829].

Xenoecista WARREN, 1897. Type species: *Xenoecista subdiversa* WARREN, 1897. Colombia.

UNAVAILABLE NAMES (incorrect subsequent spelling): *Maearia* SEYFFER, 1850 (see FLETCHER 1979).

DIVERSITY AND DISTRIBUTION: About 170 species, although deletions and additions to the list of included species are likely to be made with further research. Species of this genus are found in the Nearctic, Neotropical and the Palaearctic regions. Most species are found in the New World. If *Parosteodes* WARREN, 1895 were to be included in *Macaria*, then Australasia would be included in the distribution of the latter (SCOBLE & KRÜGER 2002).

EXTERNAL CHARACTERS AND ABDOMEN: Medium-sized moths, European species different shades of grey, brown and orange-brown. Forewing margin round or concave below apex. Females occasionally short-winged (brachypterous). Forewing often with dark spot in terminal area. Chaetosemata extended across head. Male antennae bipectinate or fasciculate. Male 3rd sternite occasionally with stout setae. Posterior margin of male 8th sternite deeply cleft.

MALE GENITALIA: Uncus horns present, rather small, occasionally absent. Sclerite below uncus absent. Valva divided. Ventral arm of valva rounded or weakly elongated, its dorsal margin base often with sclerotised, setose ridge. Vesica simple tube, with basal sclerotisation (see Table 1). SCOBLE & KRÜGER (2002) consider the shape of valva with the minimal separation between costa and sacculus a possible apomorphic feature. Male genitalia of the European species are similar. Diagnostic features are often quantitative and differences between species are minute.

FEMALE GENITALIA: Lamella antevaginalis and lamella postvaginalis fused, forming narrow ring around ostium bursae. Lamella postvaginalis roundish plate. Posterior part of corpus bursae long, rather narrow. Anterior part of corpus bursae angled or roundish. Signum large, stellate (see Table 1).

BIOLOGY: The range of food plants is very wide and includes coniferous and deciduous trees, and both monophagous and polyphagous species exist (SCOBLE & KRÜGER 2002). Majority of North American species are conifer feeders, mostly on Pinaceae, and a few species have been recorded feeding on *Acer*, *Betula* and *Populus* (FERGUSON 2008). In Europe food plants of larvae have been recorded to include for instance *Salix*, *Betula*, *Ribes*, *Vaccinium*, but also *Picea* and *Pinus* (SEPPÄNEN 1970). Few species are diurnal, for instance *Macaria carbonaria* (CLERCK), but most are nocturnal.

IMMATURE STAGES: Larvae are long, narrow, green or brown, with several longitudinal whitish stripes. Several species are illustrated in AIZPÚRUA (2002), EBERT (2003) and FERGUSON (2008). Pupation occurs in cocoon on ground. Cremaster elongated, with two short lateral hooks, occasionally with two additional minute teeth (PATOČKA & TURČANI 2005).

REMARKS: SCOBLE & KRÜGER (2002) combine the North-American *Epelis truncataria* (WALKER, 1862) to *Macaria* and mention that this Holarctic species occurs in Europe also. We are not aware that *E. truncataria* has been reported from the Palaearctic region or from Europe, nor is it mentioned by FERGUSON (2008). We agree with SCOBLE (1999), and SCOBLE & KRÜGER (2002), and consider *Speranza* CURTIS, 1828, with *M. brunneata* THUNBERG, 1784) as its type species, a junior synonym of *Macaria*. FERGUSON (2008) did not agree with this and classified the North American fauna in the genus *Speranza*. Selected literature: KRÜGER (2001), SCOBLE & KRÜGER (2002), FERGUSON (2008). Systematic position: Macariini (KRÜGER 2001, SCOBLE & KRÜGER 2002).

35. *Macaria notata* (LINNAEUS, 1758)

notata notata LINNAEUS, 1758, Syst. Nat. (Ed. 10) 1: 523, (*Phalaena* (*Geometra*)). Syntypes, including putative lectotype (without abdomen), (Linnean Society of London, London, UK), Europe (examined externally).

exustata HUFNAGEL, 1767, Berlin Mag. 4 (6): 600, (*Phalaena*). Syntype(s), lost, [Germany]: Berlin region.

hieroglyphica FOURCROY, 1785, in FOURCROY & GEOFFROY, Entomologia Parisiensis: 286, (*Phalaena*). France: Paris. [Synonymy after LERAUT (1997)].

notaria MORRIS, 1861, Nat. Hist. Br. Moths 1: 177, (*Macaria*). [Emendation of *notata* LINNAEUS].

notataria [DENIS & SCHIFFERMÜLLER], 1775, Ankündung syst. Werkes Schmett. Wienergegend: 104 (Geometra). [Emendation of *notata* LINNAEUS].

notataria ESPER, [1801], Die Schmett. 5 (3): 87, pl. 16, figs 4-6, (*Ph*[*alaena*] *Geom*[*etra*]). [Emendation of *notate* LINNAEUS].

ulsterata PEARSALL, 1913, Jour. N. Y. ent. soc. 21 (3): 188, pl. 3, fig. 1, (*Philobia*). Lectotype male (American Museum of Natural History, New York, USA), [U.S.A].: Big Indian Valley, Catskill Mountains, New York. Lectotype designated and synonymy proposed by FERGUSON (2008).

notata appalachiata FERGUSON, 2008, Moths of North America, fascicle 17.2: 180, pl. 4, figs 21–23, (*Macaria*). Holotype male (National Museum of Natural History, Smithsonian Institution, Washington, DC, USA), [U.S.A].: North Carolina, Jackson County, Waterrock Knob (examined, based on literature).

notata kirina WEHRLI, 1940, in SEITZ, Gross-Schmett. Erde 4 (Suppl.): 386, pl. 30: k, (*Semiothisa*). Syntype(s) (Zoologische Forschungsinstitut und Museum Alexander Koenig, Bonn, Germany), Manchuria (Gouv. Kirin): Maunschan; Korea: Poktussan, 2500 m. Valid at subspecies rank according to SCOBLE (1999).

INTRASUBSPECIFIC NAMES, POTENTIALLY UNAVAILABLE: *albescens* LEMPKE, 1970 (f.); *agraptus* SCHULTZ, 1930 (f.); *costijuncta* LEMPKE, 1970 (f.); *diluta* LEMPKE, 1953 (f.); *fusca* LEMPKE, 1953 (f.); *fuscomarginata* LEMPKE, 1953 (f.); *infuscaria* REBEL, 1910 (ab.); *innotata* FUCHS, 1901 (ab.); *lineata* LEMPKE, 1970 (f.); *lunulacarens* LEMPKE, 1953 (f.); *luteolaria* TENGSTRÖM, 1869 (ab.); *signata* LEMPKE, 1970 (f.); *uniformata* LEMPKE, 1953 (f.).

EXTERNAL CHARACTERS AND ABDOMEN: Wingspan 24–30 mm. Wings whitish-yellow, irrorated with grey. Medial lines narrow, undulating, grey, expanded to notable, orange-brown spots on costa. Subterminal band expanded on costa, orange-brown. Large, blackish brown spot in subterminal band, usually broken into five small spots. Subterminal band below spot inconspicuous. Terminal line narrow, continuous. Forewing margin concave below apex, fringes not uniformly dark. Elsewhere fringes weakly chequered whitish and grey. Hindwing margin elongated at M3. Discal spots small, often absent on forewings. Wings below as above, markings more striated, brown, subterminal line and discal spots more pronounced. Frons, collar, thorax and abdomen concolorous with wings, weakly irrorated with brown. Abdomen often with two rows of black dots dorsally. Antennae concolorous with wings dorsally, fasciculate in male, filiform in female. Hindtibia with 2+2 spurs. Tympanal organs large. Posterior margin of male 8th sternite deeply invaginated. Other sternites and tergites of both sexes undifferentiated.

VARIATION: Very little. Ground colour varies from whitish to whitish yellow. Density of markings and irroration somewhat variable. Numerous forsm have been described, see for instance LERAUT (2009). Melanistic specimens are rare (MIKKOLA *et al.* 1989).

MALE GENITALIA: Uncus margins slightly undulating, setose dorsally, apex semi-acute, with two horns. Gnathos arms narrow, fused ventrally, forming weakly upturned, triangular element. Juxta small plate. Valva divided. Dorsal arm of valva long, wide, straight, costa and apex setose. Ventral arm of valva slightly elongated, apex round, weakly setose. Its ventral margin curved inwards, with several submarginal setal groups. Saccus wide, shallow. Aedeagus short, stout, caecum round knob, apex sharp. Vesica opens subapically, at approximately 90 degree angle, basal part wider, weakly ribbed, with scale-like sclerotised plate. Apical part angled, narrower.

FEMALE GENITALIA: Papillae anales roundish, setose. Apophyses straight, narrow. Lamella antevaginalis membranous. Lamella postvaginalis small, irregular, posterior margin concave plate. Ductus bursae wide, short, sclerotised, weakly ribbed. Base of ductus seminalis weakly expanded. Posterior part of corpus bursae with finger-like extension, otherwise long, weakly striated. Anterior part of corpus bursae large, weakly angular with apical diverticulum. Signum large, stellate.

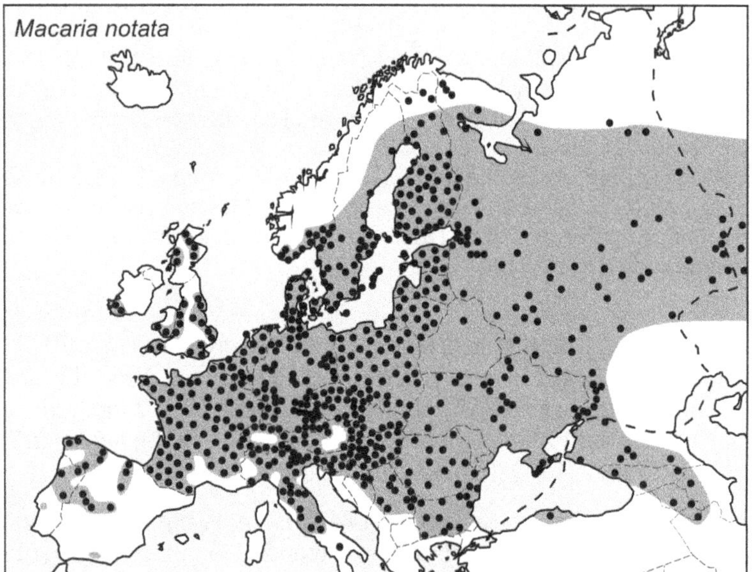

DISTRIBUTION AND ABUNDANCE: Holarctic. In Europe widely distributed from northern Fennoscandia to the Iberian Peninsula, central Italy and a large part of the Balkans and from Ireland to the Ural mountains. – Outside Europe in Turkey, Caucasus, Transcaucasus, Kazakhstan, Siberia, China, Korea, Far East Russia and Sakhalin island. In North America widely distributed in the nominate subspecies and the more local subsp. *appalachiata*. – In Europe a common species that now and then occurs in numbers. In Britain the species has increased in numbers in recent decades, and it is now much more common than it was earlier (FOX *et al.* 2013).

PHENOLOGY: Uni- or bivoltine. In Italy from April to June and in July-August (FLAMIGNI *et al.* 2007), in southern Germany in overlapping generations from late April to late August with maximum abundance mid-May to mid-June and again late July to early August (EBERT 2003). In northern Europe only in one generation from late May to late July. Depending on latitude and altitude, the larva lives from May or June to August or September. Hibernation takes place as a pupa. The adult moths are rather rarely seen by day and fly by night when they come readily to light.

BIOLOGY: Polyphagous. The larva has been found on *Betula* and *Salix* (HULDÉN *et al.* 2000), repeatedly on *Betula pendula* (H. WEGNER & D. FRITSCH pers. comm.). Also recorded on *Corylus avellana* and *Prunus spinosa* (SKOU 1984, 1986). The pupation takes place among mosses on the ground.

HABITAT: Sylvicolous. A typical species of light, open deciduous forests, but it occurs also in other habitat types such as birch moors and gardens. In Fennoscandia it is absent from the mountains. In the Alps occurring up to 1450 m (FLAMIGNI *et al.* 2007).

SIMILAR SPECIES: *Macaria alternata* (DENIS & SCHIFFERMÜLLER) is smaller, wings grey, with tint of violet (larger, wings whitish grey in *M. notata*). Dark spot in subterminal band with four spots (normally five in *M. notata*), postmedial band is wide, grey (normally weak or absent in *M. notata*), see Text-figs 69–70. In *M. alternata* ventral arm of valva is clearly protruded outwards (ventral arm of valva is shorter in *M. notata*). Female genitalia without notable differences. Shape of corpus bursae varies in both species.

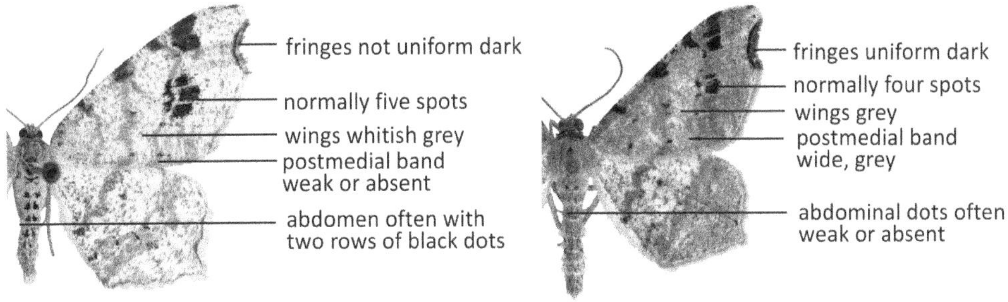

M. notata 69 ***M. alternata*** 70

Text-figs 69–70. Diagnostic characters (indicated) of *Macaria* species. Text-fig. x: *M. notata* (LINNAEUS, 1758), France: Amfreville, 22.vii.1950). Text-fig. x: *M. alternata* (DENIS & SCHIFFERMÜLLER, 1775), Hungary: 25 km W Kecskemet, 28.vii.2002.

GENETIC DATA: Genetically homogeneous from Europe (n=19, from six countries) to Mongolia, but with a separate BIN at 4.6% distance from Turkey to Iran (n=8), in Georgia both BINs present. Nearest species: North American populations identified as '*ulsterata* (PEARSALL, 1913)' (2.8%), treated as a junior synonym of *M. notata* subsp. *notata* (LINNAEUS) by FERGUSON (2008), see above. In Europe *M. liturata* (6.1%).

36. *Macaria alternata* (DENIS & SCHIFFERMÜLLER, 1775)

alternata [DENIS & SCHIFFERMÜLLER], 1775, Ankündung syst. Werkes Schmett. Wienergegend: 106, (*Geometra*). Syntype(s), lost, [Austria]: Vienna district.

alternaria HÜBNER, [1809], Samml. Eur. Schmett. 5 Geometrae (1): pl. 61, fig. 314, (*Geometra*). [Incorrect subsequent spelling of *alternata* [DENIS & SCHIFFERMÜLLER].

praenotata HAWORTH, 1809, Lepid. Br. (2): 346, (*Phalaena*). Syntype(s), [United Kingdom].

INTRASUBSPECIFIC NAMES, POTENTIALLY UNAVAILABLE: *costanuda* LEMPKE, 1953 (f.); *demaculata* HEINRICH, 1923 (ab.); *fusca* LEMPKE, 1953 (f.); *fuscomarginaria* DANNEHL, 1927 (ab.); *immacularia* DANNEHL, 1927 (ab.); *lineata* LEMPKE, 1953 (f.); *lunulacarens* LEMPKE, 1953 (f.); *lutescens* LEMPKE, 1953 (f.); *pallida* LEMPKE, 1953 (f.).

EXTERNAL CHARACTERS AND ABDOMEN: Generally as *M. notata* (LINNAEUS), see above, but with following differences. *M. alternata* is smaller, wingspan 22–29 mm. Wing markings grey to grey-violet. Blackish brown spot in subterminal band smaller, usually broken into four small spots. Subterminal line wide, grey, in forewings and hindwings, also below forewing spot. Forewing margin markedly concave below apex, fringes uniformly dark in concave area and uniform grey elsewhere. Wings below often more contrasting, wings whitish, markings orange-brown.

MALE GENITALIA: Generally as in *Macaria notata*, but with following difference: In *M. alternata* ventral arm of valva is clearly protruded outwards (ventral arm of valva is shorter in *M. notata*).

FEMALE GENITALIA: Generally as in *Macaria notata*, without notable differences. Shape of corpus bursae varies in both species. FLAMIGNI *et al.* (2007) indicate that relative lengths of ductus bursae and corpus bursae are diagnostic. Our material does not support that. Genitalia of *M.alternata* is illustrated as enlarged in the first female genitalia plate that explains terminology.

DISTRIBUTION AND ABUNDANCE: Eurasiatic. In Europe widely distributed from northern Fennoscandia to the Iberian Peninsula, Italy and most of the Balkans and from Ireland to the Ural mountains. – Outside Europe the species is known from Caucasus, Transcaucasus, central Asia, Siberia and Far East Russia. Reputedly recorded from Turkey (VIIDALEPP 1996), but no Turkish specimens have been possible to locate. – In Europe a common species that now and then occurs in numbers.

PHENOLOGY: Uni- or bivoltine. In southern Europe in two more or less overlapping generations from April to June/July and from July to September (FLAMIGNI *et al.* 2007, REDONDO *et al.* 2009). North of the Alps in southern central Europe also in two generations in warm places, otherwise in one generation from late May to mid-August in southern Germany (EBERT 2003) and in Denmark from early June to mid-July. Caterpillars can be found from May or June and until September and sometimes even in October. The species overwinters as pupa. The moths fly at night and come frequently to light, but they are rarely disturbed during daytime.

BIOLOGY: Polyphagous. The caterpillar has been found on *Quercus robur*, *Prunus padus*, *P. spinosa*, *Rhamnus*, *Hippophaes rhamnoides* (EBERT 2003), on *Salix* and *Sorbus* (HULDÉN *et al.* 2000) and on *Betula pubescens* (H. WEGNER pers. comm.). It is also recorded on *Frangula alnus* (SKOU 1984, 1986), *Populus*, *Alnus*, *Fagus*, *Malus* and *Crataegus* (FLAMIGNI *et al.* 2007). Pupation takes place in the ground.

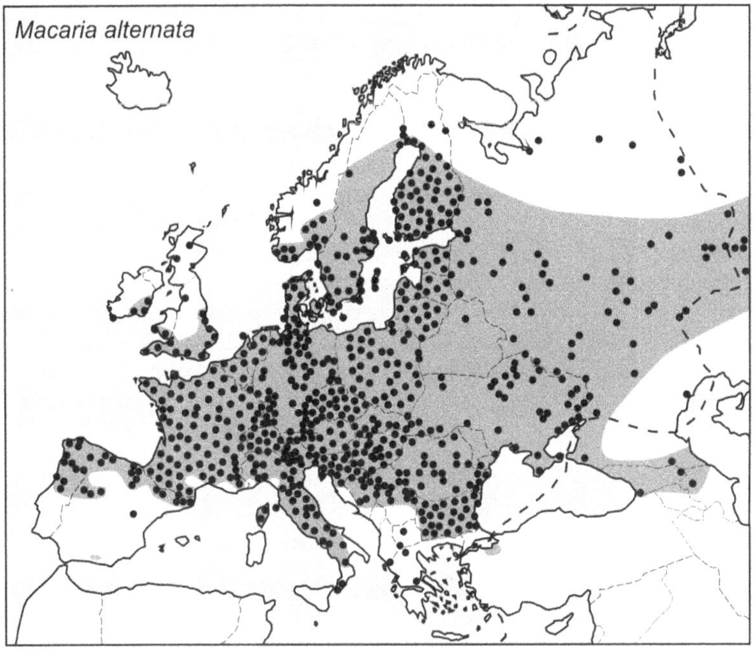

HABITAT: Sylvicolous. In light and open forests, often on moist soil and in rather dry moors with trees and bushes. In Fennoscandia not in the mountains. In the Alps up to 1800 m altitude (FLAMIGNI *et al.* 2007).

SIMILAR SPECIES: *Macaria notata*, see entry above and Text-figs 69–70.

GENETIC DATA: Genetically homogeneous from Europe (n=31, from eleven countries). Nearest species in Europe: *M. notata* (6.7%).

37. *Macaria signaria* (HÜBNER, 1809)

signaria signaria HÜBNER, [1809], Samml. Eur. Schmett. 5 Geometrae (1): pl. 61, fig. 313, (*Geometra*). Syntype(s), lost, Europe.

signaria dispuncta WALKER, 1860, List Specimens lepid. Insects Colln Br. Mus. 21: 405, (*Macaria*). Holotype female (The Natural History Museum, London, UK), [Canada]: Hudson's Bay, Albany River, St Martin's Falls (examined externally). Originally as sp., valid at subspecies rank according to SCOBLE (1999). FERGUSON (2008, p. 216) states under *dispuncta* 'revised synonymy', but the meaning is unclear. Later (p. 217) he considers the European specimens and American specimens as forms ['... the nominate European form and predominant North American form ...'], but also as subspecies ['The overall coloring of European specimens is slightly more brownish, not as grey as the American subspecies.']

exnotata WALKER, 1862, List Specimens lepid. Insects Colln. Br. Mus. 24: 1186, (*Larentia*?). Holotype female (The Natural History Museum, London, UK), [Canada]: Nova Scotia (examined externally).

fraserata Ferguson, 1974, Can. Ent. 106: 581, figs 62–70, (*Semiothisa*). Holotype male, [U.S.A].: North Carolina, Richland Balsam Mountain, 6000 ft, Jackson-Haywood County line. [Synonymy proposed by FERGUSON (2008)].

haliata WALKER, 1861, List Specimens lepid. Insects Colln. Br. Mus. 23: 885, (*Macaria*). Syntypes 1 male, 1 female (The Natural History Museum, London, UK), [USA]: New York; [Canada]: Nova Scotia (examined externally).
inordinaria WALKER, 1860, List Specimens lepid. Insects Colln Br. Mus. 21: 488, (*Boarmia*). Holotype female (Canadian National Insect Collection, Ottawa, Ontaria, Canada), Canada.
irregulata WALKER, 1861, List Specimens lepid. Insects Colln. Br. Mus. 23: 890, (*Macaria*). Holotype male (The Natural History Museum, London, UK), [U.S.A].: New York (examined externally).
quadrisignata WALKER, 1866, List Specimens lepid. Insects Colln Br. Mus. 35: 1655, (*Macaria*). Holotype male (The Natural History Museum, London, UK), North America (examined externally).
subapicaria WALKER, 1861, in Urban, Can. nat. & Geol. 6 (1): 40, (*Macaria?*). Holotype male (Canadian National Insect Collection, Ottawa, Ontario, Canada), [Canada]: Valley of the River Rouge.

INTRASUBSPECIFIC NAMES, POTENTIALLY UNAVAILABLE: *agraptus* SCHULTZ, 1930 (ab.); *cinerata* WERNEBURG, 1864 (f.); *canescens* FEICHENBERGER, 1962 (f.); *demaculata* LEMPKE, 1952 (f.); *fusca* LEMPKE, 1953 (f.); *lineata* LEMPKE, 1970 (f.); *lutescens* LEMPKE, 1952 (f.).

EXTERNAL CHARACTERS AND ABDOMEN: Wingspan 21–30 mm. Wings greyish brown, irrorated with grey and brown. Medial lines brown, serrated, widest on costa. Subterminal line expanded on costa, brown-orange. Black, suffused spot in subterminal band. Subterminal line greyish white. Terminal area often with whitish spot, ground colour more uniform brown than in basal and medial areas. Terminal line weak, interrupted. Forewing margin weakly concave, or nearly absent below apex, fringes in that area not noticeably darker than elsewhere. Hindwings lighter than forewings, terminal area darker. Hindwing margin weakly elongated at M3. Discal spots small. Wings below as above, markings weaker. Frons, collar, thorax and abdomen concolorous with wings. Antennae concolorous with wings dorsally, fasciculate in male, filiform in female. Hindtibia with 2+2 spurs. Tympanal organs medium-sized. Male 3rd sternite with setae, posterior margin of 8th sternite deeply invaginated. Other sternites and tergites of both sexes undifferentiated.

VARIATION: Rather little. Amount of brown irroration and intensity of markings varies. Terminal area often darker than basal and medial areas.

MALE GENITALIA: Generally as *Macaria notata*, without notable differences (see entry above).

FEMALE GENITALIA: Generally as *Macaria notata*, but with following difference: In *M. signaria* lamella postvaginalis is small plate, posterior margin round, lateral appendices small (lamella postvaginalis is small, irregular plate, posterior margin is concave and lateral appendices are absent in *M. notata*).

DISTRIBUTION AND ABUNDANCE: Holarctic. In Europe from central Fennoscandia to France, the Alps, and the northern Balkans and from south-eastern England through central Europe and European Russia to the Ural mountains. – Outside Europe the species occurs through Siberia to Far East Russia, Kamchatka, Sakhalin and Japan, it is also widely distributed throughout temperate North America (subsp. *dispuncta*). Reputedly recorded from Caucasus and Turkey (VIIDALEPP 1996). KOÇAK & KEMAL (2009) also list *M. signaria* from Turkey, from the southern province Kahramanmaras, but no specimens from Caucasus or Turkey and we regard records from these areas doubtful. – In Europe a rather common species that rarely appears in numbers.

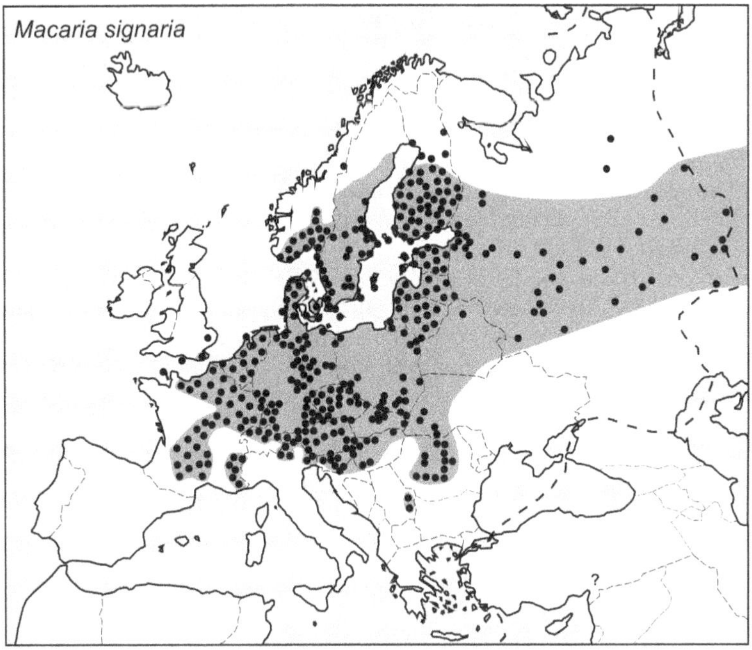

PHENOLOGY: Univoltine. In northern Europe the species flies from early June to late July. In south-western Germany also in one generation, but with an extended flight time from mid-May to early August (EBERT 2003) and in north-eastern Italy in the Alps from late May to July (FLAMIGNI *et al.* 2007). According to EBERT (2003) the larva remains unpupated in its silk web during the whole winter to pupate a few weeks before the moth emerges. The moths are rarely seen by daytime, but fly at night where they come to light.

BIOLOGY: Oligophagous on conifers. The larva is recorded on *Picea abies* (EBERT 2003), *P. excelsa* (BERGMANN 1955) and *Larix* (HULDÉN *et al.* 2000) from June to September. However, among others FLAMIGNI *et al.* (2007) also records *Abies* and *Pinus*.

HABITAT: Sylvicolous. Especially in spruce forests and plantations, but also in garden areas with sufficient density of the food plant.

SIMILAR SPECIES: *Macaria notata* (LINNAEUS) and *M. alternata* (DENIS & SCHIFFERMÜLLER) have forewing margin more concave below apex (Text-figs 69–70), subterminal line not visible and male 3rd sternite is without setae (forewing margin weakly concave below apex, subterminal line greyish white and male 3rd sternite is present in *M. signaria*). In *Macaria liturata* subsp. *liturata* (CLERCK) subterminal band is orange-brown, blackish spot is normally absent (subterminal band is brown, blackish spot present in *M. signaria*).

GENETIC DATA: Genetically homogeneous in Europe (n=11, from four countries). North American populations, on BOLD partly identified as *M. signaria*, partly as *M. marmorata* (FERGUSON, 1972), at a distance of 1.5%. Nearest species in Europe: *M. liturata* (4.3%).

REMARKS: *M. signaria* is Holarctic in distribution. It is the most abundant and widely distributed member of the *signaria*-group in North America. According to FERGUSON (2008), it is represented in North America by at least three visibly different entities, which he did not name formally because their status is uncertain. These are illustrated in FERGUSON (2008).

38. Macaria liturata (CLERCK, 1759)

liturata liturata CLERCK, 1759, Icones Insectorum Rariorum 1: pl. 6, fig. 6, ([*Phalaena*]). Syntypes, including putative lectotype male (Linnean Society of London, London, UK), [Sweden] (examined externally).

fuscata THUNBERG, 1788, D. D. Mus. nat. Acad. Upsal. 6: 75, figured, (*Phalaena*). Lectotype male (Universitets Zoologiska Institut, Uppsala, Sweden), Sweden.

igneata HUFNAGEL, 1767, Berlin Mag. 4 (6): 612, (*Phalaena*). Syntype(s), lost, [Germany]: Berlin region.

liturata deceptans STAUDINGER, 1901, in Staudinger & Rebel, Cat. Lepid. palaearct. Faunengeb. (Edn 3) 1: 333, (*Semiothisa*). Syntype(s) male(s), female(s), Graecia [Greece]; Italy: Vallombrosa. Originally as var., treated valid at subspecies level by SCOBLE (1999) and FLAMIGNI *et al.* (2007), downgraded here from subspecies rank (new synonym), quoting partially sympatric occurrence with nominal subspecies and identical COI barcode. See remarks.

liturata hispanica HERBULOT, 1957, Bull. mens. Soc. linn. Lyon 26 (7): 191, (*Semiothisa*). Holotype male (Coll. HERBULOT, in Zoologische Staatssammlung, Munich, Germany), [Spain]: Aragon, Orihuela del Tremedal (examined externally). See remarks.

lituraria HÜBNER, [1823], Verz. bekannter Schmett.: 298, (*Semiothisa*). [Emendation of *liturata* CLERCK]

liturataria ESPER, [1801], Die Schmett. 5 (4): 89, pl. 16, figs 7-9, (*Ph*[*alaena*] *Geom*[*etra*]). [Emendation of *liturata* CLERCK]

liturata pressaria CHRISTOPH, 1893, Dt. ent. Z. Iris 6 (1): 95, (*Macaria*). Syntype(s), Russia: eastern Amur (fortress Albasin). Originally as var., valid at subspecies rank according to SCOBLE (1999).

liturata depressa INOUE, 1955, Tinea 2 (1/2): 84, (*Semiothisa*). Holotype male (The Natural History Museum, London, UK), Japan: Yamanashi Prefecture, Yashajin Pass, 1400 m. Valid at subspecies rank according to SCOBLE (1999) (examined externally).

INTRASUBSPECIFIC NAMES, POTENTIALLY UNAVAILABLE: *albescens* KAUCKI, 1928 (f.); *bilineata* LEMPKE, 1970 (f.); *brunnescens* LEMPKE, 1970 (f.); *contrasta* LEMPKE, 1953 (f.); *costijuncta* LEMPKE, 1970 (f.); *debrunneata* OSTHELDER, 1931 (ab.?); *flavofasciata* LEMPKE, 1970 (f.); *fuscata* FOLTIN, 1942 (ab.); *lineata* LEMPKE, 1953 (f.); *marginata* LEMPKE, 1970 (f.); *nigra* PEERDEMAN, 1962 (f.); *nigra* REBEL, 1910 (ab.) (=synonym of *nigrofulvata* COLLINS, 1905 according to The Global Lepidoptera Names Index (ANONYMOUS 2013)); *nigrofulvata* COLLINS, 1905 (ab.); *pallida* KAUCKI, 1929 (ab.); *semivirgata* COCKAYNE, 1953 (ab.); *trexleri* SCHAWERDA, 1913 (ab.); *unicolorata* RICHARDSON, 1952 (ab.).

EXTERNAL CHARACTERS AND ABDOMEN: Wingspan 24–32 mm. Wings grey, basal and medial areas weakly irrorated with grey and brown. Subterminal band rusty brown-yellow, expanded on costa, without brownish spot in middle. Terminal area grey on costa, elsewhere brown, darker than basal and medial areas. Medial lines narrow, blackish, expanded on costa. Subterminal line present on costa, narrow, blackish, elsewhere faint. Terminal line narrow, continuous. Forewing margin weakly concave below apex, fringes slightly darker than elsewhere in fringes. Hindwing margin weakly elongated at M3. Discal spots small, more distinct on hindwings. Wings below as above, lighter, forewing apex whitish. Frons, collar, thorax and abdomen concolorous with wings, weakly irrorated with brown. Antennae concolorous with wings dorsally, fasciculate in male, filiform in female. Hindtibia with 2+2 spurs. Tympanal organs medium-sized, bullae tympani with small roundish lobe. Male

3rd sternite with setae, posterior margin of 8th sternite deeply invaginated. Other sternites and tergites of both sexes undifferentiated.

VARIATION: Ground colour rather uniform grey, density of markings and irroration somewhat variable. Numerous forms have been described, see for instance PROUT (1912–1916). f. *deceptans* (STAUDINGER) lighter, including terminal area, medial lines weak or absent. Subterminal band expanded on costa, with brownish spot in middle, occasionally band dark throughout forewing, thus general appearance more contrasting. f. *hispanica* (HERBULOT) lighter, subterminal band and terminal area light yellowish orange, including hindwings. Known from the Iberian Peninsula. See remarks.

MALE GENITALIA: Generally as *Macaria signaria*, without notable differences (see entry above).

FEMALE GENITALIA: Generally as *Macaria signaria*, without notable differences (see entry above).

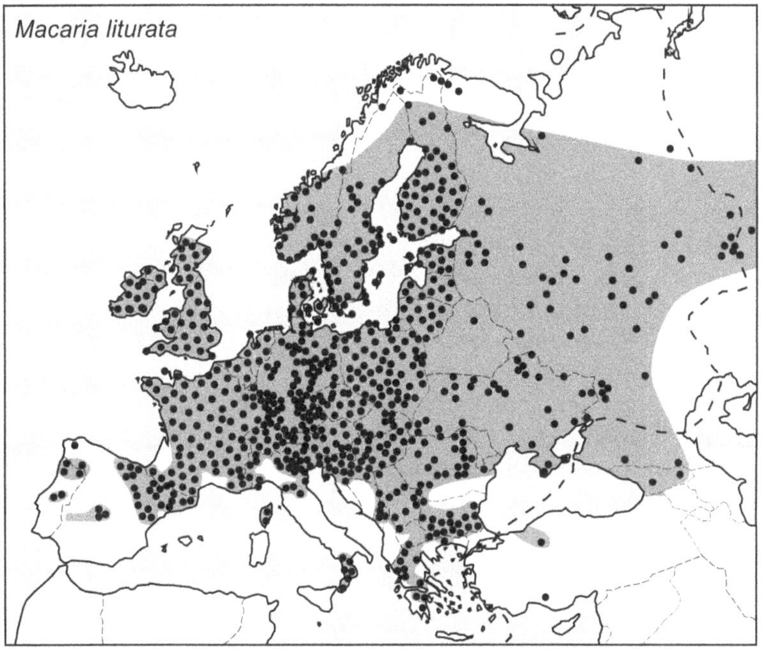

DISTRIBUTION AND ABUNDANCE: Eurasiatic. In Europe widely distributed from northernmost Fennoscandia to the Iberian Peninsula, Italy and Greece and from the British Isles to the Urals. – Outside Europe the species occurs in Turkey, the Caucasus region, Siberia and China to Japan. – In Europe a common species that does not often occurs in numbers.

PHENOLOGY: Uni- or bivoltine. In southern Fennoscandia the species flies from late May to late August, in southern Germany from late April to late August (EBERT 2003), in Italy in two generations from mid-April to June and again from July to mid-September (FLAMIGNI *et al.* 2007). Larva in northern Europe from June to September, in southern Europe in May-June and from August to October. Hibernation takes place as pupa. The moths are relative rarely found by daytime, but they fly actively at night where they abun come to light abundantly.

BIOLOGY: Oligophagous. The caterpillar has been found on *Abies alba*, *Picea abies*, *Pinus sylvestris* and other *Pinus* species (EBERT 2003). Also H. WEGNER (pers. comm.) has found caterpillars on *Pinus sylvestris*, which is possibly the main food plant in a large part of the European distribution area. The pupation takes place under moss.

HABITAT: Sylvicolous. In forests, plantations, gardens and parks where the food plants are growing. In Fennoscandia the species is absent from the mountain regions, in Italy from sea level up to 2100 m.

SIMILAR SPECIES: Rusty brown-yellow subterminal band and brown terminal area make *M. liturata* easily identifiable. *Macaria notata* (LINNAEUS) and *M. alternata* (DENIS & SCHIFFERMÜLLER) have forewing margin more concave below apex (Text-figs 69–70) and male 3rd sternite is without setae (forewing margin weakly concave below apex and male 3rd sternite is present in *M. liturata*). In *M. signaria* subterminal band is brown and blackish spot is present in forewing (subterminal band rusty brown-yellow and blackish spot is absent in forewing in subsp. *liturata*).

GENETIC DATA: Genetically somewhat heterogeneous (n=44, from nine European countries), with three homogeneous BINs at distances of 1.0–1.8%, without clear regional separation from each other and not rarely occurring in sympatry. Populations from Transcaucasia and north-eastern China identical with the widespread European haplotypes. Nearest species: *M. signaria* (4.3%).

REMARKS: Taxon *deceptans* STAUDINGER, 1901, originally described as a variation from Greece, has been considered a mountain form in Greece (PROUT 1912–1916) or a subspecies of *liturata* CLERCK, 1759 (SCOBLE 1999, FLAMIGNI *et al.* 2007). Distribution is not known in detail, but it is found sympatrically among nominotypical specimens in Eptachori in central Greece at 1400 m (our observations), in Badia della Valle in Tuscany (Adriatic side of the Apennines) and rarely in northern Apennines (C. FLAMIGNI pers. comm.). Further south in Italy (Romagna Apennines, central eastern Tuscany, Basilicata and Calabria) only *deceptans* is found (FLAMIGNI *et al.* 2007, C. FLAMIGNI pers. comm.). Sympatric occurrence in several places, lack of genitalia differences, and identical COI barcode lead us to treat *deceptans* as a form of *liturata*.

No specimens from Spain (subsp. *hispanica* (HERBULOT)) have been DNA barcoded so far. REDONDO *et al.* (2009) concluded that 'subspecies *hispanica* HERBULOT, 1957, described from Orihuela del Tremedal (Teruel), is within the range of morphological variation of this species, and therefore we are considering it as just another individual form.' SCOBLE (1999) considered *hispanica* valid at subspecies level, we follow REDONDO *et al.* (2009). We have not examined the genitalia.

Macaria bicolorata (FABRICIUS, 1798)

bicolorata FABRICIUS, 1798, Syst. Ent. (Suppl.): 454, (*Phalaena*). Syntype(s) (lost), [U.S.A].: Virginia.
praeatomata HAWORTH, 1809, Lepid. Br. (2): 345, (*Phalaena*). Syntype(s), England.

Phalaena praeatomata HAWORTH, 1809, which is a junior synonym of *Phalaena bicolorata* FABRICIUS, 1798, was originally described from the British Fauna, but later *praeatomata* was identified as North American (GUENÉE, 1858). According to FERGUSON (2008: 198): 'The two specimens that GUENÉE referred to *praeatomata* are now in the National Museum of Natural History, Smithsonian Institution, Washington D.C., USA. A coloured figure in WOOD's Index Entomologicus (1839, pl. 26, fig. 746) may represent the type of *praeatomata*. WOOD, in a later edition (1854), changed its status to that of a form of *liturata*, although his original figure does show a moth that closely resembles *bicolorata* (the figure in the 1854 edition is poor). The type had come from the well-known London dealer, JOHN FRANCILLON, who is believed to have received and distributed specimens collected by JOHN ABBOT in Georgia. It is likely that *praeatomata* was based on an American specimen with misleading or no locality data.' Therefore *Macaria bicolorata* (FABRICIUS) is excluded from the European fauna.

Macaria bicolorata, which is similar to *M. liturata*, is found in North America only, ranging in the eastern states of USA from Massachusetts to the Gulf Coast to Florida and Texas (FERGUSON 2008). See SCOBLE (1999) for other synonyms of *bicolorata*.

39. Macaria wauaria (LINNAEUS, 1758)

wauaria wauaria LINNAEUS, 1758, Syst. Nat. (Ed. 10) 1: 522, (*Phalaena* (*Geometra*)). Syntypes (Linnean Society of London, London, UK) (examined externally).
fuscaria THUNBERG, 1792, Diss. ent. sistens Insecta Suecia 4: 59, pl. 4, (*Phalaena*). Lectotype male, Sweden. [Junior primary homonym of *Phalaena fuscaria* VILLERS, 1789].
vauaria HAWORTH, 1809, Lepid. Br. (2): 283, (*Geometra*). [Emendation of *wauaria* LINNAEUS].
v-nigraria HAWORTH, 1809, Lepid. Br. (2). 282, (*Geometra*). Holotype, [England]: London.
wauaria africana ZERNY, 1934, Z. öst. Ent. Ver. 19 (7/8): 52, pl. 5, figs 11, 12, (*Itame*). Syntypes 17 males 7 females, Morocco: Tachdirt, 2300-2700 m, (syntypes are distributed at least between The Natural History Museum, London, UK, Zoologische Forschungsinstitut und Museum Alexander Koenig, Bonn, Germany, Zoologische Staatssammlung, Munich, Germany and Naturhistorisches Museum, Vienna, Austria). We have examined two syntype males (London, Munich), including genitalia. See remarks.
carolata LUCAS, 1938, Bull. Soc. ent. Fr. 1938 (13/14): 181, (*Larentia*). Holotype male, Morocco: Ifrane.
wauaria chinensis STERNECK, 1928, Dt. ent. Z. Iris 42 (2): 236, (*Itame*). Syntypes 2 males, 2 females, China (west): Sungpanting. Valid at subspecies rank according to SCOBLE (1999).
wauaria koreaebia BRYK, 1949, Ark. Zool. 41A (1): 213, (*Itame*). Holotype male, Korea: Shuotsu. Valid at subspecies rank according to SCOBLE (1999).

UNAVAILABLE NAMES (incorrect subsequent spellings): *wavaria* (numerous references, for instance HULST 1896); *vavaria* (for instance PACKARD 1876). – Intrasubspecific names, potentially unavailable: *alba* PROUT, 1915 (ab.); *dealbaria* KOLOSSOW, 1936 (ab.); *f-signum* SILBERNAGEL, 1943 (f.); *fusca* LEMPKE, 1953 (f.); *grisearia* HEQVIST, 1950 (ab.); *lineata* LEMPKE, 1953 (f.); *marginata* LUCAS, 1959 (f.); *plana* LEMPKE, 1953 (f.); *reducta* LEMPKE, 1953 (f.); *unicolor* LUCAS, 1959 (f.); *v-remotum* SCHULTZ, 1931 (ab.); *v-solum* SCHULTZ, 1931 (ab.).

EXTERNAL CHARACTERS AND ABDOMEN: Wingspan 26–31 mm. Wings grey, irrorated with darker grey and black. Terminal area brown. Medial lines and subterminal band distinct on costa, forming blackish spots, elsewhere lines marked by small dots. Forewing medial line fused with discal spot, forming hook-shaped angle. Hindwings lighter, less irrorated, without markings, medial lines present as blackish dots on anal margin, terminal area brownish. Forewing discal spot long, large, hindwing discal spot minute. Terminal line blackish, fine, discontinuous. Fringes chequered grey and brown. Wings below darker, yellowish, heavily irrorated with brown, markings weaker. Frons, collar, thorax and abdomen concolorous with wings. Antennae concolorous with wings dorsally, bipectinate in male, filiform in female. Hindtibia with 2+2 spurs. Tympanal organs large, meeting ventrally. Posterior margin of male 8th sternite deeply invaginated, with two large, sclerotised extensions. Other sternites and tergites of both sexes undifferentiated.

VARIATION: Very little. Ground colour occasionally brown, medial lines can be absent apart from distinct spots on costa. Shape of the extensions in the posterior margin of male 8th sternite varies from acute to flat.

Subsp. ***africana*** more brownish, without hook-shaped angle below discal spot in forewing. Small, quantitative differences in the genitalia structures are explained in Remarks.

MALE GENITALIA: Uncus setose dorsally, apex concave, with two minute horns. Gnathos arms narrow, fused ventrally, forming weakly upturned, triangular element. Juxta large plate. Valva divided. Dorsal arm of valva long, narrow, weakly bent, costa and apex setose. Ventral arm of valva slightly elongated, apex round, weakly setose. Its dorsal margin base with sclerotised, setose ridge. Its ventral margin curved inwards, with several setae. Saccus wide, shallow. Aedeagus narrow, straight, caecum minute, subapical area with serrate ridge. Vesica opens apically, at approximately 90 degree angle, basal part weakly constricted, otherwise straight membranous tube. Basal part with scale-like sclerotised plate. Genitalia of *M. wauaria* illustrated as enlarged in the first male genitalia plate (explanation of terminology).

FEMALE GENITALIA: Papillae anales distinctly elongated, setose. Apophyses straight, narrow. Lamella antevaginalis membranous. Lamella postvaginalis small plate, posterior margin irregular, lateral appendices small. Ductus bursae narrow, short, sclerotised, more sclerotised laterally. Posterior end of corpus bursae partly sclerotised, long, narrow, striated. Anterior part of corpus bursae large, round. Signum large, densely stellate.

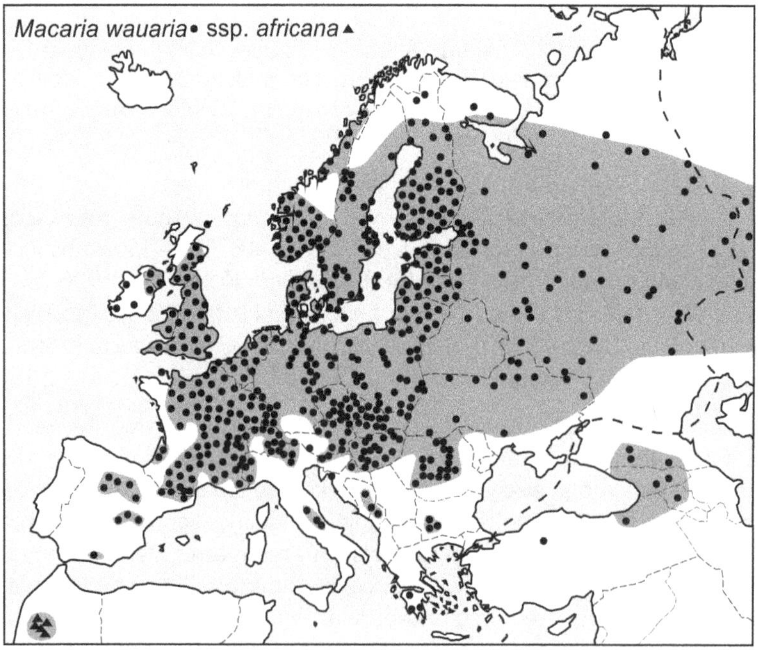

DISTRIBUTION AND ABUNDANCE: Eurasiatic (introduced in North America). In Europe from northern Fennoscandia to Spain, France, Italy and northern Balkans and from the British Isles through central Europe and European Russia to the Ural mountains. – Outside Europe in the Atlas mountains in Morocco (ssp. *africana*) and in Turkey, Transcaucasus, Caucasus, central Asia, Siberia, Far East Russia, Kamchatka and Sakhalin island. The species was accidentally introduced to North America rather early, as by the 1860's it was already common and widespread in the eastern part of North America. Nowadays it is more restricted, but still common for instance in Quebec, Ottawa and Ontario provinces. – In Europe this species is in general common but occurs rarely in larger numbers. In Britain it has in recent decades decreased by 99% of its previous abundance and it is now scarce (FOX *et al.* 2013).

PHENOLOGY: Univoltine. The moths fly in the whole of Europe from June to August, with relatively little variation depending of latitude. Larva from April to June. The species overwinters as an egg. The adult insect is not often seen by day, it flies at night and comes readily to light.

BIOLOGY: Oligophagous. The main food plant of the larva is probably *Ribes uva-crispa*, on which among others EBERT (2003) mentions it to be found, but it has also been found on other *Ribes* species such as *R. nigrum* (B. MÜLLER pers. comm.), *R. rubrum* (H. WEGNER pers. comm.) and *R. alpinum* (EBERT 2003). LHOMME (1935) mentions *Prunus spinosa* as food plant, contrary of all other sources, but this is probably incorrect. The pupation takes place in the ground

HABITAT: Forest fringes and clearings, gardens and parks, often on slightly moist ground. In Scandinavia the species is missing in the mountains, but in southern Europe it has been found up to 2000 m in Sierra Nevada in Spain (coll. P. SKOU), and up to 2200 m in Italy (FLAMIGNI *et al.* 2007).

SIMILAR SPECIES: No similar species in Europe.

GENETIC DATA: Genetically homogeneous in Europe and Transcaucasia (n=18, from four European countries), but in Bosnia diverging by 4.9%, exactly matching the haplotype found in a specimen from Mongolia. One specimen from southern Germany at a distance of 8.4%, awaiting further investigation. Subsp. *africana* (ZERNY, 1934) from North Africa diverging by 3.9%. Nearest species: North American *M. bitacta epigenata* (BARNES & MCDUNNOUGH, 1917) (3.6% from subsp. *africana*) and North American *M. quadrilinearia* (PACKARD, 1873) (5.2% from nominate subspecies). Nearest species in Europe: *M. fusca* (5.7%). See remarks.

REMARKS: LERAUT (2009) revised the status of *africana* (ZERNY, 1934), and unlike elsewhere in the recent literature (for instance SCOBLE 1999, KRÜGER 2001, REDONDO *et al.* 2009), he raised it from a subspecies of *wauria* and considered it valid at species level. He did not indicate if any type material was examined, and although a hand-drawn picture of the male genitalia was illustrated, the diagnostic features were not discussed. We have examined two syntype males of *africana* (ZERNY), including the genitalia, and compared them against numerous European specimens. We conclude that forewing discal spot and antemedial line form a hook-shape in the European specimens, hook is weak or absent in *africana* (ZERNY). The apex of ventral arm of valva is on average more rounded and longer in European specimens, apex is slightly angled and shorter in *africana* (ZERNY). Base of vesica is weakly constricted in the European specimens (the character may appear different if vesica is not fully inflated) and not in *africana* (ZERNY). Posterior margin of male 8th sternite is deeply invaginated, with two large, sclerotised extensions. Shape of the extensions varies in European material, they can be acute (see genitalia plate) or flat as in *africana* (ZERNY) (see genitalia plate). Signum is larger in the European specimens. Because *africana* has small quantitative differences and population is allopatric, we consider tentatively the taxon *africana* (ZERNY) valid at subspecies level. KRÜGER (2001) considered *africana* probably not to represent more than a local race. See genetic data. More research is needed.

RUNGS (1981) proposed *Larentia carolata* LUCAS, 1938, to be a junior synonym of *Macaria wauaria* subsp. *africana* (ZERNY, 1934). KRÜGER (2001) followed the synonymy. The holotype male is from Morocco, Ifrane, presumably in the Muséum National d'Histoire Naturelle, Paris, France. We have not examined it.

There are several closely related species in *Macaria wauaria* group in North America (FERGUSON 2008). FERGUSON (2008) classified *M. wauaria* in genus *Speranza* CURTIS, 1828, together with numerous other North American taxa, whereas SCOBLE (1999) considered *Speranza* a junior synonym of *Macaria* CURTIS.

40. *Macaria artesiaria* (DENIS & SCHIFFERMÜLLER, 1775)

artesiaria [DENIS & SCHIFFERMÜLLER], 1775, Ankündung syst. Werkes Schmett. Wienergegend: 102, (*Geometra*). Syntype(s), lost, Central Europe [surroundings of Austria: Vienna?].

festucaria HÜBNER, 1790, Beitr. Gesch. Schmett. 2 (2): 55, pl. (2)4, fig. X, (*Ph[aleana] Geom[etra]*). Syntype(s) female(s), lost, Austria.

EXTERNAL CHARACTERS AND ABDOMEN: Wingspan 23–27 mm. Wings grey, antemedial and medial lines weak, sometimes enlarged on costa, postmedial line whitish, rather straight, its outer margin often dark. Forewing medial area with yellowish, wedge-shaped area, bordered by yellowish veins. Subterminal band brown on costa, yellowish brown in middle, otherwise reduced. Terminal area often brown, darker than rest of wing. Forewing discal spots oval, brown, large. Hindwings rather uniform grey, postmedial line whitish, terminal area darker. Hindwing discal spots minute. Fringes concolorous with wings. Forewing margin weakly concave below apex. Terminal line weakly yellowish or concolorous with wings. Wings below uniform grey-brown, striated with grey, particularly hindwings. Frons, collar, thorax and abdomen concolorous with wings. Antennae concolorous with wings dorsally, bipectinate in male, filiform in female. Hindtibia with 2+2 spurs. Tympanal organs large. Posterior margin of male 8th sternite deeply invaginated. Other sternites and tergites of both sexes undifferentiated.

VARIATION: Rather little. Forewings mostly rather uniform grey, occasionally irrorated with brown or wings more uniform brown. Forewing discal spots may be absent. Typically the yellowish, wedge-shaped area bordered by yellowish veins, but other veins may be yellowish also.

MALE GENITALIA: Uncus setose dorsally, apex concave, with two minute horns. Gnathos arms narrow, fused ventrally, forming weakly upturned, triangular element. Juxta large plate. Valva divided. Dorsal arm of valva long, narrow, straight, costa and apex setose. Ventral arm of valva slightly elongated, apex round, weakly setose. Its dorsal margin base with sclerotised, setose ridge. Its ventral margin curved inwards, with several setae. Saccus wide, shallow. Aedeagus narrow, straight, caecum minute, subapical area with serrate ridge. Vesica opens apically, at approximately 90 degree angle, basal part weakly constricted, otherwise weakly bent, membranous tube. Basal part with scale-like sclerotised plate.

FEMALE GENITALIA: Papillae anales distinctly elongated, setose. 8th and 9th segments elongated. Apophyses straight, narrow. Lamella antevaginalis membranous. Lamella postvaginalis small plate, posterior margin round, lateral appendices small. Ductus bursae narrow, short, sclerotised, more sclerotised laterally. Base of ductus seminalis weakly expanded. Posterior end of corpus bursae partly sclerotised, long, narrow, striated. Anterior part of corpus bursae large, oval. Signum large, sparsely stellate.

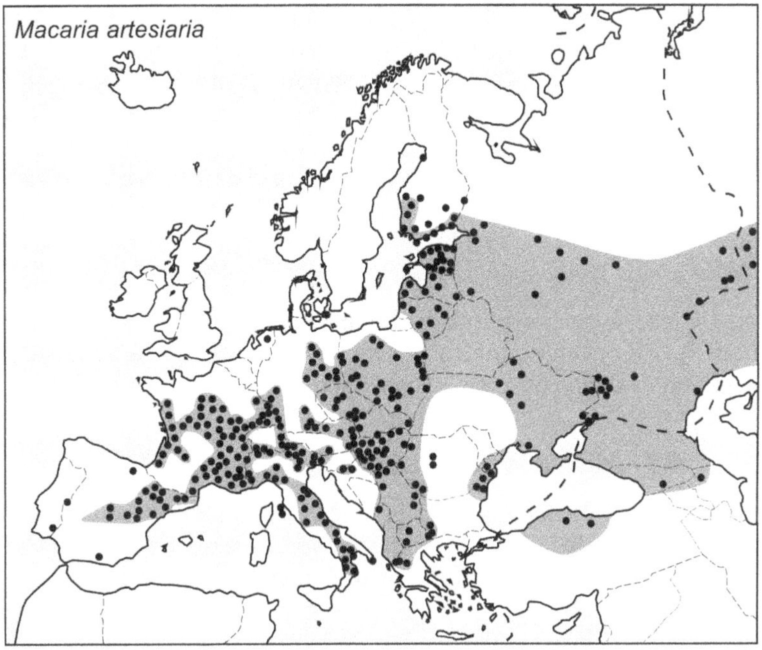

DISTRIBUTION AND ABUNDANCE: Eurasiatic. In Europe from central Finland through the Baltic countries to the Mediterranean from the Iberian Peninsula and France eastwards through southern and central Europe, parts of the Balkans, Ukraine and European Russia to the Urals. – Outside Europe in Turkey, Transcaucasus, Kazakhstan, southern parts of Siberia and Far East Russia. – In Europe less common and somewhat local and it usually occurs in limited numbers.

PHENOLOGY: Uni- or bivoltine. In Finland in one generation from mid-July to mid-August, in Poland early June to late July and from mid-August to mid-September (MALKIEWICZ 2012), and in south-western Germany from late May to late July and from August to late September, hardly with a gap between the two generations (EBERT 2003), and in Italy in two overlapping generations from late April to early October where the first generation ends and the second starts in early July (FLAMIGNI *et al.* 2007). Larva from late May to late June (BERGMANN 1955). Hibernation takes place as an egg. The imago flies at night and comes to light. In Finland most specimens are found as caterpillars.

BIOLOGY: Monophagous.The caterpillar has been found on *Salix* (EBERT 2003), in Finland found on *S. repens* and *S. phylicifolia*, in northern Europe in the spring and early summer, further to the south also later in the summer until about August. In rearings, the caterpillars have accepted narrow leaved *Salix* species as food plant (B. MÜLLER pers. comm.). Pupation takes place between spun leaves. A rearing of the species is described in detail by E. HAEGER (1965).

HABITAT: Willow scrubs and moist meadows with willow bushes, often along rivers. According to EBERT (2003) the species prefers warm localities. In northern Europe only in the lowlands, in Italy up to 1650 m (FLAMIGNI *et al.* 2007) and in Spain up to 1500 m (REDONDO *et al.* 2009).

SIMILAR SPECIES: *Macaria ichnusae* GOVI & FIUMI is smaller, forewing postmedial line is faint, discontinuous, postmedial band is wide, dark brown, fringes are chequered grey and brown and hindwings are brown (larger, forewing postmedial line whitish, rather straight, postmedial band barely noticeable, fringes concolorous with wings and hindwings are grey in *M. artesiaria*). Uncus margin is rounded and ventral arm of valva is narrower (uncus margin weakly concave and ventral arm of valva is wider in *M. artesiaria*). Female genitalia are without notable differences. According to current knowledge *M. ichnusae* occurs in Sardinia only, where *M. artesiaria* is not known to occur.

GENETIC DATA: Genetically somewhat heterogeneous from Europe and Turkey to Mongolia (n=18, from six European countries), maximum variation 1.8%. One specimen from north-eastern China diverging by 2.2%. Nearest species: *M. ichnusae* (4.6%).

41. *Macaria ichnusae* GOVI & FIUMI, 2005

ichnusae GOVI & FIUMI, 2005, Quaderno di Studi e Notizie di Storia Naturale della Romagna 21: 130, (*Macaria*). Holotype male (private collections of G. FIUMI and G. GOVI, Forli, Italy), Italy: Sardinia, Cala Liberotto (Rio Sos Alinos).

EXTERNAL CHARACTERS AND VARIATION: Wingspan 19–22 mm. Males strongly contrasting, wings brown, striated with dark brown. Antemedial and medial lines weak, stronger on costa. Subterminal band dark brown, inner and outer margins whitish, incomplete, yellowish area in middle weak, bordered by yellowish veins. Subterminal line blackish on costa. Females less contrasting, grey-brown, weakly irrorated with dark brown. Postmedial band brown, margins not whitish, yellowish area in middle weak, bordered by yellowish veins. Terminal line faint. Fringes chequered brown and dark brown. Hindwings in both sexes uniform brown, terminal area darker. Forewing discal spots brown, elongated, hindwing discal spots minute. Forewing margin weakly concave below apex. Wings below brown, hindwing basal and medial areas striated with brown, terminal area darker. Frons, collar, thorax and abdomen concolorous with wings. Antennae concolorous with wings dorsally, bipectinate in male, weakly fasciculate in female. Hindtibia with 2+2 spurs. Tympanal organs large. Posterior margin of male 8th sternite deeply invaginated, with two large, sclerotised extensions. Other sternites and tergites of both sexes undifferentiated.

VARIATION: Rather little. Subterminal band normally expanded on forewing costa to blackish spot but it may be absent. The yellowish area in forewing subterminal band may be distinct (illustrated in FIUMI *et al.* 2009) or weak (see colour plates).

MALE GENITALIA: Generally as in *M. artesiaria* (see entry above), with following minor differences: Uncus margin is rounded and ventral arm of valva is narrower (uncus margin weakly concave and ventral arm of valva is wider in *M. artesiaria*).

FEMALE GENITALIA: Generally as in *M. artesiaria* (see entry above), without notable differences.

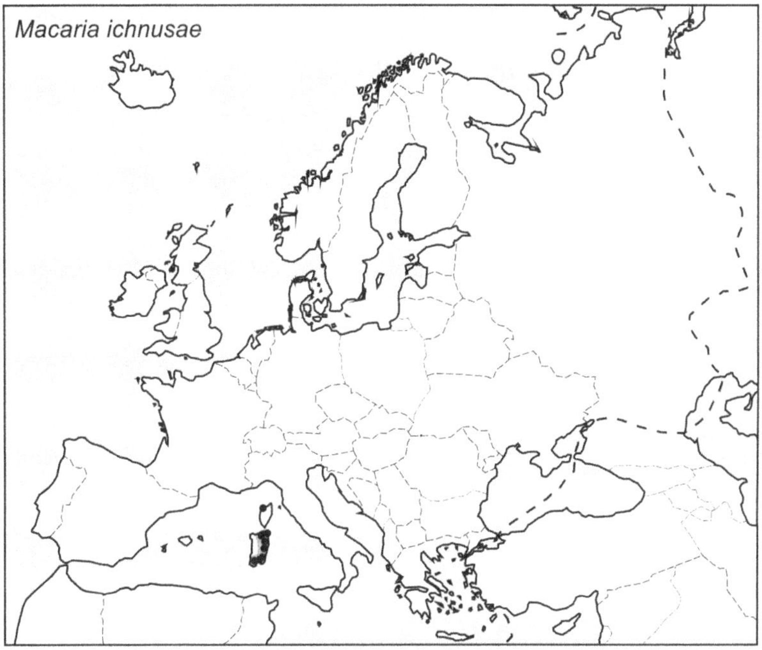

DISTRIBUTION AND ABUNDANCE: Tyrrhenian. Endemic to Sardinia and Corsica where it is local and not common, but it has on one occasion been found in numbers. The record further to the north in Corsica could indicate, that the species may occur in many suitable places also in this island. We estimate that there have so far been found 60–70 specimens in Sardinia, in Corsica only one specimen (GROS 2014). The future may very well show that *M. ichnusae* is more common than we presently know.

PHENOLOGY: Probably bivoltine. So far the adult moths have been found from early June to late July, in northern Corsica one specimen in early August (GROS 2014). From 30 specimens found in the first days of July only 8 were males and many specimens were rather worn. This could indicate that the specimens were found close to the end of the flying time. Perhaps the species has a first generation flying until early or mid-July and a second generation starting to fly in late July. More records are necessary before the flying time can be defined precisely. Larvae have been found in late May and early June (M. LEIPNITZ pers. comm.). The moths fly at night and come readily to light.

BIOLOGY: Monophagous. Larvae have been found on *Salix triandra*, *S. viminalis* and *S. purpurea* (M. LEIPNITZ pers. comm.). In rearings the caterpillars have accepted leaves of other *Salix* species.The biology of this recently described species, including illustrations of the immature stages, is given by FIUMI et al. (2009).

HABITAT: So far the species has been found from sea level up to 650 m in Mediterranean macchia landscapes along water-courses with riparian hygrophilous vegetation (FLAMIGNI *et al.* 2007). The species has been found abundantly along a small river at 140 m altitude where *Salix* bushes were growing in large numbers.

SIMILAR SPECIES: *Macaria artesiaria* (DENIS & SCHIFFERMÜLLER), see entry above.

GENETIC DATA: Nearest species: *M. artesiaria* (4.6%).

42. *Macaria carbonaria* (CLERCK, 1759)

carbonaria carbonaria CLERCK, 1759, Icones Insectorum Rariorum 1: pl. 1, fig. 11, ([*Phalaena*]). Paratype male (The Natural History Museum, London, UK), [Europe], [Sweden] (examined externally).

amnicularia ZETTERSTEDT, 1840, Insecta Lapponica: 957, (*Fidonia*). Holotype female, Southern Lapponia ['Lappland' (Lapponia Umensi?)].

carbonaria fuliginaria DANNEHL, 1928, Ent. Z., Frankf. a.M. 41 (20): 405, (*Fidonia*). Syntypes 23 males, 6 females, South Tyrol: Penegal, Vigiljoch, Ultental. Originally as var. See Variation.

carbonaria valesiaca WEHRLI, 1940, in SEITZ, Gross-Schmett. Erde 4 (Suppl.): 394, pl. 31: i (*Isturgia*). Syntype(s), Alps: Davos, 1600–2000 m. [Originally as var].

carbonaria wolfiaria WARNECKE, 1939, Mitt. münch. ent. Ges. 29 (2/3): 388, (*Isturgia*). Syntype(s), [Germany]: Schleswig Holstein. [Originally as f].

carbonarius MEYRICK, 1895, A handbook of British Lepidoptera: 266, (*Bupalus*). [Emendation of *carbonaria* CLERCK].

picearia GEYER, [1831], Samml. Eur. Schmett. 5 Geometrae (3): pl. 106, figs 552–555, (*Geometra*). Syntype(s), Europe.

roseidaria HÜBNER [1799], Samml. Eur. Schmett. 5 Geometrae (1): pl. 24, fig. 128, (*Geometra*). Syntype(s), lost, Europe.

UNAVAILABLE NAMES (incorrect subsequent spelling): *roscidaria* HÜBNER [incorrect subsequent spelling of *roseidaria* HÜBNER] (WEHRLI 1939–1954). – Intrasubspecific names, potentially unavailable: *atra* BRANDT, 1933 (ab.); *fasciaria* BURMANN, 1950 (ab.); *fumigata* STAUDER, 1928 (f.).

EXTERNAL CHARACTERS AND ABDOMEN: Wingspan 18–22 mm. Wings white, irrorated with blackish. Transverse lines and subterminal line blackish, wide, angled, also terminal area and veins blackish, creating an overall black and white appearance. Females tend to be lighter. Terminal line blackish. Fringes chequered black-and-white. Discal spots blackish. Wings below as above, lighter, occasionally yellowish. Frons, collar, thorax and abdomen concolorous with wings. Antennae chequered black and white dorsally, bipectinate in male, filiform in female. Hindtibia with 2+2 spurs. Tympanal organs very large, meeting ventrally. Posterior margin of male 8th sternite concave, with two minute, sclerotised extensions. Other sternites and tergites of both sexes undifferentiated.

VARIATION: Rather little. Ground colour varies from white to yellowish, particularly underside may be yellowish. Also frons may be yellowish. Density of irroration and width of blackish markings varies, occasionally specimens almost uniformly blackish. DANNEHL (1928) described var. *fuliginaria* from South Tyrol (Italy), the specimens being black-brown, and only outer margin of the postmedial line is white. We consider such specimens to fall within intraspecific variation, but SCOBLE (1999) considered *fuliginaria* valid at subspecies level. Several forms have been described, see for instance WEHRLI (1939–1954).

MALE GENITALIA: Uncus setose dorsally, apex concave, with two minute horns. Gnathos arms narrow, fused ventrally, forming weakly upturned, triangular element. Juxta small plate. Valva divided. Dorsal arm of valva long, narrow, curved ventrally, costa and apex weakly setose. Ventral arm of valva very wide, apex weakly angled, weakly setose. Its dorsal margin base with sclerotised, weakly setose ridge. Saccus wide, round. Aedeagus stout, straight, caecum absent, subapical area with short, narrow, weakly serrate lateral extension, apex round. Vesica opens apically at approximately 90 degree angle, basal part striated, otherwise straight membranous tube. Sclerotisations absent.

FEMALE GENITALIA: Papillae anales elongated, setose. 8th and 9th segments elongated. Apophyses straight, narrow. Lamella antevaginalis membranous, margin concave. Lamella postvaginalis membranous, striated horizontally. Antrum bowl-shaped. Ductus bursae narrow, short, more sclerotised laterally. Base of ductus seminalis weakly expanded. Posterior end of corpus bursae membranous, long, narrow, striated. Anterior part of corpus bursae wider, oval. Signum large, stellate, arms short.

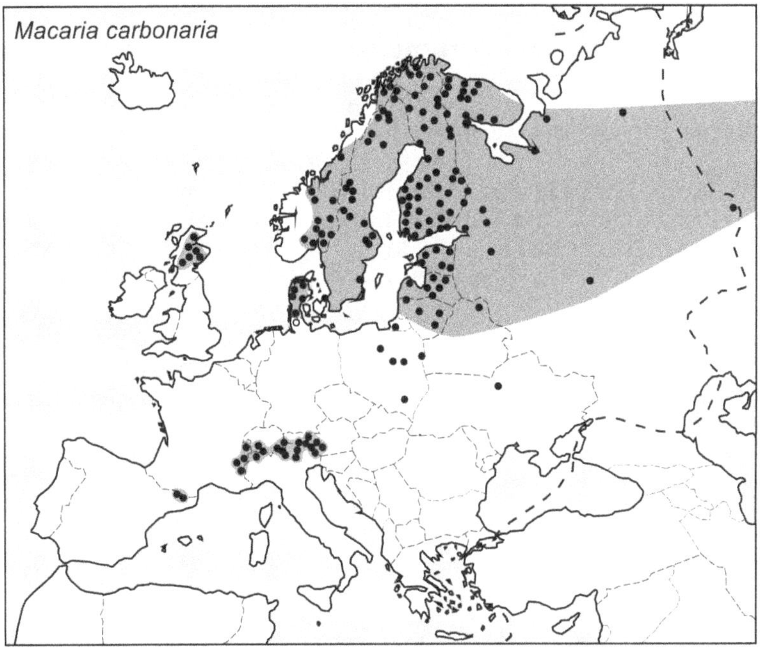

DISTRIBUTION AND ABUNDANCE: Eurasiatic. In Europe boreo-montane occurring in Scotland, Fennoscandia and in the north European lowlands across northern European Russia to the Ural mountains, further south only in the mountains such as the Alps and the Pyrenees. – Outside Europe through northern Siberia to Yakutia. – In Europe a local species that rather often appears in numbers at its localities.

PHENOLOGY: Univoltine. In northern Europe from early May to early June, in Poland from late April to late May (MALKIEWICZ 2012) and in the Italian Alps from March to June, depending on altitude. The pupa overwinters. The moths fly only by daytime when they fly rapidly in the sunshine just above the vegetation.

BIOLOGY: Oligophagous. The caterpillar lives from May or June to September on a number of food plants, in Denmark it was found on *Arctostaphylos uva-ursi* (SØNDERUP 1901), in Finland primarily on *Andromea polifolia* (O. PELTONEN pers. comm.) and on *Empetrum* (HULDÉN *et al.* 2000). BLESZYNSKI (1966) records it on *Salix* and *Betula*.

HABITAT: In Denmark exclusively on moors, in Fennoscandia also in marshes, open moorland spruce forests and in the mountain region on both dry and wet soils. In the Alps on slopes with rocky grassland. In northern Europe from sea level up to about 1200 m, in the Alps from 1800 m (ROBINEAU *et al.* 2007) up to 2300 m (P. HUEMER pers. comm.).

SIMILAR SPECIES: No similar species in Europe, but compare with *Chiasmia clathrata* (LINNAEUS), *Narraga* WALKER species, *Ematurga atomaria* (LINNAEUS) and *Heliomata glarearia* (DENIS & SCHIFFERMÜLLER). In North America *Macaria truncataria* (WALKER, 1862) is similar, sharing with *M. carbonaria* similar food plants, habitat and flight time (FERGUSON 2008).

GENETIC DATA: Genetically homogeneous in Europe (n=7, from four countries). Nearest species: *M. fusca* (6.3%).

43. *Macaria brunneata* (THUNBERG, 1784)

brunneata brunneata THUNBERG, 1784, Diss. ent. sistens Insecta Suecia 1: 9, (*Geometra*). Lectotype male (Universitets Zoologiska Institut, Uppsala, Sweden), Sweden: Uppland, Uppsala.

atomaria FABRICIUS, 1794, Ent. Syst. 3 (2): 144, (*Phalaena*). Syntype(s) (lost), [Europe].

ferruginaria PACKARD, 1873, 5th Ann. Rep. Trustees Peab. Acad. Sci.: 78, (*Eupistheria*). Syntype(s) male(s), [USA]: Maine.

flammata FABRICIUS, 1794, Ent. Syst. 3 (2): 194, (*Phalaena*). Syntype(s) (lost), Italy: province Torino; Piemonte.

flammataria ZETTERSTEDT, 1839, Insecta Lapponica, Sectio Quarta, Lepidoptera: 957, (Fidonia). Syntype(s), [Sweden/Finland]: Lapponia Tornensis, Karesuvanto.

fulvaria VILLERS, 1789, Linn. ent. 2: 329, (*Phal*[*aena*] *Geom*[*etra*]. Syntype(s), Gallia [southern France].

fulvaria coreame BRYK, 1949, Ark. Zool. 41A (1): 213, (*Itame*). Holotype male, Korea: Motojondo.

pinetaria HÜBNER, [1799], Samml. Eur. Schmett. 5 Geometrae (1): pl. 24, fig. 130, (*Geometra*). Syntype(s), lost, Europe.

quinquaria HÜBNER, [1822], Samml. Eur. Schmett. 5 Geometrae (1): pl. 100, figs. 516, 517, (*Geometra*). Syntype(s), lost, Europe.

sylvaria CURTIS, 1828, Br. Ent. 5: folio 225, pl. 225, (*Speranza*). Syntypes many males, 1 female, Scotland: Loch Rannoch, Black Wood.

brunneata imperatoria KARDAKOFF, 1928, Ent. Mitt. 17 (6): 423, pl. 9, fig. 24, (*Itame*). (Syn) type male (The Natural History Museum, London, UK), [Russia]: Ussuri, Imperatorhafen. [Originally as subsp. of *fulvaria* VILLERS]. Valid at subspecies rank according to SCOBLE (1999) (examined externally).

brunneata sordida BUTLER, 1881, Trans. ent. Soc. Lond. 1881 (3): 418, (*Cleogene*). Syntype(s) (The Natural History Museum, London, UK), Japan: Kamakura: Tokei (Kanagawa Prefecture). Originally as sp., valid at subspecies rank according to SCOBLE (1999) (examined externally).

INTRASUBSPECIFIC NAMES, POTENTIALLY UNAVAILABLE: *atomarioides* NORDSTRÖM, 1941 (ab.?); *fusca* LEMPKE, 1959 (f.); *obscurior* KLEMENSIEWICZ, 1913 (var.); *uniformis* HOFFMANN, 1919 (ab.?); *unicinctata* STRAND, 1901 (ab.?); *unilineata* LEMPKE, 1953 (f.).

EXTERNAL CHARACTERS AND ABDOMEN: Wingspan 20–27 mm. Wings rusty brown, heavily irrorated with brown. Males rather uniform rusty brown, medial lines weakly developed, except costa, where present as brown spots, terminal area often darkest. Females lighter, more yellowish, medial lines distinct, brown, often suffused. Terminal line brown, interrupted, more developed in females. Fringes concolorous with wings in male, chequered brown and rusty brown in females. Those of females often strongly chequered brown and rusty yellow. Discal spots small. Wings coloured below as above, males often more ochreous, postmedial line often most distinct. Frons, collar, thorax and abdomen concolorous with wings. Antennae concolorous with wings dorsally, bipectinate in male, weakly fasciculate in female. Hindtibia with 2+2 spurs. Tympanal organs large. Posterior margin of male 8th sternite invaginated, with two sclerotised extensions. Other sternites and tergites of both sexes undifferentiated.

VARIATION: Very little, specimens are rather uniform throughout the distribution area. Males occasionally lighter and medial lines more visible. Females occasionally darker and medial lines less visible. Also North American specimens are similar (FERGUSON 2008).

MALE GENITALIA: Uncus setose dorsally, apex concave, with two minute horns. Gnathos arms narrow, fused ventrally, forming weakly upturned, triangular element. Juxta small plate. Valva divided. Dorsal arm of valva long, narrow, straight, costa and apex setose. Ventral arm of valva elongated, apex narrow. Its dorsal margin base with small lobe. Its ventral margin curved inwards, convex, setose. Saccus wide, shallow. Aedeagus stout, straight, caecum absent, apex round. Vesica opens apically, at approximately 45 degree angle, basal part weakly constricted, otherwise straight, membranous tube. Basal part with scale-like sclerotised plate.

FEMALE GENITALIA: Papillae anales distinctly elongated, setose. 8th and 9th segments elongated. Apophyses straight, narrow. Lamella antevaginalis membranous. Lamella postvaginalis small plate, posterior margin round, lateral appendices minute. Antrum wide. Ductus bursae narrow, short, more sclerotised laterally. Base of ductus seminalis weakly expanded. Posterior end of corpus bursae weakly sclerotised, long, narrow, weakly striated. Anterior part of corpus bursae large, round. Signum large, stellate, located laterally.

DISTRIBUTION AND ABUNDANCE: Holarctic. In Europe widespread from northernmost Fennoscandia to the Pyrenees, the Alps and northern Balkan and from Great Britain through central and eastern Europe, Ukraine and European Russia to the Urals. – Outside Europe further to the east through Siberia, Mongolia, Korea, Far East Russia, to the Magadan district and Japan. The species also occurs in North America. – In Europe a local species that rather often appears in numbers where it occurs. In Denmark, where the species occurs rather locally, migrations are recorded and often with large numbers of specimens suddenly showing up outside the habitats.

PHENOLOGY: Univoltine. In Denmark (own data) and in the Pyrenees from early July to early August (REDONDO *et al.* 2009) and in the Italian Alps from May to August (FLAMIGNI *et al.* 2007). Larva from April to June. Hibernation takes place as an egg. The adult moths fly easily up by daytime, perhaps they are even dayflying, but they are also readily attracted to light.

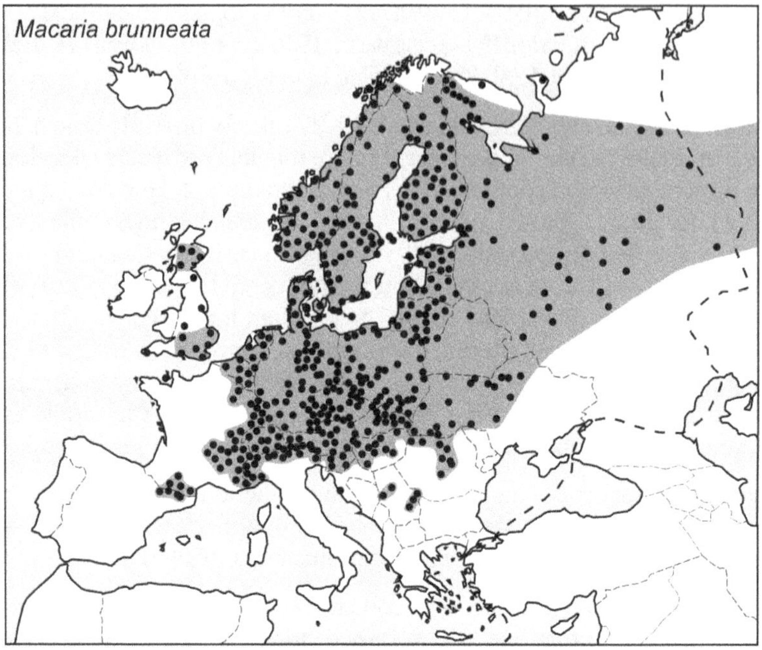

BIOLOGY: Oligophagous. The larva has been found on *Vaccinium* and *Ledum* (HULDÉN *et al.* 2000), on *V. myrtillus* (E. FRIEDRICH & H. WEGNER pers. comm.) and on *V. uliginosum* (E. FRIEDRICH pers. comm.). The pupation takes place in the ground.

HABITAT: In open forests, mountain slopes and other places with acidic soil. In Scandinavia the species occurs from sea level up to 500–600 m (N. RYRHOLM pers. comm.). In the Alps from 600 m up to 2700 m (FLAMIGNI *et al.* 2007) and in the Pyrenees from 1700 m up to 2300 m (REDONDO *et al.* 2009).

SIMILAR SPECIES: No similar species in Europe.

GENETIC DATA: Genetically homogeneous in Europe (n=12, from five countries), Canadian populations diverging by 2.5%. Nearest species: *M. wauaria africana* (7.1%; *M. wauaria wauaria* 7.4%).

REMARKS: *Macaria brunneata* has been reported to exhibit migrant behaviour (for instance HURST 1964, RYRHOLM 1994), most recently to southern England in 2009 (HIGGOTT & DAVEY 2010) and 2010–2011 (ANONYMOUS 2011b), where several thousand specimens were observed. FERGUSON (2008) classified *M. brunneata* in genus *Speranza* CURTIS, 1828, together with numerous other North American taxa, whereas SCOBLE (1999) considered *Speranza* a junior synonym of *Macaria* CURTIS.

44. *Macaria loricaria* (EVERSMANN, 1837)

loricaria EVERSMANN, 1837, Bull. Soc. imp. Nat. Moscou 1837 (6): 59, (*Fidonia*). Syntype(s), [Russia]: Kasan [Kazan].

julia HULST, 1896, Trans. Am. ent. Soc. 23 (3): 338, (*Sympherta*). Lectotype male, (National Museum of Natural History, Smithsonian Institution, Washington DC, U.S.A.), [U.S.A].:

Washington, Easton [Kitttitat County] [Originally as sp].. Lectotype designated and synonymy proposed in FERGUSON (2008).

vinctaria LIENIG & ZELLER, 1846, Isis 30: 203, (*Idaea*). Syntype(s), Lievland [Latvia]; Curland [Latvia]. [We follow GAEDIKE (2014) and consider the author to be LIENIG & ZELLER].

nubilata WARREN 1904, Novit. zool. 11 (2): 563, (*Thamnonoma*). Holotype male (The Natural History Museum, London, UK), [U.S.A].: Colorado, South Park (examined externally).

EXTERNAL CHARACTERS AND ABDOMEN: Male: wingspan 23–29 mm. Males grey, irrorated with brown, terminal area darker. Antemedial and medial lines suffused, grey-brown, weak, strongest on costa. Postmedial line wide, grey, often with minute black dots. Subterminal band grey-brown, bordered by undulating, rusty brown subterminal line, often with several oval, brown spots. Terminal line discontinuous, with distinct black dots. Fringes concolorous with wings. Hindwings lighter, weakly irrorated with grey, terminal area often darker, postmedial line grey. Forewing discal spots large, elongated, grey-brown, hindwing discal spots minute. Forewing margin below apex round. Wings below yellowish grey, irrorated with grey, subterminal line and discal spots suffused. Female: wingspan 11–12 mm. Females brachypterous, wings narrow, elongated, grey-brown, irrorated with brown. Antemedial line and postmedial band brown. Terminal area dark brown. Hindwings grey-brown, with brown postmedial line. Discal spots minute. Wings below as above, more striated. Female abdomen very large (Text-figs 71–72).

Frons, collar, thorax and abdomen of both sexes concolorous with wings. Antennae bipectinate in male, filiform in female. Hindtibia with 2+2 spurs. Tympanal organs medium-sized. Posterior margin of male 8th sternite invaginated, with two sclerotised extensions. Other sternites and tergites of both sexes undifferentiated.

Text-figs 71–72. The short-winged female of *Macaria loricaria* (EVERSMANN, 1837). When the female climbs up, it keeps the wings vertically upwards, and when it sits still, the wings are held more or less in horizontal position (Finland: Tampere, end of June, 1992). Photos: MICHAEL LEIPNITZ.

VARIATION: Very little, specimens are rather uniform throughout the distribution area, including North America (FERGUSON 2008).

MALE GENITALIA: Uncus setose dorsally, with two medium-sized horns. Gnathos arms narrow, fused ventrally, forming weakly upturned, triangular element. Juxta small plate. Valva divided. Dorsal arm of valva long, narrow, straight, costa and apex setose. Ventral arm of valva wide, apex round, weakly setose. Its dorsal margin base with sclerotised, setose ridge. Its ventral margin weakly curved inwards, with several setae. Saccus wide, elongated. Aedeagus stout, straight, caecum absent, subapical area with acute tooth, apex round. Vesica opens apically, at approximately 90 degree angle, basal part weakly constricted, otherwise straight membranous tube. Basal part with scale-like sclerotised plates, one of them enlarged to cornutus-like sclerotisation.

FEMALE GENITALIA: Papillae anales large, distinctly elongated, setose. 8th and 9th segments elongated. Apophyses straight, narrow. Margin of lamella antevaginalis sclerotised, otherwise membranous. Lamella postvaginalis small plate, posterior margin round, lateral appendices reduced. Ductus bursae short, sclerotised tube. Base of ductus seminalis weakly expanded. Posterior end of corpus bursae wide, densely scobinate, membranous apart from sclerotised ridges. Anterior part of corpus bursae large, oval, membranous. Signum densely stellate.

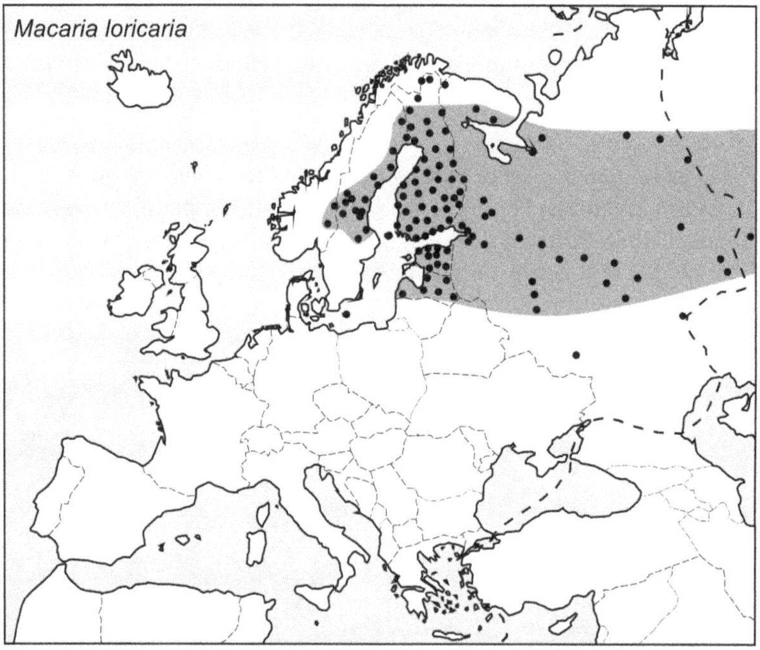

DISTRIBUTION AND ABUNDANCE: Holarctic. In Europe only in Fennoscandia, the Baltic countries and eastwards through European Russia to the Ural mountains. In Denmark a single migrant male on the island Bornholm. – Outside Europe further to the east through Siberia, the northern Far East Russia to Sakhalin. Also in North America (subsp. *julia*). – In Europe a relatively common species that mainly occurs in limited numbers.

PHENOLOGY: Univoltine. The adult moths can be found from mid-July to early August, in some years somewhat earlier. The species overwinters as an egg. The males fly at night and come to light, while the female is very rarely found, most females in collections have been reared from caterpillars.

BIOLOGY: Oligophagous. The caterpillar lives from late May until early July. It has been found on *Salix* and *Betula* (HULDÉN *et al.* 2000). The pupation takes place in the ground.

HABITAT: On moist meadows and forest edges. The species occurs mainly in the lowlands and it has been found up to 450 m (N. RYRHOLM pers. comm.).

SIMILAR SPECIES: No similar species in Europe.

GENETIC DATA: Nearest species: *M. artesiaria* (6.9%).

REMARKS: FERGUSON (2008) designated a lectotype for the North American *Sympherta julia* (HULST, 1896) and downgraded it from a subspecies of *Fidonia loricaria* (EVERSMANN, 1837) to a junior synonym. FERGUSON (2008) classified *M. loricaria* in genus *Speranza* CURTIS, 1828, together with numerous other North American taxa, whereas SCOBLE (1999) considered *Speranza* a junior synonym of *Macaria* CURTIS.

45. *Macaria fusca* (THUNBERG, 1792)

fusca THUNBERG, 1792, Diss. ent. sistens Insecta Suecica 4: 53, pl. 4, fig. 2, (*Bombyx*). Lectotype male (Universitets Zoologiska Institut, Uppsala, Sweden), Sweden: 'Lappland'.

canitiaria FREYER, 1830, Beitr. Gesch. Eur. Schmett. 3: 105, pl. 125, fig. 4, (*Psodos Geom.* [*etra*]). Syntype(s), France: Hautes-Alpes.

venetaria HÜBNER, [1809], Samml. Eur. Schmett. 5: Geometrae (1): pl. 64, fig. 329, (*Geometra*). Syntype(s), lost, Europe.

INFRASUBSPECIFIC NAMES, POTENTIALLY UNAVAILABLE: *destrigata* STRAND, 1904 (ab.); *unistrigata* STRAND, 1904 (ab.).

EXTERNAL CHARACTERS AND ABDOMEN: Male: wingspan 18–20 mm. Males uniform greyish dark brown, only antemedial line and postmedial line visible, suffused blackish-brown. Terminal line dark brown, fringes concolorous with wings. Discal spots brown, suffused. Wings below as above. Female: wingspan 15–18 mm. Females narrow-winged, lighter, irrorated with dark brown. Antemedial line and postmedial line dark brown, enlarged on costa. Discal spots small. Wings below as above.

Frons, collar, thorax and abdomen of both sexes concolorous with wings. Antennae bipectinate in male, weakly fasciculate in female. Hindtibia with 2+2 spurs. Tympanal organs very large, meeting ventrally. Posterior margin of male 8th sternite invaginated, with two sclerotised extensions. Other sternites and tergites of both sexes undifferentiated.

VARIATION: Very little. Normally postmedial visible in males but may be weak or absent.

MALE GENITALIA: Uncus setose dorsally, with two small horns. Gnathos arms narrow, fused ventrally, forming weakly upturned, triangular element. Juxta small plate. Valva divided. Dorsal arm of valva long, narrow, straight, costa and apex setose. Ventral arm of valva wide, apex round, weakly setose. Its dorsal margin base with small, sclerotised, setose ridge. Its ventral margin weakly curved inwards, with several setae. Saccus wide, elongated. Aedeagus narrow, straight, caecum absent, subapical area with serrated ridge, apex round. Vesica opens apically, at approximately 90 degree angle, basal part weakly constricted, otherwise straight membranous tube. Basal part with scale-like sclerotised plate.

FEMALE GENITALIA: Papillae anales large, distinctly elongated, setose. 8th and 9th segments elongated. Apophyses straight, narrow, long. Lamella antevaginalis membranous. Lamella postvaginalis small plate, posterior margin round, lateral appendices reduced. Ductus bursae short, more sclerotised laterally. Base of ductus seminalis expanded. Posterior end of corpus bursae narrow, scobinate, membranous, striated. Anterior part of corpus bursae large, round, membranous. Signum small, stellate, stellar arms short.

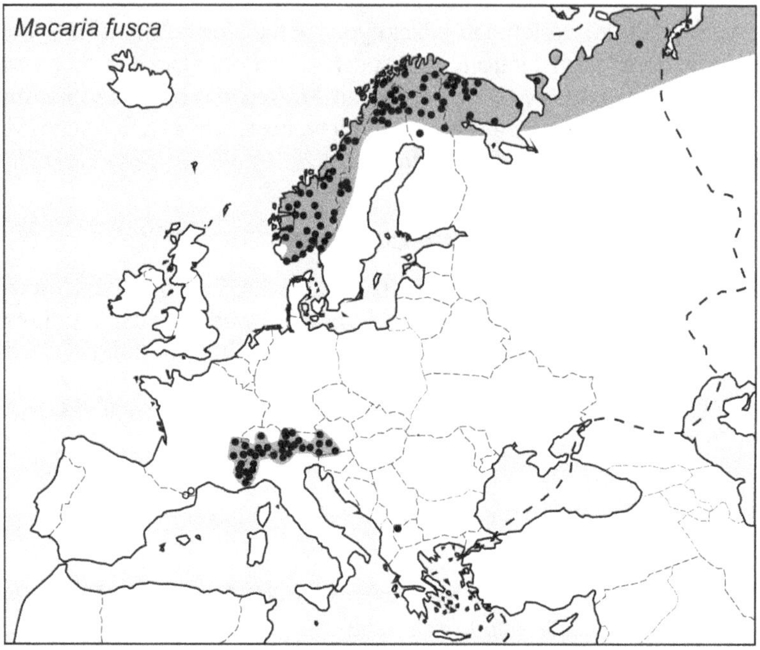

DISTRIBUTION AND ABUNDANCE: Eurasiatic. In Europe arctic-alpine, in Fennoscandia and northern European Russia, the Alps and according to ROBINEAU (2007) also in the eastern French Pyrenees where it was recorded around 1900. Further there is an isolated occurrence in north-western Macedonia which is well documented by PINKER (1968). – Outside Europe through northern Siberia to Kamchatka. – In Europe a locally abundant species.

PHENOLOGY: Univoltine. The flight time in Scandinavia is from early to late July, in the Alps also from early July and considerably longer until early September. The hibernation takes place as a small larva. The moths fly by day when especially the males are found. The females fly rarely and then only for short distances, but they can be found sitting on the ground or in the low vegetation.

BIOLOGY: Mono- or oligophagous. The caterpillar is recorded in Scandinavia on *Empetrum* and *Vaccinium* (HULDÉN *et al.* 2000), overwintering from August to June. According to FLAMIGNI *et al.* (2007), it is polyphagous on low plants, while ROBINEAU (2007) records *Viola calcarata*, *Vaccinium uliginosum* and *Erophila verna* as food plants.

HABITAT: In Scandinavia on mountain heathland just above the timber line, but locally and on warm, south-facing slopes often higher. In the Alps also above the timber line on grassland and alpine meadows up to 2700 m.

SIMILAR SPECIES: No similar species in Europe. The unrelated *Glacies coracina* (ESPER, 1805) shows some resemblance.

GENETIC DATA: Genetically homogeneous in Europe (n=19, from eight countries), but populations from Macedonia diverging by 1.2%. Nearest species: *M. occiduaria* (PACKARD, 1874) from North America (4.7%), in Europe *M. wauaria* (5.7%). The DNA barcodes of *M. fusca* specimens from Fennoscandia and Alps show a mean variation of 1.08% (min. 0.31%, max. 1.71%) (MUTANEN *et al.* 2012).

REMARKS: KULLBERG *et al.* (2001), who proposed the monotypic genus *Pygmaena* BOISDUVAL, 1840 to be a junior synonym of *Macaria* CURTIS, 1826, noted that the synonymy of *Pygmaena* and *Macaria* makes the North African species *Itame fusca* KRÜGER, 1939 (if included in *Macaria* as in SCOBLE 1999) a junior homonym of *Bombyx fusca* THUNBERG, 1792. We have failed to find the type specimen of *Itame fusca* KRÜGER. KRÜGER (1939) did not indicate where the type was deposited, and we don't know the fate of KRÜGER collection, but potentially it may have been in a Libyan museum. According to a black and white photograph in the original description (plate XVII, fig. 120), and the written description, *Itame fusca* KRÜGER is similar (or even conspecific?) to *Isturgia deerraria* (WALKER), a species that is widely distributed in Africa and in Europe only in south-western Iberian penisula, Malta and Lampedusa. According to the SCOBLE & KRÜGER's concept of *Macaria* (2002), there is only one *Macaria* species in the Palaearctic part of Africa (*M. wauaria* (LINNAEUS)), and none occurs south of the Sahara (KRÜGER 2001). For these reasons it is unlikely that *Itame fusca* KRÜGER, 1939, would belong to *Macaria*, and consequently it is unlikely that *Itame fusca* KRÜGER, 1939, would need a replacement name.

Chiasmia HÜBNER, 1823

Chiasmia HÜBNER, [1823] 1816, Verz. bekannter Schmett.: 295. Type species: *Phalaena clathrata* LINNAEUS, 1758, Europe. Genus synonyms are after SCOBLE & KRÜGER (2002).

Acadra HERRICH-SCHÄFFER, [1854]. Type species: *Acadra rectistriaria* HERRICH-SCHÄFFER, [1854]. [South Africa]: Durban. [Synonymized, as junior name, with *Godonela* BOISDUVAL by HOLLOWAY, [1994]: 162].

Allochrosis STRAND, 1912. Type species: *Allochrosis suriens* STRAND, 1912. [Tanzania]. [Synonymized, as junior name, with *Godonela* BOISDUVAL by HOLLOWAY, [1994]: 162].

Arte STEPHENS, 1829. [Replacement name for *Hercyna* STEPHENS, 1829].

Automolodes WARREN (1894). Type species: *Bociraza vacuna* DRUCE, 1889. Solomon Islands. [Synonymized, as junior name, with *Godonela* BOISDUVAL by HOLLOWAY, [1994]: 162].

Discalma MEYRICK, 1892. Type species: *Tephrina normata* WALKER, 1861. Australia: Queensland, Moreton Bay.

Evarzia WALKER, 1860. Type species: *Evarzia ozararia* WALKER, 1860. Borneo: Sarawak. [Synonymized, as junior name, with *Godonela* BOISDUVAL by HOLLOWAY, [1994]: 162].

Godonela BOISDUVAL, 1840. Type species: *Geometra aestimaria* HÜBNER, [1823] 1816. Europe.

Gonodella DESMAREST, [1858] [incorrect subsequent spelling].

Gonodela MOORE [1887] 1884–7 [incorrect subsequent spelling]. [*Godonela* BOISDUVAL is a replacement name for *Pharmacis* HÜBNER, [1820]].

Gubaria MOORE, [1887]. Type species: *Semiothisa fasciosaria* HÜBNER, [1823] 1816. India. [Synonymized, as junior name, with *Godonela* BOISDUVAL by HOLLOWAY, [1994]].

Hercyna STEPHENS, 1829. Type species: *Phalaena clathrata* LINNAEUS, 1758. [Junior homonym of *Hercyna* TREITSCHKE, 1828 (Lepidoptera, Pyralidae), replaced by *Arte* STEPHENS, 1829, and junior objective synonym of *Chiasmia* HÜBNER].

Hyostomodes WARREN, 1897. Type species: *Hyostomodes nubilata* WARREN, 1897. [Kenya].

Iulocera WARREN, 1905. Type species: *Azata variegata* WARREN, 1896. Papua New Guinea: D'Entrecasteaux Islands (= Fergusson Island). [Synonymized, as junior name, with *Godonela* BOISDUVAL by HOLLOWAY, [1994]: 162].

Osteodes GUENÉE, [1858]. Type species: *Osteodes procidata* GUENÉE, [1858]. A junior homonym of *Osteodes* CONRAD, 1855 (Mollusca). [There is no replacement name. The type species, *procidata* GUENÉE, is congeneric with *Chiasmia* HÜBNER].

Pharmacis HÜBNER, [1823] 1816. [Junior homonym of *Pharmacis* HÜBNER, [1820] 1816 (Lepidoptera: Hepialidae). Replaced by *Godonela* BOISDUVAL, 1840].

Peridela WARREN, 1897. Type species: *Peridela crassata* WARREN, 1897. Zambia.

Strenia DUPONCHEL, [1829]. Type species: *Phalaena clathrata* LINNAEUS, 1758. Europe. [Junior objective synonym of *Chiasmia* HÜBNER].

Tephrinopsis WARREN, 1896. Type species: *Aspilates parallelaria* WALKER, [1863] 1862, a junior subjective synonym of *Tephrina normata* WALKER, the type species of *Discalma* MEYRICK, see FLETCHER, 1979: 201. Ceylon [Sri Lanka].

Thyridesia WEHRLI, 1940. Type species: *Phalaena pluviata* FABRICIUS, 1798. India. [Originally proposed as a subgenus of *Semiothisa* HÜBNER [1818]].

Xenoneura WARREN, 1896. Type species: *Azata tephrinata* WARREN, 1896. Australia. [Junior homonym of *Xenoneura* SCUDDER, 1867 (Neuroptera, fossil)]. [There is no replacement name].

UNAVAILABLE NAMES (incorrect subsequent spellings): *Chiannia* STEPHENS, 1835 (FLETCHER 1979); *Chiasma* HÜBNER, 1823 (PROUT 1912–1916); *Sternia* KOLENATI, 1846 [Incorrect subsequent spelling of *Strenia* DUPONCHEL] (see FLETCHER 1979).

DIVERSITY AND DISTRIBUTION: SCOBLE & KRÜGER (2002) included in the expanded sense of the genus 265 putatively valid species, which are found in the Afrotropics, Indo-Australia and the eastern Palaearctic. The Neotropical taxa assigned to *Chiasmia* were considered misplaced.

EXTERNAL CHARACTERS AND ABDOMEN: In Europe medium-sized moths, wingspan between 20–30 mm. Wings different shades of brown, grey, yellow and white. Wings with two lines or wide single band across hindwing and most of forewing. Hindwing often with black spot near M3 (in Europe *C. aestimaria*), rarely with reticulated wings (in Europe *C. clathrata*). Outside Europe wingspan often exceeds 30 mm, occasionally wings with conspicuous broad white fascia in both wings, or red fascia on black forewing, or wings grey-violet with golden irroration (illustrated in SCOBLE & KRÜGER 2002). Male antenna typically fasciculate (not bipectinate). Male 3rd sternite often with setae, posterior margin of 8th sternite excavated.

MALE GENITALIA: Uncus horns present, spine-like, moderate to very large. Often distinct pair of setae present at base of gnathos arms. Sclerite below uncus present (Text-fig. 73). Valva divided. Dorsal arm long, narrow, occasionally with ventral process, apex setose. Ventral arm of valva rounded, or acute, or flat with downturned, elongated tip, its dorsal margin base often with sclerotised, setose ridge. Aedeagus often with apical tooth. Vesica simple tube, or with several diverticula, variably sclerotised (see Table 1). Diagnostic features are often quantitative and differences between species are minute.

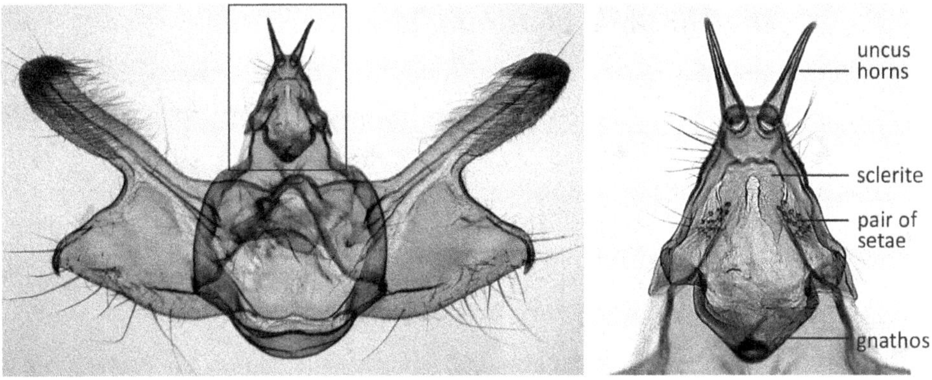

Text-fig. 73. Distinctive, horseshoe-shaped sclerite below the uncus, a potential synapomorphy of *Chiasmia* HÜBNER, 1823 (SCOBLE & KRÜGER 2002). The sclerite can only be seen if the uncus is fully bent upwards. *Chiasmia syriacaria* (STAUDINGER, 1871), Turkmenistan: Kopetdag, 15.iv–10.v.1994, slide PS1600 (coll. ZSM).

FEMALE GENITALIA: European species have short, wide, strongly sclerotised ductus bursae. Base of ductus seminalis expanded, striated. Signum stellate or absent. On wordwide scale the female genitalia are too variable to provide a useful diagnosis for the genus (SCOBLE & KRÜGER 2002).

BIOLOGY: The larva of *C. clathrata* (LINNAEUS), the type species of the genus, has been recorded feeding on numerous Fabaceae, for instance *Lathyrus pratensis* and *Trifolium*, but also on *Galium album*, *G. verum* and *Medicago sativa* (SEPPÄNEN 1970, EBERT 2003). Otherwise the range of food plants is very wide, including plant species from the families Elaeagnaceae, Tamaricaceae, Combretaceae, Anacardiaceae, Compositae (Asteraceae) and Sapindaceae. For details, see HOLLOWAY (1994), KRÜGER (2001), SCOBLE & KRÜGER (2002). Majority of the species are nocturnal, but *C. clathrata* is day-flying also. BÄNZIGER & FLETCHER (1988) reported that adults of some species of *Chiasmia* from South-East Asia feed on various mammalian body fluids, e.g. eye secretions. *Chiasmia inaequilinea* (WARREN) has been observed to visit the eyes of sheep in the Karoo region in southern Africa (du TOIT *in* BÜTTIKER & WHELLAN 1966).

IMMATURE STAGES: The larva of *C. clathrata* (LINNAEUS) has two colour forms, green and violet-brown. Both have several whitish longitudinal stripes, dorsal stripes are narrow, the lateral stripe is the most notable (EBERT 2003). Cremaster base rather wide, apex with two short lateral hooks and two additional minute teeth (PATOČKA & TURČANI 2005). AIZPÚRUA (2002) illustrated immatures stages of *C. aestimaria* (HÜBNER) and *C. clathrata*.

REMARKS: *Chiasmia* has previously included just the type species, the Palaearctic *C. clathrata* (LINNAEUS). INOUE (1986) noted that the Indo-Malayan species with uncus horns, which were earlier assigned to *Semiothisa*, were misplaced and suggested that they should be placed in *Chiasmia*. SCOBLE & KRÜGER (2002) adopted and formalised the suggestion and redefined the genus *Chiasmia* as a very extensive taxon. The genus in the expanded sense includes 265 putatively valid species which are found in the Afrotropics, Indo-Australia and the eastern Palaearctic. SCOBLE & KRÜGER (2002) discuss systematics of the genus in detail. Systematic position: Macariini (SCOBLE & KRÜGER 2002).

46. Chiasmia clathrata (LINNAEUS, 1758)

clathrata clathrata, LINNAEUS, 1758, Syst. Nat. (Ed. 10) 1: 524, (*Phalaena* (*Geometra*)). Syntypes, including putative lectotype male (Linnean Society of London, London, UK), Europe (examined externally). See remarks.

cancellaria HÜBNER, [1809], Samml. Eur. Schmett. 5 Geometrae (1 & 2): pl. 62, fig. 322, (*Geometra*). Syntype(s), lost, Europe.

clathrata nivea ROCCI, 1923, Memorie Soc. ent. Ital. 2 (1): 9, (*Phasiane*). Syntype(s), [Italy:] Genova [Genoa]. [Originally as f].

clathrata tschangkuensis WEHRLI, 1940, in SEITZ, Gross-Schmett. Erde 4 (Suppl.): 389, pl. 31: g, (*Semiothisa*). Syntype(s), China: West China [Szechwan (Sichuan)], Tschang-Kou, Sepin-Lou-Chan, Ya Tcheou.

clathrata vandarbana WEHRLI, 1940, in SEITZ, Gross-Schmett. Erde 4 (Suppl.): 389, pl. 31: k, (*Chiasmia*). Syntype(s) female, North Persia [Iran:] Elburs Mts, Vandarbantal. [Originally as var].

decussata SCHRANK, 1802, Fauna boica 2 (2): 27, (*Phalaena*). Syntype(s), [Germany]: Ingolstadt.

radiata HAWORTH, 1809, Lepid. Br. (2): 348, (*Phalaena*). Syntype(s), [United Kingdom]: Kentshire.

retialis SCOPOLI, 1763, Ent. Carniolica: 217, fig. 536, (*Phalaena*). Syntype(s), [Slovenja (north-west)]: Carniolia.

wehrlii STAUDER, 1920, Int. ent. Z. 14: 38, (*Chiasmia* (*Phasiane*)). Syntypes 4 males (The Natural History Museum, London, UK), Italy [southern:] [Campania:] Sorrento; [Calabria:] Monte Martinella. [Originally as f].. We follow FLAMIGNI *et al.* (2007), who consider taxon intrasubspecific.

clathrata albifenestra INOUE, 1942, Trans. Kansai ent. Soc. 12 (1): 16, pl. 4, fig. 5, (*Semiothisa*). Syntype(s) 3 males (The Natural History Museum, London, UK), Japan: Iwate, Morioka (examined externally). Valid at subspecies rank according to SCOBLE (1999).

clathrata almacola WEHRLI, 1934, Mitt. münch. ent. Ges. 24 (2): 54, (*Chiasmia*). Syntypes 17 males, 2 females, ‚Syria' [Turkey:] Marasch [Maras]. Valid at subspecies rank according to SCOBLE (1999).

clathrata aurata TURATI, 1905, Naturalista sicil. 18 (2–3): 45, pl. 9, figs 5–12, (*Phasiane*). Syntypes male, female, [Italy:] Sicily: Ficuzza and Busambra. Originally as var., valid at subspecies rank according to SCOBLE (1999).

clathrata azrouensis WEHRLI, 1937, Ent. Z., Frankf. a.M. 51 (12): 119, (*Semiothisa*). Holotype female (Zoologische Forschungsinstitut und Museum Alexander Koenig, Bonn, Germany), North Africa: Forest d'Azrou (examined externally). Valid at subspecies rank according to SCOBLE (1999).

clathrata centralasiae KRULIKOVSKY, 1911, Zap. ural. Obshch. Lyub. Estest. 31: 120, (*Phasiane*). Syntype(s), Central Asia: Vierny; Prshevalsk. Originally as var., valid at subspecies rank according to SCOBLE (1999).

clathrata clathraria HÜBNER, [1799], Samml. Eur. Schmett. 5 Geometrae (1): pl. 25, fig. 132, (*Geometra*). Syntype(s), lost, Europe. Originally as sp., valid at subspecies rank according to SCOBLE (1999).

clathrata djakonovi KARDAKOFF, 1928, Ent. Mitt. 17 (6): 423, pl. 9, figs. 25, 26, (*Chiasmia*). Syntypes male, female (The Natural History Museum, London, UK), [Russia:] Ussuri (examined externally). Valid at subspecies rank according to SCOBLE (1999).

clathrata hoenei SCHULTE, 1954, Bonn. zool. Beitr. 5 (1/2): 156, figs 2, 3, (*Chiasma* (sic!)). Holotype male (Zoologische Forschungsinstitut und Museum Alexander Koenig, Bonn, Germany), [China:] South Shensi, Tapaishan im Tsinling, 1700 m (examined externally). Valid at subspecies rank according to SCOBLE (1999).

clathrata hugginsi BAYNES, 1959, Entomologist 92: 21, (*Chiasmia*). Holotype male (The Natural History Museum, London, UK), [Ireland]: County Clare, Burren (examined externally). Valid at subspecies rank according to SCOBLE (1999) and SKINNER & WILSON (2009).

clathrata kurilata BRYK, 1942, Dt. ent. Z. Iris 56: 85, pl. 2, fig. 23, (*Semiothisa*). Holotype male (Naturhistoriska Riksmuseet, Stockholm, Sweden), [Russia] Kurile islands: Yeterefu, Shana. Valid at subspecies rank according to SCOBLE (1999).

clathrata punctata BANG-HAAS, 1910, Dt. ent. Z. Iris 24 (3): 49, (*Phasiane*). Syntype(s), [China]: [Xinjiang], Juldus [Yining] Territory. Originally as var., valid at subspecies rank according to SCOBLE (1999).

clathrata retata HAWORTH, 1809, Lepid. Br. (2): 348, (*Phalaena*). Syntype(s), [United Kingdom]: Kentshire. Originally as sp., valid at subspecies rank according to SCOBLE (1999).

UNAVAILABLE NAMES (incorrect subsequent authorship): *flaveola* STAUDINGER (The Global Lepidoptera Names Index (ANONYMOUS 2013)); should be *flaveola* STAUDER, 1928); *wehrli* STAUDINGER (The Global Lepidoptera Names Index (ANONYMOUS 2013)); should be *wehrli* STAUDER, 1920). – Intrasubspecific names, potentially unavailable: *aethiopissa* MEVES, 1914 (ab.); *aequifasciata* JACQUIMIN, 1931 (ab.); *albicans* OBERTHÜR, 1896 (ab.); *albissima* COCKAYNE, 1952 (ab.); *alboguttata* FETTIG, 1889 (ab.); *anastomosaria* LEMPKE, 1970 (f.); *anfaltyi* BEZSILLA, 1943 (ab.); *basibrunnea* CAPUSE, 1963 (f.); *brunnea* LERAUT, 2009 (f.); *chretieni* THIERRY-MIEG, 1910 (f.); *crassesignata* LEMPKE, 1953 (f.); *declathrata* DAHL, 1930 (ab.); *demarginata* LEMPKE, 1953 (f.); *diluta* COCKAYNE, 1950 (ab.); *disjuncta* LEMPKE, 1970 (f.); *fasciata* PROUT, 1915 (ab.); *flaveola* STAUDER, 1920 (ab.); *germanae* LE CERF, 1924 (ab.); *lactilineata* COCKAYNE, 1952 (ab.); *lativirgata* COCKAYNE, 1952 (ab.); *nigra* KÖNIG, 1975 (f.); *nigricans* OBERTHÜR, 1896 (=synonym of *nocturnata* FUCHS according to The Global Lepidoptera Names Index (ANONYMOUS 2013)); *nocturnata* FUCHS, 1898 (ab.); *obsoletissima* COCKAYNE, 1952 (ab.); *opipara* DAHL, 1930 (ab.); *ornataria* KRULIKOWSKY, 1907 (ab.); *reducta* LEMPKE, 1952 (f.); *semiaurata* CAPUSE, 1963 (f.); *wehrli* STAUDER, 1920 (f.). WEHRLI (1939–1954) lists numerous further intrasubspecific names.

EXTERNAL CHARACTERS AND ABDOMEN: Wingspan ♂ 22–28 mm, ♀ 18–24 mm. Wings white to yellowish. Medial lines and veins brown, creating reticulate pattern. Forewing costa and terminal area often yellowish, irrorated with brown. Terminal line continuous, brown. Fringes chequered brown and white. Discal spots lunular. Wings below as above. Frons, collar, thorax and abdomen concolorous with wings. Antennae weakly fasciculate in male, filiform in female. Hindtibia with 2+2 spurs. Tympanal organs large. Posterior margin of male 8th sternite excavated, with two sclerotised lobes. Other sternites and tergites of both sexes undifferentiated.

VARIATION: Ground colour varies from white to golden yellow and brownish yellow, medial lines may be weak, or wider and partly fused, or almost uniform brown. Numerous subspecies and forms have been described, see for instance WEHRLI (1939–1954). FLAMIGNI *et al.* (2007) illustrate several specimens demonstrating the variation within the species. See remarks.

MALE GENITALIA: Uncus setose dorsally, apex round, with two large horns. Gnathos arms narrow, fused ventrally, forming weakly upturned, triangular element. Juxta small plate. Valva divided. Dorsal arm of valva long, narrow, straight, costa and apex setose. Its ventral margin constricted, with medial, sclerotised element. Ventral arm of valva short, round, margin curved inwards, setose. Its dorsal margin base with sclerotised, setose ridge. Saccus rather narrow, elongated. Aedeagus stout, straight, caecum minute, subapical area with one or two acute teeth and row of dentate sclerotisations, apex round. Vesica opens apically, at approximately 90 degree angle, basal part weakly enlarged, striated, with numerous different-sized microcornuti. Medial part with weakly sclerotised plate, and small lateral diverticulum, otherwise straight membranous tube.

FEMALE GENITALIA: Papillae anales elongated, setose. Apophyses straight, narrow. Lamella antevaginalis membranous. Lamella postvaginalis narrow, sclerotised plate, posterior margin round, lateral appendices small, irregular. Ductus bursae short, wide, sclerotised. Base of ductus seminalis notably expanded. Posterior end of corpus bursae wide, scobinate, striated, partly sclerotised. Corpus bursae slightly constricted near middle. Anterior part of corpus bursae elongated, with apical diverticulum. Signum stellate, stellar arms of unequal length.

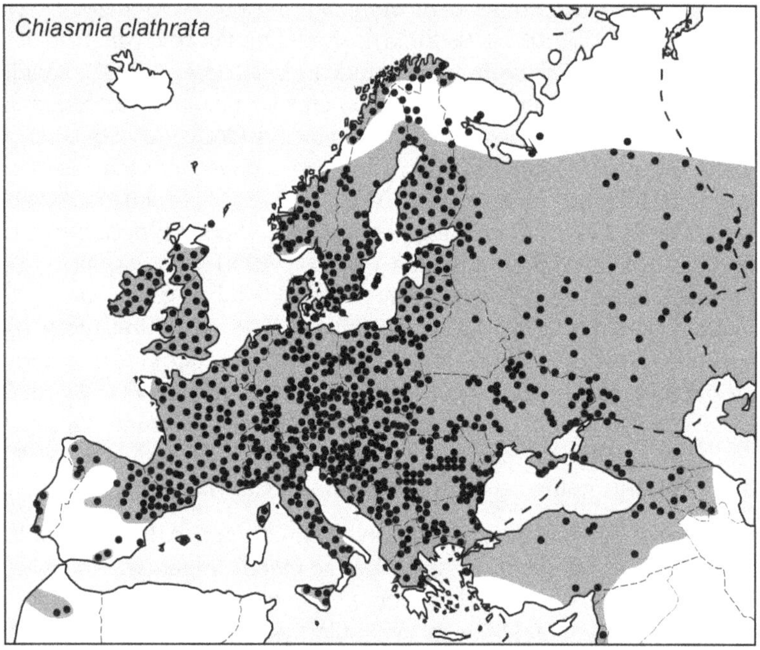

DISTRIBUTION AND ABUNDANCE: Eurasiatic-Palaearctic. Widely distributed in Europe from northern Fennoscandia to the Mediterranean and from the British Isles to the Urals. Absent only from Sardinia, Crete and Iceland. – Outside Europe the species occurs in Morocco, Lebanon, Turkey, the Caucasus region, central Asia, southern Siberia, north-western China, Korea, Far East Russia, Japan, Sakhalin and the Kurile Islands. – In Europe a common species that rather often can be found in numbers.

PHENOLOGY: Bi- or trivoltine. In northern Europe in two overlapping generations from mid-May to mid or late August, in southern central Europe from late April to late August (EBERT 2003) and in Italy from March to September, maybe in three generations (FLAMIGNI *et al.* 2007). The species overwinters as pupa. The species is easily flushed up during daytime, possibly it flies actively by day and it also flies by night when it comes to light.

BIOLOGY: Polyphagous. The larva was found on *Trifolium*, *Medicago*, *Cytisus scoparius*, *Genista sagittalis*, *Galium mollugo* and *G. sylvaticum* according to EBERT (2003). Abundant in a field with *Onobrychis viciifolia* (M. CORLEY pers. comm.). Also FLAMIGNI *et al.* (2007) mentions *Galium* as food plant. The larva has been found on *Vicia cracca* (H. WEGNER pers. comm.). Pupation takes place in the ground.

HABITAT: In meadows, both dry and more moist places, gardens, in fields with *Medicago* and *Trifolium* and in other open land habitats with herbaceous vegetation. In south-western Germany the species is found up to 1000 m (EBERT 2003), in Italy up to 2300 m (FLAMIGNI *et al.* 2007).

REMARKS: In Denmark the species has shown tendencies to migration by suddenly occurring at the same time in different localities where the species normally does not occur.

SIMILAR SPECIES: There are no similar species in Europe.

GENETIC DATA: Genetically homogeneous from Europe (n=24, from nine countries), to Middle East (n=9) and Far East Asia (subsp. *djakonovi*; n=12). Nearest species: African *C. multistrigata* (WARREN, 1897) (6.6%).

REMARKS: *C. clathrata* (LINNAEUS) is a widespread species with numerous putative subspecies, the subspecies given above are from SCOBLE (1999). These need to be evaluated critically, therefore we don't give descriptive accounts and do not recognise them in the distribution map. See genetic data. Those recognised in the recent literature valid at subspecies level in Europe and North Africa, in addition to SCOBLE (1999), include *azrouensis* (WEHRLI) in Morocco and Algeria (small, wings pale yellow) (KRÜGER 2001), *aurata* (TURATI) in Italy: Sicily (wings golden yellow, including fringes) (FLAMIGNI *et al.* 2007) and *hugginsi* BAYNES in the United Kingdom (wings white) (SKINNER & WILSON 2009, NATIONAL BIODIVERSITY NETWORK 2012).

In a population in southern Finland (Vantaa: 2 km North of Viherlaakso, sandy ruderal habitat), occasionally specimens are found which lack left or right hindwing, very rarely both hindwings are missing. The frequency of three-winged specimens reaches approximately up to 10% of specimens (SIHVONEN, personal observation).

Chiasmia saburraria (EVERSMANN, 1851)

Chiasmia saburraria is a species that could potentially be found in Europe. The nearest recorded locality, which is mentioned in the literature, is in the Ilmenskii Reserve near Miass, West of Chelyabinsk in the Chelyabinsk oblast in southern Ural, Russia (OLSHVANG *et al.* 2004). *C. saburraria* is known from southern Siberia, from southern Ural to Mongolia, northern China to Amur basin in Russia (VIIDALEPP 1996). *C. saburraria* is externally somewhat similar to *Macaria signaria* (HÜBNER, 1809), but the spot on the forewing postmedial line is broken up into small spots, extending into the medial area, small spots also present near costa (spot less broken, not extending into the medial area, small spots weak or absent near costa in *signaria*). An externally variable species, several subspecies and forms have been described, these are summarised in SCOBLE (1999). The male and female genitalia are similar, but with minor diagnostic differences, to *C. clathrata* (illustrated in Text-fig 74).

According to the label data the adults are from mid-June to mid-July. The species comes to light. Food plant of the larva is unknown. Adults have been taken in mixed forests and meadows up to 700 m.

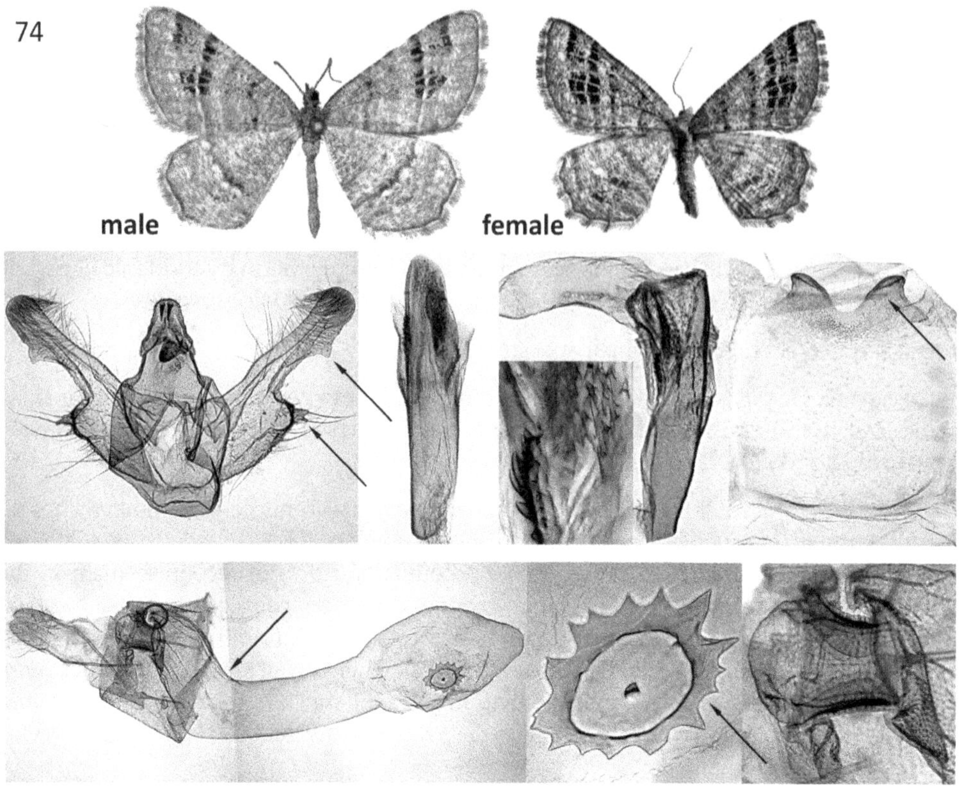

Text-fig. 74. Adults and genitalia of *Chiasmia saburraria* (EVERSMANN, 1851), a species that could potentially be found in Europe. Male adult (wingspan 21 mm): Russia: [Altai Republic] Altai Mountains, Ongodai [Ongoday], 10.vi.[18]98 (coll. BMNH). Male genitalia and aedeagus illustrated from same specimen: Russia: Buryatia, Barquzin valley, Upper Kurumkan river, 700 m, 1.vii.1996, slide PS2005 (coll. ZMH); vesica (insert: sclerotisations in aedeagus apex and microcornuti in vesica) and 8th sternite illustrated from same specimen: Russia: SW-Altai, Katun valley 10 km W Katanda, 1200 m, 22–27.vi.1983, slide PS1499 (coll. ZMH). Female adult (wingspan 20 mm): Russia: Buryatia, Barguzin valley, 600 m, 9.vii.1996 (coll. ZMH). Female genitalia (point of origin of ostium bursae indicated by circle), signum, ostium bursae and adjacent structures illustrated from same specimen: USSR: SW-Altai, Katun valley 10 km, W Katanda, 1200 m, 22–27.vi.1983, slide PS1500 (coll. ZMH).

47. *Chiasmia aestimaria* (HÜBNER, 1809)

aestimaria aestimaria HÜBNER, [1809], Samml. Eur. Schmett. 5 Geometrae (1): pl. 64, fig. 333, (*Geometra*). Syntype(s), lost, Europe.

aestimata HERRICH-SCHÄFFER, 1861, Syst. Verz. Eur. Schmett (Edn 2): 15, (*Macaria*). [Emendation of *aestimaria* HÜBNER].

contemptata GUENÉE, [1858], in BOISDUVAL & GUENÉE, Hist. nat. Insectes (Spec. gén. Lépid.) 10: 86, (*Macaria*), lectotype male (National Museum of Natural History, Smithsonian Institution, Washington DC, U.S.A.), Amérique septentrionale´. [mislabelled]

aestimaria kuldschana WEHRLI, 1940, in SEITZ, Gross-Schmett. Erde 4 (Suppl.): 384, pl. 31: d, (*Semiothisa*). Syntype(s), [China:] Kuldscha [Yining (Gulja)], Dsungarei. Valid at subspecies rank according to SCOBLE (1999).

aestimaria sareptanaria STAUDINGER, 1871, in STAUDINGER & WOCKE, Cat. Lepid. eur. Faunengeb. 1: 160, (*Macaria*). Lectotype male, [Russia:] Sarepta [Volgograd], paralectotype female [Russia?] (Zoologisches Museum der Humboldt Universität, Berlin, Germany). [Originally as var. of *aestimaria* HÜBNER]. Lectotype designated by KRÜGER (2001). See remarks.

aestimaria tunesiella LUCAS, 1949, Bull. Soc. ent. Fr. 53: 96, (*Macaria*). Lectotype female (Coll. HERBULOT in Zoologische Staatsammlung, Munich, Germany), Tunisia: Sfax (examined externally). [Originally as var. of *aestimaria* HÜBNER, treated as subspecies by HERBULOT (pers. comm. in KRÜGER (2001)), raised formally to subspecies rank by LERAUT (2009)]. See remarks.

EXTERNAL CHARACTERS AND ABDOMEN: Wingspan 20–27 mm. Wings light ochre-yellow to grey, weakly irrorated with grey or brown. Antemedial and medial lines weak. Postmedial line well developed, narrow, dark brown, oblique, angled below costa. Inner margin of postmedial line wide, pale band, reaching forewing apex and often making part of postmedial line invisible. Outer margin of postmedial line dark, particularly towards termen, with dark, elongated spots on fore and hindwings. Forewing costa with characteristic brown triangle, formed by postmedial and subterminal lines. Terminal line narrow, brown. Fringes chequered. Discal spots lunular. Forewing margin slightly concave below apex, hindwing margin crenulated, elongated at M3. Wings below heavily irrorated with brown, particularly terminal area. Postmedial line strongest. Terminal area often with whitish spots. Frons, collar, thorax and abdomen concolorous with wings. Antennae weakly fasciculate in male, filiform in female. Hindtibia with 2+2 spurs. Tympanal organs large. Posterior margin of male 8th sternite excavated, with two sclerotised lobes. Other sternites and tergites of both sexes undifferentiated.

VARIATION: Ground colour varies from light ochre-yellow to grey. Pale band on inner margin of postmedial line is more pronounced in males. Postmedial line often incomplete or weak below angle on upperside, but complete on underside.

Subsp. ***sareptanaria*** is more contrasting. Wings whitish, irrorated heavily with dark brown, outer margin of postmedial line extensively brown. Antemedial and medial lines well developed. Length of sclerotisation at base of vesica variable but structure similar with *aestimaria* subsp. *aestimaria*. See remarks.

Subsp. ***tunesiella*** is intermediate between subspecies *aestimaria* and subspecies *sareptanaria*. Wings whitish, irrorated with dark brown and outer margin of postmedial line brown, but not as extensively as in *sareptanaria*. Antemedial and medial lines weak. See remarks.

MALE GENITALIA: Base of uncus with two setal patches, enclosed in elongated sacs. Uncus setose dorsally, apex round, with two large horns. Gnathos arms narrow, fused ventrally, forming weakly upturned, triangular element. Juxta medium-sized plate. Valva divided. Dorsal arm of valva long, narrow, straight, costa and apex setose. Ventral arm of valva wide, flat, triangular, apex extended to form prominent, downturned tip. Its ventral margin convex, weakly curved inwards, with long setae. Its dorsal margin weakly setose. Saccus narrow, wide. Aedeagus straight, caecum minute. Subapical area with row of dentate sclerotisations, on opposite side long, narrow, sclerotised rod-like apex. Vesica extends beyond apex, opens at approximately 90 degree angle. Basal part extended, surface granulate laterally, with dentate sclerotisation, length variable (see Genitalia plate). Medial part with one dorsal and two ventral diverticula, small cornutus. Apex narrow, weakly curved, membranous tube.

FEMALE GENITALIA: Papillae anales setose. Apophyses straight, narrow. Lamella antevaginalis membranous. Lamella postvaginalis narrow, sclerotised plate, posterior margin round. Ductus bursae short, wide, sclerotised, particularly anterior parts laterally. Base of ductus seminalis notable expanded, elongated. Posterior part of corpus bursae wide, densely scobinate, striated, partly sclerotised. Corpus bursae slightly constricted near middle. Anterior part of corpus bursae roundish. Signum absent.

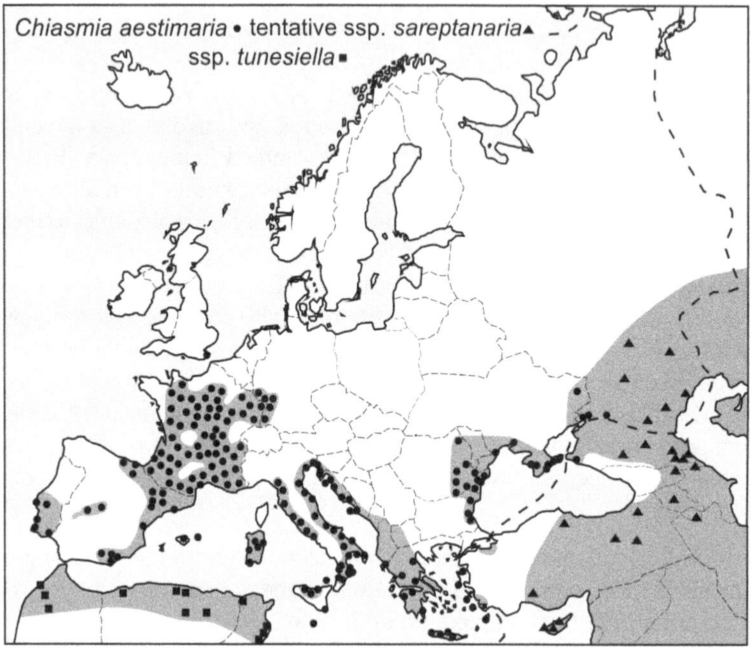

DISTRIBUTION AND ABUNDANCE: Mediterranean-Turanian. In Europe rather disjunct. Restricted to southern Europe from Portugal and France to the Balkans, Ukraine and south-eastern European Russia (here subsp. *sareptanaria*) to the southern Ural mountains. Recently recorded in south-western Germany. – Outside Europe in Morocco, Algeria and Tunisia (subsp. *tunesiella*) to Senegal (RUNGS 1981, the conspecificity of the Afrotropical populations awaiting confirmation), Cyprus, Turkey, Transcaucasus, Caucasus, near and Middle East, Yemen (A. HAUSMANN 2006, pers. comm.), central Asia, Iran and Afghanistan (VIIDALEPP 1996) (subsp. *sareptanaria*). – In Europe rather common, but usually in limited numbers.

PHENOLOGY: Bi- or trivoltine. In two or three generations from early April to September or October. In Israel in all months of the year (A. HAUSMANN pers. comm.).The caterpillar has been found from May to October. The adult moths fly by night and come to light.

BIOLOGY: Monophagous. The caterpillar has been recorded on *Tamarix gallica* and *T. africana*. Also reared on *Tamarix* (E. FRIEDRICH pers. comm.).

HABITAT: In various coastal localities and along rivers where the food plants grow, also in very warm inland and lowland localities in Spain. According to FLAMIGNI *et al.* (2007) it has been found up to 1300 m in Sicily and up to 1200 m in Spain (REDONDO *et al.* 2009).

SIMILAR SPECIES: Non-European *Chiasmia syriacaria* (STAUDINGER) is paler, rather uniform ochreous yellow, subapical triangle in forewing costa is weak or absent, postmedial spot is absent (wings darker, contrasting ochre-yellow or grey, forewing costa with prominent subapical triangle and with postmedial spot in *C. aestimaria*). Sclerotisation at base of vesica is broken up into several small sclerotisations in *C. syriacaria* (sclerotisation continuous in *C. aestimaria*). Base of ductus seminalis is short in *C. syriacaria* (elongated in *C. aestimaria*).

GENETIC DATA: Genetically homogeneous from Europe to the eastern Mediterranean (n=19), but with two other BINs in the Middle East (n=15). Nearest species: *C. syriacaria* (4.4%).

REMARKS: The taxonomic status of *Macaria sareptanaria* STAUDINGER, 1871, described from Sarepta (currently Volgograd), Russia, has remained controversial. Recently it has been considered a form of *C. aestimaria* (for instance LERAUT 2009), subspecies of *C. aestimaria* (for instance WILTSHIRE 1990b, HAUSMANN 1995a), or valid at species level (for instance SCOBLE 1999). KRÜGER (2001) designated a lectotype male from Sarepta and considered the taxon valid at species level. The diagnostic differences he describes, mainly the location and number of sclerotisations in (uneverted?) vesica, appear difficult for us to interpret. We fail to see the described differences from fully inflated vesicas. Zoological Museum of Helsinki, Finland, has several *sareptanaria* specimens from Sarepta. Based on the these materials, we consider *C. sareptanaria* a junior synonym of *C. aestimaria* (HÜBNER), tentatively valid at subspecies level. The South Russian specimens appear externally distinct from the European specimens, but we acknowledge that if more specimens are found, the differences may appear gradual and eventually the South Russian specimens, classified tentatively valid at subspecies level by us, should perhaps be treated as a synonym of the nominal subspecies. We have not examined genitalia of North African specimens (subsp. *tunesiella*). More research is needed.

Chiasmia aestimaria has been used as a model organism in DNA based identification. The unknown geometrid larvae from Israel were identified from a tissue sample (CO1 gene) as *C. aestimaria* and its food plant was identified from the gut content (chloroplast rbcL gene), the latter showing sequence homologies with the suspected host-plant *Tamarix nilotica* (MILLER *et al.* 2006).

Chiasmia syriacaria (STAUDINGER, 1871), not found in Europe

syriacaria STAUDINGER, 1871, in STAUDINGER & WOCKE, Cat. Lepid. eur. Faunengeb. 1: 160, (*Macaria*). Lectotype male, paralectotype female, two paralectotype males (Zoologisches Museum der Humboldt Universität, Berlin, Germany), Cyprus, Lebanon: Beirut, Syria. [Originally as var. of *aestimaria* HÜBNER]. (lectotype male examined, including genitalia).

venerata CHRISTOPH, 1887, in Romanoff, Mém. Lépid. 3: 94, pl. 4, fig. 12, (*Macaria*). Syntypes 4 males, (potentially in Zoological Institute, Russian Academy of Sciences, St Petersburg, Russia), [Turkmenistan:] Ashkhabad; Armenia: Kasikoparan; North Persia [Iran:] Schahrud.

EXTERNAL CHARACTERS AND ABDOMEN: Wingspan 19–24 mm. Wings rather uniform ochreous yellow-brown, weakly irrorated with brown. Normally antemedial and medial lines absent or weak, postmedial line well developed near termen, oblique, brown, reduced towards costa. Brown triangle on forewing costa absent or weak. Outer margin of postmedial line suffused with brown in forewing termen and hindwings. Terminal area often slightly darker than rest of wing. Terminal line weak, interrupted. Fringes concolorous with wings or weakly chequered. Discal spots lunular, brown. Forewing margin round, hindwing margin weakly crenulated, weakly elongated at M3. Wings below weakly striated with brown, markings weaker. Frons, collar, thorax and abdomen concolorous with wings. Antennae weakly fasciculate in male, filiform in female. Hindtibia with 2+2 spurs. Tympanal organs large. Posterior margin of male 8th sternite excavated, with two sclerotised lobes. Other sternites and tergites of both sexes undifferentiated.

VARIATION: Ground colour varies from whitish to deep sandy yellow. Occasionally markings more pronounced, antemedial and medial lines visible and subterminal band wider. Postmedial line may be visible also near costa. If visible, angled below costa.

MALE GENITALIA: Generally as in *C. aestimaria*, but with following difference: In *C. syriacaria* sclerotisation at base of vesica is broken up into several sclerotisations (sclerotisation continuous in *C. aestimaria*). Uncus and gnathos regions are shown enlarged in Text-fig 73.

FEMALE GENITALIA: Generally as in *C. aestimaria*, but with following difference: In *C. syriacaria* base of ductus seminalis is short and lamella postvaginalis is bigger (base of ductus seminalis elongated and lamella postvaginalis smaller in *C. aestimaria*).

DISTRIBUTION: Anatolian-Iranian. Outside Europe the species has been found in Turkmenistan, Uzbekistan, Kirghizstan, northern Iran (VIIDALEPP 1996) and in Turkey (F. CAN CENGIZ, pers. comm.).

PHENOLOGY: Uni- or bivoltine. The scattered collecting data indicates a flying time from mid April till late May and again in late September. In Israel and Jordan the species has been found from mid March to early June, from early to late July, from late August to mid October and in early November (HAUSMANN 1991, 1997).

BIOLOGY: Polyphagous. Only little is known. The larva is recorded in Iraq on *Prosopis farcta*, and probably also *Glycyrrhiza* and other Fabaceae (WILTSHIRE 1957).

HABITAT: In Turkey up to 1300 m altitude.

SIMILAR SPECIES: *Chiasmia aestimaria* (HÜBNER), see entry above. Outside Europe *Chiasmia tenuiata* (STAUDINGER, 1901), which is known from Lebanon and potentially Syria only, is very similar. KRÜGER (2001) illustrates the lectotype male, including its genitalia, which is the only known specimen. The minor differences in the arrangement of cornuti in the vesica between *C. syriacaria* and *C. tenuiata*, which we find difficult to see in the illustrations (Figs 661, 662 in KRÜGER 2001), raise the question of the validity of *C. tenuiata* (STAUDINGER, 1901) at the species level, being potentially a junior synonym of *C. syriacaria* (STAUDINGER, 1871).

GENETIC DATA: Nearest species: *C. aestimaria* (4.4%).

REMARKS: *C. syriacaria* (STAUDINGER) has been reported from Crete: Chania (in May) and Megali Episkopi (in October) (REISSER 1958), but no authentic European material has been found despite extensive searches. Even the Reisser collection, deposited in Staatliches Museum für Naturkunde, Karlsruhe, Germany, did not contain any European specimens (R. TRUSCH pers. comm.). Therefore *C. syriacaria* is removed from the European list.

According to LERAUT (2009): 'genitalia very similar to those of *aestimaria*; and this moth could be just a local form of that species.' We consider *C. syriacaria* (STAUDINGER) valid at species level because it is diagnosable by external characters, male and female genitalia, genetic data, and the food plant preference of larvae is different from in *C. aestimaria*. Also other recent works, for instance WILTSHIRE (1990), VIIDALEPP (1996) and KRÜGER (2001) consider *C. syriacaria* valid at species level.

Digrammia GUMPPENBERG, 1887

Digrammia GUMPPENBERG, 1887, Nova Acta Acad. Caesar. Leop. Carol. 49: 325. Type species: *Phasiane mellistrigata* GROTE, 1873. North America. Genus synonyms are after SCOBLE & KRÜGER (2002).

Digrammia GUMPPENBERG, 1887. Type species: *Phasiane mellistrigata* GROTE, 1873. North America.

Asmate GUMPPENBERG, 1887. Type species: *Phasiane rippertaria* DUPONCHEL, 1830. France.

Spinuncina WEHRLI, 1937. Type species: *Phasiane rippertaria* DUPONCHEL, 1830. [Junior objective synonym of *Asmate* GUMPPENBERG. (*Asmate* is a junior subjective synonym of *Digrammia*)].

DIVERSITY AND DISTRIBUTION: *Digrammia* contains over 50 species, the majority of which are found in North America, but also in South America and one species in Europe.

EXTERNAL CHARACTERS AND ABDOMEN: Medium-sized. Wings mostly different shades of grey or brown, often antemedial and postmedial lines strong, dark, enclosing medial area that is wider on costa. Male forewing without fovea. Male antennae ciliated-setose, female antennae filiform. Male hindtibia enlarged, without hair pencil. Male 3rd sternite with setae, posterior margin of 8th sternite excavated or U-shaped, occasionally asymmetrical.

MALE GENITALIA: Uncus beak-like, with 2–4 pairs of strong subapical setae. Sclerite below uncus absent. Valva divided. Dorsal arm of valva long, narrow. Ventral arm of valva wider, often with acute apical process. Valva often asymmetrical. Aedeagus with digitate process in numerous species. Vesica with or without cornuti (see Table 1).

FEMALE GENITALIA: Lamella postvaginalis bilobed, 'winged'. Signum absent (see Table 1).

BIOLOGY: The range of recorded food plants is very wide and includes Cupressaceae (for instance *Juniperus*, *Cupressus*, *Thuja*, *Calocedrus* and *Chamaecyparis*), Salicaceae, Fabaceae (for instance *Lotus*, *Vicia* and *Robinia*), Asteraceae, Polygonaceae, Rosaceae and Zygophyllaceae (FERGUSON 2008, DEBORAH *et al.* 2014). In North America the largest taxonomic group has been recorded feeding on Salicaceae, on willows (*Salix*) and rarely on poplars (*Populus*) (FERGUSON 2008). The only European species is recorded feeding on *Salix* species (FLAMIGNI *et al.* 2007) and according to SCOBLE (1999) also on *Alnus*.

IMMATURE STAGES: Larvae externally varied. Numerous species with fine, whitish or yellowish longitudinal stripes, or with contrasting and complex patterns, lateral markings especially prominent, with strongly oblique aspect. The latter group often often feeds on Cupressaceae. Numerous species are illustrated and described in FERGUSON (2008). Pupae with prespiracular furrows (FERGUSON 2008).

REMARKS: Systematic position: Macariini (SCOBLE & KRÜGER 2002).

48. *Digrammia rippertaria* (DUPONCHEL, 1830)

rippertaria rippertaria DUPONCHEL, 1830, in GODART & DUPONCHEL, Hist. nat. Lépid. Papillons Fr. 8 (1): 159, pl. 180, fig. 5, (*Phasiane*). Syntype(s), France: Digne.

hebetata HULST, 1881, Bull. Brooklyn ent. Soc. 4 (5/6): 34, (*Phasiane*). Syntypes 2 (National Museum of Natural History, Smithsonian Institution, Washington DC, USA), [U.S.A].: Colorado.

ponderosa BARNES & MCDUNNOUGH, 1917, Contr. nat. Hist. N. Am. 3 (4): 234, pl. 23, figs 7, 8, (*Phasiane* (*Sciagraphia*)). Lectotype male (National Museum of Natural History, Smithsonian Institution), [Canada:] Manitoba, Cartwright. Lectotype designated in FERGUSON (2008).

ponderosa demaculata BARNES & MCDUNNOUGH, 1917, Contr. nat. Hist. N. Am. 3 (4): 235, pl. 23, fig. 9, (*Phasiane* (*Sciagraphia*)). Lectotype male (National Museum of Natural History, Smithsonian Institution), [Canada:] Alberta, Calgary. Lectotype designated in FERGUSON (2008).

tulareata CASSINO & SWETT, 1923, Lepidopterist 4 (3): 22, (*Phasiane*). Holotype male, [U.S.A].: California, Monachee Meadows.

rippertaria flavularia PÜNGELER, 1902, Dt. ent. Z. Iris 15 (1): 158, pl. 6, fig. 6, (*Phasiane*). Syntype(s), [Uzbekistan]: South Aksu. [Originally as var]. Valid at subspecies rank according to SCOBLE (1999).

INTRASUBSPECIFIC NAMES, POTENTIALLY UNAVAILABLE: *analogaria* PÜNGELER, 1902 (ab.); *irrorata* CHRISTOPH, 1897 (ab.); *nigriscente* LUCAS, 1948 (ab.); *pallidaria* STAUDINGER, 1871 (ab.); *ravouxi* THIERRY-MIEG, 1910 (ab.).

EXTERNAL CHARACTERS AND ABDOMEN: Wingspan 22–26 mm. Wings grey, grey-brown or brown, irrorated sparsely with grey or black. Antemedial line black, distinct on inner margin, weak or absent on costa. Postmedial line black, distinct on inner margin, angled below costa, present as black spot on costa, its outer margin shaded with black. Medial area defined by medial lines, these are wider apart on costa. Discal spots hollow, roundish. Terminal area often with whitish-grey band. Terminal line fine, discontinuous. Fringes weakly chequered whitish and grey. Hindwings irrorated with grey, postmedial line visible on anal margin only. Wings below same colour, striated with brown, markings weak. Frons, collar, thorax and abdomen concolorous with wings. Antennae concolorous with wings dorsally, shortly fasciculate in male, filiform in female. Male hindtibia slightly enlarged, hindtibia with 2+2 spurs. Tympanal organs large. Male 3rd sternite with setae, posterior margin of 8th sternite strongly sclerotised, deeply invaginated, with two large, scobinated extensions. Other sternites and tergites of both sexes undifferentiated.

VARIATION: Ground colour varies from grey to brown. Fresh specimens are grey, these turn to grey-brown with age. Thus most specimens in the collections are grey-brown. Medial area may be lighter than rest of wings. Forewing black bands (antemedial and postmedial lines) may be brown, narrow and weak (f. *pallidaria* STAUDINGER) or black, wide and distinct. Degree of markings varies between these extremes. The proportions of black-banded to unbanded specimens seems to vary in concordance with the geographical area. In Canada and Alaska black-banded form is in the minority by a ratio 3:4, in South Dakota black-banded form is in the majority by a ratio 6:1. Black banding appears unrelated to sex (FERGUSON 2008).

MALE GENITALIA: Uncus setose dorsally, with numerous strong subapical setae, apex round. Gnathos arms wide, fused ventrally, forming small, upturned, triangular element. Juxta large plate. Valva divided. Dorsal arm of valva long, narrow, weakly curved, costa and apex setose. Ventral arm of valva wide, apex with acute tip. Its ventral margin slightly curved inwards, with several setae. Saccus wide, shallow. Aedeagus short, very wide, weakly curved, caecum enlarged, apex round. Subapical area scobinate. Vesica opens at approximately 90 degree angle, membranous, basal part large, with small lateral diverticulum, otherwise narrower.

FEMALE GENITALIA: Papillae anales large, densely setose. Apophyses short, narrow. Lamella antevaginalis narrow, concave, weakly sclerotised, with lateral lobes. Lamella postvaginalis sclerotised, large bilobed, 'winged' plate, posterior margin invaginated medially. Ductus bursae very short, wide, membranous. Posterior part of corpus bursae wide, weakly striated. Anterior part of corpus bursae wider, roundish. Signum absent.

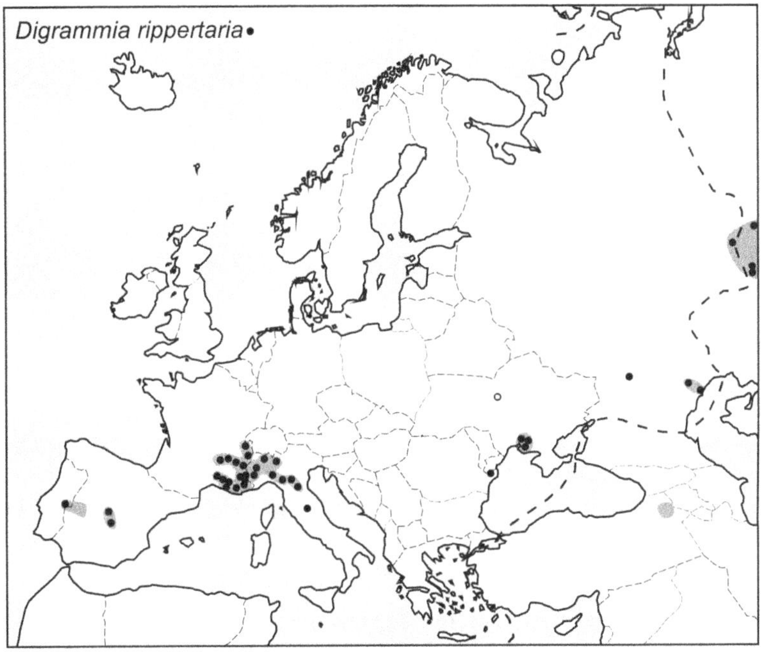

DISTRIBUTION AND ABUNDANCE: Holarctic. In Europe disjunct with a western distribution from central Iberian Peninsula through southern France to north-eastern Italy and an eastern occurrence in Romania, Ukraine, south-eastern European Russian and southern Urals. – Outside Europe across large parts of the Palaearctic region from Far East Russia through southern Siberia and central Asia. Also in North America. Recorded from Kars province in north-eastern Turkey (KOÇAK & KEMAL 2009), but no specimens from there have been traced. – In Europe apparently very local, but rather often the species occurs in numbers at its localities.

PHENOLOGY: Uni- or bivoltine. In south-eastern France in two generations from mid or late April to mid or late June and from mid-July to late September. In Romania early June until late July (L. SZÉKELY pers. comm.), apparently in only one generation. The species hibernates as a pupa. The moths fly at night and come readily to light.

BIOLOGY: Monophagous. The caterpillar lives from early summer until autumn and is recorded on *Salix*, especially *S. elaeagnos*, *S. viminalis* and *S. repens* subsp. *rosmarinifolia* (FLAMIGNI *et al.* 2007) and according to SCOBLE *et al.* (1999) (reputedly) also on *Alnus*.

HABITAT: In various habitats where the food plants grow, but especially along rivers and streams. In Italy up to 1000 m altitude in the northern Appenines (FLAMIGNI *et al.* 2007), in Turkey up to 2400 m and in Kirghizstan up to 2800 m (coll. P. SKOU).

SIMILAR SPECIES: No similar species in Europe. The North American *Digrammia hebetata* (HULST, 1881) is very similar, with small but consistent differences in the external characters and the male genitalia. The female genitalia are perhaps not reliably distinguishable (FERGUSON 2008).

GENETIC DATA: Genetically heterogeneous, populations from western Kazakhstan and Uzbekistan diverging from those of North America by 1.9%. Nearest species: North American *D. irrorata* (PACKARD, 1876) (2.8%).

Rhoptria GUENÉE, 1858

Rhoptria GUENÉE, [1858], Hist. nat. Insectes (Spec. gén. Lépid.) 10: 114. Type species: *Geometra asperaria* HÜBNER, [1817] (Europe). Genus synonyms are after SCOBLE (1999).
Pseudognophos STAUDINGER, 1900, Dt. ent. Z. Iris 12: 395. Type species: *Pseudognophos mardinata* STAUDINGER, 1900.

DIVERSITY AND DISTRIBUTION: *Rhoptria* contains three species, two of which occur in Europe and one species in Central Asia. The Neotropical species assigned to *Rhoptria* is clearly not a Macariini, thus being misplaced (PITKIN 2002).

EXTERNAL CHARACTERS AND ABDOMEN (description is based on two European species): Medium-sized moths, wingspan 20–28 mm. Wings brown-grey, markings brown or blackish. Discal spots minute. Hindwing margin weakly crenulate. Male antennae filiform, densely setose. Male hindtibia slightly enlarged, without hair pencil. Male 3rd sternite without setae, posterior margin of 8th sternite weakly concave, not otherwise modified.

MALE GENITALIA (description is based on two European species): Uncus densely setose, horns absent. Medial element of gnathos upturned, round, granulate. Valva divided. Ventral arm of valva wide, flat, slightly triangular. Juxta very large plate. Vesica narrow at base, middle part enlarged, with several long cornuti (see Table 1).

FEMALE GENITALIA (description is based on two European species): Lamella postvaginalis membranous. Ductus bursae short, sclerotised. Posterior part of corpus bursae long, curved, striated. Anterior part of corpus bursae oval, large. Signum stellate or absent (see Table 1).

BIOLOGY: Larva of *R. asperaria* has been recorded to feed on *Cistus*, particularly on *C. monspeliensis* and *C. salvifolius*, from sea level up to 2000 metres (FLAMIGNI *et al.* 2007, REDONDO *et al.* 2009). Larva of *R. dolosaria* has been found on *Pistacia lentiscus* (FRITSCH *et al.* 2014). Both European species are nocturnal.

IMMATURE STAGES: Larva of *R. asperaria* has two colour forms, yellowish and brown, which are illustrated in AIZPÚRUA (2002). We have no information on pupa, apart from superficial illustration in AIZPÚRUA (2002).

REMARKS: KRÜGER (2001) and SCOBLE & KRÜGER (2002) did not include *Rhoptria* in the Macariini. We agree with HAUSMANN (1993), VIVES MORENO (1994), LERAUT (1997) and HAUSMANN *et al.* (2011) and include it in the Macariini. HAUSMANN (1993) and VIVES MORENO (1994) classified *Rhoptria* in the Semiothisini, agreeing with the current definition of the Macariini, because the male and female genitalia have numerous, characteristic Macariini structures. We have not examined *R. mardinata* (STAUDINGER, 1900) from central Asia, but according to WEHRLI (1939–1954) it is structurally similar, including the male genitalia, to *R. asperaria* (HÜBNER) and *R. dolosaria* (HERRICH-SCHÄFFER). Systematic position: Macariini.

49. Rhoptria asperaria (HÜBNER, 1817)

asperaria HÜBNER, [1817], Samml. Eur. Schmett. 5 Geometrae (1): pl. 94, fig. 484, (*Geometra*). Syntype(s), lost, Europe.

asperaria esteparentis AGENJO, 1933, Boln R. Soc. esp. Hist. nat. 33 (8): 311, pl. 2, fig. 1, (*Rhoptria*). Holotype male (Museo Nacional de Ciencias Naturales, Madrid, Spain), [Spain]: Burgos, Estépar, 810 m. [Originally as var].

collata HÜBNER, [1826], Samml. Eur. Schmett. 5 Geometrae (1): pl. 104, fig. 524, (*Geometra*). Syntype(s), lost, Europe.

pityata RAMBUR, 1829, Annls Sci. Observ. 2 (2): 267, pl. 6, fig. 6, (*Phalena* [sic!]). Syntype(s), France (mid): near Montpellier.

UNAVAILABLE NAMES (incorrect subsequent spelling): *aspersaria* (LERAUT 2009). – Intrasubspecific names, potentially unavailable: *melanochroma* LORITZ, 1954 (f.); *obsoleta* LORITZ, 1954 (f.).

EXTERNAL CHARACTERS AND ABDOMEN: Wingspan 22–27 mm. Wings grey-brown, irrorated with brown. Forewing lines oblique, parallel with wing margin. Antemedial line distinctly angled on costa. Medial line wide, blackish. Postmedial line narrow, straight. Subterminal line wide brown band, adjoined by white wavy line. Terminal line blackish. Fringes concolorous with wings. Wing margins crenulate, hindwings more strongly. Hindwing medial lines brown, strongest on anal margin. Discal spots small, blackish. Wings below brown, often with yellowish tint, markings weaker, postmedial line strongest. Frons, collar, thorax and abdomen concolorous with wings. Antennae concolorous with wings dorsally. Male antennae filiform, densely setose. Female antennae filiform. Hindtibia with 2+2 spurs. Tympanal organs medium-sized. Posterior margin of male 8th sternite weakly concave, not otherwise modified. Other sternites and tergites of both sexes undifferentiated.

VARIATION: Specimens may be strongly contrasting, with greyish ground and markings brown-blackish, to brown ground colour and markings slightly darker brown, to almost uniform grey-brown or brown and markings nearly absent (f. *pityata* RAMBUR, 1829).

MALE GENITALIA: Uncus densely setose, without horns, apex with small knob. Gnathos arms narrow, fused ventrally, forming upturned, round, granulate element. Juxta large plate. Valva divided. Dorsal arm of valva long, rather narrow, straight, costa and apex setose. Ventral arm of valva triangular, flat, apex round, weakly setose. Its ventral margin weakly curved inwards, with narrow subapical, weakly sclerotised ridge. Saccus narrow, elongated medially. Aedeagus straight, caecum minute, apex sharp. Vesica opens at approximately 180 degree angle, basal part narrow, medial part with two diverticula, smaller with approximately 10 long, sclerotised cornuti. Ductus ejaculatorius opens from apex.

FEMALE GENITALIA: Papillae anales roundish, setose. Apophyses straight, narrow. Lamella antevaginalis membranous ridge. Lamella postvaginalis unrecognisable. Ductus bursae short, rather wide, most sclerotised laterally. Posterior part of corpus bursae long, curved, sclerotised, striated. Anterior part of corpus bursae large, elongated sac, weakly sclerotised. Signum small, stellate, located laterally.

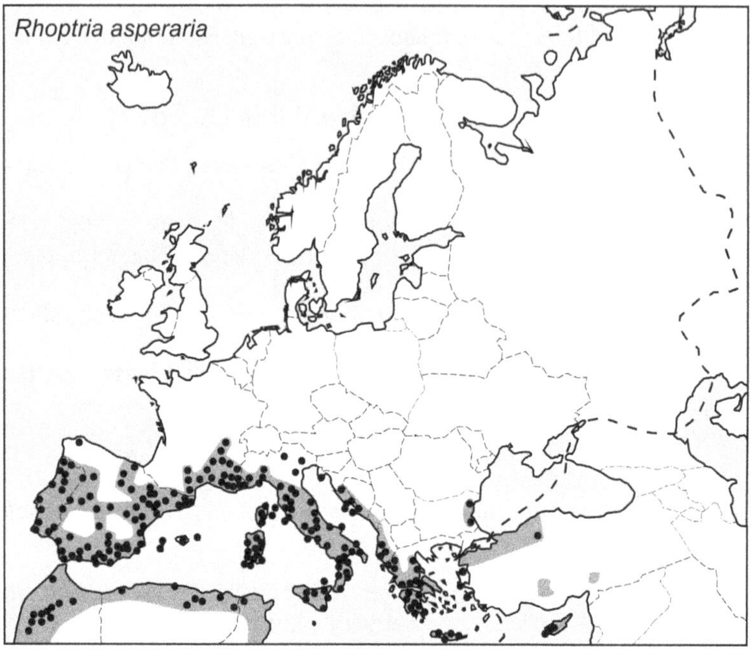

DISTRIBUTION AND ABUNDANCE: Mediterranean(-Macaronesian). In Europe only in the Mediterranean region from Greece including Crete to Portugal with an isolated occurrence in Bulgaria. – Outside Europe in Turkey, Cyprus, Tunisia, Algeria, Morocco and the Macaronesian Islands. – In Europe a common species that often occurs in numbers.

PHENOLOGY: Bi- or more likely trivoltine from mid-March to late November, in some years specimens may be found earlier or later than this. The moths are nocturnal and are strongly attracted to light.

BIOLOGY: Monophagous. The caterpillar feeds on *Cistus*, probably over most of the year. In Portugal found on *C. ladanifer* in mid-October (M. CORLEY pers. comm.).

HABITAT: The species occurs in most habitat types from sea level up to the mountains, most common, however, in the Medierranean macchia. In Sierra Nevada in southern Spain up to 2250 m above sea level (coll. SKOU).

SIMILAR SPECIES: No similar species in Europe. The uniform grey-brown specimens of *R. asperaria* (f. *pityata* RAMBUR) are somewhat similar to *R. dolosaria*. *R. dolosaria* are more uniform grey, postmedial line is broken into spots, angled inwards on costa (grey-brown, postmedial line narrow, continuous, not angled on costa in *R. asperaria*). Male and female genitalia of these species are diagnostic, see plates.

GENETIC DATA: Genetically heterogeneous (n=27), in Europe with two BINs at 1.4% distance, one from France, Portugal, central Spain and Sardinia, another from eastern Spain, Italy, Sicily, Croatia, Bulgaria and Cyprus. In Morocco with an additional BIN at 2.7% distance. Nearest species in Europe (genetic 'similarity' due to long branch attraction): *Calamodes subscudularia* (TURATI, 1919) (9.6%). Congenerics genetically less similar.

50. *Rhoptria dolosaria* (HERRICH-SCHÄFFER, 1848)

dolosaria HERRICH-SCHÄFFER, 1848, Syst. Bearbeitung Schmett. Eur. 3 (31): 74; ibidem, pl. 70, fig. 430 (non binominal), (*Gnophos*). Syntype(s), [Greece]: Creta.

EXTERNAL CHARACTERS AND ABDOMEN: Wingspan 21–24 mm. Wings grey-brownish, irrorated with brown. Medial lines reduced to blackish dots, strongest on costa, postmedial line more complete. Wavy line grey, undulate. Terminal area slightly darker than rest of wing. Terminal line weak, incomplete. Fringes concolorous with wings. Hindwing margin weakly crenulate. Discal spots small, blackish. Wings below brownish grey, postmedial line and discal spots most visible. Frons, collar, thorax and abdomen concolorous with wings. Male antennae filiform, densely setose. Female antennae filiform. Hindtibia with 2+2 spurs, male hindtibia with hairpencil. Tympanal organs large, small sclerotized plate between tympanic bullae. Posterior margin of male 8th sternite weakly concave, not otherwise modified. Other sternites and tergites of both sexes undifferentiated.

VARIATION: Very little. Medial lines usually reduced to row of blackish dots, occasionally lines are more complete, particularly the postmedial line.

MALE GENITALIA: Generally as in *Rhoptria asperaria*, but with the following differences: In *R. dolosaria* ventral arm of valva is more triangular, apex is more acute and narrow, subapical, weakly sclerotised ridge is absent (ventral arm of valva not triangular, apex round, subapical ridge present in *R. asperaria*). Vesica opens at approximately 90 degree angle, basal part narrow, medial part large, with small diverticulum, with approximately 10 long, sclerotised cornuti. Ductus ejaculatorius opens from apex.

FEMALE GENITALIA: Papillae anales roundish, setose. Apophyses straight, narrow. Lamella antevaginalis deeply embedded in sinus vaginalis, sclerotised, with round, lateral lobes. Lamella postvaginalis unrecognisable. Ductus bursae short, rather wide, sclerotised. Posterior part of corpus bursae long, curved, sclerotised, striated, with numerous minute dentate sclerotisations. Anterior part of corpus bursae large, elliptic sac, sclerotised. Signum absent.

DISTRIBUTION AND ABUNDANCE: East-Mediterranean. In Europe only in Greece. Though *R. dolosaria* has been described on the basis of a specimen reputedly from Crete, it has not been possible to locate any specimens from there (RUCKDESCHEL 2007) so the species cannot be verified from the island. – Outside Europe apparently only in Turkey. – In Europe the species is not common and rather local and it occurs only rarely in numbers.

PHENOLOGY: Univoltine. From early April until late June, in Samos slightly earlier in March and April. Larvae from late April and May (FRITSCH *et al.* 2014). Pupation took place in late May or June and the pupae hibernated (D. FRITSCH pers. comm.). The adult insects fly at night and come readily to light.

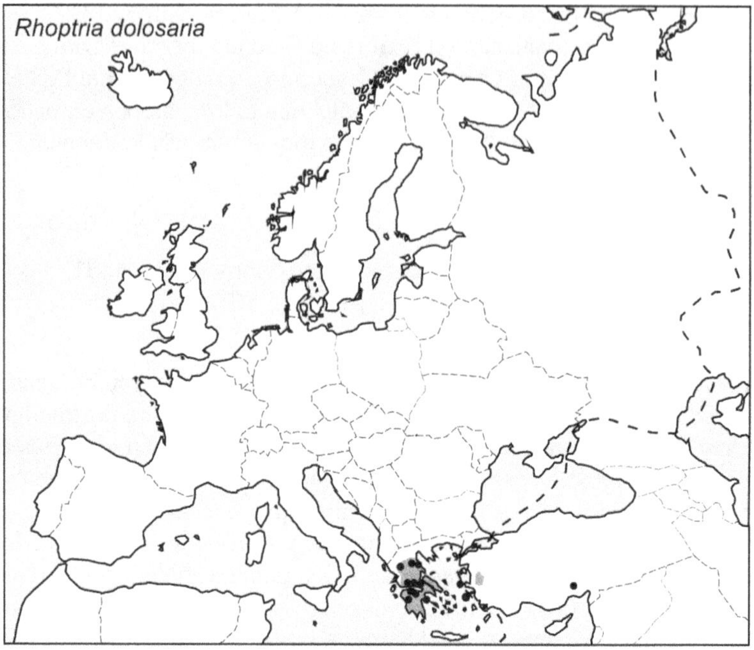

BIOLOGY: The caterpillar has been found on *Pistacia lentiscus* in Samos (FRITSCH *et al.* 2014).

HABITAT: Mainly in warm lowland habitats with macchia vegetation, both in localities with rather low and open growth, and in places where the vegetation is rather high and dense. In central Greece at 750 m (WEIDLICH 2008).

SIMILAR SPECIES: *R. asperaria*, see above.

GENETIC DATA: Nearest species in Europe (genetic 'similarity' due to long branch attraction): *Tephronia sepiaria* (HUFNAGEL, 1767) (9.6%). Congeneric *R. asperaria* genetically much less similar (14.5%). The large distance to the morphologically nearest neighbour possibly indicating a phylogenetically old, isolated lineage.

REMARKS: The morphology does not support the relationship between *Rhoptria* and *Tephronia*, as superficially suggested by the genetic data (neighbouring position being due to the long branch attraction). The male genitalia of European *Rhoptria* species, with divided valva, suggests a Macariini position. This character is not present in the morphologically rather uniform European *Tephronia* species, which have rather large setose lobe in the male valva. Also the female signa of *Tephronia* are diagnostic, not sharing the stellate shape of the Macariini.

From the technical point of view, the vesica of *R. dolosaria* was the most difficult to evert of all species covered in this book. The vesica is large, with numerous long cornuti, and the aedeagus is rather small and weakly sclerotised. Therefore the uninflated vesica completely fills the aedeagus, making it difficult to evert the vesica without breaking the structures.

Ortaliella HAUSMANN, 1993

Ortaliella HAUSMANN, 1993, Bonn. zool. Beitr. 44 (3/4): 226. Type species: *Gnophos palaestinensis* CALBERLA, 1891 (Palestine [Israel]). There are no genus synonyms.

DIVERSITY AND DISTRIBUTION: *Ortaliella* contains three species, two of which occur in Israel and Jordania, one species in Greece (HAUSMANN 1993).

EXTERNAL CHARACTERS AND ABDOMEN: Medium-sized to large moths, wingspan 29–40 mm. Wings different shades of brown to ochreous. Postmedial line distinct (2 species) or absent (1 species). Forewing discal spots distinct. Hindwing margin weakly crenulate. Male antennae filiform, densely setose. Male 3rd sternite without setae, posterior margin of 8th sternite weakly concave, not otherwise modified.

MALE GENITALIA: Uncus densely setose, horns absent. Medial element of gnathos downturned, round, granulate. Valva divided. Ventral arm of valva narrow, its dorsal margin with sclerotised process. Juxta large plate. Vesica tubular, with several short cornuti (see Table 1).

FEMALE GENITALIA: Papillae anales, apophyses anteriores, apophyses posteriores and 9th segment extremely long. Lamella antevaginalis sclerotised. Ductus bursae short. Signum stellate (see Table 1).

BIOLOGY: In *ex ovo* cultivation the larvae accepted *Rhamnus cathartica* (M. LEIPNITZ pers. comm.). Adults fly from March to June.

IMMATURE STAGES: No data.

REMARKS: SCOBLE & KRÜGER (2002) did not include *Ortaliella* in the Macariini. We include it in the Macariini because the male and female genitalia have numerous, characteristic Macariini structures. HAUSMANN (1993) proposed that *Rhoptria* GUENÉE, 1858, and *Stueningia* HAUSMANN, 1993, are closely related to *Ortaliella*. Systematic position: Macariini.

51. *Ortaliella gruneraria* (STAUDINGER, 1862)

gruneraria STAUDINGER, 1862, Stettin. ent. Ztg 23 (4-6): 266, (*Gnophos*). Syntypes 1 male, 4 females, Greece: Peloponnes, Taygetos Mts.

EXTERNAL CHARACTERS AND ABDOMEN: Wingspan ♂ 31–34 mm, ♀ 39–40 mm. Wings grey-brown, heavily irrorated with brown. Postmedial line distinct, brown, angled on costa, its outer margin often lighter. Other markings weak. Terminal area often more brown than rest of wing. Terminal line and fringes concolorous with wings. Forewing discal spot distinct, blackish. Hindwing discal spot minute. Forewing margin weakly concave below apex. Hindwing margin weakly crenulate. Wings below uniform yellowish brown, only forewing discal spots clearly visible. Frons, collar, thorax and abdomen concolorous with wings. Male antennae filiform, densely setose. Female antennae filiform. Hindtibia with 2+2 spurs. Tympanal organs large, medial corner with incurved plate. Male 2nd sternite with sclerotised plate between tympanic bullae. Male 8th sternite amphora-shaped, posterior margin weakly concave. Other sternites and tergites of both sexes undifferentiated.

VARIATION: Very little. Ground colour varies from grey-brown to brown. Degree of irroration is somewhat variable.

MALE GENITALIA: Uncus densely setose, without horns, apex round. Gnathos arms narrow, fused ventrally, forming downturned, round, granulate element. Juxta large plate. Valva divided. Dorsal arm of valva long, rather wide, weakly curved, costa weakly setose, apex with horn-like setae. Its ventral margin terminated subapically. Ventral arm of valva narrow, apex acute, setose. Its ventral margin weakly curved inwards, weakly setose. Its dorsal margin with sclerotised, round process. Saccus wide, round. Aedeagus weakly curved, caecum minute, apex sharp. Vesica opens at approximately 90 degree angle. Basal part wide, with several minute, tooth-like sclerotised cornuti above, another group of approximately six short, sclerotised cornuti below, apex narrow tube.

FEMALE GENITALIA: Papillae anales elongated, setose. Apophyses anteriores, apophyses posteriores and 9th segment extremely long. Lamella antevaginalis and lamella postvaginalis fused, forming triangular sclerotised sclerite around ostium bursae. Ductus bursae short, narrow, most sclerotised laterally. Posterior part of corpus bursae very long, narrow, weakly sclerotised and ribbed near ductus seminalis, otherwise membranous, striated. Anterior part of corpus bursae roundish, membranous sac. Signum small, stellar arms short.

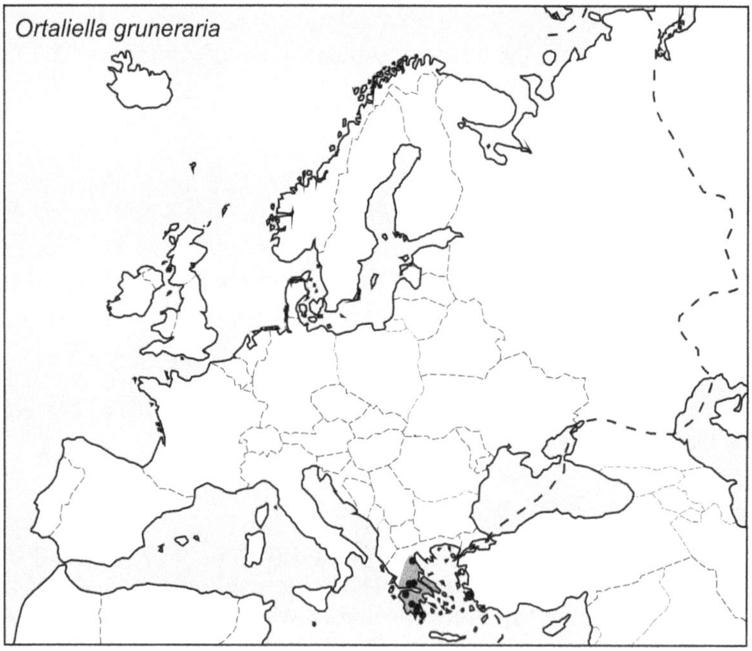

DISTRIBUTION AND ABUNDANCE: East Mediterranean. Endemic in Greece. – Outside Europe there are no records of this species and it has not yet been found in Turkey (F. CAN CENGIZ pers. comm.), but since it has been found in Samos close to Turkey, it is likely to occur also along the Turkish west coast. – In Greece a rather uncommon species with a relatively scarce occurrence, but it has by one occasion been found in numbers (Mount Parnassos, 1700 m, N. PÖLL pers. comm.). Its relative scarcity may be from a lack of collecting in suitable habitats.

PHENOLOGY: Univoltine. From mid-May until late June, in higher altitudes in late June and early July and sometimes until late July. The moths are nocturnal and are attracted to light. In rearing the species has hibernated as an egg (M. LEIPNITZ pers. comm.). The moths fly by night and come to light.

BIOLOGY: In rearing from eggs the larvae have accepted *Rhamnus cathartica* as food plant, caterpillars from late March to early May (M. LEIPNITZ pers. comm.).

HABITAT: Apparently occurring in a variety of habitats from sea level up to 1700 m such as rocky and dry slopes with bushes, gorges and valleys with Mediterranean macchia vegetation including *Cistus*, *Genista*, *Rhamnus*, *Crataegus* and others. At Mount Parnassos just above the timber line.

SIMILAR SPECIES: No similar species in Europe. *O. palaestinensis* (CALBERLA, 1891), which is known from Israel and Jordan, is yellowish brown, with tint of olive, and postmedial line is wider. HAUSMANN (1993) illustrates adult, male and female genitalia.

GENETIC DATA: Genetically homogeneous in Greece (n=3). Nearest species: *O. palaestinensis* (STAUDINGER, 1862) from the eastern Mediterranean (5.1%).

Cepphis HÜBNER, 1823

Cepphis HÜBNER, [1823] 1816, Verz. bekannter Schmett.: 294. Type species: *Geometra advenaria* HÜBNER, [1790] 1796, Ukraine; [Germany:] Leipzig. Genus synonyms are after SCOBLE (1999).

Hetrione POPPIUS, 1887, Acta Soc. Fauna Flora fenn. 3 (3): 29 (key), 56. Type species: *Geometra advenaria* HÜBNER, [1799] 1796.

Priocycla GUENÉE, [1858], in BOISDUVAL & GUENÉE, Hist. nat. Insectes (Spec. gén. Lépid.) 9: 90. Type species: *Priocycla armataria* GUENÉE, [1858].

UNAVAILABLE NAMES (incorrect subsequent spelling): *Cephis* (WEHRLI 1939–1954).

DIVERSITY AND DISTRIBUTION: *Cepphis* contains three species, one in the Palaearctic region and two in the Nearctic region. PITKIN (2002) noted that two further Neotropical taxa assigned to *Cepphis* have been misplaced.

EXTERNAL CHARACTERS AND ABDOMEN (description is based on one European and one Nearctic species): Medium-sized moths, wingspan 21–32 mm. Wings yellow-brown, red-brown or grey-brown. Postmedial line angled. Forewing margin concave below apex. Hindwing margin weakly concave or crenulate. Forewing accessory cell absent. Male antennae bipectinate, female antennae shortly bipectinate. Male 3rd sternite without setae.

MALE GENITALIA (description is based on one European and one Nearctic species): Uncus long, evenly tapering. Socii minute, setose. Gnathos arms fused ventrally, apex narrow, upcurved. Valves membranous, costa weakly sclerotised, apex round. Membranous sac with setae (coremata) present or absent. Furca arms long, narrow, symmetrical. Saccus concave. Aedeagus with several cornuti.

FEMALE GENITALIA (description is based on one European and one Nearctic species): Papillae anales round. Ductus bursae short. Signum minute, located at anterior part of corpus bursae.

BIOLOGY: Larvae have been recorded feeding on numerous deciduous trees and shrubs, that of *C. advenaria* feeds for instance on *Vaccinium myrtillus*, *Ribes nigrum*, *Melampyrum* (MIKKOLA *et al.* 1989), *Actaea* (EBERT 2003), the North American *C. armataria* (HERRICH-SCHÄFFER, 1855) is polyphagous (see MCGUFFIN 1987, SCOBLE 1999) and *C. decoloraria* (HULST, 1886) has been recorded feeding on *Rubus* (MCGUFFIN 1987). Hibernation takes place in the pupal stage. Adults are nocturnal, active from May to July.

IMMATURE STAGES: Larva of *C. advenaria* has zigzag pattern laterally, consisting of white, brown and black areas (EBERT 2003), the North American *C. armataria* (HERRICH-SCHÄFFER, 1855) is a light brown twig-mimic (MCGUFFIN 1987). Chaetotaxy of *C. armataria* is illustrated in MCGUFFIN (1987). Pupa cremaster with eight hooked setae (MCGUFFIN 1987, PATOČKA & TURČANI 2005).

REMARKS: The systematic position of the genus *Cepphis* HÜBNER, including its tribus combination, is unclear. Recently it has been classified in the Anagogini (=Hypochrosini according to HOLLOWAY (1994)) (MCGUFFIN 1987), Ourapterygini (VIIDALEPP 1996, KULLBERG *et al.* 2001 (following VIIDALEPP)), Ennomini (VIVES MORENO 1994) and to Lithinini (LERAUT 1997, HAUSMANN, MIRONOV & VIIDALEPP 2011). Only MCGUFFIN (1987) justified the tribus combination literally. We have compared *C. advenaria* against the tribus descriptions given in HOLLOWAY (1994), and conclude that it agrees with numerous diagnostic characters of the Hypochrosini but the male furcae do not originate from a pocket between the valves, a diagnostic feature of the Hypochrosini. *Cepphis* furcae have a more dorsal position, thus resembling the Ourapterygini as defined in HOLLOWAY (1994). Female genitalia, particularly shape of the corpus bursae, is similar to the Lithinini. Finally, *Cepphis* HÜBNER and *Epione* DUPONCHEL are also structurally similar. They share V-shapedsclerotised pockets at base of valva, membranous sacs with setae (coremata) and concave saccus. Also male 8th sternite is weakly sclerotised, posterior margin is narrow. Morphological differences between Hypochrosini, Ourapterygini and Lithinini, as diagnosed in HOLLOWAY (1994), are small. In a molecular, phylogenetic analysis of limited European taxa only, ÕUNAP *et al.* (2011) concluded that *Cepphis*, *Selenia*, *Plagodis* and *Pseudopanthera* grouped together. In a molecular, phylogenetic analysis of the world Geometridae the Hypochrosini, Ourapterygini and Lithinini did not group together (SIHVONEN *et al.* 2011). Systematic position: Epionini of uncertain association. Further research is needed.

As is explained in more detail in the *Ennominae systematics* chapter, the European Ennominae fauna as a whole has not previously been classified into tribes. Systematic review of tribal classifications and their constituent genera is beyond the scope of this work, mainly because it would require extensive studies on numerous non-European taxa. Therefore in several cases, including this one, we have been forced to arrange higher taxa according to the conventional and practical aspects.

52. *Cepphis advenaria* (HÜBNER, 1790)

advenaria HÜBNER, [1790], Beitr. Gesch. Schmett. 2 (3): 70, pl. (3) 3, fig. Q, (*Ph[alaena] Geom[etra]*). Syntype(s), lost, Ukraine; [Germany]: Leipzig.

advenaria kurilibia BRYK, 1942, Dt. ent. Z. Iris 56: 83, (*Cepphis*). Holotype female (Naturhistoriska Riksmuseet, Stockholm, Sweden), Kurile Islands: Yeterefu, Kamikotan.

UNAVAILABLE NAMES (incorrect subsequent spelling): *adversaria* (REUTTI 1898). – (Intrasubspecific names, potentially unavailable): *approximata* LEMPKE, 1970 (f.); *aurantia* CHALMERS-

Hunt, 1961 (ab.); *brunnea* Lempke, 1970 (f.); *fasciata* Schwingenschuss, 1917 (ab.?); *fulva* Gillner, 1904 (ab.); *lilacina* Lempke, 1951 (f.); *obsoleta* Lempke, 1970 (f.); *rectilinea* Strand, 1919 (ab.); *reducta* Lempke, 1951 (f.); *totibrunnea* Silbernagel, 1944 (f.).

External characters and abdomen: Wingspan 24–30 mm. Wings whitish, markings pale yellow, heavily irrorated with brown, veins brown. Antemedial line angled on costa. Postmedial line distinctly concave, angled outwards before costa. Medial area dark yellow. Terminal area paler, mixed with whitish and pale yellow. Hindwing postmedial line angled. Terminal line narrow, brown. Fringes chequered whitish and pale yellow. Subterminal line reduced to dark spot on costa. Forewing margin concave below apex. Hindwing margin concave at M2. Forewing discal spots absent or weak, hindwing discal spots distinct, small. Wings below more uniform, striated, contrast between medial and terminal areas distinct, forewing discal spots visible. Frons, collar, thorax and abdomen concolorous with wings. Antennae concolorous with wings dorsally. Male antennae bipectinate. Female antennae shortly bipectinate. Hindtibia with 2+2 spurs. Tympanal organs large. Posterior margin of male 8th sternite weakly elongated medially, membranous, not otherwise modified. Other sternites and tergites of both sexes undifferentiated.

Variation: Very little. The markings vary from pale yellow to deep ochre-yellow. Mikkola *et al.* (1989) report a melanistic and an albinistic specimen from Finland.

Male genitalia: Uncus long, evenly tapering, base wide, apex weakly setose. Socii minute, setose. Gnathos arms narrow, fused ventrally, apex acute, upcurved. Tegumen margins x-shaped medially. Furca base bowl-shaped, sclerotised plate. Furca arms long, curved, spinose, symmetrical. Valves membranous, weakly setose all over, costa weakly sclerotised, apex blunt-ended, densely setose. Membranous sacs with long setae dorsad of valva base (coremata). Saccus concave. Aedeagus straight, caecum minute, sclerotised laterally, apex sharp. Vesica opens at approximately 120 degree angle, straight membranous tube. Basal part wider, with small, lateral diverticulum, with about 2–5 long cornuti, apex narrower.

Female genitalia: Papillae anales round, setose. Apophyses narrow, straight. Lamella antevaginalis and postvaginalis fused, forming two round pockets, with elliptical, sclerotised plates, whole complex partially spinose. Ductus bursae short, most sclerotised laterally. Ductus bursae very long, narrow, curved, membranous. Anterior part slightly wider. Signum minute, located at anterior part of corpus bursae.

Distribution and abundance: Eurasiatic. Widely distributed in temperate Europe from southern Scandinavia to northern Spain, northern Italy and the northern part of the Balkans and from the British Isles through central and eastern Europe to European Russia and the Ural mountains. – Outside Europe a single record in northern Turkey and further the species occurs through Siberia, Korea and Far East Russia to Japan, Sakhalin and the Kurile Islands. – In Europe a local species that now and then occurs in numbers at its localities.

Phenology: Univoltine. The moths fly from late May to early July in northern Europe and in central Europe from early May to mid-July. In the Czech Republic from late May to early July (Z. Laštůvka pers. comm.). In northernmost Spain in May and June (V. Redondo pers. comm.). In Romania from late May to late July (L. Székely pers. comm.). Larva from June to August. The species hibernates as pupa in the ground. The moths fly at night when they readily come to light and are easily disturbed by daytime.

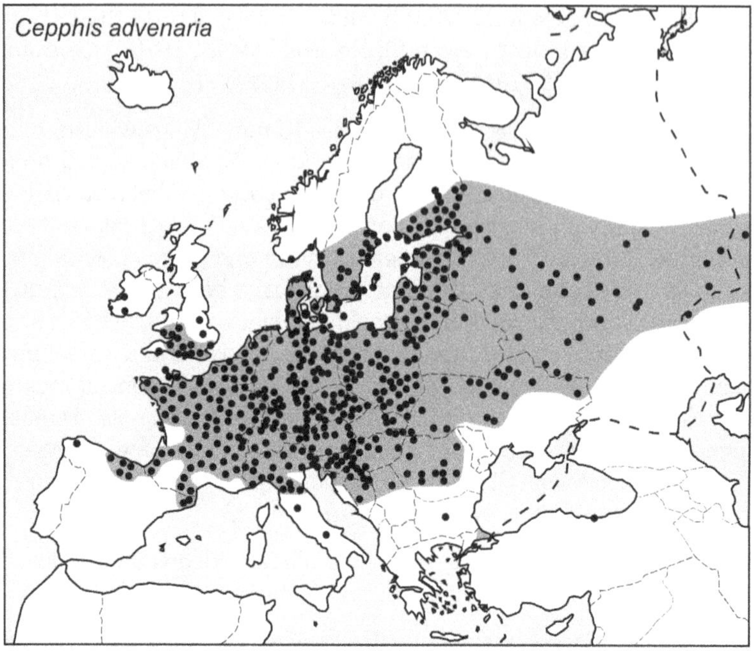

BIOLOGY: Polyphagous. The larva has been found on *Actaea spicata* (B. MÜLLER & E. FRIEDRICH pers. comm.) and on *Vaccinium myrtillus* (E. FRIEDRICH & H. WEGNER pers. comm.). Also on *Ribes* and *Salix* (HULDÉN *et al.* 2000). In the literature (e. g. REDONDO *et al.* 2009, FLAMIGNI *et al.* 2007) also other food plants are recorded, such as *Fragaria*, *Melampyrum*, *Gentiana acaulis*, *Ribes*, *Rubus*, *Crataegus*, *Prunus spinosa*, *Cornus sanguinea* and *Quercus*.

HABITAT: Sylvicolous, hygrophilous. Forests, both on acid as well as on calcareous ground.

SIMILAR SPECIES: No similar species in Europe.

GENETIC DATA: Genetically homogeneous (n=14, from five European countries). Nearest species in Europe: *Selenia lunularia* (6.4%), outside Europe *Epholca fractistriga* (ALPHÉRAKY, 1892) (4.9%). Distance of the closest congeneric species in current taxonomy, the North American *Cepphis decoloraria* (HULST, 1886) much larger (9.6%).

Petrophora HÜBNER, 1811

Petrophora HÜBNER, [1811] 1806, Samml. exot. Schmett. 1: pl. [207]. Type species: *Petrophora divisata* HÜBNER, [1811] 1806, [North America]. Genus synonyms are after SCOBLE (1999).

Lithina HÜBNER, [1825] 1816, Verz. bekannter Schmett.: 338. Type species: *Geometra petraria* HÜBNER, [1799] 1796.

Loxogramma AGASSIZ, 1847, Nomencl. zool. (Index univl.): 216 [Emendation of *Lozogramma* STEPHENS, 1829].

Lozogramma STEPHENS, 1829, Nom. Br. Insects: 44. Type species: *Geometra petraria* HÜBNER, [1799] 1796. [Junior objective synonym of *Lithina* HÜBNER].

Ortholitha HÜBNER, 1821, Index. exot. Lepid.: [3]. Type species: *Petrophora divisata* HÜBNER, [1811] 1806. [Junior objective synonym of *Petrophora* HÜBNER].

UNAVAILABLE NAMES (incorrect subsequent spelling): *Ortholita* GLASER, 1887 (see FLETCHER 1979).

DIVERSITY AND DISTRIBUTION: *Petrophora* contains four species in the Holarctic region, one species in Europe.

EXTERNAL CHARACTERS AND ABDOMEN: Medium-sized, wingspan approximately 30 mm. Wings grey-brown, medial and postmedial lines distinct, subterminal line weaker. Forewing discal spots distinct, elongated. Hindwing lighter, postmedial line weak on anal margin. Male antennae filiform. Abdomen slender. Male 3rd sternite with transverse comb of setae, 8th sternite unmodified.

MALE GENITALIA: Uncus long, curved, apex round, setose. Gnathos arms fused ventrally, upturned, spinose. Valva membranous, weakly setose. Transtilla present. Anellus with pair of setose processes. Furca long, narrow rod. Vesica with small number of of large cornuti.

FEMALE GENITALIA: Papillae anales small. Ductus bursae very short. Posterior part of corpus bursae striated, long and narrow or short and wide. Signum sclerotised, dentate plate.

BIOLOGY: Larvae of the Palaearctic *P. chlorosata* (SCOPOLI) and the Neartic *P. subaequaria* (WALKER, 1860) have been recorded to feed on *Pteridium aquilinum* (MIKKOLA *et al.* 1989, MCGUFFIN 1987). Adults are nocturnal.

IMMATURE STAGES: Egg of North American *P. subaequaria* (WALKER) oval, with longitudinal ridges (MCGUFFIN 1987). Larva of *P. chlorosata* (SCOPOLI) green, green-brown or brown with longitudinal stripes, lateral stripe whitish (EBERT 2003). Cremaster of *P. chlorosata* with two stout, apical setae and three pairs of fine, curved subapical setae (PATOČKA & TURČANI 2005).

REMARKS: VIVES MORENO (1994) included *Petrophora* in the Ennomini, but this view has not been adopted elsewhere. Systematic position: Lithinini (WEINTRAUB & SCOBLE 2004).

53. *Petrophora chlorosata* (SCOPOLI, 1763)

chlorosata SCOPOLI, 1763, Ent. Carniolica: 222, fig. 551, (*Phalaena*). Syntype(s), [Slovenja (north-west]: Carniolia.

chlorosata lignaria ZERNY, 1927, Dt. ent. Z. Iris 41 (2): 125, pl. 2, fig. 10, (*Lithina*). Syntypes 2 males, [Spain]: Southern Andalusia, Barranco del Algarrobo. [considered valid at subspecies level by VIVES MORENO (1994)].

innata FABRICIUS, 1794, Ent. Syst. 3 (2): 198, (*Phalaena*). Syntype(s) (lost), Gallia [France].

petraria HÜBNER, [1799], Samml. Eur. Schmett. 5 Geometrae (1): pl. 21, fig. 113, (*Geometra*). Syntype(s), lost, Europe.

petrata HÜBNER, [1825], Verz. bekannter Schmett.: 338, (*Lithina*). [Emendation of *petraria* HÜBNER].

romanaria DANNEHL, 1928, Ent. Zeitschr., Frankf. a. M. 41: 408, (*Phasiane*). Syntype(s), Italy: Lazio, Monti Albani, Campagna romana; Abruzzo: Monte Sirente.

virgaria BORKHAUSEN, 1794, Natur. eur. Schmett. 5: 61, (*Phalaena Geometra*). Syntype(s), Europe.

INTRASUBSPECIFIC NAMES, POTENTIALLY UNAVAILABLE: *brunnescens* LEMPKE, 1970 (f.); *fuscata* LEMPKE, 1951 (f.); *impuncta* LEMPKE, 1970 (f.); *umbrosa* LEMPKE, 1970 (f.); *unilinea* LEMPKE, 1951 (f.).

EXTERNAL CHARACTERS AND ABDOMEN: Wingspan 26–32 mm. Wings uniform brown-grey, costa slightly darker. Medial and postmedial lines oblique, straight, whitish, inner margin brown. Wavy line whitish, narrow, weakly undulating. Terminal line narrow, brown. Fringes concolorous with wings. Forewing discal spots distinct, brown, hindwing discal spots weak. Hindwings lighter, whitish, subterminal line whitish, inner margin brown, strongest on anal margin. Wings below irrorated with brown, markings weaker. Forewing margin weakly concave below apex. Frons, collar and abdomen concolorous with wings. Male and female antennae filiform. Hindtibia with 2+2 spurs, male hindtibia with hairpencil. Tympanal organs large, meeting ventrally, margin of bullae tympani with round lobe. Male 3rd sternite with stout setae. Other sternites and tergites of both sexes undifferentiated.

VARIATION: Very little. Intensity of markings varies little. Hindwing colour varies from whitish to grey-white.

MALE GENITALIA: Uncus long, curved ventrally, apex round, setose. Gnathos arms narrow, fused ventrally, upturned, spinose. Transtilla present. Furca arms rather long, straight, bare, symmetrical. Valva wide, membranous, weakly setose all over, costa weakly sclerotised, angled before apex, apex round. Ventral margin of valva evenly curved. Base of valva setose. Saccus blunt. Aedeagus curved, caecum minute, apex sclerotised projection. Vesica opens at approximately 180 degree angle. Vesica narrow at base, followed by elongated diverticulum, with row of numerous cornuti, apex narrow tube.

FEMALE GENITALIA: Papillae anales small, round, setose. Apophyses narrow, short. Lamella antevaginalis unrecognisable. Lamella postvaginalis minute, irregular sclerotisation. Sinus vaginalis funnel-shaped, most sclerotised laterally. Ductus bursae very short. Posterior part of corpus bursae narrower, striated, with two irregular small sclerotisations. Anterior part large, partly striated, membranous. Signum sclerotised, dentate plate.

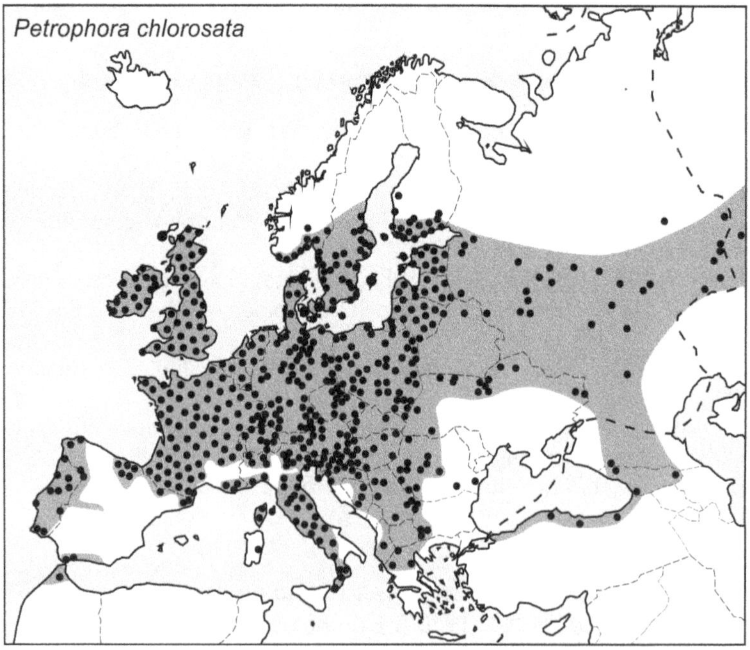

DISTRIBUTION AND ABUNDANCE: Eurasiatic. In Europe the species occurs from southern Scandinavia to the Mediterranean including Corsica and Sardinia and from Ireland in the west to the Ural mountains in the east. – Outside Europe the species occurs in Turkey and the Caucasus region and it is distributed further to the east through southern Siberia to Far East Russia, China, Korea, Japan and Sakhalin. There are also a few records from Morocco (RUNGS 1981). – In Europe a relatively common species that usually occurs in limited numbers.

PHENOLOGY: Univoltine. In southern Scandinavia from late May to late June, in Baden-Württemberg in Germany and Romania from early May to early July (EBERT 2003, L. SZÉKELY pers comm.), in Italy from April to June (FLAMIGNI *et al.* 2007), in Spain from mid-April to late June (V. REDONDO pers. comm., J. YLLA I ULLASTRE pers. comm.) and in Portugal two very early records from early and late March (M. CORLEY pers. comm.). Larva from May-June to August. The species hibernates as pupa. The adult moths fly at night and come regularly to light and are also relatively often disturbed by daytime.

BIOLOGY: Monophagous on ferns. The caterpillar has been found on *Pteridium aquilinum*, and reared on *Athyrium filix-femina* and *Dryopteris filix-mas* (EBERT 2003).

HABITAT: Sylvicolous. Forests and woodland, both moist and more dry, and in meadows and moors, from sea level up to 1500 m in Italy (FLAMIGNI *et al.* 2007) and up to 2000 m in Spain in the Pyrenees (REDONDO *et al.* 2009).

SIMILAR SPECIES: No similar species in Europe. A few *Scotopteryx* HÜBNER, 1825 species (Larentiinae) appear somewhat similar superficially.

GENETIC DATA: Genetically homogeneous (n=24, from eight European countries). Nearest species: *Toulgoetia cauteriata* (7.8%).

Perigune GUMPPENBERG, 1887

Perigune GUMPPENBERG, 1887, Nova Acta Acad. Caesar. Leop. Carol. 59: 331 (key). Type species: *Geometra partitaria* HÜBNER, 1813, Europe. *G. partitaria* HÜBNER is junior synonym of *Phalaena narbonea* LINNAEUS, 1767. There are no genus synonyms.

DIVERSITY AND DISTRIBUTION: *Perigune* contains four species in the Palaearctic region.

EXTERNAL CHARACTERS AND ABDOMEN: Medium-sized, wingspan approximately 20–28 mm. Wings grey to grey-brown. Medial and postmedial lines distinct, yellowish. Hindwings almost uniform grey. Male antennae filiform. Male 3rd sternite without setae, 8th sternite unmodified.

MALE GENITALIA: Uncus long, curved, apex acute, setose. Socii minute, setose. Gnathos arms fused ventrally, upturned, spinose. Valva membranous, costa setose. Transtilla present. Anellus with pair of setose processes. Furca short, serrate. Saccus bifurcate. Vesica wide at base, with one or two large cornuti.

FEMALE GENITALIA: Papillae anales small. Ductus bursae very short. Corpus bursae narrow, long, curved or wide, elongated. Signum absent or minute.

BIOLOGY: Larvae are oligophagous. They have been recorded feeding on *Thymus*, *Rosmarinus officinalis*, *Lavandula* and *Teucrium* (REDONDO *et al.* 2009).

IMMATURE STAGES: Larva of *P. convergata* (VILLERS) green or brown with longitudinal whitish stripes (AIZPÚRUA 2002). No data on pupa, apart from superficial illustration in Aizpúrua (2002). No data available on egg.

REMARKS: The taxa that we combine to *Perigune* GUMPPENBERG, in accordance with REDONDO *et al.* (2009) and LERAUT (2009), have often been combined to genus *Petrophora* HÜBNER. Table 2 lists diagnostic features of the European species in these genera. Type species of the genus *Liodesina* is *L. mesatlantis* WEHRLI, 1938, described from Morocco. *L. mesatlantis* is unicolorous brown, externally similar to *Isturgia homochromata* (MABILLE). WEHRLI (1938) described the male genitalia of *L. mesatlantis* (no illustration) and concluded that *Liodesina* is related to the genera *Epione* and *Lignyoptera*. We have dissected and examined both male (slide PS1267) and female genitalia (slide BMNH GEO 24001) of *L. mesatlantis* (material is in the Natural History Museum, London, UK) and conclude that these are structurally very similar to the European *Perigune* species, potentially subject to synonymisation. Male antennae of *L. mesatlantis* are bipectinate, with long pectinations, unlike in *Perigune*. Systematic position: Lithinini of uncertain association. Further research is needed.

Table 2. Diagnostic characters of genera *Petrophora* HÜBNER, 1811 and *Perigune* GUMPPENBERG, 1887.

Character	*Petrophora*	*Perigune*
male 3rd sternite setae	present	absent
male hindtibial hairpencil	present	absent
apex of uncus	round	acute
saccus	blunt	bifurcate
vesica	narrow at base	wide at base
cornutus	numerous (>10)	one or two
signum	large, sclerotised, dentate plate	absent or minute
larval food plants	ferns (e.g. *Pteridium*)	various Lamiaceae, e.g. thyme (*Thymus*), rosemary (*Rosmarinus*)
Included European species	*P. chlorosata*	*P. binaevata, P. convergata, P. narbonea*

54. *Perigune binaevata* (MABILLE, 1869)

binaevata binaevata MABILLE, 1869, Annls. Soc. ent. Fr. (4) 9 (1): 56, pl. 2, figs 1, 2, (*Tephrina*). Lectotype male (C. HERBULOT Collection in Zoologische Staatssammlung, Munich, Germany), [France: Corsica]: vallée du Fango (examined externally).

binaevata austautaria OBERTHÜR, 1881, Études ent. 6: 83, pl. 3, fig. 11, (*Tephrina*). Syntype(s), Algeria: near Oran. [Originally as sp].. Putatively valid at subspecies level, See remarks.

binaevata fraternaria ROTHSCHILD, 1914, Novit. zool. 21 (4): 352, (*Tephrina*). Syntypes 68 males, more than 1 female (The Natural History Museum, London, UK), Algeria (central): Guelt-Es-Stel (examined externally). [Originally as sp].. Putatively valid at subspecies level, See remarks.

INTRASUBSPECIFIC NAMES, POTENTIALLY UNAVAILABLE: *adustata* BYTINSKI-SALZ, 1937 (ab.?); *transversata* BYTINSKI-SALZ, 1937 (ab.?).

EXTERNAL CHARACTERS AND ABDOMEN: Wingspan 21–25 mm. Wings light grey, irrorated with blackish, with two suffused bands traversing forewings. Medial area often lightest. Medial area yellowish, interior margin blackish. Postmedial lines yellowish, weakly undulating, outer margin blackish. Subterminal line blackish, wide, suffused. Terminal area with distinct, blackish spot. Terminal line faint, blackish. Fringes concolorous with wings. Forewing discal spots distinct, blackish. Hindwing discal spots absent or minute. Hindwings uniform grey. Wings below with weaker markings, hindwings yellowish, heavily irrorated or striated with blackish. Frons light grey. Collar, thorax and abdomen concolorous with wings. Male and female antennae filiform. Hindtibia with 2+2 spurs. Tympanal organs large. Sternites and tergites of both sexes undifferentiated.

VARIATION: Density of irroration variable. Medial and postmedial lines normally yellowish, occasionally yellowish colour absent. Hindwings grey in fresh specimens, browner in old specimens. Two suffused, forewing traversing bands normally present, well-developed, particularly one near inner margin, but these may be reduced.

Putative subsp. ***austautaria*** is less contrasting (Text-figs 75–77). Medial area light grey, medial and postmedial yellow, bordered with brown areas. Terminal area with two distinct, blackish spots. Hindwings light grey. Described from Algeria, known also from Libya (our observation) and from Morocco: Oujda (D. STADIE pers. comm.). Distribution not known in detail. LERAUT (2009) considered *austautaria* a junior synonym of *binaevata* MABILLE. See remarks.

Putative subsp. ***fraternaria*** is more uniform, ground colour grey-brown (Text-figs 75–77). Postmedial line less angled, external appearance approaching *P. convergata* (VILLERS). Described from Algeria, recorded also from Tunisia and Morocco: Oujda (D. STADIE pers. comm.). Distribution not known in detail. LERAUT (2009) proposed *fraternaria* a junior synonym of *binaevata* MABILLE. See remarks.

MALE GENITALIA: Uncus long, curved ventrally, apex acute, setose. Socii minute, setose. Gnathos arms narrow, fused ventrally, upturned, spinose. Transtilla with two lobes. Furca arms short, apex weakly serrate. Valva narrow, membranous, apex round. Costa sclerotised, setose and surface granulate, particularly below apex. Base of valva weakly setose. Saccus bifurcate, wide. Aedeagus long, narrow, apex acute. Vesica opens at approximately 90 degree angle. Vesica wide at base, with one or two small cornuti, followed by small diverticulum and one long cornutus. Apex narrow tube.

FEMALE GENITALIA: Papillae anales small, round, setose. Apophyses narrow, short. Lamella antevaginalis narrow, lunular, weakly sclerotised. Lamella postvaginalis unrecognisable. Ductus bursae short, narrow, most sclerotised laterally. Corpus bursae membranous, its shape somewhat variable. Posterior part of corpus bursae very long, narrow, weakly curved, membranous, weakly striated. Anterior part weakly expanded, roundish. Signum absent or minute, weakly stellate, located at anterior end.

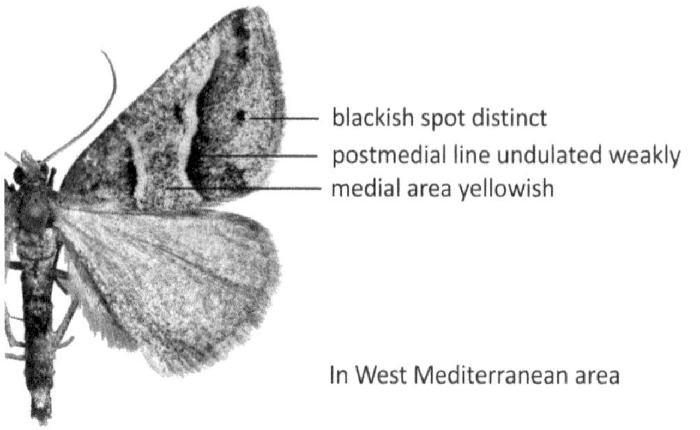

P. binaevata* subsp. *binaevata 75

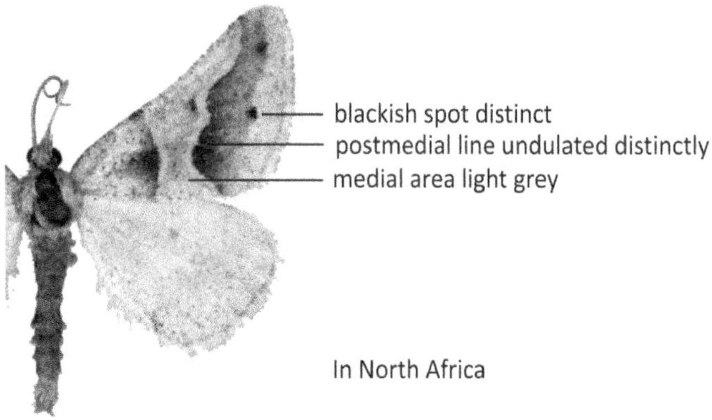

P. binaevata* subsp. *austautaria 76

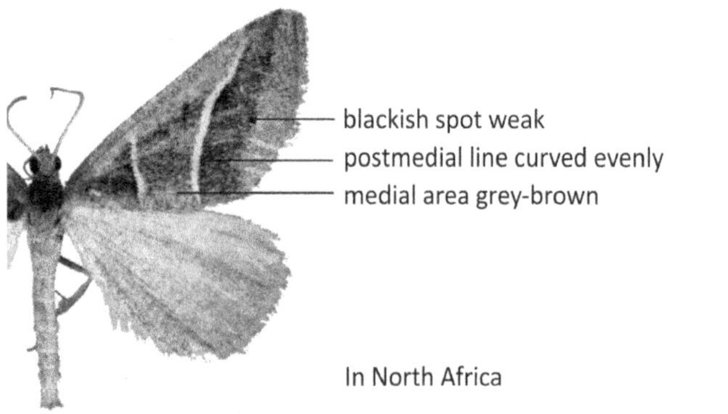

P. binaevata* subsp. *fraternaria 77

Text-figs 75–77. Diagnostic characters (indicated) of *Perigune binaevata* (MABILLE, 1869) and its subspecies. The taxonomic status of putative subspecies needs further research. Text-fig. 75. *P. binaevata binaevata* (MABILLE, 1869), Italy, 13.x.2010 (coll. SKOU). Text-fig. 76. *P. binaevata austautaria* (OBERTHÜR, 1881), Libya: Bardia, November (coll. BMNH). Text-fig. 77. *P. binaevata fraternaria* (ROTHSCHILD, 1914), Algeria, 20.x.1913 (coll. BMNH, syntype).

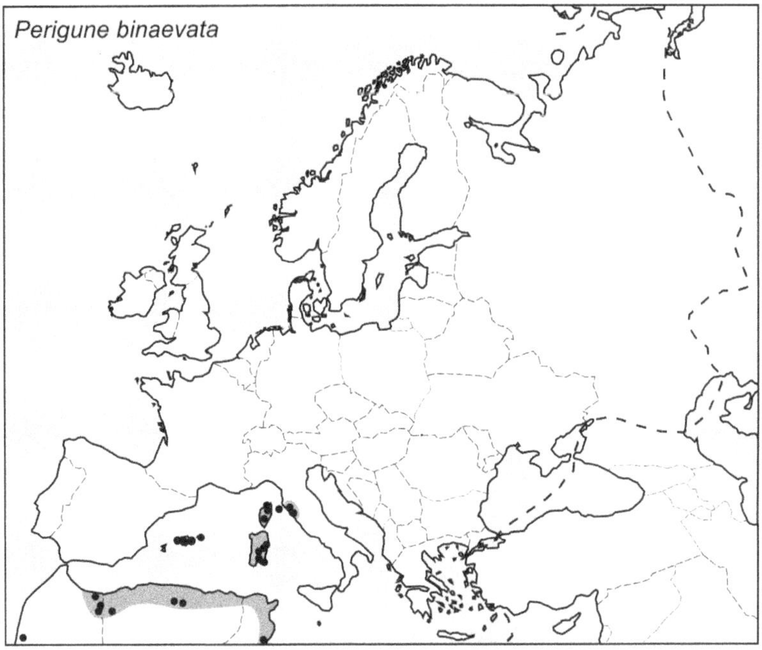

Distribution and abundance: West-Mediterranean. In Europe a rather limited distribution in the western and central Mediterranean area where it occurs in the Balearic Islands, Corsica and Sardinia, in the small islands Capráia, Elba and Montecristo, and on the Italian mainland in a limited coastal area in southernmost Tuscany. – Outside Europe the species is recorded from Libya, Tunisia, Algeria and Morocco (the putative subsp. *austautaria* and subsp. *fraternaria*). – In Europe a local and uncommon species that rarely occurs in numbers.

Phenology: Univoltine. From late September to early November. The adult moths are attracted to light. According to Rungs (1981) in Morocco from February to April and rarely in October and November. The moths fly at night and come to light.

Biology: Oligophagous. The larva is recorded on *Rosmarinus officinalis* (Flamigni *et al.* 2007) and on *Thymus* (Redondo *et al.* 2009).

Habitat: In Mediterranean macchia and scrubland, and according to Redondo *et al.* (2009), also in places with *Quercus* and *Pinus halepensis* woodland up to 900 m altitude, otherwise mainly in lowlands.

Similar species: *Perigune convergata* (Villers) has evenly curved postmedial line, without angle in middle, less sinuous and reaching costa further away from apex (postmedial line not evenly curved, with angle in middle, more sinuous, reaching costa closer to apex in *P. binaevata*). *Perigune convergata* has subterminal line less marked, normally without dark area between postmedial line and subterminal line, dark spot small (subterminal line dark, normally with dark area between postmedial line and subterminal line, dark spot distinct in *P. binaevata*). Male and female genitalia of *P. binaevata* and *P. convergata* are without qualitative differences. *P. narbonea* has medial and postmedial lines nearly parallel, hindwing subterminal line normally visible (Text-figs 78–80).

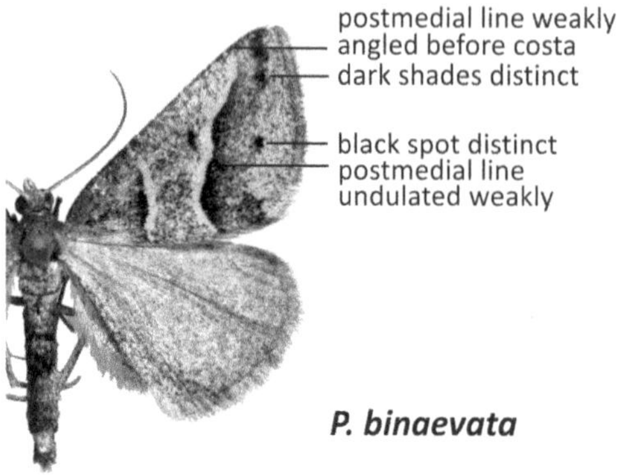

P. binaevata 78

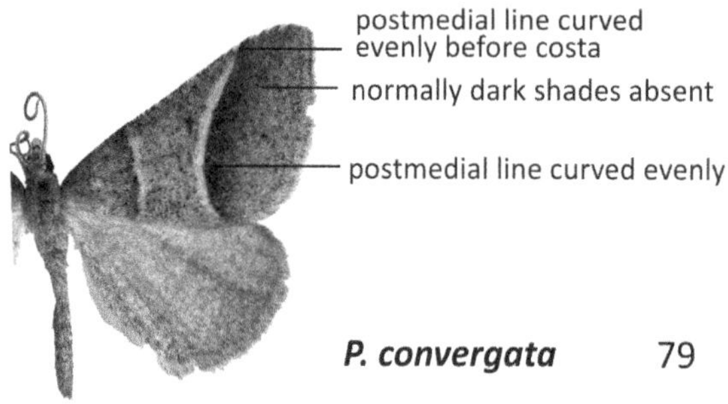

P. convergata 79

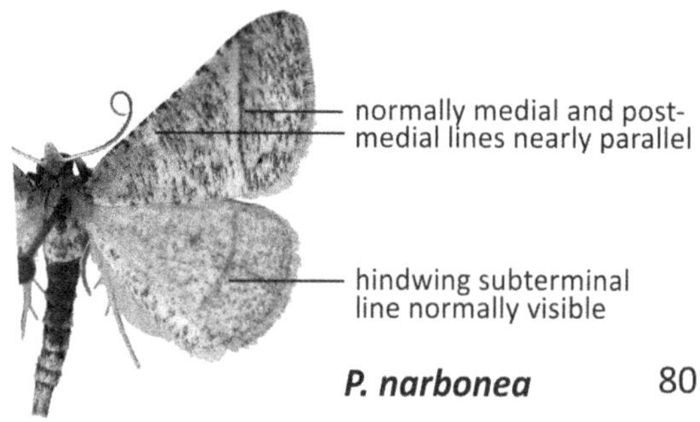

P. narbonea 80

Text-figs 78–80. Diagnostic characters (indicated) of European *Perigune* species. Text-fig. 78. *P. binaevata* (MABILLE, 1869), Italy, 13.x.2010. Text-fig. 79. *P. convergata* (VILLERS, 1789), France, 13.x.1996. Text-fig. 80. *P. narbonea* (LINNAEUS, 1767), France, 10.viii.1996.

GENETIC DATA: The *binaevata/convergata* species complex in Europe genetically splits up into four BINs. One specimen from Sardinia BIN-sharing (but at 1.2% distance) with populations from Tuscany and Calabria. All populations from North Africa currently assigned to *P. binaevata* genetically strongly diverging (minimum distance 5.0% in three different BINs from Morocco to Tunisia). Nearest species: *P. convergata* (1.4%).

REMARKS: *Perigune binaevata*, *P. convergata* and their subspecies form a species complex that needs further study. Most specimens can be identified on the basis of external characters and the taxa seem to be confined to different habitats: *P. binaevata* has so far been found close to the coasts at low altitudes while *P. convergata* has been found from coast, inland and in high altitudes (up to 2250 metres in Sierra Nevada, Spain). However, there are intermediate forms and the male and female genitalia are without qualitative differences. Non-European *P. jordanaria* (STAUDINGER, 1901) is part of the complex also.

The taxonomic status of *Tephrina austautaria* OBERTHÜR has remained controversial. It has been considered valid at species level (e.g. RUNGS 1981), subspecies level (e.g. SCOBLE 1999) and a synonym of *P. binaevata* (LERAUT 2009). Male and female genitalia are without qualitative differences when compared to other taxa in the *P. binaevata* and *P. convergata* complex. *P. binaevata* subsp. *austautaria* and *P. binaevata* subsp. *fraternaria* occur sympatrically in Oujda, Marocco (D. STADIE pers. comm.). Genetic data may indicate that populations from North Africa, currently assigned to *P. binaevata austautaria* and *P. binaevata fraternaria*, are isolated and potentially requiring consideration at species level. Further, it brings into question also the validity of *P. binaevata* and *P. convergata* at species level, as currently defined. Because the matter is unsatisfactorily examined at the moment, we do not assign specimens from different areas to different subspecies in the distribution map.

55. Perigune convergata (VILLERS, 1789)

convergata convergata VILLERS, 1789, Linn. ent. 2: 382, pl. 6, fig. 18, (*Phal*[*aena*] *Geom*[*etra*]. Syntype(s), [France]: Lyon.
peltaria BOISDUVAL, 1840, Genera Index meth. eur. Lepid.: 201, (*Eubolia*). Syntype(s), Gall. mer. [Spain].
permutaria FREYER, [1836], Neuere Beitr. Schmettkde 3: 28, pl. 210, fig. 3, (*Aspilates*). Syntype(s), France (southern).
scutularia DUPONCHEL, 1830, in GODART & DUPONCHEL, Hist. nat. Lépid. Papillons Fr. 8 (1): 161; pl. 181, fig. 6, (*Phasiane*). Syntype(s), France: near Montpellier.
convergata exoglypha PRAVEIL, 1936, Livre jubilaire de m. Eugene-Louis Bouvier. 279, pl. 11, figs 6-10, 17 (*Lithina*). Holotype male (Muséum National d´Histoire Naturelle, Paris, France), [France] Alpes-Maritimes, Valdeblore-la-Bolline, 1020 m. [Originally as subsp. of *scutularia*].

INTRASUBSPECIFIC NAMES, POTENTIALLY UNAVAILABLE: *approximata* LERAUT, 2009 (f.); *aureolineata* SCHWINGENSCHUSS, 1962 (f.); *carmen* SCHAWERDA, 1927 (f.); *flavofasciata* SCHWINGENSCHUSS, 1962 (f.); *lineacrocea* LERAUT, 2009 (f.); *petiti* LORITZ, 1946 (ab.); *semiumbrosa* LORITZ, 1946 (ab.); *unilineata* SCHWINGENSCHUSS, 1962 (f.).

EXTERNAL CHARACTERS AND ABDOMEN: Wingspan 22–27 mm. Generally as *P. binaevata*, but with the following differences. Postmedial line evenly curved, without angle in middle, less sinuous and it reaches costa further away from apex. Subterminal line less marked, normally without dark area between postmedial line and subterminal line, dark spot small.

Subsp. ***exoglypha*** with wing background ash grey, medial and postmedial lines almost without light yellow. Subspecies status needs to be confirmed, therefore not illustrated in the distribution map.

MALE GENITALIA: Generally as in *P. binaevata* (MABILLE), without notable differences. See *P. binaevata*.

FEMALE GENITALIA: Generally as in *P. binaevata* (MABILLE), but with following potential difference: posterior part of corpus bursae potentially bigger in *P. convergata*. Corpus bursae membranous, therefore its shape somewhat variable. Signum absent or minute, weakly stellate, located at anterior end.

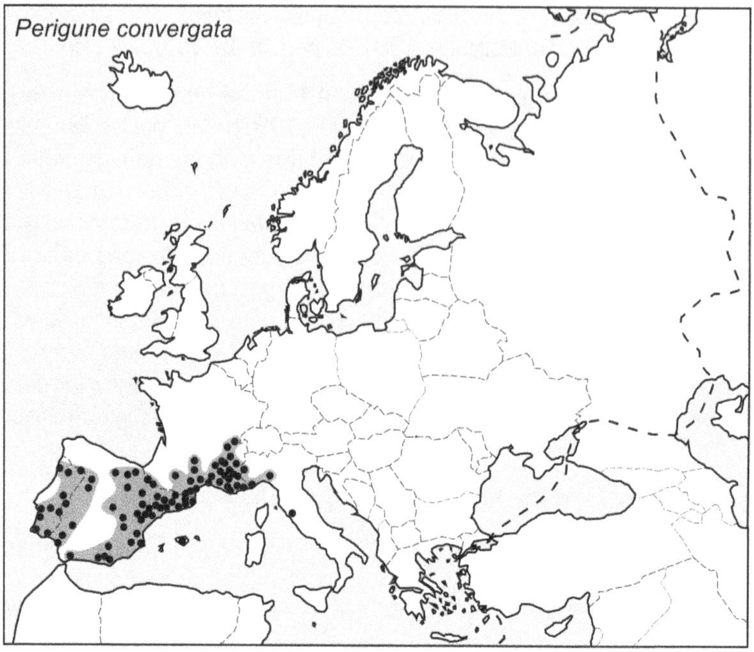

DISTRIBUTION AND ABUNDANCE: West Mediterranean. The species is only known with certainty from Europe where it occurs in most of the Iberian Peninsula, in southern France and in north-westernmost Italy. In Italy furthermore a single specimen in northern Lombardy (L. DAPPORTO pers comm.). – The species was reported from Biskra in Algeria in the early 20th century (CHRÉTIEN 1917), but never since, and we have not been able to locate any specimens from North Africa. – A rather common species that not rarely occurs in numbers.

PHENOLOGY: Univoltine. The moths fly from late August to late November. In coll. SKOU there are two specimens from early April in south-eastern Spain. In north-eastern Spain from September to October (J. YLLA I ULLASTRE pers. comm.), in southernmost Spain recorded in November and December (P. HALE pers. comm.). Caterpillars from early February to mid-May. Probably the species hibernates as an egg until late January, and then as caterpillar. The adult moths fly at night and come abundantly to light.

BIOLOGY: Oligophagous. G. KING (pers. comm.) has found numerous larvae on *Thymus lacaitae* – which may be the main food plant – and on *Lavandula*. Also found on *Lavandula stoechas* (D. FRITSCH pers. comm.). The caterpillar has been recorded on *Thymus* and *Rosmarinus officinalis* (REDONDO *et al.* 2009) and on *Ephedra major* (AIZPÚRUA 2002).

HABITAT: A large variety of open land habitats including macchia. In south-eastern Spain in semi-desert habitats. The species occurs from sea level up to higher altitudes, in central Spain up to 1400 m, in Sierra Nevada in southern Spain up to 2250 m, and abundant there at 1600 m.

SIMILAR SPECIES: *P. binaevata* (MABILLE) and *P. narbonea* (LINNAEUS), see *P. binaevata* and Text-figs. 78–80.

GENETIC DATA: The *binaevata/convergata* species-complex genetically splitting up into four BINs. *P. convergata* genetically heterogeneous and paraphyletic (n=5, Spain, central and north-western Italy), with three BINs at maximum variation 6.7% (pairwise distance). Nearest species: *P. binaevata* (1.4%).

REMARKS: See *P. binaevata* (MABILLE).

56. *Perigune narbonea* (LINNAEUS, 1767)

narbonea narbonea LINNAEUS, 1767, Syst. Nat. (Ed. 12) 1 (2): 837, (*Phalaena* (*Noctua*)). Type (without head) (The Linnean Society of London, London, UK), Lusitania [Portugal] (examined externally).

partitaria HÜBNER, [1813], Samml. Eur. Schmett. 5 Geometrae (1): pl. 72, fig. 374, (*Geometra*). Syntype(s), lost, Europe.

narbonea obliterata STAUDINGER, 1901, in STAUDINGER & REBEL, Cat. Lepid. palaearct. Faunengeb. (Edn 3) 1: 353, (*Phasiane*). Syntype(s), Mauretania (mid) [Algeria]: Lambessa; S Mauretania and [Spain]: Murcia. [Originally as var. of *partitaria*].

UNAVAILABLE NAMES (incorrect subsequent spelling): *littoralaria* (CULOT 1919–1920, WEHRLI 1940 *in* WEHRLI 1939–1954; should be *litoralaria* TURATI, 1905). Intrasubspecific names, potentially unavailable: *faeculenta* THIERRY-MIEG, 1893 (ab.); *jahandiezi* OBERTHÜR, 1922 (ab.?); *litoralaria* TURATI, 1905 (ab.); *punctiferata* REBEL, 1926 (ab.?); *ruscinonensis* CULOT, 1919 (ab.); *unilineata* LORITZ, 1950 (ab.); *versofasciata* LORITZ, 1950 (ab.). WEHRLI (1939–1954) lists further intrasubspecific names. – (invalid name): *bipartita* RAMBUR, 1866 (see LERAUT 1997).

EXTERNAL CHARACTERS AND ABDOMEN: Wingspan 22–28 mm. Wings yellowish grey to reddish brown, heavily irrorated with black. Medial and postmedial lines light yellow, rather straight, parallel. Basal, medial and terminal areas normally with similar colours. Subterminal line absent except minute black and light yellow spot in middle. Terminal line faint, blackish. Fringes concolorous with wings. Discal spots weak. Hindwings uniform grey or terminal area darker, subterminal line weak. Wings below heavily irrorated with black, discal spots larger, hindwing postmedial line often more pronounced. Frons cream-coloured laterally. Collar, thorax and abdomen concolorous with wings. Antennae cream colour dorsally. Male and female antennae filiform. Hindtibia with 2+2 spurs. Tympanal organs medium-sized, medial margin of cavus tympani enlarged. Male 8th tergite weakly triangular. Other sternites and tergites of both sexes undifferentiated.

VARIATION: Postmedial line can be curved, its outer margin with narrow, blackish border (f. *litoralaria*). Density of irroration varies. Hindwing postmedial line normally visible, faint, but can be absent.

Subsp. ***obliterata*** with narrower medial and postmedial lines. Postmedial line curved outwards, not parallel with medial line. Treated as a junior synonym of *narbonea* LINNAEUS by VIVES MORENO (1994), validated at subspecies level by SCOBLE (1999). Subspecies status needs further research.

MALE GENITALIA: Generally as in *P. binaevata* (MABILLE), but with the following differences: Vesica huge at base, with one large cornutus. Apex wider tube, opens laterally.

FEMALE GENITALIA: Generally as in *P. binaevata* (MABILLE), but with the following differences: Ductus bursae bigger. Corpus bursae large, roundish, its anterior tip with small diverticulum. Signum absent or minute, weakly stellate, located near anterior end.

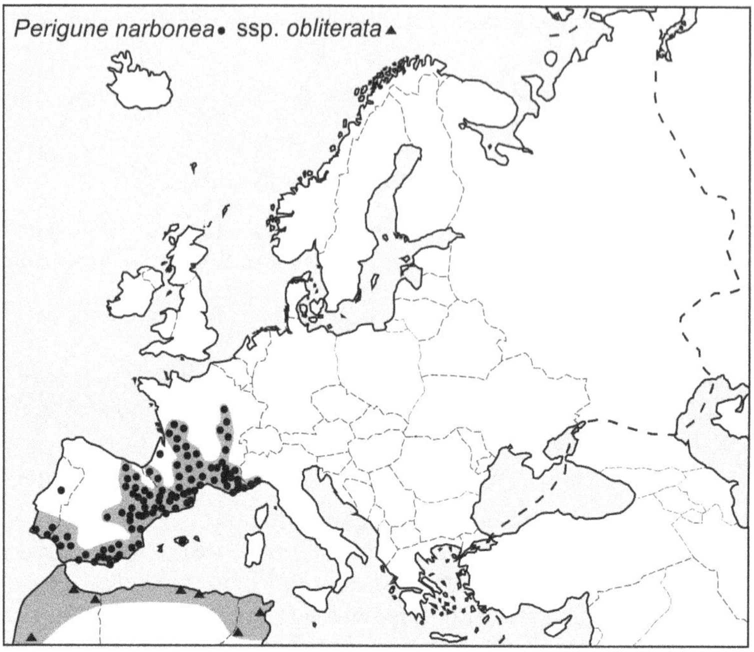

DISTRIBUTION AND ABUNDANCE: West Mediterranean. The European distribution is limited to the Iberian Peninsula, southern and central France and a small area in western Liguria in Italy. – Outside Europe the species is found in Tunisia, Algeria and Morocco (subsp. *obliterata*). – In Europe a rather common species that rather often occurs in numbers.

PHENOLOGY: Bivoltine. The moths fly from early March to early May and from early August to late October, sometimes slightly earlier or later (own data, REDONDO *et al.* 2009). Caterpillars have been recorded from October to June. The adult insects come readily to light, rather often in numbers.

BIOLOGY: Oligophagous on Lamiacae. In Spain monophagous on *Teucrium polium* (KING & VIEJO MONTESINOS 2010). In Italy reported from *Teucrium chaemaedrys*, *T. scorodonia* and *Thymus* (FLAMIGNI *et al.* 2007).

HABITAT: The species occurs in a large variety of open land habitats including macchia. In south-eastern Spain also in semi-desert localities. It occurs from sea level up to at least 1700 m, according to REDONDO *et al.* (2009) up to 2000 m.

SIMILAR SPECIES: No similar species in Europe. *P. narbonea* f. *litoralaria* (TURATI, 1905) resembles *P. convergata* in having the postmedial line curved and its outer margin dark. *P. convergata* has more grey ground colour and hindwing is usually uniform grey, without postmedial line (ground colour more brown, hindwing with postmedial line and dark terminal area in *P. narbonea* f. *litoralaria*). Numerous f. *litoralaria* specimens are illustrated in FLAMIGNI *et al.* (2007). See also Similar species section under *P. binaevata* (MABILLE) and Text-figs 78–80.

GENETIC DATA: Genetically heterogeneous (n=7) from southern Spain to Balearic islands, maximum variation 2.3% (pairwise distance). Nearest species: *P. jordanaria* (STAUDINGER, 1901) from Israel (6.7%).

Lignyoptera LEDERER, 1853

Lignyoptera LEDERER, 1853, Verh. zool. –bot. Ver. Wien 3 (Abh.): 177, 211,212. Type species: *Geometra fumidaria* HÜBNER, [1825], 1796, Europe [objective replacement name for *Acalia* GUENÉE]. Genus synonyms are after SCOBLE (1999).

Acalia GUENÉE, [1845], 1844, in DUPONCHEL, Cat. méth. Lépid. Eur.: 278. Type species: *Geometra fumidaria* HÜBNER, [1825] 1796. [Junior homonym of *Acalia* GREY, 1840 (Echinodermata)].

DIVERSITY AND DISTRIBUTION: *Lignyoptera* contains two species in the Palaearctic region.

EXTERNAL CHARACTERS AND ABDOMEN: Males medium-sized, wingspan 28–33 mm, females brachypterous. Males almost uniform grey-brown or deep reddish yellow, only traces of medial and postmedial lines present on costa. Hindwings uniform, lighter. Male and female antennae filiform. Male tympanal organs small, female tympanal organs apparently reduced. Male 3rd sternite without setae, posterior margin of 8th sternite round medially, elongated laterally. Male 8th tergite weakly triangular.

MALE GENITALIA: Uncus long, curved, apex acute, setose. Socii minute, setose. Gnathos arms fused ventrally, upturned, spinose. Valva membranous, costa setose. Transtilla present. Anellus with pair of setose processes. Furca short, serrate. Saccus weakly bifurcate. Aedeagus serrate subapically. Vesica with cornuti.

FEMALE GENITALIA: Papillae anales small, surface granulate. Lamella antevaginalis/sinus vaginalis sclerotised, funnel-shaped, finely setose. Corpus bursae membranous.

BIOLOGY: Larva of *L. fumidaria* (HÜBNER) has been found on *Achillea millefolium* and on *Gallatella linosyris* (LUNAK 1940) and larva of *L. thaumastaria* has been reared on *Artemisia vulgaris*, *Achillea millefolium* and garden *Chrysanthemum* (LUNAK 1940). Adults are active from October to December. Eggs or small caterpillars overwinter (VOJNITS 1980).

IMMATURE STAGES: We have no data, except the superficial description in VOJNITS (1980): larva of *L. fumidaria* (HÜBNER) is cylindrical, grey-green in colour, with longitudinal light stripes.

REMARKS: The systematic position of genus *Lignyoptera* LEDERER, including its tribus combination, is unclear. Recently it has been classified in the Ennomini (VIIDALEPP 1996, HAUSMANN, MIRONOV & VIIDALEPP 2011), but without any written justification. BELJAEV (2008a) did not include *Lignyoptera* in his new concept of the Ennomini. Male and female genitalia of *Lignyoptera* are structurally very similar to *Perigune* (Lithinini). As is explained in more detail in the *Ennominae systematics* chapter, the European Ennominae fauna as a whole has not previously been classified into tribes. Systematic review of tribal classifications and their constituent genera is beyond scope of this work, mainly because it would require extensive studies on numerous non-European taxa. Therefore in several cases, including this one, we have been forced to arrange higher taxa according to conventional and practical aspects. Systematic position: Uncertain, further research is needed.

57. Lignyoptera fumidaria (HÜBNER, 1825)

fumidaria HÜBNER, [1825], Samml. Eur. Schmett. 5 Geometrae (1): pl. 101, figs 520, 521, (*Geometra*). Syntype(s), lost, Europe.

INTRASUBSPECIFIC NAMES, POTENTIALLY UNAVAILABLE: *nausearia* KRULIKOWSKI, 1906 (ab.) (treated valid at subspecies level in VIIDALEPP (1996), but VIIDALEPP published the name without description, thus not available (ICZN 1999, Art. 13.1., Art. 45.5.1).), see remarks; *rubraria* KAMMEL, 1917 (ab.); *unicoloraria* KAMMEL, 1917 (ab.).

EXTERNAL CHARACTERS AND ABDOMEN: Male: wingspan 27–30 mm. Wings uniform ochreous grey, irrorated with brown, glossy. Medial and postmedial lines present on costa only, suffused, brown. Vein (M3) between medial and postmedial lines brown, these three elements form weakly rectangular, brown shape in forewing medial area. Terminal line absent or very weak. Fringes concolorous with wings. Discal spots small, suffused. Hindwing lighter, irrorated with brown, veins brown. Wings below as above, without medial lines, irrorated with brown, veins brown. Head and thorax covered with long, black hairs. Frons, collar and thorax black. Abdomen concolorous with wings. Antennae whitish dorsally, filiform, flagellomeres heart-shaped. Tympanal organs small. 8th sternite weakly sclerotised, posterior margin round medially, elongated laterally. 8th tergite weakly sclerotised, triangular. Female: brachypterous, wing length about 1 mm. Wings uniform ochreous grey, with few brown scales. Head and thorax not covered with long hairs. Frons, collar, thorax and abdomen concolorous with wings. Antennae blackish dorsally, filiform. Tympanal organs apparently reduced. Sternites and tergites undifferentiated. Hindtibia with 2+2 spurs.

VARIATION: Very little. Density of irroration variable in males, thus appearance varies from ochreous grey to brown.

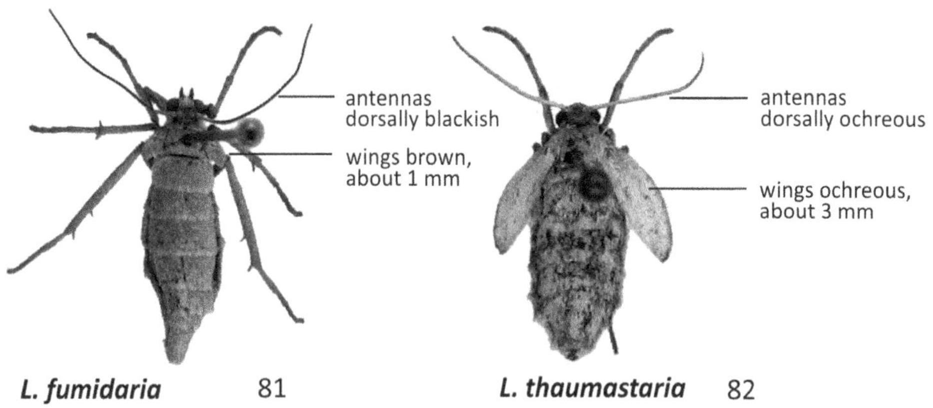

Text-figs 81–82. Diagnostic female characters (indicated) of European *Lignyoptera* LEDERER, 1853 species. Text-fig. 81. *L. fumidaria* (HÜBNER, 1825), Austria, 6.xi.1910. Text-fig. 82. *L. thaumastaria* REBEL, 1901, Bosnia, ex ovo 20.xi.1931.

MALE GENITALIA: Uncus long, curved ventrally, apex acute, setose. Socii minute, setose. Gnathos arms narrow, fused ventrally, upturned, spinose. Transtilla with two lobes. Furca arms serrate, asymmetrical, one apex acute, other apex blunt. Valva narrow, membranous, weakly setose, apex round. Costa sclerotised, more setose, particularly below apex. Base of valva weakly setose. Saccus bifurcate. Aedeagus long, narrow, caecum short, apex long, acute. Aedeagus serrate subapically. Vesica opens at approximately 90 degree angle. Vesica wider at base, with sclerotised plate/row of about 10 microcornuti. Apex narrow, acute.

FEMALE GENITALIA: Papillae anales setose, surface granulate. Apophyses narrow, short. Lamella antevaginalis/sinus vaginalis sclerotised, funnel-shaped, finely setose. Lamella postvaginalis membranous. Ductus bursae short, sclerotised. Posterior part of corpus bursae narrow, membranous, weakly striated. Anterior part of corpus bursae large, round, membranous. Signum absent.

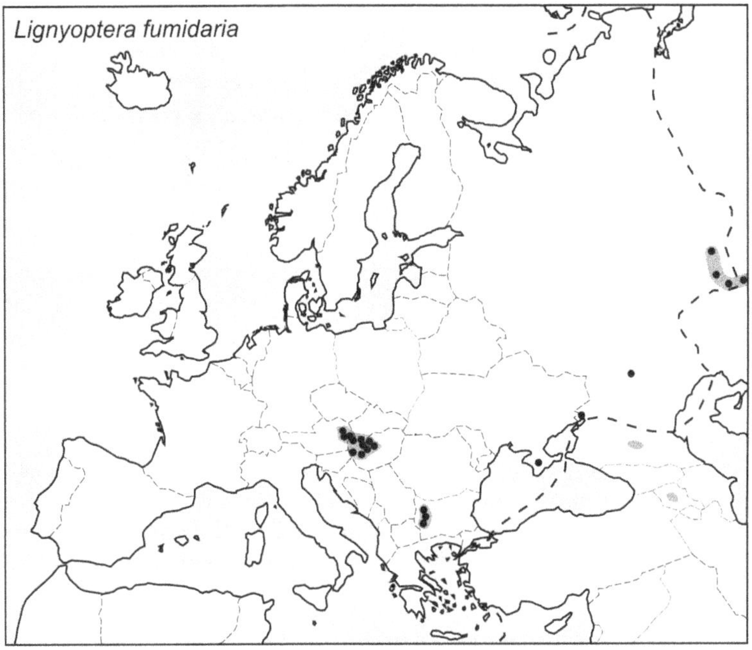

DISTRIBUTION AND ABUNDANCE: Eurasiatic. Recorded from eastern Austria, Hungary, Bulgaria, Crimea and south-eastern European Russia. – Outside Europe the species is recorded from north-west Caucasus and the Altai region (MIRONOV *et al.* 2008). – In Europe a rare and local species that now and then is found in numbers.

PHENOLOGY: Univoltine. The adults occur from late October or early November to early or mid-December. In southern Urals from late September (K. NUPPONEN pers. comm.). Hibernation takes place as an egg. The males fly by day, in sunshine or cloudy conditions, at dusk or at night when they come to light, but not that often (Z. LAŠTŮVKA pers. comm.). PETRÁNYI (2014) also reports the males as day flying, preferable on windless days with sunshine, but they may also fly actively in cloudy warm weather. According to M. LEIPNITZ (pers. comm.), the males may also fly in foggy weather. The wingless females sit half way up stems in quite tall grass vegetation and they are very difficult to find. Copulation takes place in the morning between 9 and 12 a.m.

BIOLOGY: Oligophagous on Asteraceae. The caterpillar has been found on *Achillea millefolium* in more moist meadows, and on *Gallatella linosyris* on drier slopes, but it is also mentioned that these two plant species are the most common food plants and that *L. fumidaria* is polyphagous (LUNAK 1940). KASY (1965) assumes that in eastern Austria the main food plant is *Achillea aspleniifolia*. The species has been reared on *Achillea* (PETRÁNYI 2014). Pupation takes place on the ground.

HABITAT: According to PETRÁNYI (2014), the typical habitats are wind sheltered or partially open grasslands on limestone or even dolomite and in north-western Hungary also halophilous (saline) meadows. Normally the species prefers areas with taller grass vegetation. In Austria also found on sun-exposed, steep slopes with meadow vegetation (LUNAK 1940). Restricted to lowlands, in Austria from 120 to 300 m altitude (N. PÖLL pers. comm.). In southern Urals found at 220 m (K. NUPPONEN pers. comm.).

SIMILAR SPECIES: No similar species in Europe. Female of *L. thaumastaria* REBEL has longer wings, wing length about 3.5 mm, antennae are ochreous dorsally (wings shorter, wing length about 1 mm, antennae blackish dorsally in *L. fumidaria*) (Text-figs 81–82).

GENETIC DATA: Nearest species in Europe (*L. thaumastaria* not yet barcoded; genetic 'similarity' due to long branch attraction): *Lycia hirtaria* (10.7%) and *Pseudopanthera macularia* (11.4%).

REMARKS: Taxon *nausearia* KRULIKOWSKI was described originally as a variation from Pjatigorsk, Stavropol Krai in southern Russia (KRULIKOWSKI 1906), and it occurs in Caucasus region and in Armenia (VIIDALEPP 1996). KRULIKOWSKI (1906) described the Caucasian specimens as larger than European specimens (wingspan 32–35 mm) and unicolorous grey. According to current knowledge, the Caucasian populations are allopatric, disjunct from the European populations. We have examined specimens from the Caucasus area (in The National Museum of Natural History, Kiev, Ukraine), and conclude that the Caucasian material matches the European material, both wingspan and external appearance are similar. WARDIKJAN (1985) provided a rather superficial illustration of the male genitalia of *L. fumidaria* from Armenia, which are also similar to the European material.

58. *Lignyoptera thaumastaria* REBEL, 1901

thaumastaria REBEL, 1901, Verh. zool. –bot. Ges. Wien 51 (10): 800, (*Lignyoptera*). Holotype male, North Bosnia [Bosnia and Herzegovina], South of Zepce.

INTRASUBSPECIFIC NAMES, POTENTIALLY UNAVAILABLE: *vinosa* LUNAK, 1938 (ab.).

EXTERNAL CHARACTERS AND ABDOMEN: Male: wingspan 34–36 mm (n=2). Wings deep yellow, irrorated with reddish, particularly near wing base and terminal area. Medial and postmedial lines present on costa only, suffused, reddish. Terminal line narrow, reddish. Fringes reddish. Forewing discal spot, large, suffused. Hindwing discal spot small. Hindwings pale. Wings below lighter, weakly striated with reddish. Frons, collar, thorax reddish-yellow. Abdomen yellow, irrorated with reddish. Antennae ochreous dorsally, filiform. Tympanal organs small. 8th sternite weakly sclerotised, posterior margin round medially, elongated laterally. 8th tergite weakly sclerotised, triangular. Female: brachypterous, wing length 3.5 mm (n=1). Wings uniformly pale, with few brownish scales. Frons, collar, thorax and abdomen concolorous with wings. Antennae ochreous dorsally, filiform. Tympanal organs apparently reduced. Sternites and tergites undifferentiated. Hindtibia with 2+2 spurs.

VARIATION: Very little in males (n=2). Female variation unknown, only one female was available for study.

MALE GENITALIA: Generally as in *L. fumidaria* (HÜBNER), but with following differences: Furca arms symmetrical. Aedeagus wider. Vesica opens at approximately 45 degree angle. Vesica with large diverticulum, with two long cornuti. Single cornutus at base.

FEMALE GENITALIA: Papillae anales setose, surface granulate. Apophyses narrow, short, weakly sclerotised. Lamella antevaginalis/sinus vaginalis sclerotised, funnel-shaped, finely setose. Lamella postvaginalis membranous. Ductus bursae long, membranous, striated. Corpus bursae long, narrow, striated, membranous. Anterior end with narrow extension. Signum absent.

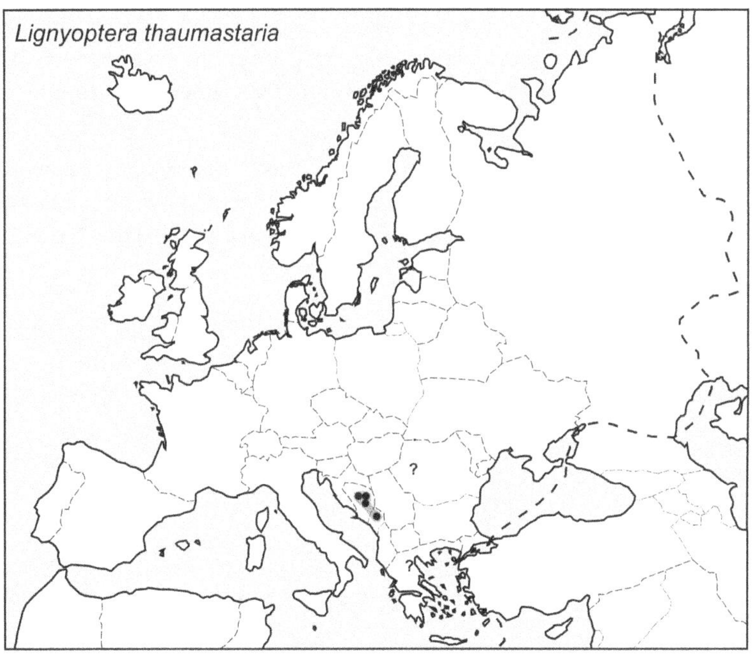

DISTRIBUTION AND ABUNDANCE: South-east European. Endemic to Bosnia-Hercegovina and Montenegro. LUNAK (1938) reports a male caught at mount Retyézat in the Transsylvanian Alps in Romania. It has not been possible to trace this specimen and the record cannot be verified. In his Macedonia fauna, PINKER (1968) mentions in a footnote, that the species had recently been found at Mount Olympus in Greece. A search for a specimen from there has been without success and therefore the record cannot be verified. *L. thaumastaria* appears in the species list of the Galicica National Park in Macedonia, but according to S. BESHKOV (pers. comm.) this is a mistake and due to a confusion with *Enanthyperythra legataria*. An extremely rare species with a very limited number of specimens, although it was described more than hundred years ago.

PHENOLOGY: Univoltine. The very few records of this species are from mid and late October. The hibernation takes place as an egg. LUNAK (1938) supposes that the species flies after the first snow has fallen in the autumn and melted away again. The putative specimen from Romania (Transslyvania) was found in September and this specimen was flying by day (LUNAK 1938). The males fly from dusk for a couple of hours and then again after midnight with increasing intensity until dawn, the females sit at the tips of grass leaves at temperatures down to minus 2 degrees Celsius (LUNAK 1938). A pupa has been found on 16th August from which a female hatched 2nd October (LUNAK 1938).

BIOLOGY: Perhaps oligophagous on Asteraceae. No records from nature available. The caterpillar has been reared on *Artemisia vulgaris*, *Achillea millefolium* and garden *Chrysanthemum*, but preferred the first mentioned (LUNAK 1940). Pupation takes place on the ground, preferably under stones.

HABITAT: Open mountain meadows from just below to above the timber line where such plants as *Dorycnium* and *Amelanchier ovalis* are growing. There are very few records about altitude. A pupa has been found at approximately 1500 m. The locality in Bosnia where the species was found in 1937 is not defined exactly, but it is apparently around 1600 m (LUNAK 1940). In Montenegro in 1300 m. Finally the assumed specimen from Mount Olympus in Greece was from 1000 m.

SIMILAR SPECIES: No similar species in Europe. Female of *L. fumidaria* (HÜBNER) has shorter wings, wing length about 1 mm, antennae are blackish dorsally (wings longer, wing length about 3.5 mm, antennae ochreous dorsally in *L. thaumastaria*) (Text-figs 81–82).

GENETIC DATA: No data available.

REMARKS: Despite the fact that *L. thaumastaria* was described more than hundred years ago, it is among the least known European Geometridae species. The Austrian entomologist ROBERT LUNAK discovered many details about this species which he described in two papers (1938, 1940).

Heterolocha LEDERER, 1853

Heterolocha LEDERER, 1853, Verh. zool. – bot. Ver. Wien 3 (Abh.): 176, 202, 207. Type species: *Hypoplectis laminaria* HERRICH-SCHÄFFER, 1852, Asia Minor. Genus synonyms are after SCOBLE (1999).

- *Hetaera* GUMPPENBERG, 1887, Nova Acta Acad. Caesar Leop. Carol. 49: 325, 342. [Unnecessary replacement name for *Heterolocha* LEDERER, and a junior homonym of *Hetaera* HOFFMANNSEGG (Lepidoptera: Satyridae)].
- *Nabla* WALKER, 1866, List Specimens lepid. Insects Colln Br. Mus. 35: 1668. Type species: *Nabla pyreniata* WALKER, 1866.
- *Symmetresia* WEHRLI, 1937, Ent. Rdsch. 54: 502. Type species: *Hyperythra aristonaria* WALKER, 1860.

DIVERSITY AND DISTRIBUTION: *Heterolocha* contains 41 species in the southern Palaearctic and Oriental regions.

EXTERNAL CHARACTERS AND ABDOMEN: Small species, wingspan about 20–30 mm. Wings various shades of yellow to yellow-brown, markings reddish to pinkish. Postmedial line wide, suffuse, prominent on both wings. Terminal area often darker. Forewing discal spot large, often with pale centre. Male antennae bipectinate. Male 3rd sternite without setae, 8th sternite unmodified.

MALE GENITALIA (diagnosis based on *H. laminaria* (HERRICH-SCHÄFFER, 1852), *H. aristonaria* (WALKER, 1860), *H. pyreniata* (WALKER, 1866) and *H. polymorphoides* HOLLOWAY, 1994) only: Uncus apex spatulate. Socii minute, setose. Gnathos arms fused, apex vertically elongated, with weak row of spines. Valva wide, membranous. Costa setose subapically. Base of valva with group of long setae. Furca arms arise from pocket between valves, often asymmetrical, dentate. Vinculum arms distinct, each containing membranous coremata laterally. Saccus absent. Vesica with or without sclerotisations.

FEMALE GENITALIA (diagnosis based on *H. laminaria* (HERRICH-SCHÄFFER) and *H. polymorphoides* HOLLOWAY only): Papillae anales elongated, setose. Apophyses anteriores reduced, about 1/5 length of apophyses posteriores. Lamella postvaginalis and lamella antevaginalis membranous. Ductus bursae short. Posterior part of corpus bursae long, striated. Corpus bursae large, round. Signum large, with three long spines.

BIOLOGY: *H. laminaria* has been recorded feeding on *Salix* and *Corylus* species (POLTAVSKY *et al.* 2007). Japanese species have been recorded feeding on *Abelia*, *Lonicera* and *Weigela* (Caprifoliaceae) (SUGI 1987, HOLLOWAY 1994). Adults are nocturnal.

IMMATURE STAGES: Two Japanese species have a bold, reticular or striped pattern of black, pale yellow and pale grey (SUGI 1987). No data available on eggs and pupae.

REMARKS: *Heterolocha* has recently been classified in the Hypochrosini (HOLLOWAY 1994) and Ourapterygini (VIIDALEPP 1996). KIM *et al.* (2001) did not recognise tribal groupings, but placed *Heterolocha* between *Plagodis* HÜBNER, 1823 and *Cepphis* HÜBNER, 1823 in the sequence of genera. Structurally *Heterolocha* males are very similar to *Cepphis advenaria* (HÜBNER, 1790), and to *Plagodis* HÜBNER, 1823 females. See the genitalia plates for structures. We follow HOLLOWAY (1994), because *Heterolocha* agrees with the diagnostic features of the Hypochrosini. Systematic position: Hypochrosini.

59. *Heterolocha laminaria* (HERRICH-SCHÄFFER, 1852)

laminaria laminaria HERRICH-SCHÄFFER, 1852, Syst. Bearbeitung Schmett. Eur. 6 (55): 71; ibidem (1851), 3: pl. 79, figs 489, 490 (non binominal), (*Hypoplectis*). Syntype(s), [Asia Minor].

laminaria euxantha WEHRLI, 1937, Ent. Rdsch. 54 (40): 502, (*Heterolocha*). Syntype(s), China: West-Tienmushan, Chekiang. Valid at subspecies rank according to SCOBLE (1999).

laminaria lungtana WEHRLI, 1940, in SEITZ, Gross-Schmett. Erde 4 (Suppl.): 364, pl. 29:i, (*Heterolocha*). Syntype(s), China: Chekiang, South East China, Lungtan, near Nanking, Mokanshan, Tien-Mu-Shan. Valid at subspecies rank according to SCOBLE (1999).

laminaria sutschanska WEHRLI, 1937, Ent. Rdsch. 54 (40): 502, (*Heterolocha*). Syntypes(s), Russia: Amur; Ussuri. Valid at subspecies rank according to SCOBLE (1999).

EXTERNAL CHARACTERS AND ABDOMEN: Wingspan 19–21 mm. Wings pale to dark yellow, basal and terminal areas irrorated and striated with pinkish brown. Medial area often less irrorated. Markings pinkish brown. Forewing antemedial line wide, suffused, curved on costa. Postmedial line approximately parallel with wing margin, wide, suffused. Terminal area irrorated with pinkish brown, most pronounced near apex on both wings. Hindwing postmedial line wide, suffused. Terminal line and fringes concolorous with wings. Forewing discal spot large, elongate, suffused. Hindwing discal spot smaller. Wings below as above, slightly paler. Frons, collar, thorax and abdomen concolorous with wings. Male antennae bipectinate. Female antennae weakly fasciculate. Hindtibia with 2+2 spurs. Tympanal organs medium-sized. Abdominal segments of both sexes undifferentiated.

VARIATION: Density of irroration, colour of basal and terminal area variable. Wings can be more or less uniformly pale to dark yellow, with well-developed postmedial line and forewing discal spot being the only distinct markings. Basal area can be pinkish brown, also terminal area can be darker than rest of wing. Forewing discal spot can have pale centre. In addition to the West Palaearctic specimens, we have examined specimens from China: Heilongjiang. Although the examined specimens are externally different, the male and female genitalia are similar. The validity of subspecies needs to be critically evaluated (see genetic data).

MALE GENITALIA: Uncus long, apex spatulate, setose. Socii minute, setose. Gnathos arms narrow, fused, apex vertically elongated, with weak row of spines. Tegumen large. Valva wide, membranous, ventral margin weakly concave, basal part weakly sclerotised. Costa setose subapically, apex roundish. Base of valva with group of long setae. Furca arms arise from pocket between valves, asymmetrical, apex dentate, left arm longer. Vinculum arms distinct, each containing membranous, round coremata laterally. Saccus absent. Aedeagus narrow, weakly curved, caecum long, apex very acute. Vesica opens at approximately 180 degree angle. Vesica straight, striated tube, with two small lateral diverticula, distal one with long, acute, sclerotised cornutus.

FEMALE GENITALIA: Papillae anales elongated, setose. Apophyses anteriores reduced, about 1/5 length of apophyses posteriores. Lamella postvaginalis and lamella antevaginalis membranous. Ductus bursae short, membranous. Base of ductus seminalis expanded, distinct laterally. Posterior part of corpus bursae long, straight, striated. Corpus bursae large, round. Signum large, with three long spines, central disc round.

DISTRIBUTION AND ABUNDANCE: Palaearctic. In Europe only known from the Rostov oblast in the southern European part of Russia (POLTAVSKY *et al.* 2007). – Outside Europe the species is known from north-eastern Turkey, Transcaucasus, Caucasus, northern Iran, northern Afghanistan (subsp. *sutschanska*) and from Far East Russia, Sakhalin Island, China and Japan. According to KOÇAK & KEMAL (2009) also found in western Turkey in the provinces Bursa and Bolu. These records are far from the known distribution area and need verification. – In Europe a newly discovered species that is only known from a few specimens.

PHENOLOGY: Univoltine or possibly bivoltine in warmer places. Recorded from mid-June to late August. Based on the rather limited material it is difficult to say if the species has one or two generations, but it may well have two generations in the south in the distribution area and at low altitudes. In the Shepkinsky forest by Rostov-on-Don abundant in early May (A.N. POLTAVSKY pers. comm.).

BIOLOGY: The caterpillar has been recorded feeding on *Salix* and *Corylus* species (POLTAVSKY *et al.* 2007).

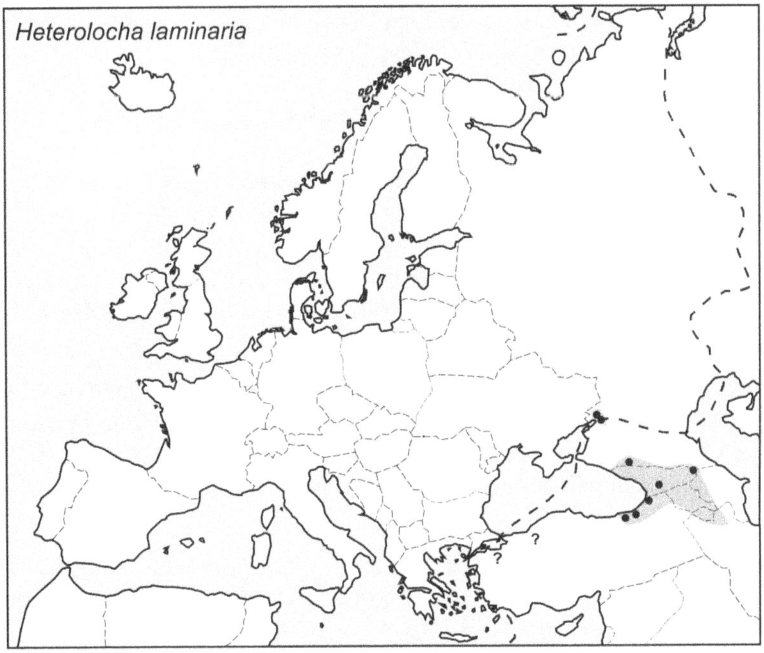

HABITAT: In south-eastern European Russia the species has been found on slopes with bushes and with steppe vegetation, in a planted deciduous forest and in a village. Here it has been found from 30 up to 80 m altitude. In Turkey up to 1700 m.

SIMILAR SPECIES: No similar species in Europe.

GENETIC DATA: Nearest species: Chinese *H. coccinea* INOUE, 1976 (4.5%). Nominotypical populations (n=7 from Turkey and Iran) genetically homogenous, but strongly divergent from East Asian *lungtana* WEHRLI, 1940, *sutschanska* WEHRLI, 1937 (6.9%) and *euxantha* WEHRLI, 1937 (4.2%), suggesting that the taxonomic status of these taxa needs re-evaluation.

Plagodis HÜBNER, 1823

Plagodis HÜBNER, [1823] 1816, Verz. bekannter Schmett.: 294. Type species: *Phalaena dolabraria* LINNAEUS, 1767, Germany. Genus synonyms are after SCOBLE (1999) unless otherwise specified.

Axinophora AGASSIZ, 1847, Nom. zool. (Index. univ.): 42 [Unjustified emendation of *Azinephora* STEPHENS, 1827]. Synonymy after VIVES MORENO (1994).

Anagoga HÜBNER, [1823] 1816, Verz. bekannter Schmett.: 294. Type species: *Phalaena pulveraria* LINNAEUS, 1758.

Apoplagodis WEHRLI, 1939, in SEITZ, Gross-Schmett. Erde 4 (Suppl.): 358. Type species: *Plagodis reticulata* WARREN, 1893.

Azinephora STEPHENS, 1827, Retrospective Review (2) 1: 245. Type species: *Phalaena pulveraria* LINNAEUS, 1758. Synonymy after VIVES MORENO (1994).

Eurymene DUPONCHEL, 1829, in GODART & DUPONCHEL, Hist. nat. Lépid. Papillons Fr. 7 (2): 105, 185. Type species: *Phalaena dolabraria* LINNAEUS, 1767.

Numeria DUPONCHEL, 1829, Hist. nat. Lépid. Fr. 7 (2) : 107. Type species: *Phalaena pulveraria* LINNAEUS, 1758. Synonymy after VIVES MORENO (1994).

UNAVAILABLE NAMES (incorrect subsequent spellings): *Anagoge* STEPHENS, 1850 (see VIVES MORENO 1994); *Eurimene* RONDANI, 1872 (see FLETCHER 1979).

DIVERSITY AND DISTRIBUTION: *Plagodis* contains 18 species in the Palaearctic, Oriental and Nearctic regions.

EXTERNAL CHARACTERS AND ABDOMEN: Medium-sized, wingspan about 30 mm. Wings different shades of yellow-brown, often with tint of violet, striated with brownish. Medial and postmedial lines distinct. Forewing margin often weakly convave below apex, distinctly concave on tornus. Hindwings with trace of postmedial line on anal margin. Male antennae bipectinate. Male 3rd sternite without setae, 8th sternite weakly sclerotised but somewhat modified.

MALE GENITALIA: Uncus long, curved. Socii minute, setose. Gnathos arms fused ventrally, upturned, weakly spinose. Valva wide at base, apex narrower, round. Costa setose subapically. Base of valva with group of long setae. Transtilla present. Furca arms arise from pocket between valves, often asymmetrical, dentate. Saccus not elongated. Coremata absent. Vesica curved or coiled, with or without cornuti.

FEMALE GENITALIA: Papillae anales densely setose, surface granulate. Lamella postvaginalis striated horizontally. Ductus bursae short, sclerotised. Posterior part of corpus bursae often sclerotised, elongated. Anterior part of corpus bursae round, membranous. Signum small, irregular plate with more definitive sclerotisation in middle.

BIOLOGY: *Plagodis* species have been recorded feeding on Aceraceae, Betulaceae, Caprifoliaceae, Fagaceae, Salicaceae, Tiliaceae, Oleaceae, Pinaceae, Rosaceae, Corylaceae, Ericaceae (MCGUFFIN 1987, SCOBLE 1999), the European species have been found for instance on *Corylus avellana, Sorbus aucuparia* and *Prunus spinosa* (EBERT 2003). Adults are nocturnal.

IMMATURE STAGES: Egg smooth, elliptical (SALKELD 1983, MCGUFFIN 1987). Larvae of *P. pulveraria* and *P. dolabraria* are twig-mimics, brown to brown-violet in colour, and they have nodular extension dorsally, apparently on A8 (EBERT 2003). Pupa cremaster with four pairs of curved setae, of similar length (PATOČKA & TURČANI 2005).

REMARKS: HOLLOWAY (1994), LERAUT (1997) and HAUSMANN *et al.* (2011) have adopted Hypochrosini position, VIIDALEPP (1996) and KULLBERG *et al.* (2001) classified the genus in the Ourapterygini and VIVES MORENO (1994) in the Ennomini. HOLLOWAY (1994) proposed that the Anagogini FORBES, 1948 are a junior synonym of the Hypochrosini GUENÉE, 1858. Type species of the Anagogini is *Phalaena pulveraria* (LINNAEUS, 1758) from Europe and type of the Hypochrosini is *Hypochrosis sternaria* GUENÉE, 1858 from India. Structurally *Plagodis* and *Hypochrosis* are dissimilar, but other taxa included in the Hypochrosini by HOLLOWAY (1994) are more similar to *Plagodis*. In a molecular, phylogenetic analysis of the world Geometridae (SIHVONEN *et al.* 2011), genus *Plagodis* grouped together with *Epione* DUPONCHEL, 1829 (Epionini) and *Apeira* GISTL, 1848 (Apeirini). The Ourapterygini were not related. In a molecular, phylogenetic analysis of limited European taxa only, ÕUNAP *et al.* (2011) concluded that *Cepphis*, *Selenia*, *Plagodis* and *Pseudopanthera* grouped together. In *P. pulveraria* the furcae originate from a pocket between the valves and their base is setose. *O. sambucaria* has similar, setose structure, but the arms are very short and easily overlooked. In addition to these, *O. sambucaria* has second pair of furca-like structures. These massive, asymmetrical structures have more dorsal origin. BELJAEV (2008) did not include *Plagodis* in his concept of the Ennomini. Finally, the triangular transtilla lobes resemble those found commonly in the Lithinini (see for instance WEINTRAUB & SCOBLE 2004). Systematic position: Hypochrosini of uncertain association.

Seasonal variation in the length of furca arms in the male genitalia has been observed for numerous species, in Europe, Japan and in North America (RUPERT 1949, SATTLER 1963, HEITZMAN & ENNS 1977, VITALIEVNA 1996).

60. *Plagodis pulveraria* (LINNAEUS, 1758)

pulveraria pulveraria LINNAEUS, 1758, Syst. Nat. (Ed. 10) 1: 521, (*Phalaena Geometra*). Syntypes (Linnean Society of London, London, UK), Sweden: Stockholm (examined externally).

diffusaria WALKER, 1861, List Specimens lepid. Insects Colln Br. Mus. 24: 1075, (*Aspilates*). Holotype male (The Natural History Museum, London, UK) (examined externally).

distinctaria BRYK, 1949, Ark. Zool. 41A (1): 214, (*Itame*). Holotype female (Swedish Museum of Natural History, Stockholm, Sweden), Korea: Shuotsu. Synonymy proposed in KIM *et al.* (2001) (examined externally).

pulveraria gadmensis RÄTZER, 1890, Mitt. schweiz. ent. Ges. 8 (6): 224, (*Numeria*). Syntype(s), Switzerland: Gadmen. Originally as var., treated valid at subspecies rank by SCOBLE (1999), downgraded here from subspecies rank, quoting absence of diagnostic features (new synonym). See remarks.

pulveraria ruforaria KARDAKOFF, 1928, Entomol. Mitteilungen 17 (4): 422, (*Anagoga*). Syntype(s), Russia: Siberia, Ussuri. Originally as subspecies, treated as ab. of *pulveraria* by WEHRLI (1939–1954).

trifasciata PILLER & MITTERPACHER, 1783, Iter per Poseganam Sclavoniae provinciam mensibus Junio, et Julio anno MDCCLXXXII: 71, (*Phal.* (*Geometra*)). Syntype(s), Slovenia: Pozhega. Synonymy after VIVES MORENO (1994) and LERAUT (1997).

pulveraria japonica BUTLER, 1881, Trans. ent. Soc. London. 1881 (3): 418, (*Numeria*). Syntype(s) (The Natural History Museum, London, UK), Japan: Kamakura: Tokei (Kanagawa Prefecture) (examined externally). Originally as sp., valid at subspecies rank according to SCOBLE (1999).

pulveraria jezoensis INOUE, 1954, Butterfl. Moths 4 (4): 25 [English translation p. 27], fig. 5, (*Anagoga*). Syntypes 7 males (lectotype male in The Natural History Museum, London, UK), Japan: Kushiro, Futatsuyama (examined externally). Valid at subspecies rank according to SCOBLE (1999).

pulveraria montana INOUE, 1954, Butterfl. Moths 4 (4): 25 [English translation p. 27], fig. 4, (*Anagoga*). Syntypes 2 males, 1 female (The Natural History Museum, London, UK), Japan: Japanese Alps, Mt Jonen, 2500 m (examined externally). Valid at subspecies rank according to SCOBLE (1999).

pulveraria occiduaria WALKER, 1861, List Specimens lepid. Insects Colln Br. Mus. 23: 1016, (*Numeria*). Syntypes 2 males (The Natural History Museum, London, UK), [U.S.A].: New York, Trenton Falls (examined externally). [Originally as sp]. The North American taxon *occiduaria* is illustrated in MCGUFFIN (1987) as *Anagoga occiduaria* (WALKER). Valid at subspecies rank according to SCOBLE (1999).

californiaria PACKARD, 1871, Proc. Boston Soc. nat. Hist. 13: 384, (*Ellopia*). Syntype(s) female (Peabody Museum, Yale University, New Haven, USA), [U.S.A].: California.

plagifasciata WALKER, [1863], List Specimens lepid. Insects Colln Br. Mus. 26: 1508, (*Ellopia*). Syntypes 1 male, 1 female (The Natural History Museum, London, UK), [U.S.A].: New York (examined externally).

pulveraria singularis VOJNITS, 1975, Annales hist.-nat. Mus. natn. hung. 67: 184, figs 1, 2, (*Plagodis*). Holotype female (Hungarian Natural History Museum, Budapest, Hungary),

Mongolia: Central Aimak, Somon Bajancogt, 1600 m. Paratype female in the Natural History Museum, London, UK (examined externally). Valid at subspecies rank according to SCOBLE (1999).

pulveraria violacearia PROUT, 1915, in SEITZ, Gross-Schmett. Erde 4: 320, pl. 15 h, (*Anagoga*). Syntypes 2 females, [Russia:] Wlad. [Vladivostok]. Valid at subspecies rank according to VIIDALEPP (1996), tentatively valid at subspecies rank according to KIM *et al.* (2001). According to KIM *et al.* (2001, p. 177): 'The authority and availability of the name *violacearia* is discussable. This name was established by GRAESER (1889, Berl. Ent. Z. 32 (2): 392) as intrasubspecific name (aberration) [*Numeria pulveraria* ab. *violacearia* GRAESER]. STAUDINGER (1897; STAUDINGER and REBEL, 1901) and HERZ (1904) treated this name in the same sense, but they proposed treatment of this name as variety. PROUT (1915) first applied a geographical sense to the name, therefore he has to be established as author of the available subspecific name *violacearia* (International Code [ICZN], 1999: 45.5.1).'

UNAVAILABLE NAMES (incorrect subsequent spelling): *pulverata* (HULST 1896; should be *pulveraria* LINNAEUS). – Intrasubspecific names, potentially unavailable: *approximata* LEMPKE, 1951 (f.); *extrema* BRETT, 1936 (ab.); *gigantea* INOUE, 1954 (f.); *grisescens* LEMPKE, 1951 (f.); *ijimai* INOUE, 1954 (f.); *linearia* LEMPKE 1951 (f.); *marginepurpuraria* BASTELBERGER, 1897 (ab.); *obscurior* DAHLSTROM, 1900 (ab.); *passetii* THIERRY-MIEG, 1884 (ab.); *rufescens* LEMPKE, 1951 (f.); *saturaria* WAGNER, 1909 (ab.) (=synonym of *passetii* THIERRY-MIEG, 1884 according to The Global Lepidoptera Names Index (ANONYMOUS 2013)); *unicolor* HIRSCHKE, 1910 (ab.); *violacearia* GRAESER, 1889 (see *pulveraria violacearia* PROUT, 1915 above).

EXTERNAL CHARACTERS AND ABDOMEN: Wingspan 29–33 mm. Wings yellow-brown, heavily irrorated or striated with brown. Medial and postmedial lines dark brown, former slightly convave, latter distinctly angled. Medial area darker, rusty brown, widest on costa. Terminal line absent. Fringes concolorous with wings. Discal spots minute, suffused. Hindwings lighter, subterminal line weak on costa. Wings below slightly darker, heavily striated with brown. Frons, collar, thorax and abdomen concolorous with wings. Male antennae bipectinate. Female antennae weakly fasciculate. Hindtibia with 2+2 spurs. Tympanal organs medium-sized. Male 8th sternite weakly sclerotised, posterior margin concave. Other sternites and tergites of both sexes undifferentiated.

VARIATION: Female forewings more elongated, medial area often lighter and postmedial line more suffused than in males. Density of irroration, colour of medial area, and to lesser degree the ground colour varies. Numerous subspecies and forms have been described, see for instance PROUT (1912–1916), WEHRLI (1939–1954) and LERAUT (2009). The validity of several subspecies, particularly in the Palaearctic region, needs to be critically evaluated.

MALE GENITALIA: Uncus long, curved ventrally, apex weakly setose. Socii short, setose. Gnathos arms narrow, fused ventrally, upturned, weakly spinose. Transtilla with two, slightly triangular, large lobes. Furca arms arise from pocket, sclerotised, long, curved, apical half spinose, sometimes branched. Valva wide at base, partly striated, apex narrower, round. Costa weakly sclerotised, setose subapically. Ventral margin round, strongly sclerotised, dentate. Base of valva with group of setae. Saccus narrow ring. Aedeagus weakly curved, apex elongated, sharp. Vesica opens at approximately 135 degree angle. Vesica membranous, with small lateral diverticulum, apex narrower, angled. Cornuti absent.

FEMALE GENITALIA: Papillae anales densely setose, surface granulate. Apophyses narrow. 8th segment strongly sclerotised. Lamella antevaginalis undifferentiated. Lamella postvaginalis striated horizontally. Ductus bursae short, sclerotised. Posterior part of corpus bursae weakly sclerotised, funnel-shaped. Anterior part of corpus bursae large, round, membranous. Signum small, irregular plate with more definitive, concave sclerotisation in middle.

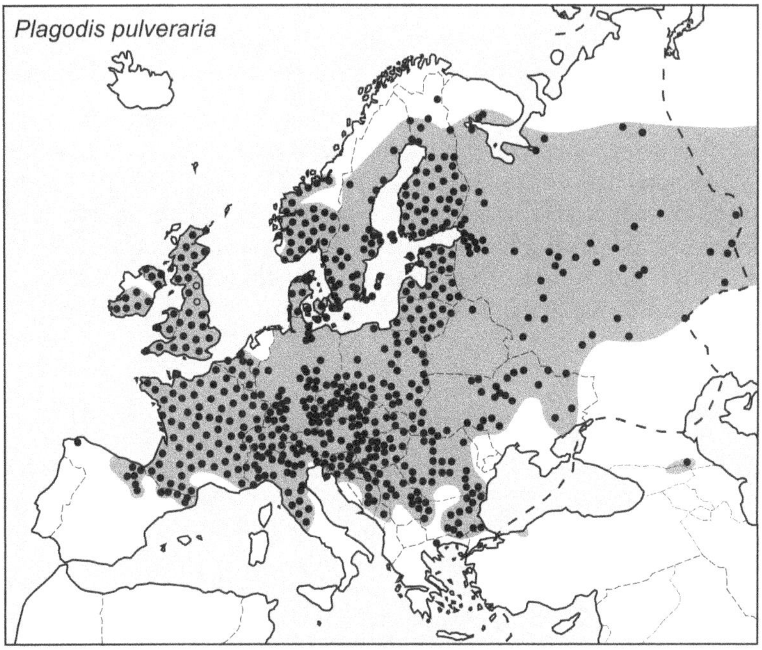

DISTRIBUTION AND ABUNDANCE: Holarctic. In Europe from central Scandinavia to northernmost Spain, central Italy, Bulgaria and northern Greece and from the British Isles through central and eastern Europe to the Ural mountains. The species has earlier been reported from Malta (BORG 1932), but its presence is unverified (P. SAMMUT pers. comm.) and most likely an error. – Outside Europe the species is known from Düzce province Turkey (KOÇAK & KEMAL 2009) and in Caucasus and it is distributed further to the east from the Ural mountains and Caucasus through western and central Siberia, Mongolia, China, Korea and Far East Russia to Japan and Sakhalin. In North America widely distributed in U.S.A. and Canada (FORBES 1948, MCGUFFIN 1987). – In Europe a common, but somewhat locally distributed species that usually is found in limited numbers. In Britain the abundance has increased in recent decades (FOX *et al.* 2013).

PHENOLOGY: Uni- or bivoltine. In northern Europe from mid-May to late June, in south-western Germany from early May to early July, in warm regions in two generations from early April to mid-June and from mid-July to late August (EBERT 2003) and in southern Europe from late April to late June and again from mid-July to late August (REDONDO *et al.* 2009). Larvae in northern Europe from June to August or early September, further to the south depending on flight times and generations. Hibernation takes place as pupa in the soil. The species flies at night and comes to light.

BIOLOGY: Polyphagous. The caterpillar has been found on *Corylus avellana*, *Sorbus aucuparia*, *S. aria*, *Prunus padus* and *Vaccinium myrtillus* (EBERT 2003). The caterpillars have been reared on *Vaccinium myrtillus* and *Betula pendula* (B. MÜLLER pers. comm.).

HABITAT: Sylvicolous. In semi-open deciduous and mixed forests and scrubland, both on moist and more dry soil, forests along rivers etc. From sea level and in the Pyrenees up to 1800 m (REDONDO *et al.* 2009) and in the Italian Alps up to 2100 m (FLAMIGNI *et al.* 2007).

SIMILAR SPECIES: No similar species in Europe.

GENETIC DATA: Genetically homogeneous in Europe (n=13, from six countries), well corresponding to populations from north-eastern China. Nearest species: *P. dolabraria* (4.4%).

REMARKS: Taxon *Numeria gadmensis* RÄTZER, 1890 is downgraded here from subspecies rank (SCOBLE 1999) to synonymy of *P. pulveraria* (LINNAEUS). PROUT (1912–1916) treated it as 'quite small, (yellow) brown form' from Switzerland. WEHRLI (1939–1954) added that the form also occurs in southern Bavaria, Poland, North-Tirol and Estonia. We have examined numerous specimens across Europe and conclude that such features are within the geographical variation of the species. Genetic and morphological information available does not support the hypothesis that there exists an allopatric, distinct population of *P. pulveraria* in central Europe.

61. *Plagodis dolabraria* (LINNAEUS, 1767)

dolabraria dolabraria LINNAEUS, 1767, Syst. Nat. (Ed. 12) 1 (2): 861, (*Phalaena* (*Geometra*)). Syntypes (Linnean Society of London, London, UK), Germany (examined externally).

dolobraria BORKHAUSEN, 1794, Natur. eur. Schmett. 5: 137, (*Phalaena Geometra*). [Emendation of *dolabraria* LINNAEUS].

dolabraria violacea WEHRLI, 1939, in SEITZ, Gross-Schmett. Erde 4 (Suppl.): 358, pl. 29: b, (*Plagodis*). Syntypes 2 males, [Switzerland:] Tessin, Maroggia. Originally as sp., downgraded from subspecies rank (for instance SCOBLE 1999) to synonymy by FLAMIGNI *et al.* (2007). See remarks.

ustulataria HUFNAGEL, 1767, Berlin Mag. 4 (5): 516, (*Phalaena*). Syntype(s), lost, [Germany]: Berlin region.

dolabraria costisignata WEHRLI, 1939, in SEITZ, Gross-Schmett. Erde 4 (Suppl.): 358, pl. 29: b, (*Plagodis*). Syntype(s), [China]: Chekiang, West Tien-Mu-Shan; Lungtan near Nanking. 2 paratype males, 1 paratype female (Natural History Museum, London, UK) (examined externally). Valid at subspecies rank according to SCOBLE (1999).

UNAVAILABLE NAMES (intrasubspecific): *aestiva* DIÓSZEGHY, 1935 (summer generation); *atrox* ZERNY, 1909 (ab.); *aurantiaca* LEMPKE, 1951 (f.); *fasciata* LEMPKE, 1970 (f.); *pallida* LEMPKE, 1970 (f.); *rufescens* LEMPKE, 1951 (f.).

EXTERNAL CHARACTERS AND ABDOMEN: Wingspan ♂ 28–33 mm, ♀ 31–36 mm. Forewings narrow, wings whitish to rusty brown, densely striated with olive-brown. Medial and postmedial lines brownish, suffused, former strongest on costa, latter strongest on inner margin. Tornus with violet patch. Hindwings more uniform whitish yellow. Violet patch and postmedial line present on anal margin. Terminal line faint. Fringes concolorous with wings. Discal spots absent. Wings below lighter, markings weaker, terminal area often dark yellow to violet. Forewing margin on tornus and hindwing margin on termen concave. Frons, collar and part of thorax brown-violet. Abdomen concolorous with wings, last segments violet. Male antennae bipectinate. Female antennae weakly fasciculate. Hindtibia with 2+2 spurs. Tympanal organs large. Male 8th sternite weakly sclerotised, posterior margin weakly elongated. Other sternites and tergites of both sexes undifferentiated.

VARIATION: Female forewings are more elongated. Density of striations, extent of violet on forewing tornus, extent of white on forewings and hindwing ground colour varies. Numerous forms have been described. Summer generation specimens smaller, forewings with less striations.

MALE GENITALIA: Uncus long, curved ventrally, widest subapically. Socii short, setose. Gnathos arms narrow, fused ventrally, upturned, weakly setose. Transtilla with two large lobes. Furca arms arise from pocket, asymmetrical, sclerotised, long, recurved, apical half densely setose/spinose. Valva setose, wide at base, apex narrower, round. Costa sclerotised, setose subapically. Ventral margin weakly concave, strongly sclerotised, ridge-like, followed by angle. Base of valva with group of setae. Saccus narrow ring. Aedeagus wide, short, apex elongated, curved. Vesica opens at 180 degree angle. Vesica with large, round diverticulum, with about 10–15 cornuti. Apex narrow, straight tube.

FEMALE GENITALIA: Papillae anales densely setose, surface granulate. Apophyses long, narrow. Lamella antevaginalis long, horizontal plate, with round pockets laterally. Lamella postvaginalis striated horizontally. Ductus bursae wide, short, sclerotised. Posterior part of ductus bursae strongly sclerotised, long, striated. Anterior part of corpus bursae membranous, round, partly striated. Signum at anterior end, small, irregular plate with pointed sclerotisation.

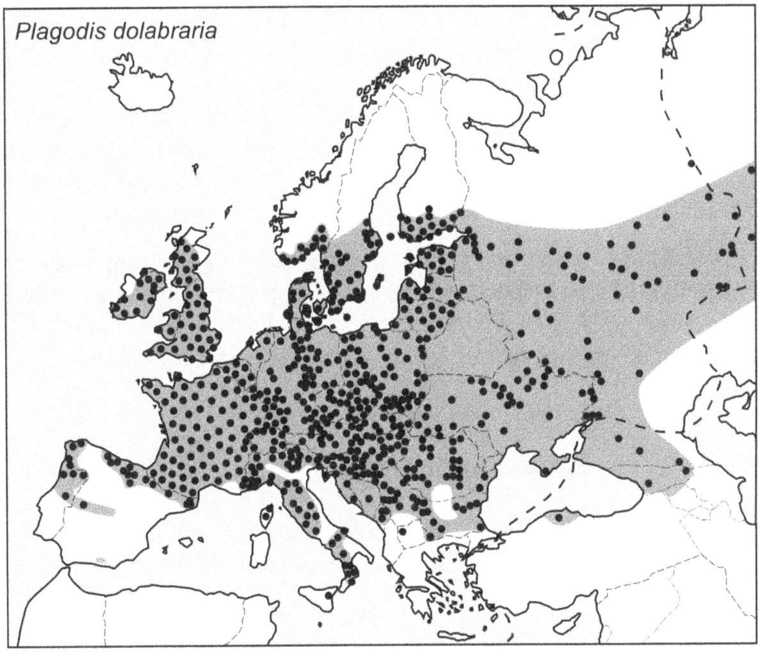

DISTRIBUTION AND ABUNDANCE: Eurasiatic. In Europe from southern Scandinavia to the Iberian Peninsula, southern France including Corsica, southern Italy including Sicily and most of the Balkans and from the British Isles through central and eastern Europe to the Ural mountains. – Outside Europe the species occurs in Turkey, Caucasus and through northern Kazakhstan, southern Siberia and China to Korea, Far East Russia, Sakhalin and the Kuriles. – A common species in Europe that now and then can occur in numbers.

PHENOLOGY: Univoltine or bivoltine. In northern Europe from late May to early July, in southern central Europe and in southern Europe in two generations from late April to late May and from from mid-July to late August or early September. In central Italy at sea level as early as late March (coll. SKOU). The larval time in northern Europe is June to August, further to the south and depending on flight time, the larval time can be both earlier and later. The species overwinters as pupa among withered leaves on the ground. The adult moths come readily to light.

BIOLOGY: Polyphagous. The larva has been found on *Quercus robur*, *Ulmus*, *Sorbus aucuparia*, *Prunus spinosa*, *Tilia* and *Lonicera xylosteum* (EBERT 2003). H. WEGNER (pers. comm.) found larvae in August on *Quercus robur* and *Q. petraea*. A large number of other deciduous food plants have been recorded in the literature.

HABITAT: Sylvicolous. Deciduous forests and scrubland from sea level and in southern Europe up to 1500 m in Spain (REDONDO *et al.* 2009) and Italy (FLAMIGNI *et al.* 2007).

SIMILAR SPECIES: No similar species in Europe.

GENETIC DATA: Genetically homogeneous in Europe (n=30, from eight countries). Nearest species: *P. pulveraria* (4.4%).

REMARKS: Taxon *Plagodis dolabraria* subspecies *violacea* WEHRLI, 1939 was downgraded from subspecies rank (for instance SCOBLE 1999) to synonymy of *P. dolabraria* subspecies *dolabraria* (LINNAEUS, 1767) by FLAMIGNI *et al.* (2007); we agree with this. WEHRLI (1939–1954) diagnosed the taxon as 'hindwings above and below with a broad, fairly demarcated purple line, which extends even to the fringes. The spot at the interior angle of the forewing also purple (underside more clear).' We have examined numerous specimens across Europe and conclude that such features are within the geographical variation of the species. Genetic and morphological information available does not support the hypothesis that there exists an allopatric, distinct population of *P. dolabraria* in central Europe.

Length of furca arms in *P. dolabraria* varies between spring and summer generations. Left arm to right arm ratio of spring specimens is on average 0.78, while in summer specimens the ratio is on average 0.94. The differences are not disjunct. It was found out in breeding experiments that furca length is determined by the day-length conditions of the caterpillar (VITALIEVNA 1996).

Pachycnemia STEPHENS, 1829

Pachycnemia STEPHENS, 1829, Nom. Br. Insects: 44. Type species: *Geometra hippocastanaria* HÜBNER, [1799], 1796, Europe. Genus synonyms are after FLETCHER (1979) and SCOBLE (1999).

Continia HARTIG, 1976, Mitt. ent. Ges. Basel (NF) 26: 18. Type species: *Phalaena tibiaria* RAMBUR, 1829.

Sthanelia BOISDUVAL, 1840, Genera Index meth. eur. Lepid.: 229. Type species: *Phalaena tibiaria* RAMBUR, 1829.

DIVERSITY AND DISTRIBUTION: *Pachycnemia* contains two species, which occur in the west Palaearctic region and in Northern Africa. Species of genus *Pachycnemoides* KRÜGER, 1999, which extends across southern Africa, are similar externally (for instance KRÜGER 1999, 2009). *Pachycnemoides* is classified in the Nacophorini (KRÜGER, 2009).

EXTERNAL CHARACTERS AND ABDOMEN: Medium-sized, wingspan about 25–30 mm. Wings grey to grey-brown, markings and veins blackish. Male and female antennae filiform. Male hindtibia swollen, with hairpencil. Male 2nd sternite with sclerotised plate between tympanal organs, posterior corners with pockets. Male 3rd sternite with setae, 8th sternite unmodified.

MALE GENITALIA: Uncus long, apex weakly bilobed, setose. Socii absent. Gnathos arms fused ventrally, upturned, apex granulate. Valva narrow, setose, membranous. Costa weakly sclerotised. Base of valva with group of setae. Transtilla present. Vinculum(?) with lateral sclerotisations, with membranous sacs (coremata) present or absent. Anellus processes long, narrow. Saccus narrow. Vesica with one large cornutus.

FEMALE GENITALIA: Papillae anales setose, surface granulate. Posterior part of corpus bursae long, narrow, striated, curved or coiled. Signum small, dentate plate.

BIOLOGY: *Pachycnemia* species have been recorded feeding for instance on *Calluna vulgaris* and *Erica* (REDONDO *et al.* 2009). Adults are nocturnal but can be flushed out quite easily during daytime.

IMMATURE STAGES: Data available on *P. hippocastanaria* only: eggs yellow, oval, laid singly on *C. vulgaris* flowers (EBERT 2003). Larva with brown and white longitudinal stripes or patches. Depending on the specimen, either white or brown can be more dominant colour (EBERT 2003, AIZPÚRUA 2002). Pupa cremaster with three, sometimes only two pairs of setae, apical pair very long. Cremaster base not extended (PATOČKA & TURČANI 2005).

REMARKS: The systematic position of genus *Pachycnemia* STEPHENS, including its tribal position, is unclear. It is the type genus of the Pachycnemiini KIRBY, 1903 (a replacement name for Ligiini GUENÉE, 1858), which has been proposed to be a junior synonym of Prosopolophini (FORUM HERBULOT 2014). Also Compsopterini HERBULOT, 1963 is considered a junior synonym of Prosopolophini (FORUM HERBULOT 2014). In addition to the Prosopolophini, *Pachycnemia* has recently been classified in the Ennomini (WOLF 1988, HAUSMANN 1993, VIVES MORENO 1994, VIIDALEPP 1996) and in the Hypochrosini (LERAUT 1997, HAUSMANN, MIRONOV & VIIDALEPP 2011). BELJAEV (2006b) proposed that the Pachycnemiini and Epionini are related, the relationship being supported by 'labides with dorsal end of basis far from base of transtilla'. Later BELJAEV (2008a) proposed that the Ennomini and Prosopolophini are related. In a molecular, phylogenetic analysis of the world Geometridae (SIHVONEN *et al.* 2011), genus *Pachycnemia* grouped together with the Oriental *Curbia* WARREN, 1894 (Baptini/Caberini in HOLLOWAY 1994, PITKIN 2002) and the New Zealand genus *Ischalis* WALKER, 1863 (Lithinini in WEINTRAUB & SCOBLE 2004). The Ennomini, Hypochrosini and Epionini were not related to *Pachycnemia*, neither were *Colotois* HÜBNER, 1823 or *Wilemania* PROUT, 1928, which have also been proposed to be members of the Prosopolophini (BELJAEV 2006b, BELJAEV 2008a). The male and female genitalia, and other structural features of *Pachycnemia* are somewhat similar to *Petrophora chlorosata* (SCOPOLI, 1763) (Lithinini) (our observation) and confirm with the Lithinini as recently defined on a global scale (RINDGE 1986, HOLLOWAY 1994, WEINTRAUB & SCOBLE 2004, PITKIN 2002, PARRA & HERNANDEZ 2010). Most notably the Lithinini association is supported (our observations, see also SIHVONEN *et al.* 2015) by the filiform male antennae; transtilla weakly triangular; anellus with pair of processes, apex weakly setose; valva with sclerotised costal band; male 3rd sternite with row of setae; female signum with horizontal ridge and dentate margins; and appendix bursae is present in corpus bursae, constricted at base. The three pairs of cremaster setae in *P. hippocastanaria* are rare among the Geometridae, typically the Lithinini have four pairs (WEINTRAUB & SCOBLE 2004). See text under Lithinini above. Systematic position: Lithinini of uncertain association.

As is explained in more detail in *Ennominae systematics* chapter, the European Ennominae fauna as a whole has not previously been classified into tribes. Systematic review of tribal classifications and their constituent genera is beyond scope of this work, mainly because it would require extensive studies on numerous non-European taxa. Therefore in several cases, including this one, we have been forced to arrange higher taxa according to the conventional and practical aspects.

62. *Pachycnemia hippocastanaria* (HÜBNER, 1799)

hippocastanaria HÜBNER, [1799], Samml. Eur. Schmett. 5 Geometrae (2): pl. 36, fig. 186, (*Geometra*). Syntype(s), lost, Europe.

anomalata HAWORTH, 1812, Ent.Trans. (old series) 1 (3): 337, (*Phalaena*?). Syntype(s), United Kingdom.

degenerata HÜBNER, [1813], Samml. Eur. Schmett. 5 Geometrae (1): pl. 78, fig. 405, (*Geometra*). Syntype(s), lost, Europe.

difformaria HÜBNER, [1825], Verz. bekannter Schmett.: 321, (*Alsophila*). [Replacement name for *degenerata* HÜBNER].

dilutior WAGNER, 1924, Verh. zool.-bot. Ges. Wien 74: 10, (*Pachycnemia*). Syntype(s), [Croatia:] South Dalmatia, Gravosa. [Originally as var. of *hippocastanaria* HÜBNER].

hippocastanaria modestaria LEMPKE, 1970, Tijdschr. Ent. 113: 153, (*Pachycnemia*). Holotype male, (Amsterdam Zoological Museum, Netherlands), Holland: Gelderland, Wiessel (examined externally, based on original description).

hippocastanata TREITSCHKE, 1827, Schmett. Eur. 6 (1): 341, (*Chesias*). [Emendation of *hippocastanaria* HÜBNER].

INTRASUBSPECIFIC NAMES, POTENTIALLY UNAVAILABLE: *aestiva* HANNEMANN, 1913 (f.); *grisescens* LERAUT, 2009 (f.); *nigrescens* LEMPKE, 1951 (f.); *rachonica* KOUTSAFTIKIS, 1973 (ab.).

EXTERNAL CHARACTERS AND ABDOMEN: Wingspan 25–32 mm. Forewings narrow, pyralid-like, grey-brown, with tint of reddish brown in medial area, veins blackish. Medial lines broken into black dots: antemedial line angled, medial line suffused, strongest on costa, postmedial line curved, its outer margin whitish. Hindwings light grey, postmedial line whitish, terminal area darker. Terminal line black, interrupted. Fringes whitish. Forewing discal spots sharp, black, white spot on costal side. Hindwing discal spots small, suffused. Wings below uniform light grey, glossy. Frons, collar, thorax and abdomen concolorous with wings. Male and female antennae filiform. Male hindtibia swollen, with hair pencil. Hindtibia with 2+2 spurs. Tympanal organs medium-sized, sclerotised plate between them, posterior corners with pockets. Male 3rd sternite with setae, 8th sternite unmodified. Other sternites and tergites of both sexes undifferentiated.

VARIATION: Forewing ground colour varies from grey to grey-brown to brown. Hindwing ground colour varies from whitish to dark grey to grey-brown, always lighter than forewings.

MALE GENITALIA: Uncus rather wide, long, apex weakly bilobed, setose. Gnathos arms straight, fused ventrally, upturned, apex granulate. Valva equally narrow, setose, membranous, apex straight. Costa weakly sclerotised, posterior end with acute extension. Base of valva with few setae. Transtilla present, very weakly sclerotised. Vinculum sclerotised. Long, narrow, membranous sac (coremata) arises between valva and vinculum, covered with apically wide setae. Anellus arms long, narrow, apical part setose. Saccus narrow. Aedeagus curved, caecum minute, apex elongated, round. Vesica opens at approximately 90 degree angle. Base weakly expanded, apical part coiled, wrinkled, with one long, curved cornutus.

FEMALE GENITALIA: Papillae anales setose, surface granulate. Apophyses short, narrow. Lamella antevaginalis lunular plate, surface spinose. Lamella postvaginalis small plate, surface granulate. Ductus bursae long, narrow, sclerotised. Posterior part of corpus bursae long, narrow, partly sclerotised and spinose, spiralled two times along its axis. Anterior end of corpus bursae larger, globular, with anterior diverticulum. Signum elongated, margins dentate.

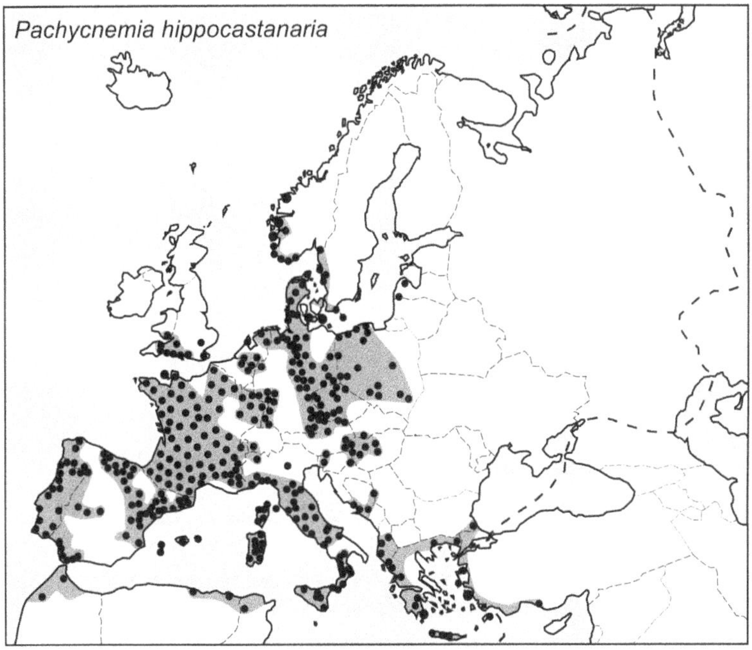

DISTRIBUTION AND ABUNDANCE: West-Palaearctic. In Europe a somewhat disjunct occurrence from southern Scandinavia to the Iberian Peninsula, Corsica, southern Italy including Sicily and Sardinia and Greece and from Portugal, France and southern England to Poland and Hungary. – Outside Europe the species has been found in Turkey, Tunisia, Algeria and in Morocco. – In northern and central Europe local, in southern Europe more commonly occurring and rather often in numbers.

PHENOLOGY: Bi- or trivoltine. In northern Europe in two well defined generations from early May to early June and from late July to early September. In Britain the second generation is regarded as partial (SKINNER 2009). In south-western Germany generally from early April to early June and from early July to early September (EBERT 2003). In the Iberian Peninsula from February to November (REDONDO *et al.* 2009). In southernmost Spain recorded in all months of the year (P. HALE pers. comm.). In southern Italy from late March to early October (A. HAUSMANN pers. comm.). Larva from late May to early July and again in September (PORTER 2010). In southern Europe probably in a large part of the year, in Samos caterpillars have been found from late March to mid-May and from late November to mid-December (FRITSCH *et al.* 2014). The species hibernates as a pupa. The moths fly at night and come to light.

BIOLOGY: Mono- or oligophagous. In northern and central Europe monophagous on *Calluna vulgaris* (EBERT 2003, H. WEGNER pers. comm.). In southern Europe found on *Erica arborea*, *E. manipuliflora*, *Rosmarinus officinalis* and *Calluna vulgaris* throughout the year (D. FRITSCH pers. comm.). Recorded on *Cistus incanus* and *Erica cinerea* (FLAMIGNI *et al.* 2007). The species overwinters as pupa among moss on the ground.

HABITAT: In northern and central Europe on heaths and other places where the food plant occurs, while in southern Europe the species occurs in most open land habitat types including Mediterranean macchia and scattered oak forests. In northern and central Europe at low altitudes, in southern Europe up to 1600 m in Italy (FLAMIGNI *et al.* 2007) and up to 1800 m in Spain (REDONDO *et al.* 2009).

SIMILAR SPECIES: No similar species in Europe.

GENETIC DATA: Genetically homogeneous from Europe (n=25, from seven countries) to Turkey but populations from Sardinia somewhat diverging (1.6%), approaching a haplotype occurring also in Tunisia. Nearest species in Europe: *Pseudopanthera macularia* (8.1%), *Pachycnemia tibiaria* (10.5%).

REMARKS: The inheritance and maintenance of the melanic form *nigrescens* LEMPKE, 1951 of *P. hippocastanaria* has been studied extensively in England (MAJERUS 1981). It was found out that f. *nigrescens* is inherited as a unifactorial recessive, but temperature also plays a role in the phenotype.

63. *Pachycnemia tibiaria* (RAMBUR, 1829)

tibiaria tibiaria RAMBUR, 1829, Annls Sci. Observ. 2 (2): 266, pl. 6, fig. 2 (as *Geometra tibiaria*), (*Phalaena*). Lectotype male (C. HERBULOT Collection in Zoologische Staatssammlung, Munich, Germany), [France]: Tours (examined externally).

benesignata uniformis BYTINSKI-SALZ, 1937, Memorie Soc. Ent. Ital. 15 (2): 212, (*Gnophos*). Holotype male, Sardinia, Aritzo. [Originally as f].

tibiaria pravieli HERBULOT, 1944, Miscnea. ent. 41: 67, (*Gnophos*). Holotype male, [France]: Saint-Tropez, Var. (synonymy after LERAUT 2009).

tibiaria aindrahamaria OBERTHÜR, 1923, Études Lépid. comp. 20: 255, pl. 556, fig. 4763, (*Liodes*). Syntype(s), Tunisia: Ain-Draham.

tibiaria benesignata BELLIER, 1861 Annls Soc. ent. Fr. (4) 1 (1): 30, pl. 2, fig. 13, (*Liodes*). Syntype(s), mountains of Corsica. (synonymy after LERAUT 2009, here confirmed).

EXTERNAL CHARACTERS AND ABDOMEN: Wingspan 25–28 mm. Wings almost uniform brown-grey, sparsely irrorated with black. Antemedial line and postmedial lines broken into small blackish dots. Outer margin of postmedial line lighter. Terminal line narrow, black, interrupted. Fringes concolorous with wings. Hindwings lighter, terminal area darker. Discal spots small, diffuse. Wings below as above, hindwing discal spot often larger. Frons, collar, thorax and abdomen concolorous with wings. Male and female antennae filiform. Male hindtibia swollen, with hair pencil. Hindtibia with 2+2 spurs. Tympanal organs medium-sized, sclerotised plate between them, posterior corners with pockets. Male 3rd sternite with setae, posterior margin of 8th sternite weakly sclerotised, concave, otherwise unmodified. Other sternites and tergites of both sexes undifferentiated.

VARIATION: Very little. Ground colour varies from pale brown to grey-brown to dark brown. Antemedial and postmedial lines may be reduced.

Subsp. ***benesignata*** darker, forewing veins blackish in medial area. Male genitalia with minor quantitative differences: Apex of gnathos more dentate. Posterior end of costa with smaller, round extension and furca arms evenly curved.

Subsp. ***aindrahamaria*** is paler, light sand-brown, veins are not black in medial area. Some specimens appear as intermediate between subspecies *tibiaria* and subspecies *benesignata*. Subspecies status needs to be confirmed, see also WEHRLI (1939–1954).

MALE GENITALIA: Uncus narrow subapically, long, apex with knob, setose. Gnathos arms narrow, fused ventrally, upturned, apex weakly dentate. Valva narrow at base, setose, membranous, apex round. Costa weakly sclerotised, posterior end with round extension. Base of valva with row of setae. Transtilla present, moderately sclerotised. Vinculum(?) with lateral sclerotisations, without membranous sac (coremata). Furca arms long, narrow, weakly angled below apex. Saccus narrow. Aedeagus curved, caecum minute. Vesica opens at approximately 45 degree angle. Base with small lateral diverticulum, otherwise weakly curved tube, with one long, rather straight cornutus.

FEMALE GENITALIA: Papillae anales setose, surface weakly granulate. Apophyses long, narrow. Lamella antevaginalis large, roundish plate. Lamella postvaginalis with few horizontal grooves. Ductus bursae short, sclerotised. Ductus seminalis arises from diverticulum. Posterior part of corpus bursae long, narrow, grooved, weakly twisted along its axis. Anterior end of corpus bursae larger, elongated, with anterior diverticulum. Signum small, elongated, margins dentate.

DISTRIBUTION AND ABUNDANCE: Mediterranean. In Europe only in the south from the Iberian Peninsula, southern France and Italy to Greece including Crete. Reported several times from Croatia, but apparently no specimens seem to exist (I. MIHOCI pers. comm.). However, a specimen was collected in Istria in 2003 (G. LASZLO pers. comm.). In Corsica and Sardinia replaced by subsp. *benesignata*. – Outside Europe the species has been found in Morocco and Tunisia (subsp. *aindrahamaria*). Reported from Algeria (RUNGS 1981), but no specimen from there has been traced. – In Europe a somewhat local and not common species that not often occurs in numbers.

PHENOLOGY: Univoltine. The adult moths occur from late August to mid-November. They are easily attracted to light, but occur usually in limited numbers.

BIOLOGY: Oligophagous. The caterpillar is recorded on *Erica scoparia* (REDONDO *et al.* 2009) and according to FLAMIGNI *et al.* (2007) also on *Calluna vulgaris* (subsp. *benesignata*). Egg deposition observed on *Erica arborea* (M. LEIPNITZ pers. comm.: subsp. *benesignata*). In rearings, the larvae have accepted *Erica carnea* as food plant, the larvae prefered the buds which they hollowed out. (B. MÜLLER pers. comm.).

HABITAT: In a variety of open land habitats with bushes and scattered trees (Mediterranean macchia). The species is found from sea level up to 1400 m.

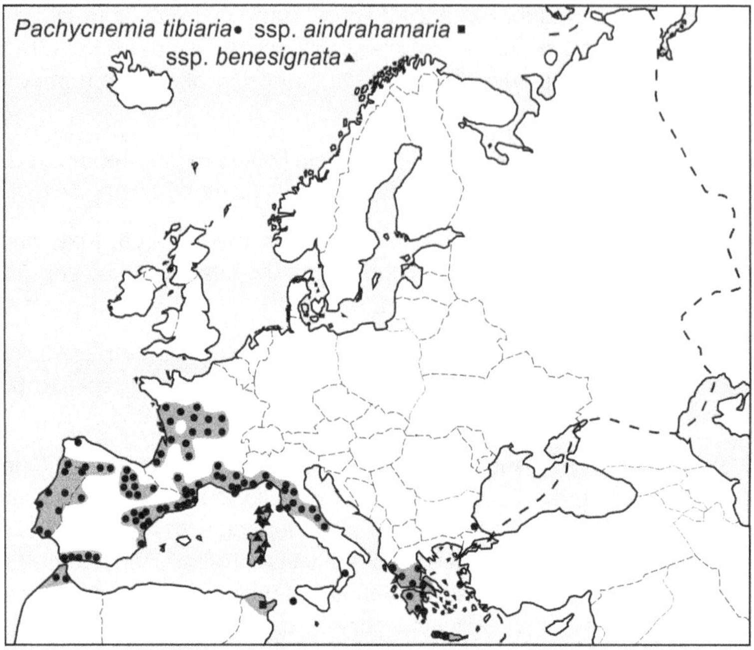

SIMILAR SPECIES: No similar species in Europe, but lack of distinct patterns and uniform brown-grey colour may cause trouble in identification. *Ortaliella gruneraria* (STAUDINGER) and few Larentiinae, Noctuidae and even Crambidae, exhibit some superficial resemblance.

GENETIC DATA: Genetically slightly heterogeneous (Sicily, France, southern Spain), subsp. *benesignata* from Sardinia diverging by 1.7% but barcode-sharing with populations from Tuscany and Liguria. Nearest species in Europe: *Selenia lunularia* (8.4%), but African *Idiodes saxaria* (GUENÉE, 1858) and Tasmanian *I. apicata* (GUENÉE, 1858) much closer (5.2% and 6.0%). *Pachycnemia hippocastanaria* at 10.5% distance.

REMARKS: *Pachycnemia benesignata* (BELLIER) has traditionally been considered valid at species level (for instance MÜLLER 1996). In recent literature it has been treated as a subspecies of *P. tibiaria* (for instance HAUSMANN, MIRONOV & VIIDALEPP 2011, LERAUT 2009). We consider the taxon valid at subspecies level also, and for the sake of clarity illustrate genitalia of both taxa (*tibiaria* and *benesignata*). The treatment of *benesignata* as a subspecies of *P. hippocastanaria* in SCOBLE (1999) is an error.

Opisthograptis HÜBNER, 1823

Opisthograptis HÜBNER, [1823] 1816, Verz. bekannter Schmett.: 292. Type species: *Phalaena crataegata* LINNAEUS, 1761 (unnecessary replacement name for *luteolata* LINNAEUS) [Europe?]. Genus synonyms are after SCOBLE (1999).

Rumia DUPONCHEL, 1829, in GODART & DUPONCHEL, Hist. nat.Lépid. Papillons Fr. 7 (2): 103, 117. Type species: *Phalaena crataegata* LINNAEUS, 1761.

UNAVAILABLE NAMES (incorrect subsequent spelling): *Opistograptis* (for instance in ECKSTEIN (1923), see FLETCHER 1979).

DIVERSITY AND DISTRIBUTION: *Opisthograptis* contains 11 species, occuring in the Palaearctic and the Oriental regions.

EXTERNAL CHARACTERS AND ABDOMEN: Medium-sized, wingspan about 30–40 mm. Wings yellow, markings mostly reddish brown or blackish. Male and female antennae filiform. Male hindtibia not swollen, without hairpencil. Male 3rd sternite without setae, 8th sternite unmodified.

MALE GENITALIA, description is based on European species: Uncus stout. Socii absent. Gnathos arms fused ventrally, upturned, weakly granulate. Valva wide, membranous, setose. Base of valva with small group of setae. Transtilla present. Furca arms absent. Saccus not elongated. Vesica with numerous microcornuti.

FEMALE GENITALIA, description is based on European species: Papillae anales weakly setose, partially granulate. Posterior part of corpus bursae very long, narrow, membranous. Anterior part of corpus bursae round, membranous. Signum absent.

BIOLOGY: *Opisthograptis luteolata* (LINNAEUS) has been recorded feeding on numerous deciduous trees and bushes, for instance *Betula*, *Salix*, *Malus* and *Ribes* (MIKKOLA *et al.* 1989), *Prunus*, *Crataegus* and *Corylus* (EBERT 2003). Adults are nocturnal.

IMMATURE STAGES: Larva of *Opisthograptis luteolata* (LINNAEUS) is a twig-mimic, brown-violet or violet in colour, and it has twig-mimicking extension dorsally, apparently on A3 (EBERT 2003). Pupa cremaster with four pairs of curved setae, apical pair biggest (PATOČKA & TURČANI 2005). Immature stages are illustrated in AIZPÚRUA (2002).

REMARKS: *Opisthograptis* has recently been classified in the Ennomini (URBAHN 1967, VIVES MORENO 1994), Epionini (LERAUT 1997, HAUSMANN, MIRONOV & VIIDALEPP 2011) and Ourapterygini (VIIDALEPP 1996). HOLLOWAY (1994) diagnosed the Epionini, and the potentially related Ennomini, Hypochrosiini and Scardamiini, by consistent absence of a setal comb on male third abdominal sternite, perhaps associated with loss of a tibial hair pencil. Further, the Epionini has extension and looping of the vinculum to support lateral coremata. In a molecular, phylogenetic analysis of the world Geometridae (SIHVONEN *et al.* 2011), *Opisthograptis* grouped together with the Eurasian *Epirranthis diversata* (DENIS & SCHIFFERMÜLLER, 1775). *Epione* (Epionini), *Ennomos* (Ennomini) and *Ourapteryx* (Ourapterygini) were not related. In a molecular, phylogenetic analysis of limited European taxa, ÕUNAP *et al.* (2011) recovered *Ennomos* as sister lineage to *Opisthograptis* and *Crocallis*.

O. luteolata, type species of genus *Opisthograptis*, does not have the characteristic adult morphology features of the Epionini, Ourapterygini and Ennomini. *O. luteolata* lacks lateral coremata (characteristic feature of the Epionini sensu HOLLOWAY (1993) and HOLLOWAY *et al.* (2001)), it lacks setal comb on male third sternite and furca in the male genitalia (characteristic features of the Ourapterygini sensu HOLLOWAY (1994) and HOLLOWAY *et al.* (2001)) and the numerous structures described in BELJAEV (2008a) that he used to define the new concept of Ennomini. The tribus name Rumiini (Rumiinae TUTT, 1896) is available, based on *Rumia* DUPONCHEL, 1829, a junior objective synonym of *Opisthograptis* HÜBNER. HOLLOWAY (1994) did not locate current usage. Systematic position: Uncertain, further research is needed.

As is explained in more detail in *Ennominae systematics* chapter, the European Ennominae fauna as a whole has not previously been classified into tribes. Systematic review of tribal classifications and their constituent genera is beyond scope of this work, mainly because it would require extensive studies on numerous non-European taxa. Therefore in several cases, including this one, we have been forced to arrange higher taxa according to the conventional and practical aspects.

64. Opisthograptis luteolata (LINNAEUS, 1758)

luteolata LINNAEUS, 1758, Syst. Nat. (Ed. 10) 1: 525, (*Phalaena* (*Geometra*)). Syntypes, including putative lectotype male (Linnean Society of London, London, UK). [Europe?] (examined externally).

crataegaria HÜBNER, [1823], Verz. bekannter Schmett.: 292, (*Opisthograptis*). [Emendation of *crataegata* LINNAEUS].

crataegata LINNAEUS, 1761, Fauna Suecica (Edn 2): 336, (*Phalaena Geometra*). [Unnecessary replacement name for *luteolata* LINNAEUS].

luteolata impunctata OSTHELDER, 1931, Mitt. münch. ent. Ges. suppl. 21 (4): 504, (*Opisthograptis*). Syntype(s), [Germany]: near Munich. [Originally as var].

emaculata GRAESER, 1892, Berl. Ent. Z. 37 (3): 317, (*Rumina*). Syntypes two males, one female. [Kyrgyzstan:] Issyk-Kul, Alexander-Gebirge, [China:] Kuldja-Distrikt. [Originally as ab. of *luteolata* LINNAEUS]. See genetic data and Remarks.

niko CHRISTOPH, 1893, Dt. ent. Z. Iris 6 (1): 95, (*Rumia*). Syntype(s) female(s), [Georgia]: Swanetia.

provincialis OBERTHÜR, 1912, Études Lépid. comp. 6: 270, pl. 155, fig. 1495, (*Rumia*). Syntype(s), [France]: Digne. [Originally as subsp. of *crataegata*]. See remarks.

INTRASUBSPECIFIC NAMES, POTENTIALLY UNAVAILABLE: *aestiva* VORBRODT, 1914 (f.); *albescens* COCKERELL, 1888 (ab.); *albicans* REBEL, 1907 (ab.); *apicolutea* COCKAYNE, 1950 (ab.); *apiconigrescens* LEMPKE, 1970 (f.); *delineata* LEMPKE, 1951 (f.); *flavissima* KRULIKOWSKY, 1908 (ab.); *immaculata* OBERTHÜR, 1912 (ab.); *intensa* LEMPKE, 1970 (f.); *intermedia* HARRISON, 1905 (ab.); *lacticolor* HARRISON, 1905 (ab.); *poskini*, DERENNE, 1928 (ab.); *quadrilineata* NORDSTRÖM, 1941 (ab.?); *ruficosta* LEMPKE, 1951 (f.); *tangens* LEMPKE, 1951 (f.).

EXTERNAL CHARACTERS AND ABDOMEN: Wingspan 27–40 mm. Wings deep yellow. Markings rusty brown, strongest on forewing costa. Medial lines weak, broken into diffuse, rusty brown spots. Forewing apex with triangular, rusty brown spot. Terminal line and fringes concolorous with wings. Forewing discal spots white, margins rusty brown, posterior margin dentate. Hindwing discal spots smaller, often with white centre. Wings below as above. Lower part of frons rusty brown, otherwise frons, collar, thorax and abdomen concolorous with wings. Male and female antennae filiform. Hindtibia with 2+2 spurs. Tympanal organs medium-sized. All sternites and tergites of both sexes undifferentiated.

VARIATION: Very little. Ground colour is mostly deep yellow, but pale yellow, and even albinistic specimens have been found (SLOT 1999). Density of rusty brown markings varies somewhat. LERAUT (2009) gives an overview of the named forms. Second generation specimens tend to be smaller and darker.

MALE GENITALIA: Uncus long, stout, apex round, weakly setose. Socii absent. Gnathos arms fused ventrally, upturned, weakly granulate. Juxta sclerotised plate. Valva wide, membranous, setose, apex round. Base of valva with small group of setae. Transtilla membranous. Furca arms absent. Saccus narrow, interior margin medially concave. Aedeagus wide, caecum minute, apex acute. Vesica opens at approximately 135 degree angle. Vesica straight, membranous tube, basal part with numerous microcornuti.

FEMALE GENITALIA: Papillae anales weakly setose, partially granulate. Lamella antevaginalis and lamella postvaginalis membranous. Ductus bursae short, wide, most sclerotised laterally. Base of ductus seminalis weakly expanded. Posterior part of corpus bursae very long, narrow, membranous. Anterior part of corpus bursae wider, round, membranous. Signum absent.

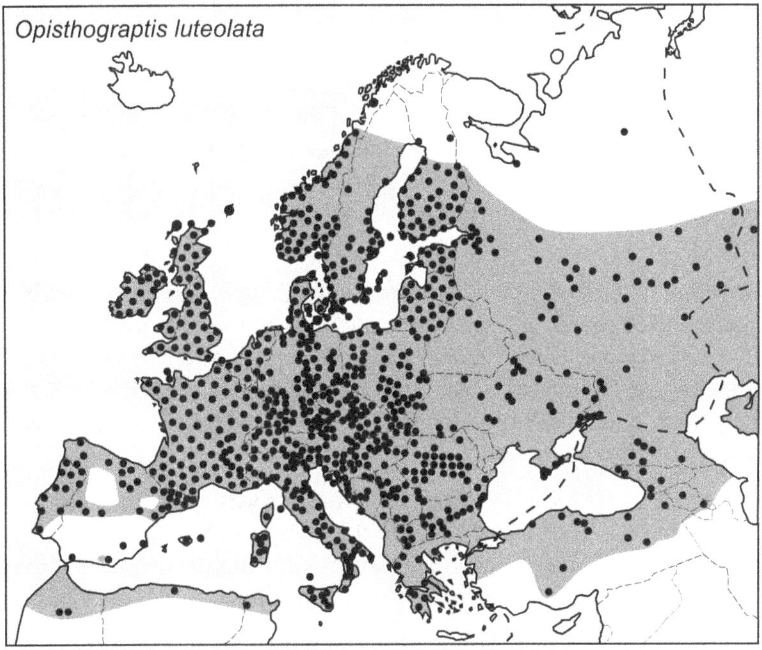

DISTRIBUTION AND ABUNDANCE: European-Westasiatic. Widespread over most of Europe, only missing in Iceland, northernmost Fennoscandia and northern Russia. – Outside Europe the species is found in Turkey and the Caucasus region and further to the east in central Asia and Afghanistan. South of the Mediterranean the species also occurs in Morocco, Algeria and Tunisia. – A common or very common species in most parts of Europe, but in most cases it occurs in limited numbers. However, in southern Italy at medium and higher altitudes it is one of the most numerous geometrid species (A. HAUSMANN pers. comm.).

PHENOLOGY: Uni- or bivoltine or even trivoltine. In northern Europe univoltine from mid-May to early July, but specimens of a partial second generation are found now and then in September. In southern central Europe and further to the south in two more or less overlapping generations from April to early October. In England there are almost certainly three generations in most years and with more and more overlap as the summer progresses (M. CORLEY pers. comm.). The overwintering takes place as pupa. In south-western Germany also larva can hibernate (EBERT 2003). The moths fly at night and are easily attracted to light, they are often seen flying at dusk and they can frequently be disturbed by day.

BIOLOGY: Polyphagous. The larva has been found on many deciduous trees and bushes. In south-western Germany it has been found on *Salix caprea*, *Corylus avellana*, *Cotoneaster horizontalis*, *Chaenomeles japonica*, *Sorbus aucuparia*, *Crataegus*, *Pyracantha submollis*, *Rosa canina*, *Prunus laurocerasus*, *P. cerasus*, *P. padus*, *P. avium*, *P. insititia*, *P. spinosa*,

Lonicera xylosteum, *Acer campestre* and *Tilia* (EBERT 2003). In northern Germany the larvae have been found on *Sorbus aucuparia* (H. WEGNER pers. comm.), *S. aria* and *Prunus cerasus* (C. KAYSER pers. comm.). Pupation takes place between leaves and plant litter on the ground or on the food plant (M. LEIPNITZ pers. comm.).

HABITAT: Sylvicolous. In most habitat types with deciduous trees and bushes. Vertically the species is distributed from sea level up to about 2100 m in the Alps and the Pyrenees.

SIMILAR SPECIES: No similar species in Europe.

GENETIC DATA: Genetically homogeneous from Europe to Turkey and Iran (n=51, from 12 countries), but populations from Lebanon and Iran with deeper divergences, requiring further study. Large genetic distance of examined specimens from Kyrgyzstan and Pakistan (minimum distance 7.4%) suggesting cryptic species diversity, See remarks. Nearest species: *O. tridentifera* (MOORE, 1888) from northern India (7.4%), in Europe *Dyscia innocentaria* (7.2%).

REMARKS: MIKKOLA *et al.* (1989) mention subspecies *emaculata* (GRAESER, 1892) from central Asian mountains and Afghanistan, while VIIDALEPP (1988) treats it as a form of *O. luteolata*. PROUT (1912–1916) also treats *emaculata*, but the name is not mentioned in SCOBLE (1999). We have not examined the matter, and do not know if *emaculata* is a relevant name in the context of Kyrgyzstan and Pakistan specimens, see genetic data. Taxon *emaculata* lacks the apical triangular, rusty brown marking.

SCOBLE (1999) treated *provincialis* (OBERTHÜR) (rusty brown markings very weak) valid at subspecies level, whereas in LERAUT (1997) and HAUSMANN *et al.* (2011) it is considered a junior synonym of *O. luteolata* (LINNAEUS). Genetic and morphological information available do not support the hypothesis that there exists an allopatric, distinct population of *O. luteolata* in southern France.

Epione DUPONCHEL, 1829

Epione DUPONCHEL, 1829, Hist. nat. Lépid. Papillons Fr. 7 (2): 105, 211. Type species: *Geometra paralellaria* [DENIS & SCHIFFERMÜLLER], 1775, [Austria:] Vienna district, which is a junior synonym of *Phalaena (Geometra) vespertaria* LINNAEUS, 1767. There are no genus synonyms.

DIVERSITY AND DISTRIBUTION: *Epione* contains six species, five of them in the Palaearctic region, one species in South Africa.

EXTERNAL CHARACTERS AND ABDOMEN: Medium-sized moths, wingspan about 25–30 mm. Wings different shades of orange, yellow, grey and white or mixture of these. European species have terminal area darker than rest of wing. Forewing margin commonly weakly concave below apex. Male antennae bipectinate, female antennae filiform. Male hindleg without hair pencil, 3rd sternite without setae, 8th sternite weakly sclerotised, posterior margin narrower.

MALE GENITALIA, description is based on European species: Uncus long, narrow, curved ventrally. Socii minute. Gnathos arms fused ventrally, upturned, weakly dentate. End of tuba analis (subscaphium?) sclerotised. Valva narrow, membranous, setose. Costa weakly sclerotised, setose subapically. Base of valva with small group of setae, with V-shaped, sclerotised pockets. Membranous sac with setae (coremata), originating from vinculum. Transtilla with two hollow rings. Saccus concave. Vesica narrow tube, without cornuti.

FEMALE GENITALIA, description is based on European species: Papillae anales striated. Lamella antevaginalis narrow, curved, sclerotised. Corpus bursae angled. Signum small, margins dentate.

BIOLOGY: Larvae of European species have been recorded to feed on numerous deciduous trees, particularly on *Salix* and *Populus*, but also on *Alnus* and *Betula* (MIKKOLA *et al.* 1989), and on *Lonicera xylosteum* (EBERT 2003). Caterpillars live between leaves, which are spun together. Pupation takes place between leaves also (MIKKOLA *et al.* 1989). Adults are nocturnal, but can be easily flushed up during the day.

IMMATURE STAGES: Larvae of European *Epione* species are twig-mimics, brown, brown-violet or green in colour, and they have twig-mimicking hump dorsally, apparently on A2 (EBERT 2003). Pupa cremaster with four pairs of curved setae, apical pair biggest (PATOČKA & TURČANI 2005).

REMARKS: *Cepphis* HÜBNER and *Epione* DUPONCHEL are structurally similar. They share for instance V-shapedsclerotised pockets at base of valva, membranous sacs with setae (coremata) and concave saccus. Also male 8th sternite is weakly sclerotised, posterior margin is narrow. In a molecular, phylogenetic analysis of the world Geometridae (SIHVONEN *et al.* 2011), *Epione* grouped together with *Apeira* and *Plagodis*. *Cepphis* was not included in the analysis. VIVES MORENO (1994) classified *Epione* in the Ennomini, but this view was not adopted by BELJAEV (2008a). Systematic position: Epionini (HOLLOWAY 1994).

65. Epione repandaria (HUFNAGEL, 1767)

repandaria HUFNAGEL, 1767, Berlin Mag. 4 (5): 508, (*Phalaena*). Syntype(s), lost, [Germany:] Berlin region.

apiciaria [DENIS & SCHIFFERMÜLLER], 1775, Ankündung syst. Werkes Schmett. Wienergegend: 104, (*Geometra*). Syntype(s), lost, [Austria:] Vienna district.

apiciaria franziscaria MÜLLER, 1962, Ent. Z., Frankf. a.M. 72 (15): 163, (*Epione*). Holotype male, [Germany:] Lechauwald, Kissingen near Augsburg. [Originally as f].

marginaria VILLERS, 1789, Linn. ent. 2: 303, (*Phal*[*aena*] *Geom*[*etra*]). Syntype(s), France: near Marseille, Bouches-du-Rhône.

INTRASUBSPECIFIC NAMES, POTENTIALLY UNAVAILABLE: *aurantiaca* REBEL, 1924 (ab.); *crasselineata* LEMPKE, 1953 (f.); *demarginaria* HELLWEGER, 1914 (ab.?); *glabra* LEMPKE, 1951 (f.); *impuncta* LEMPKE, 1970 (f.); *lateritia* RUDOLPH, 1926 (ab.); *linearecedens* LEMPKE, 1951 (f.); *maculosa* LEMPKE, 1970 (f.); *purpureomarginata* LEMPKE, 1970 (f.). Incorrect subsequent authorship: *demarginaria* WEHRLI, 1940 (SCOBLE 1999; should be *demarginaria* HELLWEGER, 1914. Also cited with incorrect authorship in OKYAR and MIRONOV (2008).

EXTERNAL CHARACTERS AND ABDOMEN: Wingspan 23–29 mm. Wings orange-yellow, heavily irrorated with brown. Veins brown, thus creating reticulated pattern. Antemedial line sharply angled. Forewing postmedial line curved in middle, terminates on costa just before apex. Terminal area darker, violet-brown. Terminal line and fringes concolorous with wings. Discal spots small, sharp. Forewing margin concave below apex, hindwing margin concave also. Wings below as above, markings more diffuse. Frons, collar, thorax and abdomen concolorous with wings. Male antennae bipectinate, female antennae filiform. Hindtibia with 2+2 spurs. Tympanal organs medium-sized. Male 8th sternite weakly sclerotised, posterior margin narrow medially. Other sternites and tergites of both sexes undifferentiated.

VARIATION: Rather little. Ground colour varies from deep orange-yellow to pale yellow. Density of irroration variable, from heavy to absent. Terminal area normally darker, violet-brown, but can be almost the same colour as rest of the wing. Discal spots can be absent.

MALE GENITALIA: Uncus long, narrow, weakly setose. Socii minute, setose. Gnathos arms fused ventrally, upturned, weakly dentate. End of tuba analis (subscaphium?) sclerotised. Valva narrow, membranous, setose. Costa weakly sclerotised, setose subapically. Base of valva with small group of setae, with V-shapedsclerotised pockets. Narrow ridge connecting base of valva and saccus medially. Membranous sac with setae (coremata) present. Transtilla with two hollow rings. Saccus concave. Aedeagus narrow, curved, caecum minute, apex sharp. Vesica opens at approximately 135 degree angle. Vesica narrow tube, with small dorsal diverticulum, without cornuti.

FEMALE GENITALIA: Papillae anales elongated, setose, surface striated. Apophyses long, straight. Lamella antevaginalis narrow, curved, sclerotised. Lamella postvaginalis membranous, weakly striated horizontally. Ductus bursae long, narrow, membranous. Posterior part of corpus bursae narrow. Anterior part of corpus bursae angled, apex elongated. Signum small, margins dentate.

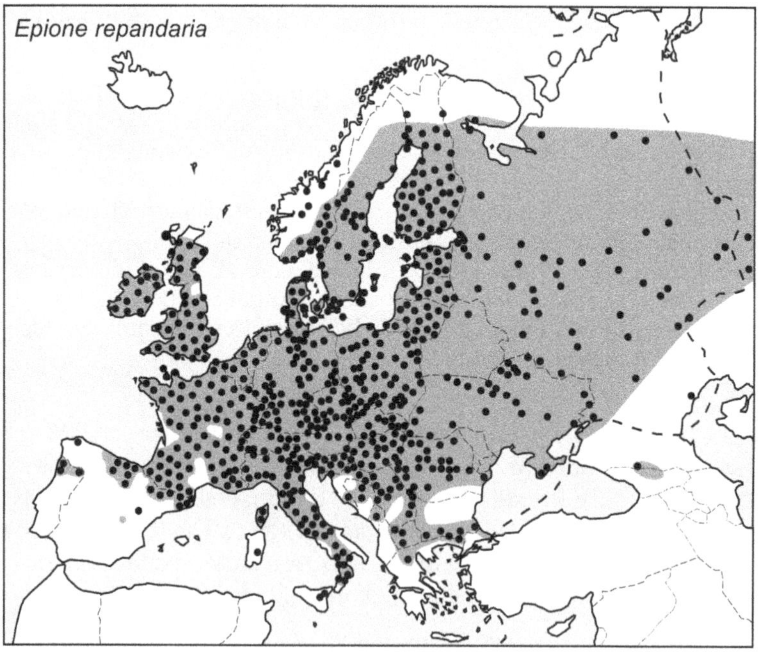

DISTRIBUTION AND ABUNDANCE: Eurasiatic. In Europe from central Scandinavia to the northern part of the Iberian Peninsula, southern France including Corsica, Italy including Sardinia and Sicily and to northern Greece and from the British Isles to the Urals. – Outside Europe the species is known from Transcaucasus, Caucasus, northern Kazakhstan and further to the east through Siberia to Far East Russia. – In Europe a relatively common species that usually is found in limited numbers.

PHENOLOGY: Uni- or bivoltine. In northern Europe from mid-July to mid-September. In southern central Europe from early June to late September. In southern Europe in two generations in Italy from May to June and July to September (FLAMIGNI *et al.* 2007), in the Iberian Peninsula from June to July and from August to September (REDONDO *et al.* 2009). Larva from late May to late June, in the south also in August. The species hibernates as egg. The moths fly at night and come to light but are rarely seen by day.

BIOLOGY: Oligophagous. The caterpillar has been found on *Salix* (HULDÉN *et al.* 2000), *Populus balsamifera*, *Lonicera xylosteum* (EBERT 2003) and on *Alnus glutinosa* (H. WEGNER pers. comm.). Larva has been found feeding on young shrubs of *Salix aurita* in Germany (A. HAUSMANN pers. comm.). In rearing the caterpillar has accepted *Salix caprea* (B. MÜLLER pers. comm.). Pupation takes place between spun leaves.

HABITAT: On moors, on meadows with *Salix*, semi-open deciduous forests, and similar places with moist soil. The species is found from sea level up to 1700 m in the Italian Alps (FLAMIGNI *et al.* 2007).

SIMILAR SPECIES: *Epione vespertaria* (LINNAEUS). In male forewing postmedial line terminates on costa, clearly before apex, and there is distinct concavity or angle in middle (postmedial line terminates just before apex, line evenly curved in middle in *E. repandaria*). In female ground colour is pale yellow, terminal area almost uniform brown and forewing postmedial line deeply concave in middle (ground colour orange-yellow, terminal area violet-brown, not uniform in colour and postmedial line evenly curved in middle in *E. repandaria*). Male and female genitalia similar, without qualitative differences.

GENETIC DATA: Genetically homogeneous in Europe (n=20 from seven countries). Nearest species: *E. vespertaria* (9.2%).

66. Epione vespertaria (LINNAEUS, 1767)

vespertaria vespertaria LINNAEUS, 1767, Syst. Nat. (Ed. 12) 1 (2): 864, (*Phalaena (Geometra)*). Syntype(s), Lusitania [Portugal].

affiniaria BORKHAUSEN, 1794, Natur. eur. Schmett. 5: 136, (*Phalaena Geometra*). Syntype(s), Europe: Teutschland [Germany].

paralellaria [DENIS & SCHIFFERMÜLLER], 1775, Ankündung syst. Werkes Schmett. Wienergegend: 104, (*Geometra*). Syntype(s), lost, [Austria]: Vienna district.

vespertaria amura WEHRLI, 1940, in SEITZ, Gross-Schmett. Erde 4 (Suppl.): 371, pl. 34: d, (*Epione*). Syntype(s) male(s), Russia: Amur.

vespertaria hespera BRYK, 1949, Ark. Zool. 41A (1): 196, (*Epione*). Holotype female, Korea: Motojondo. Valid at subspecies rank according to SCOBLE (1999).

UNAVAILABLE NAMES (incorrect subsequent spelling): *parallelaria* (frequently, for instance in LERAUT 2009). – (hybrid): *isabellae* HARRIS, 1914 (*vespertaria* LINNAEUS male × *repandaria* HUFNAGEL female). Intrasubspecific names, potentially unavailable: *alba* COCKAYNE, 1942 (ab.); *dereticularia* RUDOLPH, 1936 (ab.); *fulva* COCKAYNE, 1934 (ab.); *impuncta* LEMPKE, 1970 (f.); *reticulata* LEMPKE, 1951 (f.).

EXTERNAL CHARACTERS AND ABDOMEN: Wingspan 24–29 mm. Male: Generally as *E. repandaria*, but with following differences: Antemedial line more round. Postmedial line with distinct concavity or angle in middle, line terminates on costa. Forewing margin less concave below apex. Female: Generally as *E. repandaria*, but with following differences: Ground colour pale yellow, weakly irrorated with brown. Veins only slightly darker than ground colour, thus reticulate pattern absent or weak. Antemedial line more round. Postmedial line with distinct concavity or angle in middle. Terminal area reddish brown. Terminal line chequered pale yellow and reddish brown. Postmedial line terminates on costa just before apex as in *E. repandaria*. Other structures of both sexes as in *E. repandaria*.

VARIATION: Rather little: Density of irroration variable in male. Veins may occasionally be darker also in female, thus creating weakly reticulate pattern.

MALE GENITALIA: Generally as in *E. repandaria* (HUFNAGEL), without any qualitative differences, see entry above.

FEMALE GENITALIA: Generally as in *E. repandaria* (HUFNAGEL), without any qualitative differences, see entry above. Corpus bursae slightly less angled and signum often slightly more elongated, or broken into granulate, discontinuous plate in *E. vespertaria*, but there is variation in both species.

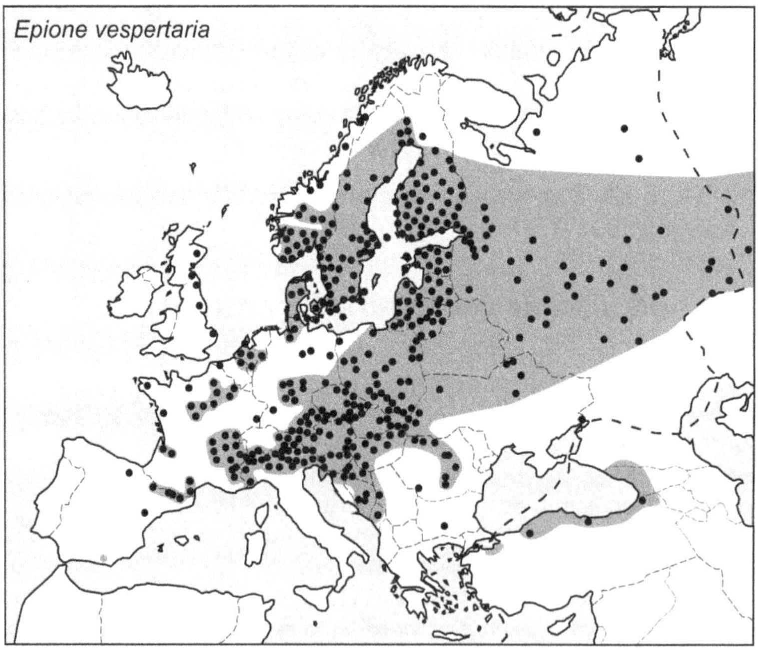

DISTRIBUTION AND ABUNDANCE: Eurasiatic. In Europe from central Scandinavia to the Alps and with a scattered occurrence in the northern Balkans and the Pyrenees and from Great Britain to the Urals. Reported from Sardinia (MÜLLER 1996), but this is apparently due to an error (B. MÜLLER pers. comm.). – Outside Europe the species occurs in Turkey, Georgia and Caucasus and further to the east from the Ural mountains and the Caucasus region through central Asia and Siberia to Far East Russia and Sakhalin. – In Europe a somewhat local species that now and then can occur in numbers.

PHENOLOGY: Univoltine. In northern Europe from late June to late August. In the Italian Alps from mid-June to August or September (FLAMIGNI *et al.* 2007). Caterpillars in May and June (BERGMANN 1955). The species overwinters as egg. The adult insects fly at night and come readily to light but are rarely seen by day.

BIOLOGY: Oligophagous. The caterpillar has been found on *Populus tremula* (H. WEGNER pers. comm.), *Salix caprea* and *Betula pendula* (EBERT 2003). In Britain also recorded on *Salix repens* and *S. phylicifolia* (SKINNER 2009).

HABITAT: In light deciduous forests on moist soil, in Britain in wet and lightly wooded heathland (SKINNER 2009). In northern Europe the species occurs from sea level, in southern Germany up to 1300 m and in the Italian Alps from 500 up to 2000 m (FLAMIGNI *et al.* 2007).

SIMILAR SPECIES: *Epione repandaria* (HUFNAGEL), see entry above.

GENETIC DATA: Genetically homogeneous in Germany, Finland and Russia/Altai (n=6), but one specimen from Turkey diverging by 2.9%. Nearest species: *E. repandaria* (9.2%).

Therapis HÜBNER, 1823

Therapis HÜBNER, [1823] 1816, Verz. bekannter Schmett.: 288. Type species: *Geometra flavicaria* [DENIS & SCHIFFERMÜLLER], 1775, Hungary. Genus synonymy is after SCOBLE (1999).

Caustoloma LEDERER, 1853, Verh. zool. –bot. Ver. Wien 3 (Abh.): 176, 203, 209. Type species: *Geometra flavicaria* [DENIS & SCHIFFERMÜLLER], 1775.

UNAVAILABLE NAMES (incorrect subsequent spelling): *Caulostoma* PACKARD, 1876 (see FLETCHER 1979).

DIVERSITY AND DISTRIBUTION: *Therapis* is a monotypic genus, the only species occurs from Eastern Europe and Turkey to south-west Russia and northern Iran.

EXTERNAL CHARACTERS AND ABDOMEN: Medium-sized, wingspan 25–29 mm. Wings yellow, markings brown, spot-like. Forewing margin concave below apex. Male and female antennae bipectinate. Male hindtibia without hair pencil, 3rd sternite without setae, 8th sternite unmodified.

MALE GENITALIA: Uncus long, narrow, curved ventrally. Socii minute. Gnathos arms fused ventrally, upturned, weakly dentate. Valva narrow, membranous, setose. Costa weakly sclerotised, setose subapically. Base of valva with small group of setae, with weak V-shaped sclerotised ridges. Membranous sac with setae (coremata), originating from vinculum. Transtilla present. Saccus concave. Vesica with numerous cornuti.

FEMALE GENITALIA: Papillae anales striated. Lamella antevaginalis and lamella postvaginalis unrecognisable. Corpus bursae long, narrow. Posterior part of corpus bursae sclerotised, striated, weakly dentate. Signum small, margins dentate.

BIOLOGY: Larva of *T. flavicaria* has been recorded feeding on various herbaceous plants, for instance on *Galium*, *Galeopsis* (SZÉKELY 2011) and on *Lamium* (LERAUT 2009). Habitats include grasslands, meadows, steppe meadows and steppe forests. Two generations per year, species overwinters as pupa (SZÉKELY 2011). Adults are partly diurnal (LERAUT 2009). Pupa of *T. flavicaria* (DENIS & SCHIFFERMÜLLER) overwinters (VOJNITS 1980).

IMMATURE STAGES: We have no data available, apart from superficial mention in VOJNITS (1980): the larva of *T. flavicaria* (DENIS & SCHIFFERMÜLLER) is short, green, with white stripes. Pupa is brown.

REMARKS: Structurally *Epione* and *Therapis* are very similar, and they could even be considered congeneric. The most notable differences are, in addition to the superficial appearance, that in *Epione* the female antennae are filiform, in *Therapis* shortly bipectinate, and in *Epione* the male vesica is without cornuti, in *Therapis* with numerous cornuti. VIIDALEPP (1996) placed *Therapis* in the Ourapterygini, but we emphasise structural similarities with the Epionini. Systematic position: Epionini.

67. *Therapis flavicaria* (DENIS & SCHIFFERMÜLLER, 1775)

flavicaria [DENIS & SCHIFFERMÜLLER], 1775, Ankündung syst. Werkes Schmett. Wienergegend: 104, (*Geometra*). Syntype(s), lost, Hungary.

UNAVAILABLE NAMES (incorrect subsequent spelling): *fiavicaria* (HERRICH-SCHÄFFER, 1863, referred to in The Global Lepidoptera Names Index (ANONYMOUS 2013).

EXTERNAL CHARACTERS AND ABDOMEN: Wingspan 25–29 mm. Wings yellow, markings brown, spot-like. Basal line, medial lines, subterminal line on forewings weak, incomplete, present as five brown spots on costa. Postmedial line brownish spot on inner margin. Terminal area with brown spot on each wing. Postmedial line on hindwings most distinct on anal margin. Forewing margin concave below apex, brown, hindwing margin concave also. Terminal line narrow, brown. Fringes weakly chequered yellow and brown. Discal spots absent. Wings below heavily irrorated with brown, postmedial line wide, forewing apex brown. Discal spots large, elongated, brown. Frons protruding. Frons, collar, thorax and abdomen concolorous with wings. Male antennae markedly bipectinate, female antennae shortly bipectinate. Hindtibia with 2+2 spurs. Tympanal organs medium-sized. Sternites and tergites 3–8 of both sexes undifferentiated.

VARIATION: Very little. Occasionally specimens weakly irrorated with brown, irroration usually more or less absent.

MALE GENITALIA: Generally as in *E. repandaria* (HUFNAGEL), with the following differences: Uncus longer. Base of valva with weak V-shapedsclerotised ridges. Transtilla irregular, without two hollow rings. Aedeagus wider, apex round. Vesica opens at approximately 180 degree angle. Vesica wide at base, with two diverticula and three groups of numerous cornuti. Apex narrow, membranous tube.

FEMALE GENITALIA: Papillae anales elongated, setose, surface striated. Apophyses rather short, straight. Lamella antevaginalis and lamella postvaginalis unrecognisable. Ductus bursae short, narrow, most sclerotised laterally. Corpus bursae long, narrow. Posterior part of corpus bursae sclerotised, striated, weakly dentate. Anterior part membranous. Signum small, margins dentate.

DISTRIBUTION AND ABUNDANCE: (East-)European-Westasiatic. In Europe in south-eastern central Europe and in south-eastern Europe including Ukraine, Crimea and southern European

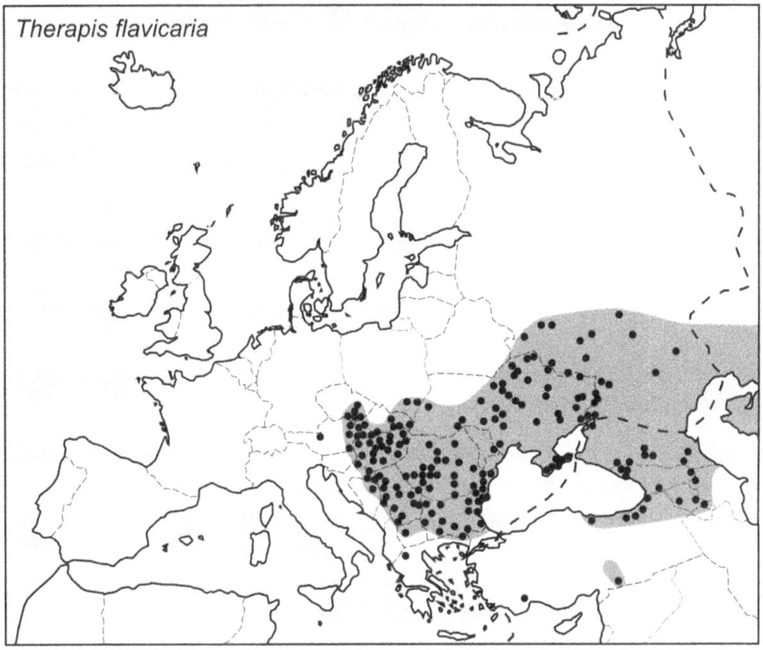

Russia. – Outside Europe in Turkey, the Caucasus region, Iran and central Asia. – In Europe a rather local species that for example in Bulgaria can be rather common, but usually it occurs in limited numbers (S. BESHKOV pers. comm.) and in Hungary where it is widespread and relatively common (G. PETRÁNYI pers. comm.).

PHENOLOGY: Bivoltine. From mid-April to mid-June and from mid-July to mid-September. In Poland from late April to early June and from mid-July to early September, occasionally somewhat earlier or later (Poland MALKIEWICZ 2012, Czech and Slovak Republics Z. LAŠTŮVKA pers. comm., Romania L. SZÉKELY pers. comm.). Three specimens from Iran from late October belong probably to the second generation (coll. SKOU). Caterpillars from June to July and from August to October. Hibernation as a pupa (SZÉKELY 2011). The moths fly at night and are attracted to light.

BIOLOGY: Oligophagous. The caterpillar is recorded feeding on *Lamium album*, *Galeopsis* ((BUSZKO 2000) and *Galium* (SZÉKELY 2011).

HABITAT: Thermophilous scrublands on sunny slopes (MALKIEWICZ 2012), grasslands, steppe with bushes, steppe meadows (SZÉKELY 2011), light lowland forests, forest edges, river bank vegetation, edges of gardens, old quarries and ruderal areas (Z. LAŠTŮVKA pers. comm.). In Europe generally distributed at low altitudes up to 800 m. In Turkey up to 1900 m and in Iran recorded up to 2500 m.

SIMILAR SPECIES: No similar species in Europe. Non-European *Pseudopanthera syriacata* (GUENÉE, 1858) from Middle-East is almost identical in wing pattern and coloration. Antennae of both sexes are ciliate-fasciculate in *P. syriacata* (markedly bipectinate in males, shortly bipectinate in females of *T. flavicaria*). *P. syriacata* is illustrated in LERAUT (2009).

GENETIC DATA: Nearest species: *Pseudopanthera syriacata* (GUENÉE, 1858) from Turkey and the eastern Mediterranean (9.9%).

Pseudopanthera HÜBNER, 1823

Pseudopanthera HÜBNER, [1823] 1816, Verz. bekannter Schmett.: 295. Type species: *Phalaena macularia* LINNAEUS, 1758, Germany. Genus synonyms are after SCOBLE (1999).

Apolema HULST, 1896, Trans. Am. ent. Soc. 23: 324, 336. Type species: *Aspilates carata* HULST, 1887. See FERGUSON (1983) for details.

Venilia DUPONCHEL, 1829, in GODART & DUPONCHEL, Hist. nat. Lépid. Papillons Fr. 7 (2): 110. Type species: *Phalaena macularia* LINNAEUS, 1758. [Junior homonym of *Venilia* RAFINESQUE, 1815 (Crustacea), and junior objective synonym of *Pseudopanthera* HÜBNER, [1823] 1816].

DIVERSITY AND DISTRIBUTION: *Pseudopanthera* contains eight species, which occur in the Palaearctic and the Oriental regions (North India).

EXTERNAL CHARACTERS AND ABDOMEN: Medium-sized, wingspan about 30 mm. Wings yellow, orange, whitish or mixture of these. Markings spot-like or wings with strong, straight medial lines. Following is based on *P. macularia* only: Male antennae shortly ciliate-fasciculate, female antennae filiform. Male hindleg without hair pencil, 3rd sternite without setae. 8th sternite weakly sclerotised, posterior margin with two weak, concavities. 8th tergite weakly sclerotised, elongated medially.

MALE GENITALIA, description based on European species: Uncus long, stout, weakly curved ventrally. Socii minute. Gnathos arms fused ventrally, upturned, strongly dentate. Valva narrow, sclerotised, apex and ventral margin setose. Subcosta strongly sclerotised. Base of valva with distinct group of setae and with membranous sac (coremata). Base of valvae(?) fused dorsally, forming wide, sclerotised, setose, ridge. Base of valva with membranous sac with setae (coremata). Furca long, sclerotised. Transtilla membranous. Saccus concave. Vesica without cornuti.

FEMALE GENITALIA, description based on European species: Papillae anales striated. Lamella antevaginalis sclerotised laterally, weakly dentate. Lamella postvaginalis with horizontal striations. Ductus bursae long, sclerotised. Corpus bursae partly sclerotised, round. Signum small, margins dentate.

BIOLOGY: Larva of *P. macularia* has been recorded feeding for instance on *Lamium album*, *Lamium purpureum*, *Stachys sylvatica*, *Mentha* (MIKKOLA *et al.* 1989) and *Salvia pratensis* (EBERT 2003, SKINNER & WILSON 2009). Habitats include for instance deciduous mixed forests and groves. Adults are partly diurnal.

IMMATURE STAGES: Larva of *P. macularia* is light green with white longitudinal stripes, and a dark green lateral stripe (EBERT 2003). Pupa cremaster with four pairs of curved setae, apical pair biggest (PATOČKA & TURČANI 2005).

REMARKS: Systematic position of genus *Pseudopanthera* has not stabilised yet. It has been placed in the Ennomini (URBAHN 1967, VIVES MORENO 1994), Ourapterygini (VIIDALEPP 1996, KULLBERG *et al.* 2001) and in the Epionini (LERAUT 1997, HAUSMANN, MIRONOV & VIIDALEPP 2011). The coremata originate from base of valva in *P. macularia*, not from vinculum as in *Epione*. Further, furcae originate from medial extension, dorsad of valva, a condition not shared with other assumed relatives. The dorsal fusion of valva(?), which forms wide, sclerotised, setose ridge is also unique in the European fauna. BELJAEV (2008) did not include *Pseudopanthera* in his concept of the Ennomini. In a molecular, phylogenetic analysis of limited European taxa only, ÕUNAP *et al.* (2011) concluded that *Cepphis*, *Selenia*, *Plagodis* and *Pseudopanthera* grouped together but were not related to the Ennomini. Systematic position: Uncertain, potentially related to Epionini.

Aspilates carata HULST, 1887, described from Florida, USA, has proven to be conspecific with *Pseudopanthera himaleyica* (KOLLAR, 1844) or something very close to it, the specimen probably originating from India or Pakistan. *A. carata* is type species of genus *Apolema* HULST, 1896. Therefore *Pseudopanthera* does not occur in the Nearctic region, although mentioned in check lists prior the check list of HODGES *et al.* (1983) and *Apolema* is a junior synonym of *Pseudopanthera* HÜBNER, [1823] 1816 (FERGUSON 1983).

68. *Pseudopanthera macularia* (LINNAEUS, 1758)

macularia macularia LINNAEUS, 1758, Syst. Nat. (Ed. 10) 1: 521, (*Phalaena* (*Geometra*)). Syntypes (Linnean Society of London, London, UK), Germany (examined externally).
maculata FABRICIUS, 1775, Syst. Ent.: 637, (*Phalaena*). [Emendation of *macularia* LINNAEUS.
macularia aureoadflata STAUDER, 1915, Z. wiss. InsektBiol. 11 (1/2): 6, (*Venilia*). Syntypes 6 males, 2 females, [Italy: Calabria]: Monte Martinella, 600–1000 m. Originally as subspecies (STAUDER 1915), valid at subspecies rank according to SCOBLE (1999), synonymy proposed by FLAMIGNI *et al.* (2007).
macularia meridionalis GALVAGNI, 1908, Verh. zool. -bot. Ges. Wien 58 (4/5): (158), (*Venilia*). Syntypes, [Italy-Slovenia border]: Görz, Mt Calvary, Grojnatal; [Italy, S Tirol]: Bozen, Virglwarte. Originally as var., valid at subspecies rank according to SCOBLE (1999), junior synonym according to FLAMIGNI *et al.* (2007). See remarks.
quadrimaculata HATCHETT, [1809, possibly 1812?], Ent. Trans. (old series). 1: 245, [pl. 7], fig. 2, (*Phalaena*). Holotype, [Great Britain].
macularia cryptica BELJAEV, 1997, Far Eastern Entomologist 51: 2, figs 1–3, 6–8, (*Pseudopanthera*). Holotype male (Zoological Institute, Russian Academy of Sciences, St Petersburg), Russia: Irkutskaya oblast, Bunbui. Valid at subspecies rank according to SCOBLE (1999).

UNAVAILABLE NAMES (incorrect subsequent spelling): *aureoadflava* (SCOBLE 1999). – (incorrect subsequent spelling and authorship): *aureoadflava* STAUDINGER (The Global Lepidoptera Names Index (ANONYMOUS 2013); should be *aureoadflata* STAUDER). – Intrasubspecific names, potentially unavailable: *albicans* OBERTHÜR, 1912 (ab.); *argillea* LORITZ, 1947 (ab.); *completa* COCKAYNE, 1950 (ab.); *elegans* COCKAYNE, 1946 (ab.?); *interligata* LEMPKE, 1970 (f.); *fuscaria* STAUDINGER, 1871 (ab.); *krombholzi* STERNECK, 1931 (ab.); *nigrescens* LEMPKE, 1951 (f.); *parvipunctaria* COCKAYNE, 1950 (ab.); *quadrimaculata* HATCHETT, 1807 (ab.); *radiata* LORITZ, 1947 (ab.); *transversaria* KRULIKOWSKY, 1908 (ab.); *viridimaculata* COCKERELL, 1889 (ab.); *xantholeuca* BUBACEK, 1924 (ab.); *zonata* COCKAYNE, 1950 (ab.).

EXTERNAL CHARACTERS AND ABDOMEN: Wingspan 25–28 mm. Wings yellow, markings brown, spot- or panther-like, wing bases irrorated with brown. Medial lines and subterminal line spot-like, incomplete, present as spots on costa. Forewing margin weakly concave below apex, hindwing margin weakly concave also. Terminal line concolorous with wings. Fringes chequered yellow and brown. Discal spots brown, elongated. Wings below as above. Frons, collar, thorax and abdomen concolorous with wings. Male antennae shortly fasciculate, female antennae filiform. Hindtibia with 2+2 spurs. Tympanal organs large. Male 8th sternite weakly sclerotised, posterior margin with two weak concavities. 8th tergite weakly sclerotised, elongated medially. Other sternites and tergites of both sexes undifferentiated.

VARIATION: Rather variable. The amount and extent of brown markings varies a lot, they can be fused together or distinctly separated, and all stages in between. The spots can occasionally be olive-green and ground colour can be whitish. Several forms have been described, which are summarised in LERAUT (2009). EBERT (2003) illustrates uniform, brownish black specimen. The non-European subspecies *cryptica* BELJAEV, 1997 has reddish yellow wings (BELJAEV 1997).

MALE GENITALIA: Uncus long, stout, weakly curved ventrally. Socii minute, setose. Gnathos arms fused ventrally, upturned, strongly dentate. Valva narrow, sclerotised, apex blunt. Valva setose, particularly near apex, medial area bare. Subcosta with strong, weakly curved, sclerotised ridge, apex weakly extending beyond apex. Base of valva with distinct group of setae and with membranous sac (coremata). Base of valvae(?) fused dorsally, forming wide, sclerotised, setose, ridge. Furca long, sclerotised. Transtilla membranous. Saccus concave. Aedeagus narrow, recurved. Vesica opens at approximately 135 degree angle. Vesica wider at base, without cornuti.

FEMALE GENITALIA: Papillae anales elongated, setose, surface striated. 9th segment elongated. Apophyses long, straight. Lamella antevaginalis sclerotised laterally, weakly dentate. Lamella postvaginalis with horizontal striations. Ductus bursae long, sclerotised, slightly curved, partially slightly striated and weakly dentate. Corpus bursae round, posterior part irregularly sclerotised. Signum small, asymmetrical, margins dentate.

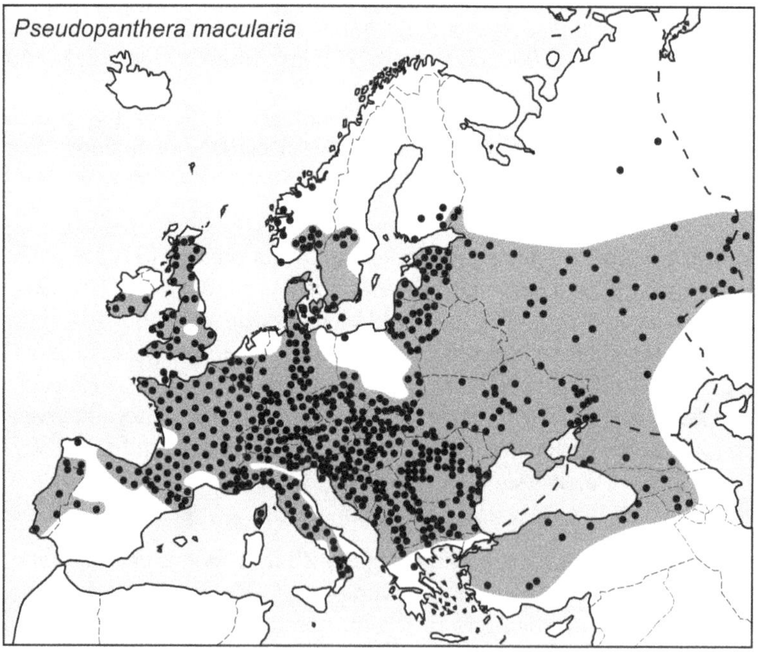

DISTRIBUTION AND ABUNDANCE: Eurasiatic. In Europe from southern Scandinavia to the Iberian Peninsula, southern France including Corsica, southern Italy and most of the Balkans and from the British Isles to the Ural mountains. – Outside Europe in Turkey, the Caucasus region and south-western Siberia. – In Europe the species is in general rather common, in northern and central Europe somewhat local and it can now and then occur in numbers. In Denmark the species has declined much in recent decades.

PHENOLOGY: Univoltine. In northern Europe from mid-May to late June, in south-western Germany from early May to early July (EBERT 2003), and in southern Europe depending on altitude from April to early August (FLAMIGNI *et al.* 2007). Caterpillars from June to August or September. The hibernation takes place as pupa. The adults fly actively and visit flowers in sunshine, but they are not often seen at light.

BIOLOGY: Polyphagous. The larva is found on *Salvia pratensis*, *Lamium album*, *Galium boreale* (EBERT 2003) and on *Teucrium scorodonia*, *Stachys* and *Lamium galeobdolon* (SKINNER 2009). Pupation between mosses on the ground.

HABITAT: In clearings in forests, between bushes, on meadows and in moors, always on moist ground. In northern Europe from sea level, in southern Europe from 500 up to 1700 m altitude (REDONDO *et al.* 2009).

SIMILAR SPECIES: No similar species in Europe.

GENETIC DATA: Genetically homogeneous in Europe (n=20 from seven countries), but in the United Kingdom with a second haplotype at a distance of 3.3%. Nearest species: *P. syriacata* (GUENÉE, 1858) from Turkey and the Eastern Mediterranean (5.5%).

REMARKS: Three subspecies have been regarded valid in the recent literature (LERAUT 1997, SCOBLE 1999, HAUSMANN, MIRONOV & VIIDALEPP 2011). We agree with FLAMIGNI *et al.* (2007) and LERAUT (2009) and do not recognise subspecies in Europe, quoting the absence of disjunct, diagnostic features. LERAUT (2009, p. 91–92) writes: 'Several subspecies have been described, but the high degree of individual variation observed means that none can be accepted.' FLAMIGNI *et al.* (2007) consider taxon *meridionalis* (GALVAGNI) a junior synonym of *macularia* (LINNAEUS) on the grounds that nominotypical, transitional and *meridionalis* specimens occur sympatrically for instance in Trentino-Alto Adige, northern Italy. The external appearance of some populations correlates with geographical locality in a potentially discontinuous manner.

The non-European subspecies *cryptica* BELJAEV from Russia, Irkutskaya oblast has reddish yellow wings. The female genitalia has, according to BELJAEV (1997), tiny quantitative differences in the corpus bursae (without membranous sacs, signum with pair of lateral spines) when compared to European specimens.

Eilicrinia HÜBNER, 1823

Eilicrinia HÜBNER, [1823] 1816, Verz. bekannter Schmett.: 287. Type species: *Phalaena cordiaria* HÜBNER, 1790, [Austria]: Vienna. Genus synonyms are after SCOBLE (1999).

Ilicrinia AGASSIZ, 1847, Nomencl. zool. (Index univl.): 135, 194. [Emendation of *Eilicrinia* HÜBNER].

Pareilicrinia WARREN, 1894, Novit. zool. 1: 462. Type species: *Noreia flava* MOORE, 1888.

UNAVAILABLE NAMES (incorrect subsequent spellings): *Elicrina* (HERRICH-SCHÄFFER, 1847 and STAUDINGER, 1859); *Ilicrina* (HERRICH-SCHÄFFER, 1855); *Elicrinia* (BREMER, 1864) (see FLETCHER 1979).

DIVERSITY AND DISTRIBUTION: *Eilicrinia* contains 11 species, which occur mainly in the Palaearctic region, also in the Oriental region, one species is known from the Neotropics.

EXTERNAL CHARACTERS AND ABDOMEN: Medium-sized, wingspan about 25–30 mm. Wings yellow, yellowish, grey, brown or mixture of these. Forewing discal spot normally large, postmedial line most distinct, yet often suffuse. Forewing margin concave below apex, often dark. Male and female antennae filiform. Male 8th sternite weakly sclerotised, posterior margin narrower, thus sclerite slightly triangular.

MALE GENITALIA, description is based on European species: Uncus long, narrow, weakly curved ventrally. Socii minute. Gnathos arms narrow, weakly sclerotised, fused ventrally, without medial elements. Valva widest subapically, weakly sclerotised, mostly setose, apex round. Valva with medial sclerotisation. Base of valva with group of setae and with reduced membranous sac. Pedunculi narrow. Saccus margin round or weakly concave. Vesica with small sclerotisations (no sharp cornuti) or sclerotisations absent.

FEMALE GENITALIA, description is based on European species: Papillae anales setose, surface weakly granulate or with very strong, curved setae. Base of ductus seminalis elongated, sclerotised, followed by expanded, membranous sac. Anterior part of corpus bursae enlarged, membranous. Signum large plate, margins serrate.

BIOLOGY: Larva of *E. cordiaria* (HÜBNER) has been recorded feeding on *Salix*, that of *E. trinotata* (METZNER) on *Silene* (SZÉKELY 2011). These two species have two generations per year, egg overwinters.

IMMATURE STAGES: No data available, apart from superficial remarks in PROUT (1912–1916), where he does not specify to which species the reference is made to: 'Caterpillar smooth, thin, flat head.'

REMARKS: *Eilicrinia* has recently been classified in the Ennomini (VIIDALEPP 1996) and the Epionini (HAUSMANN, MIRONOV & VIIDALEPP 2011). The Ennomini connection, as defined in BELJAEV (2008a), is not supported by morphology. The Epionini connection also needs verification. In Epionini (*Epione*) coremata originate from vinculum and the apex is free, whereas *Eilicrinia* has small, lateral coremata, but these originate from valva and the apex is not free. Systematic position: Uncertain, potentially related to the Epionini.

69. *Eilicrinia cordiaria* (HÜBNER, 1790)

cordiaria cordiaria HÜBNER, 1790, Beitr. Gesch. Schmett. 2 (1): 31, pl. (1) 4, fig. Z, (*Ph*[*alaena*] *Geom*[*etra*]). Syntype(s) female(s), lost, [Austria]: Vienna.

animaria HERRICH-SCHÄFFER, 1847, Syst. Bearbeitung Schmett. Eur. 3 (27): 53; ibidem (1848), pl. 59, fig. 364 (non binominal), (*Eilicrinia*). Syntype(s), [Serbia/Croatia].

animata FISCHER VON RÖSLERSTAMM, 1836, Abbildungen zur Berichtigung und Ergänzung der Schmetterlingskunde (5): 45, (*Ennom*[*os*]). Syntype(s), Syrnien; Weidenstämmen.

animata roeslerstammaria STAUDINGER, 1871, in STAUDINGER & WOCKE, Cat. Lepid. eur. Faunengeb. 1: 160, (*Eilicrinia*). Syntype(s), [Originally as var].

cordiata HERRICH-SCHÄFFER, 1861, Syst. Verz. Eur. Schmett. (Edn 2): 15, (*Ilicrinia*): [Emendation of *cordiaria* HÜBNER].

cordiaria signigera BUTLER, 1889, Illust. typical Specimens Lepid. Heterocera Colln Br. Mus. 7: 112, pl. 137, fig. 3, (*Eilicrinia*). Syntype(s) (The Natural History Museum, London, UK), [India]: Kúlú. [Originally as sp]. (examined externally). Valid at subspecies rank according to SCOBLE (1999).

Unavailable names (incorrect subsequent spelling): *cordiarii* Hübner, 1823 (see Fletcher 1979). – Intrasubspecific names, potentially unavailable: *astigmaria* Bytinski-Salz & Brandt, 1937 (f.).

External characters and abdomen: Wingspan 25–30 mm. Summer generation: Wings pale yellow, weakly irrorated with brown. Markings weak, dominated by large, brown forewing discal spot and brown, lunular area below apex. Postmedial line weak, suffuse. Hindwings with weak, suffuse discal spot and postmedial line. Terminal line and fringes concolorous with wings, except subapical area brown. Forewing margin concave below apex. Wings below more irrorated with brown, forewing discal smaller, about same size as hindwing spot. Postmedial line narrow, dentate. Frons, collar, thorax and abdomen concolorous with wings. Spring generation: Wings whitish grey, irrorated with grey. Male and female antennae filiform. Hindtibia with 2+2 spurs. Tympanal organs medium-sized. Male 8th sternite weakly sclerotised, anterior margin concave, posterior margin narrower, thus sclerite triangular. Anterior margin of male 8th tergite with two lobes, concave medially. Other sternites and tergites of both sexes undifferentiated.

Variation: Very little. Density of irroration varies little. Postmedial line occasionally grey.

Male genitalia: Uncus long, narrow, weakly curved ventrally, setose. Socii minute, setose. Gnathos arms narrow, weakly sclerotised, fused ventrally, without medial elements. Valva widest subapically, weakly sclerotised, mostly setose, apex round. Valva with medial sclerotised ridge. Costa weakly granulate. Base of valva with group of setae and with small membranous sac, its apex fused to valva. Pedunculi narrow. Saccus margin round. Aedeagus long, straight, striated, caecum distinct, apex round. Area below vesica with minute teeth. Vesica opens at approximately 90 degree angle. Vesica membranous, constricted after wider base. Medial area with small diverticulum and sclerotised plate.

Female genitalia: Papillae anales round, setose, surface weakly granulate. Apophyses long, straight. Lamella antevaginalis/posterior part of ductus bursae sclerotised, weakly serrate all over, with lateral wings, concave medially. Lamella postvaginalis with horizontal striations. Posterior part of ductus bursae narrow, sclerotised, serrate. Anterior part of ductus bursae sclerotised, smooth. Base of ductus seminalis elongated, sclerotised, followed by expanded, membranous sac. Posterior part of corpus bursae narrow, sclerotised, with few teeth. Anterior part of corpus bursae enlarged, membranous, dorsal side partly dentate. Signum large plate, margins serrate.

Distribution and abundance: West-Palaearctic. In Europe from easternmost Austria, southern Slovakia and the Balkans through Ukraine including Crimea, southern Russia and Crimea to southern Urals. – Outside Europe the species occurs in Turkey, Caucasus region, northern Iran and central Asia. – In Europe a somewhat local species that in general occurs in limited numbers.

Phenology: Bivoltine. In southern Slovakia from mid-May to mid-June and from mid-July to early August. In Romania from mid-April to late May and from early July to late August (L. Székely pers. comm.). The earliest records are from Greece. Overwintering takes place as pupa. The adult moths fly at night and are attracted to light.

Biology: Monophagous. The caterpillar is recorded feeding on *Salix* (Fajcik 2003, Székely 2011). Has been reared on *Salix matsudana* (B. Müller pers. comm.)

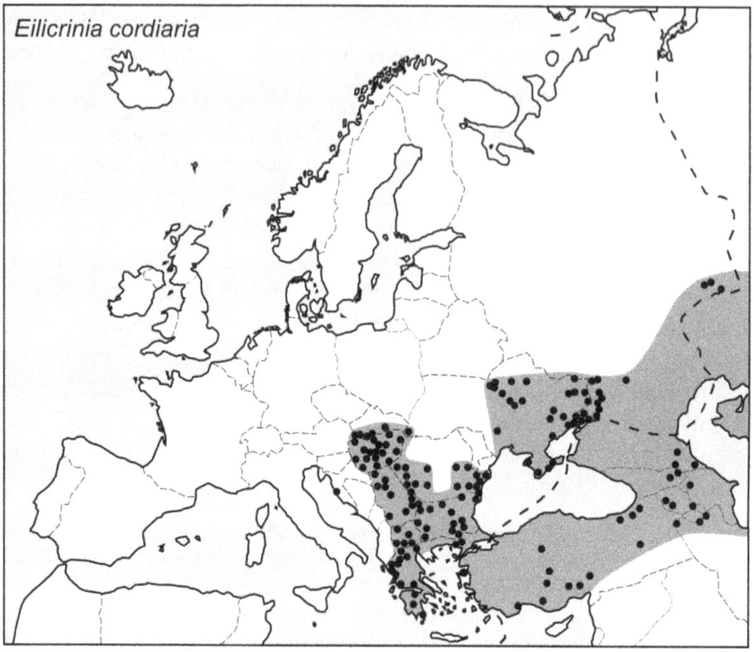

HABITAT: In Europe in meadows along rivers and forest fringes at low altitudes. In Greece the species is found up to 800 m altitude, in north-western Iran it has been found up to 2000 m, in Turkey up to 2200 m.

SIMILAR SPECIES: *Eilicrinia subcordaria* HERRICH-SCHÄFFER has ochre-yellow or greyish brown wings (wings pale yellow or whitish grey in *E. cordiaria*). Forewings with pale, suffuse forewing discal spot and lunular, subapical mark (forewing discal spot and lunular, subapical mark are brown, distinct in *E. cordiaria*). Subcosta of valva with distinct, elongated sclerotisation (valva without sclerotisation in subcosta of valva in *E. cordiaria*). Lamella antevaginalis/posterior part of ductus bursae wide, without concavity medially (lamella antevaginalis/posterior part of ductus bursae narrow, with distinct concavity medially in *E. cordiaria*).

GENETIC DATA: Genetically heterogeneous with three BINs in Turkey, Iran and central Asia (n=17). Nearest species: *E. subcordaria* (2.5%).

70. *Eilicrinia subcordaria* HERRICH-SCHÄFFER, 1852

subcordaria subcordaria HERRICH-SCHÄFFER, 1852, Syst. Bearbeitung Schmett. Eur. 6 (55): 71; ibidem (1850), 3: pl. 73, fig. 446 (non binominal), (*Eilicrinia*). Syntype(s), Turkey.

anicularia EVERSMANN, 1852, Bull. Soc. Nat. Moscou 25 (1): 159, (*Ennomos*). Syntypes, Russia: lower Volga; Ural River.

freitagaria MÖSCHLER, 1854, Stettin. ent. Ztg 15: 144, (*Pericallia*). Syntypes male, female, Russia.

subcordata HERRICH-SCHÄFFER, 1861, Syst. Verz. Eur. Schmett. (Edn 2): 15, (*Ilicrina*). [Emendation of *subcordaria* HERRICH-SCHÄFFER].

subcordaria acardia STICHEL, 1911, Z. wiss. InsektBiol. 7 (5/6): 162, fig. 12, (*Eilicrinia*). Holotype male, [Iran]: [Khuzestan], Sultanabad. Valid at subspecies rank according to SCOBLE (1999) [originally as f. of *subcordaria* HERRICH-SCHÄFFER].

UNAVAILABLE NAMES (incorrect subsequent spelling): *subcordiaria* (LERAUT 2009, HAUSMANN, MIRONOV & VIIDALEPP 2011).

EXTERNAL CHARACTERS AND ABDOMEN: Wingspan 25–31 mm. Generally as *E. cordiaria* (HÜBNER), but with the following differences. Summer generation: Wings ochre-yellow. Forewing discal spot faint, light brown. Lunular area below apex faint, light brown. Hindwing discal spots almost reduced. Spring generation: Wings light grey-brown, weakly irrorated with brown. Forewing discal spot and lunular area below apex brown, standing out more clearly than in summer generation. Antemedial and postmedial lines distinct, brown. Hindwing discal spot and postmedial line normally visible, grey-brown.

VARIATION: Very little. Medial area occasionally darker than rest of wing.

MALE GENITALIA: Generally as *E. cordiaria* (HÜBNER), but with the following differences. Subcosta of valva with distinct, elongated sclerotisation. Aedeagus with row of minute teeth on opposite side of vesica. Base of vesica with irregular, sclerotised plate. Vesica not constricted after base, rather straight.

FEMALE GENITALIA: Generally as *E. cordiaria* (HÜBNER), but with the following differences. Lamella antevaginalis/posterior part of ductus bursae wide, without concavity medially (margin irregular). Ductus bursae shorter, membranous. Posterior part of corpus bursae longer, dorsal side covered entirely with small teeth. Signum more triangular.

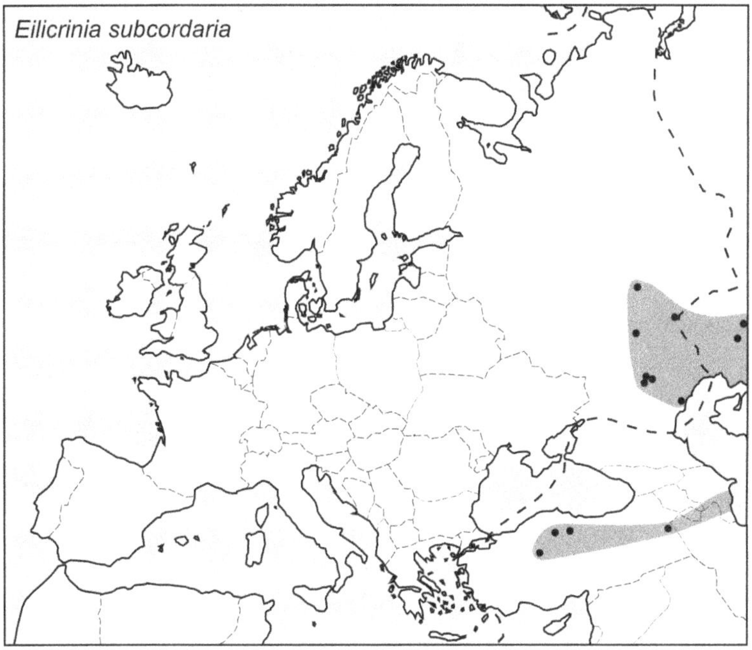

DISTRIBUTION AND ABUNDANCE: Turanian. In Europe only known from south-eastern European Russia. – Outside Europe the species is known from Turkey, the Caucasus region and central Asia. – In Europe apparently a very rare and little known species. Most, perhaps all, records are very old.

PHENOLOGY: Univoltine. From mid-May to late July. The moths fly at night and come to light.

BIOLOGY: Unknown.

HABITAT: Recorded from various habitat types, such as rocky steppe slopes, riversides and moist patches in sandy deserts (K. NUPPONEN pers. comm.). In Kazakhstan from sea level up to 1200 m (K. NUPPONEN pers. comm.). Up to 2000 m in southern central Asian mountains.

SIMILAR SPECIES: *E. cordiaria* (HÜBNER), see entry above.

GENETIC DATA: Nearest species: *E. cordiaria* (2.5%).

REMARKS: *Larentia tritomata* FISCHER VON WALDHEIM, 1840 from Russia: Lower Volga, which is an older name, is potentially conspecific with *E. subcordaria* HERRICH-SCHÄFFER, 1852. STAUDINGER & REBEL (1901) considered *tritomata* a synonym of *Ennomos anicularia* EVERSMANN, 1852, which is considered a junior synonym of *E. subcordaria* HERRICH-SCHÄFFER, 1852. PROUT (1912–1916) considered it a potential synonym of *E. subcordaria* (with a question mark), which was removed by SCOBLE (1999). We have failed to find reference in the literature where the synonymy is established. The type specimen(s) of *tritomata* is/are potentially deposited in the Zoological Museum of the Moscow University, where those of FISCHER VON WALDHEIM's specimens are kept that were collected after the 1812 Moscow fire. Despite efforts, the type specimen(s) have not been located (F. V. SVIRIDOV pers. comm.). The original, short description in latin (FISCHER VON WALDHEIM 1840, p. 88) describes *tritomata* as ‚forewings brownish grey, white-marbled, with three black spots on the anterior margin‘ and ‚hindwings irrorated with brownish at base‘. The written description fits better that of *E. trinotata* (METZNER, 1845). Because the identity of this senior name remains unconfirmed, we have not followed the synonymy of PROUT (1912–1916) and SCOBLE (1999).

71. Eilicrinia trinotata (METZNER, 1845)

Eilicrinia trinotata METZNER, 1845, Stettin. ent. Ztg 6 (6): 184, (*Geom[etra]* (*Ennomos*)). Syntypes 2, including at least 1 male, [European part(?) of] Turkey.

UNAVAILABLE NAMES: *aestiva* REBEL, 1903 (summer generation).

EXTERNAL CHARACTERS AND ABDOMEN: Wingspan 21–26 mm. Summer generation: Wings sand-yellow, with tint of olive. Forewing medial and postmedial lines distinct, brown, enlarged to spots on costa. Terminal area darker, particularly towards tornus. Forewing discal spot distinct, brown, centre sand-yellow. Terminal line undulating, sharp, brown. Fringes chequered sand-yellow and brown. Hindwings with postmedial line present only, distinct, brown. Discal spot small or absent. Forewing margin concave below apex, hindwing margin undulating. Wings below as above, markings more suffuse. Forewing discal spot uniform brown. Frons, collar, thorax and abdomen concolorous with wings. Spring generation: Wings ash-grey, heavily irrorated with brown. Male and female antennae filiform. Hindtibia with 2+2 spurs. Tympanal organs medium-sized. Male 8th sternite and tergite weakly sclerotised, posterior margins narrower, thus sclerites triangular. Other sternites and tergites of both sexes undifferentiated.

VARIATION: Very little. Density of irroration varies little. Summer generation specimens may occasionally be light yellow and greyish in colour.

MALE GENITALIA: Uncus long, narrow, weakly curved ventrally, apex with spine. Socii minute, setose. Gnathos arms narrow, weakly sclerotised, fused ventrally, without medial elements. Valva widest subapically, weakly sclerotised, mostly setose (several of the setae with expanded apex), apex round. Valva with medial sclerotised ridge, basal parts fused medially, medial area with distinct concavity, followed by several (about three) sclerotised spines. Costa weakly granulate. Base of valva with small group of setae and reduced membranous sac. Dorsal margin of valva with two dentate, artichoke-shaped, sclerotised extentions medially. Pedunculi narrow. Saccus margin round. Aedeagus almost straight, caecum minute, apex sharp. Vesica opens at approximately 90 degree angle. Vesica membranous.

FEMALE GENITALIA: Papillae anales unusual, large, almost separated medially, inner margin with row of thick, curved spines, otherwise setose (Text-fig. 83). Anterior margin sclerotised. Apophyses short, thick. Lamella antevaginalis and lamella postvaginalis fused, plate-like sclerotisation widest laterally. Ductus bursae short, narrow, most sclerotised laterally. Base of ductus seminalis weakly elongated, sclerotised, followed by slightly expanded, membranous sac. Posterior part of corpus bursae long, narrow, membranous. Anterior part of corpus bursae slightly wider, membranous. Signum large, slightly lunular, outer margin serrate.

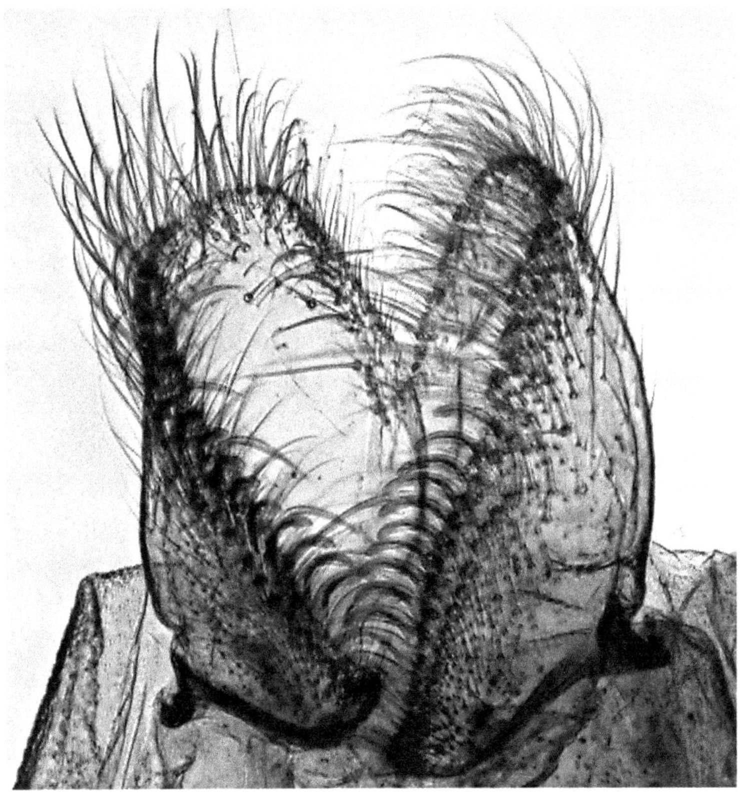

83

Text-fig. 83. Papillae anales of *Eilicrinia trinotata* (METZNER, 1845). The papillae anales are unusual, being almost separated medially, and the inner margin is with a row of thick, curved spines. Hungary: Fót Somlyóhegy, 22.v.1958, slide PS1412 (coll. ZMH).

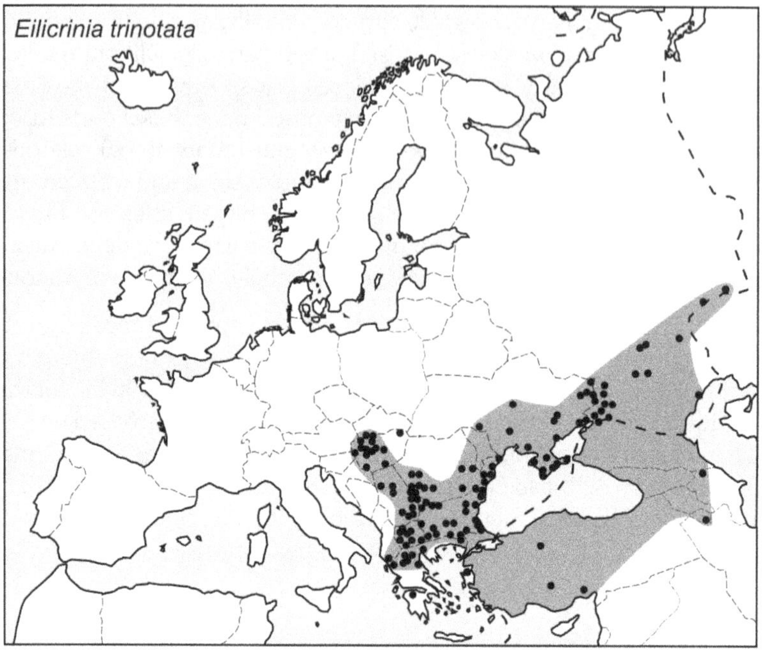

DISTRIBUTION AND ABUNDANCE: (East-)European-Westasiatic. In Europe only in the south-east, in Hungary, in southern Slovakia and further to the south-east through most of the Balkans and including southern Ukraine, south-eastern European Russia and the European part of Kazakhstan. – Outside Europe in Turkey and Caucasus. – In Europe a rather local species. However, in Bulgaria it is a common species that usually occurs in limited numbers (S. BESHKOV pers. comm.).

PHENOLOGY: Bivoltine. From mid-April to early June and again from late June to late August. In southern Urals and the lower Volga region also in early September (K. NUPPONEN pers. comm.). Larvae from May to October. The hibernation takes place as pupa. The moths fly at night and come to light.

BIOLOGY: Monophagous. The caterpillar is recorded on *Silene* (FAJCIK 2003, SZÉKELY 2011).

HABITAT: In steppe and forest steppe localities. Vertically the species is distributed from sea level up to 800 m altitude. In central Turkey up to 1500 m (K. NUPPONEN pers. comm.).

SIMILAR SPECIES: No similar species in Europe.

GENETIC DATA: Barcoded from Crimea. Nearest species: *E. subcordaria* (7.1%).

REMARKS: See *E. subcordaria* HERRICH-SCHÄFFER above.

Toulgoetia HERBULOT, 1946

Toulgoetia HERBULOT, 1946, Bull. Soc. ent. Fr. 51: 79. Type species: *Elicrina cauteriata* STAUDINGER, 1859, [Spain]: Andalusia. There are no genus synonyms.

DIVERSITY AND DISTRIBUTION: *Toulgoetia* is a monotypic genus, the only species occurs in the Iberian Peninsula and in North Africa.

EXTERNAL CHARACTERS AND ABDOMEN: Medium-sized, wingspan 22–29 mm. Wings red-brown, grey-brown or light yellow-brown, irrorated with brown. Medial lines suffuse, curved inwards. Forewing distinctly concave below apex. Hindwings greyish, medial line suffuse. Male and female antennae filiform. Male hindtibia with hair pencil. Male 3rd sternite with setae. Male 8th sternite weakly sclerotised, posterior margin concave.

MALE GENITALIA: Uncus short, stout, weakly setose. Socii minute, setose. Gnathos arms fused ventrally, upturned, weakly spinose. Valva narrow, costa extending below apex, medial area with triangular lobe. Base of valva with large group of setae. Transtilla membranous, not fused(?) dorsally. Furca long, narrow, arising from vinculum/saccus. Saccus weakly concave. Vesica with two, long cornuti.

FEMALE GENITALIA: Papillae anales round, setose. Sinus vaginalis and posterior part of ductus bursae densely setose. Corpus bursae elongated, its posterior part sclerotised, striated, densely dentate. Signum large plate, margins more sclerotised.

BIOLOGY: Larva of *T. cauteriata* has been recorded to feed on *Quercus coccifera* (STAUDINGER 1859, RIBBE 1912). Moths inhabit dry woodlands, up to 1650 metres from February to late May, at the higher altitudes until early June (REDONDO *et al.* 2009).

IMMATURE STAGES: No data available.

REMARKS: *Toulgoetia* has recently been classified in the Ennomini (VIVES MORENO 1994) and in the Epionini (HAUSMANN, MIRONOV & VIIDALEPP 2011). *Toulgoetia* does not have morphological features of the Ennomini as defined in BELJAEV (2008a) or the Epionini as defined in HOLLOWAY (1994). To make the case more concrete, the Epionini is defined by consistent absence of setal comb on male 3rd sternite (present in *Toulgoetia*), loss of tibial hair pencil (present in *Toulgoetia*) and the vinculum supports lateral coremata (coremata absent in *Toulgoetia*).

T. cauteriata was classified in genus *Eilicrinia* HÜBNER until HERBULOT (1946) proposed a monotypic genus for it. We agree, *T. cauteriata* is morphologically dissimilar to *Eilicrinia*. As is explained in more detail in *Ennominae systematics* chapter, the European Ennominae fauna as a whole has not previously been classified into tribes. Systematic review of tribal classifications and their constituent genera is beyond scope of this work, mainly because it would require extensive studies on numerous non-European taxa. Therefore in several cases, including this one, we have been forced to arrange higher taxa according to the conventional and practical aspects. Systematic position: Systematic position: Uncertain, further research is needed.

72. Toulgoetia cauteriata (STAUDINGER, 1859)

cauteriata STAUDINGER, 1859, Stettin. ent. Ztg 20 (7–9): 217, (*Elicrina* HERRICH-SCHÄFFER, 1847). Syntypes male, female, [Spain]: Granada, near Chiclana.

obtusaria J. J. WALKER, 1890, Trans. Ent. Soc. L. 1890: 387, (*Pseudosestra*?). Holotype [sex not given], [Spain]: San Roque. Junior synonym according to The Global Lepidoptera Names Index (ANONYMOUS 2013). According to the original description (WALKER 1890), the species was assumed to be conspecific with *Lozogramma obtusaria* WALKER (1861) from New Zealand. WALKER (1890) does not illustrate the Spanish specimen.

EXTERNAL CHARACTERS AND ABDOMEN: Wingspan 22–29 mm. Wings red-brown, grey-brown or light yellow-brown, irrorated with brown. Medial lines suffuse, strongest on costa, curved inwards. Medial line often widest. Antemedial and postmedial lines weaker, often broken into dots, particularly postmedial line. Forewing distinctly concave below apex. Hindwings distinctly lighter, greyish, medial line suffuse. Terminal line and fringes concolorous with wings. Discal spots small. Wings below lighter, particularly forewings. Male and female antennae filiform, male antennae wider. Hindtibia with 2+2 spurs. Male hindtibia with hair pencil. Tympanal organs large, cavi tympani touching each other medially. Male 3rd sternite with setae. Male 8th sternite weakly sclerotised, posterior margin concave. Other sternites and tergites of both sexes undifferentiated.

VARIATION: Ground colour varies from red-brown to grey-brown to light yellow-brown. Degree of brown irroration variable. Medial lines normally visible, suffused, but they may be almost absent.

MALE GENITALIA: Valvae narrow and short, fused ventrally, thus they can not be spread out laterally and genitalia capsule has compact appearance. Uncus short, stout, weakly setose. Socii minute, setose. Gnathos arms fused ventrally, upturned, weakly spinose. Valva narrow, short, weakly setose, ventral margin elongated and valvae fused ventrally. Costa weakly sclerotised, extending below apex. Medial area with triangular lobe, apex sclerotised. Base of valva with large group of setae. Transtilla membranous, not fused(?) dorsally. Furca very long, narrow, arising from vinculum/saccus. Saccus weakly concave. Aedeagus narrow, curved, caecum minute, apex sharp. Aedeagus weakly granulate below apex. Vesica opens at approximately 135 degree angle. Vesica narrow, membranous tube, basal part wider, striated, with two long cornuti.

FEMALE GENITALIA: Papillae anales round, setose. Apophyses long, narrow. Lamella antevaginalis and lamella postvaginalis fused, sclerotised, dorsal area elongated medially. Sinus vaginalis and posterior part of ductus bursae sclerotised, densely setose. Ductus bursae long, narrow, sclerotised, anterior end constricted. Corpus bursae elongated, its posterior part sclerotised, striated, densely dentate. Anterior part of corpus bursae membranous, with numerous tiny spines. Signum large plate, margins more sclerotised.

DISTRIBUTION AND ABUNDANCE: West Mediterranean. In Europe only in the southern half of the Iberian Peninsula. – Outside Europe only known from Morocco and Algeria. – In Europe not rare within the distribution area, but the species is usually only found in limited numbers.

PHENOLOGY: Univoltine. From early March to late May. In southernmost Spain also recorded in February (P. HALE pers. comm.). The moths fly at night and are attracted to light.

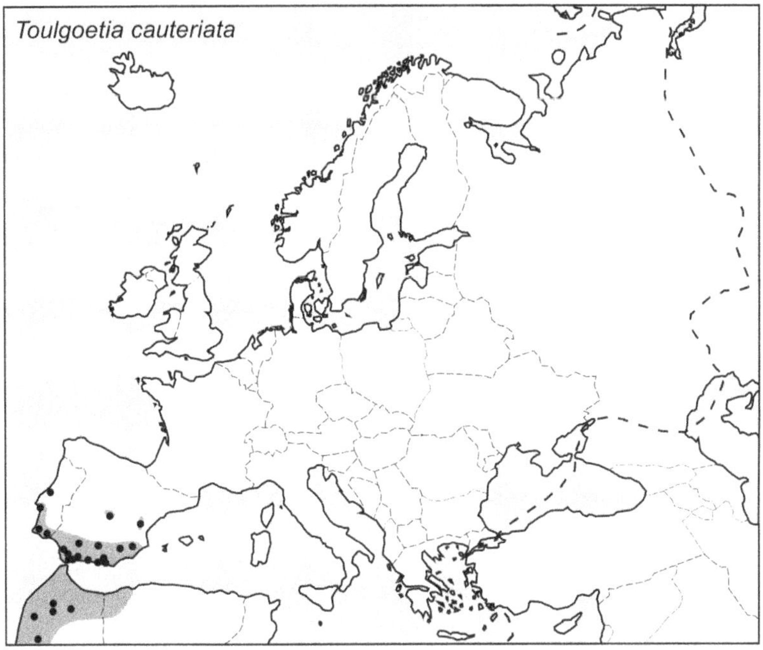

BIOLOGY: Apparently monophagous. Originally described on basis of a caterpillar found on *Quercus coccifera* (STAUDINGER 1859) which appears to be the only biological data known.

HABITAT: In warm and dry, scattered oak forests from sea level up to 1500 m, in Morocco up to 1650 m.

SIMILAR SPECIES: No similar species in Europe. The strong contrast between red-brown forewings and light greyish hindwings is characteristic.

GENETIC DATA: Nearest species: *Pachycnemia tibiaria* (7.7%).

Hypoxystis PROUT, 1915

Hypoxystis PROUT, 1915, Gross-Schmett. Erde 4: 342. Type species: *Phalaena pluviaria* FABRICIUS, 1787, [Germany]: Saxony, Halle. There are no genus synonyms.

DIVERSITY AND DISTRIBUTION: *Hypoxystis* contains three species in the Palaearctic region.

EXTERNAL CHARACTERS AND ABDOMEN: Medium-sized, wingspan about 30–35 mm. Wings pale yellow, irrorated with brown. Postmedial line strongest, often broken into brown spots. Forewing margin concave below apex. Male antennae bipectinate, female filiform. Male 3rd sternite without setae, hindleg with hair pencil. Male 8th sternite weakly sclerotised, posterior margin concave, 8th tergite elongated.

MALE GENITALIA, description is based on the European species: Uncus short, stout, weakly setose. Socii apparently absent. Gnathos arms narrow, fused ventrally, forming large granular plate. Valva complex, fused ventrally. Costa with naked, sclerotised extension (ampulla), apex setose, round. Ventral margin strongly sclerotised, with round, setose lobe. Base of valva with few stout setae. Juxta with dorsolateral extensions. Saccus margin round. Vesica with cornutus.

FEMALE GENITALIA, description is based on the European species: Papillae anales setose, surface weakly striated. Lamella antevaginalis sclerotised, lunular. Lamella postvaginalis irregular, weakly sclerotised plate. Corpus bursae large, membranous. Signum quadrangular, margins mostly sclerotised.

BIOLOGY: Larva of *H. pluviaria* has been recorded feeding on *Galium palustre*, *Vaccinium uliginosum*, *Betula nana* (MIKKOLA *et al.* 1989), *Cytisus scoparius* and *Genista* (PATOČKA & TURČANI 2005). Adults are day-active, in northern Europe also during night time. Caterpillar overwinters.

IMMATURE STAGES: Larva of *H. pluviaria* is long and narrow, distinctly bicolorous, brownish dorsally and whitish ventrally (EBERT 2003). The pupa cremaster has four pairs of curved setae, two of which are apical, the other two are more lateral (PATOČKA & TURČANI 2005).

REMARKS: *Hypoxystis* has recently been classified in the Gnophini (VIIDALEPP 1996, KULLBERG *et al.* 2001) and in the Epionini (HAUSMANN, MIRONOV & VIIDALEPP 2011). Male and female genitalia support the Gnophini association as defined in VIIDALEPP *et al.* (2007). The potential Gnophini association of *Hypoxystis* was noted late in the production process of this book, so for practical reasons it is dealt with here, although we note that its current position in the sequence of species does not reflect its phylogenetic position. Structurally *H. pluviaria* is similar to European *Aspitates* species. The association is particularly supported by the male genitalia, but also by the female genitalia. See Gnophini under *Classification of European Ennominae genera into tribes* for further details. Systematic position: Gnophini (but systematic position and extent of the Gnophini needs further research).

73. *Hypoxystis pluviaria* (FABRICIUS, 1787)

pluviaria FABRICIUS, 1787, Mantissa Insect. 2: 192, (*Phalaena*). Syntype(s), [Germany]: Saxony, Halle.

- *adspersaria* HÜBNER, 1790, Beitr. Gesch. Schmett. 2 (3): 71, pl. (3)3, fig. R, (*Ph[alaena] Geom[etra]*). Syntype(s) male(s), lost. [Junior homonym of *Phalaena adspersaria* FABRICIUS, 1787].
- *inspersaria* HÜBNER, [1823], Verz. bekannter Schmett.: 294, (*Anagoga*). [Replacement name for *adspersaria* HÜBNER].
- *irroraria* PROUT, 1915, in SEITZ, The Macrolepidoptera of the World 4: 342, (*Hypoxystis*). [Replacement name for *inspersaria* HÜBNER].
- *jacobaearia* BORKHAUSEN, 1794, Natur. eur. Schmett. 5: 253, (*Phalaena Geometra*). Syntype(s), Europe.
- *pluviaria sylvanaria* HERRICH-SCHÄFFER, 1847, Syst. Bearbeitung Schmett. Eur. 3 (27): 57; ibidem (1848), pl. 70, fig. 431 (non binominal), (*Hypoplectis*). Syntype(s), France, Germany, Russia. See remarks.

INTRASUBSPECIFIC NAMES, POTENTIALLY UNAVAILABLE: *extinctaria* KRULIKOWSKY, 1896 (ab.), *obscurata* LAMBILLION, 1905 (ab.); *pallidata* LAMBILLION, 1905 (ab.); *punctularia* LAMBILLION, 1905 (ab.).

EXTERNAL CHARACTERS AND ABDOMEN: Wingspan 25–32 mm. Wings light yellow, heavily irrorated with brown. Only medial and postmedial lines present, weak, broken into spots, brown, angled on costa. Forewing margin concave below apex. Terminal line narrow, brown. Fringes concolorous with wings, brown in forewing concavity. Discal spots small, brown. Wings below as above. Frons, collar, thorax and abdomen concolorous with wings. Male antennae bipectinate, female filiform. Hindtibia with 2+2 spurs. Tympanal organs medium-sized. Male 3rd sternite without setae, hindleg with small hair pencil. Male 8th sternite weakly sclerotised, posterior margin concave, 8th tergite elongated. Tergites 3–8 of both sexes with two minute punctures/segment.

VARIATION: Ground colour varies from dull yellow to whitish yellow. Northern European specimens are often slightly larger, ground colour whitish yellow, irrorated with light brown, and postmedial line is barely visible. Towards the south the specimens tend to be slightly smaller, more contrasting, ground colour is brownish yellow, irrorated with dark brown, and postmedial line (and often medial line also) distinct, brown.

MALE GENITALIA: Uncus short, stout, weakly setose. Socii apparently absent. Gnathos arms narrow, fused ventrally, forming large granular plate. Valva wide, complex, fused ventrally. Costa sclerotised, with naked extension (ampulla), apex setose, round. Ventral margin strongly sclerotised, setose, with round, densely setose lobe. Base of valva with few stout setae. Juxta with dorsolateral extensions. Saccus margin round. Aedeagus stout, weakly curved, apex sharp. Vesica opens at approximately 45 degree angle. Vesica wide at base, with small dentate cornutus. Apical part of vesica narrow membranous tube.

FEMALE GENITALIA: Papillae anales setose, surface weakly striated. Apophyses long, straight. Lamella antevaginalis narrow, sclerotised, lunular. Lamella postvaginalis irregular, concave medially, weakly sclerotised plate. Ductus bursae short, wide, membranous. Corpus bursae large, round, membranous. Signum quadrangular, margins mostly sclerotised.

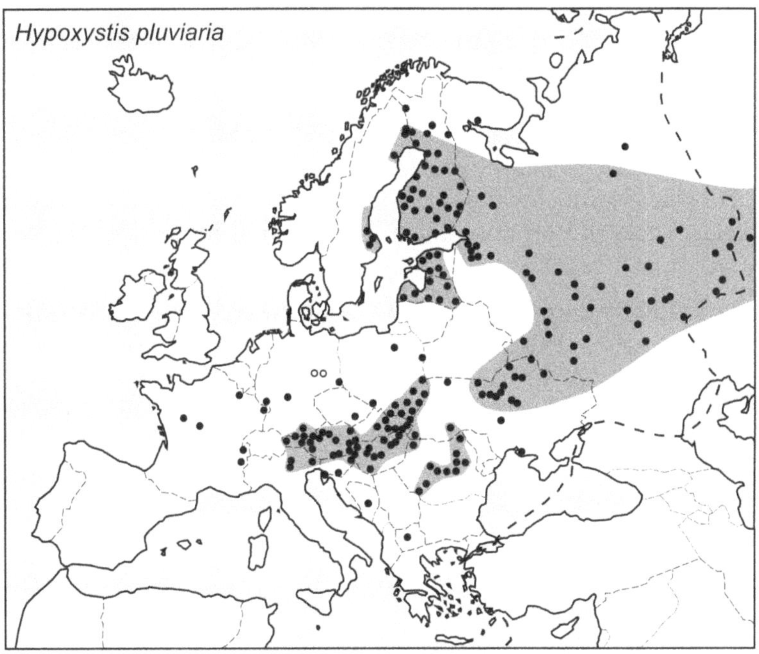

DISTRIBUTION AND ABUNDANCE: Eurasiatic. In Europe a somewhat disjunct distribution from northern Fennoscandia to southern central Europe including scattered records in France and the Balkans. To the east through Ukraine and central European Russia to the Ural mountains. The species has been recorded from Bulgaria (MÜLLER 1996). This is based on two old records, but no recent and valid records seem to exist for the occurrence in Bulgaria (S. BESHKOV pers. comm.) – Outside Europe in Kazakhstan, southern Siberia and Mongolia. Recorded from Turkey and Caucasus (VIIDALEPP 1996), but no specimens have been traced. Also not recorded from Turkey by KOÇAK & KEMAL (2009). – In most parts of Europe a local and scarce species, in central Europe, such as Hungary and Romania widespread and not rare (G. PETRÁNYI, L. SZÉKELY pers. comm.).

PHENOLOGY: Uni- or bivoltine. In northern Europe in one generation from early June to early July, further to the south in Fennoscandia from mid-May to mid-June. In Poland and Slovakia from late April to early June and from early July to early August (MALKIEWICZ 2012, Z. LAŠTŮVKA pers. comm.). In the southern part of the European distribution area from mid-April to early June and again from early July to mid-August. From Romania the species is also recorded from late September to mid-October which may represent a 3rd generation, perhaps partial. Larvae from June until the following spring. The overwintering takes place as full grown caterpillar. The moths are active by day and by night, males preferably diurnal (FLAMIGNI *et al.* 2007, Z. LAŠTŮVKA pers. comm.), in southern Urals only collected by night (K. NUPPONEN pers. comm.).

BIOLOGY: Oligophagous. The caterpillars in northern Europe are found on *Galium*, *Vaccinium*, *Filipendula* and *Betula* (HULDÉN *et al.* 2000). Further to the south they are recorded on *Galium palustre*, *Cytisus scoparius*, *Senecio jacobaeae*, *S. nemorensis* (BUSZKO 2000), *Genista* and *Coronilla* (FLAMIGNI *et al.* 2007). Reared on *Lotus* (J. GELBRECHT pers. comm.).

HABITAT: Hygrophilous. In wetlands, moist deciduous forests and moist meadows, in forest clearings and in open woodland, apparently always on moist soil. Mainly at lower altitudes, but in the Alps up to 1800 m according to FAJCIK (2003).

SIMILAR SPECIES: No similar species in Europe.

GENETIC DATA: Genetically homogeneous from Finland, Germany and Russia (n=10) to central Asia. Populations from Far East Asia, however, diverging by 5.1%. Nearest species: *Plagodis reticulata* WARREN, 1893 from China (7.1%) and *Synopsia sociaria* (HÜBNER, 1799) (7.3%).

REMARKS: SCOBLE (1999) recognised *H. pluviaria* subspecies *sylvanaria* (HERRICH-SCHÄFFER). According to PROUT (1912–1916), the taxon is found in south-east Europe (should be: France, Germany, Russia according to original description (HERRICH-SCHÄFFER 1847)) and in Finland. This view has not been adopted elsewhere in the recent literature (for instance VIIDALEPP 1996, KULLBERG *et al.* 2001, HAUSMANN, MIRONOV & VIIDALEPP 2011), and we also consider it a junior synonym of *H. pluviaria*. The DNA barcodes of *H. pluviaria* from Finland, Germany and Russia are almost identical, thus supporting the view that there is no separate subspecies of *H. pluviaria* in Finland.

Enanthyperythra WEHRLI, 1937

Enanthyperythra WEHRLI, 1937, Ent. Z., Frankf. a.M. 51: 117. Type species: *Hypoplectis legataria* HERRICH-SCHÄFFER, 1852, [Europe?]. *Enanthyperythra* was originally proposed as a subgenus of *Anthyperythra* SWINHOE, 1891. There are no genus synonyms.

DIVERSITY AND DISTRIBUTION: *Enanthyperythra* is a monotypic genus, the only species occurs in the West Palaearctic region.

EXTERNAL CHARACTERS AND ABDOMEN: Medium-sized, wingspan 25–29 mm. Wings yellow, markings and fringes pale pink. Postmedial line most pronounced. Hindwings lighter. Male and female antennae shortly fasciculate. Bullae tympani with small roundish lobe. All sternites and tergites of both sexes undifferentiated.

MALE GENITALIA: Uncus long, narrow, curved ventrally. Socii minute, setose. Gnathos arms fused ventrally, upturned, apex weakly dentate. Valva membranous, setose, apex round. Costa sclerotised, most setose subapically. Base of valva with setose, membranous sac. Furca curved, apex sharp. Saccus margin round. Vesica with two long cornuti.

FEMALE GENITALIA: Papillae anales large, setose, surface weakly striated. Lamella antevaginalis sclerotised, funnel-shaped, heavily setose. Posterior part of corpus bursae sclerotised, striated. Anterior part of corpus bursae membranous, round. Signum weakly quadrangular, margins most sclerotised.

BIOLOGY: WEHRLI (1939–1954) reports that *E. legataria* specimens from Macedonia and Turkey have been taken at rather high altitude, 1600 metres and 1000 metres, respectively. Larval food plant is unknown to us.

IMMATURE STAGES: No data available.

REMARKS: VIIDALEPP (1996) classified *legataria* (HERRICH-SCHÄFFER) in genus *Anthyperythra* SWINHOE, 1891. We have not examined type species of the *Anthyperythra*, *A. hermearia* SWINHOE, 1891 from India, but according to the superficial description of the male genitalia by WEHRLI (1939–1954), the structures are similar to their equivalents in *E. legataria*.

Recently *Enanthyperythra* has been classified in the Ourapterygini (VIIDALEPP 1996) and to the Aspitatini (HAUSMANN, MIRONOV & VIIDALEPP 2011). The male structures of *E. legataria*, including its genitalia, are very general ennomine type, therefore they do not correspond with the Aspitatini structures. *A. legataria* and *O. sambucaria* share the dorsal origin of furca. Female structures, including its genitalia, are also general ennomine type. The signum bears some similarity with the Angeronini, including the mainly Nearctic genera *Euchlaena* HÜBNER, 1823, *Lytrosis* HULST, 1896 and *Xanthotype* WARREN, 1894. In Europe *E. legataria* is structurally similar to numerous other general Ennominae genera (for instance *Lignyoptera*, *Pachycnemia* and *Opisthograptis*), which are currently placed in various tribes. *E. legataria* has membranous sac with setae (coremata), arising from valva base, similar to those in genus *Eilicrinia*. Systematic position: Uncertain.

74. *Enanthyperythra legataria* (HERRICH-SCHÄFFER 1852)

legataria HERRICH-SCHÄFFER, 1852, Syst. Bearbeitung Schmett. Eur. 6 (55); 71; ibidem (1851), 3: pl. 79, fig. 488 (non binominal), (*Hypoplectis*). Syntype(s), [Europe?].

legataria transcaucasica WEHRLI, 1937, Int. ent. Z. 51: 117, (*Enanthyperythra*). Syntypes 5 males, 2 females, [Russia]: Transcaucasus, Ordubad. [Paratype male in the Natural History Museum, London, UK (examined externally]. Originally as subspecies of *legataria*, treated valid at subspecies rank by SCOBLE (1999), downgraded here from subspecies rank (new synonym), quoting absence of diagnostic features, falling within intraspecific variation.

EXTERNAL CHARACTERS AND ABDOMEN: Wingspan 25–29 mm. Wings yellow, irrorated with grey-brown, markings pale pink. Postmedial line most pronounced, pale pink, wider on forewing inner margin and hindwing anal margin. Antemedial and medial lines reduced, remnants on costa. Costa pale pink. Hindwings lighter. Terminal line and fringes concolorous with wings or pale pink. Discal spots grey. Wings below more irrorated with grey-brown, particularly forewing basal area. Subterminal line more pronounced, grey-brown. Frons, collar, thorax and abdomen concolorous with wings. Male and female antennae shortly fasciculate. Hindtibia with 2+2 spurs. Tympanal organs large, bullae tympani with small roundish lobe. All sternites and tergites of both sexes undifferentiated.

VARIATION: Ground colour varies from pale yellow (more common in eastern Turkey and Transcaucasus, but not exclusively) to deep yellow (more common in southern Balkan and Crimea), density of irroration is variable. Fringes may be concolorous with yellow wings or pale pink. Forewing margin below apex may be round or slightly concave. In latter the terminal line is dark in concavity.

MALE GENITALIA: Uncus long, narrow, curved ventrally, weakly setose. Socii minute, setose. Gnathos arms fused ventrally, upturned, apex weakly dentate. Valva membranous, setose, apex round. Costa sclerotised, most setose subapically. Base of valva with setose, membranous sac. Furca curved, apex sharp. Transtilla membranous. Saccus margin round. Aedeagus narrow, weakly curved. Vesica opens at approximately 135 degree angle. Vesica wider at base, striated, with two long cornuti.

FEMALE GENITALIA: Papillae anales large, setose, surface weakly striated. Apophyses long, straight. Lamella antevaginalis sclerotised, funnel-shaped, heavily setose. Lamella postvaginalis unrecognisable. Ductus bursae short, most sclerotised laterally. Posterior part of corpus bursae sclerotised, weakly curved, striated. Anterior part of corpus bursae membranous, round. Signum weakly quadrangular, margins weakly dentate, sclerotised.

DISTRIBUTION AND ABUNDANCE: (East-)European-Westasiatic. In Europe the distribution is limited to the the southern Balkans, southern European Russia and Crimea. – Outside Europe the species is known from Turkey, northern Iran, Armenia and Azerbaijan. – In Europe a rare species that is apparently always found singly or as very few specimens.

PHENOLOGY: Univoltine. From mid-April to mid-June. In Crimea from late April to mid-May. The moths fly at night and come to light.

BIOLOGY: Unknown.

HABITAT: On Mount Askion by Siatista in Greece on open slopes with small scattered bushes and a few trees. In the Galicica National Park in south-western Macedonia on open mountain

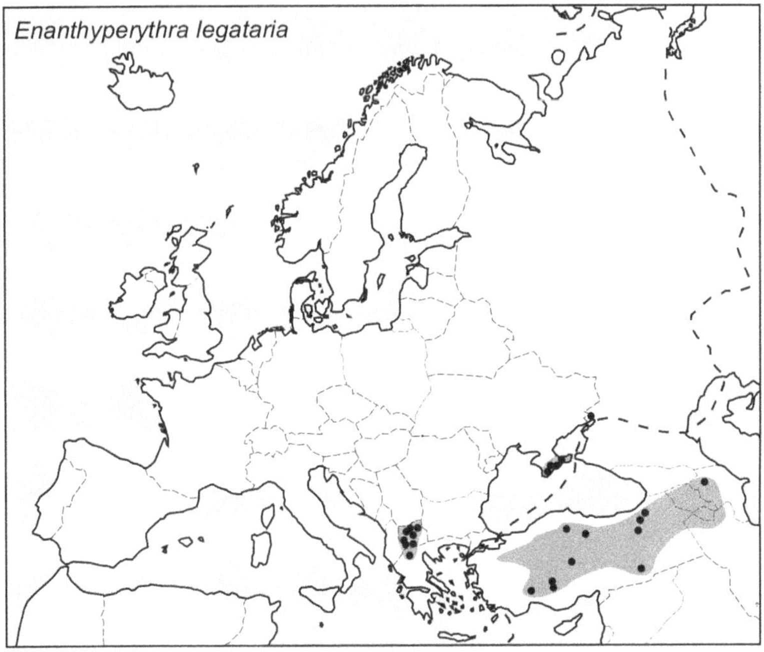

meadow slopes above the timber line at 1600 m. In central Turkey on a xerothermic chalk slope. Found from 1100 m up to 1600 m altitude. Apparently not recorded below 1100 m.

SIMILAR SPECIES: No similar species in Europe.

GENETIC DATA: No data available.

Apeira GISTL, 1848

Apeira GISTL, 1848, Naturg. Thierreichs. xi. Type species: *Phalaena syringaria* LINNAEUS, 1758, [Europe]. Genus synonyms are after SCOBLE (1999).

Cyphosea GROTE, 1896, Proc. ent. Soc. Lond. 1896: xiv. Type species: *Phalaena syringaria* LINNAEUS, 1758 [Unnecessary replacement name for *Pericallia* STEPHENS, 1828].

Pericallia STEPHENS, 1828, in KIRBY & SPENCE, Introd. Ent. (Edn 5) 3: 151. Type species: *Phalaena syringaria* LINNAEUS, 1758. [Junior homonym of *Pericallia* HÜBNER, [1820] 1816 (Lepidoptera: Arctiidae)].

DIVERSITY AND DISTRIBUTION: *Apeira* contains 10 species, which occur in the Palaearctic region.

EXTERNAL CHARACTERS AND ABDOMEN: Rather large moths, wingspan about 35 mm. Wings yellowish brown, often with whitish areas. Postmedial line oblique, angled before costa. Forewing margin concave below apex. Male and female antennae bipectinate. Male 8th sternite weakly sclerotised, posterior margin elongated medially.

MALE GENITALIA, description is based on the European species (homologies are difficult to interpret): Uncus absent. Socii sclerotised, setose. Gnathos arms massive, separate (or fused via medial sclerotisation), upturned and dentate. Valva wide, flat, setose, apex round. Two separate sclerotisations ventrad of juxta (part of valva?). Transtilla(?) with furca-like, narrow, sclerotised projections. Saccus concave. Vesica with numerous cornuti.

FEMALE GENITALIA, description is based on the European species: Lamella antevaginalis partly setose, with two large lateral cavities. Its posterolateral margins elongated, reaching papillae anales. Posterior part of corpus bursae long, sclerotised, striated. Anterior part of corpus bursae membranous, wrinkled, roundish. Signum absent.

BIOLOGY: Larva of *A. syringaria* has been recorded feeding for instance on *Lonicera*, *Syringa* (MIKKOLA *et al.* 1989), *Fraxinus excelsior* and *Ligustrum vulgare* (EBERT 2003). Small larva overwinters. The pupa is formed in a flimsy cocoon spun up on the twigs of the food plant (PORTER 2010). Adult is nocturnal.

IMMATURE STAGES: Larva of *A. syringaria* is a twig-mimic, it can be various shades of brown or yellowish, with tiny short greyish bristles and with two long, curled dorsal structures arising from A7, smaller pairs of humps on A5 and A6 (PORTER 2010). Pupa strongly expanded in middle part, thorax and abdomen dorsally with dense short hairs. Pupa cremaster short, with two curved apical setae, and with numerous (about 30) minute, curved setae (PATOČKA & TURČANI 2005).

REMARKS: Systematic position of genus *Apeira*, including its tribal position and status, are unclear. STEKOLNIKOV & KUZNETZOV (1982) described tribe Apeirini, and designated *Apeira* (GISTL) as its type genus. Afterwards the Apeirini have been recognised in two works only, by VIIDALEPP (1996) and KULLBERG *et al.* (2001), the latter adopted VIIDALEPP's system apart from few exceptions. The few works that have summarised the usage of Ennominae tribes in wider context do not mention the Apeirini (HOLLOWAY 1994, BELJAEV 2006ab, FORUM HERBULOT 2014). Elsewhere genus *Apeira* has been classified in the Ennomini (VIVES MORENO 1994, LERAUT 1997, HAUSMANN, MIRONOV & VIIDALEPP 2011), but BELJAEV (2008a) did not include it in his concept of the Ennomini. In the DNA based phylogenetic analysis of the world Geometridae tribes, genus *Apeira* (Apeirini) grouped next to *Epione* (Epionini) and *Plagodis* (potentially Hypochrosini) (SIHVONEN *et al.* 2011).

The male genitalia, larva and pupa of *Apeira syringaria* are unique within the European Ennominae fauna. In the male genitalia homologies are difficult to interpret. The absence of uncus, massive and separate gnathos arms (or derivatives of tegumen?), downwards directed valvae, two separate sclerotisations ventrad of juxta and pupa cremaster with two curved apical setae and numerous (about 30) minute, curved setae are examples of such unique structures. Numerous other tribes (for instance the Abraxini, Cassymini and Macariini group of tribes) are founded on smaller differences. Posterior margin of male 8th sternite of *A. syringaria* is elongated medially, similar condition is present in *Cepphis* (potentially Epionini) and *Epione* (Epionini). Systematic position (tentative): Apeirini, potentially related to the Epionini.

75. *Apeira syringaria* (LINNAEUS, 1758)

syringaria LINNAEUS, 1758, Syst. Nat. (Ed. 10) 1: 520, (*Phalaena (Geometra)*). Syntypes (Linnean Society of London, UK), Europe (examined externally).

circularia THUNBERG, 1792, Diss. ent. sistens Insecta Suecica 4: 58, pl. 4, (*Phalaena*). Lectotype female (Universitets Zoologiska Institut, Uppsala, Sweden), Sweden.

jaspoides FOURCROY, 1785, Entomologica Parisiensis 2: 267, (*Phalaena*). Syntype(s), [France]: Paris.

syringaria druentaria LERAUT, 2009, Moths of Europe, vol. 2, Geometrid moths: 94, pl. 20, figs 10, 12 (*Apeira*). Holotype male (Muséum National d'Histoire Naturelle, Paris, France), France: Hautes-Alpes, Pelvoux. Originally as subspecies of *syringaria*, downgraded here from subspecies rank (new synonym), quoting absence of differential features, falling within intraspecific variation.

UNAVAILABLE NAMES (incorrect subsequent spelling): *helvola* (LEMPKE 1970). – Intrasubspecific names, potentially unavailable: *feminicolor* LEMPKE, 1970 (f.); *flavescens* LEMPKE, 1951 (f.); *helvolaria* ROBSON & GARDNER, 1886 (f.); *hofmanni* SCHREIBER, 1903 (ab.); *mariscolora* LEMPKE, 1951 (f.); *obsoleta* LEMPKE, 1951 (f.); *pallidaria* COCKAYNE, 1948 (ab.); *retyezatica* DIOSZEGHY, 1930 (ab.?).

EXTERNAL CHARACTERS AND ABDOMEN: Wingspan 29–40 mm. Wings yellowish brown, with several whitish-violet areas. Medial line red-brown, oblique, meeting with partly whitish postmedial line, both are angled before costa. Costa with several whitish violet areas, lunular area just before apex. Terminal area partly yellow. Hindwings with red-brown, medial line and spotted postmedial line, strongest on anal margin. Terminal line and fringes concolorous with wings. Forewing margin concave below apex, hindwing margin undulating. Discal spots minute. Wings below more irrorated with brown, medial line brown, suffuse. Discal spots larger, brown. Frons, collar, thorax and abdomen concolorous with wings. Male and female antennae bipectinate. Hindtibia of both sexes slightly enlarged, with 2+2 spurs. Tympanal organs medium-sized. Male 8th sternite weakly sclerotised, posterior margin elongated medially. Other sternites and tergites of both sexes undifferentiated.

VARIATION: Very little. Females often with more extensive whitish-violet areas. Second generation specimens smaller. We consider taxon *druentaria* LERAUT, 2009, which has paler, olive green wings, to fall within intraspecific variation.

MALE GENITALIA (homologies are difficult to interpret, See remarks): Uncus absent. Socii round, sclerotised, setose. Gnathos arms massive, separate (or fused via medial sclerotisation), upturned, curved, dentate. Valva wide, flat, setose, apex round. Costa weakly sclerotised, less setose. Juxta concave plate. Two separate sclerotisations, pocket-shaped, ventrad of juxta (part of valva?). Transtilla(?) with furca-like, narrow, weakly dentate, sclerotised projections. Saccus concave. Aedeagus straight, caecum minute, apex blunt. Vesica opens at approximately 180 degree angle. Basal part with anteriorly directed diverticulum, second diverticulum coiled, with numerous (>10) long cornuti. Apical portion of vesica narrow, membranous tube.

FEMALE GENITALIA: Papillae anales large, setose. Apophyses long, apophyses anteriores curved. Margin of lamella antevaginalis partly setose, concave medially, with two large, sclerotised lateral cavities. Its posterolateral margins elongated, reaching papillae anales. Lamella postvaginalis membranous. Ductus bursae short, most sclerotised laterally. Posterior part of corpus bursae with dorsal diverticulum, otherwise long, sclerotised, weakly curved, striated. Anterior part of corpus bursae membranous, wrinkled, roundish. Wrinkled part concealed by membrane. Signum absent.

DISTRIBUTION AND ABUNDANCE: Eurasiatic. In Europe from southern central Scandinavia to the north-eastern Iberian Peninsula, southern France, southern Italy and the Balkans and from the British Isles through central and eastern Europe to European Russia and the Ural mountains. – Outside Europe the species is known from Turkey, the Caucasus region and

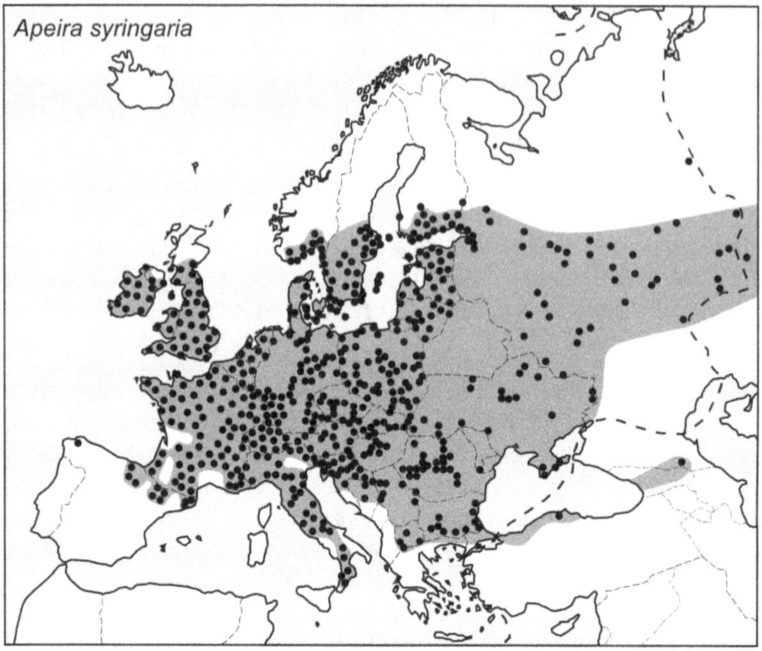

southern Siberia to Far East Russia, Japan and Sakhalin. – In Europe a common or rather common species that rarely occurs in numbers. More local and rare towards north.

PHENOLOGY: Uni- or bivoltine. In northern Europe the adult moths fly from late June to late July and rather seldom specimens of a partial second generation are found in early and mid-September. In warm places in southern Germany in two generations from mid-May to early July and again from mid-August to mid or late September (EBERT 2003). Further to the south regularly bivoltine at lower altitudes and univoltine at higher altitudes. Larva from August to the spring in the north and at higher altitudes, in the south also between the two generations of adult moths. The adult moths fly at night and come to light.

BIOLOGY: Oligophagous. The larva was found on *Syringa vulgaris* (E. FRIEDRICH pers. comm.) *Fraxinus excelsior*, *Weigela florida* (EBERT 2003), *Lonicera xylosteum* (B. MÜLLER pers. comm.) and *L. periclymenum* (H. WEGNER pers. comm.). *Ligustrum vulgare* is given as a main food-plant in Britain (SKINNER 2009). The pupation takes place in an open web on a twig on the food plant (EBERT 2003).

HABITAT: In forest edges and open deciduous forests, in gardens, parks and similar places. The species is found from sea level, in the Pyrenees up to 1400 m (REDONDO *et al.* 2009) and in the Alps up to 1790 m (FLAMIGNI *et al.* 2007).

SIMILAR SPECIES: No similar species in Europe.

GENETIC DATA: Genetically heterogeneous in Europe (n=13, from five countries), maximum variation 4.7%. Nearest species in Europe (genetic similarity due to long branch attraction): *Selidosema plumaria* (DENIS & SCHIFFERMÜLLER, 1775) (9.5%).

REMARKS: Homologies of the male genitalia in *A. syringaria* are difficult to interpret. An alternative way could be to consider the two sclerotised lobes arising from margin of the tegumen as bilobed uncus. Setae arising dorsad of massive, dentate arms could be considered socii. These are tiny, they are not visible in the photographs. This view is supported by sclerotised nature of these lobes: normally uncus is sclerotised, whereas normally socii are membranous. The massive arms could then be gnathos, which appear to be fused via a medial sclerotisation. If this interpretation is implied, the 'normal' Ennominae structures uncus, socii and fused gnathos would be present, although highly modified. The dorsal, furca-like structures are located where transtilla is normally located. We have not seen transtilla with such sclerotised extensions, so our interpretation of their origin must be considered tentative.

Ennomos TREITSCHKE, 1825

Ennomos TREITSCHKE, 1825, Schmett. Eur. 5 (2): 427, a replacement name for *Eugonia* HÜBNER, [1823] 1816. Type species: *Eugonia autumnaria* WERNEBURG, 1859, [Europe]. Genus synonyms are after SCOBLE (1999).

Deuteronomos PROUT, 1914, in PIERCE, Genitalia Group Geometridae Lepid. Br. Islands: xxvii, 8. Type species: *Phalaena alniaria* LINNAEUS, 1758.

Ennomus AGASSIZ, 1847, Nomencl. zool. (Index univl.): 139. [Emendation of *Ennomos* TREITSCHKE].

Eugonia HÜBNER, [1823] 1816, Verz. bekannter Schmett.: 291. Type species: *Eugonia autumnaria* WERNEBURG, 1859. [Junior homonym of *Eugonia* HÜBNER, [1819] 1816 (Lepidoptera: Nymphalidae)].

Odontoptera AGASSIZ, 1847, Nomencl. zool. (Index univl.): 255. [Emendation of *Odoptera* SODOFFSKY, and junior homonym of *Odontoptera* CARRENO, 1842 (Hemiptera)].

Odoptera SODOFFSKY, 1837, Bull. Soc. imp. Nat. Moscou 1837 (6): 89. [Unnecessary replacement name for *Ennomos* TREITSCHKE, 1825].

UNAVAILABLE NAMES (incorrect subsequent spellings): *Denderonomos* PROUT, 1915 in SEITZ (1912–1916); *Ennomus* WIEGMANN & RUTHE, 1832; *Eunomus* HORANINOW, 1834; *Eunomos* LUCAS, 1846; *Ennemos* MOTSCHOULSKY, 1854; *Ennom* NEAVE, 1939 (see FLETCHER 1979).

DIVERSITY AND DISTRIBUTION: *Ennomos* contains about 15 species, which occur in the Palaearctic and in the Nearctic regions.

EXTERNAL CHARACTERS AND ABDOMEN: Rather large moths, wingspan about 35–50 mm. Wings yellow-brown, colours similar to autumn leaves, occasionally with tint of olive-green, rarely uniform white. Medial and postmedial lines normally well-developed, oblique. Forewing margin concave below apex, otherwise margins undulating. Male and female antennae bipectinate, female pectinations shorter. Male hindleg often enlarged distally. Hindleg tibia of both sexes often with two apical spurs only. Male 8th sternite unmodified.

MALE GENITALIA: Uncus variable, usually short, stout. Socii minute, setose. Gnathos arms fused ventrally, ventral part variable, often strongly spinose. Valva most setose subapically. Costa sclerotised, often extending beyond apex. Furca arms often present, arising from posterior margin of juxta. Saccus often concave, membranous. Vesica with cornuti.

FEMALE GENITALIA: Papillae anales large, setose. Apophyses posteriores long, narrow. Apophyses anteriores very short, stout. Lamella postvaginalis striated horizontally. Ductus bursae sclerotised, often long. Corpus bursae membranous, roundish. Signum often small, occasionally weakly stellate.

BIOLOGY: Larvae of European and North American species have mainly been recorded feeding on leaves of deciduous trees such as *Betula*, *Quercus*, *Alnus*, *Fraxinus*, *Fagus*, *Tilia*, *Populus*, *Salix*, *Carya* and *Viburnum* and also on *Vaccinium myrtillus* (MCGUFFIN 1987, MIKKOLA *et al.* 1989, DEBORAH *et al.* 2014). Adults are nocturnal, flight-time is in autumn, egg overwinters.

IMMATURE STAGES: Eggs of North American *E. magnaria* GUENÉE, 1858 are barrel-shaped, surface is smooth (MCGUFFIN 1987), those are laid in rows. Larvae of European species are greenish brown twig-mimics, with short dorsal humps on 5th and 8th segments, occasionally laterally on 6th segment (PORTER 2010). Pupa cremaster is slightly triangular, with four pairs of curved setae, their length and width varies between species (PATOČKA & TURČANI 2005).

REMARKS: Systematic position: Ennomini. See text on Ennomini under *Classification of European Ennominae genera into tribes* and BELJAEV (2008a).

76. *Ennomos autumnaria* (WERNEBURG, 1859)

autumnaria autumnaria WERNEBURG, 1859, Stettin. ent. Ztg 20 (10–12): 361, (*Eugonia*). Syntype(s), [Europe].

autumnaria sareptana WEHRLI, 1940, in SEITZ, Gross-Schmett. Erde 4 (Suppl.): 324, pl. 24: h, (*Ennomos*). Syntype(s) (Zoologische Forschungsinstitut und Museum Alexander Koenig, Bonn, Germany), [Russia:] Sarepta [Volgograd]. [Paratype male in the Natural History Museum, London, UK (examined externally)]. Originally as subspecies, treated valid at subspecies rank also by SCOBLE (1999), downgraded here from subspecies rank (new synonym), quoting absence of differential features, falling within intraspecific variation. SATO (2005) did not recognise *sareptana* valid at subspecies rank either.

autumnaria koreennomos BRYK, 1949, Ark. Zool. 41A (1): 191, (*Ennomos*). Syntypes 1 male, 1 female, Korea: Shuotsu. Valid at subspecies rank according to SCOBLE (1999).

autumnaria pyrrosticta WEHRLI, 1940, in SEITZ, Gross-Schmett. Erde 4 (Suppl.): 324, (*Ennomos*). Syntype(s) (Zoologische Forschungsinstitut und Museum Alexander Koenig, Bonn, Germany), China: Tien-Mu-shan; Chekiang. Valid at subspecies rank according to SCOBLE (1999), but See remarks.

autumnaria sinica YANG, 1978, Moths of North China 2: 388, pl. 20, fig. 1, (*Ennomos*). Holotype male, [North-China:] inner Mongolia, Chifeng. Valid at subspecies rank according to SCOBLE (1999).

INTRASUBSPECIFIC NAMES, POTENTIALLY UNAVAILABLE: *angustaria* KRULIKOWSKY, 1897 (ab.); *apicata* LEMPKE, 1951 (f.); *approximata* LEMPKE, 1970 (f.); *aurantiaca* LEMPKE, 1951 (f.); *brunneata* COCKAYNE, 1952 (ab.); *brunnescens* LEMPKE, 1970 (f.); *clara* COCKAYNE, 1952 (ab.); *costimacula* LEMPKE, 1970 (f.); *distincta* HEINRICH, 1916 (f.); *latelineata* LEMPKE, 1970 (f.); *maculosa* LEMPKE, 1951 (f.); *nigrescens* LEMPKE, 1970 (f.); *pallida* LEMPKE, 1951 (f.); *schultzi* SIEBERT, 1907 (ab.); *tangens* LEMPKE, 1951 (f.); *triangularis* COCKAYNE, 1948 (ab.). – (hybrid): *dartfordi* TUTT, 1905 (*autumnaria* WERNEBURG male × *quercinaria* HUFNAGEL female).

EXTERNAL CHARACTERS AND ABDOMEN: Wingspan ♂ 41–47 mm, ♀ 47–52 mm. Wings deep yellow, densely marked with greyish brown irroration and striations, medial lines wide, suffuse. Medial line greyish-brown, evenly curved on costa. Postmedial line greyish brown, oblique, strongest on costa. Terminal area more uniform dark yellow-brown. Forewing margin concave below apex. Wing margins undulating, most extended on M3 on both wings. Terminal line concolorous with wings. Fringes brown or weakly chequered brown and white. Discal spots small, suffuse. Wings below with large, brown discal spots. Forewing apex and hindwing heavily irrorated with brown. Frons, collar, thorax and abdomen concolorous with wings. Male and female antennae bipectinate, female pectinations shorter. Hindleg of both sexes enlarged distally, with 2+2 spurs. Tympanal organs medium-sized. Abdominal sternites and tergites of both sexes undifferentiated.

VARIATION: Ground colour varies from light yellow to deep yellow. Irroration and striations can be absent or heavy, partly fused together. Terminal area, particularly near wing margins, is often uniform dark yellow-brown or yellow-orange in females. Forewing medial and postmedial lines can be close to each other or even touch each other. Postmedial line often reduced, diffuse, most visible on costa.

MALE GENITALIA: Uncus widest subapically, with few stout setae, apex acute. Socii minute, setose. Gnathos arms wide, fused ventrally, upturned, with few sclerotised teeth. Valva wide, most setose subapically, apex blunt. Costa sclerotised, extending beyond apex. Ventral margin weakly concave in middle, valva fused medially. Furca arms long, apex weakly curved inwards, surface smooth. Furca arms with narrow, anterior extensions. Vinculum rather wide. Saccus round, membranous medially. Aedeagus straight, caecum minute, apex blunt. Vesica opens at approximately 90 degree angle. Vesica weakly expanded at base, with small lateral diverticulum and sclerotised, dentate plate and row of short cornuti.

FEMALE GENITALIA: Papillae anales large, surface weakly granulate. 9th and 10th segments elongated. Apophyses posteriores long, straight. Apophyses anteriores short, stout. Lamella postvaginalis striated horizontally. Lamella antevaginalis sclerotised, with wide lateral extensions, margin concave medially. Ductus bursae sclerotised, long, straight. Corpus bursae membranous, round. Signum small, elongated, margins dentate.

DISTRIBUTION AND ABUNDANCE: Eurasiatic. In Europe from central Fennoscandia to central France including Corsica, parts of the Alps and the northern Balkans and from Great Britain through central and east Europe to European Russia and the Ural mountains. – Outside Europe the species is known eastwards from the Ural mountains through northern Kazakhstan and southern Siberia to Korea, Far East Russia, Sakhalin and Japan. – In Europe a common species that rarely occurs in numbers. However, in south-western Germany it has become very rare in the last 10–20 years (M. LEIPNITZ pers. comm.).

PHENOLOGY: Univoltine. In northern Europe from late August to early October while in southern central Europe the flight time is longer, from early August to early October (EBERT 2003). Caterpillar from late May to early August. The hibernation takes place as an egg. The moths fly at night and come to light while they are rarely seen by day.

BIOLOGY: Polyphagous. The caterpillar has been found on a number of species of trees and bushes, especially *Crataegus* and *Quercus* (EBERT 2003) as well as on *Betula*, *Alnus* and *Vaccinium* (HULDÉN *et al.* 2000). The caterpillar has also been found on *Prunus spinosa* (H. WEGNER pers. comm.) and it has been reared on *Tilia cordata* (E. FRIEDRICH pers. comm.). BERGMANN (1955) regards *Tilia* as the main food plant. The pupation takes place in an open web between leaves on the food plant.

HABITAT: The species occurs in a variety of open deciduous forests, forest edges, open scrubland, parks and gardens. The main part of the European distribution is lowland, but the species has been found up to 1300 m in the Italian Alps (FLAMIGNI *et al.* 2007).

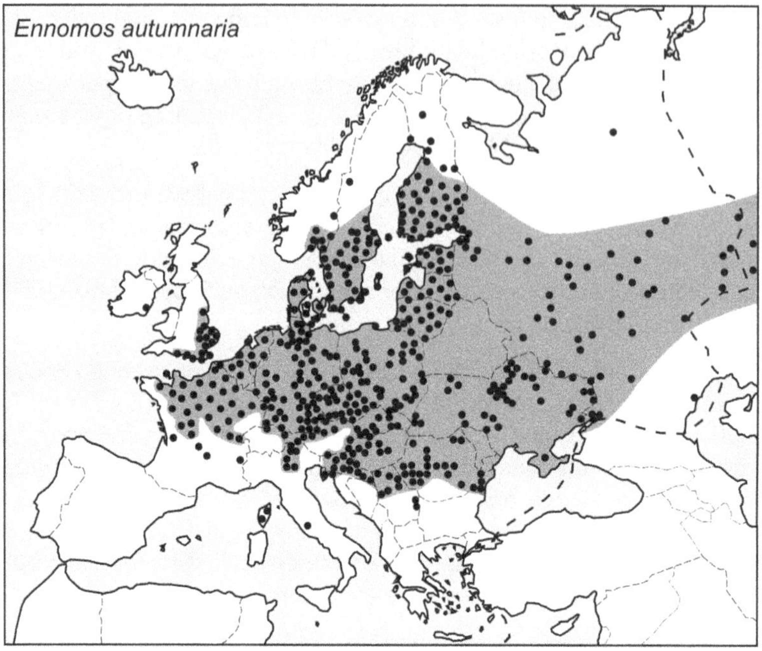

SIMILAR SPECIES: Other *Ennomos* species, see Text-figs 84–89. *E. autumnaria* is normally diagnosable by its large size, wings are densely marked with greyish brown irroration and striations. Medial lines are rather wide in *E. autumnaria*, medial line is not angled before costa (lines narrow, medial line angled before costa in *E. quercinaria*). Postmedial line is diffuse, most visible on costa. Male and female genitalia are diagnostic.

GENETIC DATA: Genetically homogeneous (n=14, from six European countries). Nearest species: *E. quercinaria* (2.5%).

REMARKS: *Ennomos autumnaria* subspecies *nephotropa* PROUT, 1930 from Japan has been raised to species status by SATO (2005). *E. autumnaria* subspecies *intermedia* INOUE, 1942, also from Japan, is considered a junior synonym of it. SATO reports also that *E. autumnaria* subspecies *pyrrosticta* WEHRLI, 1940 from China is conspecific with *E. autumnaria*.

77. *Ennomos quercinaria* (HUFNAGEL, 1767)

quercinaria HUFNAGEL, 1767, Berlin Mag. 4 (5): 520, (*Phalaena*). Syntype(s), lost, [Germany:] Berlin region.

angularia [DENIS & SCHIFFERMÜLLER], 1775, Ankündung syst. Werkes Schmett. Wienergegend: 103, *G.*[*eometra*]. Syntype(s), lost, [Austria]: Vienna district.

carpinaria HÜBNER, [1799], Samml. Eur. Schmett. 5 Geometra (1): pl. 5, fig. 27, (*Geometra*). Syntype(s), lost, Europe.

equestraria FABRICIUS, [1776], Genera Insect.: 284, (*Phalaena*). Syntype(s), [Germany:] Hamburg.

zona FOURCROY, 1785, Entomologia Parisiensis 2: 270, (*Phalaena*). Syntype(s), [France:] Paris.

INTRASUBSPECIFIC NAMES, POTENTIALLY UNAVAILABLE: *approximata* COCKAYNE, 1952 (ab.); *brunneata* COCKAYNE, 1952 (ab.); *carpiniaria* STEPHENS, 1831 (ab.) (=synonym of *carpinaria* HÜBNER, 1779 according to The Global Lepidoptera Names Index (ANONYMOUS 2013)); *clara* COCKAYNE, 1942 (ab.); *clausa* LEMPKE, 1951 (f.); *illineata* LEMPKE, 1970 (f.); *infuscata* STAUDINGER, 1904 (ab.) (=synonym of *equestraria* FABRICIUS, 1776 according to The Global Lepidoptera Names Index (ANONYMOUS 2013)); *maculosa* LEMPKE, 1951 (f.); *obsoleta* LEMPKE, 1951 (f.); *perfuscata* PROUT 1915 (ab.); *subfuscata* COCKAYNE, 1952 (ab.). The Global Lepidoptera Names Index (ANONYMOUS 2013) lists further intrasubspecific names.

EXTERNAL CHARACTERS AND ABDOMEN: Wingspan ♂ 31–41 mm, ♀ 33–47 mm. Wings deep yellow, markings and veins brown, basal and terminal areas darker. Medial lines narrow, sharp. Medial line distinctly angled before costa. Postmedial line angled inwards after discal spot, then slightly angled outwards on costa. Outer margin of postmedial line bordered with brown or terminal area more or less extensively dark. Forewing margin concave below apex. Wing margins undulating, most extended on M3 on both wings. Terminal line concolorous with wings. Fringes weakly chequered brown and white. Discal spots small, suffuse. Wings below often darker, discal spots larger. Frons, collar, thorax and abdomen concolorous with wings. Male abdomen slim. Male and female antennae bipectinate, female pectinations shorter. Hindleg of both sexes slightly enlarged distally, with 2+2 spurs. Tympanal organs medium-sized. Abdominal sternites and tergites of both sexes undifferentiated.

VARIATION: Variable species. Basal and terminal areas are often dark brown to varying degrees, compared with paler medial area. Ground colour varies from pale whitish yellow to deep brown yellow. Veins can be brown, standing out clearly and forming weakly reticulate pattern with medial lines or veins can be same colour as wings. Medial lines can be dark brown or absent altogether. Numerous forms have been described, see LERAUT (2009) for summary.

MALE GENITALIA: Uncus long, stout. Socii minute, setose. Gnathos arms weakly sclerotised, fused ventrally, upturned, with few sclerotised teeth. Valva rather narrow, with patch of long, upturned setae subapically, apex weakly round. Costa sclerotised, extending beyond apex. Ventral margin round, valva fused medially. Furca arms long, apex weakly curved inwards, surface smooth. Furca arms with narrow, anterior extensions. Vinculum rather wide. Saccus concave. Aedeagus straight, caecum bent, apex blunt. Vesica opens at approximately 135 degree angle. Vesica weakly expanded at base, with small lateral diverticulum and sclerotised, dentate plate and few short cornuti.

FEMALE GENITALIA: Papillae anales large, surface weakly granulate. 9th and 10th segments elongated. Apophyses posteriores long, straight. Apophyses anteriores short, stout. Lamella postvaginalis striated horizontally. Lamella antevaginalis sclerotised, without lateral extensions, margin weakly concave medially. Ductus bursae sclerotised, long, slightly funnel-shaped. Corpus bursae angled relative to ductus bursae. Corpus bursae membranous, elongated, anterior end densely striated. Signum small, elongated, margins dentate.

DISTRIBUTION AND ABUNDANCE: European-Westasiatic. In Europe from southern Scandinavia to the northern Iberian Peninsula, southern France including Corsica, southern Italy including Sicily and northern Greece and from the British Isles through central and eastern Europe to European Russia and the Ural mountains. The species has earlier been treated as present in Portugal, but no correctly identified specimens have been found (M. CORLEY pers. comm.). – Outside Europe the species is known from Turkey, the Caucasus region and central Asia. – In Europe a rather common species that is rarely seen in numbers. In Britain it has declined much since 1968 (FOX *et al.* 2013). Also in Denmark the species has declined much in recent decades where it was common up to the 1970's or early 1980's.

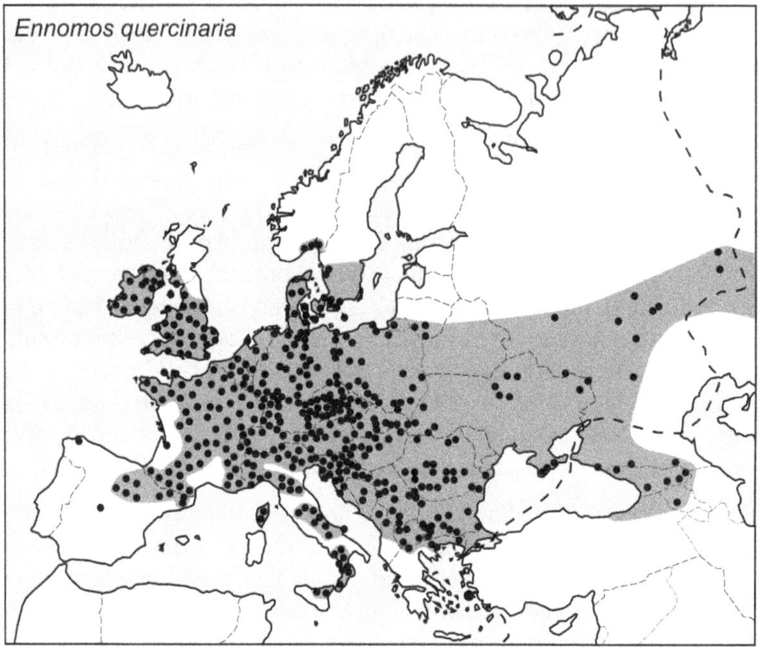

PHENOLOGY: Univoltine. In northern Europe from late July to late September, in southern central Europe from late June to early September (EBERT 2003) and in southern Europe from July to September (FLAMIGNI *et al.* 2007). Larva from May to July. The species overwinters as an egg. The moth is rarely seen by day, it flies at night and comes to light.

BIOLOGY: Polyphagous. The larva has been found on a variety of deciduous trees and bushes such as *Salix*, *Populus tremula*, *Carpinus betulus*, *Corylus avellana*, *Betula*, *Fagus*, *Quercus*, *Crataegus* and *Acer pseudoplatanus* (EBERT 2003) and on *Prunus spinosa* (E. FRIEDRICH pers. comm.). It has been found feeding on *Quercus robur* (H. WEGNER pers. comm.) and *Salix* (B. MÜLLER pers. comm.). The pupation takes place between spun leaves of the food plant.

HABITAT: Sylvicolous. Deciduous woodland. In northern Europe at low altitudes from sea level, in southern Europe it occurs from about 500 up to 1600 m altitude.

SIMILAR SPECIES: Other *Ennomos* species, see Text-figs 84–89. Male and female genitalia allow reliable identification, see the genitalia plates. Characteristically *E. quercinaria* male has slim abdomen, forewing medial line is angled before costa, postmedial line is angled inwards after discal spot, then slightly angled outwards on costa and basal and terminal areas are often darker than medial area.

GENETIC DATA: Genetically homogeneous (n=18, from six European countries). Nearest species: *E. autumnaria* (2.5%).

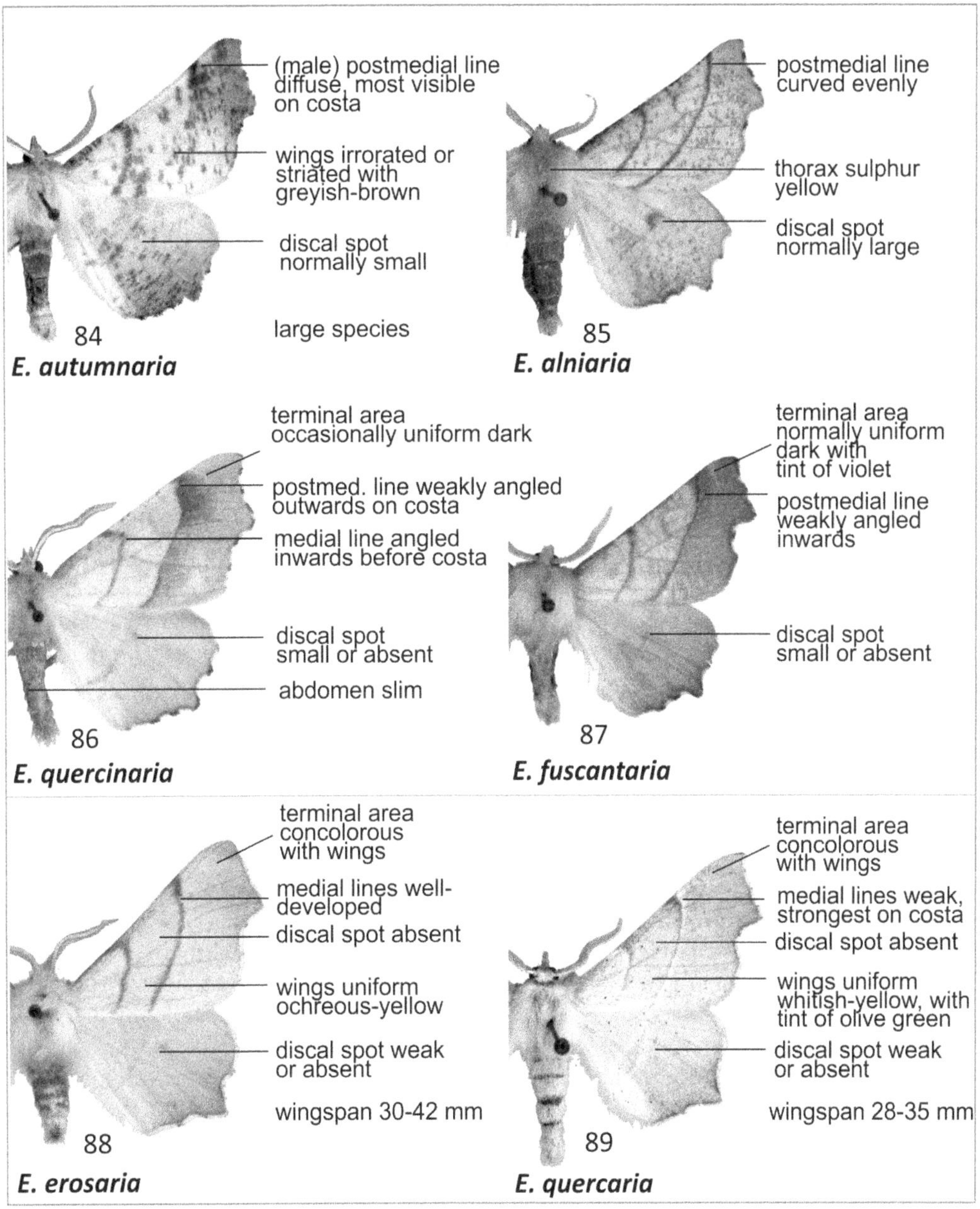

Text-figs 84–89. Diagnostic characters (indicated) of European *Ennomos* TREITSCHKE, 1825 species. Narrow grey line delimits the most similar species and first comparisons should be made against the species included in the same grey box. Specimens are not in scale relative to each other. Text-fig. 84. *E. autumnaria* (WERNEBURG, 1859), Denmark, 9.ix.1973. Text-fig. 85. *E. alniaria* (LINNAEUS, 1758), Denmark, 6.x.1972. Text-fig. 86. *E. quercinaria* (HUFNAGEL, 1767), Sweden, 16.viii.1972. Text-fig. 87. *E. fuscantaria* (HAWORTH, 1809), Denmark, 10.ix.1969. Text-fig. 88. *E. erosaria* (DENIS & SCHIFFERMÜLLER, 1775), Sweden, 2.ix.1974. Text-fig. 89. *E. quercaria* (HÜBNER, 1813), Spain: Prov. Teruel, Puerto de Orihuela, 1650 m, 23.viii.2001.

78. *Ennomos alniaria* (LINNAEUS, 1758)

alniaria LINNAEUS, 1758, Syst. Nat. (Ed. 10) 1: 519, (*Phalaena (Geometra)*). Syntypes (Linnean Society of London, UK), Europe [Sweden, Germany] (examined externally).

canaria HÜBNER, 1790, Beitr. Gesch. Schmett. 2 (4): 99, pl. (4) 4, fig. Y, 1, 2, (*Ph[alaena] Geom[etra]*). Syntypes male, female [lost].

tiliaria BORKHAUSEN, 1794, Natur. eur. Schmett. 5: 87, (*Phalaena Geometra*). Syntypes 4 males, Europe.

INTRASUBSPECIFIC NAMES, POTENTIALLY UNAVAILABLE: *approximata* LEMPKE, 1951 (f.); *aurantiaca* LEMPKE, 1951 (f.); *clausa* LEMPKE, 1951 (f.); *coarctata* COCKAYNE, 1948 (ab.); *concolor* LEMPKE, 1951 (f.); *crassestrigata* LEMPKE, 1970 (f.); *destrigata* COCKAYNE, 1948 (ab.); *fuscomarginata* LEMPKE, 1970 (f.); *glabra* LEMPKE, 1951 (f.); *grisescens* COCKAYNE, 1952 (ab.); *hannemanni* HEINRICH, 1917 (ab.); *infuscata* HANNEMANN, 1917 (ab.); *maculosa* NORDSTRÖM, 1941 (ab.?); *nigrescens* LEMPKE, 1970 (f.); *obsoleta* LEMPKE, 1970 (f.); *pallida* LEMPKE, 1951 (f.); *postfasciata* VAN WISSELINGH, 1961 (f.); *triangularis* LEMPKE, 1951 (f.).

EXTERNAL CHARACTERS AND ABDOMEN: Wingspan ♂ 33–39 mm, ♀ 34–43 mm. Wings deep yellow, marked with violet-brown irroration and striations, medial lines oblique. Medial and postmedial lines violet-brown, narrow, evenly curved, oblique. Terminal area weakly brown near apex. Forewing margin concave below apex. Wing margins undulating, most extended on M3 on both wings. Terminal line concolorous with wings. Fringes weakly chequered brown and white. Forewing discal spot small, elongated. Hindwing discal spot large, diffuse. Wings below with large, brown discal spots. Forewing apex and hindwing heavily irrorated with violet-brown. Frons, collar and thorax sulphur-yellow. Abdomen concolorous with wings. Male and female antennae bipectinate, female pectinations shorter. Hindleg of both sexes with 2 apical spurs only. Tympanal organs large. Abdominal sternites and tergites of both sexes undifferentiated.

VARIATION: Ground colour varies from yellow to deep yellow. Yellow specimens are normally less irrorated with violet-brown scales and deep yellow specimens are normally more heavily irrorated with violet-brown scales. Terminal area is often concolorous with rest of wing, but forewing apex and margins are often brownish. Darker females often have hindwing postmedial line wide, suffused, violet-brown. This is particularly strong on underside. Medial lines can be close to each other, touch each other, or be almost absent. Thorax can be concolorous with wings. Numerous forms have been described, see LERAUT (2009) for summary.

MALE GENITALIA: Uncus long, stout. Socii minute, setose. Gnathos arms sclerotised, fused ventrally, forming wide plate with stout setae. Valva wide, with patch of long, medially pointed setae subapically, apex roundish. Costa sclerotised, wide, basal extensions almost fused dorsally. Ventral margin roundish. Furca arms short, wide, apex round. Furca arms with rather wide, anterior extensions. Saccus weakly concave, membranous. Aedeagus short, stout, caecum minute, apex round. Vesica opens at approximately 90 degree angle. Vesica weakly striated at base, with row of stout cornuti. Apex narrow tube. See remarks.

FEMALE GENITALIA: Papillae anales large, surface weakly granulate. Apophyses posteriores long, straight. Apophyses anteriores short, stout. Lamella postvaginalis striated horizontally. Lamella antevaginalis sclerotised, with rounded lateral extensions, margin weakly concave. Ductus bursae sclerotised, very wide, short, weakly striated. Corpus bursae membranous, elongated. Signum minute, margins weakly dentate.

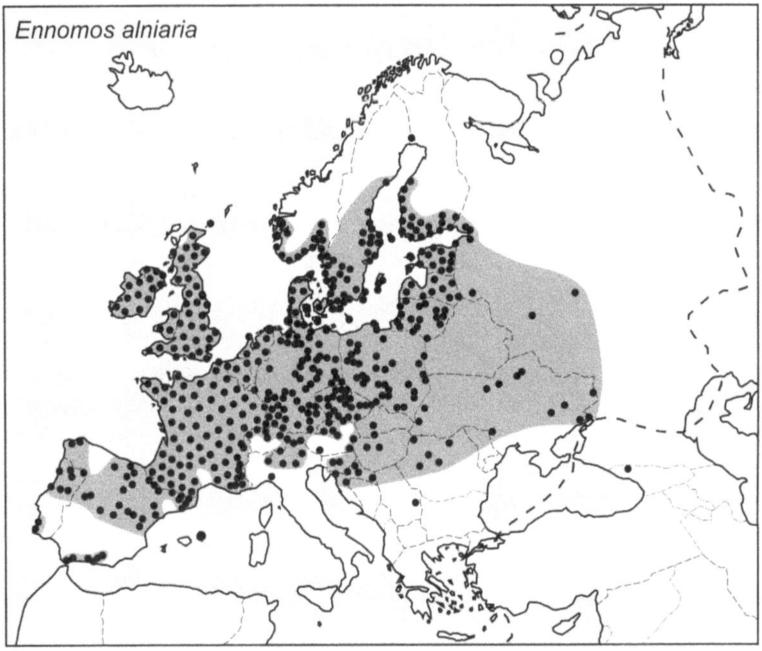

DISTRIBUTION AND ABUNDANCE: Euro-Caucasian. In Europe from central Fennoscandia to the Iberian Peninsula, southern France, northern Italy and the northern part of the Balkans and from the British Isles through central Europe to Ukraine and western European Russia. – Outside Europe the species is only known from Georgia. Further recorded from British Columbia, western North America, accidentally introduced from Europe (COVELL *et al.* 1986). – In Europe a common species that now and then can be found in numbers.

PHENOLOGY: Univoltine. In northern Europe the species flies from early August to early October, in southern central Europe from mid-July to early October (EBERT 2003) and in the Iberian Peninsula from June to November (REDONDO *et al.* 2009). Near Tarifa in southernmost Spain the species has been found as early as mid-May (coll. SKOU). Larvae from late May to July. The hibernation takes place as an egg. The moths come frequently to light but are rarely seen by day.

BIOLOGY: Polyphagous. The caterpillar feeds on *Alnus*, *Salix*, *Betula* and *Vaccinium* (HULDÉN *et al.* 2000). According to FLAMIGNI *et al.* (2007) it is recorded additionally on *Carpinus*, *Corylus*, *Quercus*, *Ulmus*, *Pyrus*, *Malus* and *Prunus* and according to AIZPÚRUA (2002) on *Alnus glutinosa*, *Populus* and *Tilia*. Reared on *Betula pendula* (B. MÜLLER pers. comm.). The caterpillar pupates between spun leaves on the food plant.

HABITAT: Sylvicolous, hygrophilous. In deciduous forests and woodland, on moors and in gardens. The species occurs from sea level and up to 900 m, in the Iberian Peninsula higher and often over 1000 m, in the Pyrenees up to 2000 m (REDONDO *et al.* 2009).

SIMILAR SPECIES: Other *Ennomos* species, see Text-figs 84–89. Male and female genitalia allow reliable identification, see the genitalia plates. Characteristically *E. alniaria* has thorax sulphur-yellow (different colour from wings) and hindwing discal spot is normally large.

GENETIC DATA: Genetically homogeneous (n=19, from seven European countries), but populations from Spain diverging by 2.7%. Nearest species: *E. quercaria* (8.2%).

REMARKS: The Linnean Society of London has in its collections six specimens under the label *Phalaena* (*Geometra*) *alniaria* LINNAEUS, 1758. There are three males (one without abdomen) and one female that appear as *E. alniaria* (LINNAEUS, 1758), and two females that appear as *E. quercinaria* (HUFNAGEL, 1767). *E. quercinaria* specimens have been clearly labelled (det. by D. S. FLETCHER, 1980), so the identity of *E. alniaria* is not in doubt. We do not designate a lectotype for *E. alniaria* because the manuscript is under preparation by MARTIN HONEY (M. HONEY pers. comm. 2011).

Due to the genetic divergence of Spanish specimens from other European specimens, we illustrate the genitalia of a Spanish male also, which was taken in May (Prov. Cádiz, 10 km NE of Tarifa, 300 m, 17.-18.v.2006, PEDER SKOU leg.). The apex of valva is narrower, otherwise the genitalia appear similar when compared to autumn specimens elsewhere in Europe. We have not had a Spanish female from spring available for study.

79. Ennomos fuscantaria (HAWORTH, 1809)

fuscantaria fuscantaria HAWORTH, 1809, Lepid. Br. (2): 295, (*Geometra*). Syntype(s), [United Kingdom].

fuscantaria algeriensis PROUT, 1929, Novit. zool. 35: 148, (*Deuteronomos*). Holotype male (The Natural History Museum, London, UK), Algeria: North side of Mt Zaccar (Miliano), Sebdou, Oran (examined externally).

INTRASUBSPECIFIC NAMES, POTENTIALLY UNAVAILABLE: *approximata* LUCAS, 1959 (f.); *aurantiaca* LEMPKE, 1951 (f.); *crassilineata* LEMPKE, 1970 (f.); *destrigaria* GALVAGNI, 1902 (ab.); *effuscaria* REBEL, 1910 (ab.); *glabra* WIZE, 1934 (ab.?); *juncta* WIZE, 1917 (ab.?); *lata* LEMPKE, 1970 (f.); *pallida* LEMPKE, 1951 (f.); *pallidior* BURRAU, 1950 (ab.?); *perfuscata* REBEL, 1910 (ab.); *romanyki* AGENJO, 1964 (f.); *triangularis* LEMPKE, 1970 (f.).

EXTERNAL CHARACTERS AND ABDOMEN: Wingspan ♂ 33–38 mm, ♀ 33–42 mm. Wings deep yellow, terminal area violet-brown, thorax concolorous with wings. Medial and postmedial lines violet-brown, narrow. Postmedial line slightly curved inwards before costa, also curved inwards on inner margin. Hindwing postmedial line narrow. Terminal area more or less uniform violet-brown, particularly near apex. Forewing margin concave below apex, distinctly curved inwards below M3. Wing margins undulating, most extended on M3 on both wings. Terminal line concolorous with wings. Fringes weakly chequered brown and white. Fore and hindwing discal spots small. Wings below with larger discal spots. Forewing apex and hindwing rather uniform violet-brown. Frons, collar, thorax and abdomen concolorous with wings. Male and female antennae bipectinate, female pectinations shorter. Hindleg of both sexes with 2 apical spurs only. Tympanal organs large. Abdominal sternites and tergites of both sexes undifferentiated.

VARIATION: Ground colour varies from pale yellow to deep yellow. Irroration often scarce, suffused. Terminal area is normally rather uniform violet-brown, particularly near apex, but violet-brown colour can be restricted to apex and wing margins only. Medial lines normally close together on inner margin, but can be wide apart, joined together or absent.

Subsp. ***algeriensis*** is paler, terminal area is only weakly darker than the rest of the wing and medial lines are widely separated and parallel on inner margin. Known from Algeria (PROUT 1929) and Morocco (RUNGS 1981), but potentially also elsewhere in North Africa.

MALE GENITALIA: Uncus wide, weakly setose, subapex with horizontal ridge, apex margins concave. Socii minute, setose. Gnathos arms narrow, fused ventrally, forming roundish, dentate plate. Valva very wide, apical part with upturned setae, apex round. Costa wide, weakly sclerotised. Ventral margin roundish. Furca arms short, wide, apex round. Saccus roundish. Aedeagus short, stout, narrow in middle, caecum minute, apex diffuse. Vesica opens at approximately 180 degree angle. Vesica weakly sclerotised at base, granulate, with 3–7 cornuti.

FEMALE GENITALIA: Papillae anales large, surface weakly granulate. Apophyses posteriores long, narrow. Apophyses anteriores short, stout. Lamella postvaginalis striated horizontally. Lamella antevaginalis with two sclerotised, separate lateral ridges, medial plate with convex margin. Ductus bursae partly sclerotised, wide, slightly twisted. Posterior part of corpus bursae narrow, striated. Anterior part of corpus bursae small, roundish. Signum very small, irregularly stellate, or absent.

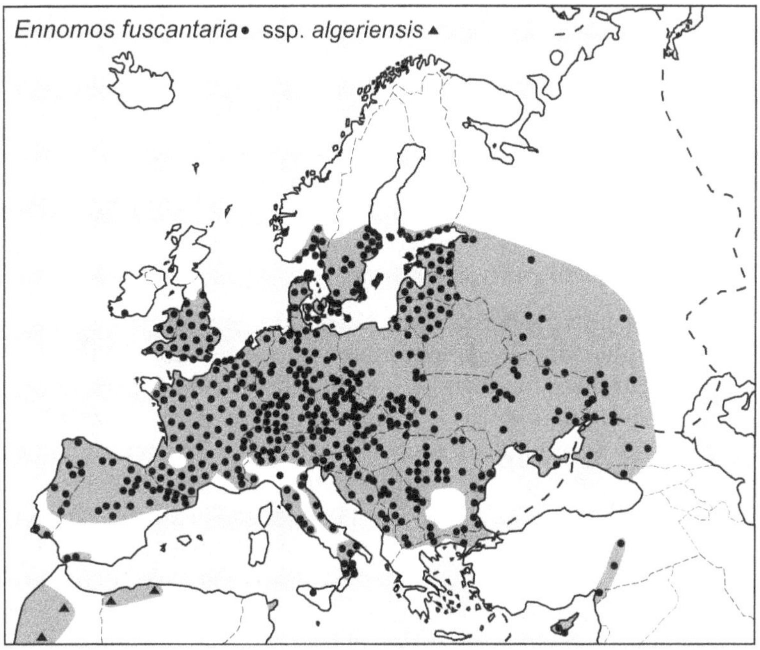

DISTRIBUTION AND ABUNDANCE: West-Palaearctic. In Europe from southern Fennoscandia to the Iberian Peninsula, southern France, Italy including Sicily and northern Greece and from Great Britain through central and eastern Europe to the central parts of European Russia. – Outside Europe the species is known from southern Turkey (first records from Turkey: Amanus [male slide PS1623, coll. ZFMK] and Malatia [female slide PS1624, coll. ZFMK]) and Cyprus (first records from Cyprus: Limasol [male slide PS1619, coll. ZFMK]) and Troodos Mountains, three specimens (E. FRIEDRICH pers. comm.)) and from Morocco, Algeria and Tunisia (subsp. *algeriensis*). – In Europe a rather common species that rarely occurs in numbers. In Britain it has in recent decades decreased by 98% of its previous abundance and it is now scarce (FOX *et al.* 2013) although this data is contrary to the experience of some British lepidopterists (M. CORLEY pers. comm.).

PHENOLOGY: Univoltine. In northern Europe from early August to mid-October, in southern Germany from mid-July to early October (EBERT 2003) and in southern Europe from June to October (FLAMIGNI *et al.* 2007). Caterpillar from spring until May, June or July depending on altitude and latitude. The species overwinters as an egg. The moths fly at night and frequently come to light.

BIOLOGY: Oligophagous. The larva has been found on the leaves of *Fraxinus excelsior* (HULDÉN *et al.* 2000) and is also recorded on *Ligustrum vulgare* (MALKIEWICZ 2012). The pupation takes place between spun leaves of the food plant.

HABITAT: In deciduous forests, in woods- and scrubland, parks and gardens. From sea level up to about 900 m in central Europe, in southern Europe up to 1400 m altitude.

SIMILAR SPECIES: Other *Ennomos* species, see Text-figs 84–89. Male and female genitalia allow reliable identification, see the genitalia plates. Characteristically *E. fuscantaria* has the terminal area normally uniform dark with a tint of violet, hindwing discal spot is small or absent and occasionally medial and postmedial lines approach each other on forewing inner margin.

GENETIC DATA: Genetically homogeneous in Europe (n=17, from six European countries), populations from Turkey and Iran slightly diverging. Nearest species: *E. quercaria* (5.4%).

Ennomos effractaria FREYER, 1841, status uncertain

effractaria FREYER, 1841, Neuere Beitr. Schmettkde 4 (59): 112, pl. 353, fig. 1, (*Ennomos*). Syntype(s), [Russia]: Sarepta [Krasnoarmeysk].

MIRONOV *et al.* (2008) reports the species from Volga-Don, Volgograd and Rostov regions in Russia. VIIDALEPP (1996) reports *E. effractaria* from southern European Russia and Turkey. We do not illustrate the taxon because its identity, and therefore distribution, is unclear. We have not managed to locate the type specimen, described from Sarepta (=Volgograd, Russia), and the original description accompanied with hand drawn picture (plate 353, fig. 1) is too superficial to unambiguously reveal its identity (FREYER 1841). FREYER considered it similar to *E. tiliaria* (BORKHAUSEN, 1794) (=junior synonym of *E. alniaria* (LINNAEUS)). WEHRLI illustrated male genitalia (1934) from a specimen from southern Turkey (Amanus, Düldül mountains) and a hand drawn picture of *E. effractaria* male (WEHRLI 1939–1954), which he considered conspecific with *E. effractaria*, and concluded that *E. effractaria* is similar to *E. fuscantaria* (HAWORTH), but different enough to be considered valid at species level. According to WEHRLI (1934), the gnathos is many times wider and uncus is higher in *fuscantaria* than in *effractaria*. We have examined WEHRLI's specimens and genitalia slides (in Zoologische Forschungsinstitut und Museum Alexander Koenig, Bonn, Germany), and conclude that the material he considered *E. effractaria*, we consider *E. fuscantaria*. The mentioned genitalia differences are quantitative and appear to fall within the variation of *E. fuscantaria*. Further, the male specimen illustrated in WEHRLI (1939–1954) is externally dissimilar to FREYER's (1841) original illustration. We have examined numerous specimens from Russia (including their genitalia), externally similar to FREYER's type with medial and postmedial lines fused together at the inner margin, but all have turned out to be *E. fuscantaria*. The identity of *E. effractaria* remains unclear, being potentially a junior synonym of *E. fuscantaria* (HAWORTH).

80. Ennomos erosaria (DENIS & SCHIFFERMÜLLER, 1775)

erosaria erosaria [DENIS & SCHIFFERMÜLLER], 1775, Ankündung syst. Werkes Schmett. Wienergegend: 103, (*Geometra*). Syntype(s), lost, [Austria]: Vienna district.

crassaria FABRICIUS, 1794, Ent. Syst. 3 (2): 135, (*Phalaena*). Syntype(s), Germania.

ochraria STEPHENS, 1831, Illust. Br. Ent. (Haustellata) 3: 165, (*Geometra*). Syntypes, [England:] Darenth.

quercinaria BORKHAUSEN, 1794, Natur. eur. Schmett. 5: 94, (*Phalaena Geometra*). Syntype(s), Europe.

tiliaria HÜBNER, 1790, Beitr. Gesch. Schmett. 2 (4): 99, pl. (4) 4, fig. 10, (*Ph[alaena] Geom[etra]*). Syntypes male, female, lost [Europe].

unicoloria ESPER, [1801], Die Schmett. 5 (4): 101, pl. 19, fig. 4, (*Ph[alena] Geom[etra]*). Syntype(s), [Europe].

erosaria ifranaria RUNGS, 1941, Bull. Soc. Sci. nat. phys. Maroc 20: 102, pl. 4, fig. 3, (*Ennomos (Deuteronomos)*). Holotype male, Morocco: Middle Atlas, Ifrane, 1650 m (original description examined, depository of the type specimen unknown). [Originally as sp]. New combination, transferred here from a subspecies of *Ennomos quercaria* (HÜBNER, 1813) (SCOBLE 1999, LERAUT 2009) to a subspecies of *erosaria* (DENIS & SCHIFFERMÜLLER, 1775).

erosaria pseudoquercaria WEHRLI, 1940, in SEITZ, Gross-Schmett. Erde 4 (Suppl.): 325, pl. 24: i, (*Ennomos*). Syntypes 12 males, 16 females (Zoologische Forschungsinstitut und Museum Alexander Koenig, Bonn, Germany), Algeria: Bone (dept. Constantine) (examined externally).

UNAVAILABLE NAMES (incorrect subsequent spellings): *angularia* (LERAUT 2009); *ifrinaria* RUNGS, 1941 (numerous authors, for instance The Global Lepidoptera Names Index (ANONYMOUS 2013). – Intrasubspecific names, potentially unavailable: *albescens* LEMPKE, 1970 (f.); *angulifera* DEMAISON, 1918 (ab.); *approximata* LEMPKE, 1951 (f.); *aurantiaca* LEMPKE, 1951 (f.); *citrina* COCKAYNE, 1952 (ab.); *clausa* LEMPKE, 1951 (f.); *fuscata* LEMPKE, 1970 (f.); *illineata* LEMPKE, 1970 (f.); *maculosa* LEMPKE, 1970 (f.); *nigrolineata* LEMPKE, 1970 (f.); *obsoleta* LEMPKE, 1951 (f.); *rufescens* LEMPKE, 1970 (f.); *saturata* SCHWINGENSCHUSS, 1953 (ab.); *semellineata* MUNSTERS, 1968 (ab.?); *tangens* LEMPKE, 1951 (f.); *triangularis* LEMPKE, 1970 (f.).

EXTERNAL CHARACTERS AND ABDOMEN: Wingspan ♂ 30–40 mm, ♀ 36–42 mm. Externally as *E. fuscantaria*, but with the following differences: Wings paler, ochreous-yellow, rather uniform in colour, terminal area not darker than the rest of the wing, only margins weakly darker. Medial lines brownish. Postmedial line not angled inwards on inner margin.

VARIATION: Ground colour often uniform pale yellow, but it can be deep brownish yellow. Terminal area often not shaded, except margins. In Denmark and in the British Isles, and potentially also elsewhere, specimens are smaller and paler from July until late August (peak in late July) and larger and darker from late August until early October (peak in early September). The phenogram from Germany (EBERT 2003) shows fewer specimens in mid-August. According to SKINNER & WILSON (2009), it is possible that two ecologically separate traits are involved. We have examined genitalia of July and September specimens from Denmark, and those are similar, without notable differences.

Subsp. ***ifranaria*** is paler, with slightly greyish tone. Male and female genitalia of *ifranaria* are slightly less sclerotised and female signum is more irregular in shape than examined European specimens (Text-figs 90–91). Known from Morocco only. See Remarks.

Subsp. ***pseudoquercaria*** is, according to the original description (WEHRLI 1940 (1939–1954)), smaller, grey, and the medial lines are similar to *E. quercaria*. WEHRLI examined the male genitalia and concluded that *pseudoquercaria* is conspecific with *E. erosaria*: processes of juxta (=furca arms) are wider and closer to each other than in *quercaria*. Known from Algeria only (WEHRLI 1939–1954). See Remarks.

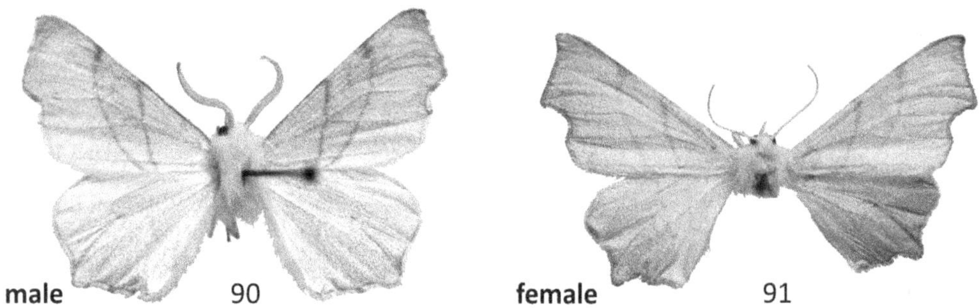

Text-figs 90–91. Male and female of *Ennomos erosaria* subsp. *ifranaria* (RUNGS, 1941). Text-fig. 90. Male, Morocco: Ifrane, 12.vii.1964, genitalia shown in genitalia plate, slide PS1725 (coll. ZSM). Text-fig. 91. Female, Morocco: Ifrane, 6.vii.1952, genitalia shown in genitalia plate, slide PS1726 (coll. ZSM). External appearance and genitalia morphology indicate that taxon *ifranaria* RUNGS, 1941 is conspecific with *E. erosaria* (DENIS & SCHIFFERMÜLLER, 1775) (new combination).

MALE GENITALIA: Generally as in *E. quercaria* (HÜBNER), see entry below, but with following differences: In *E. erosaria* furca arms are wide, weakly curved outwards and their bases are close to one another (furca arms narrow, straight or weakly curved inwards and bases far apart in *E. quercaria*). Vesicas are very similar, but apparently in *E. erosaria* base of vesica is shorter (longer in *E. quercaria*).

FEMALE GENITALIA: Generally as in *E. quercaria* (HÜBNER), see entry below, but with following differences: In *E. erosaria* lamella antevaginalis has sclerotised lateral arms and base of ductus seminalis has small lobe (lamella antevaginalis without sclerotised lateral arms, base of ductus seminalis without small lobe in *E. quercaria*).

DISTRIBUTION AND ABUNDANCE: Euro-Caucasian. In Europe from southern Fennoscandia to the Iberian Peninsula, southern France, northern Italy and the Balkans to northern Greece and from Great Britain through central and eastern Europe to European part of Russia and the Ural mountains. – Outside Europe the species is known from Morocco (subsp. *ifranaria*), Algeria (subsp. *pseudoquercaria*), Turkey and the Caucasus region. – In parts of Europe probably still a relatively common species that only rather rarely occurs in numbers. In other parts of Europe such as Great Britain (FOX *et al.* 2013) and Denmark, the species has decreased much in the recent 30–35 years.

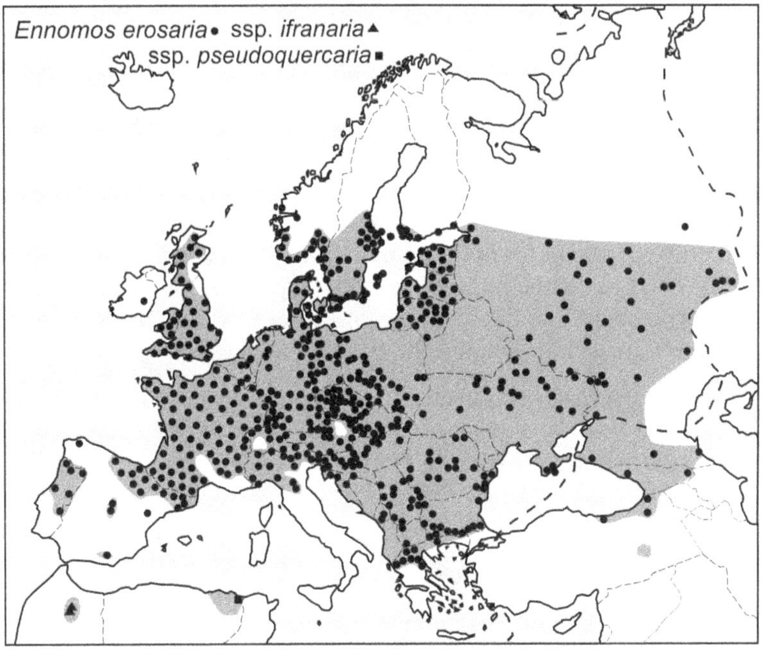

PHENOLOGY: Univoltine. In northern Europe the moths fly from mid-July to early October, in southern central Europe from late June to early October (EBERT 2003), while in southern Europe it flies from May to September or October. According to FLAMIGNI *et al.* (2007) the species flies in Italy from May to October, in some localities in two generations. Larvae from early spring until April, May or June, depending on latitude. The hibernation takes place as an egg. The moths fly at night and are easily attracted to light, but they are rarely seen by day.

BIOLOGY: Polyphagous. The caterpillar feeds especially on *Quercus*, *Betula pendula* and *Salix caprea* (EBERT 2003). According to FLAMIGNI *et al.* (2007) recorded also on *Populus*, *Carpinus*, *Castanea*, *Ulmus* and *Pyrus*. In northern Germany larvae have been found on *Quercus robur* and *Q. petraea* (H. WEGNER pers. comm.). BERGMANN (1955) regards *Quercus* as the main food plant.

HABITAT: Sylvicolous. In deciduous forests and woodland, parks and gardens. The species occurs from sea level up to 800 m in southern central Europe (EBERT 2003), in southern Europe up to 1500 m, exceptionally higher.

SIMILAR SPECIES: Other *Ennomos* species, see Text-figs 84–89. Characteristically *E. erosaria* has terminal area normally concolorous with rest of wing and hindwing discal spot is small or absent. Male and female genitalia are similar to *E. quercaria*, but allow reliable identification, see male and female genitalia entries above. FLAMIGNI & BASTIA (2003, reprinted in FLAMIGNI *et al.* 2007) illustrate male genitalia of *E. erosaria* and *E. quercaria*. In their illustration *E. erosaria* has valva apex narrow, pointed, and directed clearly outwards. The material we have seen does not support this quantitative difference, both *E. erosaria* and *E. quercaria* have similar valva apex as illustrated in the genitalia plates.

GENETIC DATA: Genetically identical all over Europe (n=25, from six European countries). Nearest species: an Iranian *Ennomos* species (2.8%) (identity between *E. fraxineti* WILTSHIRE, 1947 and *E. zandi* WILTSHIRE, 1947, and potentially other taxa, needs to examined), in Europe *E. quercaria* (3.2%).

REMARKS: External appearance and genitalia morphology indicate that *E. quercaria* subsp. *ifranaria* (RUNGS) from Morocco, Atlas Mountains, is conspecific with *E. erosaria* (DENIS & SCHIFFERMÜLLER), we therefore transfer it from subspecies of *E. quercaria* (HÜBNER) (SCOBLE 1999, LERAUT 2009) to subspecies of *E. erosaria*. We have examined the original description (RUNGS 1941) and authentic material from the type locality, Ifrane in Middle Atlas in Marocco, but have not examined the holotype *in situ* because current location of the RUNGS collection is unknown to us. See Text-figs 90–91, male and female genitalia plates.

All specimens from Morocco examined by us, including the genitalia, have turned out to be *E. erosaria*, even if the external appearance indicated them to be *E. quercaria*. Subspecies status of *ifranaria* and *pseudoquercaria* in North Africa needs further research, potentially these are synonymous.

81. Ennomos quercaria (HÜBNER, 1813)

quercaria quercaria HÜBNER, [1813], Samml. Eur. Schmett. 5 Geometrae (1): pl. 80, figs 411, 412, (*Geometra*). Syntype(s), lost, Europe.

dryadaria RAMBUR, 1866, Cat. syst. Lépid. Andalousie 2: pl. 17, fig. 1, (*Ennomos*). Holotype male (C. HERBULOT collection, in Zoologische Staatsammlung, Munich, Germany), [Spain]: Grenada (examined externally). Originally as sp., downgraded here from subspecies rank (treated as *quercaria dryadaria* in SCOBLE 1999) (new synonym), quoting absence of differential features, falling within intraspecific variation and barcode-sharing with *Ennomos quercaria* (new synonym). REDONDO *et al.* (2009) also noted that validity of *dryadaria* at subspecies rank is to be confirmed.

quercaria duercki REISSER, 1958, Z. wien. ent. Ges. 43 (7): 114, pl. 3, figs 3, 4; text-fig. 4, (*Ennomos*). Holotype male (Staatliches Museum für Naturkunde Karlsruhe, Germany), Crete: Silva Rouwa, Mons Ida, 1300 m (examined, including genitalia). Originally as sp., downgraded here from species rank (SCOBLE 1999) to a subspecies of *Ennomos quercaria* (HÜBNER, 1813), quoting allopatric occurrence (new status). See remarks.

quercaria freidbergi HAUSMANN, 1997, Entomofauna 18 (1): 12, figs. 2, 6, 7 (*Ennomos*). Holotype male (Collection of the Tel Aviv University, Israel), Northern Israel: Tivon (examined, including genitalia, see also genetic data). Originally as sp., downgraded here from species rank (HAUSMANN 1997, SCOBLE 1999) to a subspecies of *Ennomos quercaria* (HÜBNER, 1813), quoting allopatric occurrence (new status). See remarks.

quercaria lissochila PROUT, 1929, Novit. zool. 35: 149, (*Deuteronomos*). Holotype male (The Natural History Museum, London, UK), Cyprus: Platres, 4000 ft (examined externally). Originally as sp., downgraded here from species rank (SCOBLE 1999) to a subspecies of *Ennomos quercaria* (HÜBNER, 1813), quoting allopatric occurrence (new status). See remarks.

quercaria olivaria BRANDT, 1938, Ent. Rdsch. 55 (51): 597, figs 254–258, (*Ennomos*). Syntypes (Naturhistoriska Riksmuseet, Stockholm, Sweden), Iran: Fort Mian-Kotal; Sine-Sefid. Paratype male in the Natural History Museum, London, UK (examined externally). Originally as subsp., valid at subspecies rank according to SCOBLE (1999).

UNAVAILABLE NAMES (incorrect subsequent spelling): *lissochillus* (FISCHER & LEWANDOWSKI 2005). – Intrasubspecific name, potentially unavailable: *destrigaria* SCHWINGENSCHUSS, 1954 (f.).

EXTERNAL CHARACTERS AND ABDOMEN: Wingspan 28–35 mm. Small species, wings pale whitish-yellow to greenish with tint of olive-green. Wings mostly pale near base, olive tint strongest near margins. Medial and postmedial lines weak, brownish, evenly curved, expanded on costa. Anterior margin of medial line and posterior margin of postmedial line whitish. Hindwings without markings or postmedial line weak. Forewing margin concave below apex. Wing margins undulating, most extended on M3 on both wings. Terminal line concolorous with wings. Fringes concolorous with wings, white or weakly chequered brown and white. Discal spots small or absent. Wings below less marked, uniform pale. Frons, collar, thorax and abdomen concolorous with wings. Male and female antennae bipectinate, female pectinations shorter. Hindleg of both sexes with 2 apical spurs only. Tympanal organs medium-sized. Abdominal sternites and tergites of both sexes undifferentiated.

VARIATION: Ground colour pale whitish yellow to greenish in most specimens, but it can be brown or darker olive-green. Typically specimens are not irrorated, but particularly brownish specimens may be weakly irrorated with grey-brown. Medial area can be slightly darker than rest of wing, particularly on costa. Wings below normally pale greenish, almost without markings, but more contrasting specimens also have undersides well-marked.

Subsp. ***duercki*** has wings slightly darker, forewing margin concavity below vein M3 is slightly smaller than in nominate subspecies. Male and female genitalia are similar to *E. quercaria* subsp. *quercaria*. REISSER (1958) diagnosed *E. duercki* to have the dorsal margin of valva narrower and longer than in *E. quercaria*. This quantitative character is variable and appearance depends on how the coverglass has been placed on the genitalia. The caterpillar feeds on leaves of *Acer cretica*, *Quercus ilex* and other *Quercus* species, in captivity also on *Acer monspessulanum* (M. LEIPNITZ pers. comm.). The occurrence of *duercki* in southwestern Turkey confirmed by DNA barcodes.

Subsp. ***freidbergi*** is larger than nominate subspecies, more irrorated with dark scales, and length of longest branches in the male antennae is about 1.2 mm (1.1 mm in nominate subspecies). Male genitalia are larger, furca arms terminally slightly curved outwards and longer, and cornuti longer than in nominate subspecies. Female genitalia are larger than in nominate subspecies. The larva has been recorded on *Quercus ithaburensis* (BYTINSKI-SALZ & STERNLICHT 1967 [as *E. quercarius*])

Subsp. ***lissochila*** has wings darker, greenish yellow, heavily irrorated with greenish, and forewing margin concavity below vein M3 is smaller than in nominate subspecies. Fringes chequered greenish yellow. Legs, abdomen and wing base underneath have an elegant tint of violet or dark lavender colour. Ground colour somewhat variable, ranging from yellowish green to dark olive-green. Male and female genitalia are similar to those of *E. quercaria* subsp. *quercaria*. Male genitalia are illustrated also in FISCHER & LEWANDOWSKI (2005) as *E. lissochillus* (PROUT).

MALE GENITALIA: Uncus stout, weakly curved ventrally, weakly setose. Socii minute, setose. Gnathos arms wide, weakly fused ventrally, with wide patch of stout setae. Valva narrow, directed upwards, most setose medially, apex round. Costa wide, sclerotised, roundish plate, ventral margin with row of stout setae. Furca arms narrow, straight, or weakly curved inwards, bases far apart. Saccus concave. Aedeagus slightly curved. Vesica opens at approximately 135 degree angle. Vesica narrow, base long, with elongated diverticulum. Basal part with row of cornuti, diverticulum with few long cornuti.

FEMALE GENITALIA: Papillae anales large, surface weakly granulate. Apophyses posteriores long, narrow. Apophyses anteriores short, stout. Lamella postvaginalis striated horizontally. Lamella antevaginalis without sclerotised lateral arms, medial margin approximately straight. Ductus bursae short, most sclerotised laterally. Base of ductus seminalis without small lobe. Posterior part of corpus bursae narrow, weakly curved, striated. Anterior part of corpus bursae membranous, roundish. Signum small, margins roundish.

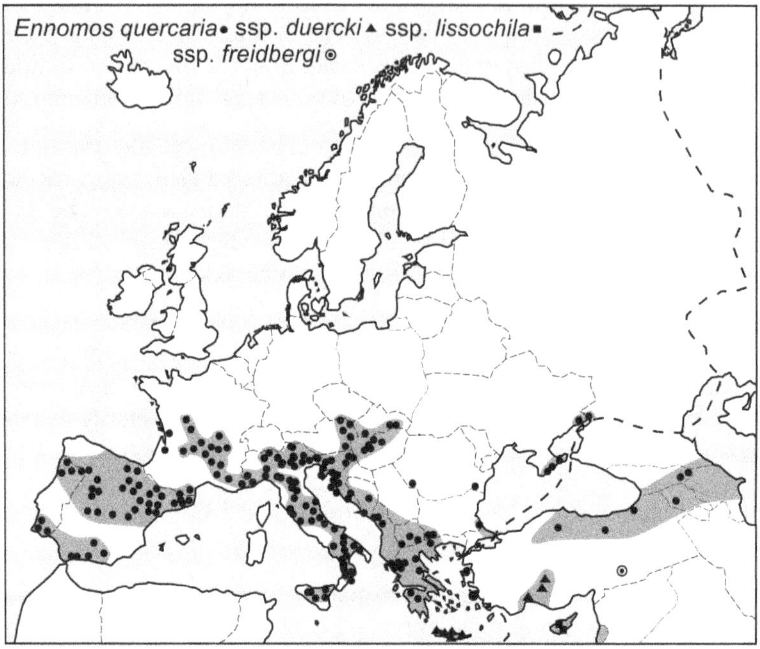

DISTRIBUTION AND ABUNDANCE: West Palaearctic. In Europe only in the Mediterranean region and south-eastern central Europe, from the Iberian Peninsula in the west to the Balkans, Crimea and eastern Ukraine in the east. Also in Crete (subsp. *duercki*). – Outside Europe in northern Turkey and the Caucasus region, in Cyprus (subsp. *lissochila*), northern Iran (subsp. *olivaria*), southern Turkey and northern Israel (subsp. *freidbergi*), probably the records in Jordan (HAUSMANN 1991) and in Lebanon (ELLISON & WILTSHIRE 1939) are also conspecific to subsp. *freidbergi* (see HAUSMANN 1997). – In southern Europe a common or rather common species that is not often found in numbers. See remarks.

PHENOLOGY: Bivoltine. From late May to early October in two more or less overlapping generations. In southern Italy from mid-April to mid-October (A. HAUSMANN pers. comm.). Larva from early May to early or mid-June. The hibernation takes place as an egg (M. LEIPNITZ pers. comm.) The moths fly at night and come readily to light.

BIOLOGY: Oligophagous, preferring *Quercus*. On Samos the larva is found on *Quercus coccifera* (D. FRITSCH pers. comm.). On Crete (subsp. *duercki*) found mainly on *Acer cretica*, but also on *Q. ilex* and in rearing the caterpillars have accepted *Acer monspessulanum* as food plant (M. LEIPNITZ pers. comm.), from Crete also mentioned on *Quercus ilex* (E. FRIEDRICH (pers. comm.). In rearings, the caterpillars have also accepted *Quercus robur* as food plant (B. MÜLLER pers. comm.) and *Q. petraea* (E. FRIEDRICH pers. comm.). The larva is recorded on *Quercus ilex*, *Q. faginea* and *Q. pyrenaica* (REDONDO *et al.* 2009), *Q. robur* (FLAMIGNI *et al.* 2007).

HABITAT: In *Quercus* forests and places with scattered growing *Quercus* trees and bushes. The species occurs from sea level up to 1650 m in central Spain.

SIMILAR SPECIES: Other *Ennomos* species, see Text-figs 84–89, particularly *E. erosaria*. Characteristically *E. quercaria* is small, wings are pale whitish yellow with tint of olive-green medial lines are weak, strongest on costa (slightly larger, wings ochreous-yellow, medial lines well-developed in *E. erosaria*). Male and female genitalia are similar to *E. erosaria*, but allow reliable identification. In *E. quercaria* furca arms are narrow, straight or weakly curved inwards and bases are far apart (furca arms wide, weakly curved outwards, bases close to one another in *E. erosaria*). Vesicas are very similar, but apparently in *E. quercaria* base of vesica is longer (shorter in *E. erosaria*). In *E. quercaria* lamella antevaginalis is without sclerotised lateral arms and base of ductus seminalis is without small lobe (lamella antevaginalis with sclerotised lateral arms, base of ductus seminalis with small lobe in *E. erosaria*).

GENETIC DATA: Genetically splitting up into three BINs: Nominotypical populations of *E. quercaria* genetically homogeneous but with constant divergence from subspecies *duercki* from Crete (and one specimen from Turkey: Antalya) by 1.8%. Subspecies *lissochila* at a distance of 1.3% from Italian populations, however, no barcodes of mainland Greek specimens included so far. *E. quercaria* subsp. *freidbergi* (HAUSMANN, 1997) from the eastern Mediterranean and *E. quercaria* subsp. *olivaria* (BRANDT, 1938) from Iran both barcode-sharing with *E. quercaria* subsp. *duercki*. Nearest species: *E. erosaria* (3.2%). See Remarks.

REMARKS: *E. quercaria* is a widespread species with several putative subspecies, requiring an extensive morphological and DNA based revision. We have not examined adequate material from Turkey and eastern Mediterranean, so the conclusions that are shown in the distribution map in those areas must be considered tentative. We treat *E. freidbergi* HAUSMANN, *E. duercki* REISSER and *E. lissochila* (PROUT) valid at subspecies level because their known populations are either limited to islands: *E. quercaria* subsp. *lissochila* is known from Cyprus, or the known populations are allopatric: *E. quercaria* subsp. *duercki* is known from Crete and southern Turkey and *E. freidbergi* is thus far recorded from Middle East only (HAUSMANN 1997). We have not examined the genitalia of the type specimen of *E. quercaria* subsp. *olivaria*, thus we retain it valid at subspecies level, despite the barcode sharing of specimens from Iran with *E. quercaria* subsp. *duercki*.

Eumera STAUDINGER, 1892

Eumera STAUDINGER, 1892, Dt. ent. Z. Iris 5: 162. Type species: *Eumera regina* STAUDINGER, 1892, Turkey. There are no genus synonyms.

DIVERSITY AND DISTRIBUTION: *Eumera* contains five species, those occur in the western Palaearctic region.

EXTERNAL CHARACTERS AND ABDOMEN: Rather large moths, wingspan about 35–40 mm. Wing coloration with different shades of orange and yellow, occasionally with light pink areas. Medial and postmedial lines distinct, rather straight. Terminal area with whitish spots in two species. Wing margins weakly undulating, particularly that of hindwings. Male and female antennae bipectinate, female pectinations shorter. Proboscis reduced. Male and female sternites and tergites 3–8 undifferentiated.

MALE GENITALIA: Uncus absent. Scaphium weakly sclerotised, bilobed, weakly setose. Socii minute, setose. Gnathos absent. Tegumen narrow. Medial part of valva membranous, weakly setose. Costa sclerotised, apex setose. Ventral margin of valva sclerotised, with hook. Juxta large, posterior margins elongated, setose. Saccus round. Vesica narrow tube, without sclerotisations. See remarks.

FEMALE GENITALIA: Papillae anales large. Apophyses anteriores shortened. Lamella antevaginalis narrow, weakly sclerotised ridge. Ductus bursae membranous, narrow. Corpus bursae membranous, thin, roundish. Signum weakly sclerotised, small. See remarks.

BIOLOGY: Larvae of *E. regina* have been found on *Acer monspessulanum* in April (H. WEGNER, pers. comm). In *ex ovo* rearing the larvae accepted *Prunus triloba* (B. MÜLLER pers. comm.), thus potentially being polyphagous. Adults are nocturnal, flying from August to early November between 600–2200 metres above sea level. Egg overwinters.

IMMATURE STAGES: No data available.

REMARKS: VIIDALEPP (1996) classified *Eumera* in the Colotoini, HAUSMANN *et al.* (2011) in the Ennomini. Externally *Eumera* are somewhat similar to *Colotois pennaria* (LINNAEUS), but the male genitalia of *Eumera* have potentially unique features. Homologies of the male genitalia are difficult to interpret, but it appears that the uncus is absent (not common in the examined Ennominae), functionally being replaced by weakly sclerotised scaphium (not common in the examined Ennominae). Socii are minute, setose (common in the examined Ennominae), gnathos is absent (not common) and the tegumen is narrow (not common), See remarks under *E. regina*. Female genitalia of *Eumera* show only general fetures of the subfamily, thus providing little additional information about its potential systematic position, except that apophyses anteriores are somewhat reduced, thus being more similar to *Ennomos* than to *Colotois* and posterior part of corpus bursae is membranous, thus being more similar to *Colotois* than to *Ennomos*. Systematic position: Uncertain.

82. Eumera regina STAUDINGER, 1892

regina STAUDINGER, 1892, Dt. ent. Z. Iris 5 (1): 162, pl. 2, fig. 8, (*Eumera*). Syntypes male, female, [Turkey]: Amasia, [Dalmatia; Croatia]: Zara.

UNAVAILABLE NAMES (incorrect subsequent authorship): *decoronata* STAUDINGER, 1913 (numerous authors, for instance The Global Lepidoptera Names Index (ANONYMOUS 2013). – Intrasubspecific names, potentially unavailable: *decoronata* STAUDER, 1913 (ab.).

EXTERNAL CHARACTERS AND ABDOMEN: Wingspan 36–40 mm. Forewing basal area, terminal area and costa yellowish pink, with tint of olive green, particularly near wing margins. Terminal area with two white spots, margins reddish brown. Medial area yellow. Medial and postmedial lines reddish brown, rather straight, distinct. Hindwings yellowish pink, more olive-green near wing margins, basal area paler. Medial and postmedial lines narrow, close to each other, strongest on anal margin. Terminal line and fringes reddish-brown. Forewing margin weakly concave below apex. Wing margins undulating, most extended on M3 on both wings. Discal spots reduced or absent. Wings below more uniform yellowish, irrorated with brown. Frons, collar, thorax and abdomen concolorous with wings. Proboscis reduced. Male and female antennae bipectinate, female pectinations shorter. Hindtibia of both sexes with 2+2 short spurs. Tympanal organs medium-sized. Abdominal sternites and tergites of both sexes undifferentiated.

VARIATION: Ground colour varies from pale yellow to deep yellow. Greyish olive-green irroration normally heaviest near wing margins. Contrast between yellow medial area and yellowish pink basal and terminal areas normally rather clear, but occasionally wings more yellow overall and less contrasting. White spots in terminal area can be reduced in size, perhaps even absent.

MALE GENITALIA: Uncus absent. Scaphium weakly sclerotised, small, bilobed, weakly setose. Socii minute, setose. Gnathos absent. Tegumen narrow. Valva costa sclerotised, small patch of setae basally, apex setose. Medial area of costa membranous, weakly setose. Ventral margin of valva sclerotised, with hook. Valva fused dorsally. Juxta large, posterior margins elongated, setose. Saccus large, round. Aedeagus narrow, caecum minute, apex blunt. Vesica opens at approximately 90 degree angle. Vesica evenly narrowing tube, without sclerotisations. See remarks.

FEMALE GENITALIA: Papillae anales large, lobes almost separate. Apophyses posteriores long, angled in near apical end. Apophyses anteriores short, stout. Lamella postvaginalis unrecognisable. Lamella antevaginalis narrow, membranous ridge with small lateral pockets. Ductus bursae membranous, long, narrow. Posterior end of corpus bursae, short, wrinkled. Anterior end of corpus bursae membranous, thin, roundish. Signum weakly sclerotised, small, roundish.

DISTRIBUTION AND ABUNDANCE: (South-east-)European-Westasiatic. In Europe only in the Balkans from Croatia to Greece. Reputedly collected in central Italy (two specimens leg. DANNEHL in coll ZSM), perhaps mislabelled, awaiting confirmation. – Outside Europe only known from Turkey and Israel. – In south-eastern Europe a rather common species, but it is not often found in numbers.

PHENOLOGY: Univoltine. From late August to mid-October, according to H. RIETZ (pers. comm.) also in late October. Larvae in April (H. WEGNER pers. comm.) and in Turkey at 1500 m altitude from mid-April to mid-June (M. LEIPNITZ pers. comm.). Hibernation takes place as an egg. The adult moths fly at night and come frequently to light.

BIOLOGY: Probably polyphagous. The caterpillar has been found on *Acer monspessulanum* (H. WEGNER pers. comm.). In Turkey on deciduous *Cotoneaster*, however, in rearing, the caterpillars did not accept evergreen *Cotoneaster* species (M. LEIPNITZ pers. comm.). In rearing, the caterpillars accepted *Prunus triloba* as food plant and they preferred to eat out the buds (B. MÜLLER pers. comm.). The eggs are laid in compact groups which resemble those of the praying mantid (M. LEIPNITZ pers. comm.). Pupation takes place in the ground (M. LEIPNITZ pers. comm.).

HABITAT: In open woodland with *Quercus* as the dominant tree species. The moth occurs from 600 up to 1700 m altitude, exceptionally up to 2200 m.

SIMILAR SPECIES: No similar species in Europe, but see remarks.

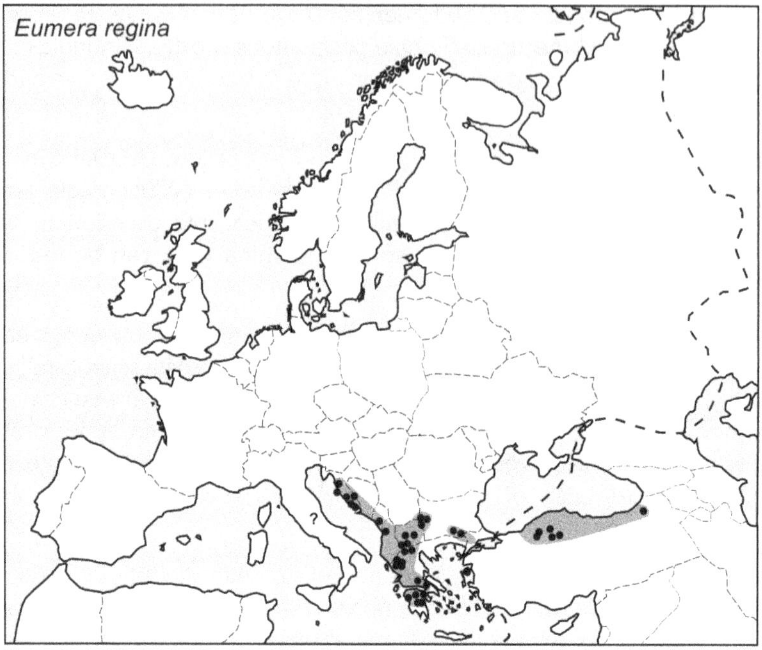

GENETIC DATA: European populations genetically close to populations from northern Turkey. *E. hoeferi* WEHRLI, 1934 from eastern Turkey diverging by 4.0%, *E. mulier* PROUT, 1929 from Cyprus by 5.7%. Nearest species: *E. turcosyrica* WEHRLI, 1934 from southern Turkey (1.6%) and *E. lewandowskii* FISCHER, 2008 from Jordan (1.7%), the validity at species rank of both these taxa may be questioned.

REMARKS: *E. turcosyrica* WEHRLI, 1932, originally described as a subspecies of *E. regina*, is externally similar to *E. regina*, only the white spots in terminal area are smaller. The (syn) type male specimen of *turcosyrica* (in Zoologische Forschungsinstitut und Museum Alexander Koenig, Bonn, Germany, from East Turkey: Marasch) and the European *regina* specimens have similar genitalia, without qualitative differences (D. STÜNING pers. comm.). Other examined *turcosyrica* specimens from the type locality have larger white spots. *E. turcosyrica* is known from the border area of Turkey and Syria and further south from the Levant area. Its status needs to be examined, see also genetic data. *E. mulier* PROUT, 1929 is endemic to Cyprus. Externally it is not similar, wings are almost uniform light rusty-brown. Postmedial line is dark rusty-brown and outer margin is whitish. Genitalia of *E. mulier* are very similar to *E. regina*: Hook on ventral margin of valva is more upturned, costa of valva is narrower and scaphium is wider in *E. mulier* (hook less upturned, costa wider and scaphium narrower in *E. regina*). Apophyses posteriores are straight, apophyses anteriores are short and narrow in *E. mulier* (apophyses posteriores angled, apophyses anteriores short, wide in *E. regina*). Otherwise female genitalia are without notable differences.

Selenia HÜBNER, 1823

Selenia HÜBNER, [1823] 1816, Verz. bekannter Schmett.: 292. Type species: *Geometra illunaria* HÜBNER, [1799] 1796, which is a junior subjective synonym of *Phalaena dentaria* (FABRICIUS, 1775), Europe. There are no genus synonyms.

UNAVAILABLE NAMES (included in work rejected for nomenclatural purposes): *Eutrapela* HÜBNER, 1806 (see FLETCHER 1979).

DIVERSITY AND DISTRIBUTION: According to SCOBLE (1999), *Selenia* contains 24 species in the Palaearctic, the Nearctic and in the Neotropical regions. PITKIN (2002) noted that the Neotropical taxa assigned to *Selenia* are unlikely even to belong in the same tribe as true *Selenia*.

EXTERNAL CHARACTERS AND ABDOMEN, description based on European and North American species: Rather large, wingspan about 30–40 mm. Wings different shades of reddish brown, yellowish and greyish. Postmedial line well developed. Forewing margin concave below apex, bordered with reddish brown lunular area, otherwise margins undulating. Discal spots often lunular. Male and female antennae bipectinate, female pectinations shorter. Dorsal part of frons with protruding hair, ventral part bare. Male 8th tergite weakly sclerotised, posterior margin narrower.

MALE GENITALIA, description based on European and North American species: Uncus variable, usually short, stout. Socii represented by setae only. Gnathos arms fused ventrally, upturned, weakly dentate. Valva wide, most setose subapically, apex round. Transtilla well-developed, with two ventral pockets. Furca large, symmetrical, bare. Vinculum with membranous sac. Saccus concave, rarely elongated. Vesica with numerous cornuti.

FEMALE GENITALIA, description based on European and North American species: Lamella postvaginalis weakly sclerotised or membranous. Lamella antevaginalis often wide, sclerotised plate. Ductus bursae narrow, sclerotised. Corpus bursae elongated or roundish. Signum small, margins weakly dentate or serrate.

BIOLOGY: Larvae of European and North American species are apparently polyphagous, commonly recorded food plants include deciduous trees such as *Betula, Alnus, Salix, Quercus, Fraxinus, Salix* and *Ulmus*, but also on other plants such as *Rosa, Calluna, Vaccinium myrtillus, V. uliginosum* (MCGUFFIN 1987, MIKKOLA *et al.* 1989). Usually *Selenia* species have two, in southern Europe occasionally even three, generations per summer, in late spring or early summer and in late summer. Late summer specimens are often different in appearance when compared to first generation specimens. According to MIKKOLA *et al.* (1989), the late summer specimens undergo one less larval stage the than the overwintering specimens, thus explaining the difference in external appearance. Hibernation is usually in the pupal stage. Adults are nocturnal.

IMMATURE STAGES: Eggs of North American *S. alciphearia* (WALKER, 1860) are rounded to elliptical; they are laid singly, but near one another (MCGUFFIN 1987). Larvae are brown or greenish brown twig-mimics with swollen structures on fifth and eighth, or seventh and eighth (PORTER 2010) or fourth and fifth abdominal segments (MCGUFFIN 1987). Pupa cremaster rounded, with four pairs of setae, apical pair strongest (MCGUFFIN 1987, PATOČKA & TURČANI 2005).

REMARKS: *Selenia* has traditionally been classified in the Ennomini (VIVES MORENO 1994, LERAUT 1997, VIIDALEPP 1997, HAUSMANN, MIRONOV & VIIDALEPP 2011), but was excluded by BELJAEV (2008a). *Selenia* are structurally dissimilar to *Ennomos* in many respects, but it includes also species (for instance *S. dentaria*) that have a potential Ennomini synapomorphy as defined by BELJAEV (2008a): presence of a membranous area on the ventral side of the vinculum. *Selenia* have membranous sac in the male genitalia, apparently originating from vinculum, similar to those in *Cepphis* HÜBNER and *Epione* DUPONCHEL. In a molecular, phylogenetic analysis of limited European taxa, ÕUNAP *et al.* (2011) concluded that *Cepphis*, *Selenia*, *Plagodis* and *Pseudopanthera* grouped together and these were not related to Ennomos. The New World *Metanema* GUENÉE, 1858, is structurally similar; North American species are illustrated in MCGUFFIN (1987). Systematic position: Ennomini of uncertain association. Further research is needed.

83. Selenia dentaria (FABRICIUS, 1775)

dentaria dentaria FABRICIUS, 1775, Syst. Ent.: 623, (*Phalaena*). Syntype(s), England.

achataea FOURCROY, 1785, Entomologia Parisiensis: 266, (*Phalaena*). Syntype(s), [Europe]. Synonymy after VIVES MORENO (1994).

alpestris WEHRLI, 1940, in SEITZ, Gross-Schmett. Erde 4 (Suppl.): 326, pl. 25: a, (*Selenia*). Syntypes male(s), female(s) (Zoologisches Forschungsinstitut und Museum Alexander Koenig, Bonn, Germany), [Switzerland]: Zermatt; Davos. [Paratype male in the Natural History Museum, London, UK (examined externally)]. [Originally as var. of *bilunaria*]. See remarks.

bilunaria ESPER, [1801], Die Schmett. 5 (3): 73, pl. 13, figs 1-10, (*Ph*[*alaena*] *Geom*[*etra*]). Syntype(s), Germany: Leipzig.

fulvolunaria ESPER, [1801], Die Schmett. 5 (3): 78, pl. 14, figs 9, 10, (*Ph*[*alaena*] *Geom*[*etra*]). Syntype(s), [Germany: Erlangen].

gagathes GOEZE, 1781, Ent. Beyträge 3: 63, (*Phalaena* (*Bombyx*)). Syntype(s), [Europe]. Synonymy after LERAUT (1997).

illunaria HÜBNER, [1799], Samml. Eur. Schmett. 5 Geometrae (1): pl. 7, figs 36, 37, (*Geometra*). Syntype(s), lost, Europe. [second generation specimens]

illunaria aestivalis GUENÉE, 1842, Annls. Soc. Ent. Fr. 11: 243, (*Ennomos*). Syntype(s), France: Chateaudun. [Originally as var].

juliaria HAWORTH, 1809, Lepid. Br. (2): 293, (*Geometra*). Syntype(s), [United Kingdom].

unilunaria ESPER, [1801], Die Schmett. 5 (3): 77, pl. 14, figs 6-8, (*Ph*[*alaena*] *Geom*[*etra*]). Syntype(s), Germany: Erlangen.

ustularia DONOVAN, 1794, The natural history of British insects 3: 25, pl. 82, (*Phalaena*). Syntype(s), [United Kingdom].

UNAVAILABLE NAMES (hybrid): *parvilunaria* BARTER, 1900 (*bilunaria* ESPER, 1801 male × *tetralunaria* HUFNAGEL, 1767 female). – Intrasubspecific names, potentially unavailable: *albicosta* BOYES, 1954 (ab.?); *bilineata* GROSSE, 1938 (ab.?); *braconieri* NORDSTRÖM, 1935 (ab.?); *brunneofasciata* SMITH, 1950 (ab.); *brunneopostulata* SMITH, 1950 (ab.); *canescens* FEICHTENBERGER, 1965 (ab.?); *centrifasciata* LEMPKE, 1970 (f.); *centrilineata* LEMPKE, 1951 (f.); *chalcescens* SMITH, 1954 (ab.?); *chinochlora* SMITH, 1954 (ab.?); *connexa* SMITH, 1950 (ab.); *costijuncta* COCKAYNE, 1948 (ab.); *deumbraria* KLEMENSIEWICZ, 1913 (ab.?); *dolichobalia* SMITH, 1954 (ab.); *eblanaria* BAYNES, 1952 (ab.); *erythrofasciata* SMITH, 1950 (ab.?); *estynensis* SMITH, 1947 (ab.?); *exquisita* SMITH, 1950 (ab.?); *extrema* SMITH, 1950 (ab.?); *flavo-marginaria* SMITH, 1950 (ab.); *fulvopustulata* SMITH, 1950 (ab.?); *fumata* SMITH, 1950 (ab.?); *garretti* WAGNER, 1928 (var.) ; *glabra* LEMPKE, 1951 (f.); *glaucescens* SMITH, 1954 (ab.); *grisescens* LEMPKE, 1951 (ab.); *harrisoni* WAGNER, 1928 (var.); *ialensis* VAUGHAN-ROBERTS, 1950 (ab.?); *infuscata* STRAND, 1901 (ab.); *infravenosa* BOYES, 1954 (ab.); *minima* STRAND, 1901 (ab.); *mixta* SMITH, 1950 (ab.); *nigra* SMITH, 1950 (ab.); *nigrobrunneata* SMITH, 1950 (ab.?); *nigrofasciata* SMITH, 1948 (ab.?); *nigrovanellata* SMITH, 1950 (ab.); *obscura* LEMPKE, 1951 (f.); *pallida* SMITH, 1950 (ab.?); *postmediojuncta* BOYES, 1954 (ab.); *radiata* BOYES, 1950 (ab.?); *rosea* SMITH, 1950 (ab.?); *rubra* SMITH, 1950 (ab.); *rubropustulata* SMITH, 1954 (ab.); *rufescens* LEMPKE, 1951 (f.); *schizomedia* VAUGHAN-ROBERTS, 1951 (ab.?); *schizo-nigra* VAUGHAN-ROBERTS, 1951 (ab.); *superba* SMITH, 1950 (ab.); *tetrafasciaria* SMITH, 1950 (ab.); *unilinearia* SMITH, 1951 (ab.); *v-signata* SMITH, 1950 (ab.?).

EXTERNAL CHARACTERS AND ABDOMEN: Wingspan 34–45 mm, second generation specimens smaller. Wings red-brown and yellowish white, heavily irrorated or striated with greyish brown. Medial lines widest on costa. Antemedial and medial lines slightly curved inwards, medial line often diffuse. Postmedial line straight, vertical, its outer margin bordered with whitish. Subterminal line whitish, weak. Medial lines diffuse on hindwings. Forewing margin concave below apex, bordered with reddish brown lunular area. Wing margins undulating, most extended on M3 on both wings. Discal spots absent or reduced. Terminal line brown. Fringes brown or weakly chequered brown and whitish. Wings below more intensely coloured, discal spots lunular with whitish centre. Frons and collar whitish. Dorsal part of frons with protruding hair, ventral part bare. Thorax and abdomen concolorous with wings. Male and female antennae bipectinate, female pectinations shorter. Male and female hindleg with 2+2 spurs. Tympanal organs medium-sized. Male 8th sternite weakly sclerotised, posterior margin concave. Male 8th tergite weakly sclerotised, posterior margin narrower. Other abdominal sternites and tergites of both sexes undifferentiated. Adult has a peculiar resting posture, wings are folded together over the abdomen.

VARIATION: High altitude and northern European specimens are darker and markings are more suffused. Females are often slightly lighter, more whitish than males. Second generation specimens are smaller (f. *illunaria* (HÜBNER), more yellow-orange, forewing postmedial line oblique, directed towards termen.

MALE GENITALIA: Uncus stout, weakly setose, apex round. Socii present as setae only. Gnathos arms wide, fused ventrally, upturned, weakly dentate. Transtilla well-developed, with two ventral pockets. Valva wide, most setose subapically, apex round. Costa sclerotised to about half of valva. Juxta roundish plate, with lateral pockets. Furca arms long, stout, symmetrical, apex curved medially. Vinculum with small membranous sac. Saccus concave. Aedeagus slightly curved, apex rounded. Vesica opens at approximately 180 degree angle. Vesica simple tube, with row of numerous cornuti.

FEMALE GENITALIA: Papillae anales large, surface weakly granulate. Apophyses long, narrow. Lamella postvaginalis weakly sclerotised, roundish plate. Lamella antevaginalis membranous, cup-shaped. Ductus bursae narrow, posterior part sclerotised, anterior part membranous. Corpus bursae elongated. Signum small, roundish.

DISTRIBUTION AND ABUNDANCE: Eurasiatic. In Europe widely distributed from northernmost Fennoscandia to the northern part of the Iberian Peninsula, southern France including Corsica, southern Italy and the Balkans excluding Greece, and from the British Isles through central and eastern Europe to European Russia and the Urals. – Outside Europe the species occurs in Turkey and the Caucasus region and eastwards of the Ural mountains through

Siberia, Mongolia, northern China to Far East Russia and Kamchatka. – In Europe a very common species that now and then can occur in numbers.

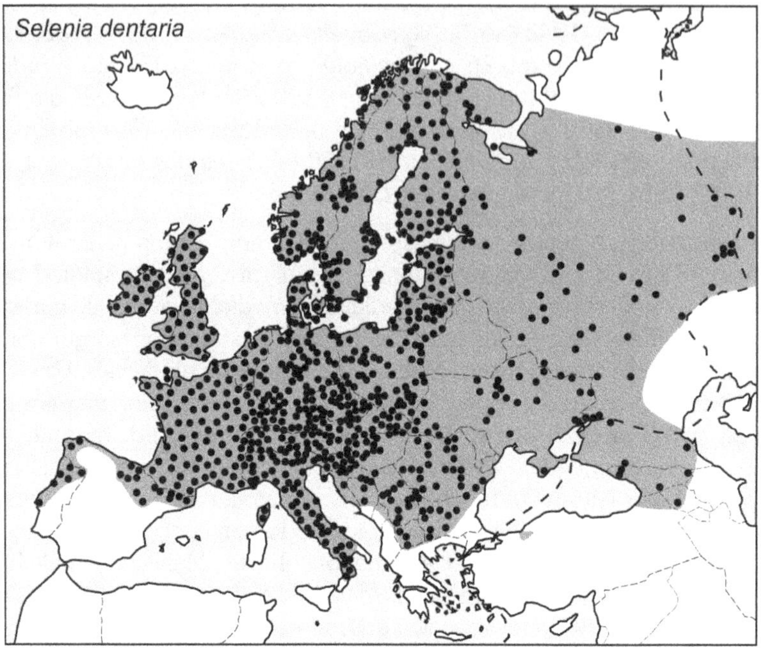

PHENOLOGY: Uni- or bivoltine. Only one generation in northern Britain in May and June and in northern Scandinavia from early June to mid-July, occasionally with second generation specimens, mainly in August, elsewhere in Europe in two generations. In southern Scandinavia from late April to early June and from mid-July to mid-August, in southern Germany from early April to late May and from early July to mid-August (EBERT 2003), in Spain from March to May and from July to September (REDONDO *et al.* 2009). In norther Portugal recorded from mid-February to mid-October, almost certainly in three generations (M. CORLEY pers. comm.). The caterpillar can be found from May to July and again from August to October, depending on latitude. Hibernation takes place as pupa. The moths fly at night and come readily to light.

BIOLOGY: Polyphagous. The caterpillar has been found on *Salix caprea*, and other *Salix* species, *Carpinus betulus*, *Corylus avellana*, *Quercus*, *Betula pendula*, *Sorbus aucuparia*, *Rosa canina*, *Cornus sanguinea*, *Acer campestre*, *Viburnum opulus* (EBERT 2003), *Crataegus* (SKINNER 2009), on *Myrica gale*, *Vaccinium myrtillus* (H. WEGNER pers. comm.) and *Rosa* (C. KAYSER pers. comm.). Further it is recorded on *Populus*, *Alnus*, *Ulmus*, *Ribes*, *Rubus*, *Prunus*, *Rhamnus*, *Frangula*, *Tilia*, *Cornus*, *Calluna*, *Vaccinium* and *Lonicera* (FLAMIGNI *et al.* 2007).

HABITAT: The species occurs in a large variety of habitat types with deciduous trees and bushes from sea level up to 1350 m in southern Germany (EBERT 2003), to 2000 m in the Pyrenees (REDONDO et al. 2009) and 2200 m in the Italian Alps (FLAMIGNI et al. 2007).

SIMILAR SPECIES: No similar species in Europe. Second generation specimens of *S. dentaria* have the postmedial line oblique, ending near tornus as in *S. lunularia* (HÜBNER). *S. dentaria* lacks the characteristic lunular discal spots of *S. lunularia*. Genitalia are similar to *S. tetralunaria*, see text below.

GENETIC DATA: Genetically comparatively homogeneous in Europe (n=28, from eight countries), populations from southern Italy diverging by 2.0%. Nearest species: *S. lunularia* (8.2%).

REMARKS: WEHRLI (1940, in WERHLI 1939–1954) described specimens from Switzerland (Zermatt and Davos) as an alpine variation, *S. bilunaria* var. *alpestris*. According to WEHRLI (1940), these high altitude specimens only occur in one generation from July to August. The subspecies status has not been considered valid in the recent literature, apart from VIIDALEPP (1996) and SCOBLE (1999). Generally, but not exclusively, the high altitude and northern European specimens are darker, similar to the specimen WEHRLI illustrated in plate 25a. The clinal change of external characters of Finnish specimens is discussed in MIKKOLA *et al.* (1989). FLAMIGNI el al. (2007) conclude that there is no genetic isolation between populations of the major alpine altitudes and differences in the habitus are not clear and consistent so that a separate subspecies should not be recognised. The preliminary DNA results do not lend support to the subspecies hypothesis either.

84. Selenia lunularia (HÜBNER, 1788)

lunularia lunularia HÜBNER, [1788], Beitr. Gesch. Schmett. 1 (3): 27, pl. (3)3, figs T, 1, 2, (*Ph*[*alaena*] *Geom*[*etra*]. Syntypes male(s), female(s), lost, [Germany]: Augsburg.

delunaria HÜBNER, [1799], Samml. Eur. Schmett. 5 Geometrae (1): pl. 7, fig. 34, (*Geometra*). Syntype(s), lost, Europe. [Summer generation].

lunaria [DENIS & SCHIFFERMÜLLER], 1775, Ankündung syst. Werkes Schmett. Wienergegend: 103, pl. 1b, fig. 4, (*Geometra*). Syntype(s), lost, [Austria]: Vienna district. [Junior secondary homonym of *Phalaena lunularia* FABRICIUS, 1775. According to LEMPKE 1970, p. 170: On page 103 of the [DENIS & SCHIFFERMÜLLER] 1775, the species was named *Geometra lunaria* and as such the name was perfectly valid. But on p. 276 of the same work DENIS & SCHIFFERMÜLLER named it *Phalaena Geometra lunaria*. By placing the species in the genus *Phalaena* they made the name a secondary homonym. As such it was invalid and must remain so according to art. 59 [Article 57 in the current code, ICZN 1999]. Mr. D. S. FLETCHER is certainly right in rejecting the name *lunaria* for the species.

nubilunaria DANNEHL, 1927, Ent. Z., Frankf. a. M. 41 (15): 318, (*Selenia*). Lectotype male (Zoologische Staatsammlung, Munich, Germany), [Austria]: [Italy: Alto Adige]: Etschtal, Terlano, Sigmundskron (examined externally). [Originally as subsp. of *lunaria* [DENIS & SCHIFFERMÜLLER]]. See remarks.

privataria DANNEHL, 1927, Ent. Z., Frankf. a. M. 41 (15): 318, (*Selenia*). Syntypes, [Austria]: South Tirol. [Originally as ab. of *delunaria* HÜBNER]. See remarks.

sublunaria STEPHENS, 1831, Illust. Br. Ent. (Haustellata) 3: 170, pl. 28, fig. 1, (*Geometra*). Syntypes 3, [England]: Derbyshire.

lunularia gamra RUNGS, 1948, Bull. Soc. Sci. nat. phys. Maroc 28: 158, pl. 5, fig. 9, pl. 6, figs 5, 6, (*Selenia*). Holotype male, Morocco: Middle Atlas, Ifrane Valley, c. 1550 m. [Originally as subsp. of *lunaria* [DENIS & SCHIFFERMÜLLER]].

lunularia kuldjana WEHRLI, 1940, in SEITZ, Gross-Schmett. Erde 4 (Suppl.): 327, pl. 25; b, (*Selenia*). Syntype(s), [China]: Kuldja [Yining (Gulja)]. Originally as subsp. of *lunaria* [DENIS & SCHIFFERMÜLLER], valid at subspecies rank according to SCOBLE (1999).

INTRASUBSPECIFIC NAMES, POTENTIALLY UNAVAILABLE: *aestivalis* RUNGS, 1981 (f.); *albilinea* WAGNER, 1919 (ab.); *basireducta* LEMPKE, 1970 (f.); *brunnea* LEMPKE, 1951 (f.); *canescens* FEICHTENBERGER, 1965 (ab.?); *centripetala* COCKAYNE, 1952 (ab.); *constricta* HÖFER, 1924 (ab.); *delineata* FEICHTENBERGER, 1965 (f.); *deumbrata* LEMPKE, 1951 (f.); *lactimarginata* COCKAYNE, 1952 (ab.); *lilacina* LEMPKE, 1970 (f.); *maculosa* LEMPKE, 1951 (f.); *nigrumbrata* LEMPKE, 1951 (f.); *ochracea* CLOSS, 1919 (f.); *pallida* LEMPKE, 1970 (f.); *roseofasciata* LEMPKE, 1951 (f.); *rubra* LEMPKE, 1970 (f.); *schizomedia* LEMPKE, 1970 (f.).

EXTERNAL CHARACTERS AND ABDOMEN: Wingspan ♂ 34–43 mm, ♀ 36–44 mm, second generation specimens smaller. Wings yellowish brown to contrasting reddish brown, irrorated with brown. Antemedial line fine, angled on costa. Medial line suffused, dark, or weak or reduced. Postmedial line almost straight, outer margin bordered with whitish. Forewing margin concave below apex, bordered with reddish brown lunular area. Wing margins undulating, most extended on M3 on both wings. Discal spots lunular, with white centre, forewing discal spot almost parallel with postmedial line. Terminal line brown. Fringes brown or weakly chequered brown and whitish. Wings below as above. Frons, collar, thorax and abdomen concolorous with wings. Dorsal part of frons with protruding hair, ventral part bare. Male and female antennae bipectinate, female pectinations shorter. Male and female hindleg with 2+2 spurs. Tympanal organs medium-sized. Male 8th sternite weakly sclerotised, posterior margin narrower. Male 8th tergite weakly sclerotised, posterior margin rounded. Other abdominal sternites and tergites of both sexes undifferentiated.

VARIATION: The ground colour varies from less contrasting pale yellow to yellowish-brown to contrasting reddish brown, even purplish specimens. MIKKOLA *et al.* (1989) note that yellowish brown and reddish brown colour forms appear rather disjunct (in northern Europe), presenting a potential case of polymorphism. Second generation specimens are smaller and more uniform yellow (f. *delunaria*). Forewing discal spots may be absent (f. *privataria*). Numerous forms have been described, those are summarised in LERAUT (2009). See remarks.

Subsp. ***gamra*** from Morocco is dark brown, with tint of purple, and the first and second generation specimens are similar. *S. lunularia* subsp. *gamra* is potentially valid at species level (D. STADIE pers. comm.).

MALE GENITALIA: Uncus shovel-shaped, apex weakly setose. Socii present as setae only. Gnathos arms wide, fused ventrally, upturned, weakly dentate. Transtilla well-developed, with two ventral pockets. Valva narrow, most setose subapically, apex round. Costa sclerotised to about 2/3 of valva. Valva base with small, acute, sclerotised spine. Juxta plate absent. Furca arms concealed partly in saccus, long, narrow, symmetrical, apex curved weakly laterally. Vinculum with elongated membranous sac. Saccus narrow, elongated. Aedeagus straight, apex rounded. Vesica opens at approximately 180 degree angle. Vesica with two large basal diverticula, with row of numerous cornuti.

FEMALE GENITALIA: Papillae anales large, surface weakly granulate. Apophyses long, narrow. Lamella postvaginalis membranous. Lamella antevaginalis sclerotised, weakly funnel-shaped, surface covered with microsetae. Ductus bursae narrow, posterior part sclerotised, anterior part membranous, wrinkled. Posterior part of corpus bursae with weakly sclerotised diverticulum, anterior part long, narrow, membranous. Signum small, weakly stellate.

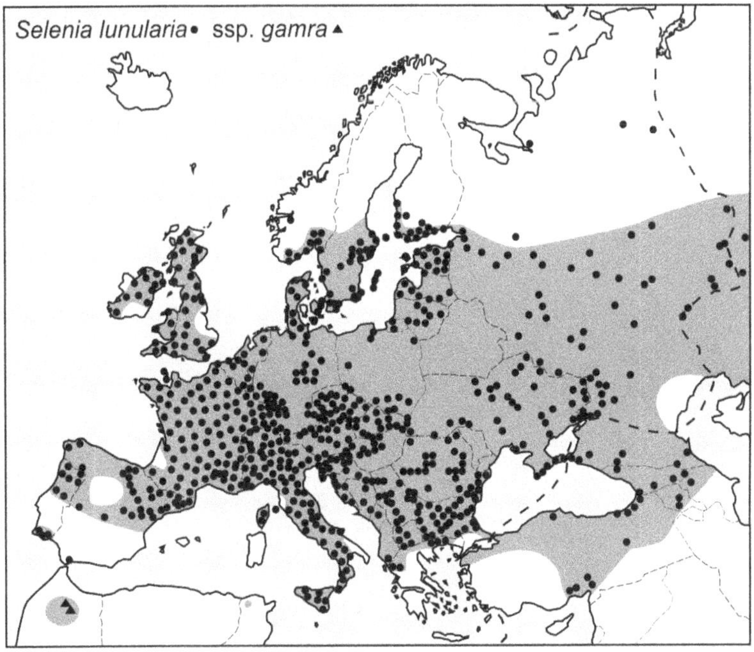

DISTRIBUTION AND ABUNDANCE: West-Palaearctic. In Europe the species occurs from southern Fennoscandia to the northern half of the Iberian Peninsula, southern France including Corsica, southern Italy including Sicily and the Balkans including northern Greece and from the British Isles through central and eastern Europe to European Russia and the Ural mountains. – Outside Europe the species is known from Turkey, the Caucasus region and central Asia and also in North Africa in Morocco and Tunisia. – In Europe a common or rather common species that is rarely found in numbers.

PHENOLOGY: Uni- or bivoltine. In southern Fennoscandia in one generation from late May to early July and in warm summers sometimes in a partial second generation in August, in warmer places in southern central Europe and in southern Europe in two generations from March or April to June and from July to late September. Based on collection material the two generations overlap, depending on altitude and latitude. Larva in June and early July and again in August and September. The species hibernates as a pupa.The moths fly at night and come frequently to light.

BIOLOGY: Polyphagous. In northern Europe the larva was found on broadleaved trees (HULDÉN *et al.* 2000). According to EBERT (2003) it has been found on *Fagus*, *Prunus spinosa*, *Vaccinium myrtillus* and *Tilia* and FLAMIGNI *et al.* (2007) record it on *Alnus*, *Ostrya*, *Corylus*, *Ulmus*, *Berberis*, *Rubus*, *Rosa*, *Pyrus*, *Malus*, *Crataegus*, *Cytisus scoparius*, *Spartium junceum*, *Acer*, *Rhamnus*, *Frangula*, *Cornus*, *Fraxinus*, *Syringa*, *Sambucus*, *Viburnum*, *Lantana* and *Lonicera*.

HABITAT: In northern Europe mainly in oak forests, further to the south in a larger variety of habitat types. The species is found from sea level up to about 1000 m north of the Alps, in the Iberian Peninsula up to 1600 m and in the Alps even up to 2100 m.

SIMILAR SPECIES: *Selenia tetralunaria* (HUFNAGEL) has forewing postmedial line angled below discal spot, forewing discal spot is not parallel with postmedial line, hindwing margin above vein M3 is less concave and hindwing terminal area is with dark spot (forewing postmedial line almost straight, discal spot almost parallel with postmedial line, hindwing margin deeply concave, hindwing terminal area without dark spot in *S. lunularia*), see Text-figs 92–93. Male and female genitalia allow reliable identification. Second generation specimens of *S. dentaria* also have postmedial line almost straight but lacks the lunular discal spots.

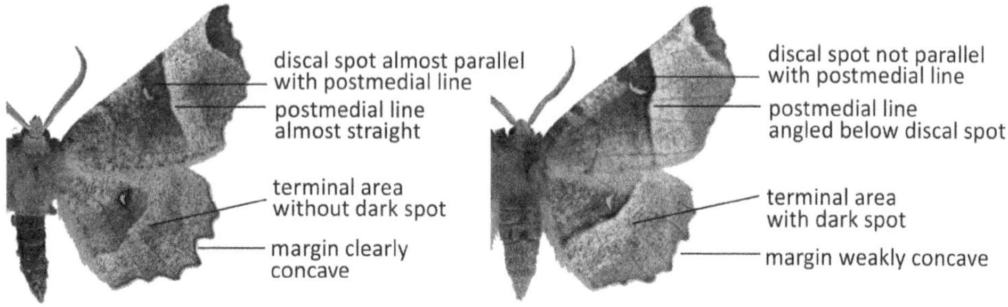

S. lunularia 92 ***S. tetralunaria*** 93

Text-figs 92–93. Diagnostic characters (indicated) of European *Selenia* HÜBNER, 1823 species. Text-fig. 92. *S. lunularia* (HÜBNER, 1788), Italy: Sicily, 5.vi.2005. Text-fig. 93. *S. tetralunaria* (HUFNAGEL, 1767), Sweden: Södermanland, 24.v.1972.

GENETIC DATA: Genetically heterogeneous, in Europe occurring in three homogeneous clusters: (1) from Finland and United Kingdom to Georgia and northern Turkey (n=18, from nine countries); (2) Germany, Austria and United Kingdom (n=7); (3) southern Italy, Sicily, Turkey, Georgia, Kazakhstan (n=17). Sympatric occurrence of (1) and (2) in the United Kingdom (at 2.7% distance), of (1) and (3) in Turkey and Georgia (at 2.7% distance). Populations from Morocco (subspecies *gamra* RUNGS, 1950) diverging by 3.1% (See remarks). Nearest species: *S. forsteri* WEHRLI, 1941 from Iran (4.4%), in Europe *S. lunularia* (8.1%) and *S. dentaria* (8.2%).

REMARKS: SCOBLE (1999) considers taxa *nubilunaria* DANNEHL (first generation specimens with yellowish brown background colour) and *privataria* DANNEHL (second generation specimens without forewing lunular discal spot) valid at subspecies rank. FLAMIGNI *et al.* (2007) report that such specimens may prevail locally, but everywhere (in Italy?) one meets other more reddish and more contrasting specimens among larger samples. The same implies to specimens without lunular discal spots. We follow FLAMIGNI *et al.* (2007) and do not recognise these taxa as valid at subspecies level. D. STADIE (pers. comm.) is in the process of evaluating the taxonomic status of *gamra* RUNGS, 1950 from Morocco.

85. Selenia tetralunaria (HUFNAGEL, 1767)

tetralunaria (HUFNAGEL, 1767), Berlin Mag. 4 (5): 506, (*Phalaena*). Syntype(s), lost, [Germany]: Berlin region.

clarilunaria DANNEHL, 1927, Ent. Z., Frankf. a.M. 41 (15): 319, (*Selenia*). Syntypes, [Austria]: South Tyrol. See remarks.

illustraria HÜBNER, [1799], Samml. Eur. Schmett. 5 Geometrae (1): pl. 7, fig. 35, (*Geometra*). Syntype(s), lost, Europe.

lunaria FABRICIUS, 1775, Syst. Ent.: 623, (*Phalaena*). Syntype(s), Germany.

phoebearia SCHRANK, 1802, Fauna boica 2 (2): 14, (*Phalaena*). Syntype(s). [Germany, Bavaria]: Ingolstadt.

quadrilunaria ESPER, [1801], Die Schmett. 5 (3): 72, pl. 12, figs 5, 6, (*Ph[alaena] Geom[etra]*). Syntype(s), [Germany]: Leipzig.

tetralunaria coreana WEHRLI, 1940, in SEITZ, Gross-Schmett. Erde 4 (Suppl.): 328, pl. 25: c, (*Selenia*). Syntype(s), [Korea (north-west)]: Pjöngjang. [Paratype male in the Natural History Museum, London, UK (examined externally)].

INTRASUBSPECIFIC NAMES, POTENTIALLY UNAVAILABLE: *aestiva* STAUDINGER, 1871 (summer generation); *aurantiaca* LEMPKE, 1970 (f.); *brunnea* SCHNAIDER, 1950 (ab.?); *clara* LEMPKE, 1951 (f.); *clarilunaria* DANNEHL, 1927 (f.); *kuhnei* KULNE, 1908 (ab.); *lacticolor* COCKAYNE, 1952 (ab.); *nigrescens* COCKAYNE & KETTLEWELL, 1949 (ab.); *nigrolineata* LEMPKE, 1951 (f.); *notabilis* THIERRY-MIEG, 1910 (ab.); *obscura* LEMPKE, 1951 (f.); *pallimarginata* LEMPKE, 1970 (f.); *postmaculata* LEMPKE, 1970 (f.); *postradiata* COCKAYNE, 1952 (ab.); *rufescens* LEMPKE, 1951 (f.); *viridobrunnea* SCHAWERDA, 1941 (ab.); *wehrlilunaria* BRYK, 1948 (f.).

EXTERNAL CHARACTERS AND ABDOMEN: Wingspan ♂ 39–45 mm, ♀ 40–47 mm, second generation specimens smaller. Generally as *S. lunularia*, but with following differences: *S. tetralunaria* (HUFNAGEL) has forewing postmedial line angled below discal spot, forewing discal spot is not parallel with postmedial line, hindwing margin above vein M3 is less concave and hindwing terminal area is often with dark spot (forewing postmedial line almost straight, discal spot almost parallel with postmedial line, hindwing margin deeply concave, hindwing dark spot absent in *S. lunularia*). Wings below more contrasting than above, often with more pronounced orange and white areas.

VARIATION: Two colour morphs are common: large, dark brown to red-brown (normally the first generation) and small, more orange with tint of violet (normally the second generation (MUTANEN & KAITALA 2006). Second generation specimens often have wing margins less undulating and hindwing discal spot is enclosed within medial and postmedial lines. In the first generation specimens the hindwing medial line is reduced and dark anterior margin of postmedial line encloses the discal spot. See remarks.

MALE GENITALIA: Generally as in *S. dentaria*, but with the following quantitative differences: Costa of valva straight, sclerotised approximately to 3/4 valva length, vesica base slightly expanded and cornuti more scattered in *S. tetralunaria* (costa valva slightly curved, sclerotised approximately to 1/2 valva length, vesica base not expanded, cornuti in row in *S. dentaria*). See remarks.

FEMALE GENITALIA: Generally as in *S. dentaria*, but with the following quantitative differences: Lamella antevaginalis more sclerotised, membranous part of ductus bursae short, posterior part of corpus bursae without diverticulum, signum irregular plate in *S. tetralunaria* (lamella antevaginalis less sclerotised, membranous part of ductus bursae longer, posterior part of corpus bursae with weakly sclerotised diverticulum, signum weakly stellate in *S. dentaria*). See remarks.

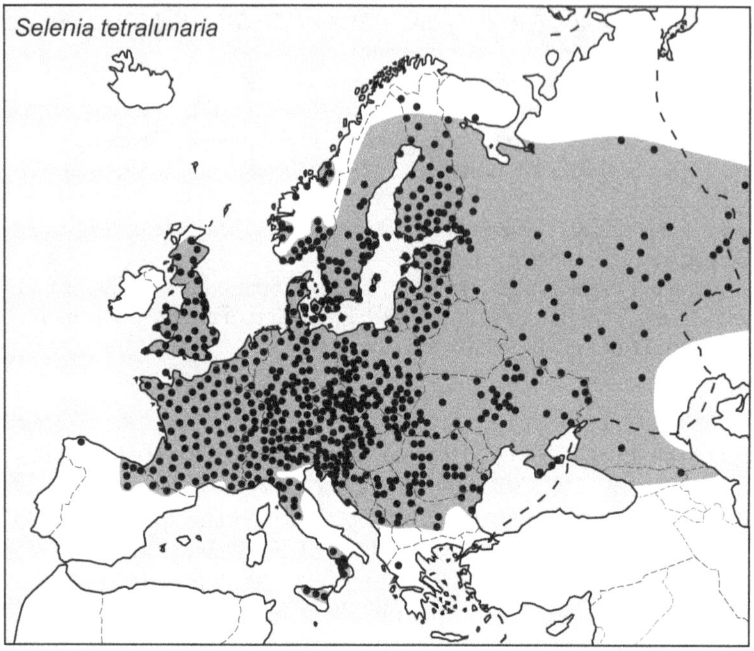

DISTRIBUTION AND ABUNDANCE: Eurasiatic. In Europe widely distributed from northern and central Fennoscandia to the northernmost part of the Iberian Peninsula, southern France, southern Italy including Sicily and the Balkans including a single record in northern Greece and from Great Britain through central and eastern Europe to European Russia and the Ural mountains. – Outside Europe the species occurs from the Caucasus region through Kazakhstan, Siberia, Mongolia and China to Korea, Far East Russia, Sakhalin and Japan. – In Europe a common species that rarely occurs in numbers.

PHENOLOGY: Uni- or bivoltine. In northern Fennoscandia mainly univoltine from early June to early July, but in warmer years a (partial) second generation may appear. In northern Europe from late April to early June and from late July to late August, in southern central Europe from early March to mid-June and from early July to mid-August (EBERT 2003), and in southern Europe from April to September with the two generations overlapping. In the northern part of the distribution area caterpillars from August and overwintering to early June. More to the south caterpillars can be found from late May and June and again in August overwintering until spring. The adult moths come frequently to light.

BIOLOGY: Polyphagous. The larva has been found on *Betula pubescens* (H. WEGNER pers. comm.) and on *Salix alba*, *S. caprea*, *Alnus glutinosa*, *Quercus petraea*, *Q. robur*, *Q. rubra*, *Prunus spinosa* and *Fraxinus excelsior* (EBERT 2003). The caterpillar has been reared on *Tilia cordata* (B. MÜLLLER pers. comm.). FLAMIGNI *et al.* (2007) record this species on *Betula*, *Corylus*, *Fagus*, *Rubus idaeus*, *Rosa*, *Malus*, *Sorbus aucuparia*, *Crataegus*, *Frangula alnus* and *Lonicera*. The pupation takes place between spun leaves of the food plant.

HABITAT: In deciduous forests and wood- and scrubland, often in slightly moist places. Also along rivers and in gardens. The species occurs from sea level up to 1600 m in southern Europe.

SIMILAR SPECIES: Externally *Selenia lunularia* (HÜBNER) is similar, see entry above and Text-figs 92–93. Genitalia are similar to *S. dentaria*, see text above.

GENETIC DATA: Genetically comparatively homogeneous in Europe (n=30, from eight countries), populations from southern Italy slightly diverging. Nearest species: *S. lunularia* (8.1%).

REMARKS: DANNEHL (1927) assumed that *clarilunaria* is a hybrid between *lunaria* (DENIS & SCHIFFERMÜLLER) (=*lunularia* (HÜBNER)) and *tetralunaria* (HUFNAGEL). The postmedial line is almost straight, as in *S. lunularia*, and discal spot is similar to *S. tetralunaria*. WEHRLI (1939–1954) did not recognise the taxon valid at subspecies level, but SCOBLE (1999) did. More research is needed.

The second generation specimens of *S. tetralunaria* are smaller in size when compared to the first generation, the males are on average 20% smaller and females on average 13.6% smaller (MUTANEN & KAITALA 2006). In a detailed morphometric analysis it was found that the male external genitalia of the second generation were on average 9.2% smaller and the female external genitalia 11.7% smaller. The internal genitalia of the first and second generation specimens were approximately the same size (second generation males 0.07% larger, females 1.26% smaller). MUTANEN & KAITALA (2006) concluded that all measured genital traits were scaled in negative allometric relation to body size, supporting previous findings on other organisms of allometric scaling of genital traits in relation to body size. Also the shape of genitalia of both sexes varied between the larger and smaller morphs.

Artiora MEYRICK, 1892

Artiora MEYRICK, 1892, Trans. ent. Soc. Lond. 1892: 101 (key), 114 [Replacement name]. Type species: *Geometra evonymaria* [DENIS & SCHIFFERMÜLLER], 1775. Genus synonymy is after SCOBLE (1999).

Pericallia HERRICH-SCHÄFFER, 1847, Syst. Bearbeitung Schmett. Eur. 3: 48. Type species: *Geometra evonymaria* [DENIS & SCHIFFERMÜLLER], 1775. [Junior homonym of *Pericallia* HÜBNER, [1820] 1816 (Lepidoptera: Arctiidae)].

DIVERSITY AND DISTRIBUTION: *Artiora* is a monotypic genus, the only species occurs in the West-Palaearctic area.

EXTERNAL CHARACTERS AND ABDOMEN: Wingspan 28–32 mm. Wings pale yellow-brown. Medial lines blackish, suffused, strongest on costa, postmedial line broken into small spots. Terminal area darker. Forewing margin deeply concave below apex, otherwise margin undulating. Male antennae bipectinate, female antennae filiform. Male 8th sternite unmodified.

MALE GENITALIA: Uncus round, setose. Socii absent or reduced to few setae. Gnathos arms fused ventrally, upturned, medial part triangular. Valva curved, weakly setose, apex round. Medial ridge sclerotised, dentate, apical half spinose. Juxta elongated medially, apex round. Saccus large, weakly concave. Vesica without cornuti.

FEMALE GENITALIA: Papillae anales weakly sclerotised, setose. 9th segment elongated. Apophyses anteriores weakly angled at 1/3 distance. Lamella postvaginalis weakly sclerotised medially. Ductus bursae most sclerotised laterally. Corpus bursae roundish, membranous. Signum small, elongated.

BIOLOGY: Larva of *A. evonymaria* has been recorded feeding on the leaves of *Euonymus europaeus* (FLAMIGNI *et al.* 2007). It has a single generation with adults on the wings from late June to September. Egg overwinters (SZÉKELY 2011). Adult has a peculiar resting posture, hindwings are under forewings thus leaving a space between the abdomen and wing margins (Text-fig. 94).

Text-fig. 94. The resting posture of *Artiora evonymaria* (DENIS & SCHIFFERMÜLLER, 1775), unusual within the Ennominae, where hindwings are held hidden under the forewings and the anal margin does not touch the abdomen, Slovakia, 1.viii.2009. Photo: FRANTIŠEK ŠARŽIK.

IMMATURE STAGES: Larva of *A. evonymaria* is white, densely covered with black reticulate pattern and a dorsal and lateral rows of black and yellow spots (LEPIFORUM 2013). Pupa cremaster short, with two straight, acute pairs of setae, the apical pair is wider (PATOČKA & TURČANI 2005).

REMARKS: *Artiora* has been classified in the Ennomini (VIIDALEPP 1996, HAUSMANN, MIRONOV & VIIDALEPP 2011). BELJAEV (2008) noted that European authors have followed the tradition founded by PIERCE (1914) and HERBULOT (1963–1964), who included in the tribe Ennomini a series of remotely related genera, united by habitus rather than strictly morphological similarity. Consequently BELJAEV excluded the majority of the European genera from the Ennomini, including *Artiora*. He did not propose where these excluded genera should be classified. The male and female structures, including the genitalia, are general Ennominae type and provide little information as regards its systematic position. The apparently aposematic caterpillar and pupa cremaster with its two acute pairs of setae in *A. evonymaria* are not similar to genus *Ennomos* TREITSCHKE. Systematic position: Ennomini of uncertain association. Further research is needed.

86. Artiora evonymaria (DENIS & SCHIFFERMÜLLER, 1775)

evonymaria [DENIS & SCHIFFERMÜLLER], 1775, Ankündung syst. Werkes Schmett. Wienergegend: 103, (*Geometra*). Syntype(s), lost, [Austria]: Vienna district.

INTRASUBSPECIFIC NAMES, POTENTIALLY UNAVAILABLE: *arrostaria* SCHAUS, 1925 (ab.); *exquisita* AIGNER-ABAFI, 1906 (ab.) (=synonym of *fuscaria* WAGNER, 1903 according to The Global Lepidoptera Names Index (ANONYMOUS 2013)); *flavescens* PROUT, 1915 (ab.); *fuscaria* WAGNER, 1903 (ab.); *obscura* AIGNER-ABAFI, 1906 (ab.?).

EXTERNAL CHARACTERS AND ABDOMEN: Wingspan 28–32 mm. Wings pale yellow-brown, irrorated with brown. Medial lines blackish, suffused, strongest on costa. Postmedial line broken into small, blackish spots, evenly curved, marked with whitish line. Terminal area darker. Forewing margin deeply concave below apex, followed by roundish extension, otherwise margins undulating. Discal spots absent or very small. Terminal line and fringes brown, darkest on forewing concavity. Wings below more orange-brown, more irrorated with brown. Frons, collar, thorax and abdomen concolorous with wings. Male antennae bipectinate, female antennae filiform. Male and female hindleg with 2+2 spurs. Tympanal organs small. Other abdominal sternites and tergites of both sexes undifferentiated.

VARIATION: Ground colour varies from pale yellow-brown to greyish brown to dark brown. Medial lines normally suffused, strongest on costa, but those may be almost absent.

MALE GENITALIA: Uncus round, setose. Socii absent or reduced to few setae. Gnathos arms fused ventrally, upturned, medial part triangular. Valva curved inwards, weakly setose, apex round. Costa sclerotised, not extending beyond apex. Medial ridge sclerotised, dentate, apical half spinose. Valva bases form small pockets. Transtilla membranous. Juxta elongated medially, apex round. Saccus large, weakly concave. Aedeagus straight, caecum long, apex with serrate ridge. Vesica opens at approximately 135 degree angle. Vesica narrow, membranous tube, without cornuti.

FEMALE GENITALIA: Papillae anales weakly sclerotised, setose. 9th segment elongated, striated vertically. Apophyses posteriores long, straight. Apophyses anteriores shorter, weakly angled at 1/3 distance. Lamella postvaginalis striated horizontally, with medial sclerotised plate and sclerotised lateral grooves. Lamella antevaginalis membranous. Ductus bursae narrow, most sclerotised laterally. Posterior part of corpus bursae narrow, wrinkled, membranous. Anterior part of corpus bursae roundish, membranous. Signum small, elongated, margins weakly dentate.

DISTRIBUTION AND ABUNDANCE: European-Westasiatic. In Europe from central Europe and north-eastern Italy to the Balkans, Ukraine and south-western European Russia. – Outside Europe the species is only found in Turkey and northern Caucasus. – A somewhat local species that is mainly found in limited numbers. The caterpillars exhibit great fluctuations in abundance. After a year with mass occurrence a year may follow where the caterpillar is completely missing in the same locality (M. LEIPNITZ pers. comm.).

PHENOLOGY: Univoltine. Usually from late July to late September (Poland, Czech Republic, Slovakia, Austria, Romania, cf. BUSZKO 2000, Z. LÁSTŮVKA, L. SZÉKELY pers. comm.), in Greece until late October. The hibernation takes place as an egg. Larval time from late April to early June (M. LEIPNITZ pers. comm.). The species flies at night and comes to light, but supposedly to a lesser degree than most other species (G. PETRÁNYI pers. comm.).

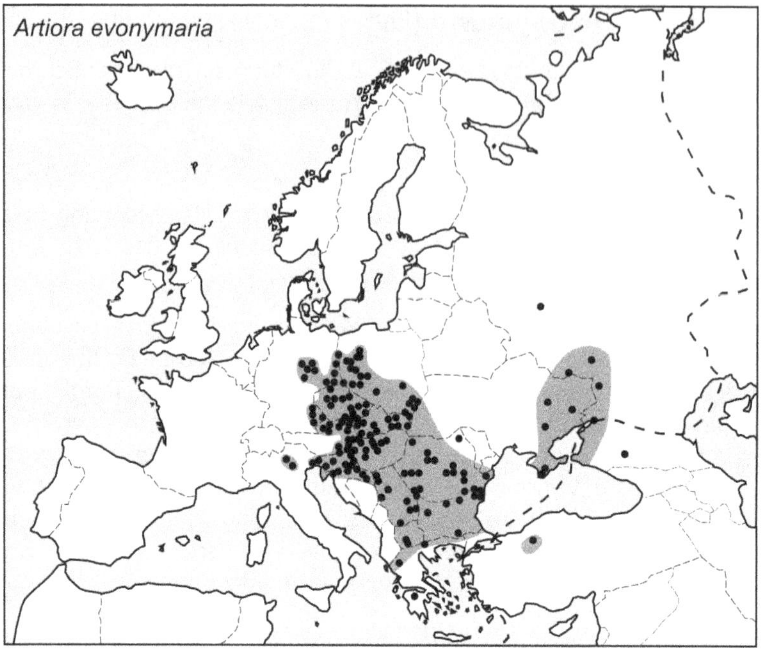

BIOLOGY: Monophagous. The larva has been recorded on *Euonymus europaeus* (FLAMIGNI *et al.* 2007, MALKIEWICZ 2012). Also reared on *E. europaeus* (B. MÜLLER pers. comm.).

HABITAT: Sylvicolous. In both more dry (FAJCIK 2003) and more moist (BUSZKO 2002) places such as forests and forest edges, along rivers and in woodland, the species prefers places where the food plant grows in the shade of trees (M. LEIPNITZ pers. comm.). In south-eastern central Europe up to 1000 m (FAJCIK 2003), in Greece up to 1700 m at Mount Chelmos (coll. SKOU), and in Turkey up to 1500 m.

SIMILAR SPECIES: No similar species in Europe.

GENETIC DATA: Nearest species in Europe: *Dyscia royaria* (8.7%) and *Ennomos quercaria* (9.1%).

Odontopera STEPHENS, 1831

Odontopera STEPHENS, 1831, Illust. Br. Ent. (Haustellata) 3: 162. Type species: [*Phalaena*] *bidentata* CLERCK, 1759, [Sweden]. Genus synonyms are after SCOBLE (1999).

Aethiopodes WARREN, 1902, Novit. zool. 9: 530. Type species: *Azelina indecoraria* WALKER, 1866.

Caripetodes WARREN, 1895, Novit. zool. 2: 139. Type species: *Colotois kametaria* FELDER & ROGENHOFER, 1875.

Cenoctenucha WARREN, 1897, Novit. zool. 4: 115. Type species: *Crocallis similaria* MOORE, 1888.

Corotia MOORE, 1868, Proc. zool. Soc. Lond. 1867: 624. Type species: *Corotia cervinaria* MOORE, 1868.

Endropia GUENÉE, [1858], in BOISDUVAL & GUENÉE, Hist. nat. Insectes (Spec. gén. Lépid.) 9: 122. Type species: *Geometra pectinaria* [DENIS & SCHIFFERMÜLLER], 1775. Synonymy established by HEPPNER (2002).

Lioptilesia WEHRLI, 1936, Ent. Rdsch. 54: 129. Type species: *Gonodontis prolita* WEHRLI, 1936.

Niphonissa BUTLER, 1878, Ann. Mag. nat. Hist. (5) 1: 394. Type species: *Niphonissa arida* BUTLER, 1878.

Paragonodontis WEHRLI, 1936, Ent. Rdsch. 54: 129. Type species: *Gonodontis postobscura* WEHRLI, 1936.

DIVERSITY AND DISTRIBUTION: *Odontopera* contains about 60 species, which occur in the Palaeartic and Oriental regions and in Africa. PITKIN (2002) noted that the Neotropical taxon assigned to *Odontopera* is unlikely to be a true member of it. One species in Europe.

EXTERNAL CHARACTERS AND ABDOMEN: Rather large moths, wingspan about 40 mm. Wings different shades of brown, often with greyish or yellowish tones. Typically postmedial line most visible and medial area darkened. Discal spots often with blackish margins and pale centre. Forewing margin typically concave below apex, otherwise wing margins undulating. Male antennae bipectinate, female antennae filiform. Male 8th sternite unmodified.

MALE GENITALIA, description is based on European species: Uncus long, stout, weakly setose. Socii minute, setose. Gnathos arms fused ventrally, upturned, apex weakly dentate. Valva narrow, with sclerotised, setose medial ridge, extends beyond apex. Furca arms long, angled (Text-fig. 95). Saccus round. Cornuti extend outside aedeagus in uneverted vesica.

FEMALE GENITALIA, description is based on European species: Papillae anales small, setose. Apophyses anteriores about 1/2 length of apophyses posteriores. Lamella postvaginalis striated horizontally. Lamella antevaginalis wide, sclerotised plate, with lateral extensions. Ductus bursae very short, wide. Posterior part of corpus bursae long, narrow, striated, sclerotised. Anterior part of corpus bursae round, membranous. Signum shortly stellate, with medial concavity.

BIOLOGY: Larva of *O. bidentata* is very polyphagous, it has been recorded feeding on numerous deciduous trees (for instance *Betula, Sorbus, Tilia*), on coniferous trees (for instance on *Pinus, Larix*), and on *Vaccinium myrtillus*, *Artemisia vulgaris* and *Arctium* (MIKKOLA *et al.* 1989). Adults are active at dusk, and come readily to light. Pupa overwinters.

IMMATURE STAGES: Larva of *O. bidentata* is of various shades of brown or grey, dark dorsal line, and small extension dorsally on the 11th segment and often a triangular-shaped dark marking on the fourth segment (PORTER 2010). Pupa cremaster rounded, with four pairs of curved setae, apical pair strongest (PATOČKA & TURČANI 2005).

REMARKS: Numerous *Odontopera* species have earlier been combined to *Gonodontis* HÜBNER, 1823, because for a long time the type species of *Gonodontis* was thought to be *Odontopera bidentata* (CLERCK). WARREN (1893) established *Phalaena* (*Geometra*) *clelia* CRAMER, 1780 from India as the type species of *Gonodontis*, as noted by FLETCHER (1979), but overlooked even in the recent literature (for instance HAUSMANN, MIRONOV & VIIDALEPP 2011). Therefore genus *Gonodontis* and tribe Gonodontini must be based on *G. clelia* and allied taxa as defined in HOLLOWAY (1994). Despite similarities in the external appearances, the genera *Gonodontis* and *Odontopera* are not closely related.

Odontopera has recently been classified in the Ennomini (VIVES MORENO 1994, VIIDALEPP 1996, LERAUT 1997, KULLBERG *et al.* 2001 (following VIIDALEPP 1996). BELJAEV (2008a) noted that the European authors have followed the tradition founded by PIERCE (1914) and HERBULOT (1963–1964), who included in the tribe Ennomini a series of remotely related genera, united by habitus rather than strictly morphological similarity. Consequently BELJAEV (2008a) excluded the majority of the European genera from the Ennomini, including *Odontopera*. He did not propose where these excluded genera should be classified. The male and female structures, including the genitalia, are general Ennominae type and provide little information as regards its systematic position. *Crocallis* TREITSCHKE and *Odontopera* STEPHENS share an unusual trait in the male aedeagus: cornuti are outside the aedeagus (not concealed within it) in an uneverted vesica. Systematic position: Ennomini of uncertain association. Further research is needed.

87. *Odontopera bidentata* (CLERCK, 1759)

bidentata bidentata CLERCK, 1759, Icones Insectorum Rariorum 1: pl. 7, fig. 2, ([*Phalaena*]). Lectotype female (Naturhistoriska Riskmuseet, Stockholm, Sweden), [Sweden].

bidentaria FABRICIUS, 1794, Ent. Syst. 3 (2): 133, (*Phalaena*). [Emendation of *bidentata* CLERCK].

dentaria HÜBNER, [1799], Samml. Eur. Schmett. 5 Geometrae (1): pl. 3, fig. 12, *Ph.*[*alaena*] *Geom.*[*etra*]. Syntype(s), lost, Germany: Bayern.

graecarius BANG-HAAS, 1910, Dt. ent. Z. Iris 24: 47, pl. 4, fig. 3, (*Gonodontis*). [Holo]type female (potentially in Senckenberg Natural History Collections of Dresden, Germany), [Greece]: Attica (examined externally, based on literature). Originally as sp., downgraded here to junior subjective synonym of *Odontopera bidentata* (CLERCK, 1759), quoting absence of disjunt, diagnostic features (new synonym). See remarks.

oreas HERBULOT, 1963, Alexanor 3 (2): 90; ibidem 3(1): 23, (*Gonodontis*). Holotype male (Coll. HERBULOT in Zoologische Staatssammlung, Munich, Germany), France: Colmars (examined externally). LERAUT (2009) considers *oreas* HERBULOT valid at subspecies level.

pectinaria [DENIS & SCHIFFERMÜLLER], 1775, Ankündung syst. Werkes Schmett. Wienergegend: 103, (Geometra), Syntype(s), lost, [Austria]: Vienna district. Synonymy noted by HEPPNER (2002), but used as such earlier, for instance in LERAUT (1997).

bidentata exsul TCHETRERIKOV, 1905, Russk. ent. Obozr. 4: 78, (*Gonodontis*). Syntype(s) male(s), [Russia]: Sajan. Originally as var., valid at subspecies rank according to SCOBLE (1999).

bidentata asiatica A. BANG-HAAS, 1907, Dt. ent. Z. Iris 20 (1/2): 85, (*Gonodontis*). Syntype(s) male(s), female(s), [Russia]: Sajan (Munku), Kentei, Amur, Saisan. [Originally as var].

bidentata harutai INOUE, 1953, Tinea 1 (1): 17, (*Gonodontis*). Syntypes 2 males, 5 females (The Natural History Museum, London, UK), Japan: Honshu, Nagano Prefecture, Azusa-yama; Nakabusaonsen; Gumma Prefecture, Marunuma; Kumanotaira; Nikko, Yumoto (examined externally). Valid at subspecies rank according to SCOBLE (1999).

bidentata kurilana BRYK, 1942, Dt. ent. Z. Iris 56: 80, (*Gonodontis*). Holotype male (Naturhistoriska Riksmuseet, Stockholm, Sweden), Kurile Islands: Yeterefu, Kamekotan. Valid at subspecies rank according to SCOBLE (1999)

bidentata rava VOJNITS, 1975, Annls hist.-nat. Mus. natn. hung. 67: 186, figs 3, 4, (*Gonodontis*). Holotype female (Hungarian Natural History Museum, Budapest, Hungary), Mongolia: Bulgan Aimak, 23 km NNE of Somon Chischig-Öndör, 1390 m. Valid at subspecies rank according to SCOBLE (1999).

UNAVAILABLE NAMES (incorrect subsequent authorship): *ochracea* FLECK, 1904 (LEMPKE 1970; should be FLECK); *trapezoides* SCHILLE, 1924 (WEHRLI 1939–1954, LEMPKE 1970; should be SCKILLE). – (homonymy): *defasciata* HANNEMANN, 1917 (junior primary homonym of *defasciata* KIEFER, 1916). – (incorrect spelling): *graecarus* (BELJAEV & VASILENKO 2014). – Intrasubspecific names, potentially unavailable: *approximata* LEMPKE, 1951 f.); *atrofasciata* FEICHTENBERGER, 1965 (f.); *bowateri* COCKAYNE, 1952 (ab.); *clausa* LEMPKE, 1951 (f.); *dealbidata* NORDSTRÖM, 1941 (ab.?); *defasciata* KIEFER, 1916 (ab.); *defasciata* HANNEMANN, 1917 (ab.?); *edentata* REBEL, 1910 (f.) (=synonym of *edentula* KRULIKOWSKY, 1908 according to The Global Lepidoptera Names Index (ANONYMOUS 2013)); *edentula* KRULIKOWSKY, 1908 (ab.); *fenestrata* COCKAYNE, 1946 (ab.?); *ferruginea* COCKAYNE, 1946 (ab.?); *fusca* LEMPKE, 1951 (f.); *fuscomarginata* LEMPKE, 1951 (f.); *hafneri* WAGNER, 1918 (ab.); *mediorufa* COCKAYNE, 1938/1939 (ab.); *nigra* PROUT, 1901 (ab.); *nigrofenestrata* COCKAYNE, 1946 (ab.?); *ochracea* FLECK, 1904 (ab.); *ochrea* DERENNE, 1927 (ab.); *pallida* LEMPKE, 1951 (f.); *paradoxa* COCKAYNE, 1946 (ab.); *semifasciata* COCKAYNE, 1948 (ab.?); *surtur* BANG-HAAS, 1907 (ab.) (=synonym of *nigra* PROUT, 1915 according to The Global Lepidoptera Names Index (ANONYMOUS 2013)); *trapezoides* SCKILLE, 1924 (ab.); *unifasciata* LEMPKE, 1970 (f.).

EXTERNAL CHARACTERS AND ABDOMEN: Wingspan 33–45 mm. Wings brown, irrorated with dark brown. Medial and postmedial line distinct, dark brown, dentate, often bordered with whitish. Medial area normally slightly darker than rest of wing. Hindwings lighter, more grey, only postmedial line well-marked. Discal spots with blackish margins and pale centre. Terminal line blackish, discontinuous. Fringes concolorous with wings, weakly chequered with white. Wing margins undulating, more pronounced on forewings. Wings below as above, markings darker. Frons, collar, thorax and abdomen concolorous with wings. Male antennae bipectinate, female antennae filiform. Male and female hindleg with 2+2 spurs. Tympanal organs medium-sized. Other abdominal sternites and tergites of both sexes undifferentiated.

VARIATION: Ground colour varies from rather pale greyish brown to yellowish brown to dark brown. Medial and postmedial lines dentate, weakly dentate or smooth. Outer margin of medial area dark, combined with smooth postmedial line, bordered with white rather commonly in (high altitude) specimens from Greece, Bulgaria (and Turkey?). Postmedial line often discontinuous on forewings, broken into row of small blackish spots. Melanistic specimens, almost uniform black-brown are also encountered (MIKKOLA *et al.* 1989). Numerous forms have been described, which are summarised in LERAUT (2009). See remarks.

MALE GENITALIA: Uncus long, stout, weakly setose. Socii minute, setose. Gnathos arms fused ventrally, upturned, apex weakly dentate. Valva narrow, with prominent, sclerotised, setose medial ridge, extending beyond apex. Costa weakly sclerotised, bases extended, almost meeting medially. Apical half of ventral margin membranous. Juxta sclerotised plate. Furca arms long, recurved (Text-fig. 95). Saccus round. Aedeagus stout, weakly curved, caecum long, apex round. Vesica opens at approximately 90 degree angle. Vesica base with about 6–8 cornuti, which lie on outside of aedeagus in uneverted vesica. Vesica base with small lateral diverticulum, medial part straight tube, apex with long, striated diverticulum, covered densely with long cornuti. See remarks.

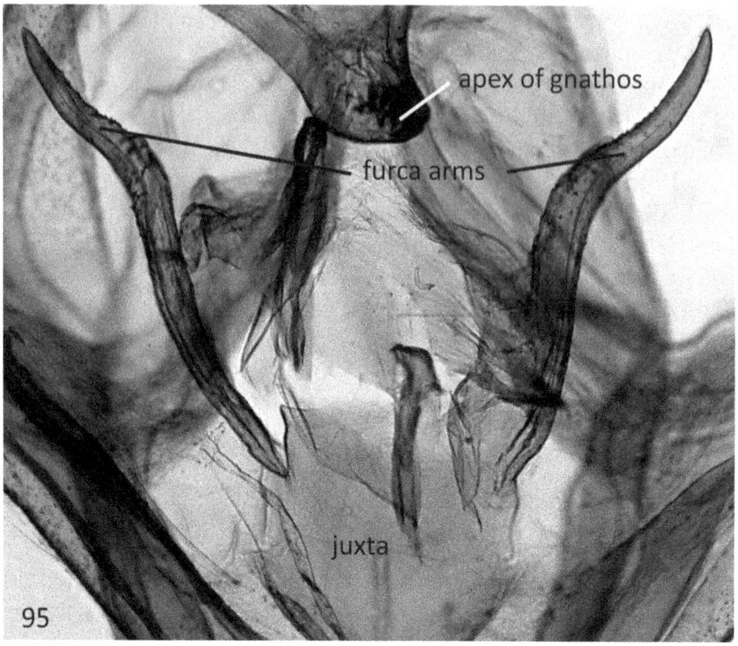

Text-fig. 95. Long, recurved furca arms of *Odontopera bidentata* (CLERCK, 1759), a diagnostic character in European *Odontopera* STEPHENS, 1831. Finland: Savonlinna, 11–14.vi.1996, slide PS1304 (coll. P. SIHVONEN).

FEMALE GENITALIA: Papillae anales small, setose. Apophyses anteriores about 1/2 length of apophyses posteriores. Lamella postvaginalis striated horizontally. Lamella antevaginalis wide, sclerotised plate, margin weakly concave to weakly convex. Lamella antevaginalis with elongated, lateral extensions. Ductus bursae very short, wide. Base of ductus seminalis weakly expanded. Posterior part of corpus bursae long, narrow, striated, sclerotised. Anterior part of corpus bursae round, rather large, membranous. Signum round, shortly stellate, with medial concavity. See remarks.

DISTRIBUTION AND ABUNDANCE: Eurasiatic. In Europe from northern central Fennoscandia to north-eastern Spain, southern France including Corsica, central Italy and the Balkans and from the British Isles through central and eastern Europe to European Russia and the Ural mountains. In Sicily only one isolated record from Castelbuono (coll. SKOU). – Outside Europe the species occurs in Turkey and Caucasus and from the Urals through southern Siberia, Mongolia and Far East Russia to Japan, Sakhalin and the Kuriles. – In Europe a common species that rather rarely occurs in numbers.

PHENOLOGY: Univoltine. In southern Scandinavia from mid-May to late June, in south-western Germany from early May to mid-June (EBERT 2003) and in Spain from late April to early July, depending on altitude (REDONDO *et al.* 2009). Larvae from June to September. The species overwinters as a pupa. The moth flies at night and comes readily to light.

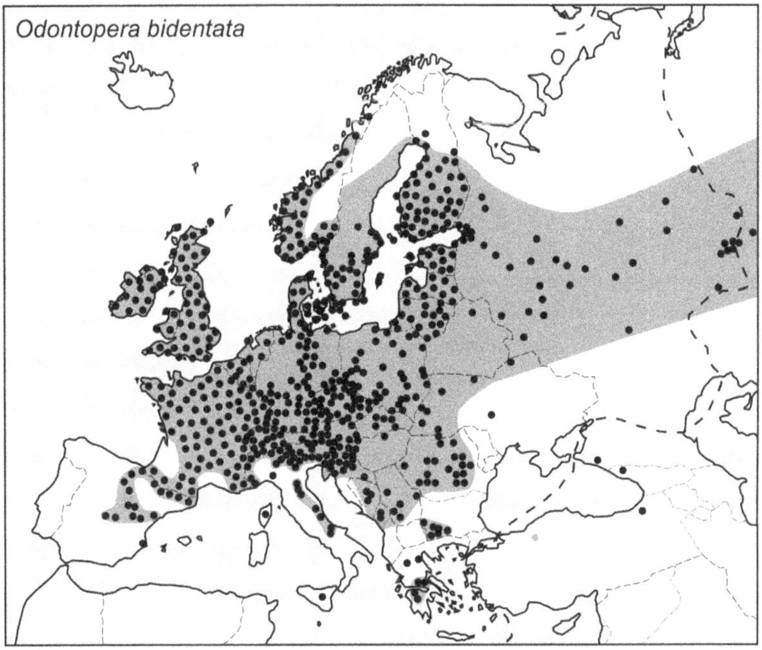

BIOLOGY: Polyphagous. The caterpillar has been found on *Picea abies*, *Salix alba*, *S. caprea*, *Sorbus aucuparia*, *Prunus spinosa* and *Cornus sanguinea* (EBERT 2003), on *Vaccinium myrtillus* and *Cytisus scoparius* (H. WEGNER pers. comm.). Further it has been recorded on *Betula*, *Populus* and *Salix* (AIZPÚRUA 2002). Also found on *Lonicera xylosteum*, *L. tatarica* and *Sorbus aucuparia* (BERGMANN 1955). Larvae have been reared on *Lonicera xylosteum* as food plant (B. MÜLLER pers. comm.). *O. bidentata* f. *graecarius* has been reared on *Vaccinium myrtillus* and *Salix* (B. MÜLLER pers. comm.). One of the most polyphagous Ennominae in Europe. Pupation among moss on the ground.

HABITAT: The species occurs in a large variety of habitats with trees and bushes. Vertically it is found from sea level up to 1200 m in the Black Forest in south-western Germany (EBERT 2003) and up to 2000 m in Spain and Italy (REDONDO *et al.* 2009, FLAMIGNI *et al.* 2007). The form *graecarius* from southern Balkan occurs in open mountain forests and woodland and it has mainly been found at higher altitudes from 1500 m up to 1800 m.

SIMILAR SPECIES: No similar species in Europe. *Odontopera muscularia* STAUDINGER, 1892, which occurs in Central Asia (VIIDALEPP 1996), is very similar. *O. muscularia* is illustrated and diagnosed in BELJAEV & VASILENKO (2014).

GENETIC DATA: *O. bidentata* (n=32, from western Europe to central Asia) and *O. graecarius* (treated as a junior subjective synonym of *O. bidentata* by us) (n=11) DNA barcode sharing, forming a homogeneous cluster. Nearest species: An unidentified *Odontopera* species from China (3.8%).

Table 3. Quantitative variation of the genitalia characters in *O. bidentata* (CLERCK, 1759). Other characters are rather uniform in size and shape. The specimens classified earlier as *O. graecarius* (BANG-HAAS, 1910), seem to represent an end of (clinal?) variation, and fall within intraspecific variation. The cornuti are easily detached during copulation, thus their number must be treated with caution. The detached cornuti leave a rupture on surface of the vesica.

Character	**Type of variation**
Corpus bursae	size: large to small [*graecarius*: small n=2]
Margin of lamella antevaginalis	shape: weakly concave to straight to convex [*graecarius*: convex, n=2]
Base of ductus seminalis	size: moderately to weakly expanded
Signum	size: concave medial area small to large
Cornuti	number and size: about 6–8, minor differences in size

REMARKS: We consider *Gonodontis graecarius* BANG-HAAS, 1910 a junior subjective synonym of *Odontopera bidentata* (CLERCK, 1759), quoting the absence of qualitative differences in the external appearance, male and female genitalia, and identical barcode sharing with *O. bidentata*. Taxon *graecarius* has recently been combined with *Odontopera*, and it has been considered valid at species level (MÜLLER 1996, SCOBLE 1999, HAUSMANN, MIRONOV & VIIDALEPP 2011). Most (all?) specimens identified as *O. graecarius* have been collected from high altitudes, from about 1100 metres and above. The external appearance of such specimens from Greece, Bulgaria and Turkey, is somewhat distinct (see plate 9, figures 87d–f), but the differences are not clearly disjunct, and fall within intraspecific variation. Furthermore, such specimens occur sympatrically with less contrasting specimens. Both male and female genitalia have minor quantitative differences, the extremes are illustrated in the genitalia plates. The differences are not qualitative and fall within intraspecific variation. Table 3 summarises results of the genitalia morphology from the examined specimens (n=8). We have not examined the (holo)type female of *Gonodontis graecarius in situ* (potentially in Senckenberg Natural History Collections of Dresden, Germany), but the identity of the taxon is not in doubt thanks to a good grey-scale photograph of the (holo)type in the original description (BANG-HAAS 1910, plate 4, fig. 3). The (holo)type is contrasting, similar to a specimen figured in 87f in plate 9. BELJAEV & VASILENKO (2014) report *O. graecarius* as new to the Caucasus fauna. The diagnostic characters given in that publication for *graecarius* (for instance the spine-like sclerotisation in the costa of valva, uncus shape and lamella antevaginalis shape), fall in our opinion within the intraspecific variation.

Odontopera bidentata subsp. *oreas* HERBULOT from France was treated valid at subspecies rank by LERAUT (1997, 2009). We follow HAUSMANN *et al.* (2011), who treated *oreas* as a synonym of *O. bidentata* subsp. *bidentata*. DNA barcodes of examined material (38 specimens from 13 countries, including two specimens from French Alps) are very similar and do not provide evidence for a genetically disjunct population in the French Alps. The ground colour of *oreas* is greyish brown, falling within variation of *O. bidentata*.

Dasycorsa Prout, 1915

Dasycorsa Prout, 1915, in Seitz, Gross-Schmett. Erde 4: 332 [Replacement name]. Type species: *Dasycephala modesta* Staudinger, 1879, Turkey. Genus synonymy is after Scoble (1999).

Dasycephala Staudinger, 1879, Horae Soc. ent. ross. 14: 445. Type species: *Dasycephala modesta* Staudinger, 1879. [Junior homonym of *Dasycephala* Swainson, [1832] 1831 (Aves)].

Diversity and distribution: *Dasycorsa* is a monotypic genus, the only species is found in Turkey and in the adjacent regions (for instance Bulgaria, Romania, Greece and Cyprus).

External characters and abdomen: Wingspan 31–34 mm. Wings uniform reddish brown, irrorated with blackish. Hindwings greyish brown. Discal spots blackish, larger on forewings. Wings narrow, elongated, giving Noctuidae-like appearance. Male antennae bipectinate, pectinations very long. Hindleg tibia of both sexes with 2+2 spurs, outer spur shorter. Male 8th sternite unmodified.

Male genitalia: Uncus evenly narrowing, weakly setose. Gnathos absent. Valva wide, weakly divided. Costa sclerotised, apex setose. Ventral margin weakly setose. Juxta sclerotised plate, dorsal margin bilobed. Aedeagus with shovel-shaped, subapical extension. Vesica with one cornutus.

Female genitalia: Papillae anales wide. Apophyses anteriores rudimentary. Ductus bursae funnel-shaped. Posterior part of corpus bursae sclerotised, narrow. Anterior part of corpus bursae elongated, membranous. Signum small, irregular plate.

Biology: Larva of *D. modesta* is polyphagous, it has been found on *Spartium junceum*, *Calicotome villosa*, *Tanacetum parthenium* and *Quercus coccifera* (D. Fritsch pers. comm., Fritsch *et al.* 2014) and on *Bituminaria bituminosa* (E. Friedrich pers. comm.) and recorded on the leaves of *Vicia sativa* (Lepiforum 2013). Most adults have been recorded between December and May. Habitats include steppe forests, bushy areas and steppe meadows (Székely 2011, Lepiforum 2013).

Immature stages: Eggs are laid in groups, and are olive-green to olive-brown (Lepiforum 2013). Larva is green, with pale yellow lines between segments, with whitish longitudinal stripe laterally and one black lateral spot per segment (Lepiforum 2013). There is no detailed information on the pupa, apart from a superficial photo in Lepiforum (2013). Pupation takes place in the ground, in a loose sandy cocoon (Lepiforum 2013).

Remarks: *Dasycorsa* has recently been classified in the Colotoini (Viidalepp 1996) and Gonodontini (Hausmann, Mironov & Viidalepp 2011). Gonodontini, as defined by Holloway (1994), and based on *Gonodontis clelia* (Cramer, 1780) from India, is different and inclusion of *Dasycorsa* in the Gonodontini is not supported by morphology. Based on the material we have examined, *Dasycorsa* Prout and *Colotois* Hübner are morphologically similar, and possibly also *Compsoptera* Blanchard, but the potential relationships should be studied in an analytical context. Systematic position: Colotoini.

As is explained in more detail in *Ennominae systematics* chapter, the European Ennominae fauna as a whole has not previously been classified into tribes. The above-mentioned similarities were noted so late in the manuscript preparation that rearrangement of taxa to reflect such affinities was not possible.

88. *Dasycorsa modesta* (STAUDINGER, 1879)

modesta modesta STAUDINGER, 1879, Horae Soc. ent. ross. 14 (4): 455, (*Dasycephala*). Syntype(s) female(s), [Turkey]: Taurus, [Toros Daglari].

modesta marerufa WILTSHIRE, 1940, Proc. R. ent. Soc. Lond. (B) 9 (5): 82, (*Dasycorsa*). Holotype male (The Natural History Museum, London, UK), Lebanon (examined externally).

INTRASUBSPECIFIC NAMES, POTENTIALLY UNAVAILABLE: *astigmatica* WILTSHIRE, 1939 (f.).

EXTERNAL CHARACTERS AND ABDOMEN: Wingspan 31–34 mm. Wings reddish to yellowish-brown, irrorated with blackish. Medial lines not visible. Hindwings greyish-brown. Discal spots blackish, larger on forewings. Wings narrow, elongated, giving Noctuidae-like appearance. Forewing margin weakly concave below apex. Terminal line blackish, fine. Fringes concolorous with wings. Wings below lighter, greyish. Frons, collar, thorax and abdomen concolorous with wings. Male antennae bipectinate, pectinations very long. Female antennae filiform, apical half very shortly fasciculate. Hindleg of both sexes with 2+2 spurs, outer spur shorter. Tympanal organs small. Other abdominal sternites and tergites of both sexes undifferentiated.

VARIATION: Ground colour varies from uniform dark, reddish brown (often males) to paler yellowish brown (often females). Wings may be almost without irroration, or heavily irrorated and partly striated with blackish. Veins may stand out lighter against darker background.

Subsp. ***marerufa*** has wings "of a rich reddish purple hue" (WILTSHIRE 1940). Described from Lebanon. HAUSMANN (1995a) has studied specimens from Israel and reports that caudal end of the juxta is slightly forked, whereas other material examined, from Macedonia and Turkey, has caudal end of the juxta blunt ended. Based on the material we have examined, the variation is clinal. HAUSMANN (1991) did not recognise subspecies *marerufa* from Jordan. Geographical extent and subspecies status of *marerufa* needs further examination.

MALE GENITALIA: Uncus wide at base, evenly narrowing, weakly setose. Socii small, setose. Gnathos absent. Valva wide, weakly divided, medial margin membranous. Costa sclerotised, wide, apex setose. Medial area with evenly curved, narrow, sclerotised ridge. Ventral margin weakly setose. Valva bases with small pockets. Juxta sclerotised plate, dorsal margin wider, bilobed. Saccus small, margin round. Aedeagus narrow, weakly curved, caecum long, apex with shovel-shaped extension. Vesica opens at approximately 135 degree angle. Vesica narrow, membranous tube, with one long, 'backwards' directed cornuti.

FEMALE GENITALIA: Papillae anales wide, setose. Apophyses posteriores long, narrow. Apophyses anteriores rudimentary. Lamella postvaginalis tiny, irregular plate. Lamella antevaginalis weakly sclerotised, margin concave medially. Accessory glands large, long, narrow. Ductus bursae funnel-shaped, anterior end with sclerotisation. Posterior part of corpus bursae sclerotised, narrow. Anterior part of corpus bursae elongated, weakly wedge-shaped, membranous. Signum minute, partly stellate or weakly sclerotised, irregular plate.

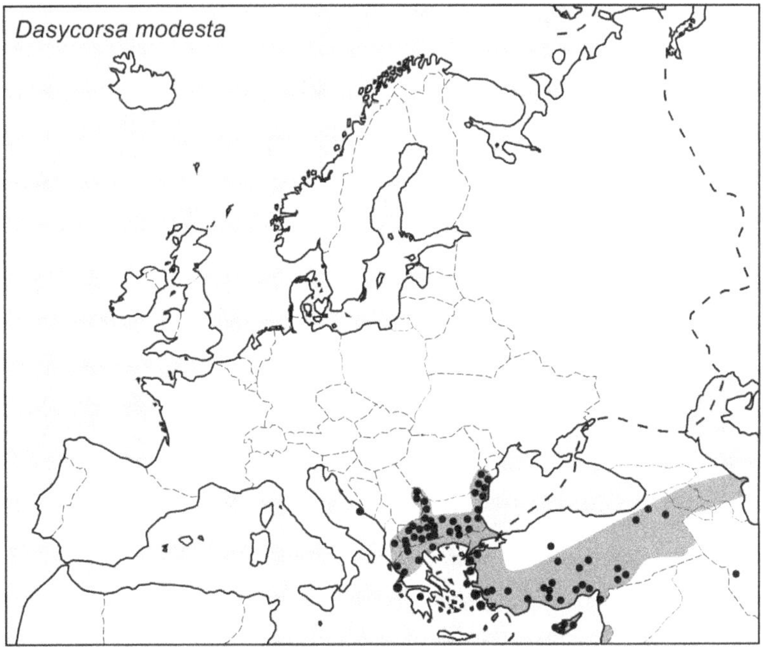

DISTRIBUTION AND ABUNDANCE: East-Mediterranean. In Europe only known from the Balkans. – Outside Europe the species has been found in Jordan, Israel, Lebanon, Cyprus, Turkey, Azerbaijan, Armenia and Iran. – In Europe apparently common, though perhaps somewhat local. It occurs not rarely in numbers.

PHENOLOGY: Univoltine. The species flies in northern Greece from late March to late May. In southern Greece from late February to late April (M. LEIPNITZ pers. comm.). In Samos from mid-February to early April (FRITSCH *et al.* 2014). In Romania from mid-April to late May (L. SZÉKELY pers. comm.). In Cyprus (coll. SKOU) and the whole Near East (A. HAUSMANN pers. comm.) flying in winter from late November to early April. Caterpillar from late April to late May (FRITSCH *et al.* 2014). The moths fly at night and come readily to light.

BIOLOGY: Polyphagous. Larvae have been found in Samos on *Spartium junceum*, *Calicotome villosa*, *Glebionis coronarium*, *Glycyrrhiza glabra* and *Quercus coccifera* (FRITSCH *et al.* 2014). Also found on *Bituminaria bituminosa* (E. FRIEDRICH pers. comm.). Recorded on *Astragalus ponticus* and *Lotus* (FAJCIK 2003).

HABITAT: In open habitats with scattered trees and bushes. The species occurs from sea level and up to 1000 m in north-western Greece and up to 1200 m in Turkey.

SIMILAR SPECIES: No similar species in Europe. *D. modesta* is externally somewhat similar to numerous Noctuidae species, for instance those in the genus *Conistra* HÜBNER, 1821.

GENETIC DATA: Genetically homogeneous in south-eastern Europe (n=4) but with three other BINs in the Middle East (n=10). Nearest species in Europe: *Ennomos quercinaria* (7.4%).

Crocallis TREITSCHKE, 1825

Crocallis TREITSCHKE, 1825, Schmett. Eur. 5 (2): 431. Type species: *Phalaena elinguaria* LINNAEUS, 1758, [Europe?]. Genus synonyms are after SCOBLE (1999).

Crocallis STEPHENS, 1829, Nom. Br. Insects: 43. Type species: *Phalaena elinguaria* LINNAEUS, 1758. [Junior homonym and junior objective synonym of *Crocallis* TREITSCHKE, 1825].

Dzhugesia WEHRLI, 1936, Ent. rdsch. 54: 145. Type species: *Crocallis rjabovi* WEHRLI, 1936.

UNAVAILABLE NAMES (included in work rejected for nomenclatural purposes): *Eusarca* HÜBNER, 1806 (see FLETCHER 1979). – (incorrect subsequent spellings): *Crocollis* HEYDENREICH, 1846; *Crocalis* LUCAS, 1846 (see FLETCHER 1979).

DIVERSITY AND DISTRIBUTION: According to SCOBLE (1999) *Crocallis* contains 18 species, most occur in the Palaearctic region. PITKIN (2002) noted that the Neotropical taxa assigned to *Crocallis* are misplaced, thus reducing the number of included species to 15. STADIE & FIEBIG (2014) revised the *Crocallis elinguaria* species lineage and described five new species, thus the current number of *Crocallis* species is 20, but potentially some species may be misplaced in closely related genera *Scodiomima* STAUDINGER, 1892, *Odontopera* STEPHENS, 1831, *Aethiopodes* WARREN, 1902.

EXTERNAL CHARACTERS AND ABDOMEN: Medium-sized, wingspan about 30–35 mm. Wings different shades of yellow and brown. Medial and postmedial lines prominent, bordering darker medial area. Forewing discal spot blackish, conspicuous. Male antennae bipectinate, female antennae filiform or shortly fasciculate. Proboscis often reduced. Male 8th sternite unmodified.

MALE GENITALIA: Uncus prominent. Gnathos arms large, separate or fused medially, upturned, apex weakly dentate. Juxta with two long extensions. Valvae elongated, evenly curved, dorsal margin variably setose, apex with prominent setae. Vesica simple tube, base often slightly expanded. Cornuti outside aedeagus (i.e. not concealed inside the aedeagus) in uneverted vesica, located at base of vesica.

FEMALE GENITALIA: Region around ostium bursae heavily sclerotised. Ductus bursae often prominent, sclerotised. Corpus bursae thin sac that often loses its shape during preparation. Signum small, weakly stellate.

BIOLOGY: Larvae of *Crocallis* species have been recorded feeding on numerous plants, for instance on *Crataegus, Prunus spinosa, Ulmus, Salix, Rosa, Vaccinium myrtillus, Lonicera xylosteum, Cupressus* (EBERT 2003, REDONDO *et al.* 2009). According to current knowledge, most species are oligo- or polyphagous (REDONDO *et al.* 2009). Adults fly in autumn, egg overwinters.

IMMATURE STAGES: Larva of *C. elinguaria* and *C. tusciaria* are twig-mimics, being various shades of brown, grey or violet, dorsally with row of diamond-shaped markings and 11th segment with small dorsal hump (EBERT 2003, PORTER 2010). Pupa cremaster of *C. elinguaria* is with four pairs of curved setae, apical pair strongest (PATOČKA & TURČANI 2005).

REMARKS: *Crocallis* species are similar in external apperance, and examination of the genitalia is often needed to identify the species. The world fauna has not been revised in a modern way. The Spanish species are reviewed by DANTART *et al.* (1993) and *C. elinguaria* complex of Asia Minor and the eastern adjacent territories are reviewed by STADIE & FIEBIG (2014).

Crocallis has recently been classified in the Ennomini (VIVES MORENO 1994, VIIDALEPP 1996, KULLBERG *et al.* 2001), Crocallini (LERAUT 1997) and in Gonodontini (HAUSMANN, MIRONOV & VIIDALEPP 2011). BELJAEV (2008a) noted that the European authors have followed the tradition founded by PIERCE (1914) and HERBULOT (1963–1964), who included in the tribe Ennomini a series of remotely related genera, united by habitus rather than strictly morphological similarity. Consenquently BELJAEV (2008a) excluded majority of the European genera from the Ennomini, including *Crocallis*. He did not propose information where these excluded genera should be classified. The male and female structures, including the genitalia, are general Ennominae type and provide little information as regards its systematic position. *Crocallis* TREITSCHKE and *Odontopera* STEPHENS share an unusual trait in the male aedeagus: cornuti are outside the aedeagus (not concealed within it) in an uneverted vesica. Gonodontini, as defined by HOLLOWAY (1994), and based on *Gonodontis clelia* (CRAMER, 1780) from India, is different and inclusion of *Crocallis* in the Gonodontini is not supported by morphology. In a molecular, phylogenetic analysis of limited European taxa, ÕUNAP *et al.* (2011) recovered *Ennomos* as sister lineage to *Opisthograptis* and *Crocallis*. Systematic position: Ennomini of uncertain association. Further research is needed.

89. *Crocallis tusciaria* (BORKHAUSEN, 1793)

tusciaria tusciaria BORKHAUSEN, 1793, in Scriba, Beitr. Ins. Gesch. 3: 217, pl. 17, fig. 10, (*Phalaena Geometra*). Syntype(s) [Italy]: [Tuscany], Toscana.

extimaria HÜBNER, [1799], Samml. Eur. Schmett. 5 Geometrae (1): pl. 4, fig. 21, (*Geometra*). Syntype(s), lost, Europe.

tusciaria bieli BANG-HAAS, 1927, Horae macrolepidopt. Reg. palaearct. 1: 96, pl. 11, fig. 33, (*Crocallis*). Holotype male, Lusitania [Portugal]: Gerez.

gaigeri STAUDINGER, 1885, Stettin. ent. Ztg 46 (7–9): 349, (*Crocallis*). Syntypes male, female, [Dalmatia, Croatia]: Zara. Originally as var., see remarks.

tusciaria cypriaca FISCHER, 2003, Entomologische Zeitschrift 113(12): 372, figs 1, 2, 3, 5, 6 (*Crocallis*). Holotype male (private coll. H. FISCHER, Rottach, Germany), Cyprus: Schlucht von Mylikouri, paratypes 3 males (two males in private coll. H. FISCHER, Rottach, Germany, one male in private coll. LEWANDOWSKI, München, Germany), Cyprus: Schlucht von Mylikouri (holotype examined from literature, the original description contains photographs, including genitalia). Originally as sp., downgraded here to a subspecies of *Crocallis tusciaria* (BORKHAUSEN, 1793), quoting allopatric occurrence (new status). See remarks.

tusciaria taurica WEHRLI, 1934, Mitt. münch. ent. Ges. 24 (1): 41, (*Crocallis*). Syntype(s), [Syria]: South-eastern Taurus [Toros Daglari], Akbés [Maydan Ikbis], 800–1000 m. [Paratype male in the Natural History Museum, London, UK (examined externally). Originally as var., tentatively valid at subspecies level, See remarks.

tusciaria transcaucasica WEHRLI, 1940, in SEITZ, Gross-Schmett. Erde 4 (Suppl.): 347, pl. 27: c, (*Crocallis*). Syntype(s), [Central Asia]: Arax River, Negram. Valid at subspecies rank according to SCOBLE (1999).

UNAVAILABLE NAMES (manuscript name): *friedrichi* MÜLLER (See remarks). – Intrasubspecific names, potentially unavailable: *aurantiaca* DURAND, 1934 (ab.); *virgata* REBEL, 1911 (ab.).

EXTERNAL CHARACTERS AND ABDOMEN: Wingspan 31–40 mm. Wings ochreous yellow, rather sparsely irrorated with blackish, medial area darker. Medial line brown, bordered with whitish, angled, meets inner margin at about 90 degree angle. Postmedial line brown, bordered with whitish, angled outwards in middle. Hindwings paler, postmedial brownish, suffuse, bordered with whitish. Discal spots narrow, blackish. Terminal line discontinuous row of black dots, strongest on upper half of forewing. Fringes concolorous with wings. Wings below more suffused, medial line reduced. Proboscis vestigial. Frons, collar, thorax and abdomen concolorous with wings. Male antennae bipectinate. Female antennae fasciculate. Hindtibia of both sexes with 2+2 spurs. Tympanal organs medium-sized. 2nd sternite sclerotised. Other abdominal sternites and tergites of both sexes undifferentiated.

VARIATION: Ground colour varies from ochreous yellow to yellow-brown to greyish brown to brown (f. *gaigeri*), which becomes more common towards South. In the Iberian Peninsula it is the most common form (DANTART *et al.* 1993). Medial area normally darker yellow than rest of wing, but it can be uniform brown. Density of blackish irroration variable. Medial and postmedial lines normally brown, weakly bordered with whitish, rarely brown is absent, thus lines whitish. Hindwings paler, colour ranging from pale ochreous yellow to greyish brown.

Subsp. ***cypriaca*** is more contrasting, especially medial area is often darker, even blackish (Text-figs 96–98). Forewing postmedial line more curved towards termen. According to MÜLLER (2004), there are minor quantitative differences in the genitalia when compared to *tusciaria*, and the pectinations in the male antennae of *cypriaca* are slightly longer in the middle part than in *tusciaria*. MÜLLER (2004) reports that the male vesica has 3–4 cornuti (*tusciaria* 3–7), but based on the examined material we have examined, *cypriaca* can have up to seven cornuti. Female genitalia are similar to *tusciaria*. Larva has been recorded feeding on *Calicotome villosa* (MÜLLER 2004). The taxon is endemic to Cyprus. See remarks.

Text-figs 96–98. Diagnostic characters (indicated) of *Crocallis tusciaria* subsp. *cypriaca* FISCHER, 2003. Text-fig. 96. Male, Cyprus: Pegeia, 22.xii.1997 (coll. MÜLLER). Text-fig. 97. Male, Cyprus: Gecitköy Göleti, 29.xi.2010 (coll. SKOU). Text-fig. 98. Female, Cyprus: Troodos Mountains, ex ovo 20.x.2009 (coll. MÜLLER). Variation in the external characters is illustrated more extensively in MÜLLER (2004).

Subsp. ***taurica*** is without clear diagnostic morphological features. It is usually bigger, more contrasting, pale straw-yellow in colour. According to WEHRLI (1939–1954) the uncus is slightly narrower and aedeagus is with four long cornuti. Tentatively valid at subspecies level, distribution unclear (Turkey, Eastern Mediterranean?), further studies are needed. See remarks.

MALE GENITALIA: Uncus wide at base, evenly narrowing, stout, weakly setose. Socii minute, setose. Gnathos arms fused ventrally, upturned, apex weakly dentate. Valva wide at base, apex setose. Sclerotised medial ridge with tooth. Costa sclerotised, wide, almost meeting dorsomedially. Juxta large sclerotised plate. Furca arms long, weakly curved. Saccus slightly angular. Aedeagus straight, caecum long, apex round. Vesica opens at approximately 45 degree angle. Vesica base with 3–7 long cornuti, lying on outside of aedeagus in uneverted vesica. Vesica base expanded, slightly striated, apical part narrower.

FEMALE GENITALIA: Papillae anales large, setose. Apophyses anteriores about 1/2 length of apophyses posteriores. Lamella antevaginalis margin with high medial ridge. Lamella post-vaginalis with two lateral sclerotised plates, medial area weakly sclerotised, slightly striated horizontally. Ductus bursae membranous, base short, narrow, then wider. Base of ductus seminalis expanded. Corpus bursae wide, elongated, very thin (easily wrinkled when dissected). Signum minute, weakly sclerotised.

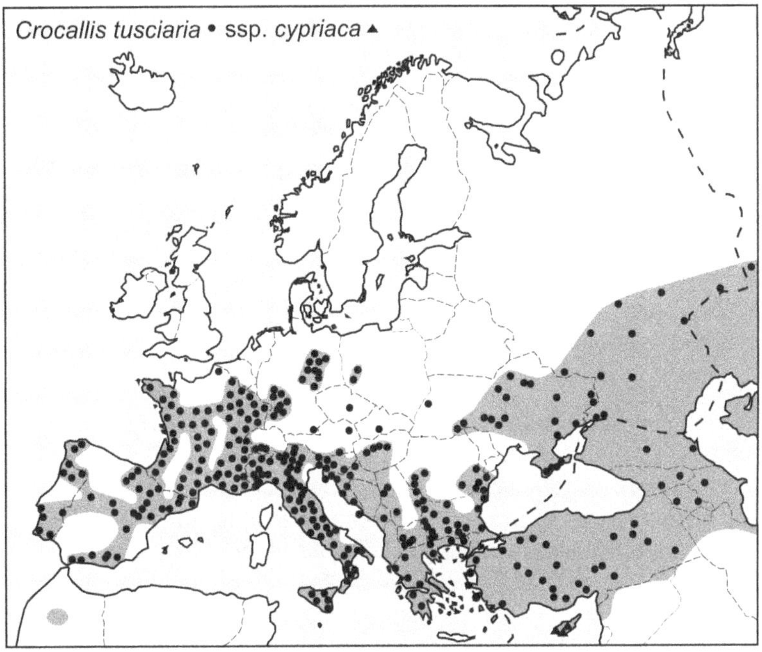

DISTRIBUTION AND ABUNDANCE: European-Westasiatic. In Europe mainly in the Mediterranean area from Portugal to Greece and in southern central Europe from France eastwards to Ukraine, European Russia and the southern Ural mountains. – Outside Europe the species occurs in Morocco, Cyprus (subsp. *cypriaca*), Turkey, Syria, Transcaucasus, Caucasus, Turkmenistan and northern Iran. – In Europe a rather common species that rather rarely occurs in numbers.

PHENOLOGY: Univoltine. In south-western Germany the species flies from early September to mid-October. In southern Europe from mid-September to mid-November according to EBERT (2003). In Spain in southern Andalusia the species is recorded in November and December (P. HALE pers. comm.). In Samos, the larva has been found from mid-March to late April (FRITSCH *et al.* 2014), in eastern Germany in May and June (BERGMANN 1955). The species overwinters as an egg. The moths fly at night and the species comes readily to light.

BIOLOGY: Polyphagous. In south-western Germany the larva has been found on *Crataegus*, *Prunus spinosa* and *Ulmus* (EBERT 2003), in eastern Germany on *Clematis* (BERGMANN 1955), in Bulgaria on *Carpinus betulus* (B. MÜLLER pers. comm.) and in Samos on *Erica arborea*, *Pistacia lentiscus*, *Prunus spinosa*, *Quercus coccifera* and *Crataegus* (FRITSCH *et al.* 2014). In Spain recorded on *Clematis vitalba*, *Berberis*, *Fraxinus*, *Salix*, *Cistus* and *Rosmarinus officinalis* (REDONDO *et al.* 2009). Caterpillars of *C. tusciaria* subsp. *cypriaca* have been found on *Calicotome spinosa* and reared on *Lonicera xylosteum* (B. MÜLLER pers. comm.). The pupation takes place in a web on the ground.

HABITAT: In open wood- and scrubland and in hedges in agricultural areas. The species occurs from sea level up to 1600 m.

SIMILAR SPECIES: Dark, greyish brown specimens (f. *gaigeri* STAUDINGER) may resemble *C. auberti* OBERTHÜR. Diagnostic characters of European *Crocallis* TREITSCHKE, 1825 are summarised in Table 4 and shown in Text-figs 99–105.

GENETIC DATA: Genetically heterogeneous with regional clusters (n=28; ten countries). Main cluster from populations from Italy and Hungary to Greece comparatively homogeneous, but populations from Sicily and Iberian Peninsula diverging by 2.7% and 4.3%. The haplotype of the cluster of subspecies *taurica* (Turkey to Israel) diverging by 1.1% and occurring also in Bulgaria. *C. tusciaria* subsp. *cypriaca*, diverging by 6.2%, see remarks.

REMARKS: We retain *taurica* WEHRLI from Syria valid at subspecies level beceause the DNA data from nearby areas, combined with geographical differences in the facies, is maybe indicative of disjunct populations. More extensive material throughout the distribution area needs to be analysed to have the issue solved.

We consider taxon *cypriaca* FISCHER, originally described as *C. cypriaca*, to be a subspecies of *C. tusciaria*. It is externally diagnosable, but the genitalia are similar to those of nominate subspecies, only minor quantitative differences have been observed (MÜLLER 2004). Because *cypriaca* is known from Cyprus only, thus being a reproductively disjunct population, it is largely a semantic (subjective) question whether it should be treated valid at species or subspecies level. We emphasize similarities with the continental populations and the connection is best conceived and communicated to the reader by linking the taxon nomenclatorially to taxon *tusciaria*. DNA barcode diverges by 6.2% from the main cluster, which can be interpreted to support isolation at species or subspecies level. In an European wide study on the Geometridae, it was found that congeneric species averaged 8.8% divergence (HAUSMANN *et al.* 2013). BERND MÜLLER, Germany, was in the process of describing the taxon from Cyprus as *C. friedrichi*, but H. FISCHER, Germany submitted his manuscript earlier. This is explained as an editorial footnote in MÜLLER (2004). Therefore *C. friedrichi* MÜLLER is an unavailable manuscript name that was never published, although it is mentioned in the literature (LERAUT 2009), without making the name available (cf Art. 16, ICZN 1999).

SCOBLE (1999) treated *gaigeri* STAUDINGER valid at subspecies level. This view has not been adopted elsewhere in the recent literature, where *gaigeri* has been treated as a dark form (for instance DANTART *et al.* 1993, FLAMIGNI *et al.* 2007, ROBINEAU 2007, LERAUT 2009). According to FLAMIGNI *et al.* (2007) f. *gaigeri* coexists everywhere in Italy with more yellow specimens. Our observations support it being a dark form, coexisting with yellow forms.

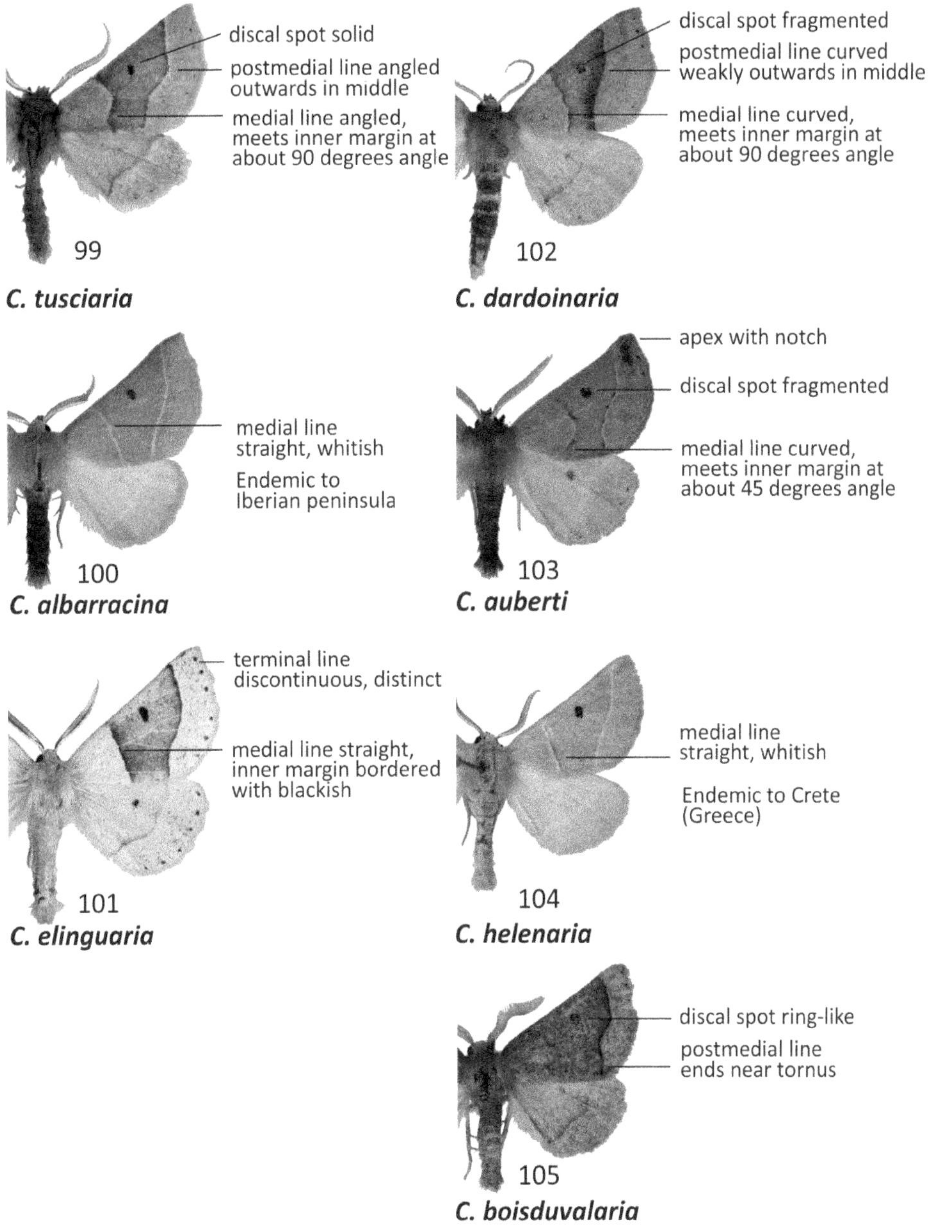

Text-figs 99–105. Diagnostic characters (indicated) of European *Crocallis* TREITSCHKE, 1825 species. Text-fig. 99. *C. tusciaria* (BORKHAUSEN, 1793), Greece: Thrakia, 23.x.1984. Text-fig. 100. *C. albarracina* WEHRLI, 1940, Spain: Teruel, 27.viii.2001. Text-fig. 101. *C. elinguaria* (LINNAEUS, 1758), Germany: Kyffhäuser, 03.viii.1984. Text-fig. 102. *C. dardoinaria* DONZEL, 1840, Spain: Teruel, 4.viii.2007). Text-fig. 103. *C. auberti* OBERTHÜR, 1883, Spain: Huesca, 13–14.ix.2002. Text-fig. 104. *C. helenaria* RUCKDESCHEL, 2006, Greece: Crete, 11–17.x.2003. Text-fig. 105. *C. boisduvalaria* (LUCAS, 1849), Italy: Sicily, 5.x.1927. *Crocallis* species are similar in external apperance, especially *C. dardoinaria - C. auberti and C. elinguaria - C. albarracina - C. helenaria*, and examination of the genitalia is often needed to identify the species.

Table 4. Diagnostic genitalia characters of European *Crocallis* TREITSCHKE, 1825 species. *Crocallis* species are similar in external apperance, especially *C. dardoinaria* - *C. auberti and C. elinguaria* - *C. albarracina* - *C. helenaria*, and examination of the genitalia is often needed to identify the species.

	Genitalia characters	Similar species
C. tusciaria	aedeagus straight medial ridge of valva with one tooth valva wide at base gnathos arms fused lamella antevaginalis margin with high medial ridge	*C. auberti*
C. albarracina	vesica with 2–4 cornuti furca arms short gnathos arms separate lamella postvaginalis not sclerotised, few striations only	*C. elinguaria,* *C. helenaria*
C. elinguaria	vesica with one cornutus furca arms long gnathos arms separate lamella postvaginalis horizontal sclerotised plate	*C. albarracina,* *C. helenaria*
C. dardoinaria	furca arms curved, almost fused gnathos arms fused lamella antevaginalis margin with medial extension signum present apophyses anteriores and apophyses posteriores equal length	*C. auberti*
C. auberti	furca arms straight, separate gnathos arms fused lamella antevaginalis margin without medial ridge signum absent apophyses anteriores about 1/2 length of apophyses posteriores	*C. dardoinaria,* *C. tusciaria*
C. helenaria	uncus short, wide costa densely setose valva wide gnathos arms separate lamella postvaginalis with two sclerotised, dentate appendices	*C. albarracina,* *C. elinguaria*
C. boisduvalaria	aedeagus stout, straight valva narrow at base medial ridge of valva without tooth gnathos arms fused lamella antevaginalis with narrow lateral projections, margin concave	No similar species

90. *Crocallis albarracina* WEHRLI, 1940

albarracina albarracina WEHRLI, 1940, in SEITZ, Gross-Schmett. Erde 4 (Suppl.): 347, pl. 27: b, (*Crocallis*). Syntype(s) (Zoologisches Forschungsinstitut und Museum Alexander Koenig, Bonn, Germany), [Spain]: Aragon, Albarracin. [Paratype male in the Natural History Museum, London, UK (examined externally)]. [Originally as subsp. of *elinguaria*].

albarracina centraliberica GASTÓN & REDONDO, 2009, Geometridae Ibericae: 57, figs 59d–f, (*Crocallis*). Holotype male (Museo Nacional de Ciencias Naturales, Madrid, Spain), [Spain]: Puerto de Pasapán (Segovia), 1700 m (examined, based on literature). Originally as subsp. of *albarracina* WEHRLI, 1940, downgraded here from subspecies rank (GASTÓN & REDONDO, 2009) to junior subjective synonym of *albarracina* WEHRLI, 1940, quoting sympatric occurrence and absence of disjunct characters (new synonym). See Remarks.

EXTERNAL CHARACTERS AND ABDOMEN: Wingspan 30–38 mm. Wings uniform ochreous yellow, with tint of brownish, medial area slightly darker. Medial line whitish, straight. Postmedial line whitish, weakly curved outwards in middle. Hindwings whitish, terminal area weakly yellowish, trace of postmedial barely visible. Forewing discal spot large, slightly elongated. Terminal line and fringes concolorous with wings. Wings below lighter, postmedial suffused, discal spots faint. Proboscis absent. Frons, collar, thorax and abdomen concolorous with wings. Antennae whitish dorsally. Male antennae bipectinate, female antennae filiform. Hindtibia of both sexes with 2+2 spurs. Tympanal organs medium-sized. Sternites and tergites 3–8 of both sexes undifferentiated.

VARIATION: Forewing medial area normally slightly darker than rest of wing, but can be concolorous with rest of wing or brownish. Specimens from the mountains of Sistema Central in central Spain (*centraliberica*) are slightly smaller than other Spanish specimens, but wingspan can reach 38 mm (see adult plate 9). Wing colour of these specimens is variable, but often irrorated with brown and medial area is darker. Legs and palpi are dark.

MALE GENITALIA: Uncus weakly setose, wide at base, slightly constricted below apex. Socii small, setose. Gnathos arms separate, angled upwards, apex acute. Valva narrow at base, subapex densely setose, costa with few small setae. Costa sclerotised, narrow. Juxta small, ventrolateral margins curved inwards. Furca arms short, vestigial. Saccus round. Aedeagus curved, caecum long. Vesica opens at approximately 135 degree angle. Vesica base with 2–4 long cornuti, those lying on outside of aedeagus in uneverted vesica. Vesica base slightly expanded, apical part narrower.

FEMALE GENITALIA: Papillae anales large, setose. Apophyses anteriores about 1/2 length of apophyses posteriores. Lamella antevaginalis funnel-shaped, margin slightly concave medially. Lamella postvaginalis not sclerotised, with few lateral striations only. Ductus bursae mostly sclerotised, wide, dorsoventrally flattened, with one hump. Posterior part of corpus bursae anteriorly narrowing, densely wrinkled. Anterior part of corpus bursae large, roundish sac. Signum small, weakly sclerotised, margins weakly stellate.

DISTRIBUTION AND ABUNDANCE: West-Mediterranean. Endemic to the Iberian Peninsula. Rather local, but rather often found in numbers.

PHENOLOGY: Univoltine. The species flies from late July or early August until late September. It flies earlier at higher than at lower altitudes. Hibernation takes place as an egg (B. MÜLLER pers. comm.). The adult moths fly at night and come to light.

BIOLOGY: Possibly polyphagous. All that is known is a caterpillar that was found on *Cupressus* in mid-April at an altitude of 175 m (leg. P. SKOU & B. SKULE), however, it can be questioned if *Cupressus* really is the food plant. In rearings, the caterpillars accepted *Lonicera xylosteum* as food plant (B. MÜLLER pers. comm.) and also *Prunus cerasifera*, *P. domestica* and *P. padus* (D. STADIE & T. DRECHSEL, pers. comm.).

HABITAT: In dry, open scrubland from 175 m up to 1400 m. Taxon *centraliberica* in open grassland with rocks and stones above the tree line, at 1400 m up to 1900 m.

SIMILAR SPECIES: *C. elinguaria* (LINNAEUS) and *C. helenaria* RUCKDESCHEL. Diagnostic characters of European *Crocallis* TREITSCHKE, 1825 are summarised in Table 4 and shown in Text-figs 99–105. We recommend that genitalia to be used for identification, particularly for south-European specimens.

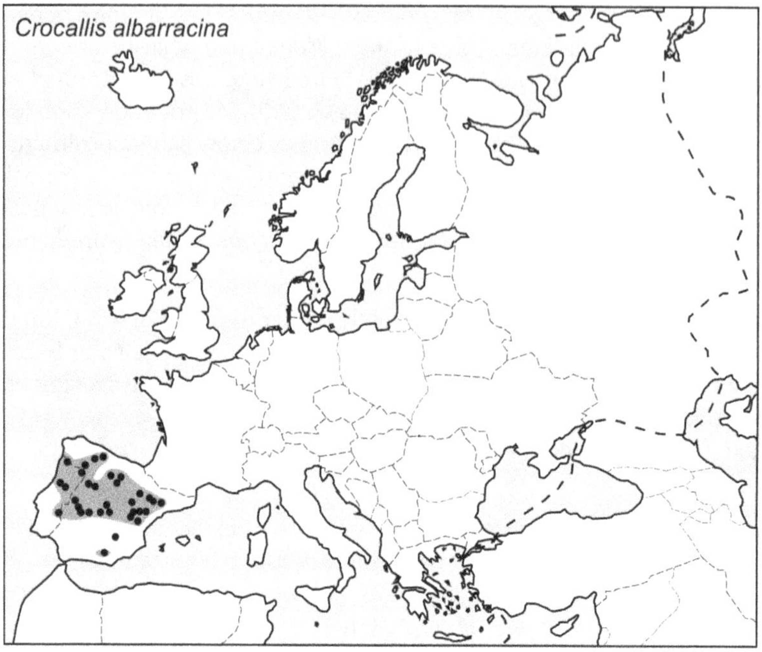

GENETIC DATA: Two specimens from southern Spain diverging by 1.4% from the main cluster of *C. elinguaria*. Further research needed to verify attribution of this cluster to *C. albarracina*. Other material from central and eastern Spain genetically barcode-sharing with *C. elinguaria*, though in habitus well corresponding to *C. albarracina*.

REMARKS: GASTÓN & REDONDO (2009) described *C. albarracina* subsp. *centraliberica* from the mountains of Sistema Central in central Spain. Normally *centraliberica* specimens are slightly smaller than other Spanish specimens, but wingspan can reach 38 mm (see adult plate 9). Wing colour is variable, but often irrorated with brown and medial area is darker. Legs and palpi are dark. The genitalia correspond to those of *C. albarracina*. We consider these specimens a mountain form of *albarracina* but note that further studies are needed, particularly because the terminal lines are discontinuous as in *elinguaria*. DNA barcodes are not available, so far, for *centraliberica* populations.

91. *Crocallis elinguaria* (LINNAEUS, 1758)

elinguaria LINNAEUS, 1758, Syst. Nat. (Ed. 10) 1: 520, (*Phalaena (Geometra)*). Syntypes (Linnean Society of London, London, UK), Germany (examined externally).

elinguaria prosapiaria ROBSON & GARDNER, 1886, Young Nat. 7 (Suppl.): 33, (Crocallis). Syntype(s), [United Kingdom]. [Originally as var].

elinguaria prunarioides HELBIG, 1940, Ent. Z. Frankf. a.M. 53 (33): 287, (*Crocallis*). Syntypes 4 males, [Italy: Trentino]: Lago di Garda, Torbole. [Originally as f].

trapezaria BOISDUVAL, 1840, Genera Index meth. eur. Lepid.: 184, (*Crocallis*). Syntype(s), [France]: Montpellier. [Originally as sp].

UNAVAILABLE NAMES (incorrect subsequent spelling): *trapecina* (CULOT 1919). – Intrasubspecific names, potentially unavailable: *aequaria* FUCHS, 1910 (ab.); *aurantiaca* LEMPKE, 1951 (f.); *brevipennis* COCKAYNE, 1946 (ab.?); *caeca* CANNAVIELLO, 1904 (ab.); *crassilineata* COCKAYNE, 1948 (ab.); *defasciata* LEMPKE, 1951 (f.); *delineata* LEMPKE, 1951 (f.); *depuncta* STEPHAN, 1925 (ab.); *fasciata* GILLMER, 1908 (ab.); *fusca* REUTTI, 1905 (ab.); *impuncta* LEMPKE, 1970 (f.); *insolitaria* FUCHS, 1901 (ab.) (=synonym of *fusca* REUTTI, 1905 according to The Global Lepidoptera Names Index (ANONYMOUS 2013)); *juncta* SCHILLE, 1918 (ab.); *marginenuda* LEMPKE, 1951 (f.); *moltrechti* OBERTHÜR, 1912 (ab.); *nigrolineata* LEMPKE, 1951 (f.); *obscura* NICKERL, 1897 (ab.); *obviaria* LJUNGDAHL, 1918 (ab.); *pallida* LEMPKE, 1951 (f.); *postlineata* LEMPKE, 1970 (f.); *radiata* COCKAYNE, 1949 (ab.); *reticulata* LEMPKE, 1951 (f.); *signatipennis* NEUSTEAD & SMITH, 1919 (ab.); *solitaria* FUCHS, 1910 (f.) (=synonym of *fusca* REUTTI, 1905 according to The Global Lepidoptera Names Index (ANONYMOUS 2013)); *triangularis* LEMPKE, 1970 (f.); *triangulata* CLOSS, 1920 (ab.); *undulata* LEMPKE, 1970 (f.); *unicolor* PROUT, 1915 (ab.).

EXTERNAL CHARACTERS AND ABDOMEN: Wingspan 32–42 mm. Forewings dark ochreous yellow, rather sparsely irrorated with blackish, medial area darker. Medial line blackish, straight, bordered with pale yellow. Postmedial line blackish, bordered with pale yellow, weakly curved outwards in middle. Hindwings lighter, postmedial line weakly visible. Forewing discal spot large, slightly elongated. Terminal line discontinous row of dark dots. Fringes concolorous with wings. Wings below lighter, markings weaker. Proboscis absent. Frons, collar, thorax and abdomen concolorous with wings. Antennae dark ochrous yellow dorsally. Male antennae bipectinate, female antennae filiform. Hindtibia of both sexes with 2+2 spurs. Tympanal organs small. Sternites and tergites 3–8 of both sexes undifferentiated.

VARIATION: Ground colour varies from pale yellow to deep ochreous yellow (see remarks). Medial area can be almost unicolorous with wings, normally slightly darker, but can be heavily irrorated with black. Medial lines can be absent, wide apart, partly joined together. Discal spots can be absent, also dark dots of terminal line. Numerous forms have been described, which are summarised in LERAUT (2009). See remarks.

MALE GENITALIA: Uncus weakly setose, wide to about 1/2 distance, apex narrow. Socii small, setose. Gnathos arms separate, angled upwards, apex blunt. Valva narrow at base, subapex densely setose, costa with row of long setae. Costa sclerotised, narrow. Juxta small, ventrolateral margins weakly curved inwards. Furca arms long, weakly curved. Saccus round. Aedeagus curved, caecum long. Aedeagus opens at approximately 180 degree angle. Vesica with one long cornutus, it lying on outside of aedeagus in uneverted vesica. Vesica base very slightly expanded, apical part narrower.

FEMALE GENITALIA: Papillae anales large, setose. Apophyses anteriores about 1/2 length of apophyses posteriores. Lamella antevaginalis funnel-shaped. Lamella postvaginalis with two large, perpendicular plates, partly fused medially and with narrow, extended, horizontal plate. Ductus bursae sclerotised, wide, dorsoventrally flattened, with one hump. Posterior part of corpus bursae narrow, membranous. Anterior part of corpus bursae large, roundish sac. Signum small, degree of sclerotisation variable, weakly stellate.

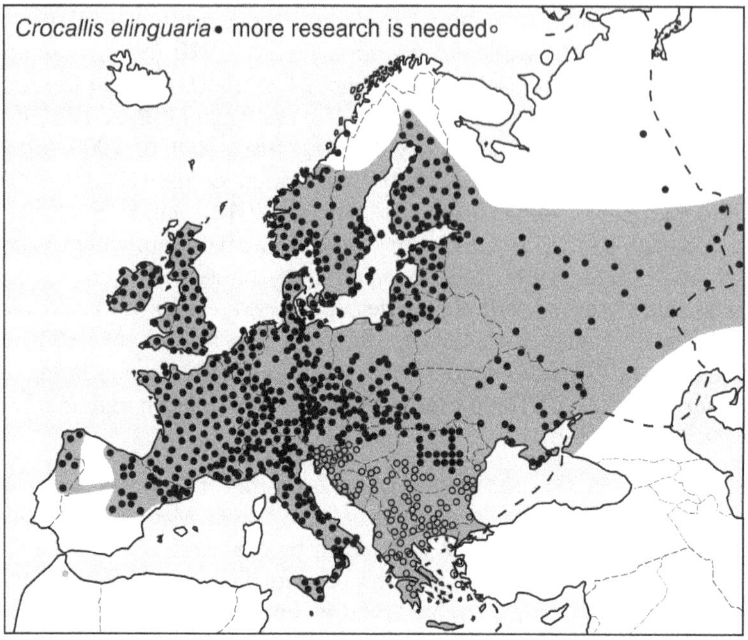

DISTRIBUTION AND ABUNDANCE: Eurasiatic. In Europe the species is widely distributed from northern Scandinavia to central Spain, southern France and Italy incl. Sicily, the Balkans, Crimea and southern European Russia and from the Urals to the British Isles. – Outside Europe the species has been found from northern central Asia and southern Siberia to Japan and toward the south-west as a single specimen from Morocco (RUNGS 1981). – In Europe a common species that rather rarely occurs in numbers. – There is potentially a sister species that is likely to occur in the Balkans, Greece, Bulgaria and in southern Romania (shown in the map as open dots). Its occurrence in Turkey and the Caucasus region is unclear.

PHENOLOGY: Univoltine. In southern Scandinavia from mid-July to late August, in southern central Europe from late June to mid-August (EBERT 2003) and in southern Europe from July to October (FLAMIGNI *et al.* 2007, REDONDO *et al.* 2009). FLAMIGNI *et al.* (2007) mention an early generation in April-May. Larva from August or September until May or June (BERGMANN 1955). The hibernation takes place as an egg or as a small larva (HULDÉN *et al.* 2000).The moths fly at night and come readily to light.

BIOLOGY: Polyphagous. The caterpillar has been found on many deciduous trees and shrubs, such as *Salix cinerea*, *S. caprea*, *Betula pendula*, *Fagus sylvatica*, *Quercus robur*, *Cotoneaster*, *Rubus idaeus*, *Rosa dumalis*, *Prunus avium*, *P. spinosa*, *Cytisus scoparius*, *Genista*, *Hippophae rhamnoides*, *Vaccinium myrtillus*, *Syringa vulgaris* and *Lonicera xylosteum* (EBERT 2003). It has in Germany been found on *Ledum palustre* and *Vaccinium myrtillus*, (H. WEGNER pers. comm.), *Prunus spinosa*, *Betula* (C. KAYSER pers. comm.) and *Lonicera periclymenum* and reared on *Genista*, *Rubus idaeus*, *Ribes*, *Salix* and *Prunus* (B. MÜLLER pers. comm.). The pupation takes place between loosely spun leaves and moss. The potential sister species has been found on Samos on *Erica arborea*, *Calicotome villosa*, *Spartium junceum* and *Crataegus* (D. FRITSCH pers. comm.).

HABITAT: The species can be found in a large variety of habitat types, however not in too warm places, such as mixed and deciduous forests, semi-open meadows with bushes, slopes with scattered trees and bushes, parks and gardens. It is found from sea level up to approximately 2000 m altitude in southern Europe.

SIMILAR SPECIES: *C. albarracina* WEHRLI and *C. helenaria* RUCKDESCHEL. Diagnostic characters of European *Crocallis* TREITSCHKE, 1825 are summarised in Table 4 and shown in Text-figs 99–105. We recommend that genitalia to be used for identification, particularly for southern-European specimens.

GENETIC DATA: Genetically comparatively homogeneous from Europe to central Asia (n=51, from 11 countries) but two other BINs in southern Italy and Sicily at distances of 2.4–2.7%. In addition there is a separate genetic cluster at a distance of 6.3% (n=23, from Slovenia, Hungary, Bulgaria, Greece, southern Russia and Turkey). Specimens of this cluster have partly been identified by specimen donors as *C. elinguaria*, partly as *C. inexpectata* WARNECKE, 1940. A separate study will be published on this question (D. STADIE, pers. comm.). Nearest species: *C. albarracina* (1.4% from the main cluster of *C. elinguaria*; see above for preliminary identification of the *C. albarracina* cluster).

REMARKS: SCOBLE (1999) considered *trapezaria* BOISDUVAL from Corsica valid at subpecies level. Elsewhere in the recent literature, for instance LERAUT (2009), pale specimens with forewing medial lines wide apart are considered to fall within variation of the species. FLAMIGNI *et al.* (2007) note that *trapezaria* represents only an extreme variation of this species, coexisting with typical *elinguaria* everywhere in Italy. DNA barcodes are not available from Corsica so far.

Specimens from the Balkans, Hungary, Bulgaria and southern Romania tend to be paler (see adult plate 10) than those normally found in Central and northern Europe, although not exclusively, a phenomenon that couples with the DNA barcode divergences and potentially with the furca length in the male genitalia. For this reason we have included records from those areas as open circles in the distribution map. More research is needed to understand if these taxa occur sympatrically. Limited, preliminary material suggests that the furca length may be discontinuous, potentially allowing grouping of the European *Crocallis elinguaria* taxa into two groups (Text-figs 106–108). In the short furca group (Finland, Spain and Germany examined so far) the length varies from 1.4–1.7 mm (n=4) and in the long furca group (Greece and Croatia examined so far) the length varies from 2.2–2.8 mm (n=6). Valva length, used as a surrogate of the genitalia size, is rather uniform in the specimens examined. We have not found diagnostic differences in the female genitalia (n=7) between North European and Mediterranean samples. If a distinct taxon exists in the Mediterranean area and eastern Europe (see genetic data), we have not yet considered what name should be applied to it. A revision of the *C. elinguaria* complex in Europe is under preparation (D. STADIE pers. comm.).

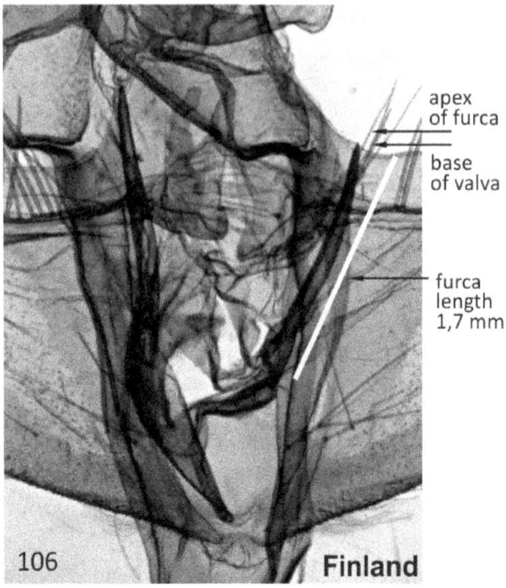

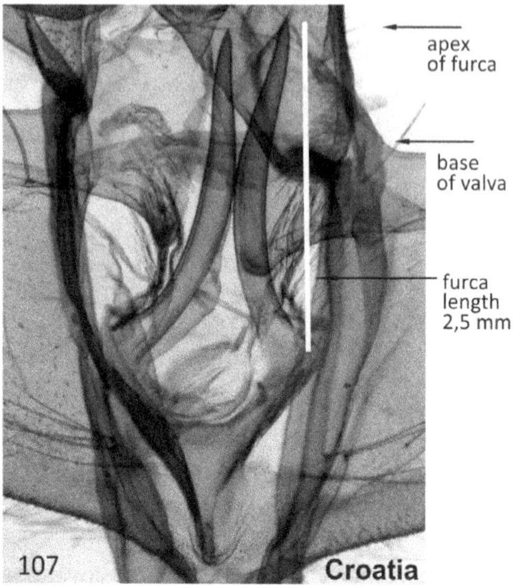

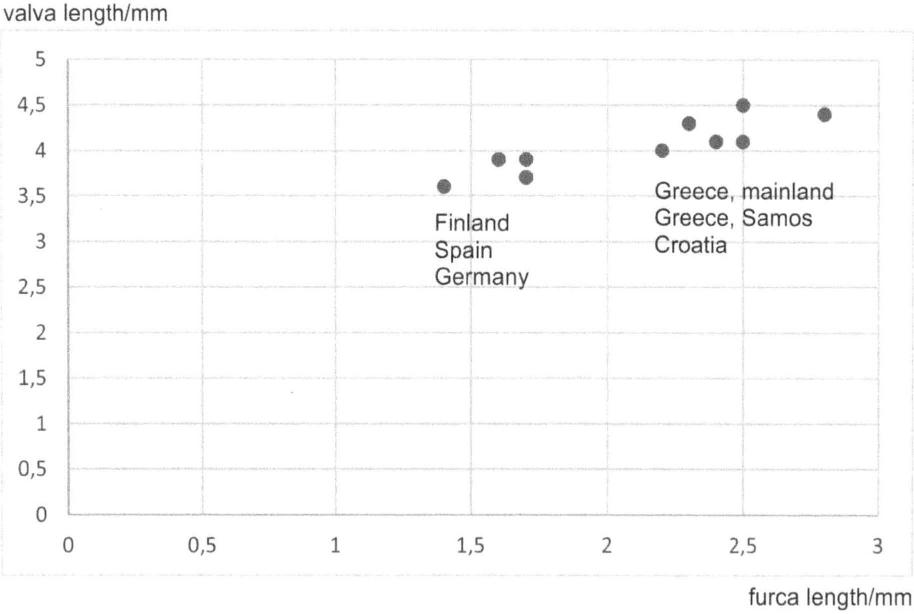

108

Text-fig. 108. Limited, preliminary material suggests that the furca length of the European *Crocallis elinguaria* (LINNAEUS, 1758) taxa may be discontinuous, potentially allowing grouping of the taxa into two groups. In the short furca group the length varies between 1,4–1,7 mm (n=4) and in the long furca group the length varies between 2,2–2,8 mm (n=6). Valva length (y-axis) was continuous in these two groups. The specimens in the long furca groups tend to have paler wings, although not exclusively. See remarks.

⇦ Text-figs 106–107. Variation in the furca length of the European *Crocallis elinguaria* (LINNAEUS, 1758) taxa (figures are in scale). Furca length is measured from the apex to the basal angle, thus omitting the basal part. Note: left arm in slide PS1910 (text-fig. 106) was distorted during the preparation. Limited, preliminary material suggests that in the examined Finnish, Spanish and German specimens the furca apex reaches the base of valva, whereas in the examined Greek and Croation specimens the furca apex exceeds the base of valva. Text-fig. 106. Finland: Savonlinna, 27.vii.1996, slide PS1910 (coll. P. SIHVONEN). Text-fig. 107. Croatia: Solta, 13–20.ix.1989, slide PS1890 (coll. ZFMK). See remarks.

92. *Crocallis dardoinaria* DONZEL, 1840

dardoinaria DONZEL, 1840, Annls Soc. ent. Fr.: 9 (1): 59, pl. 4, figs a-b, (*Crocallis*). Syntype(s), France: Marseille.

aglossaria BOISDUVAL, 1840, Genera Index meth. eur. Lepid.: 184, (*Crocallis*). Syntype(s), [France]: [Provence].

atlantica HERBULOT, 1957, Lambillionea 57 (7–8): 51, (*Crocallis*). Holotype male (Coll. HERBULOT, in Zoologische Staatssammlung, Munich, Germany), France: (Vendée), d'Olonne [Originally as subspecies of *dardoinaria*] (examined externally). See remarks.

dardouinaria FREYER, 1848, Neuere Beitr. Schmett. VI (85): 47, (*Crocallis*). Syntype(s). [Synonymy after The Global Lepidoptera Names Index (ANONYMOUS 2013)].

EXTERNAL CHARACTERS AND ABDOMEN: Wingspan 32–44 mm. Wings yellowish brown, irrorated with blackish, medial area slightly darker, particularly on distal part. Medial line brown, slightly curved, meets inner margin at about 90 degree angle. Postmedial line brown, weakly curved outwards in middle. Hindwings paler, postmedial line narrow. Forewing discal spot large, fragmented into four spots. Hindwing discal spot smaller. Terminal line discontinuous row of dark dots. Fringes concolorous with wings. Wings below lighter, less marked. Proboscis long. Frons, collar, thorax and abdomen concolorous with wings. Male antennae bipectinate, female antennae filiform. Hindtibia of both sexes with 2+2 spurs. Tympanal organs medium-sized. Sternites and tergites 3–8 of both sexes undifferentiated.

VARIATION: Ground colour varies from light rusty brown to greyish brown to pale greyish yellow. Medial area darkest distally, but whole medial area can be darker or almost same colour as rest of wing. Female is more variable than male, often more irrorated with blackish. Terminal line can be absent, or present near forewing apex only.

MALE GENITALIA: Uncus weakly setose, long, narrow. Socii minute, setose. Gnathos arms fused ventrally, elongated and upturned, apex weakly dentate. Valva narrow at base, apex weakly setose. Sclerotised medial ridge narrow, sparsely setose, with minute, weakly sclerotised tooth. Costa sclerotised, narrow. Juxta large sclerotised plate, ventrolateral margin sparsely setose. Furca arms curved, almost fused. Saccus slightly angular. Aedeagus straight, caecum short. Vesica opens at approximately 45 degree angle. Vesica base with approximately 4–5 short cornuti, those lying on outside of aedeagus in uneverted vesica. Vesica base elongated sac, apical part narrower.

FEMALE GENITALIA: Papillae anales large, setose. Apophyses anteriores and apophyses posteriores about equal in length. Lamella antevaginalis margin with medial extension and sharp, lateral projections. Lamella postvaginalis with lateral, pitted plates. Ductus bursae membranous, narrow, densely wrinkled. Posterior part of corpus bursae short, narrow. Anterior part of corpus bursae wider, elongated. Signum small, margins weakly stellate.

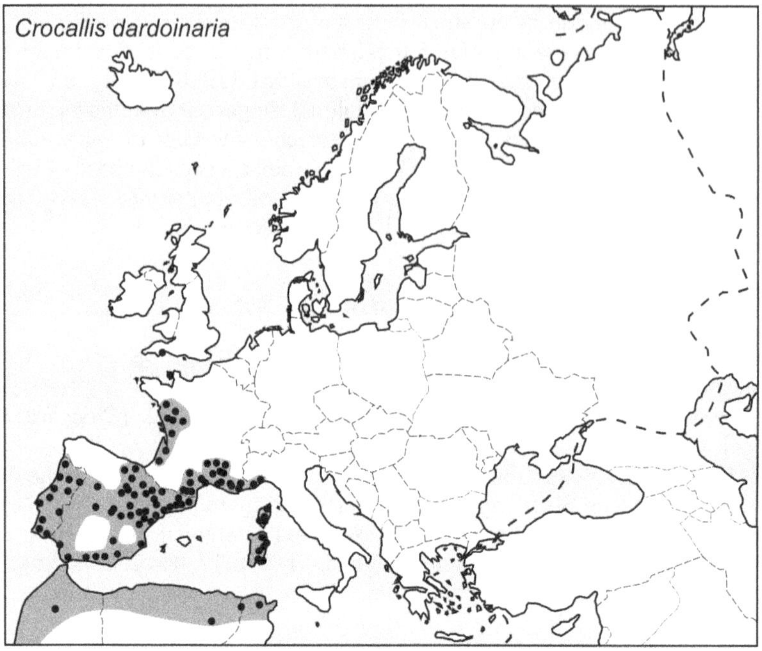

DISTRIBUTION AND ABUNDANCE: West-Mediterranean. In Europe only in Portugal, Spain through southern and western France to the British Channel islands Guernsey and Jersey and including one record from southern England (SKINNER 2009), also in north-westernmost Italy and in Corsica and Sardinia. – Outside Europe the species is known from North Africa: Morocco, Algeria and Tunisia. – In Europe a rather common species that rather rarely occurs in numbers.

PHENOLOGY: Univoltine. From late July to early or mid-October, in southernmost Spain also recorded from November and exceptionally also as early as June (P. HALE pers. comm.). Caterpillars from autumn until spring or early summer. The species hibernates as a larva (B. MÜLLER pers. comm.). The moths come readily to light.

BIOLOGY: Polyphagous. The larva is recorded on *Ulex*, *Juniperus* and *Cistus* (REDONDO *et al.* 2009). Reared on *Genista* in autumn and winter and on *Lonicera xylosteum* in spring (B. MÜLLER pers. comm.) as well as on *Ligustrum vulgare* (E. FRIEDRICH pers. comm.).

HABITAT: In open dry wood- and scrubland, often with patches of bare soil. The species occurs from sea level up to 1600 m, in Sierra Nevada in southern Spain up to 2000 m.

SIMILAR SPECIES: *C. auberti* OBERTHÜR. Diagnostic characters of European *Crocallis* TREITSCHKE, 1825 are summarised in Table 4 and shown in Text-figs 99–105. Genitalia differences are illustrated also in DANTART *et al.* (1993).

GENETIC DATA: Genetically homogeneous (Sardinia, southern Spain). Nearest species in Europe: *C. auberti* (10.8%), other congenerics genetically less similar.

REMARKS: Like LERAUT (2009), we consider taxon *atlantica* HERBULOT, described as a subspecies from the Atlantic coast of France, a lighter, greyish yellow individual form, potentially more common in females. Such specimens are not restricted to the French Atlantic coast, but have also been reported from Spain (REDONDO 1986, DANTART *et al.* 1993, REDONDO & GASTÓN 1999), and potentially the abundance varies geographically. The taxon is retained valid at subspecies level in several publications (DANTART *et al.* 1993, SCOBLE 1999, ROBINEAU 2007). More research is needed.

93. *Crocallis auberti* OBERTHÜR, 1883

auberti OBERTHÜR, 1883, Bull. Soc. ent. Fr. 6 (3): 48, (*Crocallis*). Holotype female, [Algeria]: Sebdou.

INTRASUBSPECIFIC NAMES, POTENTIALLY UNAVAILABLE: *fuliginosa* OBERTHÜR, 1883 (ab.?).

EXTERNAL CHARACTERS AND ABDOMEN: Wingspan 31–40 mm. Wings greyish brown, often with tint of yellow, irrorated with dark brown. Medial area darker than rest of wing, its margins dark brown. Forewing costa often with notch near apex, forewing rounded. Medial line yellowish brown, clearly curved, meets inner margin at about 45 degree angle. Postmedial line yellowish brown, weakly curved outwards in middle. Terminal area often with blackish area near apex. Hindwings paler, postmedial line faint, brown, bordered with whitish. Forewing discal spot large, fragmented into four spots. Hindwing discal spot smaller. Terminal line with few black dots near forewing apex, otherwise concolorous with wings. Fringes concolorous with wings. Wings below more grey, markings faint. Proboscis long. Frons, collar, thorax and abdomen concolorous with wings. Antennae cream-white dorsally. Male antennae bipectinate, female antennae filiform. Hindtibia of both sexes with 2+2 spurs. Tympanal organs medium-sized. Sternites and tergites 3–8 of both sexes undifferentiated.

VARIATION: Ground colour normally different shades of greyish brown, but wings can be brownish yellow. Medial area often darkened in its margins. Postmedial line often broken into row of black dots. Blackish area near forewing apex often absent. Density of irroration variable. Hindwing discal spots normally small, suffused, but can be absent. Specimens from North Africa (Morocco, Algeria) often more contrasting than European specimens, margins of medial area blackish (f. *fuliginosa*).

MALE GENITALIA: Generally as in *C. dardoinaria* DONZEL, but with following differences. In *C. auberti* furca arms are almost straight, clearly separate (furca arms curved, almost fused in *C. dardoinaria*), ventrolateral margin of juxta is densely setose (ventrolateral margin of juxta sparsely setose in *C. dardoinaria*), aedeagus is weakly curved, vesica with approximately 6–8 short cornuti (aedeagus straight, vesica with approximately 4–5 short cornuti in *C. dardoinaria*).

FEMALE GENITALIA: Papillae anales setose. Apophyses anteriores about 1/2 length of apophyses posteriores. Lamella antevaginalis margin rather straight, otherwise funnel-shaped. Lamella postvaginalis with lateral, recurved plates. Ductus bursae membranous, narrow, densely wrinkled. Posterior part of corpus bursae long, narrow, striated. Anterior part of corpus bursae round. Signum absent.

DISTRIBUTION AND ABUNDANCE: West-Mediterranean. In Europe only in the Iberian Peninsula and Sicily including the small islands Pantellaria and Uscita. – Outside Europe in North Africa in Morocco, Algeria, Tunisia and Libya. – In Europe a rather rare and local species.

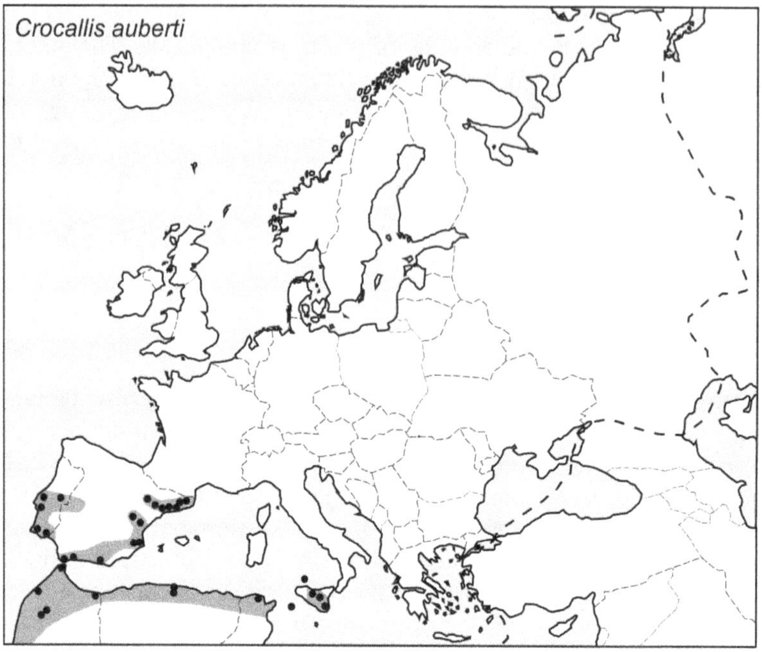

PHENOLOGY: Univoltine. From early September to mid-November, in southernmost Spain until December (P. HALE pers. comm.). In Morocco also recorded in March. The caterpillar is full grown in late January or early February (DANTART *et al.* 1993). The adult moths fly at night and the males comes regularly to light, the females more rarely.

BIOLOGY: Oligophagous. The caterpillar has been recorded on *Prunus spinosa*, *Coronilla valentina* (REDONDO *et al.* 2009). The larvae have been reared on *Genista* and *Lonicera* (B. MÜLLER pers. comm.). The pupation takes place in the ground.

HABITAT: Warm places in open scrubland from sea level and up to 1200 m in Sierra Nevada in southern Spain.

SIMILAR SPECIES: *C. dardoinaria* DONZEL. Diagnostic characters of European *Crocallis* TREITSCHKE, 1825 are summarised in Table 4 and shown in Text-figs 99–105. Genitalia differences are illustrated also in DANTART *et al.* (1993).

GENETIC DATA: Genetically heterogeneous, populations from southern Spain diverging from those of Sicily by 5.6%. Nearest species in Europe: *Dyscia royaria* (8.4%) and *Crocallis matillae* PINKER, 1974 from Canary Islands (8.8%), other congenerics genetically less similar, See remarks.

REMARKS: We have examined the genitalia of specimens from Spain, Algeria and Italy (Sicily) (n=5), but did not observe morphological differences that would correlate with the genetic data, see genetic data.

94. Crocallis helenaria RUCKDESCHEL, 2006

Crocallis helenaria RUCKDESCHEL, 2006, Entomologische Zeitschrift 116(4): 155, figs 1–6, 10e, 11e, (*Crocallis*). Holotype male (Zoologische Staatssammlung, München, Germany), Kreta, über Spili (n. Gerakari) (examined, including genitalia), allotype female (private coll. W. RUCKDESCHEL, Übersee, Germany) (examined, including genitalia), paratypes 4 males, 1 female (private coll. W. RUCKDESCHEL, Übersee, Germany), Kreta: über Spili (n. Gerakari), Kirtomados a. Xekollimenos, Fodele a. Pandomutrios, Mesa Lassithi.

UNAVAILABLE NAMES (manuscript): *insularia* REISSER (see remarks).

EXTERNAL CHARACTERS AND ABDOMEN: Wingspan 33–35 mm. Wings pale ochreous yellow, medial area concolorous with wings. Medial line whitish, straight. Postmedial line whitish, weakly curved outwards in middle. Hindwings whitish, terminal area slightly darker, trace of postmedial line barely visible. Forewing discal spot large, slightly elongated. Terminal line with few blackish dots near forewing apex. Fringes concolorous with wings. Wings below lighter, markings weaker. Proboscis absent. Frons, collar, thorax and abdomen concolorous with wings. Antennae pale ochreous yellow dorsally. Male antennae bipectinate, female antennae filiform. Thorax concolorous with wings. Hindtibia of both sexes with 2+2 spurs. Tympanal organs medium-sized. Sternites and tergites 3–8 of both sexes undifferentiated.

VARIATION: Very little. Medial area can be slightly darker than rest of wing. Medial line can be faintly brownish. Forewing discal spot can be weakly fragmented, traversed by whitish vein.

MALE GENITALIA: Uncus weakly setose, wide to about 3/4 distance, apex triangular. Socii small, setose. Gnathos arms separate, angled upwards, apex blunt. Valva wide at base, subapex and costa densely setose. Costa sclerotised, narrow. Juxta large, ventrolateral margins weakly curved inwards. Furca arms long, weakly curved. Saccus round. Aedeagus curved, caecum very long. Vesica opens at approximately 90 degree angle. Vesica with 2–4 long cornuti, lying on outside of aedeagus in uneverted vesica. Vesica base slightly expanded, apical part narrower.

FEMALE GENITALIA: Papillae anales large, setose. Apophyses anteriores about 1/2 length of apophyses posteriores. Lamella antevaginalis wide plate, margin slightly elongated medially. Lamella postvaginalis (homology tentative) with two lateral, strongly sclerotised, curved, dentate appendices. Ductus bursae wide, dorsoventrally flattened. Corpus bursae membranous, elongated. Signum small, irregular plate.

DISTRIBUTION AND ABUNDANCE: East-Mediterranean. Endemic to Crete. Apparently rather common on the island.

PHENOLOGY: Univoltine. Found from late August to mid-October, but material is still rather limited. Larvae from late March to late May or early June (WAGNER 2014). Larvae in their last instar have been found in mid-May at 1100–1200 m altitude on trees with new and small leaves (M. LEIPNITZ pers. comm.). Therefore it must be assumed, that the hibernation takes place as a full grown larva. The moths fly at night and both males and females come to light.

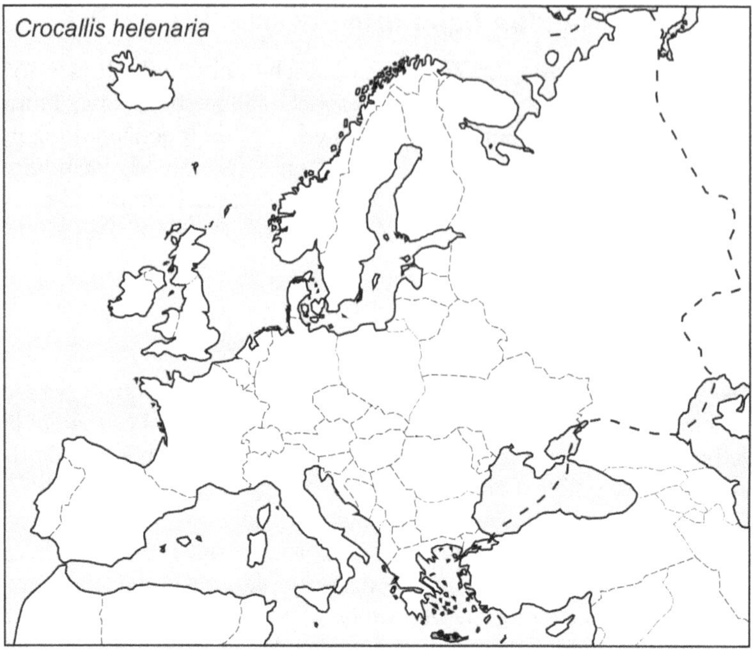

BIOLOGY: Possibly monophagous. The larva has been found feeding on *Acer sempervirens*, but possibly also other species of broadleaved trees may be suitable as host plant (WAGNER 2014).

HABITAT: The species has been found in warm and dry scrubland and open macchia, with scattered rocks, trees and bushes, meadows with many bushes from 200 m up to 1500 m.

SIMILAR SPECIES: *C. albarracina* WEHRLI and *C. elinguaria* (LINNAEUS). Diagnostic characters of European *Crocallis* TREITSCHKE, 1825 are summarised in Table 4 and shown in Text-figs 99–105. We recommend that genitalia to be used in identification, particularly for south-European specimens. The non-European *Odontopera jordanaria* (STAUDINGER, 1898), known from Jordan and Israel, has forewing postmedial line brown and it is clearly angled inwards before costa. Female genitalia have two lateral, strongly sclerotised, curved and dentate appendices similar to *C. helenaria*. Species is illustrated in STADIE & LEHMANN (2012).

GENETIC DATA: Nearest species: '*C. elinguaria*' from Balkans and Turkey (7.0%), See remarks under *C. elinguaria*.

REMARKS: Specimens from Crete have been suspected to represent a distinct *Crocallis* species for a long time (REISSER 1962), but they have been identified as *C. elinguaria* or *C. inexpectata* (RUCKDESCHEL 2006), until RUCKDESCHEL demonstrated that they represent a distinct species, *C. helenaria*. The Staatliches Museum für Naturkunde, Karlsruhe, Germany, has two females from Crete that have been dissected by H. REISSER (slide numbers 314 paratype and 322 allotype) and labelled as '*Crocallis insularia* sp. n., det. REISSER 1973'. We have examined the specimens and dissections and they are conspecific with *C. helenaria*. According to our knowledge, the name *insularia* REISSER has not been published, and it must therefore be treated as an unavailable manuscript name.

95. Crocallis boisduvalaria (LUCAS, 1849)

boisduvalaria LUCAS, 1849, Explor. Sci. Algerie, Zool. 3 (Insectes): 391; ibidem, 4: pl. 4, fig. 1, (*Boarmia*). Syntype(s), Algeria: near Constantine, in the gardens of Salah-Bey.

UNAVAILABLE NAMES (incorrect subsequent spelling): *boisduvaliaria* (numerous references, for instance PROUT 1912–1916; CULOT 1919; LERAUT 2009, HAUSMANN, MIRONOV & VIIDALEPP 2011; should be *boisduvalaria* LUCAS. See WEHRLI 1939–1954, p. 710).

EXTERNAL CHARACTERS AND ABDOMEN: Wingspan 29–33 mm. Wings dark brown, terminal area grey-brown. Medial line dark brown, barely visible, meets inner margin at about 45 degree angle. Postmedial line blackish, wide, weakly curved in middle, ending near tornus. Hindwings paler, greyish-brown, postmedial line distinct. Forewing discal spot ring-like, centre concolorous with wings. Hindwing discal spot small. Terminal line slightly undulating, brown. Fringes concolorous with wings. Wings below greyish brown, only postmedial line and discal spots distinct. Proboscis vestigial. Frons, collar, thorax and abdomen concolorous with wings. Male antennae bipectinate, female antennae filiform. Hindtibia of both sexes with 2+2 spurs. Tympanal organs small. 2nd sternite of both sexes sclerotised. Sternites and tergites 3–8 of both sexes undifferentiated.

VARIATION: Very little. Occasionally terminal area may be yellowish grey, and forewing postmedial line may be bone-white, (without blackish colour). Forewing discal spot may be fragmented (not ring-like).

MALE GENITALIA: Uncus weakly setose, long, narrow. Socii small, setose. Gnathos arms fused ventrally, elongated and upturned, apex weakly dentate. Valva narrow at base, apex weakly setose. Weakly sclerotised medial ridge very narrow, sparsely setose, without sclerotised tooth. Costa sclerotised, narrow. Juxta V-shaped, sclerotised. Furca arms angled at base, long, slightly curved. Saccus slightly angular. Aedeagus straight, stout, caecum short. Vesica opens at approximately 135 degree angle. Vesica base with approximately 8–10 short cornuti, lying on outside of aedeagus in uneverted vesica. Vesica base elongated sac, striated, apical part narrower.

FEMALE GENITALIA: Papillae anales large, setose. Apophyses anteriores slightly curved, about equal in length with apophyses posteriores. Lamella antevaginalis wide, angular, margin deeply concave, with narrow, lateral projections. Lamella postvaginalis membranous. Ductus bursae short, membranous, wrinkled. Corpus bursae long, narrow, striated, membranous. Signum minute, irregular, weakly sclerotised.

DISTRIBUTION AND ABUNDANCE: West-Mediterranean. In Europe only known from Sicily and Calabria in southern Italy. – Outside Europe in Morocco, Algeria and Tunisia. – In Europe local and rather rare.

PHENOLOGY: Univoltine. In Europe from mid-September to early November. In Morocco from late August to November (RUNGS 1981). In Morocco the larva has been found in mid-October (M. LEIPNITZ pers. comm.). In rearing the larva hibernates (M. LEIPNITZ pers. comm.). The adult males fly at night when they come to light while the females are rarely seen.

BIOLOGY: Possibly monophagous or oligophagous. The larva has been recorded on *Calicotome spinosa* and has in captivity been reared on *Asparagus albus* (FLAMIGNI *et al.* 2007) and on *Artemisia abrotanum* (M. LEIPNITZ pers. comm.).

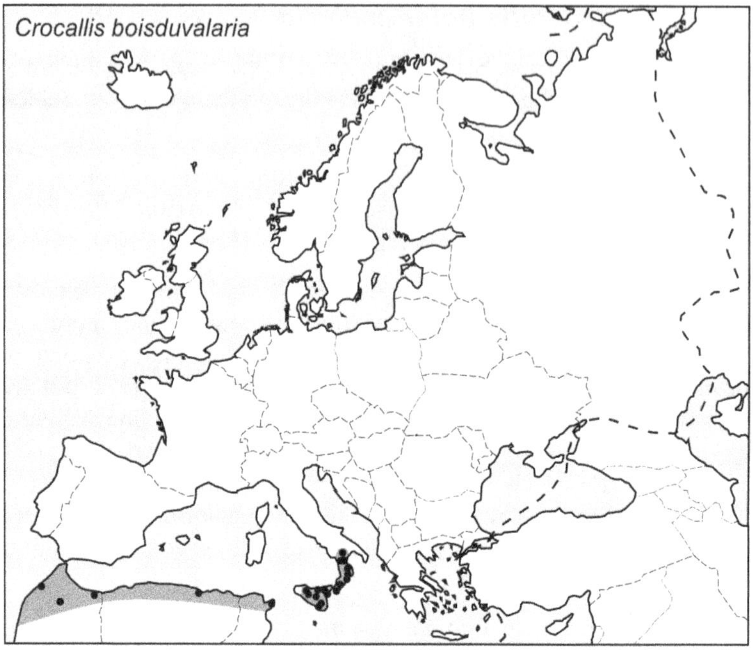

HABITAT: In open oak forests or in Mediterranean scrubland, from sea level up to 1100 m altitude.

SIMILAR SPECIES: No similar species in continental Europe. Diagnostic characters of European *Crocallis* TREITSCHKE, 1825 are summarised in Table 4 and shown in Text-figs 99–105. *C. matillae* PINKER, 1974 from Fuerteventura and *C. bacalladoi* PINKER, 1978 from La Gomera, both endemic to Canary Islands, are similar but shape of forewing postmedial line is diagnostic. Also genitalia are diagnostic (PINKER 1974, 1978). Brown-grey European *Crocallis* species show resemblance but small size, ring-like forewing discal spot and forewing medial line that ends near tornus in *C. boisduvalaria* are characteristic.

GENETIC DATA: Genetically homogeneous in Italy (n=4) but with two other BINs in North Africa. Nearest species: *C. matillae* PINKER, 1974 from Canary Islands (10.3%), other congenerics genetically less similar.

Crocallis inexpectata WARNECKE, 1940

According to our knowledge, *C. inexpectata* WARNECKE, has not been found in Europe. It is described from Turkey and subsequently it has been reported from Turkey and Dagestan (VIIDALEPP 1996) and East Caucasian region (MIRONOV *et al.* 2008). STADIE & FIEBIG (2014) revised the *C. elinguaria* species lineage, including *C. inexpectata*, concluding that it is known from the Central Anatolian Highlands, Inner Anatolian High Mountains and in south-eastern Turkey in Province Van (subspecies *inexpectata* WARNECKE), in eastern Dagestan and northern Caucasus (subspecies *caucasi* WEHRLI, 1954) and in Cappadocia, Province Nevşehir in Turkey (subspecies *cappadocica* STADIE & FIEBIG, 2014). We have examined numerous potential specimens from Europe, but none has fitted the original description (WARNECKE 1940), nor that given in RUCKDESCHEL (2006) nor the redescription and illustrations given in STADIE & FIEBIG (2014). *C. inexpectata* is similar to *C. elinguaria*, but *inexpectata* is on average smaller, paler, dark spots in forewing terminal line are absent and discal spots on hindwing upper side are absent. *C. inexpectata* belongs to the *C. elinguaria* species group having a curved aedeagus, gnathos arms are separate and the valvae are without central tooth. Wide valva base, short valva, and several cornuti (normally two) are similar to *C. helenaria*, whereas the long uncus and distribution of the valva setae are similar to *C. elinguaria*. Female, including its genitalia, of *C. inexpectata* is illustrated for the first time in STADIE & FIEBIG (2014).

Ourapteryx LEACH, 1814

Ourapteryx LEACH, 1814, in LEACH & NODDER, Zool. Miscell. 1: 79. Type species: *Phalaena sambucaria* LINNAEUS, 1758 (Sweden). Generic synonyms are after SCOBLE (1999).

Acaena TREITSCHKE, 1825, in OCHSENHEIMER, Schmett. Eur. 5 (2): 429. Type species: *Phalaena sambucaria* Linnaeus, 1758.

Energopteryx THIERRY-MIEG, 1903, Annls Soc. ent. Belg. 47: 383. Type species: *Ourapteryx nigrociliaris* LEECH, 1891.

Euctenurapteryx WARREN, 1894, Novit. zool. 1: 399. Type species: *Acaena maculicaudaria* MOTSCHULSKY, 1866.

Phrudura SWINHOE, 1906, Ann. Mag. nat. Hist. (7) 17: 554. Type species: *Baptra pura* SWINHOE, 1902.

Uraptera BILLBERG, 1820, Enumeratio Insect. Mus. G. J. Billberg. 88 [Emendation of *Ourapteryx* LEACH, 1814].

Uropteryx AGASSIZ, 1847, Nomencl. zool. (ndex univl.): 267, 384, 385. [Emendation of *Ourapteryx* LEACH, 1814, and junior homonym of *Uropteryx* AGASSIZ, 1835 (Pisces)].

UNAVAILABLE NAMES (included in work rejected for nomenclatural purposes): LARS HÜBNER, 1806 (see FLETCHER 1979). – (incorrect subsequent spellings): *Euctenurapterix* BASTELBERGER, 1911 (see VIVES MORENO 1994); *Ourapterix* GUÉRIN-MÉNEVILLE, 1827; *Urapteryx* KIRBY & SPENCE, 1828 (see FLETCHER 1979).

DIVERSITY AND DISTRIBUTION: *Ourapteryx* contains approximately 80 species, several species and species groups need to be analysed critically in modern context. Species are found in the Palaearctic region (mostly East Palaearctic) and Oriental regions, for instance in the Himalayas and Taiwan. Only one species in Europe.

EXTERNAL CHARACTERS AND ABDOMEN: Large moths, wingspan about 45–50 mm. Most species pale yellow with pale brown striae, medial and postmedial lines straight, forewing discal spot elongated, narrow and hindwing with swallowtail-like tail on M3. A few species are more colourful, different shades of reddish-brown, white, orange and black. Wings occasionally partly transparent, swallowtail-like tail rarely absent. Male and female antennae filiform, rarely pectinate. Following characters examined on *O. sambucaria* only: Male hindleg with hair pencil. Male 8th segment membrane with two minute setal groups.

MALE GENITALIA: Uncus short, stout. Socii small, setose. Gnathos arms fused ventrally, flat, medial element often dentate or scobinate. Valva wide, subapex setose. Furca arms (homologies unclear) curved, asymmetric, one apex (right) densely setose. Juxta (homologies unclear) minute, hollow, ventrolaterally curved, with asymmetric arms. Saccus margin concave. Vesica with numerous cornuti.

FEMALE GENITALIA: Papillae anales elongated. Apophyses anteriores vestigial. Lamella antevaginalis lunular, weakly sclerotised. Ductus bursae variable in width and length, in some Asian species narrow and twisted along longitudinal axis (INOUE 2002). Posterior part of corpus bursae often partly covered by spines. Anterior part of corpus bursae large, round. Signum rarely absent, normally stellate, with hollow centre.

BIOLOGY: Larva of *O. sambucaria* is polyphagous, it has been recorded feeding for instance on *Sambucus racemosa, Rosa, Prunus, Clematis vitalba, Syringa vulgaris* and *Valeriana officinalis* (MIKKOLA *et al.* 1989, EBERT 2003).

IMMATURE STAGES: Larva of *O. sambucaria* is a twig-mimic, grey-brown, long and narrow, with lateral swelling on A6 and small extension on A8 (PORTER 2010). Its pupa cremaster has four pairs of setae, subapical pair curved outwards, apical pair with elongated base and setae curved inwards (PATOČKA & TURČANI 2005).

REMARKS: Traditionally *Ourapteryx* has been classified in the Ourapterygini (for instance HOLLOWAY 1994, VIVES MORENO 1994, KULLBERG *et al.* 2001, HAUSMANN, MIRONOV & VIIDALEPP 2011). BELJAEV (2008a) synonymised the tribe with the Ennomini, the reasoning being based on two putative synapomorphic characters in the male genitalia: Vinculum with a membranous area on its ventral side and paired muscle m3 is attached distally to the medial invagination of the basal portion of the juxta. In a molecular phylogenetic analysis of the world Geometridae (SIHVONEN *et al.* 2011) *Ennomos* (Ennomini) and *Ourapteryx* (Ourapterygini) grouped next to each other. Selected literature: INOUE 1985, INOUE 1993a, INOUE 1993b, INOUE 2002, INOUE & STÜNING 1995). Systematic position: Ennomini of uncertain association. Further research is needed.

96. *Ourapteryx sambucaria* (LINNAEUS, 1758)

sambucaria LINNAEUS, 1758, Syst. Nat. (Ed. 10) 1: 519, (*Phalaena* (*Geometra*)). Syntypes (Linnean Society of London, London, UK), [Sweden: Scania [Skåne] (examined externally).

UNAVAILABLE NAMES (invalid name): *sulphureacaudata* RETZIUS, 1783 (see LERAUT 1997). – (incorrect subsequent spelling): *sulphurea* (WEHRLI 1939–1954; should be *sulfurea* SCHAWERDA). – Intrasubspecific names, potentially unavailable: *approximata* LEMPKE, 1970 (f.); *cuspidaria* BIRD, 1909 (ab.); *deflexaria* SCHULTZ, 1911 (ab.); *delineata* LEMPKE, 1951 (f.); *destrigata* LEMPKE, 1951 (f.); *ecaudata* LEMPKE, 1970 (f.); *olivacea* STANDFUSS, 1896 (ab.); *pallida* LEMPKE, 1951 (f.); *sulfurea* SCHAWERDA, 1934 (ab.); *superba* CHALMERS-HUNT, 1964 (ab.); *peterfii* DIOSZEGHY, 1935 (ab.?).

EXTERNAL CHARACTERS AND ABDOMEN: Wingspan ♂ 40–50 mm, ♀ 50–53 mm. Wings pale yellow, with pale greyish brown striae. Medial line straight, pale brown. Postmedial weakly curved, pale brown, ending near tornus. Hindwing with pale brown medial line. Forewing discal spot elongated, narrow. Forewing margin concave below apex, hindwing with swallowtail-like tail on M3, its base with two brown spots, margins blackish. Terminal line and fringes pale brown except pale yellow on forewing tornus and on anal margin. Wings below almost uniform pale yellow, hindwing without two brown spots. Frons brown. Collar, thorax and abdomen concolorous with wings. Male and female antennae filiform. Hindtibia of both sexes with 2+2 spurs, male hindleg with hair pencil. Tympanal organs medium-sized.

Other abdominal sternites and tergites of both sexes undifferentiated. Male 8th segment membranous with two minute setal groups.

VARIATION: Very little. Ground colour varies from yellowish-white to yellow with tint of olive green to deeper yellow; Portugese specimens are much more yellow than those found in the United Kingdom (M. CORLEY, pers. comm.). Medial lines can be weak, even absent, and be close to each other.

MALE GENITALIA: Uncus wide at base, short, stout, apex melanised. Socii small, setose. Gnathos arms fused ventrally, medial element dentate. Valva wide, subapex setose. Costa narrow, sclerotised. Furca arms (homologies unclear) curved, asymmetric, one apex densely setose. Valva base with group of setae. Juxta (homologies unclear) minute, hollow, ventrolaterally curved, with asymmetric arms. Saccus margin concave. Aedeagus stout, caecum short. Vesica opens at approximately 90 degree angle. Vesica base expanded, round. Medial part long, striated, apex with small lateral diverticulum, with patch of long cornuti.

FEMALE GENITALIA: Papillae anales elongated, setose. Apophyses anteriores vestigial, about 1/6 length of apophyses posteriores. Lamella antevaginalis lunular, weakly sclerotised, striated. Lamella postvaginalis unrecognisable. Ductus bursae short, posterior end sclerotised. Posterior part of corpus bursae angled, followed by long, striated, sclerotised part. Anterior part of corpus bursae large, round. Signum round, densely stellate, medial part roundish.

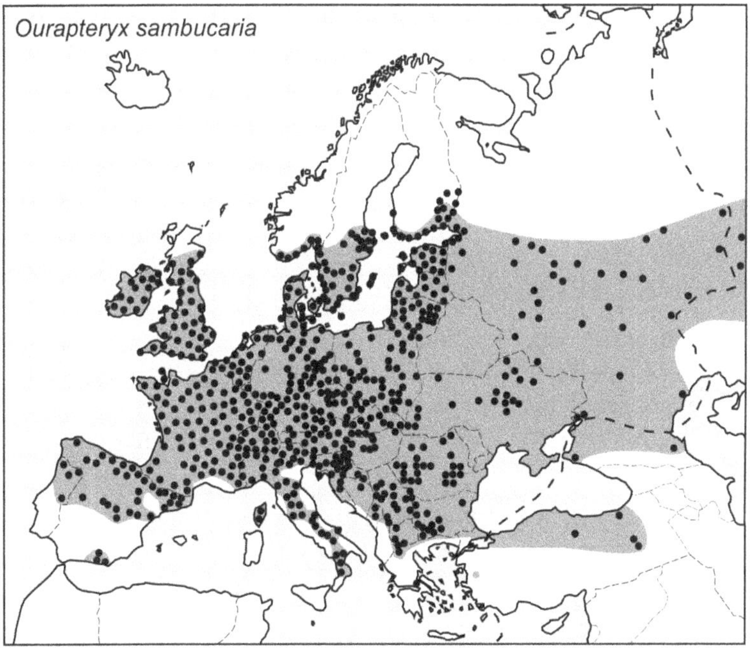

DISTRIBUTION AND ABUNDANCE: European-Westasiatic. In Europe from southern Fennoscandia to the Iberian Peninsula, southern France including Corsica, southern Italy and the Balkans including parts of Greece and from the British Isles through central and eastern Europe to European Russia and the Ural mountains. – Outside Europe in Turkey and from the Urals eastwards through northern Kazakhstan to the Altai mountains. – In Europe a common species that rather rarely occurs in numbers.

PHENOLOGY: Univoltine. In northern Europe from late June to late July or early August, in southern central Europe from mid-June to early August (EBERT 2003) and in southern Europe from June to August (REDONDO *et al.* 2009). Caterpillar from August to June. Overwintering as a small larva. The moths fly from dusk when they are easy to recognize due to size and colour and at night they readily come to light.

BIOLOGY: Polyphagous. The larva has been found on a variety of deciduous trees and bushes, such as *Clematis vitalba*, *Berberis thunbergii*, *Rosa canina*, *Prunus spinosa*, *P. domestica*, *Cornus sanguinea*, *Syringa vulgaris*, *Sambucus nigra*, *Lonicera xylosteum*, *Hedera helix* (EBERT 2003) and on *Lonicera periclymenum* (B. MÜLLER pers. comm.).

HABITAT: Sylvicolous. In woods- and scrubland, parks, gardens. The species occurs from sea level up to 900 m in south-western Germany (EBERT 2003) and up to 1600 m in southern Europe.

SIMILAR SPECIES: No similar species in Europe. *Ourapteryx falciformis* INOUE 1993 has been reported from Caucasus (INOUE 1993a, MIRONOV *et al.* 2008), but identity of the taxon and its distribution are unclear (D. STÜNING pers. comm.). More research is needed.

GENETIC DATA: Genetically homogeneous in Europe (n=25, from seven countries). Nearest species in Europe: *Ennomos autumnaria* (7.4%); outside Europe: *O. falciformis* INOUE, 1993 from Iran (1.4%).

Asovia ALPHÉRAKY, 1908

Asovia ALPHÉRAKY, 1908, Horae Soc. ent. ross. 38: 604. Type species: *Boarmia maeoticaria* ALPHÉRAKY, 1876, [Ukraine]. There are no genus synonyms.

DIVERSITY AND DISTRIBUTION: *Asovia* is a monotypic genus in the Palaearctic region.

EXTERNAL CHARACTERS AND ABDOMEN: Wingspan 28–32 mm. Wings grey, with brown striations, medial and postmedial lines black, the latter bordered with brown. Hindwings with weak postmedial line, strongest on anal margin. Male and female antennae bipectinate, female pectinations shorter. Proboscis vestigial. Sternites and tergites 3–8 of both sexes undifferentiated.

MALE GENITALIA: Uncus long, apex round. Socii small, setose. Gnathos arms fused ventrally, apex upcurved, weakly dentate. Valva evenly curved, without extensions, apex setose. Juxta large plate. Saccus elongated, narrow. Vesica very long, with numerous cornuti.

FEMALE GENITALIA: Papillae anales elongated. Lamella antevaginalis membranous ridge. Ductus bursae short, weakly sclerotised. Posterior part of corpus bursae very long, narrow, striated. Anterior part of corpus bursae round. Signum deeply forked, with triangular base.

BIOLOGY: Larva of *A. maeoticaria* has been found on *Pyrus spinosa* and *Crataegus* (D. FRITSCH pers. comm., FRITSCH *et al.* 2014), in captivity (ex ovo cultivation) on leaves of *Prunus spinosa* (LEPIFORUM 2013). Pupa overwinters.

IMMATURE STAGES: Larva of *A. maeoticaria* is a twig-mimic, pale brown, with blackish dorsal stripe behind the head, brown patches laterally, and small dorsal extension on A6. Pupa is enclosed in a loose white net. Pupa cremaster is with three pairs of curved setae, apical pair widest (LEPIFORUM 2013).

REMARKS: *Asovia* has recently been classified in the Ourapterygini (VIIDALEPP 1996) and Gonodontini (HAUSMANN, MIRONOV & VIIDALEPP 2011). Gonodontini, as defined by HOLLOWAY (1994), and based on *Gonodontis clelia* (CRAMER, 1780) from India, is different and inclusion of *Asovia* in the Gonodontini is not supported by morphology. Male morphology of *Asovia* is general, similar to numerous European Ennominae genera. Female genitalia has unique signum, deeply-forked with triangular base. These structures do not support the Ourapterygini position. WEHRLI (1939–1954), basing his judgement on the male genitalia and wing venation, considered *Asovia* to show some similarities with *Cepphis* HÜBNER, 1823. Systematic position: Uncertain. Further research is needed.

97. *Asovia maeoticaria* (ALPHÉRAKY, 1876)

maeoticaria ALPHÉRAKY, 1876, Trudy russk. ent. Obshch. 10: 38, (*Boarmia*). Syntypes male, female, [Ukraine:] Taganrog, Maji.

decoloraria ALPHÉRAKY, 1889, in ROMANOFF, Mém. Lépid. 5: 239, (*Boarmia*). Syntype(s), Asia Minor.

EXTERNAL CHARACTERS AND ABDOMEN: Wingspan 28–32 mm. Wings grey, with blackish brown striations. Medial line black, narrow, curved. Postmedial line black, with two concavities, inner margin bordered with brown. Wavy line whitish, undulating. Terminal area with one or two brown wedges. Terminal line narrow, black. Fringes chequered grey and brown. Hindwings grey, weakly irrorated with brown. Hindwing postmedial line weak, undulating, row of black dots, strongest on anal margin. Forewing discal spots black, narrow. Hindwing discal spots minute. Wings below grey, heavily irrorated with brown, markings weak. Proboscis vestigial. Frons, collar and thorax concolorous with wings. Abdomen with row of black dots dorsally. Male and female antennae bipectinate, female pectinations shorter. Hindtibia with 2+2 spurs. Tympanal organs medium-sized. Sternites and tergites 3–8 of both sexes undifferentiated.

VARIATION: Very little. Ground colour varies from whitish grey to grey-brown.

MALE GENITALIA: Uncus weakly setose, wide at base, long, apex round. Socii small, setose. Gnathos arms fused ventrally, apex upcurved, weakly dentate. Valva evenly curved, without extensions, apex setose. Costa narrow, sclerotised. Juxta large plate. Saccus elongated, narrow. Aedeagus curved, caecum very long. Vesica opens at approximately 180 degree angle. Vesica very long, narrow. Lateral diverticulum with numerous short cornuti. Apex with one long cornutus.

FEMALE GENITALIA: Papillae anales elongated, setose. Apophyses straight, long. Lamella antevaginalis membranous ridge. Lamella postvaginalis undifferentiated. Ductus bursae short, weakly sclerotised. Base of ductus seminalis expanded. Posterior part of corpus bursae very long, narrow, striated. Anterior part of corpus bursae round. Signum deeply forked, with triangular base.

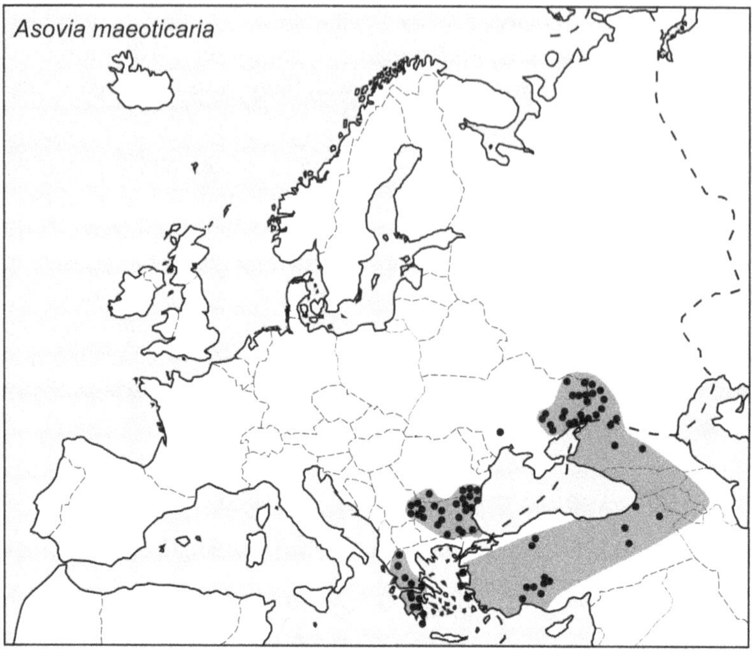

DISTRIBUTION AND ABUNDANCE: South-east-European-Westasiatic. In Europe only in the Balkans, in southern Ukraine and south-western European Russia. – Outside Europe only known from Turkey and Caucasus. – In Europe local but rather common, occasionally in numbers.

PHENOLOGY: Bivoltine or perhaps trivoltine in warm localities. In overlapping generations from late March to late September. In Romania from early May to mid-June and again from early August to late September (SZÉKELY 2011). In Samos the caterpillar has been found from mid-May to mid-June (FRITSCH *et al.* 2014). The hibernation takes place as a pupa (M. LEIPNITZ pers. comm.). The moths fly at night and both males and females come frequently to light.

BIOLOGY: Oligophagous on Rosaceae. On Samos the caterpillar has been found on *Pyrus spinosa* and *Crataegus* (FRITSCH *et al.* 2014). The caterpillar has been reared on *Prunus domestica* and *P. persica* (B. MÜLLER pers. comm.) and on *Prunus spinosa* (M. LEIPNITZ pers. comm.). Pupation takes place on and in the ground (M. LEIPNITZ pers. comm.).

HABITAT: Open woods- and scrubland. In Greece from sea level up to 1700 m altitude.

SIMILAR SPECIES: No similar species in Europe.

GENETIC DATA: Genetically heterogeneous, two specimens from Greece diverging by 3.0% (Peloponnese; central Greece). Nearest species in Europe (genetic similarity probably due to long branch attraction): *Erannis ankeraria* (STAUDINGER, 1861) (8.5%).

Colotois HÜBNER, 1823

Colotois HÜBNER, [1823] 1816, Verz. bekannter Schmett.: 288. Type species: *Phalaena pennaria* LINNAEUS, 1761, Sweden. Genus synonyms are after SCOBLE (1999).

Himera DUPONCHEL, 1829, in GODART & DUPONCHEL, Hist. nat. Lépid. Papillons Fr. 7 (2): 104, 169. Type species: *Phalaena pennaria* LINNAEUS, 1761.

Metra STEPHENS, 1827, in ANONYMOUS, Retrospective Review (2) 1: 245. Type species: *Phalaena pennaria* LINNAEUS, 1761.

DIVERSITY AND DISTRIBUTION: *Colotois* contains three species, two of those occur in the Palaearctic region and one in the Nearctic region.

EXTERNAL CHARACTERS AND ABDOMEN, description is based on Palaearctic species: Large moths, wingspan 35–48 mm. Wings ochreous brown to greyish brown. Medial lines brownish, suffuse. Discal spots brownish, suffuse. Male and female antennae bipectinate, male pectinations very long. Sternites and tergites 3–8 of both sexes undifferentiated.

MALE GENITALIA, description is based on Palaearctic species: Uncus deeply invaginated, asymmetric. Socii minute, setose. Gnathos absent. Valva divided. Costa apex setose. Ventral part wide, apex setose. Juxta sclerotised plate, dorsal margin bilobed. Vesica with cornuti.

FEMALE GENITALIA, description is based on Palaearctic species: Papillae anales wide, setose. Lamella antevaginalis narrow, sclerotised ridge. Ductus bursae short, funnel-shaped. Corpus bursae membranous. Signum absent.

BIOLOGY: Larva of *C. pennaria* has been recorded mainly feeding on the leaves of deciduous trees, for instance on *Quercus, Salix, Malus, Acer* and *Sorbus aucuparia* (MIKKOLA *et al.* 1989). Adults are nocturnal, attracted to light, flying in autumn from August to December, southern European specimens fly later. Egg overwinters.

IMMATURE STAGES: Larva of *C. pennaria* is a twig-mimic, brown, irrorated with black and with row of small white lateral spots. Two small, conical-shaped dorsal structures on A11 (PORTER 2010). Its pupa cremaster is with two pairs of curved setae, apical pair wider (PATOČKA & TURČANI 2005).

REMARKS: *Colotois* has recently been classified in the Colotoini (VIVES MORENO 1994, VIIDALEPP 1996, LERAUT 1997, HAUSMANN, MIRONOV & VIIDALEPP 2011). Based on the material we have examined, *Dasycorsa* PROUT and *Colotois* HÜBNER are morphologically similar, and potentially also *Compsoptera* BLANCHARD, but the potential relationships should be studied in an analytical context. Systematic position: Colotoini.

As is explained in more detail in *Ennominae systematics* chapter, the European Ennominae fauna as a whole has not previously been classified into tribes. The above-mentioned similarities were noted so late in the manuscript preparation that rearrangement of taxa to reflect such affinities were not possible.

98. Colotois pennaria (LINNAEUS, 1761)

pennaria pennaria LINNAEUS, 1761, Fauna Suecica (Edn 2): 324, (*Phalaena (Geometra)*). Syntypes (Linnean Society of London, UK), [Sweden]: W Gothia [Vestergötland] (examined externally).

bifidaria HAWORTH, 1809, Lepid. Br. (2): 291, (*Geometra*). Syntype(s), Anglia [England]. [The description is repeated on p. 301].

cerasi FABRICIUS, 1775, Syst. Ent.: 564, (*Bombyx*). Syntype(s), England.

zorella CAPUSE, 1960, Comunicari Zool., Bucuresti. 1957: 108, fig. 6, (*Himera*). Syntypes 5 males, [Romania]: Baile Herculane; la Mehadia [originally as f]. See remarks.

pennaria carbonii HARTIG, 1976, Mitt. ent. Ges. Basel 26 (1): 17, (*Colotois*). Syntype(s), [Italy]: Sardinia: Cagliari.

pennaria mauretanaria STÄTTERMAYER, 1930, Int. ent. Z. 24: 253, (*Himera*). Holotype male, Algeria: Guelt-es-Stel.

pennaria paupera HAUSMANN, 1995, Mitt. münch. ent. Ges. 85: 104, (*Colotois*). Holotype male (Zoologische Staatssammlung, Munich, Germany), N Zypern [Cyprus]: 8 km W Girne, 50 m (examined, including genitalia).

pennaria ussuriensis BANG-HAAS, 1927, Horae macrolepidopt. Reg. palaearct. 1: 96, pl. 11, fig. 32, (*Colotois (Himera)*). Syntypes 3 males, [Russia]: Ussuri (south), Sutschansk. Valid at subspecies rank according to SCOBLE (1999).

INTRASUBSPECIFIC NAMES, POTENTIALLY UNAVAILABLE: *alba* SCHNEIDER,]1949] 1950 (ab.); *albomaculata* SCHNEIDER, [1949] 1950 (ab.); *algiriae* STÄTTERMAYER, 1930 (ab.); *atrolineata* WARNECKE, 1944 (f.); *aurantiaca* LEMPKE, 1951 (f.); *biroi* DIOSZEGHY, 1935 (ab.); *brevipennis* LEMPKE, 1970 (f.); *castiniaria* LAMBILLION, 1905 (f.); *chobauti* TESTOUT, 1936 (ab.); *cuneata* RUDOLPH, 1935 (ab.); *cuprea* SCHWINGENSCHUSS, 1954 (ab.); *demaculata* LEMPKE, (1951 (f.); *depuncta* NITSCHE, 1924 (ab.); *flava* SCHNEIDER, [1949] 1950 (ab.); *flavescens* SCHAWERDA, 1923 (ab.); *fusca* SCHWINGENSCHUSS, 1954 (ab.); *grisea* HANNEMANN, 1920 (f.); *korbi* SCHAWERDA, 1922 (ab.); *interrupta* LEMPKE, 1970 (f.); *juncta* LEMPKE, 1970 (f.); *lugubrata* RUDOLPHI, 1935 (ab.); *mariscolora* LEMPKE, 1951 (female f.); *montanaria* STÄTTERMAYER, 1930 (ab.); *nigrata* SCHNEIDER, [1949] 1950 (ab.); *nubilosa* HEPP, 1932 (ab.); *nuda* HEPP, 1932 (ab.); *obscura* AIGNER, 1906 (ab.) (=synonym of *castiniaria* LAMBILLION, 1905 according to LEMPKE (1970); *obsoletelineata* LEMPKE, 1951 (f.); *olivacea* HOFFMANN, 1919 (ab.); *pallens* SCHNEIDER, [1949] 1950 (ab.); *pallida* LEMPKE, 1970 (female f.); *recurvata* LEMPKE, 1970 (f.); *rosea* FOLTIN, 1942 (ab.); *rufa* SCHNEIDER, [1949] 1950 (ab.); *rufescens* LEMPKE, 1951 (f.) (=synonym of *rosea* FOLTIN, 1942 according to LEMPKE (1970); *rufolineata* LEMPKE, 1951 (f.); *tangens* LEMPKE, 1951 (f.); *unilineata* LEMPKE, 1970 (f.); *vicinalis* RUDOLPH, 1935 (ab.); *virescens* SCHNEIDER, [1949] 1950 (ab.).

EXTERNAL CHARACTERS AND ABDOMEN: Wingspan 35–48 mm. Male wings ochreous brown, often heavily suffused with blackish. Female wings paler, greyish yellow-brown. Medial lines blackish brown, slightly curved on costa. Outer margin of postmedial line bordered with narrow, pale yellow area. Terminal area with white spot near forewing apex, its inner margin blackish. Hindwings paler, terminal area darker, postmedial line straight, narrow. Occasionally rounded subterminal line visible, partly fused with postmedial line. Terminal line and fringes blackish brown, often concolorous with wings on forewing tornus and on anal margin. Forewing margin weakly concave below apex. Discal spots small, blackish brown. Wings below as above, markings more suffused. Frons, collar, thorax and abdomen concolorous with wings. Male and female antennae whitish dorsally, bipectinate, male pectinations very long. Hindtibia of both sexes with 2+2 spurs. Tympanal organs medium-sized. Sternites and tergites 3–8 of both sexes undifferentiated.

VARIATION: Ground colour varies from deep reddish-brown to pale ochreous yellow, occasionally with tint of olive-green. Females are always paler. Dark suffusion can be almost absent, present as weak irroration or cover large areas of wings. Medial lines can be reddish, blackish, those can be fused for a short distance, be far apart or even absent.

Subsp. ***carbonii*** is variable in external appearance but wings always dark, with tint of purplish, irrorated with brown or black. Apparently restricted to Sardinia in Italy. Specimens are illustrated in FLAMIGNI *et al.* (2007).

Subsp. ***mauretanaria*** is variable in external in appearance but wings normally paler and females with tint of green. Known from Morocco and Algeria.

Subsp. ***paupera*** is reddish-brown, medial and postmedial lines weak, blurred, discal spots minute. Forewing margin round. Hindwings reddish. In the male genitalia uncus is shorter, caudal margin of juxta is slightly deeper than in nominal subspecies (HAUSMANN 1995b). Known from Cyprus (HAUSMANN 1995a, FISCHER & LEWANDOWSKI 2005). Female is illustrated in FISCHER & LEWANDOWSKI (2005).

MALE GENITALIA: Uncus deeply invaginated, asymmetric, left arm with acute apex, right arm with round apex. Socii minute, setose. Gnathos absent. Valva medially invaginated, membranous. Costa sclerotised, apex extended, setose. Ventral part wide, margin sclerotised, apex setose. Juxta narrow at base, sclerotised, dorsal margin bilobed. Saccus round. Aedeagus weakly curved, caecum long. Vesica opens at approximately 90 degree angle. Vescia base expanded, with small diverticulum holding one large and one small cornuti. Apex narrow tube.

FEMALE GENITALIA: Papillae anales wide, setose. Apophyses long, straight. Lamella antevaginalis narrow, concave, sclerotised ridge. Lamella postvaginalis undifferentiated. Posterior part of ductus bursae funnel-shaped, sclerotised. Anterior part of ductus bursae narrow, membranous. Corpus bursae elongated, membranous, posterior part weakly striated. Signum absent.

DISTRIBUTION AND ABUNDANCE: Holarctic. In Europe distributed from central Fennoscandia to the Mediterranean and from Ireland through central and eastern Europe to European Russia and the Ural mountains. – Outside Europe in North Africa, Turkey, Transcaucasus, Caucasus and in Far East Russia, Sakhalin and Japan (subsp. *ussuriensis*). The species also occurs in North America in eastern Canada. – In Europe a common or very common species that normally does not occur in numbers.

PHENOLOGY: Univoltine. In northern Europe from mid-September to early November, in southern central Europe from late September to mid-November (EBERT 2003) and in Italy from October to December, in southern Italy until January or even February (FLAMIGNI *et al.* 2007). In southernmost Spain recorded from October to January (P. HALE pers. comm.). Caterpillars from late April to June or possibly July. The species overwinters as an egg. The moths fly at night and come readily to light, the females less frequently than the males.

BIOLOGY: Polyphagous. The caterpillar feeds on a large variety of deciduous trees and bushes. In German Baden-Württemberg it has been found, according to EBERT (2003), on *Populus nigra*, *Salix alba*, *S. caprea*, *Corylus avellana*, *Carpinus betulus*, *Betula pendula*, *Fagus sylvatica*, *Quercus robur*, *Q. petraea*, *Quercus sp.*, *Ulmus minus*, *Ulmus sp.*, *Malus domestica*,

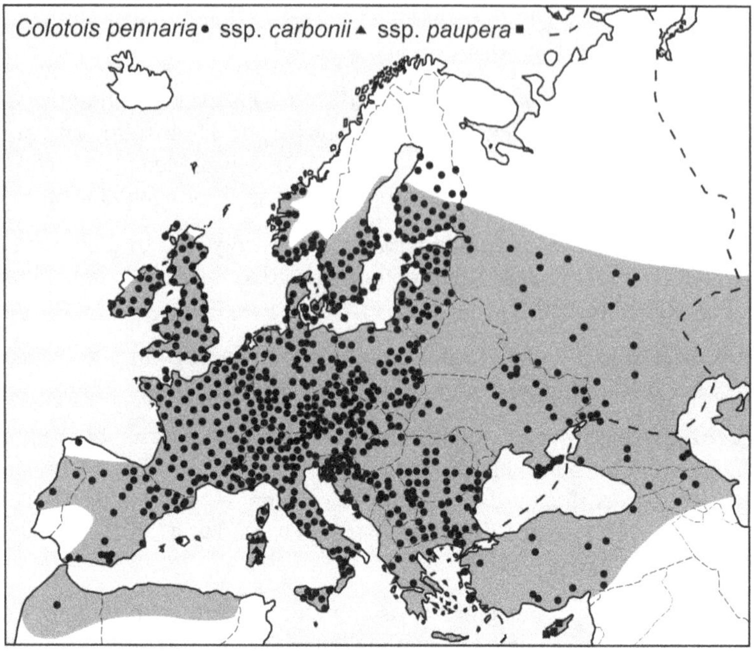

Crataegus monogyna, *Crataegus submollis*, *Crataegus sp.*, *Rosa vosagiaca*, *Rubus idaeus*, *Prunus insititia*, *P. padus*, *P. avium*, *P. spinosa*, *Acer campestre*, *A. pseudoplatanus*, *Ligustrum vulgare*, *Fraxinus excelsior* and *Fraxinus sp*. Larva found feeding on *Rhamnus frangula* in Germany (A. HAUSMANN pers. comm.). In Crete on *Prunus* species and *Acer sempervirens* (M. LEIPNITZ pers. comm.). In Sardinia, the larva has been found feeding on *Pistacia lentiscus* (B. MÜLLER pers. comm.) and *Quercus pubescens* (M. LEIPNITZ pers. comm.) and in Cyprus on *Crataegus azarolus* and *Quercus infectoria* (E. FRIEDRICH pers. comm.). The pupation takes place in the ground.

HABITAT: In a large variety of habitat types such as deciduous forests, wood- and scrubland, gardens, parks and similar bushy places. The species occurs from sea level up to 1500 m in the Alps (VORBRODT 1914) and slightly higher, up to 1600–1700 m, in southern Europe.

SIMILAR SPECIES: No similar species in Europe. *C. furcata* HAUSMANN 1995 from Israel is externally without notable differences. In the male genitalia caudal end of juxta is deeply bilobed and vesica is with one cornutus (caudal end of juxta bilobed, vesica with one large and one small cornuti in *C. pennaria*). Female genitalia are without differences (HAUSMANN 1995b).

GENETIC DATA: Genetically heterogeneous with three BINs in Europe at minimum distances of 1.9–3.1%: (1) Spain (n=2); (2) Finland, Netherlands, United Kingdom, Croatia, Greece and Turkey (n=11), and a third from United Kingdom, Germany, Italy, Austria, and France (n=21). Outside Europe with separate BINs for subspecies *paupera* HAUSMANN, 1995 from Cyprus and subspecies *ussuriensis* BANG-HAAS, 1927 from East Asia. Nearest species: *C. furcata* HAUSMANN, 1995 from Israel (4.2% from subspecies *paupera*).

REMARKS: SCOBLE (1999) recognized taxon *zorella* CAPUSE, 1960 from Romania valid at subspecies level. We follow the original intention of CAPUSE (1960), and consider it an individual form. Also SZÉKELY (2011) does not recognise *zorella* valid at subspecies level.

Angerona DUPONCHEL, 1829

Angerona DUPONCHEL, 1829, Hist. nat. Lépid. Papillons Fr. 7 (2): 104, 180. Type species: *Phalaena prunaria* LINNAEUS, 1758, [Europe]. There are no genus synonyms.

UNAVAILABLE NAMES (incorrect subsequent spelling): *Augerona* DESMAREST 1847 (see FLETCHER 1979).

DIVERSITY AND DISTRIBUTION: *Angerona* contains four species, three of those occur in the Palaearctic region. PITKIN (2002) noted that the Neotropical species assigned to *Angerona* has been misplaced.

EXTERNAL CHARACTERS AND ABDOMEN, description is based on Palaearctic species: Large moths, wingspan about 40 mm. Wings various combinations of yellow and orange, irrorated or striated with brown or blackish. Male antennae bipectinate, female antennae filiform. Male hindtibia swollen, with hair pencil. Male 3rd sternite without setae, posterior margin of 8th sternite strongly concave.

MALE GENITALIA, description is based on European species: Uncus shallow plate. Socii absent. Gnathos arms wide, fused ventrally, apex upcurved, triangular. Valva wide, apical half deeply grooved. Costa with spinose projection. Juxta granulate. Saccus wide, round. Vesica with cornutus.

FEMALE GENITALIA, description is based on European species: Papillae anales elongated. 9th and 10th segments elongated. Apophyses anteriores short, about 1/5 length of apophyses posteriores. Lamella antevaginalis elongated medially, spoon-shaped. Ductus bursae very short. Anterior part of corpus bursae elongated. Signum quadrangular, with hollow centre.

BIOLOGY: Larvae of *A. prunaria* and *A. nigrisparsa* BUTLER, 1879 are polyphagous. *A. prunaria* has been recorded feeding particularly on deciduous trees and shrubs such as *Betula, Salix, Rubus idaeus* and *Calluna* (MIKKOLA *et al.* 1989). *A. nigrisparsa* has been recorded feeding for instance on *Betula platyphylla, Viburnum plicatum, Aesculus turbinata* and *Meliosma tenuis* (SCOBLE 1999). Small larva of *A. prunaria* overwinters.

IMMATURE STAGES: Larva of *A. prunaria* is a twig-mimic, various shades of brown, with large pointed conical-shaped projections on A8, smaller dorsal structures on A4 and A11 (PORTER 2010). Pupa cremaster of *A. prunaria* is slightly elongated, with four pairs of curved setae, apical pair widest (PATOČKA & TURČANI 2005).

REMARKS: *Angerona* has recently been classified in the Angeronini (VIIDALEPP 1996, VIVES MORENO 1994, LERAUT 1997, HAUSMANN, MIRONOV & VIIDALEPP 2011). Male and female genitalia support the Gnophini association as defined in VIIDALEPP *et al.* (2007). In DNA-based analyses genus *Angerona* grouped next to *Charissa* (Gnophini) (SIHVONEN *et al.* 2011), next to *Chariaspilates* (Gnophini) (YAMAMOTO & SOTA 2007) and next to *Siona* (Gnophini) (ÕUNAP *et al.* 2011). Facies of the mainly Nearctic genera *Euchlaena* HÜBNER, *Lytrosis* HULST and *Xanthotype* WARREN are dissimilar but structurally these genera, including the genitalia, are similar to *Angerona* (illustrated for instance in MCGUFFIN 1981). The potential Gnophini association of *Angerona* was noted late in the production process of this book, so for practical reasons it is dealt with here, although we note that its current position in the sequence of species does not reflect its probable phylogenetic position. Structurally *A. prunaria* is similar to European *Aspitates* species, see the male and female genitalia plates. See

Gnophini under *Classification of European Ennominae genera into tribes* for further details. Systematic position: Gnophini (but systematic position and extent of the Gnophini needs further research).

99. Angerona prunaria (LINNAEUS, 1758)

prunaria prunaria LINNAEUS, 1758, Syst. Nat. (Ed. 10) 1: 520, (*Phalaena (Geometra)*). Syntypes (Linnean Society of London, London, UK), [Finland: Turku?] (examined) externally.

alpina KITT, 1932, Verh. zool. -bot. Ges. Wien 82 (1): 16, figured, (*Angerona*). Syntypes 1 male, 1 female, [Austria]: Ötzal, Mayerhofen in Zillertal. Originally as var., treated valid at subspecies level in SCOBLE (1999). See remarks.

corticalis SCOPOLI, 1763, Ent. Carniolica: 216, fig. 533, (*Phalaena*). Syntype(s), [Slovenia]: Carniola, Ljubljana.

corylaria THUNBERG, 1784, Diss. ent. sistens Insecta Suecica 1: 4, (*Geometra*). Lectotype male (Universitets Zoologiska Institut, Uppsala, Sweden), Sweden: Halland.

dimidiata FABRICIUS, 1775, Syst. Ent.: 630, (*Phalaena*). Syntype(s) (Zoological Museum, University of Copenhagen, Denmark), England.

fulvularia HUFNAGEL, 1767, Berlin Mag. 4 (5): 518, (*Phalaena*). Syntype(s), lost, [Germany]: Berlin region.

prunaria aquafortis BRYK, 1949, Ark. Zool. 41A (1): 194, (*Angerona*). Holotype male, Korea. Valid at subspecies rank according to SCOBLE (1999).

prunaria coreola WEHRLI, 1940, in SEITZ, Gross-Schmett. Erde 4 (Suppl.): 350, pl. 27: f, (*Angerona*). Syntype(s), Korea: South Korea, Silberbergen. Valid at subspecies rank according to SCOBLE (1999).

prunaria dominans BRYK, 1942, Dt. ent. Z. Iris 56: 82, (*Angerona*). Holotype male (Naturhistoriska Riksmuseet, Stockholm, Sweden), [Kuril Islands]: Kunashiri. Valid at subspecies rank according to SCOBLE (1999).

prunaria kentearia STAUDINGER, 1892, Dt. ent. Z. Iris 5 (2): 375, (*Angerona*). Syntype(s), [Mongolia]: Kentei. Originally as var., valid at subspecies rank according to SCOBLE (1999).

prunaria sibirica FUCHS, 1899, Jb. nassau. Ver. Naturk. 52: 145, (*Angerona*). Syntypes male, female, [Russia]: Krassnojarsk. [Originally as var].

prunaria mongoligena BRYK, 1949, Ark. Zool. 41A (1): 195, (*Angerona*). Holotype male, Mongolia (north). Valid at subspecies rank according to SCOBLE (1999).

prunaria prouterona BRYK, 1949, Ark. Zool. 41A (1): 195, (*Angerona*). Holotype male, Korea: Myokosan. Valid at subspecies rank according to SCOBLE (1999).

prunaria turbata PROUT, 1929, Novit. zool. 35 (2): 148, (*Angerona*). Holotype male (The Natural History Museum, London, UK), Japan: Asamayama (examined externally). Valid at subspecies rank according to SCOBLE (1999).

prunaria valens WEHRLI, 1940, in SEITZ, Gross-Schmett. Erde 4 (Suppl.): 350, pl. 27: g, (*Angerona*). Syntypes male, female, North Korea; [Russia]: Ussuri, [China]: Manchuria. [Paratype female in the Natural History Museum, London, UK (examined externally)]. Valid at subspecies rank according to SCOBLE (1999).

UNAVAILABLE NAMES (incorrect original spelling): *spångbergi* LAMPA, 1885 (ICZN 1999, ♂ 32.5.2, should be *spangbergi* LAMPA, 1885). – Incorrect subsequent spelling: *spångbergi* (for instance MIKKOLA *et al.* 1989). – Intrasubspecific names, potentially unavailable: *aurea* WILLIAMS, 1947 (ab.); *aureocincta* OBERTHÜR, 1912 (ab.) (=synonym of *pallidaria* PROUT, 1903 according to The Global Lepidoptera Names Index (ANONYMOUS 2013)); *bimacularia* WILLIAMS, 1947 (ab.); *bistrigata* WEBER, 1795 (ab.); *clercata* FETTIG, 1897 (ab.); *constirpataria*

FUCHS, 1899 (ab.); *diluta* WILLIAMS, 1947 (ab.); *douglasaria* WILLIAMS, 1947 (ab.); *extrema* WILLIAMS, 1947 (ab.); *feminaecoloris* VALLE, 1930 (ab.); *franckaria* LAMBILLION, 1912 (ab.) (=synonym of *fuscaria* PROUT, 1903 according to The Global Lepidoptera Names Index (ANONYMOUS 2013)); *fuscapicata* WILLIAMS, 1947 (ab.); *fuscaria* PROUT, 1903 (ab.); *griseoguttata* WILLIAMS, 1947 (ab.); *hasebroekii* KUJAN, 1933 (ab.); *infuscata* MATSUMURA, 1925 (ab.); *intensa* LEMPKE, 1970 (f.); *juncta* WILLIAMS, 1947 (ab.); *luteosordida* HÖRHAMMER, 1933 (ab.); *minor* HEIERMARK, 1942 (ab.); *nigrilineata* COCKAYNE, 1952 (ab.); *nigrolimbata* DE JOANNIS, 1908 (ab.); *ochreata* SCHAWERDA, 1923 (ab.); *pallescens* WILLIAMS, 1947 (ab.); *pallida* LEMPKE, 1970 (f.); *pallidaria* PROUT, 1903 (ab.); *parvula* ROCCI, 1914 (f.); *pickettaria* PROUT, 1903 (ab.); *pluriguttata* WILLIAMS, 1947 (ab.); *postfusca* WILLIAMS, 1947 (ab.); *postmarginata* LEMPKE, 1951 (f.); *purpurascens* LEMPKE, 1951 (f.); *selectaria* REBEL, 1914 (ab.); *smartaria* WILLIAMS, 1947 (ab.); *spangbergi* LAMPA, 1885 (ab.); *striolata* KLEMENSIEWICZ, 1913 (ab.); *subalpinaria* LAMBILLION, 1912 (ab.) (=synonym of *spangbergi* LAMPA, 1885 according to The Global Lepidoptera Names Index (ANONYMOUS 2013)); *subsordiata* INOUE, 1942 (ab.); *suffusa* LEMPKE, 1970 (f.); *turbata* PROUT, 1929 (f.); *unicolor* MATSUMURA, 1925 (ab.); *unicoloraria* HORM, 1894 (ab.); *wenzeli* SCHAWERDA, 1921 (ab.).

EXTERNAL CHARACTERS AND ABDOMEN: Wingspan 38–50mm. Wings ochre-yellow or orange-yellow in males, pale yellow in females, densely striated with brown. Discal spots narrow, elongated, brown. Fringes chequered brown and yellowish. Wings below as above, paler. Frons, collar, thorax and abdomen concolorous with wings. Male antennae bipectinate, female antennae filiform. Hindtibia of both sexes with 2+2 spurs. Male hindtibia swollen, with hair pencil. Tympanal organs large. Posterior margin of 8th sternite strongly concave.

VARIATION: Very variable. Ground colour varies from pale yellow to yellow-orange, to deep orange and to brown. Normally males are ochre-yellow or orange-yellow, females pale yellow, both are striated with brown. Basal and terminal areas can be more or less uniform brown (f. *corylaria* THUNBERG) or brown striations can be nearly absent (f. *spangbergi* LAMPA), apparently the latter being more common in females. Occasionally specimens are uniform brown. In an extensive survey in Finland the portion of f. *corylaria* specimens was about 8.5% in males and 12.5% in females (MIKKOLA *et al.* 1989) of total specimens observed. The extent of brown colour shows continuous variation, thus the definition of different forms is somewhat arbitrary.

MALE GENITALIA: Uncus weakly setose, shallow horizontal plate. Socii absent. Gnathos arms wide, fused ventrally, apex upcurved, triangular, granulate. Valva wide, apical half deeply grooved. Upper half of costa and apex densely spinose. Costa with spinose projection. Medial ridge narrow, base setose. Ventral margin sclerotised, setose. Valva bases fused together, with two pockets medially. Juxta large, granulate, dorsolateral margins extended. Saccus wide, round. Aedeagus wide, straight, caecum absent. Vesica opens at approximately 180 degree angle. Vesica almost straight tube, base weakly expanded, with long, forked cornutus.

FEMALE GENITALIA: Papillae anales elongated, setose. 9th and 10th segments elongated. Apophyses anteriores short, about 1/5 length of apophyses posteriores. Lamella antevaginalis elongated medially, spoon-shaped. Lamella postvaginalis undifferentiated. Ductus bursae very short, wide, membranous. Posterior part of corpus bursae long, narrow, membranous. Anterior part of corpus bursae wider, elongated. Signum quadrangular, with hollow centre.

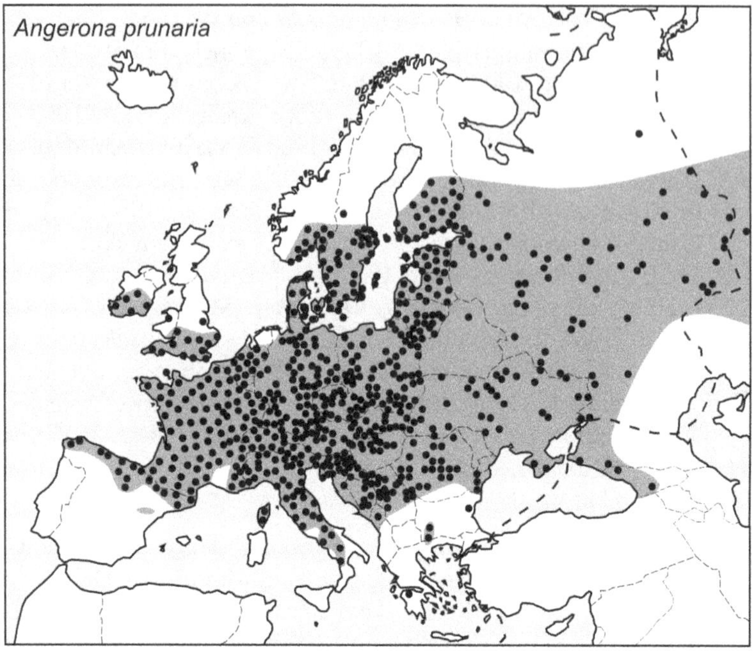

DISTRIBUTION AND ABUNDANCE: Eurasiatic. In Europe from southern central Fennoscandia to northern Iberian Peninsula, southern France including Corsica and southern Italy and the Balkans, mainly in the northern half and from the British Isles through central and eastern Europe to European Russia and the Urals. – Outside Europe the species occurs in Georgia and Caucasus and through northern Kazakhstan, southern and eastern Siberia, Mongolia, Korea and Far East Russia to Sakhalin, southern Kuriles and Japan. – In Europe rather local and not common, it rarely occurs in numbers.

PHENOLOGY: Univoltine, in the south bivoltine. In northern Europe from late June to mid or late July, in south-western Germany from early June to early August (EBERT 2003) and in Spain from mid-June to late July (REDONDO *et al.* 2009). FLAMIGNI *et al.* (2007) regard the species as being bivoltine in Italy outside the Alps, flying from May to September. In Romania from early May to mid-June and from early August to mid-September (SZÉKELY 2011). In breeding the species indoors from a Hungarian female, 100% of the reared adults appeared in the following year, which could indicate one generation only (B. MÜLLER pers. comm.). Caterpillars from August, overwintering as small larvae, then feeding until May. The moths fly at night and come frequently to light, the males more frequently than the females.

BIOLOGY: Polyphagous. The larva lives on a variety of deciduous trees and bushes. EBERT (2003) mention *Salix caprea*, *Corylus avellana*, *Quercus*, *Ribes uva-crispa*, *Rubus idaeus*, *Prunus spinosa*, *Cytisus scoparius*, *Acer campestre*, *Cornus sanguinea*, *C. mas*, *Ligustrum vulgare*, *Lonicera xylosteum* and *Sambucus nigra*. The caterpillars have been found on *Vaccinium uliginosum* (H. WEGNER pers. comm.), *Betula* (C. KAYSER pers. comm.) and *Lonicera periclymenum* and reared on *Sambucus nigra* (B. MÜLLER pers. comm.). Further it has been recorded from *Crataegus*, *Alnus glutinosa*, *Rubus idaeus* and *Populus tremula* (AIZPÚRUA 2002). The pupation takes place between loosely spun leaves.

HABITAT: In light, open deciduous forests, along forest fringes, in more open scrubland and in gardens. The species occurs from sea level up to 1600 m in southern Europe, in Spain recorded up to 1500 m (J. YLLA I ULLASTRE pers. comm.).

SIMILAR SPECIES: No similar species in Europe. Uniform brown specimens (f. *fuscaria*) may be difficult to identify and examination of the genitalia may be needed.

GENETIC DATA: Genetically homogeneous from western Europe to central Asia (n=22, from six European countries), but East Asian populations diverging by 2.3%. Nearest species: *Aplochlora pisochroa* (TURNER, 1906) from Australia (5.8%). Nearest species in Europe: *Ennomos quercinaria* (7.2%).

REMARKS: Altogether nine subspecies, in addition to the nominal subspecies, have been listed for *A. prunaria* in SCOBLE (1999). Whether this represents more a taxonomic tradition of the early 1900s or a true phenomenon of numerous disjunct populations, remains to be tested. In the recent Asian literature (for instance SHIN 1996, VIIDALEPP 1996, KIM & BELJAEV 2001) numerous subspecies are recognised. FLAMIGNI *et al.* (2007) and FORSTER & WOHLFAHRT (1981) consider the taxon *alpina* KITT an altitudinal form where brown striations are more extensive and it has been reported to fly among the specimens of nominal subspecies. SCOBLE (1999) considered it valid at subspecies level.

Cabera TREITSCHKE, 1825

Cabera TREITSCHKE, 1825 [October 18], Schmett. Eur. 5 (2): 437. Type species: *Phalaena pusaria* LINNAEUS, 1758, Europe. The objective synonyms *Cabera* TREITSCHKE and *Deilinia* HÜBNER were both proposed in 1825. The dates of publication of the relevant parts of the works concerned were discussed by FLETCHER (1966) and FLETCHER (1979). Genus synonyms are after FLETCHER (1979) and SCOBLE (1999).

- *Cabera* STEPHENS, 1829, Nom. Br. Insects. 44. Type species: *Phalaena pusaria* LINNAEUS, 1758. [Junior homonym of *Cabera* TREITSCHKE, 1825].
- *Cabira* SODOFFSKY, 1837, Bull. Soc. imp. Nat. Moscou 1837 (6): 90. [Emendation of *Cabera* TREITSCHKE].
- *Cabiro* UNGER, 1856, Arch. Ver. Freunde Naturg.-Mecklenb. 10: 61. [Emendation of *Cabera* TREITSCHKE].
- *Deilinia* HÜBNER, [1825 December 31] 1816, Verz. bekannter Schmett. (20): 310. Type species: *Phalaena pusaria* LINNAEUS, 1758.
- *Dilina* AGASSIZ, 1847, Nomencl. zool. (Index univl.): 118, 124. [Emendation of *Deilinia* HÜBNER and a junior homonym of *Dilina* DALMAN, 1817 (Lepidoptera: Sphingidae)].
- *Dilinia* HAMPSON, 1895, Fauna Br. India (Moths) 3: 216. [Emendation of *Deilinia* HÜBNER, [1825]].
- *Gyalomia* PROUT, 1913, Ann. Transv. Mus. 3: 218. Type species: *Gyalomia elatina* PROUT, 1913.
- *Thysanochilus* BUTLER, 1878, Ann. Mag. nat. Hist. (5) 1: 404. Type species: *Thysanochilus purus* BUTLER, 1878.

UNAVAILABLE NAMES (included in work rejected for nomenclatural purposes): *Sphecodes* HÜBNER, 1806 (see FLETCHER 1979). – (name used in suprageneric sense): *Sphecodes* HÜBNER, 1818 (see FLETCHER 1979). – (incorrect subsequent spelling): *Deilinea* HOLLAND, 1893 (see FLETCHER 1979).

DIVERSITY AND DISTRIBUTION: *Cabera* contains about 30 species in the Palaearctic and Nearctic regions, Africa and Madagascar. PITKIN (2002) noted that the Neotropical taxa assigned to *Cabera* have been misplaced.

EXTERNAL CHARACTERS AND ABDOMEN, description is based on Holarctic species: Wingspan approximately 20–30 mm. Wings white, whitish, grey or yellow-brown. Medial lines often straight, suffuse. Discal spots often weak or absent. Male antennae bipectinate. Female antennae fasciculate. Male 8th tergite occasionally slightly triangular, otherwise sternites and tergites 3–8 of both sexes undifferentiated.

MALE GENITALIA, description is based on Holarctic species: Uncus evenly tapering. Socii minute, setose. Gnathos absent. Valva short, apex weakly upcurved. Apex and medial part of valva setose. Valva bases almost meeting medially. Juxta small. Saccus large, elongated. Vesica with several cornuti.

FEMALE GENITALIA, description is based on Holarctic species: Papillae anales elongated, setose. Lamella antevaginalis and lamella postvaginalis often undifferentiated. Antrum sclerotised, funnel-shaped. Ductus bursae short, membranous. Corpus bursae narrow, long, densely spinose, spines often arranged in striae.

BIOLOGY: Larvae of Nearctic species have been recorded feeding on *Salix* and *Populus* (MCGUFFIN 1981), the Palaearctic species feed additionally on *Alnus*, *Betula* and *Sorbus aucuparia* (MIKKOLA *et al.* 1989). The larvae of Canadian species usually rest flat along the leaf of the host. Larva pupates either in a loose cocoon of leaves tied together with silk on the host plant or in a silken cell at or near the surface of the soil (MCGUFFIN 1981). Pupa overwinters.

IMMATURE STAGES: Larva of *C. pusaria* is green with few pinkish red markings dorsally, also brown form exists. Larva of *C. exanthemata* are green with white longitudinal and dark green dorsal stripes (PORTER 2010). Pupa cremaster is with four pairs of curved setae, the apical pair is widest (MCGUFFIN 1981, PATOČKA & TURČANI 2005).

REMARKS: *Cabera* has recently been classified unanimously in the Caberini (VIVES MORENO 1994, VIIDALEPP 1996, LERAUT 1997, HAUSMANN, MIRONOV & VIIDALEPP 2011). In a DNA-based maximum likelihood analysis *Cabera* (type genus of the Caberini) and the mainly Neotropical *Sphacelodes* GUENÉE, 1858 (type genus of Sphacelodini FORBES, 1948, synonymised with the Caberini in PITKIN (2002)) were recoved as sister taxa (SIHVONEN *et al.* 2011). See *Classification of European Ennominae genera into tribes* for further information. Systematic position: Caberini.

100. Cabera pusaria (LINNAEUS, 1758)

pusaria LINNAEUS, 1758, Syst. Nat. (Ed. 10) 1: 522, (*Phalaena* (*Geometra*)). Syntypes (Linnean Society of London, London, UK), Europe (examined) externally.

confinaria FREYER, [1832], Neuere Beitr. Schmettkde 1 (10): 115, pl. 60, fig. 2, (*Cabera*). Syntype(s), [Slovenia]: Lanbach [Laibach]. [Originally as ab]..

heyeraria HERRICH-SCHÄFFER, 1848, Syst. Bearbeitung Schmett. Eur. 3 (31): 85; ibidem (1847), pl. 41, figs 251, 252 (non binominal), (*Fidonia*). Syntype(s), Germany (northern).

rotundaria HAWORTH, 1809, Lepid. Br. (2): 289, (*Geometra*). Syntype(s), [United Kingdom].

strigata SCOPOLI, 1763, Ent. Carniolica: 218, (*Phalaena*). Syntype(s), [Europe].

UNAVAILABLE NAMES (hybrid): *fletcheri* TUTT, 1905 (*pusaria* LINNAEUS male × *exanthemata* SCOPOLI female). – Intrasubspecific names, potentially unavailable: *ablataria* FUCHS, 1899 (ab.); *bilineata* GALVAGNI, 1928 (ab.); *crassesignata* LEMPKE, 1951 (f.); *flavescens* LEMPKE, 1939 (ab.); *inornaria* MEVES, 1914 (ab.); *irrorata* LEMPKE, 1939 (ab.); *juncta* LEMPKE, 1939 (ab.); *linearia* DEBAUCHE, 1929 (ab.); *melania* OBERTHÜR, 1896 (ab.); *monotonica* STRAND, 1901 (ab.); *nivea* PETERSEN, 1902 (ab.); *posteropuncta* LEMPKE, 1939 (ab.); *quadrilineata* BOLDT, 1939 (ab.); *quadripunctata* NITSCHE, 1926 (ab.); *reducta* LEMPKE, 1939 (ab.).

EXTERNAL CHARACTERS AND ABDOMEN: Wingspan 28–33 mm. Wings silky white, costa and terminal area irrorated with greyish black. Forewing antemedial, medial and postmedial lines grey. Postmedial line almost straight. Hindwing with medial and postmedial lines grey. Hindwing postmedial line weakly curved, nearly straight. Terminal line and fringes concolorous with wings. Discal spots absent or minute. Wings below as above. Frons, collar, thorax and abdomen concolorous with wings. Male and female antennae white dorsally. Male antennae bipectinate, female antennae fasciculate. Hindtibia of both sexes with 2+2 spurs. Tympanal organs medium-sized. Male 8th tergite slightly triangular. Other abdominal sternites and tergites of both sexes undifferentiated.

VARIATION: Very little. Ground colour normally silky white, occasionally greyish specimens are found. Forewing medial lines normally grey, but can also be brown-grey or occasionally yellowish.

MALE GENITALIA: Uncus weakly setose, evenly tapering. Socii minute, setose. Gnathos absent. Valva short, narrow, apex weakly upcurved. Apex and medial part of valva setose. Costa base with small extension. Valva bases almost meeting medially. Juxta small, narrow. Saccus large, elongated. Aedeagus weakly curved, caecum short, apex round. Vesica opens at approximately 135 degree angle. Vesica large, with one diverticulum, with four long cornuti.

FEMALE GENITALIA: Papillae anales elongated, setose. Lamella antevaginalis and lamella postvaginalis undifferentiated. Antrum sclerotised (posterior part of ductus bursae), funnel-shaped. Anterior part of ductus bursae short, membranous. Corpus bursae evenly narrow, long, anterior end without diverticula. Corpus bursae sclerotised, densely spinose, spines arranged in striae.

DISTRIBUTION AND ABUNDANCE: Eurasiatic. In Europe from northern Scandinavia to the Mediterranean and from Ireland through central and eastern Europe to European Russia and the Ural mountains. – Outside Europe in Turkey, northern Iran and the Caucasus region and from the Urals eastwards through central Asia, Siberia and Mongolia to Far East Russia and Sakhalin. – In Europe a very common species that not rarely occurs in numbers.

PHENOLOGY: Usually bivoltine, but in northern Scandinavia univoltine from mid-June to late July. Further south in two overlapping generations. In southern Scandinavia from late May to late August, in south-western Germany from early May to late August (EBERT 2003) and in southern Europe from mid-May to mid-August (REDONDO *et al.* 2009). The two generations peak approximately in mid-June and early August. Caterpillars from June to September or October. The hibernation takes place as a pupa in the ground. The moths fly at night and come readily to light and are easily disturbed from trees and bushes during daytime.

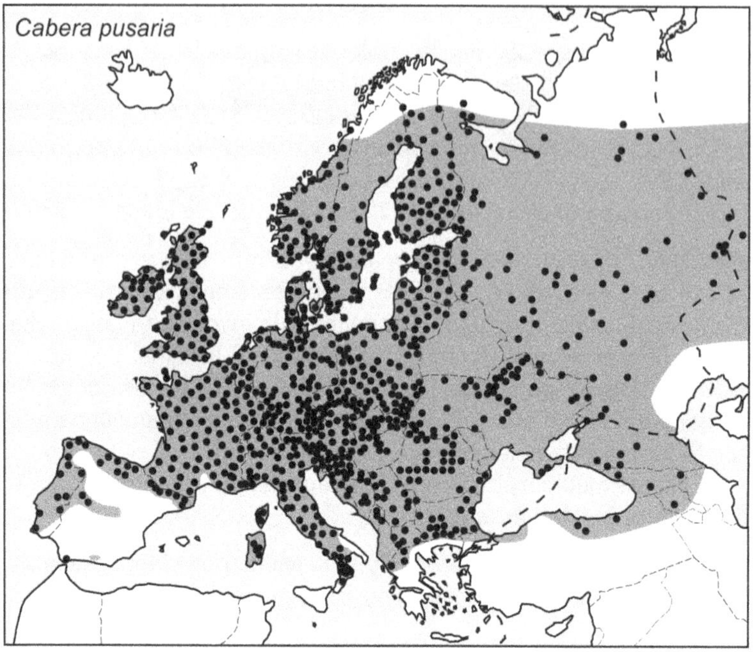

BIOLOGY: Oligophagous. The caterpillar has been found on *Salix*, *Betula*, *Alnus*, *Carpinus betulus*, *Prunus padus* and *Viburnum opulus* (EBERT 2003). H. WEGNER (pers. comm.) has found the larvae on *Salix repens*, *Betula pubescens* and *Alnus glutinosa*. Recorded on *Corylus avellana* and *Populus* (AIZPÚRUA 2002).

HABITAT: In forest fringes and bushy places, in meadows, along rivers and on moors, often on moist or slightly moist soil, but also in gardens. Vertically the species occurs from sea level up to 2100 m altitude in the Iberian Peninsula according to REDONDO *et al.* (2009) and in the western Alps up to 1750 m.

SIMILAR SPECIES: *C. leptographa* WEHRLI and *C. exanthemata* (SCOPOLI), see Text-figs 109–111. *C. leptographa* is smaller, less irrorated with greyish, irroration is often limited to costa and antemedial line is often weak (larger, more irrorated with greyish, irroration on costa and on terminal area and antemedial line often clearly visible in *C. pusaria*). Further, URBAHN illustrates (1952a, plate 3) that narrow apex of male antenna is longer in *C. leptographa* (length of narrow apex about 9 segments), shorter in *C. pusaria* (length of narrow apex about 12 segments). We have not examined adequate material to confirm the diagnostic value of this character. *C. exanthemata* is yellowish, forewing postmedial line curves inwards on costa, hindwing postmedial line is angled (silky white, forewing postmedial line almost straight, hindwing postmedial line straight in *C. pusaria*). Male and female genitalia of *C. pusaria*, *C. leptographa* and *C. exanthemata* allow reliable identification.

GENETIC DATA: Genetically homogeneous from Europe (n=42, from eight European countries) to east Asia, but two specimens from Italy and Germany diverging by 2.3%. Nearest species: *C. exanthemata* (5.5%).

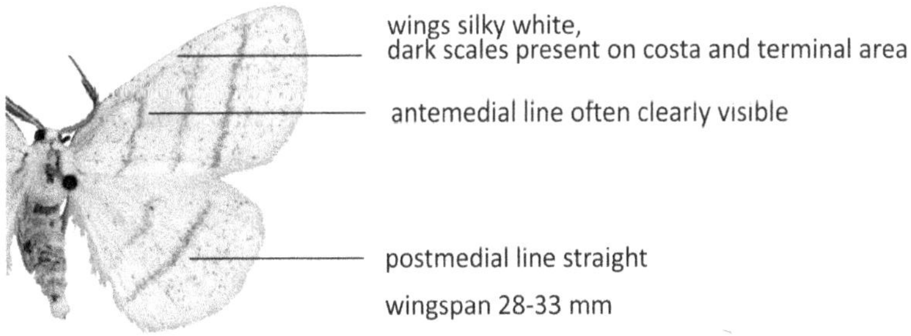

C. pusaria 109

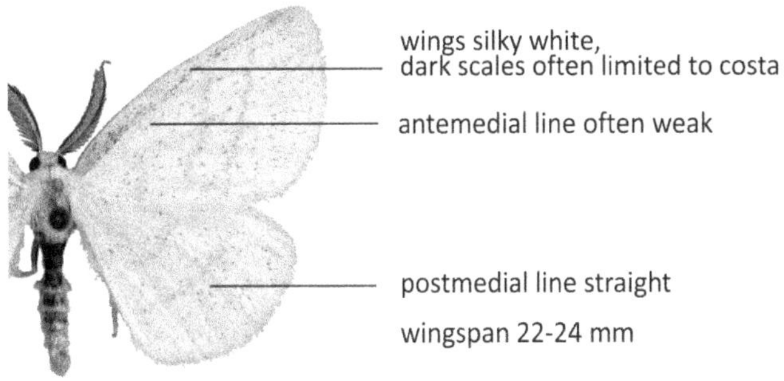

C. leptographa 110

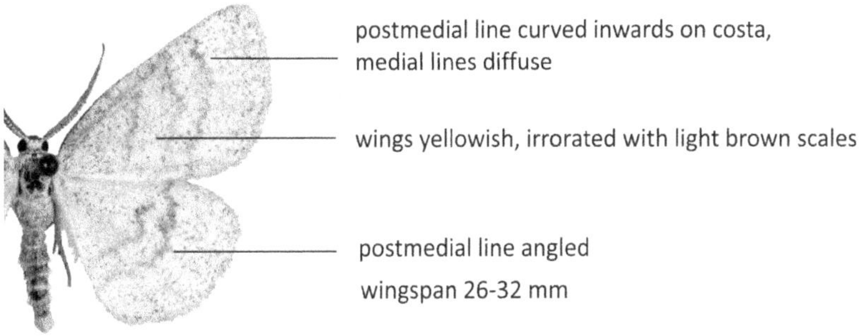

C. exanthemata 111

Text-figs 109–111. Diagnostic characters (indicated) of European *Cabera* species. Text-fig. 109. *C. pusaria* (LINNAEUS, 1758), Denmark, 4.vi.1977. Text-fig. 110. *C. leptographa* WEHRLI, 1936, Austria, ex ovo 22.vi.1981. Text-fig. 111. *C. exanthemata* (SCOPOLI, 1763), Denmark, 11–13. viii.1978.

101. Cabera leptographa WEHRLI, 1936

leptographa leptographa WEHRLI, 1936, Ent. Rdsch. 53 (40): 564, (*Cobera*). Syntypes 1 male, 1 female, Tajikistan: Southern Bucharas, Kulab. [The type locality is in Tajikistan, near Afghanistan border (WEHRLI 1939–1954), not in Afghanistan as mentioned in SCOBLE (1999).

leptographa birographa VIIDALEPP & REMM, 1982, Useful and harmful insects of Siberia: 149, figs 9a, 9b (*Cabera*). Holotype male (Estonian University of Life Sciences, Tartu, Estonia), Russia: Habarovsk, botanical garden 'Dendrarium'. [Original publication (VIIDALEPP 1982) is in Russian, therefore taxon is often overlooked in the literature].

EXTERNAL CHARACTERS AND ABDOMEN: Wingspan 22–24 mm. Generally as in *C. pusaria*, but with the following differences. In *C. leptographa* wingspan is smaller, greyish irroration is often limited to costa, antemedial line often weak (larger, greyish irroration present on costa and in terminal area, antemedial often clearly visible in *C. pusaria*).

VARIATION: Very little. Medial lines are normally weak, narrow, but occasionally they can stand out more clearly.

MALE GENITALIA: Generally as in *C. pusaria*, but with the following differences. In *C. leptographa* costa base is without extension, valva is less curved upwards, vesica opens at approximately 90 degree angle, vesica is with two diverticula (costa with small extension, valva more curved upwards, vesica opens at approximately 135 degree angle, vesica is with one diverticulum in *C. pusaria*).

FEMALE GENITALIA: Generally as in *C. pusaria*, but with the following differences. In *C. leptographa* corpus bursae is angled, less sclerotised, spines are less arranged in striae (corpus bursae straight, more sclerotised, spines arranged in striae in *C. pusaria*).

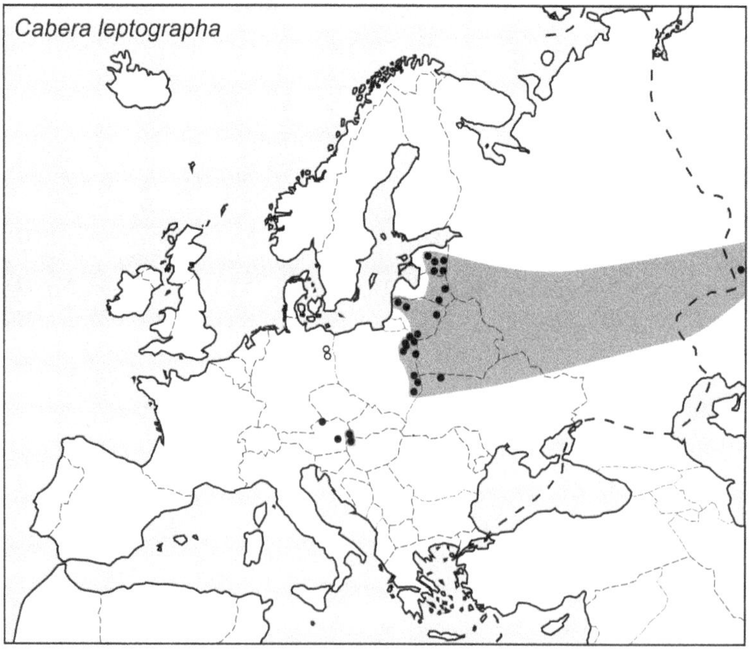

DISTRIBUTION AND ABUNDANCE: Eurasiatic. In Europe from eastern central Europe and the Baltic states across central European Russia to southern Urals. There are two records from Serbia, but one has turned out to be misidentified, and the other cannot be verified, so the species has to be removed from the species list of Serbia. – Outside Europe the species is known from Tadjikistan, south-western Siberia, Cis-Altai, Republic of Tuva, Trans-Baikal and Far East Russia. The present large gaps in the distribution are probably caused by the lack of collecting and the fact, that since its description the species has been overlooked and confused with the other two European *Cabera* species. – In Europe a local and rare species that rarely is found in numbers, possibly it remains undiscovered in many places.

PHENOLOGY: Univoltine. From mid-June to late July or early August. Larvae from July to August or September. The moths fly at night and are attracted to light.

BIOLOGY: Monophagous. The caterpillar has been recorded on *Salix rosmarinifolia*, *S. aurita*, *S. viminalis* and *S. cinerea* (MALKIEWICZ 2012). Hibernation takes place as a pupa.

HABITAT: Hygrophilous. In marshland, peat bogs, along rivers and in meadows. All known European localities are apparently at low altitude apart from two specimens (ZSM) from Seewinkel in Austria in about 900 m.

SIMILAR SPECIES: *C. pusaria* (LINNAEUS) and *C. exanthemata* (SCOPOLI). See text above on *C. pusaria* and Text-figs 109–111.

GENETIC DATA: Nearest species: *C. quadrifasciaria* (PACKARD, 1873) from North America (6.1%) and *C. exanthemata* (7.3%).

REMARKS: *C. leptographa*, originally described from central Asia, was later found in Europe (URBAHN 1952a, 1952b, URBAHN 1955, URBAHN 1960, KASY 1963). An extensive survey of diagnostic characters of *C. leptographa*, *C. pusaria* and *C. exanthemata* is given in URBAHN & URBAHN (1978).

102. Cabera exanthemata (SCOPOLI, 1763)

exanthemata exanthemata SCOPOLI, 1763, Ent. Carniolica: 218, fig. 542, (*Phalaena*). Syntype(s), [Slovenja (north-west)]: Carniolia.

approximaria HAWORTH, 1809, Lepid. Br. (2): 289, (*Geometra*). Syntype(s), [United Kingdom].

arenosaria HAWORTH, 1809, Lepid. Br. (2): 289, (*Geometra*). Syntype(s), [United Kingdom].

atomifera FOURCROY, 1785, Ent. Paris: 277, (*Phalaena*). Syntype(s), France: Paris. [Synonymy after LERAUT (1997)].

atomosa FOURCROY, 1785, Ent. Paris: 282, (*Phalaena*). Syntype(s), France: Paris. [Synonymy after LERAUT (1997)].

exanthemaria BORKHAUSEN, 1794, Natur. eur. Schmett. 5: 269, (*Phalaena Geometra*). [Emendation of *exanthemata* SCOPOLI].

striaria HÜBNER, [1799], Samml. Eur. Schmett. 5 Geometrae (1): pl. 17, fig. 88, (*Geometra*). Syntype(s), lost, Europe.

suprapunctata WEHRLI, 1925, Mitt. münch. ent. Ges. 15 (1–5): 59, pl. 1, fig. 30, (*Cabera*). Syntypes 2, [Germany]: Tessin, [Switzerland]: Maroggia; Jura: Trans. Raimeux. See remarks.

exanthemata bryantaria TAYLOR, 1906, Can. Ent. 38 (12): 402, (*Deilinia*). Holotype male (National Museum of Natural History, Smithsonian Institution, Washington DC, USA), [Canada]: British Columbia, near the Stickeen River. Originally as sp.

exanthemata hamica WEHRLI, 1939, in SEITZ, Gross-Schmett. Erde 4 (Suppl.): 307, pl. 23: e, (*Cabera*). Syntypes including 3 males, 1 female, [China]: Mongolia (north), Hami. Valid at subspecies rank according to SCOBLE (1999).

INTRASUBSPECIFIC NAMES, POTENTIALLY UNAVAILABLE: *alba* LEMPKE, 1951 (f.); *bistriaria* MEVES, 1914 (ab.); *crassesignata* LEMPKE, 1947 (f.); *glabra* LEMPKE, 1951 (f.); *holomelana* DUFAY, 1974 (f.); *inornata* LEMPKE, 1947 (f.); *irrorata* LEMPKE, 1947 (f.); *lativittata* COCKAYNE, 1952 (ab.); *linearia* LEMPKE, 1947 (f.); *nigrociliata* LEMPKE, 1970 (f.); *nogentina* THIERRY-MIEG, 1910 (f.); *pellagraria* GUENÉE, [1858] (ab.); *plumbeata* HACKRAY, 1946 (ab.); *pseudapproximaria* LEMPKE, 1947 (f.); *reducta* LEMPKE, 1947 (f); *rotundaria* MEVES, 1914 (ab.); *unicolorata* TEICH, 1892 (ab.).

EXTERNAL CHARACTERS AND ABDOMEN: Wingspan 26–32 mm. Generally as in *C. pusaria*, but with the following differences. In *C. exanthemata* wings are white with yellowish tinge, irrorated with light brown, medial lines are diffuse, forewing postmedial line curved inwards on costa, hindwing postmedial line is angled (wings silky white, weakly irrorated with greyish, medial lines are rather distinct, forewing postmedial line straight on costa, hindwing postmedial line weakly curved or nearly straight in *C. pusaria*).

VARIATION: Little. Ground colour typically white with yellowish tinge, irrorated with light brown, but occasionally white or darker specimens are found, the latter particularly in northern Europe. One, two or even all medial lines can be absent, partly fused together, and density of irroration is variable.

MALE GENITALIA: Generally as in *C. pusaria*, but with the following differences. In *C. exanthemata* costa base has notable extension, valva clearly curved upwards, vesica is without diverticula, vesica has more than 10 cornuti (costa base with small extension, valva weakly curved upwards, vesica with one diverticulum, vesica has four long cornuti in *C. pusaria*).

FEMALE GENITALIA: Generally as in *C. pusaria*, but with the following differences. In *C. exanthemata* posterior part of corpus bursae is narrow, anterior part of corpus is with small diverticulum (corpus bursae evenly narrow, anterior end is without diverticulum in *C. pusaria*).

DISTRIBUTION AND ABUNDANCE: Holarctic. In Europe from northern Fennoscandia to the Iberian Peninsula, southern France including Corsica, Italy including Sardinia and the Balkans and from Ireland through central and eastern Europe to European Russia and the Ural mountains. – Outside Europe the species is found in Turkey and the Caucasus region and eastwards from the Ural mountains through central Asia, Siberia and Far East Russia to Kamchatka, Sakhalin, the Kuriles and Japan. The species also occurs in North America. – In Europe a very common species that rarely occurs in numbers.

PHENOLOGY: Uni- or bivoltine. In northern Scandinavia in one generation from mid-June to late July. In the remaining part of the European distribution the species occurs in two overlapping generations, in southern Scandinavia from late May until late August, in southern central Europe from early May to late August (EBERT 2003) and in southern Europe from mid-April to late August (REDONDO *et al.* 2009). Caterpillars from June to September or October. The species overwinters as a pupa. The moths are easily disturbed by day. They fly at night and come readily to light.

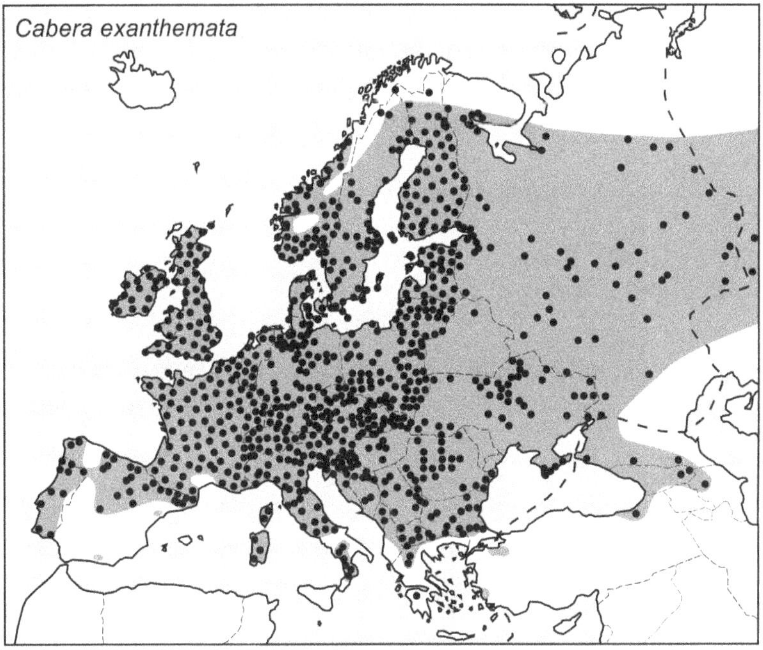

BIOLOGY: Oligophagous. The larva has been found on various *Salix* species, *Populus nigra*, *P. tremula*, *Betula pendula*, *Alnus incana* and *A. glutinosa* (EBERT 2003) and on *Salix aurita* (H. WEGNER pers. comm.). Furthermore recorded on *Corylus avellana* (AIZPÚRUA 2002).

HABITAT: In forest fringes and scrubland, especially on moist soil, along ditches, in meadows and in marshes as well as more wetter areas on moors. The species occurs from sea level in northern Europe up to 2000 m in southern Europe. In the Alps mainly found up to 1000 m, but sometimes going up to 1500 m (N. PÖLL pers. comm.) and exceptionally up to 1850 m altitude (B. MÜLLER pers. comm.).

SIMILAR SPECIES: *C. pusaria* (LINNAEUS) and *C. leptographa* WEHRLI. See text above on *C. pusaria* and Text-figs 109–111.

GENETIC DATA: Genetically homogeneous in Europe (n=39, from eight European countries), populations from eastern China slightly diverging. Nearest species: *C. pusaria* (5.5%).

REMARKS. *Cabera exanthemata insulata* INOUE, 1958, which was originally described as a subspecies of *C. exanthemata* (SCOPOLI) from Japan, was raised to species status by SATO & BELJAEV (2009). *C. exanthemata* has thus far not been recorded from Japan, but it occurs sympatrically with *C. insulata* in the Far East of Russia.

SCOBLE (1999) treated the taxon *suprapunctata* WEHRLI, 1925 as valid at subspecies level. We agree with LERAUT (2009) and consider it a form with each wing showing a small, black discal spot. Genetic and morphological information available does not support the hypothesis that there exists an allopatric, distinct population of *C. exanthemata* in central Europe.

The taxon *bryantaria* (TAYLOR) from British Columbia, Canada, which was originally described as species, was treated as subspecies of *C. exanthemata* by RINDGE (1956). Structurally *bryantaria* is very similar to *C. exanthemata* and North American *C. erythemaria*

(GUENÉE, 1858) (MCGUFFIN 1981). In an extensive DNA barcode study from British Columbia, *C. exanthemata* subsp. *bryantaria* and *C. erythemaria* were intermixed in a genetically heterogeneous cluster (DEWAARD *et al.* 2011). Further studies are needed to clarify the taxonomic position of taxon *bryantaria*.

Lhommeia WEHRLI, 1939

Lhommeia WEHRLI, 1939, Revue fr. Lépidopt. 9: 234. Type species: *Tephrina biskraria* OBERTHÜR, 1885 (Algeria: Djurjura). There are no genus synonyms.

DIVERSITY AND DISTRIBUTION: *Lhommeia* contains three species, two of those occur in North Africa and Arabian Peninsula, one of those extends into south-eastern Spain, and one occurs in South Africa.

EXTERNAL CHARACTERS AND ABDOMEN, description is based on European species: Wingspan 25–27 mm. Wings brown-yellow to green. Forewing postmedial line bordered with reddish brown suffusion. Hindwings paler. Male antennae bipectinate, female antennae shortly fasciculate. Labial palpi long, apex downturned. Posterior margin of male 8th sternite weakly concave. Anterior margin of 8th tergite and sternite with lateral pockets. Other abdominal sternites and tergites of both sexes undifferentiated.

MALE GENITALIA, description is based on the European species (see also Text-fig. 112): Uncus evenly tapering. Socii minute, setose. Gnathos absent. Valva extremely modified. Dorsal arm long, narrow, setose, membranous. Ventral arm with setose dorsal process and strongly sclerotised, narrow ventral process, apex acute. Three sclerotised arms (homologies unclear) long, narrow, medial arm curved ventrally. Saccus large, elongated. Vesica without cornuti.

FEMALE GENITALIA, description is based on the European species: Papillae anales large. Apophyses anteriores short, rudimentary. Lamella antevaginalis wide, quadrangular plate. Ductus bursae short, narrow. Corpus bursae thin, membranous. Signum absent.

BIOLOGY: Larvae of *L. biskraria* have been recorded to feed on the leaves *Ziziphus lotus* (CHRÉTIEN 1917). *L. biskraria* occurs in deserts, semi-deserts and similar dry places with sparse vegetation. In Morocco adults are found throughout the year (RUNGS 1981).

IMMATURE STAGES: According to CHRÉTIEN (1917) the elliptic egg of *L. biskraria* is green, with irregular polygonal depression. Larva is smooth, ash-grey to reddish brown, with brown longitudinal lines, ornamented with black diamond-shaped pattern laterally. No data is available on pupa.

REMARKS: WILTSHIRE (1990) noted that *Lhommeia* might be a Caberini. Absence of gnathos is the most apparent character supporting the Caberini association, in addition to the elongated saccus and superficial similarity. The extreme valva modifications, massive furca arms, wide antrum and downwards curved labial palpi are features which are difficult to associate with any other European taxa. Superficially similar dorsal arm of valva is seen in the unrelated *Lythria cruentaria* (HUFNAGEL, 1767) (Sterrhinae). Systematic position: Uncertain.

103. Lhommeia biskraria (OBERTHÜR, 1885)

biskraria biskraria OBERTHÜR, 1885, Annales Soc. ent. France, ser. 6, IV (1884): CXXXIV, (*Tephrina*). Holotype female, Algeria: Biskra.

biskraria bleusei THIERRY-MIEG, 1905, Naturaliste 27: 192, (*Tephrina*). Holotype male, [Algeria]: Oran, Ain-Séfra. Originally as sp., valid at subspecies rank according to SCOBLE (1999).

biskraria cinnamomaria ROTHSCHILD, 1914, Novit. zool. 21: 353, (*Tephrina*). Holotype female (The Natural History Museum, London, UK), Algeria (central): Guelt-Es-Stel (examined externally). Originally as sp., valid at subspecies rank according to SCOBLE (1999).

biskraria illiturata WARREN, 1897, Novit. zool. 4: 255, (*Anthyperythra*?). Holotype male (The Natural History Museum, London, UK), South-East Africa (examined externally). Originally as sp., valid at subspecies rank according to SCOBLE (1999).

biskraria viridaria ROTHSCHILD, 1914, Novit. zool. 21: 353, (*Tephrina*). Holotype female (The Natural History Museum, London, UK), Algeria (central): Guelt-Es-Stel (examined externally). Originally as sp., valid at subspecies rank according to SCOBLE (1999).

EXTERNAL CHARACTERS AND ABDOMEN: Wingspan 25–28 mm. Wings ochreous-brown. Antemedial and postmedial lines blackish brown, latter curved inwards on costa and bordered with reddish brown suffusion and small blacks spots. Medial line weak, suffuse. Hindwings paler, only postmedial line clearly visible, terminal area darker. Terminal line and fringes concolorous with wings. Discal spots minute or absent. Wings below lighter, postmedial line and discal spots suffuse. Frons, collar, thorax and abdomen concolorous with wings. Male antennae bipectinate, female antennae shortly fasciculate. Labial palpi long, apex downturned. Hindtibia of both sexes with 2+2 spurs. Tympanal organs medium-sized. Posterior margin of male 8th sternite weakly concave. Anterior margin of 8th tergite and sternite with lateral pockets. Other abdominal sternites and tergites of both sexes undifferentiated.

VARIATION: Ground colour varies from ochreous brown to greenish brown. Medial lines visible to varying degrees, forewing postmedial line normally strongest, occasionally present as few blackish spots only.

We do not re-describe subspecies here, see remarks, apart from superficially mentioning that taxon *viridaria* (ROTHSCHILD) is typically pale greenish grey, *bleusei* (THIERRY-MIEG) is typically ochre-yellow with weak markings and *cinnamomaria* (ROTHSCHILD) is typically greyish cinnamon with three dark grey spots on costa.

MALE GENITALIA (see also Text-fig. 112): Uncus weakly setose, evenly tapering. Socii minute, setose. Gnathos absent. Tegumen narrow. Valva extremely modified. Dorsal arm long, narrow, setose, membranous. Ventral arm with setose dorsal process and strongly sclerotised, narrow ventral process, apex acute. Ventral arms of valva fused medially. Three sclerotised arms (homologies unclear) long, narrow, medial arm curved ventrally. Saccus large, elongated. Aedeagus widest in middle, caecum minute, apex elongated, curved. Vesica opens at approximately 90 degree angle, apical part at 45 degree angle. Vesica wide at base, with weak sclerotisation in middle, cornuti absent.

FEMALE GENITALIA: Papillae anales large, setose. Apophyses posteriores long, straight. Apophyses anteriores short, rudimentary. Lamella antevaginalis wide, sclerotised, quadrangular plate. Its margin with narrow horizontal ridge. Lamella postvaginalis weakly sclerotised, large plate. Ductus bursae short, narrow. Corpus bursae thin, elongated, membranous. Signum absent.

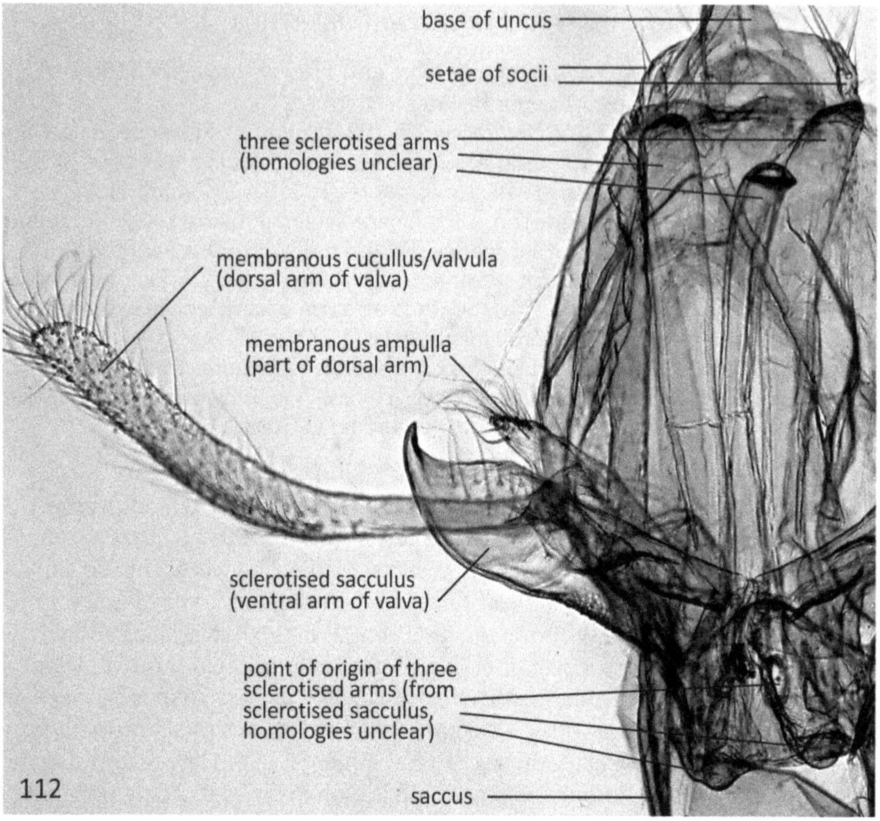

Text-fig. 112. Potential homologies of the male genitalia in *L. biskraria* (OBERTHÜR, 1885). The valva structure and the point of origin of the three sclerotised arms are unusual within the European Ennominae, making the interpretation of the homologies difficult. Morocco: High-Atlas, Asni-Ouirgane, 6–9.x.1964, slide PS1646 (coll. ZSM).

DISTRIBUTION AND ABUNDANCE: Afrotropical- North African. In Europe the species has been found in several places in the Almeria province in south-eastern Spain since the first record in Europe in 1995. Possibly the species is more widely distributed in this area since there appear to be many suitable localities for it. – Outside Europe the species is widely distributed throughout Africa including Morocco, Algeria, Tunisia, Mauretania and from Ethiopia to Botswana, Namibia and South Africa. – In Europe local and rare, but the species has on one occasion been found in numbers.

PHENOLOGY: Bi- or trivoltine. The relatively few European specimens have been found in early June, mid-July, mid-October and early November. In Morocco the species flies through the whole year, but it is most common from October to March (RUNGS 1981). Possibly the species might also be found as adult most of the year in Europe. The moths fly at night and come readily to light. Especially in hot weather they can also be flushed up by day (K. NUPPONEN pers. comm.).

BIOLOGY: Monophagous. Eggs have been laid on *Zizyphus lotus* and larvae have been reared on this plant species (CHRÉTIEN 1916).

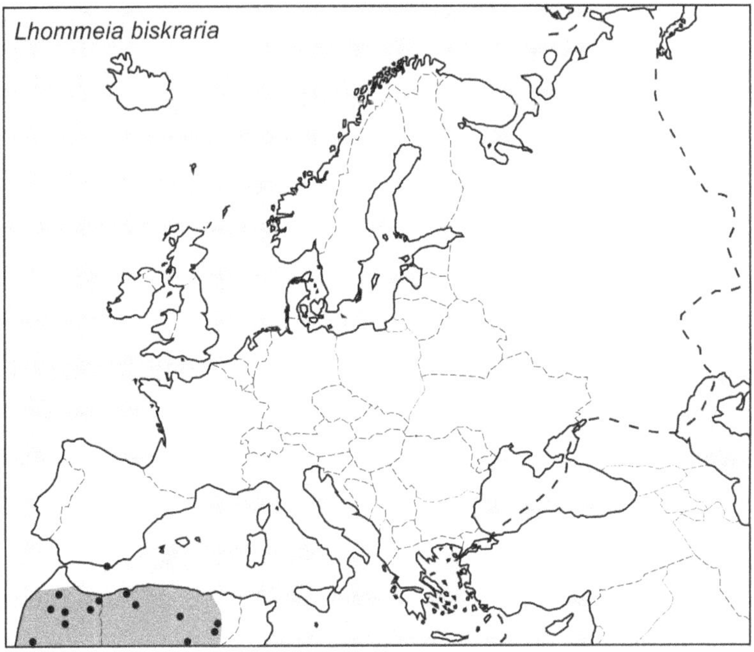

HABITAT: In deserts, semi-deserts and similar dry places with sparse vegetation. In southern Spain on an open sandy area with sparse bushes and grasses at sea level. In southern Spain up to 300 m altitude, in the Anti-Atlas in Morocco up to 500 m.

SIMILAR SPECIES: No similar species in Europe.

GENETIC DATA: Two specimens examined from Morocco. Nearest species: African *Lhommeia subapicata* (WARREN, 1899) (4.1%).

REMARKS: We have examined externally the *L. biskraria* type specimens deposited in the Natural History Museum, London, UK, but we have not had access to adequate amount of material to judge the status of various subspecies as presented in SCOBLE (1999). Therefore we repeat the list as it is shown in the mentioned reference. *L. biskraria* species complex needs a thorough, modern revision, because there is a possibility that some African taxa may be synonyms of *L. biskraria*.

Lomographa HÜBNER, 1825

Lomographa HÜBNER, [1825] 1816, Verz. bekannter Schmett.: 311. Type species: *Geometra taminata* [DENIS & SCHIFFERMÜLLER], 1775 [December 8], which is a junior synonym of *Phalaena bimaculata* FABRICIUS, 1775. Genus synonyms are after SCOBLE (1999).

Akrobapta WEHRLI, 1924, Mitt. münch. ent. Ges. 14: 136. Type species: *Bapta perapicata* WEHRLI, 1924.

Anhibernia STAUDINGER, 1892, Dt. ent. Z. Iris 5: 170. Type species: *Hybernia orientalis* STAUDINGER, 1892.

Bapta STEPHENS, 1829, Nom. Br. Insects: 45. Type species: *Phalaena bimaculata* FABRICIUS, 1775.

Cirretaera WEHRLI, 1939, in SEITZ, Gross-Schmett. Erde 4 (Suppl.): 298. Type species: *Somatina simplicior* BUTLER, 1881.

Corycia DUPONCHEL, 1829, in GODART & DUPONCHEL, Hist. nat. Lépid. Papillons Fr. 7 (2): 110. Type species: *Geometra temerata* [DENIS & SCHIFFERMÜLLER], 1775. [Junior homonym of *Corycia* HÜBNER, [1823] 1816 (Lepidoptera: Noctuidae)].

Cyrtesia WEHRLI, 1937, Lambillionea 37: 188. Type species: *Bapta ochrilinea* WARREN, 1894.

Eudjanonovia FLETCHER, 1979, The Generic Names of Moths of the World 3: 81. Type species: *Stegania pulverata* BANG-HAAS, 1910. [Unnecessary replacement name for *Neobapta* DJAKONOV, 1936].

Faroxyptera DJAKONOV, 1936, Trudy zool. Inst. Leningr. 3: 492, 515, 517. Type species: *Anhibernia buraetica* STAUDINGER, 1892.

Leucetaera WARREN, 1894, Novit. zool. 1: 405. Type species: *Acidalia inamata* WALKER, 1861.

Lomatographa AGASSIZ, 1847, Nomencl. zool. (Index univl.): 215. [Emendation of *Lomographa* HÜBNER].

Neobapta DJAKONOV, 1936, Trudy zool. Inst. Leningr. 3: 486, 505. Type species: *Stegania pulverata* BANG-HAAS, 1910. [Junior homonym of *Neobapta* WARREN, 1904 (Geometridae: Ennominae)].

DIVERSITY AND DISTRIBUTION: *Lomographa* contains about 70 species in the Nearctic, the Neotropics, the Palaearctic, the African and the Oriental regions, extending to New Guinea and the Solomon Islands.

EXTERNAL CHARACTERS AND ABDOMEN, description is based on Holarctic species: Wingspan about 25 mm. Wings white, occasionally grey or brownish, markings often blackish. Male and female antennae filiform. Male 8th sternite and tergite variably sclerotised, modified. Anterior margin of 8th tergite and sternite often with two pockets (Text-figs 113–114).

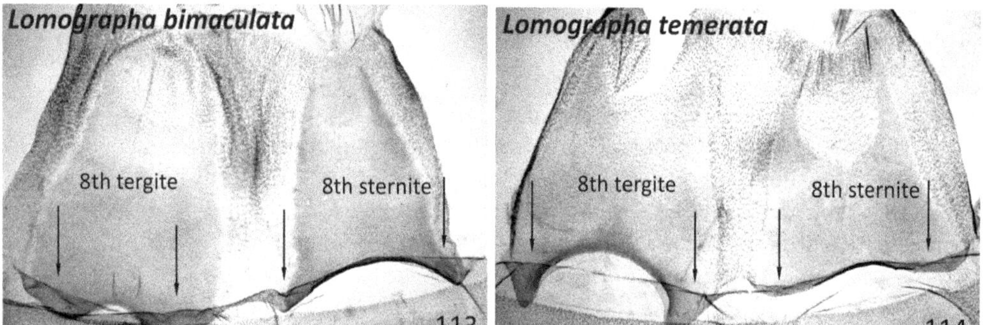

Text-figs 113–114. Eighth segment of European *Lomographa* HÜBNER, 1825 species. Arrows indicate the pockets on their anterior margins that are diagnostic in the European species. Text-fig. 113. *L. bimaculata* (FABRICIUS, 1775), Finland: Savonlinna, 1–2.vi.1996, slide PS1337 (coll. P. SIHVONEN). Text-fig. 114. *L. temerata* (DENIS & SCHIFFERMÜLLER, 1775), Finland: Lammi, 30.v–7.vi.1996, slide PS1340 (coll. P. SIHVONEN).

MALE GENITALIA, description is based on Holarctic species: Uncus evenly tapering, occasionally with small dorsal extensions at base. Socii small, setose. Gnathos weakly developed, often partly membranous. Tuba analis often sclerotised. Valva wide, with setose medial ridge. Transtilla often with roundish extensions. Coremata sacs arise from valva base and vinculum. Vesica often with numerous cornuti.

FEMALE GENITALIA, description is based on Holarctic species: Lamella antevaginalis and lamella postvaginalis small, often weakly sclerotised. Posterior part of corpus bursae very long, curved, or coiled, often striated. Anterior part of corpus bursae enlarged. Signum present, varied in shape, often spinose.

BIOLOGY: Numerous species have been recorded feeding on *Prunus, Sorbus, Crataegus, Malus, Betula* and *Tilia* (MCGUFFIN 1981, MIKKOLA *et al.* 1989). According to HOLLOWAY (1994), the larvae are most frequently recorded from Rosaceae, but not exclusively. Pupa overwinters.

IMMATURE STAGES: Larva of *L. bimaculata* is stout, green, with varying amount of reddish dorsally. Larva of *L. temerata* is stout, green, with whitish or red-edged pale dashes along the dorsal area (PORTER 2010). Pupa cremaster is with three pairs of small, curved setae, whereas apical pair is with two large, laterally curved spines (MCGUFFIN 1981, PATOČKA & TURČANI 2005).

REMARKS: *Lomographa* has recently been classified in the Baptini (HOLLOWAY 1994, LERAUT 1997, HAUSMANN, MIRONOV & VIIDALEPP 2011), in the Caberini (VIVES MORENO 1994, VIIDALEPP 1996) and in the Caberini/Baptini complex (PITKIN 2002, FORUM HERBULOT 2014). In a DNA analysis of the world Geometridae tribes, *Lomographa* (type genus of the Baptini) and *Theria* HÜBNER, 1825 (Theriini) grouped together and these two taxa were not related to the Caberini (SIHVONEN *et al.* 2011). Weakly developed gnathos is the most apparent character supporting the Caberini/Baptini association. Several other characters, such as coremata sacs, which arise from valva base and vinculum, densely setose medial ridge of valva, modified male 8th segment and spinose signum, are features which are difficult to associate with any other European taxa. Systematic position: Baptini.

104. *Lomographa bimaculata* (FABRICIUS, 1775)

bimaculata bimaculata FABRICIUS, 1775, Syst. Ent.: 635, (*Phalaena*). Syntype(s) (Zoological Museum, University of Copenhagen, Denmark), [Germany]: Leipzig.

taminaria HÜBNER, [1825], Verz. bekannter Schmett.: 311, (*Lomographa*). [Emendation of *taminata* [DENIS & SCHIFFERMÜLLER]].

taminata [DENIS & SCHIFFERMÜLLER], 1775, Ankündung syst. Werkes Schmett. Wienergegend: 116, (*Geometra*). Syntype(s), lost, [Austria]: Vienna district.

bimaculata subnotata WARREN, 1895, Novit. zool. 2: 121, (*Bapta*). Holotype male (The Natural History Museum, London, UK), Japan: Hakodate (examined externally). Originally as sp., valid at subspecies rank according to SCOBLE (1999).

bimaculata pallidiola MATSUMURA, 1931, 6000 illust. Insects Japan-Empire: 868, figured, (*Bapta*). Syntype(s), including female, Japan? [Originally as f].

INTRASUBSPECIFIC NAMES, POTENTIALLY UNAVAILABLE: *bipunctata* FUCHS, 1900 (f.); *brunneimargo* PROUT, 1915 (ab); *costijuncta* COCKAYNE, 1950 (ab.); *suffusa* COCKAYNE, 1950 (ab.); *unimaculata* COCKAYNE, 1950 (ab.).

EXTERNAL CHARACTERS AND ABDOMEN: Wingspan 23–29 mm. Wings satiny white, forewing costa with two blackish brown spots. Medial and postmedial line weak, slightly dentate, brown. Terminal area and fringes brownish grey. Terminal line narrow, brownish grey.

Hindwings with weak postmedial line, terminal area slightly suffused with brownish grey. Discal spots minute. Wings below as above, paler. Frons pale brown. Collar, thorax and abdomen concolorous with wings. Male and female antennae filiform. Hindtibia of both sexes with 2+2 spurs. Tympanal organs medium-sized. Male 8th sternite slightly triangular, anterior margin concave. Anterior margin of male 8th tergite and sternite with two pockets (Text-fig. 113). Other abdominal sternites and tergites of both sexes undifferentiated.

VARIATION: Very little. Markings are occasionally weakly developed, terminal area can be white and costal spots can be reduced in size.

MALE GENITALIA: Uncus sparsely setose, evenly tapering. Socii small, setose. Gnathos arms narrow, fused ventrally, medial part membranous. Tuba analis (subscaphium?) sclerotised. Tegumen horse-shoe-shaped. Valva wide, with two sclerotised ridges, dorsal ridge setose. Valva apex granulate. Transtilla wide, with large roundish extensions. Large coremata sacs arise from valva base and vinculum. Saccus narrow, round. Aedeagus straight, caecum long. Vesica opens at approximately 90 degree angle. Vesica base narrow, followed by large, elongated sac with numerous, weakly sclerotised, minute microcornuti. Vesica apex with numerous cornuti. Ductus ejaculatorius opens laterally from medial area.

FEMALE GENITALIA: Papillae anales soft, setose. Apophyses long, straight. Lamella antevaginalis and lamella postvaginalis undifferentiated. Sterigma sclerotised, y-shaped. Ductus bursae very short. Posterior part of corpus bursae very long, coiled, its anterior part striated. Anterior part of corpus bursae large, elongated. Signum dentate band.

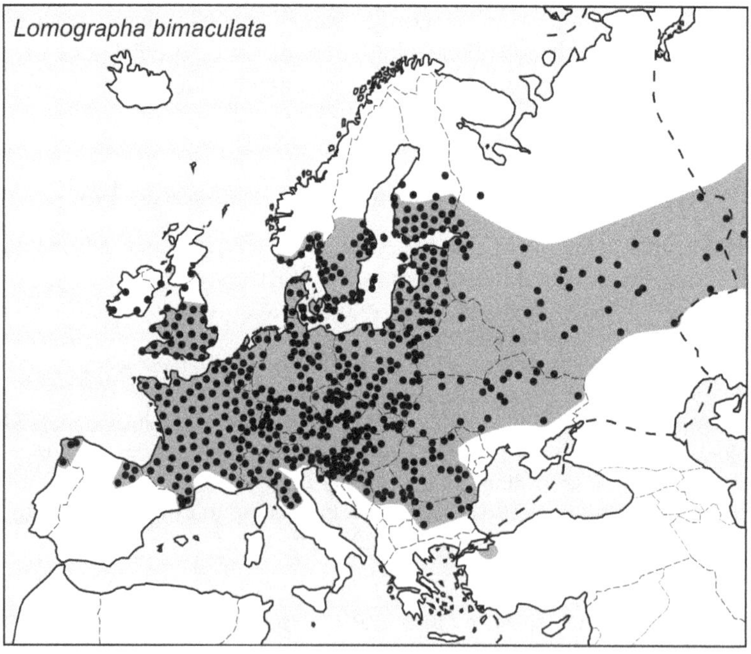

DISTRIBUTION AND ABUNDANCE: Eurasiatic. In Europe from southern central Scandinavia to northern Spain and Portugal, southern France, northern Italy and the northern half of the Balkans and from Ireland eastwards to the Ural mountains. – Outside Europe in north-western Turkey (KOÇAK & KEMAL 2009) and further it is distributed from the Ural mountains through southern Siberia to Trans-Baikal, Far East Russia, Sakhalin, Korea and Japan. – In Europe a common species that rarely occurs in numbers.

PHENOLOGY: Univoltine. In southern Scandinavia from mid-May to early July, in southwestern Germany from late April or early May to early July (EBERT 2003) and in northern Spain from April to June (REDONDO *et al.* 2009). Larva from June to August. The species hibernates as a pupa. The adult moths come to light and are easily disturbed from trees and bushes during daytime.

BIOLOGY: Polyphagous. The caterpillar has been found on various deciduous trees and bushes such as *Betula*, *Crataegus*, *Tilia*, *Prunus spinosa*, *P. avium*, *P. padus*, *Cornus sanguinea*, *Rosa canina* and *Viburnum opulus* (EBERT 2003). Recorded on *Fagus sylvatica* (AIZPÚRUA 2002).

HABITAT: In hedgerows, groves, gardens, woods- and scrubland. The species occurs from sea level up to 1200 m altitude in southern Europe.

SIMILAR SPECIES: No similar species in Europe.

GENETIC DATA: Genetically homogeneous in Europe (n=22, from five countries). Nearest species in Europe: *Dyscia innocentaria* (7.3%), outside Europe Australian *Casbia rectaria* WALKER, 1866 (6.9%) and New Guinean *Lomographa lomographata* (WARREN, 1907) (7.5%).

105. Lomographa temerata (DENIS & SCHIFFERMÜLLER, 1775)

temerata [DENIS & SCHIFFERMÜLLER], 1775, Ankündung syst. Werkes Schmett. Wienergegend: 116, (*Geometra*). Syntype(s), lost, [Austria]: Vienna district.
nitidula FOURCROY, 1785, Ent. Paris: 278, (*Phalaena*). Syntype(s), France: Paris. [Synonymy after LERAUT (1997)].
nubeculata HAWORTH, 1809, Lepid. Br. (2): 355, (*Phalaena*). Syntype(s), [United Kingdom].
punctata FABRICIUS, 1775, Syst. Ent.: 637, (*Phalaena*). Syntype(s), England.
sylvestrata HÜBNER, 1786, Beitr. Gesch. Schmett.: 1 (27), pl. IV, fig. S, (*Ph.[alaena] Geom.[etra]*. Syntype(s), lost, Germany.

INTRASUBSPECIFIC NAMES, POTENTIALLY UNAVAILABLE: *basinotata* COCKAYNE, 1950 (ab.); *obscura* LEMPKE, 1951 (f.); *obsoleta* LEMPKE, 1970 (f.); *pauper* HOFFMANN, 1918 (ab.); *sericeata* HÖRHAMMER, 1933 (ab.); *tangens* LEMPKE, 1951 (f.).

EXTERNAL CHARACTERS AND ABDOMEN: Wingspan 24–29 mm. Wings white. Forewing terminal area suffused with blackish grey-brown, traversed by weak, white wavy line. Forewing medial line blackish grey, reduced, strongest on inner margin. Postmedial line grey, suffused. Hindwings with grey medial and postmedial lines. Terminal line narrow, blackish grey, wider on forewing. Fringes white, grey-brown on forewing. Discal spots minute, acute, black. Wings below almost uniform white, discal spots as above. Frons, collar, thorax and abdomen concolorous with wings. Male and female antennae filiform. Hindtibia of both sexes with 2+2 spurs. Tympanal organs large. Posterior margin of male 8th sternite bifurcate, anterior margin concave. Anterior margin of male 8th tergite with two elongated pockets, 8th sternite with small pockets (Text-fig. 114). Other abdominal sternites and tergites of both sexes undifferentiated.

VARIATION: Very little. Extent and colour of terminal area dark suffusion varies, it can be almost absent. Medial lines can be almost absent.

MALE GENITALIA: Uncus sparsely setose, evenly tapering. Socii small, setose. Gnathos arms narrow, fused ventrally, medial part vestigial. Tuba analis (subscaphium?) weakly sclerotised. Tegumen narrow. Valva wide, with one weakly sclerotised, setose ridge. Valva apex granulate. Transtilla narrow, without roundish extensions. Large coremata sacs with very long setae arise from valva base and vinculum (Text-fig. 115). Saccus narrow, slightly elongated medially. Aedeagus straight, caecum short, apical area striated. Vesica opens at approximately 135 degree angle. Vesica base narrow, followed by large diverticulum, apex narrow, curved. Vesica with several long cornuti.

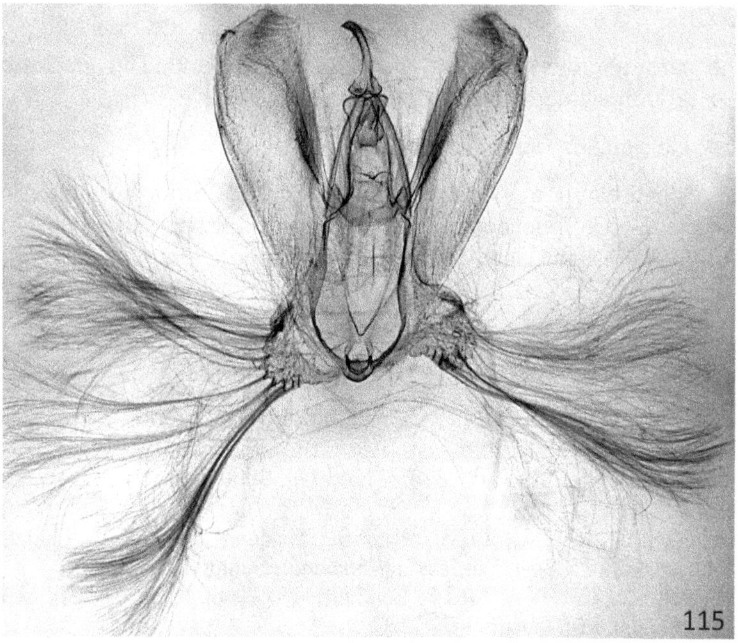

Text-fig. 115. Male genitalia and coremata with very long setae of *Lomographa temerata* (DENIS & SCHIFFERMÜLLER, 1775). Finland: Lammi, 30.v–7.vi.1996, slide PS1340 (coll. P. SIHVONEN).

FEMALE GENITALIA: Papillae anales small, setose. Apophyses long, straight. Lamella antevaginalis and lamella postvaginalis undifferentiated. Sterigma membranous. Ductus bursae very short. Posterior part of corpus bursae very long, narrow, curved, its anterior part striated, dentate. Anterior part of corpus bursae large, roundish. Signum with concave base, apex nail-shaped.

DISTRIBUTION AND ABUNDANCE: Eurasiatic. In Europe from southern central Scandinavia to northernmost Spain, southern France, northern Italy and the northern and central part of the Balkans and from the British Isles through central and eastern Europe to the Ural mountains. – Outside Europe the species occurs in Turkey and the Caucasus region and further eastwards from the Urals through central Asia and southern Siberia to Far East Russia, Korea, Japan, southern Kuriles and Sakhalin. – Common in Europe, sometimes in numbers.

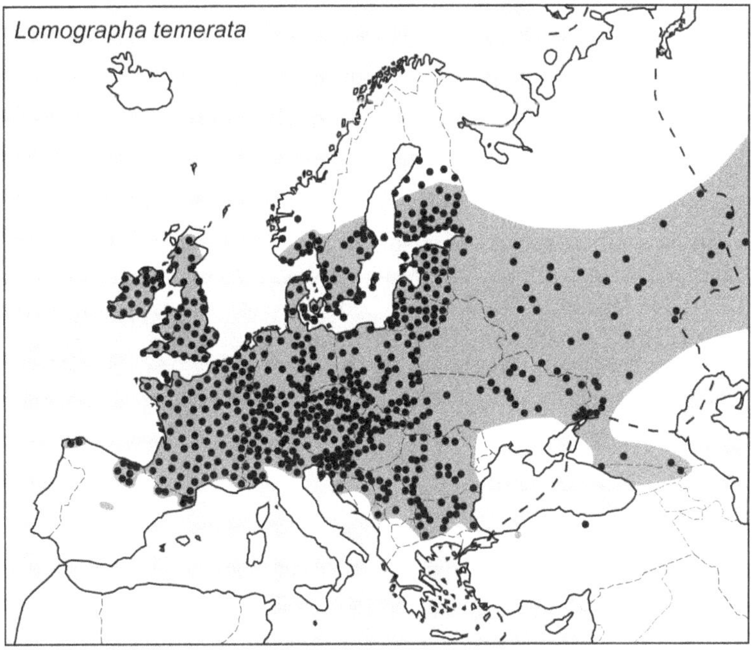

PHENOLOGY: Univoltine. In southern Scandinavia the adult moths fly from late May to early July, in southern central Europe from early May to mid-July (EBERT 2003) and in northern Spain from late May to early August (REDONDO *et al.* 2009). The late flying time so far in the south is probably due to the high altitude. Larva from June to August. The species hibernates as a pupa. The moths fly at night and come frequently to light and are less often disturbed by day.

BIOLOGY: Oligophagous. The larva has been found on *Prunus spinosa* and *Quercus* (EBERT 2003) as well as on *Sorbus* and *Malus* (HULDÉN *et al.* 2000). It has also been found feeding on *P. padus* (H. WEGNER pers. comm.). Further it has been recorded on *Rosa*, *Prunus cerasus* and *Salix* (AIZPÚRUA 2002).

HABITAT: In hedgerows, groves, wood- and scrubland, parks and in gardens. Vertically the species is distributed from sea level up to 2000 m altitude in northern Spain (REDONDO *et al.* 2009).

SIMILAR SPECIES: No similar species in Europe.

GENETIC DATA: Genetically homogeneous in Europe but both populations of Germany and United Kingdom deeply diverging into two homogeneous (intrapopulational) lineages at 3.2% distance, requiring investigation. Nearest species: An unidentified Chinese *Lomographa* species (4.9%) and East Asian *L. distans* (WARREN, 1894) (7.9%).

Aleucis GUENÉE, 1845

Aleucis GUENÉE, 1845, Cat. méth. Lépid. Eur.: 270. Type species: *Erannis distinctata* HERRICH-SCHÄFFER, 1839. Remark from FLETCHER (1979, p. 7): 'Type species of [*Aleucis*] is *distinctata* HERRICH-SCHÄFFER, 1839, by original designation, but cited as *Ephyra pictaria* CURTIS, 1833. *Geometra pictaria* THUNBERG, 1784, Diss. ent. sistens Insecta Suecica (1): 6, was misidentified and figured by CURTIS, 1833, Br. Ent. 10: 447 as *Ephyra pictaria* THUNBERG. *Geometra pictaria* THUNBERG, 1784, is a junior subjective synonym of *Phalaena lichenaria* HUFNAGEL, 1767, Berl. Mag. 4: 512. *Geometra pictaria* THUNBERG sensu CURTIS is *Erannis distinctata* HERRICH-SCHÄFFER, [1839]. Under article 70(a) of the Code, the case of a misidentified type-species should be referred to the Commission.' Genus synonymy is after SCOBLE (1999).

Aleucis STEPHENS, 1850, List Specimens Br. Anim. Colln Br. Mus. 5: 178. Type species: *Erannis distinctata* HERRICH-SCHÄFFER, [1839]. [Junior homonym and junior objective synonym of *Aleucis* GUENÉE].

DIVERSITY AND DISTRIBUTION: *Aleucis* contains three species [*distinctata* HERRICH-SCHÄFFER, 1839; *orientalis* STAUDINGER, 1892; *mimetes* WEHRLI, 1932] in the West Palaearctic region, possibly with additional species misplaced in other genera, e. g. *Epitherina* WEHRLI, 1938 and potentially in *Lomographa* HÜBNER, 1816. There is a further taxon similar to *A. orientalis*, in the process of being described from the Middle East (STÜNING & HAUSMANN pers. comm. *in* STADIE & LEHMANN 2012).

EXTERNAL CHARACTERS AND ABDOMEN: Wingspan about 18–25 mm, females also fully winged. Wings grey-brown, antemedial and postmedial lines dentate. Hindwings grey-whitish with blackish postmedial line. Male and female antennae weakly fasciculate/filiform. Tympanal organs of both sexes medium-sized. Male 8th sternite weakly sclerotised, posterior end narrower.

MALE GENITALIA: Uncus evenly curved, without ventral extension. Socii small, setose. Gnathos vestigial. Juxta granulate (laterally), dorsal margin with lobes. Valva simple, weakly setose. Saccus elongated. Vesica with one cornutus.

FEMALE GENITALIA: Papillae anales large, with scoop-shaped setae (floricomous ovipositor), covering most of ventral area. Apophyses short, stout. Lamella postvaginalis with setae laterally. Sterigma weakly sclerotised. Anterior part of corpus bursae with striated sclerotisation.

BIOLOGY: Larva of *A. distinctata* is monophagous, it has been recorded to feed on the leaves of *Prunus spinosa* (SKOU 1986, EBERT 2003, WILSON 2010), that of *A. orientalis* is oligophagous, it has been found also for instance on *Crataegus* and *Pyrus spinosa* (D. FRITSCH pers. comm., FRITSCH *et al.* 2014). Pupa overwinters, and nocturnal adults are attracted to light. Adults are active in February-April.

IMMATURE STAGES: Larva of *A. distinctata* is light green, violet-grey or greyish brown, with pale green or whitish areas on A7 and A8. In both colour forms there are indistinct V- or X-shaped darker markings dorsally (EBERT 2003, WILSON 2010). No data available on morphology of the pupa.

REMARKS: *Aleucis*, if considered valid at genus level, has recently been classified in the Caberini (VIVES MORENO 1994) and Baptini (HAUSMANN, MIRONOV & VIIDALEPP 2011). *Aleucis* and *Lomographa* have been considered closely related, and often *Aleucis* is considered a junior synonym of *Lomographa* (for instance VIIDALEPP 1996, LERAUT 1997) or the taxa that are currently classified in *Aleucis*, have earlier been classified in *Lomographa* (for instance MIKKOLA *et al.* 1989, SCOBLE 1999, EBERT 2003). The genitalia morphology does not lend support to the hypothesis that *Aleucis* and *Lomographa* are closely related, see the genitalia plates. Genitalia morphology shows that *Aleucis* and *Theria* are structurally similar, but this potential relationship is not shown in the literature because these two genera are classified in different tribes (for instance VIVES MORENO 1994, VIIDALEPP 1996, LERAUT 1997, HAUSMANN, MIRONOV & VIIDALEPP 2011). Diagnostic genus-level characters of *Aleucis* and *Theria* are summarised in Table 5. Systematic position: Theriini.

Table 5. Diagnostic genus-level characters of *Aleucis* GUENÉE, 1845 and *Theria* HÜBNER, 1825 (Theriini), see also Text-figs 116–121.

Selected diagnostic characters	***Aleucis***	***Theria***
Male antennae	Weakly fasciculate/filiform	Bipectinate
Female wings	Fully developed	Vestigial
Uncus	Single, evenly tapering	Forked (with ventral extension)
Scoop-shaped setae of papillae anales	Cover most of ventral area	Arranged in two horizontal areas
Sclerotisations of corpus bursae	Cover large area	Definitive signum

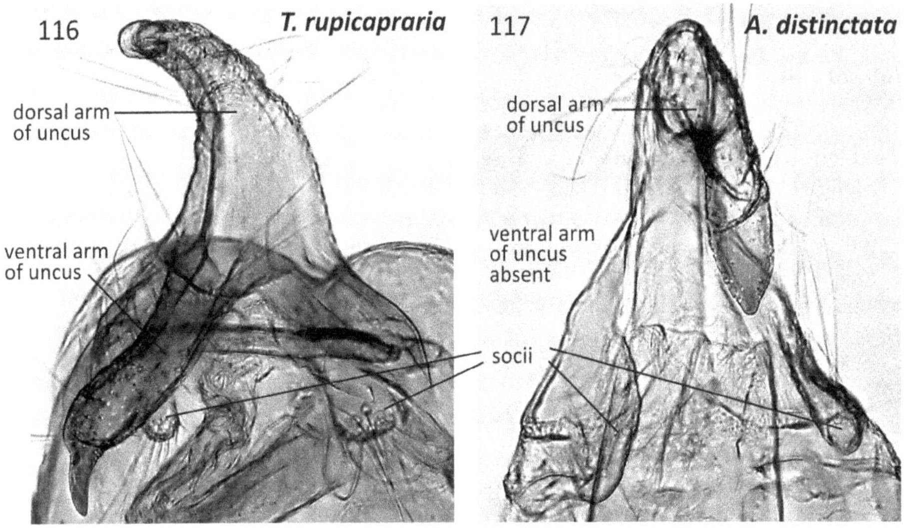

Text-figs 116–117. Uncus structures in genera *Theria* HÜBNER, 1825 (forked, with ventral extension) and *Aleucis* GUENÉE, 1845 (single, evenly tapering). Text-fig. 116. *T. rupicapraria* (DENIS & SCHIFFERMÜLLER, 1775), Germany: Wolgast, 5.iv.1941, slide PS1416 (coll. ZMH). Text-fig. 117. *A. distinctata* (HERRICH-SCHÄFFER, 1839), Germany: Bornich, 31.iii.1912, slide PS1421 (coll. ZMH).

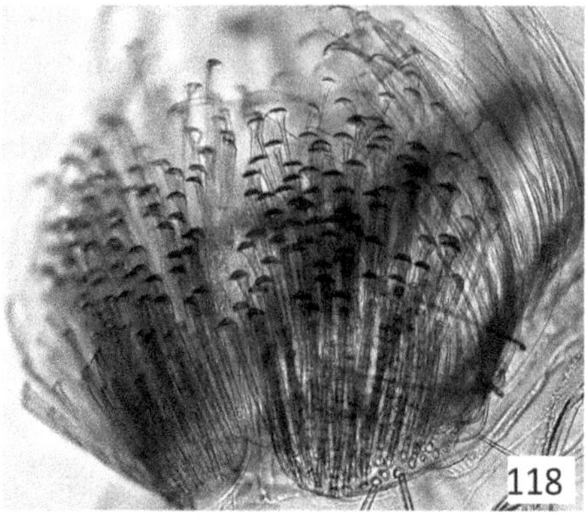

Aleucis distinctata

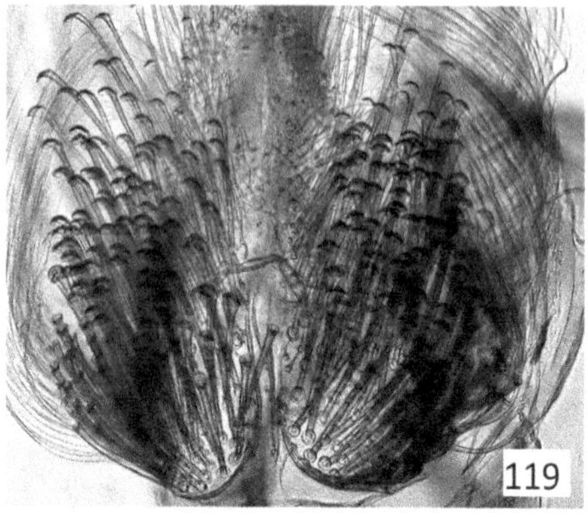

Aleucis orientalis

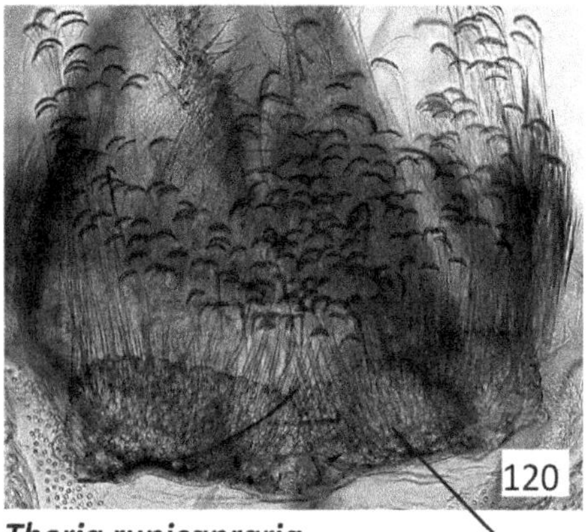

Theria rupicapraria

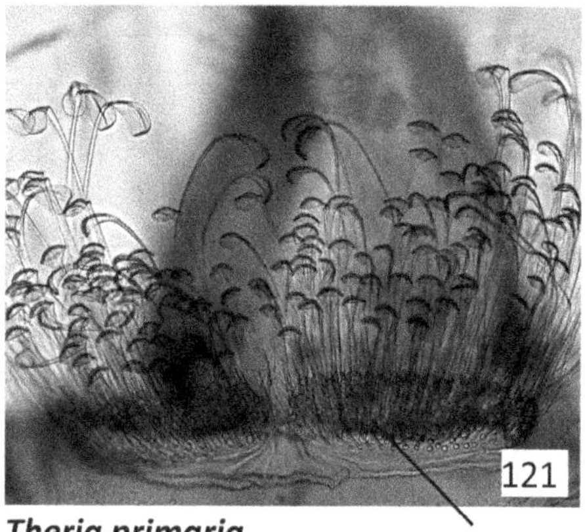

Theria primaria

Text-figs 118–121. Arrangement of papillae anales' setae in genera *Aleucis* GUENÉE, 1845 (not concentrated, covering large area) and *Theria* HÜBNER, 1825 (concentrated on anterior margin). Text-fig. 118. *A. distinctata* (HERRICH-SCHÄFFER, 1839), France: Digne, without date, slide PS1615 (coll. ZFMK). Text-fig. 119. *A. orientalis* (STAUDINGER, 1892), Greece: Kastoria, 800 m, 19.iv.1987, slide PS1540 (coll. Skou). Text-fig. 120. *T. rupicapraria* (DENIS & SCHIFFERMÜLLER, 1775), coll. DUSKE, without locality data, 8.3.[19?]67, slide PS1417 (coll. ZMH). Text-fig. 121. *T. primaria* (HAWORTH, 1809), Denmark: Hofmansgave, ex ovo 27.iii.1979, slide PS1541 (coll. SKOU).

106. Aleucis distinctata (HERRICH-SCHÄFFER, 1839)

distinctata HERRICH-SCHÄFFER, [1839], in Panzer, Faunae Insect. Germ. initia (Dtl. Insect.) (Heft 166): 81, (*Erannis*). Syntype(s), [Germany]. See text under *Aleucis* GUENÉE, 1845 for the misidentification of *Geometra pictaria* THUNBERG, 1784.

INTRASUBSPECIFIC NAMES, POTENTIALLY UNAVAILABLE: *contrastaria* FUCHS, 1901 (ab.); *pallescens* LEMPKE, 1951 (ab.); *variegata* LEMPKE, 1951 (f.).

EXTERNAL CHARACTERS AND ABDOMEN: Wingspan 20–25 mm. Forewings greyish brown, medial area slightly darker than rest of wing, terminal area often paler. Medial and postmedial lines dark brown, dentate. Hindwings lighter, grey, with weakly dentate postmedial line.

Forewing discal spots large, diffuse. Terminal line black, discontinuous. Fringes concolorous with wings. Wings below grey, irrorated with brown, markings sharper. Frons, collar, thorax and abdomen concolorous with wings. Male and female antennae weakly fasciculate/filiform. Hindtibia of both sexes with 2+2 spurs. Tympanal organs medium-sized. Male 8th sternite weakly sclerotised, posterior end narrower. Other abdominal sternites and tergites of both sexes undifferentiated.

VARIATION: Very little. Ground colour varies from dark brown to greyish brown.

MALE GENITALIA: Uncus not forked, weakly setose, evenly curved (Text-fig. 117). Socii small, setose. Gnathos vestigial, arms do not meet ventrally. Juxta granulate laterally, dorsal margin with two lobes. Valva weakly setose, rather membranous. Saccus elongated. Aedeagus straight, caecum minute. Vesica opens at approximately 135 degree angle. Vesica curved with one apical cornutus. Cornutus wide at base, apex narrower.

FEMALE GENITALIA: Papillae anales large, scoop-shaped setae (floricomous ovipositor) cover most of ventral area. Apophyses short, stout. Lamella postvaginalis with setae laterally. Sterigma minute, weakly sclerotised. Ductus bursae short, membranous. Posterior part of corpus bursae wide, weakly striated, membranous. Anterior part of corpus bursae narrow, with striated sclerotisation.

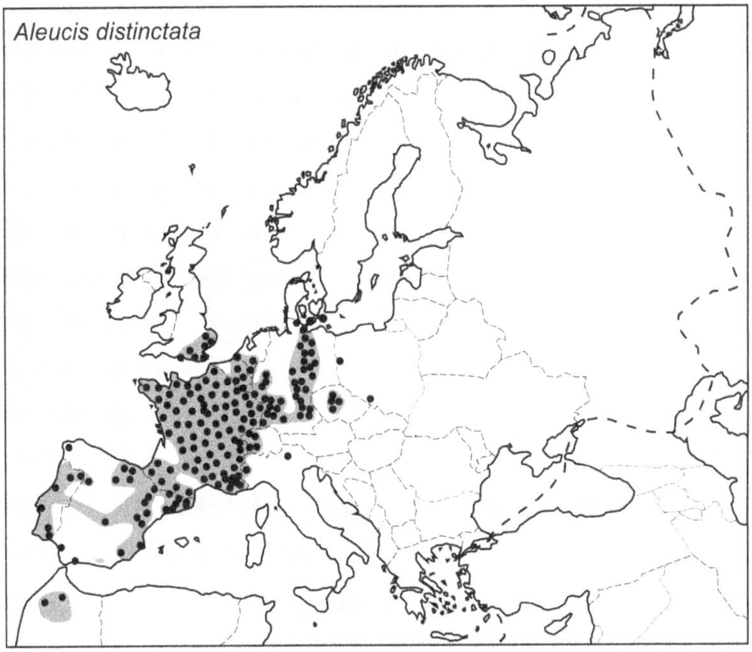

DISTRIBUTION AND ABUNDANCE: Sub-Mediterranean. In Europe from the Iberian Peninsula through France to southern England and northern Germany and to western Czech Republic and south-western Poland. According to HAUSMANN *et al.* (2011) *distinctata* is found in Slovenia, but there are no confirmed records from there (S. GOMBOC pers. comm.). In Denmark two migrant specimens. – Outside Europe the species is only known from Morocco. All records outside the mentioned distribution area appear to be *A. orientalis*, in Turkey and the Caucasus region maybe also another species in the genus. In the most recent Turkish check list (KOÇAK & KEMAL 2009), *A. distinctata* is recorded from seven provinces. This is undoubtedly due to confusion with the sister species *A. orientalis*. – The species is in Europe rather common, but somewhat local, and it can now and then be found in numbers.

PHENOLOGY: Univoltine. The adult moths fly from early March to mid-May, the latest records are from higher altitudes. In southernmost Spain recorded from January to March (P. HALE pers. comm.). In Portugal from late January in the south to beginning of June in mountainous areas in the north (M. CORLEY pers. comm.). In Poland from early April to mid-May (MALKIEWICZ 2012). Larva in May and June in central Europe (BERGMANN 1955). The species overwinters as a pupa. The adult moths fly at night and come frequently to light.

BIOLOGY: Monophagous. The larva has been found on *Prunus spinosa* (EBERT 2003). Pupation takes place in a web on the surface of the ground.

HABITAT: In forest fringes, hedgerows and scrubland. The species occurs from sea level up to 750 m in southern Germany (EBERT 2003) and up to 1300 m in Spain (REDONDO *et al.* 2009).

SIMILAR SPECIES: *Aleucis orientalis* (STAUDINGER) is very similar, without distinct qualitative differences in external appearance. *A. orientalis* is paler, more greyish, and medial lines are more dentate (darker, more brown, medial lines less dentate in *A. distinctata*). In *A. orientalis* cornutus is evenly narrow, vesica with basal diverticulum (cornutus wide at base, vesica without basal diverticulum in *A. distinctata*), corpus bursae is with medial, indistinct sclerotisation (corpus bursae with striated sclerotisation, it is located in its anterior part in *A. distinctata*).

GENETIC DATA: Genetically identical in Spain, France and Germany (n=11). Populations from Morocco somewhat diverging. Nearest species: *A. orientalis* from the eastern Mediterranean (6.4%) and *Epitherina kuznetzovi* VIIDALEPP, 1992 from Turkmenistan (6.4%).

REMARKS: *A. distinctata* has often been classified in *Lomographa* (for instance MIKKOLA *et al.* 1989, LERAUT 1997, EBERT 2003), see remarks under genus-level text. LERAUT (2009) mistakenly claims that female of *A. distinctata* is brachypterous (explained in text and illustrated in plate).

107. *Aleucis orientalis* (STAUDINGER, 1892)

orientalis STAUDINGER, 1892, Dt. ent. Z. Iris 5: 170, (*Hybernia* (*Anhibernia*)). Syntype(s) male, [Israel]: Jerusalem; [Turkey]: Amasia.

EXTERNAL CHARACTERS AND ABDOMEN: Generally as *A. distinctata* (HERRICH-SCHÄFFER), see above, but with following differences. *A. orientalis* is paler, more greyish, and medial lines are more dentate.

VARIATION: Very little. Ground colour varies from dark brown to greyish brown.

MALE GENITALIA: Generally as *A. distinctata* (HERRICH-SCHÄFFER), see above, but with following differences. In *A. orientalis* cornutus is evenly narrow (base not widened), vesica is with basal diverticulum.

FEMALE GENITALIA: Generally as *A. distinctata* (HERRICH-SCHÄFFER), see above, but with following differences. In *A. orientalis* corpus bursae is with medial, partly striated, indistinct sclerotisation.

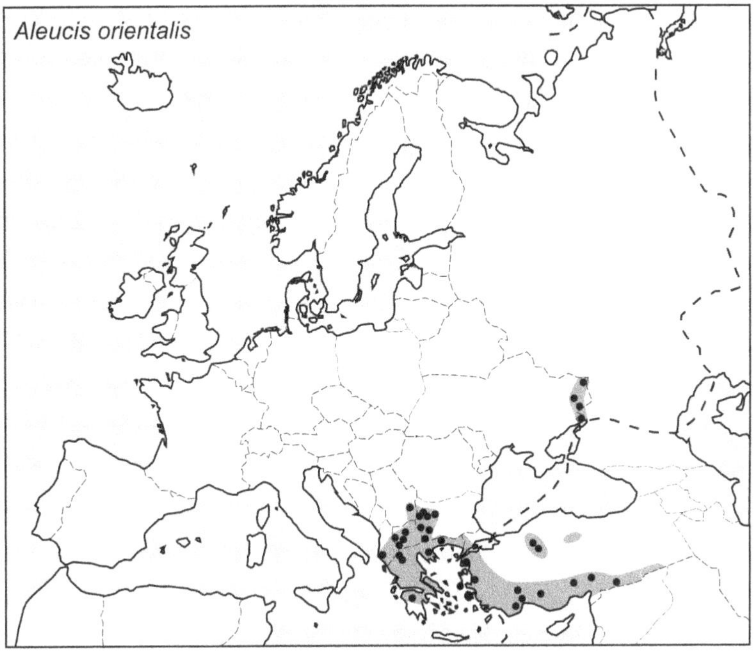

DISTRIBUTION AND ABUNDANCE: (East-)European-Westasiatic. In Europe only in the southern half of the Balkans and in eastern Ukraine. – Outside Europe in Turkey (as *distinctata* in KOÇAK & KEMAL 2009), Azerbaijan (as *distinctata* in VIIDALEPP 1996), Turkmenistan, Israel and Jordan. The distribution outside Europe and Turkey is doubtful as *A. orientalis* is possibly confused with other species of the genus or even with an undescribed species. – Local in Europe and rather rarely in numbers.

PHENOLOGY: Univoltine. In northern Greece and in Bulgaria in early and mid-April. In Samos from early March to early April (D. FRITSCH pers. comm.). In Samos the larvae have been found from late April to late May (D. FRITSCH pers. comm.) and in south-eastern Turkey in late May and early June (M. LEIPNITZ pers. comm.). Hibernation takes place as a pupa.The moths fly at night and come regularly to light.

BIOLOGY: Oligophagous on Rosaceae. In Samos the caterpillar has been found on *Prunus spinosa*, *Crataegus* and *Pyrus spinosa* (D. FRITSCH pers. comm.) and on Rosaceae and *Crataegus* bushes in Turkey (M. LEIPNITZ pers. comm.).

HABITAT: Open scrubland on mountain slopes, meadows with bushes and hedges. The vertical distribution is poorly known. In northern Greece the species has been found at 1000 m and in Turkmenistan at 1600 m altitude.

SIMILAR SPECIES: *A. distinctata* (HERRICH-SCHÄFFER), see entry above. There is a further taxon similar to *A. orientalis*, in the process of being described from the Middle East (STÜNING & HAUSMANN pers. comm. *in* STADIE & LEHMANN 2012).

GENETIC DATA: Nearest species: *A. distinctata* (6.4%) and Israeli populations of *A. orientalis* (STAUDINGER, 1892) at the same genetic distance of 6.4%. Three examined Turkmenian specimens of *Epitherina kuznetzovi* VIIDALEPP, 1992 at 1.0% distance only from a specimen examined from the Greek Aegean islands.

REMARKS: *A. orientalis* has often been classified in *Lomographa* (for instance SCOBLE 1999), see remarks under genus-level text. Validity of *A. orientalis* at species level has often been questioned, especially in the older literature it was treated as a subspecies or as a synonym of *A. distinctata*, but genitalia morphology and DNA barcodes clearly support the hypothesis of it being a separate species. WEIDLICH (2008) provides an illustration of the species from Greece.

Theria HÜBNER, 1825

Theria HÜBNER, [1825] 1816, Verz. bekannter Schmett.: 321. Type species: *Geometra rupicapraria* [DENIS & SCHIFFERMÜLLER], 1775, Austria, Vienna district. Genus synonyms are after SCOBLE (1999).

Cheimatobia STEPHENS, 1829, Nom. Br. Insects: 44. Type species: *Geometra rupicapraria* [DENIS & SCHIFFERMÜLLER], 1775. [Junior objective synonym of *Theria* HÜBNER].

Chimatobia AGASSIZ, 1847, Nomencl. zool. (Index univl.): 78, 80. [Emendation of *Cheimatobia* STEPHENS].

DIVERSITY AND DISTRIBUTION: *Theria* contains two species in the Palaearctic region.

EXTERNAL CHARACTERS AND ABDOMEN: Male wingspan about 25–30 mm, female wingspan about 10–12 mm (brachypterous). Male forewings grey-brown, antemedial and postmedial lines weakly undulating. Discal spots large. Hindwings pale, with blackish postmedial line. Female wings with dark brown medial area, basal and terminal areas pale brown. Male antennae bipectinate, female antennae filiform. Male tympanal organs medium-sized, female tympanal organs very small. Male 8th sternite weakly sclerotised, posterior end narrower.

MALE GENITALIA: Uncus forked, with ventral extension (Text-fig. 116). Socii small, setose. Gnathos vestigial. Dorsal margin of juxta with lobes. Base of valva (homology unclear) with sclerotised plate, margin serrate. Valva simple, weakly setose. Saccus elongated. Caecum of aedeagus very long. Vesica without cornutus.

FEMALE GENITALIA: Papillae anales large, with scoop-shaped setae (floricomous ovipositor), arranged in two horizontal areas (Text-figs 120–121). Apophyses short, stout. Lamella antevaginalis and lamella postvaginalis variably sclerotised. Sterigma weakly sclerotised. Corpus bursae membranous sac. Signum small, stellate.

BIOLOGY: Larvae of *T. rupicapraria* and *T. primaria* have been recorded feeding on *Crataegus* and on *Prunus spinosa* (SKOU 1986, EBERT 2003) but not exclusively (LERAUT 2009). Both species overwinter as pupa in the soil. Adults of *T. primaria* are active from December to March and those of *T. rupicapraria* from January to April (EBERT 2003). It is assumed that females use the floricomous, scoop-shaped setae of the papillae anales to scratch particles from the surface of the bark to cover and camouflage eggs. Similar structures occur also in the non-related *Gnopharmia* and *Neognopharmia* (RAJAEI 2010, RAJAEI *et al.* 2012).

IMMATURE STAGES: Colour of *T. primaria* and *T. rupicapraria* larvae varies from whitish green to darker greyish green to greyish brown. Dorsal area often distinct with white lines, dorsolateral lines widest, or with darker squarish, longitudinal markings (SKOU 1986, EBERT 2003, WILSON 2010). Pupa cremaster of *T. primaria* is with two stout, acute setae (PATOČKA & TURČANI 2005).

REMARKS: *Theria* has recently been classified in the Theriini (VIVES MORENO 1994, VIIDALEPP 1996, LERAUT 1997, HAUSMANN, MIRONOV & VIIDALEPP 2011). Genitalia morphology shows that *Aleucis* and *Theria* are structurally similar, but this potential relationship is not shown in the literature because these two genera are classified in different tribes, *Aleucis* in the Caberini and the Baptini and *Theria* in the Theriini (for instance VIVES MORENO 1994, VIIDALEPP 1996, LERAUT 1997, HAUSMANN, MIRONOV & VIIDALEPP 2011). Diagnostic genus-level characters of *Aleucis* and *Theria* are summarised in Table 5. *Theria* includes two species, but the existence of those were not noted until recently (KUHNA 1977). Systematic position: Theriini.

108. Theria rupicapraria (DENIS & SCHIFFERMÜLLER, 1775)

rupicapraria rupicapraria [DENIS & SCHIFFERMÜLLER], 1775, Ankündung syst. Werkes Schmett. Wienergegend: 105, (*Geometra*). Syntype(s), lost, [Austria]: Vienna district.

rupicapraria culminaria LERAUT, 1993, Entomol. Gall. 4 (4): 235, (*Theria*). Holotype male (Muséum National d'Histoire Naturelle, Paris, France), France: Hautes-Alpes, L'Argentière-la-Bessée. Originally as subspecies of *rupicapraria* DENIS & SCHIFFERMÜLLER. See remarks.

rupicapraria crypta WEHRLI, 1940, in SEITZ, Gross-Schmett. Erde 4 (Suppl.): 410, pl. 33: d, (*Theria*). Holotype male (Zoological Institute, Russian Academy of Sciences, St Petersburg, Russia), [Russia: Stavropol Kari (Northern Caucasus)]: Mashuk (examined externally).

INTRASUBSPECIFIC NAMES, POTENTIALLY UNAVAILABLE: *fasciata* LEMPKE, 1970 (f.); *nigrescens* LEMPKE, 1970 (f.); *pallida* LEMPKE, 1952 (f.).

EXTERNAL CHARACTERS AND ABDOMEN: Male: wingspan 26–30 mm. Forewings light grey-brown, medial area slightly darker than rest of wing. Antemedial and postmedial lines brown, weakly undulating. Postmedial line curved inwards on inner margin, medial lines rather close to each other on inner margin. Terminal line discontinuous, with distinct, slightly elongated black dots. Fringes concolorous with wings. Discal spot large, lunular. Hindwings pale grey, subterminal line and discal spots weak, blackish. Wings below uniform grey, weakly irrorated with blackish, markings blackish. Frons, collar, thorax and abdomen concolorous with wings. Antennae bipectinate. Hindtibia with 2+2 spurs. Tympanal organs medium-sized. 8th sternite weakly sclerotised, posterior end narrower. Female: wingspan 11–12 mm. Forewings grey-brown, medial area rather wide, dark brown, wider on costa. Antemedial line blackish. Hindwings paler, postmedial line blackish brown. Antennae filiform. Hindtibia with 2+2 spurs. Tympanal organs very small.

VARIATION: Very little. Medial area normally slightly darker than rest of wing but it can be almost concolorous with the rest of the wing. Dark brown medial area in female forewings often widest on costa, but not exclusively.

Subsp. ***crypta*** has the forewing medial lines less light and dots on wing margins are bigger, especially on the hindwings, where they can be more elongated. Known from North Caucasus. Taxon *crypta* WEHRLI, 1940 was considered a junior synonym by SCOBLE (1999), but treated as a valid species by VIIDALEPP (1996). We consider the allopatric Caucasus specimens valid at subspecies level.

MALE GENITALIA: Dorsal part of uncus weakly setose, evenly curved. Ventral arm of uncus weakly curved ventrally, apex with spine (Text-fig. 116). Socii small, setose. Gnathos vestigial, arms do not meet ventrally. Juxta granulate laterally, dorsal margin with two lobes. Base of valva (homology unclear) with sclerotised plate, apex acute, its margin granulate. Valva weakly setose, with narrow median ridge, otherwise membranous. Saccus elongated. Aedeagus curved, caecum very long. Subapical extension distinct, apex elongated. Vesica opens at approximately 90 degree angle. Vesica wider at base, without sclerotisations.

FEMALE GENITALIA: Papillae anales large, roundish, scoop-shaped setae (floricomous ovipositor) arranged in two, nearly fused, horizontal areas in ventral margin (Text-fig. 120). Apophyses short, stout. Lateral sclerotisations of sterigma (lamella postvaginalis) indistinct, dorsad of ostium bursae. Sterigma weakly sclerotised. Ductus bursae short, membranous. Posterior part of corpus bursae short, membranous. Anterior part of corpus bursae round, membranous. Signum small, weakly stellate.

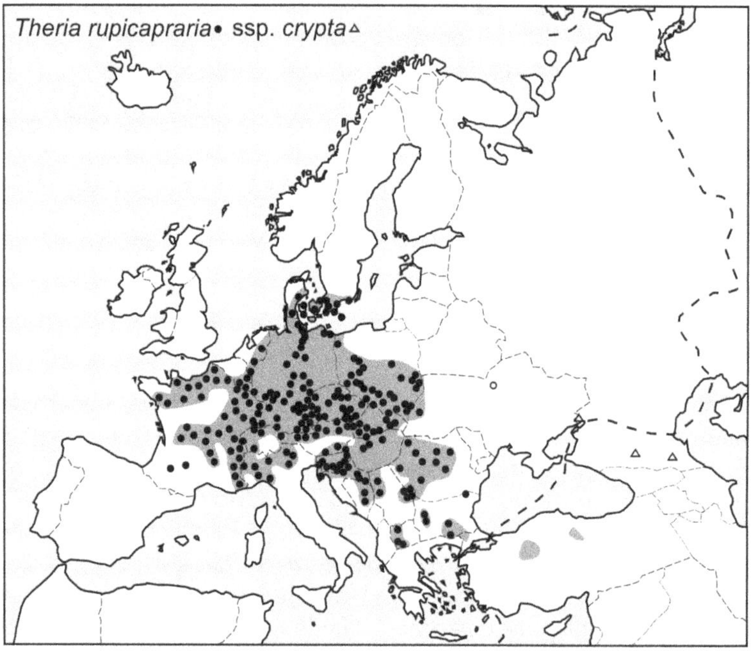

DISTRIBUTION AND ABUNDANCE: European. From southern Sweden and Denmark through central Europe to France, northern Italy and parts of the Balkans. From Ukraine one old record near Kiev and one by Taganrog in south-westernmost European Russia, the last mentioned belonging to subsp. *crypta*. – Outside Europe in Turkey and two old records from Caucasus (subsp. *crypta*). – In Europe common and rather often found in numbers.

PHENOLOGY: Univoltine. In southern Scandinavia from late February to late March and in southern central Europe from early February to late March or early April (EBERT 2003). Due to the early flight time this may vary quite much from year to year, depending on the length of the winter. Larva in May and June. The species overwinters as a pupa. The males come to light in limited numbers, but they can be found at night when they are sitting on or flying around the food plants. The wingless female can also be found at night sitting on twigs and branches of the food plants.

BIOLOGY: Oligophagous. The caterpillar has been found on *Crataegus* and *Prunus spinosa* (EBERT 2003). FLAMIGNI *et al.* (in prep.) also records *Prunus domestica* and *Mespilus germanicus*. Pupation in the ground.

HABITAT: In forest fringes, hedgerows, open scrubland and gardens. The species occurs from sea level up to about 850 m altitude.

SIMILAR SPECIES: *Theria primaria* (HAWORTH) male has forewings darker brown, postmedial line is less curved inwards on inner margin, thus medial lines are further apart on inner margin (forewings light grey-brown, medial lines rather close to each other on inner margin in *T. rupicapraria*). *T. primaria* female has forewing medial area of equal width, margins are parallel (medial area wider on costa, outer margin curves weakly towards apex in *T. rupicapraria*) (Text-figs 122–125). In the literature other quantitative external characters are

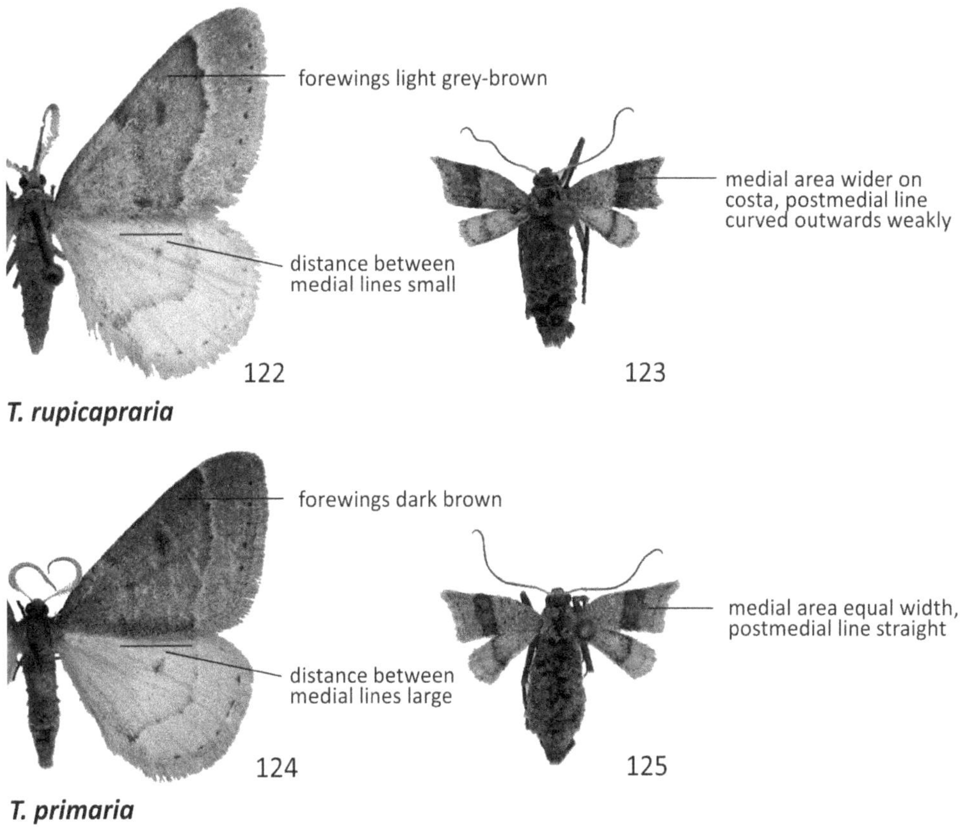

Text-figs 122–125. Diagnostic characters (indicated) of *Theria* HÜBNER, 1825 species. Text-fig. 122. *T. rupicapraria* (DENIS & SCHIFFERMÜLLER, 1775) male, France, 15.iii.1938. Text-fig. 123. *T. rupicapraria* female, Denmark, 14.iv.1979, e.o. Text-fig. 124. *T. primaria* (HAWORTH, 1809) male, Denmark, 16.ii.2002. Text-fig. 125. *T. primaria* female, Denmark, 27.iii.1979, e.o. We recommend that identity of females, particularly collection specimens, is checked from the genitalia.

described also, but often they do not allow reliable identification. For instance in male forewing terminal line dots are generally more extended and larger and forewing discal spot is lunular in *T. rupicapraria* whereas terminal dots are generally more rounded and smaller and discal spot is roundish in *T. primaria* (LEPERTEL *et al.* 2007). LERAUT (2009) mentions the shape of male forewing apex, it being generally more rounded in *T. rupicapraria*. *T. primaria* is without subapical extension in male aedeagus, basal lobe of valva is with round apex, its margin is serrate and dorsal margin of juxta is with two large lobes (subapical extension of aedeagus is present, basal lobe of valva is with acute apex, its margin granulate and dorsal margin of juxta is with two small lobes in *T. rupicapraria*) (Text-figs 126–127). Papillae anales are elongated and lateral sclerotisations of sterigma are in more ventral position in *T. primaria* (papillae anales roundish, lateral sclerotisation of sterigma in more dorsal position in *T. rupicapraria*).

GENETIC DATA: Genetically homogeneous in Europe (n=5, from Germany, France and Hungary). Nearest species: *T. primaria* (5.4%).

REMARKS: Taxon *culminaria* LERAUT, 1993, originally described as a subspecies of *rupicapraria* (DENIS & SCHIFFERMÜLLER) from southern Alps, was later considered an individual form (LERAUT 2009).

109. Theria primaria (HAWORTH, 1809)

primaria HAWORTH, 1809, Lepid. Br. (2): 305, (*Geometra*). Syntype(s), [United Kingdom].
ibicaria HERRICH-SCHÄFFER, 1856, Syst. Bearbeitung Schmett. Eur. 6 (69): 11 (Index Universalis. Macrolepidoptera. Vol. 1–3); ibidem (1851), 3: pl. 83, fig. 511 (non binominal), (*Hibernia*). Syntype(s), [France].

INTRASUBSPECIFIC NAMES, POTENTIALLY UNAVAILABLE: *illataria* FUCHS, 1900 (ab.).

EXTERNAL CHARACTERS AND ABDOMEN: Generally as *T. rupicapraria* (DENIS & SCHIFFERMÜLLER), see above, but with following differences. Male: *T. primaria* has forewings darker brown, postmedial line is less curved inwards on inner margin, thus medial lines are further apart on inner margin. Female: *T. primaria* often has forewing medial area of equal width, margins are parallel. Character is variable and does not always allow reliable identification.

VARIATION: Very little. Male forewing ground colour varies from light brown to dark brown, occasionally it may be with tint of reddish brown. Dark brown medial area in female forewings is often of equal width but not exclusively.

MALE GENITALIA: Generally as *T. rupicapraria* (DENIS & SCHIFFERMÜLLER), see above, but with following differences. *T. primaria* is without subapical extension in male aedeagus, aedeagus is almost straight and apex is less elongated, basal lobe of valva is with round apex, its margin is serrate and dorsal margin of juxta is with two large lobes (Text-figs 126–127).

FEMALE GENITALIA: Generally as *T. rupicapraria* (DENIS & SCHIFFERMÜLLER), see above, but with following differences. *T. primaria* has elongated papillae anales and lateral sclerotisations of sterigma are in more ventral position.

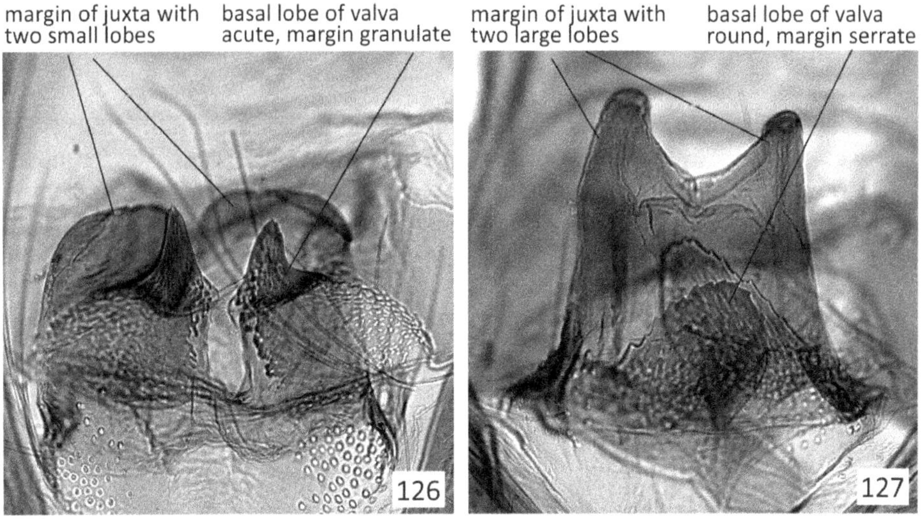

T. rupicapraria ***T. primaria***

Text-figs 126–127. Diagnostic characters (indicated) of *Theria* HÜBNER, 1825 species, showing basal lobe of valva and dorsal margin of juxta. Text-fig. 126. *T. rupicapraria* (DENIS & SCHIFFERMÜLLER, 1775), Germany, 5.iv.1941, slide PS1416 (coll. ZMH). Text-fig. 127. *T. primaria* (HAWORTH, 1809), Denmark, 11.iii.1980, slide PS1418 (coll. ZMH).

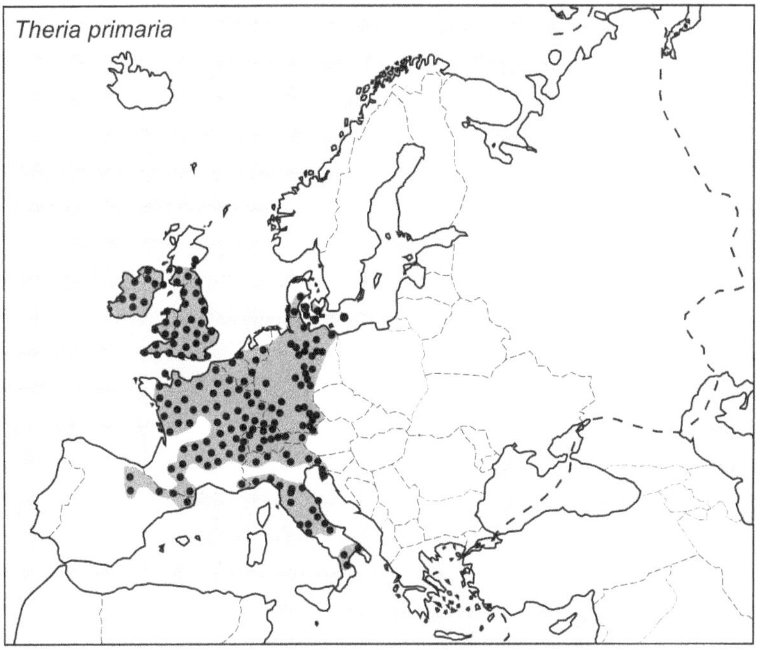

DISTRIBUTION AND ABUNDANCE: European. The species has a more western distribution than *T. rupicapraria* and due to its relatively recent separation from this species (KUHNA 1977), the distribution is possibly not yet completely known. It occurs from Denmark to Italy, southern France and the northern part of Spain and from the British Isles to central Europe. – The species is not known from outside Europe. – Local, but rather common and not rarely in numbers.

PHENOLOGY: Univoltine. In southern Scandinavia from mid- or late February to late March, in south-western Germany from late December to late February or early March. In central Italy also from late December (S. SCIVOLA pers. comm.). Due to the early flying time this may vary quite much from year to year, depending on the development of the winter weather. Caterpillar in May and June. The hibernation takes place as a pupa. The males come to light in rather small numbers, but they can be found at night when they are sitting on or flying around the food plants. The wingless female may also be found at night sitting on twigs and branches of the food plants.

BIOLOGY: Oligophagous. The larva has been found feeding on *Prunus spinosa* and *Crataegus*. FLAMIGNI *et al.* (in prep.) also mention *P. domestica* and *Malus pumila*. Pupation takes place in the ground.

HABITAT: In hedgerows and forest fringes. In Denmark the species mainly occurs in dense and windblown *Prunus spinosa* scrub directly on the coast and few meters from the sea. Vertically the species has been found from sea level up to 800 m in southern central Europe (EBERT 2003), in northern Spain from 400 up to 1200 m (REDONDO *et al.* 2009).

SIMILAR SPECIES: *Theria rupicapraria* (DENIS & SCHIFFERMÜLLER), see entry above.

GENETIC DATA: Genetically homogeneous in Europe (n=8, from Germany, France and United Kingdom). Nearest species: *T. rupicapraria* (5.4%).

Campaea LAMARCK, 1816

Campaea LAMARCK, 1816, Hist. Nat. Animaux sans Vertèbres 3: 568. Type species: *Phalaena margaritata* LINNAEUS, 1767, Sweden. Genus synonyms are after SCOBLE (1999).

- *Eratopis* [REICHENBACH], 1817, Jenaische allg. Lit.-Ztg 1 (36): 288. Type species: *Phalaena margaritata* LINNAEUS, 1767.
- *Metrocampa* BRUAND, 1846, Mém. Soc. Emul. Doubs. (1) 2 (2, livr. 3, 4): 76. Type species: *Phalaena margaritata* LINNAEUS, 1767.
- *Metrocampus* BERTHOLD, 1827, in LATREILLE, Nat. Familien des Thierreichs: 484. Type species: *Phalaena margaritata* LINNAEUS, 1767.
- *Tribacis* BILLBERG, 1820, Enumeratio Insect. Mus. G. J. BILLBERG: 89. Type species: *Phalaena margaritata* LINNAEUS, 1767.

UNAVAILABLE NAMES (incorrect subsequent spellings): *Metrocampos* THIENEMANN, 1828; *Metrocampe* DUPONCHEL, 1829; *Eratopsis* SHERBORN, 1926 (see FLETCHER 1979).

DIVERSITY AND DISTRIBUTION: *Campaea* contains four species in the Palaearctic and Nearctic regions. We have not examined the genitalia of two species, *C. biseriata* (MOORE, 1888) from India and *C. parallela* (WEHRLI, 1936) from China, thus the inclusion of these taxa in *Campaea* is based on the literature (SCOBLE 1999).

EXTERNAL CHARACTERS AND ABDOMEN, description is based on Palaearctic and Nearctic species: Wingspan about 35–40 mm. Wings pale green, antemedial and medial lines white, inner margins darker. Forewing margin concave below apex. Discal spots absent. Male antennae bipectinate, female antennae weakly fasciculate. Male hindtibia with small hair pencil. Male and female abdominal segments 3–8 unmodified.

MALE GENITALIA, description is based on Palaearctic and Nearctic species: Uncus setose, evenly tapering to apex. Socii absent. Gnathos absent (or present as weak, vestigial membranous ridge). Valva narrow, apex wider, spoon-shaped, dorsal margin densely setose. Valva with subapical spine in ventral margin, base with prominent, asymmetric, sclerotised dorsal lobe/extension. Juxta with two setose patches. Saccus elongated. Vesica with row of microcornuti.

FEMALE GENITALIA, description is based on Palaearctic and Nearctic species: Papillae anales small, setose. Apophyses posteriores long, straight. Apophyses anteriores absent. Ductus bursae short. Posterior part of corpus bursae long, narrow. Anterior part of corpus bursae large, roundish, with large patch of minute spines.

BIOLOGY: The Palaearctic *C. margaritaria* and Nearctic *C. perlata* (GUENÉE, 1858) are polyphagous species that have been recorded feeding mainly on deciduous trees (MCGUFFIN 1981, MIKKOLA *et al.* 1989, EBERT 2003). Larva overwinters in third or fouth instar on twigs and branches. *C. margaritaria* adults are nocturnal, *C. perlata* adults are diurnal in Yukon, where there is 24 hours daylight in its flight time, and nocturnal in southern parts of its distribution area (MCGUFFIN 1981, MIKKOLA *et al.* 1989, EBERT 2003).

IMMATURE STAGES: Larva of *C. margaritaria* is a twig-mimic, usually greyish brown, with hair-like structures on its ventral parts and it has an additional, rudimentary pair of legs on the eighth segment (PORTER 2010, HAUSMANN & MILLER 2000). Larva of *C. perlata* is described in MCGUFFIN (1981). Pupa cremaster is with four pairs of curved setae, apical pair is most pronounced (MCGUFFIN 1981, PATOČKA & TURČANI 2005).

REMARKS: *Campaea* has recently been classified in the Campaeini (VIVES MORENO 1994, VIIDALEPP 1996, LERAUT 1997, HAUSMANN, MIRONOV & VIIDALEPP 2011). The dorsal lobe at the base of valva, one of the putative Campaeini synapomorphies, is in a more dorsal position in *Campaea* than in other putative European Campaeini genera, apparently arising from fusion of valva and vinculum. MCGUFFIN (1981) suggests that these originate from the ends of transtilla. In the male genitalia *Campaea* and *Hylaea* HÜBNER, 1822 share a row of microcornuti in the vesica whereas in the female genitalia the structurally similar corpus bursae spines are found in *Campaea* and *Pungeleria* ROUGEMONT, 1903. In a DNA-based analysis no other putative Campaeini genera besides *Campaea* were included; as a result *Campaea* grouped next to a group of genera containing *Alsophila* HÜBNER, 1825, *Colotois* HÜBNER, 1823 and *Wilemania* PROUT, 1928 (SIHVONEN *et al.* 2011). Systematic position: Campaeini.

110. Campaea margaritaria (LINNAEUS, 1761)

margaritaria LINNAEUS, 1761, Fauna Suecica (Edn 2): 328 (*Phalaena (Geometra)*). Syntype(s), Holmiae [Sweden: Stockholm] (examined).

margaritata LINNAEUS, 1767, Syst. Nat. (Ed. 12) 1 (2): 865, (*Phalaena Geometra*). [Emendation of *margaritaria* LINNAEUS].

nayas FOURCROY, 1785, Ent. Paris: 276, (*Phalaena*). Syntype(s), France: Paris. [Synonymy after LERAUT (1997)].

sesquistriataria KNOCH, 1781, Beitr. Insektengesch. 1: 1, pl. 1, fig. 1, (*Phalaena Geometra*). Syntypes male, female, [Germany]: Leipzig region.

straminea RETZIUS, 1783, Caroli Lib. Bar. De Geer ... Genera et Species Insectorum: 43, (*Phalaena*). Syntype(s), [Europe]. [Synonymy after LERAUT (1997)].

vernaria HUFNAGEL, 1767, Berlin Mag. 4 (5): 506, (*Phalaena*). Syntype(s), lost, [Germany]: Berlin region.

vitriolata CYRILLO, 1787, Entomologias Neapolitanae specimen primium: pl. 4, fig. 4, (*Phalaena*). Syntype(s), Italy: Campania, 'Appenninis del Napoletano'.

UNAVAILABLE NAMES (incorrect authorship): *margaritaria* FABRICIUS (LAMARCK 1816); *margaritaria* TREITSCHKE (DUPONCHEL 1829) (see FLETCHER 1979); *margaritaria* DENIS & SCHIFFERMÜLLER (DENIS & SCHIFFERMÜLLER 1775); *sesquistriga* BRAHM (BRAHM 1791); *stramineraria* DE VILLERS (DE VILLERS 1789). – Intrasubspecific names, potentially unavailable: *approximata* LEMPKE, 1951 (f.); *clausa* LEMPKE, 1970 (f.); *dulcinaria* DANNEHL, 1927 (ab.); *olivata* BURRAU, 1950 (ab.); *rubrociliata* SCHAWERDA, 1932 (f.); *triangularis* COCKAYNE, 1948 (ab.); *zawiszae* WIZE, 1917 (ab.).

EXTERNAL CHARACTERS AND ABDOMEN: Wingspan 36–50 mm (first generation), 30–39 mm (second generation), females are bigger. Wings pale green. Antemedial line white, straight, outer margin olive-green. Postmedial line white, oblique, slightly curved inwards on costa, inner margin olive-green. On hindwings only postmedial line visible. Terminal line narrow, olive-green, forewing apex with reddish olive-green streak. Fringes concolorous with wings. Forewing margin concave below apex, hindwing margin weakly elongated at M3. Discal spots absent. Wings below as above, paler. Ventral margin of frons white, dorsal margin brownish orange. Collar, thorax and abdomen concolorous with wings. Male antennae bipectinate, female antennae weakly fasciculate. Hindtibia of both sexes with 2+2 spurs, male tibia with small hair pencil. Tympanal organs medium-sized, bullae tympani with roundish lobe. Sternites and tergites 3–8 of both sexes undifferentiated.

VARIATION: Little. Wings almost always uniformly pale green, but fringes may be reddish. Position and angle of medial lines somewhat variable, in individual aberrant specimens these may even anastomose (see HAUSMANN 2001, figs. 61–62). Pale green colour fades away as time goes by, thus old specimens are whitish and medial lines barely recognisable.

MALE GENITALIA: Uncus setose, evenly tapering to apex, apex round. Socii absent. Gnathos vestigial, narrow membranous ridge, fused ventrally. Valva narrow, apex wider, spoon-shaped, its dorsal margin densely setose. Valva with subapical spine in ventral margin. Valva base (apparently originating from fusion of valva and vinculum) with prominent, curved and asymmetric, sclerotised dorsal lobe/extension. Juxta with two setose patches. Saccus elongated. Aedeagus narrow, slightly curved, caecum short. Vesica opens at approximately 90 degree angle. Vesica evenly narrowing tube, base with row of microcornuti.

FEMALE GENITALIA: Papillae anales soft, setose. Apophyses posteriores long, straight. Apophyses anteriores absent. Lamella antevaginalis large, sclerotised flap. Sterigma with two finger-like extensions. Lamella postvaginalis membranous. Ductus bursae short, narrow, membranous. Posterior part of corpus bursae long, narrow, partly striated, sclerotised. Anterior part of corpus bursae large, roundish, with large patch of minute spines.

DISTRIBUTION AND ABUNDANCE: Euro-Caucasian. In Europe from central Fennoscandia to the Iberian Peninsula, southern France including Corsica, Italy including Sardinia and Sicily and the Balkans including northern Greece and from the British Isles through central and eastern Europe to southern European Russia. – Outside Europe the species occurs in Turkey and the Caucasus region and in North Africa also in Morocco. – Common or very common in Europe, now and then in numbers.

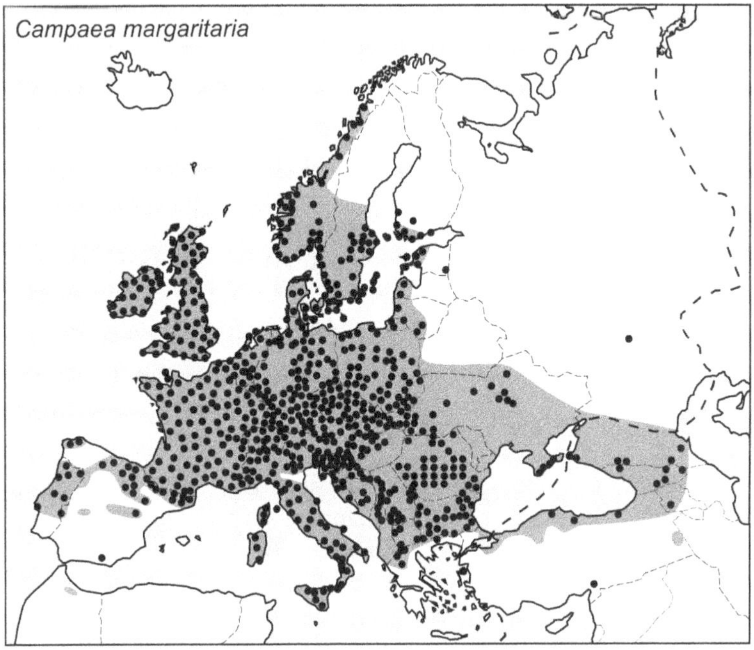

PHENOLOGY: Uni- or bivoltine. In Scandinavia in general in one generation from late June to late July. Further south from late May to early July and again from early August to late September. The larval stage in the north from July to May and further south in June and July and again from August overwintering until April. The adult moths fly at night and come frequently to light, but they are rarely flushed by day.

BIOLOGY: Polyphagous. The caterpillar has been found on a number of deciduous trees and bushes such as *Salix*, *Carpinus betulus*, *Quercus*, *Malus domestica*, *Sorbus aucuparia*, *Prunus padus*, *Acer pseudoplatanus*, *Fraxinus excelsior* (EBERT 2003), *Ulmus laevis*, *Betula pendula*, *Fagus sylvatica* (H. WEGNER pers. comm.) and *Alnus glutinosa* (C. KAYSER pers. comm.). The larva has been reared on *Quercus robur* (B. MÜLLER pers. comm.). The pupation takes place in an open web on the ground.

HABITAT: In forests and scrubland, in hedgerows, parks and gardens. The species occurs from sea level up to 1700 m in Greece, up to 2100 m in the Pyrenees in Spain (REDONDO *et al.* 2009) and up to 2300 m in the Alps in Italy.

SIMILAR SPECIES: No similar species in Europe.

GENETIC DATA: Genetically homogeneous from western Europe to Transcaucasia (n=47, from 12 countries), in the United Kingdom and the Netherlands with a second intrapopulational haplotype, diverging by 4.4% (n=5). Nearest species: *C. perlata* (GUENÉE, 1858) from North America (8.2%).

Gerinia Leraut, 2009

Gerinia Leraut, 2009, Moths of Europe, vol. 2, Geometrid Moths: 209. Type species: *Geometra honoraria* [Denis & Schiffermüller, 1775]. There are no genus synonyms.

Diversity and distribution: *Gerinia* is a monotypic genus in the West Palaearctic (Sub-Mediterranean) region.

External characters and abdomen: Wingspan 35–48 mm. Wings reddish brown, antemedial and medial lines white, inner margins darker. Forewing margin concave below apex, hindwing margin weakly undulating. Discal spots small. Male antennae bipectinate, female antennae weakly fasciculate. Male hindtibia without hair pencil. Male and female abdominal segments 3–8 unmodified.

Male genitalia: Uncus setose, wide. Socii minute. Gnathos absent. Valva asymmetrical, wide, ventral margin and apex densely spinose. Valva without subapical spine in ventral margin, base with prominent, asymmetric, sclerotised extension. Juxta with two setose patches. Saccus slightly elongated. Vesica without sclerotisations.

Female genitalia: Papillae anales large, setose. Apophyses posteriores long, straight. Apophyses anteriores vestigial. Sterigma sclerotised, funnel-shaped. Ductus bursae short. Posterior part of corpus bursae short, narrow, membranous. Anterior part of corpus bursae large, roundish, without sclerotisations.

Biology: Larva of *G. honoraria* (Denis & Schiffermüller) has been recorded feeding on *Quercus* and *Olea europaea* (Fajčík 2003, Patočka & Turčani 2005, Leraut 2009). Larva is a twig-mimic, usually greyish brown, with hair-like structures on its ventral parts and it has an additional, rudimentary pair of legs on the eighth segment (Lepiforum 2013). Eggs are laid in single rows (Lepiforum 2013). Adults are nocturnal. Pupa overwinters.

Immature stages: Larva of *G. honoraria* is a twig-mimic, usually greyish brown, with branching hair-like structures on its ventral parts and minute, white-centered spots dorsolaterally (Lepiforum 2013). Larva of *G. honoraria* has an additional, rudimentary pair of legs on the eighth segment, sharing this character with *C. margaritaria* (Hausmann & Miller 2000). Pupa cremaster is with four pairs of curved setae, apical pair most pronounced (Patočka & Turčani 2005).

Remarks: Leraut (2009) described genus *Gerinia* and combined *Geometra honoraria* (Denis & Schiffermüller, 1775) to it. External appearance of taxon *honoraria* is similar to that of *C. margaritaria* (Linnaeus), but we tentatively agree with Leraut (2009) that structurally *honoraria* is somewhat dissimilar from the other European Campaeini species (Table 6). Therefore it may warrant being classified in a genus of its own. However, presence of an additional proleg in the larvae of both genera may question this classification. A thorough analytical analysis on its relationships to other potentially related taxa must be carried out to find out whether the putatively diagnostic characters are homoplastic or synapomorphic. Homoplasy is a common phenomenon in the Lepidoptera and different combination of structures are often found in rather closely related species (e.g. Miller 1991, Kristensen 1991, Minet & Scoble 1999). Further, if the genus categories are too narrowly or widely delimited, they may fail to adequately convey the relatedness of species. A classification with very narrowly delimited genera can lead to the frequent addition of new genera when new species with slightly differing character combinations are discovered. Systematic position: Campaeini.

Table 6. Selected, diagnostic genus-level characters of *Campaea* LAMARCK, 1816 and *Gerinia* LERAUT, 2009. See remarks under *Gerinia*.

Selected diagnostic characters	***Campaea***	***Gerinia***
Labial palpi	Narrow	Thick
Male hindtibia	With small hair pencil	Without hair pencil
Uncus	Narrow	Wide
Valvae and valva base extension	Symmetric, asymmetric	Asymmetric, asymmetric
Subapical spine of valva	Present	Absent
Cornuti	Present	Absent
Signum	Large patch of minute spines	Absent

111. *Gerinia honoraria* (DENIS & SCHIFFERMÜLLER, 1775)

honoraria [DENIS & SCHIFFERMÜLLER], 1775, Ankündung syst. Werkes Schmett. Wienergegend: 315, (*Geometra*). Syntype(s), lost, [Austria]: Vienna district.

adsociaria BORKHAUSEN, 1794, Natur. eur. Schmett. 5: 78, (*Phalaena Geometra*). Holotype male, Europe.

excisaria ESPER, [1801], Die Schmett. 5 (4): 101, pl. 19, fig. 5, (*Ph*[*alaena*] *Geom*[*etra*]. Syntype(s), [Germany].

honoraria pictarorum OBERTHÜR, 1910, Études Lépid. comp. 4: 677, pl. 51, fig. 436, (*Metrocampa*). Syntypes 2 females, [France]: Charroux.

honoraria reisseri BYTINSKI-SALZ, 1937, Memorie Soc. ent. ital. 15 (2): 210, (*Metrocampa*). Holotype male, Sardinia: Teulada. [Originally as var].

honorifica ESPER, [1801], Die Schmett. 5 (7): 185, pl. 33, figs 1, 2, (*Ph*[*alaena*] *Bomb*[*yx*]). [Emendation of *honoraria* [DENIS & SCHIFFERMÜLLER]?]

ilicaria VILLERS, 1789, Linn. ent. 2: 302, pl. 6, fig. 2, (*Phal*[*aena*] *Geom*[*etra*]). Syntype(s).

INTRASUBSPECIFIC NAMES, POTENTIALLY UNAVAILABLE: *pictavorum* OBERTHÜR, 1910 (ab.); *reisseri* BYTINSKY-SALZ, 1937 (ab.?); *virescens* LUCAS, 1911 (ab.).

EXTERNAL CHARACTERS AND ABDOMEN: Wingspan 35–48 mm. Wings reddish brown, antemedial and medial lines white, inner margins darker. On hindwings only postmedial line visible. Terminal line narrow, dark reddish brown. Fringes concolorous with wings, hindwing fringes often partly white. Forewing margin concave below apex, hindwing margin weakly undulating. Discal spots small. Wings below irrorated with black, antemedial line absent, postmedial line broken into row of black dots. Frons, thorax and abdomen concolorous with wings. Collar whitish. Antennae whitish dorsally, male antennae bipectinate, female antennae weakly fasciculate. Hindtibia of both sexes with 2+2 spurs, male hindtibia without hair pencil. Tympanal organs medium-sized. Sternites and tergites 3–8 of both sexes undifferentiated.

VARIATION: Rather little. Females are paler. Antemedial and postmedial lines can be fused together on inner margin and ground colour can occasionally be greenish.

MALE GENITALIA: Uncus setose, wide, slightly quadrangular. Socii minute. Gnathos absent. Valva asymmetrical, wide, ventral margin and apex densely spinose. Valva without subapical spine in ventral margin. Valva base with prominent, asymmetric, sclerotised extensions, left extension larger, apex spinose, right apex setose. Juxta with two setose patches. Saccus slightly elongated. Aedeagus narrow, caecum minute. Vesica opens at approximately 135 degree angle. Vesica evenly narrowing tube, without sclerotisations.

FEMALE GENITALIA: Papillae anales large, setose. Apophyses posteriores very long, straight. Apophyses anteriores vestigial. Lamella antevaginalis medially narrow, sclerotised ridge, with membranous pockets laterally. Lamella postvaginalis membranous. Sterigma sclerotised, funnel-shaped. Ductus bursae short. Posterior part of corpus bursae short, narrow, membranous. Anterior part of corpus bursae large, roundish, without sclerotisations.

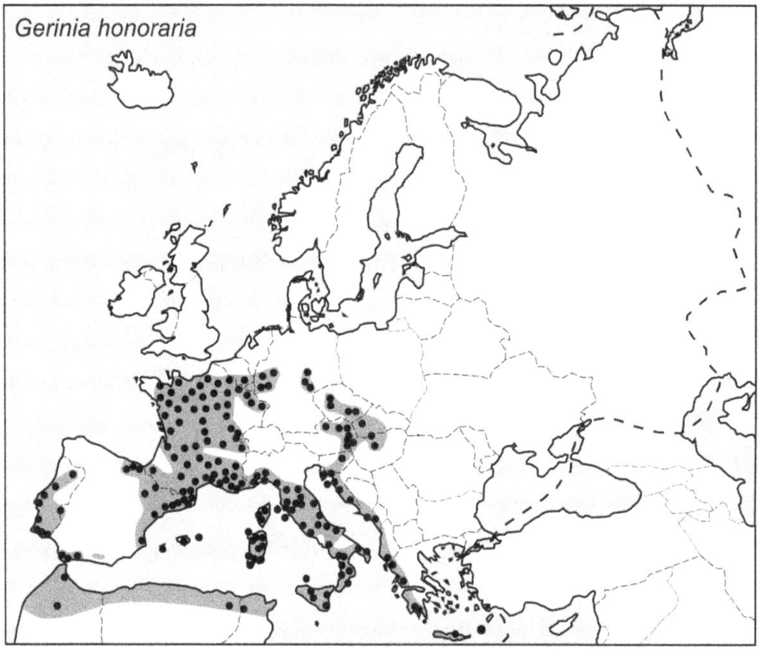

DISTRIBUTION AND ABUNDANCE: Sub-Mediterranean. In Europe from northern France, southern Belgium and central Germany to the Iberian Peninsula, southern France, Italy and on the Balkans along the Adriatic coast to Greece and including the Mediterranean islands Corsica, Sardinia, Sicily and the Balearic Islands. – Outside Europe the species occurs in Tunisia, Algeria and Morocco. – In Europe common, but not often in numbers.

PHENOLOGY: Uni-, bi- or trivoltine. The moths fly in central Europe from mid-May to late June (SCHÖNBORN 2011), in southern Europe in general from early April to late September in two overlapping generations, in northern Spain from May to September (J. YLLA I ULLASTRE pers. comm.) and in southern Spain also in November, probably in a third generation. In southernmost Spain recorded in all months of the year except August (P. HALE pers. comm.). In southern Italy until mid-November (A. HAUSMANN pers. comm.). The moths fly at night and the males come frequently to light, the females in smaller numbers. The adults are rarely seen by day.

BIOLOGY: Oligophagous. The larva is found on *Quercus pubescens* (M. LEIPNITZ pers. comm.) and recorded on *Quercus*, especially *Q. ilex* but also on *Q. cerris* and *Olea europaea* (FLAMIGNI *et al.*, in prep.). Reared on *Q. petraea* (E. FRIEDRICH pers. comm.). Reared on *Ligustrum vulgare* (southern Italy, A. HAUSMANN pers. comm.), feeding nocturnal. In rearings, some caterpillars produced moths already in November, the others hibernated (A. HAUSMANN pers. comm.).

HABITAT: In wood- and scrubland, preferring warm, but not too dry places, in southern Europe often, but not always near the coast. In coastal lowlands sometimes abundant in Mediterranean macchia and scattered oak forest (A. HAUSMANN pers. comm.). The species is found from sea level up to 1200 m altitude in the Iberian Peninsula (REDONDO *et al.* 2009) and up to 1450 m in Sardinia (FLAMIGNI *et al.*, in prep.), but it is most often found at lower altitudes.

SIMILAR SPECIES: No similar species in Europe.

GENETIC DATA: Genetically homogeneous in Europe and North Africa (n=20, from six countries). Nearest species in Europe: *Hylaea fasciaria* (9.4%).

REMARKS: Taxon *ilicaria* VILLERS, 1789, was considered valid at subspecies rank by LERAUT (1997), but not in LERAUT (2009). In other recent publications it has been considered a junior synonym of *honoraria* (DENIS & SCHIFFERMÜLLER).

Hylaea HÜBNER, 1822

Hylaea HÜBNER, 1822, Syst.-alphab. Verz.: 39, 43, 47, 50. Type species: *Phalaena fasciaria* LINNAEUS, 1758, Europe. Incorrect type species: *Phalaena prosapiaria* LINNAEUS, 1758 (see FLETCHER 1979). Genus synonyms are after SCOBLE (1999).

- *Ellopia* STEPHENS, 1829, Nom. Br. Insects: 43. Type species: *Phalaena fasciaria* LINNAEUS, 1758. [Junior homonym of *Ellopia* TREITSCHKE, 1825, and junior objective synonym of *Hylaea* HÜBNER, 1822].
- *Ellopia* TREITSCHKE, 1825, in Ochsenheimer, Schmett. Eur. 5 (2): 429. Type species: *Phalaena fasciaria* LINNAEUS, 1758.
- *Therina* HÜBNER, [1823] 1816, Verz. bekannter Schmett.: 283. Type species: *Phalaena fasciaria* LINNAEUS, 1758.

UNAVAILABLE NAMES (included in work rejected for nomenclatural purposes): *Hylaea* HÜBNER, 1806 (see FLETCHER 1979). – (incorrect subsequent spelling): *Terina* HÜBNER, 1823 (see FLETCHER 1979).

DIVERSITY AND DISTRIBUTION: According to SCOBLE (1999), *Hylaea* contains 13 species in the Palaearctic, the Nearctic and in the Neotropical regions. PITKIN (2002) noted that all Neotropical species placed in *Hylaea* appear not to belong there. LERAUT (2009) transferred *Pungeleria poeymiraui* (OBERTHÜR, 1922) to *Hylaea*, and recently *H. compararia* (STAUDINGER, 1895) was raised to species status and one new *Hylaea* species was described (SIHVONEN *et al.* 2014), thus the number of *Hylaea* species is five, and true members of the genus are Palaearctic.

EXTERNAL CHARACTERS AND ABDOMEN: Wingspan about 30–40 mm. Wings various shades of green, red-brown to greyish to yellow. Forewing medial and postmedial lines whitish, rather straight in European species, dentate in *H. poeymiraui* (OBERTHÜR, 1922). Hindwing with whitish postmedial line. Discal spots absent in European species, present in *H. poeymiraui*. Male antennae bipectinate, female antennae weakly fasciculate. Bullae tympani often with small roundish lobe. Male and female abdominal segments 3–8 unmodified.

MALE GENITALIA: Uncus setose, narrow subapically. Socii small, setose. Gnathos absent or present as vestigial, membranous ridge. Valva narrow, apex wider, sparsely setose. Valva with subapical spine in ventral margin, base with narrow, symmetric, sclerotised extension. Juxta with two setose patches. Saccus very elongated. Aedeagus with additional arm. Vesica with row of microcornuti.

FEMALE GENITALIA: Papillae anales wide, setose. Apophyses posteriores long, straight. Apophyses anteriores about 1/3 length of apophyses posteriores.Lamella postvaginalis large, horizontally striated, partly sclerotised. Sterigma with membranous, flower-like frill. Ductus bursae short. Posterior part of corpus bursae narrow, sclerotised, surface granulate. Anterior part of corpus bursae round, membranous. Signum small, roundish.

BIOLOGY: *Hylaea* species have been recorded feeding on conifer trees, *H. fasciaria* mostly on *Pinus sylvestris*, *Picea abies*, *Abies alba* and on *Larix* (MIKKOLA *et al.* 1989, EBERT 2003) and *H. pinicolaria* on *Pinus nigra* ssp. *laricio* (BELLIER 1861, ROBINEAU 2007). *H. fasciaria* overwinters as small larva on the food plant, pupation takes place in plant litter (MIKKOLA *et al.* 1989, PORTER 2010). Adults are nocturnal.

IMMATURE STAGES: Eggs of *H. fasciaria* are white or red, round, with smooth surface, laid in rows on conifer needles (LEPIFORUM 2013). Larva of *H. fasciaria* is a twig or needle-mimic, greyish, with brown mottlings and paler markings dorsally, or reddish brown with diamond-shaped patterns dorsally, and pairs of small raised wart-like structures along the back (EBERT 2003, PORTER 2010, LEPIFORUM 2013). Larva of *H. pinicolaria* is a also a twig or needle-mimic, green, but with transverse yellowish stripes or reddish brown with diamond-shaped patterns dorsally (LEPIFORUM 2013). Pupa cremaster of *H. fasciaria* is with four pairs of curved setae, apical pair is most pronounced (PATOČKA & TURČANI 2005).

REMARKS: *Hylaea* has recently been classified in the Campaeini (VIVES MORENO 1994, VIIDALEPP 1996, LERAUT 1997, HAUSMANN, MIRONOV & VIIDALEPP 2011). Systematic position: Campaeini. The descriptions of European *Hylaea* species, given below, largely follow SIHVONEN *et al.* (2014).

112. Hylaea fasciaria (LINNAEUS, 1758)

fasciaria fasciaria LINNAEUS, 1758, Syst. Nat. (Ed. 10) 1: 521, (*Phalaena (Geometra)*). Europe. (Holo)type female (Linnean Society of London, UK) (examined externally).

biliosata VILLERS, 1789, Linn. ent. 2: 386, pl. 6, fig. 22 (female), (*Phal[aena] Geom[etra]*). Syntype(s), [France]: Bressia [=Bresse, near Lyon]. SCOBLE (1999) gave the type locality as [Italy]: Brescia, but this is incorrect.

fasciaria cleui LERAUT, 1994, Entomol. Gall. 4 [1993] (4): 235, (*Hylaea*). Holotype male (Muséum National d'Histoire Naturelle, Paris, France), France: Hautes-Alpes, L'Argentiére-la-Bessée (examined externally) [originally as subspecies of *fasciaria* (LINNAEUS), downgraded from subspecies rank (SCOBLE 1999, LERAUT 2009] to synonymy by SIHVONEN *et al.* (2014), quoting the absence of distinct external features. See remarks.

neustriaria HUFNAGEL, 1767, Berlin Mag. 4 (5): 520, (*Phalaena*). Syntype(s), lost, [Germany]: Berlin region.

prasinaria [DENIS & SCHIFFERMÜLLER], 1775, Ankündung syst. Werkes Schmett. Wienergegend: 96, (*Geometra*). Syntype(s), lost, [Austria]: Vienna district. [Junior primary homonym of *Phalaena Geometra prasinaria* HUFNAGEL, 1767].

prosapiaria LINNAEUS, 1758, Syst. Nat. (Ed. 10) 1: 522, (*Phalaena (Geometra)*). Syntypes (Linnean Society of London, UK), Europe [probably near Åbo (=Turku), Finland] (examined externally).

rufofasciosa ESPER, 1794, Die Schmett. 3 Suppl. (5–6): 58, pl. 90, pl. 4, 5, (*Ph[alaena] Bomb[yx]*). Syntype(s), Sweden, Germany.

viridifasciosa ESPER, 1794, Die Schmett. 3 Suppl. (5–6): 58, pl. 90, figs 6, 7, (*Ph[alaena] Geom[etra]*). Syntypes male, female, [Germany: Erlangen; France: Brest].

fasciaria cedricola WEHRLI, 1929, Mitt. Münch. Ent. Ges. 19 (10–12): 319, pl. 24, fig. 3; pl. 25, fig. 9, 10, (*Ellopia*). Syntypes 9 males, 2 females (Zoologisches Forschungsinstitut und Museum Alexander Koenig, Bonn, Germany), Turkey (former Syria): Achyr Dagh, Bertiz Jaila, 1800 m (examined, including genitalia). [Originally as sp., downgraded from species rank (SCOBLE 1999)] to subspecies rank by SIHVONEN *et al.* (2014), quoting the absence of differential features in the genitalia and the exact barcode-sharing with *H. fasciaria.* See subspecies description below.

fasciaria flavella WEHRLI, 1940 (WEHRLI 1939–1954 in SEITZ, Gross-Schmett. Erde 4 (Suppl.): 322, pl. 24: g, (*Ellopia*), Armenia. [Originally as var, but raised to subspecies rank by WEHRLI 1954 (in WEHRLI 1939–1954)]. See subspecies description below.

INTRASUBSPECIFIC NAMES, POTENTIALLY UNAVAILABLE: *anastomosaria* HÖFER, 1920 (ab.); *angustata* FEICHTENBERGER, 1964 (f.); *approximata* LEMPKE, 1961 (f.); *aurantia* LERAUT, 2009 (f.); *conjuncta* POKORNY, 1918 (ab.); *constricta* SCHNAIDER, 1950 (f. of *prasinaria* DENIS & SCHIFFERMÜLLER); *cinereostrigaria* KLEMENSIEWICZ, 1892 (ab.); *contrasta* LERAUT, 2009 (f.); *extincta* WEHRLI, 1913 (ab.); *grisearia* FUCHS, 1877 (ab.); *intermedia* GUMPPENBERG, 1895 (ab.); *intermediaria* PROUT, 1915 (ab.) [unjustified emendation of subsp. *intermedia* GUMPPENBERG, 1895]; *interrupta* COCKAYNE, 1952 (ab.); *jacoviaci* LERAUT, 2009 (f.); *manitiaria* HERRICH-SCHÄFFER, 1848 (ab.); *ochracearia* REBEL, 1910 (ab.) (=synonym of *ochrearia* DE JOANNIS, 1921 according to PROUT (1912–1916, p. 322) and The Global Lepidoptera Names Index (ANONYMOUS 2013)); *ochrearia* DE JOANNIS, 1903 (ab.); *pallescens* FEICHTENBERGER, 1964 (f.); *reducta* ZWEIGELT, 1918 (ab.); *rosea* LERAUT, 2009 (f.); *rubronervata* ALBERS, 1941 (ab.); *rufostrigaria* LEMPKE, 1951 (f.); *unicolor* HEINRICH, 1917 (f.); *unifasciata* FEICHTENBERGER, 1964 (f.).

EXTERNAL CHARACTERS AND ABDOMEN: Wingspan ♂ 27–35 mm, ♀ 34–39 mm. Ground colour variable (see Variation), dominant colours being different shades of reddish brown and green. Medial lines often whitish (see Variation). Medial line angled before costa, basal part moves away from costa. Postmedial line angled before costa, weakly angled outwards on inner margin. Medial area often slightly darker than rest of wing, narrowest in middle. Hindwings with postmedial line visible only. Terminal line and fringes near forewing apex normally concolorous with wings. Hindwing postmedial line distinct, curved. Discal spots absent. Wings below as above, but paler. Frons pale brown to brown-red, thorax and abdomen concolorous with wings. Area between antennae (vertex) white. Antennae white dorsally, male antennae bipectinate, female antennae weakly fasciculate. Hindtibia of both sexes with 2+2 spurs. Tympanal organs medium-sized. Sternites and tergites 3–8 of both sexes undifferentiated.

VARIATION: Very variable. Ground colour varies from reddish grey to dark grey, to yellowish-green and to various shades of green. Various shades of reddish brown are dominant in northern Europe, and in northern parts of Scandinavia only these colours exist, whereas in southern Europe various shades of green (f. *prasinaria*) are dominant. In many areas both colour morphs coexist. In rearing from eggs from Brandenburg, Germany, all specimens belonged to the red form and gave a complete 2nd generation while when rearing from eggs from Black Forest, Germany, all larvae hibernated and produced the green form (B. MÜLLER pers. comm.). Position, width and colour of medial lines variable, they often stand out weakly in reddish brown specimens, being almost concolorous with wings, grey or blueish grey. In green specimens the medial lines are usually white, thinner in females. Numerous intrasubspecific forms have been described, those are summarised in PROUT (1912–1916), WEHRLI (1939–1954) and in LERAUT (2009).

Subsp. ***cedricola*** has wings rather dark green, medial lines are whitish. Forewing margin weakly concave below apex. Only green specimens are known. We follow SIHVONEN *et al.* (2014), and retain taxon valid at subspecies level, due to the concave forewing margin, the narrow medial area and the conspicuous transverse lines not reaching the forewing costa. The taxon is, according to current knowledge, allopatric and restricted to southern Turkey.

Subsp. ***flavella*** has wings yellowish grey, and forewing medial line is not visible near costa. We follow SIHVONEN *et al.* (2014), and retain the taxon valid at subspecies level. The taxon is, according to current knowledge, allopatric and restricted to Armenia (Transcaucasus). SCOBLE (1999) did not mention the taxon at all, VIIDALEPP (1996) consided it valid at subspecies level.

MALE GENITALIA: Uncus setose, subapical part rather wide, apex short, roundish. Socii small, setose. Gnathos absent. Valva narrow, apex wider dorsally, sparsely setose. Valva with subapical spine (occasionally with two spines) in ventral margin. Valva base with narrow, symmetric, sclerotised extension. Transtilla wide plate, anterior margin with two concavities. Juxta small, with two setose patches. Saccus very elongated, weakly curved laterally. Aedeagus narrow, caecum long. Aedeagus with straight additional arm, distance between aedeagus and additional arm narrow, apex not expanded, weakly dentate. Vesica opens at approximately 90 degree angle. Vesica evenly narrowing tube, base with straight row of microcornuti.

FEMALE GENITALIA: Papillae anales wide, setose. Apophyses posteriores long, straight. Apophyses anteriores about 1/4 length of apophyses posteriores. Lamella postvaginalis large, horizontally striated, partly sclerotised. Lamella antevaginalis often large, margin roundish, weakly sclerotised ridge. Sterigma with membranous, flower-like frill. Ductus bursae short, weakly sclerotised laterally. Posterior part of corpus bursae narrow, rather long, sclerotised, surface granulate. Anterior part of corpus bursae round, membranous. Signum absent or minute, roundish.

DISTRIBUTION AND ABUNDANCE: Eurasiatic. In Europe from northern Scandinavia to the central Iberian Peninsula, southern France, northern Italy and the Balkans including Greece and from the British Isles to the Ural mountains. – Outside Europe the distribution continues eastwards through southern Siberia to the Dahuria and Yakutia regions east of Lake Baikal, and to the south-east the species is found in the Caucasus region (subsp. *flavella*) and in Turkey (partly subsp. *cedricola*). – In Europe a common, but somewhat local species that now and then is found in numbers.

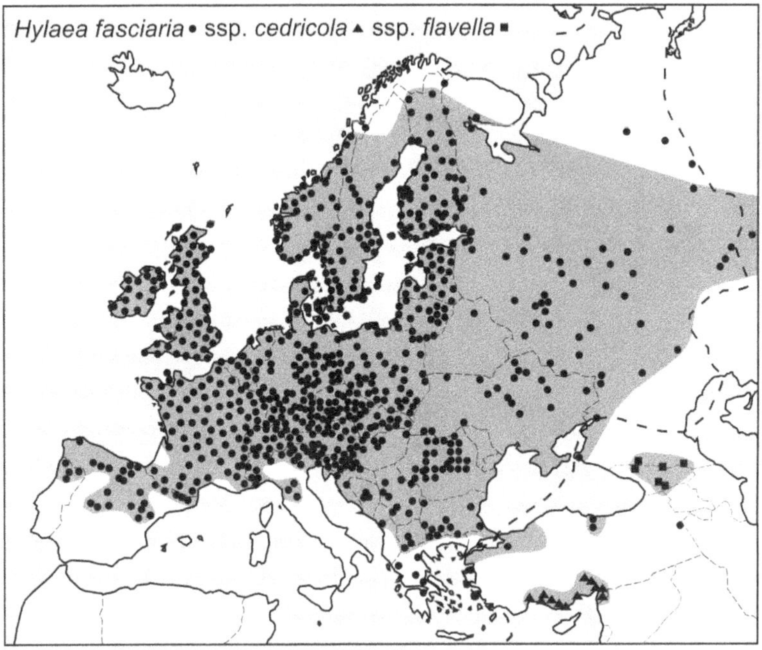

PHENOLOGY: Uni- or bivoltine. In northern Europe the moths fly in one generation from late June to early August, in south-western Germany from late May or early June to late August (EBERT 2003) and in Spain in two overlapping generations from May to early September (REDONDO *et al.* 2009). In Slovenia from late May to late August in two overlapping generations with peaks around 20th June and the first days of August (S. GOMBOC pers. comm.). Larva from August to May, in southern Europe probably also earlier from June or July. The hibernation takes place as a small larva. The moths fly at night and come to light, but they are rarely seen by day.

BIOLOGY: Oligophagous. The caterpillar has been found on *Picea abies*, *Abies alba*, *Pinus sylvestris* and also on *Larix* (EBERT 2003) and on *Pinus sylvestris* (C. KAYSER pers. comm.). Subspecies *cedricola* has been reared on *Pseudotsuga menziesii* (B. MÜLLER pers. comm.).

HABITAT: Sylvicolous. In coniferous woodland, occasionally also in gardens. In northern Europe restricted to lowlands. In Hungary from 100 m up to 1000 m, being local and rare on the Great Hungarian Plain (I. FAZEKAS pers. comm.). In the Alps usually up to 1000 m and more rarely up to 1800 m (N. PÖLL pers. comm.). In the Pyrenees up to 2300 m (REDONDO *et al.* 2009).

SIMILAR SPECIES: Species in Palaearctic *Hylaea fasciaria* species group are similar. The diagnostic, external characters shown in Text-figs 128–130 are somewhat tentative and should not be used in isolation, but should be combined with other information including biology, collecting locality, male and female genitalia and DNA barcodes. An overview of diagnostic morphological features is given in Table 7. North African *H. compararia* (STAUDINGER, 1895) is small, wings are dull green and medial lines are weak. Genitalia differences are illustrated in SIHVONEN *et al.* (2014).

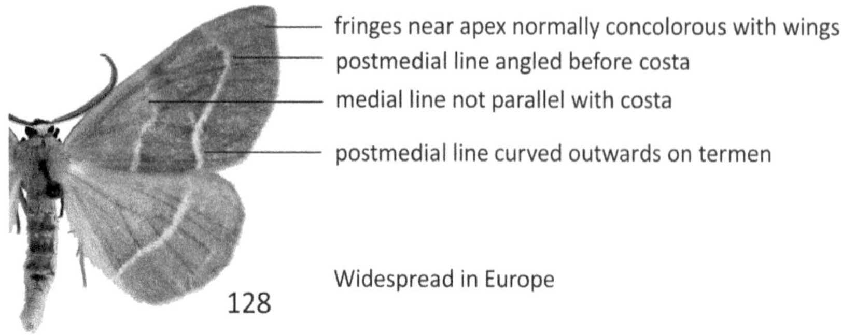

H. fasciaria

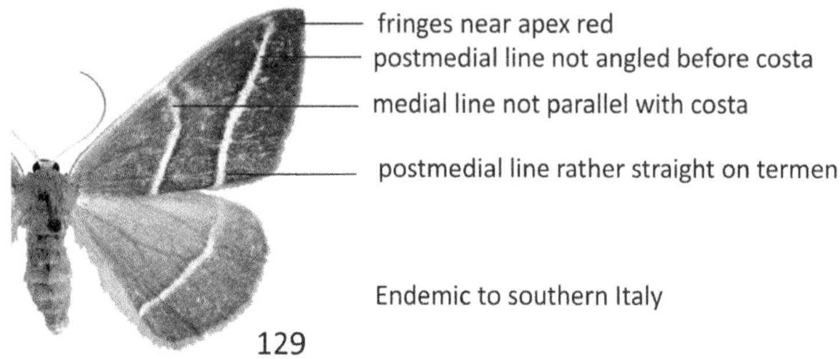

H. mediterranea

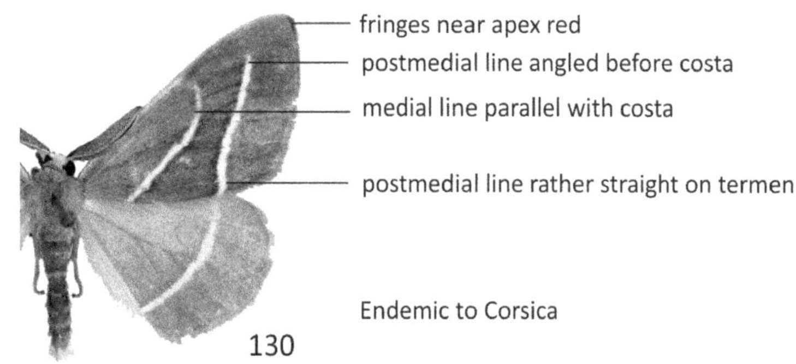

H. pinicolaria

Text-figs 128–130. Diagnostic characters (indicated) of European *Hylaea* HÜBNER, 1822 species. Text-fig. 128. *H. fasciaria* (LINNAEUS, 1758), Denmark, 27.vi.1977. Text-fig. 129. *H. mediterranea* SIHVONEN *et al.* 2014, Italy, Sicily, 5.vi.2005 (paratype). Text-fig. 130. *H. pinicolaria* (BELLIER, 1861), France, Corsica, 20–21.vii.2004. North African *H. compararia* (STAUDINGER, 1894) is also similar, see SIHVONEN *et al.* (2014) for details.

GENETIC DATA: Genetically comparatively homogeneous in Europe and Transcaucasia (n=40, from 11 countries), mean intraspecific variation 0.21%, maximum variation 1.71%. The taxon *cedricola* from Turkey exactly barcode-sharing (n=10), mean intraspecific variation 0.13%, maximum variation 0.46%. Nearest species: *Hylaea mediterranea* SIHVONEN *et al.* (minimum pairwise distance 3.3%).

Table 7. An overview of diagnostic morphological features in the Palaearctic *Hylaea fasciaria* LINNAEUS, 1758 species group after SIHVONEN *et al.* (2014). The shape and size of the lamella antevaginalis in the female genitalia is variable, and should be treated with caution.

Character	*H. fasciaria*	*H. mediterranea*	*H. pinicolaria*
forewing medial line	angled, not parallel with costa	curved, not parallel with costa	curved, parallel with costa
forewing postmedial line	angled at costa, angled outwards near inner margin	weakly angled at costa, evenly curved near inner margin	angled at costa, evenly curved near inner margin
hindwing postmedial line	distinct, curved	distinct, curved	distinct, curved
Reddish specimens	yes	no	no
Fringes at forewing apex	usually concolorous with wings	red	red
Uncus shape	subapical part rather wide	subapical part narrow	subapical part narrow
Apex of aedeagus arm	not enlarged	not enlarged	not enlarged
Cornuti	straight row	straight row	straight row
Signum	small/minute	large	small/minute
Lamella antevaginalis	often large, margin round	often large, margin concave	often large, margin concave
Distribution	Pan-European, eastwards to Russia: Yakutia	Italy: Sicily, Calabria, Molise	France: Corsica

REMARKS: *H. fasciaria* subsp. *cleui* LERAUT, illustrated in LERAUT (2009) and HAUSMANN (2001, fig. 66), was downgraded from subspecies rank to junior synonym to the nominal subspecies of *H. fasciaria* (LINNAEUS) by SIHVONEN *et al.* (2014). Wings are purple-pink to crimson-red, and medial lines are ash grey. Taxon is known from southern French Alps. According to SIHVONEN *et al.* (2014), in the adjacent Valesia (southern Switzerland, >200 specimens examined in the ZSM) such forms are dominant but mixed with green and red forms, potentially supporting the existence of a cline. This indicates that *cleui*, although locally dominant phenotype, does not constitute a subspecies because it lacks disjunct external features, and it is questionable whether the southern French Alps populations are geographically isolated from other populations in the Alps. DNA barcodes are not available, so far, for French Alps populations.

113. Hylaea mediterranea SIHVONEN ET AL., 2014

mediterranea SIHVONEN, SKOU, FLAMIGNI, FIUMI & HAUSMANN, 2014, Zootaxa 3768 (4): 478, figs. 2, 7, 11, 15, (*Hylaea*). Holotype male (coll. Skou, Denmark, to be deposited at the Zoological Museum, University of Copenhagen, Denmark), Italy: Sicily, 5.7 km ESE San Stefano Quisquina, near Pizzo della Rondine, 1000 m, 9.-10.x.2010, Prep. number 1738 (examined, including genitalia).

EXTERNAL CHARACTERS AND ABDOMEN: Wingspan ♂ 31 mm (n=4), ♀ 37–41 mm (n=4). Wings light green, medial lines white. Medial line curved before costa, basal part not parallel with costa. Postmedial line rather straight, only weakly curved, barely angled before it reaches costa near apex and evenly curved outwards on inner margin. Medial area concolorous with rest of wing. Terminal line and fringes concolorous with wings, forewing apex dark red. Hindwing postmedial line distinct, curved. Discal spots absent. Wings below as above, but paler. Frons red-brown, thorax and abdomen concolorous with wings. Area between antennae (vertex) white. Antennae white dorsally, male antennae bipectinate, female antennae fasciculate. Hindtibia of both sexes with 2+2 spurs. Tympanal organs medium-sized. Sternites and tergites 3–8 of both sexes undifferentiated.

VARIATION: Little. Forewing postmedial line is straight or weakly curved outwards on inner margin. Specimens from Calabria (Italy) often have the forewing postmedial line clearly angled before it reaches costa. Only light green specimens are known.

MALE GENITALIA: Generally as in *H. fasciaria* (LINNAEUS), see description above, and *H. pinicolaria* (BELLIER). Uncus relatively narrower before wide apex in *H. mediterranea* (uncus relatively wider before wide apex in *H. fasciaria* and *H. pinicolaria*, but the differences are not clear-cut). See also Table 7.

FEMALE GENITALIA: Generally as in *H. fasciaria* (LINNAEUS), see description above, and *H. pinicolaria* (BELLIER), but with following quantitative difference: signum large in *H. mediterranea* (signum absent or minute in *H. fasciaria* and *H. pinicolaria*).

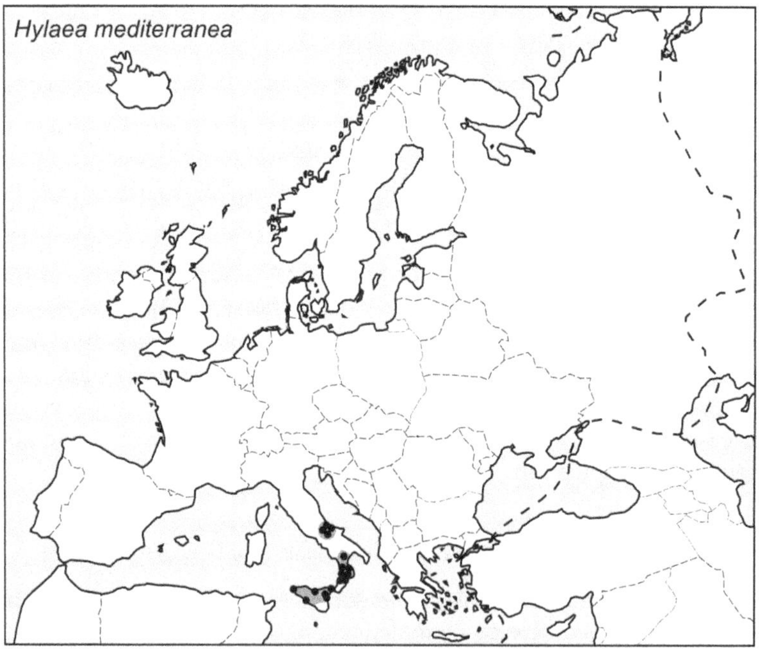

DISTRIBUTION AND ABUNDANCE: Central Mediterranean. Endemic in central and southern Italy and Sicily. A local, but not rare species which seems to appear at suitable habitats where it can occur in limited numbers.

PHENOLOGY: Bivoltine. In Sicily the species flies from late May to early July and from late August to late October (SIHVONEN *et al.* 2014). The moths fly at night and come to light.

BIOLOGY: Probably oligophagous on conifers. Reared on *Pinus sylvestris* and *Picea abies* (SIHVONEN *et al.* 2014). *Pinus sylvestris* and *Picea abies* are not present in Sicily, in the collecting localities *Pinus nigra* ssp. *laricio* and *P. halepensis* are common.

HABITAT: Sylvicolous. In pine forests and places with more scattered pine trees. Altitudinal range from sea level up to 1780 m (SIHVONEN *et al.* 2014).

SIMILAR SPECIES: Species in Palaearctic *Hylaea fasciaria* species group are similar. The diagnostic, external characters shown in Text-figs 128–130 are somewhat tentative and should not be used in isolation, but should be combined with other information including biology, collecting locality, male and female genitalia and DNA barcodes. An overview of diagnostic morphological features is given in Table 7. North African *H. compararia* (STAUDINGER, 1895) is small, wings are dull green and medial lines are weak. Genitalia differences are illustrated in SIHVONEN *et al.* (2014).

GENETIC DATA: Genetically homogeneous in Calabria, Molise and Sicily (n=8), mean intraspecific variation 0.19%, maximum variation 0.46%. Nearest species: *Hylaea fasciaria* (minimum pairwise distance 3.3%).

114. *Hylaea pinicolaria* (BELLIER, 1861)

pinicolaria BELLIER, 1861, Annls Soc. ent. Fr. (4) 1 (1): 29, pl. 2, fig. 12, (*Ellopia*). Syntype(s), Corsica (mountains of). BELLIER's collection was apparently included in the C. OBERTHÜR collection, which is currently housed in the Natural History Museum, London, UK. Despite searches, we have not located the type.

EXTERNAL CHARACTERS AND ABDOMEN: Wingspan ♂ 32–34 mm, ♀ 39–42 mm. Wings light green, medial lines white. Medial line curved before costa, basal part parallel with costa. Postmedial line rather straight, only weakly curved, angled before it reaches costa well before apex and evenly curved outwards on inner margin. Medial area concolorous with rest of wing. Terminal line and fringes concolorous with wings, forewing apex dark red. Hindwing postmedial line distinct, curved. Discal spots absent. Wings below as above, but paler. Frons red-brown, thorax and abdomen concolorous with wings. Area between antennae (vertex) white. Antennae white dorsally, male antennae bipectinate, female antennae fasciculate. Hindtibia of both sexes with 2+2 spurs. Tympanal organs medium-sized. Sternites and tergites 3–8 of both sexes undifferentiated.

VARIATION: Little. Ground colour varies from light green to light yellowish green. Only green specimens are known.

MALE GENITALIA: Generally as in *H. fasciaria* (LINNAEUS), see description above, and *H. mediterranea* SIHVONEN *et al.* Aedeagus with weakly curved additional arm in *H. pinicolaria* (additional arm straight in *H. fasciaria*).

FEMALE GENITALIA: Generally as in *H. fasciaria* (LINNAEUS), see description above, and *H. mediterranea*, but with following quantitative difference: signum large in *H. mediterranea* (signum absent or minute in *H. fasciaria* and *H. pinicolaria*).

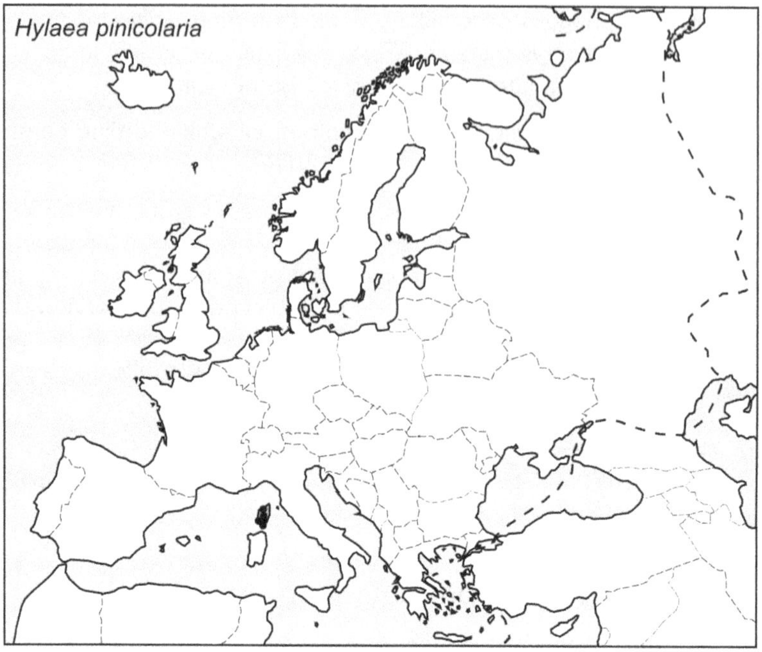

DISTRIBUTION AND ABUNDANCE: Tyrrhenian. Endemic to Corsica. Rather common, but usually seen in limited numbers.

PHENOLOGY: Uni- or bivoltine. The species has been recorded from mid-May until early October. Autumn specimens might represent a partial second generation. A single specimen as early as 19th April (C. TAUTEL pers. comm.). The earliest and latest records are from lower altitudes and indicate bivoltinism. The moths fly at night and come readily to light.

BIOLOGY: Monophagous. The larva has been found on *Pinus pinaster* (MABILLE 1867) and was recorded on *P. nigra* ssp. *laricio* (ROBINEAU 2007). In captivity it has been reared on *Pinus sylvestris* (REISER 1929) and *P. nigra* (E. FRIEDRICH pers. comm.).

HABITAT: Sylvicolous. In pine forests and places with more scattered pine trees from approximately 700 m altitude and upwards to at least 1400 m.

SIMILAR SPECIES: Species in the Palaearctic *Hylaea fasciaria* species-group are similar. The diagnostic, external characters shown in Text-figs 128–130 are somewhat tentative and should not be used in isolation, but should be combined with other information including biology, collecting locality, male and female genitalia and DNA barcodes. An overview of diagnostic morphological features is given in Table 7. North African *H. compararia* (STAUDINGER, 1895) is small, wings are dull green and medial lines are weak. Genitalia differences are illustrated in SIHVONEN *et al.* (2014).

GENETIC DATA: Genetically homogeneous (n=4), mean and maximum intraspecific variation 0.0%. Nearest species: *Hylaea fasciaria* (minimum pairwise distance 3.9%).

Hemithea squalidaria O. G. COSTA, 1848, taxonomic combination and status uncertain

squalidaria O. G. COSTA, 1848, Fauna Regno Napoli (Ent.): [331], pl. (Geom.) 2, fig. 4, (*Hemithea*). Syntype(s), [Italy]: Adriatic coast: San Cataldo, near Lecce; Tyrrhenian coast: [Lago di] Patria [near Naples].

Hemithea squalidaria O. G. COSTA, 1848 from southern Italy is a problematic taxon due to the lack of type material. Only the original, hand-drawn colour illustration on the taxon exists (COSTA 1848, reprinted in greyscale in Text-fig. 131). Prout (1912–1916) combined *squalidaria* with *Ellopia* STEPHENS (=*Hylaea*), due to its similarity with North African *H. compararia* (STAUDINGER, 1895). WEHRLI (1939–1954), in the absence of material, treated *squalidaria* putatively valid at species level, as did SCOBLE (1999), combining it to *Hylaea*.

SIHVONEN *et al.* (2014) concluded that if *Hemithea squalidaria* belonged to the genus *Hylaea*, in theory it could be conspecific with *H. mediterranea*. However, the conspecifity was excluded by mismatch of external characters and considerably smaller size of *Hemithea squalidaria*, as given in the original description. *Hylaea compararia*, *H. pinicolaria* and *H. fasciaria* are externally different and, according to current knowledge, they do not occur in southern Italy. Further, SIHVONEN *et al.* (2014) were unable to exclude the possibility that *H. squalidaria* may be a species of the subfamily Geometrinae. There are potentially several Geometrinae species, which live at sea level in southern Italy that may be relevant in this context.

Due to the unavailability of type material, we follow SIHVONEN *et al.* (2014), who returned the taxon to its original combination, leaving its taxonomic combination and status uncertain, being potentially valid at species level and potentially belonging to another genus in the subfamily Geometrinae. See SIHVONEN *et al.* (2014) for details.

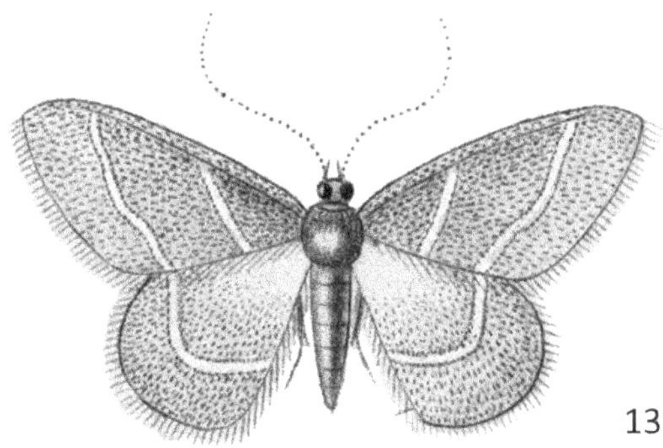

131

Text-fig. 131. The original illustration of *Hemithea squalidaria* O. G. COSTA, 1848, a taxon that has often been combined with *Hylaea* HÜBNER, 1822, but its taxonomic combination and status are unclear, see SIHVONEN *et al.* (2014) and text for details. Figure is reproduced under a licence from the Picture Library of the Natural History Museum, London, UK.

Adalbertia WEHRLI, 1931

Adalbertia WEHRLI, 1931, Ent. Z., Frankf. a. M. 44: 338. Type species: *Numeria castiliaria* STAUDINGER, 1900, [Spain]. *Adalbertia* was proposed as the objective replacement name for *Seitzia* WEHRLI, 1930. Genus synonymy is after SCOBLE (1999).

Seitzia WEHRLI, 1930, Ent. Z., Frankf. a. M. 44: 75. Type species: *Numeria castiliaria* STAUDINGER, 1900. [Junior homonym of *Seitzia* SCRIBA, 1919 (Lepidoptera: Lasiocampidae.)]

DIVERSITY AND DISTRIBUTION: *Adalbertia* is a monotypic genus, the only species is found in the Iberian Peninsula.

EXTERNAL CHARACTERS AND ABDOMEN: Wingspan 21–37 mm. Wings reddish grey-brown, medial lines dentate, often faint. Hindwings uniform greyish, often with tint of reddish. Forewing discal spots large, elongated. Male antennae bipectinate, female antennae filiform. Male hindtibia without hair pencil, 3rd sternite without setae, 8th sternite unmodified.

MALE GENITALIA: Uncus with ventral suture, basal portion densely setose. Socii minute, setose. Gnathos absent. Tegumen large, sclerotised. Valva strongly sclerotised, apex wider, spoon-shaped, densely setose. Valva with subapical spine in ventral margin, costa base with roundish, setose lobe. Medial ridge of valva microdentate. Transtilla arch-shaped. Saccus roundish. Vesica small, without sclerotisations.

FEMALE GENITALIA: Papillae anales large, setose. Apophyses posteriores long, straight. Apophyses anteriores about 1/3 length of apophyses posteriores. Lamella antevaginalis small, raised, sclerotised plate. Ductus bursae short, membranous. Corpus bursae roundish, membranous, without sclerotisations.

BIOLOGY: Larva of *A. castiliaria* has been recorded to feed on the needles of *Pinus sylvestris, P. halepensis* and *P. nigra*. Common species in pine woodlands. Two generations per year, from mid-March to early July and from early August to late October (REDONDO *et al.* 2009). Both caterpillar and pupae may overwinter (AIZPÚRUA 2002).

IMMATURE STAGES: Larva of *A. castiliaria* is grey-brown, with zig-zag pattern dorsally, with hair-like, white structures on its ventral parts (AIZPÚRUA 2002). We have no data on pupa. AIZPÚRUA (2002) provides a superficial habitus photograph.

REMARKS: *Adalbertia* has recently been classified in the Campaeini (VIVES MORENO 1994, LERAUT 1997, HAUSMANN, MIRONOV & VIIDALEPP 2011). LERAUT (2009) concluded, after examination of male and female genitalia, that *Numeria castiliaria* STAUDINGER, recently classified in the monotypic *Adalbertia* WEHRLI, 1931, is similar with *Pungeleria capreolaria* (DENIS & SCHIFFERMÜLLER) and combined *A. castiliaria* to *Pungeleria* ROUGEMONT, 1903. We agree with structural similarities, but do not follow the proposed classification until a thorough, phylogenetic analysis of the Campaeini and potentially related genera is carried out. While certain genitalia characters support the *Pungeleria* association, other characters, for instance the corpus bursae spines in the female genitalia, may be indicative of other relationships. The mentioned character is found in *Adalbertia* and in *Campaea*. Systematic position: Campaeini.

115. Adalbertia castiliaria (STAUDINGER, 1900)

castiliaria STAUDINGER, 1900, Dt. ent. Z. Iris 12: 392, (*Numeria*). Holotype male, [Spain, Castille]: near San Ildefonso.

dumonti MABILLE, 1906, Annls Soc. ent. Fr. 75 (1): 32, pl. 3, fig. 2, (*Ellopia*?). Holotype male (Muséum National d´Histoire Naturelle, Paris, France), Spain: La Granja.

cazorlensis EXPÓSITO, 1984, Revta Lepid. 12 (3): 204, (*Adalbertia*). Holotype male, [Spain: Castilla], [Sierra de] Cazorla, Fuente del Oso. See remarks.

INTRASUBSPECIFIC NAMES, POTENTIALLY UNAVAILABLE: *cortes* AGENJO, 1948 (2nd/summer generation specimens).

EXTERNAL CHARACTERS AND ABDOMEN: Wingspan 21–37 mm, second generation specimens smaller. Wings reddish grey-brown. Antemedial and postmedial lines dentate, partly blackish. Medial area often slightly reddish, particularly on its outer part. Terminal line and fringes concolorous with wings. Hindwings uniform greyish, often with tint of reddish near margin. Forewing discal spots large, blackish, elongated. Wings below uniform grey, often with tint of reddish. Frons, collar, thorax and abdomen concolorous with wings. Male antennae bipectinate, female antennae filiform. Hindtibia of both sexes with 2+2 spurs, male hindtibia without hair pencil. Tympanal organs medium-sized. Sternites and tergites 3–8 of both sexes undifferentiated.

VARIATION: Ground colour normally grey-brown, the degree of reddish variable, and it may be absent. Medial lines often partly blackish, but blackish colour can be absent.

MALE GENITALIA, see remarks: Uncus with ventral suture, basal portion densely setose, apical portion sparsely setose, apex acute. Socii minute, setose. Gnathos absent. Tegumen large, sclerotised. Valva strongly sclerotised, apex wider, spoon-shaped, densely setose. Valva with subapical spine in ventral margin, costa base with roundish, setose lobe, another sclerotised ridge at halfway. Medial ridge of valva microdentate. Transtilla arch-shaped. Saccus roundish. Aedeagus wedge-shaped, caecum small, apex acute. Vesica opens at approximately 45 degree angle. Vesica narrow tube, without sclerotisations.

FEMALE GENITALIA: Papillae anales large, setose, surface granulate. Apophyses posteriores long, straight. Apophyses anteriores about 1/3 length of apophyses posteriores. Lamella antevaginalis small, raised, sclerotised plate. Ductus bursae short, membranous. Corpus bursae roundish, membranous, without sclerotisations.

DISTRIBUTION AND ABUNDANCE: West-Mediterranean. Endemic to southernmost France and to the Iberian Peninsula where the species only occurs in the eastern and central parts. Rather common, but somewhat local, sometimes in limited numbers.

PHENOLOGY: Bivoltine. Moths of the first generation fly from mid-March to early July, of the second from early August to late October (REDONDO *et al.* 2009). In coll. SKOU from early May to mid-September with no real break between the two generations, but the first generation seems to stop in early July and the second to start with fresh specimens in mid-July. The hibernation takes place as caterpillar (ROBINEAU *et al.* 2007), according to AIZPÚRUA (2002) both larva and pupa can hibernate. The moths fly at night and both males and females come readily to light.

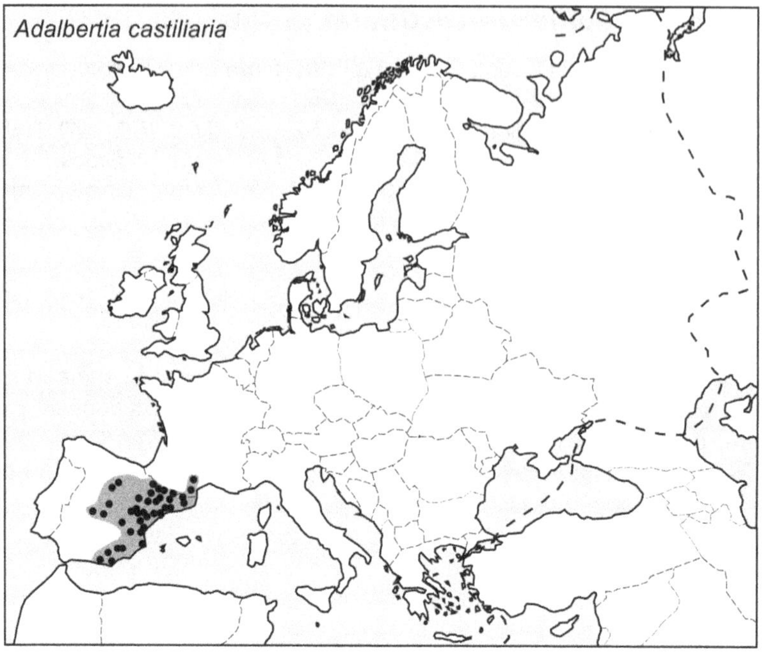

BIOLOGY: In nature probably oligophagous on *Pinus*. The caterpillar has been recorded on *Pinus sylvestris*, *P. halepensis* and *P. nigra* (REDONDO *et al.* 2009). Caterpillars have been reared on *Pseudotsuga menziesii* (B. MÜLLER pers. comm.).

HABITAT: Sylvicolous. The species occurs in most types of pine woodland and it is found from sea level up to 2000 m according to REDONDO *et al.* (2009).

SIMILAR SPECIES: No similar species in Europe.

GENETIC DATA: Genetically slightly heterogeneous in Europe: The two barcoded specimens from southern France and central Spain diverging by 1.6%. Nearest species (genetic similarity due to long branch attraction): South American *Leuciris fimbriaria* (STOLL, 1781) (7.7%), in Europe *Calamodes subscudularia* (TURATI, 1919) (8.5%).

REMARKS. The putative subspecies *cazorlensis* EXPÓSITO, 1984 from Sierra de Cazorla, Spain, has not been accepted or its validity has been questioned in the recent literature (LERAUT 2009, REDONDO *et al.* 2009, HAUSMANN, MIRONOV & VIIDALEPP 2011). No DNA barcodes have been analysed for the taxon *cazorlensis*.

Homologies of the uncus and socii in the male genitalia are tentative. We have interpreted the large setose structure to be uncus, whereas the socii are minute, barely notable, weakly setose extensions on its ventral margin. Another interpretation could be that the large setose structure is socii, a paired structure whose arms have fused together to form an organ that has functionally replaced uncus. This is supported by the paired structure, with ventral suture, and the fact that the setose structures do not arise from the apex of tegumen, where uncus normally originates, but from more lateral position. If this interpretation is correct, then the minute, barely notable, weakly setose extensions are simply a further modified part of the socii.

Pungeleria ROUGEMONT, 1903

Pungeleria ROUGEMONT, 1903, Bull. Soc. neuchâtel. Sci. nat. 31: 178. Type species: *Geometra capreolaria* [DENIS & SCHIFFERMÜLLER], 1775, [Austria]. There are no genus synonyms.

UNAVAILABLE NAMES (incorrect subsequent spellings): *Puengeleria* (found frequently in the literature, for instance KENIS *et al.* 2005, SUMPICH & CERNY 2010, BESHKOV 2011); *Püngeleria* (found frequently in the literature, for instance PROUT 1912–1916).

DIVERSITY AND DISTRIBUTION: *Pungeleria* is currently a monotypic genus in the western Palaearctic region, after LERAUT'S (2009) proposal to combine *Pungeleria poeymiraui* (OBERTHÜR) with *Hylaea* HÜBNER.

EXTERNAL CHARACTERS AND ABDOMEN: Wingspan 26–31 mm. Wings yellowish brown, dark brown medial lines bordered with whitish. Forewing apex with streak. Male antennae bipectinate, female antennae filiform. Male hindtibia slightly swollen, with hair pencil. Abdominal sternites and tergites 3–8 of both sexes undifferentiated.

MALE GENITALIA: Uncus densely setose basally. Socii and gnathos absent. Tegumen large, sclerotised. Valva narrow, weakly setose, apex wider, with longer setae. Valva with subapical spine in ventral margin, costa base/lateral end of transtilla with elongated, setose lobe. Saccus elongated. Vesica enlarged at base, without sclerotisations.

FEMALE GENITALIA: Papillae anales small, weakly setose. Apophyses posteriores long, straight. Apophyses anteriores vestigial. Lamella antevaginalis and lamella postvaginalis fused, forming sclerotised ring around ostium bursae. Corpus bursae with patch of minute spines.

BIOLOGY: Larva of *P. capreolaria* has been recorded feeding on *Picea abies* and *Abies alba*. Adults fly from July to September in subalpine conifer forests. Caterpillar overwinters (ROBINEAU 2007, REDONDO *et al.* 2009).

IMMATURE STAGES: Egg of *P. capreolaria* is smooth, reddish-beige, laid in rows on needles. Larva is green, with wide, whitish lateral and green-blue dorsal stripes (LEPIFORUM 2013). Pupa cremaster is constricted basally, with four pairs of curved setae, apical pair is most pronounced (PATOČKA & TURČANI 2005).

REMARKS: *Pungeleria* has recently been classified in the Campaeini (VIVES MORENO 1994, VIIDALEPP 1996, LERAUT 1997, HAUSMANN, MIRONOV & VIIDALEPP 2011). LERAUT (2009) concluded, after examination of male and female genitalia, that *Numeria castiliaria* STAUDINGER, recently classified in the monotypic *Adalbertia* WEHRLI, 1931, is similar with *Pungeleria capreolaria* (DENIS & SCHIFFERMÜLLER) and combined *A. castiliaria* into *Pungeleria* ROUGEMONT, 1903. We agree with structural similarities, but do not follow the proposed classification until a thorough, phylogenetic analysis of the Campaeini genera is carried out. Numerous characters in the Campaeini taxa are intermixed in the current classification, for instance the corpus bursae spines are found in *Campaea* and in *Pungeleria*, thus raising the question about their relationships. Systematic position: Campaeini.

116. Pungeleria capreolaria (DENIS & SCHIFFERMÜLLER 1775)

capreolaria [DENIS & SCHIFFERMÜLLER], 1775, Ankündung syst. Werkes Schmett. Wienergegend: 105, (*Geometra*). Syntype(s), lost, [Austria]: Vienna district.

donzelaria DUPONCHEL, 1829, in GODART & DUPONCHEL, Hist. nat. Lépid. Papillons Fr. 7 (2): 478, pl. 170, fig. 6, (*Numeria*). Syntype(s) female(s), [France]: Auvergne, Mont d'Or.

valesiaria VORBRODT & MÜLLER-RUTZ, 1914, Schmett. Schweiz 2 (2): 143, (*Pungeleria*). Syntypes, Switzerland: Mt Chemin; Mayens de Sion; Sierre; Niouc; Ponchette. See remarks.

INTRASUBSPECIFIC NAMES, POTENTIALLY UNAVAILABLE: *nigrofasciata* FELTIN, 1938 (ab.).

EXTERNAL CHARACTERS AND ABDOMEN: Wingspan 26–31 mm. Wings yellowish brown, densely irrorated with blackish brown. Antemedial and postmedial lines dark brown, latter clearly curved outwards in middle, bordered with whitish. Medial area often slightly darker than rest of wing. Forewing apex with streak. Terminal line faint, interrupted. Fringes concolorous with wings. Hindwings grey, weakly irrorated with blackish, postmedial line weak, dentate. Forewing discal spots distinct, blackish. Hindwing discal spots weak or absent. Wings below yellowish brown, postmedial line broken into row of blackish dots on hindwing. Frons, collar, thorax and abdomen concolorous with wings. Male antennae bipectinate, female antennae filiform. Male hindtibia slightly swollen, with hair pencil. Hindtibia of both sexes with 2+2 spurs. Tympanal organs medium-sized. Abdominal sternites and tergites 3–8 of both sexes undifferentiated.

VARIATION: Ground colour varies from pale whitish yellow to whitish grey to grey-brown. Density of brown irroration is variable, occasionally forming striations.

MALE GENITALIA: Uncus densely setose basally, apex elongated. Socii and gnathos absent. Tegumen large, sclerotised. Valva narrow, weakly setose, apex wider, with longer setae. Valva with subapical spine in ventral margin, costa base/lateral end of transtilla with elongated, setose lobe. Juxta setose, with medial, weakly sclerotised structure. Saccus elongated. Aedeagus short, curved, caecum short. Vesica opens at approximately 90 degree angle. Vesica enlarged at base, without sclerotisations.

FEMALE GENITALIA: Papillae anales small, weakly setose. Apophyses posteriores long, straight. Apophyses anteriores vestigial. Lamella antevaginalis and lamella postvaginalis fused, forming narrow, sclerotised ring around ostium bursae. Ductus bursae short, wide. Posterior part of corpus bursae narrow. Anterior part of corpus bursae wider, elongated, with patch of minute spines.

DISTRIBUTION AND ABUNDANCE: European-Westasiatic. In Europe from western France including Corsica and the Pyrenees through central Europe and Italy to the Balkans including Greece. Most recently also two records from the British Channel Islands and southern England (B. GOATER, pers comm.). – Outside Europe the species is only found in Turkey. Reported from European Russia (VIIDALEPP 1996), but no specimens from there have been traced. – In Europe relatively common, but somewhat local and not rarely in numbers.

PHENOLOGY: Usually univoltine, in the southernmost part of the distribution area bivoltine. In south-western Germany the species has a rather long flight period from late May to late September, peaking from late July to late August, but only in one generation (EBERT 2003). In the Czech Republic and Slovakia from early August to mid-September (Z. LAŠTŮVKA pers. comm.). In Poland from mid-July to early September (BUSZKO 2000). In Greece from late

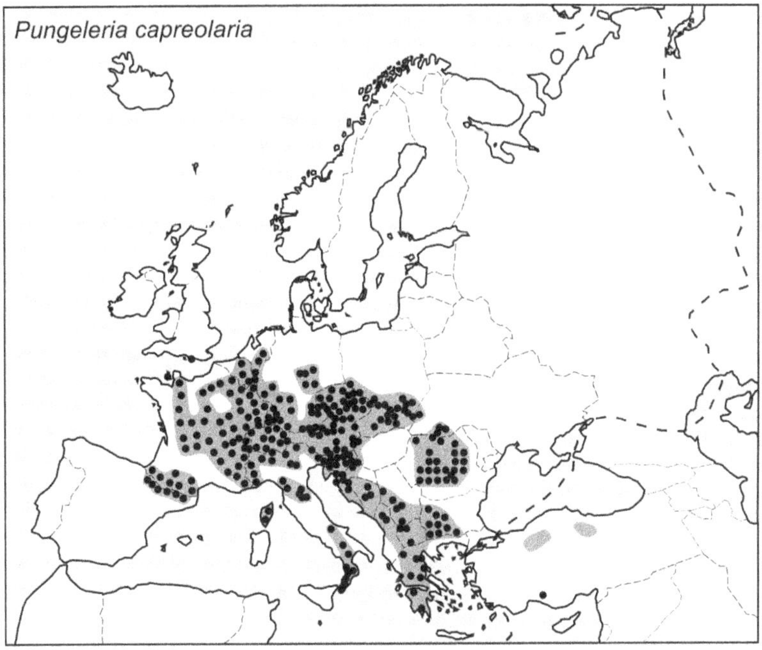

May until late October, probably in two generations. Larva from July and overwintering until May or June. The moths are active at night and both males and females come frequently to light.

BIOLOGY: Oligophagous on conifers, preferring *Abies*. The larva lives on *Abies alba* (EBERT 2003). Reared on *A. alba*, *A. cephalonica* and *Pseudotsuga menziesii* (B. MÜLLER pers. comm.), as well as on *Picea abies* (EBERT 2003). In Greek mountains the species probably feeds on *Abies cephalonica*, which is abundant there. Possibly also on *Abies borisii-regis* which occurs further to the north in Bulgaria, Macedonia, Albania and Serbia. The pupation takes place between bark and lichens near the base of the tree trunks.

HABITAT: Sylvicolous. The species occurs in coniferous and mixed forest with conifers as well as in more open woodland with conifers. In southern central Europe from low altitudes up to over 1000 m, in the central Alps up to 1500 m, in Greece up to 1700 m and in Spain in the Pyrenees from 1000 m up to 1800 m (REDONDO *et al.* 2009).

SIMILAR SPECIES: No similar species in Europe.

GENETIC DATA: Genetically slightly heterogeneous in Europe (n=11, from four countries) but without clear regional patterns. Nearest species in Europe: *Selenia tetralunaria* (8.7%).

REMARKS: Taxon *valesiaria* VORBRODT & MÜLLER-RUTZ has not been recognised as valid at subspecies level in the recent literature apart from SCOBLE (1999). Elsewhere it has been considered an individual form (LERAUT 2009) or a junior synonym (HAUSMANN, MIRONOV & VIIDALEPP 2011). We consider *valesiaria*, whose wings are almost uniform grey-brown, to fall within the intraspecific variation and do not recognise it valid at subspecies level.

Siona DUPONCHEL, 1829

Siona DUPONCHEL, 1829, Hist. nat. Lépid. Papillons Fr. 7 (2): 112. Type species: *Phalaena dealbata* LINNAEUS, 1767, which is a junior synonym of *Phalaena lineata* SCOPOLI, 1763, [north-west Slovenja]: Carniolia.) There are no genus synonyms.

UNAVAILABLE NAMES (incorrect subsequent spelling): *Scoria* STEPHENS, 1831 (see FLETCHER 1979).

DIVERSITY AND DISTRIBUTION: *Siona* contains two species, both occur in the Palaearctic region.

EXTERNAL CHARACTERS AND ABDOMEN: Wingspan 35–46 mm. Wings white to whitish yellow, veins dark brown, stronger on underside. Hindwing margin slightly concave below apex. Abdomen with longitudinal dark brown stripes. Male and female antennae filiform. Male hindtibia not enlarged. Male 8th sternite unmodified.

MALE GENITALIA, description is based on *S. lineata*: Uncus weakly setose, with acute subapex dorsally, and with narrow, granulate ventral apex. Socii and gnathos absent. Valva upturned, densely setose, costa base with long projection. Valva fused dorsally. Juxta large plate. Saccus wide, round. Vesica with row of small cornuti.

FEMALE GENITALIA, description is based on *S. lineata*: Apophyses posteriores and apophyses anteriores long, straight. Lamella antevaginalis and lamella postvaginalis fused, forming narrow ring around ostium bursae. Posterior part of corpus bursae with sclerotised striae, with large diverticulum. Signum elongated, margins dentate.

BIOLOGY: Larva of *S. lineata* is polyphagous, it has been recorded feeding for instance on *Galium, Hypericum, Achillea millefolium, Solidago virgaurea, Salix aurita, Dactylis glomerata* (MIKKOLA *et al.* 1989), *Plantago* sp., *Lembotropis nigricans, Galatella linosyris* and *Artemisia campestris* (EBERT 2003). Adults fly in daytime if disturbed, they also come readily to light. Adults often rest with wings folded together like butterflies. About half grown larvae overwinter.

IMMATURE STAGES: Eggs of *S. lineata* are quadrangular, yellow, laid in rows (EBERT 2003, LEPIFORUM 2013). Larvae is ochreous, with darker dorsal line and black speckling dorsally (PORTER 2010). Pupa is enclosed in an elongated, pale cocoon, resembling in appearance *Zygaena* (Zygaenidae) cocoons. Pupa cremaster is with an elongated base, apex with two straight, stout setae (PATOČKA & TURČÁNI 2005).

REMARKS: *Siona* has recently been classified in the Gnophini (VIVES MORENO 1994, VIIDALEPP 1996, LERAUT 1997, HAUSMANN, MIRONOV & VIIDALEPP 2011). Male and female genitalia support the Gnophini association as defined in VIIDALEPP *et al.* (2007). Systematic position: Gnophini.

117. Siona lineata (SCOPOLI, 1763)

lineata SCOPOLI, 1763, Ent. Carniolica: 218, fig. 540, (*Phalaena*). Syntype(s), [north-west Slovenja:] Carniolia.

dealbata LINNAEUS, 1767, Syst. Nat. (Ed. 12) 1 (2): 870, (*Phalaena* (*Geometra*)). Syntypes, including putative lectotype male (Linnean Society of London, London, UK), Germany (examined externally).

oenotriensis STAUDER, 1915, Z. wiss. InsektBiol. 11 (1/2): 6, (*Scoria*). Syntypes 28 males, 16 females (The Natural History Museum, London, UK), [Italy]: near San Fili, 900 m (examined externally). Originally as subspecies. See remarks.

UNAVAILABLE NAMES (incorrect subsequent authorship): *oenotriensis* STAUDINGER (SCOBLE 1999, HAUSMANN, MIRONOV & VIIDALEPP 2011). – Intrasubspecific names, potentially unavailable: *obsoleta* STEPHENS, 1925 (ab.); *panfasciata* HOEGH-GULDBERG, 1958 (f.).

EXTERNAL CHARACTERS AND ABDOMEN: Wingspan 35–40 mm. Wings white. Dark brown veins from below visible on upperside also. Terminal line fine, continuous, dark brown. Fringes white, weakly chequered dark brown and white below forewing apex. Hindwing margin slightly concave below apex. Wings below slightly suffused with dark brown, veins distinctly dark brown. Postmedial line and discal spots suffused on forewings below. Apex of labial palpi dark brown. Frons, collar and thorax concolorous with wings. Abdomen long, narrow, white, with longitudinal dark brown stripes. Male and female antennae filiform. Male hindtibia not enlarged. Hindtibia of both sexes with 2+2 spurs. Tympanal organs medium-sized. Abdominal segments 3–5 elongated. Abdominal sternites and tergites 3–8 of both sexes undifferentiated, except posterior margin of male 8th sternite weakly rounded.

VARIATION: Ground colour varies from white to cream-yellow. Forewing costa can be heavily suffused with dark brown.

MALE GENITALIA: Uncus absent. Socii potentially represented by minute setae. Tuba analis weakly sclerotised. Gnathos arms narrow, upturned, apex upturned, granulate. Valva upturned, ventral margin and apex densely setose. Costa base with long projection, apex with long spine. Valva fused ventrally via sclerotised plate. Juxta large plate. Saccus wide, round. Aedeagus curved, medial part with two dentate rows, caecum long, apex round. Vesica opens at approximately 90 degree angle. Vesica with basal diverticulum, followed by row of small cornuti.

FEMALE GENITALIA: Papillae anales setose. Apophyses posteriores and apophyses anteriores long, straight. Lamella antevaginalis and lamella postvaginalis fused, forming narrow ring around ostium bursae. Ductus bursae short, sclerotised laterally. Posterior part of corpus bursae with sclerotised striae, with large lateral diverticulum. Anterior part of corpus bursae elongated, membranous. Signum elongated, margins dentate.

DISTRIBUTION AND ABUNDANCE: Eurasiatic. In Europe from central Scandinavia to central Spain, southern France including Corsica, Italy including Sicily and the Balkans to northern Greece and from southern England through central and eastern Europe to European Russia and the Ural mountains. – Outside Europe the distribution continues east- and south-eastwards to Turkey, the Caucasus region, central Asia, southern Siberia, Mongolia and the Far East of Russia. – A common species in Europe that usually is seen in limited numbers.

PHENOLOGY: Univoltine. In northern Europe from late May to early July, in southern central Europe from mid-May to early or mid-July (EBERT 2003) and in Italy from mid-May to mid-July (FLAMIGNI *et al.*, in prep.). Larva from July or August to May, hibernation takes place as a small larva. The moths come to light, but most specimens are seen by day when they fly actively or are very easily disturbed. The moths are often found by day, resting in long grass.

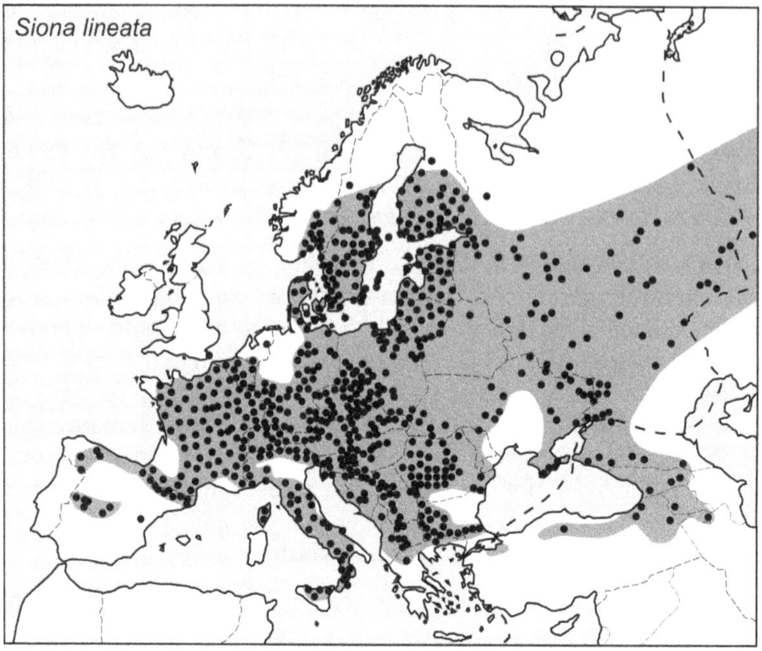

BIOLOGY: Polyphagous. The caterpillar has been found on *Plantago*, *Lembotropis nigricans* and *Artemisia campestris* (EBERT 2003), *Thymus serpyllum* (BERGMANN 1955) and on *Hypericum perforatum* (H. WEGNER pers. comm.). It has been recorded on *Rumex*, *Vicia*, *Trifolium*, *Hypericum*, *Pimpinella*, *Peucedanum*, *Galium*, *Thymus*, *Origanum*, *Lamium*, *Knautia arvensis*, *Campanula*, *Solidago*, *Aster*, *Achillea*, *Tanacetum vulgare*, *Centaurea*, *Taraxacum* and *Dactylis* (FLAMIGNI *et al.*, in prep.). Pupa enclosed in a spun whitish grey cocoon on the host plant, resembling a *Zygaena* (Zygaenidae) cocoon.

HABITAT: Open and sunny areas with low vegetation such as meadows and scrubland, in northern Europe often, but not always, on slightly dry soil. In southern Europe mainly in moist places. The species in northern Europe is restricted to lowlands, and occurs up to 1000 m in south-western Germany (EBERT 2003), and up to 2000 m in the Alps (VORBRODT 1914). In central Asia up to at least 2100 m altitude.

SIMILAR SPECIES: No similar species in Europe. *S. galactica* PROUT, 1929, which is known from Algeria, is very similar. According to the original description: 'Forewing with costa appreciably less arched near base than in *lineata*, very pale cream-yellow, entirely without markings, only the strongest veins faintly browner. Hindwing white, unmarked' (PROUT 1929). We have not examined original material.

GENETIC DATA: Genetically homogeneous in Europe (n=18, from seven countries), but populations from Sicily (1.9%), Armenia and central Asia slightly diverging. Nearest species: Central American *Neoselenia banasa* (DRUCE, 1892) (5.8%) and East Asian *Ctenognophos grandinaria* (MOTSCHULSKY, 1861) (6.5%).

REMARKS: Taxon *oenotriensis* (STAUDER) from San Fili, Calabria, Italy, was originally described as subspecies and was considered as such by SCOBLE (1999). Specimens are larger, wings cream-yellow to pale brown, with blackish brown costa, and forewing postmedial line on the underside is darker than in typical specimens. WEHRLI (1939–1954) considered the specimens as individual aberrations and noted that the above-mentioned, diagnostic features are not distinct. Further research is needed.

Chariaspilates WEHRLI, 1953

Chariaspilates WERHLI, 1953, in SEITZ, Gross-Schmett. Erde 4 (Suppl.): 674. Type species: *Aspilates formosaria* EVERSMANN, 1837, [Russia:]: Kasanischen [Kazan]. There are no genus synonyms.

DIVERSITY AND DISTRIBUTION: *Chariaspilates* contains two species in the Palaearctic region.

EXTERNAL CHARACTERS AND ABDOMEN, description is based on European species: Wingspan 32–36 mm. Wings golden brown-yellow, irrorated with brown. Postmedial line oblique, inner margin white. Forewing margin rounded on tornus. Wings below with dark terminal area. Male antennae bipectinate, female antennae fasciculate. Male hindtibia unmodified. Abdominal segments 3–8 of both sexes undifferentiated.

MALE GENITALIA, description is based on European species: Uncus evenly tapering. Socii absent. Gnathos arms fused, with large medial plate. Valva deeply divided, with a large membrane between the arms. Dorsal arm of valva with projections on costa and on ventral margin, apex with thick setae. Ventral arm of valva with basal projection, apex margin spinose. Vesica with patch of microcornuti.

FEMALE GENITALIA, description is based on European species: Papillae anales striated. Apophyses anteriores about 1/2 length of apophyses posteriores. Lamella antevaginalis large, lunular, sclerotised. Ductus bursae very short. Anterior part of corpus bursae slightly triangular. Signum elongated, with sclerotised margins.

BIOLOGY: Larva of *C. formosaria* has been recorded feeding on *Caltha palustris, Myrica gale* and *Lysimachia vulgaris* (ROBINEAU 2007, SZÉKELY 2011). Adults inhabit marshes, and fly in June and July, with partial second generation in August and September. Larva overwinters.

IMMATURE STAGES: Larva of *C. formosaria* is smooth, grey-brown, with dark brown stripe laterally (Lepiforum 2013). Pupa cremaster is stout, with four pairs of short, recurved setae (PATOČKA & TURČANI 2005).

REMARKS: *Chariaspilates* has recently been classified in the Gnophini (VIIDALEPP 1996) and Aspitatini (LERAUT 1997, HAUSMANN, MIRONOV & VIIDALEPP 2011). Based on the morphology of examined European taxa it seems apparent to us that the Angeronini, which also includes the Aspitatini DUPONCHEL, 1845 (see MCGUFFIN 1981, HOLLOWAY 1994) could potentially be brought into synonymy with the Gnophini. YAMAMOTO & SOTA (2007), whose analysis was based on DNA, recovered *Angerona* and *Chariaspilates* as sister taxa. *Angerona prunaria, Hypoxystis pluviaria* and *C. formosaria* share the quadrangular/elongated female signum with sclerotised margins, also the male valvae show some similarities. Systematic position: Gnophini.

118. Chariaspilates formosaria (EVERSMANN, 1837)

formosaria formosaria EVERSMANN, 1837, Bull. Soc. imp. Nat. Moscou 1837 (6): 54, (*Aspilates*). Syntype(s), [Russia:]: Kasanischen [Kazan].

andriana DANNEHL, 1921, Ent. Z., Frankf. a.M. 41 (20): 411, (*Aspilates*). Syntypes 4 males, 4 females, [Austria]: South Tirol, Terlan-Andrian. [Originally as sp., treated valid at subspecies level by SCOBLE (1999)]. We agree with WEHRLI (1939–1954) and consider *andriana* DANNEHL a form of *formosaria* EVERSMANN, quoting absence of constant differential features and falling within intraspecific variation.

formosaria kurilaria BRYK, 1942, Dt. ent. Z. Iris 56: 87, (*Aspilates*). Holotype male (Naturhistoriska Riksmuseet, Stockholm, Sweden), Kurile Islands: Kunashiri.

formosaria pannonicus VOJNITS, 1977, Folia ent. hung. 30 (1): 167, figs 2, 4, (*Chariaspilates*). Holotype male (Hungarian Natural History Museum, Budapest, Hungary), Hungary (east): Bátorliget (examined, including genitalia). Originally as subspecies, treated valid at subspecies rank by SCOBLE (1999), downgraded here from subspecies rank (new synonym), quoting absence of disjunct differential features, falling within intraspecific variation. See remarks.

gloriosaria BOISDUVAL, 1840, Genera Index meth. eur. Lepid.: 188, (*Aspilates*). Syntype(s), [France?]: Palud; Somonae.

niponaria FELDER & ROGENHOFER, 1875, Reise öst. Fregatte Novara (Zool.) 2 (Abt. 2): pl. 123, fig. 29, (*Aspilates*). Syntype male (The Natural History Museum, London, UK), Japan (examined externally).

INTRASUBSPECIFIC NAMES, POTENTIALLY UNAVAILABLE: *autumnalis* PFAU, 1925 (summer generation); *brunneata* KARDAKOFF, 1928 (ab.).

EXTERNAL CHARACTERS AND ABDOMEN: Wingspan 32–40 mm (first generation), 22–26 mm (second generation). Robust species. Wings golden brown-yellow, irrorated with brown. Postmedial line oblique, brown, inner margin white. Costa and terminal area whitish. Veins on terminal area brown. Forewing margin rounded on tornus. Terminal line narrow, brown. Fringes concolorous with wings. Hindwings whitish, postmedial line narrow, brown. Terminal area irrorated with brown. Discal spots small, brown. Wings below whitish, irrorated with brown, particularly on terminal area. Discal spots more pronounced than above. Frons brown. Collar, thorax and abdomen concolorous with wings. Male antennae bipectinate, female antennae fasciculate. Hindtibia of both sexes with 2+2 spurs. Tympanal organs medium-sized. Abdominal segments 3–8 of both sexes undifferentiated.

VARIATION: Ground colour varies from golden brown-yellow to pale brown-yellow. Forewing postmedial line can be almost straight or curved inwards on inner margin. Hindwing terminal area varies from whitish to heavily irrorated with brown.

MALE GENITALIA: Uncus evenly tapering, weakly setose. Socii absent. Gnathos arms fused ventrally, with large, granulate medial plate. Valva deeply divided, with large, weakly setose membrane between arms. Dorsal arm of valva narrow, apex with thick setae. Its costa with triangular extension, apex with few stout spines. Its ventral margin with curved extension, apex with row of stout spines. Ventral arm of valva narrow, apex margin with numerous stout spines. Base with long, acute, sclerotised projection. Juxta small. Saccus wide, round. Aedeagus stout, subapical part with row of small teeth, apex acute. Vesica opens at approximately 90 degree angle. Vesica base wide, with patch of microcornuti.

FEMALE GENITALIA: Papillae anales soft, setose, surface striated. Apophyses anteriores about 1/2 length of apophyses posteriores. Lamella antevaginalis large, lunular, sclerotised, adjoined with wrinkled membrane. Lamella postvaginalis small, sclerotised plate, adjoined with horizontal, membranous striations. Ductus bursae very short. Posterior part of corpus bursae narrow, membranous. Anterior part of corpus bursae wide, slightly triangular. Signum elongated, with sclerotised margins.

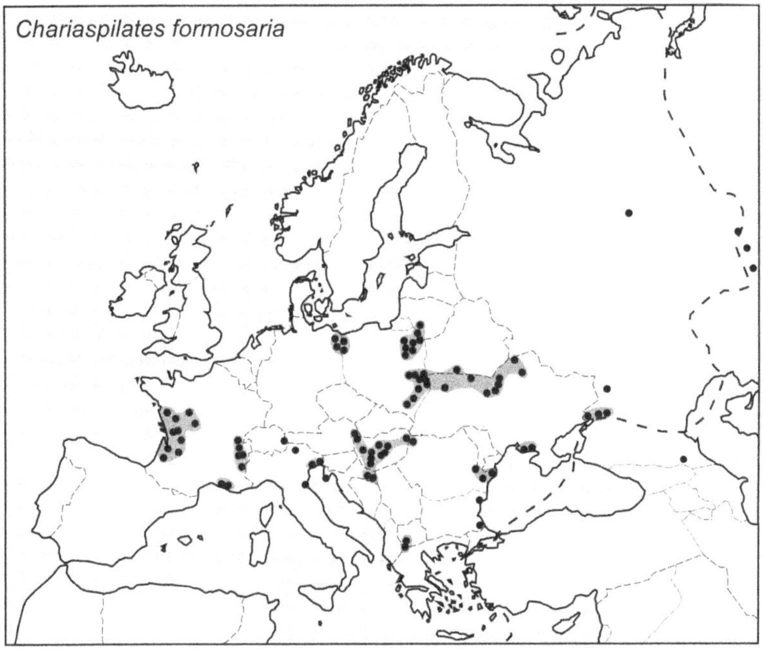

DISTRIBUTION AND ABUNDANCE: Eurasiatic. In Europe disjunct and scattered from France to central and south-eastern Europe including European Russia to the Ural mountains. – Outside Europe the species is found in Caucasus and from Kazakhstan through southern Siberia, Mongolia, China and Korea to Far East Russia, the Kuriles and Japan. – In Europe rather local, but sometimes the species occurs in numbers at its localities. It has declined, probably due to habitat destruction.

PHENOLOGY: Uni- or bivoltine. In general from late May to mid-July. In Poland from mid-May to early July (MALKIEWICZ 2012), but also recorded in mid-July (H. RIETZ pers. comm). In Pommern, (north-eastern Germany and north-western Poland), URBAHN & URBAHN (1939) recorded the species from late June until late July. In Romania bivoltine from early June to mid-July and from mid-August to early September (L. SZÉKELY pers. comm.). In southern France in two generations from late May to late June and from mid-July to mid-August while in the remaining part of the species distribution area in France in one generation in July (C. TAUTEL pers. comm.). Larva from July or August, overwintering to April or May. Especially the males are attracted to light, the females in less degree, but these can be found when flying at dusk.

BIOLOGY: Oligophagous. The larva is recorded on *Lysimachia vulgaris*, *Myrica gale* and *Caltha palustris* (MALKIEWICZ 2012). In rearing on *Lysimachia vulgaris*, the caterpillars preferred to feed on the flowers, but could also eat the leaves. After hibernation rearing continued on *Rumex* and *Doronicum* (B. MÜLLER pers. comm.).

HABITAT: Hygrophilous. Moist meadows, peat bogs and other types of wetlands. Usually restricted to lowlands, in the southern Balkans up to 900 m. In Far East Russia up to 800 m.

SIMILAR SPECIES: No similar species in Europe.

GENETIC DATA: Genetically identical in France and Liechtenstein (n=5), populations from eastern Russia only diverging by 0.5%. Nearest species: *Synopsia sociaria* (HÜBNER, 1799) (7.6%) and *Dyscia fagaria* (THUNBERG, 1784) (8.4%).

REMARKS: Taxon *pannonicus* VOJNITS, 1977, was described on the basis of small quantitative differences on wings (yellow colour of forewings more intense and brown irroration finer) and on genitalia (male capsule larger, more robust, valva spines thicker, shorter and more numerous, and female antrum more u-shaped) (VOJNITS 1977). The author admitted that there is variation within the Hungarian populations in these features, but due to the isolated occurrence of Hungarian populations a subspecies was described. Based on the material we have examined, the mentioned characters fall within the intraspecific variation. Although the general distribution pattern of *formosaria* is fragmented, the current knowledge shows that Hungarian populations are surrounded by adjacent populations in Romania, Austria and Croatia. Finally, because the species is likely to be genetically rather homogeneous (see genetic data), we do not recognise *pannonicus* valid at subspecies level.

Aspitates TREITSCHKE, 1825

Aspitates TREITSCHKE, 1825, Schmett. Eur. 5 (2): 431. Type species: *Geometra gilvaria* [DENIS & SCHIFFERMÜLLER], 1775, (Austria): Vienna district. Genus synonyms are after SCOBLE (1999).

Apoaspilates WEHRLI, 1954, in SEITZ, Gross-schmett. Erde 4 (Suppl.): 718. Type species: *Aspilates tristrigaria* BREMER & GREY, 1853.

Aspilata UNGER, 1856, Arch. ver. Freunde. Naturg.-Mecklenb. 10: 61. [Emendation of *Aspitates* TREITSCHKE].

Aspilates TREITSCHKE, 1827, in OCHSENHEIMER, Schmett. Eur. 6: 126. [Emendation of *Aspitates* TREITSCHKE].

Conchia HÜBNER, [1825] 1816, Verz. bekannter Schmett. 337. Type species: *Phalaena mundataria* STOLL, 1782. [Junior homonym of *Conchia* HÜBNER, 1821 (Lepidoptera: Pyralidae)]. See *Megaspilates*.

Erschoffia SWINHOE, 1900, Cat. east. and Aust. Lepid. Heterocera Colln Oxf. Univ. Mus. 2: 221. Type species: *Ennomos acuminaria* EVERSMANN, 1851. [Unnecessary replcament name for *Eversmannia* ERSCHOFF, 1871 [post February]].

Eversmannia ERSCHOFF, 1871 [post February], Bull. Soc. imp. Nat. Moscou 43 (3/4): 316. Type species: *Ennomos acuminaria* EVERSMANN, 1851. [Junior homonym of *Eversmannia* STAUDINGER, 1871 January 31 (Lepidoptera: Epilemidae)].

Megaspilates WARREN, 1894, Novit. zool. 1: 437 (replacement name for *Conchia* HÜBNER, [1825] 1816), considered a junior subjective synonym of *Aspitates* TREITSCHKE, 1825, quoting the structural similarities (new synonym). Type species: *Phalaena mundataria* STOLL, 1782. See remarks.

Napuca WALKER, [1863] 1862, List Specimens lepid. Insects Colln Br. Mus. 26: 1693. Type species: *Napuca orciferaria* WALKER, [1863] 1862.

Nyctiphanta HULST, 1896, Trans. Am. ent. Soc. 23: 320, 336. Type species: *Nyctiphanta laetula* HULST, 1896. See FERGUSON (1983) for details.

Parallela SODOFFSKY, 1837, Bull. Soc. imp. Nat. Moscou 1837 (6): 90. [Unnecessary replacement name for *Aspilates* TREITSCHKE].

Pseudosiona BUTLER, 1893, Ann. Mag. nat. Hist. (6) 12: 18. Type species: *Pseudosiona taylori* BUTLER, 1893.

Semiaspilates WEHRLI, 1953, in SEITZ, Gross-Schmett. Erde 4 (Suppl.): 678. Type species: *Aspilates conspersaria* WEHRLI, 1953.

Sinope GUMPPENBERG, 1887, Nova Acta Acad. Caesar. Leop. Carol. 49: 332 (key). Type species: *Phalaena ochrearia* ROSSI, 1794.

Toxogrammia GUMPPENBERG, 1887, Nova Acta Acad. Caesar. Leop. Carol. 49: 332 (key). Type species: *Aspilates curvaria* EVERSMANN, 1852.

UNAVAILABLE NAMES (incorrect subsequent spelling): *Aspilades* HEYDENREICH, 1846 (see FLETCHER 1979).

DIVERSITY AND DISTRIBUTION: *Aspitates* contains about 20 species, mostly in the Palaearctic and in the Nearctic regions, several of those have a Holarctic distribution, and a few species in Africa and one in the Neotropics. TREITSCHKE (1825: 431) named the genus originally as *Aspitates*, but emended it later (TREITSCHKE 1827: 126) to *Aspilates*. Both names have been used in the recent literature: in North American literature (for instance MCGUFFIN 1981) and in older Palaearctic literature (for instance WEHRLI 1939–1954) *Aspilates* appears to have been in more frequent use (but *Aspitates* in HODGES *et al.* 1983), whereas in the more recent Palaearctic literature *Aspitates* may have been more frequently used (for instance VIVES MORENO 1994, VIIDALEPP 1996, LERAUT 1997, 2009, KULLBERG *et al.* 2001, HAUSMANN, MIRONOV & VIIDALEPP 2011).

EXTERNAL CHARACTERS AND ABDOMEN: Wingspan about 30–40 mm, females often smaller than males. Wings various shades of yellow and white, often irrorated with brown. Forewing postmedial line oblique, often reddish brown. Hindwing below often with distinct, almost straight postmedial line that terminates near apex. Male antennae bipectinate, female antennae fasciculate or shortly bipectinate. Male foreleg tibia often with weakly developed epiphysis. Male 8th sternite unmodified.

MALE GENITALIA: Uncus weakly sclerotised, sparsely setose. Socii absent. Gnathos arms fused ventrally, with large medial plate. Valva wide, apex round, setose. Costa base with projection, apex with spines. Medial area often with roundish lobe. Valva often fused dorsally. Juxta often concave plate. Saccus wide, round. Aedeagus with apical projection(s). Vesica mostly without sclerotisations.

FEMALE GENITALIA: Apophyses anteriores about 1/2 length of apophyses posteriores. Lamella antevaginalis often sclerotised, horseshoe-shaped. Lamella postvaginalis often narrow, horizontal sclerotisation. Corpus bursae small, anterior end slightly wider. Signum absent.

BIOLOGY: Larva of *A. gilvaria* has been recorded feeding on *Andromeda polifolia, Rubus chamaemorus, Vaccinium uliginosum* (SEPPÄNEN 1970, MIKKOLA *et al.* 1989), *Lotus corniculatus, Genista sagittalis, Lonicera xylosteum* (EBERT 2003) and on *Thymus, Potentilla, Biscutella* and *Pastinaca sativa* (REDONDO *et al.* 2009). Northern species inhabit marshes and wetlands, southern species show preference towards open and arid terrains, up to 2000 metres. *A. gilvaria* overwinters as small larva in northern Europe (MIKKOLA *et al.* 1989).

IMMATURE STAGES: Larva of *A. gilvaria* is ochreous brown, with a few tiny black spots, with darker lateral lines. Posterior end is with two pointed extensions. Larva rests often in peculiar O-shaped, looped position (PORTER 2010, LEPIFORUM 2013). Pupa cremaster base is elongated, apex is with four pairs of curved setae, apical pair is longest (PATOČKA & TURČANI 2005).

REMARKS: *Aspitates* has recently been classified in the Aspitatini (LERAUT 1997, HAUSMANN, MIRONOV & VIIDALEPP 2011), in the Gnophini (VIVES MORENO 1994, VIIDALEPP 1996) and Angeronini (MCGUFFIN 1981, HODGES *et al.* 1983). Male and female genitalia support the Gnophini association as defined in VIIDALEPP *et al.* (2007). The potential Gnophini association was noted late in the production process of this book, so for practical reasons it is dealt with here, altough we note that its current position in the sequence of species does not reflect its probable phylogenetic position. See Gnophini under *Classification of European Ennominae genera into tribes* for further details. Systematic position: Gnophini (but systematic position and extent of the Gnophini needs further research).

Examination of the morphology, including the genitalia, suggests that *Megaspilates* WARREN, 1894 is a junior synonym of *Aspitates* TREITSCHKE, 1825. *Megaspilates* has been considered a monotypic genus because the female antennae are shortly, but distinctly bipectinate, whereas in *Aspitates* the female antennae are fasciculate or indistinctly bipectinate. The boundary between these two states is quantitative. PROUT (1912–1916) and WEHRLI (1939–1954) considered *Megaspilates* (as *Conchia*) a subgenus of *Aspilates*. VIIDALEPP (1996) classified several taxa to *Napuca* WALKER 1863 (=*Semiaspilates* WEHRLI, 1953), whereas HAUSMANN *et al.* (2011) considered it a subgenus of *Aspitates*. We agree with SCOBLE (1999) and consider both taxa as junior synonyms of *Aspitates* TREITSCHKE, 1825.

119. Aspitates mundataria (STOLL, 1782), new combination

mundataria mundataria STOLL, 1782, in CRAMER, Uitlandsche Kapellen (Papillons exot.) 4: 243, 250 (index), pl. 400, fig. H. (*Phal*[*aena*] *Geomet*[*ra*]. Syntype(s), [Russia]: Sarepta [Volgograd].

nitidaria FABRICIUS, 1794, Ent. Syst. 3 (2): 158, (*Phalaena*). Syntype(s), [southern Russia].

mundataria tonghata FELDER & ROGENHOFER, 1875, Reise öst. Fregatte Novara (Zool.). 2 (Abt. 2): pl. 129, fig. 12, (*Aspilates*). Syntype male (The Natural History Museum, London, UK), Japan (examined externally). Originally as sp., valid at subspecies rank according to SCOBLE (1999).

mundataria uncinataria VOJNITS, 1975, Annales hist.-nat. Mus. natn. hung. 67: 199, figs 10–11, (*Conchia*). Holotype male (Hungarian Natural History Museum, Budapest, Hungary), Mongolia: Bulgan Aimak, 9 km E of Somon Abzaga, 1300 m. Valid at subspecies rank according to SCOBLE (1999).

INTRASUBSPECIFIC NAMES, POTENTIALLY UNAVAILABLE: *tangens* WARNECKE, 1937 (ab.).

EXTERNAL CHARACTERS AND ABDOMEN: Wingspan 29–37 mm, robust species. Wings white. Forewing costa, medial line, postmedial line and terminal line golden brown. Medial line angled weakly inwards before inner margin. Hindwings white, with narrow, golden brown terminal line. Discal spots absent. Wings below white, weakly irrorated with golden brown. Wings below with forewing postmedial line near apex, hindwing postmedial line and terminal line distinct, goldenbrown. Wings below with elongated hindwing discal spot. Frons, collar and thorax white. Abdomen mixed with white and golden brown. Male and female antennae bipectinate, male pectinations longer. Hindtibia of both sexes with 2+2 spurs. Tympanal organs medium-sized. Abdominal segments 3–8 of both sexes undifferentiated.

VARIATION: Very little. Amount of golden brown irroration somewhat variable on wings below. The non-European subspecies *tonghata* (FELDER & ROGENHOFER) is larger, wingspan up to 46 mm, the golden brown markings are wider and darker. The non-European subspecies *uncinataria* (VOJNITS) is without any qualitative differences. According to the original description (VOJNITS 1975), the apex of male valva is potentially more roundish (blunt in nominal subspecies). Taxonomic status of both taxa at subspecies level needs to be confirmed. VIIDALEPP (1996) did not recognise *uncinataria* as valid at subspecies level.

MALE GENITALIA: Uncus small, sparsely setose, only apex sclerotised, triangular. Socii absent. Gnathos arms narrow, fused ventrally, apex granulate. Valva wide, apex blunt, setose. Costa base with long projection, apex with one long spine. Medial area with large, roundish lobe. Valva fused dorsally. Juxta V-shaped. Saccus wide. Aedeagus medially constricted, apex with long, curved, dentate projection. Vesica opens at approximately 90 degree angle. Vesica weakly curved, base with dentate plate.

FEMALE GENITALIA: Papillae anales large, elongated, setose. Apophyses anteriores slightly shorter than apophyses posteriores. Lamella antevaginalis/sterigma/ductus bursae heavily sclerotised, wide, with coiled diverticulum. Lamella postvaginalis narrow, lunular, sclerotised. Corpus bursae narrow, membranous. Signum absent.

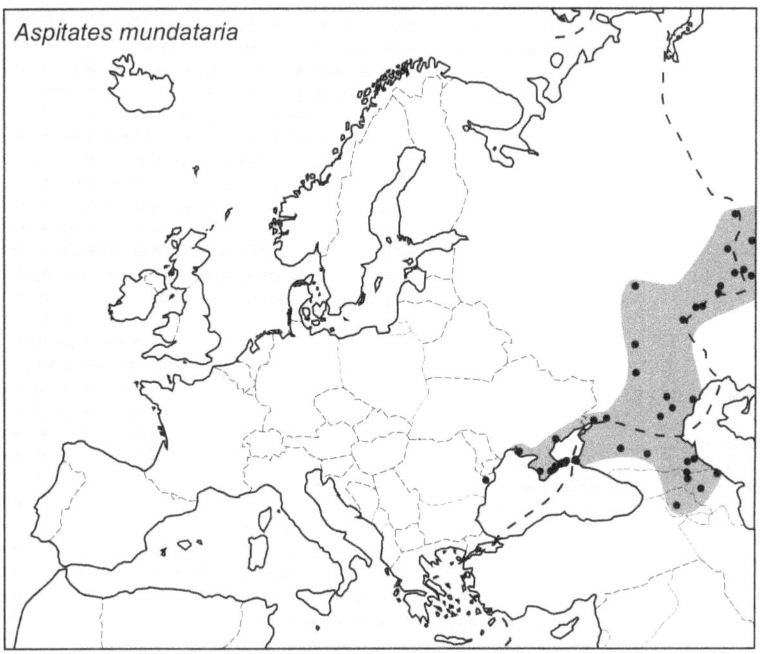

DISTRIBUTION AND ABUNDANCE: Eurasiatic. In Europe only in the south-east in Romania, Ukraine, and southern Russia. – Outside Europe the species occurs in the Caucasus region and it is distributed further eastwards from southern Urals through central Asia, southern

Siberia, Mongolia, northern China and Korea to Japan. – In Europe a very locally distributed and not common species that is rarely found in numbers (L. SZÉKELY pers. comm.).

PHENOLOGY: Univoltine. From late May to July. In Romania from mid-June to late July (L. SZÉKELY pers. comm.). In southern Urals from early June to early July and most frequent in late June (K. NUPPONEN pers. comm.). The moths are active at night and are attracted to light (K. NUPPONEN pers. comm.).

BIOLOGY: Oligophagous. Larvae have been found on *Artemisia austriaca*, *A. fragrans* and *A. nutans* (BECKER 1858).

HABITAT: Xerophilous species of open habitats. In Romania on salty steppe (L. SZÉKELY pers. comm.). On steppe, up to 400 m in southern Urals (K. NUPPONEN pers. comm.). Often on dry steppe of the *Artemisia* type (J. KULLBERG pers. comm.). In central Asia up to 2000 m.

SIMILAR SPECIES: No similar species in Europe.

GENETIC DATA: Genetically heterogeneous (n=7), populations from western Kazakhstan diverging from those of Mongolia and eastern China by 3.3%. Nearest species: *Gnophos burmesteri* (GRAESER, 1889) from Far East Russia (7.0%) and *Aspitates gilvaria* (DENIS & SCHIFFERMÜLLER) (7.7%).

120. Aspitates gilvaria (DENIS & SCHIFFERMÜLLER, 1775)

gilvaria gilvaria [DENIS & SCHIFFERMÜLLER], 1775, Ankündung syst. Werkes Schmett. Wienergegend: 102, (*Geometra*). Type(s), lost, [Austria]: Vienna district.

gilvaria fenica FUCHS, 1899, Jahrbücher des Nassauischen Vereins für Naturkunde 52: 154, (*Aspilates*). Syntype(s), Finland: Helsinki. [Originally as var].

gilvaria burrenensis COCKAYNE, 1951, Entomologists's Gaz. 2: 100, (*Aspitates*). Holotype male, Ireland (west): County Clare, The Burren. Paratype male, Ireland (west): County Clare, Burren (The Natural History Museum, London, UK, in the COCKAYNE collection) (examined externally).

gilvaria kukunorensis WEHRLI, 1953, in SEITZ, Gross-Schmett. Erde 4 (Suppl.): 677, pl. 53: d, (*Aspilates*). Holotype male, China: Szetschwan [Sichuan], Sungpanting, paratype male (The Natural History Museum, London, UK) (examined externally).

gilvaria minimus VOJNITS, 1975, Annales hist.-nat. Mus. natn. hung. 67: 201, figured, (*Aspilates*). Holotype male, Mongolia: Central Aimak, Ulaan Chodag, 16 km S of Somon Öndörschireet, 1500 m. Valid at subspecies rank according to SCOBLE (1999).

gilvaria orientaria ALPHÉRAKY, 1883, Horae Soc. ent. ross. 17 (3): 184, (*Aspilates*). Syntypes, [Central Asia]: Bermamoude, Caucasus; Tian-Chian; Charkhode; Saisan-noor. Originally as var., valid at subspecies rank according to SCOBLE (1999).

INTRASUBSPECIFIC NAMES, POTENTIALLY UNAVAILABLE: *crassata* LERAUT, 2009 (f.); *fuscedinaria* FUCHS, 1899 (ab.); *marginata* LERAUT, 2009 (f.); *nigricans* JOURDHOUILLE, 1883 (ab.); *sublataria* FUCHS, 1900 (ab.); *suffusa* COCKAYNE, 1953 (ab.); *variegata* LERAUT, 2009 (f.); *vittata* COCKAYNE, 1953 (ab.).

EXTERNAL CHARACTERS AND ABDOMEN: Wingspan 28–33 mm, males often slightly larger, forewing more roundish. Wings whitish yellow to yellow, irrorated with brown. Forewing postmedial line oblique, brown, wide, reaching costa just before apex. Terminal line narrow, brown. Fringes concolorous with wings. Forewing discal spot small, elongated. Hindwings whitish, postmedial line and discal spot weak. Terminal area occasionally darker. Wings below irrorated with brown, especially on forewing costa and terminal area. Postmedial line wide, suffused. Discal spots large, suffused. Frons, collar, thorax and abdomen concolorous with wings. Male antennae bipectinate, female antennae fasciculate/shortly bipectinate. Hindtibia of both sexes with 2+2 spurs. Tympanal organs small. Posterior margin of male 8th sternite narrower. Other abdominal segments of both sexes undifferentiated.

VARIATION: Forewing ground colour varies from very pale yellow, with brown-purple terminal area (North Europe) to whitish yellow to yellow, with concolorous terminal area (central and southern Europe). Hindwing terminal area whitish or weakly purple. Wings below normally irrorated with brown to various degrees, but wings can be almost uniform brown (northern Europe). Wings can be almost without markings.

Subsp. ***burrenensis*** is darker and hindwing postmedial line is more oblique, ending almost on apex. Taxon is known from the Burren, County Clare, Ireland (SKINNER & WILSON 2009).

MALE GENITALIA: Uncus small, sparsely setose, only apex sclerotised, elongated. Socii absent. Gnathos arms large, fused ventrally, triangular medial plate granulate. Valva wide, apex roundish, setose. Costa base with long projection, apex with two, rarely with three long spines. Ventral margin of valva with two round lobes. Valva fused dorsally. Juxta large, weakly V-shaped. Saccus wide, slightly elongated. Aedeagus rather straight, apex with several sclerotised teeth. Vesica opens at approximately 135 degree angle. Vesica base with two small, lateral diverticula. Vesica without sclerotisations.

FEMALE GENITALIA: Papillae anales large, setose. Apophyses anteriores slightly shorter than apophyses posteriores. Lamella antevaginalis/sterigma/ductus bursae sclerotised, wide. Lamella postvaginalis narrow, straight, sclerotised, dorsal margin with two small lobes. Posterior part of corpus bursae narrow, membranous. Anterior part of corpus bursae wider, membranous. Signum absent.

DISTRIBUTION AND ABUNDANCE: Eurasiatic. In Europe from southern Finland to southern Spain, France, parts of Italy and the Balkans and from south-eastern England through central Europe to European Russia and the Ural mountains. Subsp. *burrenensis* rather isolated in western Ireland – Outside Europe the species occurs from Turkey, Iran, the Caucasus region and the Urals through central Asia and southern Siberia to Mongolia and northern China. – In Europe a rather common species that now and then is found in numbers.

PHENOLOGY: Univoltine. In northern Europe from late June to early August, in southern central Europe from early August to early September (EBERT 2003) and in southern Europe the flight period is extended from late June to early September. In Finland larva from August and overwintering to June (SEPPÄNEN 1970), in Romania from October to June (SZÉKELY 2011). Hibernation takes place as a small larva. The adult moths come to light but are also readily disturbed during the day.

BIOLOGY: Polyphagous. The larva has been found in Finland on *Andromeda polifolia* and it has been reared on *Empetrum nigrum*, *Vaccinium uliginosum* and *Rubus chamaemorus* (SEPPÄNEN 1970). According to EBERT (2003), the caterpillar has been found on *Lotus corniculatus*, *Genista sagittalis* and *Lonicera xylosteum*. BERGMANN (1955) records *Achillea millefolium*, *Onobrychis*, *Hypericum perforatum*, *Hippocrepis comosa* and *Artemisia campestris* and REDONDO *et al.* (2009) *Thymus*, *Potentilla*, *Biscutella* and *Pastinaca sativa*. The pupation takes place in an open web among moss on the ground.

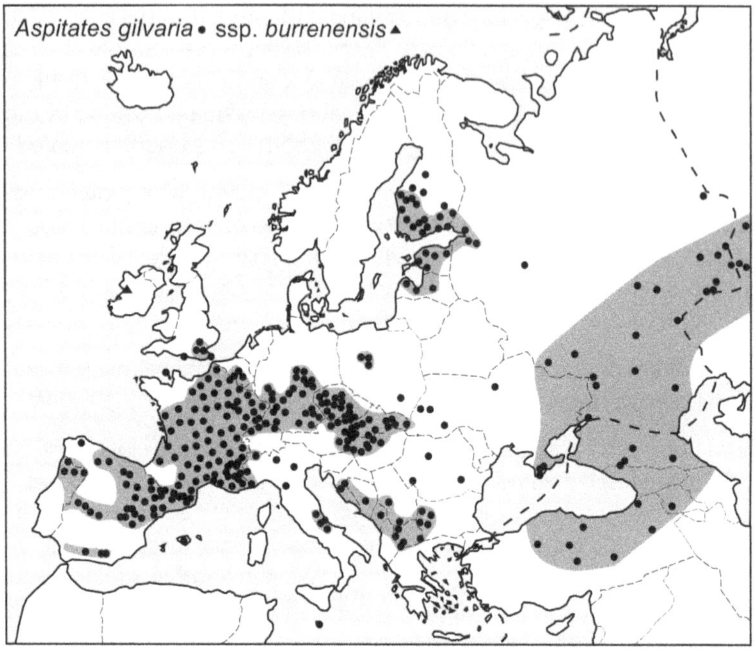

HABITAT: On low nutrient soils. In Fennoscandia in open, wet meadows and peat bogs. Further south in Europe the species prefers considerably dryer habitats and it occurs here in many different open land habitats including south-facing warm slopes, many of them with scattered trees and bushes. The species occurs from sea level up to 1000 m in the Alps and up to 2000 m in the Pyrenees.

SIMILAR SPECIES: *A. gilvaria* is usually diagnosable by yellow colour and brown, oblique forewing postmedial line. *A. gilvaria* is the most widespread species in *A. gilvaria* complex, the other species have restricted distribution areas. Diagnostic characters are summarised in Table 8. External characters are indicative but do not always allow reliable identication. Genitalia allow reliable identification.

GENETIC DATA: Populations from Germany, United Kingdom and France deeply diverging from those of Finland (5.3%; n=16), the latter genetically identical with examined populations from Central Asia (Altai mountains), see remarks. Nearest species: East Asian *Ctenognophos grandinaria* (MOTSCHULSKY, 1861) (7.4%) and central Asian *Aspitates obscurata* (WEHRLI, 1953) (7.6%).

REMARKS: The Holarctic *Aspitates gilvaria* group is a complex of taxa that requires a thorough, modern revision. Several taxa may be present both in the Palaearctic and in the Nearctic regions and in numerous instances morphology alone is apparently not adequate to solve the taxonomical and nomenclatorial questions. Data from other sources, including DNA and biological data, should be included also.

North European pale yellow and central European bright yellow specimens have distinct external appearance but the genitalia do not show significant differences (genitalia studied from Finland, Germany and Austria). More material is needed to draw conclusions, particularly from the Baltic states and Poland, to understand whether the variation is clinal or discontinous.

Table 8. Comparison of selected diagnostic characters in European species in the *Aspitates gilvaria* (DENIS & SCHIFFERMÜLLER, 1775) complex. Male genitalia allow the most reliable identification, see the genitalia plates.

	A. gilvaria	*A. orciferaria*	*A. taylorae*	*A. forbesi*	*A. albaria*
Wings	yellow	yellowish grey-brown	whitish-yellow, irrorated with brown	greyish yellow	pale yellow
Valva costa projection	long, narrow, 2–3 spines	wide at apex, >10 spines	wide at apex, >10 spines	wide, 5–10 spines	wide at apex, 5–10 spines
Valva ventral margin lobe(s)	two lobes: both round	one lobe: round	one lobe: outer margin concave, medial margin straight	one lobe: elongated	two lobes: medial indistinct, basal triangular
Aedeagus apex	no projections: apex dentate	two projections: short, one curved	two projections: long, both curved	one projection: long, angled, dentate	one projection: long, curved
Lamella antevaginalis	roundish	horseshoe: margin equally wide, flat	horseshoe: margin equally wide, tubular	horseshoe: margin equally narrow	horseshoe: margin narrow medially
Distribution	widespread	NE-Europe	NE-Europe	NE-Europe	S-Russia

121. Aspitates taylorae (BUTLER, 1893)

taylori taylori BUTLER, 1893, Ann. Mag. nat. Hist. (6) 121: 18, (*Pseudosiona*). Syntype(s) (The Natural History Museum, London, UK), [Canada, North-west Territories]: Pulo River, near Mackenzie River (examined externally). The original spelling *taylori* is incorrect, and the suffix must be '-ae' as the species is named after a woman (MUNROE 1963). *A. taylorae* is applied in HODGES *et al.* (1983) and in The Global Lepidoptera Names Index (ANONYMOUS 2013).

taylori sibirica DJAKONOV, 1955, Zool. Zh. 34: 557, (*Aspilates*). Type(s), [Russia]: Siberia. [Originally as sp].

conspersaria WEHRLI, 1953, in SEITZ, Gross-Schmett. Erde 4 (Suppl.): 678, pl. 53 e, (*Aspilates* (*Semiaspilates*)) [mistakenly as *insignis* ALPHÉRAKY, 1883]. Syntype male(s, female(s), [Russia:] Irkutsk, Sajan, Tunkish-Weissgebirge, 2000 m. [Synonymy after The Global Lepidoptera Names Index (ANONYMOUS 2013), see also WEHRLI (1939–1954, p. 678)].

EXTERNAL CHARACTERS AND ABDOMEN, based on European subsp. ***sibirica***: Wingspan ♂ 29–36 mm, ♀ 30–33 mm. Wings whitish to pale straw-yellow, heavily irrorated with brown. Forewing postmedial line wide, suffused, ending on costa before apex. Hindwing postmedial line oblique. Terminal area darker. Terminal line fine, brown. Fringes concolorous with wings. Discal spots brown, suffuse. Forewing margin rounded. Wings below more whitish, postmedial lines and discal spots darker, sharper. Frons, collar, thorax and abdomen concolorous with wings. Male antennae bipectinate, female antennae fasciculate/shortly bipectinate. Hindtibia of both sexes 2+2 spurs. Tympanal organs small. Posterior margin of male 8th sternite clearly narrower. Other abdominal segments of both sexes undifferentiated.

VARIATION: Ground colour varies from whitish to pale straw-yellow. Degree of brown irroration variable. Forewing postmedial line oblique or straight. Male forewing margin normally rounded, but it can be also rather straight.

MALE GENITALIA: Many species in the *A. gilvaria* complex are similar, see *A. gilvaria* (DENIS & SCHIFFERMÜLLER for general description). *A. taylorae* has the following unique features (see also Table 8): Valva costa projection wide at apex, with more than 10 spines. Valva ventral margin with one lobe, outer margin concave, medial margin straight. Aedeagus apex with two long, curved projections.

FEMALE GENITALIA: Many species in the *A. gilvaria* complex are similar. *A. taylorae* has the following unique feature (see also Table 8): Lamella antevaginalis horseshoe shaped, margin is equally wide and tubular.

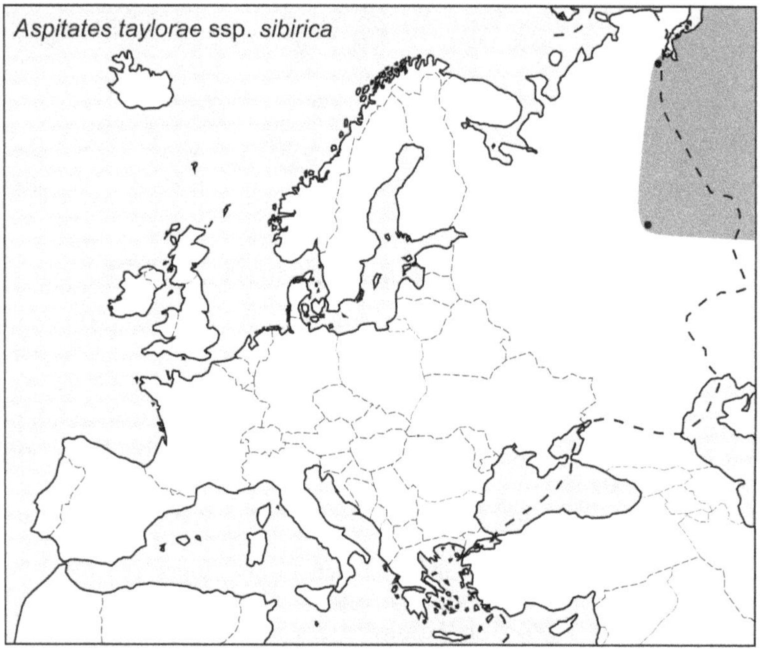

DISTRIBUTION AND ABUNDANCE: Holarctic. In Europe only known from polar Urals and Udmurtia. – Outside Europe the species occurs throughout most of Siberia to Far East Russia and Kamchatka, and in Mongolia. Also in northern North America. – In Europe a rare species that is poorly known.

PHENOLOGY: Univoltine. In polar Urals in early July. Recorded in Buryatia in mid-June (K. NUPPONEN pers. comm.). In Transbaikalia and Far East Russia from early June to early July. In northern Canada from mid-June to late July (MCGUFFIN 1981). The moths fly at night and are attracted to light (K. NUPPONEN pers. comm.). In polar Urals the species flies in the afternoon.

BIOLOGY: Unknown.

HABITAT: On low nutrient soils. In polar Urals found in peat bogs and river sides in humid meadows from 200 m up to 400 m altitude (J. KULLBERG pers. comm.). In steppe habitats from 700 up to 1450 m in Buryatia (K. NUPPONEN pers. comm.).

SIMILAR SPECIES: See Table 8. *A. taylorae* is diagnosable by whitish to pale straw-yellow wings, which are heavily irrorated with brown. The general appearance is somewhat reminiscent of *Timandra rectistrigaria* (EVERSMANN, 1851, Sterrhinae). *A. orciferaria* has slightly shorter extension on male valva costa, ventral margin lobe of valva has one round lobe, aedeagus apex projections are short (valva costa extension slightly longer, ventral margin lobe of valva has one lobe, outer margin is concave and medial margin is straight and aedeagus apex lobes are long in *A. taylorae*). *A. orciferaria* has flat margin on horseshoe-shaped lamella antevaginalis (tubular margin in *A. taylorae*). External characters allow identification in most cases. Male and female genitalia of *A. orciferaria* and *A. taylorae* are very similar, and often do not allow reliable identification.

GENETIC DATA: According to the DNA barcodes of North American specimens *A. orciferaria* differs from *A. aberrata* (EDWARDS, 1884) by 5.67% (min. 4.94%, max. 7.18%, not in Europe), from *A. forbesi* by 5.56% (min. 5.11%, max. 6.14%) and from *A. taylorae* by 5.42% (min. 4.78%, max. 6.04%) (DEWAARD *et al.* 2011, J. R. DEWAARD pers. comm.).

122. *Aspitates orciferaria* (WALKER, 1863)

orciferaria orciferaria WALKER, [1863], List Specimens lepid. Insects Colln Br. Mus. 26: 1693, (*Napuca*). Lectotype male (The Natural History Museum, London, UK), Arctic America [Canada] (examined externally).

orciferata HULST, [1903], in DYAR, A list of North American Lepidoptera and key to the literature ...: 332, (*Dyscia*). [Emendation of *orciferaria* WALKER]

orciferaria baffinensis MUNROE, 1963, Can. Ent. 95 (3): 276, figs 45, 46, (*Aspilates*). Holotype male (Canadian National Insect Collection, Ottawa, Ontario, Canada), [Canada]: Baffin Island, Frobisher Bay. [Originally as subsp. of *orciferarius*], paratype male (The Natural History Museum, London, UK) (examined externally).

orciferaria churchillensis MUNROE, 1963, Can. Ent. 95 (3): 278, figs 49, 50, (*Aspilates*). Holotype male (Canadian National Insect Collection, Ottawa, Ontario, Canada), [Canada]: Manitoba, Fort Churchill. [Originally as subsp. of *orciferarius*], paratype male, paratype female (The Natural History Museum, London, UK) (examined externally).

orciferaria occidentalis MUNROE, 1963, Can. Ent. 95 (3): 278, fig. 51, (*Aspilates*). Holotype male (Canadian National Insect Collection, Ottawa, Ontario, Canada), Yukon, Dawson, 1100 ft. [Originally as subsp. of *orciferarius*], paratype male (The Natural History Museum, London, UK) (examined externally).

EXTERNAL CHARACTERS AND ABDOMEN: Wingspan ♂ 27–29 mm, ♀ 25 mm (n=1). Generally as *A. gilvaria* (DENIS & SCHIFFERMÜLLER), see above, but with following differences: Wings yellowish grey-brown, heavily suffused, having smoky appearance, without white. Forewing postmedial line dark grey-brown. Fringes whitish. Hindwings weakly whitish, heavily irrorated grey. Wings below whitish, heavily suffused with dark grey-brown. Postmedial line wide, suffused. Discal spots large, dark grey.

VARIATION: Only a few European specimens have been available for study. Wings variably suffused with dark grey, those can be almost uniform dark grey. Wings below mostly with whitish background, but white can be almost absent and wings can be nearly uniform dark grey.

MALE GENITALIA: Many species in the *A. gilvaria* complex are similar, see *A. gilvaria* (DENIS & SCHIFFERMÜLLER for general description). *A. orciferaria* has the following unique features (see also Table 8): Valva costa projection wide at apex, with more than 10 spines, valva ventral margin with one round lobe. Aedeagus apex with two short projections, one being curved, other rather straight.

FEMALE GENITALIA: Many species in the *A. gilvaria* complex are similar. *A. orciferaria* has the following unique feature (see also Table 8): Lamella antevaginalis horseshoe-shaped, margin equally wide and flat.

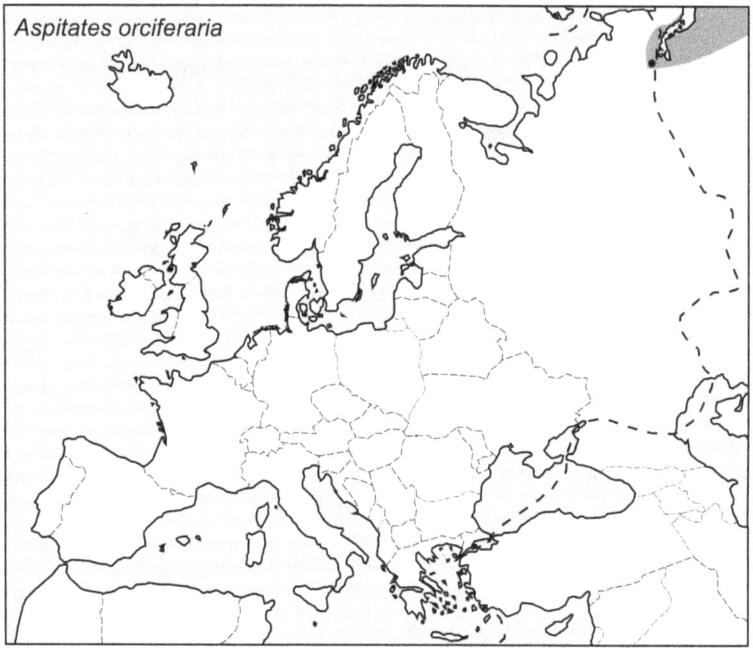

DISTRIBUTION AND ABUNDANCE: Holarctic. In Europe only known from polar Urals. – Outside Europe the species has been found in the lower Ob river region in the Yamalo-Nenets autonomous district, in the Northern-Yenisei region in Taimyr, Chukotka region and Magadan district (MIRONOV *et al.* 2008). Also in northern North America. – In Europe a rare species that is poorly known.

PHENOLOGY: Univoltine. In polar Urals in early July. In northern Canada from mid-June to late July (MCGUFFIN 1981). In polar Urals the moths fly in the afternoon.

BIOLOGY: Unknown.

HABITAT: In polar Urals on scree slopes and mountain tundra from 400 m up to 800 m altitude (J. KULLBERG pers. comm.).

SIMILAR SPECIES: *A. orciferaria* is usually diagnosable by yellowish grey-brown colour, with smoky appearance, which is darker than in other species in *A. gilvaria* complex, and by weak, dark grey brown forewing postmedial line. Externally *A. orciferaria* is the most similar to *A. forbesi*, but the genitalia are diagnostic: aedeagus with two, short and curved projections in *A. orciferaria* (aedeagus with one, long and dentate projection in *A. forbesi*). *A. taylorae* is whitish-yellow, irrorated with brown (wings yellowish grey brown in *A. orciferaria*). The male and female genitalia of *A. orciferaria* and *A. taylorae* are very similar (see, Table 8 and genitalia plates). In Europe *A. orciferaria* is known from north-eastern Europe (Polar Urals) only. Diagnostic characters are summarised in Table 8. External characters are somewhat indicative and do not always allow reliable identication. Combination of external and genitalia characters allows reliable identification.

GENETIC DATA: According to the DNA barcodes of North American specimens *A. orciferaria* differs from *A. aberrata* (EDWARDS, 1884) by 5.67% (min. 4.94%, max. 7.18%, not in Europe), from *A. forbesi* by 5.56% (min. 5.11%, max. 6.14%) and from *A. taylorae* by 5.42% (min. 4.78%, max. 6.04%) (DEWAARD *et al.* 2011, J. R. DEWAARD pers. comm.).

REMARKS: Based on the specimens collected from the Polar Urals, most *A. orciferaria* specimens were taken from around 500 metres altitude and majority of *A. taylorae* and *A. forbesi* specimens were taken at 200 metres. The genitalia of *A. taylorae* and *A. orciferaria* are similar, without qualitative differences, as was noted by MCGUFFIN (1981) also. These taxa are diagnosable by external characters. The Holarctic *Aspitates gilvaria* group is a complex of taxa that requires a thorough, modern revision. Several taxa may be present both in the Palaearctic and in the Nearctic regions and in numerous instances morphology alone is apparently not adequate to solve the taxonomical and nomenclatorial questions. Data from other sources, including DNA and biological data, should be included also.

123. Aspitates forbesi (MUNROE, 1963)

forbesi MUNROE, 1963, Can. Ent. 95 (3): 270, figs 10, 22, 41, 42, (*Aspilates*). Holotype male (Canadian National Insect Collection, Ottawa, Ontario, Canada), [Canada]: Manitoba, Churchill, paratype female (The Natural History Museum, London, UK) (examined, including genitalia).

EXTERNAL CHARACTERS AND ABDOMEN: Wingspan ♂ 29 mm (n=1) (North American specimens 30–34 mm (MCGUFFIN 1981)), ♀ 28–32 mm (North American specimens, based on MCGUFFIN (1981), no European specimens examined). Generally as *A. gilvaria* (DENIS & SCHIFFERMÜLLER), see above, but with the following differences: Wings straw yellow, finely irrorated with greyish. Forewing postmedial line weak. Fringes whitish. Wings below pale straw, forewings heavily suffused with brown. Postmedial line rather narrow, brown. Discal spots large, brown.

VARIATION: Unknown, only one European male was available for study. In North America females are lighter in colour (MCGUFFIN 1981).

MALE GENITALIA: Many species in the *A. gilvaria* complex are similar, see *A. gilvaria* (DENIS & SCHIFFERMÜLLER for general description). *A. forbesi* has the following unique features (see also Table 8): Valva costa projection is wide, with 5–10 spines, valva ventral margin is with one elongated lobe. Aedeagus apex is with one long, angled dentate projection.

FEMALE GENITALIA: Many species in the *A. gilvaria* complex are similar. *A. forbesi* has the following unique feature (see also Table 8): Lamella antevaginalis is horseshoe-shaped, margin is equally narrow.

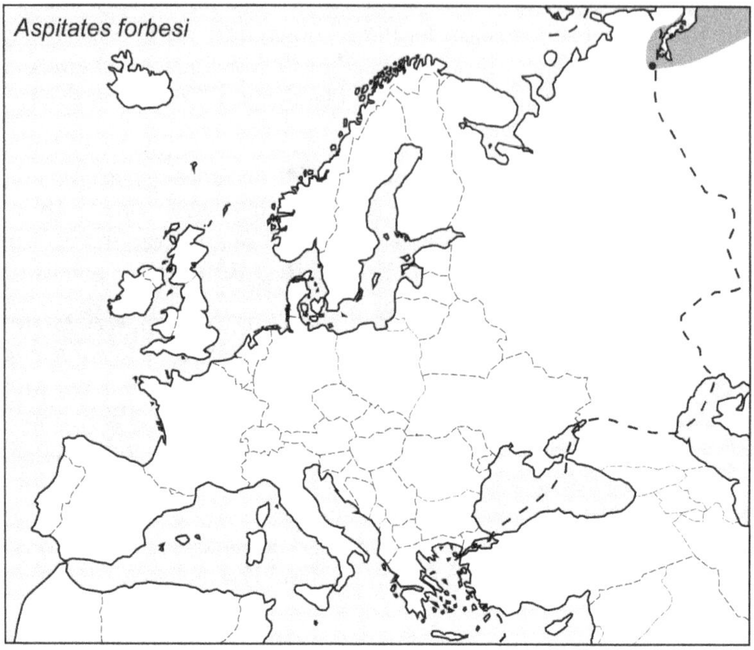

DISTRIBUTION AND ABUNDANCE: Holarctic. In Europe the species only occurs in polar Urals. – Outside Europe it has been found in the Siberian regions Eastern Yakutiya and Southern Yakutiya and also in northern North America. – In Europe a rare species that is poorly known.

PHENOLOGY: Univoltine. In polar Urals found in early July. In northern Canada from early July to mid-August (MCGUFFIN 1981). In polar Urals the species flies in the afternoon.

BIOLOGY: Unknown.

HABITAT: In polar Urals found in a peat bog at 200 m altitude.

SIMILAR SPECIES: *A. forbesi* is usually diagnosable by pale yellow colour, finely irrorated with greyish. Externally *A. forbesi* is the most similar with *A. orciferaria*, but the genitalia are diagnostic, see *A. orciferaria*. In Europe *A. forbesi* is known from north-eastern Europe (Polar Urals) only. Diagnostic characters are summarised in Table 8. External characters are somewhat indicative and do not always allow reliable identication. Combination of external and genitalia characters allows reliable identification.

GENETIC DATA: According to the DNA barcodes of North American specimens *A. orciferaria* differs from *A. aberrata* (EDWARDS, 1884) by 5.67% (min. 4.94%, max. 7.18%, not in Europe), from *A. forbesi* by 5.56% (min. 5.11%, max. 6.14%) and from *A. taylorae* by 5.42% (min. 4.78%, max. 6.04%) (DEWAARD *et al.* 2011, J. R. DEWAARD pers. comm.).

124. Aspitates albaria (BARTEL, 1903)

albaria BARTEL, 1903, Dt. ent. Z. Iris 15 (2): 221, (*Aspilates*). Syntype(s) (Institut Royal des Sciences Naturelles, Brussels, Belgium), Russia: Orenburg.

EXTERNAL CHARACTERS AND ABDOMEN: Wingspan 26–30 mm. Generally as *A. gilvaria* (DENIS & SCHIFFERMÜLLER), see above, but with following differences: Wings pale yellow, very weakly irrorated. Forewing postmedial line very weak, weakly curved inwards on inner margin. Wings below as above but postmedial line and discal spots wider, brown. Tympanal organs medium-sized.

VARIATION: Wings normally very weakly irrorated with grey-brown, but irroration can be widespread, particularly in females. Hindwings normally slightly paler than forewings but can be same colour as forewings.

MALE GENITALIA: Many species in the *A. gilvaria* complex are similar, see *A. gilvaria* (DENIS & SCHIFFERMÜLLER for general description). *A. albaria* has the following unique features (see also Table 8): Valva costa projection is wide at apex, with 5–10 spines. Valva ventral margin is with two lobes, medial lobe is indistinct, basal lobe triangular. Aedeagus apex is with one long, curved, weakly dentate projection.

FEMALE GENITALIA: Many species in the *A. gilvaria* complex are similar. *A. albaria* has the following unique feature (see also Table 8): Lamella antevaginalis is horseshoe-shaped, margin is narrow medially.

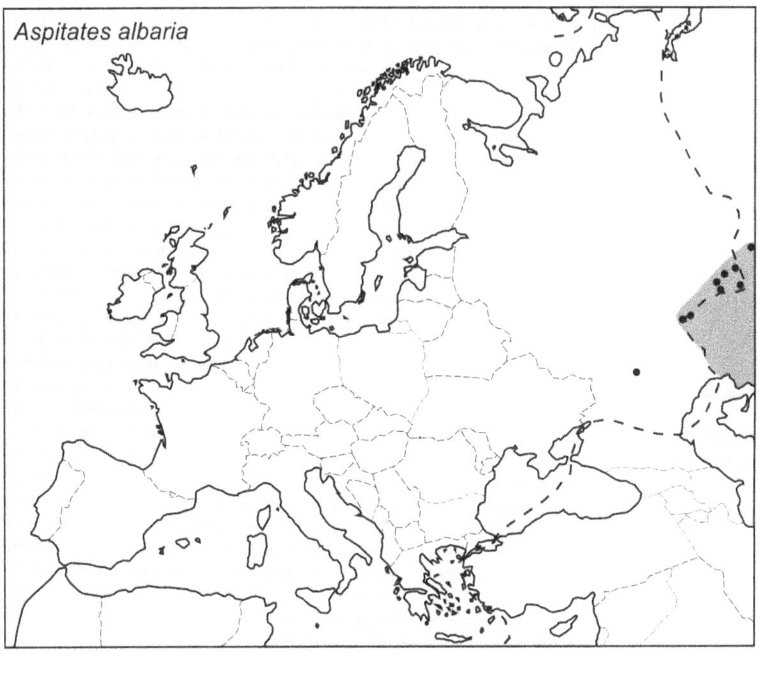

DISTRIBUTION AND ABUNDANCE: Central Asian. In Europe only in the southern Urals and in south-eastern European Russia. – Outside Europe in northern Kazakhstan (VIIDALEPP 1996)

and in Siberia in the southern west Siberian region, Cis-Altai region, Mountain-Altai region (MIRONOV *et al.* 2008). – In Europe a rather rare and poorly known species.

PHENOLOGY: Univoltine. In southern Urals and lower Volga region from mid-May to early June (K. NUPPONEN pers. comm.). The moths fly at night and come to light (K. NUPPONEN pers. comm.).

BIOLOGY: Unknown.

HABITAT: Xerophilous. Steppe, up to 250 m (K. NUPPONEN pers. comm.).

Similar species: *A. albaria* is usually diagnosable by pale yellow colour and very weak forewing postmedial line that is weakly curved inwards on inner margin. In Europe *A. albaria* is known from South Russia only. Diagnostic characters are summarised in Table 8. External characters are somewhat indicative and do not always allow reliable identication. Combination of external and genitalia characters allows reliable identification.

GENETIC DATA: No data available.

125. Aspitates ochrearia (ROSSI, 1794)

ochrearia ROSSI, 1794, Mantissa Insectorum Etrusia 2: 33, number 429, (*Phalaena*). Syntype(s), Italy: Tuscany.

citraria HÜBNER, [1799], Samml. Eur. Schmett. 5 Geometrae (1): pl. 103, fig. 536, (*Geometra*). Syntype(s), lost, Europe.

laetula HULST, 1896, Trans. Am. ent. Soc. 23 (3): 336, (*Nyctiphanta*). Holotype male (American Museum of Natural History, New York, USA), [U.S.A].: Arizona [incorrect locality]. See remarks, detailed in FERGUSON (1983).

parvularia LEMPKE, 1952, Tijdschr. Ent. 95: 264, fig. 55 (4–6,) (*Aspitates*). Holotype male (Instituut voor Taxonomisch Zoölogie, Universiteit van Amsterdam, The Netherlands), Netherlands: Hoek van Holland (examined externally). [originally as subspecies]. See remarks.

sulphuraria FABRICIUS, 1794, Ent. Syst. 3 (2): 142, (*Phalaena*). Syntype(s) male (Zoological Museum, University of Copenhagen, Denmark), [North Africa]: Barbaria [Barbary]. See remarks.

UNAVAILABLE NAMES (summer generation): *aestiva* SCHAWERDA, 1913. – (incorrect subsequent author): *crassesignata* LERAUT (LERAUT 2009, p. 570 as 'nov.'; should be *crassesignata* LEMPKE, 1953). – Intrasubspecific names, potentially unavailable: *alba* KRAUSSE, 1912 (ab.); *aurantiaca* LUCAS, 1960 (f.); *clausa* LEMPKE, 1953 (f.); *conjuncta* LEMPKE, 1953 (f.); *crassesignata* LEMPKE, 1953 (f.); *fuscata* COCKAYNE, 1953 (ab.); *glabra* LEMPKE, 1953 (f.); *impuncta* LEMPKE, 1953 (f.); *lativittata* COCKAYNE, 1953 (ab.); *obsoleta* LEMPKE, 1953 (f.); *pallidaria* RAGUSA, 1905 (spring generation); *unicolorata* SEEBOLD, 1895 (ab.).

EXTERNAL CHARACTERS AND ABDOMEN: Wingspan 27–30 mm (first generation), 23–29 mm (second generation). Forewings ochreous yellow, irrorated with grey, particularly on costa. Antemedial grey, angled inwards on costa. Postmedial line grey, suffuse, slightly angled inwards in middle, reaching costa well before apex. Terminal line absent. Fringes concolorous with wings or chequered ochreous yellow and grey. Hindwings paler, whitish. Postmedial line grey, suffused, often narrower in middle. Discal spots roundish, grey. Wings below more contrasting, markings darker, hindwings also ochreous yellow. Frons, collar, thorax and abdomen concolorous with wings. Male antennae bipectinate, female antennae filiform.

Hindtibia of both sexes with 2+2 spurs. Tympanal organs small, larger in female. Abdominal segments of both sexes undifferentiated.

VARIATION: Ground colour varies from pale whitish-yellow to deep ochreous yellow. Markings normally grey, even silvery grey, often with tint of violet or brown. Hindwings normally paler than forewings, whitish, but can also be yellow. Fringes can be concolorous with wings or chequered with ochreous yellow and grey. Numerous forms have been described, which are summarised in LERAUT (2009).

MALE GENITALIA: Uncus small, sparsely setose, apex spatulate. Socii absent. Gnathos arms narrow, fused ventrally, apex weakly granulate. Valva wide, densely setose, apex round. Costa base with long projection, apex with two long spines. Ventral margin of valva with one round lobe, surface densely spinose. Valva fused dorsally. Juxta large, weakly V-shaped. Saccus wide, quandrangular. Aedeagus weakly curved, caecum absent, subapical part with dentate ridge. Vesica opens at approximately 90 degree angle. Vesica slightly expanded at base, otherwise tubular. Vesica without sclerotisations.

FEMALE GENITALIA: Papillae anales small, setose. Apophyses anteriores less than half length of apophyses posteriores. Lamella antevaginalis narrow, large, laterally elongated. Lamella postvaginalis tiny, sclerotised. Sterigma/ductus bursae sclerotised, wide. Posterior part of corpus bursae wide, sclerotised. Anterior part of corpus bursae elongated, membranous. Signum absent.

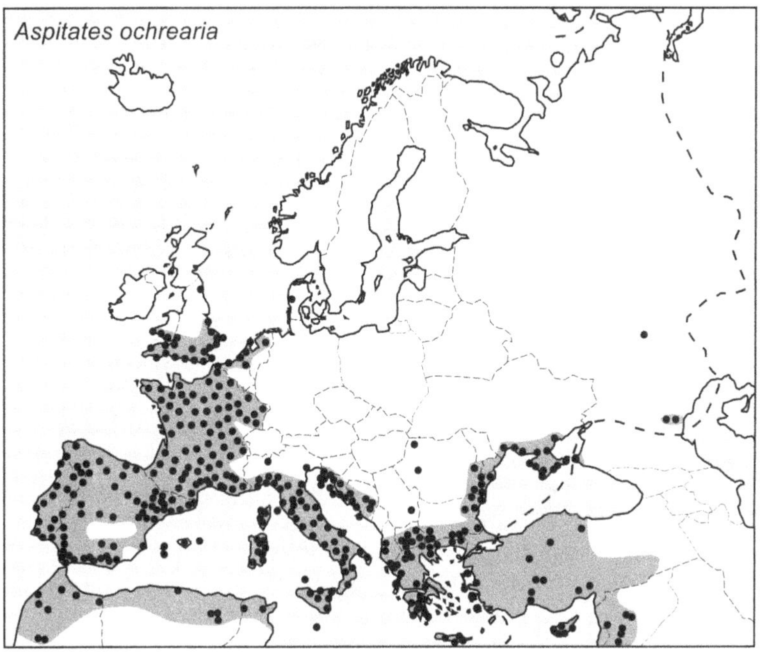

DISTRIBUTION AND ABUNDANCE: Sub-Mediterranean. In Europe only in the south and west, in the east to southern Ukraine and southern European Russia and westwards through the Balkans including Crete, Italy including Sicily and Sardinia to the Iberian Peninsula, France including Corsica and southern England. In Denmark only one migrant specimen. Towards north of the European distribution area, the species occurs mainly close to the coasts. – Outside Europe the species occurs in Cyprus, Turkey, the Levant down to Israel, Libya, Tunisia, Algeria and Morocco. – Common in Europe and not rarely in numbers.

PHENOLOGY: Bivoltine. Moths of the first generation fly from late March to late June, those of the second generation from mid-August to mid-October. Larva from September to April or May and from June to July or August. Hibernation takes place as a larva. The moths fly at night and come frequently to light and can be flushed up by day.

BIOLOGY: Polyphagous. The larva has been found on *Daucus carota*, *Plantago* and *Crepis vesicaria* (SKINNER 2009). KING & VIEJO MONTESINOS (2010) have found larvae on *Centaurea hyssopifolia*, *Lepidium subulatum*, *Thymus lacaitae*, *Helianthemum hirtum* and *Gypsophila struthium*. Recorded on *Rumex*, *Lavendula* and *Cistus* (REDONDO *et al.* 2009). AIZPÚRUA (2002) also records *Plantago* and *Daucus carota*. The caterpillar has been reared on *Vicia* (B. MÜLLER pers. comm.) and on *Achillea millefolium* (E. FRIEDRICH pers. comm.).

HABITAT: The species occurs in a large variety of open land habitats, especially dry places, along coasts and in sand dunes. Vertically the species is found from sea level up to 2000 m in Sierra Nevada in southern Spain (REDONDO *et al.* 2009), but it is most common at low altitudes.

SIMILAR SPECIES: No similar species in Europe.

GENETIC DATA: Genetically comparatively homogeneous from United Kingdom and France to southern Spain, southern Italy and Israel (n=29), with slight regional divergences of up to 1.0%. Nearest congeneric Central Asian *Aspitates stschurovskyi* (ERSCHOFF, 1874) (7.8%). North African *Dyscia nobilaria* (BANG-HAAS, 1906) diverging by 7.4% (long branch attraction).

REMARKS: We follow KUCHLEIN & VOS (1999) and HAUSMANN *et al.* (2011), and do not recognise *parvularia* LEMPKE (1952) from the Netherlands valid at subspecies level. LERAUT (2009) considers the North African taxa, i.e. *sulphuraria* (FABRICIUS, 1794), valid at subspecies level. We have not examined original material, neither have been specimens been available for DNA study. Contrary to the indication in LERAUT (2009), *A. ochrearia* is not present in North America (FERGUSON 1983).

Aspitates acuminaria (EVERSMANN, 1851)

Aspitates acuminaria is a species that could potentially be found in Europe. The nearest localities include Georgia and Dagestan in Transcaucasus region, Krasnodarsk and Stavropol regions in Russia and areas between the above-mentioned localities (VIIDALEPP 1996, MIRONOV *et al.* 2008, V. MIRONOV pers. comm.). *A. acuminaria* is known from the Transcaucasus region to Kazakhstan, Kyrgyzstan and north-west China (VIIDALEPP 1996). Wings are pale yellow, basal area and outer margin of postmedial lines are brown, occasionally with tint of violet. Females are smaller, wing markings are sharper, forewing apex is acute and antennae are filiform. Male valvae are asymmetrical in that a dense group of setae is present on left valva only (ventral view) (Text-fig. 132).

According to the label data the adults are found from late June to early August. They can be found in the vegetation during daytime, they fly at night also and come readily to light. The species can be locally abundant. Food plant of the larva is unknown. Adults occur in meadows, often with scattered trees, up to 2750 m.

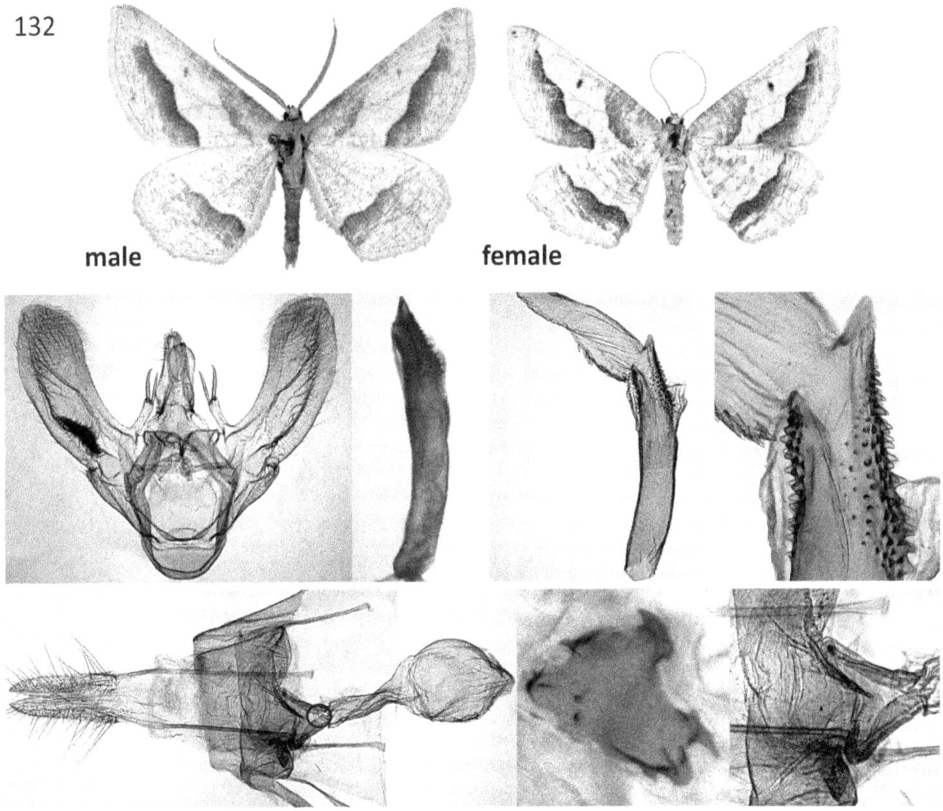

Text-fig. 132. Adults and genitalia of *Aspitates acuminaria* (EVERSMANN, 1851), a species that could potentially be found in Europe. Male adult (wingspan 41 mm): [China?], Sanka, Tian-Shan Mts., 2500–3500 m, July–August 1906 (coll. BMNH). Male genitalia, aedeagus, vesica, sclerotisations in aedeagus apex and cornuti illustrated from same specimen: Kazakhstan, Zauliskiy Alatau, Alma-Atinskij Nature Park, 1700 m, 22.6.vi.1990, slide PS1490 (coll. ZMH). Female adult (wingspan 36 mm): Kyrgyzstan, East-side of Lake Issyk-kul, 19–24.vi.1905 (coll. BMNH). Female genitalia (point of origin of ostium bursae indicated by circle), signum, ostium bursae and adjacent structures illustrated from same specimen: Kazakhstan, Zauliskiy Alatau, Alma-Atinskij Nature Park, 2750 m, meadow, 30.vi.1990, slide PS1491 (coll. ZMH).

Dyscia HÜBNER, 1825

Dyscia HÜBNER, [1825] 1816, Verz. bekannter Schmett.: 314. Type species: *Geometra conspersaria* [DENIS & SCHIFFERMÜLLER], 1775. Genus synonyms are after TRUSCH & ERLACHER (2002).

Calodyscia WEHRLI, 1950, Ent. Ber., Amst. 13: 78. Type species: *Scodiona sicanaria* OBERTHÜR, 1923.

Catadyscia WEHRLI, 1950, Ent. Ber., Amst. 13: 78. Type species: *Dyscia atlantica* REISSER, 1933.

Eudyscia WEHRLI, 1950, Ent. Ber., Amst. 13: 77. Type species: *Geometra fagaria* THUNBERG, 1784.

Iberafrina WEHRLI, 1950, Ent. Ber., Amst. 13: 78. Type species: *Geometra penulataria* HÜBNER, [1819] 1796.

Maesia STEPHENS, 1831, Illustr. Br. Ent. (Haustellata) 3: 150. Type species: *Geometra favillacearia* HÜBNER, [1799] 1796.
Psednothrix HÜBNER, [1825] 1816, Verz. bekannter Schmett.: 320. Type species: *Geometra belgiaria* HÜBNER, [1799] 1796.
Rjabovana WEHRLI, 1950, Ent. Ber., Amst. 13: 80. Type species: *Dyscia negrama* WEHRLI, 1950.
Scodiona BOISDUVAL, 1840, Genera Index meth. eur. Lepid.: 185. Type species: *Geometra conspersaria* [DENIS & SCHIFFERMÜLLER], 1775.
Warneckeella WEHRLI, 1950, Ent. Ber., Amst. 13: 77. Type species: *Dyscia malatyana* WEHRLI, 1934.
Zuleika BANG-HAAS, 1906, Dt. ent. Z. Iris 19: 142. Type species: *Scodiona nobiliaria* BANG-HAAS, 1906.

UNAVAILABLE NAMES (incorrect subsequent spellings): *Scodionia* GERSTAECKER, 1863; *Sciadiona* GLASER, 1887; *Scodonia* GLASER, 1887; *Scoidiona* GLASER, 1887 (see FLETCHER 1979).

DIVERSITY AND DISTRIBUTION: *Dyscia* contains about 20 species, which occur on steppe land and along the margins of deserts of North-Africa, Arabia, and Asia, and in the south of Europe. Only one species, *Dyscia fagaria* (THUNBERG), occurs further north to north-western Europe as a species of open sandy heaths (TRUSCH & ERLACHER 2002). PITKIN (2002) noted that the two Neotropical species placed in *Dyscia* are unlikely to be true members of it, one of those was transferred to *Thysanopyga* HERRICH-SCHÄFFER, 1855 (Caberini/Baptini). The taxon was also transferred to a new genus combination in TRUSCH & ERLACHER (2002).

EXTERNAL CHARACTERS AND ABDOMEN: Wingspan approximately 35–40 mm. Wings pale grey to beige, irrorated with dark scales. Forewing costa weakly concave. Antemedial and subterminal lines often broken into dots. Discal spots distinct. Proboscis reduced. Male antennae bipectinate, female antennae filiform. Abdomen stout. Tympanal organs small. Abdominal segments of both sexes undifferentiated.

MALE GENITALIA: Uncus weakly sclerotised, sparsely setose. Socii absent. Gnathos arms narrow, fused, often upturned ventrally. Valva wide or narrow, apex round, setose. Valva costa base with projection, apex often spinose or setose. Valva fused dorsally. Juxta large, sclerotised. Saccus with upright extension medially. Aedeagus often with teeth or other projections. Vesica small, mostly without sclerotisations.

FEMALE GENITALIA: Apophyses anteriores reduced in length, about one-third length of apophyses posteriores. Lamella antevaginalis and lamella postvaginalis variable: often large, sclerotised, or without sclerotisations. Accessory gland often very large (Text-fig. 133), larger than corpus bursae. Corpus bursae membranous, very thin, often inner surface covered with minute spines. Signum absent.

BIOLOGY: *Dyscia* larvae have been recorded on numerous low-growing plants including for instance *Salvia pratensis*, *Artemisia* spp., *Lavandula* spp., *Genista monspessulana*, *Cytisus* spp. and *Helianthemum* spp. *Dyscia fagaria* has been recorded feeding on *Calluna vulgaris* in north-western Europe (TRUSCH & ERLACHER 2001). Numerous species occur on steppes, along the margins of deserts and on open sandy heaths and sandy pine forests (TRUSCH & ERLACHER 2001, PATOČKA & TURČANI 2005). *D. fagaria* overwinters as larva in north-western Europe (SKOU 1986).

IMMATURE STAGES: Eggs of *D. royaria* TAUTEL & BILLI are elongated, laid in groups on larval hostplant (TAUTEL *et al.* 2008). Larva of *D. fagaria* is greyish-brown, mottled with reddish-

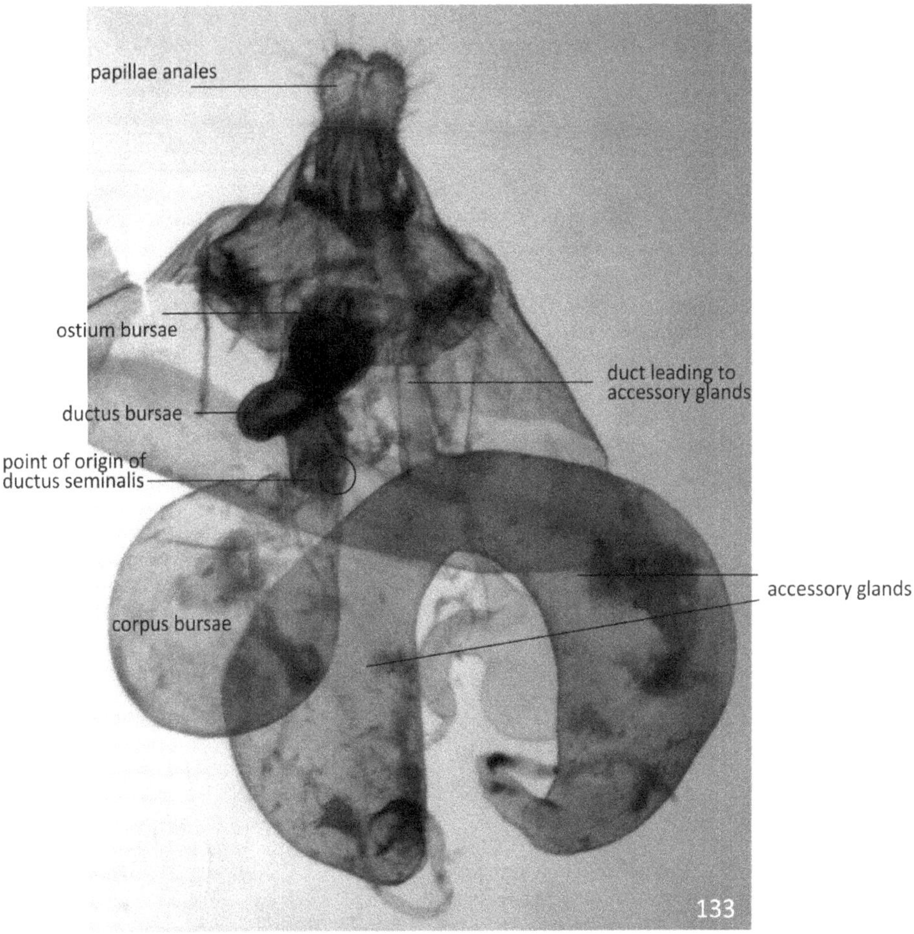

Text-fig. 133. Female genitalia and accessory glands of *Dyscia lentiscaria* (DONZEL, 1837). Accessory glands are often very large in genus *Dyscia*, even larger than the corpus bursae. Spain: Prov. Cadiz, Roche, 6.iii.2005, slide PS1517 (coll. SKOU). Photographed in ethanol.

brown and whitish (PORTER 2010). The larvae of all *Dyscia* species have a dorsal process on 9th segment (on 11th segment according to PORTER (2010)), except *D. nobiliaria* (BANG-HAAS, 1906), which occurs in North Africa (TRUSCH & ERLACHER 2001, 2002). Pupa cremaster of *D. fagaria* is wide at base, weakly elongated. Apex is with three pairs of curved setae, apical pair is longest, setae Sd1 is absent (PATOČKA & TURČANI 2005). *D. royaria* TAUTEL & BILLI has four pairs of curved setae (TAUTEL *et al.* 2008).

REMARKS: *Dyscia* has recently been classified in the Gnophini (VIVES MORENO 1994, VIIDALEPP 1996, TRUSCH & ERLACHER 2002) and in the Aspitatini (LERAUT 1997, HAUSMANN, MIRONOV & VIIDALEPP 2011). *Dyscia* has the characteristic Gnophini features as diagnosed in VIIDALEPP *et al.* (2007). The genitalia are similar to genera such as *Siona* and *Aspitates*. VIVES MORENO (1994) classified the Iberian *Dyscia* species into four subgenera, but these have been considered junior synonyms in other recent works (SCOBLE 1999, TRUSCH & ERLACHER 2001, TRUSCH & ERLACHER 2002). Systematic position: Gnophini.

Table 9. Diagnostic genitalia characters of European *Dyscia* HÜBNER, 1825 species. *Dyscia* species are similar in external apperance, and examination of the genitalia is often needed to identify the species.

	Genitalia characters	Similar species
D. raunaria	valva base extensions far apart lamella postvaginalis margins weakly sclerotised	*royaria*
D. royaria	valva base extensions close to each other lamella postvaginalis margins strongly sclerotised	*raunaria*
D. innocentaria	valva divided: dorsal arm massive, ventral arm with spines aedeagus apex with elongated projection lamella antevaginalis with two large, tubular sclerotisations	–
D. malatyana	costa base with curved extension valva long, narrow lamella postvaginalis with narrow, wing-like sclerotisations	–
D. crassipunctaria	costa extensions blunt aedeagus apex with one projection lamella antevaginalis with medial extension	*conspersaria*
D. conspersaria	costa extensions narrow aedeagus apex with two dentate projections lamella antevaginalis with lateral sclerotisations	*crassipunctaria*
D. fagaria	costa extensions acute, curved aedeagus apex with elongated projections vesica with row of minute cornuti sterigma weakly sclerotised anterior part of ductus bursae striated	–
D. penulataria	costa base with spoon-shaped extensions, with sclerotised spines sterigma weakly sclerotised corpus bursae large, very thin	–
D. lentiscaria	gnathos evenly narrowing towards apex vesica extends into aedeagus apex ductus ejaculatorius opens from membranous vesica base posterior part of ductus bursae wide, anterior part narrow	*distinctaria*
D. distinctaria	gnathos strongly curved vesica does not extend into aedeagus apex ductus ejaculatorius opens from subapical part posterior part of ductus bursae narrow, anterior part wide	*lentiscaria*

Text-figs 134–143. Diagnostic characters (indicated) of European *Dyscia* HÜBNER, 1825 species. Narrow grey lines delimit the most similar species and first comparisons should be made against the species included in the same grey box. Specimens are not in scale relative to each other. Text-fig. 134. *D. raunaria* (FREYER, 1851), Italy: Province Verona, 18.vi.1985. Text-fig. 135. *D. royaria* TAUTEL & BILLI, 2006, France: Tende, 1300 m, 7.viii.2008. Text-fig. 136. *D. innocentaria* (CHRISTOPH, 1885), European part of Kazakstan: 15 km E from Bisen, 28-29.v.2000. Text-fig. 137. *D. malatyana* subsp. *clara* PUGAEV & ZOLOTUHIN, 2006 (paratype), Russia: Republic Kalmykia, Komsomolskiy village, 21.v.2005. Text-fig. 138. *D. conspersaria* (DENIS & SCHIFFERMÜLLER, 1775), Austria: Altenburg, 14.vi.1935. Text-fig. 139. *D. crassipunctaria* (REBEL, 1916), Greece: Crete, 750 m, 6.vi.2004. Text-fig. 140. *D. penulataria* (HÜBNER, 1819), contrasting specimen, Spain: Province Huelva, 17.iv.2007. Text-fig. 141. *D. fagaria* (THUNBERG, 1784), Denmark: Oksby, 27.v.1972. Text-fig. 142. *D. lentiscaria* (DONZEL, 1837), Spain: Province Cadiz, 6.iii.2005. Text-fig. 143. *D. distinctaria* (BANG-HAAS, 1910), Spain: Province Sevilla, 150 m, 2.iv.2003. ⇨

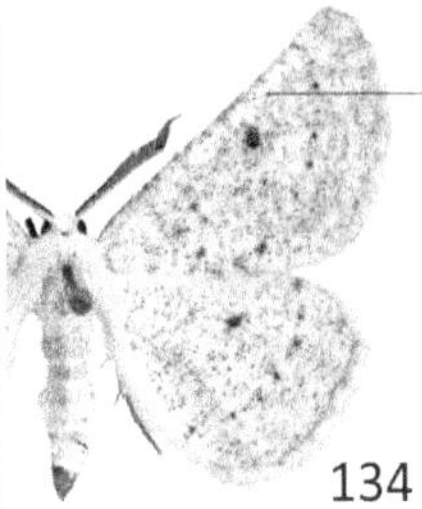

134 ***D. raunaria***

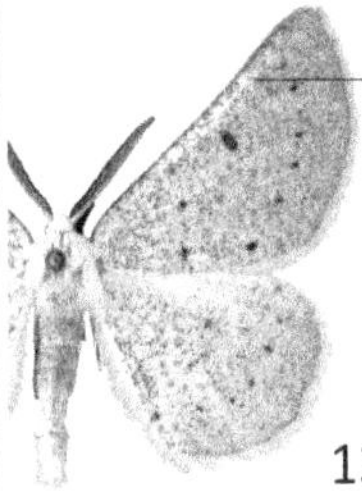

135 ***D. royaria***

136 ***D. innocentaria***

137 ***D. malatyana***

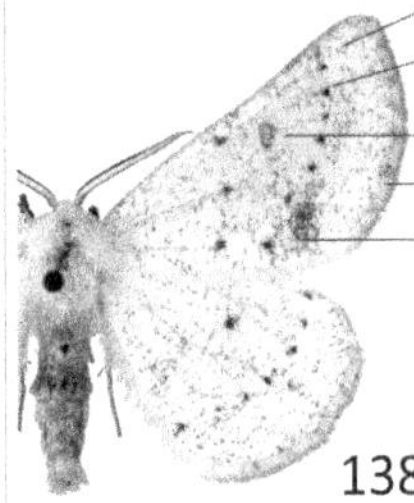

138 ***D. conspersaria***

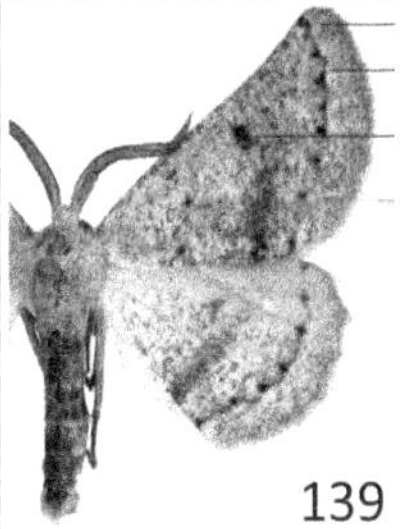

139 ***D. crassipunctaria***

140 ***D. penulataria***

141 ***D. fagaria***

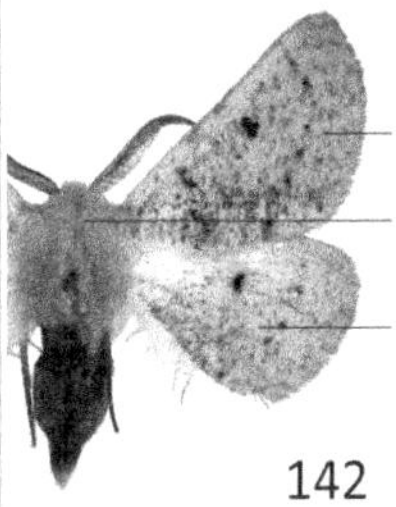

142 ***D. lentiscaria***

143 ***D. distinctaria***

TRUSCH & ERLACHER (2002) noted that a diagnostic feature of *Dyscia* is the loss of the uncus. We note that *Dyscia*, *Siona* and *Aspitates* have a similar, weakly setose and mostly membranous structure arising from dorsal margin of the tegumen. Due to its origin, dorsad of tuba analis, and because it is a single structure, not paired as typical socii, we have interpreted it as uncus. Our treatment of *Dyscia* largely follows the recent revisions (TRUSCH & ERLACHER 2001, 2002).

126. Dyscia raunaria (FREYER, 1851)

raunaria FREYER, 1851, Neuere Beitr. Schmettkde 6 (97): 160, pl. 582, figs 3, 4, (*Cabera*). Syntype(s), [Slovenia]: Karst near Raunach [Ravne near Gorizia].

INTRASUBSPECIFIC NAMES, POTENTIALLY UNAVAILABLE: *integeraria* SCHAWERDA, 1920 (ab.); *obfuscaria* SCHAWERDA, 1920 (ab.); *splichali* HAFNER, 1911 (ab.); *vernalis* WARNECKE, 1941 (f.)

EXTERNAL CHARACTERS AND ABDOMEN: Wingspan 25–33 mm, second generation specimens slightly smaller. Wings grey, with tint of brown, irrorated weakly with blackish grey-brown. Antemedial and subterminal lines broken into dots, latter evenly curved. Terminal line fine, brownish. Fringes slightly paler than wings. Discal spots distinct, blackish brown, solid or circled. Wings below as above, less irrorated with blackish grey-brown. Frons, collar and thorax uniform white. Abdomen concolorous with wings. Proboscis reduced. Male antennae bipectinate, female antennae filiform. Hindtibia of both sexes with 2+2 spurs. Tympanal organs small, bullae tympani with small roundish lobes. Other abdominal segments of both sexes undifferentiated.

VARIATION: Very little. Amount of irroration somewhat variable.

MALE GENITALIA: Uncus small, membranous, weakly setose. Socii absent. Gnathos arms narrow, fused ventrally. Valva wide, upright, apex round, with stout setae. Costa base with wide extension, these are wide distance apart. Margin between these extensions almost horizontal. Ventral margin of valva with concavity. Valva fused dorsally. Juxta large, sclerotised. Saccus wide, roundish, with upright extension medially. Aedeagus weakly curved, caecum long, with keel-shaped extension. Apex with short, dentate projection. Vesica opens at approximately 135 degree angle. Vesica small, tubular, without sclerotisations.

FEMALE GENITALIA: Papillae anales small, setose. Apophyses anteriores about 1/3 length of apophyses posteriores. Lamella antevaginalis tubular, weakly sclerotised, margin constricted medially. Lamella postvaginalis margins weakly sclerotised. Sterigma/ductus bursae funnel-shaped, sclerotised. Corpus bursae large, membranous, roundish, inner surface covered with minute spines.

DISTRIBUTION AND ABUNDANCE: Central-Mediterranean. Distributed only in northern Italy, westernmost Balkans and northern and central Greece. The species has not been found outside Europe. A local and not common species that is usually found in limited numbers where it occurs.

PHENOLOGY: Bivoltine. First generation from early May to mid-July and second generation from August to mid-October. At higher altitudes probably univoltine from late June or early July to late July or early August. The hibernation takes place as a small larva. The species flies at night when both males and females come to light.

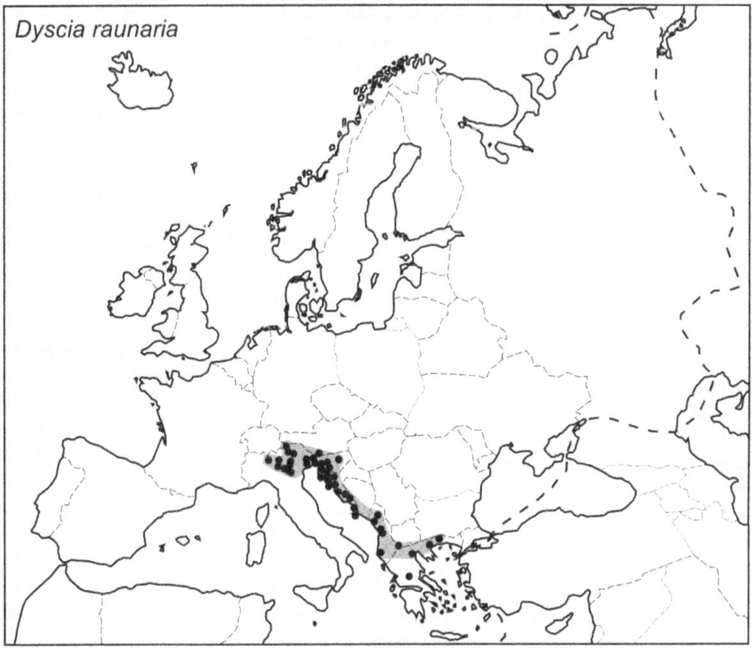

BIOLOGY: Monophagous. The caterpillar has been found on *Artemisia alba* in August and September (TRUSCH & ERLACHER 2001).

HABITAT: Xerothermophilous. Rocky slopes and similar warm places, often with sparse vegetation. The species has been found from sea level and up to 1700 m altitude (TRUSCH & ERLACHER 2001.

SIMILAR SPECIES: *D. raunaria* and *D. royaria* TAUTEL & BILLI are very similar, *D. royaria* is on average slightly larger and wings often have fewer darker scales, giving smoother appearance, but otherwise these two taxa are without reliable external differences. In *D. royaria* the valva base extensions are close to each other, margin between extensions without horizontal area (valva base extensions further apart, margin between extensions with almost horizontal area in *D. raunaria*). In *D. royaria* lamella antevaginalis is tubular, strongly sclerotised, lamella postvaginalis margins are strongly sclerotised (lamella antevaginalis tubular, weakly sclerotised, lamella postvaginalis margins weakly sclerotised in *D. raunaria*). *D. conspersaria* is tinted with ochreous, irrorated with brown, forewing discal spot often circular (wings whitish grey, irrorated with blackish, discal spot solid in *D. raunaria* and *D. royaria*). Male and female genitalia of *D. conspersaria* allow reliable identification. See Text-figs 134–143 and Table 9.

GENETIC DATA: The *raunaria/royaria* species complex (n=34) genetically splitting up into three BINs. Populations from South Tyrol to Slovenia and Croatia (*D. raunaria*) comparatively homogeneous and clustering into one single BIN. Nearest species: *D. royaria* (1.5%).

REMARKS: The genitalia differences in *D. raunaria* and *D. royaria* are quantitative: in male genitalia the distance between the valva base extensions and angle of margin between those, and in the female genitalia the degree of sclerotisations. The similarity of structures and DNA barcodes could lead one to consider *D. royaria* a subspecies or a junior synonym of *D. raunaria*. There may be differences in the ecology and in various life stages. Those of *D. royaria* are described in TAUTEL *et al.* (2008).

127. *Dyscia royaria* TAUTEL & BILLI, 2006

royaria TAUTEL & BILLI, 2006, Bulletin de la Societe Entomologique de France 111(4): 525, figs 1, 2, gen. figs. 7, 8, (*Dyscia*). Holotype female (Muséum National d'Histoire Naturelle, Paris, France), [France]: Alpes-Maritimes, vallée de la Roya, Tende, 1080 m (examined, including genitalia).

EXTERNAL CHARACTERS AND ABDOMEN: Wingspan 33–35 mm. Generally as *D. raunaria*, see above, but with following differences: *D. royaria* is on average slightly larger, wings have fewer dark scales, giving smoother appearance, and fringes are paler.

VARIATION: Very little. Amount of irroration somewhat variable.

MALE GENITALIA: Generally as in *D. raunaria*, but with following differences: In *D. royaria* the valva base extensions are close to each other, margin between these extensions is without horizontal area and ventral margin of valva is less concave (valva base extensions further apart, margin between extensions with almost horizontal area and ventral margin of valva is concave in *D. raunaria*).

FEMALE GENITALIA: Generally as in *D. raunaria*, but with following differences: In *D. royaria* lamella antevaginalis is tubular, strongly sclerotised and lamella postvaginalis margins are strongly sclerotised (lamella antevaginalis tubular, weakly sclerotised, lamella postvaginalis margins weakly sclerotised in *D. raunaria*).

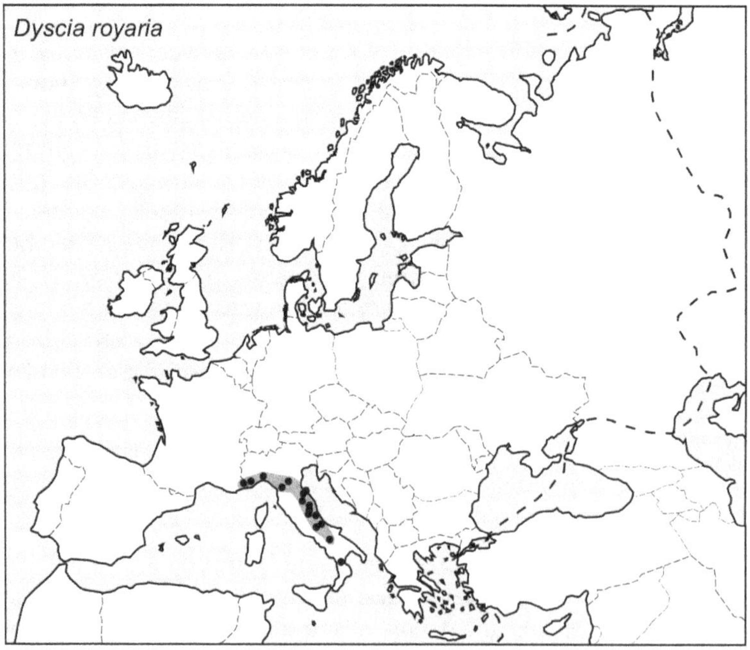

DISTRIBUTION AND ABUNDANCE: Central-Mediterranean. Endemic in the Alps in south-easternmost France through Liguria and the Apennines in Italy. Since this species is newly described, its abundance is still rather poorly known, but it appears that it is rather local. In south-eastern France sometimes in numbers (C. TAUTEL pers. comm.), in central and southern Italy mainly in limited numbers (C. FLAMIGNI & P. PARENZAN pers. comm.).

PHENOLOGY: Uni- or bivoltine. From late June to mid-August in the south-western Alps. Further to the south in two generations from late May to late June and from late August to early September. In late July at 2000 m in the central Apennines indicates univoltinism there. Larvae are found from August and overwintering until April or May (C. TAUTEL pers. comm.). The moths fly at night. A few males and females can be seen flying at dusk, but the main flight activity is in the night. The moths fly only in places with the food plant and they are attacted to light (TAUTEL *et al.* 2007).

BIOLOGY: Mono- or oligophagous. The caterpillar has been found on *Artemisia alba*. Larva reared on various Apiaceae, e. g. *Daucus carota* (southern Italy A. HAUSMANN pers. comm.). Pupation takes place in a light spun web under a stone (C. TAUTEL pers. comm.).

HABITAT: Xerophilous. Open areas on south-facing mountain slopes, warm and dry meadows with sandy, uncultivated areas, ruderal vegetation and along old walls. In France and Liguria around 1000 m altitude and at Monte Pollino in southern Italy at 2050 m (FLAMIGNI *et al.*, in prep.). At Campo Imperatore in central Italy up to 2000 m.

SIMILAR SPECIES: *D. raunaria* and *D. conspersaria*, see *D. raunaria*. See Text-figs 134–143 and Table 9.

GENETIC DATA: The *raunaria/royaria* species complex (n=34) genetically splitting up into three BINs. Populations from France and north-westernmost Italy diverging from those of central and southern Italy by 1.7%. Nearest species: *D. raunaria* (1.5%).

REMARKS: See *D. raunaria*. TAUTEL & BILLI (2006) describe the diagnostic adult features and TAUTEL *et al.* (2008) describe the early stages of *D. royaria*. Due to only minor morphological differences between *D. raunaria* and *D. royaria*, a thorough, modern taxonomic study needs to be carried out, to examine in detail whether *D. royaria* is a distinct species (or two?, see the genetic data) or if it represents, for instance, an end of a cline of *D. raunaria*.

128. Dyscia innocentaria (CHRISTOPH, 1885)

innocentaria innocentaria CHRISTOPH, 1885, in ROMANOFF, Mém. Lépid. 2: 125, pl. 6, fig. 5, (*Aspilates*). Lectotype female (Zoological Institute,St.-Petersburg, Russia), [Turkmenistan], Aschchabad, Achal-Tekke district, near Nuchur. Only one female is left of the original type material (1 male, 2 females) (TRUSCH & ERLACHER, 2001).

innocentaria govii FIUMI & FLAMIGNI, 2013, Quad. Studi Nat. Romagna 38: 192, figs. 1a-f, gen. figs. 3a, 4a, 5ab, 6bd, (*Dyscia*). Holotype male (Museo Civico di Storia Naturale 'G. Doria' di Genova, Italy), [France]: Corsica, Omessa (Haute-Corse), Francardo, 370 m, 28.V.2012 (examined, including genitalia, based on the original illustrations). Originally as sp., dowgraded here from species rank to a subspecies of *Dyscia innocentaria* (CHRISTOPH, 1885), quoting minor quantitative differences in the external appearance and genitalia characters and allopatric occurrence (new status).

innocentaria osmanica WAGNER, 1931, Int. ent. Z. 24 (47): 491, pl. 1, figs 9, 10 (*Scodiona*). Lectotype male, paralectotypes 1 male, 1 female (Naturhistorisches Museum, Wien, Austria), [Turkey]: Aksehehir.

innocentaria sicanaria OBERTHÜR, 1923, Études Lépid. comp. 20: 261, pl. 557, figs 4780, 4781, (*Scodiona*). Lectotype female (Zoologisches Forschungsinstitut und Museum Alexander Koenig, Bonn, Germany), Italy, Sicily, paralectotypes 1 male, 2 females (Zoologisches Forschungsinstitut und Museum Alexander Koenig, Bonn, Germany). [Originally as subsp. of *conspersaria* [DENIS & SCHIFFERMÜLLER]]. See remarks.

scannaria DANNEHL, 1933, Ent. Z., Frankf. a. M. 47 (18): viii, 146, (*Scodiona*). Lectotype male (Zoologische Staatssammlung, München, Germany), [Italy]: Apenn. Central, M[on] t[a]gna Grande, Scanno (examined externally), paralectotypes 4 males (Zoologische Staatssammlung, München, Germany) (examined externally).

sicula REISSER, 1940 in ALBERS & WARNECKE, Zeitschrift des Wiener Entomologen-Vereines 25: 119 (*Dyscia*). Lectotype male (Staatliches Museum für Naturkunde, Karlsruhe, Germany), [Italy, Sicily]: Mistretta, paralectotypes 34 males, 14 females (Staatliches Museum für Naturkunde, Karlsruhe, Germany), [Italy, Sicily]: Mistretta. [Originally as subsp. of *osmanica* WAGNER]

EXTERNAL CHARACTERS AND ABDOMEN, based on European specimens (see Remarks): Wingspan 26–36 mm. Generally as *D. raunaria*, see above, but with following differences: *D. innocentaria* is slightly paler, fringes chequered brown and grey, more pronounced in females, forewing subterminal line near tornus often heavily irrorated with blackish brown. Tympanal organs large, bullae tympani with small roundish lobes.

VARIATION: Amount of irroration somewhat variable. Fringes can be concolorous with wings. Forewing postmedial line near tornus can be with or without irroration. The Asian specimens (nominotypical subspecies) are more uniform brownish grey-white, slightly smaller, and fringes are less chequered.

Subsp. ***govii*** has wingspan between 35–37 mm. Antemedial and postmedial lines are more pronounced, broken into dots and discal spots are distinct. Male and female genitalia show minor quantitative differences: Aedeagus is symmetrically angled at middle and ductus bursae is larger in subsp. *govii* (FIUMI & FLAMIGNI 2013). These structures vary within the distribution area of *Dyscia innocentaria* and those of *govii* fall in our opinion within intraspecific variation. The lack of qualitative differences, both in the external appearance and genitalia characters, and allopatric occurrence in Corsica and Sardinia, lead us to conclude that taxon *govii* FIUMI & FLAMIGNI is a subspecies of *Dyscia innocentaria* (CHRISTOPH, 1885) (new status), though the DNA barcodes generated immediately before the printing of this book again questioned this view (see genetic data).

MALE GENITALIA: Uncus large, margins sclerotised, weakly setose. Socii absent. Gnathos arms narrow, fused ventrally, upturned. Valva divided. Dorsal arm of valva massive, narrow at base, apex round, setose. Costa base with elongated, sharp projections. Ventral arm of valva sclerotised, tapering towards apex. Ventral margin with row of acute spines. Valva fused dorsally. Juxta large, sclerotised. Saccus wide, weakly quadrangular, with upright extension medially. Aedeagus weakly angled, caecum long, apex with elongated projection. Vesica opens at approximately 90 degree angle. Vesica wide at base, with roundish diverticulum, followed by 90 degree angle, without sclerotisations.

FEMALE GENITALIA: Papillae anales elongated, setose. Apophyses anteriores about 1/4 length of apophyses posteriores. Ostium bursae region complex, heavily sclerotised. Lamella antevaginalis with two large, tubular sclerotisations. Lamella postvaginalis with two pairs of sclerotisations, anterior pair almost fused (or fused) ventrally. Sterigma/ductus bursae funnel-shaped, margins most sclerotised. Corpus bursae large, membranous, round, inner surface covered with minute spines.

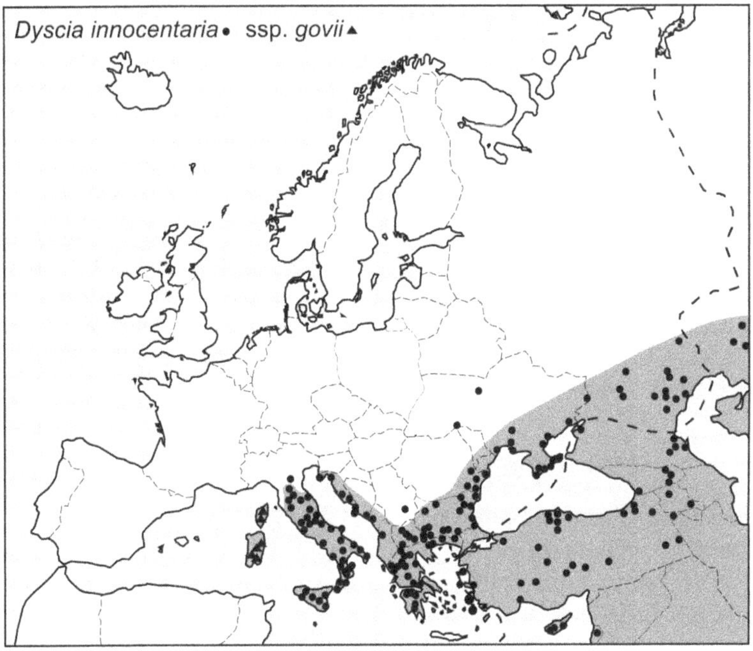

DISTRIBUTION AND ABUNDANCE: Southeuropean-Westasiatic. In Europe from Sicily and Italy eastwards through, the Balkans, Ukraine and southernmost European Russia to southern Urals. In Corsica and Sardinia subsp. *govii*. – Outside Europe the species has been found in the Caucasus region, Turkey, Cyprus, Lebanon, Syria, Israel, Jordan, northern Iran, central Asia, Afghanistan and China (TRUSCH & ERLACHER 2001). – In Europe a common species that is usually found in limited numbers.

PHENOLOGY: Bivoltine. From early May to late June or early July and from late August or early September to early October. Hibernation takes place as a small larva (M. LEIPNITZ pers. comm.), larvae were found in mid-February and mid-December (D. FRITSCH pers. comm.). The adult moths fly at night when they come readily to light.

BIOLOGY: Oligophagous, possibly polyphagous, often on Lamiaceae. On Samos caterpillars have been found on *Thymbra capitata* (Lamiaceae) and *Sarcopoterium spinosum* (Rosaceae) (FRITSCH *et al.* 2014). Reared on *Artemisia campestris* (Asteraceae) (TRUSCH & ERLACHER 2001), *A. vulgaris* and *Origanum vulgare* (Lamiaceae) (B. MÜLLER pers. comm.). Larvae of subsp. *govii* have been found on *Teucrium marum* (Lamiaceae) (ANONYMOUS 2014).

HABITAT: Xerothermophilous. In a variety of warm places with scattered bushes and trees. The species occurs from sea level and up to 1600 m (TRUSCH & ERLACHER 2001). In Turkey up to 2000 m.

SIMILAR SPECIES: Most *Dyscia* species are externally similar. *D. raunaria* and *D. royaria* are darker grey and fringes are concolorous with wings. *D conspersaria* is normally more brown and fringes are concolorous with wings. Male and female genitalia of *D. innocentaria* allow reliable identification. See Text-figs 134–143 and Table 9.

GENETIC DATA: Genetically homogeneous in the main cluster from Balkan countries to Turkey and the Eastern Mediterranean (n=24), but populations from Italy diverging by 2.2% (n=8), subsp. *govii* by 2.7% (n=2), the latter diverging by 2.0% from *D. royaria*. Populations of northern Iran belonging to a fourth BIN. Nearest species: *D. conspersaria* (2.8%).

REMARKS: We do not recognise subspecies in the distribution map, apart from from subspecies *govii* in Corsica and Sardinia, because the homogeneous genetic data from Europe to Turkey does support the morphologically defined subspecies concepts.The taxon *sicanaria* (OBERTHÜR) was originally described as a geographical variation of *Scodiona conspersaria* (DENIS & SCHIFFERMÜLLER) and raised to species rank by WARNECKE (1941). Synonymy was established by TRUSCH & ERLACHER (2002), who concluded that 'because of its more bright basal colour and some differences in the markings of the wings, *Dyscia innocentaria sicanaria* (OBERTHÜR, 1923) can be treated as a subspecies of *Dyscia innocentaria*, its range is South Europe.' More research is needed.

129. Dyscia malatyana WEHRLI, 1934

malatyana malatyana WEHRLI, 1934, Int. ent. Z. 27 (47): 535, fig. 19 (*Dyscia*). Holotype male, 2 paratype males (Zoologisches Forschungsinstitut und Museum Alexander Koenig, Bonn, Germany), [Turkey:] Malatya-Tecde (holotype examined externally, illustrated in TRUSCH & ERLACHER 2001).

plebejaria dagestana WEHRLI, 1934, Int. ent. Z. 27 (47): 536, fig. 20 (*Dyscia*). Lectotype female (Zoologisches Forschungsinstitut und Museum Alexander Koenig, Bonn, Germany), [Russia:] Dagestan, Kumtorkale (synonymy proposed in TRUSCH & ERLACHER 2002).

malatyana albersaria WARNECKE, 1940, Mitt. münch. ent. Ges. 30 (3): 1048 (*Dyscia*). Holotype male (coll. FIXSEN in Zoologisches Institut und Zoologisches Museum, Universität von Hamburg, Germany, destroyed in 1943), [Turkmenistan:] Achal Tekke (synonymised with *Dyscia malatyana* by WARNECKE (1941: 244), treated as a synonym of *malatyana* WEHRLI in SCOBLE (1999), new status proposed in TRUSCH & ERLACHER 2002).

malatyana nachadira BRANDT, 1941, Mitt. münch. ent. Ges. 31 (3): 883, pl. 30, fig. 47 (*Dyscia*). Lectotype male (Swedish Museum of Natural History, Stockholm, Sweden), Iran: Baloutchistan, street Khach-Zahedan, Fort Sengan, 1800 m (treated valid at subspecies level in SCOBLE (1999), synonymy proposed in TRUSCH & ERLACHER 2002).

malatyana clara PUGAEV & ZOLOTUHIN, 2006, Eversmannia 6: 9, figs. 25–28 (*Dyscia*). Holotype female (Zoological Institute, St.-Petersburg, Russia), [Russia:] Republic of Kalmykia, St. Andrew Sands (examined externally).

malatyana senecai WILTSHIRE, 1990, Nota lepid. 12 (4): 355, figs. 1, 4, 12, 15 (*Dyscia*). Holotype male (Zoological Museum, University of Copenhagen, Denmark), Libya: Gharian, Wadi el Hira (originally as species, subspecies-level synonymy proposed in TRUSCH & ERLACHER 2002).

EXTERNAL CHARACTERS AND ABDOMEN, based on European subsp. ***clara***: Wingspan 29–34 mm. Wings chalk-white to cream, weakly irrorated with blackish. Antemedial line, subterminal line and terminal line blackish, creating strong black and white contrast. Terminal line broken into dots, distinct, blackish. Fringes concolorous with wings. Forewing discal spot distinct, blackish, elongated. Hindwing discal spots smaller, roundish. Wings below whitish, almost without markings, except discal spots well-developed. Frons, collar, thorax and abdomen concolorous with wings. Proboscis reduced. Male antennae bipectinate, female antennae filiform. Hindtibia of both sexes with 2+2 spurs. Tympanal organs very small, bullae tympani with small roundish lobes. Other abdominal segments of both sexes undifferentiated.

VARIATION: Very variable. Ground colour varies from brown to brownish yellow (subsp. *malatyana*) to pure cream (subsp. *albersaria*) to sand-yellow (subsp. *senecai*) to pure or chalk-white (subsp. *clara*). Antemedial and postmedial lines continous or broken into spots. The external appearance and geographical location correlate to some degree, but the external differences between the putative subspecies are not clear-cut. Brownish specimens often have the subterminal and terminal line brownish, and these do not stand out from the ground colour. Whitish and white specimens often have the subterminal and terminal line blackish, creating contrasting black-and-white appearance. Amount of irroration somewhat variable, often amount of irroration is reduced.

Subsp. ***clara*** (see description above) occurs in the extreme south-eastern part of Europe, in Astrakhan Region and Republic of Kalmykia in Russia. Native to desert biotopes (PUGAEV & ZOLOTUHIN 2006). We illustrate the European subspecies only (Plate 14).

MALE GENITALIA: Uncus small, membranous, weakly setose. Socii absent. Gnathos arms wide, fused ventrally, forming prominent, cone-shaped, upturned medial projection. Costa base with long, evenly curved extension, apex with small sclerotised tooth. Valva narrow and long, ventral margin with concavity, apex round, weakly setose. Valva fused dorsally. Juxta large, sclerotised, ventrolateral margins folded inwards, forming pocket. Saccus wide, roundish, with upright extension medially. Aedeagus curved, wide at base, caecum long. Apex sharp, sclerotised. Vesica opens at approximately 90 degree angle. Vesica narrow, tubular, without sclerotisations.

FEMALE GENITALIA: Papillae anales wide, setose, surface weakly granulate. Apophyses anteriores about 1/3 length of apophyses posteriores. Lamella antevaginalis weakly sclerotised medially. Lamella postvaginalis with narrow, horizontal, wing-like sclerotisations, with further lunular plate posteriorly. Sterigma/ductus bursae sclerotised, anterior part narrower. Corpus bursae large, membranous, round, inner surface covered with very minute spines.

DISTRIBUTION AND ABUNDANCE: Westpalaearctic. In Europe only known from the south-easternmost part of European Russia and European part of Kazakhstan (subsp. *clara*) (PUGAEV & ZOLOTUHIN 2006). – Outside Europe widely distributed in Tunisia (outside map area), Libya, Turkey, southern and eastern Levant, the Caucasus region, Iran, Afghanistan, Pakistan, central Asia and Transbaikalia. In Europe a rare and local species. Outside Europe the species is not rarely found in numbers and this may also apply to the limited European occurrence.

PHENOLOGY: Bivoltine. From early April to mid-June and again from late September to mid-November (TRUSCH & ERLACHER 2001). The moths fly at night and come abundantly to light, now and then with more females than males.

BIOLOGY: No natural food plants are known. In rearings accepting *Artemisia campestris* (TRUSCH & ERLACHER 2001). In North Africa possibly on *Artemisia herba-alba* (WILTSHIRE 1990a).

HABITAT: Xerothermophilous. A desert and semi-desert species that occurs in oases and also in steppe habitats.

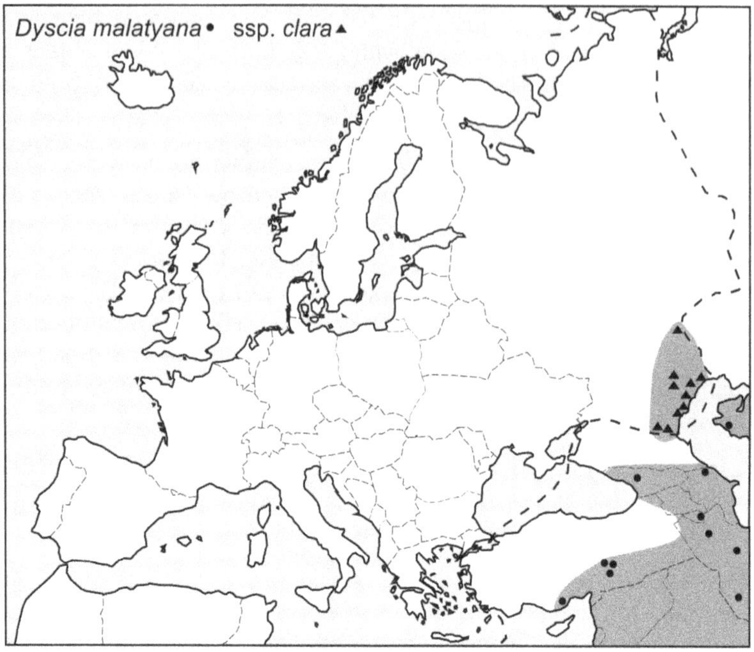

SIMILAR SPECIES: Strong black-and-white contrast, white wings, chequered blackish and white fringes, distinct subterminal line, which is angled inwards on costa, and elongated forewing discal spot of *D. malatyana* allow reliable identification. Male and female genitalia allow reliable identification, diagnostic characters include for instance: Valva long and narrow, costa base with large, weakly inwards curved extension. Aedeagus curved and apex acute. Lamella postvaginalis is with two, narrow, lateral sclerotisations. See Text-figs 134–143 and Table 9.

GENETIC DATA: Genetically heterogeneous from Jordan, Iran, Kazakhstan and Uzbekistan (n=11), maximum variation 3.5%, but all assigned to the same BIN. Nearest species: *D. negrama* WEHRLI, 1953 from Iran (5.2%) and *D. penulataria* (5.4%).

130. *Dyscia crassipunctaria* (REBEL, 1916)

crassipunctaria REBEL, 1916, Annln naturh. Mus. Wien 30: 140, pl. 4, fig. 5, (*Scodiona*). Lectotype female (Naturhistorisches Museum, Wien, Austria), Hagia, Varvára, paralectotypes 3 males, 1 female (Naturhistorisches Museum, Wien, Austria), [Greece, Crete]: Sphakia, Mèlabes, Asitaes, Neapolis [Originally as spp. of *conspersaria* [DENIS & SCHIFFERMÜLLER]].

INTRASUBSPECIFIC NAME, POTENTIALLY UNAVAILABLE: *phthinopora* REISSER, 1962 (summer generation).

EXTERNAL CHARACTERS AND ABDOMEN: Wingspan 34–36 mm, second generation specimens smaller, wingspan 27–28 mm. Wings brownis -grey, irrorated with brown, veins yellowish in medial area. Antemedial line weak. Medial line represented by brown shade, most distinct near inner margin. Subterminal line evenly curved, broken into blackish brown dots.

Hindwings paler in basal area. Terminal line fine. Fringes concolorous with wings. Discal spots large, blackish brown, solid or circled. Wings below as above, less irrorated with brown and markings weaker. Frons, collar and thorax uniform whitish. Abdomen concolorous with wings. Proboscis reduced. Male antennae bipectinate, female antennae filiform. Hindtibia of both sexes with 2+2 spurs. Tympanal organs medium-sized, bullae tympani with small roundish lobes. Other abdominal segments of both sexes undifferentiated.

VARIATION: Amount of irroration somewhat variable. Ground colour varies from pale grey to grey-brown to brown-grey. Subterminal line can be reduced to line of weak dots.

MALE GENITALIA: Uncus small, sclerotised, weakly setose. Socii absent. Gnathos arms narrow, fused ventrally, acute apex downturned. Valva wide, apex round, weakly setose. Costa base with wide, blunt extension, base with shorter pair of acute extensions, apex weakly setose. Ventral margin of valva with concavity. Valva fused. Juxta large, sclerotised. Saccus round, with upright extension medially. Aedeagus narrow, caecum long, apex angled, with acute projection. Vesica opens at approximately 135 degree angle. Vesica narrow, tubular, without sclerotisations.

FEMALE GENITALIA: Papillae anales large, weakly setose. Apophyses anteriores about 1/3 length of apophyses posteriores. Ostium bursae surrounded by large sclerotised element. Ventral margin narrow sclerotised ridge, with long medial extension. Posterior margin narrower. Corpus bursae elongated, membranous, inner surface covered with minute spines.

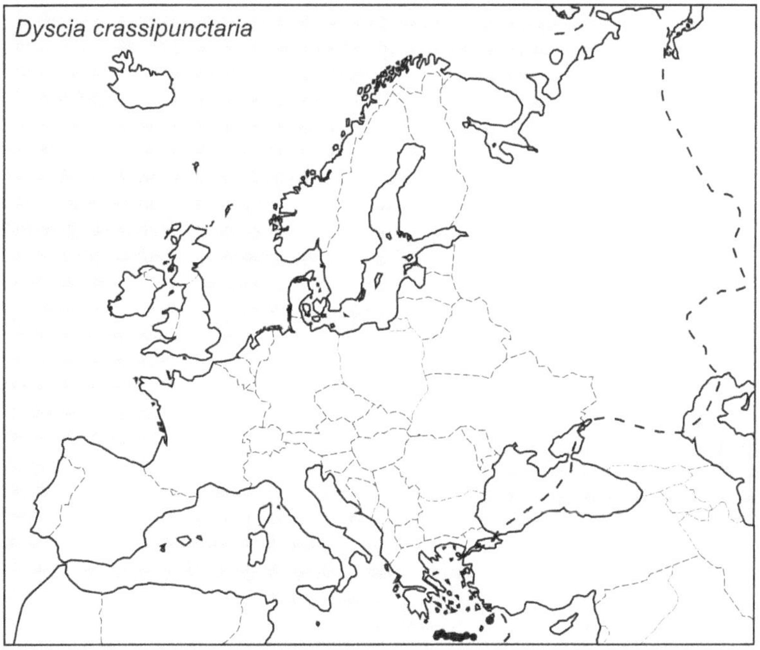

DISTRIBUTION AND ABUNDANCE: East-Mediterranean. Endemic to Crete and Rhodes, from Rhodes only a single specimen is known (coll. ZMUC), perhaps a migrant from the population on Crete. In Crete a rather common species that rather often occurs in numbers.

PHENOLOGY: Bivoltine. From late March to early June and from mid-September to late October (TRUSCH & ERLACHER 2001). At higher altitudes the species has also been found in mid-June and it is probably flying later here, perhaps in only one generation. The hibernation takes place as a caterpillar. The species flies at night and comes readily to light.

BIOLOGY: Mono- or oligophagous. The caterpillar has been found in early and mid-May on *Genista monspessulana* (M. LEIPNITZ pers. comm.). In rearing accepting *Lonicera tatarica* and *Artemisia campestris* (TRUSCH & ERLACHER 2001) and *Achillea millefolium* and *Artemisia vulgare* (E. FRIEDRICH pers. comm.).

HABITAT: In rather open macchia localities mainly on various kinds of limestone ground. The species has been found from sea level up to 1150 m.

SIMILAR SPECIES: Most *Dyscia* species are externally similar. *D fagaria* has chequered fringes and forewing subterminal line is angled, weakly concave, before it reaches costa (fringes concolorous with wings, postmedial line evenly curved in *D. crassipunctaria*). *D. fagaria* and *D. crassipunctaria* do not occur sympatrically. Male and female genitalia allow reliable identification. See Text-figs 134–143 and Table 9.

GENETIC DATA: Genetically homogeneous (n=4). Nearest species: *D. conspersaria* (4.6%) and *D. simplicaria* REBEL, 1939 from Cyprus (4.9%).

131. Dyscia conspersaria (DENIS & SCHIFFERMÜLLER, 1775)

conspersaria conspersaria [DENIS & SCHIFFERMÜLLER], 1775, Ankündung syst. Werkes Schmett. Wienergegend: 106, (*Geometra*). Neotype male, (Zoologisches Forschungsinstitut und Museum Alexander Koenig, Bonn, Germany), Austria: Wien, Mödling.
cunicularia ESPER, 1803, Die Schmett. 5 (9): 239, pl. 43, figs 5, 6, (*Phal*[*aena*] *Geom*[*etra*]). Syntype(s), [Austria]: Vienna district, [Italy]: Florence.
cuniculina HÜBNER, 1790, Beitr. Gesch. Schmett. 2 (2): 38, pl. (2) 1, fig. E, (*Ph*[*alaena*] *Noct*[*ua*]). Holotype (lost), [Austria]: Vienna district.
conspersaria sultanica WEHRLI, 1936, Mitt. münch. ent. Ges. 26 (1): 36, pl. 2, fig. 5, (*Dyscia*). Lectotype male, paralectotypes 10 males (Zoologisches Forschungsinstitut und Museum Alexander Koenig, Bonn, Germany), [Turkey]: Asia min., Aksehir, Sultan Dagh. See remarks.
conspersaria turturaria BOISDUVAL, 1840, Genera Index meth. eur. Lepid.: 185, (*Scodiona*). Syntype(s), [France]: Digne. See remarks.

UNAVAILABLE NAMES (incorrect subsequent authorship): *conspersaria* FABRICIUS (see FLETCHER 1979).

EXTERNAL CHARACTERS AND ABDOMEN: Wingspan 31–38 mm. Generally as *D. raunaria*, see above, but with following differences: *D conspersaria* is yellowish brown, khaki-coloured, finely irrorated with greyish. Forewing discal spot often hollow. Fringes weakly rusty brown. Wings below as above, forewings heavily suffused with brown.

VARIATION: Amount of irroration somewhat variable. Fringes may be concolorous with wings. Forewing subterminal line near tornus often irrorated with brown, this irroration normally lacking in females. Specimens in France and Lower Volga area in Russia are paler, the other European specimens are more light khaki-coloured, the Turkish specimens distinctly yellowish-brown.

Subsp. ***turturaria*** is paler, whitish, postmedial line broken into small, separate dots. No difference observed in the genitalia compared to other European specimens. Restricted to south-eastern France, occurring mostly in xeric rocky slopes between 800–1100 m altitude.

MALE GENITALIA: Uncus small, membranous, weakly setose, margins tubular. Socii absent. Gnathos arms narrow, fused ventrally. Valva wide, upright, apex round, setose. Costa base with sclerotised, upright extension, apex with acute tooth. Margin between these extensions with round lobe. Ventral margin of valva with concavity. Valva fused ventrally. Juxta large, sclerotised. Saccus round, with upright extension medially. Aedeagus rather straight, caecum long. Apex with two dentate projections. Vesica opens at approximately 135 degree angle. Vesica wider at base, tubular, without sclerotisations.

FEMALE GENITALIA: Papillae anales elongated, setose. Apophyses anteriores about 1/3 length of apophyses posteriores. Ostium bursae surrounded by several sclerotised striations laterally. Lamella postvaginalis with two narrow, wrinkled sclerotisations and with minute, elongated sclerotisation medially. Ductus bursae narrow, sclerotised. Corpus bursae large, membranous, round, inner surface covered with minute spines.

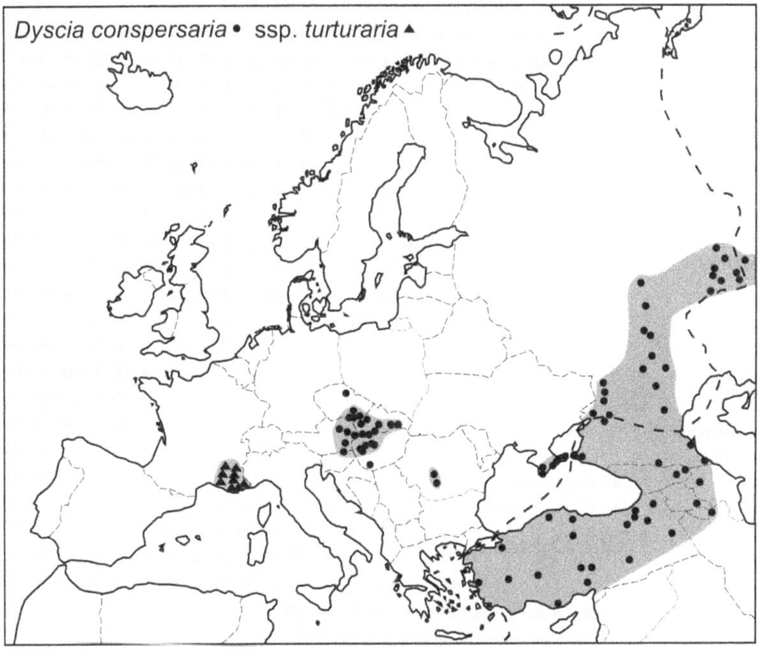

DISTRIBUTION AND ABUNDANCE: Euro-Caucasian. In Europe isolated populations in south-eastern France, south-eastern central Europe, Romania and from southern Ukraine through south-eastern European Russia to southern Urals. – Outside Europe known from Turkey and the Caucasus region. – Uncommon and local in Europe, rarely in numbers.

PHENOLOGY: Univoltine. From late May to early August (TRUSCH & ERLACHER 2001). In south-eastern France from mid-June to mid-July (C. TAUTEL pers. comm.). Caterpillars have been found in late summer or early autumn and until spring. The overwintering takes place as a larva. The moths fly at night and come to light.

BIOLOGY: Polyphagous. In Crimea the larva has been found on *Hedysarum tauricum* (BUDASHKIN & SAVCHUK 2010). Larva recorded on *Salvia, Artemisia absinthium* (VOJNITS 1980) *Salvia pratensis, S. officinalis, Artemisia, Lavandula* and *Dorycnium* in France (LHOMME 1935). Reared on *Artemisia campestris* (TRUSCH & ERLACHER 2001) and *A. vulgaris* (B. MÜLLER pers. comm.).

HABITAT: Xerothermophilous. On dry steppe localities, often on lime- or sandstone soil or south exposed slopes from sea level up to at least 1100 m altitude (TRUSCH & ERLACHER 2001). In south-eastern France in sunny, rocky places in the lower montane zone from mainly 800 m to 1000 m.

SIMILAR SPECIES: Most *Dyscia* species are externally similar. *D. innocentaria* is paler, fringes are chequered brown and grey (yellowish-brown, fringes not chequered in *D. conspersaria*). Pale specimens of *D. penulataria* have forewing subterminal line angled before costa (forewing postmedial line evenly curved on costa in *D. conspersaria*). Male and female genitalia allow reliable identification. See Text-figs 134–143 and Table 9.

GENETIC DATA: Genetically heterogeneous in Europe, populations of France diverging from those of Hungary and Austria by 1.9%. Two further BINs in Turkey. Nearest species: *D. raunaria* (2.1%).

REMARKS: Taxon *sultanica* WEHRLI is considered valid at subspecies level by TRUSCH & ERLACHER (2002). They noted that the ventral margin of the male valva bears a small 'lateral step', which has not so far been seen in the European material, but all other diagnostic male genitalia characters, as described in WEHRLI (1936, 1939–1954) are gradual and not clearly definable. We have not examined adequate amount of material to conclude where the potential subspecies boundary lies between nominotypical subspecies and *sultanica*, and further, how the Turkish material relates to those from Transcaucaus and Lower Volga region in Russia. Therefore subspecies *sultanica* is not recognised in the distribution map.

Taxon *turturaria* (BOISDUVAL) is considered valid at subspecies level by SCOBLE (1999) and TRUSCH & ERLACHER (2002), and as a junior synonym of *D. conspersaria* by LERAUT (1997, 2009). According to TRUSCH & ERLACHER (2002), the type material should be in the Zoologisches Forschungsinstitut und Museum Alexander Koenig, Bonn, Germany, but the type is missing there.

132. Dyscia fagaria (THUNBERG, 1784)

Synonymies and subspecies of *D. fagaria* are after TRUSCH & ERLACHER (2002), who provide valuable information on the taxa and the type material, and after TRUSCH & OCHSE (2004).

fagaria fagaria THUNBERG, 1784, Diss. ent. sistens Insecta Suecica 1: 7, (*Geometra*). Lectotype male (Universitets Zoologiska Institut, Uppsala, Sweden), Sweden: Halland.

alvarensis WAHLGREN, 1913, Ent. Tid. 34: 166, (*Scodiona*). Syntypes 2 males (probably in Zoological Institute University of Lund, Lund, Sweden), [Sweden:] Öland Island. [Originally as var.of *fagaria* THUNBERG].

belgaria HÜBNER, 1790, Beitr. Gesch. Schmett. 2 (4): 92, pl. (4) 2, fig. N, (*Ph[alaena] Geom[etra]*). Syntype(s) [type material lost], Holland [Netherlands].

belgiaria BORKHAUSEN, 1794, Natur. eur. Schmett. 5: 195, (*Phalaena Geometra*). Syntype(s), [Belgium].

belgaria HÜBNER, [1799], Samml. Eur. Schmett. 5 Geometrae (1): pl. 26, fig. 140, (*Geometra*). Syntype(s), lost, Europe.

mediopunctaria DONOVAN, 1808, The natural history of British insects 13: 59, pl. 461, fig. 1, (*Phalaena*). Syntype(s), (The Natural History Museum, London, UK), [United Kingdom]. (examined externally)

fagaria albidaria STAUDINGER, 1871, in STAUDINGER & WOCKE, Cat. Lepid. eur. Faunengeb. 1: 173, (*Scodiona*). Syntype(s), Gallia [France (central, south and western], [Originally as var. of *fagaria* THUNBERG]. Lectotype male (Museum für Naturkunde der Humboldt-Universität), [Western France]: Vendée.

nigerrima DURAND, 1934, Lambillionea 34: 173, (*Dyscia*?). Holotype male, [Western France]: Longeville. [Originally as var. of *fagaria* THUNBERG].

fagaria emucidaria HÜBNER, [1813], Samml. Eur. Schmett. 5 Geometrae (1): pl. 82, fig. 425, (*Geometra*). Neotype male (Zoologisches Forschungsinstitut und Museum Alexander Koenig, Bonn, Germany), [South France]: Basses Alpes, Env. de Digne.

fagaria favillacearia HÜBNER, [1799], Samml. Eur. Schmett. 5 Geometrae (1): pl. 26, fig. 139, (*Geometra*). Neotype male (Zoologisches Forschungsinstitut und Museum Alexander Koenig, Bonn, Germany), Austria inf., [Vienna district], Mödling. [Originally as sp].

duanjiao YANG, 1978, Moths of North China 2: 392, pl. 20, fig. 13, (*Dyscia*). Holotype male, (Agricultural University Beijing, Beijing, China), [China] (north): Beijing, Huai-rou district.

fagaria favillacearius WOOD, [1834], Index Ent.: 78, pl. 18, fig. 458a, (*Bupalus*). [Emendation of *favillacearia* HÜBNER?]

fleischmanni REBEL, 1910, Verh. zool-bot. Ges. Wien 60: (17), fig. 4 (*Scodiona*). Holotype male, [Lower Austria]: Sollenau, Steinfeld (probably in Naturhistorisches Museum, Wien, Austria). [Originally as var. of *fagaria favillacearia* HÜBNER].

ilivolans WEHRLI, 1953, in SEITZ, Gross-Schmett. Erde 4 (Suppl.): 662, (*Dyscia*). Lectotype female, [China]: Kuldja [Yining (Gulja)], (paralectotype male, [China/Kazakhstan]: Ili region, (both type specimens in Zoologisches Forschungsinstitut und Museum Alexander Koenig, Bonn, Germany).

psoricaria EVERSMANN, 1848, Bull. Soc. imp. Nat. Moscou 21 (3): 221, (*Boarmia*). Lectotype female (Zoological Institute, St.-Petersburg, Russia), [Russia]: Lower Volga, Calmna. [Originally as sp].

INTRASUBSPECIFIC NAMES, POTENTIALLY UNAVAILABLE: *albescens* LEMPKE, 1952 (var.); *fusca* LEMPKE, 1952 (var.); *postdelineata* LEMPKE, 1952 (var.); *signata* COCKAYNE, 1942 (var.).

EXTERNAL CHARACTERS AND ABDOMEN: Wingspan ♂ 23–33 mm, ♀ 23–28 mm, second generation specimens smaller. Wings grey(-brownish), irrorated with brown, markings contrasting. Antemedial line weak or broken into row of blackish brown dots. Subterminal line broken into row of blackish brown dots, curved outwards in middle, weakly concave before costa. Terminal line row of blackish dots, often extending to fringes. Fringes concolorous with wings. Discal spots blackish, distinct, elongated. Wings below as above, antemedial line often weak or absent. Frons, collar, thorax and abdomen concolorous with wings. Proboscis reduced. Male antennae bipectinate, female antennae filiform. Hindtibia of both sexes with 2+2 spurs. Tympanal organs small, bullae tympani with small roundish lobes. Other abdominal segments of both sexes undifferentiated.

VARIATION: Ground colour varies from whitish-grey to grey-brown to pale brown-grey. Markings normally contrasting, standing out clearly, but can be weaker and medial lines can be almost absent. Medial lines near inner margin often widely irrorated with brown. Terminal line can be concolorous with wings.

Subsp. ***albidaria*** has wings with whitish ground colour, markings are less extensive, yet clearly contrasting. Restricted to western coastal region in France: Gironde, Charente-Maritime, Vendée and Ile d'Olèron. Not detailed as a subspecies in the distribution map.

Subsp. ***emucidaria*** has wings uniform grey-brown, markings finer. Described from southern France: Basses Alpes, Env. de Digne. Not detailed as a subspecies in the distribution map.

Subsp. ***favillacearia*** has wings with brownish grey ground colour, markings are less extensive. Widespread, from Central Europe to East Asia. Not detailed as a subspecies in the distribution map.

MALE GENITALIA: Uncus narrow, acute, weakly setose. Socii absent. Gnathos arms narrow, fused ventrally, forming curved plate. Valva narrow, upright, apex round, weakly setose. Costa base with long, narrow, curved extension, these nearly fused medially. Ventral margin of valva with minor concavity. Juxta large, sclerotised. Saccus wide, slightly quadrangular, with upright extension medially. Aedeagus curved, caecum long. Apex with acute, sclerotised projections dorsally and ventrally. Vesica opens at approximately 90 degree angle. Vesica developed as narrow tube. Apex with rounded apical diverticulum, with row of minute cornuti.

FEMALE GENITALIA: Papillae anales large, setose. Apophyses anteriores about 1/2 length of apophyses posteriores. Lamella antevaginalis membranous. Lamella postvaginalis weakly sclerotised, posterior margin with acute sclerotisation. Posterior part of ductus bursae partly sclerotised, funnel-shaped. Anterior part of ductus bursae membranous, narrow, striated. Corpus bursae small, membranous, inner surface without minute spines.

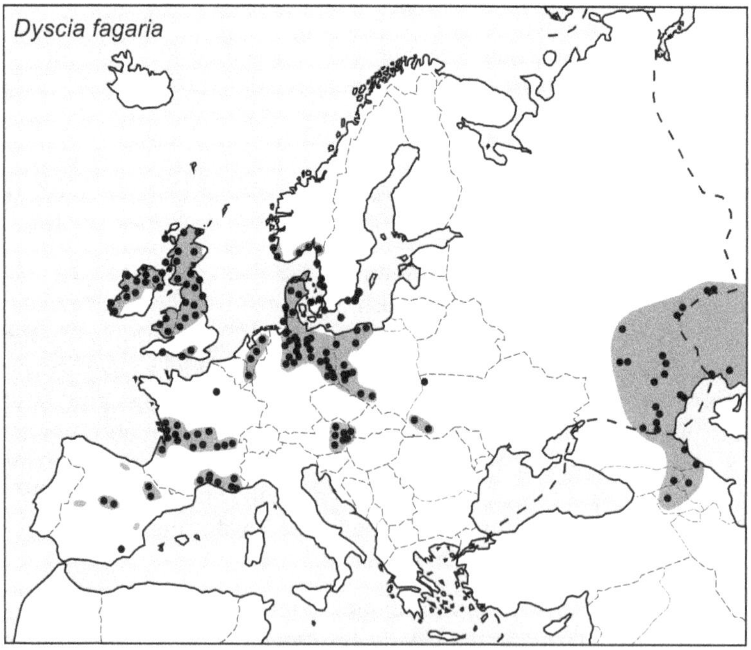

DISTRIBUTION AND ABUNDANCE: Eurasiatic. Occurrence in Europe scattered and disjunct. The species is found around the North Sea and the southern Baltic Sea, in south-west Europe and further the species is recorded in various parts of central Europe and in southern European Russia and the southern Urals. – Outside Europe there are scattered records from Caucasus, central Asia, Transbaikal, Far East Russia and China. – Though widely distributed, this species is local and not common in Europe and mainly found in limited numbers.

PHENOLOGY: Uni- or bivoltine. In southern Denmark and southern Scandinavia in one generation from mid-May to late June, in the British Isles from late May to early July (SKINNER 2009), in the Iberian Peninsula from April to May and from August to September. In north-western Europe the larva lives from July to April, in south-west Europe also between the two flying periods. The males come now and then to light, but they are probably more often found at night sitting on the vegetation in their localities. The females are rarely seen.

BIOLOGY: Oligophagous. In northern Germany the larva has been found on *Calluna vulgaris* (H. WEGNER pers. comm.), probably the main food plant in central and northern Europe. In Britain it has also been found on *Erica tetralix* (SKINNER 2009). In Spain recorded on *Erica* and *Globularia* (REDONDO *et al.* 2009). Further recorded on *Artemisia maritima* and *Globularia vulgaris* and reared on *Taraxacum officinale* (TRUSCH & ERLACHER 2001).

HABITAT: Xerophilous. In north-western Europe on heaths, in the island Öland in Sweden also on the habitat type called "alvar", sandstone plains with sparse vegetation. In Spain in a variety of open land habitat types (REDONDO *et al.* 2009). In north-western Europe at low altitudes, in Spain from 300 up to 900 m.

SIMILAR SPECIES: Most *Dyscia* species are externally similar. *D. penulataria* is also contrasting, but it is often with tint of pink and forewing postmedial line is parallel with wing margin or curved inwards on costa (grey-brown, forewing postmedial weakly concave before costa in *D. fagaria*). Male and female genitalia of *D. fagaria* allow reliable identification. See Text-figs 134–143 and Table 9.

GENETIC DATA: Genetically homogeneous in Europe (n=10), populations of central and eastern Asia belonging to two other BINs. Nearest species: *D. lentiscaria* (5.9%).

REMARKS: LERAUT (2009) considers *belgaria* (HÜBNER) and *psoricaria* (EVERSMANN) valid at subspecies level. We follow TRUSCH & ERLACHER (2002), who consider them to be junior synonyms and clearly indicate, unlike LERAUT, on which material their conclusions are based. We have not shown subspecies in the distribution map because their boundaries need further research.

133. *Dyscia penulataria* (HÜBNER, 1819)

Synonymies and subspecies of *D. penulataria* are after TRUSCH & ERLACHER (2002).

penulataria penulataria HÜBNER, [1819], Samml. Eur. Schmett. 5 Geometrae (1): pl. 98, figs 507, 508, (*Geometra*). Syntype(s), lost, Europe.

eisenbergi HÖRHAMMER, 1959, Opusc. zool. 34: 2 (*Dyscia* (*Iberafrina*)). Holotype male (coll. WARNECKE in coll. Zool. Mus. Hamburg, Germany), Spain: Villamanrique. [Originally as var. of *penulataria hispanaria* MILLIÈRE].

hispanaria MILLIÈRE, 1866, Iconogr. Descr. Chenilles Lépid. inédits 2: 373 (265), pl. 79, figs 5–9, (*Scodiona*). Syntype(s), Spain.

naevata WEHRLI, 1953, in SEITZ, Gross-Schmett. Erde 4 (Suppl.): 664, (*Dyscia*). Syntype(s), Algier: Oran; Morocco: Hammam Rirha; Nemours; Mrassine. [Originally as var]. See remarks.

rubentaria RAMBUR, 1866, Cat. syst. Lépid. Andalousie 2: pl. 17, fig. 4 (non binominal). Holotype male (Coll. HERBULOT, in Zoologische Staatssammlung, Munich, Germany), [Spain]: Granada. Validated by GUMPPENBERG, 1896: 236, 243 (examined externally).

trabucaria OBERTHÜR, 1923, Études Lépid. comp. 20: 265, pl. 560, figs 4815–18, (*Scodiona*). Syntypes 4 (males, females), Pyrenees (east): Ille-sur-la-Tet; St-Paul-de-Fenouillet; Grotte de Galamus.

penulataria atlasica LERAUT, 2009, Moths of Europe, vol. 2, Geometrid Moths: 575 (*Dyscia*). Holotype female (Muséum National d´Histoire Naturelle, Paris, France), Morocco, Atlas Mountains: Sidi Oueddar.

penulataria combustaria OBERTHÜR, 1923, Ètudes Lépid. comp. 20: 265, pl. 560, figs 4811–4814, (*Scodiona*). Lectotype male, paralectotypes 3 females (Zoologisches Forschungsinstitut und Museum Alexander Koenig, Bonn, Germany, [Algeria]: Lambèése; Andalusia [Spain] (examined externally).

EXTERNAL CHARACTERS AND ABDOMEN: Wingspan 27–36 mm. Wings often with tint of pink, ground colour varying from pale beige to brown to grey, and to reddish brown, sparsely or heavily irrorated with brown. Markings vary from almost absent to blackish brown and contrasting. Antemedial visible to varying degrees, curved inwards on costa. Subterminal line often broken into row of dots, curved outwards in middle, angled, about straight before costa. Terminal line fine, or row of dots, often extending to fringes. Fringes concolorous with wings. Discal spots large, elongated, brown to blackish. Wings below weakly marked, often postmedial line and discal spots visible. Frons, collar, thorax and abdomen concolorous with wings. Proboscis reduced. Male antennae bipectinate, female antennae filiform. Hindtibia of both sexes with 2+2 spurs. Male tympanal organs large, bullae tympani with small roundish lobes. Other abdominal segments of both sexes undifferentiated.

VARIATION: Very variable. Ground colour often ochreous beige, with tint of violet, but it varies from pale beige to various shades of grey and reddish brown. Markings can be almost absent and wings whitish (f. *hispanaria*), but normally subterminal line developed to varying degrees, occasionally being blackish and very contrasting. Medial area can be partly shaded, most often near inner margin. Amount of irroration variable.

Subsp. ***combustaria*** is similar to f. *hispanaria*, but more yellowish, weakly irrorated with brown, markings darker and more distinct. Subterminal line present as distinct row of blackish dots. In addition to the original description, the taxon is illustrated also in WEHRLI (1939–1954). Type specimens are from Algeria. Considered valid at subspecies level by TRUSCH & ERLACHER (2002).

Subsp. ***atlasica*** is, according to LERAUT (2009) whitish, with tint of light pink. Subterminal line brown, with brown spot near inner margin. Discal spots brown. Thorax and abdomen whitish. Known from Morocco and Algeria, and externally similar to North African *D. holli* (OBERTHÜR, 1910). Validity of the subspecies needs further research.

MALE GENITALIA: Uncus small, sclerotised, weakly setose. Socii absent. Gnathos arms narrow, fused ventrally, wide, ventral margin concave, dorsal margin rounded extension. Valva narrow, upright, apex round, weakly spinose. Costa base with spoon-shaped extension, normally with two inwardly directed, sclerotised spines. Ventral margin of valva with small concavity. Juxta large, weakly sclerotised. Saccus narrow, wide, slightly quadrangular, with small upright extension medially. Aedeagus weakly curved, stout, caecum short. Apex evenly narrowing, acute. Vesica opens at approximately 90 degree angle. Vesica narrow tube, without sclerotisations.

FEMALE GENITALIA: Papillae anales large, setose, surface weakly granulate. Apophyses anteriores about 1/2 length of apophyses posteriores. Lamella antevaginalis membranous. Lamella postvaginalis weakly sclerotised, quadrangular, extending ventrad of ostium bursae, dorsal margin with small extension medially. Ductus bursae short, membranous. Corpus bursae large, membranous, very thin, inner surface without minute spines.

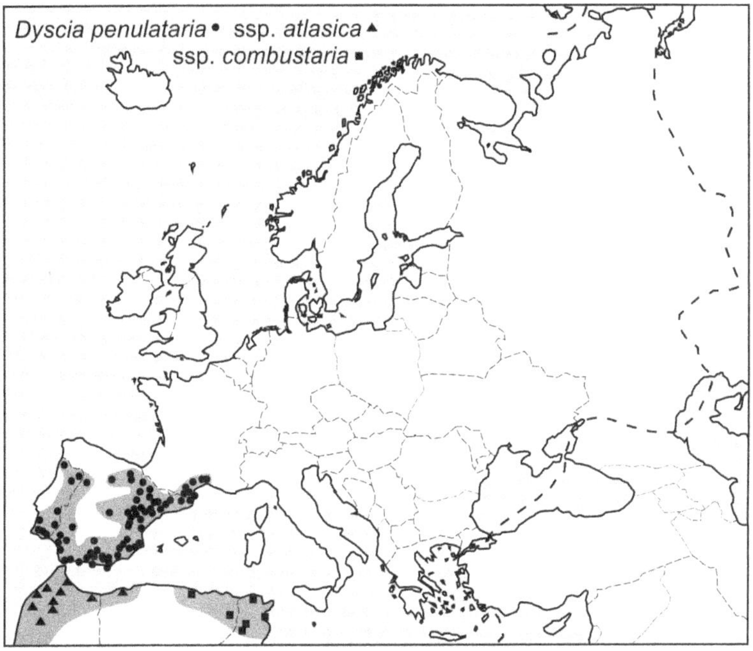

DISTRIBUTION AND ABUNDANCE: West-Mediterranean. In Europe the species is restricted to the Iberian Peninsula and southern France. – Outside Europe in North Africa from Morocco (subsp. *atlasica*), Algeria (partly subsp. *atlasica*, partly subsp. *combustaria*) and Tunisia (subsp. *combustaria*). – In Europe a common or very common species that often occurs in numbers.

PHENOLOGY: Bivoltine. From early April to late October in two apparently overlapping generations. In southernmost Spain in two distinct generations from March to April and from September to October (P. HALE pers. comm.). In southern France in two generations in May and from September to October (BACHELARD *et al.* 2007). The hibernation takes place as a small caterpillar (M. LEIPNITZ pers. comm.). The moths fly at night and come readily to light.

BIOLOGY: Oligophagous. The caterpillar is recorded on *Genista*, *Foeniculum*, *Dorycnium* and *Artemisia* (REDONDO *et al.* 2009). In rearing on *Artemisia* as food plant (B. MÜLLER pers. comm.).

HABITAT: Xerothermophilous. In a variety of open land habitats with low and often sparse vegetation, also often in warm and dry places. The species has been found from sea level up to about 2000 m in Sierra Nevada in southern Spain.

SIMILAR SPECIES: Most *Dyscia* species are externally similar. *D. fagaria* is grey(-brownish) and forewing postmedial line is weakly concave before costa (often with tint of pink, forewing postmedial line about straight near costa in *D. penulataria*). Male and female genitalia of *D. penulataria* allow reliable identification. See Text-figs 134–143 and Table 9.

GENETIC DATA: Genetically heterogeneous (n=7), with intraspecific divergence of 2.7%, both haplotypes occurring in eastern and southern Spain. Tunisian populations belonging to a third BIN. Nearest species: *D. conspersaria* (5.4%).

REMARKS: We have illustrated numerous specimens of *D. penulataria* in order to demonstrate the amount of external variation. LERAUT (2009) considered *naevata* WEHRLI, 1953 from North Africa valid at subspecies rank, being diagnosable by reddish-brown wings that are heavily irrorated with blackish brown. We follow TRUSCH & ERLACHER (2002) and consider such specimens to fall within the variation of the species.

134. Dyscia lentiscaria (DONZEL, 1837)

lentiscaria lentiscaria DONZEL, 1837, Annls Soc. ent. Fr. 6: 13, pl. 1, figs 1, 2, (*Crocallis*). Syntypes male, female, (Type material has been in coll. TESTOUT (Lyon), present location unknown), [France]: Provence, Hyères [Hières].

lentiscaria agacles OBERTHÜR, 1923, Études Lépid. comp. 20: 261, pl. 558, fig. 4782, (*Scodiona*). Lectotype male (Zoologisches Forschungsinstitut und Museum Alexander Koenig, Bonn, Germany), [Algeria]: Guelt-es-Stel. [Originally as var]. See remarks.

EXTERNAL CHARACTERS AND ABDOMEN: Wingspan 27–35 mm. Robust species, wings short. Wings ochreous, irrorated with brown, particularly on forewing inner margin. Antemedial line weak. Subterminal line evenly curved, slightly curved outwards on inner margin, often broken into row of blackish brown dots. Terminal line fine. Fringes concolorous with wings or weakly chequered brown and ochreous. Discal spots large, blackish brown. Hindwings about same colour as forewings, base paler. Wings below less irrorated with brown, hindwings almost pure white. Frons, collar and thorax hairy, concolorous with wings. Abdomen concolorous with wings. Male antennae bipectinate, female antennae filiform. Hindtibia of both sexes with 2+2 spurs. Tympanal organs large, bullae tympani of both sexes with elongated, roundish lobes. Other abdominal segments of both sexes undifferentiated.

VARIATION: Ground colour varies from pale whitish ochreous to rusty ochreous. Wings often rather heavily irrorated with blackish brown, particularly on forewing inner margin, but irroration can be weak. Discal spots often large, distinct, but they can be almost absent or with pale centre.

Subsp. ***agacles*** is, according to WEHRLI (1939–1954), larger, hindwings are pronounced white and weakly marked. Described from Algeria. See remarks.

MALE GENITALIA: Uncus rounded, weakly sclerotised, weakly setose. Dorsal margin of tegumen setose. Socii absent. Gnathos arms large, fused ventrally, evenly narrowing towards acute apex, apex upright, round, surface granulate. Valva narrow, medially constricted, apex round, setose. Costa base with weakly inwards curved, sclerotised wide extension, apex with few stout spines. Ventral margin of valva with small concavity, basal part with concavity. Juxta large, sclerotised. Saccus wide, quadrangular, with upright extension medially. Aedeagus curved, narrow, caecum long. Subapical part with acute extension. Apex long, narrow, outer margin weakly dentate. Vesica opens parallel to aedeagus stem. Vesica base elongated, extending into aedeagus apex. Ductus ejaculatorius opens from membranous vesica base.

FEMALE GENITALIA: Papillae anales elongated, setose. Apophyses anteriores about 1/2 length of apophyses posteriores. Lamella postvaginalis weakly sclerotised medially. Lamella antevaginalis/posterior part of ductus bursae cup-shaped (inside ductus bursae), its posterior margin rather straight or concave medially. Ductus bursae twisted, posterior part short, funnel-shaped, curved, sclerotised. Anterior part of ductus bursae narrow, membranous. Corpus bursae elongated, membranous, inner surface with few minute spines.

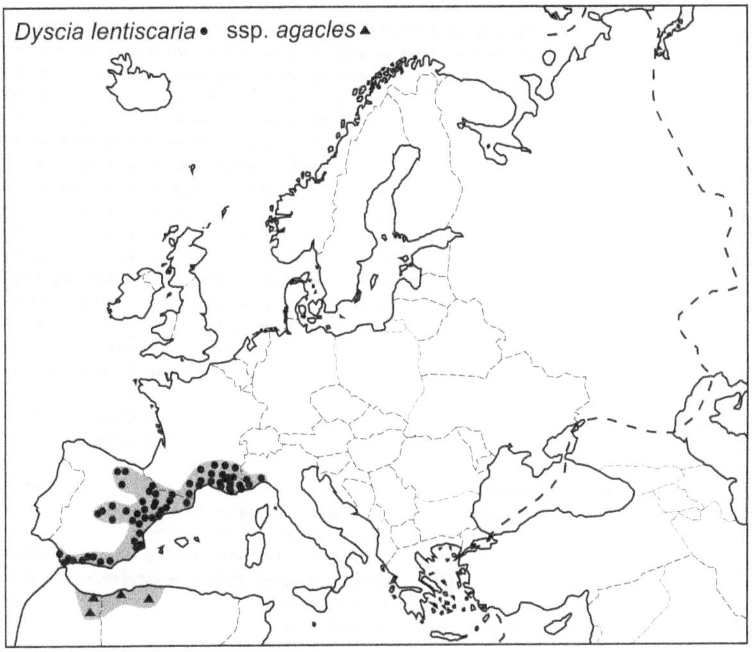

DISTRIBUTION AND ABUNDANCE: West-Mediterranean. In Europe restricted to the Iberian Peninsula, southern France and north-westernmost Italy. – Outside Europe in North Africa in Morocco and Algeria. The species is listed in the most recent Turkish checklist (KOÇAK & KEMAL 2009) from the Sivas province, but this record must be wrong. – In Europe the species is less common and somewhat local and is rather rarely seen in numbers.

PHENOLOGY: Univoltine. From early March in south-western Spain to late May, at higher altitudes and in southern France to mid-June. In southernmost Spain also in February (P. HALE pers. comm.). Larva from April to November (LHOMME 1935). The species hibernates as a pupa. Both females and especially males come to light.

BIOLOGY: Oligophagous. The caterpillar has been recorded on *Helianthemum vulgare*, *H. polifolium* and *Cistus* (LHOMME 1935). The caterpillar has been reared on *Artemisia campestris* as food plant (TRUSCH & ERLACHER 2001).

HABITAT: The species lives in a variety of open scrubland localities. It occurs from sea level up to 1700 m.

SIMILAR SPECIES: Most *Dyscia* species are externally similar. *D. distinctaria* has subterminal line often unbroken, weakly undulating, hindwings slightly paler than forewings (subterminal line often broken into row of blackish brown dots, hindwings about same colour as forewings in *D. lentiscaria*). *D. distinctaria* has wide, strongly curved gnathos with medial projection, costa base extension is narrow, vesica does not extend into aedeagus apex and ductus ejaculatorius opens from subapical part (gnathos evenly narrowing towards apex, costa base extension is wide, vesica extends into aedeagus apex and ductus ejaculatorius opens from membranous vesica base in *D. lentiscaria*). *D. distinctaria* has posterior part of ductus bursae long, anterior part of ductus bursae wide (posterior part short, anterior part narrow in *D. lentiscaria*). Finally, there are a number of quantitative differences in the female genitalia between these species, but we have not examined the diagnostic value of such structures on larger samples: Ductus bursae are slightly differently twisted, anterior end of ductus bursae are of different width, and shape and size of ductus bursae's cup-shaped sclerotisations are slightly different. See Text-figs 134–143 and Table 9.

GENETIC DATA: Genetically heterogeneous, populations from southern France and southern Spain diverging by 4.4% (n=6). Nearest species: *D. distinctaria* (1.3% from French populations; 4.0% from Spanish populations of *D. lentiscaria*). See remarks.

REMARKS: LERAUT (2009) upgraded taxon *agacles* OBERTHÜR from Algeria to species level. He did not explain the diagnostic features, but instead he provided an illustration of the female genitalia. Based on that, it apparently has the posterior part of the corpus bursae more elongated than in the European *lentiscaria*. LERAUT did not indicate how the identity of *agacles* was established, or on which material the conclusion was based. For that reason we follow TRUSCH & ERLACHER (2002), and consider *agacles* tentatively valid at subspecies level awaiting future studies.

Genetic data show that populations from southern France and southern Spain are considerably divergent. We have not studied material from southern Spain, but the genitalia of the specimens (both sexes examined) from northern Spain (Zarazoga) and southern France (Marseille) are without notable differences. Further research is needed.

135. *Dyscia distinctaria* (BANG-HAAS, 1910)

distinctaria BANG-HAAS, 1910, Dt. ent. Z. Iris 24 (3): 49, (*Scodiona*). Lectotype male, paralectotype female [Museum für Naturkunde der Humboldt-Universität, Berlin, Germany], [Spain]: Castilia, San Ildefonso. [Originally as var. of *lentiscaria* DONZEL].

distinctaria perdistincta HERBULOT, 1957, Bull. mens. Soc. linn. Lyon 26 (7): 191, (*Dyscia*). Holotype male (Coll. HERBULOT, in Zoologische Staatssammlung, Munich, Germany), [Spain]: Andalusia, Hinojos (examined externally).

EXTERNAL CHARACTERS AND ABDOMEN: Wingspan 31–36 mm. Generally as in *D. lentiscaria* (DONZEL), see above, but with following differences: Wings often more whitish, particularly hindwings. Subterminal line often unbroken, weakly undulating (subterminal line often broken into row of brown-blackish spots in *D. lentiscaria*).

VARIATION: Ground colour varies from pale whitish grey to ochreous, occasionally with weak tint of reddish. Medial area on inner margin occasionally heavily irrorated with blackish brown, also veins may be blackish brown. Fringes concolorous with wings or chequered blackish brown and ground colour.

MALE GENITALIA: Generally as in *D. lentiscaria* (DONZEL), see above, but with the following differences: *D. distinctaria* has strongly curved gnathos, costa base extension is narrow, inner margin of aedeagus apex weakly dentate, vesica does not extend into aedeagus apex and ductus ejaculatorius opens from subapical part (gnathos evenly narrowing towards apex, costa base extension is wide, outer margin of aedeagus apex weakly dentate, vesica extends into aedeagus apex and ductus ejaculatorius opens from membranous vesica base in *D. lentiscaria*).

FEMALE GENITALIA: Generally as in *D. lentiscaria* (DONZEL), see above, but with the following differences: *D. distinctaria* has posterior part of ductus bursae narrow, and anterior part of ductus bursae wide (posterior part wide, anterior part narrow in *D. lentiscaria*).

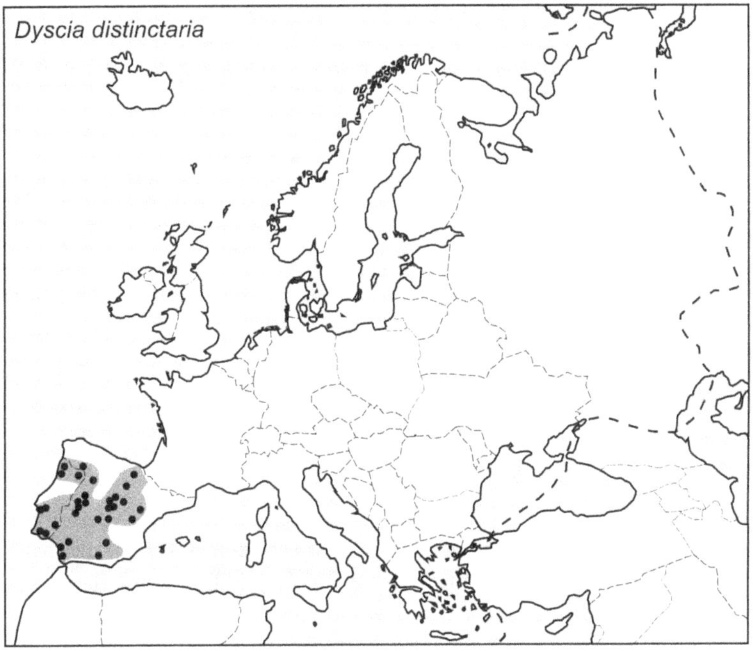

DISTRIBUTION AND ABUNDANCE: West-Mediterranean. Endemic to the Iberian Peninsula. An uncommon and local species that usually occurs in limited numbers.

PHENOLOGY: Univoltine. From early March to late May. Few specimens also found in early June (J. YLLA I ULLASTRE pers. comm.). Overwintering takes place as larva, or perhaps sometimes as a pupa, by rearing as a small larva (M. LEIPNITZ pers. comm.). Larvae have been found from May to March (ZERKOWITZ 1946) and in early December (KING & VIEJO MONTESINOS 2010). The moths fly at night, females and especially males come to light.

BIOLOGY: Oligophagous. The larva has been found on *Tuberaria guttata* and *Cistus albidus* (ZERKOWITZ 1946). KING & VIEJO MONTESINOS (2010) have found a caterpillar on *Helianthemum hirtum*. JAMBRINA PÉREZ (2010) suggest that *Lavandula stoechas* is the main larval food plant. In rearing accepting *Artemisia campestris* (TRUSCH & ERLACHER 2001, many caterpillars died, however) as well as on *Cistus* (M. CORLEY pers. comm.).

HABITAT: Xerothermophilous. In central Spain in rocky open scrubland. In southern Portugal in an area with gardens, scattered trees with grassland in between and citrus plantations (C. HVIID pers. comm.). The species occurs from sea level up to 1600 m.

SIMILAR SPECIES: Most *Dyscia* species are externally similar. *D. lentiscaria* (DONZEL) is similar, see the preceding entry. Male and female genitalia of *D. distinctaria* allow reliable identification. See Text-figs 134–143 and Table 9.

GENETIC DATA: Genetically homogeneous (n=8, including the holotype of *perdistincta*). Nearest species: *D. lentiscaria* (1.3%), distance larger in syntopic populations, see under preceding species.

Perconia HÜBNER, 1823

Perconia HÜBNER, [1823] 1816, Verz. bekannter Schmett: 296. Type species: *Phalaena strigillaria* HÜBNER, [1787]. There are no genus synonyms.

DIVERSITY AND DISTRIBUTION: *Perconia* contains two species in the Palaearctic region.

EXTERNAL CHARACTERS AND ABDOMEN: Wingspan approximately 30–40 mm. Wings whitish grey, irrorated with brown. Medial lines oblique, parallel. Forewing postmedial line on underside curved inwards on costa. Discal spots weak. Male antennae bipectinate, female antennae weakly fasciculate. Abdominal segments of both sexes undifferentiated.

MALE GENITALIA: Uncus sclerotised, tapering, weakly setose. Socii minute. Gnathos rudimentary, arms separate. Valva long, narrow, weakly setose. Valva costa base with sclerotised projection. Ventral margin of juxta V-shaped. Saccus small, roundish. Aedeagus with curved extension. Vesica without sclerotisations.

FEMALE GENITALIA: Apophyses anteriores about 1/2 length of apophyses posteriores. Lamella antevaginalis with two large, curved sclerotisations. Lamella postvaginalis with two large, vertically directed sclerotisations. Ostium bursae with additional sclerite. Corpus bursae large, membranous. Signum solid, anterior margin curved inwards.

BIOLOGY: Larva of *P. strigillaria* (HÜBNER) has been recorded feeding on *Calluna vulgaris* in northern Europe (MIKKOLA *et al.* 1989, in more southern areas food plants include also *Erica* spp., *Vicia, Lavandula, Rubus, Cytisus scoparius* and the flowers of *Ulex* (EBERT 2003, REDONDO *et al.* 2009, SKINNER & WILSON 2009). *P. baeticaria* (STAUDINGER) has been recorded feeding on *Genista* and *Cytisus* (REDONDO *et al.* 2009). Nearly fully-grown larva of *P. strigillaria* overwinters in northern Europe. Pupa is in a spun cocoon, which is attached to the food plant (MIKKOLA *et al.* 1989).

IMMATURE STAGES: Larva of *P. strigillaria* is brown-grey, with irregular black markings dorsally and laterally. Each abdominal segment 5–8 bears two wart-like structures dorsally, most prominent on 8th segment (PORTER 2010, LEPIFORUM 2013). Larva of *P. baeticaria* is longitudinally striped white, ochre and black, without dorsal structures (AIZPÚRUA 2002). Pupa cremaster of *P. strigillaria* wide at base and stout. Apex with four pairs of curved setae, apical pair longest (PATOČKA & TURČANI 2005).

REMARKS: *Perconia* has recently been classified in the Gnophini (VIVES MORENO 1994, VIIDALEPP 1996) and in the Aspitatini (LERAUT 1997, HAUSMANN, MIRONOV & VIIDALEPP 2011).

Perconia has the characteristic Gnophini features as diagnosed in VIIDALEPP *et al.* (2007). The Gnophini association was noted late in the production process of this book, so for practical reasons it is dealt with here, altough we note that its current position in the sequence of species does not reflect its probable phylogenetic position. See Gnophini under *Classification of European Ennominae genera into tribes* for further details. Systematic position: Gnophini.

136. *Perconia strigillaria* (HÜBNER, 1787)

strigillaria strigillaria HÜBNER, 1787, Beitr. Gesch. Schmett. 1 (2): 13, pl. 2, fig. I (as fig. F), (*Ph*[*alaena*] *Geom*[*etra*]). Syntype(s) male, lost, [Germany]: Augsburg.

fuscosignaria LEMPKE, 1952, Tijdschr. Ent. 95: 265, fig. 56: 4–6, (*Perconia*). Holotype male (Instituut voor Taxonomisch Zoölogie, Universiteit van Amsterdam, The Netherlands), Netherlands: Roermond. [originally as subspecies]. See remarks.

herpeticaria GUMPPENBERG, 1887, Nova Acta Acad. Caesar Leop. Carol. 49 (4): 381, 386, (*Fidonia*). Holotype female (C. HERBULOT Collection in Zoologische Staatssammlung, Munich, Germany), [Spain]: Andalusia (examined externally).

inaequaria HAWORTH, 1809, Lepid. Br. (2): 288, (*Geometra*). Syntype(s), [United Kingdom].

respersaria HÜBNER, [1799], Samml. Eur. Schmett. 5 Geometrae (1): pl. 23, fig. 125, (*Geometra*). Syntype(s), lost, Europe.

strigillaria cretaria EVERSMANN, 1844, Fauna Lepidopterologica Volgo-Uralensis 395, (*Cabera*). Syntype(s), [Russia]: Ural Mountains. Originally as sp., treated valid at subspecies rank by SCOBLE (1999), downgraded here from subspecies rank, quoting absence of diagnostic features, falling within intraspecific variation (new synonym). Type specimen(s) not designated in the original description. We have examined specimens from South Ural (Russia: S-Ural, Orenburg district, 25–30.v.2004, coll. Nupponen, Finland, male genitalia PS1570, female genitalia PS1592), assumably near the type locality. See Variation.

INTRASUBSPECIFIC NAMES, POTENTIALLY UNAVAILABLE: *armoris* OBERTHÜR, 1923 (ab.); *atra* COCKAYNE, 1953 (ab.); *basiapproximata* LEMPKE, 1953 (f.); *centrifasciata* LEMPKE, 1953 (ab.); *contrasta* LEMPKE, 1953 (f.); *crassesignata* LEMPKE, 1953 (f.); *dilatata* STRAND, 1912 (f.); *diluta* LEMPKE, 1953 (f.); *discoapproximata* LEMPKE, 1953 (f.); *fuscata* HANNEMANN, 1917 (f.); *fumosa* FJELDBORG, 1945 (ab.); *fuscomarginata* LEMPKE, 1953 (f.); *grisearia* STAUDINGER, 1871 (f.); *obscura* LEMPKE, 1953 (f.); *quadripunctata* WARNECKE, 1944 (f.); *reducta* LEMPKE, 1952 (f.); *semifasciata* COCKAYNE, 1953 (ab.).

EXTERNAL CHARACTERS AND ABDOMEN: Wingspan 28–36 mm. Wings whitish grey, heavily irrorated with different shades of brown and grey, particularly on forewing costa. Females slightly paler. Forewing antemedial, medial and postmedial lines oblique, parallel. Hindwing antemedial line weak or absent. Terminal line fine. Fringes weakly chequered brown or grey and whitish grey. Hindwing margin weakly undulating on M veins. Discal spots minute, more distinct on hindwings. Wings below as above, postmedial line most pronounced, curved inwards on forewing costa. Discal spots larger. Frons, collar, thorax and abdomen concolorous with wings. Male antennae bipectinate, female antennae weakly fasciculate. Hindtibia of both sexes with 2+2 spurs. Tympanal organs medium-sized. Posterior margin of male 8th sternite weakly elongated medially, other abdominal segments of both sexes undifferentiated.

VARIATION: Ground colour varies from dark grey to almost pure white. Markings and irroration vary from brown to pale grey, and from very extensive, covering most of wings to almost non-existent. Females tend to be paler, less irrorated and markings finer. Northern European specimens appear darker, or more heavily irrorated, than specimens from southern Europe. Position and density of medial lines variable, numerous forms have been described, which are summarised in LERAUT (2009). Specimens from South Ural are almost pure white (f. *cretaria*), transverse lines barely visible or absent altogether. Male genitalia of those specimens are similar to European specimens. In the female genitalia the sclerotised plate of the ostium bursae is more elongated than in the European *strigillaria* specimens, similar to that of *P. baeticaria*. VIIDALEPP (1996) does not mention taxon *cretaria*. Further research is needed.

MALE GENITALIA: Uncus sclerotised, evenly tapering, weakly setose. Socii minute. Gnathos rudimentary, arms separate. Valva long, narrow, weakly setose. Valva costa base with sclerotised, weakly inwards curved projection. Ventral margin of valva with minute, round extension on basal half. Transtilla narrow, weakly dentate. Juxta small, sclerotised, dorsal margin with minute extension, ventral margin V-shaped. Saccus small, roundish. Aedeagus stout, straight, caecum short. Subapical part with curved extension. Vesica opens at approximately 90 degree angle. Vesica narrow tube, base expanded. Vesica without sclerotisations.

FEMALE GENITALIA: Papillae anales round, setose. Apophyses anteriores about 1/2 length of apophyses posteriores. Lamella antevaginalis with two large, curved sclerotisations, interrupted ventromedially. Lamella postvaginalis with two large, vertically directed, smooth sclerotisations. Ostium bursae with additional, lunular sclerite. Ductus bursae short, wide, partly sclerotised. Corpus bursae large, membranous. Signum solid, anterior margin curved inwards, weakly striated.

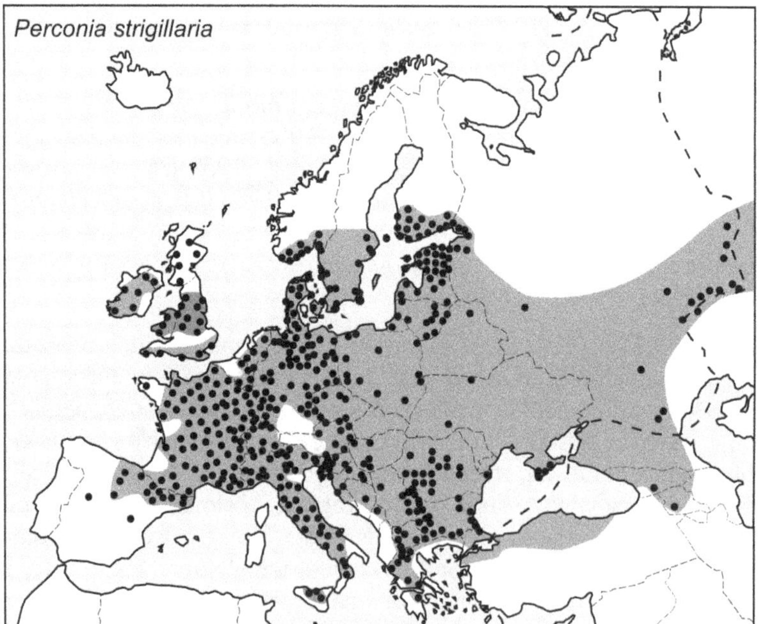

DISTRIBUTION AND ABUNDANCE: Palaearctic. In Europe from southern Fennoscandia to northern Spain, southern France, Italy including Sicily, the Balkans and from Ireland through central and eastern Europe to southern European Russia and the Urals. – Outside Europe the species has been found in Turkey, the Caucasus region and eastwards from the Urals through southern and central Siberia and Mongolia to Far East of Russia. Reputedly also from south-western Morocco (two specimens leg. CERNY in coll ZSM), perhaps mislabelled, awaiting confirmation. – In Europe a relatively common species. In central and northern Europe local, in the northern part of the Mediterranean area relatively more common. It is now and then seen in numbers.

PHENOLOGY: Univoltine, probably bivoltine in the southernmost part of the European distribution area. In southern Scandinavia the moths fly from late May until mid-July, in south-western Germany from mid-May to early July (EBERT 2003) and in northern Spain from May to September (REDONDO *et al.* 2009). Specimens flying late in the season in southern Europe probably represent a partial second generation. Caterpillars from June or July overwintering to May. The species flies at night and comes frequently to light, but it is also readily disturbed during the day.

BIOLOGY: Oligophagous. The larva has been found on *Calluna vulgaris* (B. MÜLLER pers. comm.), in south-western Germany it is found only on *Cytisus scoparius* (EBERT 2003) and in Britain also on the flowers of *Ulex* (PORTER 2010). REDONDO *et al.* (2009) record it on *Vicia*, *Lavandula*, *Rubus* and *Erica*. The pupation takes place in a web on the host plant.

HABITAT: In northern Europe in heaths, moors and places with heath vegetation, further south in a larger variety of habitats where the potential food plants grow. The species occurs from sea level, in the Alps up to 1300 m, in Spain up to 1700 m (REDONDO *et al.* 2009), in Greece up to 1550 m and in Sicily up to 1350 m.

SIMILAR SPECIES: *Perconia baeticaria* (STAUDINGER) is slightly bigger, irroration is more yellowish brown, forewing antemedial line is angled inwards on costa, and forewing postmedial line is often most pronounced (slightly smaller, irroration brown to grey, forewing antemedial line oblique, forewing antemedial line as strong as other medial lines in *P. strigillaria*). Male and female genitalia are without qualitative differences, and the diagnostic value of the mentioned structures may become obscure when larger samples, also from various geographical localities, are examined. *P. baeticaria* has extension on valva costa roundish and aedeagus extension is potentially straighter (valva extension weakly curved inwards, aedeagus extension potentially weakly curved in *P. strigillaria*). *P. baeticaria* has lamella postvaginalis horizontally striated, sclerotisation of ostium bursae is rounded and signum is smooth (lamella postvaginalis smooth, sclerotisation of ostium bursae lunular, signum weakly striated in *P. strigillaria*). The larvae of *P. strigillaria* and *P. baeticaria* feed on different host plants and the external appearance of larvae is different. Larva of *P. strigillaria* is illustrated for instance in PORTER (2010) and LEPIFORUM (2013) and *P. baeticaria* in AIZPÚRUA (2002).

GENETIC DATA: Genetically very heterogeneous (n=17), requiring further study, with four BINs in Europe and another in central Asia, at distances of 2.0 up to 5.7%. BIN (1) from Germany, France, United Kingdom and Morocco (perhaps mislabelled, awaiting confirmation); (2) from Germany, Finland and United Kingdom; (3) from southern Italy; (4) from Finland, comparatively close to the populations of Altai mountains (2.0%). Nearest species: *P. baeticaria* (6.4%).

REMARKS: We follow KUCHLEIN & VOS (1999) and HAUSMANN, MIRONOV & VIIDALEPP (2011), and do not recognise *fuscosignaria* LEMPKE, 1952 from the Netherlands valid at subspecies level. LERAUT (2009) considered *fuscosignaria* a junior synonym of f. *inaequaria* (HAWORTH, 1809).

137. Perconia baeticaria (STAUDINGER, 1871)

baeticaria STAUDINGER, 1871, in STAUDINGER & WOCKE, Cat. Lepid. eur. Faunengeb. 1: 175, (*Aspilates*). Syntype(s), [Spain]: Andalusia.

baeticaria RAMBUR, 1866, Cat. syst. Lépid. Andalousie 2: pl. 19, fig. 1, (non binominal), Holotype male (C. HERBULOT Collection in Zoologische Staatssammlung, Munich, Germany), [Spain]: Andalusia (examined externally).

castiliaria STAUDINGER, 1901, in STAUDINGER & REBEL, Cat. Lepid. palaearct. Faunengeb. (Edn 3): 1: 357, (*Perconia*). Syntypes 2 males, 3 females, [Spain]: Castilia, San Ildefonso. [Originally as var].. See remarks.

EXTERNAL CHARACTERS AND ABDOMEN: Wingspan 29–39 mm. Generally as in *P. strigillaria* (HÜBNER), see above, but with the following differences: wingspan is slightly bigger, irroration is more yellowish-brown, forewing antemedial line is angled inwards on costa, and forewing postmedial line is often most pronounced (slightly smaller, irroration brown to grey, forewing antemedial line oblique, forewing antemedial line as strong as other medial lines in *P. strigillaria*).

VARIATION: Density and extent of yellowish brown irroration variable. Markings often rather faint, but particularly the postmedial line can be dark brown and contrasting.

MALE GENITALIA: Generally as in *P. strigillaria* (HÜBNER), see above, without qualitative differences: *P. baeticaria* has extension on valva costa roundish and aedeagus extension is potentially straighter (valva extension weakly curved inwards, aedeagus extension potentially weakly curved in *P. strigillaria*).

FEMALE GENITALIA: Generally as in *P. strigillaria* (HÜBNER), see above, without qualitative differences: *P. baeticaria* has lamella postvaginalis horizontally striated, sclerotisation of ostium bursae is rounded and signum is smooth (lamella postvaginalis smooth, sclerotisation of ostium bursae lunular, signum weakly striated in *P. strigillaria*).

DISTRIBUTION AND ABUNDANCE: West-Mediterranean. Endemic to the Iberian Peninsula. Local and not common, usually in limited numbers.

PHENOLOGY: Univoltine. The moths fly from mid-May to late July (M. CORLEY pers. comm.). The moths fly at night, the males come readily to light, the females more rarely.

BIOLOGY: Oligophagous on Fabaceae. The caterpillar is recorded on *Genista florida* and *Cytisus balansae* (AIZPÚRUA 2002).

HABITAT: In open woods and scrubland and also above the timber line in places with rocks and grassland or very low growing bushes. The species occurs from low altitudes up to 1700 m in Sierra de Gredos in central Spain and up to 2400 in Sierra Nevada in southern Spain.

SIMILAR SPECIES: *P. strigillaria* (HÜBNER), see the preceding entry. Male and female genitalia of *P. baeticaria* and *P. strigillaria* are without qualitative differences and do not appear to allow reliable identification.

GENETIC DATA: Nearest species: *P. strigillaria* (6.4%).

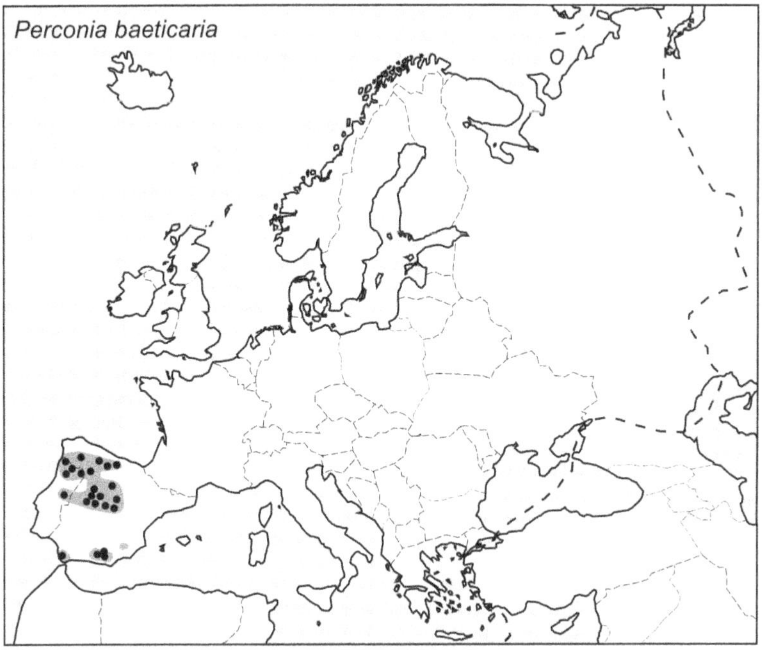

REMARKS: SCOBLE (1999) and LERAUT (2009) consider taxon *castiliaria* STAUDINGER valid at subspecies rank. We follow REDONDO *et al.* (2009) and consider these paler, more contrasting specimens to fall within the variation of the species because the diagnostic features appear not to be disjunct between different populations.

Onychora MEYRICK, 1892

Onychora MEYRICK, 1892, Trans. ent. Soc. London 1892: 101 (key). Type species: *Numeria agaritharia* DARDOIN, 1842, [France]. There are no genus synonyms.

DIVERSITY AND DISTRIBUTION: *Onychora* is a monotypic genus occuring in the western Mediterranean and North West Africa.

EXTERNAL CHARACTERS AND ABDOMEN: Wingspan 32–45 mm. Wings brown, grey, or mixture of these, often without any markings, except blackish discal spots. Antemedial and postmedial lines weak or absent. Proboscis reduced. Male antennae bipectinate, female antennae weakly fasciculate. Male bullae tympani with elongated, flat lobes (Text-fig. 144). Other abdominal segments of both sexes undifferentiated.

MALE GENITALIA: Uncus long, weakly setose. Socii large, with long setae. Gnathos arms fused ventrally, forming upturned, sclerotised plate. Valva narrow, setose, with sclerotised medial ridge. Juxta large, V-shaped, dorsolateral margins serrate. Saccus elongated. Vesica partly granulate, with several cornuti.

FEMALE GENITALIA: Papillae anales weakly granulate. Apophyses anteriores about 1/2 length of apophyses posteriores. Lamella antevaginalis weakly sclerotised, bowl-shaped. Corpus bursae narrow, elongated, posterior end striated, signum absent.

BIOLOGY: Larva of *O. agaritharia* has been recorded feeding on *Genista*, *Spartium* and *Ulex*, particularly on flowers and young shoots (AIZPÚRUA 2002, REDONDO *et al.* 2009). Adults are active from September until November.

IMMATURE STAGES: Larva of *O. agaritharia* is longitudinally striped black, whitish yellow and brown (AIZPÚRUA 2002). We have no data on pupa, apart from a superficial photograph (AIZPÚRUA 2002), and apparently pupa cremaster is weakly elongated, apex is forked, with two stout setae.

REMARKS: *Onychora* has recently been classified in the Onychorini (HERBULOT 1962–63, VIVES MORENO 1994, LERAUT 1997, HAUSMANN, MIRONOV & VIIDALEPP 2011). In a molecular phylogenetic analysis of the world Geometridae (SIHVONEN *et al.* 2011) *Onychora* was affiliated with the 'Ennominae' branch (sensu WAHLBERG *et al.* 2010). *Onychora* MEYRICK did not group closely with any other genera, but it was included in an unresolved trichotomy, containing for instance taxa from the Campaeini, Colotoini, Prosopolophini, Angeronini, Gnophini, Nephodiini and Azelinini. The support values of the branches were low.

Morphogically *O. agaritharia* shows a general Ennominae built, similar to several European genera, which are placed in this monograph in uncertain tribus association. Particularly similar are the males in the genera *Enanthyperythra* WEHRLI, *Asovia* ALPHÉRAKY, *Lignyoptera* LEDERER and to some degree also *Opisthograptis* HÜBNER. Female genitalia of *O. agaritharia* are also of rather general build, thus not lending straightforward information about potential systematic associations. Against this background we do not agree that *O. agaritharia* is so different from other European Ennominae taxa that it merits recognition in a monotypic tribus. Systematic position: Uncertain tribus association.

138. Onychora agaritharia (DARDOIN, 1842)

agaritharia agaritharia DARDOIN, 1842, Annls Soc. ent. Fr. 11 (2): 201, pl. 8, figs 3, 4, (*Numeria*). Syntypes male, female, [France]. Valid at subspecies level according to Scoble (1999).

agaritharia atlasica LERAUT, 2009, Moths of Europe, vol. 2, Geometrid Moths: 579, pl. 75, fig. 11, (*Onychora*). Holotype male (Muséum National d´Histoire Naturelle, Paris, France), Morocco: High Atlas, Oukaimeden.

agaritharia hepaticaria SCHWINGENSCHUSS, 1930, Verh. zool. -bot. Ges. Wien 80 (1/2): 26, (*Enconista*). Syntypes, [Spain]: Andalusia, Algearas; Chiclana. Valid at subspecies rank according to SCOBLE (1999).

agaritharia mauretanicaria STAUDINGER, 1901, in STAUDINGER & REBEL, Cat. Lepid. palaearct. Faunengeb. (Edn 3) 1: 355, (*Enconista*). (Holo)type male, 'Mauritania' [Algeria: Guelma] (Zoologisches Museum der Humboldt Universität, Berlin, Germany). Originally as var. of *agaritharia* DARDOIN, valid at subspecies rank according to SCOBLE (1999).

agaritharia rungsi LERAUT, 2009, Moths of Europe, vol. 2, Geometrid Moths: 580, pl. 75, figs 12, 13, (*Onychora*). Holotype male (Muséum National d´Histoire Naturelle, Paris, France), Morocco: Ifrane, 1650 m.

EXTERNAL CHARACTERS AND ABDOMEN: Wingspan 32–45 mm, wingspan varies considerably in both sexes. Wings brown, grey, or mixture of these, often without any markings, except blackish discal spots. Forewing antemedial and postmedial lines weak, blackish brown or absent, latter weakly undulating, curved inwards on costa. Hindwings paler, greyish, terminal area darker. Terminal line fine. Fringes concolorous with wings. Discal spots distinct, blackish. Wings below as above, markings weaker. Proboscis reduced. Frons, collar and thorax concolorous with wings. Male antennae bipectinate, female antennae weakly fasciculate. Hindtibia of both sexes with 2+2 spurs. Tympanal organs medium-sized, male organs with elongated, flat lobes (Text-fig. 144). Other abdominal segments of both sexes undifferentiated.

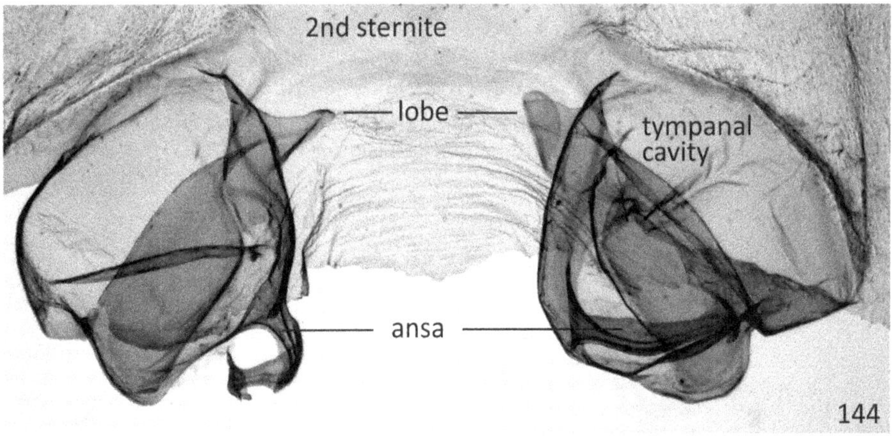

Text-fig 144. Male tympanal organs of *O. agaritharia* (DARDOIN, 1842) with elongated, flat lobes (France: Nice, without date, slide PS1487 (coll. ZMH)). Similar structures, often roundish, are found across the European Ennominae, for instance in genera *Isturgia* HÜBNER, 1823 (see Text-fig. 3, *Isturgia deerraria* (WALKER, 1861)) and *Dyscia* HÜBNER 1825. Function of the lobes is unknown.

VARIATION: Ground colour varies from grey-brown to reddish brown to yellowish brown to blackish brown. Specimens can be almost without any markings, or medial area can be darkened. Forewing medial lines normally weakly visible or absent but can be blackish and rather distinct. Size of discal spots somewhat variable.

Subsp. ***atlasica*** from the High Atlas (Morocco) is ochre-brown, discal spots extremely small. Cornuti are smaller than in the nominate subspecies. Female is unknown (LERAUT 2009). Validity at subspecies level needs further research. Not detailed as a subspecies in the distribution map.

Subsp. ***hepaticaria*** from southern Spain is chestnut-brown (LERAUT 2009). Validity at subspecies level needs further research. Not detailed as a subspecies in the distribution map.

Subsp. ***mauretanicaria*** from eastern Algeria is russet-brown, discal spots quite large, but the external characters are variable. Cornuti are smaller than in the nominate subspecies and the ductus bursae is more elongate and sclerotised over longer distance (LERAUT 2009). SCOBLE (1999) considers *mauretanicaria* valid at species level, whereas LERAUT (2009) considers it a subspecies of *agaritharia*. Validity at subspecies level needs further research. Not detailed as a subspecies in the distribution map.

Subsp. ***rungsi*** from the Middle Atlas (Morocco) is beige-grey, discal spots clearly visible. Cornuti are smaller than in the nominate subspecies (LERAUT 2009). Validity at subspecies level needs further research. Not detailed as a subspecies in the distribution map.

MALE GENITALIA: Uncus long, sclerotised, weakly setose. Socii large, with long setae. Gnathos arms fused ventrally, forming upturned, sclerotised plate. Valva narrow, setose, with sclerotised medial ridge. Juxta large, V-shaped, dorsolateral margins serrate. Saccus elongated. Aedeagus straight, caecum minute, apex with small lateral extension. Vesica opens at approximately 135 degree angle. Vesica straight tube, basal part granulate, with several cornuti.

FEMALE GENITALIA: Papillae anales elongated, surface weakly granulate, setose. Apophyses anteriores about 1/2 length of apophyses posteriores. Lamella antevaginalis weakly sclerotised, narrow, bowl-shaped. Lamella postvaginalis membranous. Ductus bursae wide, partly sclerotised. Posterior end of corpus bursae narrow, weakly curved, elongated, striated. Anterior end of corpus bursae slightly rounded. Signum absent.

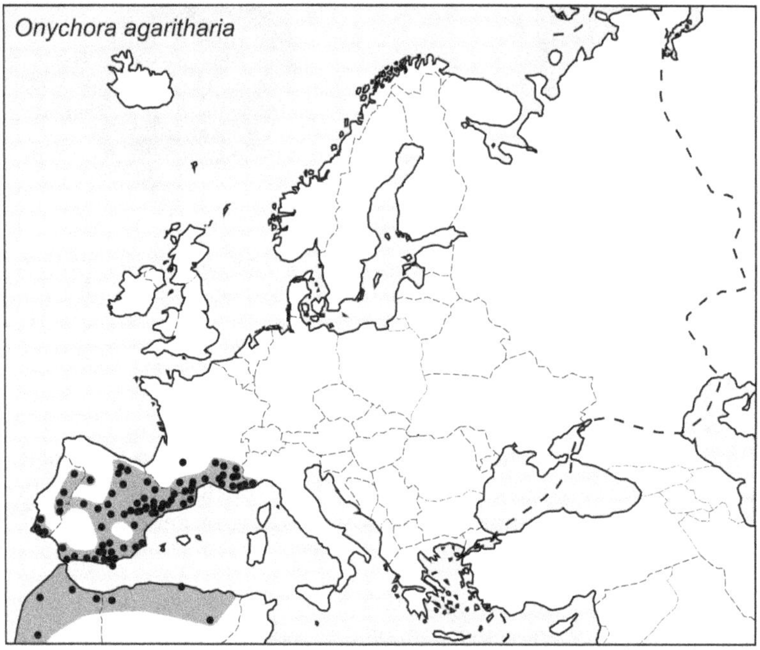

DISTRIBUTION AND ABUNDANCE: West-Mediterranean. In Europe restricted to the south-west from Portugal and Spain through southern France and into north-westernmost Italy. – Outside Europe the species occurs in northern Africa where it has been found in Algeria and Morocco. – In Europe a common or rather common species that now and then is found in numbers.

PHENOLOGY: Univoltine. From late August to late November in the Iberian Peninsula. The overwintering takes place as an egg (M. LEIPNITZ pers. comm.). The moths fly at night and both males and females come to light.

BIOLOGY: Oligophagous on Fabaceae. The larva is recorded on *Genista*, *Spartium* and *Ulex* (REDONDO *et al.* 2009). Larvae from the High Atlas mountains in Morocco have been reared on *Genista* (B. MÜLLER pers. comm.).

HABITAT: The species occurs in a large variety of habitats from Mediterranean macchia to open mountain scrubland and in the mountains above the timber line. Vertically the species occurs from sea level up to 1700 m in Sierra de Gredos in central Spain and up to 2250 m in Sierra Nevada in southern Spain.

SIMILAR SPECIES: No similar species in Europe.

GENETIC DATA: Populations from Andalusia and North Africa deeply diverging from those of northern Spain and France, with an additional BIN in Tunisia. All three BINs homogeneous and at distances of 5.6–6.8%. Nearest species (genetic similarity due to long branch attraction): Australian *Syneora symatomita* (TURNER, 1947) (8.1%).

REMARKS: Due to the scarcity of material available to us from North Africa, we have repeated the subpecies of *O. agaritharia* as they are presented in LERAUT (2009). We acknowledge that such fine division of taxa may be artificial when more extensive materials are examined, and the named subspecies may fall within gradual geographical variation of the species. A detailed analysis on *O. agaritharia*, including mtDNA, is under preparation (A. HAUSMANN & H. RIETZ pers. comm.).

Compsoptera BLANCHARD, 1845

Compsoptera BLANCHARD, 1845, Hist. Insectes 2: 394, 397 (replacement name). Type species: *Geometra jourdanaria* SERRES, 1826, [France]: near Montpellier. Genus synonyms are after SCOBLE (1999).

Endymia GISTL, 1848, Naturg. Thierreichs: ix. Type species: *Geometra jourdanaria* SERRES, 1826. [Unnecessary replacement name for *Ligia* DUPONCHEL].

Ligia DUPONCHEL, 1829, in GODART & DUPONCHEL, Hist. nat. Lépid. Papillons Fr. 7 (2): 107. Type species: *Geometra jourdanaria* SERRES, 1826. [Junior homonym of *Ligia* WEBER, 1795 (Crustacea)].

Prosopolopha LEDERER, 1853, Verh. zool.-bot. Ver. Wien 3 (Abh.): 180, 227, 238. [Unnecessary replacement name for *Ligia* DUPONCHEL].

UNAVAILABLE NAMES (incorrect subsequent authorship): *Ligia* BOISDUVAL (see FLETCHER 1979).

DIVERSITY AND DISTRIBUTION: *Compsoptera* contains three species in the western Mediterranean region.

EXTERNAL CHARACTERS AND ABDOMEN: Wingspan about 35–44 mm. Forewings narrow, elongated, female apex more acute. Forewings grey-brown to reddish brown to ochreous. Forewing postmedial line oblique, terminating below apex. Hindwings pale. Male antennae bipectinate, female antennae weakly fasciculate. Frons protruding, collar with dorsally raised setae, like Mohican hairstyle (Text-fig. 145). Abdominal segments of both sexes undifferentiated.

MALE GENITALIA: Uncus long, curved ventrally, weakly setose. Socii round lobes, weakly setose. Gnathos absent. Valva divided. Dorsal arm narrow at base, apex blunt or roundish. Ventral arm wider, apex acute. Pedunculi long, narrow. Saccus quadrangular. Vesica with backwards directed cornuti.

Text-fig. 145. Potential synapomorphies of genus *Compsoptera* BLANCHARD, 1845, shown here on *C. jourdanaria* (SERRES, 1826) female, Spain, 19.ix.1987. Frons protruding and collar with dorsally raised setae, giving Mohican-like appearance, are diagnostic in all *Compsoptera* species.

FEMALE GENITALIA: Papillae anales weakly granulate. Apophyses anteriores about same length as apophyses posteriores. Lamella antevaginalis/sterigma sclerotised, wide. Lamella postvaginalis membranous. Signum roundish, ventral margin dentate.

BIOLOGY: Larva of *C. opacaria* HÜBNER and *C. jourdanaria* (SERRES) are polyphagous, they have been recorded feeding for instance on *Retama, Genista scorpius, Ulex, Dorycnium, Thymus, Rosmarinus officinalis, Juniperus, Calluna* and *Artemisia herba-alba* (AIZPÚRUA 2002). Adults are active in autumn.

IMMATURE STAGES: Eggs of *C. jourdanaria* are whitish, laid in a group on the food plant (LEPIFORUM 2013). Larva of *C. opacaria* HÜBNER is ochreous brown, with faint whitish line dorsally and small blackish dots laterally. Larva of *C. jourdanaria* (SERRES) is greyish brown, with two dark brown, undulating lines dorsally (AIZPÚRUA 2002). We have no data on pupa, apart from superficial photographs (AIZPÚRUA 2002), and apparently pupa cremaster is very short, and apex is with two short, inwards curved, stout setae.

REMARKS: *Compsoptera* has recently been classified in the Prosopolophini (LERAUT 1997, HAUSMANN, MIRONOV & VIIDALEPP 2011) and Compsopterini (VIVES MORENO 1994, misspelled as Comsopterini). Compsopterini is considered a junior synonym of Prosopolophini

(FORUM HERBULOT 2014). Based on the material we have examined, *Dasycorsa* PROUT, *Colotois* HÜBNER and *Compsoptera* BLANCHARD are morphologically similar but the potential relationships should be studied in an analytical context. Systematic position: Prosopolophini.

As is explained in more detail in *Ennominae systematics* chapter, the European Ennominae fauna as a whole has not previously been classified into tribes. The above-mentioned similarities were noted so late in the manuscript preparation that rearrangement of taxa to reflect such potential affinities were not possible.

139. *Compsoptera opacaria* (HÜBNER, 1819)

opacaria HÜBNER, [1819], Samml. Eur. Schmett. 5 Geometrae (1): pl. 96, figs 493–496, (*Geometra*). Syntype(s), lost, Europe.

aemiliorum RAINERI, 1994, Revta Lepid. 22 (88): 318, figs 1, 2, 5–7, (*Compsoptera*). Holotype male (Museo Civico di Storia Naturale, Genoa, Italy), Italy: Sardinia, Soleminis. See remarks.

INTRASUBSPECIFIC NAMES, POTENTIALLY UNAVAILABLE: *huebneri* SCHWINGENSCHUSS, 1952 (f. '*hübneri*'); *intermediaria* BYTINSKY-SALZ, 1937 (ab.?); *rubra* STAUDINGER, 1871 (ab.).

EXTERNAL CHARACTERS AND ABDOMEN: Wingspan 36–44 mm. Forewings narrow, elongated, grey-brown to red-brown, female apex more acute. Postmedial line very oblique, white, with brown inner margin, terminating below apex. Hindwings wide, whitish, terminal area darker. Terminal line fine. Fringes concolorous with wings. Forewing discal spot acute, blackish, hindwing discal spot absent. Wings below uniform greyish, postmedial line absent. Frons, collar and thorax concolorous with wings. Frons protruding, collar with dorsally raised setae, Mohican-like. Male antennae white dorsally, bipectinate. Female antennae weakly fasciculate. Hindtibia of both sexes with 2+2 spurs. Tympanal organs small. Other abdominal segments of both sexes undifferentiated.

VARIATION: Forewing ground colour varies from greyish brown to deep red-brown (f. *rubra*). Red colour can cover the entire forewing or be limited to costa and reddish irroration. Hindwings vary from almost pure white to greyish. Postmedial line can be straight or curved, inner margin variably brown. Forewing discal spots can be absent.

MALE GENITALIA: Uncus long, curved ventrally, weakly setose. Socii round lobes, weakly setose. Gnathos absent. Valva divided. Dorsal arm with group of setae at narrow base, apex blunt. Ventral arm wider, apex acute, straight, ventral margin weakly setose. Juxta hourglass-shaped. Pedunculi long, narrow. Saccus quadrangular. Aedeagus curved, caecum short, apex with minute tooth. Vesica opens at approximately 45 degree angle. Vesica wide at base, with row of backwards directed cornuti.

FEMALE GENITALIA: Papillae anales wide, weakly granulate. Apophyses anteriores about same length as apophyses posteriores. Lamella antevaginalis/sterigma sclerotised, long, wide. Margin invaginated medially. Lamella postvaginalis membranous, with few horizontal striae. Corpus bursae round, membranous. Signum roundish, ventral margin dentate.

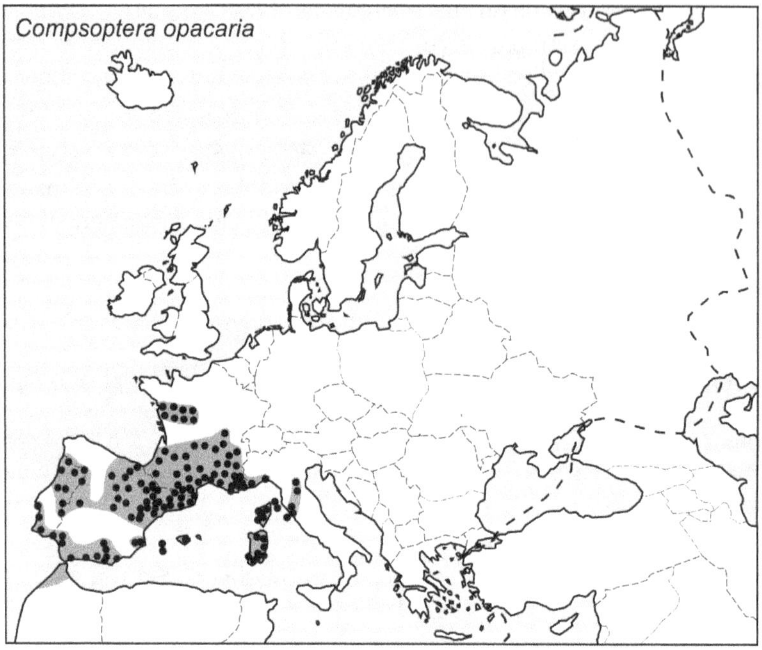

Distribution and abundance: West-Mediterranean. In Europe from Portugal to the Italian mainland including the Balearic Islands, southern and central France, Corsica and Sardinia. – Outside Europe the species occurs in Morocco in northern Africa. – In Europe a common or rather common species, but somewhat local, that now and then is found in numbers.

Phenology: Univoltine. From early September to late November, in southernmost Spain from October to December (P. Hale pers. comm.). The species overwinters as caterpillar from September or October to April. In rearing, the egg overwinters (M. Leipnitz pers. comm.). The moths fly at night and come readily to light, especially the males.

Biology: Polyphagous. The caterpillar has been found on *Cytisus* (own data). It has been recorded on *Retama*, *Genista scorpius*, *Ulex*, *Dorycnium*, *Thymus*, *Rosmarinus officinalis*, *Juniperus*, *Calluna* and *Cytisus* (Redondo *et al.* 2009) as well as on *Arbutus unedo* and *Adenocarpus hispanicus* (Aizpúrua 2002). In rearing accepting *Genista* and *Erica carnea* as food plants (B. Müller pers. comm.).

Habitat: In a large variety of habitat types, often dry and warm places, such as macchia, scrubland and open wood land. The species occurs from sea level up to 2000 m.

Similar species: No similar species in Europe.

Genetic data: Genetically homogeneous from southern France to southern Spain. Nearest species: *C. jourdanaria* (5.9%).

Remarks: Synonymy of *C. aemiliorum* Raineri, 1994 with *C. opacaria* (Hübner) was proposed by Dapporto (2004), see also (Murria Beltran 2008, Redondo *et al.* 2009), and subspecies status was also rejected because both colour forms occur sympatrically. The deep brick -ed individuals of *C. opacaria* are to be considered as a chromatic form, referable to ab. *rubra* (Staudinger, 1871).

140. Compsoptera jourdanaria (SERRES, 1827)

jourdanaria jourdanaria SERRES, 1827, in VILLIERS, Mém. Soc. linn. Paris 5: 480, pl. 11, figs h-n, (*Geometra*). Syntype(s), [France]: near Montpellier.

acutaria HERRICH-SCHÄFFER, 1852, Syst. Bearbeitung Schmett. Eur. 6 (55): 74; ibidem (1850), 3: pl. 73, fig. 447 (non binominal), (*Ligia*). Syntype(s), Southern Russia?

caesaraugustanus REDONDO, 1995, Boln. Soc. ent. aragon. 9: 14, figs 1a, 2, 2a, b, (*Compsoptera*). Holotype male, Spain: Aragón, provincia de Zaragoza, Alfocea, 221 m. See remarks.

jourdanaria anargyra TURATI, 1913, Atti Soc. ital. Sci. nat. 51: 335, figs 17–19, (*Prosolopha*). Syntypes 18 males, 2 females, [Italy]: Sardinia: 'Monti' Gennargentu.

UNAVAILABLE NAMES (incorrect subsequent authorship): *jourdanaria* TREITSCHKE; *jourdanaria* VILLIERS (see FLETCHER 1979). – Intrasubspecific names, potentially unavailable: *obscura* BETHUNE-BAKER, 1885 (ab.)

EXTERNAL CHARACTERS AND ABDOMEN: Wingspan 35–43 mm. Forewings narrow, elongated, olive-brown with brown irroration, markings white. Postmedial line oblique, undulating, white, terminating below apex. Rather straight white line crosses forewing, parallel with costa, terminates in apex. Veins white in terminal area. Hindwings wide, whitish, terminal area darker. Terminal line fine, blackish. Forewing fringes chequered white and olive-brown, hindwing fringes concolorous with wings. Forewing discal spot dark brown, hindwing discal spot absent. Wings below uniform grey-brown, forewing fringes chequered brown and greyish. Frons, collar and thorax concolorous with wings. Frons protruding, collar with dorsally raised setae, Mohican-like (Text-fig. 145). Male antennae white dorsally, bipectinate. Female antennae weakly fasciculate. Hindtibia of both sexes with 2+2 spurs. Tympanal organs medium-sized, smaller in females. Other abdominal segments of both sexes undifferentiated.

VARIATION: Ground colour varies from pale olive-brown to dark olive-brown. Width and extent of brown areas and irroration variable, white markings can be very weak in Portugese specimens.

Subsp. ***anargyra*** is, on average, slightly darker brown with weaker white postmedial line. The differences are minor, and somewhat variable. LERAUT (2009) illustrates minor quantitative differences in the male genitalia (uncus apparently wider and ventral margin of valva perhaps slightly more concave). According to the material we have examined, these potential minute differences are continuous, open to interpretation and may be an artefact of the dissection techniques. Despite the lack of clear-cut differences, we retain taxon valid at subspecies level due to its allopatric occurrence in Corsica and Sardinia. More research is needed.

MALE GENITALIA: Generally as in *C. opacaria* (HÜBNER), see above, but with following differences: In *C. jourdanaria* dorsal arm of valva is with round apex, ventral arm of valva is with slightly upright apex, juxta narrow, dorsal margin round (dorsal arm with blunt apex, ventral arm with straight apex, juxta hourglass-shaped in *C. opacaria*). Aedeagus straight. Vesica opens at approximately 135 degree angle. Vesica weakly expanded at base, with few backwards directed cornuti.

FEMALE GENITALIA: Papillae anales wide, weakly granulate. Apophyses anteriores about same length as apophyses posteriores. Lamella antevaginalis/sterigma sclerotised, narrow. Margin evenly concave, with minute invagination medially. Lamella postvaginalis membranous. Ductus bursae short, membranous. Corpus bursae narrow, elongated, membranous, weakly wrinkled. Signum roundish, ventral margin dentate.

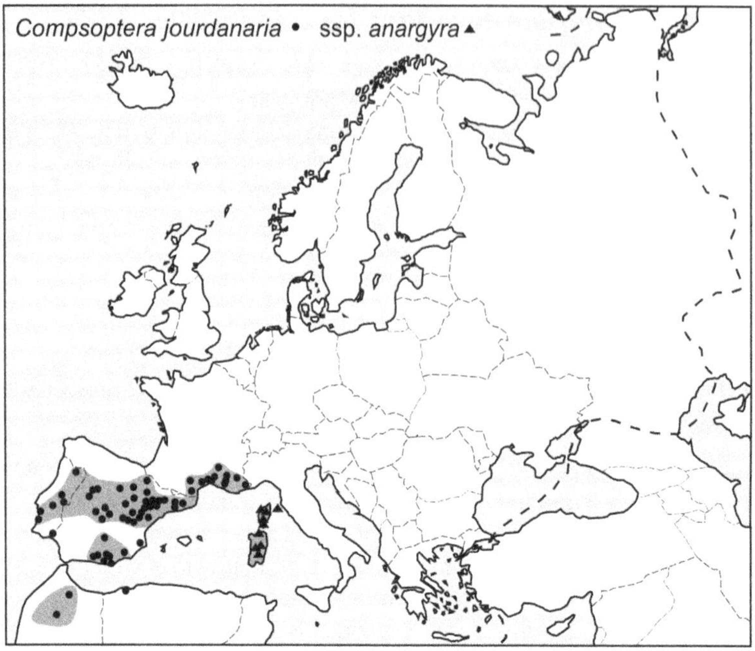

DISTRIBUTION AND ABUNDANCE: West-Mediterranean. In Europe only recorded in the Iberian Peninsula, southern France, Elba, Corsica and Sardinia. – Outside Europe only in North Africa where the species occurs in Morocco and Algeria. In Europe a common or rather common species, but somewhat local, that is now and then found in numbers.

PHENOLOGY: Univoltine. From early September to late November. Overwintering takes place in central Spain as a caterpillar from late December or January to March and pupation takes place in April (G. KING pers. comm.). In rearing the egg overwintered (M. LEIPNITZ pers. comm.). The moths fly at night, the females and especially the males come frequently to light.

BIOLOGY: Polyphagous. KING & VIEJO MONTESINOS (2010) have found the caterpillars on *Thymus lacaitae*, *Helianthemum hirtum*, *Artemisia campestris*, *A. herba-alba* and *Santolina rosmarinifolia*. Also recorded on *Helichrysum*, *Dorycnium* and *Salsola vermiculata* (REDONDO *et al.* 2009).

HABITAT: Xerothermophilous. The species inhabits various warm and dry habitats of scrubland and open woodland. It occurs from sea level up to 2000 m altitude in Sierra Nevada in southern Spain.

SIMILAR SPECIES: No similar species in Europe.

GENETIC DATA: Nearest species: *C. opacaria* (5.9%).

REMARKS: Taxon *caesaraugustanus* REDONDO, 1995 is currently considered a hybrid of *C. jourdanaria* (SERRES) and *C. opacaria* (HÜBNER) (REDONDO & GASTON 1999, REDONDO *et al.* 2009). Only the holotype is known (illustrated also in REDONDO & GASTON 2009), and it is somewhat intermediate between the mentioned taxa: forewing postmedial line is dentate as in *C. jourdanaria*, but otherwise wings are uniform greyish-brown as in *C. opacaria*.

141. Compsoptera argentaria (HERRICH-SCHÄFFER, 1839)

argentaria HERRICH-SCHÄFFER, [1839], in Panzer, Faunae Insect. Germ. initia (Dtl. Insect.) (Heft 165): 57, (*Aspilates*). Syntype(s), Sicily.

yaminaria OBERTHÜR, 1881, Études ent. 6: 84, pl. 11, fig. 4, (*Ligia*). Syntype(s) male(s), Algeria: Ben-Youb.

EXTERNAL CHARACTERS AND ABDOMEN: Wingspan 40–46 mm. Male: forewings narrow, elongated, pale ochreous to yellow-brown. Forewing postmedial line oblique, almost horizontal, yellow-brown, distinctly angled, then almost parallel with wing margin, terminating below apex. Outer margin whitish. Rather straight white line crosses forewing, parallel with costa, terminates in apex. Parallel brown line present in wing base. Veins white in medial area. Hindwings whitish, terminal area darker. Female: forewing apex more acute, wings more grey, forewing postmedial line more evenly curved. Hindwings paler, almost white. Both sexes: Terminal line very fine. Fringes concolorous with wings. Discal spots absent. Wings below uniform pale grey-ochreous. Frons, collar, thorax and abdomen concolorous with wings. Frons protruding, collar with dorsally raised setae, Mohican-like. Male antennae white dorsally, bipectinate. Female antennae weakly fasciculate. Hindtibia of both sexes with 2+2 spurs. Tympanal organs medium-sized, smaller in females. Other abdominal segments of both sexes undifferentiated.

VARIATION: Male ground colour varies from pale ochreous to yellow-brown, white areas can be ash grey. More contrasting specimens may have forewing postmedial line visible throughout wing, it is distinctly angled before inner margin. Female ground colour varies very little. Hindwing of both sexes normally greyish, darker on terminal area, but wings can be almost uniform white.

MALE GENITALIA: Generally as in *C. opacaria* (HÜBNER), see above, but with following differences: In *C. argentaria* dorsal arm of valva is with blunt apex, but with ventral extension (dorsal arm with blunt apex, without ventral extension in *C. opacaria*).

FEMALE GENITALIA: Generally as in *C. opacaria* (HÜBNER), see above, without qualitative differences, but with following minor differences: In *C. argentaria* papillae anales are narrower, lamella antevaginalis/sterigma shorter and corpus bursae more elongated (papillae anales wide, lamella antevaginalis/sterigma longer and corpus bursae round in *C. opacaria*).

DISTRIBUTION AND ABUNDANCE: West-Mediterranean. In Europe the species only occurs in Sicily. – Outside Europe it is distributed in North Africa in Tunisia, Algeria and Morocco. – In Sicily a rare, local and poorly known species of which probably less than 70 specimens have been recorded so far.

PHENOLOGY: Univoltine. In Sicily from mid-September to early November. In Morocco in September and October (RUNGS 1981). In Tunisia the species has been found in mid-November (K. NUPPONEN pers. comm.). Caterpillars in Morocco in early April at 1000 m altitude (D. STADIE pers. comm.). Hibernation takes place as a larva (FLAMIGNI *et al.* in press). The moths fly at night and the males come to light, the females less abundantly.

BIOLOGY: The natural food plant appears to be unknown, but should belong to Fabaceae. In north-eastern Morocco, caterpillars have been found on *Anthyllis cytisoides* (D. STADIE pers. comm.). Reared on *Cytisus* and *Calicotome spinosa* (FLAMIGNI *et al.*, in prep.).

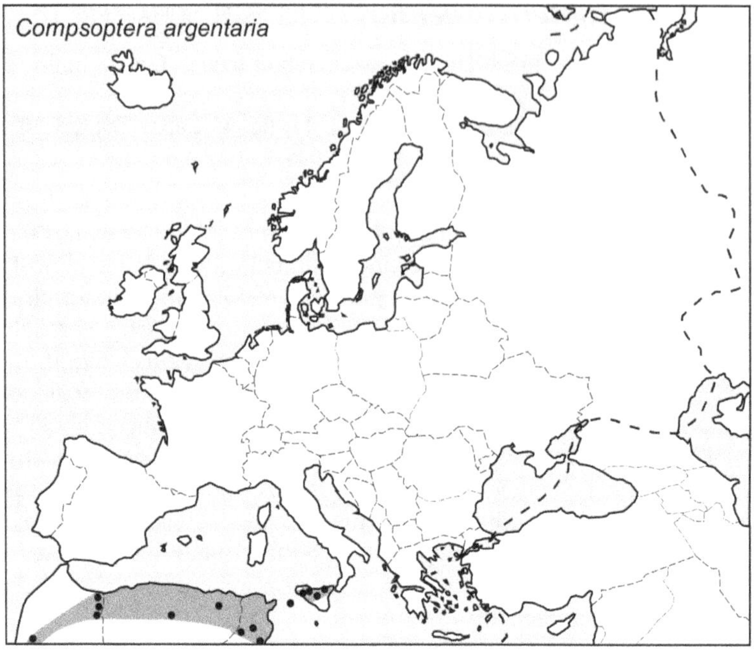

HABITAT: In Sicily in open forest where there had been a fire 10–15 years earlier. At lower altitudes mainly in *Quercus pubescens* and *Q. ilex* forests, at higher levels forests with *Pinus halepensis*, *P. pinea*, *Cupressus sempervirens*, *C. arizonica*, *Cedrus atlantica*, *Fraxinus ornus*, *Acer pseudoplatanus*, *Celtis australis* and *Ulmus minor* (P. PARENZAN pers. comm.). In Tunisia in an open and xerothermic rocky slope (K. NUPPONEN pers. comm.). In Sicily the species has been found at altitudes from 500 up to 1200 m (P. PARENZAN pers. comm.).

SIMILAR SPECIES: No similar species in Europe.

GENETIC DATA: No data available.

Ennominae taxa treated in volume 1 of the Geometrid Moths of Europe

Unlike the traditional views, recent molecular and morphological studies have provided evidence that several taxa are better placed in the Ennominae.

Subfamily Orthostixinae MEYRICK, 1892, has Palaearctic and Indo-Pacific distribution containing 17 species, classified in four genera. Traditionally classified in the Oenochrominae *sensu lato* (see SCOBLE & EDWARDS 1989), but in recent works it has been considered a subfamily of its own (HAUSMANN 2001, FORUM HERBULOT 2014) or a tribe within the Ennominae (HOLLOWAY 1996). In molecular phylogenetic analyses by YAMAMOTO & SOTA (2007), the Orthostixinae were recovered as sister to the Desmobathrinae (most analyses with mitochondrial or nuclear data) or it formed a monophyletic clade together with the Ennominae and other taxa (Bayesian and ML analyses). In another molecular phylogenetic analysis (SIHVONEN *et al.* 2011) the Orthostixinae were placed within the Ennominae, being sister taxon to the Baptini/Caberini lineage. The single European species, *Orthostixis cribraria* (HÜBNER, 1799) is dealt with in volume 1 of the *Geometrid Moths of Europe* (HAUSMANN 2001).

Subfamily Alsophilinae HERBULOT, 1962, has Palaearctic distribution containing 26 species classified in four genera. Often classified in the Oenochrominae *sensu lato* (see SCOBLE & EDWARDS 1989), recently in a subfamily of its own (SCOBLE 1999, HAUSMANN 2001), but pupal morphology is similar to the Theriini (Ennominae) and to the Operophterini (Larentiinae) (PATOČKA 1978, PATOČKA & ZACH 1994). In molecular analyses the Alsophilinae have grouped within the Ennominae (YAMAMOTO & SOTA 2007, WAHLBERG *et al.* 2010, SIHVONEN *et al.* 2011). The European fauna contains two species in genus *Alsophila* HÜBNER, 1825, which are dealt with in volume 1 of the *Geometrid Moths of Europe* (HAUSMANN 2001).

Genus *Epirranthis* HÜBNER, 1823 has two species in the Holarctic region (Nearctic species illustrated in MCGUFFIN 1981, Palaearctic species illustrated in HAUSMANN 2001), whereas SCOBLE (1999) classified the Nearctic species in *Spodolepis* HULST, 1896. Systematic position has remained controversial, recently it has been classified in the Desmobathrinae (HAUSMANN 2001), Oenochrominae *sensu lato* (MÜLLER 1996, LERAUT 1997) or in the Ennominae (MCGUFFIN 1981, VIIDALEPP 1996). The Ennominae, tribe Ennomini, relationship is supported by larval and pupal morphology (PATOČKA 1992, KULLBERG *et al.* 2001). In a molecular analysis (SIHVONEN *et al.* 2011), *Epirranthis* was nested within the Ennominae, next to genus *Opisthograptis* HÜBNER, 1823. The single European species, *Epirranthis diversata* (DENIS & SCHIFFERMÜLLER, 1775) is dealt with in volume 1 of the *Geometrid Moths of Europe* (HAUSMANN 2001).

Systematic checklist of the Geometridae of Europe (vol. 5)

The checklist shows, in a putative systematic order, the Ennominae taxa of the European Geometridae treated in vol. 5. The checklist reflects the taxonomic changes noted during preparation of the manuscript, but which were not possible to carry out in the main text and plates. Numerous subspecies level taxa need further research, those are mentioned in the text. Non-European species and subspecies, which occur adjacent to Europe (as defined in the current context), are inserted at the end of relevant genera because in numerous cases their systematic position was not examined. Those are marked with *, and their distribution area is given in square brackets.

Taxonomic note: *Itame aspersaria* KRÜGER, 1939, from Libya, classified as *Macaria aspersaria* (KRÜGER, 1939) in SCOBLE (1999), is likely to be an *Isturgia* species because currently only one *Macaria* species (*M. wauaria* (LINNAEUS)) is known to occur in the Palaearctic part of Africa (KRÜGER 2001). We have failed to find the type specimen of *Itame aspersaria* KRÜGER. KRÜGER (1939) did not indicate where the type was deposited, and we don't know the fate of KRÜGER collection, but potentially it may have been in a Libyan museum. According to a black and white photograph in the original description (plate XVII, fig. 116), and the written description, *Itame aspersaria* KRÜGER is similar (or even conspecific?) to *Isturgia deerraria* (WALKER), a species that is found throughout Africa and in southern Europe also.

Ennominae DUPONCHEL, 1845

Abraxini WARREN, 1893
Abraxas LEACH, 1815
grossulariata (LINNAEUS, 1758)
* subsp. *conspurcata* BUTLER, 1878 [Japan]
* subsp. *dsungarica* WEHRLI, 1939 [Central Asia]
* subsp. *minor* HERZ, 1905 [East Asia]
* subsp. *notabilis* WEHRLI, 1939 [Europe – China]
* subsp. *ribesata* STAUDINGER, 1892 [Central Asia]
sylvata (SCOPOLI, 1763)
* subsp. *ijimai* INOUE, 1955 [Japan]
* subsp. *intensa* COCKERELL, 1906 [Japan]
* subsp. *ischna* WEHRLI, 1939 [Russia: Sajan Mountains]
* subsp. *microtate* WEHRLI, 1931 [East Asia]
pantaria (LINNAEUS, 1867)

Ligdia GUENÉE, 1858
adustata (DENIS & SCHIFFERMÜLLER, 1775)
* *lassulata* (ROGENHOFER, 1873) [Turkey]. Type locality is Gülek, Bulgar-Dagh (Bolkar Mountains) on Taurus Mountains in Turkey (ROGENHOFER 1873: 572), not Bulgaria as written in SCOBLE (1999).

Cassymini HOLLOWAY, 1994
Lomaspilis HÜBNER, 1825
marginata (LINNAEUS, 1758)
* *opis* BUTLER, 1878 [Japan]
* subsp. *amurensis* (HEDEMANN, 1881) [Russia: Amur]
subsp. *nigrita* HEYDEMANN, 1936
bithynica WEHRLI, 1954

Stegania GUENÉE, 1845
cararia (HÜBNER, 1790)
* subsp. *lungtanensis* (WEHRLI, 1939) [South-East China]
trimaculata (VILLERS, 1789)
* subsp. *ochrearia* (BANG-HAAS, 1910) [North Africa]
dilectaria (HÜBNER, 1790)
* subsp. *trimaculoides* (WEHRLI, 1931) [Turkey]
dalmataria GUENÉE, 1858
* subsp. *altaica* (WEHRLI, 1939) [Russia: Altai]

* ***Zamarada*** MOORE, 1887
* *torrida* FLETCHER, 1974 [Israel, Jordan, Egypt, Algeria, widely spread in Africa]

Macariini GUENÉE, 1858
Acanthovalva KRÜGER, 2001
inconspicuaria (HÜBNER, 1819)
* subsp. *tsekubia* (WEHRLI, 1940) [China: Sichuan]

Narraga WALKER, 1861
fasciolaria (HUFNAGEL, 1767)
* subsp. *fumipennis* PROUT, 1915 [China]
* subsp. *turkestanica* MOUCHA & POVOLNY, 1957 [Turkestan]
* *nelvae* (ROTHSCHILD, 1912) [northern Africa]
* subsp. *cappadocica* HERBULOT, 1943 [Turkey, Armenia]
subsp. *catalaunica* HERBULOT, 1943
tessularia (METZNER, 1945)
* subsp. *ilia* WEHRLI, 1940 [Central Asia]

Heliomata GROTE & ROBINSON, 1866
glarearia (DENIS & SCHIFFERMÜLLER, 1775)

Isturgia HÜBNER, 1823
famula (ESPER, 1787)
* subsp. *brunnea* (LE CERF, 1923) [Morocco]
limbaria (FABRICIUS, 1775)
subsp. *anzascaria* (STAUDINGER, 1892)
subsp. *delimbaria* (STAUDINGER, 1892)
subsp. *nigrostriata* (HEYDEMANN, 1938)
subsp. *pedemontaria* (STAUDINGER, 1892)
subsp. *rablensis* (ZELLER, 1868)
subsp. *styriaca* (SCHWINGENSCHUSS, 1911)
roraria (FABRICIUS, 1776)
homochromata (MABILLE, 1869)
murinaria (DENIS & SCHIFFERMÜLLER, 1775)
catalaunaria (GUENÉE, 1858)
deerraria (WALKER, 1861)
disputaria (GUENÉE, 1858)
spodiaria (LEFÈBVRE, 1832)
assimilaria (RAMBUR, 1833)
miniosaria (DUPONCHEL, 1829)
* subsp. *rungsi* LERAUT, 2009 [Morocco]
berytaria (STAUDINGER, 1892)
sparsaria (HÜBNER, 1809)
messapiaria (SOHN-RETHEL, 1929)
arenacearia (DENIS & SCHIFFERMÜLLER, 1775)
* subsp. *mongolica* (VOJNITS, 1974) [Mongolia]
punctistrigaria (CHRISTOPH, 1897)
* *aspersaria* (KRÜGER, 1939) [Libya, see taxonomic note in the beginning of checklist]
* *exustaria* (STAUDINGER, 1897) [Israel, Jordan, Sinai]
* *hausmanni* KRÜGER, 2001 [Tunisia]
* *hedemanni* (CHRISTOPH, 1885) [Iran, Near East, Turkmenistan, Uzbekistan]
* *hopffearia* (STAUDINGER, 1878) [Turkey, Transcaucasus]
* *perviaria* (LEDERER, 1855) [Middle East, southern India]
* *rubrior* (HAUSMANN, 1990) [Morocco]
* *tennoa* (PINKER, 1978) [Canary Islands]
* *terminipuncta* KRÜGER, 2001 [Tunisia]
* *tozeurensis* LERAUT, 2009 [Tunisia]

Itame HÜBNER, 1823
vincularia (HÜBNER, 1813)
* subsp. *latefasciata* ROTHSCHILD, 1914 [Algeria]
* subsp. *lycioidaria* HERBULOT, 1957 [Tunisia]
* subsp. *mrassinaria* (OBERTHÜR, 1923) [Morocco]
* *teknaria* POWELL & RUNGS, 1942 [North Africa, see text for additional information]

Neognopharmia STAUDINGER, 1892
stevenaria (BOISDUVAL, 1840)
* *hoerhammeri* (BRANDT, 1941) [Iran, see text for additional information]

Macaria CURTIS, 1826
notata (LINNAEUS, 1758)
* subsp. *kirina* (WEHRLI, 1940) [East Asia]
alternata (DENIS & SCHIFFERMÜLLER, 1775)
signaria (HÜBNER, 1809)
* subsp. *dispuncta* WALKER, 1860 [North America]
liturata (CLERCK, 1759)
subsp. *deceptans* (STAUDINGER, 1901)
subsp. *hispanica* (HERBULOT, 1957)
* subsp. *pressaria* CHRISTOPH, 1893 [Russia: Amur]
* subsp. *depressa* (INOUE, 1955) [Japan]
wauaria (LINNAEUS, 1758)
* subsp. *africana* (ZERNY, 1934) [Morocco, Algeria]
* subsp. *chinensis* (STERNECK, 1928) [China]
* subsp. *koreaebia* (BRYK, 1949) [Korea]
artesiaria (DENIS & SCHIFFERMÜLLER, 1775)
ichnusae GOVI & FIUMI, 2005
carbonaria (CLERCK, 1759)
brunneata (THUNBERG, 1784)
* subsp. *imperatoria* (KARDAKOFF, 1928) [Russia: Ussuri]
* subsp. *sordida* (BUTLER, 1881) [Japan]
loricaria (EVERSMANN, 1837)
fusca (THUNBERG, 1792) [see text for additional details as concerns *Macaria fusca* (KRÜGER, 1939) from Libya]

* **Boarmioides** LUCAS, 1932 [*Boarmioides* is potentially a junior synonym of *Macaria* CURTIS, 1826, see also SCOBLE & KRÜGER, 2002]
* *colpias* (PROUT, 1928) [Morocco, Algeria]

Chiasmia HÜBNER, 1823
clathrata (LINNAEUS, 1758)
* subsp. *albifenestra* (INOUE, 1942) [Japan]
* subsp. *almacola* WEHRLI, 1934 [Turkey]
subsp. *aurata* (TURATI, 1905)
* subsp. *azrouensis* (WEHRLI, 1937) [northern Africa]
* subsp. *centralasiae* (KRULIKOWSKY, 1911) [Central Asia, south Kazakhstan]
subsp. *clathraria* (HÜBNER, 1799)
* subsp. *djakonovi* KARDAKOFF, 1928 [Russia: Ussuri]
* subsp. *hoenei* SCHULTZE, 1954 [China]
subsp. *hugginsi* BAYNES, 1959
* subsp. *kurilata* (BRYK, 1942) [Russia: Kuril Islands]
* subsp. *punctata* (BANG-HAAS, 1910) [China]
subsp. *retata* (HAWORTH, 1809)
aestimaria (HÜBNER, 1809)
* subsp. *kuldschana* (WEHRLI, 1940) [China]
subsp. *sareptanaria* (STAUDINGER, 1871)
* subsp. *tunesiella* (LUCAS, 1949) [Tunisia, perhaps elsewhere in northern Africa also]
* *syriacaria* (STAUDINGER, 1871) [Cyprus]
* *latimarginaria* (REBEL, 1907) [northern Africa]
* *saburraria* (EVERSMANN, 1851) [Russia: southern Siberia, see text for additional details]

* *tenuiata* (STAUDINGER, 1901) [Lebanon, potentially Syria. Potentially a junior synonym of *C. syriacaria* (STAUDINGER, 1871), see text for details].

'Semiothisa' HÜBNER, 1818
* *erevanica* (VARDIKJAN, 1985) [Armenia. Systematic combination uncertain. Originally combined to *Semiothisa* HÜBNER, 1818, but *Semiothisa* sensu stricto is restricted to the Neotropics (SCOBLE & KRÜGER 2002)].

Digrammia GUMPPENBERG, 1887
rippertaria (DUPONCHEL, 1830)
* subsp. *flavularia* (PÜNGELER, 1902) [Uzbekistan]

Rhoptria GUENÉE, 1858
asperaria (HÜBNER, 1817)
dolosaria (HERRICH-SCHÄFFER, 1848)

Ortaliella HAUSMANN, 1993
gruneraria (STAUDINGER, 1862)
* *palaestinensis* (CALBERLA, 1891) [Israel, Jordan, Lebanon]
* *stonei* (HAUSMANN, 1991) [Jordan]

* ***Stueningia*** HAUSMANN, 1993
* *poggearia* (LEDERER, 1855) [Lebanon, NW Israel]
* subsp. *meyi* HAUSMANN, 1993 [Syria, C Israel]
* *wolfi* HAUSMANN, 1993 [Turkey]
* subsp. *taurica* HAUSMANN, 1993 [S Turkey]

Lithinini FORBES, 1948
Petrophora HÜBNER, 1811
chlorosata (SCOPOLI, 1763)

Lithinini of uncertain association
Perigune GUMPPENBERG, 1887
binaevata (MABILLE, 1869)
* subsp. *austautaria* (OBERTHÜR, 1881) [Algeria, Libya]
* subsp. *fraternaria* (ROTHSCHILD, 1914) [northern Africa]
convergata (VILLERS, 1789)
subsp. *exoglypha* (PRAVIEL, 1936)
narbonea (LINNAEUS, 1767)
subsp. *obliterata* (STAUDINGER, 1901)
* *jordanaria* (STAUDINGER, 1901) [Israel, Jordan]

Pachycnemia STEPHENS, 1829
hippocastanaria (HÜBNER, 1799)
tibiaria (RAMBUR, 1829)
* subsp. *aindrahamaria* (OBERTHÜR, 1923) [northern Africa]
subsp. *benesignata* (BELLIER, 1861)

Hypochrosini GUENÉE, 1858
Heterolocha LEDERER, 1853
laminaria (HERRICH-SCHÄFFER, 1852)
* subsp. *euxantha* WEHRLI, 1937 [East China]
* subsp. *lungtana* WEHRLI, 1940 [South East China]
* subsp. *sutschanska* WEHRLI, 1937 [Russia: Amur]

Hypochrosini of uncertain association
Plagodis HÜBNER, 1823
pulveraria (LINNAEUS), 1758
* subsp. *japonica* (BUTLER, 1881) [Japan: Honshu]
* subsp. *jezoensis* (INOUE, 1954) [Japan: Hokkaido]
* subsp. *montana* (INOUE, 1954) [Japan: Honshu]
* subsp. *occiduaria* (WALKER, 1861) [USA]
* subsp. *singularis* VOJNITS, 1975 [Mongolia]
dolabraria (LINNAEUS, 1767)
* subsp. *costisignata* WEHRLI, 1939 [China]

Epionini BRUAND, 1846
Epione DUPONCHEL, 1829
repandaria (HUFNAGEL, 1767)
vespertaria (LINNAEUS, 1767)
* subsp. *hespera* BRYK, 1949 [Korea]

Therapis HÜBNER, 1823
flavicaria (DENIS & SCHIFFERMÜLLER, 1775)

Epionini of uncertain association
Cepphis HÜBNER, 1823
advenaria (HÜBNER, 1790)

Pseudopanthera HÜBNER, 1823
macularia (LINNAEUS, 1758)
subsp. *aureoadflava* (STAUDER, 1915)
* subsp. *cryptica* BELJAEV, 1997 [Russia: Irkutsk]
subsp. *meridionalis* (GALVAGNI, 1908)
* *syriacata* (GUENÉE, 1858) [Turkey to Israel and Jordan]

* ***Pynthanosis*** TURATI, 1922 [Systematic position of the genus not examined by us, we place it near *Pseudopanthera* HÜBNER, 1823, following WEHRLI (1939–1954)]
* *henricaria* (OBERTHÜR, 1881) [Morocco, Algeria]

Eilicrinia HÜBNER, 1823
cordiaria (HÜBNER, 1790)
* subsp. *signigera* BUTLER, 1889 [Caucasus, Armenia, Turkey, India]
subcordaria HERRICH-SCHÄFFER, 1852
* subsp. *acardia* STICHEL, 1911 [Iran]
trinotata (METZNER, 1845)

Apeirini STEKOLNIKOV & KUZNETZOV, 1982
Apeira GISTL, 1848
syringaria (LINNAEUS, 1758)

Ennomini DUPONCHEL, 1845
Ennomos TREITSCHKE, 1825
autumnaria (WERNEBURG, 1859)

* subsp. *koreennomos* BRYK, 1949 [Korea]
* subsp. *pyrrosticta* WEHRLI, 1940 [China]
* subsp. *sinica* YANG, 1978 [North China]
quercinaria (HUFNAGEL, 1767)
alniaria (LINNAEUS, 1758)
fuscantaria (HAWORTH, 1809)
* subsp. *algeriensis* PROUT, 1929 [North Africa]
effractaria FREYER, 1841 [southern European part of Russia]. Identity unclear, potentially junior synonym of *E. fuscantaria* (HAWORTH). See text for additional information.
erosaria (DENIS & SCHIFFERMÜLLER, 1775)
* subsp. *ifranaria* RUNGS, 1941 [Morocco]
* subsp. *pseudoquercaria* WEHRLI, 1940 (Algeria)
quercaria (HÜBNER, 1813)
subsp. *duercki* REISSER, 1958
* subsp. *freidbergi* HAUSMANN, 1997 [Eastern Mediterranean]
* subsp. *lissochila* (PROUT, 1929) [Cyprus]
* subsp. *olivaria* BRANDT, 1938 [Iran]
* *fraxineti* WILTSHIRE, 1947 [South Turkmenistan, Iran]
Ennomini of uncertain association
Selenia HÜBNER, 1823
dentaria (FABRICIUS, 1775)
lunularia (HÜBNER, 1788)
* subsp. *gamra* RUNGS, 1950 [Morocco]
* subsp. *kuldjana* WEHRLI, 1940 [China]
tetralunaria (HUFNAGEL, 1767)

Artiora MEYRICK, 1892
evonymaria (DENIS & SCHIFFERMÜLLER, 1775)

Odontopera STEPHENS, 1831
bidentata (CLERCK, 1759)
* subsp. *exsul* (TCHETRERIKOV, 1905) [Russia: Sajan]
* subsp. *harutai* (INOUE, 1953) [Japan: Honshu]
* subsp. *kurilana* (BRYK, 1942) [Russia: Kuril Islands]
* subsp. *rava* (VOJNITS, 1975) [Mongolia]

Crocallis TREITSCHKE, 1825
tusciaria (BORKHAUSEN, 1793)
* subsp. *cypriaca* FISCHER, 2003 [Cyprus]
* subsp. *taurica* WEHRLI, 1934 [Turkey, Eastern Mediterranean]
* subsp. *transcaucasica* WEHRLI, 1940 [Turkey, Caucasus]
albarracina WEHRLI, 1940
elinguaria (LINNAEUS, 1758)
dardoinaria DONZEL, 1840
auberti OBERTHÜR, 1883
helenaria RUCKDESCHEL, 2006
boisduvalaria (LUCAS, 1849)
* *bacalladoi* PINKER, 1978 [Canary Islands]
* *jordanaria* STAUDINGER, 1898 [Jordan, Israel]
* *inexpectata* WARNECKE, 1940 [Turkey, see text for additional information]
* *matillae* PINKER, 1974 [Canary Islands]

* *rjabovi* WEHRLI, 1936 [Transcaucasus, Armenia]

Ourapteryx LEACH, 1814
sambucaria (LINNAEUS, 1758)
* *falciformis* INOUE, 1993 [Caucasus, N. Iran. See text for additional information].

*** *Coenina*** WALKER, 1860 [systematic position not examined by us, we place it in the Ennomini after WILTSHIRE (1990) and HACKER & HAUSMANN (1999)].
* *dentataria* SWINHOE, 1904 [Israel, Arabian Peninsula, widespread in Africa]
* *islamitica* AMSEL, 1935 [Palestine]
* *paulusi* (REBEL, 1906) [Israel, Arabian Peninsula]

Colotoini WEHRLI, 1940
Colotois HÜBNER, 1823
pennaria (LINNAEUS, 1761)
subsp. *carbonii* HARTIG, 1976
* subsp. *mauretanaria* (STÄTTERMAYER, 1930) [Morocco, Algeria]
* subsp. *paupera* HAUSMANN, 1995 [Cyprus]
subsp. *ussuriensis* BANG-HAAS, 1927 [Russia: Ussuri, Japan]
* *furcata* HAUSMANN, 1995 [Israel]

Dasycorsa PROUT, 1915
modesta (STAUDINGER, 1879)
* subsp. *marerufa* WILTSHIRE, 1940 [Eastern Mediterranean]

Caberini DUPONCHEL, 1845
Cabera TREITSCHKE, 1825
pusaria (LINNAEUS, 1758)
leptographa WEHRLI, 1939
* subsp. *biographa* VIIDALEPP & REMM, 1982 (Russia: Habarovsk)
exanthemata (SCOPOLI, 1763)
* subsp. *bryantaria* (TAYLOR, 1906) [Canada]
* subsp. *hamica* WERHLI, 1939 [Mongolia]

Baptini FORBES, 1948
Lomographa HÜBNER, 1825
bimaculata (FABRICIUS, 1775)
* subsp. *subnotata* (WARREN, 1895) [Japan]
temerata (DENIS & SCHIFFERMÜLLER, 1775)

Theriini HERBULOT, 1963
Aleucis GUENÉE, 1845
distinctata (HERRICH-SCHÄFFER, 1839)
orientalis (STAUDINGER, 1892)

Theria HÜBNER, 1825
rupicapraria (DENIS & SCHIFFERMÜLLER, 1775)
primaria (HAWORTH, 1809)

Campaeini FORBES, 1948

Campaea LAMARCK, 1816
margaritaria (LINNAEUS, 1761)
Gerinia LERAUT, 2009
honoraria (DENIS & SCHIFFERMÜLLER, 1775)

Hylaea HÜBNER, 1822
fasciaria (LINNAEUS, 1758)
* subsp. *cedricola* (WEHRLI, 1929) [Turkey, potentially Near East]
* subsp. *flavella* (WEHRLI, 1940) [Transcaucasus]
mediterranea SIHVONEN *et al.*, 2014
pinicolaria (BELLIER, 1861)
* *compararia* (STAUDINGER, 1895) [North Africa, see SIHVONEN *et al.* (2014) for additional information].
* *poeymiraui* (OBERTHÜR, 1922) [Morocco]

Adalbertia WEHRLI, 1931
castiliaria (STAUDINGER, 1900)

Pungeleria ROUGEMONT, 1903
capreolaria (DENIS & SCHIFFERMÜLLER 1775)

Gnophini DUPONCHEL, 1845
Hypoxystis PROUT, 1915
pluviaria (FABRICIUS, 1787)

Angerona DUPONCHEL, 1829
prunaria (LINNAEUS, 1758)
* subsp. *aquafortis* BRYK, 1949 [Korea]
* subsp. *coreola* WEHRLI, 1940 [Korea]
* subsp. *dominans* BRYK, 1942 [Russia: Kuril Islands]
* subsp. *kentearia* STAUDINGER, 1892 [Mongolia]
* subsp. *mongoligena* BRYK, 1949 [northern Mongolia]
* subsp. *prouterona* BRYK, 1949 [Korea]
* subsp. *turbata* PROUT, 1929 [Japan]
* subsp. *valens* WEHRLI, 1940 [Russia: Ussuri, Korea]

Siona DUPONCHEL, 1829
lineata (SCOPOLI, 1763)

Chariaspilates WEHRLI, 1953
formosaria (EVERSMANN, 1837)

Aspitates TREITSCHKE, 1825
mundataria (STOLL, 1782)
* subsp. *tonghata* (FELDER & ROGENHOFER, 1875) [Japan]
* subsp. *uncinataria* (VOJNITS, 1975) [Mongolia]
gilvaria (DENIS & SCHIFFERMÜLLER, 1775)
subsp. *burrenensis* COCKAYNE, 1951
* subsp. *kukunorensis* (WEHRLI, 1953) [China]
* subsp. *minimus* (VOJNITS, 1975) [Mongolia]
* subsp. *orientaria* (ALPHÉRAKY, 1883) [Central Asia]
taylorae (BUTLER, 1893) [Canada]
subsp. *sibirica* (DJAKONOV, 1955)
orciferaria (WALKER, 1862)
* subsp. *baffinensis* (MUNROE, 1963) [Canada]
* subsp. *churchillensis* (MUNROE, 1963) [Canada]
* subsp. *occidentalis* (MUNROE, 1963) [Canada]
forbesi (MUNROE, 1963)
albaria (BARTEL, 1903)
ochrearia (ROSSI, 1794)
* *acuminaria* (EVERSMANN, 1851) [Caucasus, Georgia, Dagestan]
* *collinaria* (HOLT-WHITE, 1894) [Canary Islands]

Dyscia HÜBNER, 1825
raunaria (FREYER, 1851)
royaria TAUTEL & BILLI 2006
* *innocentaria* (CHRISTOPH, 1885) [Turkmenistan]
* subsp. *osmanica* (WAGNER, 1931) [Turkey, Cyprus to Israel]
subsp. *sicanaria* (OBERTHÜR, 1923)
* *malatyana* WEHRLI, 1934 [Turkey to Jordan and Israel]
* subsp. *albersaria* WARNECKE, 1940 [Turkmenistan]
subsp. *clara* PUGAEV & ZOLOTUHIN, 2006
* subsp. *senecai* WILTSHIRE, 1990 [Libya]
crassipunctaria (REBEL, 1916)
conspersaria (DENIS & SCHIFFERMÜLLER, 1775)
* subsp. *sultanica* WEHRLI, 1936 [Turkey]
subsp. *turturaria* (BOISDUVAL, 1840)
fagaria (THUNBERG, 1784)
subsp. *albidaria* (STAUDINGER, 1871)
subsp. *emucidaria* (HÜBNER, 1813)
subsp. *favillacearia* (HÜBNER, 1799)
penulataria (HÜBNER, 1819)
* subsp. *atlasica* LERAUT, 2009 [Morocco, Algeria]
* subsp. *combustaria* (OBERTHÜR, 1923) [Algeria]
lentiscaria (DONZEL, 1837)
* subsp. *agacles* (OBERTHÜR, 1923) [Algeria]
distinctaria (BANG-HAAS, 1910)
* *atlantica* (REISSER, 1933) [Morocco]
* *galactaria* TURATI, 1934 [North Africa]
* *holli* (OBERTHÜR, 1910) [North-West Africa]
* *negrama* WEHRLI, 1953 [Turkey, Caucasus]
* *nobiliaria* (BANG-HAAS, 1906) [North Africa]
* *plebejaria* (OBERTHÜR, 1910) [North-West Africa]
* *rungsi* HERBULOT, 1981 [Morocco]
* *simplicaria* (REBEL, 1939) [Cyprus]

Perconia HÜBNER, 1823
strigillaria (HÜBNER, 1787)
baeticaria (STAUDINGER, 1871)

Prosopolophini WARREN, 1894
Compsoptera BLANCHARD, 1845
opacaria HÜBNER, (1819)
jourdanaria (SERRES, 1826)
subsp. *anargyra* (TURATI, 1913)
argentaria (HERRICH-SCHÄFFER, 1839)

Genera of uncertain tribus association (see remarks under each genus)

Lignyoptera LEDERER, 1853
fumidaria (HÜBNER, 1825)
thaumastaria REBEL, 1901

Opisthograptis HÜBNER, 1823
luteolata (LINNAEUS, 1758)

Toulgoetia HERBULOT, 1946
cauteriata (STAUDINGER, 1859)

Enanthyperythra WEHRLI, 1937
legataria (HERRICH-SCHÄFFER, 1852)

Eumera STAUDINGER, 1892
regina STAUDINGER, 1892
* *hoeferi* WEHRLI, 1934 [Turkey]
* subsp. *transcaucasica* WEHRLI, 1940 [Transcaucasus, Armenia]
* *mulier* PROUT, 1929 [Cyprus]
* *turcosyrica* WEHRLI, 1932 [Turkey, Eastern Mediterranean]

* ***Apocolotois*** WEHRLI, 1936
* *smirnovi* (ROMANOFF, 1885) [Caucasus, Armenia]

Asovia ALPHÉRAKY, 1908
maeoticaria (ALPHÉRAKY, 1876)

Lhommeia WERHLI, 1939
biskraria (OBERTHÜR, 1885)
* subsp. *bleusei* (THIERRY-MIEG, 1905) [Algeria]
* subsp. *cinnamomaria* (ROTHSCHILD, 1914) [Algeria]
* subsp. *illiturata* (WARREN, 1897) [South-East Africa]
* subsp. *viridaria* (ROTHSCHILD, 1914) [Algeria]

Onychora MEYRICK, 1892
agaritharia (DARDOIN, 1842)
* subsp. *atlasica* LERAUT, 2009 [Morocco]
subsp. *hepaticaria* (SCHWINGENSCHUSS, 1930)
* subsp. *mauretanicaria* (STAUDINGER, 1901) [Mauretania?, Algeria]
* subsp. *rungsi* LERAUT, 2009 [Morocco]

Colour plates

PLATE 1 (all specimens natural size)

1a. ***Abraxas grossulariata*** ♂, Stengade skov, Langeland, Fyn, Denmark, 28.vii.1977 (coll. SKOU)
1b. *A. grossulariata* ♀, Nebbeskov, Langeland, Fyn, Denmark, 23.vii.1977 (coll. SKOU)
1c. *A. grossulariata* ♂, Bejershamn, Öland, Sweden, 10.vii.1973 (coll. SKOU)
1d. *A. grossulariata* ♂, Stengade skov, Langeland, Fyn, Denmark, 11.viii.1975 (coll. SKOU)
2a. ***Abraxas sylvata*** ♂, Fenyőfő, Nyugat-Dunántúl, Hungary, 8.vii.1984 (coll. SKOU)
2b. *A. sylvata* ♂, Ræveskov, Fyn, Denmark, 6.viii.1969 (coll. SKOU)
2c. *A. sylvata* ♀, Allested, Fyn, Denmark, 17.vi.1973 (coll. SKOU)
2d. *A. sylvata* ♂, Lillering, East Jutland, Denmark, 17-18.vii.1979 (coll. SKOU)
3a. ***Abraxas pantaria*** ♂, Musei, Sardinia, Italy, 26.v.1975 (coll. SKOU)
3b. *A. pantaria* ♀, El Escorial, Prov. Madrid, Spain, 2.vii.1980 (coll. SKOU)
3c. *A. pantaria* ♂, Lozoya, Prov. Madrid, Spain, 20-21.vii.1995 (coll. SKOU)
3d. *A. pantaria* ♀, Almoraima, Prov. Cádiz, Spain, 24.ix.1987 (coll. SKOU)
4a. ***Ligdia adustata*** ♂, Nørresø, Fyn, Denmark, 20.v.1978 (coll. SKOU)
4b. *L. adustata* ♂, Allarpsberg, Skåne, Sweden, 17.vi.1972 (coll. SKOU)
4c. *L. adustata* ♀, Preci, Mt. Sibillini, Umbria, Italy, 20.iv.2009 (coll. SKOU)
4d. *L. adustata* ♂, Preci, Mt. Sibillini, Umbria, Italy, 20.iv.2009 (coll SKOU)
4e. *L. adustata* ♂, Monodendri, Prov. Ioannina, Greece, 25.viii.2008 (coll. SKOU)
4f. *L. adustata* ♀, Milo, Mt. Etna, Sicily, Italy, 7.vi.2005 (coll. SKOU)
5a. ***Lomaspilis marginata*** ♂, les Vigneaux, Dept. Hautes-Alpes, France, 31.vii.2006 (coll. SKOU)
5b. *L. marginata* ♂, Komárom, Komárom-Esztergom County, Hungary, 29.vi.1999 (coll. SKOU)
5c. *L. marginata* ♀, Kirkeby Sand, Fyn, Denmark, 14.vii.2002 (coll. SKOU)
5d. *L. marginata* ♂, Tranekær, Langeland, Fyn, Denmark, 26.vii.1977 (coll. SKOU)
5e. *L. marginata* ♂, Melsomvik, Vestfold, Norway, 28.vi.2000 (coll. SKOU)
5f. *L. marginata* ♀, Snarup Mose, Fyn, Denmark, 12.vi.1976 (coll. SKOU)
6a. ***Lomaspilis opis nigrita*** ♂, Liperi, North Karelia, Finland, 12.vi.1973 (coll. SKOU)
6b. *L. opis nigrita* ♂, Liperi, North Karelia, Finland, 26.vi.1958 (ZMH)
6c. *L. opis nigrita* ♂, Kiwiõli, Estonia, 1.vi.1943 (ZSM
6d. *L. opis nigrita* ♀, Kiwiõli, Estonia, 16.vi.1939 (ZSM)
6e. *L. opis nigrita* ♂, Kiwiõli, Estonia, 13.vi.1942 (ZSM)
6f. *L. opis nigrita* ♀, Tyrma, Bureinskyi Mountains, Amur Region, Russia, 19.vi.1997 (coll. SKOU)
7a. ***Lomaspilis bithynica*** ♀ Aliakmon River, Ag. Georgios, Grevena, Greece, 6.vi.1992 (ZSM)
7b. *L. bithynica* ♀, Aliakmon River, Ag. Georgios, Grevena, Greece, 6.vi.1992 (ZSM)
7c. *L. bithynica* ♀, ♂, Yaylalar, Kaçkar Dağları, N. E. Turkey, 4-5.vii.1983 (ZSM)
7d. *L. bithynica* ♂, Yaylalar, Kaçkar Dağları, N. E. Turkey, 4-5.vii.1983 (ZSM)
7e. *L. bithynica* ♀, Mengen, Bolu, N. W. Turkey, 18.vi.1979 (ZSM)
7f. *L. bithynica* ♀, Mengen, Bolu, N. W. Turkey, 18.vi.1979 (ZSM)
8a. ***Stegania cararia*** ♀, Dagneux, Dept. Ain, France, 2.vii.1967 (coll. SKOU)
8b. *S. cararia* ♂, Bialowiska Puszcza, Poland, 10.vi.1975 (coll. SKOU)
8c. *S. cararia* ♀ Kołaczyce, Jasło, Poland, 6.vii.1938 (coll. SKOU)
8d. *S. cararia* ♂, Amstetten, Niederösterreich, Austria, 26.vi.1992 (coll. SKOU)
8e. *S. cararia* ♂, Ft. d'Armainvilliers, Dept. Seine-et-Marne, France, 3.vi.1945 (ZSM)
8f. *S. cararia* ♂, Ft. d'Armainvilliers, Dept. Seine-et-Marne, France, 10.vi.1945 (ZSM)
9a. ***Stegania trimaculata*** ♂, Tosos, Prov. Zaragoza, Spain, 24.vi.1992 (coll. SKOU)
9b. *S. trimaculata* ♀, Brihuega, Prov. Guadalajara, Spain, 29.vi.1992 (coll. SKOU)
9c. *S. trimaculata* ♂, Uña, Prov. Cuenca, Spain, 28.vi.1992 (coll. SKOU)
9d. *S. trimaculata* ♂, Marbella, Prov. Málaga, Spain, 18.iii.1988 (coll. SKOU)
9e. *S. trimaculata* ♂, Fluminimaggiore, Sardinia, Italy, 7-8.iv.2002 (coll. SKOU)
9f. *S. trimaculata* ♂, la-Roche-de-Rame, Dept. Hautes-Alpes, France, 1.viii.2006 (coll. SKOU)
10a. ***Stegania dilectaria*** ♂, Dragalevtzi, Sofia, Bulgaria, 15.vi.1987 (coll. SKOU)
10b. *S. dilectaria* ♀, Dragalevtzi, Sofia, Bulgaria, 15.vi.1987 (coll. SKOU)
10c. *S. dilectaria* ♂, Inzersdorf, Wien, Austria, 17.viii.1960 (coll. SKOU)
10d. *S. dilectaria* ♂, Lobau, Wien, Austria, 18.v.1946 (coll. SKOU)
10e. *S. dilectaria* ♀, Stara Kresna, Kresna Gorge, Distr. Blagoevgrad, Bulgaria, 8.v.1988 (coll. SKOU)
10f. *S. dilectaria* ♂, Kavála, Prov. Kavála, Greece, 1.ix.2008 (coll. SKOU)
11a. ***Stegania dalmataria*** ♂, Dalmatia, Croatia, without date (ZSM)
11b. *S. dalmataria* ♀, Sarepta, S. European Russia, without date (ZSM)
11c. *S. dalmataria* ♂, Yeruslan River Valley, Saratov District, Russia, 23.v.2001 (coll. SKOU)
11d. *S. dalmataria* ♂, Yeruslan River Valley, Saratov District, Russia, 23.v.2001 (coll. SKOU)
11e. *S. dalmataria* ♀, Saur Mountains, E. Kazakhstan, 14.vii.1997 (coll. SKOU)
11f. *S. dalmataria* ♀, near Bisen village, Ryn-Kum Steppe, Kazakhstan, 28–29.v.2000 (coll. SKOU)

1a
1b
1c
1d
2a
2b
2c
2d
3a
3b
3c
3d
4a
4b
4c
4d
4e
4f
5a
5b
5c
5d
5e
5f
6a
6b
6c
6d
6e
6f
7a
7b
7c
7d
7e
7f
8a
8b
8c
8d
8e
8f
9a
9b
9c
9d
9e
9f
10a
10b
10c
10d
10e
10f
11a
11b
11c
11d
11e
11f

PLATE 2 (all specimens natural size)

12a. ***Acanthovalva inconspicuaria*** ♂, La Rahige, Sedella, Prov. Málaga, Spain, 25.vii.2003 (coll. SKOU)
12b. *A. inconspicuaria* ♂, La Rahige, Sedella, Prov. Málaga, Spain, 26.iv.2007 (coll. SKOU)
12c. *A. inconspicuaria* ♂, Matalascañas, Prov. Huelva, Spain, 16.iv.2007 (coll. SKOU)
12d. *A. inconspicuaria* ♀, El Pinar, Los Gallardos, Prov. Almería, Spain, 19-27.iv.2001 (coll. SKOU)
12e. *A. inconspicuaria* ♀, Rambla de Tabernas, Prov. Almería, Spain, 18.iv.2001 (coll. SKOU)
12f. *A. inconspicuaria* ♀, El Campello, Busot, Prov. Alicante, Spain, 2.v.1997 (coll. SKOU)
13a. ***Narraga fasciolaria*** ♂, Kiskunsági Nemzeti Park, Bács-Kiskun County, Hungary, 28.vii.2002 (coll. SKOU)
13b. *N. fasciolaria* ♂, Kiskunsági Nemzeti Park, Bács-Kiskun County, Hungary, 2.vii.1999 (coll. SKOU)
13c. *N. fasciolaria* ♀, Chotin, Ukraine, 17.viii.1987 (coll. SKOU)
13d. *N. fasciolaria* ♂, near Bisen village, Ryn-Kum Steppe, Kazakhstan, 28-29.v.2000 (coll. SKOU)
13e. *N. fasciolaria* ♂, Püspökladány, Hajdú-Bihar County, Hungary, 1.vii.1999 (coll. SKOU)
13f. *N. fasciolaria* ♀, N. E. Wien, Austria, 21.v.1914 (ZSM)
14a. ***Narraga nelvae catalaunica*** ♂, Cabo de Gata, Prov. Almería, Spain, 9-11.iv.1997 (coll. SKOU)
14b. *N. nelvae catalaunica* ♂, Bolnuevo, Mazarrón, Prov. Murcia, Spain, 15.iv.2001 (coll. SKOU)
14c. *N. nelvae catalaunica* ♂, Bolnuevo, Mazarrón, Prov. Murcia, Spain, 25.v.1998 (coll. SKOU)
14d. *N. nelvae catalaunica* ♀, Barranco de Valcuerna, Candasnos, Prov. Huesca, Spain, 5.vii.2002 (coll. SKOU)
14e. *N. nelvae catalaunica* ♀, Bolnuevo, Mazarrón, Prov. Murcia, Spain, 1.vi.1998 (coll. SKOU)
14f. *N. nelvae catalaunica* ♀, Bolnuevo, Mazarrón, Prov. Murcia, Spain, 25.v.1998 (coll. SKOU)
15a. ***Narraga tessularia*** ♂, Arkaim Nature Reserve, South Ural, Russia, 23.vii.1998 (coll. SKOU)
15b. *N. tessularia* ♂, Tchepun Hill, Dragoman, Distr. Sofia, Bulgaria, 11.vi.2007 (coll. SKOU)
15c. *N. tessularia* ♂, Hortobágy National Park, Hungary, 24.v.1978 (coll. SKOU)
15d. *N. tessularia* ♂, Sivash Lake, Kherson Region, Ukraine, 8.v.1987 (ZMKU)
15e. *N. tessularia* ♀, Illmitz, Burgenland, Austria (coll. SKOU), reared
15f. *N. tessularia* ♀, Berezovka village, Pavlodar Region, N. E. Kazakhstan, 1.vi.1984 (coll. SKOU)
16a. ***Heliomata glarearia*** ♂, Tubre, Bolzano, Italy, 27.vi.2008 (coll. SKOU)
16b. *H. glarearia* ♂, Murgia San Elia, Altamura, Puglia, Italy, 27-29.v.2005 (coll. SKOU)
16c. *H. glarearia* ♂, Kara-Dagh, Crimea, Ukraine, 11-12.vi.1987 (coll. SKOU)
16d. *H. glarearia* ♂, Martonyi, Szendrő, Borsod-Abaúj-Zemplén County, Hungary, 26.vi.2002 (coll. SKOU)
16e. *H. glarearia* ♀, Hundsheimerberg, Hainburg, Niederösterreich, Austria, 22.vi.1981 (coll. SKOU)
16f. *H. glarearia* ♂, Vatachorion, Prov. Florina, Greece, 26.vi.1985 (coll. SKOU)
17a. ***Isturgia famula*** ♂, Rio de Baza, Baza, Prov. Granada, Spain, 6.v.1997 (coll. SKOU)
17b. *I. famula* ♂, El Escorial, Prov. Madrid, Spain, 2.vi.1983 (coll. SKOU)
17c. *I. famula* ♂, Banyuls, Pyrénées-Orientales, France, 6.v.1948 (coll. SKOU)
17d. *I. famula* ♀, Villiers-Sous-Grez, Dept. Seine-et-Marne, France, 12.vi.1925 (ZSM)
17e. *I. famula* ♀, Vallée, St. Etienne, Dept. Loire, France, 19.vi.1921 (ZSM)
18a. ***Isturgia limbaria*** ♂, Laubert, Massif Central, Dept. Lozère, France, 9.vi.2007 (coll. SKOU)
18b. *I. limbaria* ♂, Queralbs, Ribes de Freser, Prov. Girona, Spain, 12.vii.2001 (coll. SKOU)
18c. *I. limbaria* ♂, Col de Perty, Montagne de Chamouse, Provence, Dept. Rhône-Alpes, France, 13-14.vi.1980 (coll. SKOU)
18d. *I. limbaria* ♀, Orpierre les Bagnes, Dept. Hautes-Alpes, France, 26.iv.1982 (coll. SKOU)
18e. *I. limbaria* ♀, Col de Perty, Montagne de Chamouse, Provence, Dept. Rhône-Alpes, France, 13-14.vi.1980 (coll. SKOU)
19a. ***Isturgia roraria*** ♂, Remetské Hámre, Slovakia, 17-18.vii.1983 (coll. SKOU)
19b. *I. roraria* ♂, Javorník, Bile Karpaty, Moravia, Czech Republic (coll. SKOU), reared
19c. *I. roraria* ♀, Harz Mts, Niedersachsen, Germany, without date (ZSM)
19d. *I. roraria* ♀, R. Osogovo, Kyustendil, Bulgaria, 2.vii.1981 (coll. SKOU)
20. ***Isturgia homochromata*** ♂, near Bastia, Corsica, France (Herbulot/ZSM), reared, holotype
21a. ***Isturgia murinaria*** ♂, Vallouise, Dept. Hautes-Alpes, France, 29.vii.2006 (coll. SKOU)
21b. *I. murinaria* ♂, Kuvandyk Village, South Urals, Russia, 15-30.vi.1996 (coll. SKOU)
21c. *I. murinaria* ♂, Val de Vecar, Albarracín, Prov. Teruel, Spain, 22.v.1998 (coll. SKOU)
21d. *I. murinaria* ♀, Retz, Niederösterreich, Austria, without date (ZSM)
21e. *I. murinaria* ♀, Bruck an der Leitha, Niederösterreich, Austria, 30.iv.1916 (ZSM)
22a. ***Isturgia catalaunaria*** ♂, Mojacar, Prov. Almería, Spain, 9.iv.2007 (coll. SKOU)
22b. *I. catalaunaria* ♂, Rambla de Tabernas, Prov. Almería, Spain, 31.vii.2003 (coll. SKOU)
22c. *I. catalaunaria* ♀, Rambla de Tabernas, Prov. Almería, Spain, 18.iv.2001 (coll. SKOU)
22d. *I. catalaunaria* ♂, Rambla de Tabernas, Prov. Almería, Spain, 5.v.1997 (coll. SKOU)
22e. *I. catalaunaria* ♀, Rambla de Tabernas, Prov. Almería, Spain, 30.x.2005 (coll. SKOU)
23a. ***Isturgia deerraria*** ♂, Marfa, Ridge, Rdum tal-Madonna, Malta, 16.viii.2007 (coll. SAMMUT)
23b. *I. deerraria* ♂, Torre la Higuera, Prov. Huelva, Spain, 21.ix.1981 (coll. SKOU)
23c. *I. deerraria* ♀, Mellieha, Rdum tal-Madonna, Malta (coll. SAMMUT), reared
23d. *I. deerraria* ♂, Mellieha, Rdum tal-Madonna, Malta (coll. SAMMUT), reared
24a. ***Isturgia disputaria*** ♂, Biskra, Algeria, 13.v.1907 (Herbulot/ZSM)
24b. *I. disputaria* ♀, Tozeur, Tunisia, v.1909 (Herbulot/ZSM)

12a 12b 12c 12d 12e 12f
13a 13b 13c 13d 13e 13f
14a 14b 14c 14d 14e 14f
15a 15b 15c 15d 15e 15f
16a 16b 16c 16d 16e 16f
17a 17b 17c 17d 17e
18a 18b 18c 18d 18e
19a 19b 19c 19d 20
21a 21b 21c 21d 21e
22a 22b 22c 22d 22e
23a 23b 23c 23d 24a 24b

PLATE 3 (all specimens natural size)

25a. **Isturgia spodiaria** ♂, Puerto de Conil, Cabo Roche, Prov. Cádiz, Spain, 14.iv.2007 (coll. Skou)
25b. *I. spodiaria* ♂, Otivar, Prov. Granada, Spain, 7.v.2000 (coll. Skou)
25c. *I. spodiaria* ♂, Civita, Monte Pollino, Calabria, Italy, 31.v-1.vi.2005 (coll. Skou)
25d. *I. spodiaria* ♀, El Cuartón, Tarifa, Prov. Cádiz, Spain, 17-18.v.2006 (coll. Skou)
25e. *I. spodiaria* ♀, Almoraima, Prov. Cádiz, Spain, 24.ix.1987 (coll. Skou)
26a. **Isturgia assimilaria** ♂, Monte Limbara, Sardinia, Italy, 22.v.2004 (coll. Skou)
26b. *I. assimilaria* ♂, Monte Spada, Sardinia, Italy, 19-20.v.2004 (coll. Skou)
26c. *I. assimilaria* ♀, Haute Asco, Corsica, France, 23.vi.1994 (coll. Skou)
26d. *I. assimilaria* ♀, Monte Limbara, Sardinia, Italy, 22.v.2004 (coll. Skou)
26e. *I. assimilaria* ♂, underside, Monte Spada, Sardinia, Italy, 19-20.v.2004 (coll. Skou)
27a. **Isturgia miniosaria** ♂, La Plataforma, Sierra de Gredos, Prov. Ávila, Spain, 27-28.ix.2009 (coll. Skou)
27b. *I. miniosaria* ♂, Rambla de Tabernas, Prov. Almería, Spain, 30.x.2005 (coll. Skou)
27c. *I. miniosaria* ♂, La Plataforma, Sierra de Gredos, Prov. Ávila, Spain, 27-28.ix.2009 (coll. Skou)
27d. *I. miniosaria* ♂, Val de Vecar, Albarracín, Prov. Teruel, Spain, 1.x.2009 (coll. Skou)
27e. *I. miniosaria* ♂, Penalba, Prov. Huesca, Spain, 17.x.1984 (coll. Skou)
27f. *I. miniosaria* ♂, Albarracín, Prov. Teruel, Spain, 30.ix.2009 (coll. Skou)
27g. *I. miniosaria* ♂, Lagoa de Santo André, Prov. Alentejo, Portugal, 26.ix.2009 (coll. Skou)
27h. *I. miniosaria* ♂, La Plataforma, Sierra de Gredos, Prov. Ávila, Spain, 27-28.ix.2009 (coll. Skou)
27i. *I. miniosaria* ♀, La Plataforma, Sierra de Gredos, Prov. Ávila, Spain, 27-28.ix.2009 (coll. Skou)
27j. *I. miniosaria* ♀, Rambla de Tabernas, Prov. Almería, Spain, 30.x.2005 (coll. Skou)
27k. *I. miniosaria* ♀, La Plataforma, Sierra de Gredos, Prov. Ávila, Spain, 27-28.ix.2009 (coll. Skou)
27l. *I. miniosaria* ♀, La Plataforma, Sierra de Gredos, Prov. Ávila, Spain, 27-28.ix.2009 (coll. Skou)
28a. **Isturgia berytaria** ♂, Vlahokerasia, Tripoli, Peloponnese, Prov. Arcadia, Greece, 3.x.1984 (coll. Skou)
28b. *I. berytaria* ♂, Kalavryta, Chelmos, Peloponnese, Prov. Achaia, Greece, 3.x.1984 (coll. Skou)
28c. *I. berytaria* ♀, Vlahokerasia, Tripoli, Peloponnese, Prov. Arcadia, Greece, 3.x.1984 (coll. Skou)
28d. *I. berytaria* ♂, underside, Hisarköy, Northern Cyprus, 27.xi-3.xii.2010(coll. Skou)
29a. **Isturgia sparsaria** ♂, Santuaria di Macereto, Marche, Italy, 19.vii.1979 (coll. Skou)
29b. *I. sparsaria* ♀, Monte Capra, Emilia-Romagna, Italy, 29.vi.1987 (coll. Flamigni)
29c. *I. sparsaria* ♂, Monte Capra, Emilia-Romagna, Italy, 26.vii.1987 (coll. Flamigni)
29d. *I. sparsaria* ♀, Monte Capra, Emilia-Romagna, Italy, 11.v.1985 (coll. Flamigni)
29e. *I. sparsaria* ♂, Monte Capra, Emilia-Romagna, Italy, 11.v.1985 (coll. Flamigni)
30a. **Isturgia messapiaria** ♂, Fossiata, Sila Mountains, Calabria, Italy, 6.viii.1921 (MSNM)
30b. *I. messapiaria* ♂, Fossiata, Sila Mountains, Calabria, Italy, 6.viii.1921 (MSNM)
30c. *I. messapiaria* ♂, Fossiata, Sila Mountains, Calabria, Italy, 27-28.vii.2011 (coll. Skou)
30d. *I. messapiaria* ♀, Fossiata, Sila Mountains, Calabria, Italy, 6.viii.1921 (MSNM)
30e. *I. messapiaria* ♀, Fossiata, Sila Mountains, Calabria, Italy, 6.viii.1921 (MSNM)
31a. **Isturgia arenacearia** ♂, Kiskunsági at Kecskemét, Bács-Kiskun County, Hungary, 2.vii.1999 (coll. Skou)
31b. *I. arenacearia* ♂, Lanterna Camp, Istria, Croatia, 20-30.v.1987(coll. Skou)
31c. *I. arenacearia* ♀, Kiskunsági at Kecskemét, Bács-Kiskun County, Hungary, 28.vii.2002 (coll. Skou)
31d. *I. arenacearia* ♀, Vivaro, Friuli-Venezia Giulia, Italy, 28.vi.2008 (coll. Skou)
31e. *I. arenacearia* ♀, Kara-Dagh, Crimea, Ukraine, 25-27.vi.1987 (coll. Skou)
32a. **Isturgia punctistrigaria** ♀, Atyrau District, European part of Kazakhstan, 30.v.2007 (LSSU)
32b. *I. punctistrigaria* ♀, Repetek, Turkmenistan, 12.v.1988 (coll. Skou)
32c. *I. punctistrigaria* ♀, Askhabad, Turkmenistan, without date (ZISP), lectotype
32d. *I. punctistrigaria* ♂, Askhabad, Turkmenistan, without date (ZISP), paralectotype
32e. *I. punctistrigaria* ♂, Askhabad, Turkmenistan, without date (ZISP), paralectotype
32f. *I. punctistrigaria* ♂, Kara-Chokat, Turgaiskaya obl., Kazakhstan, 22.vii.1908 (ZISP)

25a
25b
25c
25d
25e
26a
26b
26c
26d
26e
27a
27b
27c
27d
27e
27f
27g
27h
27i
27j
27k
27l
28a
28b
28c
28d
29a
29b
29c
29d
29e
30a
30b
30c
30d
30e
31a
31b
31c
31d
31e
32a
32b
32c
32d
32e
32f

PLATE 4 (all specimens natural size)

33a. ***Itame vincularia*** ♂, El Pinar, Los Gallardos, Prov. Almería, Spain, 19-27.iv.2001 (coll. SKOU)
33b. *I. vincularia* ♂, Forêt de Malmont, Pont du Gard, Dept. Gard, France, 17.v.1998 (coll. SKOU)
33c. *I. vincularia* ♀, El Pinar, Los Gallardos, Prov. Almería, Spain, 19-27.iv.2001 (coll. SKOU)
33d. *I. vincularia* ♀, Val de Vecar, Albarracín, Prov. Teruel, Spain, 7.vii.1950 (HERBULOT/ZSM)
34a. ***Neognopharmia stevenaria*** ♂, Stara Baska, Krk Island, Croatia, 1.v.2008 (coll. SKOU)
34b. *N. stevenaria* ♂, Kara-Dagh, Crimea, Ukraine, 11-12.vi.1987 (coll. SKOU)
34c. *N. stevenaria* ♂, Itea, Prov. Fokidos, Greece, 20.vii.1987 (coll. SKOU)
34d. *N. stevenaria* ♀, Arginoda, Kalymnos, Prov. Dodecanese, Greece, 29.iv-1.v.2004 (coll. SKOU)
35a. ***Macaria notata*** ♂, Kruglik Village by Kiev, Ukraine, 7.viii.1994 (coll. SKOU)
35b. *M. notata* ♂, Mätorhusen, Skåne, Sweden, 17.vii.1973 (coll. SKOU)
35c. *M. notata* ♂, Snarup, Fyn, Denmark, 14.vi.1974 (coll. SKOU)
35d. *M. notata* ♀, Amfreville, Dept. Eure, France, 22.vii.1950 (HERBULOT/ZSM)
35e. *M. notata* ♀, Snarup, Fyn, Denmark, 14.vi.1974 (coll. SKOU)
36a. ***Macaria alternata*** ♂, la Roche-de-Rame, by river Durance, Dept. Hautes-Alpes, France, 1.viii.2006 (coll. SKOU)
36b. *M. alternata* ♂, Duilhac-sous-Peyrepertuse, Mayry, Dept. Aude, France, 5.viii.2003 (coll. SKOU)
36c. *M. alternata* ♂, Kiskunsági at Kecskemét, Bács-Kiskun County, Hungary, 28.vii.2002 (coll. SKOU)
36d. *M. alternata* ♂, Yeruslan River Valley, South Saratov District, South East European Russia, 23.v.2001 (coll. SKOU)
36e. *M. alternata* ♂, Vejlen, Tåsinge, Fyn Denmark, 23-25.vi.1977 (coll. SKOU)
37a. ***Macaria signaria*** ♂, Frøslev, South Jutland, Denmark, 19.vii.1974 (coll. SKOU)
37b. *M. signaria* ♂, Stengade Skov, Langeland, Fyn, Denmark, 20.vi.1978 (coll. SKOU)
37c. *M. signaria* ♀, Forêt de St. Germain, Dept. Yvelines, France, 22.v.1948 (HERBULOT/ZSM)
37d. *M. signaria* ♂, Eteignières, Dept. Ardennes, France, 31.v.1964 (HERBULOT/ZSM)
37e. *M. signaria* ♀, Les Hauts Buttés, Dept. Ardennes, France, 20-21.vi.1953 (HERBULOT/ZSM)
38a. ***Macaria liturata*** f. *hispanica* ♂, Biel-Fuencalderas, Prov. Huesca, Spain, 17.vi.1999 (coll. SKOU)
38b. *M. liturata* ♂, Eptachori, Prov. Kastoria, Greece, 13.vii.1998 (coll. SKOU)
38c. *M. liturata* ♀, Tranekær, Langeland, Fyn, Denmark, 26.vii.1977 (coll. SKOU)
38d. *M. liturata* ♀, Komárom, at Duna River, Komárom-Esztergom County, Hungary, 29.vi.1999 (coll. SKOU)
38e. *M. liturata* f. *deceptans* ♀, Mount Parnassos, Prov. Voiotia, Greece, 28.vi.1985 (coll. SKOU)
39a. ***Macaria wauaria*** ♂, Tubre, Trentino-Alto Adige, Italy, 27.vi.2008 (coll. SKOU)
39b. *M. wauaria* ♂, Puerto de Ragua, Prov. Granada, Spain, 28.vii.2003 (coll. SKOU)
39c. *M. wauaria* ♂, Vejlen, Tåsinge, Fyn, Denmark, 16-19.vii.1979 (coll. SKOU)
39d. *M. wauaria* ♂, Puisieux, Dept. Aisne, France, 27.vi.1964 (HERBULOT/ZSM)
39e. *M. wauaria* ♀, Sälen, Dalarna, Sweden, 5.viii.1978 (coll. SKOU)
40a. ***Macaria artesiaria*** ♂, Villar d'Arène, la Grave, Dept. Hautes-Alpes, France, 26.vii.2000 (coll. SKOU)
40b. *M. artesiaria* ♂, Uña, Prov. Cuenca, Spain, 28.vi.1992 (coll. SKOU)
40c. *M. artesiaria* ♀, Janakkala, EH, Finland, 9.viii.1982 (ZMH)
40d. *M. artesiaria* ♀, Odenas, Dept. Rhône-Alpes, France, 12.vii.1956 (HERBULOT/ZSM)
40e. *M. artesiaria* ♀, les Vigneaux, Dept. Hautes-Alpes, France, 31.vii.2006 (coll. SKOU)
41a. ***Macaria ichnusae*** ♂, Capoterra, Rio Gutturu, Mannu, Sardinia, Italy, 18.vi.2009 (coll. GOVI)
41b. *M. ichnusae* ♂, Capoterra, Rio Gutturu, Mannu, Sardinia, Italy, 18.vi.2009 (coll. GOVI)
41c. *M. ichnusae* ♀, Capoterra, Rio Gutturu, Mannu, Sardinia, Italy, 18.vi.2009 (coll. GOVI)
41d. *M. ichnusae* ♀, Rio Isalle, NE Nuoro, Sardinia, Italy, 3-4.vii.2011 (coll. SKOU)
41e. *M. ichnusae* ♀, Rio Isalle, NE Nuoro, Sardinia, Italy, 3-4.vii.2011 (coll. SKOU)
42a. ***Macaria carbonaria*** ♂, Veda, Upland, Sweden, 27.v.1956 (coll. SKOU)
42b. *M. carbonaria* ♂, Outrup Hede, North East Jylland, Denmark, 9.v.1975 (coll. SKOU)
42c. *M. carbonaria* ♀, Abisko, Lapland, Sweden, 29.vi.1982 (coll. SKOU)
42d. *M. carbonaria* ♂, Abisko, Lapland, Sweden, 29.vi.1982 (coll. SKOU)
42e. *M. carbonaria* ♂, Abisko, Lapland, Sweden, 29.vi.1982 (coll. SKOU)
42f. *M. carbonaria* ♀, Outrup Hede, North East Jutland, Denmark, 9.v.1975 (coll. SKOU)
43a. ***Macaria brunneata*** ♂, Prestseter, Vestfold, Norway, 2.vii.2000 (coll. SKOU)
43b. *M. brunneata* ♂, Vejlen, Tåsinge, Fyn, Denmark, 21-24.vi.1979 (coll. SKOU)
43c. *M. brunneata* ♂, Højkol, East Jutland, Denmark, 5.vii.1975 (coll. SKOU)
43d. *M. brunneata* ♀, Højkol, East Jutland, Denmark, 5.vii.1975 (coll. SKOU)
43e. *M. brunneata* ♀, Prestseter, Vestfold, Norway, 2.vii.2000 (coll. SKOU)
43f. *M. brunneata* ♀, Mont Flassin, Jovencan, Aosta, Italy, 18.viii.2000 (coll. SKOU)
44a. ***Macaria loricaria*** ♂, Färila, Hälsingland, Sweden, 8.vii.1988 (coll. SKOU)
44b. *M. loricaria* ♂, Koski HL [Hämeenkoski], Finland, 13.vii.1981 (coll. SKOU)
44c. *M. loricaria* ♂, Koski HL [Hämeenkoski], Finland, 13.vii.1981 (coll. SKOU)
44d. *M. loricaria* ♂, Hanko, Tvärminne, Finland, 13.vii.1970 (ZMH)
44e. *M. loricaria* ♀, Färila, Hälsingland, Sweden, 8.vii.1988 (coll. SKOU)

33a
33b
33c
33d
34a
34b
34c
34d
35a
35b
35c
35d
35e
36a
36b
36c
36d
36e
37a
37b
37c
37d
37e
38a
38b
38c
38d
38e
39a
39b
39c
39d
39e
40a
40b
40c
40d
40e
41a
41b
41c
41d
41e
42a
42b
42c
42d
42e
42f
43a
43b
43c
43d
43e
43f
44a
44b
44c
44d
44e

PLATE 5 (all specimens natural size)

45a. ***Macaria fusca*** ♂, Col de l'Iseran, Dept. Savoie, France, 4.viii.2009 (coll. SKOU)
45b. *M. fusca* ♂, Col de l'Iseran, Dept. Savoie, France, 4.viii.2009 (coll. SKOU)
45c. *M. fusca* ♂, Col de l'Iseran, Dept. Savoie, France, 4.viii.2009 (coll. SKOU)
45d. *M. fusca* ♂, Graddis, Saltdal, NSI, Norway, 9.vii.1976 (coll. SKOU)
45e. *M. fusca* ♂, Karigasnniemi, Utsjoki, Inari Lapland, Finland, 17.vii.1985 (coll. SKOU)
45f. *M. fusca* ♀, Jebrenjokk, Torne Lapmark, Sweden, 9.vii.1966 (coll. SKOU)
45g. *M. fusca* ♀, Torne Lapmark, Sweden, 7.vii.1980 (coll. SKOU)
46a. ***Chiasmia clathrata*** ♂, Troserum, Småland, Sweden, 4.vii.1976 (coll. SKOU)
46b. *C. clathrata* ♂, Martonyi at Szendrő, Borsod-Abaúj-Zemplén County, Hungary, 26.vi.2002 (coll. SKOU)
46c. *C. clathrata* ♂, Ruta del Veleta, Sierra Nevada, Prov. Granada, Spain, 5.vii.1987 (coll. SKOU)
46d. *C. clathrata* ♀, Svedmyra, Södermanland, Sweden, 31.vii.1974 (coll. SKOU)
46e. *C. clathrata* ♂, Kuvandyk District, Orenburg Region, South Ural, Russia, 15.vi.1998 (coll. SKOU)
46f. *C. clathrata* ♀, Frejus, Dept. Var, France, 5.vii.1999 (coll. SKOU)
47a. ***Chiasmia aestimaria*** ♂, Kos, Prov. Dodecanese, Greece, 7.viii.1975 (coll. SKOU)
47b. *C. aestimaria* ♂, Kamena Vourla, Prov. Fthiotis, Greece, 5.vii.1985 (coll. SKOU)
47c. *C. aestimaria* ♀, Skines, Prov. Chania, Crete, Greece, 8.vi.2004 (coll. SKOU)
47d. *C. aestimaria* ♀, Rambla de Tabernas, Almería, Spain, 23.iv.2001 (coll. SKOU)
47e. *C. aestimaria sareptanaria* ♂, Sarepta, Russia, 1886 (ZMKU)
47f. ***Chiasmia syriacaria*** ♂, Marrasch, Taurus, Turkey, without date (ZSM)
48a. ***Digrammia rippertaria*** ♀, Padurea Letea, Tulcea, Romania, 20.viii.1981 (ZSM)
48b. *D. rippertaria* ♂, Ryn-Kum sand steppe, European part of Kazakhstan, 29.v-1.vi.2001 (coll. SKOU)
48c. *D. rippertaria* ♂, la Roche-de-Rame, by river Durance, Dept. Hautes-Alpes, France, 1.viii.2006 (coll. SKOU)
48d. *D. rippertaria* ♀, la Roche-de-Rame, by river Durance, Dept. Hautes-Alpes, France, 1.viii.2006 (coll. SKOU)
48e. *D. rippertaria* ♂, la Roche-de-Rame, by river Durance, Dept. Hautes-Alpes, France, 1.viii.2006 (coll. SKOU)
48f. *D. rippertaria* ♂, Ryn-Kum sand steppe, European part of Kazakhstan, 12-15.ix.2000 (coll. SKOU)
49a. ***Rhoptria asperaria*** ♀, Matalascañas, Prov. Huelva, Spain, 21.iv.2007 (coll. SKOU)
49b. *R. asperaria* ♀, Rethymno, Argiroupoli, Prov. Rethymno, Crete, Greece, 25-29.vi.2000 (coll. SKOU)
49c. *R. asperaria* ♂, El Pinar, Los Gallardos, Prov. Almería, Spain, 19-27.iv.2001 (coll. SKOU)
49d. *R. asperaria* ♂, El Pinar, Los Gallardos, Prov. Almería, Spain, 19-27.iv.2001 (coll. SKOU)
49e *R. asperaria* ♀, Noguera, Monte Universales, Prov. Teruel, Spain, 7.vii.1980 (coll. SKOU)
49f. *R. asperaria* ♀, Sierra de Espuña, Prov. Murcia, Spain, 20.iii.1994 (coll. SKOU)
50a. ***Rhoptria dolosaria*** ♂, Thermopylae, Prov. Fthiotida, Greece, 23.iv.1973 (ZSM)
50b. *R. dolosaria* ♂, Lithochoron, Mt. Olympus, Province Pieria, Greece, 10-23.vi.1957 (ZSM)
50c. *R. dolosaria* ♂, Lithochoron, Mt. Olympus, Province Pieria, Greece, 10-23.vi.1957 (ZSM)
51a. ***Ortaliella gruneraria*** ♂, Tripolis, Province Arkadia, Greece, 14.vi.1954 (coll. SKOU)
51b. *O. gruneraria* ♀, Anogia, Mt. Taygetos, Province Lakonia, Greece, 6.vi.1964 (ZSM)
52a. ***Cepphis advenaria*** ♂, Gærup Skov, Fyn, Denmark, 4.vi.1972 (coll. SKOU)
52b. *C. advenaria* ♂, Ouville-la-Rivière, Dept. Seine-Maritime, France, 29.v.1939 (ZSM)
52c. *C. advenaria* ♀, Gærup Skov, Fyn, Denmark, 4.vi.1972 (coll. SKOU)
52d. *C. advenaria* ♀, Les Hauts Buttés, Dept. Ardennes, France, 20-21.vi.1953 (ZSM)
52e. *C. advenaria* ♀, Vita Pochtovaja, Kievskaja obl., Ukraine, 23.vi.2002 (ZMKU)
53a. ***Petrophora chlorosata*** ♀, Snarup Mose, Fyn, Denmark, 12.vi.1976 (coll. SKOU)
53b. *P. chlorosata* ♂, Snarup Mose, Fyn, Denmark, 3.v.1974 (coll. SKOU)
53c. *P. chlorosata* ♂, Stensund, Södermanland, Sweden, 18.v.1974 (coll. SKOU)
53d. *P. chlorosata* ♂, Voropaev, Vyshgorod District, Kievskaja obl., Ukraine, 29.iv.1983 (ZMKU)
53e. *P. chlorosata* ♂, Novije Bezradychi, Obuchov District, Kievskaja obl., Ukraine, 11.v.1983 (ZMKU)
54a. ***Perigune binaevata*** ♀, S' Albufera, Es Comu Baix, Mallorca, Spain, 28.x.2003 (coll. HONEY)
54b. *P. binaevata* ♀, S' Albufera, Es Comu Baix, Mallorca, Spain, 28.x.2003 (coll. HONEY)
54c. *P. binaevata* ♂, S' Albufera, Es Comu Baix, Mallorca, Spain, 18.x.2005 (coll. HONEY)
54d. *P. binaevata* ♂, S' Albufera, Es Comu Baix, Mallorca, Spain, 28.x.2003 (coll. HONEY)
54e. *P. binaevata* ♂, Tombolo di Giannella, Grosseto, Toscana, Italy, 13.x.2010 (coll. SKOU)
54f. *P. binaevata* ♂, Tombolo di Giannella, Grosseto, Toscana, Italy, 13.x.2010 (coll. SKOU)
55a. ***Perigune convergata*** ♂, El Pozo del Esparto, Prov. Almería, Spain, 7-8.iv.2007 (coll. SKOU)
55b. *P. convergata* ♂, El Pozo del Esparto, Prov. Almería, Spain, 1.xi.2005 (coll. SKOU)
55c. *P. convergata* ♂, Tramacastilla, Prov. Teruel, Spain, 17.ix.2002 (coll. SKOU)
55d. *P. convergata* ♂, Mornas, Dept. Vaucluse, France, 13.x.1998 (coll. SKOU)
55e. *P. convergata* ♀, Albarracín, Prov. Teruel, Spain, 2.xi.2005 (coll. SKOU)
55f. *P. convergata* ♀, El Pozo del Esparto, Prov. Almería, Spain, 1.xi.2005 (coll. SKOU)
56a. ***Perigune narbonea*** ♂, Albuñol, Prov. Granada, Spain, 5.v.2000 (coll. SKOU)
56b. *P. narbonea* ♂, Orpierre, les Bègues, Dept. Hautes-Alpes, France, 26.iv.1982 (coll. SKOU)
56c. *P. narbonea* ♂, Gorges de la Meouge, Dept. Hautes-Alpes, France, 10.viii.1996 (coll. SKOU)
56d. *P. narbonea* ♂, El Pinar, Los Gallardos, Prov. Almería, Spain, 19-27.iv.2001 (coll. SKOU)
56e. *P. narbonea* ♀, Cucugnan, Dept. Pyrénées-Orientales, France, 5.iv.2007 (coll. SKOU)
56f. *P. narbonea* ♀, El Pozo del Esparto, Prov. Almería, Spain, 1.xi.2005 (coll. SKOU)

45a 45b 45c 45d 45e 45f 54g
46a 46b 46c 46d 46e 46f
47a 47b 47c 47d 47e 47f
48a 48b 48c 48d 48e 48f
49a 49b 49c 49d 49e 49f
50a 50b 50c 51a 51b
52a 52b 52c 52d 52e
53a 53b 53c 53d 53e
54a 54b 54c 54d 54e 54f
55a 55b 55c 55d 55e 55f
56a 56b 56c 56d 56e 56f

PLATE 6 (all specimens natural size)

57a. ***Lignyoptera fumidaria*** ♂, Zitzmannsdorfer Wiesen, Burgenland, Austria, 1.xi.1971 (coll. Skou)
57b. *L. fumidaria* ♂, Tihany, Veszprém, Hungary, uncertain date (coll. Skou)
57c. *L. fumidaria* ♂, Budapest, Pest, Hungary, 13.xi.1938 (coll. Skou)
57d. *L. fumidaria* ♂, Wien, Austria, without date (ZMKU)
57e. *L. fumidaria* ♀, Laxenburg, Niederösterreich, Austria, 6.xi.1910 (ZSM)
58a. ***Lignyoptera thaumastaria*** ♂, Vlasic planina, Travnik, Bosnia, (ZSM), reared
58b. *L. thaumastaria* ♂, Vlasic planina, Travnik, Bosnia, 20.x.1937 (ZFMK)
58c. *L. thaumastaria* ♀, Vlasic planina, Travnik, Bosnia, (ZSM), reared
59a. ***Heterolocha laminaria*** ♂, Targim, Ingushetia, Russia, 27.vi.1991 (ZMKU)
59b. *H. laminaria* ♂, Targim, Ingushetia, Russia, 27.vi.1991 (ZMKU)
59c. *H. laminaria* ♀, Konygin, Ust-Donetzk dist., Rostov-on-Don, Russia, 21.vii.2006 (coll. Poltavsky)
60a. ***Plagodis pulveraria*** ♂, Bialowieza, District Podlaskie, Poland, 28.v.2000 (coll. Skou)
60b. *P. pulveraria* ♂, Møns Klint, Møn, Denmark, 4.vi.1979 (coll. Skou)
60c. *P. pulveraria* ♂, Snarup Mose, Fyn, Denmark, 30.v.1973 (coll. Skou)
60d. *P. pulveraria* ♀, Snarup Mose, Fyn, Denmark, 5.vi.1978 (coll. Skou)
61a. ***Plagodis dolabraria*** ♂, Martonyi at Szendrő, Borsod-Abaúj-Zemplén County, Hungary, 26.vi.2002 (coll. Skou)
61b. *P. dolabraria* ♂, Stengade Skov, Langeland, Fyn, Denmark, 3.vi.1977 (coll. Skou)
61c. *P. dolabraria* ♂, Bejershamn, Öland, Sweden, 8.vi.1972 (coll. Skou)
61d. *P. dolabraria* ♀, Tranekær, Langeland, Fyn, Denmark, 14.vi.1977 (coll. Skou)
62a. ***Pachycnemia hippocastanaria*** ♀, Kavala, Prov. Kavala, Greece, 1.ix.2008 (coll. Skou)
62b. *P. hippocastanaria* ♀, Sant Marti de Tous, Prov. Barcelona, Spain, 16.viii.2001 (coll. Skou)
62c. *P. hippocastanaria* ♀, Monte Limbara, Sardinia, Italy, 22.v.2004 (coll. Skou)
62d. *P. hippocastanaria* ♂, Eslida, Sierra Espadán, Prov. Castellón de la Plana, Spain, 1.v.1997 (coll. Skou)
62e. *P. hippocastanaria* ♂, Nebbeskov, Langeland, Fyn, Denmark, 15.v.1979 (coll. Skou)
63a. ***Pachycnemia tibiaria*** ♂, Sierra del Chaparral, Prov. Granada, Spain, 26.viii.2005 (coll. Skou)
63b. *P. tibiaria* ♂, Noli, Liguria, Italy, 12-25.ix.1951 (ZSM)
63c. *P. tibiaria* ♀, Gravosa-Zaton, Dalmatia, Croatia, 14.x.1934 (ZSM)
63d. *P. tibiaria benesignata* ♂, Col de Bacinu, Porto Vecchio, Corsica, France, 22.ix.2003 (coll. Skou)
63e. *P. tibiaria benesignata* ♂, Col de Bacinu, Porto Vecchio, Corsica, France, 22.ix.2003 (coll. Skou)
64a. ***Opisthograptis luteolata*** ♂, Saint-Laurent-du-Pape, Dept. Ardèche, France, 9.ix.2001 (coll. Skou)
64b. *O. luteolata* ♂, Monte Spada, Fonni, Sardinia, Italy, 19-20.v.2004 (coll. Skou)
64c. *O. luteolata* ♀, Stengade Skov, Langeland, Fyn, Denmark, 6.vi.1974 (coll. Skou)
64d. *O. luteolata* ♀, Castelbuono, Sicily, Italy, 5.vi.2005 (coll. Skou)
65a. ***Epione repandaria*** ♂, Storelung, Fyn, Denmark, 19.viii.1972 (coll. Skou)
65b. *E. repandaria* ♂, Vejlen, Tåsinge, Fyn, Denmark, 11-13.viii.1978 (coll. Skou)
65c. *E. repandaria* ♂, Ljusvattnet, Västerbotten, Sweden, 6.ix.1974 (coll. Skou)
65d. *E. repandaria* ♂, Moscardón, Prov. Teruel, Spain, 6.viii.2007 (coll. Skou)
65e. *E. repandaria* ♀, Borlänge, Grängshammar, Dalarna, Sweden, 1.ix.1979(coll. Skou)
66a. ***Epione vespertaria*** ♂, Als Odde, Northeast Jylland, Denmark, 17.vii.1979 (coll. Skou)
66b. *E. vespertaria* ♂, Torö, Södermanland, Sweden, 25.vii.1972 (coll. Skou)
66c. *E. vespertaria* ♂, Grugnay (Chamoson), Valais, Switzerland, 2.viii.1983 (coll. Skou)
66d. *E. vespertaria* ♀, Rohrwald, Baden-Württemberg, Germany, 15.vii.1900 (ZSM)
66e. *E. vespertaria* ♀, Forst Fritzen, Ostpreussen, Poland, vii.1914 (ZSM)
67a. ***Therapis flavicaria*** ♂, Židlochovice, Moravia, Czech Republic, 10.vi.1991 (coll. Skou)
67b. *T. flavicaria* ♂, Skakavitza, Distr. Kyustendil, Bulgaria, 10.vi.1962 (coll. Skou)
67c. *T. flavicaria* ♂, Volsk, Saratov Region, Southeast Russia, 23.v.1987 (coll. Skou)
67d. *T. flavicaria* ♂, Rostov by Don, Southwest Russia, 25.v.1989 (coll. Skou)
67e. *T. flavicaria* ♀, Prater, Wien, Wien, Austria, (ZSM), reared
68a. ***Pseudopanthera macularia*** ♂, Col de Perty, Provence, Dept. Rhône-Alpes, France, 13.vi.1980 (coll. Skou)
68b. *P. macularia* ♂, Häckeberga, Skåne, Sweden, 4.vi.1972 (coll. Skou)
68c. *P. macularia* ♀, Brændegård, Fyn, Denmark, 20.vi.1972 (coll. Skou)
68d. *P. macularia* ♀, Häckeberga, Skåne, Sweden, 4.vi.1972 (coll. Skou)
68e. *P. macularia* ♀, Akritochori, Prov. Serres, Greece, 25.vi.2004 (coll. Skou)

57a
57b
57c
57d
57e
58a
58b
58c
59a
59b
59c
60a
60b
60c
60d
61a
61b
61c
61d
62a
62b
62c
62d
62e
63a
63b
63c
63d
63e
64a
64b
64c
64d
65a
65b
65c
65d
65e
66a
66b
66c
66d
66e
67a
67b
67c
67d
67e
68a
68b
68c
68d
68e

PLATE 7 (all specimens natural size)

69a. ***Eilicrinia cordiaria*** ♂, Kuvandyk Village, South Urals, Russia, 15-30.vi.1996 (coll. SKOU)
69b. *E. cordiaria* ♂, Aliakmonas River, Grevena, Prov. Grevena, Greece, 4.vii.2004 (coll. SKOU)
69c. *E. cordiaria* ♀, Monemvasia, Prov. Lakonia, Greece, 7.vii.1984 (coll. SKOU)
69d. *E. cordiaria* ♀, Vatachorion, Prov. Florina, Greece, 26.vi.1985 (coll. SKOU)
69e. *E. cordiaria* ♀, Monemvasia, Prov. Lakonia, Greece, 2.iv.1979 (coll. SKOU)
70a. ***Eilicrinia subcordiaria*** ♀, Astrakhan Oblast, Southeast European Russia, 11.vii.1965 (coll. SKOU)
70b. *E. subcordiaria* ♀, South Urals, Russia, 13.vii.1906 (ZSM)
70c. *E. subcordiaria* ♀, South Urals, Russia, 12.vii.1906 (ZSM)
70d. *E. subcordiaria* Vicinity of Kashka-Sun, Kyrgyzstan, 29.vi-1.vii.1999 (coll. SKOU)
70e. *E. subcordiaria* ♀, Sarepta, S. European Russia, 1889 (ZSM)
71a. ***Eilicrinia trinotata*** ♂, Matka, Prov. Skopje, Macedonia, 22.vii.1971 (coll. SKOU)
71b. *E. trinotata* ♀, Avandas Gorge, Prov. Evros, Greece, 29.vi.2004 (coll. SKOU)
71c. *E. trinotata* ♀, Micurin, Distr. Burgas, Bulgaria, 1.vii.1972 (coll. SKOU)
71d. *E. trinotata* ♂, Kara-Dagh, Crimea, Ukraine, 27.v.1992 (ZSM)
71e. *E. trinotata* ♀, Csákberény, Fejér County, Hungary, 11.v.2007 (coll. SKOU)
72a. ***Toulgoetia cauteriata*** ♂, Novo Sancti Petri, Prov. Cádiz, Spain, 4-7.iii.2005 (coll. SKOU)
72b. *T. cauteriata* ♂, El Cuartón, Tarifa, Prov. Cádiz, Spain, 22-24.iv.2007 (coll. SKOU)
72c. *T. cauteriata* ♂, Novo Sancti Petri, Prov. Cádiz, Spain, 4-7.iii.2005 (coll. SKOU)
72d. *T. cauteriata* ♂, Novo Sancti Petri, Prov. Cádiz, Spain, 4-7.iii.2005 (coll. SKOU)
72e. *T. cauteriata* ♀, Burgau, Prov. Algarve, Portugal, 18.iv.2007 (coll. SKOU)
73a. ***Hypoxystis pluviaria*** ♂, Veszprém, Veszprém County, Hungary, 4.vii.1999 (coll. SKOU)
73b. *H. pluviaria* ♂, Isokyrö, Etelä-Pohjanmaa, Finland, 29.v.1976 (coll. SKOU)
73c. *H. pluviaria* ♂, Kengis, Norrbotten, Sweden, 16.vi.1980 (coll. SKOU)
73d. *H. pluviaria* ♀, Opponitz, Niederösterreich, Austria, 6.vi.1996 (coll. SKOU)
74a. ***Enanthyperythra legataria*** ♂, Karadagh, Crimea, Ukraine, 12.iv.1989 (ZMKU)
74b. *E. legataria* ♂, Coruh Valley, Mescit Daglari, Prov. Erzerum, Turkey, 23.iv.2001 (coll. SKOU)
74c. *E. legataria* ♀, Drenovo, Macedonia, 11.iv.1959 (ZSM)
74d. *E. legataria* ♀, Beysehir, Konya, Turkey, 4.vi.1964 (ZSM)
75a. ***Apeira syringaria*** ♂, Nebbeskov, Langeland, Fyn Denmark, 5-9.vii.1976 (coll. SKOU)
75b. *A. syringaria* ♀, Säftekärr, Öland, Sweden, 21.vii.1971 (coll. SKOU)
75c. *A. syringaria* ♂, Vic, Gurb, Prov. Barcelona, Spain, 15.viii.2001 (coll. SKOU)
75d. *A. syringaria* ♂, Kirkeby Sand, Fyn, Denmark, 24.viii.2002 (coll.SKOU)
76a. ***Ennomos autumnaria*** ♀, Svanninge Bakker, Fyn, Denmark, 30.viii.1975 (coll. SKOU)
76b. *E. autumnaria* ♂, Vejlen, Tåsinge, Fyn, Denmark, 9.ix.1973 (coll. SKOU)
76c. *E. autumnaria* ♂, Lerbäck, Nyckelhult, Närke, Sweden, 28.viii.1973 (coll. SKOU)
77a. ***Ennomos quercinaria*** ♂, Kullaberg, Skåne, Sweden, 10.viii.1972 (coll. SKOU)
77b. *E. quercinaria* ♂, Grugnay (Chamoson), Valais, Switzerland, 2.viii.1983 (coll. SKOU)
77c. *E. quercinaria* ♀, Vejlen, Tåsinge, Fyn, Denmark, 2-5.ix.1979 (coll. SKOU)
77d. *E. quercinaria* ♂, Troense, Tåsinge, Fyn, Denmark, 13.viii.1973 (coll. SKOU)
77e. *E. quercinaria* ♀, Kallithea, Prov. Xanthi, Greece, 6.vii.1985 (coll. SKOU)
77f. *E. quercinaria* ♂, Germany, 1887 (ZSM)
77g. *E. quercinaria* ♀, Weidling by Wien, Niederösterreich, Austria, 15.vii.1880 (ZSM)

69a
69b
69c
69d
69e
70a
70b
70c
70d
70e
71a
71b
71c
71d
71e
72a
72b
72c
72d
72e
73a
73b
73c
73d
74a
74b
74c
74d
75a
75b
75c
75d
76a
76b
76c
77a
77b
77c
77d
77e
77f
77g

PLATE 8 (all specimens natural size)

78a. ***Ennomos alniaria*** ♂, Ågersta, Södermanland, Sweden, 23.ix.1971 (coll. SKOU)
78b. *E. alniaria* ♂, Odense, Fyn, Denmark, 6.x.1972 (coll. SKOU)
78c. *E. alniaria* ♀, Snarup Mose, Fyn, Denmark, 27.ix.1970 (coll. SKOU)
78d. *E. alniaria* ♀, Asserbo, Northeast Zealand, Denmark, 6.x.1973 (coll. SKOU)
79a. ***Ennomos fuscantaria*** ♂, Angistro, Prov. Serres, Greece, 30.viii.2008 (coll. SKOU)
79b. *E. fuscantaria* ♂, Odense, Fyn, Denmark, 10.ix.1969 (coll. SKOU)
79c. *E. fuscantaria* ♀, Fyns Hoved, Fyn, Denmark, 11.ix.1969 (coll. SKOU)
79d. *E. fuscantaria* ♀, Lozoya, Prov. Madrid, Spain, 20-21.vii.1995 (coll. SKOU)
80a. ***Ennomos erosaria*** ♂, Lysemose, Fyn, Denmark, 5.ix.1972 (coll. SKOU)
80b. *E. erosaria* ♂, Svedmyra, Södermanland, Sweden, 2.ix.1974 (coll. SKOU)
80c. *E. erosaria* ♀, Kullaberg, Skåne, Sweden, 17.viii.1973 (coll. SKOU)
80d. *E. erosaria* ♀, Tranekær, Langeland, Fyn, Denmark, 26.vii.1977 (coll. SKOU)
81a. ***Ennomos quercaria*** ♂, Manfredónia, Puglia, Italy, 26.v.2005 (coll. SKOU)
81b. *E. quercaria* ♂, Puerto de Orihuela, Prov. Teruel, Spain, 23.viii.2001 (coll. SKOU)
81c. *E. quercaria* ♀, Lozoya, Prov. Madrid, Spain, 20-21.vii.1995 (coll. SKOU)
81d. *E. quercaria* ♀, Oštarije, Karlovačka Županija, Croatia, 9.viii.1972 (coll. SKOU)
81e. *E. quercaria duercki* ♂, Zourva, Prov. Chania, Crete, Greece, 29.vii-2.viii.2001 (coll. SKOU)
81f. *E. quercaria duercki* ♀, Mount Ida, Crete, Greece, 15-31.vii.1938 (SMNK)
81g. *E. quercaria lissochila* ♂, Foini, Troodos Mountains, Cyprus, (coll. MÜLLER), reared
81h. *E. quercaria lissochila* ♀, Foini, Troodos Mountains, Cyprus, (coll. MÜLLER), reared
82a. ***Eumera regina*** ♂, Monodendri, Prov. Ioannina, Greece, 25.viii.2008 (coll. SKOU)
82b. *E. regina* ♂, Ohrid, Macedonia, 8.ix.1936 (ZSM)
82c. *E. regina* ♀, Zara, Dalmatia, Croatia, (ZSM), reared
83a. ***Selenia dentaria*** ♂, Wellington, Somerset, England, 12.iv.1992 (coll. SKOU)
83b. *S. dentaria* ♂, Saratov, S. European Russia, 8.v.1988 (coll. SKOU)
83c. *S. dentaria* ♂, Lysemose, Fyn, Denmark, 26.vii.1972 (coll. SKOU)
83d. *S. dentaria* ♀, Stensjö, Småland, Sweden, 6.vi.1972 (coll. SKOU)
84a. ***Selenia lunularia*** ♂, Castelbuono, Sicily, Italy, 5.vi.2005 (coll. SKOU)
84b. *S. lunularia* ♂, Remsfeld by Kassel, Hessen, Germany, 22.vi.1985 (coll. SKOU)
84c. *S. lunularia* ♀, Kallithea, Prov. Xanthi, Greece, 6.vii.1985 (coll. SKOU)
84d. *S. lunularia* ♀, Allarpsberg, Skåne, Sweden, 17.vi.1972 (coll. SKOU)

78a
78b
78c
78d
79a
79b
79c
79d
80a
80b
80c
80d
81a
81b
81c
81d
81e
81f
81g
81h
82a
82b
82c
83a
83b
83c
83d
84a
84b
84c
84d

PLATE 9 (all specimens natural size)

85a. ***Selenia tetralunaria*** ♂, Torö Brygga, Södermanland, Sweden, 24.v.1972 (coll. SKOU)
85b. *S. tetralunaria* ♂, Chandlers Ford, Hampshire, England, 16.iv.1992 (coll. SKOU)
85c. *S. tetralunaria* ♂, Klagshamn, Skåne, Sweden, 9.viii.1972 (coll. SKOU)
85d. *S. tetralunaria* ♀, Nagyegyháza at Bicske, Fejér County, Hungary, 25.vi.2002 (coll. SKOU)
86a. ***Artiora evonymaria*** ♂, Ried a. Riederberg, Niederösterreich, Austria, 5.viii.1947 (ZSM)
86b. *A. evonymaria* ♂, Greece, 20.x.1984 (coll. SKOU)
86c. *A. evonymaria* ♂, Wola Roźwienicka, Jarosław, Poland, 28.viii.1954 (ZSM)
86d. *A. evonymaria* ♀, Balatonalmádi, Veszprém County, Hungary, 8-9.ix.1990 (coll. SKOU)
87a. ***Odontopera bidentata*** ♂, Stensgade Skov, Langeland, Fyn, Denmark, 3.vi.1977 (coll. SKOU)
87b. *O. bidentata* ♀, Castelbuono, Sicily, Italy, 5.vi.2005 (coll. SKOU)
87c. *O. bidentata* ♀, Tubre, Trentino-Alto Adige, Italy, 27.vi.2008 (coll. SKOU)
87d. *O. bidentata* f. *graecarius* ♂, Lagkadia, Prov. Arkadia, Greece (ZSM) reared
87e. *O. bidentata* f. *graecarius* ♀, Lagkadia, Prov. Arkadia, Greece (ZSM) reared
87f. *O. bidentata* f. *graecarius* ♀, Lagkadia, Prov. Arkadia, Greece (ZSM) reared
88a. ***Dasycorsa modesta*** ♂, Kristalopigi, Prov. Florina, Greece, 17.iv.1987 (coll. SKOU)
88b. *D. modesta* ♂, Skakavitza, Distr. Kyustendil, Bulgaria, 29.iv.1953 (coll. SKOU)
88c. *D. modesta* ♂, Karadja Bey, Ajtay-Kovács, Turkey (ZSM) 8.iv.1920
88d. *D. modesta* ♀, Kastoria, Prov. Kastoria, Greece, 19.iv.1987 (coll. SKOU)
89a. ***Crocallis tusciaria*** ♂, Albarracín, Prov. Teruel, Spain, 29.ix.1987 (coll. SKOU)
89b. *C. tusciaria* ♂, Alexandroupolis, Thrakia, Prov. Evros, Greece, 23.x.1984 (coll. SKOU)
89c. *C. tusciaria* ♂, Gradechka banja by Kresna, Distr. Blagoevgrad, Bulgaria,15.x.1987 (coll. SKOU)
89d. *C. tusciaria* ♂, Pampaneira, Sierra Nevada, Prov. Granada, Spain, 19.xi.2006 (coll. SKOU)
89e. *C. tusciaria* ♀, Sykorrachi, Prov. Evros, Greece, 22.x.1984 (coll. SKOU)
89f. *C. tusciaria* ♀, Sykorrachi, Prov. Evros, Greece, 22.x.1984 (coll. SKOU)
89g. *C. tusciaria* ♀, Tosos, Prov. Zaragoza, Spain, 3.xi.2005 (coll. SKOU)
89h. *C. tusciaria* ♀, Pampaneira, Sierra Nevada, Prov. Granada, Spain, 19.xi.2006 (coll. SKOU)
90a. ***Crocallis albarracina*** ♂, Tramacastilla, Prov. Teruel, Spain, 27.viii.2001 (coll. SKOU)
90b. *C. albarracina* ♂, Tramacastilla, Prov. Teruel, Spain, 27.viii.2001 (coll. SKOU)
90c. *C. albarracina* ♀, Tramacastilla, Prov. Teruel, Spain, 27.viii.2001 (coll. SKOU)
90d. *C. albarracina* ♀, Barranco de Valcuerna, Candasnos, Prov. Huesca, Spain, 13-14.ix.2002 (coll. SKOU)
90e. *C. albarracina* f. *centraliberica* ♂, La Plataforma, Sierra de Gredos, Prov. Ávila, Spain, 18.viii.1988 (coll. SKOU)
90f. *C. albarracina* f. *centraliberica* ♂, La Plataforma, Sierra de Gredos, Prov. Ávila, Spain, 18.viii.1988 (coll. SKOU)
90g. *C. albarracina* f. *centraliberica* ♂, La Plataforma, Sierra de Gredos, Prov. Ávila, Spain, 18.viii.1980 (coll. SKOU)

85a
85b
85c
85d
86a
86b
86c
86d
87a
87b
87c
87d
87e
87f
88a
88b
88c
88d
89a
89b
89c
89d
89e
89f
89g
89h
90a
90b
90c
90d
90e
90f
90g

PLATE 10 (all specimens natural size)

91a. ***Crocallis elinguaria*** ♂, Vita Pochtovaja, Kievskaja obl., Ukraine, 10.viii.2002 (ZMKU)
91b. *C. elinguaria* ♂, Kyffhäuser, Thüringen, Germany, 3.viii.1984 (coll. FIEBIG)
91c. *C. elinguaria* ♀, Arnager, Bornholm, Denmark,13.viii.1979 (coll. SKOU)
91d. *C. elinguaria* ♀, Sondershausen, Thüringen, Germany (coll. FIEBIG), reared
91e. *C. elinguaria* ♀, Lytkarino, Moskovskaya obl., Russia, 27.vii.1975 (ZMKU)
91f. ***Crocallis*** sp. near *elinguaria* ♂, Pigon Lake, Prov. Ioannina, Greece, 24.viii.2008 (coll. SKOU)
91g. *C.* sp. near *elinguaria* ♂, Lithochoron, Mt. Olympus, Prov. Pieria, Greece, 4.ix.2008 (coll. SKOU)
91h. *C.* sp. near *elinguaria* ♀, Pigon Lake, Prov. Ioannina, Greece, 24.viii.2008 (coll. SKOU)
92a. ***Crocallis dardoinaria*** ♂, Montalbán, Prov. Teruel, Spain, 4.viii.2007 (coll. SKOU)
92b. *C. dardoinaria* ♂, Barranco de Valcuerna, Candasnos, Prov. Huesca, Spain, 13-14.ix.2002 (coll. SKOU)
92c. *C. dardoinaria* ♂, Tosos, Prov. Zaragoza, Spain, 15.ix.2002 (coll. SKOU)
92d. *C. dardoinaria* ♂, Val de Vecar, Albarracín, Prov. Teruel, Spain, 21.viii.2001 (coll. SKOU)
92e. *C. dardoinaria* ♀, Sant Marti de Tous, Prov. Barcelona, Spain, 16.viii.2001 (coll. SKOU)
92f. *C. dardoinaria* ♀, Duilhac-sous-Peyrepertuse, Maury, Dept. Aude, France, 5.viii.2003 (coll. SKOU)
92g. *C. dardoinaria* ♂, Val de Vecar, Albarracín, Prov. Teruel, Spain, 1.viii.2003 (coll. SKOU)
92h. *C. dardoinaria* ♂, Aguarón, Cariñena, Prov. Zaragoza, Spain, 3.viii.2007 (coll. SKOU)
93a. ***Crocallis auberti*** ♂, Barranco de Valcuerna, Candasnos, Prov. Huesca, Spain, 13-14.ix.2002 (coll. SKOU)
93b. *C. auberti* ♂, Barranco de Valcuerna, Candasnos, Prov. Huesca, Spain, 13-14.ix.2002 (coll. SKOU)
93c. *C. auberti* ♀, Casares, Estepona, Prov. Málaga, Spain, 28.x.2003 (coll. SKOU)
93d. *C. auberti* ♂, Vicinity of Algiers, Algeria, 25.x.1913 (coll. SKOU)
93e. *C. auberti* ♂, El Ksiba, Haute Atlas, Morocco, 3-8.x.1973 (ZSM)
94a. ***Crocallis helenaria*** ♂, Maroulas, Prov. Rethymno, Crete, Greece, 11-17.x.2003 (coll. SKOU)
94b. *C. helenaria* ♂, Maroulas, Prov. Rethymno, Crete, Greece, 11-17.x.2003 (coll. SKOU)
94c. *C. helenaria* ♀, Vasilakion, Crete, Greece, 12.x.1972 (SMNK)
95a. ***Crocallis boisduvalaria*** ♂, Castelluccio, Sicily, Italy, 19.x.2001 (coll. SKOU)
95b. *C. boisduvalaria* ♂, Sicily, Italy, 5.x.1927 (ZSM)
95c. *C. boisduvalaria* ♂, Palermo, Sicily, Italy, ix.1928 (ZSM)
95d. *C. bosduvalaria* ♀, Casteldaccia, Palermo, Sicily, Italy, x.1928 (ZSM)
96a. ***Ourapteryx sambucaria*** ♂, Uña, Prov. Cuenca, Spain, 8-9.vii.2002 (coll. SKOU)
96b. *O. sambucaria* ♂, Midskov, Fyn, Denmark, 12.vii.1975 (coll. SKOU)
96c. *O. sambucaria* ♀, Dellach, Kärnten, Austria, 2.viii.1986 (coll. SKOU)

91a
91b
91c
91d
91e
91f
91g
91h
92a
92b
92c
92d
92e
92f
92g
92h
93a
93b
93c
93d
93e
94a
94b
94c
95a
95b
95c
95d
96a
96b
96c

PLATE 11 (all specimens natural size)

97a. ***Asovia maeoticaria*** ♂, Kalavrita, Mount Chelmos, Prov. Achaia, Greece, 17-18.vi.1982 (coll. SKOU)
97b. *A. maeoticaria* ♂, Kalavrita, Mount Chelmos, Prov. Achaia, Greece, 17-18.vi.1982 (coll. SKOU)
97c. *A. maeoticaria* ♂, Mount Oros, Prov. Arkadia, Greece, 24.vii.1998 (coll. SKOU)
97d. *A. maeoticaria* ♀, Monemvaria, Prov. Lakonia, Greece, 9.iv.1981 (coll. SKOU)
98a. ***Colotois pennaria*** ♂, Ågeta, Södermanland, Sweden, 21.ix.1973 (coll. SKOU)
98b. *C. pennaria* ♂, Odense, Fyn, Denmark, 10.x.1972 (coll. SKOU)
98c. *C. pennaria* ♂, Tosos, Prov. Zaragoza, Spain, 3.x.2005 (coll. SKOU)
98d. *C. pennaria* ♀, Tosos, Prov. Zaragoza, Spain, 3.x.2005 (coll. SKOU)
98e. *C. pennaria* ♀, Greece,20.x.1984 (coll. SKOU)
98f. *C. pennaria* ♀, Vejlen, Tåsinge, Fyn, Denmark, 14-20.x.1979 (coll. SKOU)
99a. ***Angerona prunaria*** ♂, Galgbacken, Öland, Sweden, 25.vi.1972 (coll. SKOU)
99b. *A. prunaria* ♂, Vinsløv, Skåne, Sweden, 18.vi.1972 (coll. SKOU)
99c. *A. prunaria* ♂, Eger, Heves County, Hungary, 30.vi.1999 (coll. SKOU)
99d. *A. prunaria* ♀, Tranekær, Langeland, Fyn, Denmark, 22.vi.1978 (coll. SKOU)
99e. *A. prunaria* ♂, Stengade Skov, Langeland, Fyn, Denmark, 28.vi.1976 (coll. SKOU)
99f. *A. prunaria* ♀, Dept. Seine-et-Marne, France, 11.vi.1943 (ZSM)
100a. ***Cabera pusaria*** ♂, Snarup Mose, Fyn, Denmark, 4.vi.1977 (coll. SKOU)
100b. *C. pusaria* ♂, Prestseter, Vestfold, Norge, 2.vii.2000 (coll. SKOU)
100c. *C. pusaria* ♂, Vejlen, Tåsinge, Fyn, Denmark, 30.vi.-2.vii.1978 (coll. SKOU)
100d. *C. pusaria* ♀, Allarpsberg, Skåne, Sweden, 3.vii.1972 (coll. SKOU)
100e. *C. pusaria* ♀, Pljuty, Kievskaja obl., Ukraine, 10.vii.1901 (ZMKU)
101a. ***Cabera leptographa*** ♂, Zitzmannsdorfer Wiesen, Burgenland, Austria (coll. SKOU), reared
101b. *C. leptographa* ♂, Zitzmannsdorfer Wiesen, Burgenland, Austria (coll. SKOU), reared
101c. *C. leptographa* ♂, Zitzmannsdorfer Wiesen, Burgenland, Austria (coll. SKOU), reared
101d. *C. leptographa* ♀, Zitzmannsdorfer Wiesen, Burgenland, Austria (coll. SKOU), reared
101e. *C. leptographa* ♂, Svalovichi, Volynskaja obl., Ukraine, 2-4.vii.2005 (ZMKU)
102a. ***Cabera exanthemata*** ♂, Kaunertal, Tirol, Austria, 2.vii.2008 (coll. SKOU)
102b. *C. exanthemata* ♀, Terre Rouge, Briançon, Dept. Hautes-Alpes, France, 28.vii.2000 (coll. SKOU)
102c. *C. exanthemata* ♂, Stensund, Södermanland, Sweden, 22.vi.1971 (coll. SKOU)
102d. *C. exanthemata* ♂, Vejlen, Tåsinge, Fyn, Denmark, 11-13.viii. 1978 (coll. SKOU).
102e. *C. exanthemata* ♀, Troense, Tåsinge, Fyn, Denmark, 31.v.1973 (coll. SKOU)
103a. ***Lhommeia biskraria*** ♂, Moulay Idriss, Morocco, 23.xi.1970 (ZSM)
103b. *L. biskraria* ♂, Asni-Quirgane, Haut-Atlas, Morocco, 6-9.xi.1964 (ZSM)
103c. *L. biskraria* ♂, Marrakech, Morocco, 11.xi.1964 (ZSM)
103d. *L. biskraria* ♀, Wadi El Hira, Gharian, Libya, 11.v.1983 (coll. SKOU)
103e. *L. biskraria* ♀, Tizi n'Test, Haut-Atlas, Morocco, 14.vii.1978 (ZSM)

97a
97b
97c
97d
98a
98b
98c
98d
98e
98f
99a
99b
99c
99d
99e
99f
100a
100b
100c
100d
100e
101a
101b
101c
101d
101e
102a
102b
102c
102d
102e
103a
103b
103c
103d
103e

PLATE 12 (all specimens natural size)

104a. ***Lomographa bimaculata*** ♀, Snarup Mose, Fyn, Denmark, 16.v.1979 (coll. SKOU)
104b. *L. bimaculata* ♂, Nørresø, Fyn, Denmark, 30.v.1978 (coll. SKOU)
104c. *L. bimaculata* ♀, Donskoe, Bylyaevka District, Orenburg Region, South Ural, Russia, 1-4.vi.1998 (coll. SKOU)
104d. *L. bimaculata* ♀, Gejershamn, Öland, Sweden8.vi.1972 (coll. SKOU)
104e. *L. bimaculata* ♀, Spiazzi, Monte Baldo, Veneto, Italy, 22.iv.2009 (coll. SKOU)
105a. ***Lomographa temerata*** ♀, Donskoje, Beljajewka Distr., Orenburg, South Ural, Russia, 1-4.vi.1998, (coll. SKOU)
105b. *L. temerata* ♀, Trosa, Södermanland, Sweden, 7.vi.1973 (coll. SKOU)
105c. *L. temerata* ♂, Besalú, Figueras, Prov. Girona, Spain, 17.iv.1979 (coll. SKOU)
105d. *L. temerata* ♂, Stengade Skov, Langeland, Fyn, Denmark, 5.vi.1976 (coll. SKOU)
105e. *L. temerata* ♂, Nordenhuse, Fyn, Denmark, 12.v.1971 (coll. SKOU)
106a. ***Aleucis distinctata*** ♂, Amfreville-sous-les-Monts, Dept. Eure, France, 30.iii.1959 (HERBULOT/ZSM)
106b. *A. distinctata* ♀, Dept. Essonne, France, 3.iv.1909 (HERBULOT/ZSM)
106c. *A. distinctata* ♀, Lorch-Nollig, Hessen, Germany, 10.iv.1952 (ZSM)
106d. *A. distinctata* ♀, Corsavy, Arles-sur-Tech, Dept. Pyrénées Orientales, France, 26.iii.2002 (coll. SKOU)
106e. *A. distinctata* ♀, Lorch-Nollig, Hessen, Germany, 10.iv.1952 (ZSM)
107a. ***Aleucis orientalis*** ♂, Kastoria, Prov. Kastoria, Greece, 19.iv.1987 (coll. SKOU)
107b. *A. orientalis* ♀, Monemvasia, Prov. Lakonia, Greece, 2.iv.1983 (coll. SKOU)
107c. *A. orientalis* ♀, Kastoria, Prov. Kastoria, Greece, 19.iv.1987 (coll. SKOU)
107d. *A. orientalis* ♀, Matka, Prov. Skopje, Macedonia, 9.iv.1973 (ZSM)
107e. *A. orientalis* ♂, Marasch (Kahranmaras), Central Taurus, Turkey, iii.1930 (ZSM)
108a. ***Theria rupicapraria*** ♂, Lehnskov, Fyn, Denmark, 8.iii.1980 (coll. SKOU)
108b. *T. rupicapraria* ♂, Mölle Kyrka, Skåne, Sweden, 25.iii.1972 (coll. SKOU)
108c. *T. rupicapraria* ♂, Nordenhuse, Fyn, Denmark, 26.iii.1979 (coll. SKOU)
108d. *T. rupicapraria* ♂, Bois du Mont Briseuil , Darnétal, Seine-Maritime, France, 15.iii.1938 (ZSM)
108e. *T. rupicapraria* ♀, Snarup Mose, Fyn, Denmark (coll. SKOU), reared
108f. *T. rupicapraria* ♀, Snarup Mose, Fyn, Denmark (coll. SKOU), reared
109a. ***Theria primaria*** ♂, Lehnskov Strand, Fyn, Denmark, 16.ii.2002 (coll. SKOU)
109b. *T. primaria* ♂, Lehnskov Strand, Fyn, Denmark, 16.ii.2002 (coll. SKOU)
109c. *T. primaria* ♂, Nordenhuse, Fyn, Denmark, 21.ii.1972 (coll. SKOU)
109d. *T. primaria* ♂, Lehnskov Strand, Fyn, Denmark (coll. SKOU)
109e. *T. primaria* ♀, Hofmansgave, Fyn, Denmark (coll. SKOU), reared
109f. *T. primaria* ♀, Conches, Switzerland (ZSM) reared
110a. ***Campaea margaritaria*** ♂, Milo, Monte Etna, Sicily, Italy, 7.vi.2005 (coll. SKOU)
110b. *C. margaritaria* ♂, Val d'Ese, Bastelica, Corsica; France, 15-16.vii.2004 (coll. SKOU)
110c. *C. margaritaria* ♀, Mount Phalakron, Prov. Drama, Greece, 8.vii.1985 (coll. SKOU)
110d. *C. margaritaria* ♂, Queralbs, Ribes de Freser, Prov. Girona, Spain, 10.ix.2002 (coll. SKOU)
111a. ***Gerinia honoraria*** ♂, Forêt de Bonifatu, Calvi, Corsica, France, 22.vi.1994 (coll. SKOU)
111b. *G. honoraria* ♂, Forêt de Malmont, Pont du Gard, Dept. Gard, France, 17.v.1998 (coll. SKOU)
111c. *G. honoraria* ♂, San Roque, Almoraima, Prov. Cádiz, Spain, 24.xi.1980 (coll. SKOU)
111d. *G. honoraria* ♂, Fluminimaggiore, Sardinia, Italy, 1.iv.2002 (coll. SKOU)
111e. *G. honoraria* ♀, Ciamanacce, Palneca, Corsica, France, 10.x.1972 (HERBULOT/ZSM)
111f. *G. honoraria* ♀, Duilhac-sous-Peyrepertuse, Dept. Aude, France, 8.ix.2002 (coll. SKOU)

104a
104b
104c
104d
104e
105a
105b
105c
105d
105e
106a
106b
106c
106d
106e
107a
107b
107c
107d
107e
108a
108b
108c
108d
108e
108f
109a
109b
109c
109d
109e
109f
110a
110b
110c
110d
111a
111b
111c
111d
111e
111f

PLATE 13 (all specimens natural size)

112a. ***Hylaea fasciaria*** ♂, Volsk, Saratov Region, South East Russia13.vi.1987 (coll. SKOU)
112b. *H. fasciaria* ♂, Skanør, Falsterbo, Skåne, Sweden, 6.vii.1972 (coll. SKOU)
112c. *H. fasciaria* ♂, Skallingen, West Jylland, Denmark, 27.vi.1977 (coll. SKOU)
112d. *H. fasciaria* ♂, Byerum, Öland, Sweden, 1.viii.1972 (coll. SKOU)
112e. *H. fasciaria* ♂, Dueodde, Bornholm, Denmark, 18.vii.1970 (coll. SKOU)
112f. *H. fasciaria* ♂, Mount Parnassos, Central Greece, 10.vi.2010 (coll. SKOU)
112g. *H. fasciaria* ♀, Skanør, Falsterbo, Skåne, Sweden, 6.vii.1972 (coll. SKOU)
112h. *H. fasciaria* ♀, Karpenision, Prov. Aitolia kai Akarnania, Greece, 28.vi.1981 (coll. SKOU)
112i. *H. fasciaria* ♀, Terre Rouge, Briançon, Dept. Hautes-Alpes, France, 29.vii.2000 (coll. SKOU)
113a. ***Hylaea mediterranea*** ♂, Santo Stefano Quisquina, Sicily, Italy, 9-10.x.2010 (coll. SKOU) holotype
113b. *H. mediterranea* ♂, Santo Stefano Quisquina, Sicily, Italy, 9-10.x.2010 (coll. SKOU) paratype
113c. *H. mediterranea* ♀, Castelbuono, Sicily, Italy, 5.vi.2005 (coll. SKOU) paratype
113d. *H. mediterranea* ♀, Fosso Cucolo, Donnici, Calabria, Italy, 30.x.1996 (ZSM)
114a. ***Hylaea pinicolaria*** ♂, Camping Monte Cinto, Haut-Asco, Corsica, France, 20-21.vii.2004 (coll. SKOU)
114b. *H. pinicolaria* ♂, Camping Monte Cinto, Haut-Asco, Corsica, France, 20-21.vii.2004 (coll. SKOU)
114c. *H. pinicolaria* ♂, La Foce, Vizzavona, Corsica, France, 6.vii.1951 (HERBULOT/ZSM)
114d. *H. pinicolaria* ♀, Haut Asco, Corsica, France, 23.vi.1994 (coll. SKOU)
115a. ***Adalbertia castiliaria*** ♂, Tosos, Prov. Zaragoza, Spain, 20.viii.2001 (coll. SKOU)
115b. *A. castiliaria* ♂, Alhama de Murcia, Sierra de Espuña, Prov. Murcia, Spain, 28.v.1998 (coll. SKOU)
115c. *A. castiliaria* ♂, ♂, Puerto de Navafría, Sierra de Guadarrama, Prov. Madrid, Spain, 18-19.vii.2003 (coll. SKOU)
115d. *A. castiliaria* ♂, Solsona, Prov. Lérida, Spain, 3.vii.1991 (coll. SKOU)
115e. *A. castiliaria* ♀, Puerto de Navafría, Sierra de Guadarrama, Prov. Madrid, Spain, 18-19.vii.2003 (coll. SKOU)
116a. ***Pungeleria capreolaria*** ♂, Il Castagno d'Andrea, Foreste Casentinesi, Prov. Firenze, Italy, 6.ix.1999 (coll. SKOU)
116b. *P. capreolaria* ♂, Ski Centre, Mount Parnassos, Prov. Fokida, Greece, 6.ix.2008 (coll. SKOU)
116c. *P. capreolaria* ♀, Eptachori, Prov. Ioannina, Greece, 23.vi.2004 (coll. SKOU)
116d. *P. capreolaria* ♀, Eptachori, Prov. Ioannina, Greece, 23.vi.2004 (coll. SKOU)
116e. *P. capreolaria* ♀, Il Castagno d'Andrea, Foreste Casentinesi, Prov. Firenze, Italy, 6.ix.1999 (coll. SKOU)
117a. ***Siona lineata*** ♂, Eplány, Veszprém County, Hungary, 2-8.vi.2001 (coll. SKOU)
117b. *S. lineata* ♂, Kolartorp, Södermanland, Sweden, 10.vi.1971 (coll. SKOU)
117c. *S. lineata* ♀, Ebernburg, Nahegebiet, Rheinland-Pfalz, Germany, 12.vi.1965 (ZSM)
117d. *S. lineata* ♂, Troense, Tåsinge, Fyn, Denmark, 6.vi.1973 (coll. SKOU)
118a. ***Chariaspilates formosaria*** ♂, Moosbrunn, Niederösterreich, Austria, 17.vi.1979 (coll. SKOU)
118b. *C. formosaria* ♂, Epannes, Dept. Deux-Sèvres, France, 21.vii.1924 (ZSM)
118c. *C. formosaria* ♀, Romieu, Dept. Bouches du Rhône, France, 18.vi.1984 (ZSM)
118d. *C. formosaria* ♀, Peenemoor, Anklam, Mecklenburg-Vorpommern, Germany, 14-29.v.1960 (ZSM)
119a. ***Aspitates mundataria*** ♂, Berezovka, Pavlodar Region, North East Kazakhstan1.vii.1984 (coll. SKOU)
119b. *A. mundataria* ♀, Elton Salt Lake, South East European Russia, 18-23.v.2001 (coll. SKOU)
119c. *A. mundataria* ♀, Elton Salt Lake, South East European Russia, 18-23.v.2001 (coll. SKOU)
119d. *A. mundataria* ♂, Kurortnoje, Crimea, Ukraine, 25.vi.1987 (ZMKU)

112a
112b
112c
112d
112e
112f
112g
112h
112i
113a
113b
113c
113d
114a
114b
114c
114d
115a
115b
115c
115d
115e
116a
116b
116c
116d
116e
117a
117b
117c
117d
118a
118b
118c
118d
119a
119b
119c
119d

PLATE 14 (all specimens natural size)

120a. ***Aspitates gilvaria*** ♂, Kalvola, Etelä-Häme, Finland, 11.vii.1961 (SMNK)
120b. *A. gilvaria* ♂, Ribes de Freser, Prov. Girona, Spain, 14.viii.2001 (coll. Skou)
120c. *A. gilvaria* ♂, Moscardón, Prov. Teruel, Spain, 6.viii.2007 (coll. Skou)
120d. *A. gilvaria* ♂, Kassel, Hessen, Germany, 18.vii.1946 (ZSM)
120e. *A. gilvaria* ♀, Kalvola, Etelä-Häme, Finland, 11.vii.1961 (SMNK)
120f. *A. gilvaria* ♀, Dzhanibek, European part of Kazakhstan, 7.ix.2001 (coll. Skou)
120g. *A. gilvaria* ♀, Prelles, Dept. Hautes-Alpes, France, 2.viii.2009 (coll. Skou)
120h. *A. gilvaria* ♀, Kelheim, Bavaria, Germany, 27.viii.1918 (ZSM)
121a. ***Aspitates taylorae sibirica*** ♂, Artyk Village, Jakutia, Far East Russia, 20.vi.-2.vii.2000 (coll. Skou)
121b. *A. taylorae sibirica* ♂, Ust-Nukzha Village, Tynda District, Amur, Far East Russia, 22-24.vi.1996 (coll. Skou)
121c. *A. taylorae sibirica* ♂, Tonnelnyi Village, Buryatia, Russia, 15.vi.2000 (coll. Skou)
121d. *A. taylorae sibirica* ♀, Ust-Nukzha Village, Tynda District, Amur, Far East Russia, 22-24.vi.1996 (coll. Skou)
122a. ***Aspitates orciferaria*** ♂, Krasnyi Kamen, Polar Urals, Russia, 3.vii.1994 (ZMH)
122b. *A. orciferaria* ♀, Krasnyi Kamen, Polar Urals, Russia, 8.vii.1994 (ZMH)
123a. ***Aspitates forbesi*** ♂, Krasnyi Kamen, Polar Urals, Russia, 7.vii.1994 (ZMH)
123b. *A. forbesi* ♀, Churchill, Manitoba, Canada, 12.vii.1952 (NHM) paratype
124a. ***Aspitates albaria*** ♂, Belyaevsky District, Orenburg Region, Russia, 1-4.vi.1996 (coll. Skou)
124b. *A. albaria* ♂, Belyaevsky District, Orenburg Region, Russia, 1-4.vi.1996 (coll. Skou)
124c. *A. albaria* ♂, Kuvandyk Village, Orenburg Region, Russia, 5-10.vi.1996 (coll. Skou)
124d. *A. albaria* ♀, Beryozovka village, Pavlodarskaja obl., N. W. Kazakhstan, 4.vi.1984 (coll. Skou)
125a. ***Aspitates ochrearia*** ♂, Urb. Madroñal, Andalucía, Prov. Málaga, Spain,20.ix.1987 (coll. Skou)
125b. *A. ochrearia* ♂, Sorgono, Ortueri, Sardinia, Italy, 18.v.2004 (coll. Skou)
125c. *A. ochrearia* ♂, Drapanias, Kissamos, Prov. Chania, Crete, Greece, 27.iii.-2.iv.1999 (coll. Skou)
125d. *A. ochrearia* ♀, Monemvasia, Prov. Lakonia, Greece, 5.x.1979 (coll. Skou)
125e. *A. ochrearia* ♀, Kaifuan, D. Djelloula, Tunisia, 12.iv.1983 (coll. Skou)
126a. ***Dyscia raunaria*** ♂, Monte. Prov. Verona, Veneto, Italy, 18.vi.1985 (ZSM)
126b. *D. raunaria* ♂, Mt. Phalakron, Prov. Drama, Greece, 8.vii.1985 (coll. Skou)
126c. *D. raunaria* ♀, Garda, Prov. Verona, Veneto, Italy, 17.vi.1981 (ZSM)
126d. *D. raunaria* ♂, Gravosa, Dalmatia, Croatia, early vi.1939 (ZSM)
127a. ***Dyscia royaria*** ♂, Col de Tende, Dept. Alpes-Maritimes, France,2.viii.2007 (coll. Skou)
127b. *D. royaria* ♂, Col de Tende, Dept. Alpes-Maritimes, France, 7.viii.2008 (coll. Tautel)
127c. *D. royaria* ♂, Col de Tende, Dept. Alpes-Maritimes, France (coll. Tautel), reared
127d. *D. royaria* ♀, Col de Tende, Dept. Alpes-Maritimes, France (coll. Tautel), reared
128a. ***Dyscia innocentaria*** ♂, Castel di Ieri, Prov. L'Aquila, Abruzzo, Italy, 9-11.vi.2005 (coll. Skou)
128b. *D. innocentaria* ♂, Kara-Dagh, Crimea, Ukraine, 25-27.vi.1987 (coll. Skou)
128c. *D. innocentaria* ♂, Ryn-Kum Sand Steppe, European part of Kazakhstan, 28-29.v.2000 (coll. Skou)
128d. *D. innocentaria* ♀, Blagoevgrad, Kupita, Distr. Blagoevgrad, Bulgaria, 14.vi.1982 (coll. Skou)
128e. *D. innocentaria* ♀, Manfredonia, Puglia, Italy, 26.v.2005 (coll. Skou)
128f. *D. innocentaria govii* ♂, road to Haut Asco, Corsica, France,13-14.viii.2012 (coll. Skou)
129a. ***Dyscia malatyana clara*** ♂, Komsomolskiy, Chernozemelskiy Region, Kalmykia, Russia, 21.v.2005 (coll. Zolotuhin)
129b. *D. malatyana clara* ♀, Andreevskie Sands, Kalmykia, Russia, 29.iv.2000 (ZISP)

120a
120b
120c
120d
120e
120f
120g
120h
121a
121b
121c
121d
122a
122b
123a
123b
124a
124b
124c
124d
125a
125b
125c
125d
125e
126a
126b
126c
126d
127a
127b
127c
127d
128a
128b
128c
128d
128e
128f
129a
129b

PLATE 15 (all specimens natural size)

130a. ***Dyscia crassipunctaria*** ♂, Lakki, Omalos, Prov. Chania, Crete, Greece, 6.vi.2004 (coll. SKOU)
130b. *D. crassipunctaria* ♂, Drapanias by Kissamos, Prov. Chania, Crete, Greece, 27.iii.-2.iv.1999 (coll. SKOU)
130c. *D. crassipunctaria* ♀, Lakki, Omalos, Prov. Chania, Crete, Greece, 6.vi.2004 (coll. SKOU)
130d. *D. crassipunctaria* ♂, Agia Triada, Prov. Rethymno, Crete, Greece, 14.x.2003 (coll. SKOU)
130e. *D. crassipunctaria* ♀, Agia Triada, Prov. Rethymno, Crete, Greece, 14.x.2003 (coll. SKOU)
131a. ***Dyscia conspersaria*** ♂, Altenburg, Niederösterreich. Austria14.vi.1936 (ZSM)
131b. *D. conspersaria* ♂, Baden, Niederösterreich, Austria, vi.1909 (ZSM)
131c. *D. conspersaria* ♀, Altenburg, Niederösterreich, Austria, 13.v.1934 (ZSM)
131d. *D. conspersaria* ♀, Mödling, Niederösterreich, Austria, 15.vii.1899 (ZSM)
132a. ***Dyscia fagaria*** ♂, Oksby, West Jylland, Denmark, 27.v.1972 (coll. SKOU)
132b. *D. fagaria* ♂, Weisswasser, Nochten, Jungfernberge, Sachsen, Germany, 16.v.1996 (ZSM)
132c. *D. fagaria* ♂, Beelitz-Borkwalde, Brandenburg, Germany, 11.v.1995 (ZSM)
132d. *D. fagaria* ♀, Beelitz-Borkwalde, Brandenburg, Germany (ZSM), reared
133a. ***Dyscia penulataria*** ♂, Ruta del Veleta, Sierra Nevada, Prov. Granada, Spain, 19.ix.1987 (coll. SKOU)
133b. *D. penulataria* ♂, El Pinar de Bédar, Prov. Almería, Spain, 19-27.iv.2001 (coll. SKOU)
133c. *D. penulataria* ♂, Tabernas, Prov. Almería, Spain, 28.v.1998 (coll. SKOU)
133d. *D. penulataria* ♂, El Campello, Busot, Prov. Alicante, Spain, 2.v.1997 (coll. SKOU)
133e. *D. penulataria* ♂, El Ronquillo, Prov. Huelva, Spain, 2.iv.2003 (coll. SKOU)
133f. *D. penulataria* ♂, Playa del Rompeculos, Mazagón, Prov. Huelva, Spain, 17.iv.2007 (coll. SKOU)
133g. *D. penulataria* ♂, Marbella, Prov. Málaga, Spain, 22.ix.1987 (coll. SKOU)
133h. *D. penulataria* ♂, Lagoa de Santo André, Prov. Alentejo, Portugal, 25.ix.2009 (coll. SKOU)
133i. *D. penulataria* ♀, Lagoa de Santo André, Prov. Alentejo, Portugal, 25.ix.2009 (coll. SKOU)
133j. *D. penulataria* ♀, El Curatón, Tarifa, Prov. Cádiz, Spain, 22-24.iv.2007 (coll. SKOU)
133k. *D. penulataria* ♀, El Ronquillo, Prov. Huelva, Spain, 2.iv.2003 (coll. SKOU)
133l. *D. penulataria* ♀, El Pozo del Esparto, Prov. Almería, Spain, 1.xi.2005 (coll. SKOU)
133m. *D. penulataria* ♀, Salema, Prov. Algarve, Portugal,19.iv.2007 (coll. SKOU)
133n. *D. penulataria* ♀, El Pozo del Esparto, Prov. Almería, Spain, 22.iv.2001 (coll. SKOU)
133o. *D. penulataria* ♀, Tosos, Prov. Zaragoza, Spain, 20.viii.2001 (coll. SKOU)
133p. *D. penulataria* ♀, Matalascañas, Prov. Huelva, Spain, 16.iv.2007 (coll. SKOU)
134a. ***Dyscia lentiscaria*** ♂, Saint-Michel-l'Observatoire, Dept. Alpes-de-Haute-Provence, France, 25.iv.1973 (coll. SKOU)
134b. *D. lentiscaria* ♂, Roche, Prov. Cádiz, Spain, 6.iii.2005 (coll. SKOU)
134c. *D. lentiscaria* ♀, Roche, Prov. Cádiz, Spain, 6.iii.2005 (coll. SKOU)
134d. *D. lentiscaria* ♀, Embid, Prov. Guadalajara, Spain, 9.iv.2009 (coll. SKOU)
135a. ***Dyscia distinctaria*** ♂, Pobes, Prov. Álava, Spain,6.iv.1997 (coll. SKOU)
135b. *D. distinctaria* ♂, El Ronquillo, Prov. Huelva, Spain, 2.iv.2003 (coll. SKOU)
135c. *D. distinctaria* ♀, Serra de Monchique, Prov. Algarve, Portugal, 1-6.iv.2005 (coll. SKOU)
135d. *D. distinctaria* ♀, El Ronquillo, Prov. Huelva, Spain, 2.iv.2003 (coll. SKOU)

130a
130b
130c
130d
130e
131a
131b
131c
131d
132a
132b
132c
132d
133a
133b
133c
133d
133e
133f
133g
133h
133i
133j
133k
133l
133m
133n
133o
133p
134a
134b
134c
134d
135a
135b
135c
135d

PLATE 16 (all specimens natural size)

136a. ***Perconia strigillaria*** ♂, Stråsø Plantage, West Jylland, Denmark, 14.vi.1997 (coll. Skou)
136b. *P. strigillaria* ♀, Tyresta, Södermanland, Sweden, 6.vii.1961 (coll. Skou)
136c. *P. strigillaria* ♂, Malles Venosta, Prov. Bolzano, Italy, 26.vi.2008 (coll. Skou)
136d. *P. strigillaria* ♀, Malles Venosta, Prov. Bolzano, Italy, 26.vi.2008 (coll. Skou)
136e. *P. strigillaria* ♂, Castelbuono, Sicily, Italy, 5.vi.2005 (coll. Skou)
136f. *P. strigillaria* ♂, Donskoe Village, Bylyaevka District, South Ural, Russia, 1-4.vi.1998 (coll. Skou)
136g. *P. strigillaria* ♀, les Begues, Dept. Hautes-Alpes, France,15.vi.1980 (coll. Skou)
136h. *P. strigillaria* ♀, Col d'Ey, Buis-les-Baronnies, Provence, Dept. Drôme, France, 13.vi.1980 (coll. Skou)
137a. ***Perconia baeticaria*** ♂, La Granja, Prov. Segovia, Spain, late vi.1911 (ZSM)
137b. *P. baeticaria* ♂, Puerto de Navacerrada, Prov. Madrid, Spain30.vi.-8.vii.1959 (ZSM)
137c. *P. baeticaria* ♀, San Ildefonso, Prov. Segovia, Spain,7.vi.1902 (ZSM)
137d. *P. baeticaria* ♀, Penhas da Saúde, Estrela, Prov. Beira, Portugal, 23.vi.2007 (coll. Skou)
138a. ***Onychora agaritharia*** ♂, Almunecar, Prov. Málaga, Spain, 22-27.x.2000 (coll. Skou)
138b. *O. agaritharia* ♂, Tarifa, Prov. Cádiz, Spain, 12-17.x.2006 (coll. Skou)
138c. *O. agaritharia* ♂, Lagoa de Santo André, Prov. Alentejo, Portugal, 25.ix.2009 (coll. Skou)
138d. *O. agaritharia* ♂, La Plataforma, Sierra de Gredos, Prov. Ávila, Spain, 27-28.ix.2009 (coll. Skou)
138e. *O. agaritharia* ♂, La Plataforma, Sierra de Gredos, Prov. Ávila, Spain, 27-28.ix.2009 (coll. Skou)
138f. *O. agaritharia* ♀, Uña, Prov. Cuenca, Spain, 29.ix.2009 (coll. Skou)
138g. *O. agaritharia* ♀, Canjayar-Fondon, Sierra Nevada, Prov. Almería, Spain,22.xi.1980 (coll. Skou)
139a. ***Compsoptera opacaria*** ♂, Col de Vergio, Corsica, France, 25.ix.2003 (coll. Skou)
139b. *C. opacaria* ♂, Tabernas, Prov. Almería, Spain, 30.x.2005 (coll. Skou)
139c. *C. opacaria* ♂, Sedella, La Rahige, Prov. Málaga, Spain, 18.ix.2009 (coll. Skou)
139d. *C. opacaria* ♂, Col de Vergio, Corsica, France, 25.ix.2003 (coll. Skou)
139e. *C. opacaria* ♀, El Pozo del Esparto, Almería, Spain, 30.x.2005 (coll. Skou)
139f. *C. opacaria* ♀, Albarracín, Prov. Teruel, Spain, 17-18.ix.2002 (coll. Skou)
140a. ***Compsoptera jourdanaria*** ♂, Val de Vecar, Albarracín, Prov. Teruel, Spain, 18.ix.2002 (coll. Skou)
140b. *C. jourdanaria* ♂, Ruta del Veleta, Sierra Nevada, Prov. Granada, Spain, 19.ix.1987 (coll. Skou)
140c. *C. jourdanaria anargyra* ♂, Castirla, Corsica, France, 26.ix.2003 (coll. Skou)
140d. *C. jourdanaria* ♀, Val de Vecar, Albarracín, Prov. Teruel, 1.x.2009 (coll. Skou)
141a. ***Compsoptera argentaria*** ♂, Sicily, Italy, x.1920 (ZSM)
141b. *C. argentaria* ♂, Santo Stefano Quisquina, Sicily, 9-10.x.2010 Italy (coll. Skou)
141c. *C. argentaria* ♂, Guelt es Stel, Algeria, 11.x.1913 (coll. Skou)
141d. *C. argentaria* ♀, Col de Djerada, East Morocco, 9.x.1974 (ZSM)

136a
136b
136c
136d
136e
136f
136g
136h
137a
137b
137c
137d
138a
138b
138c
138d
138e
138f
138g
139a
139b
139c
139d
139e
139f
140a
140b
140c
140d
141a
141b
141c
141d

Genitalia plates

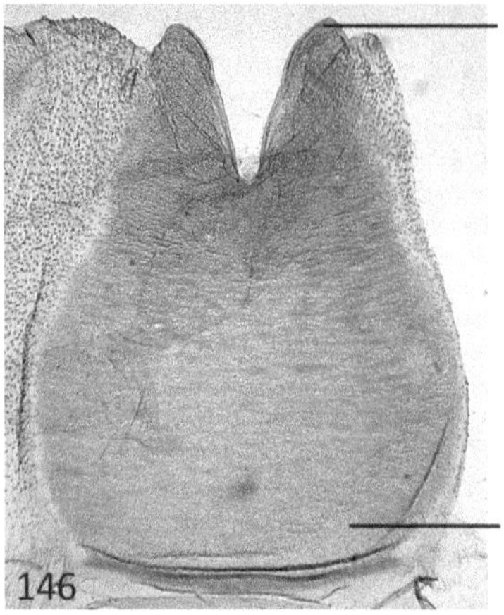

8th sternite

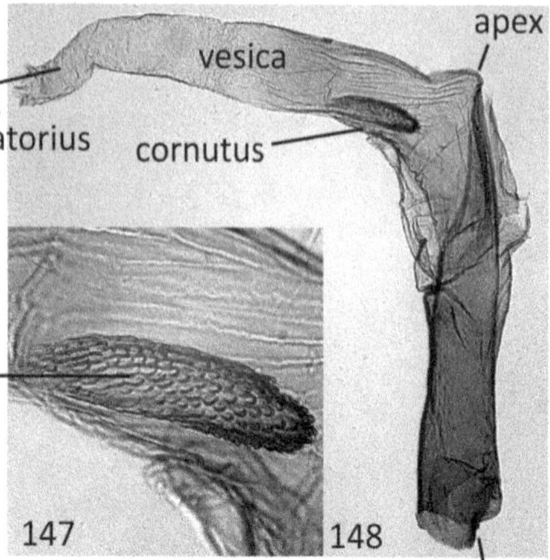

Aedeagus (cornutus enlarged)

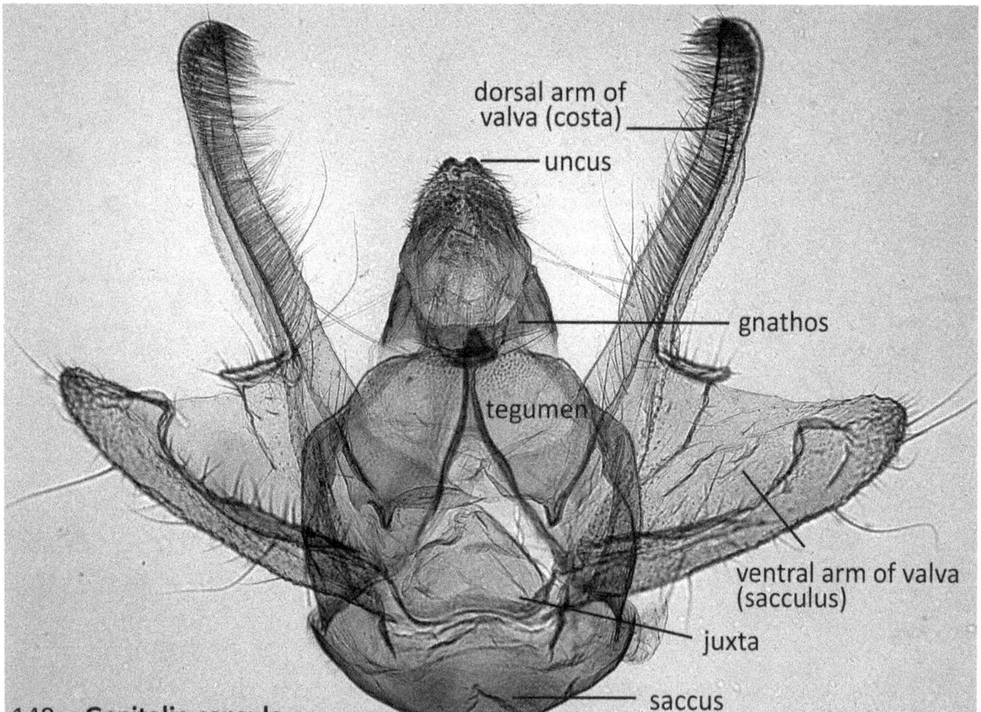

Text-figs 146–149. Terminology of the male genitalia, shown on *Macaria wauaria* (LINNAEUS, 1758), Finland; Savonlinna, 17.vii.1993, slide PS1148 (coll. P. SIHVONEN). Text-fig. 146. 8th sternite. Text-fig. 147. Cornutus enlarged. Text-fig. 148. Aedeagus. Text-fig. 149. Genitalia.

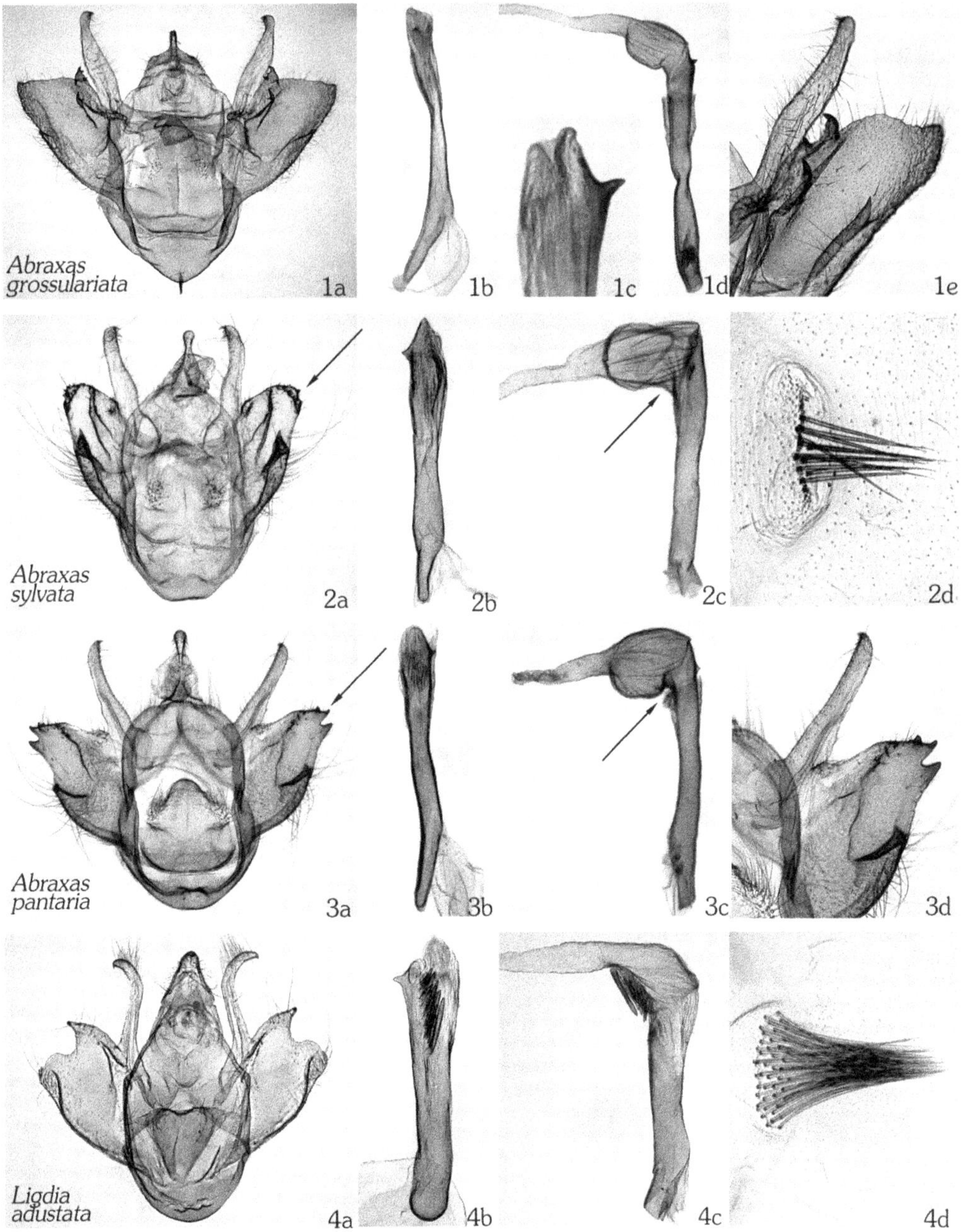

MALE GENITALIA (Figures are not in scale, specimen details are shown in Appendix 1.)

1. *Abraxas grossulariata* (LINNAEUS). a. genitalia, b. aedeagus, c. apex of aedeagus, d. vesica, e. apex of valva
2. *Abraxas sylvata* (SCOPOLI). a. genitalia, b. aedeagus, c. vesica, d. 3rd sternite setae
3. *Abraxas pantaria* (LINNAEUS). a. genitalia, b. aedeagus, c. vesica, d. apex of valva
4. *Ligdia adustata* (DENIS & SCHIFFERMÜLLER). a genitalia, b. aedeagus, c. vesica, d. 3rd sternite setae

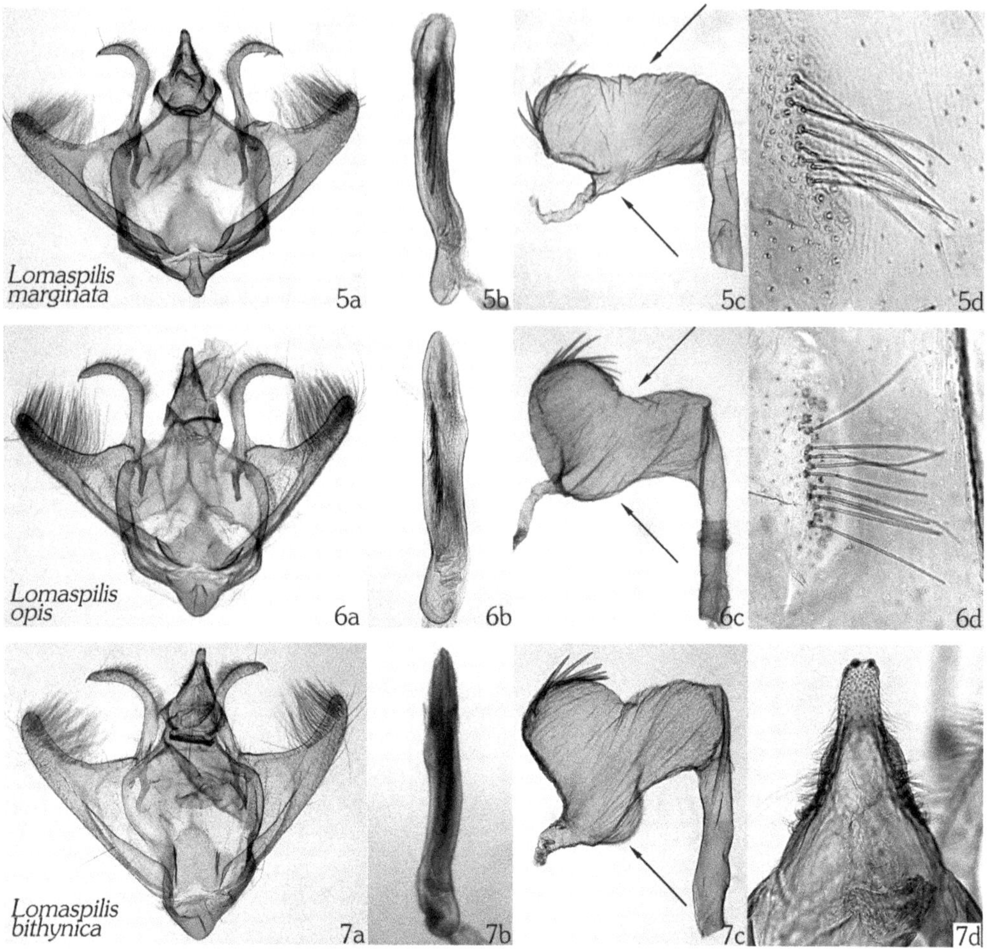

MALE GENITALIA (Figures are not in scale, specimen details are shown in Appendix 1.)

5. *Lomaspilis marginata* (LINNAEUS). a. genitalia, b. aedeagus, c. vesica, d. 3rd sternite setae
6. *Lomaspilis opis* BUTLER. a. genitalia, b. aedeagus, c. vesica, d. 3rd sternite setae
7. *Lomaspilis bithynica* WEHRLI. a. genitalia, b. aedeagus, c. vesica, d. uncus

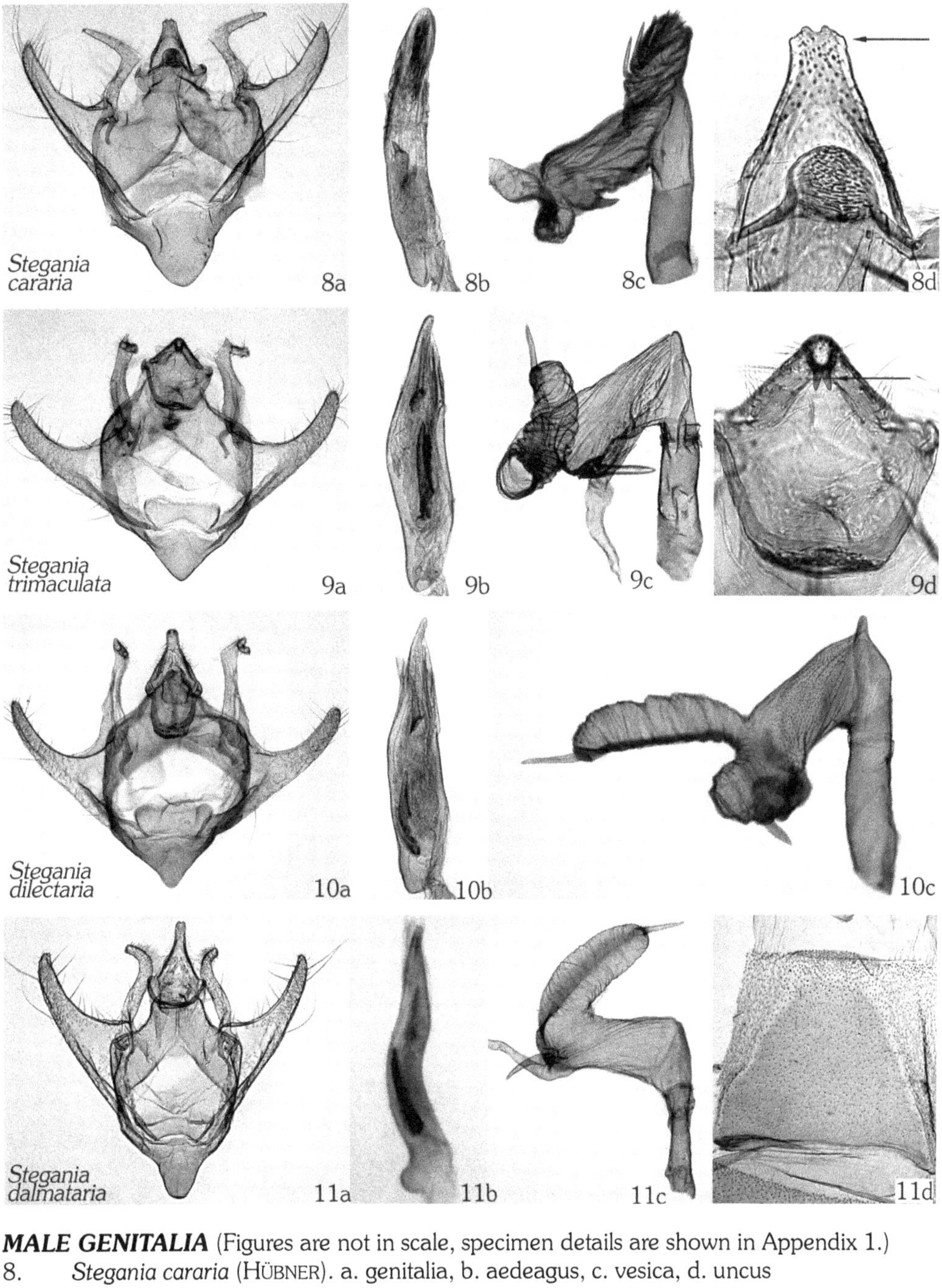

MALE GENITALIA (Figures are not in scale, specimen details are shown in Appendix 1.)

8. *Stegania cararia* (HÜBNER). a. genitalia, b. aedeagus, c. vesica, d. uncus
9. *Stegania trimaculata* (VILLERS). a. genitalia, b. aedeagus, c. vesica, d. uncus horns
10. *Stegania dilectaria* (HÜBNER). a. genitalia, b. aedeagus, c. vesica
11. *Stegania dalmataria* GUENÉE. a. genitalia, b. aedeagus, c. vesica, d. 8th tergite

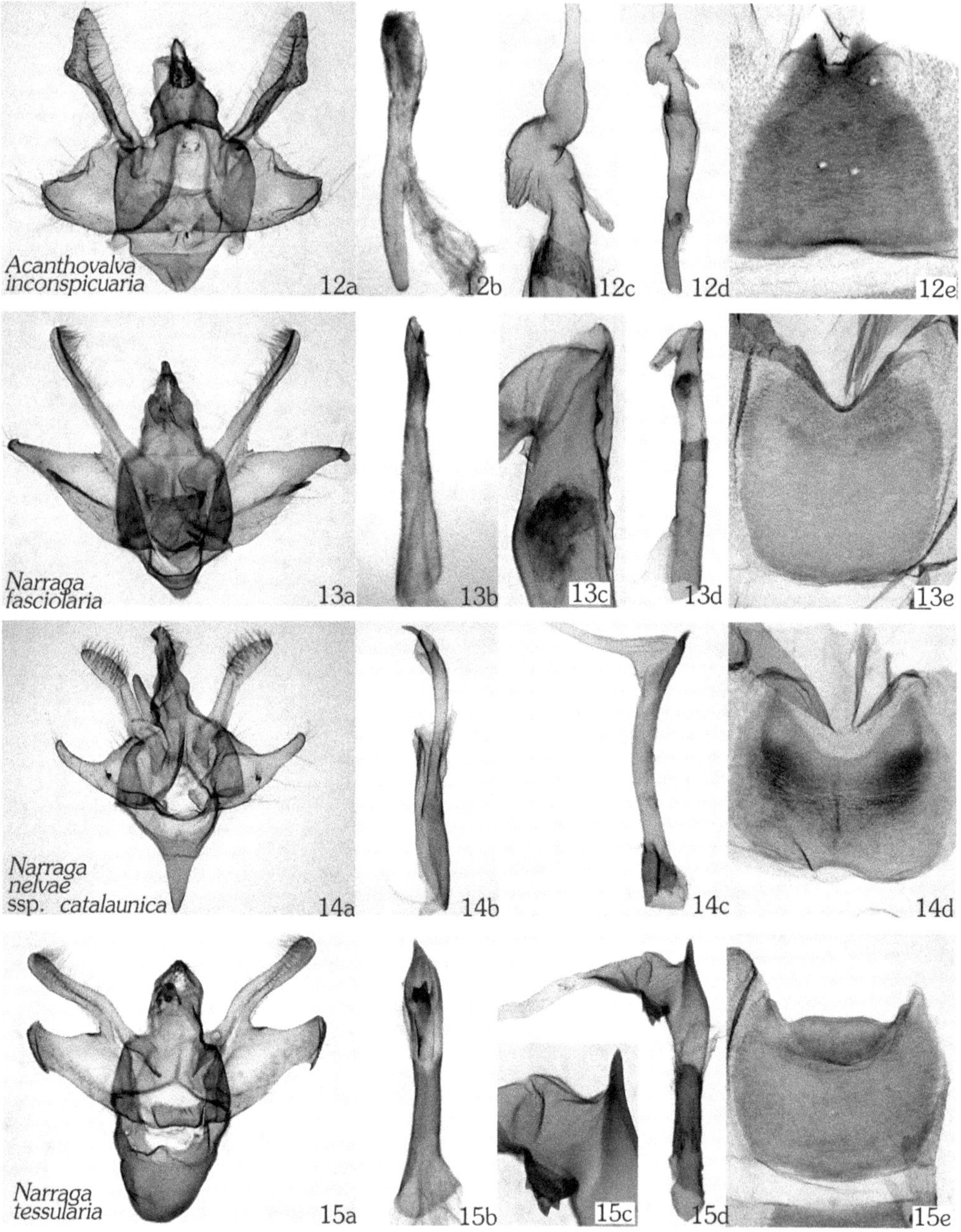

MALE GENITALIA (Figures are not in scale, specimen details are shown in Appendix 1.)

12. *Acanthovalva inconspicuaria* (HÜBNER). a. genitalia, b. aedeagus, c. vesica enlarged, d. vesica, e. 8th sternite
13. *Narraga fasciolaria* (HUFNAGEL). a. genitalia, b. aedeagus, c. apex of aedeagus, d. vesica, e. 8th sternite
14. *Narraga nelvae* subsp. *catalaunica* HERBULOT. a. genitalia, b. aedeagus, c. vesica, d. 8th sternite
15. *Narraga tessularia* (METZNER). a. genitalia, b. aedeagus, c. cornuti, d. vesica, e. 8th sternite

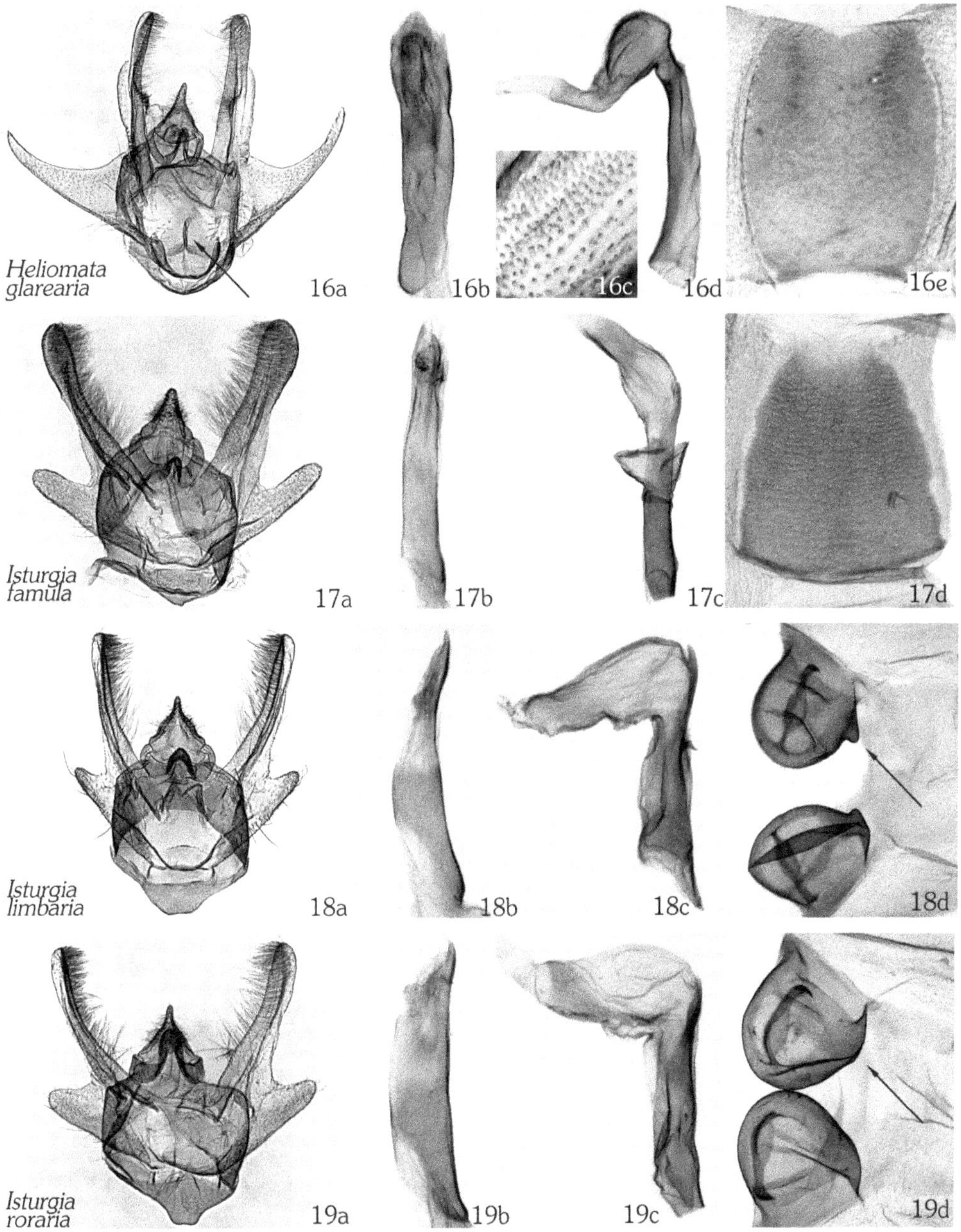

MALE GENITALIA (Figures are not in scale, specimen details are shown in Appendix 1.)

16. *Heliomata glarearia* (DENIS & SCHIFFERMÜLLER). a. genitalia, b. aedeagus, c. microcornuti, d. vesica, e. 8th sternite
17. *Isturgia famula* (ESPER). a. genitalia, b. aedeagus, c. vesica, d. 8th sternite
18. *Isturgia limbaria* (FABRICIUS). a. genitalia, b. aedeagus, c. vesica, d. tympanal organs
19. *Isturgia roraria* (FABRICIUS). a. genitalia, b. aedeagus, c. vesica, d. tympanal organs

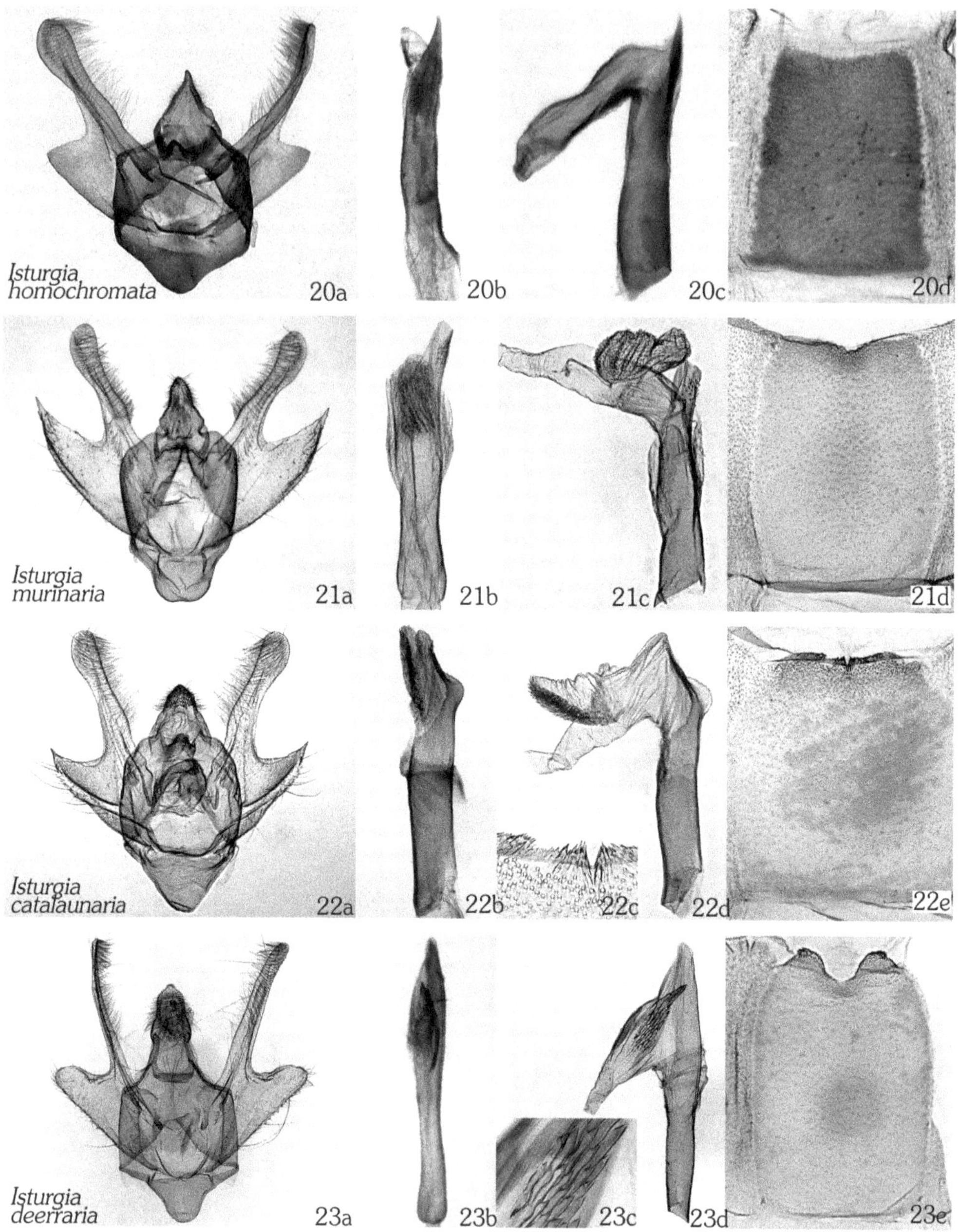

MALE GENITALIA (Figures are not in scale, specimen details are shown in Appendix 1.)

20. *Isturgia homochromata* (MABILLE). a. genitalia, b. aedeagus, c. vesica, d. 8th sternite
21. *Isturgia murinaria* (DENIS & SCHIFFERMÜLLER). a. genitalia, b. aedeagus, c. vesica, d. 8th sternite
22. *Isturgia catalaunaria* (GUENÉE). a. genitalia, b. aedeagus, c. posterior margin of 8th sternite, d. vesica, e. 8th sternite
23. *Isturgia deerraria* (WALKER). a. genitalia, b. aedeagus, c. microcornuti, d. vesica, e. 8th sternite

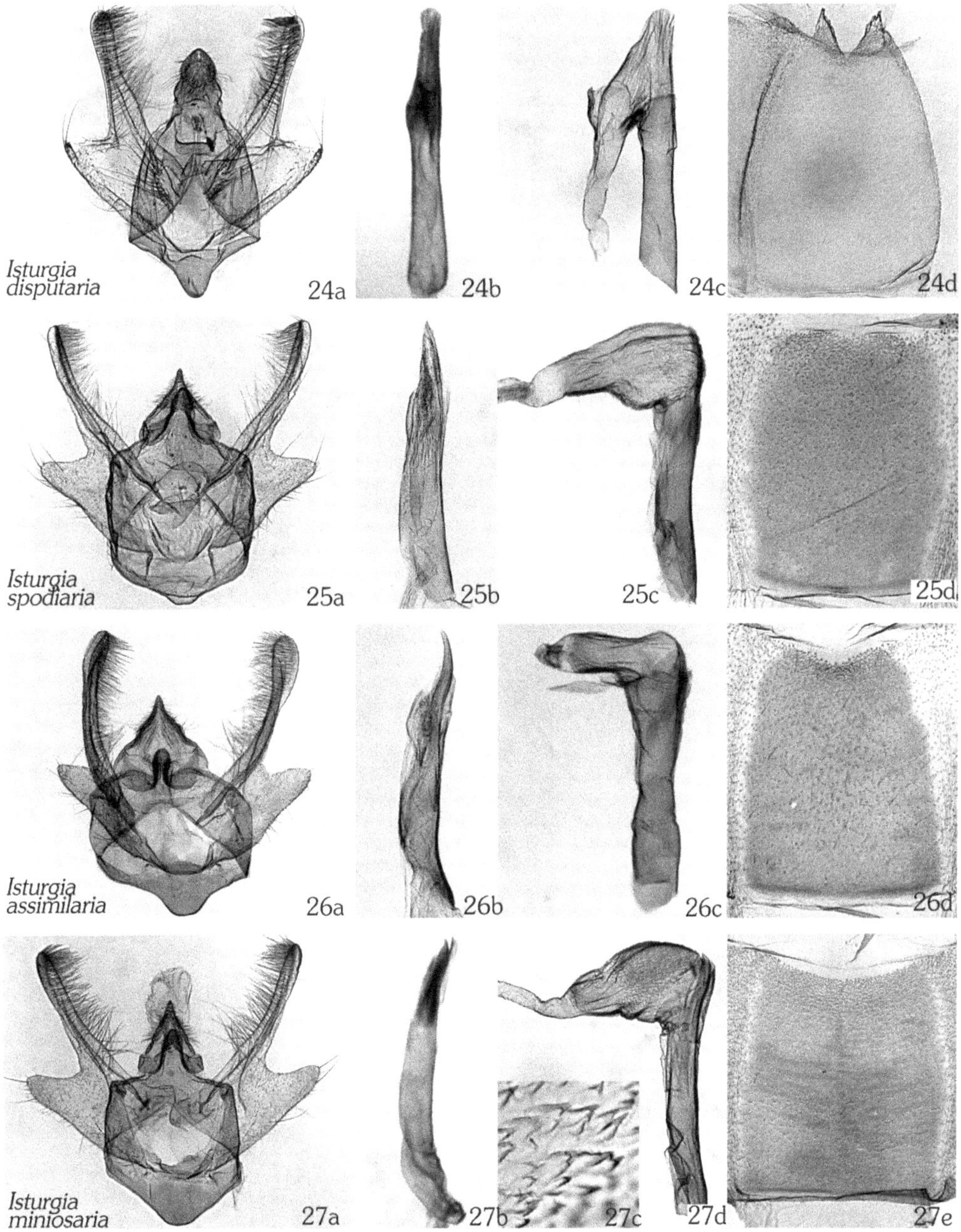

MALE GENITALIA (Figures are not in scale, specimen details are shown in Appendix 1.)

24. *Isturgia disputaria* (GUENÉE). a. genitalia, b. aedeagus, c. vesica, d. 8th sternite
25. *Isturgia spodiaria* (LEFÈBVRE). a. genitalia, b. aedeagus, c. vesica, d. 8th sternite. See also Text-figs 47–52.
26. *Isturgia assimilaria* (RAMBUR). a. genitalia, b. aedeagus, c. vesica, d. 8th sternite
27. *Isturgia miniosaria* (DUPONCHEL). a. genitalia, b. aedeagus, c. microcornuti, d. vesica, e. 8th sternite

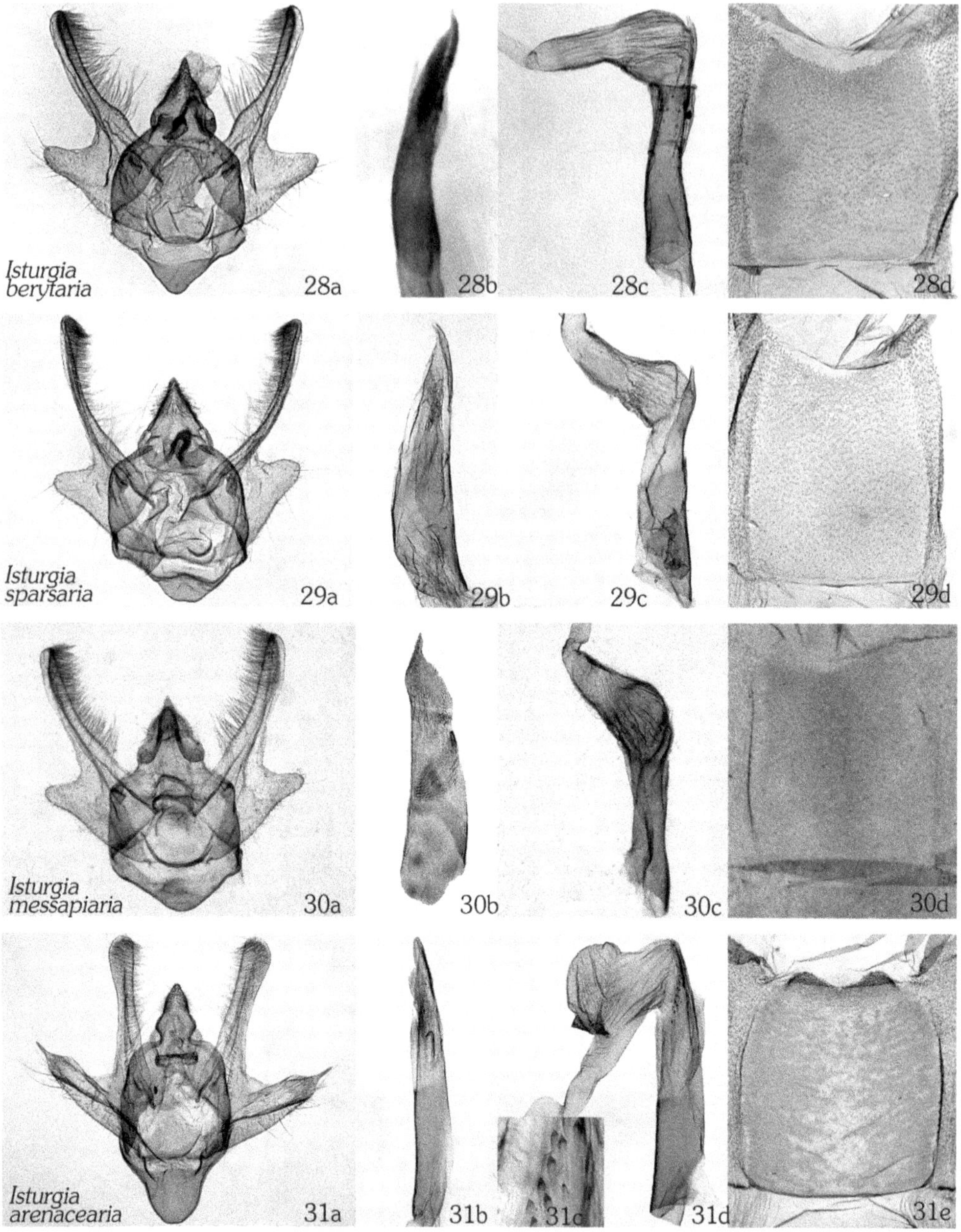

MALE GENITALIA (Figures are not in scale, specimen details are shown in Appendix 1.)

28. *Isturgia berytaria* (STAUDINGER). a. genitalia, b. aedeagus, c. vesica, d. 8th sternite
29. *Isturgia sparsaria* (HÜBNER). a. genitalia, b. aedeagus, c. vesica, d. 8th sternite
30. *Isturgia messapiaria* (SOHN-RETHEL). a. genitalia, b. aedeagus, c. vesica, d. 8th sternite
31. *Isturgia arenacearia* (DENIS & SCHIFFERMÜLLER). a. genitalia, b. aedeagus, c. spines in aedeagus apex, d. vesica, e. 8th sternite

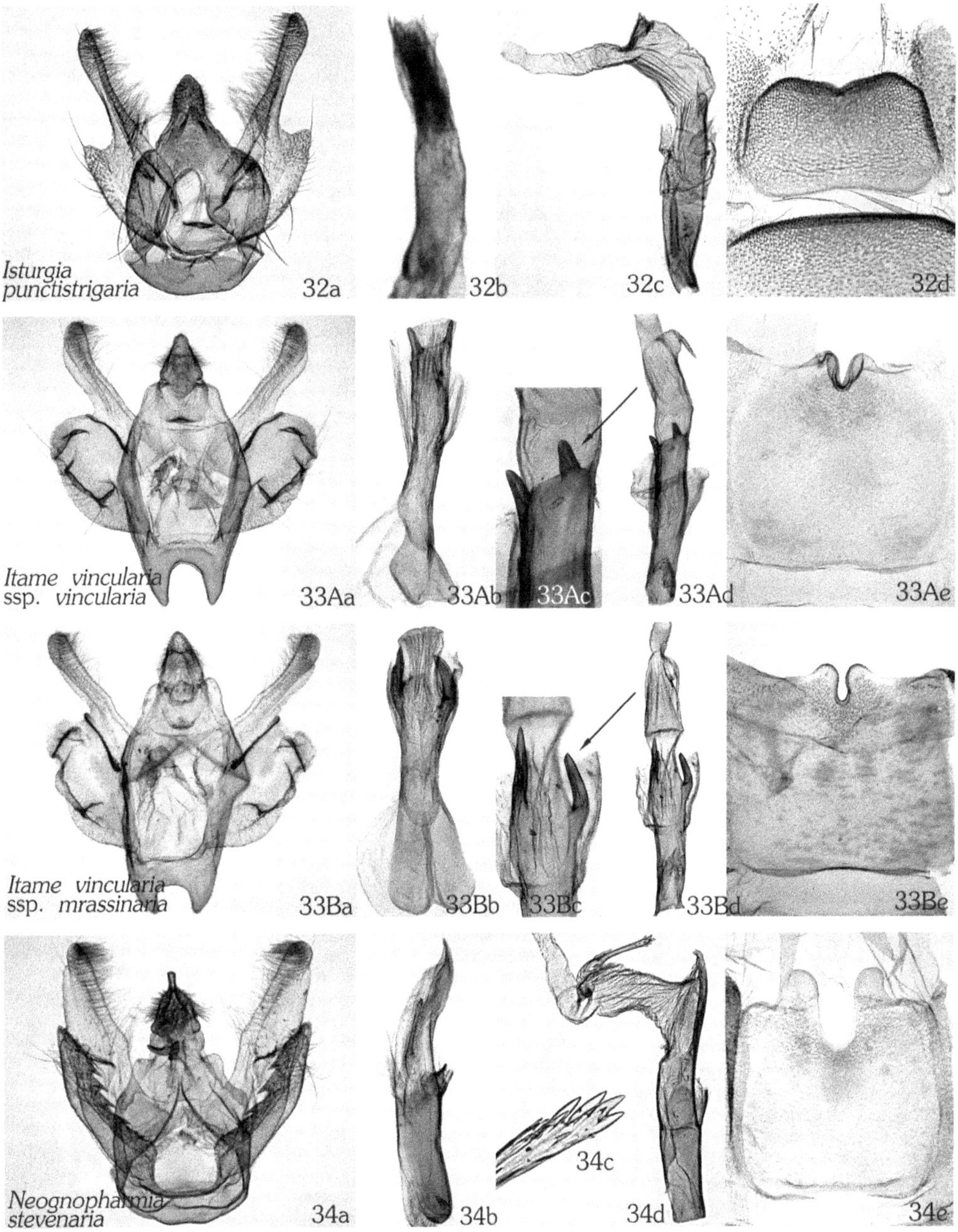

MALE GENITALIA (Figures are not in scale, specimen details are shown in Appendix 1.)

32. *Isturgia punctistrigaria* (CHRISTOPH). a. genitalia, b. aedeagus, c. vesica, d. 8th sternite

33A. *Itame vincularia* subsp. *vincularia* HÜBNER. a. genitalia, b. aedeagus, c. apex of aedeagus, d. vesica, e. 8th sternite

33B. *Itame vincularia* subsp. *mrassinaria* (OBERTHÜR). a. genitalia, b. aedeagus, c. apex of aedeagus, d. vesica, e. 8th sternite

34. *Neognopharmia stevenaria* (BOISDUVAL). a. genitalia, b. aedeagus, c. apex of cornutus, d. vesica, e. 8th sternite. See also Text-figs 60–63.

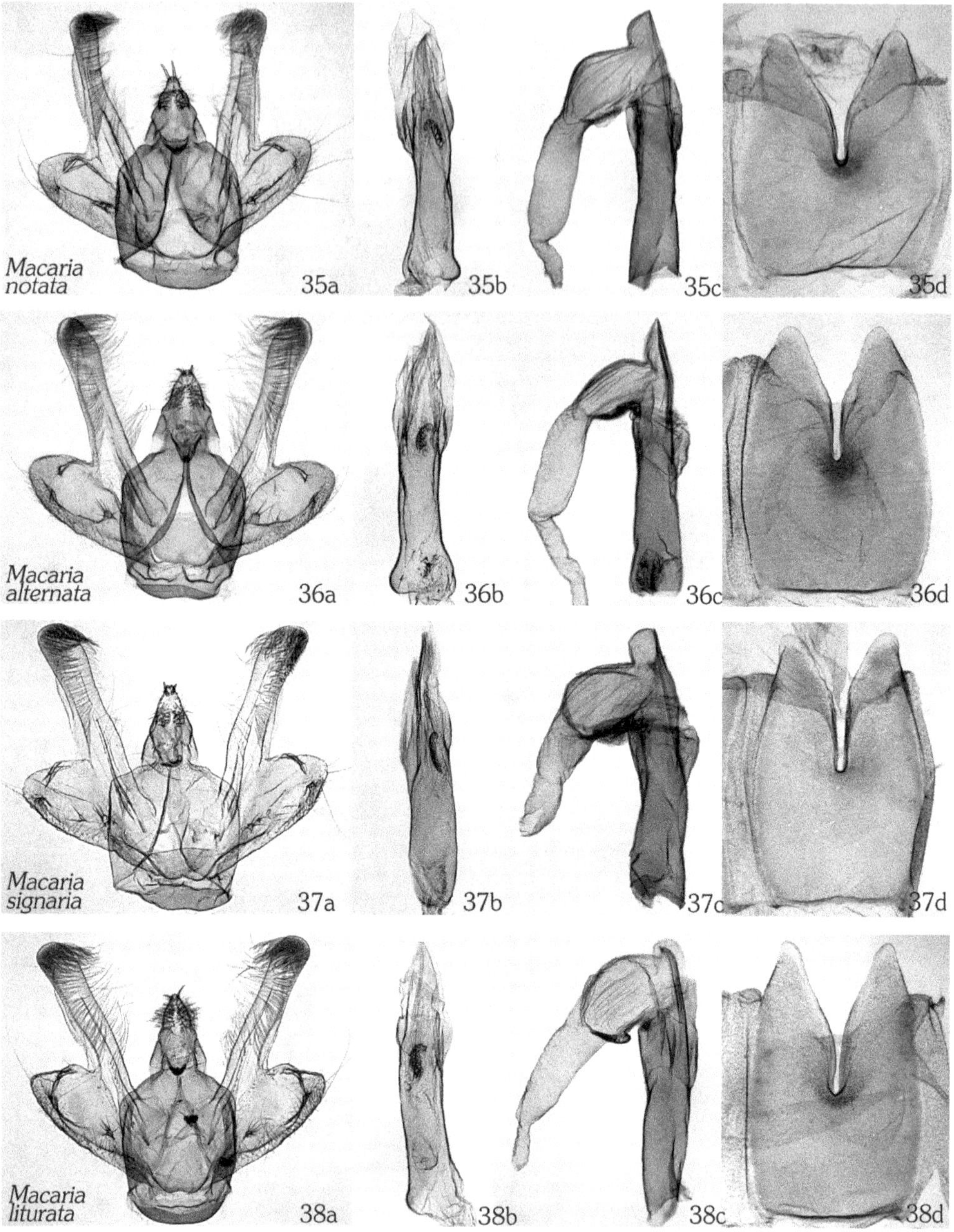

MALE GENITALIA (Figures are not in scale, specimen details are shown in Appendix 1.)

35. *Macaria notata* (LINNAEUS). a. genitalia, b. aedeagus, c. vesica, d. 8th sternite
36. *Macaria alternata* (DENIS & SCHIFFERMÜLLER). a genitalia, b. aedeagus, c. vesica, d. 8th sternite
37. *Macaria signaria* (HÜBNER). a. genitalia, b. aedeagus, c. vesica, d. 8th sternite
38. *Macaria liturata* (CLERCK). a. genitalia, b. aedeagus, c. vesica, d. 8th sternite

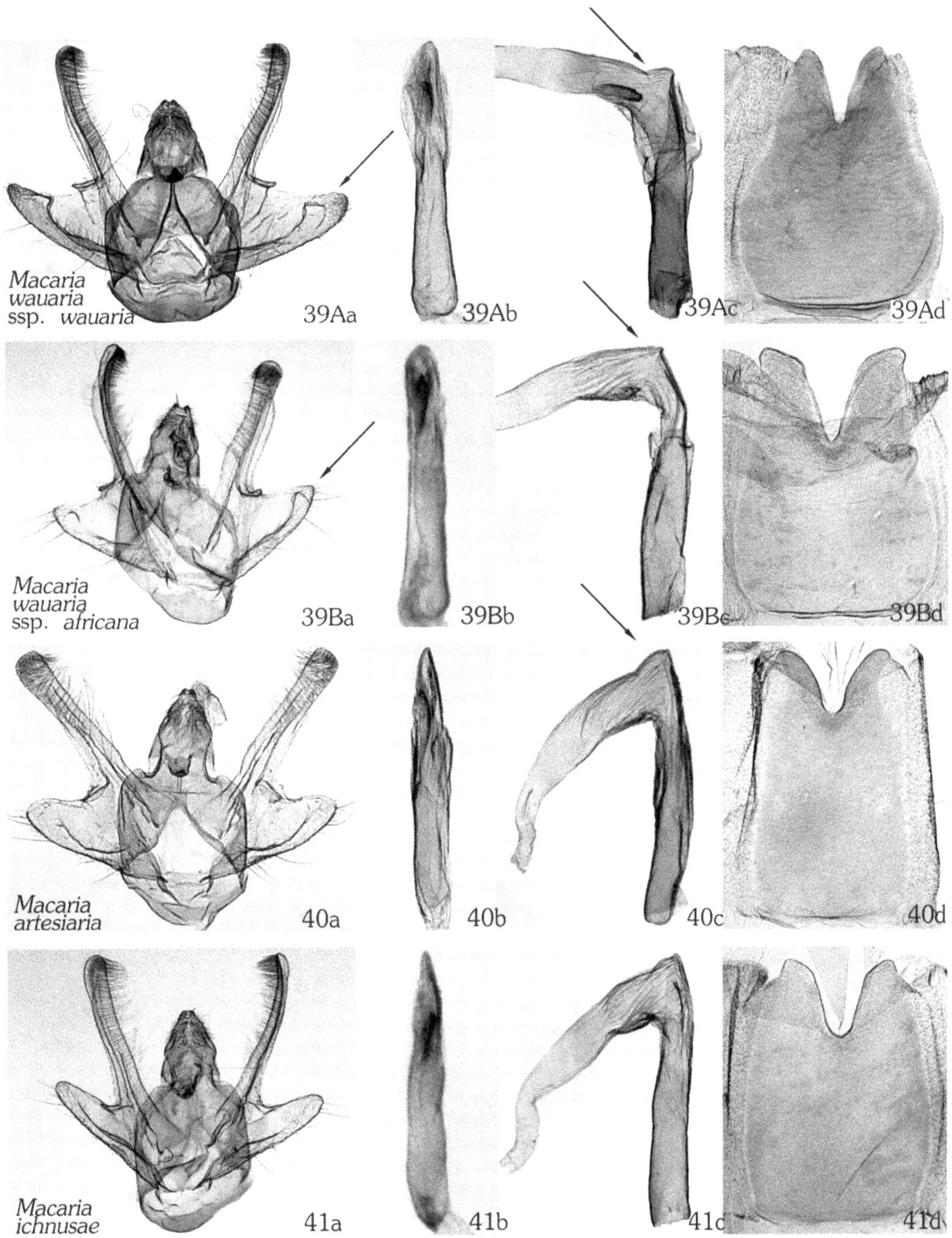

MALE GENITALIA (Figures are not in scale, specimen details are shown in Appendix 1.)

39A. *Macaria wauaria* subsp. *wauaria* (LINNAEUS). a. genitalia, b. aedeagus, c. vesica, d. 8th sternite

39B. *Macaria wauaria* subsp. *africana* (ZERNY). a. genitalia, b. aedeagus, c. vesica, d. 8th sternite

40. *Macaria artesiaria* (DENIS & SCHIFFERMÜLLER). a. genitalia, b. aedeagus, c. vesica, d. 8th sternite

41. *Macaria ichnusae* GOVI & FIUMI. a. genitalia, b. aedeagus, c. vesica, d. 8th sternite

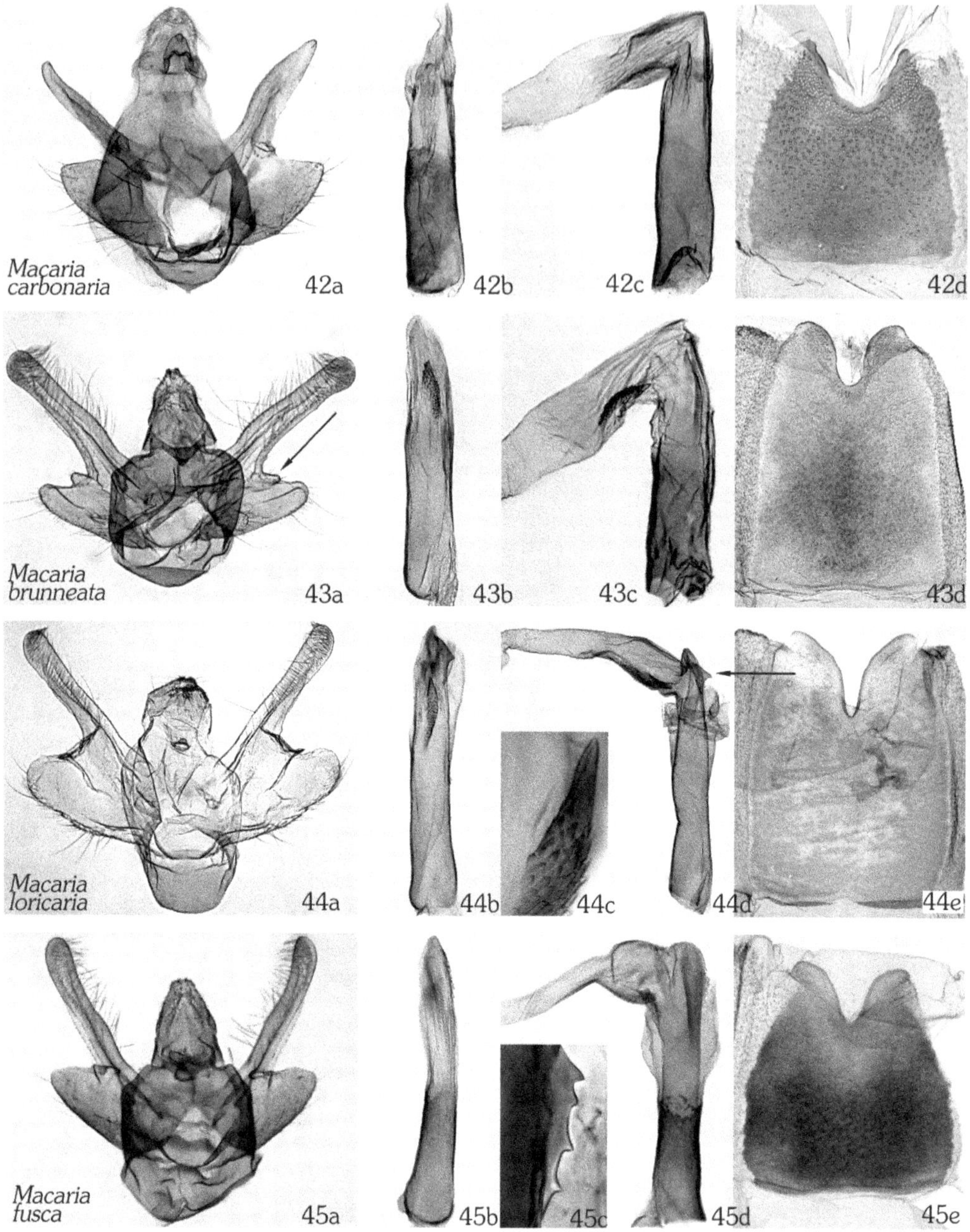

MALE GENITALIA (Figures are not in scale, specimen details are shown in Appendix 1.)

42. *Macaria carbonaria* (CLERCK). a. genitalia, b. aedeagus, c. vesica, d. 8th sternite
43. *Macaria brunneata* (THUNBERG). a. genitalia, b. aedeagus, c. vesica, d. 8th sternite
44. *Macaria loricaria* (EVERSMANN). a. genitalia, b. aedeagus, c. cornuti, d. vesica, e. 8th sternite
45. *Macaria fusca* (THUNBERG). a. genitalia, b. aedeagus, c. apex of aedeagus, d. vesica, e. 8th sternite

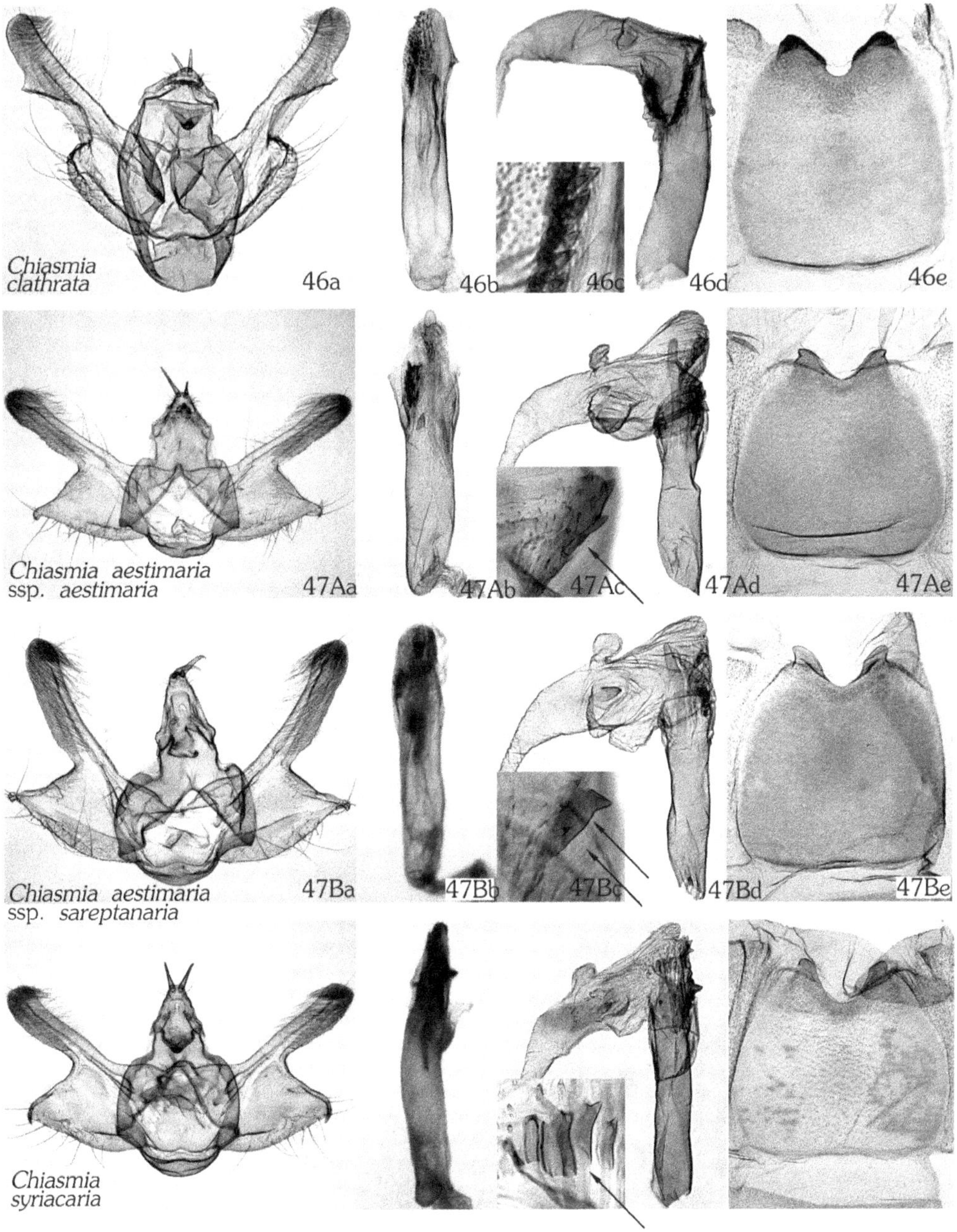

MALE GENITALIA (Figures are not in scale, specimen details are shown in Appendix 1.)

46. *Chiasmia clathrata* (LINNAEUS). a. genitalia, b. aedeagus, c. apex of aedeagus, d. vesica, e. 8th sternite

47A. *Chiasmia aestimaria* subsp. *aestimaria* (HÜBNER). a. genitalia, b. aedeagus, c. cornutus, d. vesica, e. 8th sternite

47B. *Chiasmia aestimaria* subsp. *sareptanaria* (STAUDINGER). a. genitalia, b. aedeagus, c. cornutus, d. vesica, e. 8th sternite

Chiasmia syriacaria (STAUDINGER) Non-European, see text for details. a. genitalia, b. aedeagus, c. cornutus, d. vesica, e. 8th sternite

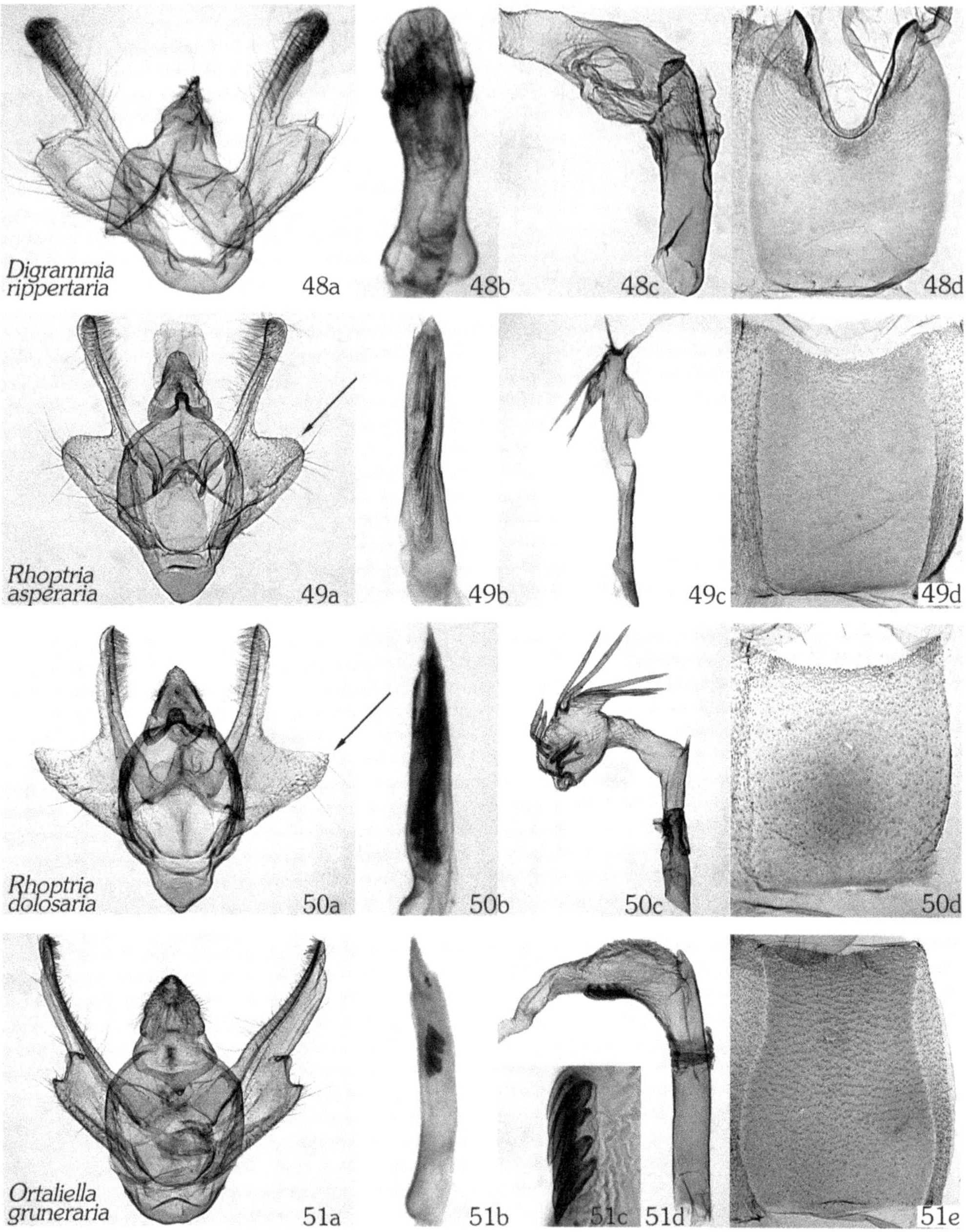

MALE GENITALIA (Figures are not in scale, specimen details are shown in Appendix 1.)

48. *Digrammia rippertaria* (Duponchel). a. genitalia, b. aedeagus, c. vesica, d. 8th sternite
49. *Rhoptria asperaria* (Hübner). a. genitalia, b. aedeagus, c. vesica, d. 8th sternite
50. *Rhoptria dolosaria* (Herrich-Schäffer). a. genitalia, b. aedeagus, c. vesica, d. 8th sternite
51. *Ortaliella gruneraria* (Staudinger). a. genitalia, b. aedeagus, c. cornuti, d. vesica, e. 8th sternite

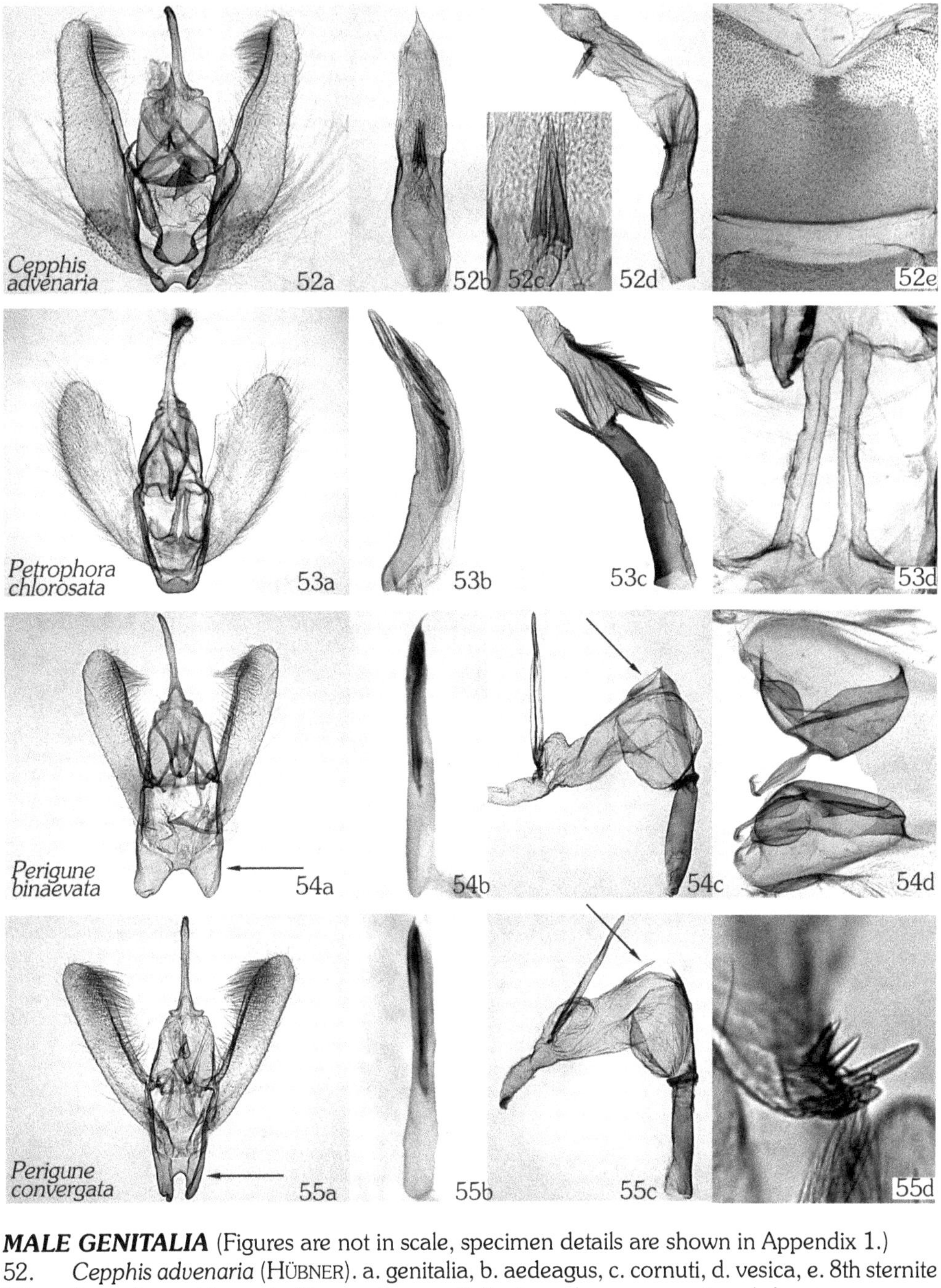

MALE GENITALIA (Figures are not in scale, specimen details are shown in Appendix 1.)

52. *Cepphis advenaria* (HÜBNER). a. genitalia, b. aedeagus, c. cornuti, d. vesica, e. 8th sternite
53. *Petrophora chlorosata* (SCOPOLI). a. genitalia, b. aedeagus, c. vesica, d. furca arms
54. *Perigune binaevata* (MABILLE). a. genitalia, b. aedeagus, c. vesica, d. tympanal organs
55. *Perigune convergata* (VILLERS). a. genitalia, b. aedeagus, c. vesica, d. apex of gnathos

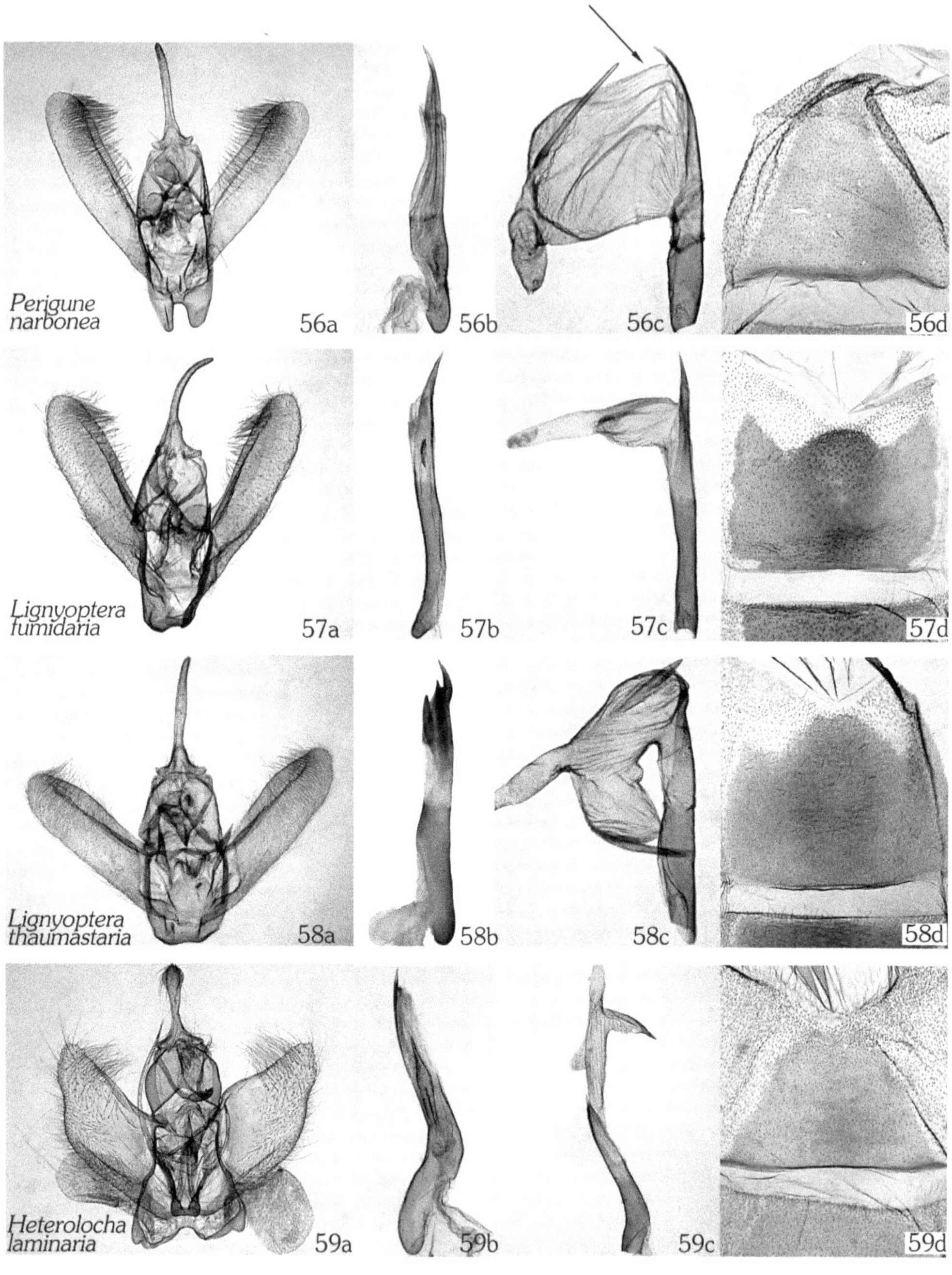

MALE GENITALIA (Figures are not in scale, specimen details are shown in Appendix 1.)

56. *Perigune narbonea* (LINNAEUS). a. genitalia, b. aedeagus, c. vesica, d. 8th tergite
57. *Lignyoptera fumidaria* (HÜBNER). a. genitalia, b. aedeagus, c. vesica, d. 8th sternite
58. *Lignyoptera thaumastaria* REBEL. a. genitalia, b. aedeagus, c. vesica, d. 8th sternite
59. *Heterolocha laminaria* (HERRICH-SCHÄFFER). a. genitalia, b. aedeagus, c. vesica, d. 8th sternite

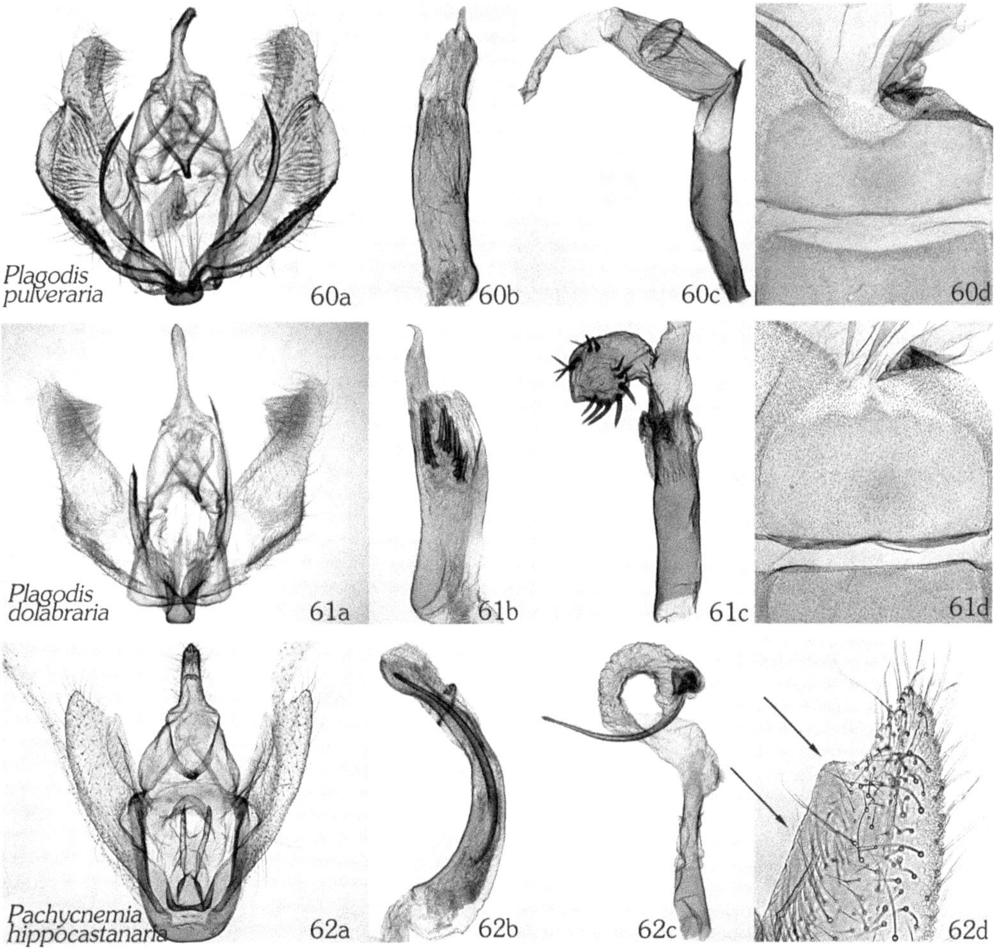

MALE GENITALIA (Figures are not in scale, specimen details are shown in Appendix 1.)

60. *Plagodis pulveraria* (LINNAEUS). a. genitalia, b. aedeagus, c. vesica, d. 8th sternite
61. *Plagodis dolabraria* (LINNAEUS). a. genitalia, b. aedeagus, c. vesica, d. 8th sternite
62. *Pachycnemia hippocastanaria* (HÜBNER). a. genitalia, b. aedeagus, c. vesica, d. apex of valva

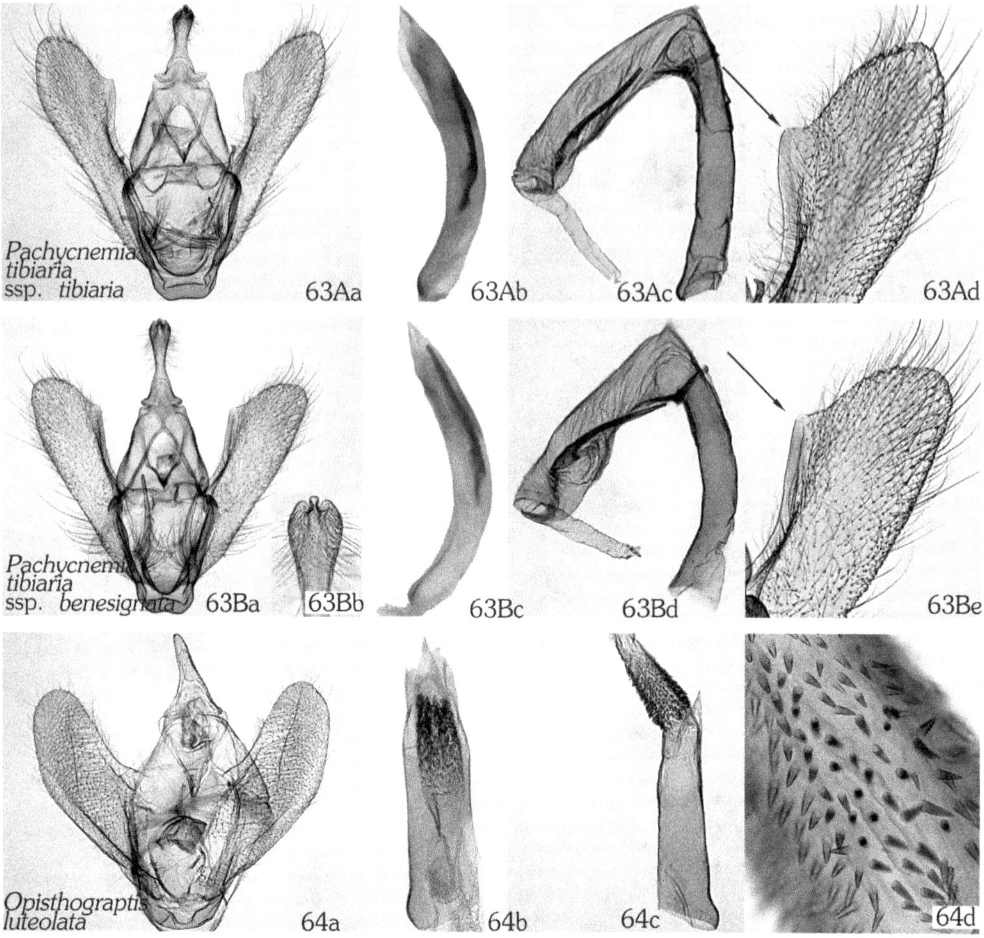

MALE GENITALIA (Figures are not in scale, specimen details are shown in Appendix 1.)

63A. *Pachycnemia tibiaria* subsp. *tibiaria* (RAMBUR). a. genitalia, b. aedeagus, c. vesica, d. apex of valva

63B. *Pachycnemia tibiaria* subsp. *benesignata* (BELLIER). a. genitalia, b. apex of uncus, c. aedeagus, d. vesica, e. apex of valva

64. *Opisthograptis luteolata* (LINNAEUS). a. genitalia, b. aedeagus, c. vesica, d. cornuti

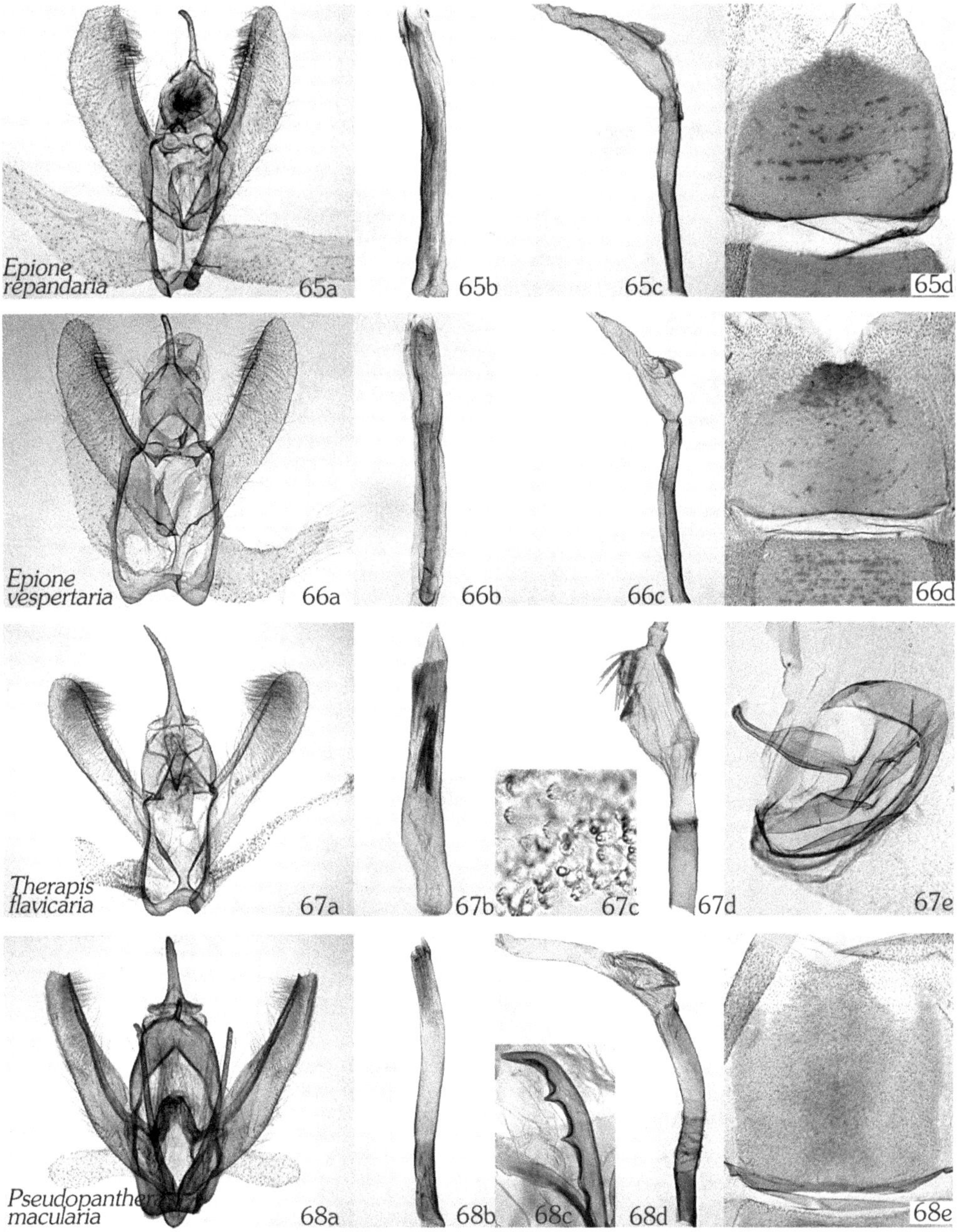

MALE GENITALIA (Figures are not in scale, specimen details are shown in Appendix 1.)

65. *Epione repandaria* (HUFNAGEL). a. genitalia, b. aedeagus, c. vesica, d. 8th sternite
66. *Epione vespertaria* (LINNAEUS). a. genitalia, b. aedeagus, c. vesica, d. 8th sternite
67. *Therapis flavicaria* (DENIS & SCHIFFERMÜLLER). a. genitalia, b. aedeagus, c. surface of membranous coremata, d. vesica, e. tympanal organ
68. *Pseudopanthera macularia* (LINNAEUS). a. genitalia, b. aedeagus, c. apex of furca arm, d. vesica, e. 8th sternite

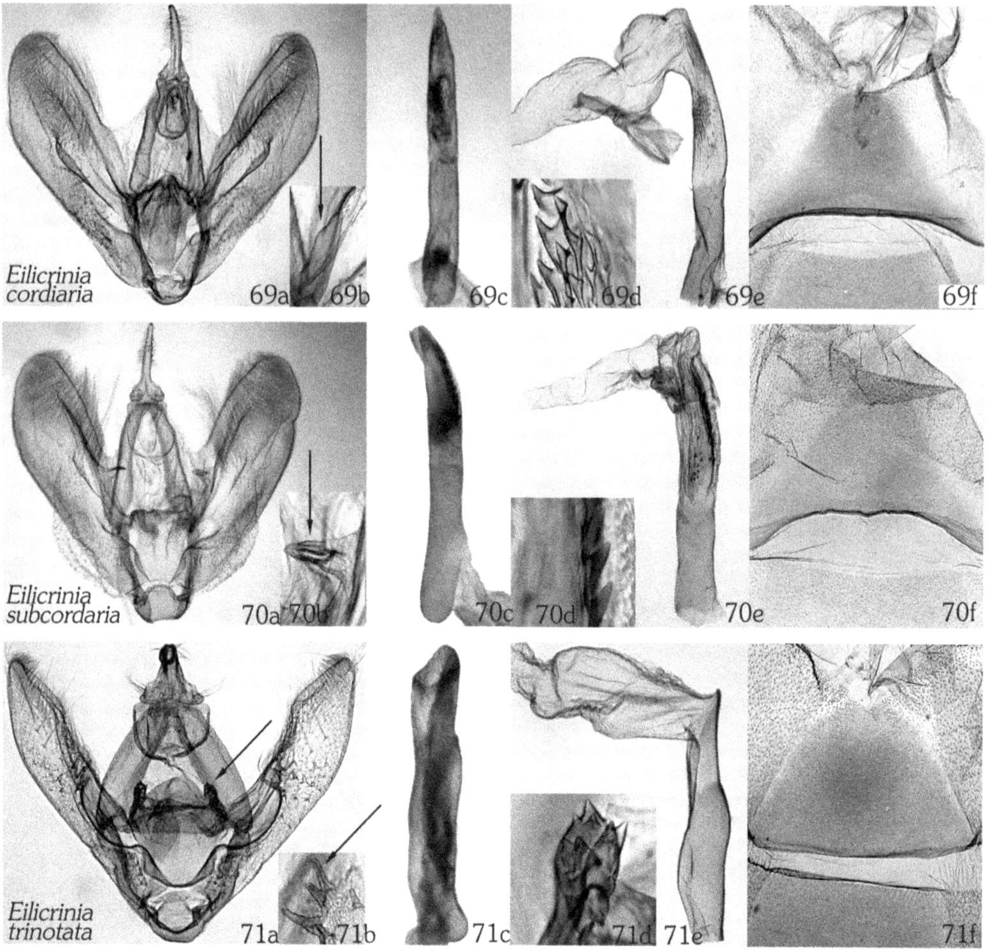

MALE GENITALIA (Figures are not in scale, specimen details are shown in Appendix 1.)

69. *Eilicrinia cordiaria* (HÜBNER). a. genitalia, b. valva margin (without extension), c. aedeagus, d. spines of aedeagus, e. vesica, f. 8th sternite
70. *Eilicrinia subcordaria* HERRICH-SCHÄFFER. a. genitalia, b. valva margin (with single spine), c. aedeagus, d. spines of aedeagus, e. vesica, f. 8th sternite
71. *Eilicrinia trinotata* (METZNER). a. genitalia, b. valva margin (with several spines), c. aedeagus, d. base of valva, e. vesica, f. 8th sternite

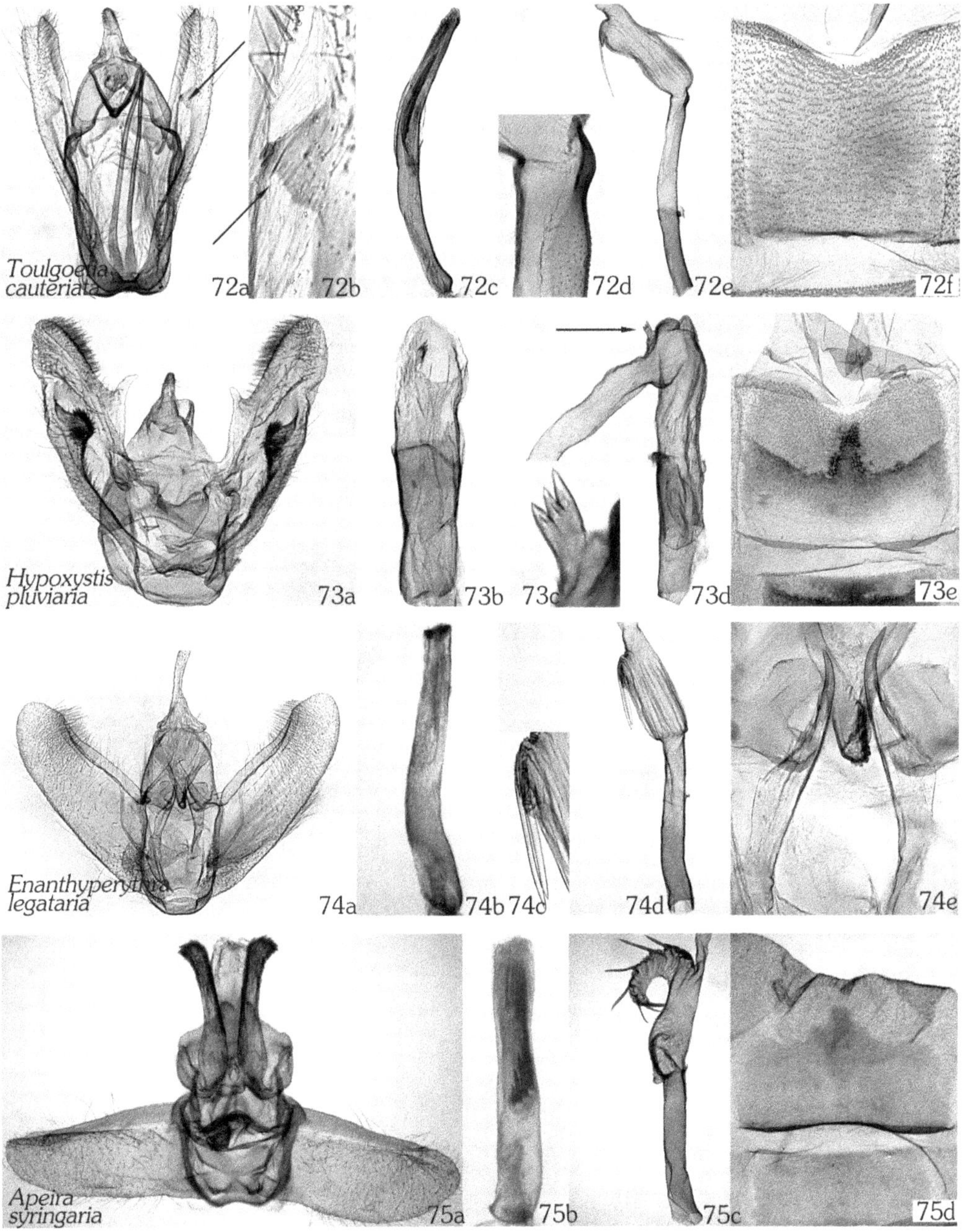

MALE GENITALIA (Figures are not in scale, specimen details are shown in Appendix 1.)

72. *Toulgoetia cauteriata* (STAUDINGER). a. genitalia, b. inner margin of valva, c. aedeagus, d. apex of aedeagus, e. vesica, f. 8th sternite
73. *Hypoxystis pluviaria* (FABRICIUS). a. genitalia, b. aedeagus, c. cornutus, d. vesica, e. 8th sternite
74. *Enanthyperythra legataria* (HERRICH-SCHÄFFER). a. genitalia, b. aedeagus, c. cornuti, d. vesica, e. furca arms
75. *Apeira syringaria* (LINNAEUS). a. genitalia, b. aedeagus, c. vesica, d. 8th sternite

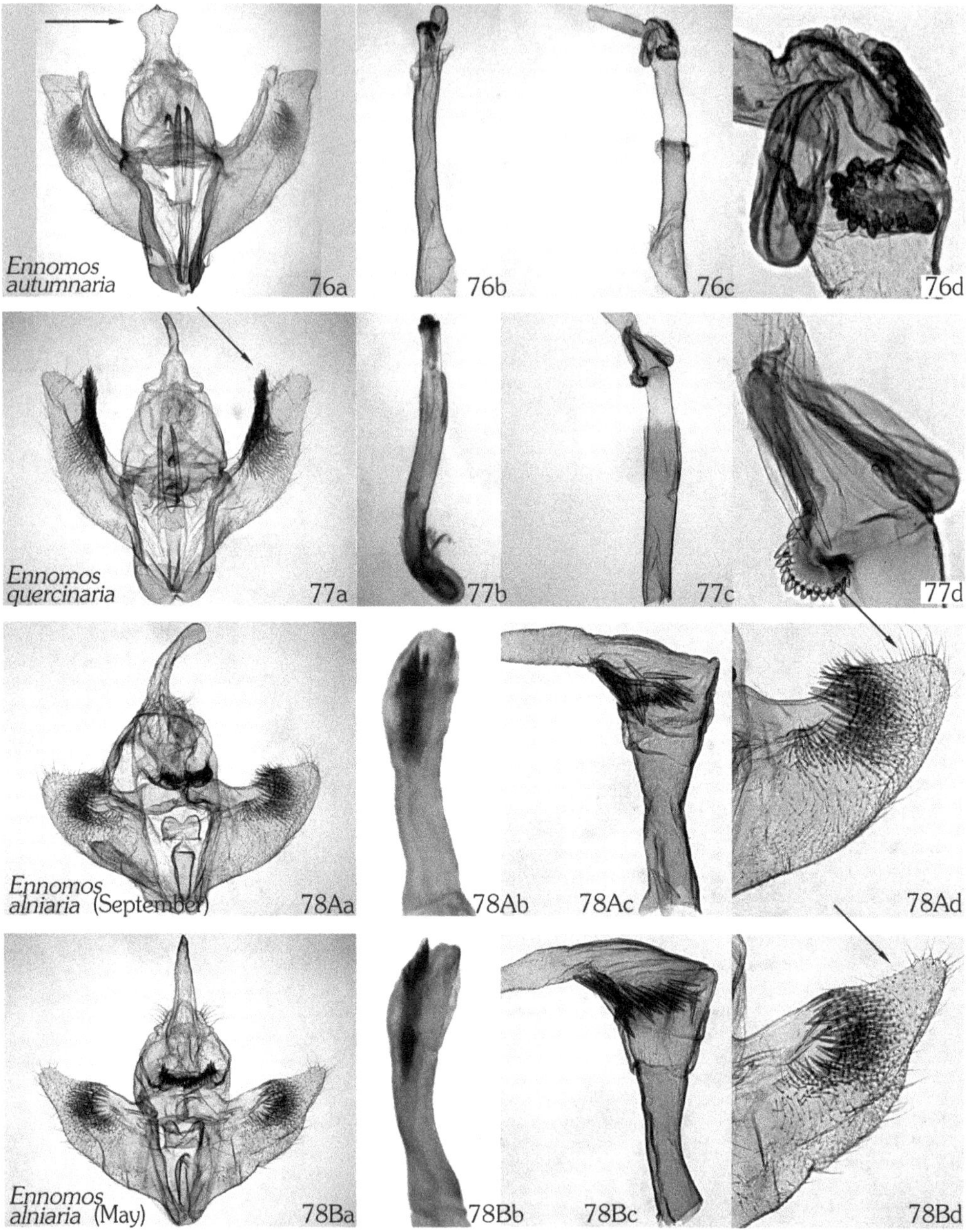

MALE GENITALIA (Figures are not in scale, specimen details are shown in Appendix 1.)

76. *Ennomos autumnaria* (WERNEBURG). a. genitalia, b. aedeagus, c. vesica, d. apex of aedeagus and base of vesica
77. *Ennomos quercinaria* (HUFNAGEL). a. genitalia, b. aedeagus, c. vesica, d. cornuti

78A. *Ennomos alniaria* (LINNAEUS), specimen taken in September. a. genitalia, b. aedeagus, c. vesica, d. valva

78B. *Ennomos alniaria* (LINNAEUS), specimen taken in May. a. genitalia, b. aedeagus, c. vesica, d. valva

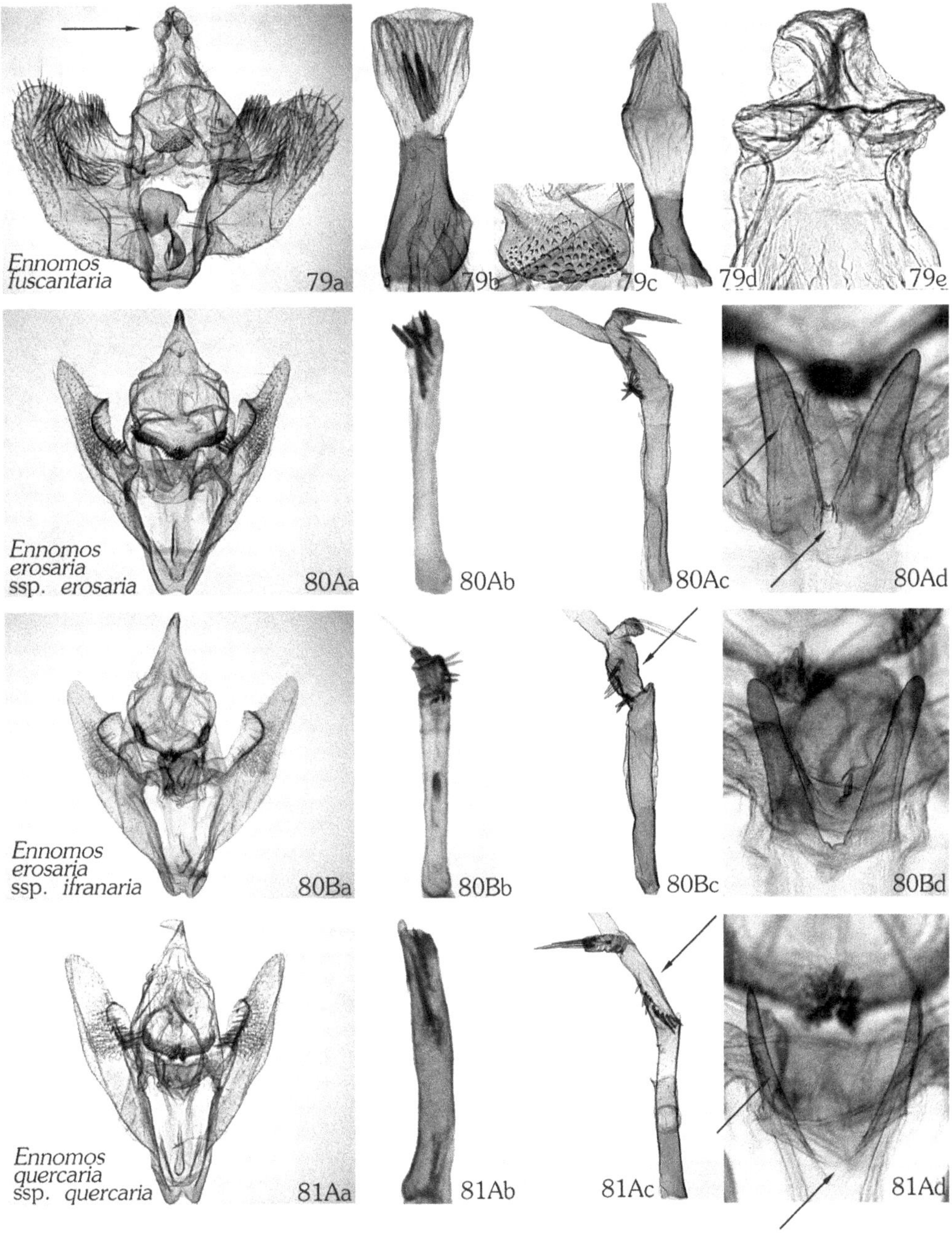

MALE GENITALIA (Figures are not in scale, specimen details are shown in Appendix 1.)

79. *Ennomos fuscantaria* (HAWORTH). a. genitalia, b. aedeagus, c. apex of gnathos, d. vesica, e. uncus

80A. *Ennomos erosaria* subsp. *erosaria* (DENIS & SCHIFFERMÜLLER). a. genitalia, b. aedeagus, c. vesica, d. furca arms

80B. *Ennomos erosaria* subsp. *ifranaria* (RUNGS). a. genitalia, b. aedeagus, c. vesica, d. furca arms

81A. *Ennomos quercaria* subsp. *quercaria* (HÜBNER). a. genitalia, b. aedeagus, c. vesica, d. furca arms

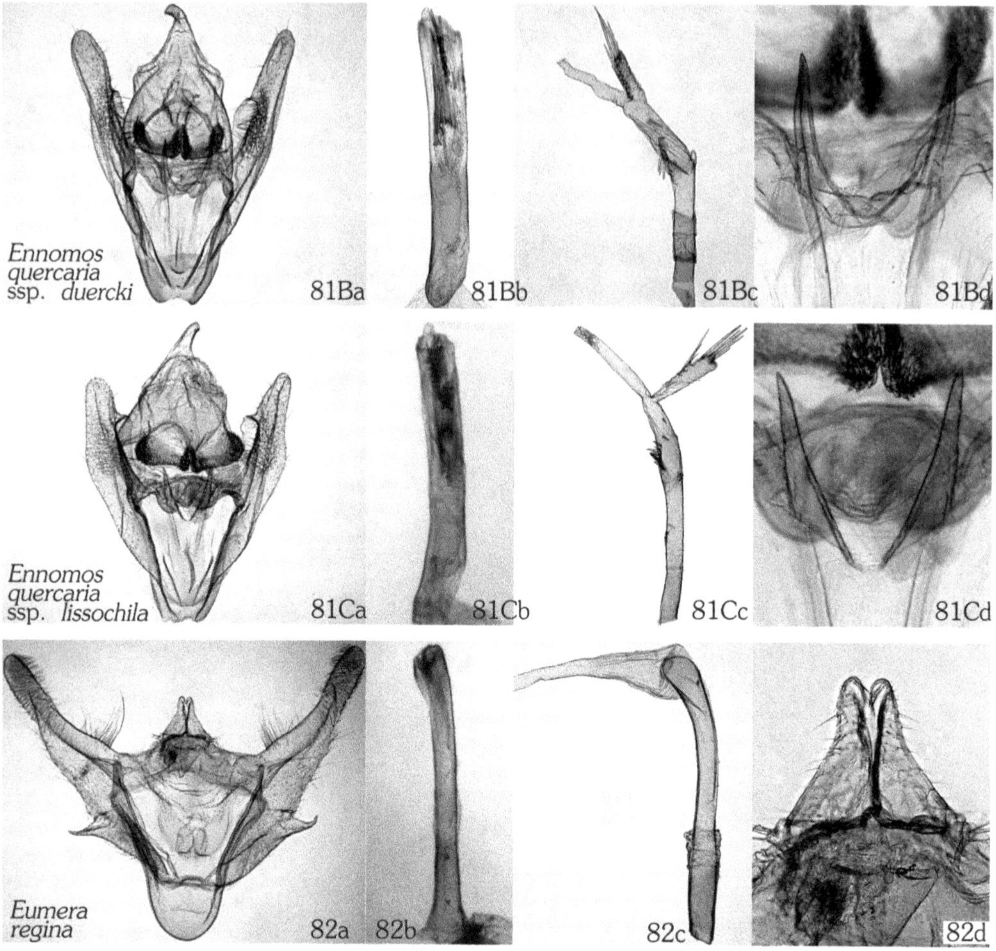

MALE GENITALIA (Figures are not in scale, specimen details are shown in Appendix 1.)

81B. *Ennomos quercaria* subsp. *duercki* REISSER. a. genitalia, b. aedeagus, c. vesica, d. furca arms

81C. *Ennomos quercaria* subsp. *lissochila* (PROUT). a. genitalia, b. aedeagus, c. vesica, d. furca arms

82. *Eumera regina* STAUDINGER. a. genitalia, b. aedeagus, c. vesica, d. scaphium and socii

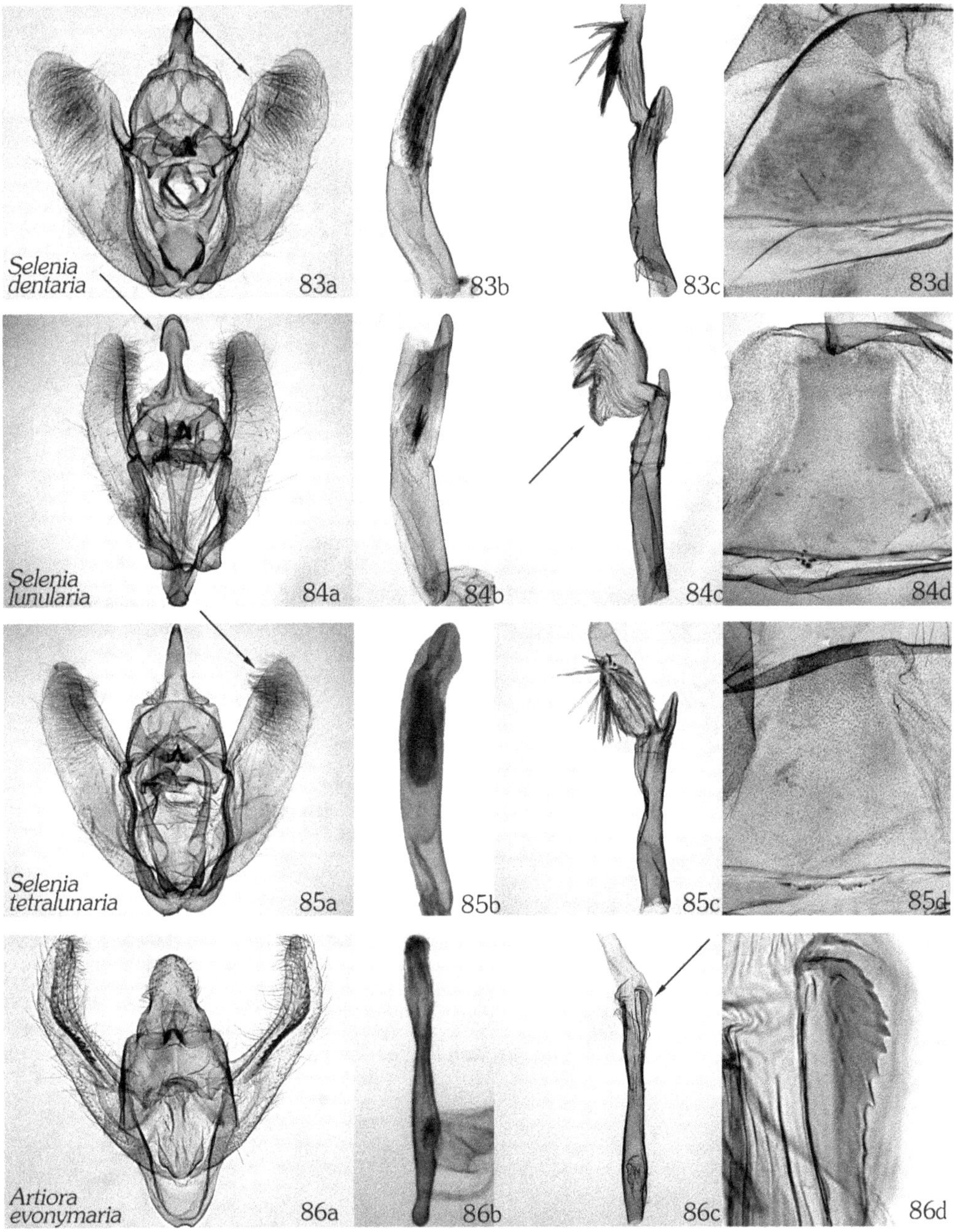

MALE GENITALIA (Figures are not in scale, specimen details are shown in Appendix 1.)

83. *Selenia dentaria* (FABRICIUS). a. genitalia, b. aedeagus, c. vesica, d. 8th tergite
84. *Selenia lunularia* (HÜBNER). a. genitalia, b. aedeagus, c. vesica, d. 8th tergite
85. *Selenia tetralunaria* (HUFNAGEL). a. genitalia, b. aedeagus, c. vesica, d. 8th tergite
86. *Artiora evonymaria* (DENIS & SCHIFFERMÜLLER). a. genitalia, b. aedeagus, c. vesica, d. apex of aedeagus

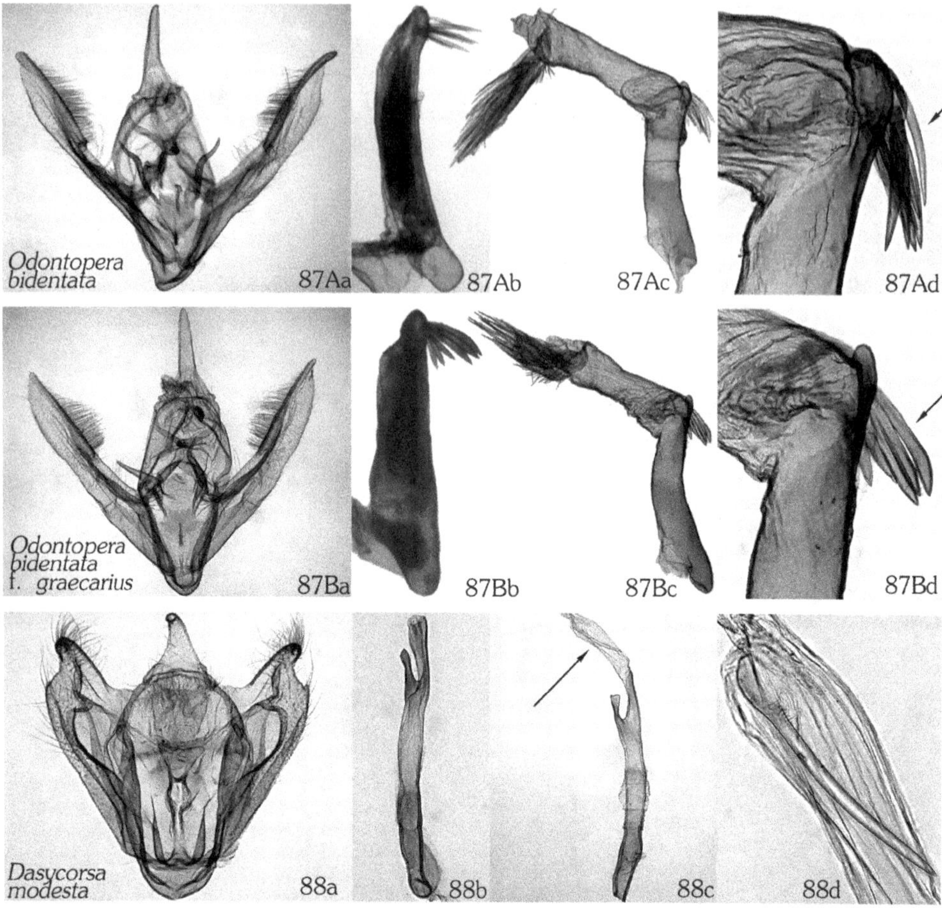

MALE GENITALIA (Figures are not in scale, specimen details are shown in Appendix 1.)

87A. *Odontopera bidentata* (CLERCK), specimens from Finland. See text for details. a. genitalia, b. aedeagus, c. vesica, d. cornuti

87B. *Odontopera bidentata* (CLERCK) f. *graecarius*, specimens from Greece. See text for details. a. genitalia, b. aedeagus, c. vesica, d. cornuti

88. *Dasycorsa modesta* (STAUDINGER). a. genitalia, b. aedeagus, c. vesica, d. cornutus

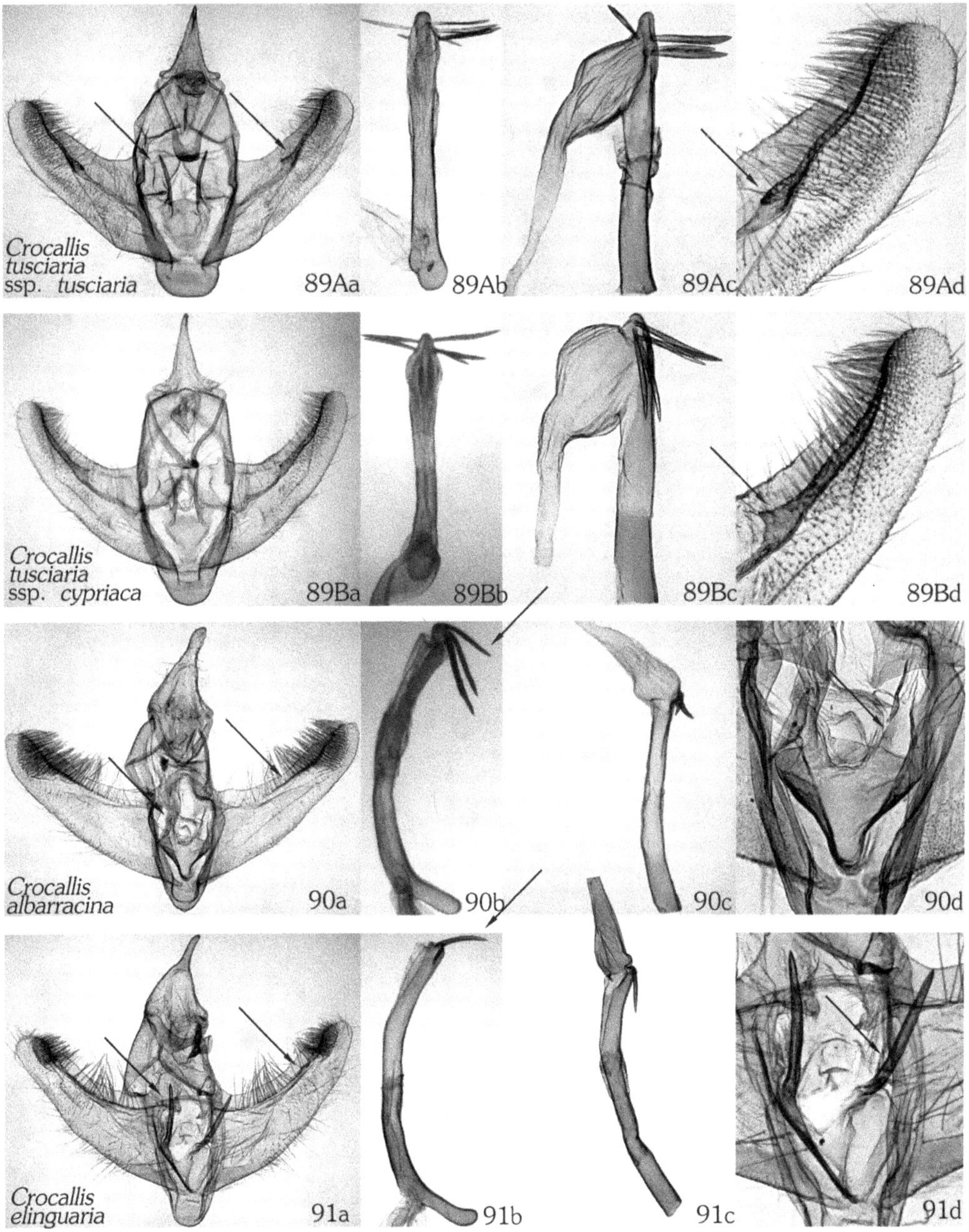

MALE GENITALIA (Figures are not in scale, specimen details are shown in Appendix 1.)

89A. *Crocallis tusciaria* subsp. *tusciaria* (BORKHAUSEN). a. genitalia, b. aedeagus, c. vesica, d. apex of valva

89B. *Crocallis tusciaria* subsp. *cypriaca* FISCHER. a. genitalia, b. aedeagus, c. vesica, d. apex of valva

90. *Crocallis albarracina* WEHRLI. a. genitalia, b. aedeagus, c. vesica, d. lateral appendices of juxta

91. *Crocallis elinguaria* (LINNAEUS). a. genitalia, b. aedeagus, c. vesica, d. lateral appendices of juxta

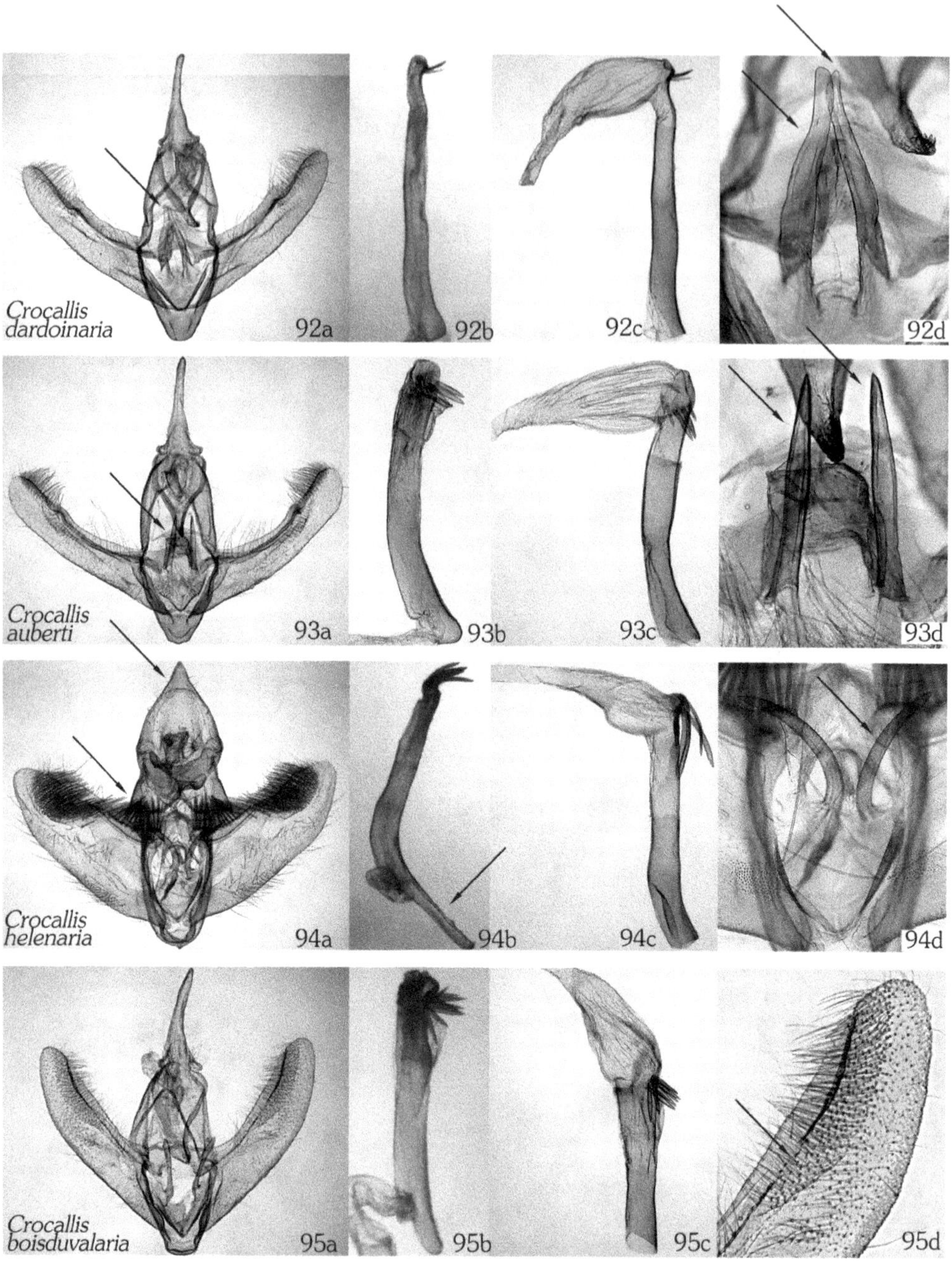

MALE GENITALIA (Figures are not in scale, specimen details are shown in Appendix 1.)

92. *Crocallis dardoinaria* Donzel. a. genitalia, b. aedeagus, c. vesica, d. lateral appendices of juxta
93. *Crocallis auberti* Oberthür. a. genitalia, b. aedeagus, c. vesica, d. lateral appendices of juxta
94. *Crocallis helenaria* Ruckdeschel. a. genitalia, b. aedeagus, c. vesica, d. lateral appendices of juxta
95. *Crocallis boisduvalaria* (Lucas). a. genitalia, b. aedeagus, c. vesica, d. apex of valva

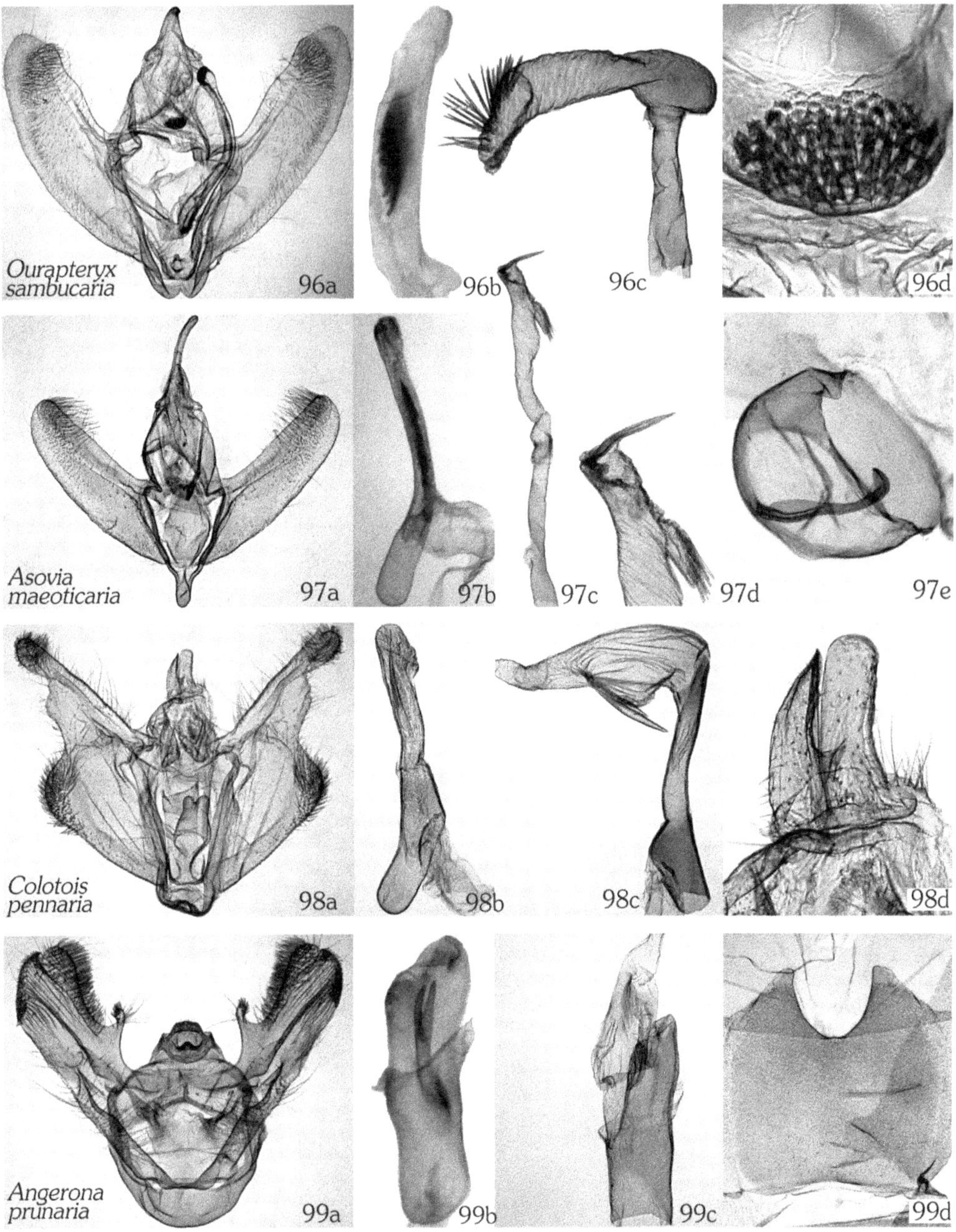

MALE GENITALIA (Figures are not in scale, specimen details are shown in Appendix 1.)

96. *Ourapteryx sambucaria* (Linnaeus). a. genitalia, b. aedeagus, c. vesica, d. apex of gnathos
97. *Asovia maeoticaria* (Alphéraky). a. genitalia, b. aedeagus, c. vesica, d. cornuti, e. tympanal organ
98. *Colotois pennaria* (Linnaeus). a. genitalia, b. aedeagus, c. vesica, d. uncus
99. *Angerona prunaria* (Linnaeus). a. genitalia, b. aedeagus, c. vesica, d. 8th sternite

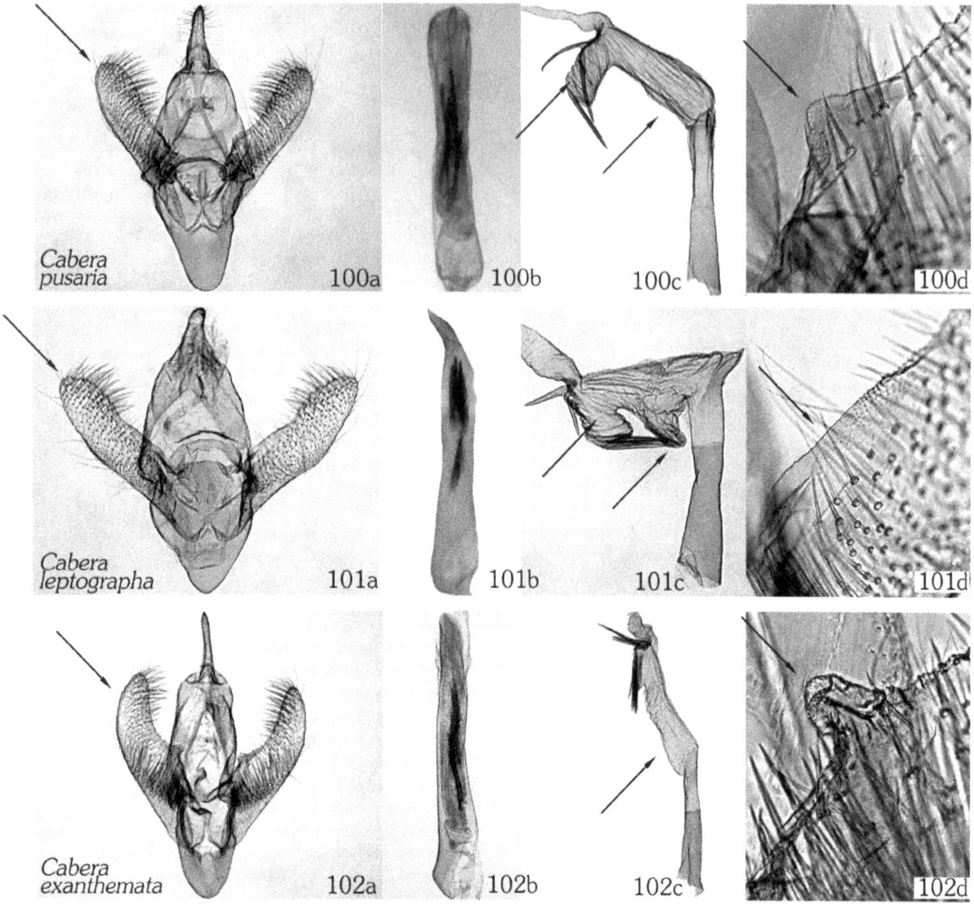

MALE GENITALIA (Figures are not in scale, specimen details are shown in Appendix 1.)

100. *Cabera pusaria* (LINNAEUS). a. genitalia, b. aedeagus, c. vesica, d. base of valva

101. *Cabera leptographa* WEHRLI. a. genitalia, b. aedeagus, c. vesica, d. base of valva

102. *Cabera exanthemata* (SCOPOLI). a. genitalia, b. aedeagus, c. vesica, d. base of valva

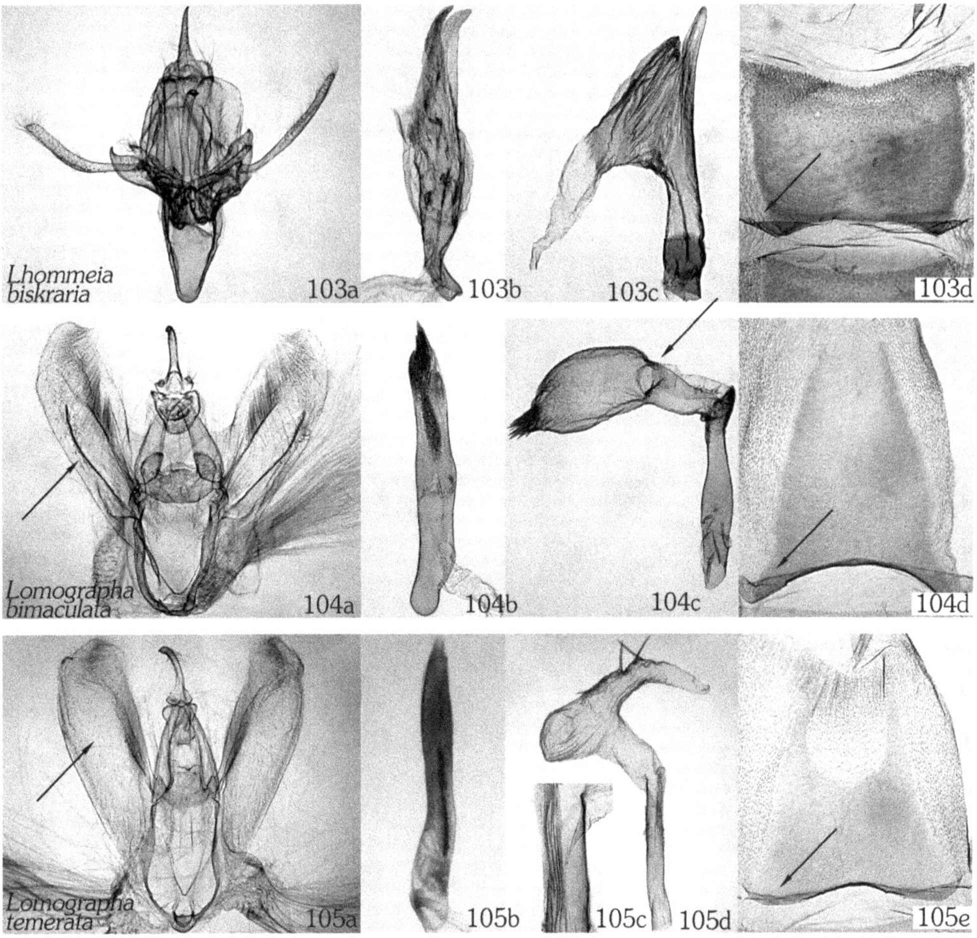

MALE GENITALIA (Figures are not in scale, specimen details are shown in Appendix 1.)

103. *Lhommeia biskraria* (OBERTHÜR). a. genitalia, b. aedeagus, c. vesica, d. 8th sternite
104. *Lomographa bimaculata* (FABRICIUS). a. genitalia, b. aedeagus, c. vesica, d. 8th sternite
105. *Lomographa temerata* (DENIS & SCHIFFERMÜLLER). a. genitalia, b. aedeagus, c. striations in aedeagus apex, d. vesica, e. 8th sternite

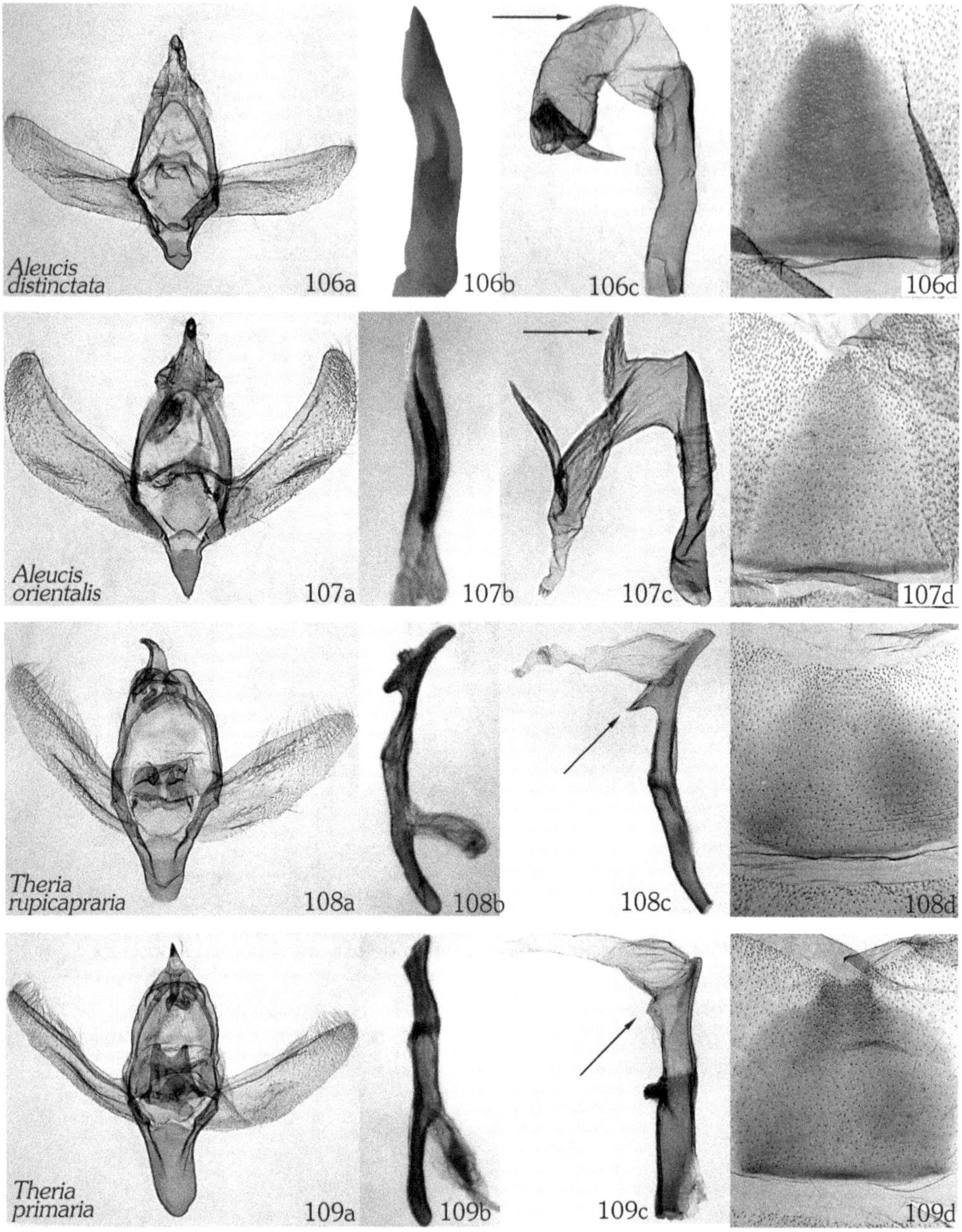

MALE GENITALIA (Figures are not in scale, specimen details are shown in Appendix 1.)

106. *Aleucis distinctata* (HERRICH-SCHÄFFER). a. genitalia, b. aedeagus, c. vesica, d. 8th sternite
107. *Aleucis orientalis* (STAUDINGER). a. genitalia, b. aedeagus, c. vesica, d. 8th sternite
108. *Theria rupicapraria* (DENIS & SCHIFFERMÜLLER). a. genitalia, b. aedeagus, c. vesica, d. 8th sternite
109. *Theria primaria* (HAWORTH). a. genitalia, b. aedeagus, c. vesica, d. 8th sternite

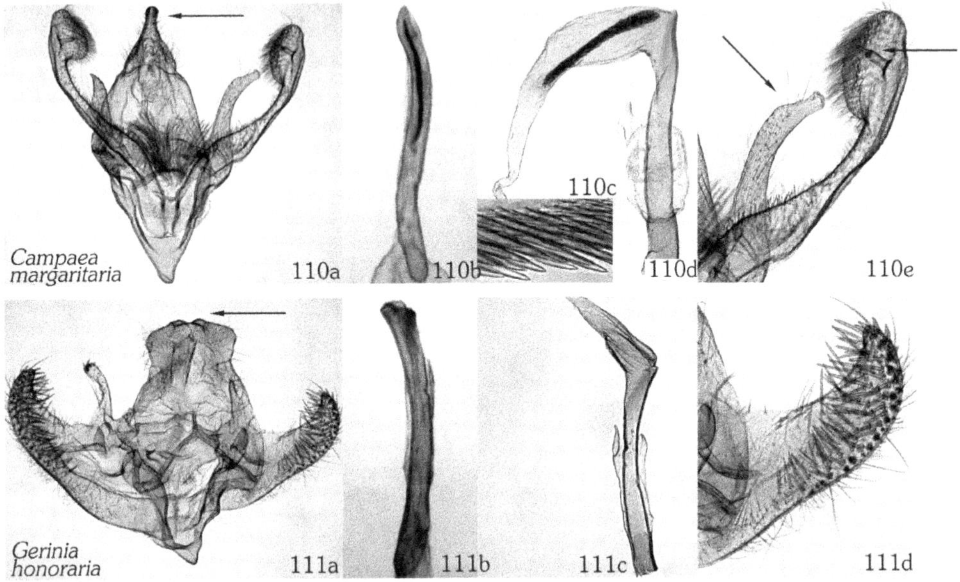

MALE GENITALIA (Figures are not in scale, specimen details are shown in Appendix 1.)

110. *Campaea margaritaria* (LINNAEUS). a. genitalia, b. aedeagus, c. cornuti, d. vesica, e. valva

111. *Gerinia honoraria* (DENIS & SCHIFFERMÜLLER). a. genitalia, b. aedeagus, c. vesica, d. valva

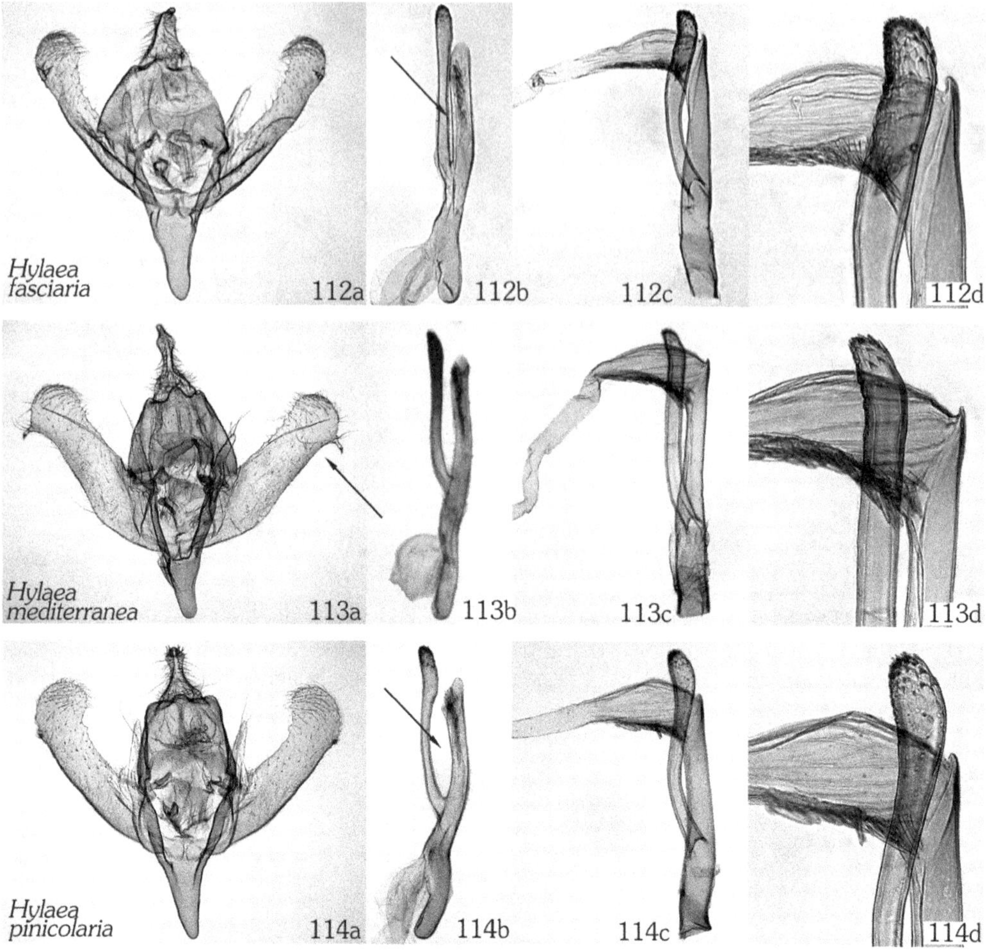

MALE GENITALIA (Figures are not in scale, specimen details are shown in Appendix 1.)
112. *Hylaea fasciaria* (LINNAEUS). a. genitalia, b. aedeagus, c. vesica, d. base of vesica
113. *Hylaea mediterranea* SIHVONEN *et al.* a. genitalia, b. aedeagus, c. vesica, d. base of vesica
114. *Hylaea pinicolaria* (BELLIER). a. genitalia, b. aedeagus, c. vesica, d. base of vesica

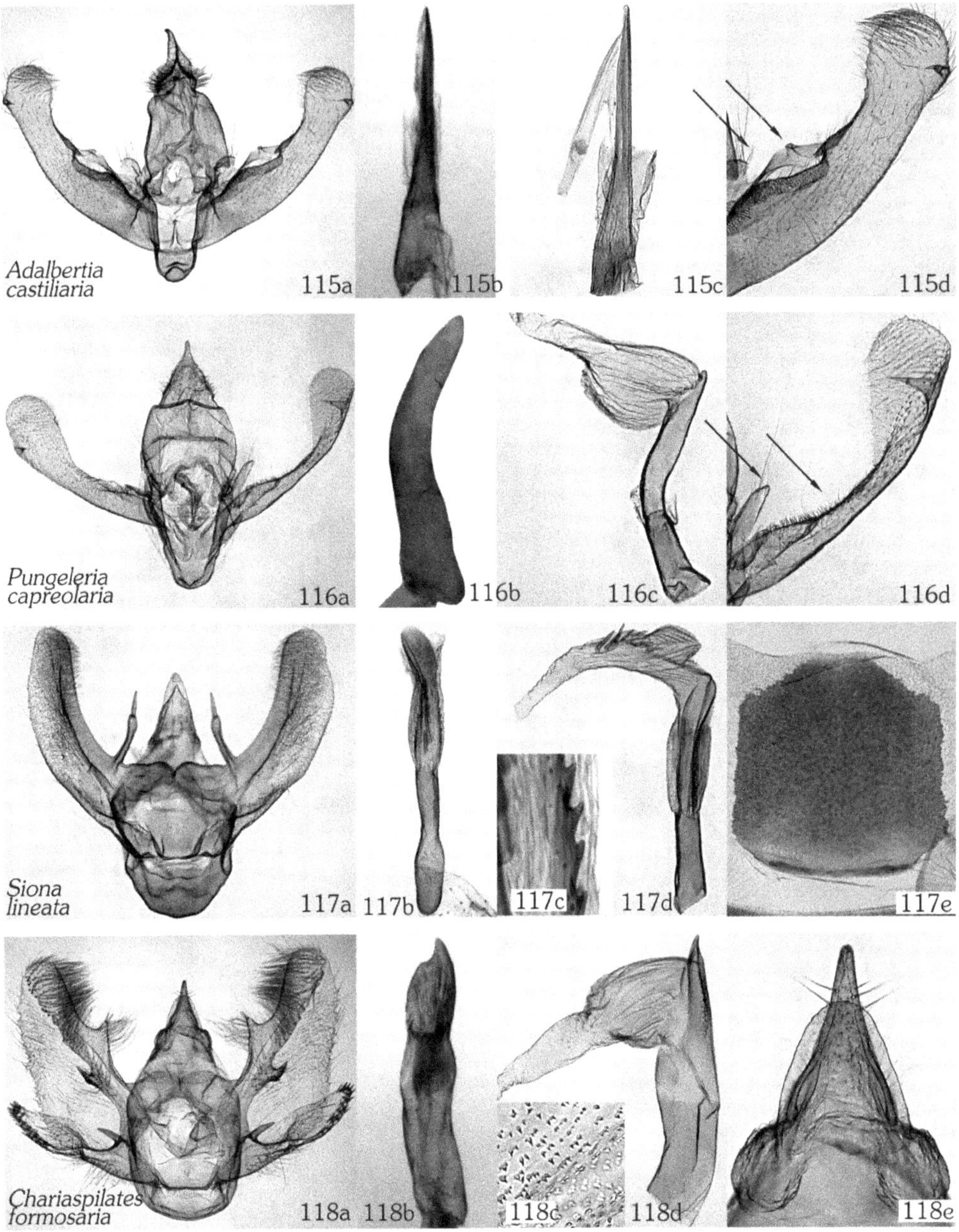

MALE GENITALIA (Figures are not in scale, specimen details are shown in Appendix 1.)

115. *Adalbertia castiliaria* (Staudinger). a. genitalia, b. aedeagus, c. vesica, d. valva
116. *Pungeleria capreolaria* (Denis & Schiffermüller). a. genitalia, b. aedeagus, c. vesica, d. valva
117. *Siona lineata* (Scopoli). a. genitalia, b. aedeagus, c. spines of aedeagus, d. vesica, e. 8th sternite
118. *Chariaspilates formosaria* (Eversmann). a. genitalia, b. aedeagus, c. microcornuti, d. vesica, e. uncus and gnathos

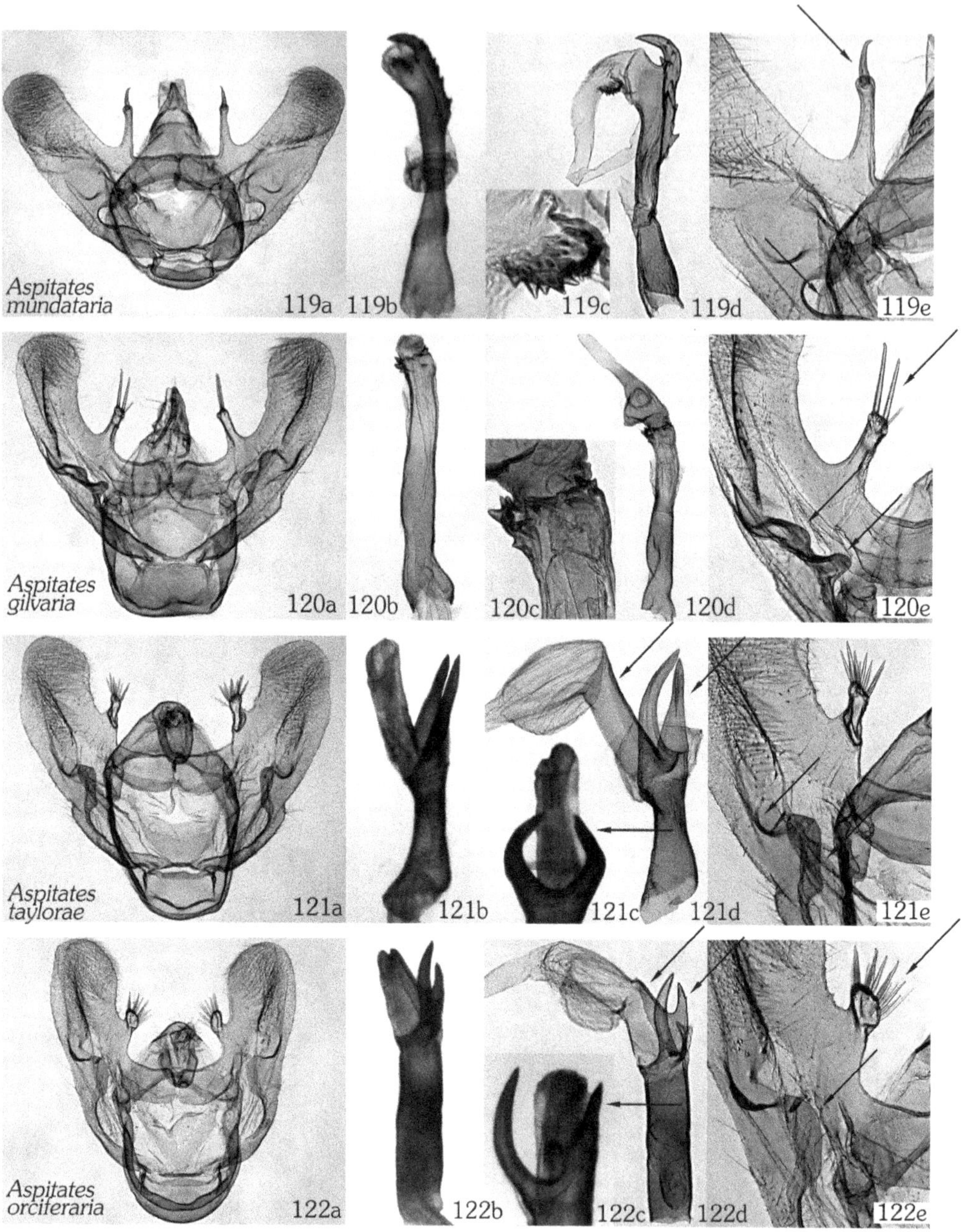

MALE GENITALIA (Figures are not in scale, specimen details are shown in Appendix 1.)

119. *Aspitates mundataria* (STOLL). a. genitalia, b. aedeagus, c. cornutus, d. vesica, e. base of valva

120. *Aspitates gilvaria* (DENIS & SCHIFFERMÜLLER). a. genitalia, b. aedeagus, c. apex of aedeagus, d. vesica, e. base of valva

121. *Aspitates taylorae* (BUTLER). a. genitalia, b. aedeagus, c. apex of aedeagus, ventral view, d. vesica, e. base of valva

122. *Aspitates orciferaria* (WALKER). a. genitalia, b. aedeagus, c. apex of aedeagus, ventral view, d. vesica, e. base of valva

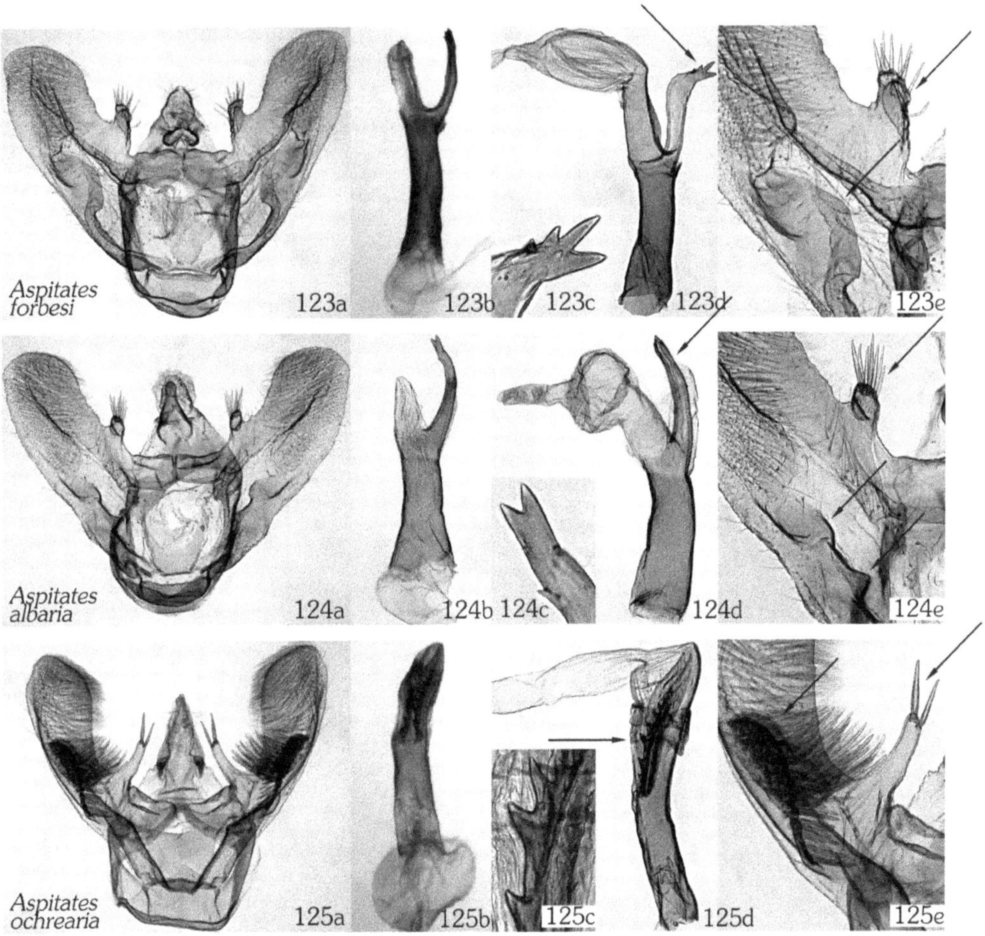

MALE GENITALIA (Figures are not in scale, specimen details are shown in Appendix 1.)

123. *Aspitates forbesi* (MUNROE). a. genitalia, b. aedeagus, c. apex of aedeagus arm, d. vesica, e. base of valva

124. *Aspitates albaria* (BARTEL). a. genitalia, b. aedeagus, c. apex of aedeagus arm, d. vesica, e. base of valva

125. *Aspitates ochrearia* (ROSSI). a. genitalia, b. aedeagus, c. spines of aedeagus, d. vesica, e. base of valva

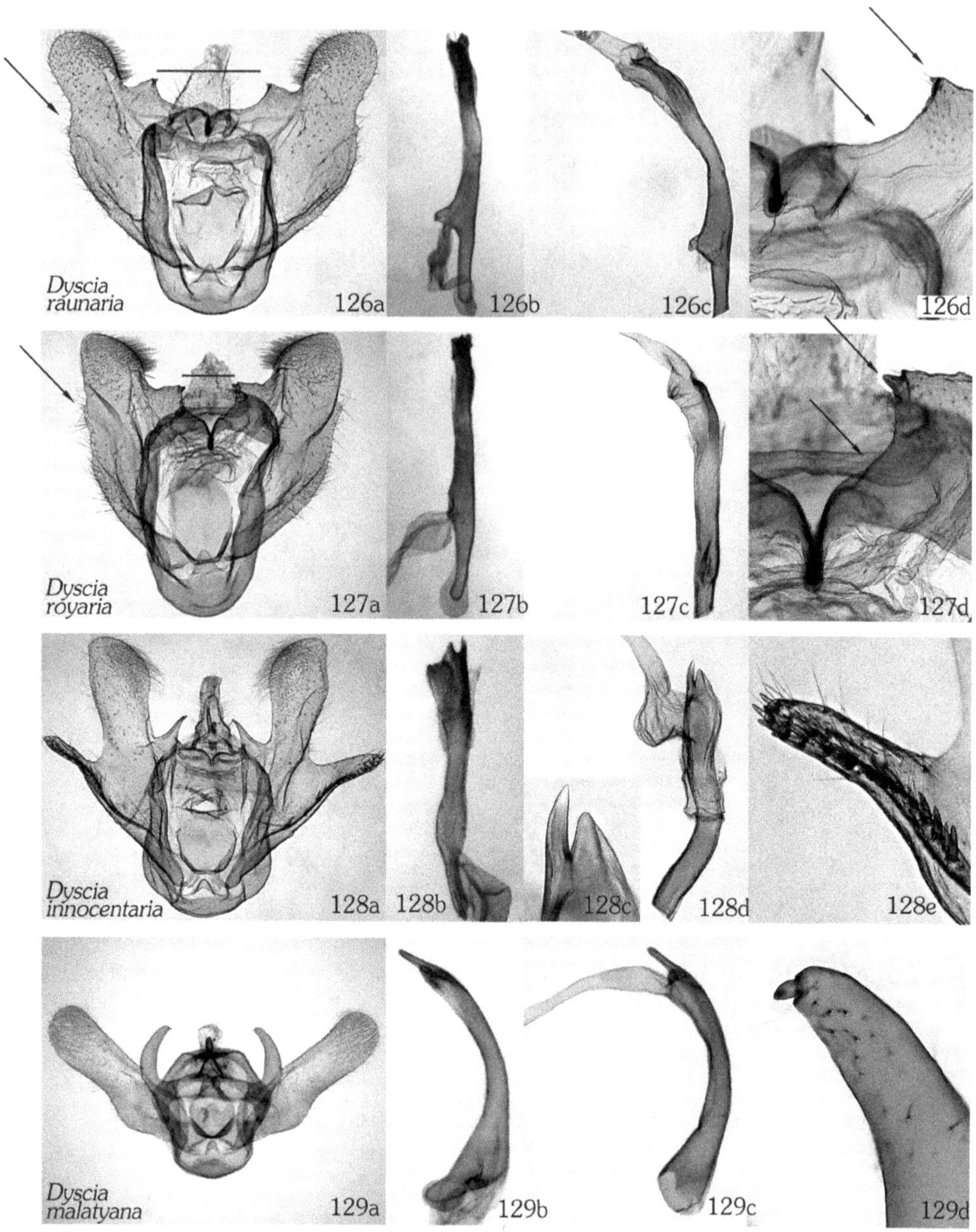

MALE GENITALIA (Figures are not in scale, specimen details are shown in Appendix 1.)

126. *Dyscia raunaria* (FREYER). a. genitalia, b. aedeagus, c. vesica, d. base of valva

127. *Dyscia royaria* TAUTEL & BILLI. a. genitalia, b. aedeagus, c. vesica, d. base of valva

128. *Dyscia innocentaria* (CHRISTOPH). a. genitalia, b. aedeagus, c. apex of aedeagus, d. vesica, e. apex of valva

129. *Dyscia malatyana* WEHRLI. a. genitalia, b. aedeagus, c. vesica, d. extension at base of valva

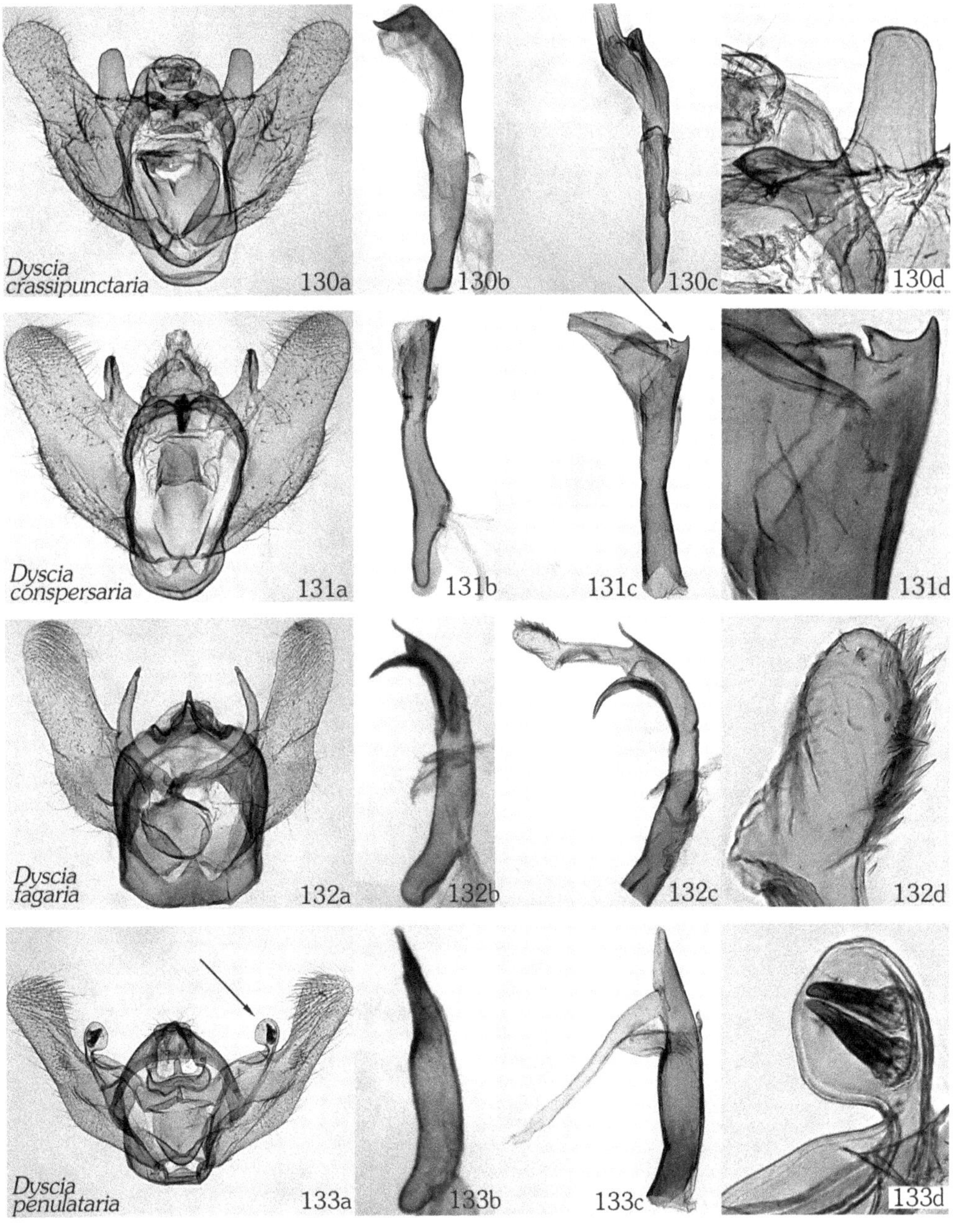

MALE GENITALIA (Figures are not in scale, specimen details are shown in Appendix 1.)

130. *Dyscia crassipunctaria* (REBEL). a. genitalia, b. aedeagus, c. vesica, d. extension at base of valva
131. *Dyscia conspersaria* (DENIS & SCHIFFERMÜLLER). a. genitalia, b. aedeagus, c. vesica, d. apex of aedeagus
132. *Dyscia fagaria* (THUNBERG). a. genitalia, b. aedeagus, c. vesica, d. sclerotisations of vesica
133. *Dyscia penulataria* (HÜBNER). a. genitalia, b. aedeagus, c. vesica, d. extension at base of valva

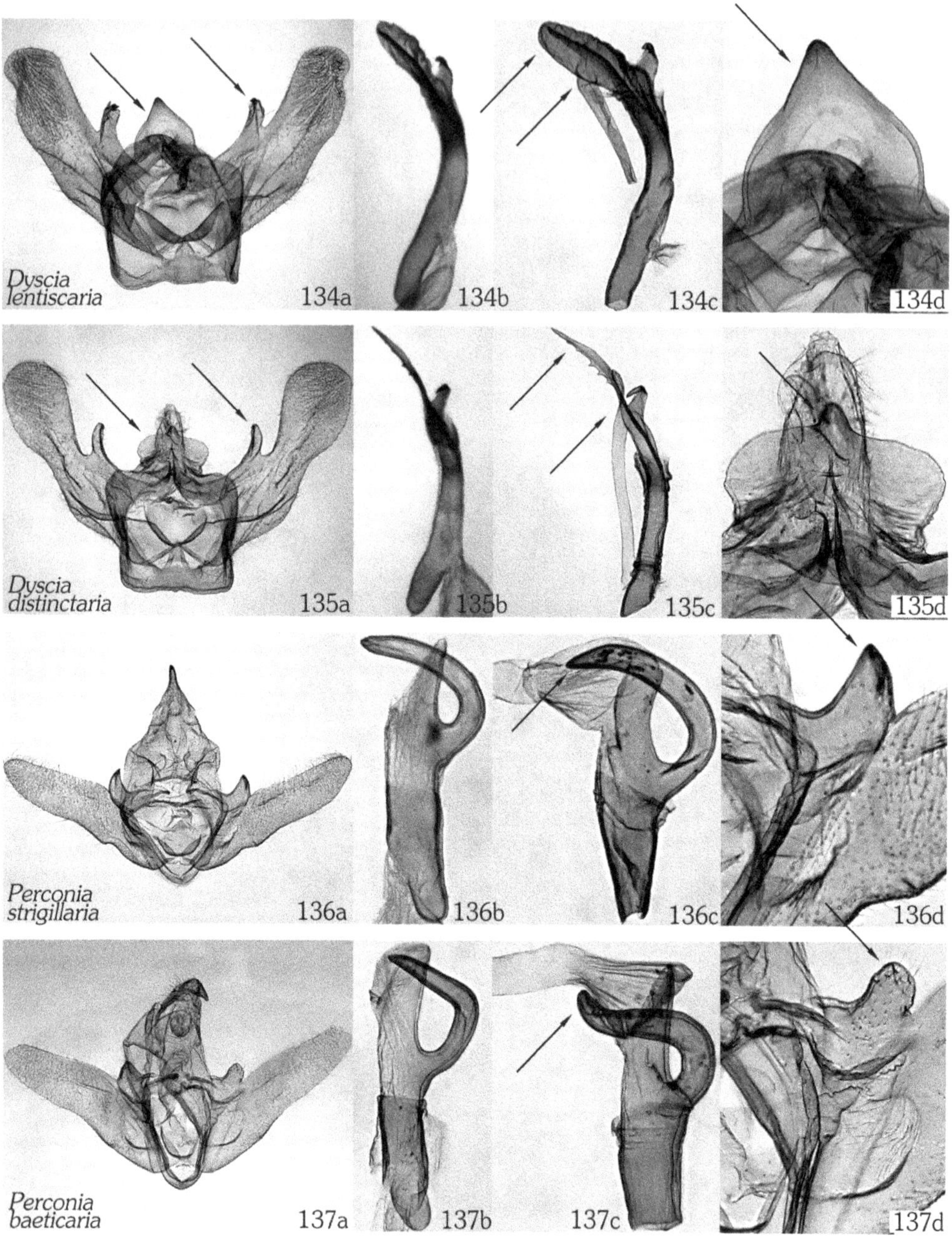

MALE GENITALIA (Figures are not in scale, specimen details are shown in Appendix 1.)
134. *Dyscia lentiscaria* (DONZEL). a. genitalia, b. aedeagus, c. vesica, d. gnathos
135. *Dyscia distinctaria* (BANG-HAAS). a. genitalia, b. aedeagus, c. vesica, d. gnathos
136. *Perconia strigillaria* (HÜBNER). a. genitalia, b. aedeagus, c. vesica, d. base of valva
137. *Perconia baeticaria* (STAUDINGER). a. genitalia, b. aedeagus, c. vesica, d. base of valva

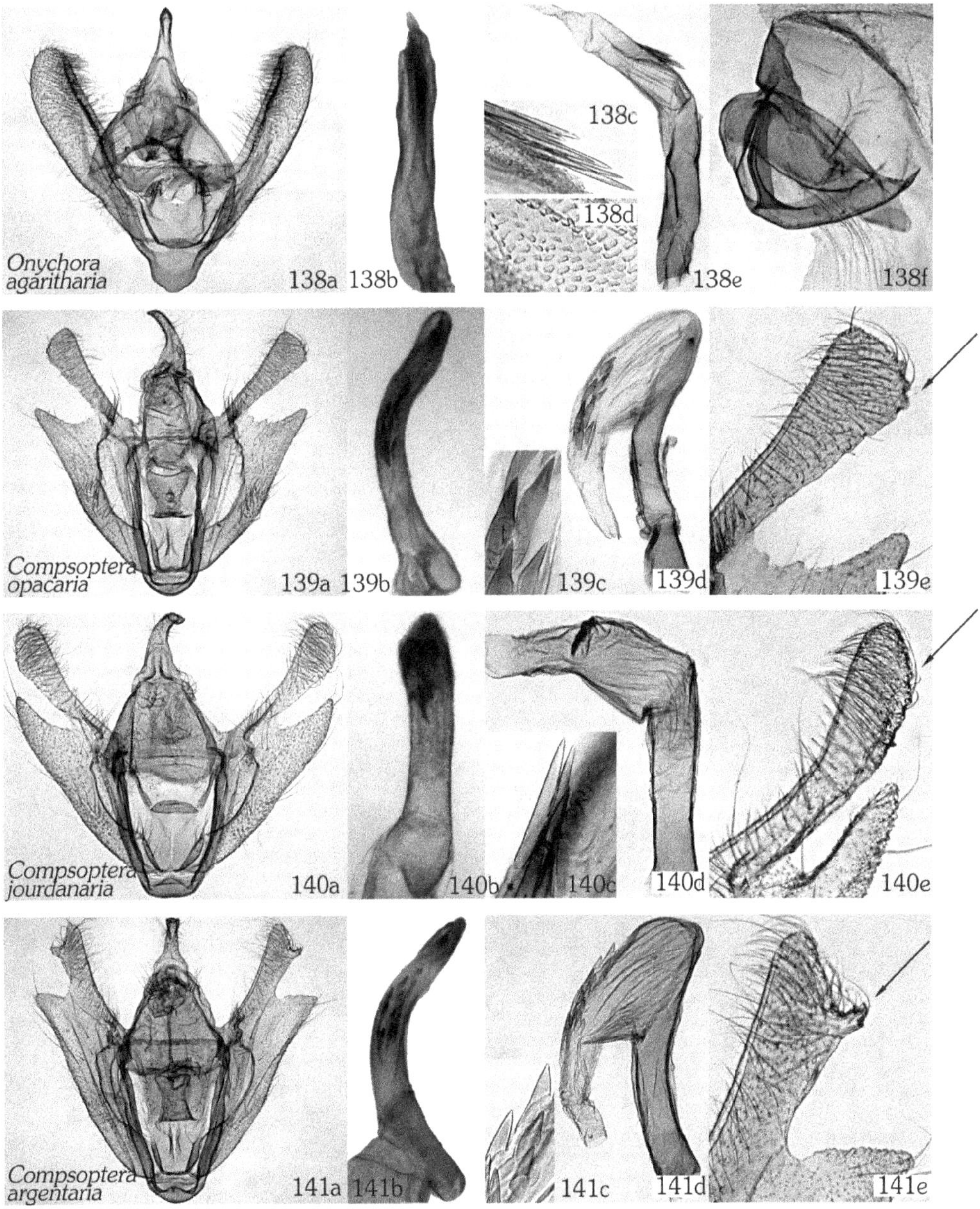

MALE GENITALIA (Figures are not in scale, specimen details are shown in Appendix 1.)

138. *Onychora agaritharia* (DARDOIN). a. genitalia, b. aedeagus, c. cornuti, d. granulate surface of vesica, e. vesica, f. tympanal organ
139. *Compsoptera opacaria* HÜBNER. a. genitalia, b. aedeagus, c. cornuti, d. vesica, e. dorsal arm of valva
140. *Compsoptera jourdanaria* (SERRES). a. genitalia, b. aedeagus, c. cornuti, d. vesica, e. dorsal arm of valva
141. *Compsoptera argentaria* (HERRICH-SCHÄFFER). a. genitalia, b. aedeagus, c. cornuti, d. vesica, e. dorsal arm of valva

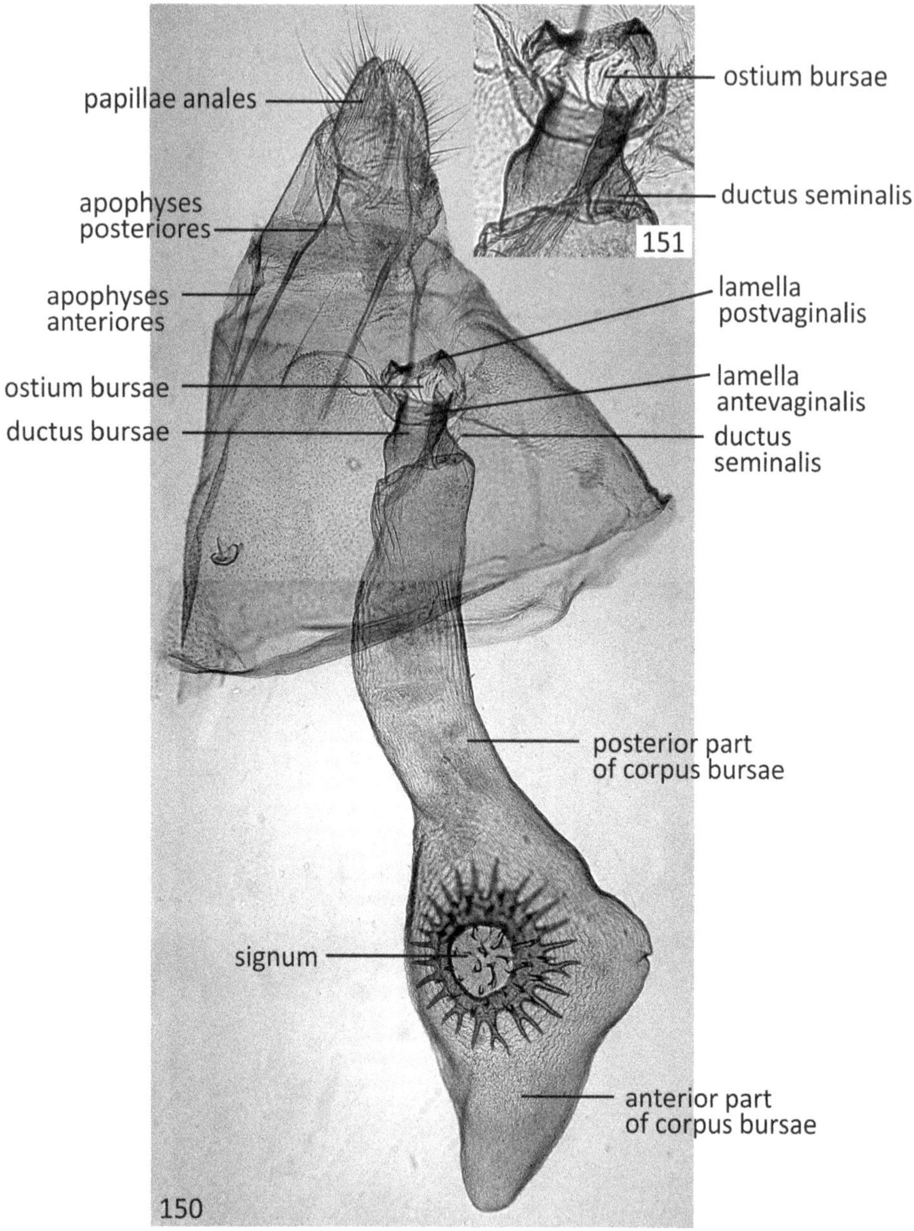

Text-figs 150–151. Terminology of the female genitalia, *Macaria alternata* (DENIS & SCHIFFERMÜLLER, 1775), Finland, Kirkkonummi, 5–19.viii.1998, slide PS1138 (coll. P. SIHVONEN). Text-fig. 150. Genitalia. Text-fig. 151. Region around the ostium bursae.

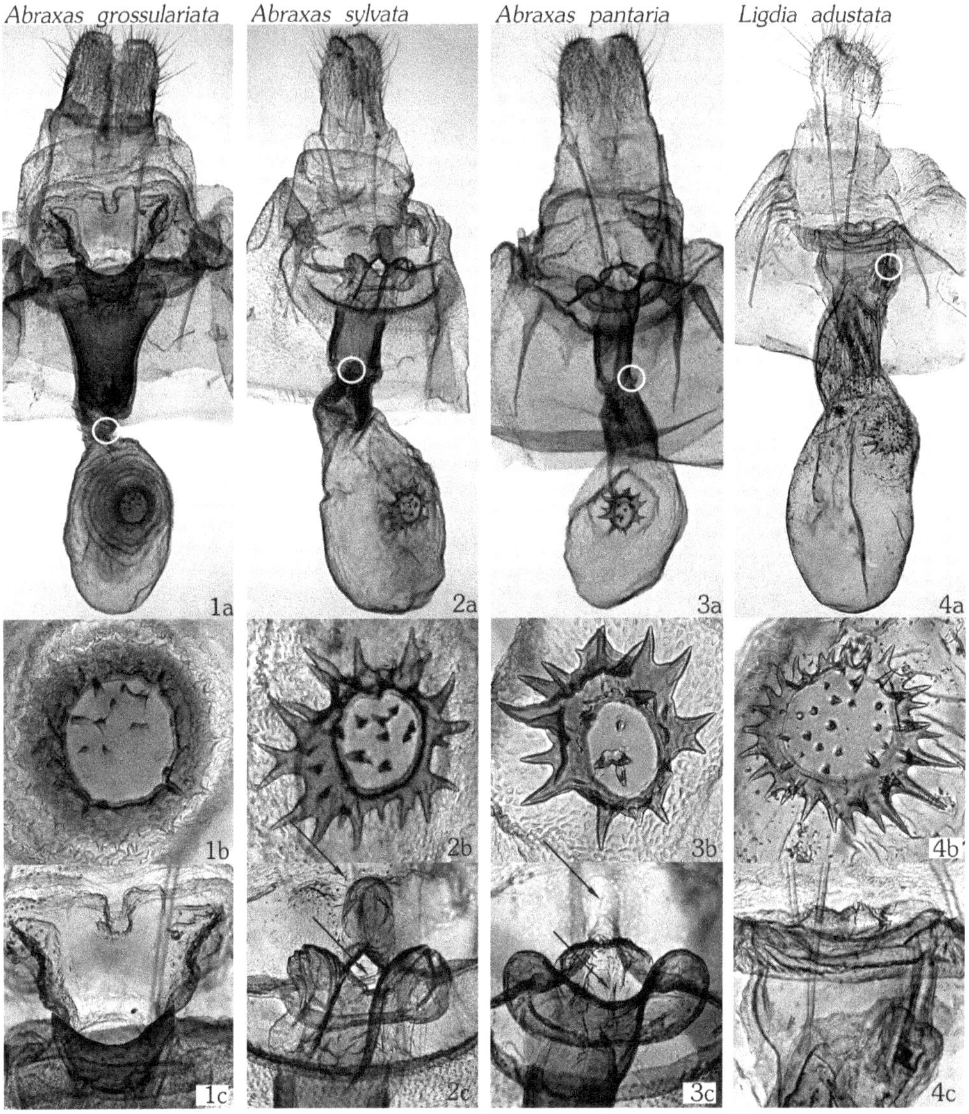

FEMALE GENITALIA (Figures are not in scale, specimen details are shown in Appendix 2.)

1. *Abraxas grossulariata* (LINNAEUS). a. genitalia, b. signum, c. ostium bursae and adjacent structures. Circle indicates the point of origin of the ductus seminalis.
2. *Abraxas sylvata* (SCOPOLI). a. genitalia, b. signum, c. ostium bursae and adjacent structures.
3. *Abraxas pantaria* (LINNAEUS). a. genitalia, b. signum, c. ostium bursae and adjacent structures.
4. *Ligdia adustata* (DENIS & SCHIFFERMÜLLER). a. genitalia, b. signum, c. ostium bursae and adjacent structures

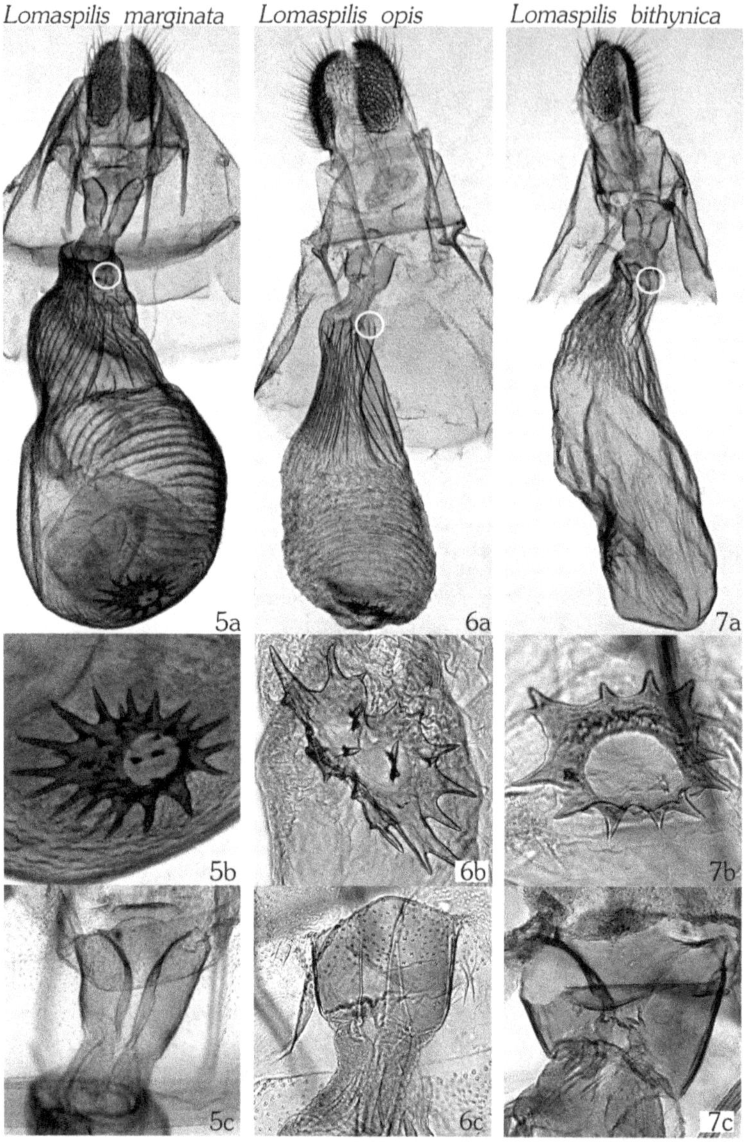

FEMALE GENITALIA (Figures are not in scale, specimen details are shown in Appendix 2.)

5. *Lomaspilis marginata* (LINNAEUS). a. genitalia, b. signum, c. ostium bursae and adjacent structures
6. *Lomaspilis opis* BUTLER. a. genitalia, b. signum, c. ostium bursae and adjacent structures
7. *Lomaspilis bithynica* WEHRLI. a. genitalia, b. signum, c. ostium bursae and adjacent structures

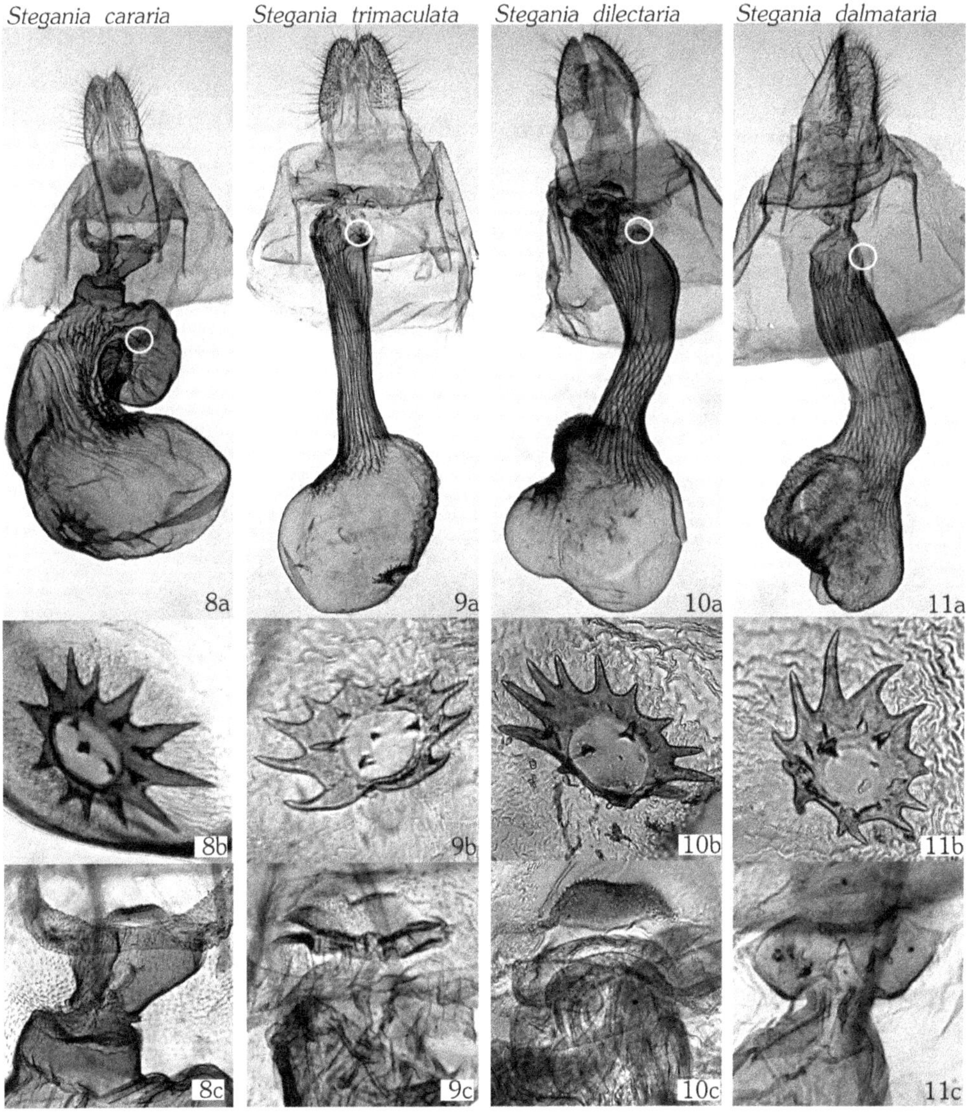

FEMALE GENITALIA (Figures are not in scale, specimen details are shown in Appendix 2.)

8. *Stegania cararia* (HÜBNER). a. genitalia, b. signum, c. ostium bursae and adjacent structures
9. *Stegania trimaculata* (VILLERS). a. genitalia, b. signum, c. ostium bursae and adjacent structures
10. *Stegania dilectaria* (HÜBNER). a. genitalia, b. signum, c. ostium bursae and adjacent structures
11. *Stegania dalmataria* GUENÉE. a. genitalia, b. signum, c. ostium bursae and adjacent structures

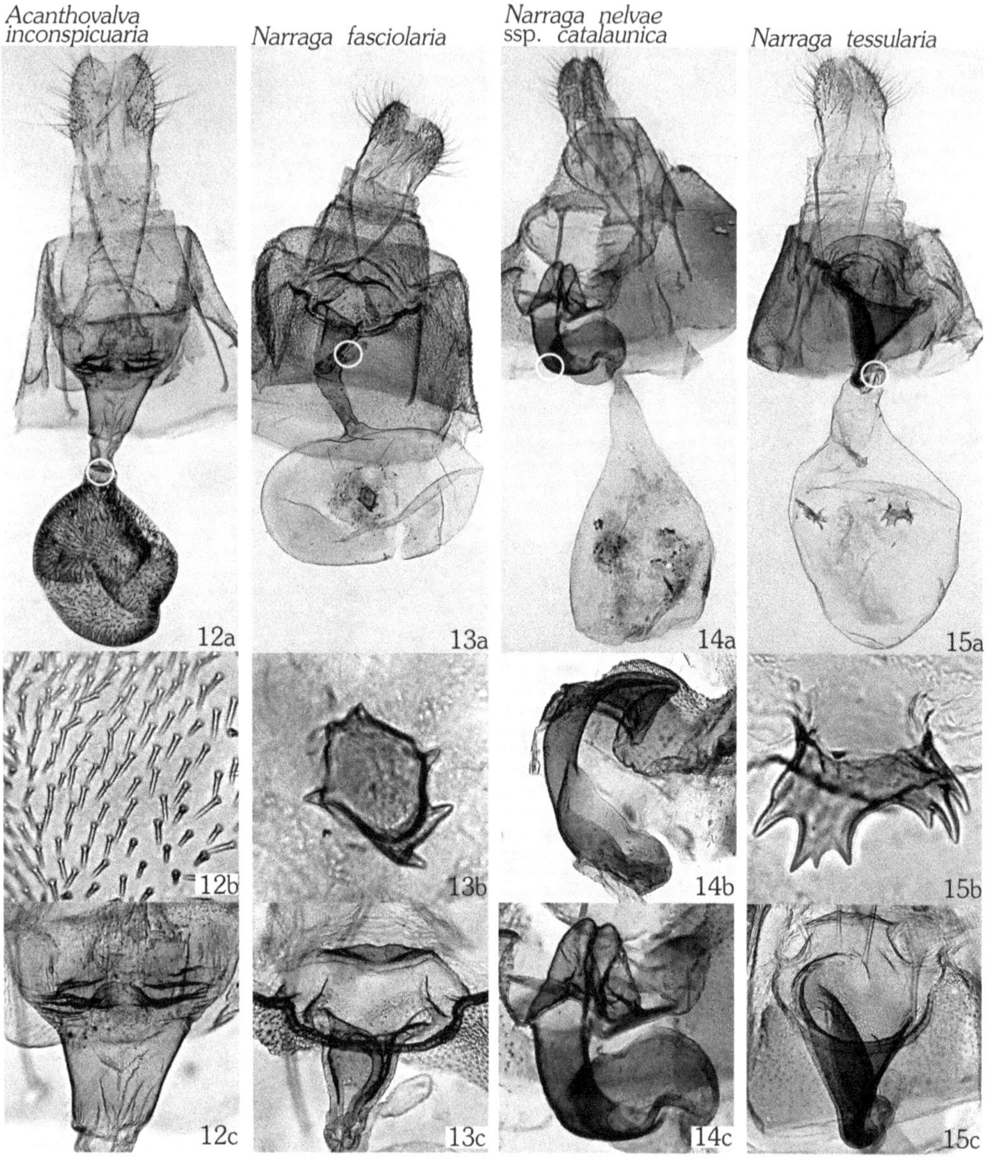

FEMALE GENITALIA (Figures are not in scale, specimen details are shown in Appendix 2.)

12. *Acanthovalva inconspicuaria* (HÜBNER). a. genitalia, b. spines of corpus bursae, c. ostium bursae and adjacent structures
13. *Narraga fasciolaria* (HUFNAGEL). a. genitalia, b. signum, c. ostium bursae and adjacent structures
14. *Narraga nelvae* subsp. *catalaunica* HERBULOT. a. genitalia, b. ductus bursae (lateral view), c. ostium bursae and adjacent structures
15. *Narraga tessularia* (METZNER). a. genitalia, b. signum, c. ostium bursae and adjacent structures

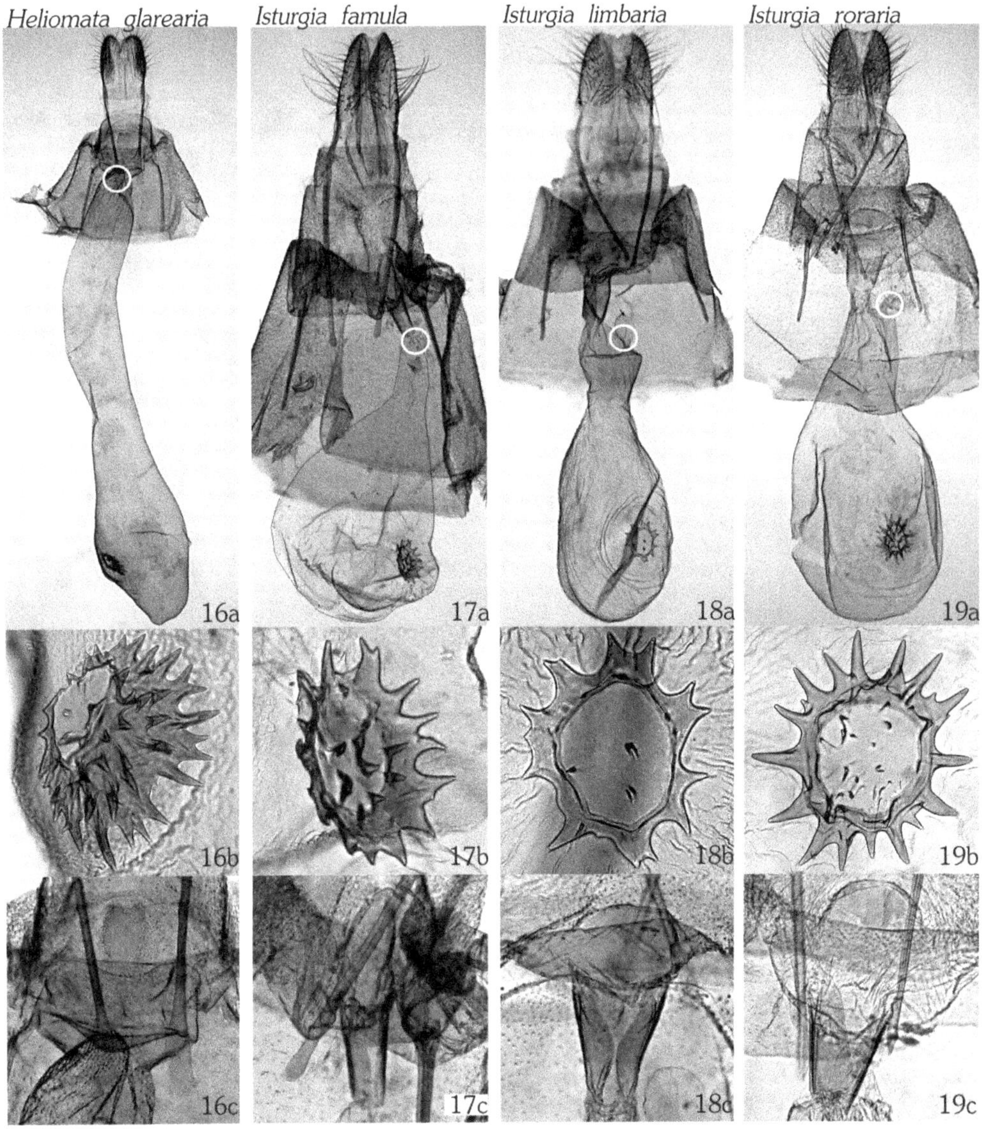

FEMALE GENITALIA (Figures are not in scale, specimen details are shown in Appendix 2.)

16. *Heliomata glarearia* (DENIS & SCHIFFERMÜLLER). a. genitalia, b. signum, c. ostium bursae and adjacent structures
17. *Isturgia famula* (ESPER). a. genitalia, b. signum, c. ostium bursae and adjacent structures
18. *Isturgia limbaria* (FABRICIUS). a. genitalia, b. signum, c. ostium bursae and adjacent structures
19. *Isturgia roraria* (FABRICIUS). a. genitalia, b. signum, c. ostium bursae and adjacent structures

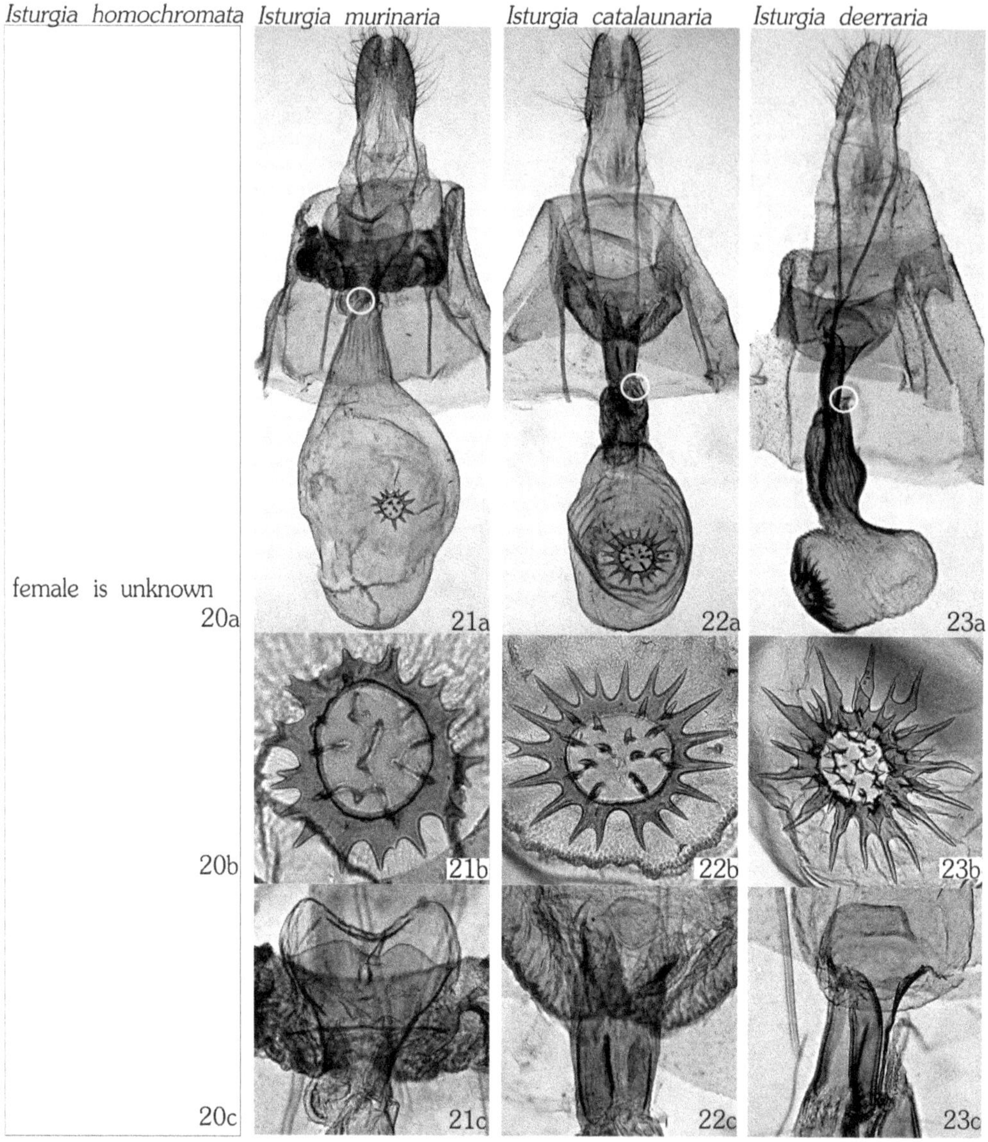

FEMALE GENITALIA (Figures are not in scale, specimen details are shown in Appendix 2.)

20. *Isturgia homochromata* (MABILLE). Not illustrated, female is unknown.
21. *Isturgia murinaria* (DENIS & SCHIFFERMÜLLER). a. genitalia, b. signum, c. ostium bursae and adjacent structures
22. *Isturgia catalaunaria* (GUENÉE). a. genitalia, b. signum, c. ostium bursae and adjacent structures
23. *Isturgia deerraria* (WALKER). a. genitalia, b. signum, c. ostium bursae and adjacent structures

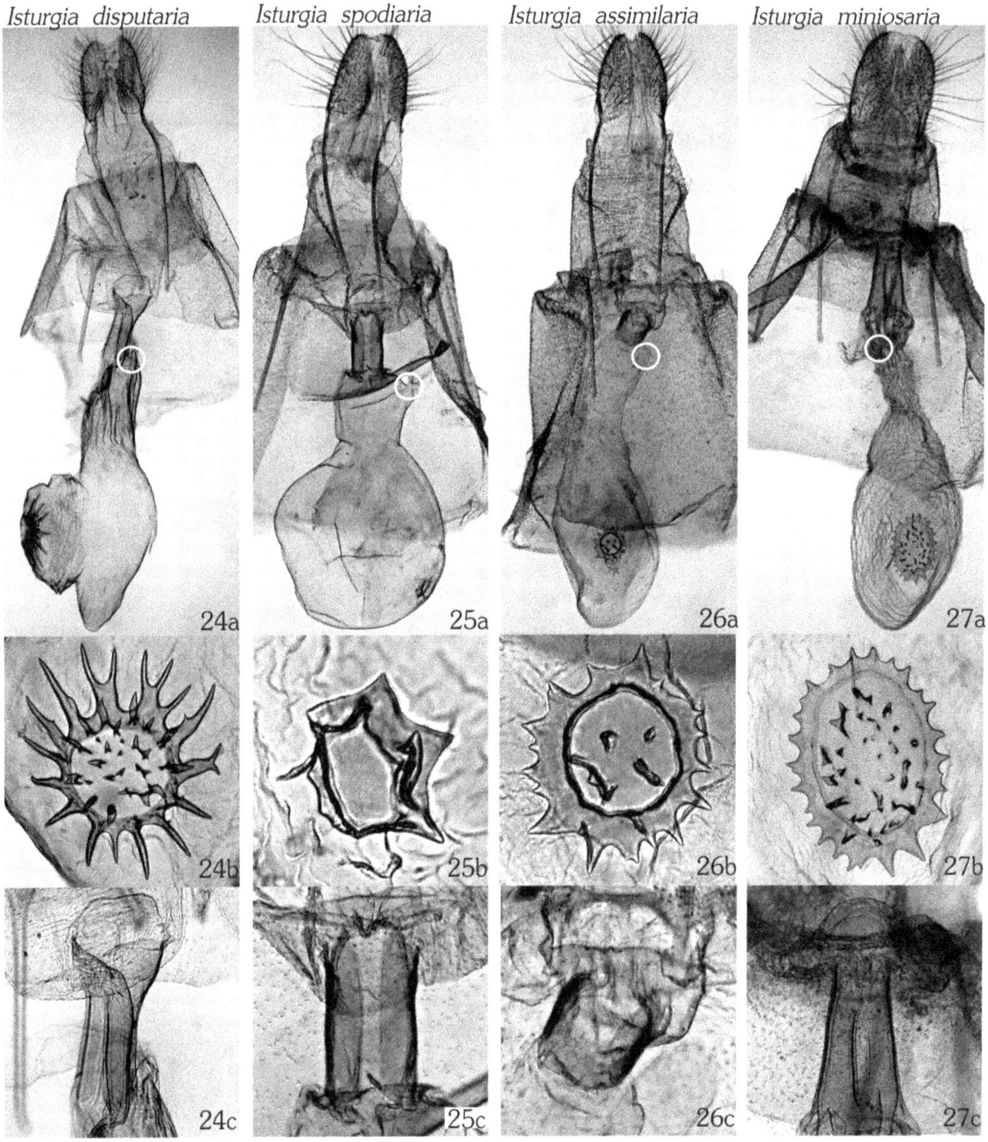

FEMALE GENITALIA (Figures are not in scale, specimen details are shown in Appendix 2.)

24. *Isturgia disputaria* (GUENÉE). a. genitalia, b. signum, c. ostium bursae and adjacent structures
25. *Isturgia spodiaria* (LEFÈBVRE), *see* also Text-figs 53–58. a. genitalia, b. signum, c. ostium bursae and adjacent structures
26. *Isturgia assimilaria* (RAMBUR). a. genitalia, b. signum, c. ostium bursae and adjacent structures
27. *Isturgia miniosaria* (DUPONCHEL). a. genitalia, b. signum, c. ostium bursae and adjacent structures

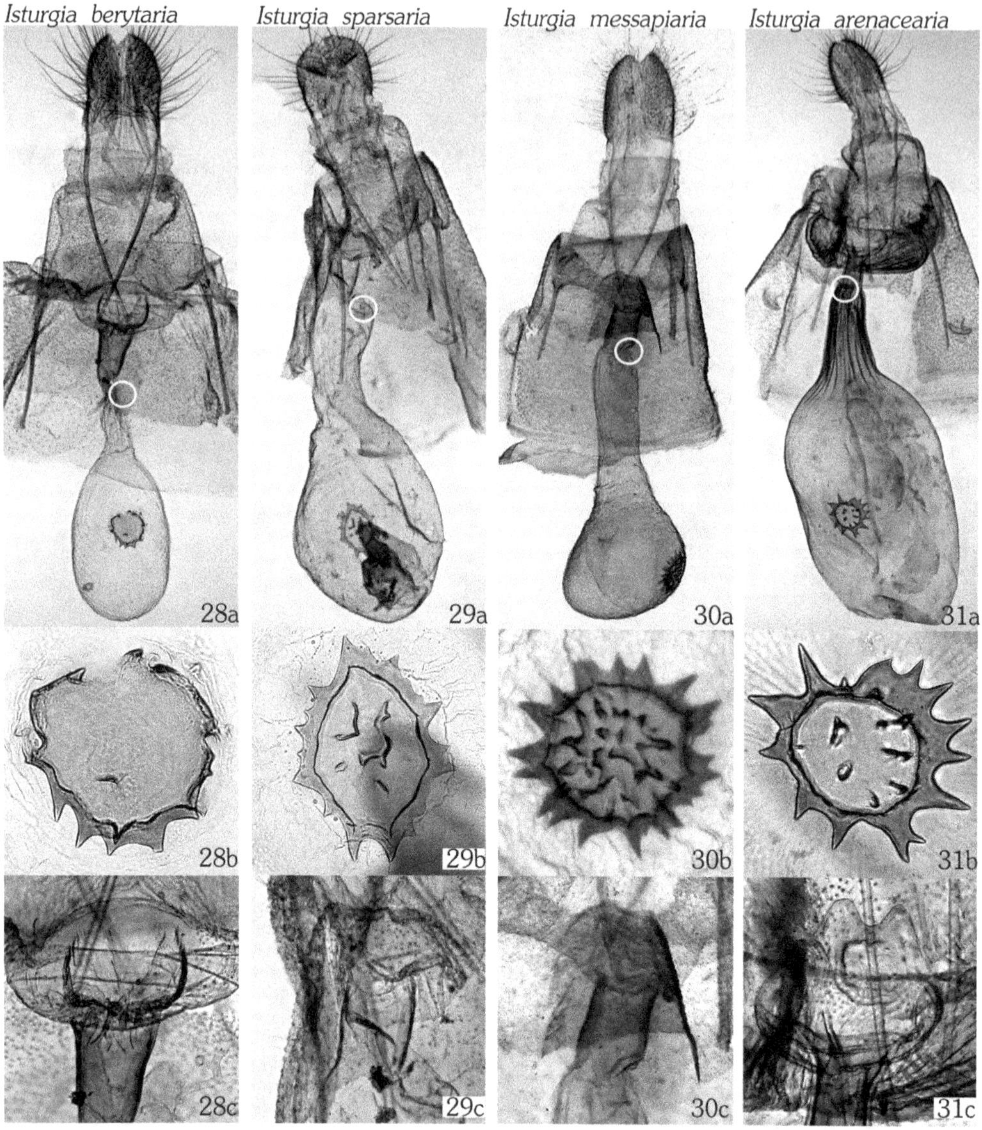

FEMALE GENITALIA (Figures are not in scale, specimen details are shown in Appendix 2.)

28. *Isturgia berytaria* (STAUDINGER). a. genitalia, b. signum, c. ostium bursae and adjacent structures
29. *Isturgia sparsaria* (HÜBNER). a. genitalia, b. signum, c. ostium bursae and adjacent structures
30. *Isturgia messapiaria* (SOHN-RETHEL). a. genitalia, b. signum, c. ostium bursae and adjacent structures
31. *Isturgia arenacearia* (DENIS & SCHIFFERMÜLLER). a. genitalia, b. signum, c. ostium bursae and adjacent structures

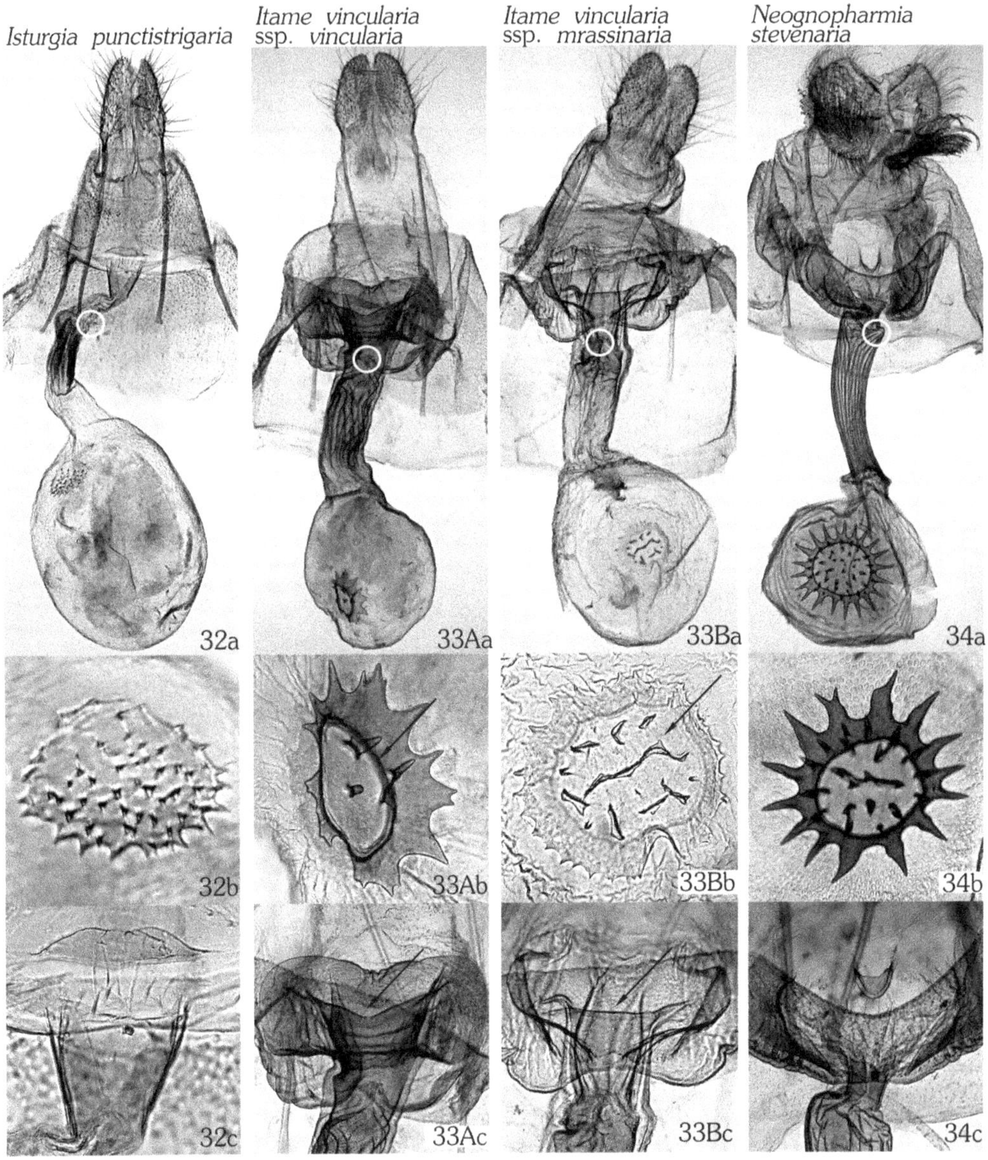

FEMALE GENITALIA (Figures are not in scale, specimen details are shown in Appendix 2.)

32. *Isturgia punctistrigaria* (CHRISTOPH). a. genitalia, b. signum, c. ostium bursae and adjacent structures

33A. *Itame vincularia* subsp. *vincularia* HÜBNER. a. genitalia, b. signum, c. ostium bursae and adjacent structures

33B. *Itame vincularia* subsp. *mrassinaria* (OBERTHÜR). a. genitalia, b. signum, c. ostium bursae and adjacent structures

34. *Neognopharmia stevenaria* (BOISDUVAL). a. genitalia, b. signum, c. ostium bursae and adjacent structures

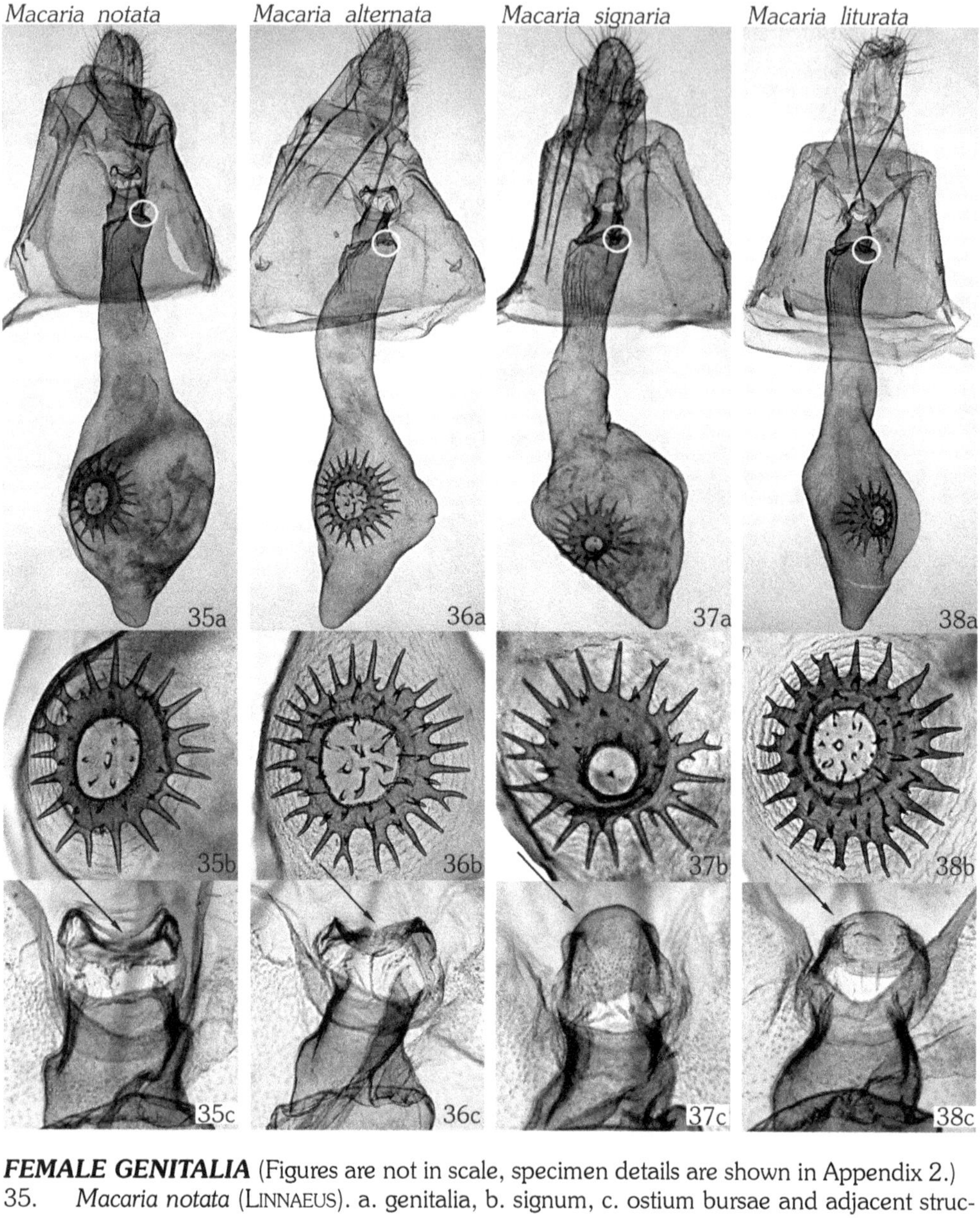

FEMALE GENITALIA (Figures are not in scale, specimen details are shown in Appendix 2.)

35. *Macaria notata* (LINNAEUS). a. genitalia, b. signum, c. ostium bursae and adjacent structures

36. *Macaria alternata* (DENIS & SCHIFFERMÜLLER). a. genitalia, b. signum, c. ostium bursae and adjacent structures

37. *Macaria signaria* (HÜBNER). a. genitalia, b. signum, c. ostium bursae and adjacent structures

38. *Macaria liturata* (CLERCK). a. genitalia, b. signum, c. ostium bursae and adjacent structures

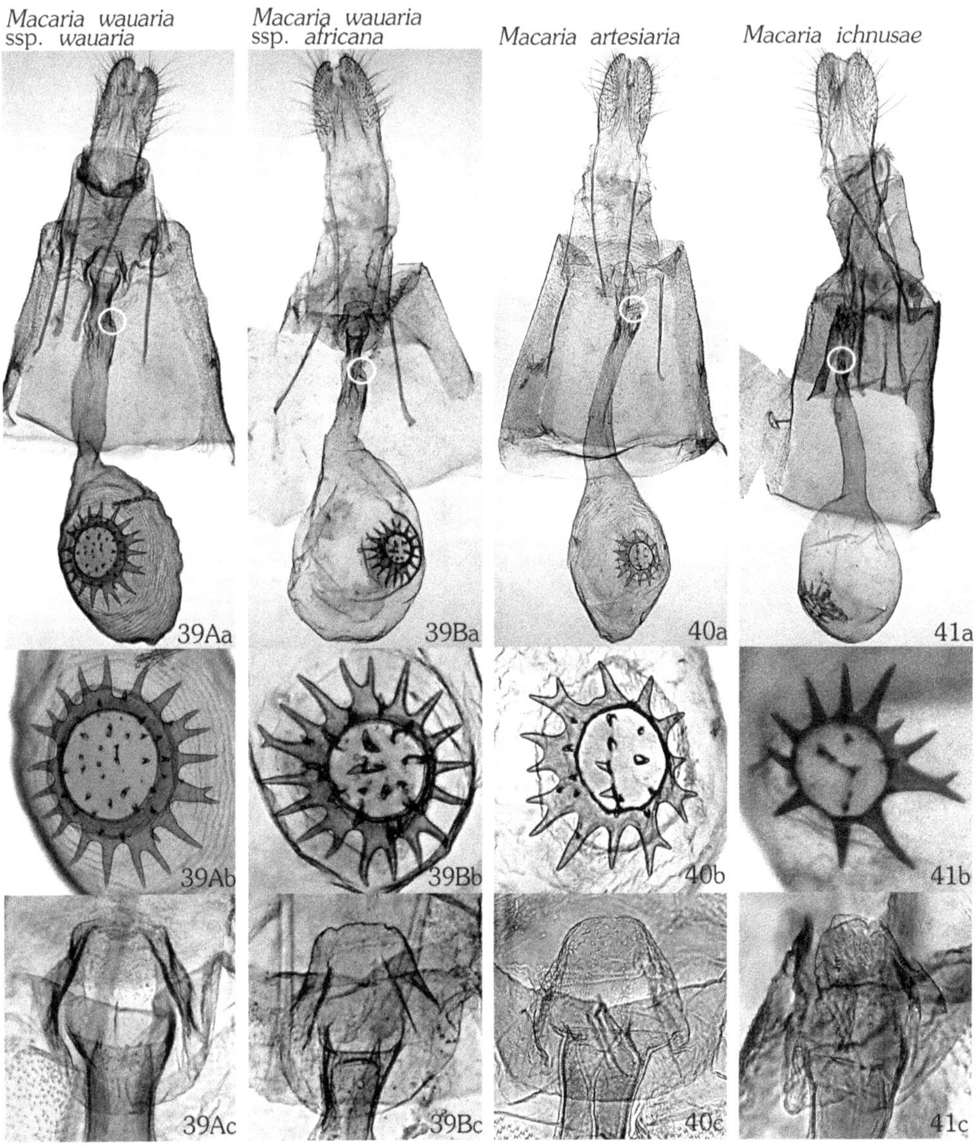

FEMALE GENITALIA (Figures are not in scale, specimen details are shown in Appendix 2.)

39A. *Macaria wauaria* subsp. *wauaria* (LINNAEUS). a. genitalia, b. signum, c. ostium bursae and adjacent structures

39B. *Macaria wauaria* subsp. *africana* (ZERNY). a. genitalia, b. signum, c. ostium bursae and adjacent structures

40. *Macaria artesiaria* (DENIS & SCHIFFERMÜLLER). a. genitalia, b. signum, c. ostium bursae and adjacent structures

41. *Macaria ichnusae* GOVI & FIUMI. a. genitalia, b. signum, c. ostium bursae and adjacent structures

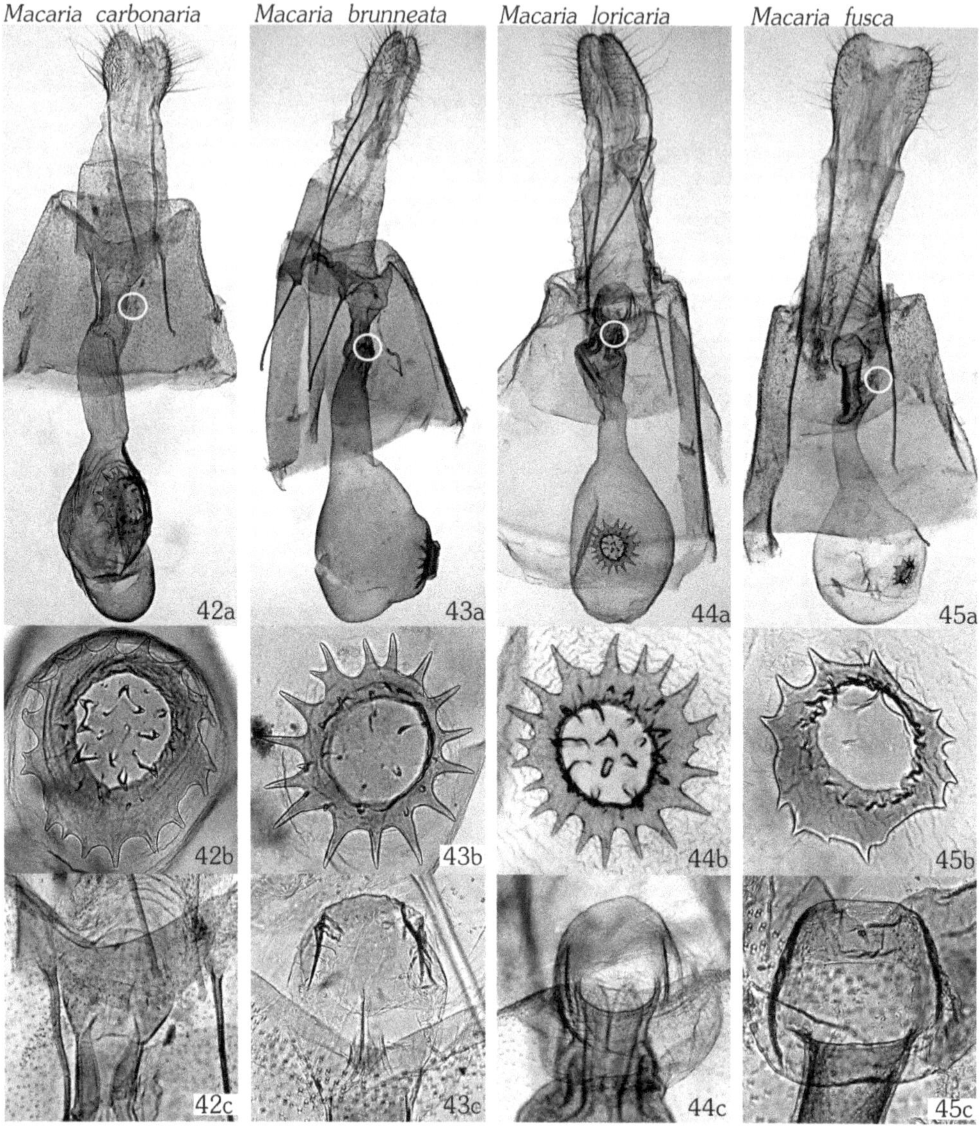

FEMALE GENITALIA (Figures are not in scale, specimen details are shown in Appendix 2.)

42. *Macaria carbonaria* (CLERCK). a. genitalia, b. signum, c. ostium bursae and adjacent structures
43. *Macaria brunneata* (THUNBERG). a. genitalia, b. signum, c. ostium bursae and adjacent structures
44. *Macaria loricaria* (EVERSMANN). a. genitalia, b. signum, c. ostium bursae and adjacent structures
45. *Macaria fusca* (THUNBERG). a. genitalia, b. signum, c. ostium bursae and adjacent structures

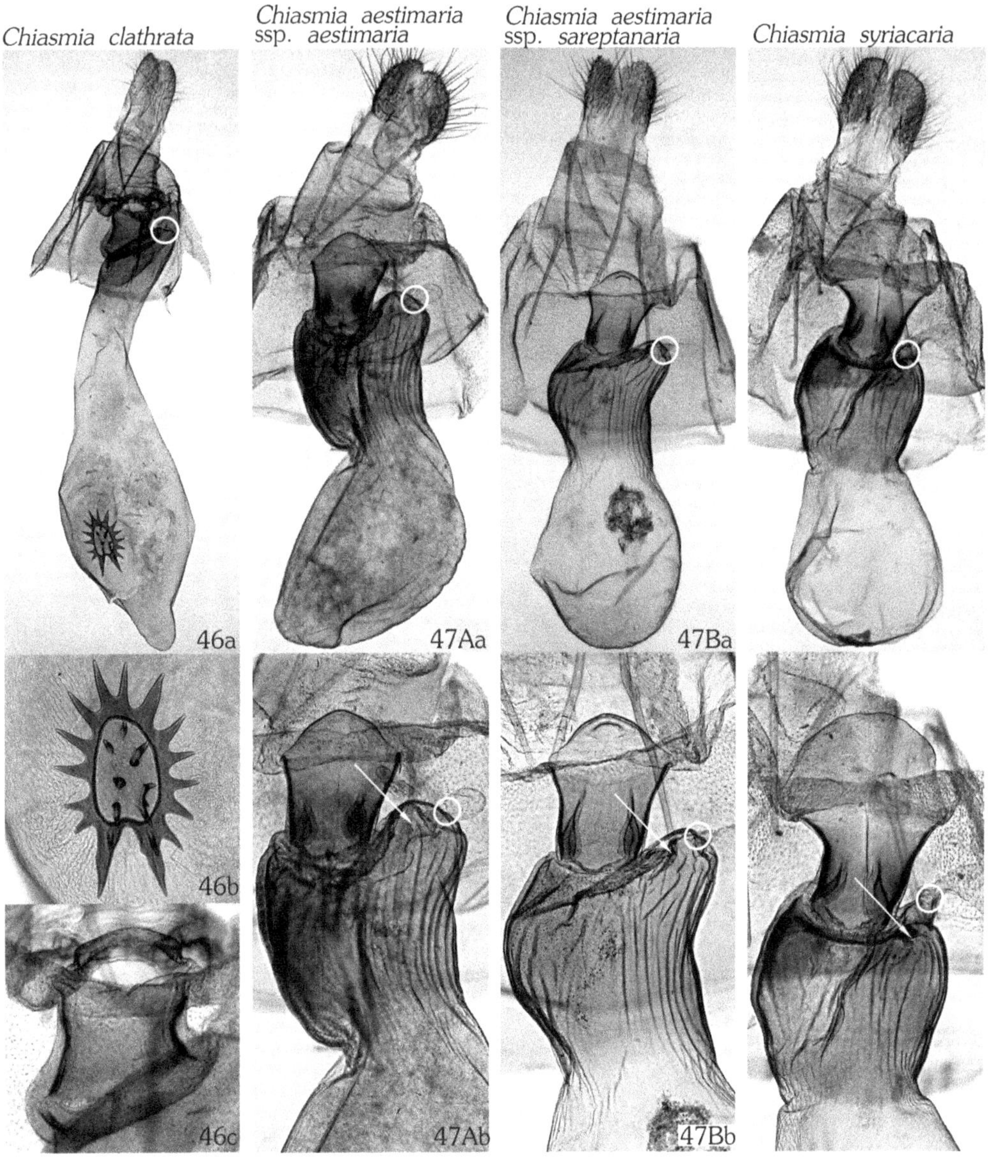

FEMALE GENITALIA (Figures are not in scale, specimen details are shown in Appendix 2.)

46. *Chiasmia clathrata* (LINNAEUS). a. genitalia, b. signum, c. ostium bursae and adjacent structures

47A. *Chiasmia aestimaria* subsp. *aestimaria* (HÜBNER). a. genitalia (signum is absent), b. ostium bursae and adjacent structures and posterior part of corpus bursae

47B. *Chiasmia aestimaria* subsp. *sareptanaria* (STAUDINGER). a. genitalia (signum is absent), b. ostium bursae and adjacent structures and posterior part of corpus bursae.

Chiasmia syriacaria (STAUDINGER) Non-European, see text for details. a. genitalia (signum is absent), b. ostium bursae and adjacent structures and posterior part of corpus bursae

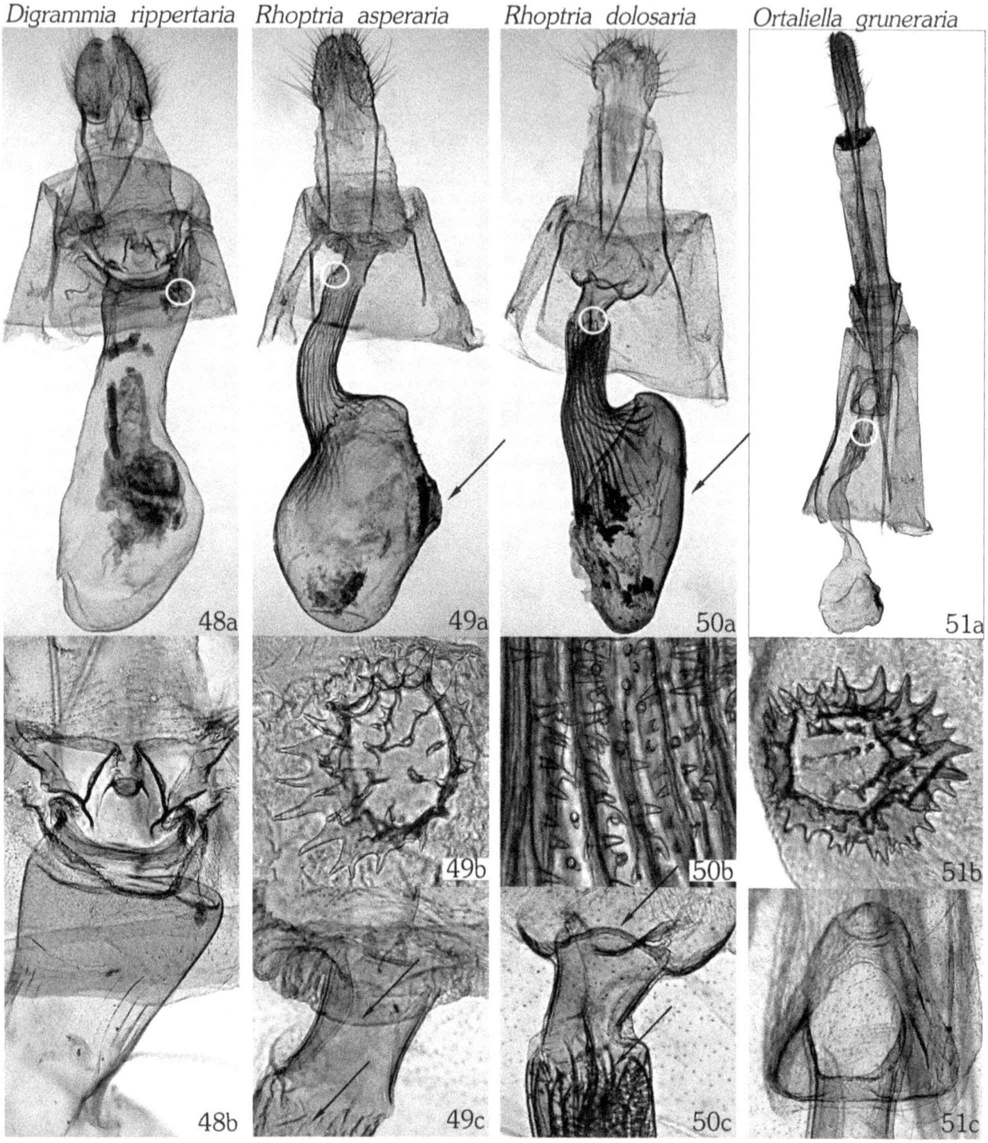

FEMALE GENITALIA (Figures are not in scale, specimen details are shown in Appendix 2.)

48. *Digrammia rippertaria* (DUPONCHEL). a. genitalia (signum is absent), b. ostium bursae and adjacent structures
49. *Rhoptria asperaria* (HÜBNER). a. genitalia, b. signum, c. ostium bursae and adjacent structures
50. *Rhoptria dolosaria* (HERRICH-SCHÄFFER). a. genitalia, b. spines in posterior part of corpus bursae, c. ostium bursae and adjacent structures
51. *Ortaliella gruneraria* (STAUDINGER). a. genitalia, b. signum, c. ostium bursae and adjacent structures

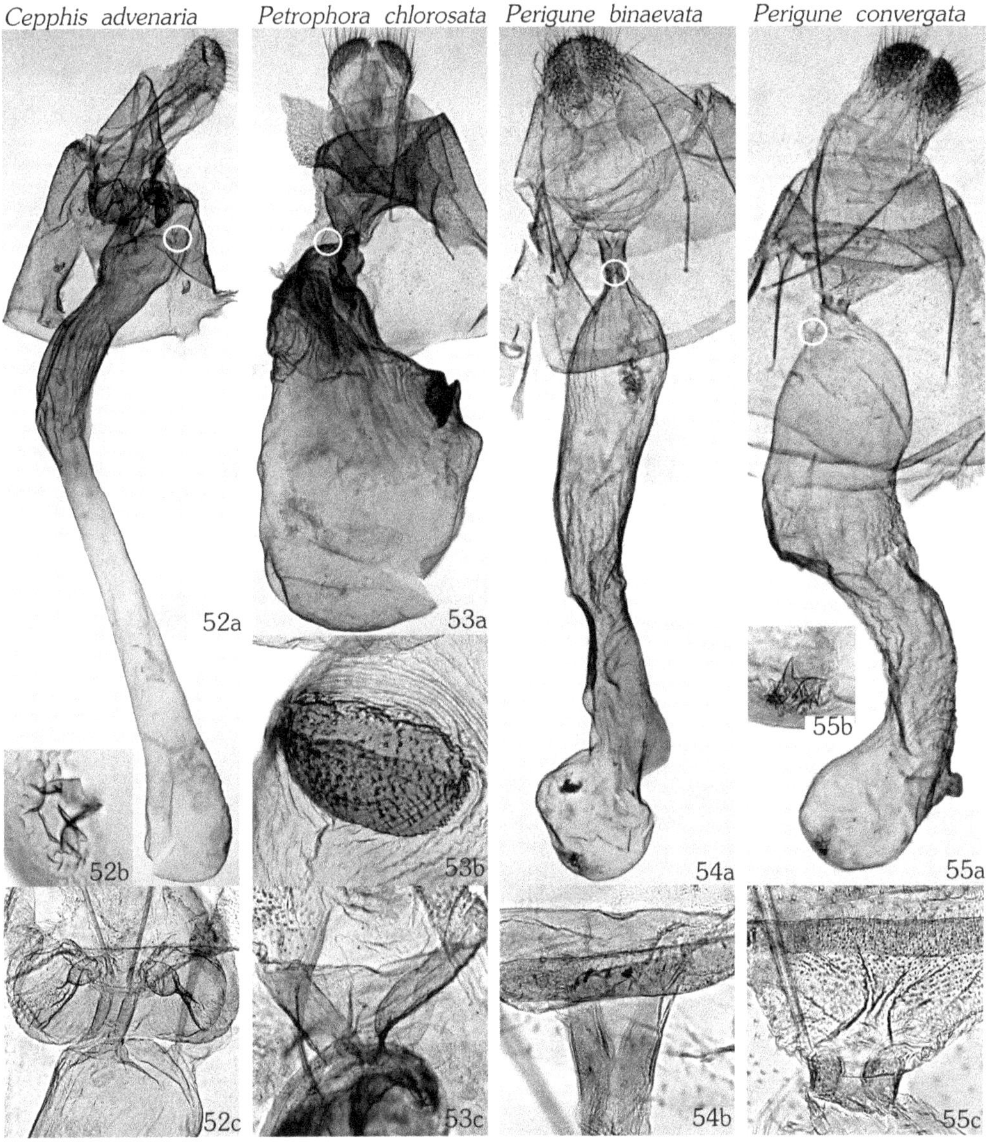

FEMALE GENITALIA (Figures are not in scale, specimen details are shown in Appendix 2.)

52. *Cepphis advenaria* (HÜBNER). a. genitalia, b. signum, c. ostium bursae and adjacent structures
53. *Petrophora chlorosata* (SCOPOLI). a. genitalia, b. signum, c. ostium bursae and adjacent structures
54. *Perigune binaevata* (MABILLE). a. genitalia (signum is often absent or if present, minute), b. ostium bursae and adjacent structures
55. *Perigune convergata* (VILLERS). a. genitalia, b. signum (often signum is absent), c. ostium bursae and adjacent structures

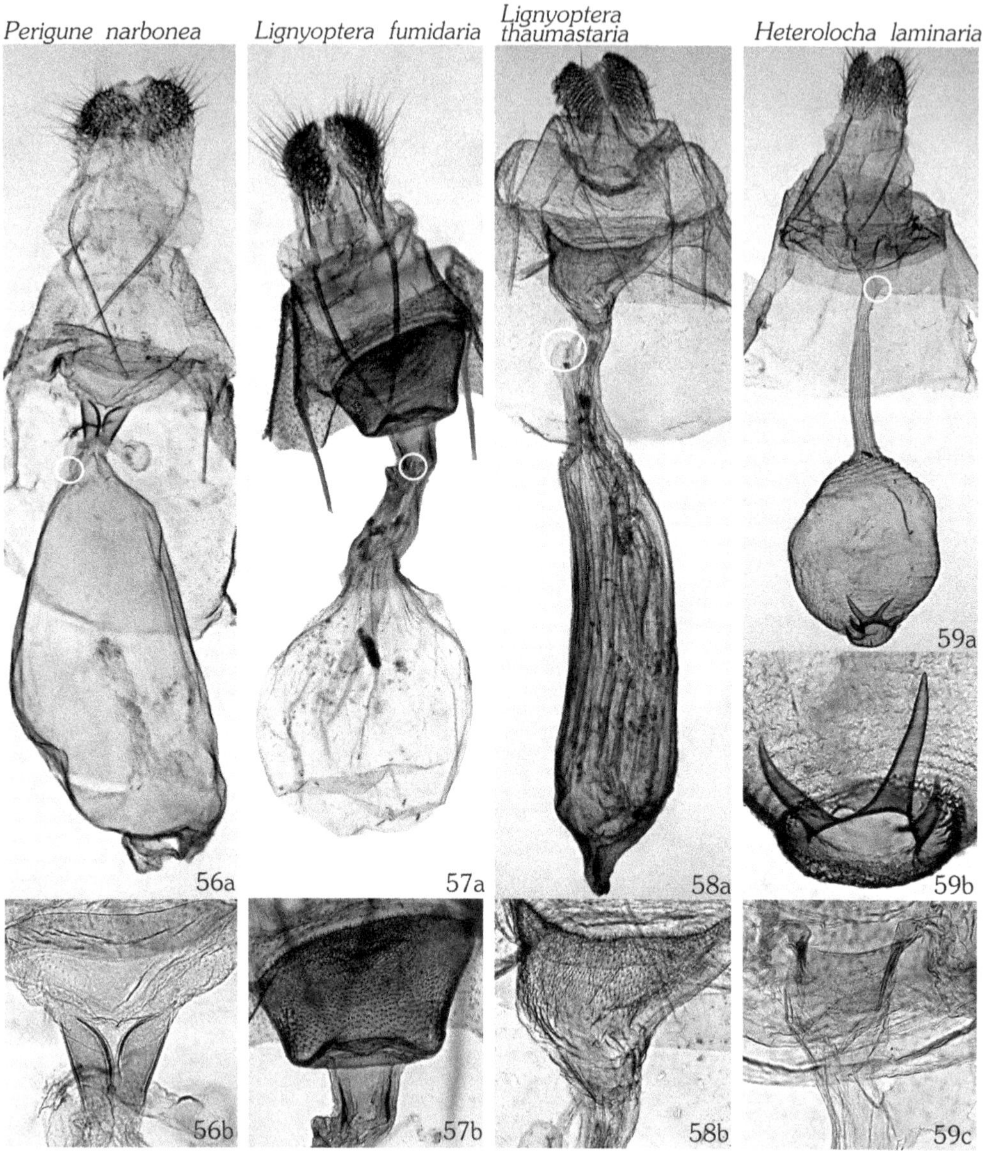

FEMALE GENITALIA (Figures are not in scale, specimen details are shown in Appendix 2.)

56. *Perigune narbonea* (LINNAEUS). a. genitalia (signum is often absent and if present, minute), b. ostium bursae and adjacent structures
57. *Lignyoptera fumidaria* (HÜBNER). a. genitalia (signum is absent), b. ostium bursae and adjacent structures
58. *Lignyoptera thaumastaria* REBEL. a. genitalia (signum is absent), b. ostium bursae and adjacent structures
59. *Heterolocha laminaria* (HERRICH-SCHÄFFER). a. genitalia, b. signum, c. ostium bursae and adjacent structures

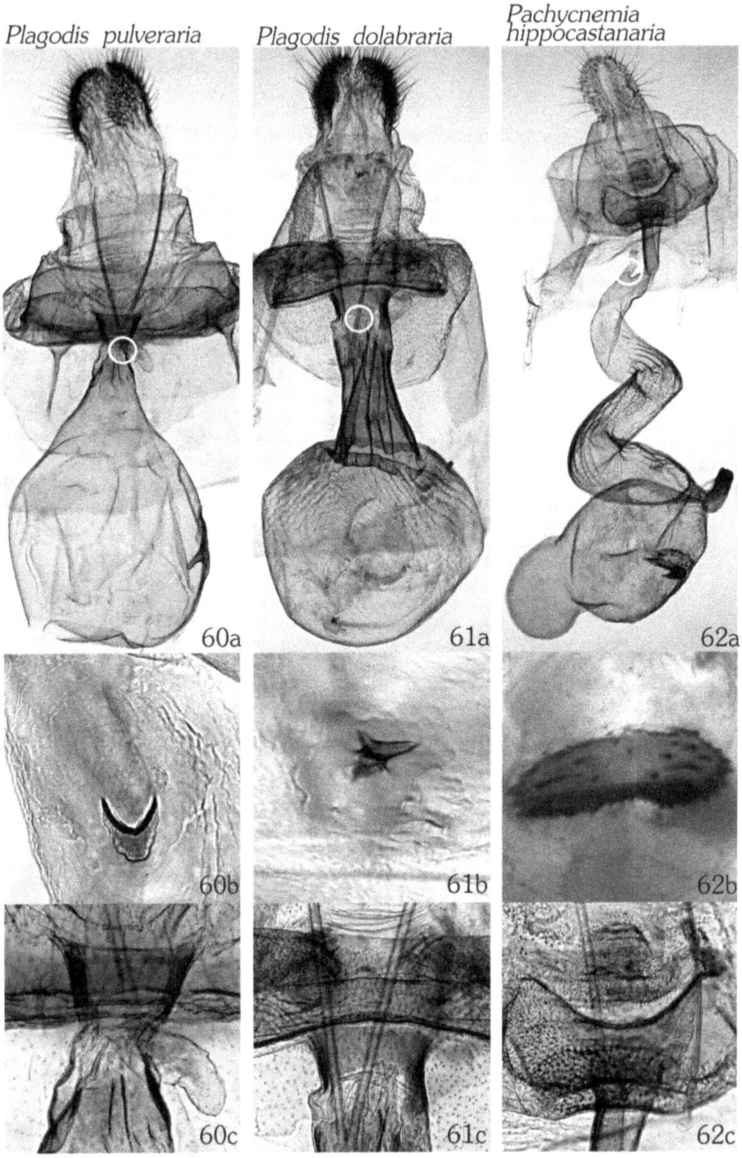

FEMALE GENITALIA (Figures are not in scale, specimen details are shown in Appendix 2.)

60. *Plagodis pulveraria* (LINNAEUS). a. genitalia, b. signum, c. ostium bursae and adjacent structures
61. *Plagodis dolabraria* (LINNAEUS). a. genitalia, b. signum, c. ostium bursae and adjacent structures
62. *Pachycnemia hippocastanaria* (HÜBNER). a. genitalia, b. signum, c. ostium bursae and adjacent structures

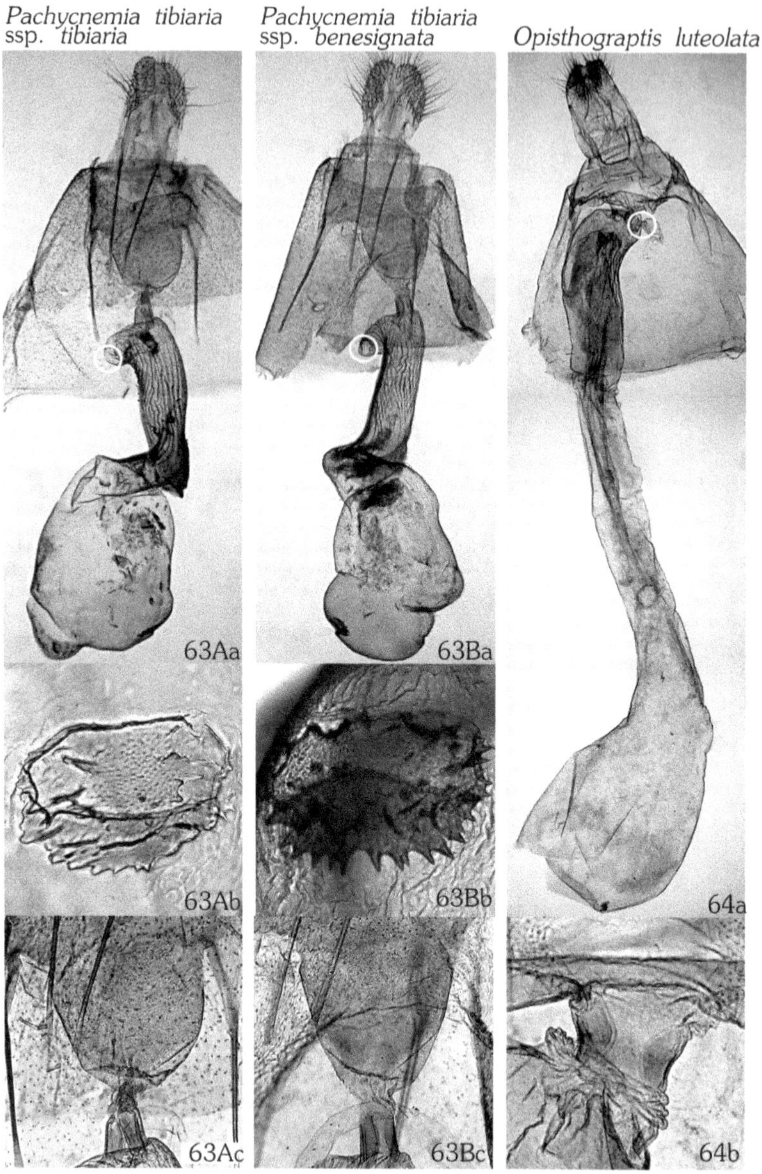

FEMALE GENITALIA (Figures are not in scale, specimen details are shown in Appendix 2.)

63A. *Pachycnemia tibiaria* subsp. *tibiaria* (RAMBUR). a. genitalia, b. signum, c. ostium bursae and adjacent structures

63B. *Pachycnemia tibiaria* subsp. *benesignata* (BELLIER). a. genitalia, b. signum, c. ostium bursae and adjacent structures

64. *Opisthograptis luteolata* (LINNAEUS). a. genitalia (signum is absent), b. ostium bursae and adjacent structures

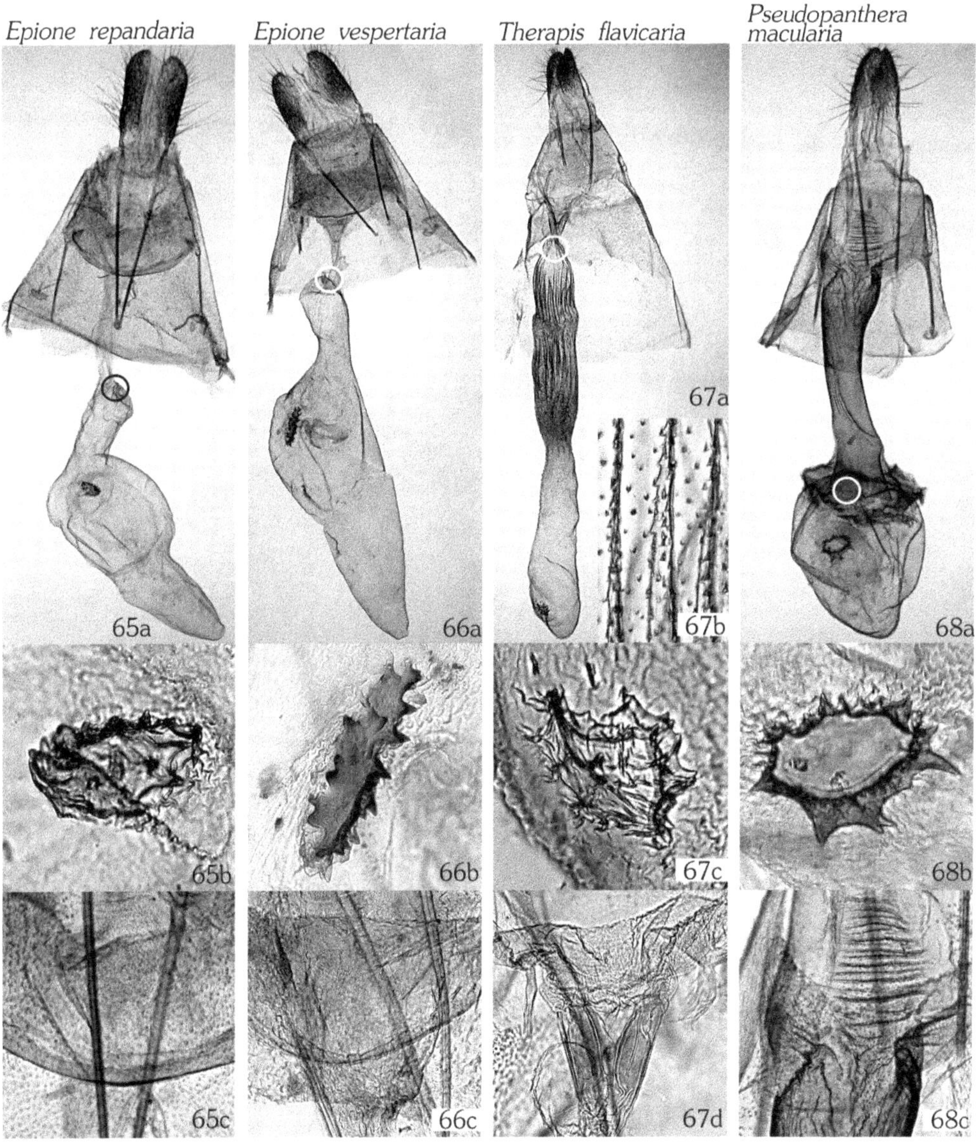

FEMALE GENITALIA (Figures are not in scale, specimen details are shown in Appendix 2.)

65. *Epione repandaria* (HUFNAGEL). a. genitalia, b. signum, c. ostium bursae and adjacent structures
66. *Epione vespertaria* (LINNAEUS). a. genitalia, b. signum, c. ostium bursae and adjacent structures
67. *Therapis flavicaria* (DENIS & SCHIFFERMÜLLER). a. genitalia, b. corpus bursae spines, c. signum, d. ostium bursae and adjacent structures
68. *Pseudopanthera macularia* (LINNAEUS). a. genitalia, b. signum, c. ostium bursae and adjacent structures

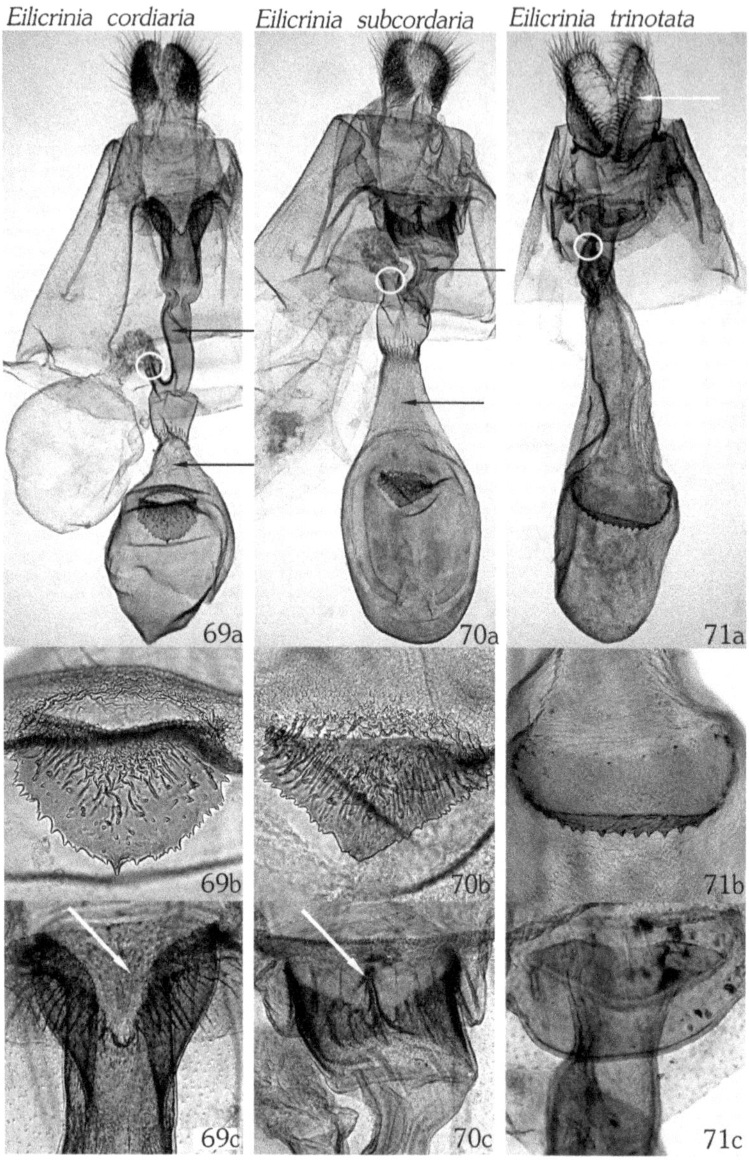

FEMALE GENITALIA (Figures are not in scale, specimen details are shown in Appendix 2.)

69. *Eilicrinia cordiaria* (HÜBNER). a. genitalia, b. signum, c. ostium bursae and adjacent structures
70. *Eilicrinia subcordaria* HERRICH-SCHÄFFER. a. genitalia, b. signum, c. ostium bursae and adjacent structures
71. *Eilicrinia trinotata* (METZNER). a. genitalia, b. signum, c. ostium bursae and adjacent structures

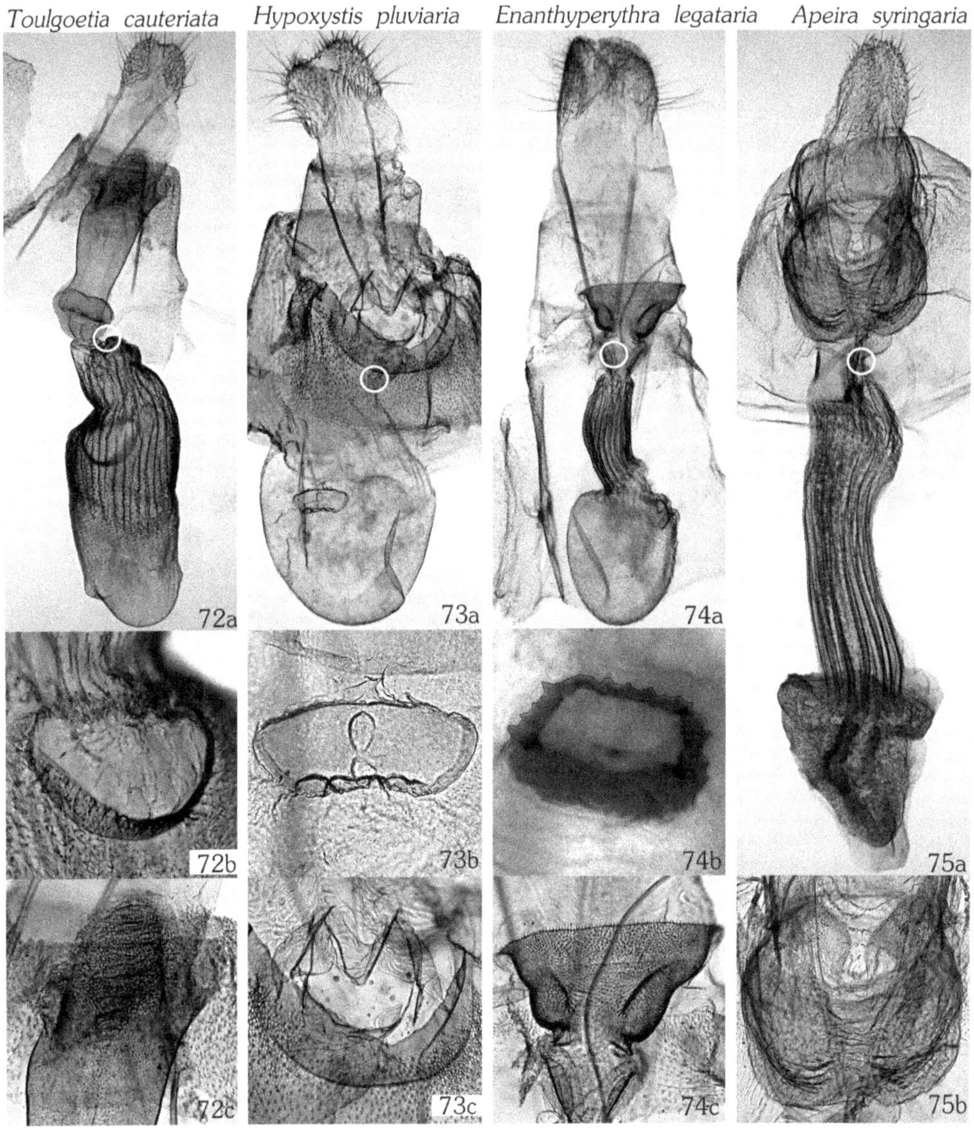

FEMALE GENITALIA (Figures are not in scale, specimen details are shown in Appendix 2.)

72. *Toulgoetia cauteriata* (STAUDINGER). a. genitalia, b. signum, c. ostium bursae and adjacent structures
73. *Hypoxystis pluviaria* (FABRICIUS). a. genitalia, b. signum, c. ostium bursae and adjacent structures
74. *Enanthyperythra legataria* (HERRICH-SCHÄFFER). a. genitalia, b. signum, c. ostium bursae and adjacent structures
75. *Apeira syringaria* (LINNAEUS). a. genitalia (signum is absent), b. ostium bursae and adjacent structures

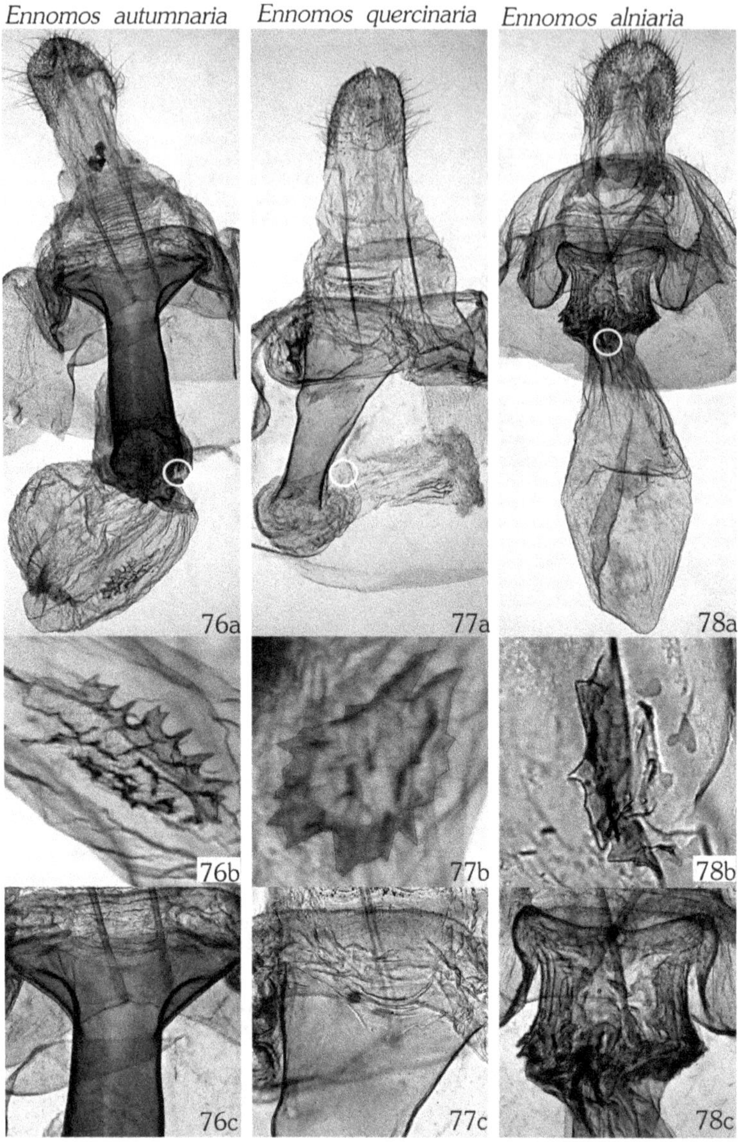

FEMALE GENITALIA (Figures are not in scale, specimen details are shown in Appendix 2.)

76. *Ennomos autumnaria* (WERNEBURG). a. genitalia, b. signum, c. ostium bursae and adjacent structures
77. *Ennomos quercinaria* (HUFNAGEL). a. genitalia, b. signum, c. ostium bursae and adjacent structures
78. *Ennomos alniaria* (LINNAEUS). a. genitalia, b. signum, c. ostium bursae and adjacent structures

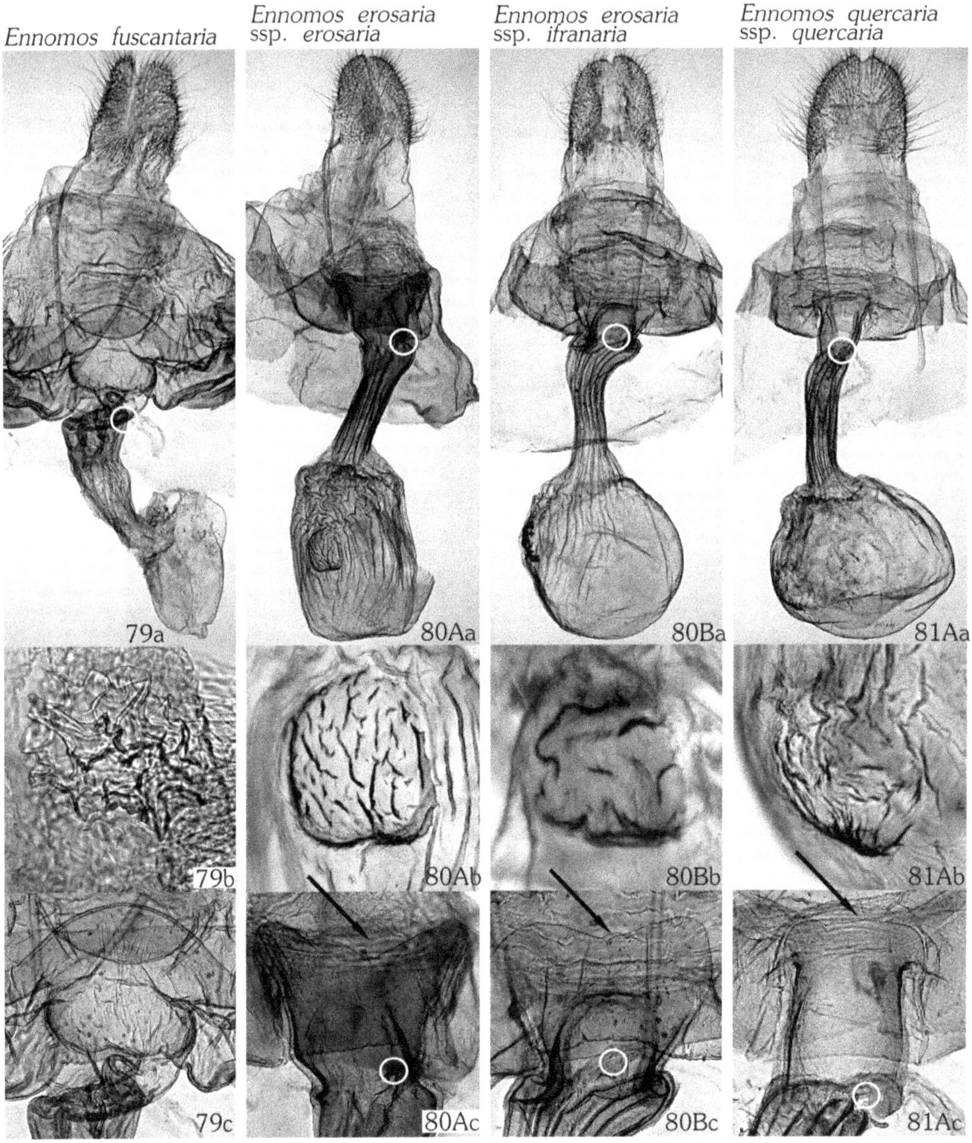

FEMALE GENITALIA (Figures are not in scale, specimen details are shown in Appendix 2.)

79. *Ennomos fuscantaria* (HAWORTH). a. genitalia, b. signum, c. ostium bursae and adjacent structures

80A. *Ennomos erosaria* subsp. *erosaria* (DENIS & SCHIFFERMÜLLER). a. genitalia, b. signum, c. ostium bursae and adjacent structures

80B. *Ennomos erosaria* subsp. *ifranaria* (RUNGS). a. genitalia, b. signum, c. ostium bursae and adjacent structures

81A. *Ennomos quercaria* subsp. *quercaria* (HÜBNER). a. genitalia, b. signum, c. ostium bursae and adjacent structures

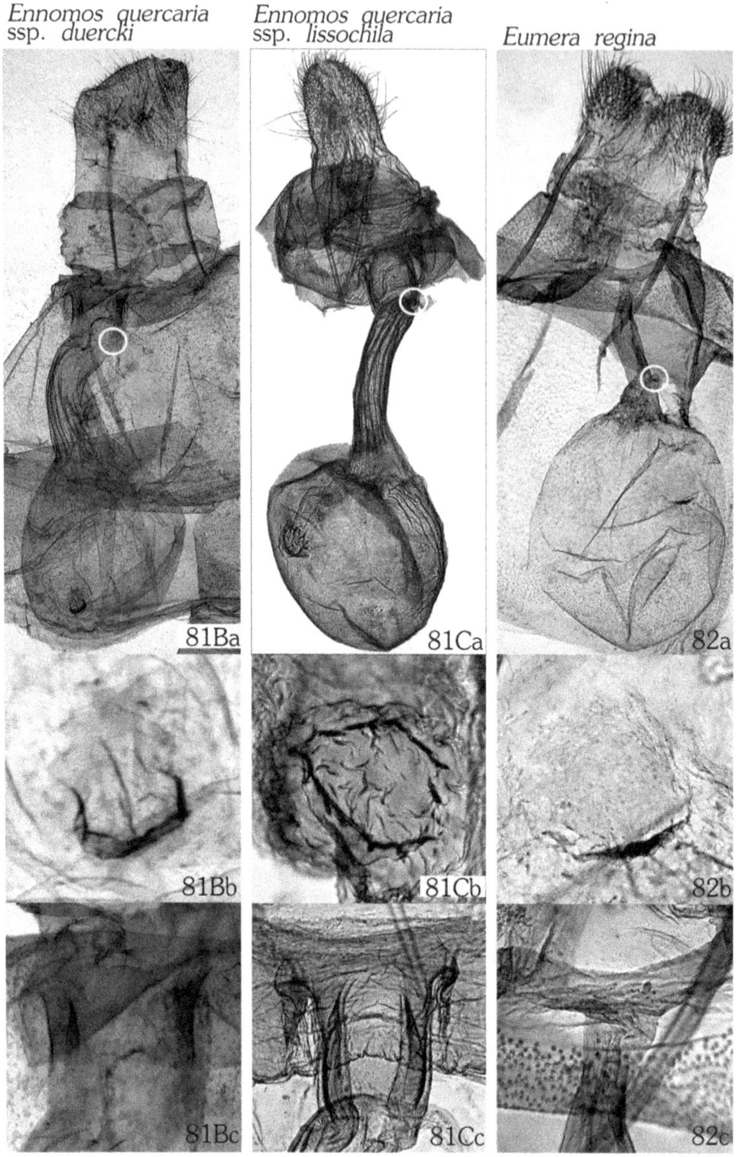

FEMALE GENITALIA (Figures are not in scale, specimen details are shown in Appendix 2.)

81B. *Ennomos quercaria* subsp. *duercki* REISSER. a. genitalia, b. signum, c. ostium bursae and adjacent structures

81C. *Ennomos quercaria* subsp. *lissochila* (PROUT). a. genitalia, b. signum, c. ostium bursae and adjacent structures

82. *Eumera regina* STAUDINGER. a. genitalia, b. signum, c. ostium bursae and adjacent structures

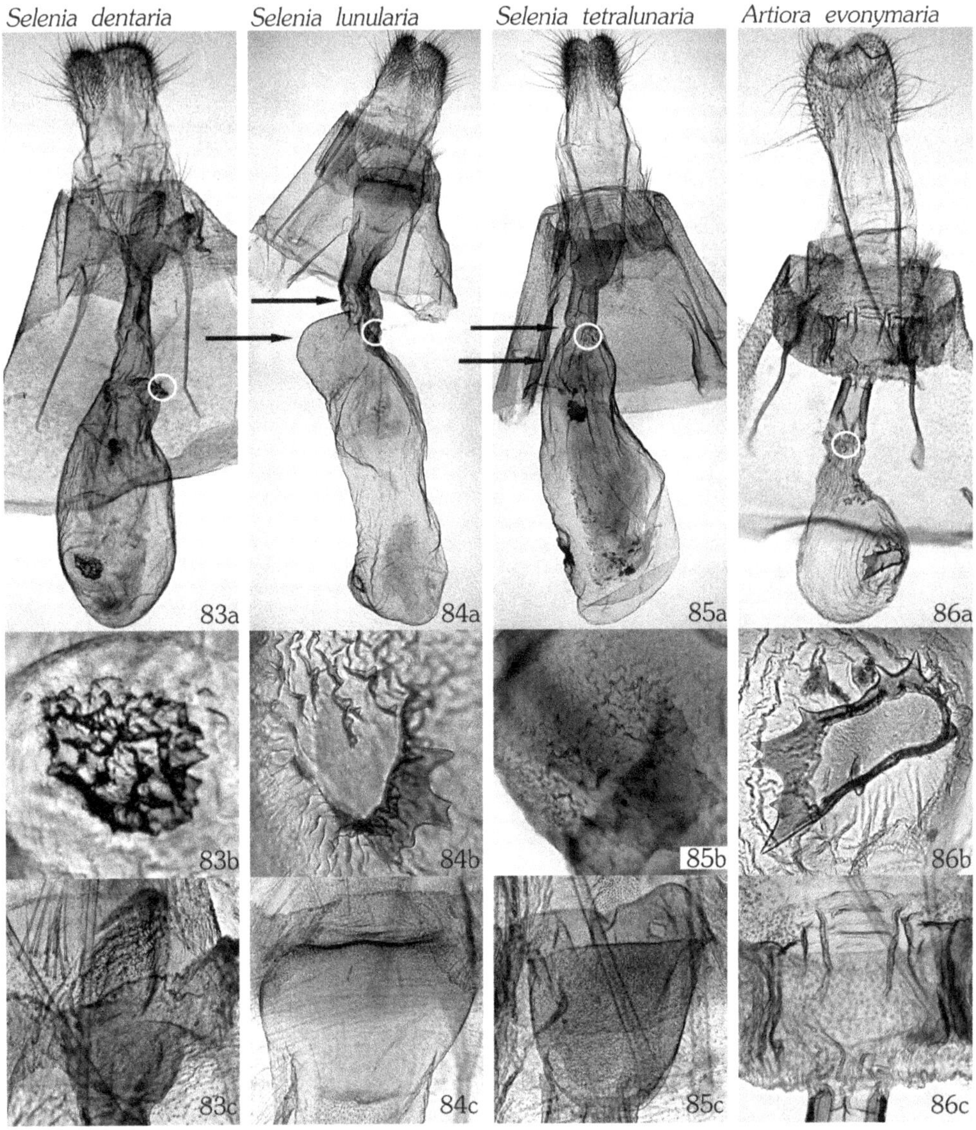

FEMALE GENITALIA (Figures are not in scale, specimen details are shown in Appendix 2.)

83. *Selenia dentaria* (FABRICIUS). a. genitalia, b. signum, c. ostium bursae and adjacent structures
84. *Selenia lunularia* (HÜBNER). a. genitalia, b. signum, c. ostium bursae and adjacent structures
85. *Selenia tetralunaria* (HUFNAGEL). a. genitalia, b. signum, c. ostium bursae and adjacent structures
86. *Artiora evonymaria* (DENIS & SCHIFFERMÜLLER). a. genitalia, b. signum, c. ostium bursae and adjacent structures

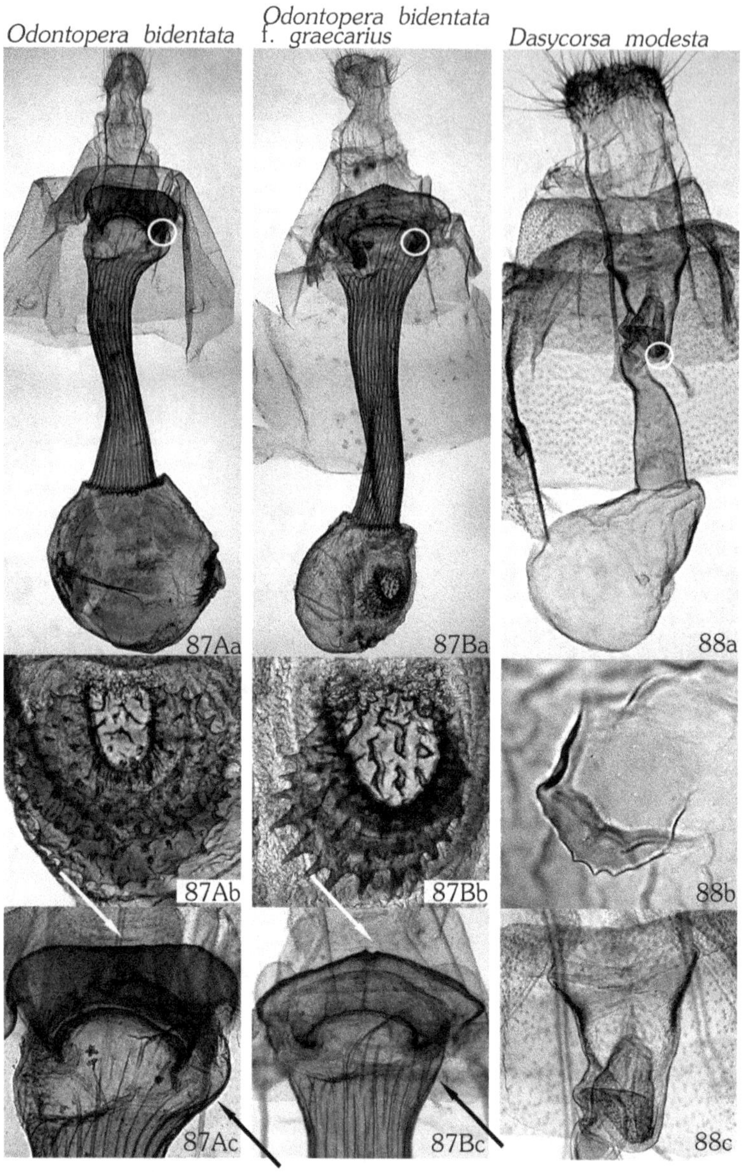

FEMALE GENITALIA (Figures are not in scale, specimen details are shown in Appendix 2.)

87A. *Odontopera bidentata* (CLERCK), structures from Finnish specimens. See text for details. a. genitalia, b. signum, c. ostium bursae and adjacent structures

87B. *Odontopera bidentata* (CLERCK) f. *graecarius*, structures from Greek specimens. See text for details. a. genitalia, b. signum, c. ostium bursae and adjacent structures

88. *Dasycorsa modesta* (STAUDINGER). a. genitalia, b. signum, c. ostium bursae and adjacent structures

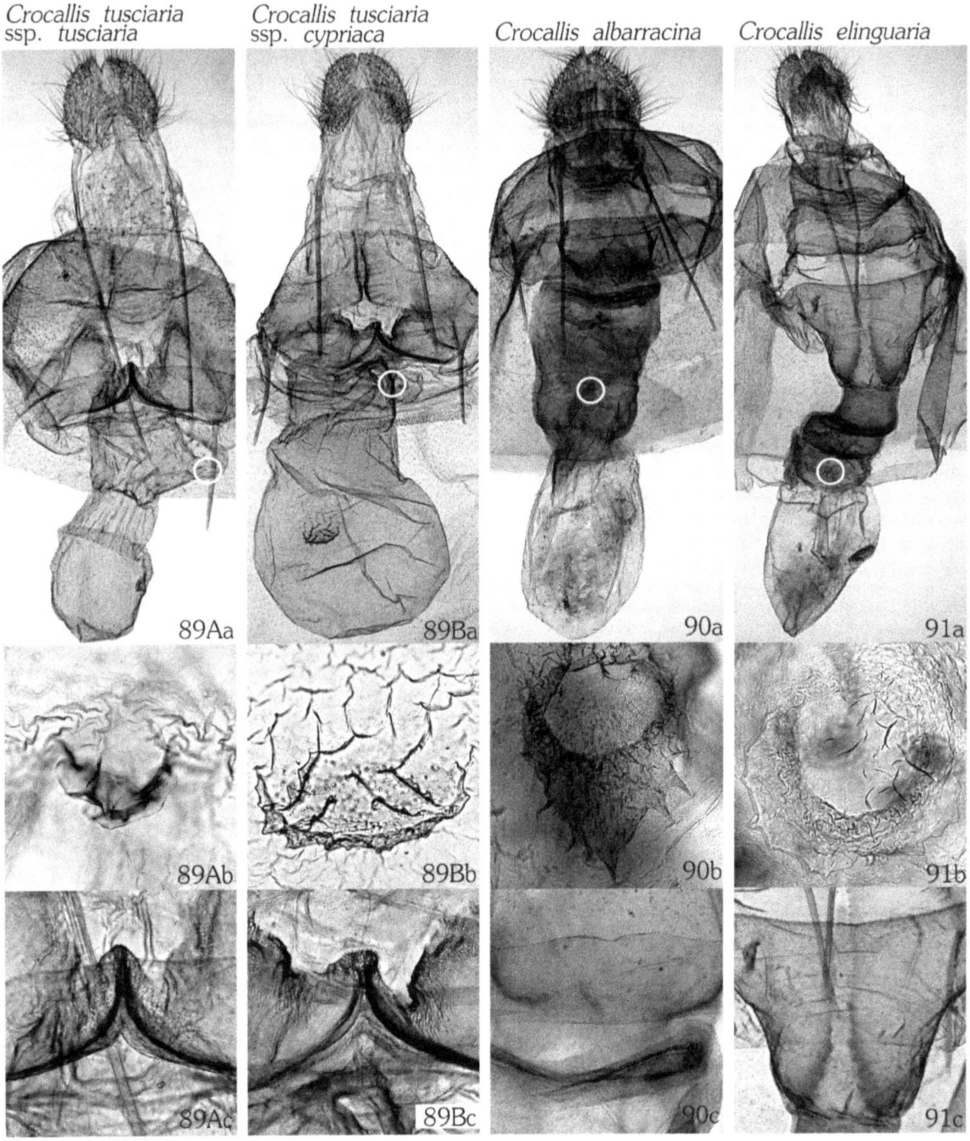

FEMALE GENITALIA (Figures are not in scale, specimen details are shown in Appendix 2.)

89A. *Crocallis tusciaria* subsp. *tusciaria* (BORKHAUSEN). a. genitalia, b. signum, c. ostium bursae and adjacent structures

89B. *Crocallis tusciaria* subsp. *cypriaca* FISCHER. a. genitalia, b. signum, c. ostium bursae and adjacent structures

90. *Crocallis albarracina* WEHRLI. a. genitalia, b. signum, c. ostium bursae and adjacent structures

91. *Crocallis elinguaria* (LINNAEUS). a. genitalia, b. signum, c. ostium bursae and adjacent structures

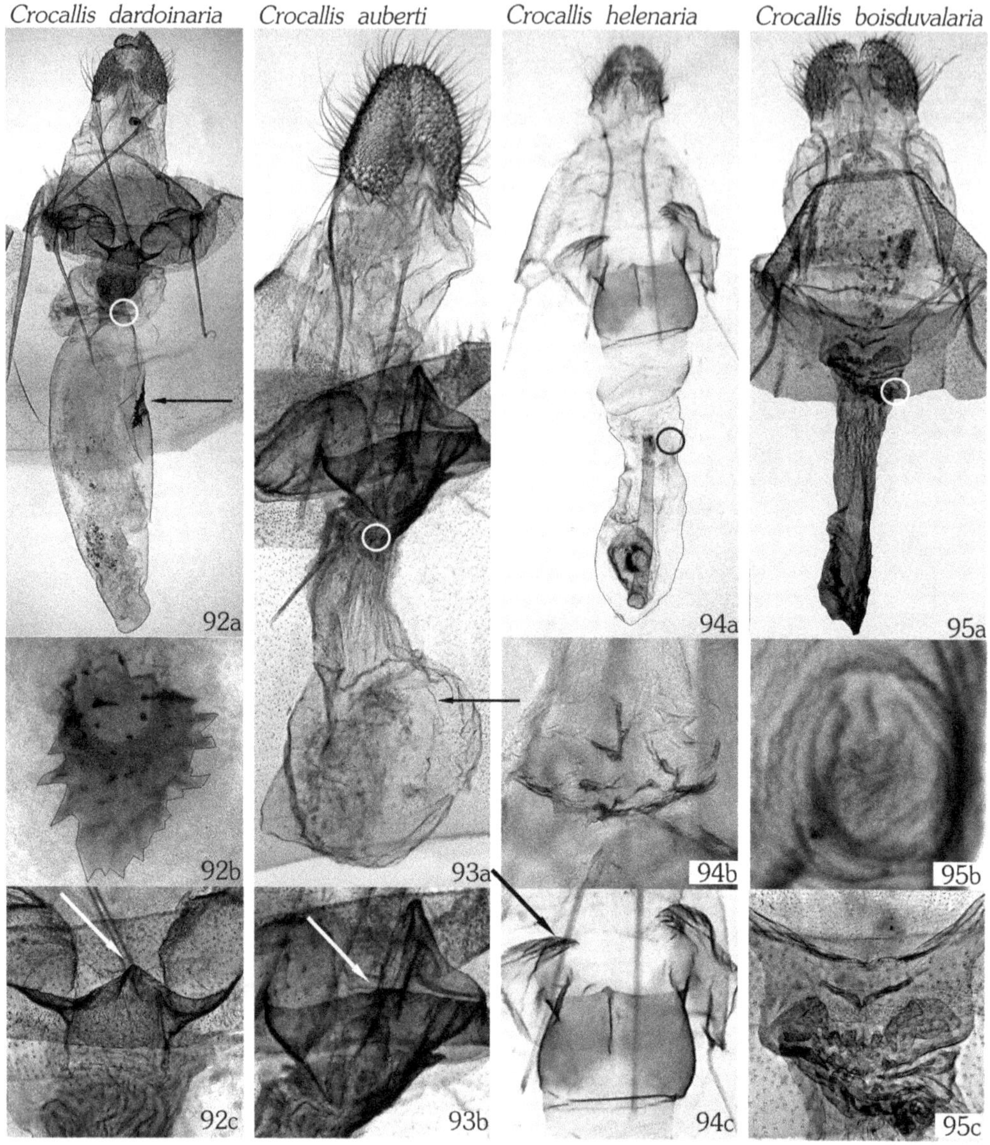

FEMALE GENITALIA (Figures are not in scale, specimen details are shown in Appendix 2.)

92. *Crocallis dardoinaria* DONZEL. a. genitalia, b. signum, c. ostium bursae and adjacent structures
93. *Crocallis auberti* OBERTHÜR. a. genitalia (signum is absent), b. ostium bursae and adjacent structures
94. *Crocallis helenaria* RUCKDESCHEL. a. genitalia, b. signum, c. ostium bursae and adjacent structures
95. *Crocallis boisduvalaria* (LUCAS). a. genitalia, b. signum, c. ostium bursae and adjacent structures

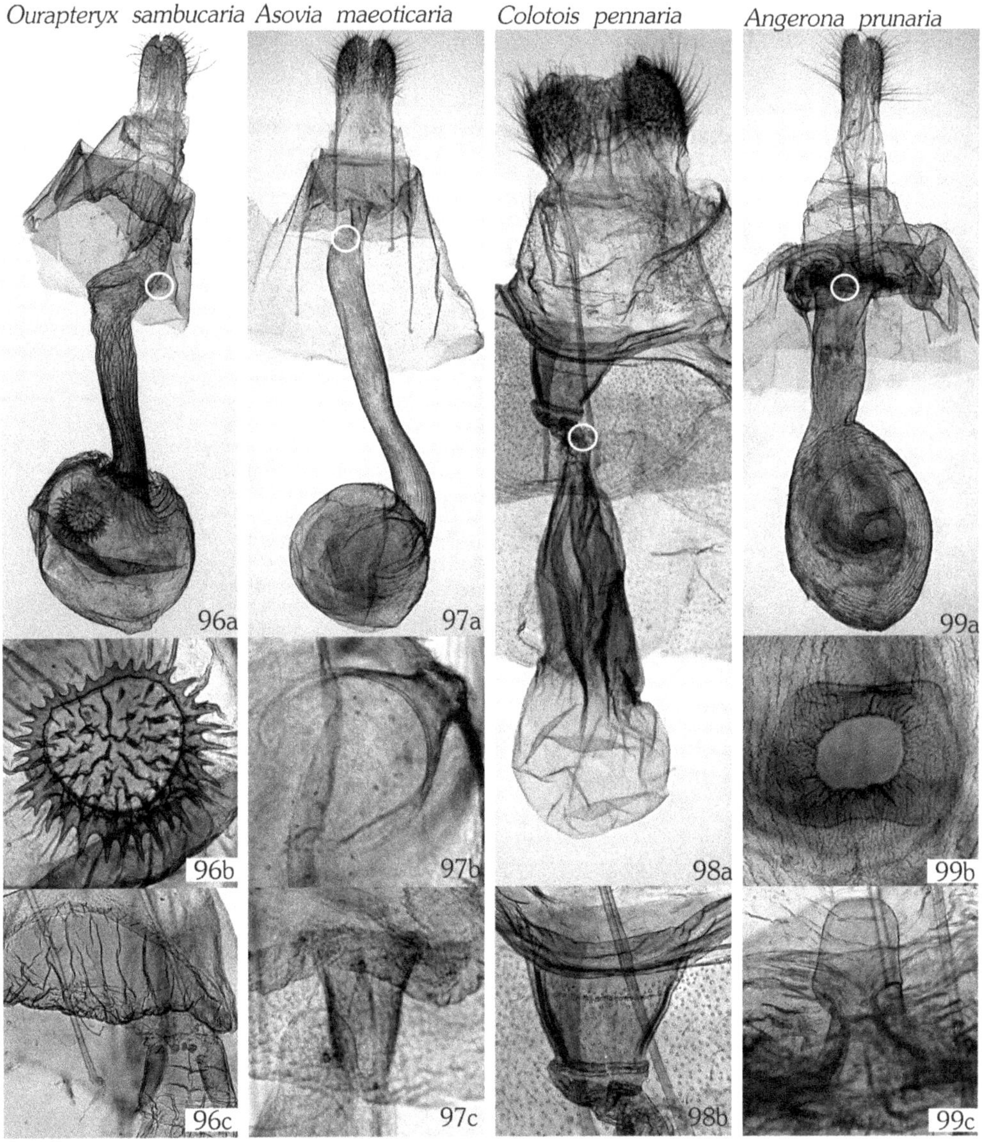

FEMALE GENITALIA (Figures are not in scale, specimen details are shown in Appendix 2.)

96. *Ourapteryx sambucaria* (LINNAEUS). a. genitalia, b. signum, c. ostium bursae and adjacent structures
97. *Asovia maeoticaria* (ALPHÉRAKY). a. genitalia, b. signum, c. ostium bursae and adjacent structures
98. *Colotois pennaria* (LINNAEUS). a. genitalia (signum is absent), b. otium bursae and adjacent structures
99. *Angerona prunaria* (LINNAEUS). a. genitalia, b. signum, c. ostium bursae and adjacent structures

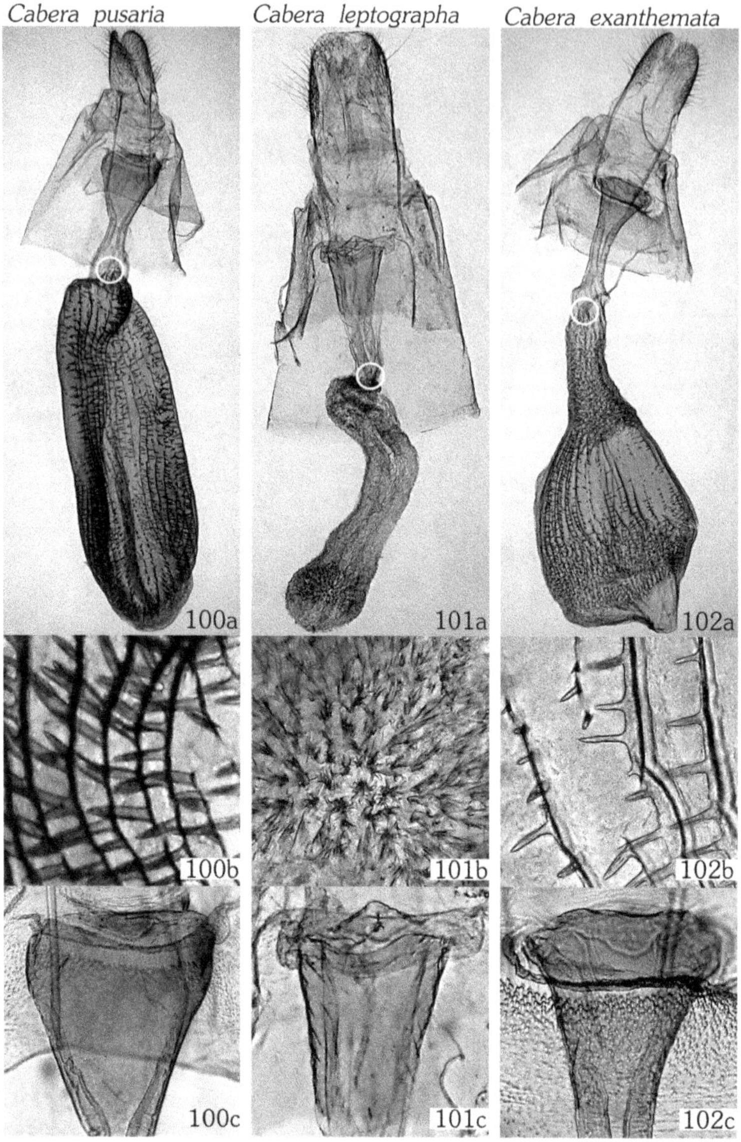

FEMALE GENITALIA (Figures are not in scale, specimen details are shown in Appendix 2.)

100. *Cabera pusaria* (LINNAEUS). a. genitalia, b. corpus bursae spines/signa, c. ostium bursae and adjacent structures
101. *Cabera leptographa* WEHRLI. a. genitalia, b. corpus bursae spines/signa, c. ostium bursae and adjacent structures
102. *Cabera exanthemata* (SCOPOLI). a. genitalia, b. corpus bursae spines/signa, c. ostium bursae and adjacent structures

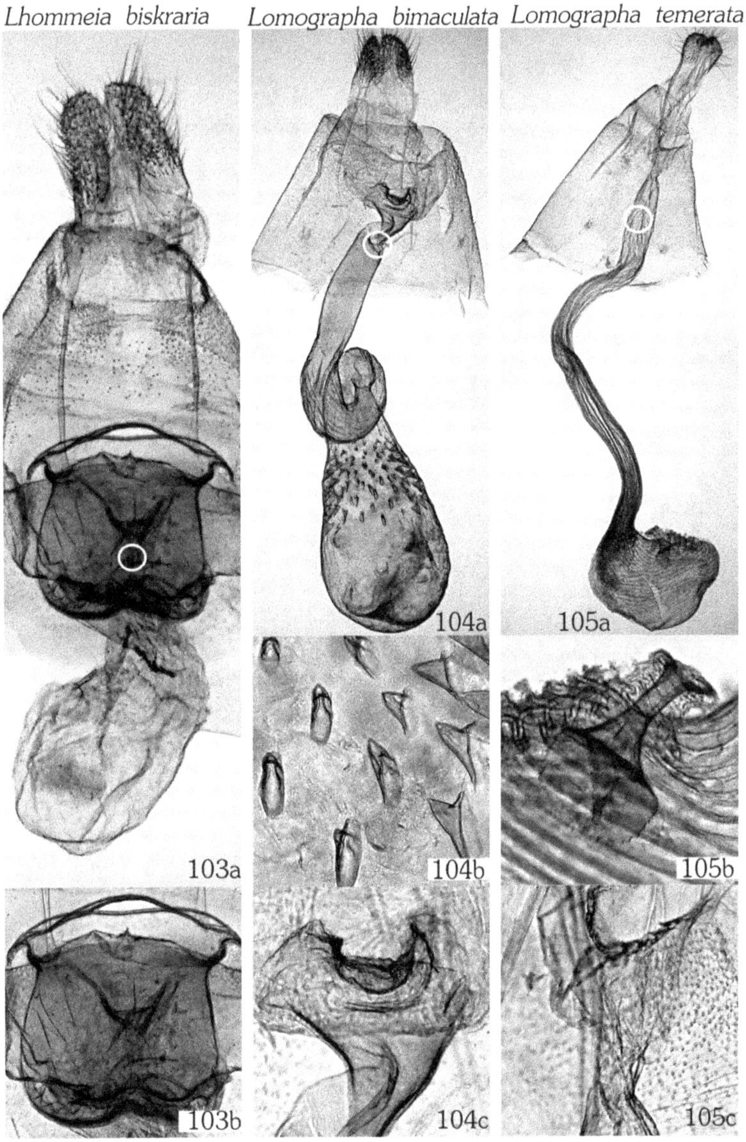

FEMALE GENITALIA (Figures are not in scale, specimen details are shown in Appendix 2.)

103. *Lhommeia biskraria* (Oberthür). a. genitalia (signum is absent), b. ostium bursae and adjacent structures
104. *Lomographa bimaculata* (Fabricius). a. genitalia, b. corpus bursae spines/signa, c. ostium bursae and adjacent structures
105. *Lomographa temerata* (Denis & Schiffermüller). a. genitalia, b. signum, c. ostium bursae and adjacent structures

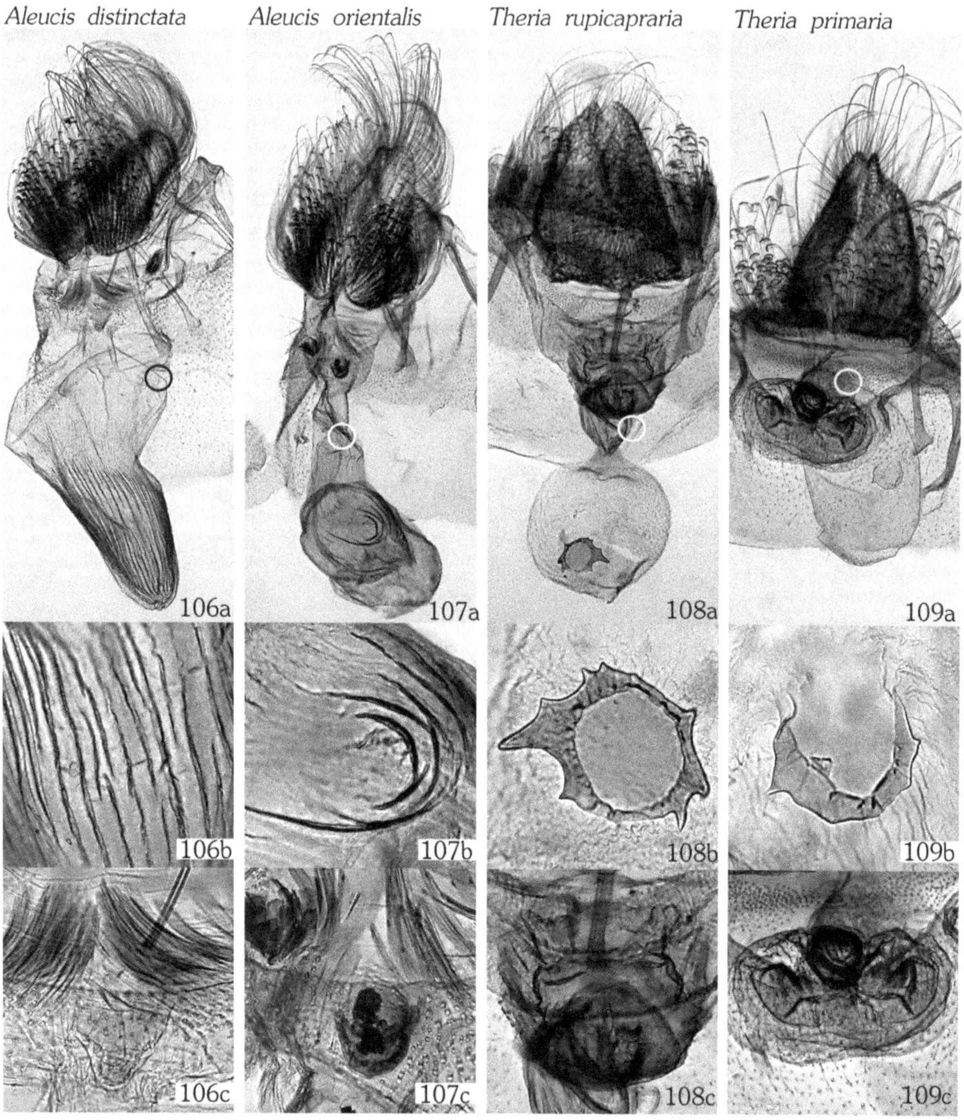

FEMALE GENITALIA (Figures are not in scale, specimen details are shown in Appendix 2.)

106. *Aleucis distinctata* (HERRICH-SCHÄFFER). a. genitalia, b. sclerotisation of corpus bursae, c. ostium bursae and adjacent structures

107. *Aleucis orientalis* (STAUDINGER). a. genitalia, b. sclerotisation of corpus bursae, c. ostium bursae and adjacent structures

108. *Theria rupicapraria* (DENIS & SCHIFFERMÜLLER). a. genitalia, b. signum, c. ostium bursae and adjacent structures

109. *Theria primaria* (HAWORTH). a. genitalia, b. signum, c. ostium bursae and adjacent structures

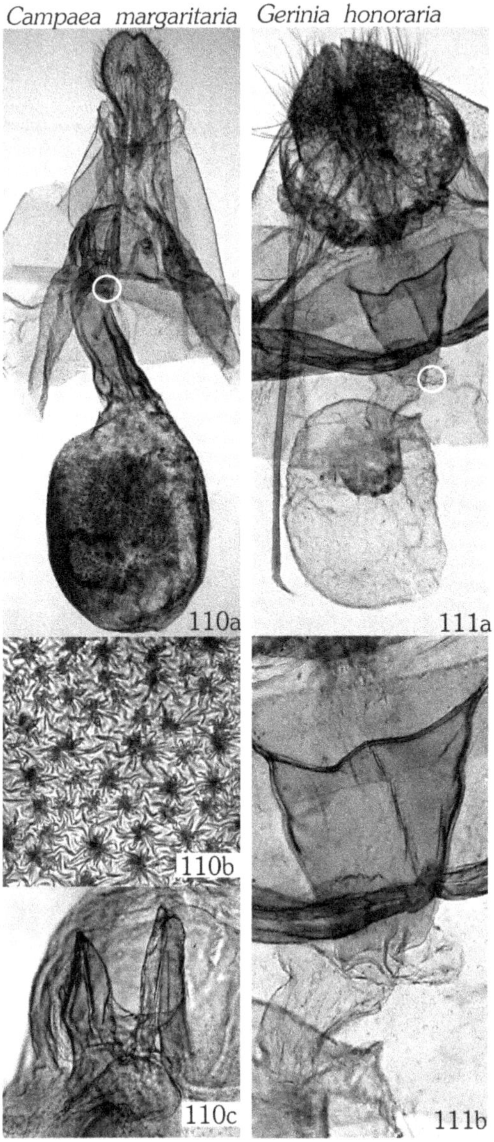

FEMALE GENITALIA (Figures are not in scale, specimen details are shown in Appendix 2.)

110. *Campaea margaritaria* (LINNAEUS). a. genitalia, b. corpus bursae spines/signa, c. ostium bursae and adjacent structures
111. *Gerinia honoraria* (DENIS & SCHIFFERMÜLLER). a. genitalia (signum is absent), b. ostium bursae and adjacent structures

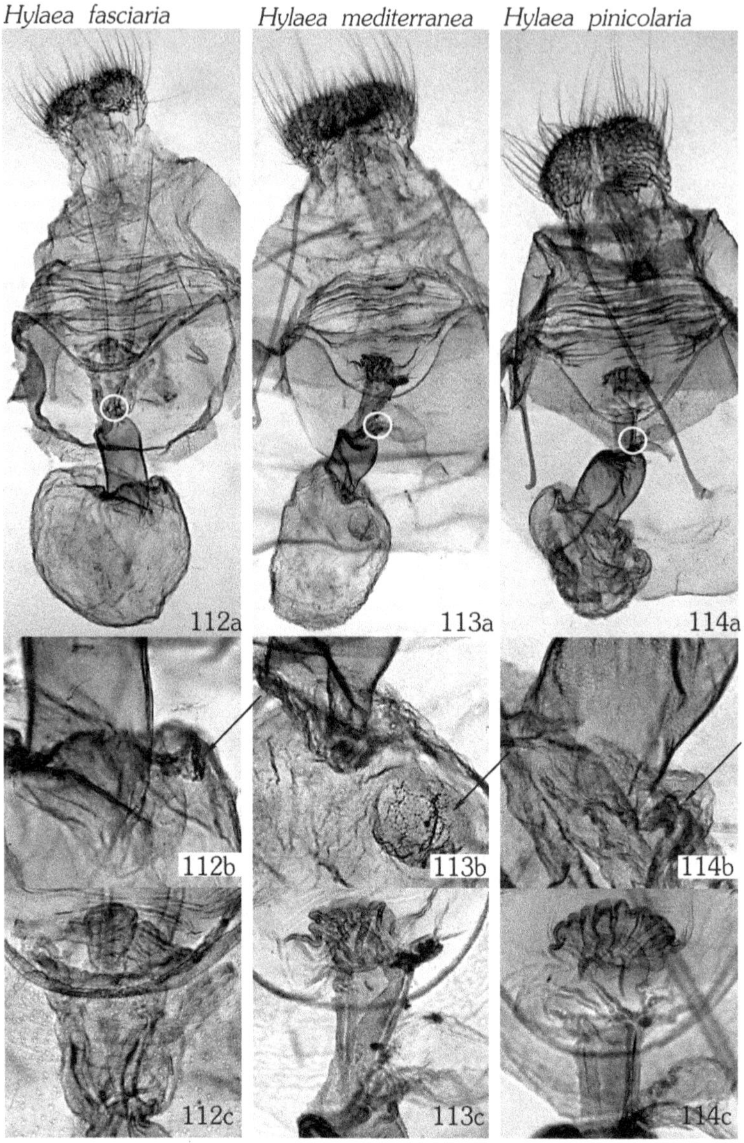

FEMALE GENITALIA (Figures are not in scale, specimen details are shown in Appendix 2.)

112. *Hylaea fasciaria* (LINNAEUS). a. genitalia, b. reduced signum, c. ostium bursae and adjacent structures
113. *Hylaea mediterranea* SIHVONEN *et al.* a. genitalia, b. signum, c. ostium bursae and adjacent structures
114. *Hylaea pinicolaria* (BELLIER). a. genitalia, b. reduced signum, c. ostium bursae and adjacent structures

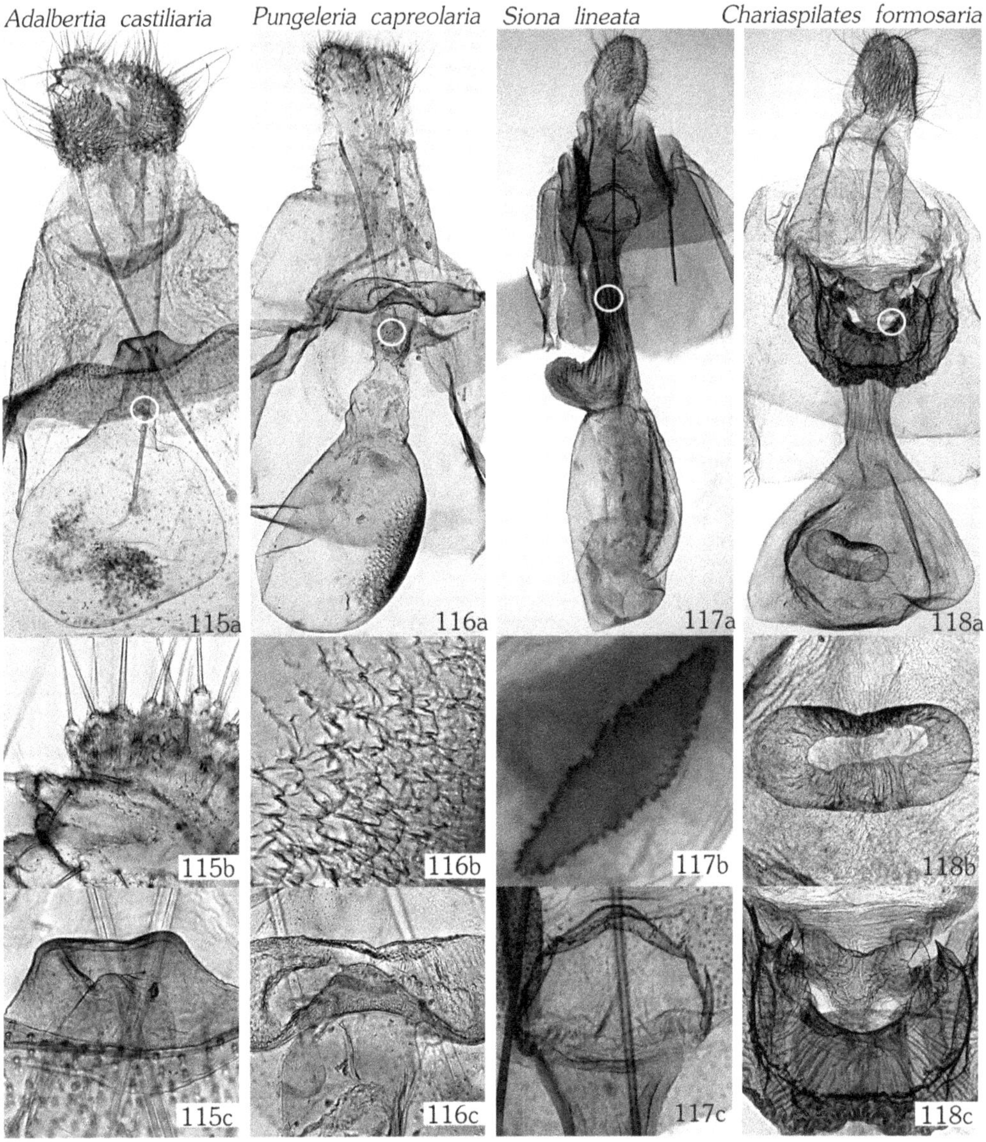

FEMALE GENITALIA (Figures are not in scale, specimen details are shown in Appendix 2.)

115. *Adalbertia castiliaria* (STAUDINGER). a. genitalia (signum is absent), b. detail of papillae anales, c. ostium bursae and adjacent structures
116. *Pungeleria capreolaria* (DENIS & SCHIFFERMÜLLER). a. genitalia, b. corpus bursae spines/signa, c. ostium bursae and adjacent structures
117. *Siona lineata* (SCOPOLI). a. genitalia, b. signum, c. ostium bursae and adjacent structures
118. *Chariaspilates formosaria* (EVERSMANN). a. genitalia, b. signum, c. ostium bursae and adjacent structures

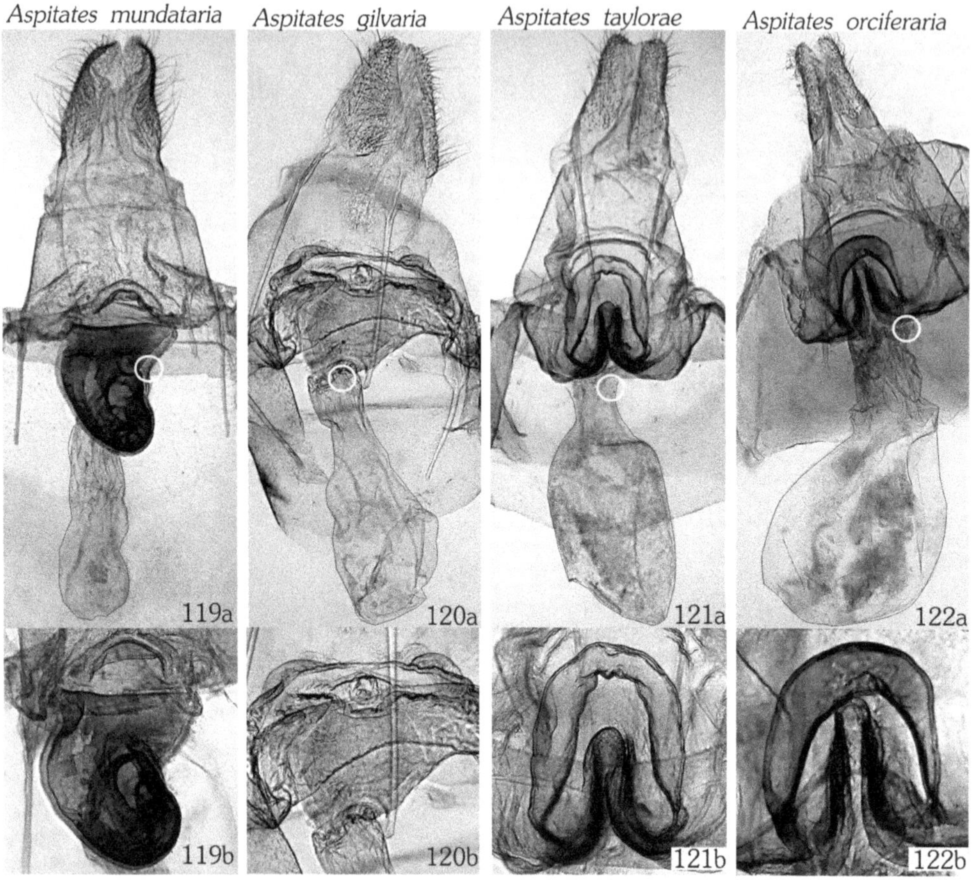

FEMALE GENITALIA (Figures are not in scale, specimen details are shown in Appendix 2.)

119. *Aspitates mundataria* (STOLL). a. genitalia (signum is absent), b. ostium bursae and adjacent structures
120. *Aspitates gilvaria* (DENIS & SCHIFFERMÜLLER). a. genitalia (signum is absent), b. ostium bursae and adjacent structures
121. *Aspitates taylorae* (BUTLER). a. genitalia (signum is absent), b. ostium bursae and adjacent structures
122. *Aspitates orciferaria* (WALKER). a. genitalia (signum is absent), b. ostium bursae and adjacent structures

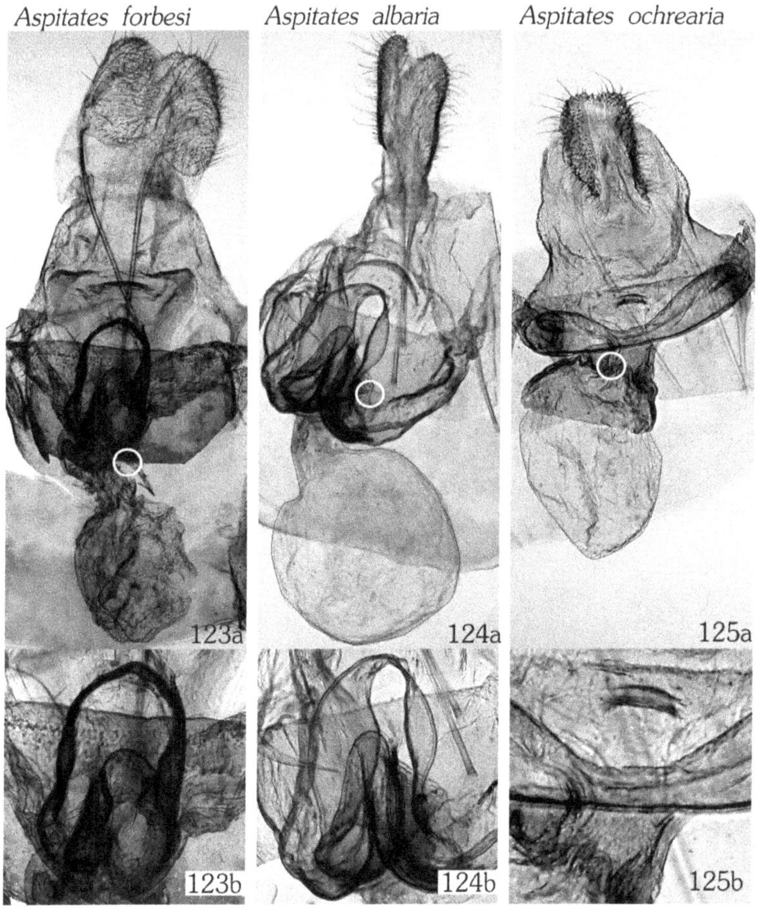

FEMALE GENITALIA (Figures are not in scale, specimen details are shown in Appendix 2.)

123. *Aspitates forbesi* (MUNROE). a. genitalia (signum is absent), b. ostium bursae and adjacent structures
124. *Aspitates albaria* (BARTEL). a. genitalia (signum is absent), b. ostium bursae and adjacent structures
125. *Aspitates ochrearia* (ROSSI). a. genitalia (signum is absent), b. ostium bursae and adjacent structures

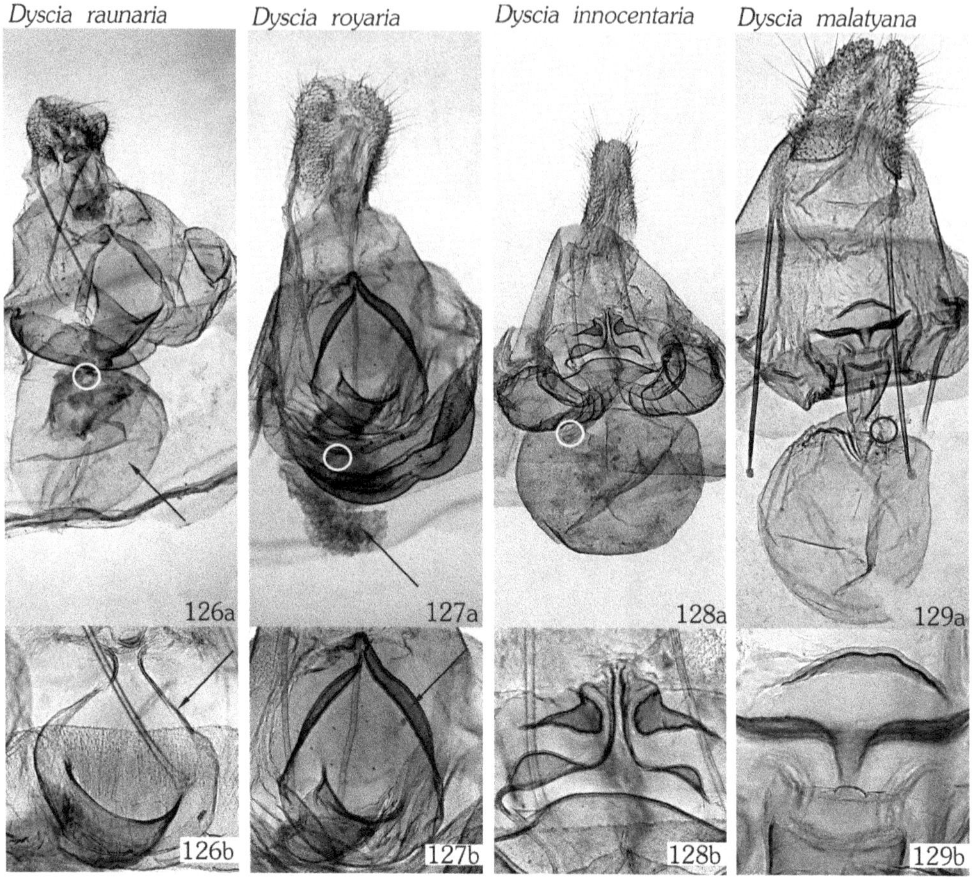

FEMALE GENITALIA (Figures are not in scale, specimen details are shown in Appendix 2.)

126. *Dyscia raunaria* (FREYER). a. genitalia (signum is absent), b. ostium bursae and adjacent structures
127. *Dyscia royaria* TAUTEL & BILLI. a. genitalia (signum is absent), b. ostium bursae and adjacent structures
128. *Dyscia innocentaria* (CHRISTOPH). a. genitalia (signum is absent), b. ostium bursae and adjacent structures
129. *Dyscia malatyana* WEHRLI. a. genitalia (signum is absent), b. ostium bursae and adjacent structures

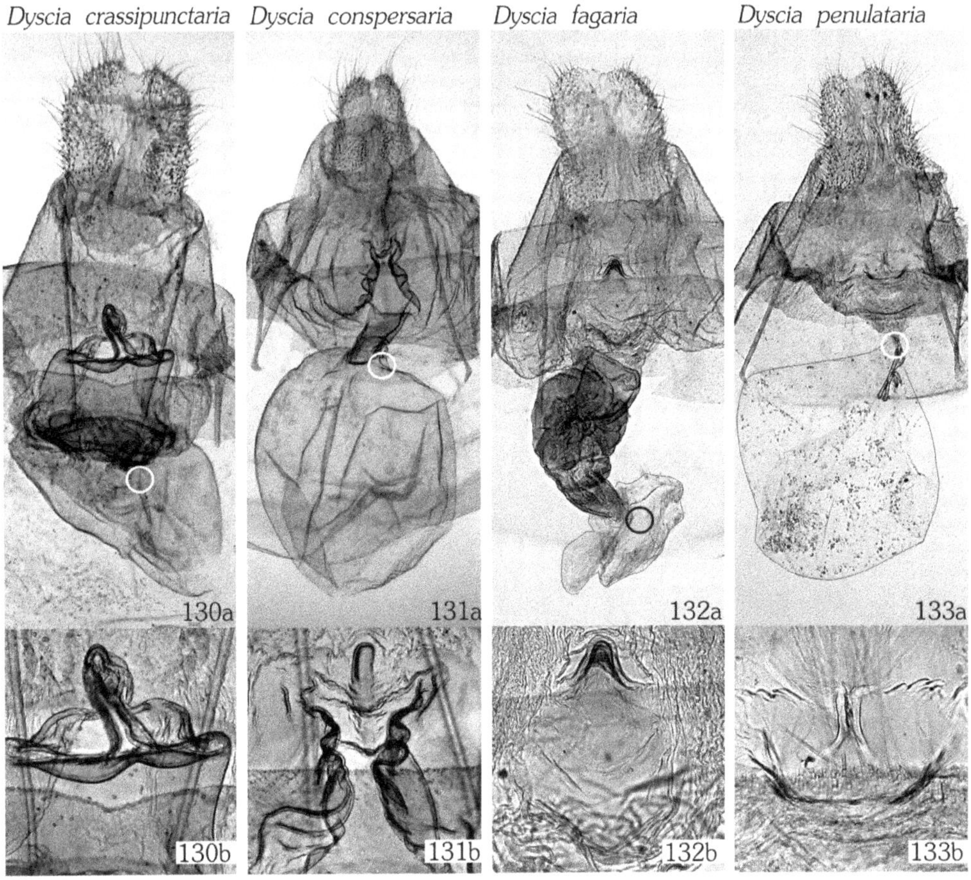

FEMALE GENITALIA (Figures are not in scale, specimen details are shown in Appendix 2.)

130. *Dyscia crassipunctaria* (REBEL). a. genitalia (signum is absent), b. ostium bursae and adjacent structures
131. *Dyscia conspersaria* (DENIS & SCHIFFERMÜLLER). a. genitalia (signum is absent), b. ostium bursae and adjacent structures
132. *Dyscia fagaria* (THUNBERG). a. genitalia (signum is absent), b. ostium bursae and adjacent structures
133. *Dyscia penulataria* (HÜBNER). a. genitalia (signum is absent), b. ostium bursae and adjacent structures

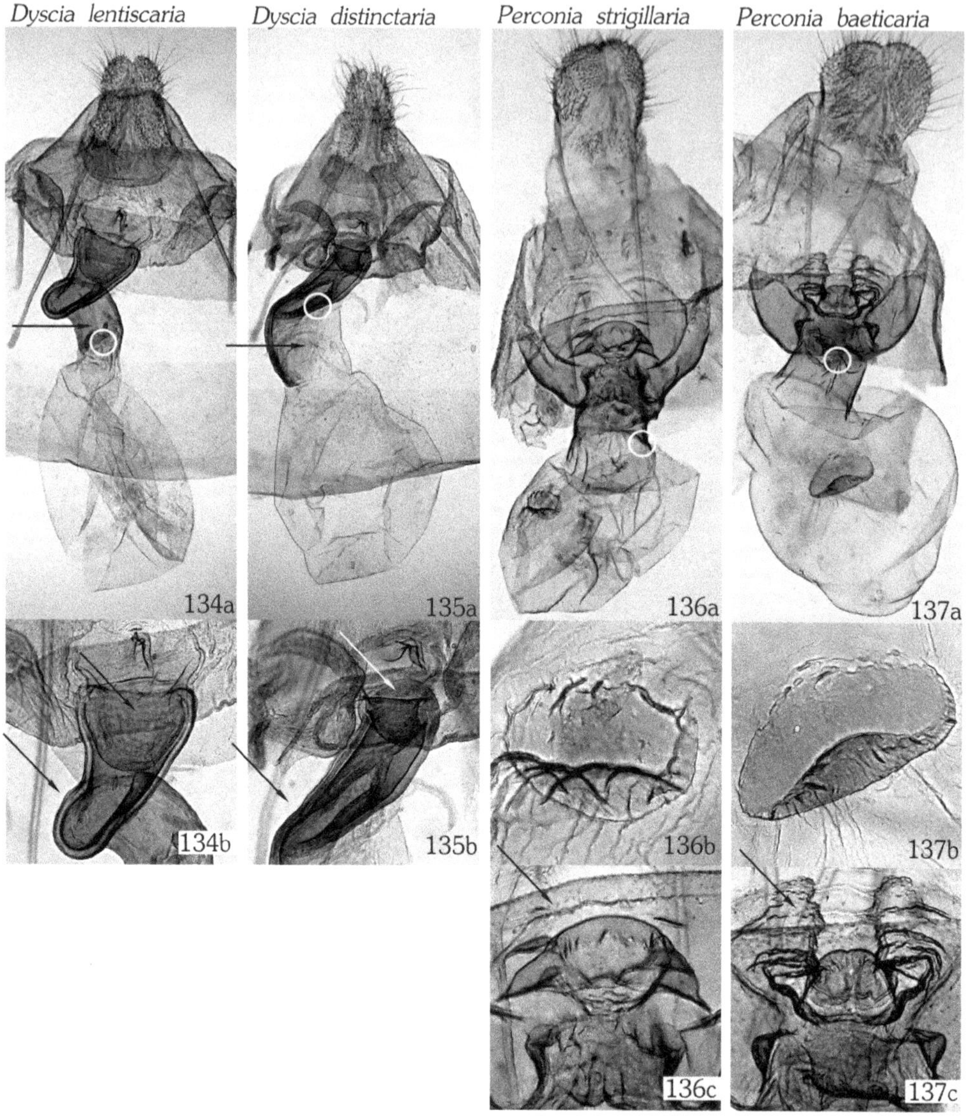

FEMALE GENITALIA (Figures are not in scale, specimen details are shown in Appendix 2.)

134. *Dyscia lentiscaria* (DONZEL). a. genitalia (signum is absent), b. ostium bursae and adjacent structures
135. *Dyscia distinctaria* (BANG-HAAS). a. genitalia (signum is absent), b. ostium bursae and adjacent structures
136. *Perconia strigillaria* (HÜBNER). a. genitalia, b. signum, c. ostium bursae and adjacent structures
137. *Perconia baeticaria* (STAUDINGER). a. genitalia, b. signum, c. ostium bursae and adjacent structures

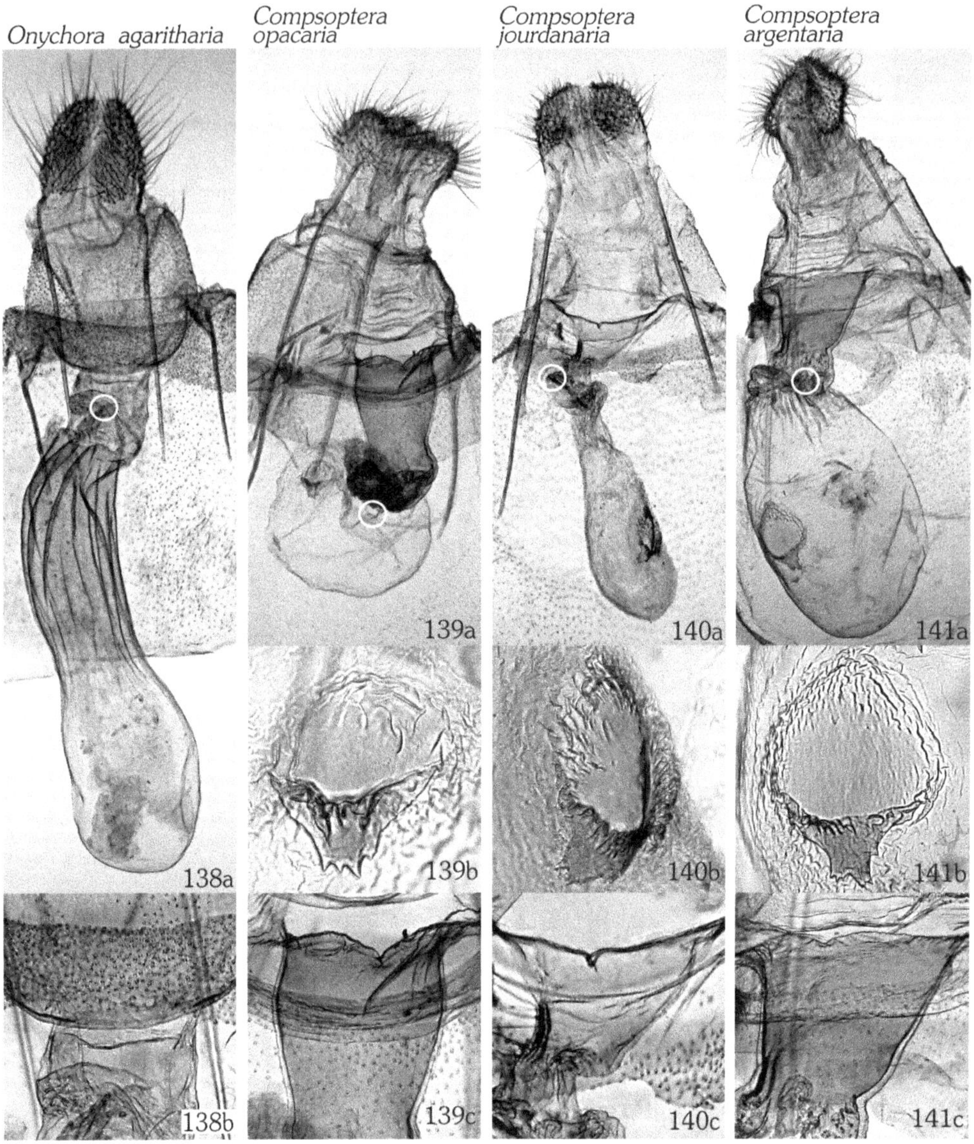

FEMALE GENITALIA (Figures are not in scale, specimen details are shown in Appendix 2.)

138. *Onychora agaritharia* (DARDOIN). a. genitalia (signum is absent), b. ostium bursae and adjacent structures
139. *Compsoptera opacaria* HÜBNER. a. genitalia, b. signum, c. ostium bursae and adjacent structures
140. *Compsoptera jourdanaria* (SERRES). a. genitalia, b. signum, c. ostium bursae and adjacent structures
141. *Compsoptera argentaria* (HERRICH-SCHÄFFER). a. genitalia, b. signum, c. ostium bursae and adjacent structures

Appendix 1. Data of the microscopic preparations shown in male genitalia plates

Abraxas grossulariata (LINNAEUS)
a. genitalia, Finland: Kökar, 10–15.vii.1993, slide PS1127 (coll. P. SIHVONEN)
b. aedeagus, Finland: Föglö, 24.vii.1995, slide PS1453 (coll. P. SIHVONEN)
c. apex of aedeagus, Finland: Föglö, 18.vii.1995, slide PS1876 (coll. H. SIHVONEN)
d. vesica, Finland: Kökar, 10–15.vii.1993, slide PS1127 (coll. P. SIHVONEN)
e. apex of valva, Finland: Föglö, 18.vii.1995, slide PS1876 (coll. H. SIHVONEN)

Abraxas sylvata (SCOPOLI)
a. genitalia, Finland: Kirkkonummi, 11–21.vi.1999, slide PS1454 (coll. P. SIHVONEN)
b. aedeagus, Finland: Kirkkonummi, 11–21.vi.1999, slide PS1454 (coll. P. SIHVONEN)
c. vesica, Finland: Kirkkonummi, 11–21.vi.1999, slide PS1128 (coll. P. SIHVONEN)
d. 3rd sternite setae, Finland: Kirkkonummi, 11–21.vi.1999, slide PS1128 (coll. P. SIHVONEN)

Abraxas pantaria (LINNAEUS)
a. genitalia, Italy: Sardinia, 15 km E Putififari, 24.v.2009, slide PS1568 (coll. NUPPONEN)
b. aedeagus, Italy: Sardinia, 15 km E Putififari, 24.v.2009, slide PS1568 (coll. NUPPONEN)
c. vesica, Dalmatia [a historical region mostly in present-day Croatia], without date, slide PS1161 (coll. ZMH)
d. apex of valva, Italy: Sardinia, 15 km E Putififari, 24.v.2009, slide PS1568 (coll. NUPPONEN)

Ligdia adustata (DENIS & SCHIFFERMÜLLER)
a. genitalia, Böhmen [Germany], without date, slide PS1163 (coll. ZMH)
b. aedeagus, Hungary: Balaton, Abrahámhegy, 11.vii.1992, slide PS1468 (coll. P. SIHVONEN)
c. vesica, Böhmen [Germany], without date, slide PS1163 (coll. ZMH)
d. 3rd sternite setae, Hungary: Balaton, Abrahámhegy, 11.vii.1992, slide PS1468 (coll. P. SIHVONEN)

Lomaspilis marginata (LINNAEUS)
a. genitalia, Finland: Enonkoski, 26.vi.1996, slide PS1136 (coll. P. SIHVONEN)
b. aedeagus, Finland: Enonkoski, 23.vi.1996, slide PS1236 (coll. P. SIHVONEN)
c. vesica, Finland: Enonkoski, 21.vi.1996, slide PS1611 (coll. P. SIHVONEN)
d. 3rd sternite setae, Finland: Enonkoski, 23.vi.1996, slide PS1236 (coll. P. SIHVONEN)

Lomaspilis opis BUTLER
a. genitalia, Finland: Vantaa, without date, slide PS1134 (coll. P. SIHVONEN)
b. aedeagus, Finland: Vantaa, 4.vi.1989, slide PS1455 (coll. P. SIHVONEN)
c. vesica, Finland: Vantaa, without date, slide PS1134 (coll. P. SIHVONEN)
d. 3rd sternite setae, Finland: Vantaa, without date, slide PS1134 (coll. P. SIHVONEN)

Lomaspilis bithynica WEHRLI
a. genitalia, Turkey: Erzurum, 2100 m, 3.vii.2001, slide PS1505 (coll. SKOU)
b. aedeagus, Turkey: Erzurum, 2100 m, 3.vii.2001, slide PS1505 (coll. SKOU)
c. vesica, Turkey: Erzurum, 2100 m, 3.vii.2001, slide PS1505 (coll. SKOU)
d. uncus, apex with reduced setae, Turkey: Erzurum, 2100 m, 3.vii.2001, slide PS1505 (coll. SKOU)

Stegania cararia (HÜBNER)

a. genitalia, Russia: Novosibirskaja oblast, Akademogorodok, 2.vii.1984, slide PS1704 (coll. ZMH)
b. aedeagus, Russia: Novosibirskaja oblast, Akademogorodok, 2.vii.1984, slide PS1469 (coll. ZMH)
c. vesica, Spain: Gerona, Ossar, 15.vi.1982, slide PS1165 (coll. ZMH)
d. uncus, apex without setae, Russia: Novosibirskaja oblast, Akademogorodok, 2.vii.1984, slide PS1704 (coll. ZMH)

Stegania trimaculata (VILLERS)

a. genitalia, France: Corsica, Porticcio, 23.vii.1992, slide PS1579 (coll. P. SIHVONEN)
b. aedeagus, France: Corsica, Porticcio, 25.vii.1992, slide PS1290 (coll. P. SIHVONEN)
c. vesica, France: Corsica, Porticcio, 25.vii.1992, slide PS1126 (coll. P. SIHVONEN)
d. uncus horns, France: Corsica, Porticcio, 23.vii.1992, slide PS1579 (coll. P. SIHVONEN)

Stegania dilectaria (HÜBNER)

a. genitalia, Hungary, without date, slide PS1470 (coll. ZMH)
b. aedeagus, Hungary, without date, slide PS1470 (coll. ZMH)
c. vesica, Hungary: Felsõpeszér, 5.v.1948, slide PS1167 (coll. ZMH)

Stegania dalmataria GUENÉE

a. genitalia, Kirghizstan: Kazarman, 1260 m, 2.viii.1983, slide PS1221 (coll. SKOU)
b. aedeagus, Kirghizstan: Kazarman, 1260 m, 2.viii.1983, slide PS1221 (coll. SKOU)
c. vesica, Russia: South Ural, without date, slide PS1172 (coll. ZMH)
d. 8th tergite, Kirghizstan: Kazarman, 1260 m, 2.8.1983, slide PS1221 (coll. SKOU)

Acanthovalva inconspicuaria (HÜBNER)

a. genitalia, Spain: Prov. Almeria, 3 km W of Tabernas, 400 m 3.v.2000, slide PS1244 (coll. SKOU)
b. aedeagus, Spain: province Alicante, Campello, 100 m, 26.viii.1984, slide PS1243 (coll. SKOU)
c. vesica enlarged, Spain: province Alicante, Campello, 100 m, 26.viii.1984, slide PS1243 (coll. SKOU)
d. vesica, Spain: province Alicante, Campello, 100 m, 26.viii.1984, slide PS1243 (coll. SKOU)
e. 8th sternite, Spain: province Alicante, Campello, 100 m, 26.viii.1984, slide PS1243 (coll. SKOU)

Narraga fasciolaria (HUFNAGEL)

a. genitalia, Kazakhstan: Zauliskiy Alatau, 1750, 14.vii.1990, slide PS1352 (coll. ZMH)
b. aedeagus, Kazakhstan: Zauliskiy Alatau, 1750, 14.vii.1990, slide PS1352 (coll. ZMH)
c. apex of aedeagus, Kazakhstan: Zauliskiy Alatau, 1750, 14.vii.1990, slide PS1352 (coll. ZMH)
d. vesica, Kazakhstan: Zauliskiy Alatau, 1750, 14.vii.1990, slide PS1352 (coll. ZMH)
e. 8th sternite, Kazakhstan: Zauliskiy Alatau, 1750, 14.vii.1990, slide PS1352 (coll. ZMH)

Narraga nelvae subsp. *catalaunica* HERBULOT
a. genitalia, Spain: Granada, Órgiva, 17.vii.1980, slide PS1481 (coll. SKOU)
b. aedeagus, Spain: Granada, Órgiva, 17.vii.1980, slide PS1481 (coll. SKOU)
c. vesica, Spain: province Granada, 10 km N of Baza, 800 m, 27.v.1988, slide PS1204 (coll. SKOU)
d. 8th sternite, Spain: province Granada, 10 km N of Baza, 800 m, 27.v.1988, slide PS1204 (coll. SKOU)

Narraga tessularia (METZNER)
a. genitalia, Austria: Stinlaersee(?), 22.vi.1958, slide PS1180 (coll. ZMH)
b. aedeagus, Hungary: Kiskunság national park, 20.vi.1998, slide PS1179 (coll. ZMH)
c. cornuti, Austria: Stinlaersee(?), 22.vi.1958, slide PS1180 (coll. ZMH)
d. vesica, Austria: Stinlaersee(?), 22.vi.1958, slide PS1180 (coll. ZMH)
e. 8th sternite, Austria: Stinlaersee(?), 22.vi.1958, slide PS1180 (coll. ZMH)

Heliomata glarearia (DENIS & SCHIFFERMÜLLER)
a. genitalia, Austria, 26.vi.1958, slide PS1471 (coll. ZMH)
b. aedeagus, Austria, 26.vi.1958, slide PS1471 (coll. ZMH)
c. microcornuti, Ukraine: Ivano-Frankivsk, Tlumach raion, 8.vi.2004, slide PS1182 (coll. ZMH)
d. vesica, Ukraine: Ivano-Frankivsk, Tlumach raion, 8.vi.2004, slide PS1182 (coll. ZMH)
e. 8th sternite, Ukraine: Ivano-Frankivsk, Tlumach raion, 8.vi.2004, slide PS1182 (coll. ZMH)

Isturgia famula (ESPER)
a. genitalia, Poland: Breslau [Wroclaw], without date, slide PS1474 (coll. ZMH)
b. aedeagus, Poland: Breslau [Wroclaw], without date, slide PS1474 (coll. ZMH)
c. vesica, Portugal: Beira Alto, 22–23.vi.1971, slide PS1194 (coll. ZMH)
d. 8th sternite, Portugal: Beira Alto, 22–23.vi.1971, slide PS1194 (coll. ZMH)

Isturgia limbaria (FABRICIUS)
a. genitalia, Germany: Lüneburg, ex larva, slide PS1193 (coll. ZMH)
b. aedeagus, Germany: Lüneburg, ex larva, slide PS1193 (coll. ZMH)
c. vesica, France: Cannes, without date, slide PS1189 (coll. ZMH)
d. tympanal organs, France: Cannes, without date, slide PS1189 (coll. ZMH)

Isturgia roraria (FABRICIUS)
a. genitalia, Austria: Wachau, ex ovo, slide PS1598 (coll. ZSM)
b. aedeagus, [Poland?]: Silesia, without date, slide PS1473 (coll. ZMH)
c. vesica, Austria: Wachau, ex ovo, slide PS1598 (coll. ZSM)
d. tympanal organs, [Poland?]: Silesia, without date, slide PS1473 (coll. ZMH)

Isturgia homochromata (MABILLE)
a. genitalia, France: Corsica, without date, slide PS1226/ ZSM G 11668, holotype (coll. ZSM)
b. aedeagus, France: Corsica, without date, slide PS1226/ ZSM G 11668, holotype (coll. ZSM)
c. vesica, France: Corsica, without date, slide PS1226/ ZSM G 11668, holotype (coll. ZSM)
d. 8th sternite, France: Corsica, without date, slide PS1226/ ZSM G 11668, holotype (coll. ZSM)

Isturgia murinaria (DENIS & SCHIFFERMÜLLER)
a. genitalia, Ukraine: Crimea, Karadagh, 26.v.2000, slide PS1472 (coll. ZMH)
b. aedeagus, Ukraine: Crimea, Karadagh, 26.v.2000, slide PS1472 (coll. ZMH)
c. vesica, Russia: S-Ural, near Moskovo village, 26.v.1998, slide PS1571 (coll. P. SIHVONEN)
d. 8th sternite, Russia: Novosibirsk Oblast, Karasuk steppe, 25–28.8.1982, slide PS1187 (coll. ZMH)

Isturgia catalaunaria (GUENÉE)
a. genitalia, Spain: prov. Almeria, Almeria 20 km N, 22.iv.2009, slide PS1581 (coll. NUPPONEN)
b. aedeagus, Spain: prov. Almeria, Almeria 20 km N, 22.iv.2009, slide PS1911 (coll. NUPPONEN)
c. posterior margin of 8th sternite, Spain: prov. Almeria, Almeria 20 km N, 22.iv.2009, slide PS1581 (coll. NUPPONEN)
d. vesica, Spain: prov. Almeria, Almeria 20 km N, 22.iv.2009, slide PS1581 (coll. NUPPONEN)
e. 8th sternite, Spain: prov. Almeria, Almeria 20 km N, 22.iv.2009, slide PS1581 (coll. NUPPONEN)

Isturgia deerraria (WALKER)
a. genitalia, Spain: province Huelva, Torre la Higuera, 21.ix.1981, slide PS1834 (coll. SKOU)
b. aedeagus, Malta: Mellieha, ex larva 13.x.2007, slide PS1836 (coll. SAMMUT)
c. microcornuti, Spain: province Huelva, Torre la Higuera, 21.ix.1981, slide PS1834 (coll. SKOU)
d. vesica, Spain: province Huelva, Torre la Higuera, 21.ix.1981, slide PS1834 (coll. SKOU)
e. 8th sternite, Malta: Mellieha, ex larva 13.x.2007, slide PS1836 (coll. SAMMUT)

Isturgia disputaria (GUENÉE)
a. genitalia, Libya: Gharian, Wadi El Hira, 17.vi.1983, slide PS1634 (coll. SKOU)
b. aedeagus, Libya: Gharian, Wadi El Hira, 17.vi.1983, slide PS1634 (coll. SKOU)
c. vesica, Libya: Gharian, Wadi El Hira, 17.vi.1983, slide PS1634 (coll. SKOU)
d. 8th sternite, Libya: Gharian, Wadi El Hira, 17.vi.1983, slide PS1634 (coll. SKOU)

Isturgia spodiaria (LEFÈBVRE)
a. genitalia, Tunisia: Medenine, 16.xi.2009, slide PS1580 (coll. NUPPONEN)
b. aedeagus, Italy: Sicily, without date, slide PS1478 (coll. ZMH)
c. vesica, Tunisia: Hammamet, 12–18.iii.1973, slide PS1212 (coll. ZMH)
d. 8th sternite, Tunisia: Hammamet, 12–18.iii.1973, slide PS1212 (coll. ZMH)

Isturgia assimilaria (RAMBUR)
a. genitalia, France: Corsica, without date, slide PS1196 (coll. ZMH)
b. aedeagus, Italy: Sardinia, Gennargentu 1800 m, 20.v.2009, slide PS1573 (coll. P. SIHVONEN)
c. vesica, France: Corsica, without date, slide PS1196 (coll. ZMH)
d. 8th sternite, France: Corsica, without date, slide PS1196 (coll. ZMH)

Isturgia miniosaria (DUPONCHEL)
a. genitalia, Spain: prov. Almeria, Sierra de Cabrera, 680 m, 22.xi.2006, slide PS1253 (coll. SKOU)
b. aedeagus, Spain: prov. Almeria, Sierra de Cabrera, 680 m, 22.xi.2006, slide PS1253 (coll. SKOU)
c. microcornuti, Spain: prov. Almeria, Sierra de Cabrera, 680 m, 22.xi.2006, slide PS1253 (coll. SKOU)
d. vesica, Spain: prov. Almeria, Sierra de Cabrera, 680 m, 22.xi.2006, slide PS1253 (coll. SKOU)
e. 8th sternite, Spain: prov. Almeria, Sierra de Cabrera, 680 m, 22.xi.2006, slide PS1253 (coll. SKOU)

Isturgia berytaria (STAUDINGER)
a. genitalia, Greece: Peleponnisos, Vlahokerasia, 3.x.1984, slide PS1254 (coll. SKOU)
b. aedeagus, Greece: Peleponnisos, Vlahokerasia, 3.x.1984, slide PS1254 (coll. SKOU)
c. vesica, Greece: Peleponnisos, Vlahokerasia, 3.x.1984, slide PS1254 (coll. SKOU)
d. 8th sternite, Greece: Peleponnisos, Vlahokerasia, 3.x.1984, slide PS1254 (coll. SKOU)

Isturgia sparsaria (HÜBNER)
a. genitalia, Italy: Lazio, 30.v.1952, slide PS1825 (coll. SKOU)
b. aedeagus, Italy: Lazio, 30.v.1952, slide PS1825 (coll. SKOU)
c. vesica, Italy: Basilicata, Trecchina 400 m, 17.ix.1990, slide PS1239 (coll. P. SIHVONEN)
d. 8th sternite, Italy: Basilicata, Trecchina 400 m, 17.ix.1990, slide PS1239 (coll. P. SIHVONEN)

Isturgia messapiaria (SOHN-RETHEL)
a. genitalia, Italy: Calabria, Fossiata, 6.viii.1921, slide PS1705 (coll. MSNM)
b. aedeagus, Italy: Calabria, Fossiata, 6.viii.1921, G. Fiumi prep. (coll. MSNM)
c. vesica, Italy: Calabria, Fossiata, 6.viii.1921, slide PS1705 (coll. MSNM)
d. 8th sternite, Italy: Calabria, Fossiata, 6.viii.1921, slide PS1705 (coll. MSNM)

Isturgia arenacearia (DENIS & SCHIFFERMÜLLER)
a. genitalia, Hungary: Budapest, 10.vii.1992, slide PS1466 (coll. P. SIHVONEN)
b. aedeagus, Hungary: Budapest, 10.vii.1992, slide PS1466 (coll. P. SIHVONEN)
c. spines in aedeagus apex, Ukraine: Ivano-Frankivsk, 800 m, Yaremcha, 28-29.v.2003, slide PS1184 (coll. ZMH)
d. vesica, Ukraine: Ivano-Frankivsk, 800 m, Yaremcha, 28-29.v.2003, slide PS1184 (coll. ZMH)
e. 8th sternite Ukraine: Ivano-Frankivsk,800 m, Yaremcha, 28-29.v.2003, slide PS1184 (coll. ZMH)

Isturgia punctistrigaria (CHRISTOPH)
a. genitalia, Turkmenistan: Repetek, 16–17.iv.1977, slide PS1242 (coll. IPPE)
b. aedeagus, Turkmenistan: Repetek, 16–17.iv.1977, slide PS1242 (coll. IPPE)
c. vesica, Turkmenistan: Repetek, 16–17.iv.1977, slide PS1242 (coll. IPPE)
d. 8th sternite, Turkmenistan: Repetek, 16–17.iv.1977, slide PS1242 (coll. IPPE)

Itame vincularia subsp. *vincularia* HÜBNER
a. genitalia, Portugal: Algarve, 11.v.1971, slide PS1206 (coll. ZMH)
b. aedeagus, Spain: Aragon, without date, slide PS1476 (coll. ZMH)
c. apex of aedeagus, Portugal: Algarve, 11.v.1971, slide PS1206 (coll. ZMH)
d. vesica, Portugal: Algarve, 11.v.1971, slide PS1206 (coll. ZMH)
d. 8th sternite, Portugal: Algarve, 11.v.1971, slide PS1206 (coll. ZMH)

Itame vincularia subsp. *mrassinaria* (OBERTHÜR)
a. genitalia, Morocco: Ouirgane, 4.v.1965, slide PS1647 (coll. ZSM)
b. aedeagus, Morocco: Essaouira, 21.iv.1969, slide PS1666 (coll. ZSM)
c. apex of aedeagus, Morocco: Ouirgane, 4.v.1965, slide PS1647 (coll. ZSM)
d. vesica, Morocco: Ouirgane, 4.v.1965, slide PS1647 (coll. ZSM)
e. 8th sternite, Morocco: Ouirgane, 4.v.1965, slide PS1647 (coll. ZSM)

Neognopharmia stevenaria (BOISDUVAL)
a. genitalia, Spain: province Huelva, Bielsa, 1900 m, 22.vii.1992, slide PS1562 (coll. ZMH)
b. aedeagus, Spain: province Huelva, Bielsa, 1900 m, 22.vii.1992, slide PS1562 (coll. ZMH)
c. apex of cornutus, Turkey: Konya, Toros Daglari, 1700 m, 14.vii.1986, slide PS1954 (coll. SKOU)
d. vesica, Greece: Lakonia, 30.vi.–3.vii.1985, slide PS1953 (coll. SKOU)
e. 8th sternite, Spain: province Huelva, Bielsa, 1900 m, 22.vii.1992, slide PS1562 (coll. ZMH)

Macaria notata (LINNAEUS)
a. genitalia, Finland: Savonlinna, 21–23.vi.1993, slide PS1459 (coll. P. SIHVONEN)
b. aedeagus, Finland: Savonlinna, 21–23.vi.1993, slide PS1459 (coll. P. SIHVONEN)
c. vesica, Finland: Savonlinna, 21–23.vi.1993, slide PS1146 (coll. P. SIHVONEN)
d. 8th sternite, Finland: Savonlinna, 21–23.vi.1993, slide PS1459 (coll. P. SIHVONEN)

Macaria alternata (DENIS & SCHIFFERMÜLLER)
a. genitalia, Finland: Savonlinna, 20.v.1993, slide PS1139 (coll. P. SIHVONEN)
b. aedeagus, Finland: Savonlinna, 6–7.vii.1997, slide PS1456 (coll. P. SIHVONEN)
c. vesica, Finland: Savonlinna, 20.v.1993, slide PS1139 (coll. P. SIHVONEN)
d. 8th sternite, Finland: Savonlinna, 20.v.1993, slide PS1139 (coll. P. SIHVONEN)

Macaria signaria (HÜBNER)
a. genitalia, Finland: Lammi, 14.vi.1995, slide PS1462 (coll. P. SIHVONEN)
b. aedeagus, Finland: Lammi, 14.vi.1995, slide PS1462 (coll. P. SIHVONEN)
c. vesica, Finland: Espoo, 13.vii.1998, slide PS1151 (coll. P. SIHVONEN)
d. 8th sternite, Finland: Espoo, 13.vii.1998, slide PS1151 (coll. P. SIHVONEN)

Macaria liturata (CLERCK)
a. genitalia, Finland: Savonlinna, 20–25.ix.1993, slide PS1125 (coll. P. SIHVONEN)
b. aedeagus, Finland: Enonkoski, 5.vii.1996, slide PS1452 (coll. P. SIHVONEN)
c. vesica, Finland: Kirkkonummi, 15.vi.2008, slide PS1826 (coll. P. SIHVONEN)
d. 8th sternite, Finland: Savonlinna, 20–25.ix.1993, slide PS1125 (coll. P. SIHVONEN)

Macaria wauaria subsp. *wauaria* (LINNAEUS)
a. genitalia, Finland: Savonlinna, 17.vii.1993, slide PS1148 (coll. P. SIHVONEN)
b. aedeagus, Finland: Savonlinna, 29–31.vii.1995, slide PS1460 (coll. P. SIHVONEN)
c. vesica, Finland: Savonlinna, 17.vii.1993, slide PS1148 (coll. P. SIHVONEN)
d. 8th sternite, Finland: Savonlinna, 17.vii.1993, slide PS1148 (coll. P. SIHVONEN)

Macaria wauaria subsp. *africana* (ZERNY)
a. genitalia, Morocco: Oukaïmeden, 2700 m, 5.–21.vii.1972, slide PS1650 (coll. ZSM)
b. aedeagus, Morocco: Oukaïmeden, 2700 m, 5.–21.vii.1972, slide PS1650 (coll. ZSM)
c. vesica, Morocco: Oukaïmeden, 2700 m, 5.–21.vii.1972, slide PS1650 (coll. ZSM)
d. 8th sternite, Morocco: Oukaïmeden, 2700 m, 5.–21.vii.1972, slide PS1650 (coll. ZSM)

Macaria artesiaria (DENIS & SCHIFFERMÜLLER)
a. genitalia, Finland: Loimaa, 3.viii.1982, slide PS1477 (coll. ZMH)
b. aedeagus, Finland: Loimaa, 3.viii.1982, slide PS1477 (coll. ZMH)
c. vesica, Finland: Paattinen, without date, slide PS1208 (coll. ZMH)
d. 8th sternite, Finland: Loimaa, 3.viii.1982, slide PS1477 (coll. ZMH)

Macaria ichnusae GOVI & FIUMI
a. genitalia, Italy: Sardinia, Rio Gutturu Mannu, 18.vi.2009, slide PS1659 (coll. GOVI)
b. aedeagus, Italy: Sardinia, Rio Gutturu Mannu, 18.vi.2009, slide PS1659 (coll. GOVI)
c. vesica, Italy: Sardinia, Rio Gutturu Mannu, 18.vi.2009, slide PS1659 (coll. GOVI)
d. 8th sternite, Italy: Sardinia, Rio Gutturu Mannu, 18.vi.2009, slide PS1659 (coll. GOVI)

Macaria carbonaria (CLERCK)
a. genitalia, Finland: Rantasalmi, 31.v.1996, slide PS1827 (coll. P. SIHVONEN)
b. aedeagus, Finland: Rantasalmi, 8.vi.1994, slide PS1289 (coll. P. SIHVONEN)
c. vesica, Finland: Rantasalmi, 31.v.1996, slide PS1141 (coll. P. SIHVONEN)
d. 8th sternite, Finland: Rantasalmi, 31.v.1996, slide PS1141 (coll. P. SIHVONEN)

Macaria brunneata (THUNBERG)
a. genitalia, Finland: Vantaa, 18–29.vi.1995, slide PS1828 (coll. P. SIHVONEN)
b. aedeagus, Finland: Vantaa, 18–29.vi.1995, slide PS1464 (coll. P. SIHVONEN)
c. vesica, Finland: Vantaa, 18–29.vi.1995, slide PS1828 (coll. P. SIHVONEN)
d. 8th sternite, Finland: Vantaa, 18–29.vi.1995, slide PS1464 (coll. P. SIHVONEN)

Macaria loricaria (EVERSMANN)
a. genitalia, Finland: Savonlinna, 28.vii.1993, slide PS1461 (coll. P. SIHVONEN)
b. aedeagus, Finland: Savonlinna, 28.vii.1993, slide PS1461 (coll. P. SIHVONEN)
c. cornuti, Finland: Savonlinna, 28–30.vii.1996, slide PS1150 (coll. P. SIHVONEN)
d. vesica, Finland: Savonlinna, 28–30.vii.1996, slide PS1150 (coll. P. SIHVONEN)
e. 8th sternite, Finland: Savonlinna, 28–30.vii.1996, slide PS1150 (coll. P. SIHVONEN)

Macaria fusca (THUNBERG)
a. genitalia, Norway: Vestlige Finnmark, c. 300 m, 20.vii.1998, slide PS1143 (coll. P. SIHVONEN)
b. aedeagus, Norway: Vestlige Finnmark, c. 300 m, 20.vii.1998, slidePS1458 (coll. P. SIHVONEN)
c. apex of aedeagus, Norway: Vestlige Finnmark, c. 300 m, 20.vii.1998, slide PS1143 (coll. P. SIHVONEN)
d. vesica, Norway: Vestlige Finnmark, c. 300 m, 20.vii.1998, slide PS1143 (coll. P. SIHVONEN)
e. 8th sternite, Norway: Vestlige Finnmark, c. 300 m, 20.vii.1998, slide PS1143 (coll. P. SIHVONEN)

Chiasmia clathrata (LINNAEUS)
a. genitalia, Finland: Savonlinna, 5.vii.1993 slide PS1457 (coll. P. SIHVONEN)
b. aedeagus, Finland: Savonlinna, 5.vii.1993, slide PS1457 (coll. P. SIHVONEN)
c. apex of aedeagus, Finland: Enonkoski, 9.vi.1996, slide PS1145 (coll. P. SIHVONEN)
d. vesica, Finland: Enonkoski, 9.vi.1996, slide PS1145 (coll. P. SIHVONEN)
e. 8th sternite, Finland: Savonlinna, 4.vii.1993, slide PS1140 (coll. P. SIHVONEN)

Chiasmia aestimaria subsp. *aestimaria* (HÜBNER)
a. genitalia, Greece: Crete, 25–29.vi.2000, slide PS1607 (coll. SKOU)
b. aedeagus, Bulgaria: Družbe, 6–11.ix.1976, slide PS1479 (coll. ZMH)
c. cornutus, Bulgaria: Družbe, 6–11.ix.1976, slide PS1215 (coll. ZMH)
d. vesica, Bulgaria: Družbe, 6–11.ix.1976, slide PS1215 (coll. ZMH)
e. 8th sternite, Bulgaria: Družbe, 6–11.ix.1976, slide PS1215 (coll. ZMH)

Chiasmia aestimaria subsp. *sareptanaria* (STAUDINGER)
a. genitalia, Russia: Sarepta [Volgograd], 31.vii.1895, slide PS1699 (coll. ZMH)
b. aedeagus, Russia: Sarepta [Volgograd], 31.vii.1895, slide PS1699 (coll. ZMH)
c. cornutus, Russia: Sarepta [Volgograd], 31.vii.1895, slide PS1699 (coll. ZMH)
d. vesica, Russia: Sarepta [Volgograd], 31.vii.1895, slide PS1699 (coll. ZMH)
e. 8th sternite, Russia: Sarepta [Volgograd], 31.vii.1895, slide PS1699 (coll. ZMH)

Chiasmia syriacaria (STAUDINGER) Non-European, see text for details
a. genitalia, Turkmenistan: Kopetdag, Garygala, 15.iv–10.v.1994, slide PS1600 (coll. ZSM)
b. aedeagus, Turkestan: Ashchabad, 8.v.1990, slide PS1225 (coll. SKOU)
c. cornutus, Turkmenistan: Kopetdag, Garygala, 15.iv–10.v.1994, slide PS1600 (coll. ZSM)
d. vesica, Turkmenistan: Kopetdag, Garygala, 15.iv–10.v.1994, slide PS1600 (coll. ZSM)
e. 8th sternite, Turkestan: Ashchabad, 8.v.1990, slide PS1225 (coll. SKOU)

Digrammia rippertaria (DUPONCHEL)
a. genitalia, France: department Loire, St-Étienne, 15.vi.1921, slide PS1599 (coll. ZSM)
b. aedeagus, France: department Loire, St-Étienne, 15.vi.1921, slide PS1599 (coll. ZSM)
c. vesica, France: department Loire, St-Étienne, 15.vi.1921, slide PS1599 (coll. ZSM)
d. 8th sternite, France: department Loire, St-Étienne, 15.vi.1921, slide PS1599 (coll. ZSM)

Rhoptria asperaria (HÜBNER)
a. genitalia, Spain: province Gerona, Massanet de la Selva, 10.vi.1982, slide PS1480 (coll. ZMH)
b. aedeagus, Spain: province Gerona, Massanet/ de la Selva, 10.vi.1982, slide PS1480 (coll. ZMH)
c. vesica, Spain: [province Cadiz], Castellar [de la Frontera], 26.v.1989, slide PS1217 (coll. ZMH)
d. 8th sternite, Spain: [province Cadiz], Castellar [de la Frontera], 26.v.1989, slide PS1217 (coll. ZMH)

Rhoptria dolosaria (HERRICH-SCHÄFFER)
a. genitalia, Greece: Olympia, 15.iv.1979, slide PS1613 (coll. ZFMK)
b. aedeagus, Greece: Olympia, 15.iv.1979, slide PS1613 (coll. ZFMK)
c. vesica, Greece: Olympia, 15.iv.1979, slide PS1830 (coll. ZFMK)
d. 8th sternite, Greece, Olympia, 15.iv.1979, slide PS1613 (coll. ZFMK)

Ortaliella gruneraria (STAUDINGER)
a. genitalia, Greece: Lakonia, Mt. Taygetos, 1000 m, 28–29.vi.1982, slide PS1256 (coll. SKOU)
b. aedeagus, Greece: Lakonia, Mt. Taygetos, 1000 m, 28–29.vi.1982, slide PS1256 (coll. SKOU)
c. cornuti, Greece: Lakonia, Mt. Taygetos, 1000 m, 28–29.vi.1982, slide PS1256 (coll. SKOU)
d. vesica, Greece: Lakonia, Mt. Taygetos, 1000 m, 28–29.vi.1982, slide PS1256 (coll. SKOU)
e. 8th sternite, Greece: Lakonia, Mt. Taygetos, 1000 m, 28–29.vi.1982, slide PS1256 (coll. SKOU)

Cepphis advenaria (HÜBNER)
a. genitalia, Finland: Savonlinna, 20.vi.1993, slide PS1153 (coll. P. SIHVONEN)
b. aedeagus, Finland: Savonlinna, 12–19.vi.1997, slide PS1463 (coll. P. SIHVONEN)
c. cornuti, Finland: Savonlinna, 12–19.vi.1997, slide PS1463 (coll. P. SIHVONEN)
d. vesica, Finland: Savonlinna, 11–14.vi.1996, slide PS1829 (coll. P. SIHVONEN)
e. 8th sternite, Finland: Savonlinna, 20.vi.1993, slide PS1153 (coll. P. SIHVONEN)

Petrophora chlorosata (SCOPOLI)
a. genitalia, Finland: Kirkkonummi, 14.vi.1997, slide PS1465 (coll. P. SIHVONEN)
b. aedeagus, Finland: Kirkkonummi, 14.vi.1997, slide PS1465 (coll. P. SIHVONEN)
c. vesica, Finland: Kirkkonummi, 16.vi.1998, slide PS1159 (coll. P. SIHVONEN)
d. furca arms, Finland: Kirkkonummi, 14.vi.1997, slide PS1465 (coll. P. SIHVONEN)

Perigune binaevata (MABILLE)
a. genitalia, without locality data, specimen from coll. P. Mabille, slide PS1601 (coll. ZSM)
b. aedeagus, Spain: Mallorca, 1979, slide PS1743 (coll. SKOU)
c. vesica, without locality data, specimen from coll. P. Mabille, slide PS1601 (coll. ZSM)
d. tympanal organs, without locality data, specimen from coll. P. Mabille, slide PS1601 (coll. ZSM)

Perigune convergata (VILLERS)
a. genitalia, Spain: Sierra Nevada, 1600 m, 19.ix.1987, slide PS1276 (coll. SKOU)
b. aedeagus, Spain: Sierra Nevada, 1600 m, 19.ix.1987, slide PS1276 (coll. SKOU)
c. vesica, Tunisia: Medenine, 16.xi.2009, slide PS1583 (coll. NUPPONEN)
d. apex of gnathos, Tunisia: Medenine, 16.xi.2009, slide PS1583 (coll. NUPPONEN)

Perigune narbonea (LINNAEUS)
a. genitalia, Spain: province Almeria, Almeria 20 km N, 10.xi.2007, slide PS1591 (coll. NUPPONEN)
b. aedeagus, Spain: prov. Girona, at Sant Marti Sesserres, 300 m, 10.vii.1992, slide PS1899 (coll. SKOU)
c. vesica, Spain: province Almeria, Almeria 20 km N, 10.xi.2007, slide PS1591 (coll. NUPPONEN)
d. 8th tergite, Spain: province Almeria, Almeria 20 km N, 10.xi.2007, slide PS1591 (coll. NUPPONEN)

Lignyoptera fumidaria (HÜBNER)
a. genitalia, Austria: Braunsberg, 23.xi.1980, slide PS1280 (coll. SKOU)
b. aedeagus, Austria: Gramatneusiedl, 23.xi.1978, , slide PS1898 (coll. SKOU)
c. vesica, Austria: Braunsberg, 23.xi.1980, slide PS1280 (coll. SKOU)
d. 8th sternite, Austria: Braunsberg, 23.xi.1980, slide PS1280 (coll. SKOU)

Lignyoptera thaumastaria REBEL
a. genitalia, Bosnia and Herzegovina? [label: Nova-Bosnien], 1550 m, 20.x.1937, slide PS1626 (coll. ZFMK)
b. aedeagus, Bosnia and Herzegovina? [label: Nova-Bosnien], 1550 m, 20.x.1937, slide PS1626 (coll. ZFMK)
c. vesica, Bosnia and Herzegovina? [label: Nova-Bosnien], 1550 m, 20.x.1937, slide PS1626 (coll. ZFMK)
d. 8th sternite, Bosnia and Herzegovina? [label: Nova-Bosnien], 1550 m, 20.x.1937, slide PS1626 (coll. ZFMK)

Heterolocha laminaria (HERRICH-SCHÄFFER)
a. genitalia, Azerbaijan: Lerik, 27.vi.1984, slide PS1894 (coll. ZMH)
b. aedeagus, Caucasus (no other data), slide PS1895 (coll. ZMH)
c. vesica, Azerbaijan: Lerik, 27.vi.1984, slide PS1894 (coll. ZMH)
d. 8th sternite, Azerbaijan: Lerik, 27.vi.1984, slide PS1894 (coll. ZMH)

Plagodis pulveraria (LINNAEUS)
a. genitalia, Finland: Enonkoski, 5–6.vi.1996, slide PS1228 (coll. P. SIHVONEN)
b. aedeagus, Finland: Savonlinna, 19.v.1993, slide PS1900 (coll. P. SIHVONEN)
c. vesica, Finland: Enonkoski, 5–6.vi.1996, slide PS1228 (coll. P. SIHVONEN)
d. 8th sternite, Finland: Enonkoski, 5–6.vi.1996, slide PS1228 (coll. P. SIHVONEN)

Plagodis dolabraria (LINNAEUS)
a. genitalia, Finland: Espoo, 20.vi.1991, slide PS1565 (coll. P. SIHVONEN)
b. aedeagus, Finland: Espoo, 20.vi.1991, slide PS1565 (coll. P. SIHVONEN)
c. vesica, Finland: Vantaa, 14–17.vi.1995, slide PS1230 (coll. P. SIHVONEN)
d. 8th sternite, Finland: Vantaa, 14–17.vi.1995, slide PS1230 (coll. P. SIHVONEN)

Pachycnemia hippocastanaria (HÜBNER)

a. genitalia, Italy: region Tuscany, Coreglia Antelminelli (Lucca), 600 m, 28.vi–10.vii.1999, slide PS1315 (coll. P. SIHVONEN)
b. aedeagus, Italy: 100 km NE from Florence, Alpi Apuane, 800 m, 6.vii.1999, slide PS1907 (coll. H. SIHVONEN)
c. vesica, Italy: region Tuscany, Coreglia Antelminelli (Lucca), 600 m, 28.vi–10.vii.1999, slide PS1315 (coll. P. SIHVONEN)
d. apex of valva, Italy: 100 km NE from Florence, Alpi Apuane, 800 m, 6.vii.1999,, slide PS1907 (coll. H. SIHVONEN)

Pachycnemia tibiaria subsp. *tibiaria* (RAMBUR)

a. genitalia, France: Dept. Var, Frejus, 5–7.ix.2006, slide PS1261 (coll. SKOU)
b. aedeagus, France: Dept. Var, Frejus, 5–7.ix.2006, slide PS1261 (coll. SKOU)
c. vesica, France: Dept. Var, Frejus, 5–7.ix.2006, slide PS1261 (coll. SKOU)
d. apex of valva, France: Dept. Var, Frejus, 5–7.ix.2006, slide PS1261 (coll. SKOU)

Pachycnemia tibiaria subsp. *benesignata* (BELLIER)

a. genitalia (insert: uncus apex), France: Corsica, sea level, 27.ix.2003, slide PS1265 (coll. SKOU)
b. aedeagus, France: Corsica, 22.ix.2003, 800 m, slide PS1263 (coll. SKOU)
c. vesica, France: Corsica, 27.ix.2003, sea level, slide PS1265 (coll. SKOU)
d. apex of valva, France: Corsica, sea level, 27.ix.2003, slide PS1265 (coll. SKOU)

Opisthograptis luteolata (LINNAEUS)

a. genitalia, Finland: Enonkoski, 27.vi.1996, slide PS1281 (coll. P. SIHVONEN)
b. aedeagus, Finland: Enonkoski, 27.vi.1996, slide PS1841 (coll. P. SIHVONEN)
c. vesica, Finland: Enonkoski, 27.vi.1996, slide PS1281 (coll. P. SIHVONEN)
d. cornuti, Finland: Enonkoski, 27.vi.1996, slide PS1281 (coll. P. SIHVONEN)

Epione repandaria (HUFNAGEL)

a. genitalia, Finland: Savonlinna, 29–31.vii.1995, slide PS1285 (coll. P. SIHVONEN)
b. aedeagus, Finland: Savonlinna, 5–8.viii.1995, slide PS1842 (coll. P. SIHVONEN)
c. vesica, Finland: Savonlinna, 29–31.vii.1995, slide PS1285 (coll. P. SIHVONEN)
d. 8th sternite, Finland: Savonlinna, 29–31.vii.1995, slide PS1285 (coll. P. SIHVONEN)

Epione vespertaria (LINNAEUS)

a. genitalia, Finland: Savonlinna, 6–8.viii.1993, slide PS1574 (coll. P. SIHVONEN)
b. aedeagus, Finland: Savonlinna, 6–8.viii.1993, slide PS1574 (coll. P. SIHVONEN)
c. vesica, Finland: Savonlinna, 2–4.viii.1993, slide PS1287 (coll. P. SIHVONEN)
d. 8th sternite, Finland: Savonlinna, 2–4.viii.1993, slide PS1287 (coll. P. SIHVONEN)

Therapis flavicaria (DENIS & SCHIFFERMÜLLER)

a. genitalia, Russia: North Caucasus, Kislovodsk 800 m, 30.vii.1990, slide PS1909 (coll. ZMH)
b. aedeagus, Russia: North Caucasus, Kislovodsk 800 m, 30.vii.1990, slide PS1909 (coll. ZMH)
c. surface of membranous coremata, Russia: North Caucasus, Kislovodsk 800 m, 30.vii.1990, slide PS1364 (coll. ZMH)
d. vesica, Russia: North Caucasus, Kislovodsk 800 m, 30.vii.1990, slide PS1364 (coll. ZMH)
e. tympanal organ, Russia: North Caucasus, Kislovodsk 800 m, 30.vii.1990, slide PS1364 (coll. ZMH)

Pseudopanthera macularia (LINNAEUS)
a. genitalia, Finland: Kitee, 15.vi.1996, slide PS1283 (coll. P. SIHVONEN)
b. aedeagus, Finland: Kitee, 16.vi.1993, slide PS1901 (coll. P. SIHVONEN)
c. apex of furca arm, Finland: Kitee, 16.vi.1993, slide PS1901 (coll. P. SIHVONEN)
d. vesica, Finland: Kitee, 15.vi.1996, slide PS1283 (coll. P. SIHVONEN)
e. 8th sternite, Finland: Kitee, 15.vi.1996, slide PS1283 (coll. P. SIHVONEN)

Eilicrinia cordiaria (HÜBNER)
a. genitalia, Hungary: Tolna, 16.vii.1964, slide PS1387 (coll. ZMH)
b. valva margin (without extensions), Hungary: Tolna, 16.vii.1964, slide PS1387 (coll. ZMH)
c. aedeagus, Hungary: Tolna, 16.vii.1964, slide PS1387 (coll. ZMH)
d. spines of aedeagus, Hungary: Tolna, 16.vii.1964, slide PS1387 (coll. ZMH)
e. vesica, Hungary: Tolna, 16.vii.1964, slide PS1387 (coll. ZMH)
f. 8th sternite, Hungary: Tolna, 16.vii.1964, slide PS1387 (coll. ZMH)

Eilicrinia subcordaria HERRICH-SCHÄFFER
a. genitalia, Russia: Sarepta [Volgograd], 23.vi.1900, slide PS1409 (coll. ZMH)
b. valva margin (with single spine), Russia: Sarepta [Volgograd], 23.vi.1900, slide PS1409 (coll. ZMH)
c. aedeagus, Russia: Sarepta [Volgograd], 23.vi.1900, slide PS1409 (coll. ZMH)
d. spines of aedeagus, Russia: Sarepta [Volgograd], 23.vi.1900, slide PS1409 (coll. ZMH)
e. vesica, Russia: Sarepta [Volgograd], 23.vi.1900, slide PS1409 (coll. ZMH)
f. 8th sternite, Russia: Sarepta [Volgograd], 23.vi.1900, slide PS1409 (coll. ZMH)

Eilicrinia trinotata (METZNER)
a. genitalia, Hungary: Budapest, 5.v.1949, slide PS1411 (coll. ZMH)
b. valva margin (with several spines), Hungary: Budapest, 5.v.1949, slide PS1411 (coll. ZMH)
c. aedeagus, Hungary: Budapest, 5.v.1949, slide PS1411 (coll. ZMH)
d. base of valva, Hungary: Budapest, 5.v.1949, slide PS1411 (coll. ZMH)
e. vesica, Hungary: Budapest, 5.v.1949, slide PS1411 (coll. ZMH)
f. 8th sternite, Hungary: Budapest, 5.v.1949, slide PS1411 (coll. ZMH)

Toulgoetia cauteriata (STAUDINGER)
a. genitalia, Spain: Province Cádiz, 8 km NE Tarifa, 300 m, 22–24.iv.2007, slide PS1518 (coll. SKOU)
b. inner margin of valva, Spain: Province Cádiz, 8 km NE Tarifa, 300 m, 22–24.iv.2007, slide PS1518 (coll. SKOU)
c. aedeagus, Spain: Province Cádiz, Algeciras, 27.iii.1901, slide PS1966/BMNH GEO 24894 (coll. BMNH)
d. apex of aedeagus, Spain: Province Cádiz, 8 km NE Tarifa, 300 m, 22–24.iv.2007, slide PS1518 (coll. SKOU)
e. vesica, Spain: Province Cádiz, 8 km NE Tarifa, 300 m, 22–24.iv.2007, slide PS1518 (coll. SKOU)
f. 8th sternite, Spain: Province Cádiz, 8 km NE Tarifa, 300 m, 22–24.iv.2007, slide PS1518 (coll. SKOU)

Hypoxystis pluviaria (FABRICIUS)
a. genitalia, Finland: Kerimäki, 2.vi.2001, slide PS1291 (coll. P. SIHVONEN)
b. aedeagus, Finland: Malax [Maalahti], 30.v.1995, slide PS1902 (coll. P. SIHVONEN)
c. cornutus, Finland: Kerimäki, 2.vi.2001, slide PS1291 (coll. P. SIHVONEN)
d. vesica, Finland: Kerimäki, 2.vi.2001, slide PS1291 (coll. P. SIHVONEN)
d. 8th sternite, Finland: Malax [Maalahti], 30.v.1995, slide PS1902 (coll. P. SIHVONEN)

Enanthyperythra legataria (HERRICH-SCHÄFFER)
a. genitalia, Macedonia: Drenovo, 12.iv.1959, slide PS1502 (coll. ZMH)
b. aedeagus, Macedonia: Drenovo, 12.iv.1959, slide PS1502 (coll. ZMH)
c. cornuti, Macedonia: Drenovo, 12.iv.1959, slide PS1502 (coll. ZMH)
d. vesica, Macedonia: Drenovo, 12.iv.1959, slide PS1502 (coll. ZMH)
e. furca arms, Macedonia: Drenovo, 12.iv.1959, slide PS1502 (coll. ZMH)

Apeira syringaria (LINNAEUS)
a. genitalia, Finland: Sipoo, 8.vii.1989, slide PS1303 (coll. P. SIHVONEN)
b. aedeagus, Finland: Sipoo, 8.vii.1989, slide PS1303 (coll. P. SIHVONEN)
c. vesica, Finland: Sipoo, 8.vii.1989, slide PS1303 (coll. P. SIHVONEN)
d. 8th sternite, Finland: Sipoo, 8.vii.1989, slide PS1303 (coll. P. SIHVONEN)

Ennomos autumnaria (WERNEBURG)
a. genitalia, Finland: Vantaa, 13–17.viii.1995, slide PS1293 (coll. P. SIHVONEN)
b. aedeagus, Finland: Vantaa, 13–17.viii.1995, slide PS1293 (coll. P. SIHVONEN)
c. vesica, Finland: Vantaa, 13–17.viii.1995, slide PS1124 (coll. P. SIHVONEN)
d. apex of aedeagus and base of vesica, Finland: Vantaa, 13–17.viii.1995, slide PS1124 (coll. P. SIHVONEN)

Ennomos quercinaria (HUFNAGEL)
a. genitalia, Germany: Berlin region, without date, slide PS1367 (coll. ZMH)
b. aedeagus, Germany: Berlin region, without date, slide PS1367 (coll. ZMH)
c. vesica, Germany: Berlin region, without date, slide PS1367 (coll. ZMH)
d. cornuti, Germany: Berlin region, without date, slide PS1367 (coll. ZMH)

Ennomos alniaria (LINNAEUS), specimen taken in September
a. genitalia, Finland: Lohja, 14–19.ix.1998, slide PS1275 (coll. P. SIHVONEN)
b. aedeagus, Finland: Lohja, 14–19.ix.1998, slide PS1275 (coll. P. SIHVONEN)
c. vesica, Finland: Lohja, 14–19.ix.1998, slide PS1275 (coll. P. SIHVONEN)
d. valva, Finland: Lohja, 14–19.ix.1998, slide PS1275 (coll. P. SIHVONEN)

Ennomos alniaria (LINNAEUS), specimen taken in May
a. genitalia, Spain: Province Cádiz, 10 km NE of Tarifa, 300 m, 17–18.v.2006, slide PS1269 (coll. SKOU)
b. aedeagus, Spain: Province Cádiz, 10 km NE of Tarifa, 300 m, 17–18.v.2006, slide PS1269 (coll. SKOU)
c. vesica, Spain: Province Cádiz, 10 km NE of Tarifa, 300 m, 17–18.v.2006, slide PS1269 (coll. SKOU)
d. valva, Spain: Province Cádiz, 10 km NE of Tarifa, 300 m, 17–18.v.2006, slide PS1269 (coll. SKOU)

Ennomos fuscantaria (HAWORTH)
a. genitalia, Finland: Föglö, 22–25.viii.1971, slide PS1903 (coll. P. SIHVONEN)
b. aedeagus, Finland: Föglö, 22–25.viii.1971, slide PS1903 (coll. P. SIHVONEN)
c. apex of gnathos, Finland: Föglö, 22–25.viii.1971, slide PS1903 (coll. P. SIHVONEN)
d. vesica, Russia: Rostov, vii.1984, slide PS1620 (coll. ZFMK)
e. uncus, Cypros: Limassol, without date, slide PS1619 (coll. ZFMK)

Ennomos erosaria subsp. *erosaria* (DENIS & SCHIFFERMÜLLER)
a. genitalia, Denmark: Vejlen, 27.vii.1973, slide PS1271 (coll. SKOU)
b. aedeagus, Denmark: Lysemose, 3.ix.1972, slide PS1270 (coll. SKOU)
c. vesica, Denmark: Vejlen, 27.vii.1973, slide PS1271 (coll. SKOU)
d. furca arms, Denmark: Vejlen, 27.vii.1973, slide PS1271 (coll. SKOU)

Ennomos erosaria subsp. *ifranaria* (RUNGS)
a. genitalia, Morocco: Atlas Mountains, Ifrane, 4–12.vii.1964, slide PS1725 (coll. ZSM)
b. aedeagus, Morocco: Atlas Mountains, Ifrane, 4–12.vii.1964, slide PS1725 (coll. ZSM)
c. vesica, Morocco: Atlas Mountains, Ifrane, 4–12.vii.1964, slide PS1725 (coll. ZSM)

d.furca arms, Morocco: Atlas Mountains, Ifrane, 4–12.vii.1964, slide PS1725 (coll. ZSM)

Ennomos quercaria subsp. *quercaria* (HÜBNER)
a. genitalia, Greece: Nom. Ioannina, 4 km SW Geroplatanos, 800 m, 12.vii.1998, slide PS1272 (coll. SKOU)
b. aedeagus, Austria: Wien, ex larva 23.vi.1934,, slide PS1612 (coll. ZSM)
c. vesica, Greece: Nom. Ioannina, 4 km SW Geroplatanos, 800 m, 12.vii.1998, slide PS1272 (coll. SKOU)
d. furca arms, Greece: Nom. Ioannina, 4 km SW Geroplatanos, 800 m, 12.vii.1998, slide PS1272 (coll. SKOU)

Ennomos quercaria subsp. *duercki* REISSER
a. genitalia, Greece: Crete, Topolia, 300 m, 25–29.vi.2000, slide PS1273 (coll. SKOU)
b. aedeagus, Greece: Crete, Zourva, 800 m, 27–30.vi.2000, slide PS1274 (coll. SKOU)
c. vesica, Greece: Crete, Topolia, 300 m, 25–29.vi.2000, slide PS1273 (coll. SKOU)
d. furca arms, Greece: Crete, Topolia, 300 m, 25–29.vi.2000, slide PS1273 (coll. SKOU)

Ennomos quercaria subsp. *lissochila* (PROUT)
a. genitalia, Cyprus: Lamela, 550 m, 23–29.vi.1997, slide PS1524 (coll. SKOU)
b. aedeagus, Cyprus: Troodos mts., 1020 m, ex ovo 12.vi.2009, slide PS1718 (coll. MÜLLER)
c. vesica, Cyprus: Lamela, 550 m, 23–29.vi.1997, slide PS1524 (coll. SKOU)
d. furca arms, Cyprus: Lamela, 550 m, 23–29.vi.1997, slide PS1524 (coll. SKOU)

Eumera regina STAUDINGER
a. genitalia, Greece: Kastoria, 1200 m, ix.1983, slide PS1382 (coll. ZMH)
b. aedeagus, Greece: Kastoria, 1200 m, ix.1983, slide PS1382 (coll. ZMH)
c. vesica, Greece: Kastoria, 1200 m, ix.1983, slide PS1382 (coll. ZMH)
d. scaphium and socii, Greece: Kastoria, 1200 m, ix.1983, slide PS1382 (coll. ZMH)

Selenia dentaria (FABRICIUS)
a. genitalia, Finland: Savonlinna, 12–13.v.1994, slide PS1298 (coll. P. SIHVONEN)
b. aedeagus, Italy: 100 km NE from Florence, Alpi Apuane, 800 m, 6.vii.1999, slide PS1904 (coll. P. SIHVONEN)
c. vesica, Finland: Savonlinna, 12–13.v.1994, slide PS1298 (coll. P. SIHVONEN)
d. 8th tergite, Finland: Savonlinna, 12–13.v.1994, slide PS1298 (coll. P. SIHVONEN)

Selenia lunularia (HÜBNER)
a. genitalia, Finland: Kirkkonummi, 11–21.vi.1999, slide PS1300 (coll. P. SIHVONEN)
b. aedeagus, Finland: Kirkkonummi, 26.v–12.vi.1993, slide PS1905 (coll. P. SIHVONEN)
c. vesica, Finland: Kirkkonummi, 11–21.vi.1999, slide PS1300 (coll. P. SIHVONEN)
d. 8th tergite, Finland: Kirkkonummi, 11–21.vi.1999, slide PS1300 (coll. P. SIHVONEN)

Selenia tetralunaria (HUFNAGEL)
a. genitalia, Finland: Savonlinna, 28–29.v.1996, slide PS1301 (coll. P. SIHVONEN)
b. aedeagus, Finland: Savonlinna, 28–29.v.1996, slide PS1301 (coll. P. SIHVONEN)
c. vesica, Finland: Savonlinna, 28–29.v.1996, slide PS1301 (coll. P. SIHVONEN)
d. 8th tergite, Finland: Savonlinna, 28–29.v.1996, slide PS1301 (coll. P. SIHVONEN)

Artiora evonymaria (DENIS & SCHIFFERMÜLLER)
a. genitalia, Czech Republic: Moravia, Palava mountains, 5.ix.1994, slide PS1384 (coll. ZMH)
b. aedeagus, Czech Republic: Moravia, Palava mountains, 5.ix.1994, slide PS1384 (coll. ZMH)
c. vesica, Czech Republic: Moravia, Palava mountains, 5.ix.1994, slide PS1384 (coll. ZMH)
d. apex of aedeagus, Czech Republic: Moravia, Palava mountains, 5.ix.1994, slide PS1384 (coll. ZMH)

Odontopera bidentata (CLERCK), specimens from Finland. See text for details.
a. genitalia, Finland: Savonlinna, 11–14.vi.1996, slide PS1304 (coll. P. SIHVONEN)
b. aedeagus, Finland: Savonlinna, 11–14.vi.1996, slide PS1304 (coll. P. SIHVONEN)
c. vesica, Finland: Savonlinna, 26.vi–2.vii.1996, slide PS1657 (coll. P. SIHVONEN)
d. cornuti, Finland: Savonlinna, 11–14.vi.1996, slide PS1304 (coll. P. SIHVONEN)

Odontopera bidentata (CLERCK) f. *graecarius*, specimens from Greece. See text for details.
a. genitalia, Greece: Lagkádia, 1200 m, 19.v.1969, slide PS1656 (coll. ZSM)
b. aedeagus, Greece: Peleponnese, Mount Chelmos, 1600–1700 m, 15.vi.2010, slide PS1838 (coll. SKOU)
c. vesica, Greece: Peleponnese, Mount Chelmos, 1600–1700 m, 15.vi.2010, slide PS1838 (coll. SKOU)
d. cornuti, Greece: Peleponnese, Mount Chelmos, 1600–1700 m, 15.vi.2010, slide PS1838 (coll. SKOU)

Dasycorsa modesta (STAUDINGER)
a. genitalia, Romania: Agigea, 26.iv.1964, slide PS1386 (coll. ZMH)
b. aedeagus, Cyprus: Pergamoth, January 1983, slide PS1929 (coll. SKOU)
c. vesica, Romania: Agigea, 26.iv.1964, slide PS1386 (coll. ZMH)
d. cornutus, Romania: Agigea, 26.iv.1964, slide PS1386 (coll. ZMH)

Crocallis tusciaria subsp. *tusciaria* (BORKHAUSEN)

a. genitalia, Italy: region Alto Adige (Südtirol), 800 m, 16.x.1979, slide PS1567 (coll. ZMH)
b. aedeagus, Italy: region Alto Adige (Südtirol), 800 m, 16.x.1979, slide PS1567 (coll. ZMH)
c. vesica, Macedonia: Ohrid, ex ovo 19.ix.1955, slide PS1369 (coll. ZMH)
d. apex of valva, Austria: region Alto Adige (Südtirol), 800 m, 16.x.1979, slide PS1567 (coll. ZMH)

Crocallis tusciaria subsp. *cypriaca* FISCHER

a. genitalia, Cyprus: Platres, 1150 m, 14–20.x.1996, slide PS1525 (coll. SKOU)
b. aedeagus, Cyprus: Platres, 1150 m, 14–20.x.1996, slide PS1525 (coll. SKOU)
c. vesica, Cyprus: Platres, 1150 m, 14–20.x.1996, slide PS1525 (coll. SKOU)
d. apex of valva, Cyprus: Platres, 1150 m, 14–20.x.1996, slide PS1525 (coll. SKOU)

Crocallis albarracina WEHRLI

a. genitalia, Spain: Avila, Sierra de Gredos, 1700 m, 18.viii.1988, slide PS1526 (coll. SKOU)
b. aedeagus, Spain: Province Teruel, 1 km E of Tramacastilla, 1250 m, 27.viii.2001, slide PS1528 (coll. SKOU)
c. vesica, Spain: Avila, Sierra de Gredos, 1700 m, 18.viii.1988, slide PS1526 (coll. SKOU)
d. lateral appendices of juxta, Spain: Avila, Sierra de Gredos, 1700 m, 18.viii.1988, slide PS1526 (coll. SKOU)

Crocallis elinguaria (LINNAEUS)

a. genitalia, Finland: Savonlinna, 29–31.vii.1995, slide PS1308 (coll. P. SIHVONEN)
b. aedeagus, Finland, Savonlinna, 27.vii.1996, slide PS1910 (coll. P. SIHVONEN)
c. vesica, Finland: Savonlinna, 29–31.vii.1995, slide PS1308 (coll. P. SIHVONEN)
d. lateral appendices of juxta, Finland: Savonlinna, 29–31.vii.1995, slide PS1308 (coll. P. SIHVONEN)

Crocallis dardoinaria DONZEL

a. genitalia, Spain, Granada, Sierra Nevada, 1600 m, 19.ix.1987, slide PS1530 (coll. SKOU)
b. aedeagus, Spain, Granada, Sierra Nevada, 1600 m, 19.ix.1987, slide PS1530 (coll. SKOU)
c. vesica, Spain, Granada, Sierra Nevada, 1600 m, 19.ix.1987, slide PS1530 (coll. SKOU)
d. lateral appendices of juxta, Spain, Granada, Sierra Nevada, 1600 m, 19.ix.1987, slide PS1530 (coll. SKOU)

Crocallis auberti OBERTHÜR

a. genitalia, Spain: province Cadiz, [Castellar de la Frontera] Almoraima, 17.xi.1980, slide PS1531 (coll. SKOU)
b. aedeagus, Algeria: Algêrie, 8.xi.1918, slide PS1728 (coll. ZSM)
c. vesica, Spain: province Cadiz, [Castellar de la Frontera] Almoraima, 17.xi.1980, slide PS1531 (coll. SKOU)
d. lateral appendices of juxta, Spain: province Cadiz, [Castellar de la Frontera] Almoraima, 17.xi.1980, slide PS1531 (coll. SKOU)

Crocallis helenaria RUCKDESCHEL
a. genitalia, Greece: Crete, 7 km ESE of Rethimno, 11–17.x.2003, slide PS1534 (coll. SKOU)
b. aedeagus, Greece: Crete, 7 km ESE of Rethimno, 11–17.x.2003, slide PS1534 (coll. SKOU)
c. vesica, Greece: Crete, 7 km ESE of Rethimno, 11–17.x.2003, slide PS1534 (coll. SKOU)
d. lateral appendices of juxta, Greece: Crete, 7 km ESE of Rethimno, 11–17.x.2003, slide PS1534 (coll. SKOU)

Crocallis boisduvalaria (LUCAS)
a. genitalia, Italy: Calabria, Civita, 7.x.1982, slide PS1533 (coll. SKOU)
b. aedeagus, Italy: Calabria, Civita, 7.x.1982, slide PS1533 (coll. SKOU)
c. vesica, Italy: Calabria, Civita, 7.x.1982, slide PS1533 (coll. SKOU)
d. apex of valva, Italy: Calabria, Civita, 7.x.1982, slide PS1533 (coll. SKOU)

Ourapteryx sambucaria (LINNAEUS)
a. genitalia, Russia: Leningrad Oblast, Kingisepp, Kurgulovo, 6–13.vii.1996, slide PS1306 (coll. P. SIHVONEN)
b. aedeagus, Russia: Leningrad Oblast, Kingisepp, Kurgulovo, 6–13.vii.1996, slide PS1306 (coll. P. SIHVONEN)
c. vesica, Russia: Leningrad Oblast, Kingisepp, Kurgulovo, 6–13.vii.1996, slide PS1306 (coll. P. SIHVONEN)
d. apex of gnathos, Russia: Leningrad Oblast, Kingisepp, Kurgulovo, 6–13.vii.1996, slide PS1306 (coll. P. SIHVONEN)

Asovia maeoticaria (ALPHÉRAKY)
a. genitalia, Greece: Voiotis, Mt. Parnassos, 1400 m, 28.vi.1985, slide PS1537 (coll. SKOU)
b. aedeagus, Greece: Voiotis, Mt. Parnassos, 1400 m, 28.vi.1985, slide PS1537 (coll. SKOU)
c. vesica, Greece: Voiotis, Mt. Parnassos, 1400 m, 28.vi.1985, slide PS1537 (coll. SKOU)
d. cornuti, Greece: Voiotis, Mt. Parnassos, 1400 m, 28.vi.1985, slide PS1537 (coll. SKOU)
e. tympanal organ, Greece: Voiotis, Mt. Parnassos, 1400 m, 28.vi.1985, slide PS1537 (coll. SKOU)

Colotois pennaria (LINNAEUS)
a. genitalia, Finland: Vantaa, 2–9.ix.1995, slide PS1862 (coll. P. SIHVONEN)
b. aedeagus, Finland: Vantaa, 2–9.ix.1995, slide PS1862 (coll. P. SIHVONEN)
c. vesica, Finland: Savonlinna, 6–16.x.1993, slide PS1310 (coll. P. SIHVONEN)
d. uncus, Finland: Vantaa, 2–9.ix.1995, slide PS1862 (coll. P. SIHVONEN)

Angerona prunaria (LINNAEUS)
a. genitalia, Finland: Savonlinna, 17–18.vi.1995, slide PS1312 (coll. P. SIHVONEN)
b. aedeagus, Finland: Savonlinna, 26.xi–2.vii.1996, slide PS1572 (coll. P. SIHVONEN)
c. vesica, Finland: Savonlinna, 26.xi–2.vii.1996, slide PS1572 (coll. P. SIHVONEN)
d. 8th sternite, Finland: Savonlinna, 17–18.vi.1995, slide PS1312 (coll. P. SIHVONEN)

Cabera pusaria (LINNAEUS)
a. genitalia, Italy: Sardinia, Gennargentu 1600 m, 2.vii.2007, slide PS1569 (coll. NUPPONEN)
b. aedeagus, Finland: Savonlinna, 12–19.vi.1997, slide PS1333 (coll. P. SIHVONEN)
c. vesica, Finland: Savonlinna, 12–19.vi.1997, slide PS1333 (coll. P. SIHVONEN)
d. base of valva, Italy: Sardinia, Gennargentu 1600 m, 2.vii.2007, slide PS1569 (coll. P. SIHVONEN)

Cabera leptographa WEHRLI
a. genitalia, Austria: Burgenland, Zitzmannsdorfer Wiesen by Neusiedlersee, 22.vi.1981, slide PS1362 (coll. ZMH)
b. aedeagus, Austria: Burgenland, Zitzmannsdorfer Wiesen by Neusiedlersee, 22.vi.1981, slide PS1362 (coll. ZMH)
c. vesica, Austria: Burgenland, Zitzmannsdorfer Wiesen by Neusiedlersee, 22.vi.1981, slide PS1362 (coll. ZMH)
d. base of valva, Austria: Burgenland, Zitzmannsdorfer Wiesen by Neusiedlersee, 22.vi.1981, slide PS1362 (coll. ZMH)

Cabera exanthemata (SCOPOLI)
a. genitalia, Finland: Enonkoski, 19.xi.1996, slide PS1578 (coll. P. SIHVONEN)
b. aedeagus, Finland: Enonkoski, 19.xi.1996, slide PS1578 (coll. P. SIHVONEN)
c. vesica, Finland: Enonkoski, 3.vii.1996, slide PS1335 (coll. P. SIHVONEN)
d. base of valva, Finland: Enonkoski, 3.vii.1996, slide PS1335 (coll. P. SIHVONEN)

Lhommeia biskraria (OBERTHÜR)
a. genitalia, Morocco: High-Atlas, Asni-Ouirgane, 6–9.x.1964, slide PS1646 (coll. ZSM)
b. aedeagus, Morocco: High-Atlas, Asni-Ouirgane, 6–9.x.1964, slide PS1646 (coll. ZSM)
c. vesica, Libya: Gharian, Wadi El Hira, 1.vi.1983, slide PS1251 (coll. Skou)
d. 8th sternite, Morocco: High-Atlas, Asni-Ouirgane, 6–9.x.1964, slide PS1646 (coll. ZSM)

Lomographa bimaculata (FABRICIUS)
a. genitalia, Finland: Savonlinna, 3.vi.1996, slide PS1577 (coll. P. SIHVONEN)
b. aedeagus, Finland: Savonlinna, 3.vi.1996, slide PS1577 (coll. P. SIHVONEN)
c. vesica, Finland: Savonlinna, 1–2.vi.1996, slide PS1337 (coll. P. SIHVONEN)
d. 8th sternite, Finland: Savonlinna, 1–2.vi.1996, slide PS1337 (coll. P. SIHVONEN)

Lomographa temerata (DENIS & SCHIFFERMÜLLER)
a. genitalia, Finland: Lammi, 30.v–7.vi.1996, slide PS1340 (coll. P. SIHVONEN)
b. aedeagus, Finland: Lammi, 30.v–7.vi.1996, slide PS1340 (coll. P. SIHVONEN)
c. striations in aedeagus apex, Finland: Lammi, 30.v–7.vi.1996, slide PS1340 (coll. P. SIHVONEN)
d. vesica, Finland: Lammi, 30.v–7.vi.1996, slide PS1340 (coll. P. SIHVONEN)
e. 8th sternite, Finland: Lammi, 30.v–7.vi.1996, slide PS1340 (coll. P. SIHVONEN)

Aleucis distinctata (HERRICH-SCHÄFFER)
a. genitalia, Germany: Land Rheinland-Pfalz, Bornich, 31.iii.1912, slide PS1421 (coll. ZMH)
b. aedeagus, Germany: Land Rheinland-Pfalz, Bornich, 31.iii.1912, slide PS1421 (coll. ZMH)
c. vesica, Germany: Land Rheinland-Pfalz, Bornich, 31.iii.1912, slide PS1421 (coll. ZMH)
d. 8th sternite, Germany: Land Rheinland-Pfalz, Bornich, 31.iii.1912, slide PS1421 (coll. ZMH)

Aleucis orientalis (STAUDINGER)
a. genitalia, Greece: Florini, Kristalopigi, 1000 m, 17.iv.1987, slide PS1539 (coll. SKOU)
b. aedeagus, Greece: Florini, Kristalopigi, 1000 m, 17.iv.1987, slide PS1539 (coll. SKOU)
c. vesica, Greece: Florini, Kristalopigi, 1000 m, 17.iv.1987, slide PS1539 (coll. SKOU)
d. 8th sternite, Greece: Florini, Kristalopigi, 1000 m, 17.iv.1987, slide PS1539 (coll. SKOU)

Theria rupicapraria (DENIS & SCHIFFERMÜLLER)
a. genitalia, Germany: Land Mecklenburg, Wolgast, 5.iv.1941, slide PS1416 (coll. ZMH)
b. aedeagus, Germany: Land Mecklenburg, Wolgast, 5.iv.1941, slide PS1416 (coll. ZMH)
c. vesica, Germany: Land Mecklenburg, Wolgast, 5.iv.1941, slide PS1416 (coll. ZMH)
d. 8th sternite, Germany: Land Mecklenburg, Wolgast, 5.iv.1941, slide PS1416 (coll. ZMH)

Theria primaria (HAWORTH)
a. genitalia, Denmark: Lehnskov, 11.iii.1980, slide PS1418 (coll. ZMH)
b. aedeagus, Denmark: Lehnskov, 11.iii.1980, slide PS1418 (coll. ZMH)
c. vesica, Denmark: Lehnskov, 11.iii.1980, slide PS1418 (coll. ZMH)
d. 8th sternite, Denmark: Lehnskov, 11.iii.1980, slide PS1418 (coll. ZMH)

Campaea margaritaria (LINNAEUS)
a. genitalia, Finland: Lemland, 20–27.vii.1996, slide PS1342 (coll. P. SIHVONEN)
b. aedeagus, Finland: Lemland, 20–27.vii.1996, slide PS1342 (coll. P. SIHVONEN)
c. cornuti, Finland: Lemland, 20–27.vii.1996, slide PS1342 (coll. P. SIHVONEN)
d. vesica, Finland: Lemland, 20–27.vii.1996, slide PS1342 (coll. P. SIHVONEN)
e. valva, Finland: Lemland, 20–27.vii.1996, slide PS1342 (coll. P. SIHVONEN)

Gerinia honoraria (DENIS & SCHIFFERMÜLLER)
a. genitalia, Portugal: Algarve, 13.v.1971, slide PS1438 (coll. ZMH)
b. aedeagus, Portugal: Algarve, 13.v.1971, slide PS1438 (coll. ZMH)
c. vesica, Portugal: Algarve, 13.v.1971, slide PS1438 (coll. ZMH)
d. valva, Portugal: Algarve, 13.v.1971, slide PS1438 (coll. ZMH)

Hylaea fasciaria (LINNAEUS)
a. genitalia, Finland: Enonkoski, 16–18.vii.1996, slide PS1344 (coll. P. SIHVONEN)
b. aedeagus, Germany: Bad Harzburg, 2.vii.1988, slide 1741 (coll. P. SIHVONEN)
c. vesica, Finland: Enonkoski, 16–18.vii.1996, slide PS1344 (coll. P. SIHVONEN)
d. base of vesica, Finland: Enonkoski, 16–18.vii.1996, slide PS1344 (coll. P. SIHVONEN)

Hylaea mediterranea SIHVONEN *et al.*
a. genitalia, Italy: Sicily, near Santo Stefano Quisquina, 1000 m, 9–10.x.2010, slide PS1738, holotype (coll SKOU)
b. aedeagus, Italy: Sicily, near Santo Stefano Quisquina, 1000 m, 9–10.x.2010, slide PS1738, holotype (coll SKOU)
c. vesica, Italy: Sicily, near Santo Stefano Quisquina, 1000 m, 9–10.x.2010, slide PS1738, holotype (coll SKOU)
d. base of vesica, Italy: Sicily, near Santo Stefano Quisquina, 1000 m, 9–10.x.2010, slide PS1738, holotype (coll SKOU)

Hylaea pinicolaria (BELLIER)
a. genitalia, France: Corsica, 3 km E of Haut-Asco, 1100 m, 20–21.vii.2004, slide PS1542 (coll. SKOU)
b. aedeagus, France: Corsica, 3 km E of Haut-Asco, 1100 m, 20–21.vii.2004, slide PS1742 (coll. SKOU)
c. vesica, France: Corsica, 3 km E of Haut-Asco, 1100 m, 20–21.vii.2004, slide PS1542 (coll. SKOU)
d. base of vesica, France: Corsica, 3 km E of Haut-Asco, 1100 m, 20–21.vii.2004, slide PS1542 (coll. SKOU)

Adalbertia castiliaria (Staudinger)
a. genitalia, Spain: Lerida, 1650 m, 7.vii.1991, slide PS1501 (coll. ZMH)
b. aedeagus, Spain: Lerida, 1650 m, 7.vii.1991, slide PS1501 (coll. ZMH)
c. vesica, Spain: Lerida, 1650 m, 7.vii.1991, slide PS1501 (coll. ZMH)
d. valva, Spain: Lerida, 1650 m, 7.vii.1991, slide PS1501 (coll. ZMH)

Pungeleria capreolaria (Denis & Schiffermüller)
a. genitalia, Austria, without date, slide PS1419 (coll. ZMH)
b. aedeagus, Austria, without date, slide PS1419 (coll. ZMH)
c. vesica, Austria, without date, slide PS1419 (coll. ZMH)
d. valva, Austria, without date, slide PS1419 (coll. ZMH)

Siona lineata (Scopoli)
a. genitalia, Finland: Imatra, 6–14.vii.1996, slide PS1346 (coll. P. Sihvonen)
b. aedeagus, Finland: Imatra, 15–23.vi.1996, slide PS1906 (coll. P. Sihvonen)
c. spines of aedeagus, Finland: Imatra, 15–23.vi.1996, slide PS1906 (coll. P. Sihvonen)
d. vesica, Finland: Imatra, 6–14.vii.1996, slide PS1346 (coll. P. Sihvonen)
e. 8th sternite, Finland: Imatra, 6–14.vii.1996, slide PS1346 (coll. P. Sihvonen)

Chariaspilates formosaria (Eversmann)
a. genitalia, Russia: Orenburg oblast, Burtinskij Step, 12–17.vi.2001, slide PS1426 (coll. ZMH)
b. aedeagus, Russia: Orenburg oblast, Burtinskij Step, 12–17.vi.2001, slide PS1426 (coll. ZMH)
c. microcornuti, Russia: Orenburg oblast, Burtinskij Step, 12–17.vi.2001, slide PS1426 (coll. ZMH)
d. vesica, Russia: Orenburg oblast, Burtinskij Step, 12–17.vi.2001, slide PS1426 (coll. ZMH)
e. uncus and gnathos, Russia: Orenburg oblast, Burtinskij Step, 12–17.vi.2001, slide PS1426 (coll. ZMH)

Aspitates mundataria (Stoll)
a. genitalia, Russia: Volgograd, 27.vi.1894, slide PS1492 (coll. ZMH)
b. aedeagus, Russia: Volgograd, 27.vi.1894, slide PS1492 (coll. ZMH)
c. cornutus, Russia: Volgograd, 27.vi.1894, slide PS1492 (coll. ZMH)
d. vesica, Russia: Volgograd, 27.vi.1894, slide PS1492 (coll. ZMH)
e. base of valva, Russia: Volgograd, 27.vi.1894, slide PS1492 (coll. ZMH)

Aspitates gilvaria (Denis & Schiffermüller)
a. genitalia, Finland: Hyvinkää, 7.vii.2001, slide PS1348 (coll. P. Sihvonen)
b. aedeagus, Finland: Kalvola, 11.vii.1961, slide PS1712 (coll. SMNK)
c. apex of aedeagus, Finland: Hyvinkää, 7.vii.2001, slide PS1348 (coll. P. Sihvonen)
d. vesica, Finland: Hyvinkää, 7.vii.2001, slide PS1348 (coll. P. Sihvonen)
e. base of valva, Finland: Hyvinkää, 7.vii.2001, slide PS1348 (coll. P. Sihvonen)

Aspitates taylorae (BUTLER)
a. genitalia, Russia: Polar Ural, Krasnyi Kamen 200 m, 1.vii.1994, slide PS1423 (coll. ZMH)
b. aedeagus, Russia: Polar Ural, Krasnyi Kamen 200 m, 1.vii.1994, slide PS1423 (coll. ZMH)
c. apex of aedeagus, ventral view, Russia: Polar Ural, Krasnyi Kamen 200 m, 1.vii.1994, slide PS1423 (coll. ZMH)
d. vesica, Russia: Polar Ural, Krasnyi Kamen 200 m, 1.vii.1994, slide PS1423 (coll. ZMH)
e. base of valva, Russia: Polar Ural, Krasnyi Kamen 200 m, 1.vii.1994, slide PS1423 (coll. ZMH)

Aspitates orciferaria (WALKER)
a. genitalia, Russia: Polar Ural, Krasnyi Kamen 500 m, 3.vii.1994, slide PS1428 (coll. ZMH)
b. aedeagus, Russia: Polar Ural, Krasnyi Kamen 500 m, 3.vii.1994, slide PS1428 (coll. ZMH)
c. apex of aedeagus, ventral view, Russia: Polar Ural, Krasnyi Kamen 500 m, 3.vii.1994, slide PS1428 (coll. ZMH)
d. vesica, Russia: Polar Ural, Krasnyi Kamen 500 m, 3.vii.1994, slide PS1428 (coll. ZMH)
e. base of valva, Russia: Polar Ural, Krasnyi Kamen 500 m, 3.vii.1994, slide PS1428 (coll. ZMH)

Aspitates forbesi (MUNROE)
a. genitalia, Russia: Polar Ural, Krasnyi Kamen, 200 m, 7.vii.1994, slide PS1425 (coll. ZMH)
b. aedeagus, Russia: Polar Ural, Krasnyi Kamen, 200 m, 7.vii.1994, slide PS1425 (coll. ZMH)
c. apex of aedeagus arm, Russia: Polar Ural, Krasnyi Kamen, 200 m, 7.vii.1994, slide PS1425 (coll. ZMH)
d. vesica, Russia: Polar Ural, Krasnyi Kamen, 200 m, 7.vii.1994, slide PS1425 (coll. ZMH)
e. base of valva, Russia: Polar Ural, Krasnyi Kamen, 200 m, 7.vii.1994, slide PS1425 (coll. ZMH)

Aspitates albaria (BARTEL)
a. genitalia, Russia: South Ural, Orenburg region, Donskoe Village, 1–4.vi.1998, slide PS1545 (coll. SKOU)
b. aedeagus, Russia: Chellabinsk district, Kizilskoye, 27.v.1998, slide PS1593 (coll. P. SIHVONEN)
c. apex of aedeagus arm, Russia: Chellabinsk district, Kizilskoye, 27.v.1998, slide PS1593 (coll. P. SIHVONEN)
d. vesica, Russia: South Ural, Orenburg region, Donskoe Village, 1–4.vi.1998, slide PS1545 (coll. SKOU)
e. base of valva, Russia: South Ural, Orenburg region, Donskoe Village, 1–4.vi.1998, slide PS1545 (coll. SKOU)

Aspitates ochrearia (ROSSI)
a. genitalia, Portugal: Estremadura, 18.v.1971, slide PS1430 (coll. ZMH)
b. aedeagus, Portugal: Estremadura, 18.v.1971, slide PS1430 (coll. ZMH)
c. spines of aedeagus, Portugal: Estremadura, 18.v.1971, slide PS1430 (coll. ZMH)
d. vesica, Portugal: Estremadura, 18.v.1971, slide PS1430 (coll. ZMH)
e. 8th sternite, Portugal: Estremadura, 18.v.1971, slide PS1430 (coll. ZMH)

Dyscia raunaria (FREYER)

a. genitalia, Croatia: Kraljevica, 15.vi.1972, slide PS1434 (coll. ZMH)
b. aedeagus, Croatia: Kraljevica, 15.vi.1972, slide PS1434 (coll. ZMH)
c. vesica, Croatia: Kraljevica, 15.vi.1972, slide PS1434 (coll. ZMH)
d. base of valva, Croatia: Kraljevica, 15.vi.1972, slide PS1434 (coll. ZMH)

Dyscia royaria TAUTEL & BILLI

a. genitalia, France: Tende, Cagnourina, 900m, 3.viii.2007, slide PS1637 (coll. SKOU)
b. aedeagus, France: Tende, Cagnourina, 900m, 3.viii.2007, slide PS1637 (coll. SKOU)
c. vesica, France: Tende, Cagnourina, 900m, 3.viii.2007, slide PS1637 (coll. SKOU)
d. base of valva, France: Tende, Cagnourina, 900m, 3.viii.2007, slide PS1637 (coll. SKOU)

Dyscia innocentaria (CHRISTOPH)

a. genitalia, Italy: Region Púglia, 5 km SW of Manfredónia, 26.v.2005, slide PS1510 (coll. SKOU)
b. aedeagus, Russia: Astrahan district, Sasykoli village, 29.viii.2006, slide PS1582 (coll. NUPPONEN)
c. apex of aedeagus, Russia: Astrahan district, Sasykoli village, 29.viii.2006, slide PS1582 (coll. NUPPONEN)
d. vesica, Russia: Astrahan district, Sasykoli village, 29.viii.2006, slide PS1582 (coll. NUPPONEN)
e. apex of valva, Russia: Astrahan district, Sasykoli village, 29.viii.2006, slide PS1582 (coll. NUPPONEN)

Dyscia malatyana WEHRLI

a. genitalia, Kazakstan: 25 km W from Oporniy settlement, 7.v.2000, slide PS1892 (coll. SKOU)
b. aedeagus, Kazakstan: 25 km W from Oporniy settlement, 7.v.2000, slide PS1892 (coll. SKOU)
c. vesica, Kazakstan: 25 km W from Oporniy settlement, 7.v.2000, slide PS1891 (coll. SKOU)
d. extension at base of valva, apex with single tooth, Kazakstan: 25 km W from Oporniy settlement, 7.v.2000, slide PS1892 (coll. SKOU)

Dyscia crassipunctaria (REBEL)

a. genitalia, Greece: Crete, Gonies, 28.v.1986, slide PS1512 (coll. SKOU)
b. aedeagus, Greece: Crete, Gonies, 26.v.1986, slide PS1555 (coll. SKOU)
c. vesica, Greece: Crete, Gonies, 28.v.1986, slide PS1512 (coll. SKOU)
d. extension at base of valva, Greece: Crete, Gonies, 28.v.1986, slide PS1512 (coll. SKOU)

Dyscia conspersaria (DENIS & SCHIFFERMÜLLER)

a. genitalia, Austria: Niederösterreich, Hundsheimer Berge, 200–500 m, 22.vi.1981, slide PS1509 (coll. SKOU)
b. aedeagus, Turkey: Aksehir, Sultan Dag, 1750 m, 12.vii.1986, slide PS1882 (coll. SKOU)
c. vesica, Austria: Niederösterreich, Hundsheimer Berge, 200–500 m, 22.vi.1981, slide PS1509 (coll. SKOU)
d. apex of aedeagus, Austria: Niederösterreich, Hundsheimer Berge, 200–500 m, 22.vi.1981, slide PS1509 (coll. SKOU)

Dyscia fagaria (THUNBERG)
a. genitalia, Denmark: Nord Sjælland, Asserbo, 4.vi.1969, slide PS1432 (coll. ZMH)
b. aedeagus, Denmark: Nord Sjælland, Asserbo, 4.vi.1969, slide PS1432 (coll. ZMH)
c. vesica, Denmark: Nord Sjælland, Asserbo, 4.vi.1969, slide PS1432 (coll. ZMH)
d. sclerotisations of vesica, Denmark: Nord Sjælland, Asserbo, 4.vi.1969, slide PS1432 (coll. ZMH)

Dyscia penulataria (HÜBNER)
a. genitalia, Spain: Sevilla, Rivera de Huelva, 150 m, 2.iv.2003, slide PS1522 (coll. SKOU)
b. aedeagus, Spain: Sevilla, Rivera de Huelva, 150 m, 2.iv.2003, slide PS1522 (coll. SKOU)
c. vesica, Spain: Sevilla, Rivera de Huelva, 150 m, 2.iv.2003, slide PS1522 (coll. SKOU)
d. extension at base of valva, Spain: Sevilla, Rivera de Huelva, 150 m, 2.iv.2003, slide PS1522 (coll. SKOU)

Dyscia lentiscaria (DONZEL)
a. genitalia, France: Marseille, 27.v.1913, slide PS1967/BMNH GEO 24895 (coll. BMNH)
b. aedeagus, Spain: Zaragoza, Tosos, 400 m, 28.iv.1997, slide PS1516 (coll. Skou)
c. vesica, Spain: Zaragoza, Tosos, 400 m, 28.iv.1997, slide PS1516 (coll. Skou)
d. gnathos, France: Marseille, 27.v.1913, slide PS1967/BMNH GEO 24895 (coll. BMNH)

Dyscia distinctaria (BANG-HAAS)
a. genitalia, Spain: province Segovia, San Ildefonso, without date, slide PS1436 (coll. ZMH)
b. aedeagus, Spain: province Sevilla, Villamanrique, 7.iv.1956, slide PS1645 (coll. ZSM)
c. vesica, Spain: province Sevilla, Villamanrique, 7.iv.1956, slide PS1645 (coll. ZSM)
d. gnathos, Spain: province Segovia, San Ildefonso, without date, slide PS1436 (coll. ZMH)

Perconia strigillaria (HÜBNER)
a. genitalia, Finland: Kirkkonummi, 21.vi.1999, slide PS1350 (coll. P. SIHVONEN)
b. aedeagus, Russia: Orenburg district, Pokrovka village, 25.v.2004, slide PS1570 (coll. NUPPONEN)
c. vesica, Finland: Kirkkonummi, 21.vi.1999, slide PS1350 (coll. P. SIHVONEN)
d. base of valva, Finland: Kirkkonummi, 21.vi.1999, slide PS1350 (coll. P. SIHVONEN)

Perconia baeticaria (STAUDINGER)
a. genitalia, Portugal: Estrela, Penhaz, 23.vi.2007, slide PS1547 (coll. SKOU)
b. aedeagus, Spain: province Granada, Aldeire, 8.vi.1983, slide PS1597 (coll. SKOU)
c. vesica, Portugal: Estrela, Penhaz, 23.vi.2007, slide PS1547 (coll. SKOU)
d. base of valva, Portugal: Estrela, Penhaz, 23.vi.2007, slide PS1547 (coll. SKOU)

Onychora agaritharia (DARDOIN)
a. genitalia, France: Alpes Maritimes, Nice-Cimiez, without date, slide PS1487 (coll. ZMH)
b. aedeagus, France: Alpes Maritimes, Nice-Cimiez, without date, slide PS1487 (coll. ZMH)
c. cornuti, France: Alpes Maritimes, Nice-Cimiez, without date, slide PS1487 (coll. ZMH)
d. granulate surface of vesica, France: Alpes Maritimes, Nice-Cimiez, without date, slide PS1487 (coll. ZMH)
e. vesica, France: Alpes Maritimes, Nice-Cimiez, without date, slide PS1487 (coll. ZMH)
f. tympanal organ, France: Alpes Maritimes, Nice-Cimiez, without date, slide PS1487 (coll. ZMH)

Compsoptera opacaria HÜBNER

a. genitalia, Spain: Barcelona, without date, slide PS1440 (coll. ZMH)
b. aedeagus, Spain: Barcelona, without date, slide PS1440 (coll. ZMH)
c. cornuti, Spain: Barcelona, without date, slide PS1440 (coll. ZMH)
d. vesica, Spain: Barcelona, without date, slide PS1440 (coll. ZMH)
e. dorsal arm of valva, Spain: Barcelona, without date, slide PS1440 (coll. ZMH)

Compsoptera jourdanaria (SERRES)

a. genitalia, France: Corsica, 1 km NE of Castirla, 350 m, 26.ix.2003, slide PS1550 (coll. SKOU)
b. aedeagus, Spain: province Segovia, San Ildefonso, 1899, without other date data, slide PS1488 (coll. ZMH)
c. cornuti, France: Corsica, 1 km NE of Castirla, 350 m, 26.ix.2003, slide PS1550 (coll. SKOU)
d. vesica, Spain: province Segovia, San Ildefonso, 1899, without other date data, slide PS1488 (coll. ZMH)
e. dorsal arm of valva, Spain: province Segovia, San Ildefonso, 1899, without other date data, slide PS1488 (coll. ZMH)

Compsoptera argentaria (HERRICH-SCHÄFFER)

a. genitalia, Italy: Sicily, October, without other date data, slide PS1632 (coll. ZFMK)
b. aedeagus, Italy: Sicily, October, without other date data, slide PS1632 (coll. ZFMK)
c. cornuti, Italy: Sicily, October, without other date data, slide PS1632 (coll. ZFMK)
d. vesica, Italy: Sicily, October, without other date data, slide PS1632 (coll. ZFMK)
e. dorsal arm of valva, Italy: Sicily, October, without other date data, slide PS1632 (coll. ZFMK)

Appendix 2. Data of the microscopic preparations shown in female genitalia plates

Abraxas grossulariata (LINNAEUS)
a. genitalia, Finland: Föglö, 29.vii.1995, slide PS1132 (coll. P. SIHVONEN)
b. signum, Finland: Föglö, 29.vii.1995, slide PS1132 (coll. P. SIHVONEN)
c. ostium bursae and adjacent structures, Finland: Föglö, 29.vii.1995, slide PS1132 (coll. P. SIHVONEN)

Abraxas sylvata (SCOPOLI)
a. genitalia, Finland: Kirkkonummi, 11–21.vi.1999, slide PS1130 (coll. P. SIHVONEN)
b. signum, Finland: Kirkkonummi, 11–21.vi.1999, slide PS1130 (coll. P. SIHVONEN)
c. ostium bursae and adjacent structures, Finland: Kirkkonummi, 11–21.vi.1999, slide PS1130 (coll. P. SIHVONEN)

Abraxas pantaria (LINNAEUS)
a. genitalia, Spain, without date, slide PS1162 (coll. ZMH)
b. signum, Spain, without date, slide PS1162 (coll. ZMH)
c. ostium bursae and adjacent structures, Spain, without date, slide PS1162 (coll. ZMH)

Ligdia adustata (DENIS & SCHIFFERMÜLLER)
a. genitalia, Italy: region Tuscany, Coreglia Antelminelli (Lucca), 600 m, 28.vi–10.vii.1999, slide PS1749 (coll. P. SIHVONEN)
b. signum, Italy: region Tuscany, Coreglia Antelminelli (Lucca), 600 m, 28.vi–10.vii.1999, slide PS1749 (coll. P. SIHVONEN)
c. ostium bursae and adjacent structures, Italy: Tuscany, Coreglia Antelminelli (Lucca), 600 m, 28.vi–10.vii.1999, slide PS1749 (coll. P. SIHVONEN)

Lomaspilis marginata (LINNAEUS)
a. genitalia, Finland: Enonkoski, 24.vi.1996, slide PS1133 (coll. P. SIHVONEN)
b. signum, Finland: Enonkoski, 24.vi.1996, slide PS1133 (coll. P. SIHVONEN)
c. ostium bursae and adjacent structures, Finland: Enonkoski, 24.vi.1996, slide PS1133 (coll. P. SIHVONEN)

Lomaspilis opis BUTLER
a. genitalia, Finland: Helsinki, 8.vi.1988, slide PS1174 (coll. ZMH)
b. signum, Finland: Pyhäselkä, 22.vi.1988, slide PS1560 (coll. ZMH)
c. ostium bursae and adjacent structures, Finland: Helsinki, 8.vi.1988, slide PS1174 (coll. ZMH)

Lomaspilis bithynica WEHRLI
a. genitalia, Turkey: Paphlagon, June 1955, slide PS1614 (coll. ZFMK)
b. signum, Turkey: Gümüshane, 1850 m, 16.vii.2000, slide PS1506 (coll. SKOU)
c. ostium bursae and adjacent structures, Turkey: Gümüshane, 1850 m, 16.vii.2000, slide PS1506 (coll. SKOU)

Stegania cararia (HÜBNER)
a. genitalia, Russia: South Ussuri region, Rjazanovka, 5–7.vii.1982, slide PS1166 (coll. ZMH)
b. signum, Russia: South Ussuri region, Rjazanovka, 5–7.vii.1982, slide PS1166 (coll. ZMH)
c. ostium bursae and adjacent structures, Russia: South Ussuri region, Rjazanovka, 5–7.vii.1982, slide PS1166 (coll. ZMH)

Stegania trimaculata (VILLERS)
a. genitalia, France: Corsica, Ajaccio, 25.vii.1992,slide PS1446 (coll. P. SIHVONEN)
b. signum, France: Corsica, Ajaccio, 24.vii.1992, slide PS1559 (coll. P. SIHVONEN)
c. ostium bursae and adjacent structures, Italy: region Tuscany, Coreglia Antelminelli (Lucca), 600 m, 28.vi–10.vii.1999, slide PS1135 (coll. P. SIHVONEN)

Stegania dilectaria (HÜBNER)
a. genitalia, Austria: without date, slide PS1168 (coll. ZMH)
b. signum, Austria: Niederosterreich, [Klosterneuburg-]Weidling, 10.viii.1905, slide PS1959/BMNH GEO 24887 (coll. BMNH)
c. ostium bursae and adjacent structures, Austria: without date, slide PS1168 (coll. ZMH)

Stegania dalmataria GUENÉE
a. genitalia, Russia: Volgograd, 9.viii.1895, slide PS1173 (coll. ZMH)
b. signum, Kirghistan: Kazarman, 1260 m, 2.viii.1983, slide PS1222 (coll. SKOU)
c. ostium bursae and adjacent structures, Russia: Volgograd, 9.viii.1895, slide PS1173 (coll. ZMH)

Acanthovalva inconspicuaria (HÜBNER)
a. genitalia, Spain: province Malaga, Casares, 350 m, 26–27.vii.2003, slide PS1245 (coll. SKOU)
b. spines of corpus bursae, Spain: province Granada, Albunol, 600 m, 5.v.2000, slide PS1246 (coll. Skou)
c. ostium bursae and adjacent structures, Spain: province Malaga, Casares, 350 m, 26–27. vii.2003, slide PS1245 (coll. SKOU)

Narraga fasciolaria (HUFNAGEL)
a. genitalia, Germany: Lüneburger Heide, Höpen, 20.v.1949, slide PS1176 (coll. ZMH)
b. signum, Germany: Lüneburger Heide, Höpen, 20.v.1949, slide PS1176 (coll. ZMH)
c. ostium bursae and adjacent structures, Poland: Poznan, ex larva 29.vi.1958, slide PS1202 (coll. SKOU)

Narraga nelvae subsp. *catalaunica* HERBULOT
a. genitalia, combined picture (signum is absent): upper part (excluding corpus bursae), Spain: province Huesca, Candasnos, 300 m, 15.vi.1999, slide PS1205 (coll. SKOU), lower part (corpus bursae), Spain: province Zaragoza, Tosos, 600 m, 16.vi.1999, slide PS1238 (coll. SKOU)
b. ductus bursae (lateral view), Spain: province Zaragoza, Tosos, 600 m, 16.vi.1999, slide PS1238 (coll. SKOU)
c. ostium bursae and adjacent structures, Spain: province Huesca, Candasnos,300 m, 15.vi.1999, slide PS1205 (coll. SKOU)

Narraga tessularia (METZNER)
a. genitalia (corpus bursae margins highlighted), Russia: Tuva republic, Tannu-Ola mountains, 1250–2000 m, 17.vi.1995, slide PS1354 (coll. ZMH)
b. signum, Russia: Tuva republic, Tannu-Ola mountains, 1250–2000 m, 17.vi.1995, slide PS1354 (coll. ZMH)
c. ostium bursae and adjacent structures, Austria: Burgenland, Illmitz am Neusiedlersee, ex larva July 1956, slide PS1181 (coll. ZMH)

Heliomata glarearia (DENIS & SCHIFFERMÜLLER)
a. genitalia, Ukraine: Ivano-Frankivsk, 8.vi.2004, slide PS1183 (coll. ZMH)
b. signum, Austria: Niederösterreichs, Marchfeld, 26.vi.1958, slide PS1926 (coll. ZMH)
c. ostium bursae and adjacent structures, Ukraine: Ivano-Frankivsk, 8.vi.2004, slide PS1183 (coll. ZMH)

Isturgia famula (ESPER)
a. genitalia (corpus bursae margins highlighted), Poland: Wroclaw, without date, slide PS1195 (coll. ZMH)
b. signum, Poland: Wroclaw, without date, slide PS1195 (coll. ZMH)
c. ostium bursae and adjacent structures, Poland: Wroclaw, without date, slide PS1195 (coll. ZMH)

Isturgia limbaria (FABRICIUS)
a. genitalia, France: department Basses-Alpes, Digne, 21.v–8.vi.1956, slide PS1927 (coll. ZSM)
b. signum, France: department Basses-Alpes, Digne, 21.v–8.vi.1956, slide PS1927 (coll. ZSM)
c. ostium bursae and adjacent structures, Germany: Rheinland, without date, slide PS1190 (coll. ZMH)

Isturgia roraria (FABRICIUS)
a. genitalia, Austria: Langegg [bei Graz?], 6.v.[19]28, slide PS2007 (coll. ZFMK)
b. signum, Germany: Nürnberg, 1912 [no exact date], slide PS2006 (coll. ZFMK)
c. ostium bursae and adjacent structures, Germany: Nürnberg, 1912 [no exact date], slide PS2006 (coll. ZFMK)

Isturgia homochromata (MABILLE). Not illustrated, female of this species is unknown.

Isturgia murinaria (DENIS & SCHIFFERMÜLLER)
a. genitalia, Austria: Burgenland, Neusiedlersee, 22.vi.1958, slide PS1188 (coll. ZMH)
b. signum, Russia: S-Ural, Cheliabinsk district, near Nikolaevka village, 3.vii.1997, slide PS1912 (coll. NUPPONEN)
c. ostium bursae and adjacent structures, Austria: Burgenland, Neusiedlersee, 22.vi.1958, slide PS1188 (coll. ZMH)

Isturgia catalaunaria (GUENÉE)
a. genitalia, Spain: province Almeria, El Pinar, 19–27.iv.2001, slide PS1840 (coll. SKOU)
b. signum, Spain: province Almeria, El Pinar, 19–27.iv.2001, slide PS1840 (coll. SKOU)
c. ostium bursae and adjacent structures, Spain: province Almeria, El Pinar, 19–27.iv.2001, slide PS1840 (coll. SKOU)

Isturgia deerraria (WALKER)
a. genitalia, Malta: Mellieha, ex larva 12.x.2007, slide PS1837 (coll. SAMMUT)
b. signum, Yemen: Province Al. Hudaydah, Tihama, 4.xi.1996, slide PS1248 (coll. SKOU)
c. ostium bursae and adjacent structures, Spain: Province Huelva, Matalascanas, 20.ix.2009, slide PS1835 (coll. SKOU)

Isturgia disputaria (GUENÉE)
a. genitalia, Oman: Al Mintirib, 400m, 11–12.iv.2005, slide PS1635 (coll. SKOU)
b. signum, Libya: Gharian, Wadi El Hira, 17.vi.1983, slide PS1821 (coll. SKOU)
c. ostium bursae and adjacent structures, Oman: Al Mintirib, 400m, 11–12.iv.2005, slide PS1635 (coll. SKOU)

Isturgia spodiaria (LEFÈBVRE)
a. genitalia, Tunisia: Sousse, 11–14.iv.1984, slide PS1213 (coll. ZMH)
b. signum, Spain: prov. Malaga, San Pedro de Alcantara, 16.xi.1980, PS1854 (coll. SKOU)
c. ostium bursae and adjacent structures, Tunisia: Sousse, 11–14.iv.1984, slide PS1213 (coll. ZMH)

Isturgia assimilaria (RAMBUR)
a. genitalia (corpus bursae margins highlighted partly), Italy: Sardinia, Soleminis, May 1987, slide PS1197 (coll. ZMH)
b. signum, Italy: Sardinia, Soleminis, May 1987, slide PS1197 (coll. ZMH)
c. ostium bursae and adjacent structures, Italy: Sardinia, Soleminis, May 1987, slide PS1197 (coll. ZMH)

Isturgia miniosaria (DUPONCHEL)
a. genitalia, Spain: Granada, without date, slide PS1201 (coll. ZMH)
b. signum, Spain: Granada, without date, slide PS1201 (coll. ZMH)
c. ostium bursae and adjacent structures, Spain: Granada, without date, slide PS1201 (coll. ZMH)

Isturgia berytaria (STAUDINGER)
a. genitalia, Greece: Peleponnisos, Tripoli, 3.x.1984, slide PS1255 (coll. SKOU)
b. signum, Greece: Peleponnisos, Tripoli, 3.x.1984, slide PS1255 (coll. SKOU)
c. ostium bursae and adjacent structures, Greece: Peleponnisos, Tripoli, 3.x.1984, slide PS1255 (coll. SKOU)

Isturgia sparsaria (HÜBNER)
a. genitalia, Italy: Calabria, Fme Argentino, 290 m, 4.ix.1991, slide PS1240 (coll. P. SIHVONEN)
b. signum, Italy: Calabria, Fme Argentino, 290 m, 4.ix.1991, slide PS1240 (coll. P. SIHVONEN)
c. ostium bursae and adjacent structures, Italy: Calabria, Fme Argentino, 290 m, 4.ix.1991, slide PS1240 (coll. P. SIHVONEN)

Isturgia messapiaria (SOHN-RETHEL)
a. genitalia, Italy: Calabria, Fossiata, 6.viii.1921, slide PS1706 (coll. MSNM)
b. signum, Italy: Calabria, Fossiata, 6.viii.1921, slide PS1706 (coll. MSNM)
c. ostium bursae and adjacent structures, Italy: Calabria, Fossiata, 6.viii.1921, slide PS1706 (coll. MSNM)

Isturgia arenacearia (DENIS & SCHIFFERMÜLLER)
a. genitalia, Turkey: Kare, 5 km. NE Igdir, 850 m, 22.vii.1989, slide PS1186 (coll. ZMH)
b. signum, Turkey: Kare, 5 km. NE Igdir, 850 m, 22.vii.1989, slide PS1186 (coll. ZMH)
c. ostium bursae and adjacent structures, Turkey: Kare, 5 km. NE Igdir, 850 m, 22.vii.1989, slide PS1186 (coll. ZMH)

Isturgia punctistrigaria (CHRISTOPH)
a. genitalia, Turkmenistan(?): Tschardschui, without date, slide PS1413 (coll. ZMH)
b. signum, Turkmenistan(?): Tschardschui, without date, slide PS1413 (coll. ZMH)
c. ostium bursae and adjacent structures, Turkmenistan(?): Tschardschui, without date, slide PS1413 (coll. ZMH)

Itame vincularia subsp. *vincularia* (HÜBNER)
a. genitalia, Spain: prov. Granada, Granada, without date, slide PS1207 (coll. ZMH)
b. signum, Spain: prov. Granada, Granada, without date, slide PS1207 (coll. ZMH)
c. ostium bursae and adjacent structures, Spain: prov. Granada, Granada, without date, slide PS1207 (coll. ZMH)

Itame vincularia subsp. *mrassinaria* (OBERTHÜR)
a. genitalia, Morocco: High Atlas, Ouirgane, 4.v.1965, slide PS1648 (coll. ZSM)
b. signum, Morocco: High Atlas, Ouirgane, 4.v.1965, slide PS1648 (coll. ZSM)
c. ostium bursae and adjacent structures, Morocco: High Atlas, Ouirgane, 4.v.1965, slide PS1648 (coll. ZSM)

Neognopharmia stevenaria (BOISDUVAL)
a. genitalia, Ukraine: Crimea, Karadagh [nature] res.[erve]10.vi.1996, slide PS1356 (coll. ZMH)
b. signum, Greece: Epiros, Parga, 17–31.v.1999, slide PS1951 (coll. P. SIHVONEN)
c. ostium bursae and adjacent structures, Greece: Epiros, Parga, 17–31.v.1999, slide PS1951 (coll. P. SIHVONEN)

Macaria notata (LINNAEUS)
a. genitalia, Finland: Savonlinna, 13.vii.1993, slide PS1147 (coll. P. SIHVONEN)
b. signum, Finland: Savonlinna, 13.vii.1993, slide PS1147 (coll. P. SIHVONEN)
c. ostium bursae and adjacent structures, Finland: Savonlinna, 13.vii.1993, slide PS1147 (coll. P. SIHVONEN)

Macaria alternata (DENIS & SCHIFFERMÜLLER)
a. genitalia, Finland: Kirkkonummi, 5–19.viii.1999, slide PS1138 (coll. P. SIHVONEN)
b. signum, Finland: Kirkkonummi, 5–19.viii.1999, slide PS1138 (coll. P. SIHVONEN)
c. ostium bursae and adjacent structures, Finland: Kirkkonummi, 5–19.viii.1999, slide PS1138 (coll. P. SIHVONEN)

Macaria signaria (HÜBNER)
a. genitalia, Finland: Kerimäki, 3.vii.1997, slide PS1152 (coll. P. SIHVONEN)
b. signum, Finland: Kerimäki, 3.vii.1997, slide PS1152 (coll. P. SIHVONEN)
c. ostium bursae and adjacent structures, Finland: Kerimäki, 3.vii.1997, slide PS1152 (coll. P. SIHVONEN)

Macaria liturata (CLERCK)
a. genitalia, Finland: Enonkoski, 5.vii.1996, slide PS1137 (coll. P. SIHVONEN)
b. signum, Finland: Kirkkonummi, 15.vi.2008, slide PS1931 (coll. P. SIHVONEN)
c. ostium bursae and adjacent structures, Finland: Enonkoski, 5.vii.1996, slide PS1137 (coll. P. SIHVONEN)

Macaria wauaria subsp. *wauaria* (LINNAEUS)
a. genitalia, Finland: Savonlinna, 16–22.vii.1995, slide PS1149 (coll. P. SIHVONEN)
b. signum, Finland: Savonlinna, 16–22.vii.1995, slide PS1149 (coll. P. SIHVONEN)
c. ostium bursae and adjacent structures, Finland: Savonlinna, 16–22.vii.1995, slide PS1149 (coll. P. SIHVONEN)

Macaria wauaria subsp. *africana* (ZERNY)
a. genitalia, Morocco: High Atlas, Oukaïmeden 2400–2700 m, 5–21.vii.1972, slide PS1651 (coll. ZSM)
b. signum, Morocco: High Atlas, Oukaïmeden 2400–2700 m, 5–21.vii.1972, slide PS1651 (coll. ZSM)
c. ostium bursae and adjacent structures, Morocco: High Atlas, Oukaïmeden 2400–2700 m, 5–21.vii.1972, slide PS1651 (coll. ZSM)

Macaria artesiaria (DENIS & SCHIFFERMÜLLER)
a. genitalia, Finland: Paattinen, without date, slide PS1449 (coll. ZMH)
b. signum, Finland: Paattinen, without date, slide PS1449 (coll. ZMH)
c. ostium bursae and adjacent structures, Finland: Paattinen, without date, slide PS1209 (coll. ZMH)

Macaria ichnusae GOVI & FIUMI
a. genitalia, Italy: Sardinia, Rio Gutturu Mannu, 100 m, 18.vi.2009, slide PS1660 (coll. GOVI)
b. signum, Italy: Sardinia, Rio Gutturu Mannu, 100 m, 18.vi.2009, slide PS1660 (coll. GOVI)
c. ostium bursae and adjacent structures, Italy: Sardinia, Rio Gutturu Mannu, 100 m, 18.vi.2009, slide PS1660 (coll. GOVI)

Macaria carbonaria (CLERCK)
a. genitalia, Finland: Rantasalmi, 31.v.1996, slide PS1142 (coll. P. SIHVONEN)
b. signum, Finland: Kerimäki, 27.v.2001, slide PS1932 (coll. P. SIHVONEN)
c. ostium bursae and adjacent structures, Finland: Rantasalmi, 31.v.1996, slide PS1142 (coll. P. SIHVONEN)

Macaria brunneata (THUNBERG)
a. genitalia, Finland: Enonkoski, 8.viii.1996, slide PS1158 (coll. P. SIHVONEN)
b. signum, Finland: Enonkoski, 3.viii.1996, slide PS1447 (coll. P. SIHVONEN)
c. ostium bursae and adjacent structures, Finland: Enonkoski, 3.viii.1996, slide PS1447 (coll. P. SIHVONEN)

Macaria loricaria (EVERSMANN)
a. genitalia, Finland: Kärkölä, ex larva 1960, slide PS1210 (coll. ZMH)
b. signum, Finland: Kärkölä, ex larva 1960, slide PS1210 (coll. ZMH)
c. ostium bursae and adjacent structures, Finland: Kärkölä, ex larva 1960, slide PS1210 (coll. ZMH)

Macaria fusca (THUNBERG)
a. genitalia, Finland: Kaunispää, 6.vii.1970, slide PS1211 (coll. ZMH)
b. signum, Finland: Viibustunturi, without date, slide PS1670 (coll. ZMH)
c. ostium bursae and adjacent structures, Finland: Kaunispää, 6.vii.1970, slide PS1211 (coll. ZMH)

Chiasmia clathrata (LINNAEUS)
a. genitalia, Finland: Savonlinna, 19.vi.1994, slide PS1155 (coll. P. SIHVONEN)
b. signum, Finland: Savonlinna, 19.vi.1994, slide PS1155 (coll. P. SIHVONEN)
c. ostium bursae and adjacent structures, Finland: Savonlinna, 19.vi.1994, slide PS1155 (coll. P. SIHVONEN)

Chiasmia aestimaria subsp. *aestimaria* (HÜBNER)
a. genitalia (signum is absent), France: Alpes-Maritimes, Nice, without date, slide PS1216 (coll. ZMH)
b. ostium bursae and adjacent structures and posterior part of corpus bursae, France: Alpes-Maritimes, Nice, without date, slide PS1216 (coll. ZMH)

Chiasmia aestimaria subsp. *sareptanaria* (STAUDINGER)
a. genitalia (signum is absent), Russia: Volgograd, 18.vii.1903, slide PS1483 (coll. ZMH)
b. ostium bursae and adjacent structures and posterior part of corpus bursae, Russia: Volgograd, 18.vii.1903, slide PS1483 (coll. ZMH)

Chiasmia syriacaria (STAUDINGER) Non-European, see text for details
a. genitalia (signum is absent), Turkmenistan: Kara-Kala, 12.v.1974, slide PS1241 (coll. IPPE)
b. ostium bursae and adjacent structures and posterior part of corpus bursae, Turkmenistan: Kara-Kala, 12.v.1974, slide PS1241 (coll. IPPE)

Digrammia rippertaria (DUPONCHEL)
a. genitalia (signum is absent), Tajikistan: Hissarskij mountains, 1150 m, 25.vii.1990, slide PS1220 (coll. ZMH)
b. ostium bursae and adjacent structures: France: departement Loire, St-Etienne Vallée, 27.vi.1921, slide PS1595 (coll. ZSM)

Rhoptria asperaria (HÜBNER)
a. genitalia, Spain: prov. Malaga, San Pedro, 24.v.1989, slide PS1218 (coll. ZMH)
b. signum , Portugal: Algarve, Serra de Monchique Foia, 13.v.1971, slide PS1450 (coll. ZMH)
c. ostium bursae and adjacent structures, Spain: prov. Malaga, San Pedro, 24.v.1989, slide PS1218 (coll. ZMH)

Rhoptria dolosaria (HERRICH-SCHÄFFER)
a. genitalia (definitive signum is absent), Greece: Korinthia, 29.vi.1985, slide PS1258 (coll. SKOU)
b. spines in posterior part of corpus bursae, Greece: Korinthia, 29.vi.1985, slide PS1258 (coll. SKOU)
c. ostium bursae and adjacent structures, Greece: Korinthia, 29.vi.1985, slide PS1258 (coll. SKOU)

Ortaliella gruneraria (STAUDINGER)
a. genitalia, Greece: Tripolis, 14.vi.1954, slide PS1257 (coll. SKOU)
b. signum, Greece: Peloponnese, 4.vi.1972, slide PS1627 (coll. ZFMK)
c. ostium bursae and adjacent structures, Greece: Tripolis, 14.vi.1954, slide PS1257 (coll. SKOU)

Cepphis advenaria (HÜBNER)
a. genitalia, Finland: Rantasalmi, 19.vi.1998, slide PS1154 (coll. P. SIHVONEN)
b. signum, Finland: Rantasalmi, 25.vi.1998, slide PS1563 (coll. P. SIHVONEN)
c. ostium bursae and adjacent structures, Finland: Rantasalmi, 25.vi.1998, slide PS1563 (coll. P. SIHVONEN)

Petrophora chlorosata (SCOPOLI)
a. genitalia, Finland: Tammisaari, 7.vi.1999, slide PS1160 (coll. P. SIHVONEN)
b. signum, Finland: Parainen, 13.vi.1998, slide PS1564 (coll. P. SIHVONEN)
c. ostium bursae and adjacent structures, Finland: Parainen, 13.vi.1998, slide PS1564 (coll. P. SIHVONEN)

Perigune binaevata (MABILLE)
a. genitalia (signum is often absent or if present, minute), Italy: Toscany, 3 km SSW Albinia, 13.x.2010, slide PS1745 (coll. SKOU)
b. ostium bursae and adjacent structures, Mabille collection, without locality data, slide PS1602 (coll. ZSM). Signum is often absent. If present, it is very small and weakly stellate, located at anterior end.

Perigune convergata (VILLERS)
a. genitalia, Spain: province Huelva, Valverde del Camino, 11.xi.1980, slide PS1277 (coll. SKOU).
b. signum (often signum is absent), Spain: province Huelva, Valverde del Camino, 11.xi.1980, slide PS1277 (coll. SKOU).
b. ostium bursae and adjacent structures, Spain: province Huelva, Valverde del Camino, 11.xi.1980, slide PS1277 (coll. SKOU). Signum is often absent. If present, it is very small and weakly stellate, located at anterior end.

Perigune narbonea (LINNAEUS)
a. genitalia (signum is often absent and if present, minute), Spain: province Murcia, Sierra Espuna, 800 m, 12.iv.2000,slide PS1451 (coll. SKOU)
b. ostium bursae and adjacent structures, Spain: province Murcia, Sierra Espuna, 800 m, 12.iv.2000,slide PS1451 (coll. SKOU). Signum is often absent. If present, it is very small and weakly stellate, located at anterior end.

Lignyoptera fumidaria (HÜBNER)
a. genitalia (signum is absent), Austria: Niedesösterreich, Hainburger, ex ovo 11.xi.1940, slide PS1604 (coll. ZSM)
b. ostium bursae and adjacent structures, Austria: Niederösterreich, Hainburger, ex ovo 11.xi.1940, slide PS1604 (coll. ZSM)

Lignyoptera thaumastaria REBEL
a. genitalia (signum is absent), Bosnia [and Herzegovina]: Travnik, 1500 m, ex ovo 20.xi.1938, slide PS1658 (coll. ZSM)
b. ostium bursae and adjacent structures, Bosnia [and Herzegovina], Travnik, 1500 m, ex ovo 20.xi.1938, slide PS1658 (coll. ZSM)

Heterolocha laminaria (HERRICH-SCHÄFFER)
a. genitalia, Azerbaijan: Talyskije Mountains, 27.vi.1984, slide PS1896 (coll. ZMH)
b. signum, Coll. Duske. (no locality or date data; potentially from Russia: Volgograd), slide PS1897 (coll. ZMH)
c. ostium bursae and adjacent structures, Azerbaijan: Talyskije Mountains, 27.vi.1984, slide PS1896 (coll. ZMH)

Plagodis pulveraria (LINNAEUS)
a. genitalia, Finland: Enonkoski, 27.vi.1996, slide PS1229 (coll. P. SIHVONEN)
b. signum, Finland: Enonkoski, 23.vi.1996, slide PS1669 (coll. P. SIHVONEN)
c. ostium bursae and adjacent structures, Finland: Enonkoski, 27.vi.1996, slide PS1229 (coll. P. SIHVONEN)

Plagodis dolabraria (LINNAEUS)
a. genitalia, Finland: Lemland, 5.vii.1991, slide PS1231 (coll. P. SIHVONEN)
b. signum, Finland: Lemland, 5.vii.1991, slide PS1231 (coll. P. SIHVONEN)
c. ostium bursae and adjacent structures, Finland: Lemland, 5.vii.1991, slide PS1231 (coll. P. SIHVONEN)

Pachycnemia hippocastanaria (HÜBNER)
a. genitalia, Spain: province Cadiz, Castellar [de la Frontera], 24.v.1972, slide PS1366 (coll. ZMH)
b. signum, Spain: province Cadiz, Castellar [de la Frontera], 24.v.1972, slide PS1366 (coll. ZMH)
c. ostium bursae and adjacent structures, Spain: province Cadiz, Castellar [de la Frontera], 24.v.1972, slide PS1366 (coll. ZMH)

Pachycnemia tibiaria subsp. *tibiaria* (RAMBUR)
a. genitalia, France: departement Var, Frejus, 15–20.ix.2005, slide PS1262 (coll. SKOU)
b. signum, France: departement Aquitanie, Gironde, 22.ix.1935, slide PS1603 (coll. ZSM)
c. ostium bursae and adjacent structures, France: departement Var, Frejus, 15–20.ix.2005, slide PS1262 (coll. Skou)

Pachycnemia tibiaria subsp. *benesignata* (BELLIER)
a. genitalia, France: Corsica, Etang de Biguglia,27.ix.2003, sea level, slide PS1266 (coll. SKOU)
b. signum, France: Corsica, 8 km E of Col de Vergio, 25.ix.2003, 950 m, slide PS1928 (coll. SKOU)
c. ostium bursae and adjacent structures, France: Corsica, Etang de Biguglia, 27.ix.2003, sea level, slide PS1266 (coll. SKOU)

Opisthograptis luteolata (LINNAEUS)
a. genitalia (signum is absent), Finland: Kirkkonummi, 2–23.x.1999, slide PS1282 (coll. P. SIHVONEN)
b. ostium bursae and adjacent structures, Finland: Enonkoski, 20.vi.1996, slide PS1933 (coll. P. SIHVONEN)

Epione repandaria (HUFNAGEL)
a. genitalia, Finland: Kirkkonummi, 24.viii.1996, slide PS1286 (coll. P. SIHVONEN)
b. signum, Finland: Kirkkonummi, 24.viii.1996, slide PS1286 (coll. P. SIHVONEN)
c. ostium bursae and adjacent structures, Finland: Kirkkonummi, 24.viii.1996, slide PS1286 (coll. P. SIHVONEN)

Epione vespertaria (LINNAEUS)
a. genitalia, Finland: Savonlinna, 29–31.vii.1995, slide PS1288 (coll. P. SIHVONEN)
b. signum, Finland: Savonlinna, 29–31.vii.1995, slide PS1288 (coll. P. SIHVONEN)
c. ostium bursae and adjacent structures, Finland: Rantasalmi, 10.viii.1996, slide PS1846 (coll. P. SIHVONEN)

Therapis flavicaria (DENIS & SCHIFFERMÜLLER)
a. genitalia, Bulgaria: Black Sea coast, ex ovo, 20.viii.1984, slide PS1365 (coll. ZMH)
b. corpus bursae spines, Bulgaria: Black Sea coast, ex ovo, 20.viii.1984, slide PS1365 (coll. ZMH)
c. signum, Bulgaria: Black Sea coast, ex ovo, 20.viii.1984, slide PS1365 (coll. ZMH)
d. ostium bursae and adjacent structures, Bulgaria: Black Sea coast, ex ovo, 20.viii.1984, slide PS1365 (coll. ZMH)

Pseudopanthera macularia (LINNAEUS)
a. genitalia, Finland: Kitee, 16.vi.1993, slide PS1284 (coll. P. SIHVONEN)
b. signum, Finland: Kitee, 16.vi.1993, slide PS1284 (coll. P. SIHVONEN)
c. ostium bursae and adjacent structures, Finland: Kitee, 16.vi.1993, slide PS1284 (coll. P. SIHVONEN)

Eilicrinia cordiaria (HÜBNER)
a. genitalia, specimen from coll. Duske [handwritten label, locality unclear to read, perhaps Lagod?], 30.vi.1888, slide PS1388 (coll. ZMH)
b. signum, specimen from coll. Duske [handwritten label, locality unclear to read, perhaps Lagod?], 30.vi.1888, slide PS1388 (coll. ZMH)
c. ostium bursae and adjacent structures, specimen from coll. Duske [handwritten label, locality unclear to read, perhaps Lagod?], 30.vi.1888, slide PS1388 (coll. ZMH)

Eilicrinia subcordaria HERRICH-SCHÄFFER
a. genitalia, Russia: Volgograd, 16.vii.1902, slide PS1410 (coll. ZMH)
b. signum, Russia: Volgograd, 16.vii.1902, slide PS1410 (coll. ZMH)
c. ostium bursae and adjacent structures, Russia: Volgograd, 16.vii.1902, slide PS1410 (coll. ZMH)

Eilicrinia trinotata (METZNER)
a. genitalia, Hungary: Fót Somlyóhegy, 22.v.1958, slide PS1412 (coll. ZMH)
b. signum, Hungary: Fót Somlyóhegy, 22.v.1958, slide PS1412 (coll. ZMH)
c. ostium bursae and adjacent structures, Hungary: Fót Somlyóhegy, 22.v.1958, slide PS1412 (coll. ZMH)

Toulgoetia cauteriata (STAUDINGER)
a. genitalia, Portugal: Algarve, Salema, 19.iv.2007, slide PS1519 (coll. SKOU)
b. signum, Portugal: Algarve, Salema, 19.iv.2007, slide PS1519 (coll. SKOU)
c. ostium bursae and adjacent structures, Portugal: Algarve, Salema, 19.iv.2007, slide PS1519 (coll. SKOU)

Hypoxystis pluviaria (FABRICIUS)
a. genitalia, Russia: Mari republic, Krasnogorsk, 9–17.vii.1998, slide PS1292 (coll. P. SIHVONEN)
b. signum, Russia: Mari republic, Krasnogorsk, 9–17.vii.1998, slide PS1292 (coll. P. SIHVONEN)
c. ostium bursae and adjacent structures, Russia: Mari republic, Krasnogorsk, 9–17.vii.1998, slide PS1292 (coll. P. SIHVONEN)

Enanthyperythra legataria WEHRLI
a. genitalia, Turkey: province Izmir, Ilica, 28.iv.19[?]92, slide PS1616 (coll. ZFMK)
b. signum (margin highlighed partly), Turkey: province Izmir, Ilica, 28.iv.19[?]92, slide PS1616 (coll. ZFMK)
c. ostium bursae and adjacent structures, Turkey: province Izmir, Ilica, 28.iv.19[?]92, slide PS1616 (coll. ZFMK)

Apeira syringaria (LINNAEUS)
a. genitalia (signum is absent), Finland: Helsinki, ex larva 1959, slide PS1358 (coll. ZMH)
b. ostium bursae and adjacent structures, Finland: Helsinki, ex larva 1959, slide PS1358 (coll. ZMH)

Ennomos autumnaria (WERNEBURG)
a. genitalia, Finland: Espoo, 2–11.ix.1999, slide PS1294 (coll. P. SIHVONEN)
b. signum, Finland: Espoo, 2–11.ix.1999, slide PS1294 (coll. P. SIHVONEN)
c. ostium bursae and adjacent structures, Finland: Espoo, 2–11.ix.1999, slide PS1294 (coll. P. SIHVONEN)

Ennomos quercinaria (HUFNAGEL)
a. genitalia, Poland: Oswifz, July 1900, slide PS1368 (coll. ZMH)
b. signum (margin highlighted partly), Germany: Bavaria, Kelheim, 11.viii.1920, slide PS1644 (coll. ZSM)
c. ostium bursae and adjacent structures, Poland: Oswifz, July 1900, slide PS1368 (coll. ZMH)

Ennomos alniaria (LINNAEUS)
a. genitalia, Finland: Kirkkonummi, 1–17.ix.2000, slide PS1295 (coll. P. SIHVONEN)
b. signum, Finland: Kirkkonummi, 1–17.ix.2000, slide PS1295 (coll. P. SIHVONEN)
c. ostium bursae and adjacent structures, Finland: Kirkkonummi, 1–17.ix.2000, slide PS1295 (coll. P. SIHVONEN)

Ennomos fuscantaria (HAWORTH)
a. genitalia, Finland: Jomala, 18–19.viii.1982, slide PS1357 (coll. ZMH)
b. signum (very small, margin highlighted), Finland: Jomala, 18–19.viii.1982, slide PS1357 (coll. ZMH)
c. ostium bursae and adjacent structures, Finland: Jomala, 18–19.viii.1982, slide PS1357 (coll. ZMH)

Ennomos erosaria subsp. *erosaria* (DENIS & SCHIFFERMÜLLER)
a. genitalia, Finland: Föglö, 2.ix.1996, slide PS1297 (coll. P. SIHVONEN)
b. signum, Finland: Föglö, 2.ix.1996, slide PS1297 (coll. P. SIHVONEN)
c. ostium bursae and adjacent structures, Finland: Föglö, 2.ix.1996, slide PS1297 (coll. P. SIHVONEN)

Ennomos erosaria subsp. *ifranaria* (RUNGS)
a. genitalia, Morocco: Atlas Mountains, Ifrane, 6.vii.1952, slide PS1726 (coll. ZSM)
b. signum, Morocco: Atlas Mountains, Ifrane, 6.vii.1952, slide PS1726 (coll. ZSM)
c. ostium bursae and adjacent structures, Morocco: Atlas Mountains, Ifrane, 6.vii.1952, slide PS1726 (coll. ZSM)

Ennomos quercaria subsp. *quercaria* (HÜBNER)
a. genitalia, France: department Var, St-Tropez, 30.xi.1948, slide PS1606 (coll. ZSM)
b. signum, France: department Var, St-Tropez, 30.xi.1948, slide PS1606 (coll. ZSM)
c. ostium bursae and adjacent structures, Italy: Sicily, 5,3 km SE Collesano, 1100 m, 8.x.2010, slide PS1740 (coll. SKOU)

Ennomos quercaria subsp. *duercki* REISSER
a. genitalia, Greece: Crete, Silwa Rouwa, 1300 m, 15–31.vii.1938, slide Reisser 340, allotype (coll. SMNK)
b. signum, Greece: Crete, Silwa Rouwa, 1300 m, 15–31.vii.1938, slide Reisser 340, allotype (coll. SMNK)
c. ostium bursae and adjacent structures, Greece: Crete, Silwa Rouwa, 1300 m, 15–31.vii.1938, slide REISSER 340, allotype (coll. SMNK)

Ennomos quercaria subsp. lissochila (PROUT)
a. genitalia, Cyprus: Platres, 1500–1800 m, September 1921, slide PS1715/ BMNH GEO 24006 (coll. BMNH)
b. signum, Cyprus: Troodos mountains, 1020 m, ex ovo 1.xi.2009, slide PS1719 (coll. MÜLLER)
c. ostium bursae and adjacent structures, Cyprus: Troodos mountains, 1020 m, ex ovo 1.xi.2009, slide PS1719 (coll. MÜLLER)

Eumera regina STAUDINGER
a. genitalia, Dalmatia [a historical region mostly in present-day Croatia], 1940, slide PS1497 (coll. ZMH)
b. signum, Dalmatia [a historical region mostly in present-day Croatia], 1940, slide PS1497 (coll. ZMH)
c. ostium bursae and adjacent structures, Dalmatia [a historical region mostly in present-day Croatia], 1940, slide PS1497 (coll. ZMH)

Selenia dentaria (FABRICIUS)
a. genitalia, Finland: Savonlinna, 28–29.v.1995, slide PS1299 (coll. P. SIHVONEN)
b. signum, Finland: Savonlinna, 28–29.v.1995, slide PS1299 (coll. P. SIHVONEN)
c. ostium bursae and adjacent structures, Finland: Savonlinna, 28–29.v.1995, slide PS1299 (coll. P. SIHVONEN)

Selenia lunularia (HÜBNER)
a. genitalia, Italy: region Tuscany, Coreglia Antelminelli (Lucca), 600 m, 28.vi–10.vii.1999, slide PS1314 (coll. P. SIHVONEN)
b. signum, Italy: region Tuscany, Coreglia Antelminelli (Lucca), 600 m, 28.vi–10.vii.1999, slide PS1950 (coll. P. SIHVONEN)
c. ostium bursae and adjacent structures, Italy: region Tuscany, Coreglia Antelminelli (Lucca), 600 m, 28.vi–10.vii.1999, slide PS1314 (coll. P. SIHVONEN)

Selenia tetralunaria (HUFNAGEL)
a. genitalia, Finland: Savonlinna, 3–11.vi.1997, slide PS1302 (coll. P. SIHVONEN)
b. signum, Germany: Saxony, Leipzig, 9.ix.1926, slide PS1960/BMNH GEO 24888 (coll. BMNH)
c. ostium bursae and adjacent structures, Finland: Savonlinna, 3–11.vi.1997, slide PS1302 (coll. P. SIHVONEN)

Artiora evonymaria (DENIS & SCHIFFERMÜLLER)
a. genitalia, Hungary, without date data, slide PS1498 (coll. ZMH)
b. signum, Hungary, without date data, slide PS1498 (coll. ZMH)
c. ostium bursae and adjacent structures, Hungary, without date data, slide PS1498 (coll. ZMH)

Odontopera bidentata (CLERCK), structures from Finnish specimens. See text for details.
a. genitalia, Finland: Savonlinna, 20–25.vi.1996, slide PS1305 (coll. P. SIHVONEN)
b. signum, Finland: Savonlinna, 19–21.vi.1998, slide PS1641 (coll. P. SIHVONEN)
c. ostium bursae and adjacent structures, Finland: Savonlinna, 20–25.vi.1996, slide PS1305 (coll. P. SIHVONEN)

Odontopera bidentata (CLERCK), f. *graecarius*, structures from Greek specimens. See text for details.
a. genitalia, Greece: Peleponnese, Mount Chelmos, 1600–1700 m, 15.vi.2010, slide PS1839 (coll. SKOU)
b. signum, Greece: Peleponnese, Mount Chelmos, 1600–1700 m, 15.vi.2010, slide PS1839 (coll. SKOU)
c. ostium bursae and adjacent structures, Greece: Lagkádia, 1200 m, 19.v.1969, slide PS1605 (coll. ZSM)

Dasycorsa modesta (STAUDINGER)
a. genitalia, Greece: Strymon, 4.iv.1985, slide PS1520 (coll. SKOU)
b. signum, Cyprus: Limassol, 19.xii (no year given), slide PS1961/BMNH GEO 24889 (coll. BMNH)
c. ostium bursae and adjacent structures, Greece: Strymon, 4.iv.1985, slide PS1520 (coll. SKOU)

Crocallis tusciaria subsp. *tusciaria* (BORKHAUSEN)
a. genitalia, no locality data, slide PS1374 (coll. ZMH)
b. signum, Germany: Hesse, Wiesbaden, without date, slide PS1625 (coll. ZFMK)
c. ostium bursae and adjacent structures, no locality data, slide PS1374 (coll. ZMH)

Crocallis tusciaria subsp. *cypriaca* FISCHER
a. genitalia, Cyprus: Troodos Mountains, 1020, ex ovo 1.xi.2008, emerged 20.x.2009, slide PS1717 (coll. MÜLLER)
b. signum, Cyprus: Troodos Mountains, 1020, ex ovo 1.xi.2008, emerged 20.x.2009, slide PS1717 (coll. MÜLLER)
c. ostium bursae and adjacent structures, Cyprus: Troodos Mountains, 1020, ex ovo 1.xi.2008, emerged 20.x.2009, slide PS1717 (coll. MÜLLER)

Crocallis albarracina WEHRLI
a. genitalia, Spain: province Teruel, Tramacastilla, 24.viii.2001, slide PS1930 (coll. SKOU)
b. signum, Spain: province Teruel, Tramacastilla, 24.viii.2001, slide PS1930 (coll. SKOU)
c. ostium bursae and adjacent structures, Spain: province Avila, Sierra de Gredos, 1700 m, 18.viii.1988, slide PS1527 (coll. SKOU)

Crocallis elinguaria (LINNAEUS)
a. genitalia, Finland: Savonlinna, 12–15.viii.1996, slide PS1309 (coll. P. SIHVONEN)
b. signum (margin highlighted partly), Finland: Imatra, 10–16.viii.1996, slide PS1668 (coll. P. SIHVONEN)
c. ostium bursae and adjacent structures, Finland: Savonlinna, 12–15.viii.1996, slide PS1309 (coll. P. SIHVONEN)

Crocallis dardoinaria DONZEL
a. genitalia, France: department Var, Frejus, 15–20.ix.2005, slide PS1504 (coll. SKOU)
b. signum (margin highlighted partly), France: department Var, Frejus, 15–20.ix.2005, slide PS1504 (coll. SKOU)
c. ostium bursae and adjacent structures, France: department Var, Frejus, 15–20.ix.2005, slide PS1504 (coll. SKOU)

Crocallis auberti OBERTHÜR
a. genitalia (signum is absent), Algeria: Alger, 31.x.1918, slide PS1532 (coll. SKOU)
b. ostium bursae and adjacent structures, Algeria: Alger, 31.x.1918, slide PS1532 (coll. SKOU)

Crocallis helenaria RUCKDESCHEL
a. genitalia (margins of corpus bursae highlighted), Greece: Crete, Mesa Lassithi, 850 m, 7.x.2001, slide 0612/W. RUCKDESCHEL, allotype (coll. RUCKDESCHEL)
b. signum, Greece: Crete, Mesa Lassithi, 850 m, 7.x.2001, slide 0612/W. Ruckdeschel, allotype (coll. RUCKDESCHEL)
c. ostium bursae and adjacent structures, Greece: Crete, Mesa Lassithi, 850 m, 7.x.2001, slide 0612/W. RUCKDESCHEL, allotype (coll. RUCKDESCHEL)

Crocallis boisduvalaria (LUCAS)
a. genitalia, Morocco: Zerhoun, without date, slide PS1628 (coll. ZFMK)
b. signum, Morocco: Zerhoun, without date, slide PS1628 (coll. ZFMK)
c. ostium bursae and adjacent structures, Morocco: Zerhoun, without date, slide PS1628 (coll. ZFMK)

Ourapteryx sambucaria (LINNAEUS)
a. genitalia, Finland: Imatra, 25.vii.1997, slide PS1307 (coll. P. SIHVONEN)
b. signum, Finland: Imatra, 25.vii.1997, slide PS1307 (coll. P. SIHVONEN)
c. ostium bursae and adjacent structures, Finland: Imatra, 25.vii.1997, slide PS1307 (coll. P. SIHVONEN)

Asovia maeoticaria (ALPHÉRAKY)
a. genitalia, Greece: Volotia, Mt. Parnassos, 1800 m, 28.vi.1985, slide PS1538 (coll. SKOU)
b. signum, Greece: Volotia, Mt. Parnassos, 1800 m, 28.vi.1985, slide PS1538 (coll. SKOU)
c. ostium bursae and adjacent structures, Greece: Volotia, Mt. Parnassos, 1800 m, 28.vi.1985, slide PS1538 (coll. SKOU)

Colotois pennaria (LINNAEUS)
a. genitalia (signum is absent), Finland: Vantaa, 8.x.1995, slide PS1311 (coll. P. SIHVONEN)
b. ostium bursae and adjacent structures, Finland: Vantaa, 8.x.1995, slide PS1311 (coll. P. SIHVONEN)

Angerona prunaria (LINNAEUS)
a. genitalia, Finland: Enonkoski, 6.vii.1996, slide PS1313 (coll. P. SIHVONEN)
b. signum, Finland: Enonkoski, 6.vii.1996, slide PS1313 (coll. P. SIHVONEN)
c. ostium bursae and adjacent structures, Finland: Enonkoski, 6.vii.1996, slide PS1313 (coll. P. SIHVONEN)

Cabera pusaria (LINNAEUS)
a. genitalia, Finland: Savonlinna, 27–30.vi.1997, slide PS1334 (coll. P. SIHVONEN)
b. corpus bursae spines/signa, Finland: Savonlinna, 27–30.vi.1997, slide PS1334 (coll. P. SIHVONEN)
c. ostium bursae and adjacent structures, Finland: Savonlinna, 27–30.vi.1997, slide PS1334 (coll. P. SIHVONEN)

Cabera leptographa WEHRLI
a. genitalia, Austria: Burgenland, 22.vi.1981, slide PS1363 (coll. ZMH)
b. corpus bursae spines/signa, Austria: Burgenland, 22.vi.1981, slide PS1363 (coll. ZMH)
c. ostium bursae and adjacent structures, Austria: Burgenland, 22.vi.1981, slide PS1363 (coll. ZMH)

Cabera exanthemata (SCOPOLI)
a. genitalia, Finland: Enonkoski, 20–24.vi.1997, slide PS1336 (coll. P. SIHVONEN)
b. corpus bursae spines/signa, Finland: Enonkoski, 20–24.vi.1997, slide PS1336 (coll. P. SIHVONEN)
c. ostium bursae and adjacent structures, Finland: Enonkoski, 20–24.vi.1997, slide PS1336 (coll. P. SIHVONEN)

Lhommeia biskraria (OBERTHÜR)
a. genitalia (signum is absent), Libya: Wadi El Hira, 10.vi.1983, slide PS1252 (coll. SKOU)
b. ostium bursae and adjacent structures, Libya: Wadi El Hira, 10.vi.1983, slide PS1252 (coll. SKOU)

Lomographa bimaculata (FABRICIUS)
a. genitalia, Finland: Savonlinna, 6–7.vi.1996, slide PS1338 (coll. P. SIHVONEN)
b. corpus bursae spines/signa, Finland: Savonlinna, 6–7.vi.1996, slide PS1338 (coll. P. SIHVONEN)
c. ostium bursae and adjacent structures, Finland: Savonlinna, 6–7.vi.1996, slide PS1338 (coll. P. SIHVONEN)

Lomographa temerata (DENIS & SCHIFFERMÜLLER)
a. genitalia, Finland: Savonlinna, 11.vi.1996, slide PS1341 (coll. P. SIHVONEN)
b. signum, Finland: Savonlinna, 11.vi.1996, slide PS1341 (coll. P. SIHVONEN)
c. ostium bursae and adjacent structures, Finland: Savonlinna, 11.vi.1996, slide PS1341 (coll. P. SIHVONEN)

Aleucis distinctata (HERRICH-SCHÄFFER)
a. genitalia (corpus bursae margins highlighted partly), France: department Basses-Alpes, Digne, without date, slide PS1615 (coll. ZFMK)
b. sclerotisation of corpus bursae, France: department Basses-Alpes, Digne, without date, slide PS1615 (coll. ZFMK)
c. ostium bursae and adjacent structures, France: department Basses-Alpes, Digne, without date, slide PS1615 (coll. ZFMK)

Aleucis orientalis (STAUDINGER)
a. genitalia, Greece: Kastoria, Kastoria, 800 m, 19.iv.1987, slide PS1540 (coll. SKOU)
b. sclerotisation of corpus bursae, Greece: Kastoria, Kastoria, 800 m, 19.iv.1987, slide PS1540 (coll. SKOU)
c. ostium bursae and adjacent structures, Greece: Kastoria, Kastoria, 800 m, 19.iv.1987, slide PS1540 (coll. SKOU)

Theria rupicapraria (DENIS & SCHIFFERMÜLLER)
a. genitalia, coll. Duske, without locality data [potentially from Russia: Volgograd], 8.3.[19?]67, slide PS1417 (coll. ZMH)
b. signum, coll. Duske, without locality data [potentially from Russia: Volgograd], 8.3.[19?]67, slide PS1417 (coll. ZMH)
c. ostium bursae and adjacent structures, coll. Duske, without locality data [potentially from Russia: Volgograd], 8.3.[19?]67, slide PS1417 (coll. ZMH)

Theria primaria (HAWORTH)
a. genitalia (margins of corpus bursae partly highlighted), Denmark: Hofmansgave, ex ovo 27.iii.1979, slide PS1541 (coll. SKOU)
b. signum, Denmark: Hofmansgave, ex ovo 27.iii.1979, slide PS1541 (coll. SKOU)
c. ostium bursae and adjacent structures, Denmark: Hofmansgave, ex ovo 27.iii.1979, slide PS1541 (coll. SKOU)

Campaea margaritaria (LINNAEUS)
a. genitalia, Finland: Föglö, 20.vii.1995, slide PS1343 (coll. P. SIHVONEN)
b. corpus bursae spines/signa, Finland: Föglö, 20.vii.1995, slide PS1343 (coll. P. SIHVONEN)
c. ostium bursae and adjacent structures, Finland: Föglö, 20.vii.1995, slide PS1343 (coll. P. SIHVONEN)

Gerinia honoraria (DENIS & SCHIFFERMÜLLER)
a. genitalia (signum is absent), Austria, without date, slide PS1439 (coll. ZMH)
b. ostium bursae and adjacent structures, Austria, without date, slide PS1439 (coll. ZMH)

Hylaea fasciaria (LINNAEUS)
a. genitalia, Finland: Föglö, 28.vii.1995, slide PS1448 (coll. P. SIHVONEN)
b. reduced signum, Finland: Föglö, 28.vii.1995, slide PS1448 (coll. P. SIHVONEN)
c. ostium bursae and adjacent structures, Finland: Föglö, 28.vii.1995, slide PS1448 (coll. P. SIHVONEN)

Hylaea mediterranea SIHVONEN *et al.*
a. genitalia, Italy: Sicily, 5,7 km ESE San Stefano Quisquina, 1000 m, 9–10.x.2010, slide PS1739, paratype (coll. SKOU)
b. signum, Italy: Sicily, 5,7 km ESE San Stefano Quisquina, 1000 m, 9–10.x.2010, slide PS1739, paratype (coll. SKOU)
c. ostium bursae and adjacent structures, Italy: Sicily, 5,7 km ESE San Stefano Quisquina, 1000 m, 9–10.x.2010, slide PS1739, paratype (coll. SKOU)

Hylaea pinicolaria (BELLIER)
a. genitalia, France: Corsica, 3 km E of Haut-Asco, 1100 m, 20–21.vii.2004, slide PS1543 (coll. SKOU)
b. reduced signum, France: Corsica, 3 km E of Haut-Asco, 1100 m, 20–21.vii.2004, slide PS1543 (coll. SKOU)
c. ostium bursae and adjacent structures, France: Corsica, 3 km E of Haut-Asco, 1100 m, 20–21.vii.2004, slide PS1543 (coll. SKOU)

Adalbertia castiliaria (STAUDINGER)
a. genitalia (corpus bursae margin highlighted, signum is absent), Spain: region Murcia, Sierra de Espuna, 800 m, 8.v.1995, slide PS1544 (coll. SKOU)
b. detail of papillae anales, Spain: region Murcia, Sierra de Espuna, 800 m, 8.v.1995, slide PS1544 (coll. SKOU)
c. ostium bursae and adjacent structures, Spain: region Murcia, Sierra de Espuna, 800 m, 8.v.1995, slide PS1544 (coll. SKOU)

Pungeleria capreolaria (DENIS & SCHIFFERMÜLLER)
a. genitalia, Romania: Mt. Bucegi, 1300 m, 16.viii.1969, slide PS1420 (coll. ZMH)
b. corpus bursae spines/signa, Romania: Mt. Bucegi, 1300 m, 16.viii.1969, slide PS1420 (coll. ZMH)
c. ostium bursae and adjacent structures, Romania: Mt. Bucegi, 1300 m, 16.viii.1969, slide PS1420 (coll. ZMH)

Siona lineata (SCOPOLI)
a. genitalia, Finland: Helsinki, 22.vi.2001, slide PS1347 (coll. P. SIHVONEN)
b. signum, Finland: Helsinki, 22.vi.2001, slide PS1347 (coll. P. SIHVONEN)
c. ostium bursae and adjacent structures, Finland: Helsinki, 22.vi.2001, slide PS1347 (coll. P. SIHVONEN)

Chariaspilates formosaria (EVERSMANN)
a. genitalia, Russia: Ussuri, Rjazanovka, 6.vii.1982, slide PS1427 (coll. ZMH)
b. signum, Russia: Ussuri, Rjazanovka, 6.vii.1982, slide PS1427 (coll. ZMH)
c. ostium bursae and adjacent structures, Russia: Ussuri, Rjazanovka, 6.vii.1982, slide PS1427 (coll. ZMH)

Aspitates mundataria (STOLL)
a. genitalia (signum is absent, corpus bursae margins highlighted), Russia: Stavropol'skij Kraj, Kuma River, 8.vi.1997, slide PS1576 (coll. SKOU)
b. ostium bursae and adjacent structures, Russia: Volgograd, 3.vi.1894, slide PS1493 (coll. ZMH)

Aspitates gilvaria (DENIS & SCHIFFERMÜLLER)
a. genitalia (signum is absent, corpus bursae margins highlighted), Finland: Hyvinkää, 7.vii.2001, slide PS1349 (coll. P. SIHVONEN)
b. ostium bursae and adjacent structures, Finland: Hyvinkää, 7.vii.2001, slide PS1349 (coll. P. SIHVONEN)

Aspitates taylorae (BUTLER)
a. genitalia (signum is absent, corpus bursae margins highlighted), Russia: Polar Ural, Krasnyi Kamen, 200 m, 1.vii.1994, slide PS1424 (coll. ZMH)
b. ostium bursae and adjacent structures, Russia: Polar Ural, Krasnyi Kamen, 200 m, 1.vii.1994, slide PS1424 (coll. ZMH)

Aspitates orciferaria (WALKER)
a. genitalia (signum is absent, corpus bursae margins highlighted), Russia: Polar Ural, Krasnyi Kamen 500 m, 8.vii.1994, slide PS1429 (coll. ZMH)
b. ostium bursae and adjacent structures, Russia: Polar Ural, Krasnyi Kamen 500 m, 8.vii.1994, slide PS1429 (coll. ZMH)

Aspitates forbesi (MUNROE)
a. genitalia (signum is absent), Canada: Manitoba, Churchill, 12.vii.1952, slide BMNH GEO 24007, paratype (coll. BMNH)
b. ostium bursae and adjacent structures, Canada: Manitoba, Churchill, 12.vii.1952, slide BMNH GEO 24007, paratype (coll. BMNH)

Aspitates albaria (BARTEL)
a. genitalia (signum is absent, corpus bursae margins highlighted), Russia: Transbaikal, Chara, June 2000, slide PS1546 (coll. SKOU)
b. ostium bursae and adjacent structures, Russia: Transbaikal, Chara, June 2000, slide PS1546 (coll. SKOU)

Aspitates ochrearia (ROSSI)
a. genitalia (signum is absent, corpus bursae margins highlighted), Portugal: Algarve, Lagos, 12.v.1971, slide PS1431 (coll. ZMH)
b. ostium bursae and adjacent structures, Portugal: Algarve, Lagos, 12.v.1971, slide PS1431 (coll. ZMH)

Dyscia raunaria (FREYER)
a. genitalia (signum is absent), Croatia: Plomin, 6.ix.1975, slide PS1503 (coll. SKOU)
b. ostium bursae and adjacent structures, Croatia: Kraljevica, 15.vi.1972, slide PS1435 (coll. ZMH)

Dyscia royaria TAUTEL & BILLI
a. genitalia (signum is absent), France: department Alpes-Maritimes, Tende, Ancienne piste du col de Tende, 1300 m, ex ovo 5.viii.2007, emerged 3.viii.2008, slide PS1638 (coll. SKOU)
b. ostium bursae and adjacent structures, France: department Alpes-Maritimes, Tende, Ancienne piste du col de Tende, 1300 m, ex ovo 5.viii.2007, emerged 3.viii.2008, slide PS1638 (coll. SKOU)

Dyscia innocentaria (CHRISTOPH)
a. genitalia (signum is absent), Italy: region Púglia, Manfredónia, 26.v.2005, slide PS1511 (coll. SKOU)
b. ostium bursae and adjacent structures, Italy: regioin Púglia, Manfredónia, 26.v.2005, slide PS1511 (coll. SKOU)

Dyscia malatyana WEHRLI

a. genitalia (signum is absent), Russia: Volgograd oblast, Elton salt lake, 18–23.v.2001,slide PS1893 (coll. SKOU)
b. ostium bursae and adjacent structures, Russia: Volgograd oblast, Elton salt lake, 18–23.v.2001, SLIDE PSPS1893 (coll. Skou)

Dyscia crassipunctaria (REBEL)

a. genitalia (signum is absent), Greece: Crete, Chersonisos, 26.v.1986, slide PS1513 (coll. SKOU)
b. ostium bursae and adjacent structures, Greece: Crete, Chersonisos, 26.v.1986, slide PS1513 (coll. SKOU)

Dyscia conspersaria (DENIS & SCHIFFERMÜLLER)

a. genitalia (signum is absent), Austria: Niederösterreich, Wien, 1902, slide PS1631 (coll. ZFMK)
b. ostium bursae and adjacent structures, Russia: Volgograd oblast, Olhovka village, 5.vi.2001, slide PS1590 (coll. NUPPONEN)

Dyscia fagaria (THUNBERG)

a. genitalia (signum is absent, corpus bursae margins highlighted partly), Austria: Tirol, Heinfeld, 5.v.1912, slide PS1433 (coll. ZMH)
b. ostium bursae and adjacent structures, Austria: Tirol, Heinfeld, 5.v.1912, slide PS1433 (coll. ZMH)

Dyscia penulataria (HÜBNER)

a. genitalia (signum is absent, corpus bursae margins highlighted), Spain: province Cadiz, Tarifa Punta Paloma, 30.iii–4.iv.1994, slide PS1523 (coll. SKOU)
b. ostium bursae and adjacent structures, Spain: province Cadiz, Tarifa Punta Paloma, 30.iii–4.iv.1994, slide PS1523 (coll. SKOU)

Dyscia lentiscaria (DONZEL)

a. genitalia (signum is absent), Spain: province Cadiz, Roche, 6.iii.2005, slide PS1517 (coll. SKOU)
b. ostium bursae and adjacent structures, Spain: province Cadiz, Roche, 6.iii.2005, slide PS1517 (coll. SKOU)

Dyscia distinctaria (BANG-HAAS)

a. genitalia (signum is absent, corpus bursae margins highlighted partly), Spain: Prov. Sevilla, El Ronquillo at Rivera de Huelva, 2.iv.2003, slide PS1636 (coll. SKOU)
b. ostium bursae and adjacent structures, Spain: Prov. Sevilla, El Ronquillo at Rivera de Huelva, 2.iv.2003, slide PS1636 (coll. SKOU)

Perconia strigillaria (HÜBNER)

a. genitalia, Finland: Lammi, 21.vi.1995, slide PS1351 (coll. P. SIHVONEN)
b. signum, Finland: Lammi, 21.vi.1995, slide PS1351 (coll. P. SIHVONEN)
c. ostium bursae and adjacent structures, Finland: Lammi, 21.vi.1995, slide PS1351 (coll. P. SIHVONEN)

Perconia baeticaria (STAUDINGER)
a. genitalia, Portugal: Serra da Estrela Meteorological station, 18.vi.2006, slide PS1548 (coll. SKOU)
b. signum, Portugal: Serra da Estrela Meteorological station, 18.vi.2006, slide PS1548 (coll. SKOU)
c. ostium bursae and adjacent structures, Portugal: Serra da Estrela Meteorological station, 18.vi.2006, slide PS1548 (coll. SKOU)

Onychora agaritharia (DARDOIN)
a. genitalia (signum is absent), Spain: Prov. Teruel, Albarracin, 1200 m, 25–26.viii.2001, slide PS1549 (coll. SKOU)
b. ostium bursae and adjacent strucures, Spain: Prov. Teruel, Albarracin, 1200 m, 25–26.viii.2001, slide PS1549 (coll. SKOU)

Compsoptera opacaria HÜBNER
a. genitalia, France: department Basses-Alpes, Digne, 1898, slide PS1441 (coll. ZMH)
b. signum, France: department Basses-Alpes, Digne, 1898, slide PS1441 (coll. ZMH)
c. ostium bursae and adjacent structures, France: department Basses-Alpes, Digne, 1898, slide PS1441 (coll. ZMH)

Compsoptera jourdanaria (SERRES)
a. genitalia, Spain: province Segovia, Ildefonso, 1900, slide PS1489 (coll. ZMH)
b. signum, Spain: province Segovia, Ildefonso, 1900, slide PS1489 (coll. ZMH)
c. ostium bursae and adjacent structures, Spain: province Segovia, Ildefonso, 1900, slide PS1489 (coll. ZMH)

Compsoptera argentaria (HERRICH-SCHÄFFER)
a. genitalia, Algeria: Lambèse, October 1918, slide PS1633 (coll. ZFMK)
b. signum, Algeria: Lambèse, October 1918, slide PS1633 (coll. ZFMK)
c. ostium bursae and adjacent structures, Algeria: Lambèse, October 1918, slide PS1633 (coll. ZFMK)

References

AARVIK, L., BERGGREN, K. & HANSEN, L. O. 2000: Catalogus Lepidopterorum Norvegiae. Nordberg Aksidenstryggeri AS, Oslo. 192 p.

ABRAHAM, D., RYRHOLM, N., WITTZELL, H., HOLLOWAY, J.D., SCOBLE, M.J. & LÖFSTEDT, C. 2001: Molecular phylogeny of the subfamilies in Geometridae (Geometroidea: Lepidoptera). Molecular Phylogenetics and Evolution 20: 65–77.

AGENJO, R. 1974: Ocho géneros y veinte especies de Geometridae nuevos para España. Graellsia 27: 3–21.

AIZPÚRUA, C. G. 2002: Orugas y Mariposas de Europa. Tomo III, Heterocera, Lepidópteros de actividad nocturna. Lerko Print, Madrid. 321 p.

ANONYMOUS 2011a. List of colors. A Wikipedia article on the colours and their names arranged by shade. http://en.wikipedia.org/wiki/List_of_colors. Page visited 30 October 2011.

ANONYMOUS 2011b: Atropos migrant insect review - end May/early June 2011. http://atropos.info/news.html. Page visited 30 April 2012.

ANONYMOUS 2012: Information source on Dutch macrolepidoptera. http://www.vlindernet.nl. Page visited 25 December 2012.

ANONYMOUS 2013: The Global Lepidoptera Names Index. http://www.nhm.ac.uk/research-curation/research/projects/lepindex/search/index.dsml Page visited numerous times in 2013.

ANONYMOUS 2014: Lépidoptères de France méridionale et de Corse. http://pathpiva.wifeo.com/dyscia-innocentaria-1.php and http://pathpiva.wifeo.com/dyscia-innocentaria-l5.php Page visited 16 February 2014.

ATANASOVA, D. I, TOSHOVA, T. B. & SUBCHEV, M. A. 2010: Seasonal Monitoring of the Yellow Alfalfa Geometrid, *Isturgia arenacearia* (Denis and Schiffermuller, 1775) (Lepidoptera: Geometridae) by Pheromone Traps in Three Regions of Bulgaria. Acta Phytopathologica et Entomologica Hungarica 45: 305–311.

AXMACHER, J. C., HOLTMANN, G., SCHEUERMANN, L., BREHM, G., MÜLLER-HOHENSTEIN, K. & FIEDLER, K. 2004: Diversity of geometrid moths (Lepidoptera: Geometridae) along an Afrotropical elevational rainforest transect. Diversity and Distributions 10: 293–302.

BANG-HAAS, A. 1910: Neue oder wenig bekannte palaearctische Macrolepidopteran. Deutsche entomologische Zeitschrift Iris 24: 27–51.

BECKER, A. 1858: Naturhistorische Mittheilungen von den Jahren 1856 und 1857; über die dem Entomologen wichtigsten Gewächse der Sareptaêr Umgegend und noch einige Bemerkungen über das Fangen und Tödten der Insekten. Bulletin de la Société Impériale des Naturalistes de Moscou XXXI (III): 159–187.

BELJAEV, E. A. 1997: A new subspecies of the *Pseudopanthera macularia* L. (Lepidoptera: Geometridae, Ennominae) from South Siberia. Far Eastern Entomologist 51: 1–7.

BELJAEV, E. A. 2006a: A morphological approach to the Ennominae phylogeny (Lepidoptera: Geometridae). Spixiana 29: 215–216.

BELJAEV, E. A. 2006b: Alternative order, tentative tribal system of Ennominae based on current family-group names. Personal communication, published in Forum Herbulot www pages without peer-review http://www.herbulot.de/famgroup2.htm (Page visited 1.1.2014)

BELJAEV, E. A. 2008a: A New Concept of the Generic Composition of the Geometrid Moth Tribe Ennomini (Lepidoptera, Geometridae) Based on Functional Morphology of the Male Genitalia. Entomological Review 88: 50–60.

BELJAEV, E. A. 2008b: Phylogenetic relationships of the family Geometridae and its subfamilies (Lepidoptera). Meetings in memory of N. A. Cholodkovsky 60: 1–238. In Russian with English summary.

BELJAEV, E. A. & VASILENKO, S. V. 2014: A little-known geometrid moth *Odontopera graecarus* (A. Bang-Haas, 1911) (Lepidoptera: Geometridae) - new to the fauna of the Caucasus. Euroasian Entomological Journal 13: 346–350.

BELLIER, C. 1861: Description de trois lépidoptères nouveaux de l'ile de Corse. Annales de la Société entomologique de France 1: 29–30.

BERGMANN, A. 1955: Die Gross-Schmetterlinge Mitteldeutschlands. Band 5/2: Spanner. 561–1267. Urania-Verlag, Leipzig/Jena.

BESHKOV S. 2011. Biodiversity of butterflies and moths (Insecta: Lepidoptera: Macrolepidoptera) in chestnut forests, Belassitsa Mountain (Preliminary report). *In*: Zlatanov, T., Velichkov, I. & Nikolov, B. (eds.) State and prospects of the *Castanea sativa* population in Belasitsa mountain: climate change adaptation; maintenance of biodiversity and sustainable ecosystem management. Project BG 0031 EEA report. 18 p.

BLESZYNSKI, S. 1966: Motyle - Lepidoptera. Zesnyt 46c. Geometridae. Klucze Oznacz. Owad. Pol. Panstwowe Wydawnictwo Naukowe - Warszawa. 122 p.

BLACKWELDER, R. E. 1967: Taxonomy: A text and reference book. John Wiley & Sons, New York. 698 p.

BLAMEY, M. & GREY-WILSON, C. 1993: Mediterranean Wild Flowers. HarperCollins Publishers. 560 p.

BLESZYNSKI, S. 1960: Klucze do Oznaczania Owadów Polski Cz. XXVII, Zeszyt 46a, Miernikowce Geometridae: 1–149. Polski Zwiazek Ent., Warszawa.

BLESZYNSKI, S. 1965: Klucze do Oznaczania Owadów; Polski XXVII Lepidoptera 46b, Geometridae. Polski Zwiazek Ent., Warszawa, 305 p.

BLESZYNSKI, S. 1966: Klucze do Oznaczania Owadów; Polski XXVII Lepidoptera 46c, Geometridae. Polski Zwiazek Ent., Warszawa, 122 p.

BORG, P. 1932: The Lepidoptera of the Maltese Islands. Government Printing Office. 25 p.

BRAHM, N. J. 1790–1791: Insektenkalender für Sammler und Oekonomen. Mainz, 2 vols. 558 p.

BREHM, G. 2002: Diversity of geometrid moths in a montane rainforest in Ecuador. Academic dissertation. Universität Bayreuth. 203 p.

BREHM G, & FIEDLER, K. 2003: Faunal composition of geometrid moths changes with altitude in an Andean montane rain forest. Journal of Biogeography 30: 431–440.

BREHM, G. SUESSENBACH, D. & FIEDLER, K. 2003: Unique elevational diversity patterns of geometrid moths in an Andean montane rainforest. Ecography 26: 456–466.

BROOKS, M. 1991: A complete Guide to British Moths. Jonathan Cape, London. 248 p.

BUDASHKIN, Yu. I. & SAVCHUK, V. V. 2010: Additions to fauna and bionomy of the Crimean Lepidoptera. Optimization and Protection of Ecosystems. Scientific Notes of Taurida National V.I. Vernadsky University 3: 50–68. In Russian.

BUSZKO, J. 2000: Atlas motyli Polski, part III: Thyatiridae, Drepanidae, Geometridae. Grupa Image, Warszawa. 518 p.

BÜTTIKER, W. & WHELLAN, J. A. 1966: Records of eye-frequenting moths from Rhodesia. Bulletin of the Rhodesia Agriculture Journal 63: 24–27.

BYTINSKI-SALZ, H. 1934: Ein Beitrag zur Kenntnis der Lepidopterenfauna Sardiniens. Internationale Entomologische Zeitschrift Guben, 28: 41–47, 56–62, 89–94, 97–106, 133–137, 165–169, 182–185, 213–216.

BYTINSKI-SALZ, H. & STERNLICHT, M. 1967: Insects associated with Oaks (*Quercus*) in Israel. Israel Journal of Entomology 2: 107–143.

BÄNZIGER, H. & FLETCHER, D. S. 1988: Description of five new lachryphagous and zoophilous *Semiothisa* moths from SE Asia, with five new synonymies (Lepid., Geometridae). Revue Suisse de Zoologie 94: 933–952.

CATANIA, A. 2010: *Isturgia disputaria* (Guenee, 1858), a species new to the Lepidoptera fauna of the Maltese Islands (Lepidoptera: Geometridae, Ennominae). SHILAP Revista de Lepidopterologia 38: 287–289.

CCDB (Canadian Centre for DNA Barcoding) 2013: Description of DNA barcode protocols. http://www.dnabarcoding.ca/pa/ge/research/protocols. Page visited 15 February 2014.

CHOI, S.-W., KIM, S.-S. & SHIN, Y.-H. 1998: Systematics of the tribe Abraxini (Lepidoptera: Geometridae, Ennominae) in South Korea. Insecta Koreana 15: 41–76.

CHRÉTIEN, P. 1917: Contribution à la connaissance des Lépidoptères du Nord de l'Afrique. Notes biologiques et eritiques. Annales de la Societé entomologique de France 85 [pages 369–400 published on January 1917): 369–502.

COOK, M. A. & SCOBLE M. J. 1992: Tympanal organs of geometrid moths: a review of their morphology, function, and systematic importance. Systematic Entomology 17: 219–232.

COSTA, O. G. [1848]: Lepidotteri. Geometre. In Fauna del Regno di Napoli. 1–314 [1836], 315–370 [1848], 371–402 [1849], 403–418 [1850], 419-422 [1849].

COVELL, C.V., FERGUSON, D. C. & STRALEY, G. B. 1986: *Ennomos alniaria* (Lepidoptera: Geometridae), a European moth recently discovered in British Columbia. Canadian Entomologist 118: 499–501.

CULOT, J. 1917: Noctuelles et Géomètres d'Europe. Vol. 3, Géometres I. Genéve, 269 p.

CULOT, J. 1919: Noctuelles et Géomètres d'Europe. Vol. 4, Géometres II. Genéve, 269 p.

DANNEHL, F. 1928: Beiträge zur Lepidopterenfauna Südtirols, Fortsetzung. Entomologische Zeitschrift Frankfurt am Main 41: 393–412.

DANTART, J., DOMINGUEZ, M. & BAIXERAS J. 1993: Contribution to the knowledge of the genus *Crocallis* Treitschke, 1925: Iberian species (Lepidoptera: Geometridae). Nota lepidopterologica 15: 195–216.

DAPPORTO, L. 2004: Biometric evidence of the synonymy of *Compsoptera aemiliorum* Raineri, 1994 with C. *opacaria* (Hübner, 1819) (Lepidoptera Geometridae). Redia 87: 73–75.

DENIS, M. & SCHIFFERMÜLLER, I. 1775: Ankündung eines systematischen Werkes von den Schmetterlingen der Wienergegend. Vienna. 324 pp., 3 plates.

DEWAARD, J. R., IVANOVA, N. V., HAJIBABAEI, M. & HEBERT, P. D. N. 2008: Assembling DNA barcodes: analytical protocols. In: Martin, C. (ed.), Methods in molecular biology: environmental genetics, pp. 275–293. Humana Press Inc., Totowa.

DEWAARD, J. R., HEBERT, P. D. N. & HUMBLE, L. M. 2011: A Comprehensive DNA Barcode Library for the Looper Moths (Lepidoptera: Geometridae) of British Columbia, Canada. PLoS ONE 6(3): e18290. doi:10.1371/journal.pone.0018290.

DISTANT, W. 1892: A Naturalist in the Transvaal. London. 277 pp.

DJAKONOV, A. M. 1936: Die Geometriden des Amur-Ussuri-Gebietes. II. Tribus Caberini, nebst Revision einiger Gattungen dieser Gruppe. Proceedings of the Zoological Institute 3: 475–531. In Russian and in German.

DUGDALE, J.D. 1961: Larval characters of taxonomic significance of New Zealand Ennomines (Lepidoptera: Geometridae). Transactions of the Royal Society of New Zealand, 16: 215–233.

DUMONT, C. 1932: Les Lépidopteres des Crommiers du Nord de l'Afrique. Société Entomologique de France. Livre du Centennaire 689–719.

EBERT G, (ed.) 2001: Die Schmetterlinge Baden-Württenbergs. Band 8, Nachtfalter VI (Geometridae). Stuttgart, E. Ulmer. 541 p.

EBERT, G. (ed.) 2003: Die Schmetterlinge Baden-Württenbergs. Band 9, Nachtfalter VII (Geometridae). Stuttgart, E. Ulmer. 609 p.

ECKSTEIN, K. 1923: Die Schmetterlinge Deutschlands mit besonderer Berücksichtigung ihrer Biologie und wirtschaftlichen Bedeutung. 4. Band: Spezieller Teil. Schluss (4). Die Spanner und die bärenartigen Falter. K. G. Lutz Verlag, Stuttgart. 88 p. 16 colour plates.

ELLISON, R. E. & WILTSHIRE, E. P. 1939: The Lepidoptera of the Lebanon with notes on their season and distribution. Transactions of the Royal Entomological Society, London 88: 1–56.

EXPÓSITO, A. 2005: Nueva sinonimia de *Narraga isabel* Agenjo, 1956 (Lepidoptera: Geometridae). SHILAP Revista de Lepidopterologia 33: 504.

FAJČIK, J. 2003: Die Schmetterlinge Mittel- und NordEuropas. Drepanidae, Geometridae, Lasiocampidae, Endromidae, Lemoniidae, Sphingidae, Notodontidae, Lymantriidae, Arctiidae. ER-Print, Bratislava. 172 p. + 22 black-and-white plates + 38 colour plates.

FERGUSON, D. C. 1983: The identity of two monotypic geometrid genera wrongly attributed to the Nearctic fauna (Geometridae). Journal of the Lepidopterists' Society 37: 146–147.

FERGUSON, D. C. 2008: Geometroidea, Geometridae (part): Ennominae (part): Abraxini, Cassymini, Macariini. In R.W. Hodges *et al.* (eds.), The Moths of North America, fasc. 17 (2): 1–431. The Wedge Entomological Research Foundation, Washington.

FINNISH ENTOMOLOGICAL DATABASE 2013. Database accessed 9 September 2013. http://hyonteiset.luomus.fi/insects/main/EntDatabase.html

FISCHER von WALDHEIM, G. 1840: Adnotationes de Lepidopteris. Bulletin de la Société impériale des naturalistes de Moscou 1840: 81–89.

FISCHER, H. & LEWANDOWSKI, S. 2005: Die Geometriden-Fauna von Zypern - eine Überarbeitung aller bisher bekannter Arten (1. Teil). (Lepidoptera, Geometridae, Geometrinae, Ennominae). Atalanta 36: 291–310.

FIUMI, G., GOVI, G. & SECHI, D. 2009: Notes on the immatures stages of *Macaria ichnusae* Govi & Fiumi, 2005 (Insecta Lepidoptera Geometridae Ennominae). Quaderno di Studi e Notizie di Storia Naturale della Romagna 29: 189–196. (in Italian with English summary).

FIUMI, G. & FLAMIGNI, C. 2013: *Dyscia* (*Dyscia*) *govii* n. sp. della regione sardo-corsa. Quaderno di Studi e Notizie di Storia Naturale della Romagna 38: 191–199.

FLAMIGNI, C. & BASTIA, G. 2003: I Geometridi del bolognese. I parte (Insecta Lepidoptera, Geometridae: Ennominae, Alsophilinae, Geometrinae, Sterrhinae). Quaderno di Studi e Notizie di Storia Naturale della Romagna 17 (suppl.): 87–118.

FLAMIGNI, C., FIUMI, G. & PARENZAN, P. 2007: Lepidotteri Eteroceri D'Italia – Geometridae Ennominae I. Natura Edizioni Scientifiche, Bologna. 382 p.

FLAMIGNI, C., FIUMI, G. & PARENZAN, P. (*in press*): Lepidotteri Eteroceri d'Italia. Geometridae Ennominae II. Natura Edizioni Scientifiche di Alfonso Iorio. Bologna.

FLETCHER, D. S. 1966: Some changes in the nomenclature of British Lepidoptera. Entomologist's Gazette 17: 9–18.

FLETCHER, D. S. 1974: A revision of the Old World genus *Zamarada* (Lepidoptera: Geometridae). Bulletin of the British Museum of Natural History (Entomology), supplement 22. 498 p.

FLETCHER, D. S. 1978: Geometridae (Lepidoptera) collected by Dr. J. Szunyoghy in Tanzania. Acta Zoologica Academiae Scientiarum Hungaricae 24: 41–105.

FLETCHER, D. S. 1979: In NYE, I.W.B. (ed.), The generic names of moths of the world, vol. 3. London. 243 p.

FORBES, W. T. M. 1948: Lepidoptera of New York and neighboring states: Geometridae, Notodontidae, Sphingidae, Lymantriidae, part II. Memoirs of the Cornell University Agricultural Experiment Station 274: 1–263.

FORUM HERBULOT 2014: World list of family-group names in Geometridae. http://www.herbulot.de/famgroup.htm Page visited 29 November 2014.

FORSTER, W. & WOHLFAHRT, T.A. 1981: Die Schmetterlinge Mitteleuropas, 5. Band, Spanner. Franckh'sche Verlagsbuchhandlung, Stuttgart. 312 p.

FOX, R., PARSONS, M. S., CHAPMAN, J. W., WOIWOD, I. P., WARREN, M. S. & BROOKS, D.R. 2013: The State of Britain's Larger Moths 2013. Butterfly Conservation and Rothamsted Research, Wareham, Dorset, UK. 30 p.

FREYER, C. H. 1841: Neuere Beiträge zur Schmetterlingskunde mit Abbildungen nach der Natur 4 (49–64): 167 p, plates 289–384. Augsburg.

FRITSCH, D., STANGELMAIER, G., TOP-JENSEN, M. & BECH, K. 2014: Die nachtaktive Groß-Schmetterlingsfauna von Samos (Griechenland, Östliche Ägäis) (Lepidoptera: Cossoidea, Lasiocampoidea, Bombycoidea, Drepanoidea, Geometroidea, Noctuoidea). Esperiana 19: 7–100.

GAEDIKE, R. 2014: The authorship of the taxa in the paper of Lienig & Zeller, 1846. Nota lepidopterologica 37: 107–111.

GASTON, K. J., SCOBLE, M. J. & CROOK, A. 1995: Patterns in species description: a case study using the Geometridae (Lepidoptera). Biological Journal of the Linnean Society 55: 225–237.

GELBRECHT, J. 1998: Aktuelle Verbreitung und ökologische Ansprüche von *Narraga fasciolaria* (Hufnagel, 1767) in Deutschland (Lep. Geometridae). Entomologische Nachrichte, Berlin 42: 25–29.

GELBRECHT, J., MALKIEWICZ, A. & STELMASZCZYK, R. 2004: *Lomaspilis bithynica* Wehrli, 1954 stat. rev., a distinct species new for Europe (Geometridae). Nota lepidopterologica 27: 179–186.

GRAESER, L. 1892: Neue Lepidopteren aus Central-Asien. Berliner Entomologische Zeitschrift 37: 299–318. Available on https://archive.org/details/berlinerentomolo3718berl

GRAESER, L. 1889 [1888]: Beiträge zur Kenntnis der Lepidopteren-Fauna des Amurlandes. Berliner Entomologische Zeitschrift 32: 309–414. Available on www.biodiversitylibrary.org

HACKER, H. H. & HAUSMANN, A. 1999: Systematic list of the Lepidoptera of the Arabian Peninsula with a survey of the spread with special reference to the fauna of Yemen. Esperiana Buchreihe zur Entomologie 7: 15–237.

HADLEY, A. 2012: CombineZP, released 6 June 2010, open-source image processing software for creating extended depth of field images. http://www.hadleyweb.pwp.blueyonder.co.uk/CZP/Installation.htm

HAEGER, E. 1965: *Semiothisa artesiaria* SCHIFF. (Lep., Geom.), Erkenntnisse einer ex ovo-Zucht. Entomologische Nachrichten 9: 4–5.

HARDWICK, D.F. 1950: Preparation of slide mounts of Lepidopterous genitalia. Canadian Entomologist 82: 231–235.

HAUSMANN, A. 1990: *Enconista rubrior* sp. n. aus West-Marokko mit weiteren Anmerkungen zur Gattung *Enconista* Lederer. Nachrichtenblatt der Bayerischen Entomologen 39: 104–110.

HAUSMANN, A. 1991: Beitrag zur Geometridenfauna Palästinas: Die Spanner der Klapperich-Ausbeute aus Jordanien (Lepidoptera: Geometridae). Mitteilungen der Münchner Entomologischen Gesellschaft 81: 111–163.

HAUSMANN, A. 1993: Beitrag zur Taxonomie und Systematik des *Gnophos poggearia* Lederer, 1855 Verwandtschaftskreises (Lepidoptera: Geometridae, Ennominae). Bonner zoologische Beiträge 44: 225–240.

HAUSMANN, A. 1995a: Neue Geometriden-Funde aus Zypern und Gesamtübersich über die Fauna. Mitteilungen der Münchner Entomologischen Gesellschaft 85: 79–111.

HAUSMANN, A. 1995b: Zur Systematik einiger ostmediterraner Geometridenarten (Lepidoptera, Geometridae). Mitteilungen der Münchner Entomologischen Gesellschaft 85: 73–78.

HAUSMANN, A. 1997: The Geometrid Moths of various entomological collections in Israel (Lepidoptera, Geometridae). Entomofauna 18: 1–20.

HAUSMANN, A. 2001: Introduction. Archiearinae, Orthostixinae, Desmobathrinae, Alsophilinae, Geometrinae. In A. HAUSMANN (ed.), The Geometrid Moths of Europe 1: 1–282. Apollo Books, Stenstrup.

HAUSMANN, A. 2004: Sterrhinae. In A. HAUSMANN (ed.), The Geometrid Moths of Europe 2: 1–600. Apollo Books, Stenstrup.

HAUSMANN, A. & LENZ, J. 1993: Zum Status von *Enconista miniosaria gabriellae* da Silva Cruz 1978 (Lepidoptera: Geometridae). Entomologische Zeitschrift 103: 376–379.

HAUSMANN, A. & MILLER, M. 2000 (eds.): Atlas der Raupen europäischer und kleinasiatischer Schmetterlinge. Verlag Dr. Friedrich Pfeil, München, 95 p.

HAUSMANN, A., HASZPRUNAR, G. & HEBERT, P. D. N. 2011: DNA Barcoding the Geometrid Fauna of Bavaria (Lepidoptera): Successes, Surprises, and Questions. PLoS ONE 6(2): e17134. doi:10.1371/journal.pone.0017134

HAUSMANN, A., MIRONOV, V. & VIIDALEPP, J. 2011: Geometridae. In KARSHOLT, O. & VAN NIEUKERKEN, E. J., Fauna Europaea: Lepidoptera, Moths. Fauna Europaea version 2.4, released 27 January 2011, http://www.faunaeur.org

HAUSMANN, A. & VIIDALEPP, J. 2012: Larentiinae I. In HAUSMANN, A. (ed.), The Geometrid Moths of Europe 3: 1–743. Apollo Books, Vester Skerninge.

HAUSMANN, A., GODFRAY, H. C. J., HUEMER, P., MUTANEN, M., ROUGERIE, R., VAN NIEUKERKEN, E. J., RATNASINGHAM, S. & HEBERT, P. D. N. 2013: Genetic patterns in European Geometrid Moths revealed by the Barcode Index Number (BIN) system. PLoS ONE 8(12): e84518. doi:10.1371/journal.pone.0084518

HEITZMAN, R. L. 1985: Systematic study of the mature larvae of the Nearctic Ennomini and related tribes (Lepidoptera: Geometridae), PhD Thesis. University of Maryland, Maryland. 720 p.
HEITZMAN, R. L. & ENNS, W. R. 1977: An analysis of the variation in male genitalia between broods of *Plagodis alcolaria* Gn. (Lepidoptera: Geometridae). Journal of the Kansas Entomological Society 50: 256–258.
HEPPNER, J. B. 2002: Notes on *Euchlaena 'pectinaria'* in the United States (Lepidoptera: Geometridae: Ennominae) Lepidoptera News 3–4: 6–7.
HERBULOT, C. 1943: Note sur deux nouvelles Formes de *Narraga* (Lep. Geometridae). Bulletin de la Société entomologique de France 48: 83–84.
HERBULOT, C. 1946: Un nouveau genre de Boarmiinae paléarctique (Lep. Geometridae). Bulletin de la Société entomologique de France 51: 79.
HERBULOT, C. 1962–1963: Mise à jour de la liste des Geometridae de France. Alexanor 2: 117–124, 147–154; 3: 17–24, 85–93.
HERZ, O. 1904: Lepidoptera von Korea. Noctuidae and Geometridae. Ezhegodnik Zoologicheskago Muzeya Imperatorskoi Academii Nauk., St. Petersburg. 9: 332–375.
HEYDEMANN, F. 1936: *Lomaspilis opis* Btlr. subsp. *nigrita* f. nov. eine für Europa neue Geometride. Entomologische Zeitschrift 49: 510–512, 522–524.
HIGGOTT, J.B. & DAVEY, P. 2010: The Rannoch Looper *Itame brunneata* (Thunb.) in southern England, 2009. Atropos 40: 26–34.
HILL, L., RANDLE, Z., FOX, R., & PARSONS, M. 2010: Provisional Atlas of the UK's Larger Moths. Butterfly Conservation, Wareham, Dorset. 454 p.
HODGES, R. W., DOMINICK, T., DAVIS, D. R., FERGUSON, D. C., FRANCLEMONT, J. G., MUNROE, E. G. & POWELL, J. A. (eds.) 1983: Check List of Lepidoptera of America North of Mexico. E.W. Classey & The Wedge Entomological Research Foundation, London. 284 p.
HOLLOWAY, J. D. 1984: The larger moths of Gunung Mulu National Park; a preliminary assessment of their distribution, ecology and potential as environmental indicators. In: JERMY, A. C. & KAVANAGH, K. P. (eds.), Gunung Mulu National Park, Sarawak, Part II: 149–190. Sarawak Museum Journal 30, special issue 2.
HOLLOWAY, J. D. 1994 (dated 1993): The moths of Borneo, part 11: Family Geometridae, Subfamily Ennominae. 309 p. Malayan Nature Journal 47: 1–309.
HOLLOWAY, J. D. 1996: The moths of Borneo: Family Geometridae (incl. Orthoxtixini), subfamilies Oenochrominae, Desmobathrinae and Geometrinae. Malayan Nature Journal 49: 147–326.
HOLLOWAY, J.D. 1997: The moths of Borneo: Family Geometridae, subfamilies Sterrhinae and Larentiinae. Malayan Nature Journal 51: 1–242.
HOLLOWAY, J. D. ROBINSON, G. S. & TUCK, K. R. 1990: Zonation in the Lepidoptera of northern Sulawesi. In: KNIGHT, W. J. & HOLLOWAY, J. D. (eds.), Insects and the Rain Forests of South-East Asia (Wallacea): 153–166. Royal Entomological Society, London.
HOLLOWAY J.D., KIBBY G. & PEGGIE D. 2001: The families of Malesian moths and butterflies. Brill, Leiden. 455 p.
HOLLOWAY, J. D., BARLOW, H. S., LOONG, H. K. & KHEN, C. V. 2013: Sweet or savoury? Adult feeding preferences of Lepidoptera attracted to banana and prawn baits in the oriental tropics. The Raffles Bulletin of Zoology, supplement No. 29: 71–90.
HORSFIELD, D. & MACDONALD, A. J. 2004: Recent large outbreaks of magpie moth *Abraxas grossulariata* L. (Lep.: Geometridae) on heather *Calluna vulgaris* (L.) Hull on the mainland of north-west Scotland. Entomologist's Record and Journal of Variation 116: 81–83.
HUEMER, P. & MALICKY, M. (eds.) 2009: Verbreitungsatlas der Tierwelt Österreichs: Lepidoptera, Geometridae. Denisia 28: 3–192.
HULDÉN, L. (ed.), ALBRECHT, A., ITÄMIES, J., MALINEN, P. & WETTENHOVI, J. 2000: Atlas of Finnish Macrolepidoptera. Helsinki. 328 p.
HULST, G. D. 1896: A classification of the Geometrina of North America, with descriptions of new genera and species. Transactions of the American Entomological Society 23: 245–386.

HURST, G. W. 1964: Meteorological aspects of the migration to Britain of *Lapnygma exigua* and certain other moths on specific occasions. Agricultural Meteorology 1: 271–281.

ICZN (International Commission on Zoological Nomenclature) 1999 [with effect from 1 January 2000]: International Code of Zoological Nomenclature, 4th ed. The Code Online http://iczn.org/iczn/index.jsp

INOUE, H. 1972: The genus *Abraxas* of Japan, Korea, Saghalien, and Manchuria (Lepidoptera: Geometridae). Bulletin of the Faculty of Domestic Sciences, Otsuma Women's University 8: 141–163.

INOUE, H., 1977: Catalogue of the Geometridae of Japan (Lepidoptera). Bulletin of the Faculty Of Domestic Sciences, Otsuma Women's University 13: 227–346.

INOUE, H. 1984: The genus *Abraxas* of Taiwan. Bulletin of the Faculty of Domestic Sciences, Otsuma Women's University 20: 93–128.

INOUE, H. 1985: The genera *Ourapteryx* and *Tristophis* of Taiwan (Lepidoptera: Geometridae). Bulletin of the Faculty of Domestic Sciences, Otsuma Women's University 21: 75–124.

INOUE, H. 1986: Further new and unrecorded species of the Geometridae from Taiwan with some synonymic notes (Lepidoptera). Bulletin of the Faculty of Domestic Sciences, Otsuma Women's University 22: 211–267.

INOUE, H. 1993a: On *Ourapteryx persica*-complex, with descriptions of three new species (Geometridae, Ennominae). Tyo to Ga (currently known as Transactions of the lepidopterogical Society of Japan) 44: 81–88.

INOUE, H. 1993b: The genus *Ourapteryx* Leach (Geometridae, Ennominae) from Thailand, with descriptions of six new species. Tinea 13: 75–124.

INOUE,H. 1994: The genus *Abraxas* Leach (Geometridae, Ennominae) from Thailand. Tinea 14:20–26.

INOUE, H. 1995. The genera *Abraxas* and *Ourapteryx* from Nepal (Geometridae, Ennominae). In Haruta, T. (ed.), Moths of Nepal. Part 4. Tinea 14 Supplement 2: 119–139.

INOUE, H. 2002: Four new species and one new subspecies of the genus *Ourapteryx* Leach (Geometridae, Ennominae) from South-east Asia. Transactions of the lepidopterological Society of Japan 53: 185–192.

INOUE, H. & STÜNING, D. 1995: A new species of the genus *Ourapteryx* Leach, 1814 from central China. Transactions of the lepidopterological Society of Japan 46: 255–259.

IVANOVA, N. V., DEWAARD, J. R. & HEBERT, P. D. N. 2006: An inexpensive, automation-friendly protocol for recovering high-quality DNA. Molecular Ecology Notes 6: 998–1002.

JAMBRINA PÉREZ, J. A., 2010: Contribución al conocimiento de la biología y corología de *Dyscia distinctaria* (A. Bang-Haas, 1910) (Lepidoptera: Geometridae). Arquivos Entomolóxicos, Revista Galega de Entomoloxía 3: 107–110.

JANSE, A. J. T. 1932: The Moths of South Africa, vol. 1. Transvaal Museum, Pretoria. 376 p.

KAMEDA, M. 2004: Abnormal female genitalia of *Abraxas sylvata* (Scopoli) (Geometridae). Yugato 177: 112.

KARSHOLT, O. & RAZOWSKI, J. 1996: The Lepidoptera of Europe. A Distributional Checklist. Apollo Books, Stenstrup. 380 p.

KASY, F. 1959: Halophile Lepidopteren des Neusiedlerseegebietes. Verhandlungen der Zoologischen-Botanischen Gesellschaft in Wien 98/99: 13–25.

KASY, F. 1963: Die ‚asiatische' Geometridae *Cabera leptographa* Wehrli auch am Neusiddlersee! (Lep., Geom.). Zeitschrift der Wiener Entomologischen Gesellschaft 48: 11–13.

KASY, F. 1965: Zur Kenntnis der Schmetterlingsfauna de östlichen Neusiedlersee-Gebietes. Wissenschaftliche Arbeiten aus dem Burgendland 34: 75–211.

KENIS, M., HERZ, K., WEST, R. J, & SHOW, M. R. 2005: Parasitoid assemblages reared from geometrid defoliators (lepidoptera: geometridae) of larch and fir in the alps. Agricultural and Forest Entomology 7: 307–318.

KERSCHBAUM, W. & PÖLL, N. 2010: Die Schmetterlinge Oberösterreichs, Teil 5, Geometridae (Spanner). Beiträge Naturkunde Oberösterreichs 20: 1–469.

KIM, S.-S. & BELJAEV, E. A. 2001: Lepidoptera (Geometridae: Ennominae). Economic insects of Korea 8 (Insecta Koreana Supplement 15): 1–245.

KIM, S.- S., BELJAEV, E. A. & OH, S. H. 2001: Illustrated catalogue of Geometridae in Korea (Lepidoptera: Geometrinae, Ennominae). Economic Insects of Korea 8: 1–278.

KIMURA, M. 1980: A simple method for estimating evolutionary rate of base substitution through comparative studies of nucleotide sequences. Journal of Molecular Evolution 16: 111–120.

KING, G. E. & VIEJO MONTESINOS, J. L. 2010: Geometrid and their food plants in the south of Madrid (Central Spain) (Geometridae). Nota Lepidopterologica 33: 155–171.

KLOTS, A. B. 1970: Lepidoptera. In Tuxen, S. L. (ed.), Taxonomists' glossary of genitalia in insects, pp. 115–130. Munksgaard, Copenhagen.

KOÇAK, A. Ö. 1990: Some notes on the nomenclature of Lepidoptera. Communications de la Faculté des Sciences de l'Université d'Ankara, serie C3: Zoologie, 24: 7–25.

KOÇAK, A. Ö. & KEMAL, M. 2009: Revised Checklist of the Lepidoptera of Turkey. Centre for Entomological Studies, Priamus Supplement 17: 1-253.

KOÇAK, A. Ö. & SEVEN, S. 2001: Tentative Checklist of the Turkish Lepidoptera (except Papilionoidea and Hesperioidea), Bombycoidea, Geometroidea, Drepanoidea, Sphingoidea, Noctuoidea. http://www.members.tripod.com/entlep/Checklist3.htm. Page visited 12th November 2013.

KOCH, M. 1984: Wir bestimmen Schmetterlinge. Teil IV Spanner (Geometridae). Verlag J. Neumann-Neudamm, Leipzig. 531–760.

KRISTENSEN, N. P. 1991: Phylogeny of extant hexapods. In Naumann, I. D. (ed.), Insects of Australia. Vol. 1, pp. 125–140. Melbourne University Press, Melbourne.

KRISTENSEN, N. P. 2003: Skeleton and muscles: adults. In KRISTENSEN, N. P. (ed.), Lepidoptera, Moths and Butterflies, vol. 2: Morphology, Physiology, and Development. Handbook of Zoology, volume IV, Arthropoda: Insecta, part 36: 39–122. Walter de Gruyter, Berlin.

KRULIKOWSKY, L. 1906: Neue Varietäten und Aberrationen der palaearktischen Lepidopteren. Societas Entomologica 21: 49–51.

KRÜGER, G. C. 1939: Notizie sulla fauna della Sirtica occidentale: Lepidotteri. Annali del Museo Libico di Storia Naturale 1: 317–357.

KRÜGER, M. 1999: *Pachycnemoides* n.gen., eine neue Ennominen-Gattung aus dem südlichen Afrika. Entomologische Zeitschrift 109: 192–205.

KRÜGER, M. 2001: A revision of the tribe Macariini (Lepidoptera: Geometridae: Ennominae) of Africa, Madagascar and Arabia. Bulletin of the Natural History Museum (Entomology) 70: 1–502.

KRÜGER, M. 2009: New Afrotropical Nacophorini in the genera *Pachycnemoides* Krüger, *Argyrophora* Guenée, *Microligia* Warren, *Pseudomaenas* Prout and *Drepanogynis* Guenée (Lepidoptera: Geometridae: Ennominae). Annals of the Transvaal Museum 46: 1–35.

KUCHLEIN, J. H. & VOS, R. de 1999: Annotated Checklist of the Dutch Lepidoptera. Backhuys. 304 pp.

KUDLER, J. 1978: Geometroidea. In SCHWENKE, W. (ed.), Die Forstschädlinge Europas, Vol. 3, Schmetterlinge, pp. 218–264. Verlag Paul Parey, Hamburg und Berlin.

KUHNA, P. 1977: Über den systematischen Status von *Theria ibicaria*. Jahresberischte des Naturwissenschaflischen Vereins in Wuppertal 30: 129–132.

KULLBERG, J., ALBRECHT, A., KAILA, L. & VARIS, V. 2001: Checklist of Finnish Lepidoptera. Sahlbergia 6: 45–190. An updated version (without taxonomic comments) is available at http://www.luomus.fi/elaintiede/hyonteiset/perhoset/ Page visited 30 May 2012.

LAFONTAINE, J.D. & MIKKOLA, K. 1998: Lås-och-nyckel systemen i de inre genitalierna av Noctuidae (Lepidoptera) som taxonomiska kännetecken. Entomologiske Meddelelser 55: 161–167.

LAMPA, S. 1885: Förteckning öfver Skandinaviens och Finlands Macrolepidoptera. Entomologisk Tidskrift 6: 1–137.

LEDERER, J. 1855: Beitrag zur Schmetterlings-Fauna von Cypern, Beirut und einem Theile Klein-Asiens. Verhandlungen des Zoologisch-Botanischen Vereins in Wien 5: 177–254.

LEMPKE, B. J. 1952: Catalogus der Nederlandse Macrolepidoptera, XI. Tijdschrift voor Entomologie 95: 197–319.

LEMPKE, B. J. 1970: Catalogus der Nederlandse Macrolepidoptera (Zestiende Supplement). Tijdschrift voor Entomologie 113: 125–260. Available on http://www.biodiversitylibrary.org/itemdetails/89561

LEPERTEL, N., QUINETTE, J.-P. & GUÉRARD, P. 2007: Connaissance et reconnaissance de deux espèces proches: *Theria primaria* (Haworth, 1809) et *Theria rupicapraria* (Denis & Schiffermüller, 1775) (Lepidoptera, Geometridae Ennominae). Invertébrés Armoricains 1: 25–26.

LEPIFORUM e.V. 2013: Bestimmung von Schmetterlingen (Lepidoptera) und ihren Präimaginalstadien. http://www.lepiforum.de/ Page visited in 2013.

LERAUT, P. 1997: Systematic and synonymic list of the Lepidoptera of France, Belgium and Corsica. Supplement a Alexanor, Universa, Wetteren, Belgique. 526 p. 2nd ed.

LERAUT, P. 2009: Moths of Europe, vol. 2, Geometrid moths. N.A.P. Editions, Verrières le Buisson, France, 808 p.

LHOMME, L. [1935]: XVIII. Geometridae. In LHOMME, L. (ed.) 1923–1935, Catalogue des Lépidoptères de France et de Belgique. Vol. 1: Macrolépidoptères, pp. 378–622. Le Carriol, Douelle (Lot).

LUNAK, R. 1938: Im Hochgebirge Bosniens wegen *Lignyoptera thaumastaria* Rbl. Zeitschrift des Österreichischen Entomologen-Vereines 23: 1–7.

LUNAK, R. 1940: Die Biotope und Biologie von *Lignyoptera thaumastaria* Rbl. im Vergleich mit *L. fumidaria* Hb. Zeitschrift des Österreichischen Entomologischen Vereines, Wien 25: 94–98.

MABILLE, P. 1867: Lépidoptères de la Corse. I. re notice. Annales de la Société entomologique de France, 4° ser., VI (1866) 4: 545–564.

MAJERUS, M. E. N. 1981: The inheritance and maintenance of the melanic form nigrescens of *Pachycnemia hippocastanaria* (Lepidoptera: Ennominae). Ecological Entomology 6: 417–422.

MALKIEWICZ, A. 2012: The Geometrid Moths of Poland, Vol. 1. Ennominae (Lepidoptera: Geometridae). Polish Taxonomical Monographs Vol. 19. Polish Taxonomical Society, Wroclaw. 269 p.

MATTHEWS, D. L., COVELL, C. V., LANE, K. M. & MILLER, J. Y. 2014: Larval hostplants of Geometridae (Lepidoptera) collected by Dale H. Habeck in Florida. Proceedings of the Entomological Society of Washington 116: 36–68. DOI: 10.4289/0013-8797.116.1.36

MAYR, E. 1942: Systematics and the origin of species. Columbia University Press, New York. 334 p.

MAYR, E. 1963: Animal species and evolution. Harvard University Press, Cambridge. 797 p.

MAYR, E. & Ashlock, P. D. 1991: Principles of Systematic Zoology, 2nd ed. McGraw-Hill, New York. 475 p.

MCGUFFIN, W. C. 1972: Guide to the Geometridae of Canada (Lepidoptera) II. Subfamily Ennominae 1. Memoirs of the Entomologcal Society of Canada 86: 1–159. Le Carriol.

MCGUFFIN, W. C. 1977: Guide to the Geometridae of Canada (Lepidoptera), II, Subfamily Ennominae 2. Memoirs of the entomological Society of Canada 101: 1–191. Ottawa.

MCGUFFIN, W.C. 1981: Guide to the Geometridae of Canada (Lepidoptera) II. Subfamily Ennominae. 3. Memoirs of the Entomological Society of Canada, 117: 1–153.

MCGUFFIN, W.C. 1987: Guide to the Geometridae of Canada (Lepidoptera) II. Subfamily Ennominae. 4. Memoirs of the Entomological Society of Canada 138: 1–182.

MCQUILLAN, P. B. 1985: A taxonomic revision of the Australian autumn gum moth genus *Mnesampela* Guest (Lepidoptera: Geometridae, Ennominae). Entomologica Scandanavica 16: 175–202.

MCQUILLAN, P. B. & EDWARDS, E. D. 1996: Geometridae. In Nielsen, E. S., Edwards, E. D. & Rangsi, T. V. (eds.), Checklist of the Lepidoptera of Australia. Monographs on Australian Lepidoptera 4, pp. 200–228.

MENDES, C. DE AZEVEDO 1909: Lepidopteros mais variaveis de Portugal da fam. Geometridae. *Brotéria, Série zoological, Lisboa (: 67-73.*

MIKKOLA, K. JALAS, I. & PELTONEN, O. 1985: Suomen perhoset, mittarit 1. Tampereen Kirjapaino Oy Tamprint, Tampere. 260 p. [Lepidoptera of Finland, Geometridae 1]. In Finnish.

MIKKOLA, K., JALAS, I. & PELTONEN, O. 1989: Suomen perhoset, mittarit 2. Hangon Kirjapaino, Hanko. 280 p. [Lepidoptera of Finland, Geometridae 2]. In Finnish.

MILLER, J. S. 1991: Cladistics and classification of the Notodontidae (Lepidoptera: Noctuioidea) based on larva and adult morphology. Bulletin of the American Museum of Natural History 204: 1–230.

MILLER, M. A., MÜLLER, G. C., KRAVCHENKO, V. D., JUNNILA, A., VERNON, K. K., MATHESON, C. D. & HAUSMANN, A. 2006: DNA-based identification of Lepidoptera larvae and plant meals from their gut contents. Russian Entomological Journal 15: 427–432.

MINET, J. & SCOBLE, M.J. 1999: The Drepanoid/Geometroid Assemblage. In Kristensen, N. P. (ed.), Lepidoptera: Evolution, Systematics and Biogeography, pp. 301–320. Walter de Gruyter, Berlin.

MIRONOV, V. 2003: Larentinae II. In Hausmann, A. (ed.), The Geometrid Moths of Europe 4: 1–464. Apollo Books, Stenstrup.

MIRONOV, V. G., BELJAEV, E. A. & VASILENKO, S. V. 2008: Geometridae. In SINEV, S. YU. (ed.), Catalogue of the Lepidoptera of Russia, pp. 190–226. KMK Scientific Press, Saint Petersburg.

MÜLLER, B. 1996: Geometridae. In KARSHOLT, O. & RAZOWSKI, J. (eds.), The Lepidoptera of Europe, a distributional checklist, pp. 218–249. Apollo Books, Stenstrup.

MÜLLER, B. 2004: Zwei Geometridenausbeuten aus Zypern nebst weiteren Erkenntnissen zu *Crocallis cypriaca* Fischer, 2003 (Lepidoptera: Geometridae). Entomologische Zeitschrift 114: 53–59.

MUNROE, E. G. 1963: The *gilvarius* group of *Aspilates* Treitschke (Lepidoptera: Geometridae). Canadian Entomologist 95: 260–287.

MURRIA BELTRAN, E. 2008: *Compsoptera aemiliorum* Raineri, 1994, a synonym of *C. opacaria* (Hübner, (1819)) (Lepidoptera: Geometridae, Ennominae). Boletin de la SEA 42: 297–303.

MUTANEN, M. & KAITALA, A. 2006: Genital variation in a dimorphic moth *Selenia tetralunaria* (Lepidoptera, Geometridae). Biological Journal of the Linnean Society 87: 297–307.

MUTANEN, M., HAUSMANN, A., HEBERT, P. D. N., LANDRY, J-F., DEWAARD, J. & HUEMER, P. 2012: Allopatry as a Gordian Knot for Taxonomists: Patterns of DNA Barcode Divergence in Arctic-Alpine Lepidoptera. PLoS ONE 7(10): e47214. doi:10.1371/journal.pone.0047214.

MUTANEN, M., WAHLBERG, N. & KAILA, L. 2010: Comprehensive gene and taxon coverage elucidates radiation patterns in moths and butterflies. Proceedings of the Royal Society B 277: 2839–2848.

NATIONAL BIODIVERSITY NETWORK 2012: NBN Gateway version 4.2. http://nbn.org.uk/Home.aspx. Page visited 2 May 2012.

NIXON, K. & WHEELER, Q. 1990: An amplification of the phylogenetic species concept. Cladistics 6: 211–223.

OKYAR, Z. & MIRONOV, V. 2008: Checklist of the Geometridae of European Turkey, with new records (Lepidoptera). Zootaxa 1789: 1–56.

OLSHVANG, V. N., NUPPONEN, K. T., LAGUNOV, A. V. & GORBUNOV, P. Y. 2004: The Lepidoptera of Ilmen Reserve. Yekaterinburg IGZ UB RAS. 288 pp. (in Russian).

ÕUNAP, E., JAVOIS, J., VIIDALEPP, J. & TAMMARU, T. 2011: Phylogenetic relationships of selected European Ennominae (Lepidoptera: Geometridae). European Journal of Entomology 108: 267–273.

PARRA, L. E. & HERNANDEZ, C. E. 2010: Estudio filogenético de los géneros de Lithinini de Sudamérica Austral (Lepidoptera, Geometridae): una nueva clasificación. Revista Brasileira de Entomologia 54: 1–27.

PATOČKA, J. 1978: Zur Puppenmorphologie und -taxonomie der Unterfamilie Ennominae, insbesondere der Tribus Bistonini (Lepidoptera, Geometridae). Acta Societatis Zoologicae Bohemoslovenicae 42: 143–151.

PATOČKA, J. 1992: Über einige Puppen der Spanner aus der Tribus Ennomini (Lepidoptera: Geometridae, Ennominae). Entomologische Berichten Amsterdam 52: 171–176.

PATOČKA, J. 2003: The pupae of central European geometrids (Lepidoptera, Geometridae). Subfamily Ennominae, part 1. Linzer Biologische Beitraege 35: 1001–1075.

PATOČKA, J. 2004: The pupae of the Central European geometrids, subfamily Ennominae (Lepidoptera: Geometridae), part 2. Linzer Biologische Beitraege 36: 315–388.

PATOČKA, J. 2006: Contribution to the knowledge of the pupae of Westpalaearctic geometrids (Lepidoptera, Geometridae, subfamily Ennominae). Linzer Biologische Beitraege 38: 779–801.

PATOČKA, J. & ZACH, P. 1994: On the pupae of central European Geometridae (Lepidoptera). Subfamilies Archiearinae, Oenochrominae and Ennominae, tribe Theriini. Biologia 49: 739–745.

PATOČKA, J. & TURČANI, M. 2005: Lepidoptera pupae, Central European species. Text and plate volumes. Apollo Books, Stenstrup. 321 p. (text) + 542 p (plates).

PETRÁNYI, G. 2014: Füstös ősziaraszoló *Lignyoptera fumidaria* Hübner, 1825. In: HARASZTHY L. (ed.): Natura 2000 fajok és élőhelyek Magyarországon, pp. 290–293. Pro Vértes Természetvédelmi Közalapítvány, Csákvár. 955 p.

PIERCE, F. N. 1914: The genitalia of the Group Geometridae of the Lepidoptera of the British Islands. Liverpool: Northern Publishing Co.

PINKER, R. 1968: Die Lepidopterenfauna Mazedoniens. III. Geometridae. Prirod. Muz. Skopje, Pos. Izdanie 4. Skopje. 72 p.

PINKER, R. 1974: Interessante und neue funde und erkenntnisse für die Lepidopterenfauna der Kanaren. V. Zeitschrift der Arbeitsgemeinschaft Österreichischer Entomologen 25 (1/2) (1973): 2–11.

PINKER, R. 1978: Zwei neue Spanner von den Kanaren (Lep., Geometridae). VI. Kanarenbeitrag. Nachrichtenblatt der Bayerischen Entomologen 27: 17–20.

PITKIN, L. M. 2002: Neotropical ennomine moths: a review of the genera (Lepidoptera: Geometridae). Zoological Journal of the Linnean Society 135: 121–401.

POLTAVSKY A. N., ARTOKHIN K.S. & SILKIN Y. A. 2007: To the fauna of Geometrid moths (Lepidoptera: Geometridae) of Rostov-on-Don Area. Caucasian Entomological Bulletin 3: 63–69 (in Russian).

POLTAVSKY, A. N., MATOV, A. YU & SILKIN, A. Yu 2007: Records of some rare moths (Lepidoptera) in the Rostov-on-Don area (Russia) in 2006. Phegea 35: 135–140.

POOLE, R. W. 1987: A taxonomic revision of the New World moth genus *Pero* (Lepidoptera: Geometridae). U.S. Department of Agriculture Technical Bulletin.

PORTER, J. 2010: Colour identification guide to caterpillars of the British Isles. Apollo Books, Stenstrup. 275 p.

PROUT L. B. 1912–1916: Die Spanner des Palaearktischen faunengebietes. In SEITZ, A. (ed.), Die Gross-Schmetterlinge der Erde, Vol. 4., pp. 1–479. A. Kernen, Stuttgart.

PROUT, L. B. 1929: New Palaearctic Geometridae. Novitates Zoologicae 35: 142–149.

PROUT, L. B. 1931: New Geometridae from the Indo-Australian region. Novitates Zoologicae 37: 18–34.

PROUT L. B. 1934–1939: Die Spanner des Palaearktischen faunengebietes. In SEITZ, A. (ed.), Die Gross-Schmetterlinge der Erde, Supplement zu Band 4, pp. 1–253. A. Kernen, Stuttgart.

PUGAEV, S. N. & ZOLOTUHIN, V. V. 2006: Geometrid moths of the genus *Dyscia* Hübner, [1825] (Lepidoptera: Geometridae) from the Lower and Middle Volga. Eversmannia 6: 3–12.

RAJAEI SH., H. 2010: Life-history of *Gnopharmia kasrunensis* Wehrli, 1939 and *G. colchidaria* Lederer, 1870 (Geometridae, Ennominae) and their distribution in Iran, with first host-plant records for the genus. Bonn zoological Bulletin 57: 65–73.

RAJAEI, H., GREVE, C., LETSCH, H., STÜNING, D., WAHLBERG, N., MINET, J. & MISOF, B. 2015: Advances in Geometroidea phylogeny, with characterization of a new family based on *Pseudobiston pinratanai* (Lepidoptera, Glossata). Zoologica Scripta 2015. Early view available on doi:10.1111/zsc.12108.

RAJAEI SH., H., STÜNING, D. & TRUSCH, R. 2012: Taxonomic revision and zoogeographical patterns of the species of *Gnopharmia* Staudinger, 1892 (Geometridae, Ennominae). Zootaxa 3360: 1–52.

RAJAEI SH., H., STRUWE, J.-F., RAUPACH, M. J., AHRENS, D. & WÄGELE, J. W. 2013: Integration of cytochrome c oxidase I barcodes and geometric morphometrics to delimit species in the genus *Gnopharmia* (Lepidoptera: Geometridae, Ennominae). Zoological Journal of the Linnean Society 169: 70–83.

RATNASINGHAM S. & HEBERT, P. D. N. 2007: BOLD: The Barcode of Life Data System (http://www.barcodinglife.org). Molecular Ecology Notes 7: 355–364.

RATNASINGHAM S. & HEBERT, P. D. N. 2013: A DNA-Based Registry for All Animal Species: The Barcode Index Number (BIN) System. PLoS ONE 8(7): e66213. doi:10.1371/journal.pone.0066213

REDONDO, V. M. 1986: Introducción al conocimiento de los macrolepidópteros del Valle medio del Ebro en Aragón (2a parte). SHILAP Revista de Lepidopterologia 14: 17–31.

REDONDO, V. M. & GASTÓN, F. J. 1999: Los Geometridae (Lepidoptera) de Aragón (Espana). Monografias Sociedad Entomológica Aragonesa (S.E.A.) 3. Gorfisa, Zaragoza. 131 p.

REDONDO, V. M., GASTÓN, F. J. & GIMENO, R. 2009: Geometridae Ibericae. Apollo Books, Stenstrup. 361 p.

REGIER, J. C., MITTER, C., ZWICK, A., BAZINET, A. L., CUMMINGS, M. P., *et al.* 2013: A large-scale, higher-level, molecular phylogenetic study of the insect order Lepidoptera (moths and butterflies). PLoS ONE 8(3): e58568. doi:10.1371/journal.pone.0058568

REISSER H. 1929: Beschreibung der unbekannten ersten Stände einiger corsischer Geometriden. Zeitschrift des Österreichischen Entomologischen Vereines, Wien 14: 3–6, 31–33, 42–44, 79–81.

REISSER, H. 1958: Neue Heteroceren aus Kreta. Zeitschrift der Wiener Entomologischen Gesellschaft 43: 105–128.

REISSER, H. 1962: Weitere neue Heteroceren aus Kreta. Zeitschrift der Wiener Entomologischen Gesellschaft 47: 193–216.

REUTTI, C. H. 1898: Übersicht der Lepidopteren-Fauna des Grossherzogtums Baden (und der anstossenden Länder). 2. Ausgabe. Überarbeitet und herausgegeben von Meess, A. und Spuler, A., Berlin. 361 pp.

RIBBE, C., 1912: Beiträge zu einer Lepidopteren-Fauna von Andalusien (Sud-Spanien). Macrolepidopteren (Geometridae). Deutsche Entomologische Zeitschrift Iris 23: 299–343.

RIEMIS, A. 1994: Geometridae of Turkey 3. A provisional list of the Geometridae of Turkey (Lepidoptera). Phegea 22: 15–22.

RINDGE, F. H. 1956: A revision of the American species of *Deilinia* (Lepidoptera: Geometridae). American Museum Novitates 1810: 1–32.

RINDGE, F. H. 1986: Generic descriptions of New World Lithinini (Lepidoptera, Geometridae). American Museum Novitates 2838: 1–68.

ROBERTI, D. 1964: Contributo alla conoscenza dell'entomofauna dell'erba dedica nel territoria di Piacenza. Annali della Facoltà di Agraria IV (III): 489–527.

ROBINEAU, R. 2007: Guide des papillons nocturnes de France. Delachaux et Niestlé, Paris. 288 p.

ROGENHOFER, A. & MANN, J. J. 1873: Neue Lepidopteren gesammelt von Herrn J. Haberhauer. Verhandlungen der Zoologisch-Botanischen Gesellschaft in Wien 23: 569-574. Available online at http://www.landesmuseum.at/biophp/de/band_det.php?litnr=28018 Accessed 2 February 2014.

RUCKDESCHEL, W. 2006: Eine neue Art von *Crocallis* TREITSCHKE, 1825 von Kreta: *Crocallis helenaria* sp. nov. (Lepidoptera: Geometridae). Entomologische Zeitschrift 116: 155–160.

RUCKDESCHEL, W. 2007: Die Geometridae Kretas (Lepidoptera, Geometriae). Nachrichtenblatt der Bayrischen Entomologen 56: 2–13.
RUNGS, C. 1941 [dated 1940, published in 1941]: Notes de Lépidoptérologie marocaine (9 note). Description de formes nouvelles, discussions. Bulletin de la Societe Naturelles (et Physiques) du Maroc 20: 96–105.
RUNGS, C. 1981: Catalogue raisonné des Lépidoptères du Maroc. Inventaire Faunistique et observations écologiques. Tome II. Travaux de l'Institut Scientifique, Série Zoologie, 40: 1–365.
RUNGS, C. 1982: Notes de lépidoptérologie corse (II). Alexanor XII (5): 221–229; (6): 265–271.
RUPERT, L. R. 1949: A revision of the North American species of the genus *Plagodis* (Lepidoptera: Geometridae, Ennominae). Journal of the New York Entomological Society 57: 19–50.
RYRHOLM, N. 1994: Remarkable records of Macrolepidoptera in Sweden 1993. Entomologisk Tidskrift 115: 37–44.
RÖDEL, I. & TRUSCH, R. 1997: Zur Biologie, Ökologie und Verbreitung von *Bichroma famula* (Esper, 1787) in Deutschlands. Entomologische Nachrichten und Berichte 41: 19–26.
SAITOU, N. & NEI, M. 1987: The neighbour-joining method: a new method for reconstructing evolutionary trees. Molecular Biology and Evolution 4: 406–425.
SALKELD, E. H. 1983: A catalogue of the eggs of some Canadian Geometridae (Lepidoptera), with comments. Memoirs of the Entomological Society of Canada 126: 1–271.
SATO, R. 2005: *Ennomos nephotropa* Prout, a distinct species, separated from *E. autumnaria* (Werneburg) (Geometridae, Ennominae). Japan Heterocerists' Journal 234: 161–164.
SATO, R. & BELJAEV, E. A. 2009: *Cabera insulata* Inoue, a distinct species, separated from *C. exanthemata* (Scopoli) (Geometridae, Ennominae). Tinea 20: 299–306.
SATTLER, K. 1963: Zur Kenntnis der Gattung Plagodis HÜBNER, 1823 (Lepidoptera: Geometridae). Zeitschrift der Wiener Entomologischen Gesellschaft 48: 25–32.
SCHÖNBORN, C. 2011: Schmetterlingsfauna Sachen-Anhalts, Vol. 1, Spanner. Weissdorn-Verlag, Jena. 352 p.
SCOBLE, M. J. 1995: A review of the moth tribe Palyadini with the description of a new genus (Geometridae: Ennominae). Systematic Entomology 20: 35–58.
SCOBLE, M. J. 1995: The Lepidoptera: Form, Function and Diversity. Oxford University Press, Oxford. 404 p.
SCOBLE, M. J. 1999: Geometrid Moths of the World: a catalogue (Lepidoptera, Geometridae). Vol 1 and 2. CSIRO Publishing and Apollo Books, Stenstrup. 1016 pp. (+ 129 pp. of Index).
SCOBLE, M. J. & EDWARDS, E. D. 1989: *Parepisparis* Bethune-Baker and the composition of the Oenochrominae (Lepidoptera: Geometridae). Entomologica Scandinavica 20: 371–399.
SCOBLE, M. J., GASTON, K. J. & CROOK A. 1995: Using taxonomic data to estimate species richness in Geometridae. Journal of the Lepidopterists' Society 49: 136–147.
SCOBLE, M. J. & KRÜGER, M. 2002: A review of the genera of Macariini with a revised classification of the tribe (Geometridae: Ennominae). Zoological Journal of the Linnean Society 134: 257–315.
SCOBLE, M. J. & HAUSMANN, A. [updated 2007]: Online list of valid and available names of the Geometridae of the World, http://www.herbulot.de/globalspecieslist.htm Page visited 2 February 2013.
SEPPÄNEN, E. J. 1970: Suurperhostoukkien ravintokasvit (The food plants of the larvae of the Macrolepidoptera of Finland). Animalia Fennica 14.
SHIN, Y-H. 1996: Synonymic list and distribution of the Geometridae of Korea (Lepidoptera). Center for insect systematics, Korea. 153 p.
SIHVONEN, P. 2001: Everted vesicae of the *Timandra griseata* group: methodology and differential features (Geometridae, Sterrhinae). Nota Lepidopterologica 24: 57–63.
SIHVONEN, P. 2009: Book review of LERAUT, P. 2009: Moths of Europe, vol. 2, Geometrid moths. N.A.P. Editions, Verrières le Buisson, France, 808 pp. Nota Lepidopterologica 32: 158–162.

SIHVONEN, P., MUTANEN, M., KAILA, L., BREHM, G., HAUSMANN, A. & STAUDE, H. S. 2011: Comprehensive molecular sampling yields a robust phylogeny for geometrid moths (Lepidoptera: Geometridae). PLoS ONE 6(6): e20356. doi:10.1371/journal.pone.0020356.

SIHVONEN, P., SKOU, P., FLAMIGNI, C., FIUMI, G. & HAUSMANN, A. 2014: Revision of the *Hylaea fasciaria* (Linnaeus, 1758) species group in the West Palaearctic (Lepidoptera: Geometridae, Ennominae). Zootaxa 3768: 469–486. http://dx.doi.org/10.11646/zootaxa.3768.4.5

SIHVONEN, P., STAUDE, H. S. & MUTANEN, M. 2015: Systematic position of the enigmatic African cycad moths: an integrative approach to a nearly century old problem (Lepidoptera: Geometridae, Diptychini). Systematic Entomology. Early view available on DOI 10.1111/syen.12125.

SKINNER, B. & WILSON, D. 2009: Colour identification guide to moths of the British Isles. 3rd revised and updated edition. Apollo Books, Stenstrup. 325 p.

SKOU, P. 1984: Nordens målere. Håndbog over de danske og fennoskandiske arter af Drepanidae og Geometridae (Lepidoptera). Danmarks Dyreliv 2. Fauna Bøger & Apollo Bøger, København & Svendborg. 332 p.

SKOU, P. 1986: The geometroid moths of North Europe (Lepidoptera: Drepanidae and Geometridae). Entomonograph 6: 1–348. E. J. Brill & Scandinavian Science Press, Leiden & Copenhagen.

SLOT, J. 1999: An albinistic form of *Opisthograptis luteolata* (Linnaeus, 1758) (Lepidoptera: Geometridae). Phegea 27: 33.

SOHN-RETHEL, O. 1929: Beiträge zur Heteroceren-Fauna Italiens. Deutsche Entomologische Zeitschrift ‚Iris' 43: 1–23.

SOUTH, R. 1961: The Moths oft he British Isles. II. London & New York. 379 pp.,14 pls.

SPULER, A. 1904: Die Raupen der Schmetterlinge Europas. Schweizerbart'sche Verlagsbuchhandlung, Stuttgart. 38 p. + 60 colour plates.

SPULER, A. 1903–1910: Die Schmetterlinge Europas, vol. 2. Schweizerbart'sche Verlagsbuchhandlung, Stuttgart. 523 p.

STADIE, D. & LEHMANN, L. 2012: On the winter and spring aspect of the Macrolepidoptera fauna of Jordan with remarks on the biology of some species (Lepidoptera: Tineoidea, Cossoidea, Bombycoidea, Papilionoidea, Geometroidea, Noctuoidea). Mitteilungen der Münchner Entomologischen Gesellschaft 102: 65–93.

STADIE, D. & FIEBIG, R. 2014: Taxonomic notes on the morphology, ecology and distribution of the *Crocallis elinguaria* (Linnaeus, 1758) species lineage of Asia Minor and the eastern adjacent territories with description of five new species and one new subspecies (Lep. Geometridae). Esperiana 19: 225–254.

STANESCU, M. 2002: The systematic position of *Crocota ostrogovichi* (Caradja, 1930) (Lepidoptera: Geometridae: Ennominae). Travaux du Muséum National d'Histoire Naturelle "Grigore Antipa" 44: 253–260.

STAUDE, H.S. 2001: A revision of the genus *Callioratis* Felder (Lepidoptera: Geometridae: Diptychinae). Metamorphosis 12: 125–156.

STAUDE, H. S. & SIHVONEN, P. 2014: The African geometrid genus *Zerenopsis* C. &. R. Felder revised – moths with peculiar life histories and mating behaviours (Geometridae: Ennominae: Diptychini). Metamorphosis 25: 11–55.

STAUDER, H. 1913: Weitere Beiträge zur Kenntnis der Makrolepidopterenfauna der adriatischen Küstengebiete. Bollettino della Società Adriatica di Scienze Naturali in Trieste 27: 105–166 + 3 plates.

STAUDER, H. 1915: Eine Sammelreise nach Unteritalien. Beitrag zur Kenntnis der Lepidopterenfauna der sorrentinischen Halbinsel und des Cocuzzo-Massivs in Calabrien. Zeitschrift für wissenschaftliche Insektenbiologie Berlin 11 (1/2): 1–7, (3/4): 71–75, (5/6): 132–137, (7/8): 175–180.

STAUDER, H. 1920: Neue mediterrane Geometridae. Internationale Entomologische Zeitschrift Guben 14: 34–39.

STAUDINGER, O. 1859: Diagnosen nebst kurzen Beschreibungen neuer andalusischer Lepidopteren. Entomologische Zeitung (Stettin entomologische Zeitung) 20: 211–259.
STAUDINGER, O. 1897: Die Geometriden des Amurgebiets. Deutsche Entomologische Zeitschrift Iris 10: 1–122, plates 1–4.
STAUDINGER, O. & REBEL, H. 1901: Catalog der Lepidopteren des Palaearktischen Faunengebietes. Teil 1. Verlag Friedländer & Sohn, Berlin, 411 pp.
STEKOLNIKOV, A. A. & KUZNETZOV, V. I. 1982: Functional morphology of the male genitalia and arrangement of the new tribes of the geometrid moths subfamily Ennominae (Lepidoptera, Geometridae). Entomologicheskoe Obozrenie 61: 344–374 (in Russian). [Translated to English in Stekolnikov, A. A. & Kuznetsov, V. I. 1982: Entomological Review 61: 92–122].
STOLL, C., 1782: Beschryving van Uitlandsche Kapellen, *in* CRAMER, P., De Uitlandsche Kapellen, Papillons exotiques des trois parties du Monde L'Asie, L'Afrique et L'Amerique, vol. 4, description des planches CCLXXXIX - CCCC. S. J. Baalde, Amsterdam. 252 + 29 p. Available at Open Library https://openlibrary.org/books/OL24196307M Page visited 25 January 2014.
STÜNING, D. & HAUSMANN, A. 2002: *Abraxas breueri*, spec. nov., a new Geometrid moth species from the Philippines (Lepidoptera, Geometridae, Ennominae). Spixiana 25: 157–161.
SUGI, S. (ed.) 1987: Larvae of larger moths in Japan. Kodansha, Tokyo. 453 pp.
SUMPICH, J. & CERNY, K. 2010: Schmetterlinge des Truppenübungsplatzes Boletice: bewaldeter Berg Špičák (Tschechische Republik, Šumava) [Butterflies and moths of the Military Training Area Boletice: the forested Špičák Mt. (Czech Republic, Bohemian Forest)]. Silva Gabreta 16: 93–126.
SZEKELY, L. 2011: Moths of Romania 2. Geometridae 1. Archiearinae, Ennominae, Alsophilinae, Orthostixinae, Geometrinae. Disz-Tipo, Săcele/Brasov. 196 p. + 7 plates.
SØNDERUP, H. S. P. 1901: Nye Sommerfugle. Flora og Fauna 3: 56.
TAUTEL, C. & BILLI, F. 2006: Découverte en France d'une nouvelle espéce du genre *Dyscia* (Hübner, 1825) (Lepidoptera, Geometridae, Ennominae). Bulletin de la Société entomologique de France 111: 525–527.
TAUTEL, C., BILLI, F., COLOMB, C., DUTHEIL, M., RYMARCZYK, F. & VARENNE, T. 2008: Nouvelles observations de *Dyscia royaria* Tautel & Billi, 2007, et description de ses premiers états (Lep. Geometridae, Ennominae). Oreina 2008: 17–23.
TREITSCHKE, F. 1825: Die Schmetterlinge von Europa (Fortsetzung des Ochsenheimer'schen Werks). Vol. 5. Leiptzig. 414 p.
TREITSCHKE, F. 1827: Die Schmetterlinge von Europa (Fortsetzung des Ochsenheimer'schen Werks). Vol. 6. Leiptzig. 416 p.
TROFIMOVA, T. A. 2009: *Atomorpha punctistrigaria* (Christoph, 1893) (Geometridae: Ennominae) – a new geometrid species for the fauna of Europe. Nota lepidopterologica 32: 55–62.
TRUSCH, R. & MÜLLER, B. 2001: Zur Geometridenfauna des Dahar-Berglandes in Süd-Tunesien (Lepidoptera: Geometridae). Entomologische Zeitschrift 111: 133–140.
TRUSCH, R. & ERLACHER, S. 2001: Zur Morphologie, Verbreitung, Bionomie und Identifikation der *Dyscia*-Arten (Lepidoptera, Geometridae: Ennominae). Bonner Zoologische Monographien 49: 1–116.
TRUSCH, R. & ERLACHER, S. 2002: Taxonomic changes in the genus *Dyscia* (Lepidoptera: Geometridae: Ennominae). European Journal of Entomology 99: 529–541.
TRUSCH, R. & OCHSE, M. 2004: Neue Vorkommen von *Dyscia fagaria* (THUNBERG, 1784) und Anmerkungen zur Taxonomie der Art (Lep., Geometridae). Entomologische Nachrichten und Berichte 48: 11–14.
TURATI, E. 1911: Lepidotteri del Museo Zoologico Università della R. di Napoli. Descrizioni di forme nuove e note critiche. Annuario del Museo Zoologico R. della Università di Napoli (N. S.) 3 (18): 1–31 .
URBAHN, E. & URBAHN, H. 1939: Die Schmetterlinge Pommerns mit einem vergleichenden Überblick über den Ostseeraum. Stettiner Entomologische Zeitung 100: 187–826.

URBAHN, E. 1952a: Die asiatische Geometride *Cabera leptographa* Wehrli in der Mark Brandenburg gefunden. Zeitschrift für Lepidopterologie 2: 25–35.
URBAHN, E. 1952b: Die Jugendstände und das Weibchen von *Cabera leptographa* Wehrli. Zeitschrift für Lepidopterologie 2: 97–104.
URBAHN, E. 1955: Weiteres über *Cabera leptographa* Wehrli. Mitteilungen der Deutschen Entomologische Gesellschaft 14: 5–6.
URBAHN, E. 1960: *Cabera leptographa* Wehrli aus Österreich nachgeweisen (Lep., Geom.). Zeitschrift der Wiener Entomologischen Gesellschaft 45: 167–168.
URBAHN, E. 1967: Das neue Goemetridensystem von C. Herbulot und Anmerkungen dazu (Lep. Geometridae). Entomologische Berichte 11: 33–52.
URBAHN, E. & URBAHN, H. 1978: *Cabera leptographa* Wehrli 1939. Entomologische Nachricten 22: 33–43.
VIIDALEPP, J. & REMM, H. 1982: New materials concerning higher lepidopterans (Macrolepidoptera) from Sakhalin. In Zolotarenko, G. S. (ed.), Useful and harmful insects of Siberia, pp. 137–151. Izdatelstvo "Nauka", Sibirskoje otdelenie, Novosibirsk.
VIIDALEPP, J. 1988: Fauna pyadenits gor Srednei Azii [Geometridae fauna of the Central Asian mountains]. Moscov. 240 p. (in Russian).
VIIDALEPP, J. 1996: Checklist of the Geometridae (Lepidoptera) of the former U.S.S.R. Apollo Books, Stenstrup. 111 p.
VIIDALEPP, J., TAMMARU, T., SNÄLL, N., RUOHOMÄKI, K. & WAHLBERG, N. 2007: *Cleorodes* Warren, 1894 does not belong in the tribe Boarmiini (Lepidoptera: Geometridae). European Journal of Entomology 104: 303–309.
VILLERS DE, C. 1789: Caroli Linnaei Entomologica, Faunae Sueciae descriptionibus aucta ... generum specierumque rariorum iconibus ornata. Lyon, 4 vols.
VITALIEVNA, N. M. 1996: Seasonal variation in the male genitalia of *Plagodis dolabraria* (Linnaeus, 1758). Atalanta 26: 311–313.
VIVES MORENO, A. 1994: Catálogo sistemático y sinonímico de los Lepidopteros de la Península Ibérica y Troodos (Insecta: Lepidoptera). Minsterio de Agricultura Pesca y Alimentación, Madrid. Part 2: x + 775 p.
VOJNITS, A. 1975: Ennominae species from Mongolia, II. (Lepidoptera, Geometridae). Annales Historico-Naturales Musei Nationalis Hungarici 67: 183–206.
VOJNITS, A. 1977: New geometrid subspecies from Hungary (Lepidoptera: Geometridae). Folia Entomologica Hungarica 30: 167–171.
VOJNITS, A. 1980: Araszolólepkék I., Geometridae I. Fauna Hungariae 137. Akadémiai Kiadó, Budapest. 157 p.
VORBRODT, K. 1914: Geometridae, Arctiidae, Syntomidae, Zygaenidae, Limacodidae, Psychidae, Thyrididae, Cossidae, Aegeriidae, Hepialidae. In Vorbrodt, K. & Müller-Rutz, J.: Die Schmetterlinge der Schweiz. II. Band. K. J. Wyss, Bern. 727 p.
WAGNER, W. 2014: Lepidoptera and their ecology. http://www.pyrgus.de/index_en.php Visited 12th February 2014.
WAHLBERG, N., SNÄLL, N., VIIDALEPP, J., RUOHOMÄKI, K. & TAMMARU, T. 2010: The evolution of female flightlessness among Ennominae of the Holarctic forest. Molecular Phylogenetics and Evolution 55: 929–938.
WAHLBERG, N., WHEAT, C. W. & PENÃ, C. 2013: Timing and Patterns in the Taxonomic Diversification of Lepidoptera (Butterflies and Moths). PLoS ONE 8(11): e80875. doi:10.1371/journal.pone.0080875
WALKER, J. J. 1890: Notes on Lepidoptera from the region of the Straits of Gibraltar. Transactions of the Entomological Society of London 361–391. Available on http://www.biodiversitylibrary.org
WARDIKJAN, S. A. 1985: Atlas of the genital apparatus of geometrid moths (Geometridae, Lepidoptera) of Armenian SSR. Akademia Nauk Armenian Institut of Zoology, Yerevan. 136 p.
WARNECKE, G. 1940: Zur Kenntnis der *Crocallis elinguaria* L.-Gruppe. (Lep. Geom.). Mitteilungen der Müncher Entomologischen Gesellschaft 30: 1052–1057.

WARNECKE, G. 1941: Ein weiterer Beitrag zur Kenntnis der *Dyscia conspersaria* F. Gruppe (Lep. Geometr.). Zeitschrift des Wiener Entomologische Gesellschaft 26: 208–211, 231–235, 241–246, pls 27,28.

WARREN, W. 1893: On new genera and species of moths of the family Geometridae from India in the collection of H. J. Elwes. Proceedings of the Zoological Society of London 1893: 341–433.

WEHRLI, E. 1929: Ueber die paläarktischen Arten der Gattung *Ellopia* Tr. (Lepidopt. Geometr.). Mitteilungen der Münchner Entomologische Gesellschaft 19: 311–323.

WEHRLI, E. 1934: Lepidopteren-Fauna von Marasch in türkish Nordsyrien. Mitteilungen der Münchner Entomologischen Gesellschaft 24: 25–55 + 6 plates.

WEHRLI, E. 1936: Einige neue Arten und Rassen aus den Ausbeuten des Herrn Ernst Pfeiffer, München. Mitteilungen der Münchner Entomologischen Gesellschaft 26: 33–37.

WEHRLI, E. 1937: Sur d'anciens et de nouveaux genres, especes et sous-especes de Geometridae. Amateur de Papillons, Paris 8 (16/17): 244–250.

WEHRLI, E. 1938: Neue Gattungen, Untergattungen, Arten und Rassen von Geometriden (Lep.). Entomologische Rundschau 55: 354–360.

WEHRLI, E. 1951: Une nouvelle classification du genre *Gnophos* Tr. Lambillionea 51: 6–11.

WEHRLI, E. 1939–1954: Die Spanner des Palaearktischen faunengebietes. In SEITZ, A. (ed.), Die Gross-Schmetterlinge der Erde, Vol. 4, supplement, pp. 254–766. A. Kernen, Stuttgart.

WEIDLICH, M. 2008: Beitrag zur Lepidopteren-Fauna des Notia Pindos (Tringia-Massiv, Lákmos-Gebirge und Athamáno-Gebirge) in Griechenland mit Beschreibung von zwei neuen Arten sowie Angaben zur Köcherfliegen- (Trichoptera) und Schnakenfauna (Diptera: Tipulidae). Entomofauna 29: 469–504.

WEINTRAUB, J. D. & SCOBLE, M. J. 2004: Lithinini (Insecta: Lepidoptera: Geometridae: Ennominae). Fauna of New Zealand 49: 1–47.

WENCK, E. 1950: Un phenomene lepidopterologique. Bulletin de la Société Entomologique de Mulhouse (1950): 87–89.

WILTSHIRE, E. P. 1940: The Lepidoptera of the Lebanon. Proceedings of the Royal Entomological Society of London. Series B, taxonomy 9: 79–82.

WILTSHIRE, E. P. 1948: Middle East Lepidoptera, IX: Two new forms or species and thirty-five new Records from Cyprus. Part II: Other notes on Cyprian Lepidoptera. The Entomologist's Record and Journal of Variation, LX: 79–87, pl. 3.

WILTSHIRE, E. P. 1957: The Lepidoptera of Iraq. Government of Iraq (Ministry of Agriculture). 162 p.

WILTSHIRE, E. P., 1990a: *Dyscia senecai* sp. n. from Libya, with notes on some other N. African *Dyscia* (Geometridae, Ennominae). Nota Lepidopterologica 12: 354–365.

WILTSHIRE, E. P. 1990b: An illustrated, annotated catalogue of the macro-Heterocera of Saudi Arabia. In BÜTTIKER, W. KRUPP, F. (eds.), Fauna of Saudi Arabia, vol. 11: 91–250. National Commission for Wildlife Conservation and Development, Riyadh. 381 p.

WOLF, W. 1988: Systematische und synonymische Liste der Spanner Deutschlands unter besonderer Berücksichtigung der Denis & Schiffermüller'schen Taxa (Lep., Geometridae). Neue Entomologische Nachrichten 22: 1–78.

WOLFSBERGER, J. 1960: Wo hat *Tephrina arenacearia* Schiff. Die Westgrenze ihrer Verbreitung? (2. Beitrag zur Kenntnis der Lepidopterenfauna der Südalpen). Nachrichtenblatt der Bayrischen Entomologen 9: 41–45.

YAMAMOTO, S. & SOTA, T. 2007: Phylogeny of the Geometridae and the evolution of winter moths inferred from a simultaneous analysis of mitochondrial and nuclear genes. Molecular Phylogenetics and Evolution 44: 711–723.

YOUNG, C. J. 2004: Characterisation of the Austalian Nacophorini and a phylogeny for the Geometridae from molecular and morphological data. Doctor of Philosophy Dissertation. University of Tasmania. 783 p.

Young, C.J. 2006: Molecular relationships of the Australian Ennominae (Lepidoptera: Geometridae) and implications for the phylogeny of the Geometridae from molecular and morphological data. Zootaxa 1264: 1-147.

Young, C. J. 2008: Characterisation of the Australian Nacophorini using adult morphology, and phylogeny of the Geometridae based on morphological characters. Zootaxa 1736: 1–141.

Zandigiacomo, P. & Dalla Montà, L. 1982: Prime osservazioni sull'entomofauna della Soia (*Glycine max* (L.) Merril) ne Veneto e nel Friuli. Redia 65: 215–227.

Zerkowitz, A. 1946: The Lepidoptera of Portugal. Journal of the New York Entomological Society 54: 51–261.

Zhang, B.-C. 1994: Index of economically important Lepidoptera. CAB International, Wallingford. 599 p.

Özbek, H. & Calmasur, Ö. 2010: Spottet ash looper, *Abraxas pantaria* (L.) (Lepidoptera: Geometridae), a new ash pest in Turkey. Turkish Journal of Zoology 34: 351–358.

Index to scientific names

The index gives page reference to the main treatment. Numbers in bold refer to species numbers and figures.

www.ingramcontent.com/pod-product-compliance
Lightning Source LLC
LaVergne TN
LVHW081516300926
849161LV00006B/1194
* 9 7 8 9 0 0 4 2 5 2 2 0 2 *